```
INTERNATIONAL TECHNOLOGICAL UNIVERSITY
This Book is Donated by:
Christina Bate
Date: 6/22/1999
```

though *Magnetism in the Nineties*

Magnetism in the Nineties

Edited by: A. J. Freeman
K. A. Gschneidner, Jr.

North-Holland
1991

© 1991 Elsevier Science Publishers B.V. All rights reserved

No part of this publication may be reproduced, stored in a retrieval system, or transmitted, in any form or by any means, electronic, mechanical photocopying, recording or otherwise, without the prior written permission of the publisher, Elsevier Science Publishers B.V., P.O. Box 211, 1000 AE Amsterdam, The Netherlands.

Special regulations for readers in the USA: This publication has been registered with the Copyright Clearance Center Inc. (CCC), Salem, Massachusetts. Information can be obtained from the CCC about conditions under which photocopies of parts of this publication may be made in the USA.

All other copyright questions, including photocopying outside of the USA, should be referred to the publisher.

No responsibility is assumed by the Publisher for any injury and / or damage to persons or property as a matter of products liability, negligence or otherwise, or from any use or operation of any methods, products, instructions or ideas contained in the material herein.

This book is printed on acid-free paper.

ISBN: 0 444 89357 1

Published by:

North-Holland
Elsevier Science Publishers B.V.
P.O. Box 211
1000 AE Amsterdam
The Netherlands

Sole distributors for the USA and Canada:

Elsevier Science Publishing Company, Inc.
655 Avenue of the Americas
New York, NY 10010
USA

Reprinted from:
JOURNAL OF MAGNETISM AND MAGNETIC MATERIALS 100

Library of Congress Cataloging-in-Publication Data

Printed in The Netherlands

Preface

Magnetism in the Nineties

It was recognized some time ago that the publication of the 100th volume of the Journal of Magnetism and Magnetic Materials is both an important milestone in the life of the Journal and a special opportunity to commemorate this auspicious occasion. A number of options were explored; the most exciting and challenging one was presented in August 1990 – namely, to put together a special tome composed of about 30 reviews written by experts. The purpose was to provide an historical perspective of the developments that had occurred since the first volume of the Journal was published in late 1975, to identify the most important issues and to provide a perspective for developments anticipated in the 1990s.

In September, the Editorial Advisory Board of the Journal and others were solicited for their suggestions and advice for both topics and possible authors. Nearly 45 topics were suggested and over 70 potential authors were proposed. After reviewing these recommendations and narrowing down our final choices, invitations were sent out and the authors agreed to supply their manuscripts by early summer.

In the process, a number of clear conclusions emerged: magnetism is in a period of renaissance – literally – a new golden age; studies of magnetism – both basic and applied – are in an almost explosive rate of growth driven by newly developed sophisticated experimental and theoretical techniques and novel discoveries; perhaps more than any other field, magnetism has a direct impact on the well-being of the people of the world; the future is bright for both the science and technology.

The thirty-two papers in this special tome cover many of the scientifically exciting and technologically important aspects of magnetism. These reviews cover materials, such as: amorphous magnetic materials, high strength permanent magnets, rare earth intermetallic phases, valence fluctuation and heavy fermion phases, Kondo lattices, magnetostrictive materials, high T_c superconductors, magnetic semiconductors, magnetic recording and magneto-optic materials, garnet bubble films, and magnetic thin films. Other papers cover various experimental methods for unraveling magnetic structures and behaviors, such as: NMR, neutron scattering and high energy spectroscopies; while others deal with certain phenomena, such as: nuclear magnetic ordering at extremely low subkelvin temperatures, the quenching of spin fluctuations, the Invar problem, surface magneto-optic Kerr effect, surface and low dimensional magnetism, and itinerant electron magnetism. A few reviews have a strong theoretical bent and these include: ab-initio calculations of the electronic structure of impurities in ferromagnetic transition metals, the random field Ising model, the Kondo lattice, itinerant electron magnetism, surface magnetism and magnetic multilayers.

We hope these reviews will inspire others and serve as the foundation and the jumping-off point for many of the advances and exciting discoveries in the field of magnetism which will surely occur in the 1990s.

Arthur J. Freeman, Editor
Karl A. Gschneidner, Jr., Guest Editor

August 1991

Contents

Preface	v
Contents	vii

Hasegawa, R.
  Amorphous magnetic materials – a history — 1

Zhang, Y.D., J.I. Budnick, J.C. Ford and W.A. Hines
  Some application of NMR to the study of magnetically-ordered materials with emphasis on the short-range order in (Fe–B)-based crystalline and amorphous alloys — 13

Strnat, K.J. and R.M.W. Strnat
  Rare earth–cobalt permanent magnets — 38

Herbst, J.F. and J.J. Croat
  Neodymium–iron–boron permanent magnets — 57

Buschow, K.H.J.
  Permanent magnet materials based on tetragonal rare earth compounds of the type $RFe_{12-x}M_x$ — 79

Lacroix, C.
  Magnetic properties of the Kondo lattice — 90

Gignoux, D. and D. Schmitt
  Rare earth intermetallics — 99

Adroja, D.T. and S.K. Malik
  Valence fluctuation and heavy fermion behaviour in rare earth and actinide based compounds — 126

Moon, R.M. and R.M. Nicklow
  Neutron scattering of lanthanide materials — 139

Lander, G.H. and G. Aeppli
  Neutron scattering studies of magnetic properties of actinide systems — 151

Koon, N.C., C.M. Williams and B.N. Das
  Giant magnetostriction materials — 173

Steglich, F.
  Experimental study of Ce-based heavy-fermion compounds — 186

De Visser, A. and J.J.M. Franse
  Uranium-based heavy-fermion superconductors: an experimental survey — 204

Johnston, D.C.
  Normal state magnetism of the high T_c cuprate superconductors — 218

Dederichs, P.H., R. Zeller, H. Akai and H. Ebert
  Ab-initio calculations of the electronic structure of impurities and alloys of ferromagnetic transition metals — 241

Moriya, T.
  Theory of itinerant electron magnetism — 261

Belanger, D.P. and A.P. Young
  The random field Ising model — 272

Ikeda, K., S.K. Dhar, M. Yoshizawa and K.A. Gschneidner Jr.
  Quenching of spin fluctuations by high magnetic fields — 292

De Jonge, W.J.M. and H.J.M. Swagten
  Magnetic properties of diluted magnetic semiconductors — 322

Wassermann, E.F.
  The Invar problem — 346

Kappert, R.J.H., H.R. Borsje and J.C. Fuggle
  High energy spectroscopies and magnetism — 363

Hakonen, P., O.V. Lounasmaa and A. Oja
  Spontaneous nuclear magnetic ordering in copper and silver at nano- and picokelvin temperatures — 394

Bate, G.
  Magnetic recording materials since 1975 — 413

Dillon Jr., J.F.
  Magnetooptics — 425

Bader, S.D.
  SMOKE — 440

Han, B.S
  Behavior of vertical-Bloch-line chains of hard domains in garnet bubble films — 455

Prinz, G.A.
  Metastability in epitaxial magnetic metal films — 469

Gradmann, U.
  Surface magnetism — 481

Freeman, A.J. and R.-Q. Wu
  Electronic structure theory of surface, interface and thin-film magnetism 497

Mills, D.L.
  The ferromagnetism of ultrathin films; from two to three dimensions 515

Mathon, J.
  Theory of magnetic multilayers. Exchange interactions and transport properties 527

Ozhogin, V.I. and V.L. Preobrazhenskii
  Nonlinear dynamics of coupled systems near magnetic phase transitions of the "order–order" type 544

Swartzendruber, L.J.
  Properties, units and constants in magnetism 573

Author index 577

Subject index 579

Amorphous magnetic materials – a history

Ryusuke Hasegawa

Metglas Products, Allied-Signal Inc., 6 Eastmans Road, Parsippany, NJ 07054, USA

The discovery of Au–Si and subsequent Pd–Si systems in glassy states had lead to the formation of a number of stable amorphous magnetic materials. A brief history of this development and its scientific and technological significance is given in a somewhat chronological order.

1. Introduction

Formation of a magnetic moment in a metallic system is closely related to the electronic property of the system as shown by Friedel [1] and Anderson [2]. The first amorphous or glassy alloy quenched directly from the melt was obtained by Duwez and his collaborators [3]. This amorphous alloy, $Au_{80}Si_{20}$, was not stable at room temperature and was not suited to examine the Friedel–Anderson criteria for local magnetic moment formation. Soon after the first discovery of a liquid-quenched amorphous metallic system, Cohen and Turnbull pointed out that a low eutectic temperature was in part responsible for glass formation in metallic systems [4]. On the other hand, the search for stable amorphous structures continued, using improved and new liquid-quenching apparata [5,6]. A new glassy phase was found in Pd–Si alloys containing 15–23 at% Si [7]. The discovery of this system is important for the following reasons: (1) The Pd–Si glassy system is stable at room temperature with crystallization temperature near 400°C and (2) The element Pd is a transition metal and can be partially replaced by other transition metals.

Some indication of magnetic ordering was indeed observed when part of Pd was replaced by Fe and Co in the binary Pd–Si glassy system, although details of the magnetic states were not clear [8]. While the search for well-defined magnetic states in glassy metals was on-going, the existence of ferromagnetic ordering in amorphous structures had been already predicted by Gubanov [9]. His model takes into consideration only the exchange interactions among neighboring atoms and the radial distribution function for the atoms; this is basically not different from the crystalline cases where Heisenberg–Dirac direct exchange interactions are usually considered up to the first or second nearest neighbors to calculate the magnetization and the Curie temperature of a system.

Before Gubanov's prediction became known, some thin films were found to have amorphous structures with ferromagnetic properties. These include Ni–P and Co–P films by electrodeposition [10]. These materials had little technological significance and did not receive much attention. It is instructive to note that a similar technique has been found useful to produce metallic superlattices which have more significance technologically [11]. Clear experimental evidence for the existence of ferromagnetism in amorphous solids was provided by the observation of magnetic domains in Fe [12], Co–P [13] and Co–Au [14] films obtained by vapor-evaporation. Some of the unstable glassy structures obtained by vapor-evaporation on cold substrates were somewhat stabilized by introducing some impurities [15]. This, however, did not improve the situation to a level in which these films were stable at room temperature.

Having reviewed briefly the situation in the

mid 1960's and with hindsight, one can conclude that the realization of amorphous magnetic materials was initiated when Au–Si and hence Pd–Si were made glassy from the liquids. This article is intended to summarize the subsequent development and its significance in somewhat chronological order.

2. Magnetic ordering in glassy metals (1965–1975)

2.1. Local moment formation

To clarify the magnetic states of Pd–Si base amorphous alloys, an extensive study was made by replacing Pd by 3d transition metals [16,17]. It was found that well-defined local magnetic moments are formed for Cr, Mn, Fe and Co atoms, but not for Ni atoms. The Fermi level of about 3.5 eV for the host Pd–Si is responsible for the observation, confirming that the Friedel–Anderson criteria for the local moment formation is appliable to non-crystalline solids. Fig. 1 illustrates localized Fe d-states in glassy Pd–Si alloys [17]. Other notable effects associated with the localized d-states such as the Kondo effect were also observed.

It was noticed that 3d element concentration levels for the Kondo effect were much higher in glassy metals than in crystalline hosts such as Cu. This implied reduced d–d exchange interactions in glassy metals. Electron transport properties also suggested low electron-mean-free-path of the order of interatomic distances in glassy metals. One such consequence is the interference of the scattering from localized d-states (Kondo type) and from non-magnetic atoms [18].

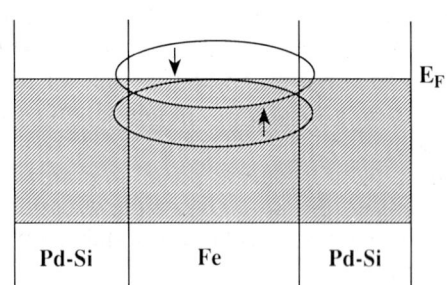

Fig. 1. Fe d-states in glassy Pd–Si alloy (taken from ref. [17]).

2.2. Long range magnetic ordering

Before experimental examination of the Friedel–Anderson criteria in glassy metals was thoroughly performed, curiosity to attain some kind of magnetic ordering led to replacement, in the glassy Pd–Si system, of as many Pd atoms as possible by 3d elements. As pointed out above, Co and Fe replacement for Pd indicated some feature of magnetic ordering [8]. More detailed work supported the finding and showed some features of long range magnetic ordering in noncrystalline solids [17]. These included reduced bulk magnetic transition temperatures and relatively unchanged magnetic states for individual atoms upon transition from crystalline to amorphous states.

The most convincing evidence for magnetic ordering was provided by the successful quenching of a liquid Fe–P–C alloy [19]. Although long-range magnetic ordering was evident in glassy metals [8], Duwez searched for a more convincing case, suggesting that one of his graduate students, Lin, try the Fe–P system. Lin's efforts were unsuccessful until he inadvertently used a carbon crucible to melt an alloy having a composition close to $Fe_{80}P_{20}$. This resulted in an well-defined glassy ferromagnet, $Fe_{80}P_{13}C_7$, with a Curie temperature of 586 K [19,20]. Fig. 2 depicts Mössbauer spectra for this glassy ferromagnet at different temperatures covering its paramagnetic (597 K) and ferromagnetic states (299 K) [20]. Handrich proposed a simple molecular field model to explain the observation in which structural fluctuations were considered [21]. This model with an experimental radial distribution function was appplied to an amorphous Fe–Pd–Si alloy with some reasonable results [17].

Thus, by the end of 1960's, clear evidence for ferromagnetism above room temperature in glassy metals was provided by Mössbauer spectra, ferromagnetic domains, coercivity and remanent magnetization. Yet, skepticism existed: one of the participants was challenged at a topical conference after he demonstrated that a glassy Fe–P–C foil was attracted to a permanent magnet.

Soon after the successful quenching of glassy $Fe_{80}P_{13}C_7$, it was found out that $Fe_{75}P_{15}C_{10}$ was

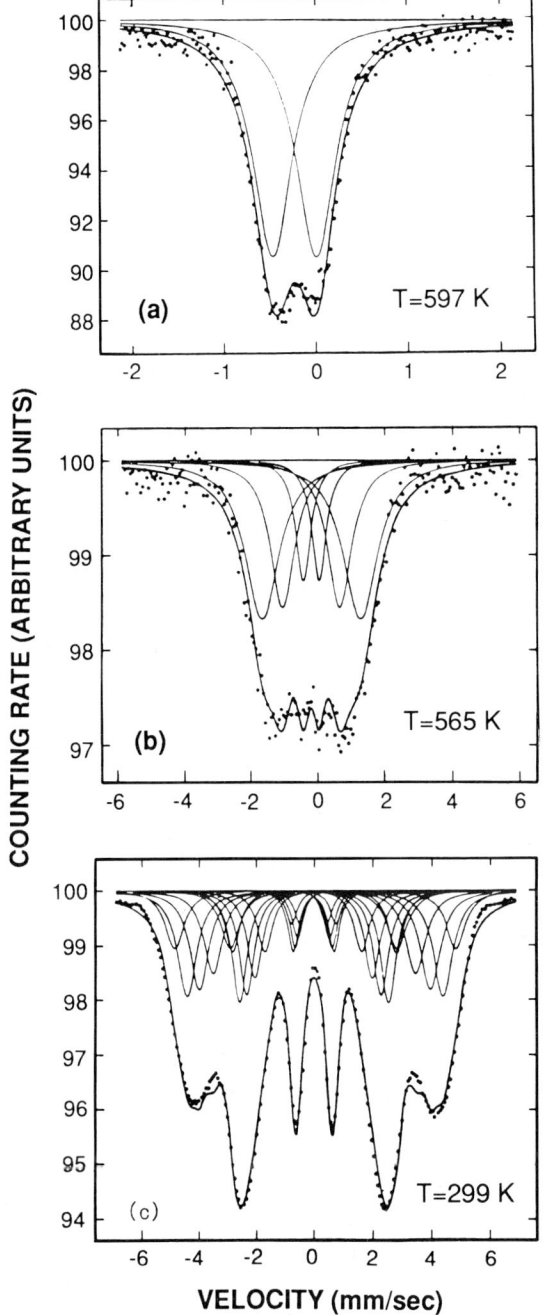

Fig. 2. Mössbauer spectra above (a), near (b) and below (c) the Curie temperature of glassy $Fe_{80}P_{13}C_7$ (taken from ref. [20]).

then thought that the local structure of the glassy Fe–P–C alloy was based on the Fe_3P structure which is ferromagnetic with a Curie temperature of 716 K [24]. Around this time, an analogous alloy, $Mn_{75}P_{15}C_{10}$, was obtained in a glassy state [25]. The local structure of this alloy was argued to be Mn_3P-like and since crystalline Mn_3P is antiferromagnetic with a Néel temperatrure of 115 K [26], the Mn base glassy alloy should exhibit antiferromagnetism. This was indeed the case [27]. Moreover, two additional antiferromagnetic states were found in the material, indicating that a mixture of local glassy states similar to those found in silicate glasses [28] is possible in glassy metals. Molecular field models [29,30] for antiferromagnetism in glassy materials were proposed along the line similar to Handrich [21]. While the search for new magnetic states in glassy metals was on-going, various properties unique to amorphous metals were of much interest [31]. Some of these are summarized below.

2.3. Unique magnetic properties

2.3.1. Resistance minima in amorphous ferromagnets

Dilute solution of 3d elements in amorphous non-magnetic materials led to observation of resistance minima at low temperatures. This was attributed to the Kondo effect [16]. However, similar findings were obtained in well-defined amorphous ferromagnets [16,32]. This was intriguing and generated some theoretical interest [33,34]. Whatever the origin is, the phenomenon implies that conduction electrons interact with some kind of local states with internal freedom. More work on this subject is mentioned below.

2.3.2. Localized states and magnetization behavior

While magnetism in amorphous metals was attracting attention, the magnetic states of Fe and Co in crystalline Pd or Pt were also generating considerable interest. Spin polarization extending further than several atomic distances was responsible for the giant magnetic moment as high as $12\mu_B$/Fe atom in dilute Fe–Pd system. The situation was found to be somewhat different

much more easily obtained in glassy states [22]. It was shown that the major equilibrium phase of the crystallized Fe–P–C was Fe_3P [23]. It was

in amorphous Pd–Si base alloys containing Fe [17]. However, a relatively high paramagnetic moment of about $5\mu_B$/Fe atom was observed in these amorphous alloys, which was attributted to "superparamagnetic" clusters consisting of Fe and Pd atoms. The atoms in these clusters were exchange-coupled and their size was temperature-dependent, the concept being different from the classical superparamagnetic clusters [35].

Although paramagnetism and ferromagnetism were observed in the glassy Fe–Pd–Si system, wider concentration coverage for Fe was not realized until the glassy Fe–Pd–P system was discovered [36]. This system, $Fe_xPd_{80-x}P_{20}$, when rapidly quenched from the liquid state, exhibits glassy states for $x = 13$–44. The magnetic states were characterized using Mössbauer techniques [37]. Some of the significant results are summarized as follows; (i) Electronic structure in a glassy metal is not significantly different from its crystalline counterparts. (ii) Hyperfine field distribution is very broad, increasing with the Fe content. (iii) The large quadrupole splitting, which is expected in glassy materials, decreases with the Fe content. (iv) The critical concentration of Fe for the appearance of ferromagnetism is much higher in the amorphous alloy. For example, the concentration is about 25 at% Fe for Fe–Pd–P and about 16 at% Fe for crystalline Au–Fe [38]. (v) The magnetization behavior of an amorphous ferromagnet can be given by a simple molecular field approximation of Handrich [21]. (vi) A calculation for disordered Heisenberg ferromagnet by Montgomery et al. is consistent with the reduced Curie temperatures observed in glassy ferromagnets.

The above-listed properties reflect some of the early findings for amorphous magnetic alloys. Within the framework of the level of understanding, some attempts were made to explain experimental results. These include resistivity minima in amorphous ferromagnets [33,34] and reduced magnetic transition temperatures in conjunction with the low energy peak in the magnon density of states [39] and d-band filling by s–p elements affecting magnetism [40]. Some of these are still controversial and require further clarification.

2.3.3. Soft ferromagnetism

Duwez and Lin found not only well-defined ferromagnetism in a glassy Fe–P–C alloy but also magnetic softness in the material [19]. This alloy was based on Fe and, therefore, potentially attractive from application stand points. Unfortunately this alloy was brittle, having limited practical applications. A breakthrough to correct this took place when Chen and Polk came up with the idea that glass formability of transition metal base alloys is enhanced by introducing such elements as Si, Al, etc. to the system [41]. As pointed out by Turnbull [42], the elements such as Si tend to stabilize, for example, the bcc structure of Fe-base alloys and, therefore, one would not expect them to enhance glass formability of metallic alloys. In addition to the effort toward basic understanding of glass formability in metals, the findings of Chen and Polk opened up a number of technologically important possibilities. Improvement of the ductility of Fe-base glassy alloys was one of the major advancements. This led to several glassy metals of technological significance [43]. A remarkable combination of soft ferromagnetic properties were reported: they included high magnetic permeabilities and low ac core losses. The advancement made by Chen and Polk has been the driving force for further developments in new glassy metals with a variety of properties, some of which are mentioned in the latter part of this article.

2.3.4. Ferrimagnetism

Conceptually ferrimagnetism is more likely the case in a disordered material in which two kinds of atoms interact antiferromagnetically. In retrospect, amorphous Gd–Co films obtained by Chaudhari, Gambino and Cuomo [44] to be candidate material for bubble domain devices were a fortunate entry into the list of amorphous magnetic materials; Gd is in a S-state and the local anisotropy at the Gd site is small compared with the exchange interaction, and therefore the ferrimagnetic interaction between Gd and Co atoms are collinear. After the discovery, other rare-earth elements in non-S-states in which random anisotropy is comparable to exchange interaction, have been found to form amorphous materials with 3d transition elements. Thus terms such as "sperimagnetism", etc. have been introduced to describe the magnetic interactions involved [45]. These magnetic states are fundamentally not different from collinear ferrimagnetism found in Gd–Co films [44] and some useful examples utilizing existing models are shown in ref. [46]. In this reference, it is instructive to notice that, within a framework of a relatively well-defined system, molecular field approximation used for many crystalline magnetic materials is applicable to glassy materials. This means that short-range magnetic interactions assumed in dealing with crystalline materials may be justified. Admittedly, mean-field approach is a crude approximation, but is a powerful tool to predict unknown properties which may be useful in practical applications. As an example, fig. 3 illustrates the point. Here the system is a disordered structure of Gd–Co films, and one would assume that the distribution of molecular fields would assimilate the system. However, without any introduction of "disorderedness" into the magnetic system, the classical "two-sublattice" approach is sufficient to describe the actual system. The model can even predict the physics of such devices as bubble memory devices. Various possible cases were considered in ref. [46], but the technology did not materialize. The efforts made in this area, however, have opened up yet other technological opportunities in magneto-optics which are mentioned later.

3. Developments of magnetic glassy metals (1975 – present)

The soft ferromagnetism recognized by Duwez and Lin [20] became more realistic when the Chen–Polk efforts had resulted in more readily available glassy ferromagnets. The technological significance has been further enhanced by another landmark invention by Narasimhan [47]. His process allows rapid solidification of liquid in

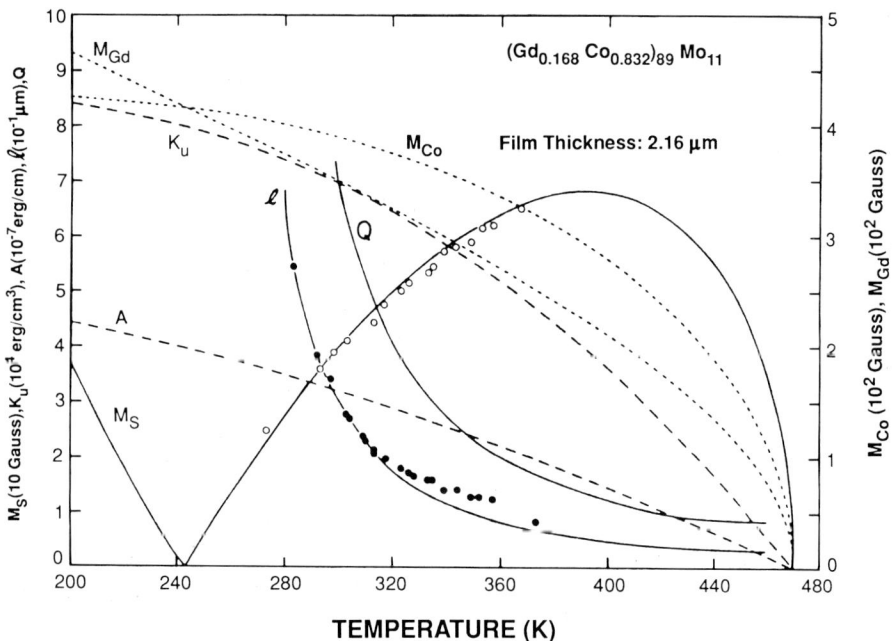

Fig. 3. Temperature dependence of the magnetic bubble device parameters determined from molecular-field fitting of the measured magnetization data. K_u and A are the anisotropy and exchange energy, respectively. Q is defined as $K_u/2\pi M_s^2$ and l is related to the domain width (taken from ref. [46]).

a "sheet" form directly into solid. The process, therefore, has essentially no limitation of width of the rapidly-quenched material, which was perceived necessary as the next step toward commercialization of the material [48]. This certainly encouraged further development of new glassy magnetic materials for various applications. To better understand the basic properties of glassy magnetic alloys, efforts to produce binary systems were made, resulting in successful quenching of Fe–B [49], Fe–P [50], Co–B [51] and Ni–B [52] alloys. These were and have been studied extensively. Some relevant works are mentioned below.

3.1. Structure and magnetism

It was thought that diffraction studies would clarify the atomic arrangements of glassy metals and a number of important works performed before 1980 are summarized in ref. [53]. It has become clear that this approach has limitations: For example, the method is not sensitive enough to probe subtle changes in the atomic arrangements upon heat-treatment. On the other hand, simple magnetization measurements in conjunction with density data can reflect structures more effectively. The above mentioned Fe–B system is one such example, which shows some kind of local structural changes around 18 at%B [54]. Mössbauer and NMR techniques have added more useful information on this system [55]. A comprehensive study of Fe-based glassy systems using Mössbauer techniques summarizes the present state of understanding as shown fig. 4 [56]. This indicates that local structures are more complex in Fe–semimetal based glassy metals than in Fe-metal based ones, indicating possibilities of more interesting magnetic states in the former materials.

Compared to Fe–B, the Co–B system is less interesting in the sense that magnetic properties are predictable [51,57]. However, as described later, this system is an important system from a technological standpoint. The wide concentration range for the glassy Ni–B system ($Ni_{100-x}B_x$; $x = 16–37$) has led to a systematic study of local structure–composition relationship [58] and general glass-formability of alloys [59]. The former

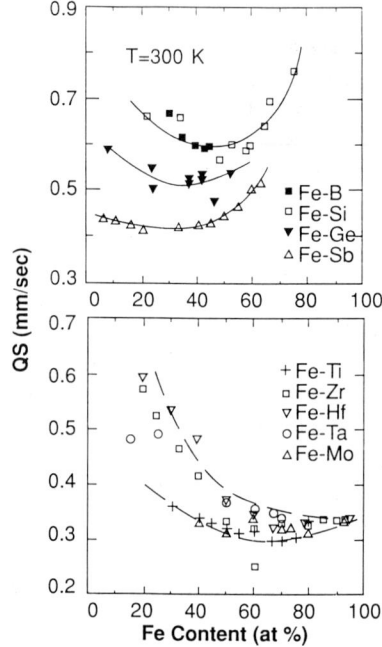

Fig. 4. Quadrupole splitting for Fe-base glassy metals (taken from ref. [56]).

work is an extention of a previous NMR study concerning atomic short-range order in crystalline and glassy materials [60]. The latter one relates to the electronic states calculation made earlier [61]. These studies would help us to understand available materials and undoubtedly help us to develop new glassy systems for various applications mentioned later.

In the early stage of glassy alloy research, there was interest in the magnetization behavior toward $T = 0$ K and at the ferromagnetic Curie temperature (Θ_f). Data on crystalline materials were well-documented and comparison with the similar data on glassy ferromagnets would give insight into fundamental understanding of magnetic materials in general. The first magnetization data toward $T = 0$ K was reported on glassy Co–P [62], followed by an inelastic neutron scattering experiment on glassy Fe–P–C [63]. These results showed that, as in crystalline ferromagnets, spin wave excitations governed the low temperature magnetization behavior obeying the $T^{3/2}$ law. Within a few years after these observations, it was found that the spin wave stiffness constant D (determined from neutron experiments) scales with Θ_f. The ratio, D/Θ_f, related to the range of exchange interaction, is close to 0.2 meV $\text{Å}^2\text{K}^{-1}$ for most glassy metals. This implies that the exchange interactions are more short-ranged than in crystalline counterparts.

The magnetization behavior near Θ_f in glassy ferromagnets attracted attention because of the interest in the critical phenomena in disordered materials. It turned out that glassy structure essentially did not affect the phenomena. A sharp magnetic transition was already observed in the first Fe-base ferromagnet [20]. Experiments covering diversified materials [65,66] indicate that the critical exponents are close to those predicted for a three-dimensional Heisenberg ferromagnet. This situation essentially has not changed at the present moment.

3.2. Glassy ferromagnets for low frequency applications

The present application requires low hysteresis and eddy current losses and high saturation inductions (B_s). The existing materials are grain-oriented silicon steels of various grades. The first Fe-based glassy material, Fe–P–C, was too brittle as mentioned earlier and a modified one, $Fe_{80}P_{16}B_1C_3$ (Metglas® 2615) [67] was introduced. The relatively ductile material caught considerable attention as mentioned earlier [43]. This Fe–P based alloy has a room temperature saturation induction of about 1.5 T and a squareness ratio (after heat-treatment) of about 0.4 [68], which are too low for present applications. This led to development of new alloys based on Fe–B; a B_s value of about 1.6 T was obtained for glassy $Fe_{80}B_{20}$ at room temperature [49]. The efforts

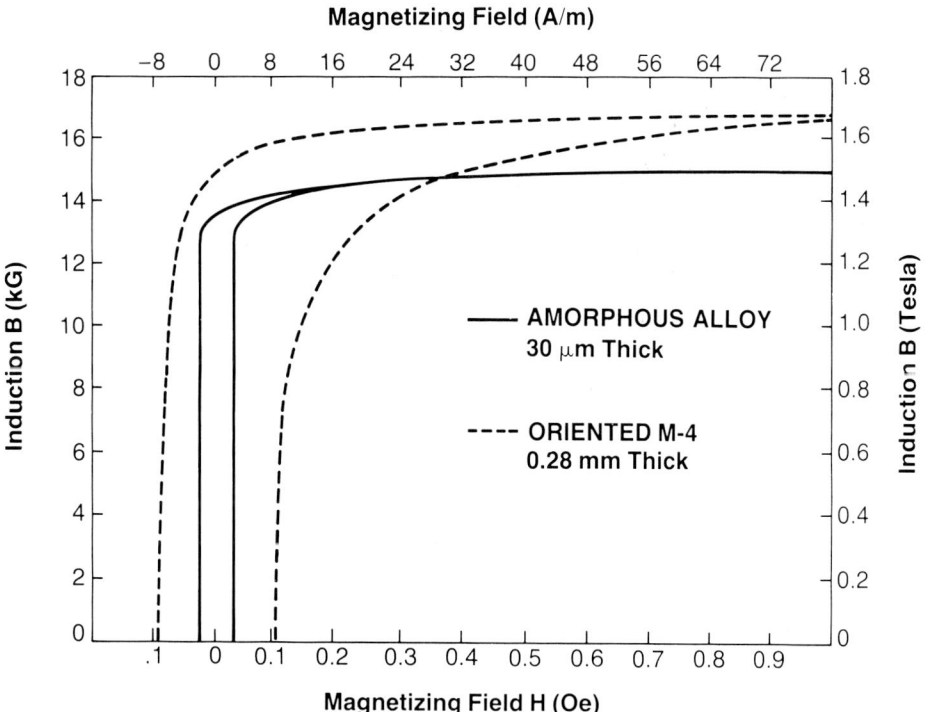

Fig. 5. $B-H$ behaviors for glassy Fe–B–Si alloy and grain-oriented silicon steel (taken from ref. [74]).

resulted in Fe–B–Si [69,70] and Fe–B–Si–C [71–73] alloys. Although the saturation induction level did not improve over that for $Fe_{80}B_{20}$, considerable improvements were made, including increase of the Curie and crystallization temperatures and decrease of magnetic losses. One such example is shown in fig. 5 in which B–H behavior is compared between a conventional Si–Fe and a typical glassy Fe-base alloy [74]. It is evident from this figure that considerable reduction can be expected in the magnetic losses in devices such as transformers and motors. This is indeed the driving force to move the industry toward more energy efficient systems. The effect is two-fold: improvements in material properties and device fabrication. This has been amply proven as indicated in fig. 6 (which also includes data on silicon steels for comparison) [75] and table 1 [76].

Table 1
Trend of core loss reduction for 25 kVA amorphous metal core distribution transformer [76]

Year	Core loss (W)
1982	35
1985	28
1986	18
1989	16

3.3. Glassy metals for high frequency applications

Although high magnetic induction is preferred, important magnetic properties in high frequencies include high permeability and low ac core loss. Conventional materials are represented by thin gauge silicon steel, permalloys, ferrites, etc.

One of the earlier glassy metals reported in ref. [43], $Fe_{40}Ni_{40}P_{14}B_6$ (Metglas® 2826) showed properties similar to those of molypermalloy. Because of the attractive properties and lack of other glassy metals, this material was studied by many investigators. During the course of the study, it was found that phosphorous atoms tend to migrate out of the material, resulting in a brittle material. A glassy alloy with the composition $Fe_{40}Ni_{38}Mo_4B_{18}$ (Metglas® 2826MB) [77] was then introduced to replace the phosphorous-containing alloy. Beside improved magnetic properties [78], this alloy has better thermal stability [79].

The above-mentioned Fe–Ni base glassy alloys have saturation magnetostriction values (λ_s) near 10 ppm. Lower values of λ_s are preferred because of reduced magnetomechanical interactions affecting the performance of the device utilizing the material. Efforts to obtain near-zero magnetostriction in glassy metals led to a number of Co-base glassy alloys. The earlier work on $(Co_{0.96}Fe_{0.04})_{75}P_{16}B_6Al_3$ resulted in magnetic properties similar to those of supermalloys [80]. In conjunction with the search for zero-magnetostrictive glassy metals, the basic mechanism for the effect was investigated [81] in light of the Hamiltonian introduced previously [82]. Some clarification was made through this study, which was helpful in the effort to search for a new

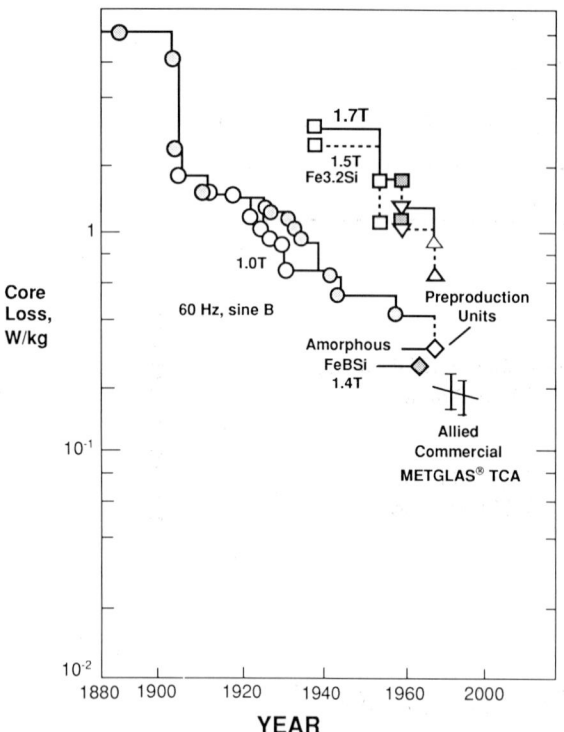

Fig. 6. Core loss reduction of transformer core materials with time (taken from ref. [75] for the data up to about 1985).

zero-magnetostrictive glassy metal. An example for this is to create a bcc-like local structure in glassy Fe and Fe–Ni base alloys, which was partially successful [83]. Thus far, this approach has not been successful in obtaining zero-magnetostriction in Fe-base glassy metals.

Along the line similar to ref. [80], on the other hand, a number of Co-base glassy metals with zero-magnetostriction have been studied. It has been found the magnetostriction value passes through zero at about the same Co/Fe ratio for which crystalline materials have the same result [84]. The situation has not been changed significantly and currently available Co-base commercial glassy alloys have similar compositions. They include Metglas®2705M ($B_s \approx 1.3$ T) and 2714A ($B_s \approx 0.5$ T); Vitrovac 6030 ($B_s \approx 0.82$ T) and 6025 ($B_s \approx 0.55$ T) [85]. These materials are widely used as saturable cores in switch-mode power supplies and various electronic devices.

An effort was made to realize, in glassy Fe-based alloys, properties attained in Co-base glassy alloys or in crystalline permalloys. It was found that a small amount of crystalline precipitates in a glassy matrix considerably reduce ac magnetic losses reaching the level of supermalloy [86,87]. The latter work clarified that the precipitates with a bcc structure are effective while those with fcc or orthorhombic structures are not [88]. Based on this work, two commercial alloys for this application, have been introduced: They are Metglas® 2605S3A ($Fe_{76.5}Cr_2B_{16}Si_5C_{0.5}$) and 2605SM ($Fe_{75}Ni_4Mo_3B_{16}Si_2$). These materials are suited for chokes in electronic filter circuits and for components which utilize ferrites.

3.4. Glassy magnetic alloys for specialty applications

When some kind of unique magnetic properties are discovered, more efforts are usually made to develop improved materials. Glassy ferromagnets are not exceptions and some examples are given below.

3.4.1. Electronic article surveillance

Low magnetic losses and high permeabilities of glassy ferromagnets attracted the attention of the investigators searching for new magnetic materials useful as identification markers. Earlier work around 1980 was done on available glassy materials and utilized higher harmonic responses of the materials excited at a given fundamental frequency [89]. With the advancement of materials processing and understanding of the required properties, better glassy materials were invented for this application [90]. Currently, near-zero magnetostrictive Co-base alloys are used for this purpose.

While the above efforts were on-going, a new method of identification was devised in 1985 [91]. As mentioned above, Fe and Fe–Ni base alloys are magnetostrictive. When these ribbons are excited by an external ac field, they vibrate at a frequency determined by the length of the strip and the acoustic velocity of the material. The magneto-mechanical resonance persists after the termination of the external field, which can be used as an identification marker.

3.4.2. Sensors

The combination of low coercivity ($H_c \approx 5$ A/m), relatively low magnetic anisotropy and high magnetostriction ($\lambda_s \approx 30$ ppm) result in high magneto-mechanical coupling constants over 0.9 for some glassy ferromagnets [92]. These properties in conjunction with high mechanical strength and ductility are suited for magneto-mechanical devices such as pressure and torque sensors. Some Fe and Fe–Ni base glassy alloys have been examined and suitable alloy modification is expected in the future. One of the interesting applications concerns digitizers for computers [93]. A narrow strip of a magnetostrictive glassy metal is pulsed at one end of the strip, resulting in a magneto-elastic wave travelling along the strip. When a small permanent magnet is placed at a given point along the strip, the acoustic wave is modified which is detected by a sensing coil. Thus a position sensor results. By forming a grid-like structure made of glassy metal strips and sensing coils, a digitizer can be constructed with a permanent magnet working as a mouse. It has been found that a medium level of magnetostriction ($\lambda_s \approx 20$ ppm) is required for this application. Thus Fe-base glassy alloys already developed for other applications (e.g. Metglas® 2605SM or S3A) are useful for this application.

3.4.3. Pulse power source

It has been known that pulse compression can be achieved by cascading L–C circuits in which L contains a saturable magnetic core [94]. However, due to high magnetic losses of conventional crystalline ferromagnets at high frequencies, actual devices using these materials have been limited to specific applications. Low ac magnetic losses of glassy metals revived interest in the field and a number of pulse power sources for accelerators have been constructed using glassy $Fe_{67}Co_{18}B_{14}Si_1$ (Metglas® 2605CO) and $Fe_{81}B_{13.5}Si_{3.5}C_2$ (Metglas® 2605SC). The former material was developed around 1984 and has one of the highest B_s values (≈ 1.7 T) among glassy ferromagnets [95]. As industries requiring pulse power sources such as those related to laser-technology grow, modification to the above alloy chemistry may be needed in future.

3.4.4. Magneto-optics

As mentioned earlier, glassy ferrimagnets have given us added insight into magnetic interactions in non-crystalline solids. It is unfortunate that envisaged development of glassy metal based magnetic bubble devices did not take place. However, a lot of efforts made to develop suitable glassy materials for the devices have been helpful in formulating materials for magneto-optics. Around 1980, glassy Tb–Fe and Tb–Fe–Gd films were found to be useful in magneto-optic recording [96]. Thanks to earlier works, by 1982 a lot of basic characterization and new materials emerged [97]. As of this writing, it appears that glassy Tb–Fe–Co film is the most promising material for magneto-optical storage system [98].

3.4.5. Glassy alloys as precurser for new crystalline ferromagnets

During 1981–84, three groups came up with Co-less rare-earth permanent magnets [99–101]. All of them used the rapid-solidification method first to obtain glassy rare-earth–transition metal alloys which were then heat-treated to produce microstructures, resulting in hard magnets. The significance was the observation of the energy product of about 14 MGOe which is the highest of all Co-less light-rare-earth–transition metal permanent magnets. It should be added that rapid-solidification of liquid metal is usually performed in air and therefore rare-earth containing metals require additional efforts and insight to avoid oxidation. In addition, use of La in ref. [99] is considered quite noteworthy because La was considered to inhibit formation of rare-earth intermetallic compounds. By 1984, investigators in the field agreed that the structure of the material having the largest energy products is based on $R_2Fe_{14}B$ [101]. This material is now a commercial product.

Another similar situation took place recently. Glassy Fe-based alloys are heat-treated until local crystalline structures of the order of 10 nm in their sizes are achieved. Some of these materials show low magnetostriction (indicating the fine crystallites are in bcc structures) and low ac magnetic losses at high frequencies [102].

4. Conclusion

A number of new magnetic materials were found and investigated in the past. Glassy magnetic materials were different from conventional materials and, when first introduced, were viewed by few as a new class of novel magnetic material. As one who has spent much of his career pursuing clarification of the nature of magnetic states in non-crystalline solids, it is satisfying to see the development of the field from the beginning to the present level in which a wide variety of applications of the material are making an impact on magnetics in general. The important point is the flow of the historical events from the first synthesis of glassy Au–Si followed by the effort to create ferromagnetic states in glassy materials to the landmark patents concerning chemical composition and production process for commercial use of the material.

References

[1] J. Friedel, Can. J. Phys. 34 (1956) 1190; Nuovo Cimento 7 (1958) 287.

[2] P.W. Anderson, Phys. Rev. 124 (1961) 41.
[3] W. Klement, Jr., R.H. Willens and P. Duwez, Nature 187 (1960) 869.
[4] M.H. Cohen and D. Turnbull, Nature 189 (1961) 131.
[5] P. Duwez, R.H. Willens and W. Klement, Jr., J. Appl. Phys. 31 (1960) 36.
[6] P. Pietrokowsky, J. Sci. Instr. 34 (1962) 445.
[7] P. Duwez, R.H. Willens and R.C. Crewdson, J. Appl. Phys. 36 (1965) 2267.
[8] C.C. Tsuei and P. Duwez, J. Appl. Phys. 37 (1966) 435.
[9] A.I. Gubanov, Fiz. Tver. Tela 2 (1960) 502.
[10] A. Brenner, D.E. Couch and E.K. Williams, J. Res. Natl. Bur. Std. 44 (1950) 109.
[11] D.S. Lashmore, R. Oberle, M.P. Dariel, L.H. Bennett and L. Swartzendruber, Mater. Res. Soc. Symp. Proc. 132 (1989) 1219.
[12] C.W.B. Grigson, D.B. Dove and G.R. Stilwell, Nature 204 (1964) 173.
[13] B.G. Bagley and D. Turnbull, Bull. Am. Phys. Soc. 10 (1965) 1101.
[14] S. Mader and A.S. Nowick, Appl. Phys. Lett. 7 (1965) 57.
[15] W. Felsch, Z. Phys. 195 (1966) 201.
[16] C.C. Tsuei and R. Hasegawa, Solid State Commun. 7 (1969) 1581.
R. Hasegawa and C.C. Tsuei, Phys. Rev. B 2 (1970) 1631, B3 (1971) 214.
[17] R. Hasegawa, J. Appl. Phys. 41 (1970) 4096; J. Phys. Chem. Sol. 32 (1971) 2487.
[18] R. Hasegawa, Phys. Rev. Lett. 28 (1972) 1376.
[19] P. Duwez and S.C.H. Lin, J. Appl. Phys. 38 (1967) 4096.
[20] C.C. Tsuei, G. Longworth and S.C.H. Lin, Phys. Rev. 170 (1968) 603.
[21] K. Handrich, Phys. Stat. Sol. 32 (1969) K55.
[22] P. Duwez, private communication (1968).
[23] P.K. Rastogi and P. Duwez, J. Non-Cryst. Solids 5 (1970) 1.
[24] A.J.P. Meyer and M.C. Cadeville, J. Phys. Soc. (Japan) 17 Suppl. (1962) B1.
[25] S.C.H. Lin, private communication (1968).
[26] R.J. Gambino, T.R. McGuire and Y. Nakamura, J. Appl. Phys. 38 (1967) 1253.
[27] R. Hasegawa, Phys. Rev. B 3 (1971) 1631.
[28] S.C.H. Lin and M. Joshi, J. Electrochem. Soc. 116 (1969) 1740.
[29] A.W. Simpson, Phys. Stat. Sol. 40 (1970) 207.
[30] R. Hasegawa, Phys. Stat. Sol. (b) 44 (1971) 613.
[31] See, for example, Amorphous Magnetism, eds. H.O. Hooper and A.M. de Graaf (Plenum Press, New York, 1976).
[32] S.C.H. Lin, J. Appl. Phys. 40 (1969) 2173.
[33] R. Hasegawa, Phys. Lett. A 36 (1971) 207.
R. Hasegawa and J.A. Dermon, Phys. Lett. A 42 (1973) 407.
A. Madhukar and R. Hasegawa, Solid State Commun. 14 (1974) 61.

[34] R.W. Cochrane, R. Harris, J.O. Strom-Olsen and M.J. Zuckermann, Phys. Rev. Lett. 35 (1975) 676.
[35] J. Crangle and W.R. Scott, J. Appl. Phys. 36 (1965) 921.
[36] P.L. Maitrepierre, J. Appl. Phys. 40 (1969) 4826.
[37] T.E. Sharaon and C.C. Tsuei, Phys. Rev. B 5 (1972) 1047.
[38] See, for example, U. Gonser, R.W. Grant, C.J. Meecham, A.H. Muir, Jr. and H. Wiedersich, J. Appl. Phys. 36 (1965) 2124.
[39] C.G. Montgomery, J.I. Kruger and R.M. Stubbs, Phys. Rev. Lett. 25 (1970) 669.
[40] T. Mizoguchi, K. Yamauchi and H. Miyajima, ref. [31], p. 325.
[41] H.S. Chen and D.E. Polk, US Patent No. 3856513 (1974); US Reissue Patent RE 32925.
[42] D. Turnbull, private communication (1983).
[43] T. Egami, P.J. Flanders and C.D. Grahams, Jr., AIP Conf. Proc. No. 24, eds. C.D. Grahams, Jr., G.H. Lander and J.J. Rhyne (1975) p. 697.
[44] P. Chaudhari, R.J. Gambino and J.J. Cuomo, Appl. Phys. Lett 22 (1973) 337.
[45] J.M.D. Coey, J. Appl. Phys. 49 (1978) 1646.
[46] R. Hasegawa, J. Appl. Phys. 46 (1975) 5263.
[47] M.C. Narasimhan, private communication (1975); US Patent No. 4142571 (1979).
[48] R.W. Cahn, Nature 259 (1976) 271.
[49] R. Hasegawa, R.C. O'Handley, L.E. Tanner, R. Ray and S. Kavesh, Appl. Phys. Lett. 29 (1976) 219.
[50] J. Durand and M. Yung, Amorpous Magnetism II, eds. R.A. Levy and R. Hasegawa (Plenum Press, New York, 1977) p. 725.
[51] R. Hasegawa and R. Ray, J. Appl. Phys. 50 (1979) 1586.
[52] J. Briggs and R. Hasegawa, US Patent No. 4338131 (1982).
[53] D.S. Boudreaux, Glassy Metals: Magnetic. Chemical and Structural Properties, ed. R. Hasegawa (CRC Press, Boca Raton, FL, 1983) p. 1.
[54] R. Hasegawa and R. Ray, J. Appl. Phys. 49 (1978) 4174.
[55] Y.D. Zhang, J.I. Budnick, J.C. Ford, W.A. Hines, F.H. Sanchez and R. Hasegawa, J. Appl. Phys. 61 (1987) 3231.
[56] G. Xiao and C.L. Chien, J. Appl. Phys. 61 (1987) 3245.
[57] J. Durand, B. Lemius, R. Hasegawa, D. Aliaga-Guerra and P. Panissod, J. Magn. Magn. Mater. 15–18 (1980) 1373.
[58] P. Panissod, I. Bakonyi and R. Hasegawa, Phys. Rev. B 28 (1983) 2374.
[59] K. Tanaka, T. Saito, K. Suzuki and R. Hasegawa, Phys. Rev. B 32 (1985) 6853.
[60] P. Panissod, D. Aliaga-Guerra, A. Amamou, J. Durand, W.L. Johnson, W.L. Carter and S.J. Poon, Phys. Rev. Lett. 44 (1980) 1465.
[61] T. Fujiwara, J. Phys. F 12 (1982) 661.
[62] R.W. Cochrane and G.S. Cargill III, Phys. Rev. Lett. 32 (1794) 476.

[63] J.D. Axe, L. Passell and C.C. Tsuei, AIP Conf. Proc. No. 24 (1975) 119.
[64] J.J. Rhyne, J.W. Lynn, F.E. Luborsky and J.L. Walter, J. Appl. Phys. 50 (1979) 583.
[65] T. Mizoguchi, N. Ueda, K. Yamauchi and H. Miyajima, J. Phys. Soc. Jpn. 34 (1973) 1691.
[66] J. Durand, K. Raj, S.J. Poon and J.I. Budnick, IEEE Trans. Magn. MAG-14 (1978) 722.
[67] Metglas® is a registered trade name of Allied-Signal Inc. for amorphous metal.
[68] R. Hasegawa and R.C. O'Handley, Proc. 2nd Intern. Conf. on Rapidly Quenched Metals, part I, eds. B.C. Giessen and N.J. Grant p. 459 (1976).
[69] R.C. O'Handley, C.P. Chou and N. DeCristofaro, J. Appl. Phys. 50 (1979) 3603.
[70] N. DeCristofaro, A. Datta, L.A. Davis and R. Hasegawa, Proc. 4th Intern. Conf. Rapidly Quenched Metals, eds. T. Masumoto and K. Suzuki (The Japan Inst. Metal., 1982) p. 1031.
[71] N. DeCristofaro, A. Freilich and D. Nathasingh, US Patent No. 4219355 (1980).
[72] M. Mitera, T. Masumoto and N.S. Kazama, J. Appl. Phys. 50 (1979) 7609.
[73] F.E. Luborsky and J. Walter, IEEE Trans. Magn. MAG-16 (1980) 572.
[74] R. Hasegawa, J. Non-Cryst. Solids 61 & 62 (1984) 725.
[75] F.E. Luborsky, Proc. NATO Advanced Study Institute on Glasses, Current Issues, eds. A.F. Wright and J. Dupuy (Martinus Nijhoff, Dordrecht, 1985) p. 139.
[76] L.A. Lowdermilk, M.P. Sampat and W.D. Nagel, presented at the Amorphous Transformer Symposium (CRIEPI/EPRI, 22–23 March 1989, Tokyo) p. 174.
[77] R. Hasegawa and C.P. Chou, US Patent No. 4152144 (1979).
[78] R. Hasegawa, M.C. Narasimhan and N. DeCristofaro, J. Appl. Phys. 49 (1978) 1712.
[79] R. Hasegawa, J. Appl. Phys. 49 (1978) 5610.
[80] R.C. Sherwood, E.M. Geory, H.S. Chen, S.D. Ferris, G. Norman and H.J. Leamy, AIP Conf. Proc. No. 24 (1975) 743.
[81] R.C. O'Handley, Phys. Rev. B 18 (1978) 930.
[82] E. Callen, J. Appl. Phys. 39 (1968) 519.
[83] R. Hasegawa, J. de Phys. 41 (1980) C8–701.
[84] R.C. O'Handley, Amorphous Magnetism II (Plenum Press, New York, 1977) p. 379.
[85] VITROVAC is a trade name of Vacuumschmelze (Germany).
[86] A. Datta, N. DeCristofaro and L.A. Davis, Proc. 4th Intern. Conf. on Rapidly Quenched Metals, eds. T. Masumoto and K. Suzuki (1981) p. 1007.
[87] R. Hasegawa, G.E. Fish and V.R.V. Ramanan, ref. [86], p. 929.
[88] R. Hasegawa, V.R.R. Ramanan and G.E. Fish, J. Appl. Phys. 53 (1982) 2276.
[89] J.A. Gregor and G.J. Sellers, US Patent No. 4298862 (1982); Reissue Patent RE 32427 and 32428.
[90] P.M. Anderson, R. Hasegawa and R.M. von Hoene, US Patent No. 4553136 (1985).
[91] P.M. Anderson, G.R. Bretts and J.E. Kearney, US Patent No. 4510489 and 4510490 (1985).
[92] C. Modzelewski, H.T. Savage, L.T. Kabakoff and A.E. Clark, IEEE Trans. Magn. MAG-17 (1981) 2837.
[93] WACOM Corporations, Japan (1985).
[94] W.S. Melville, Proc. IEE (London) 98 (1951) 1985.
[95] A. Datta and C.H. Smith, Proc. 5th Intern. Conf. on Rapidly Quenched Metals, eds. S. Steeb and H. Warlimont (North-Holland, Amsterdam, 1985) p. 1315.
[96] N. Imamura and C. Ohta, Jpn. J. Appl. Phys. 19 (1980) L731.
[97] See, for example, J. Appl. Phys. 53 (1982) 2336.
[98] W.A. McGaham, H. Ping, L.Y. Chen, S. Bonafede, J.A. Woolam, F. Sequeda, T. McDanial and H. Do, J. Appl. Phys. 69 (1991) 4568.
[99] N.C. Koon and B.N. Das, Appl. Phys. Lett. 39 (1981) 840.
[100] G.C. Hadjipanayis, R.C. Hazelton and K.R. Lawless, Appl. Phys. Lett. 43 (1983) 797; J. Appl. Phys. 55 (1984) 2073.
[101] J.J. Croat, J.F. Herbst, R.W. Lee and F.E. Pinkerton, Appl. Phys. Lett. 44 (1984) 148; J. Appl. Phys 55 (1984) 2078.
[102] Y. Yoshizawa, S. Oguma and K. Yamauchi, J. Appl. Phys. 64 (1988) 6044.

Some applications of NMR to the study of magnetically-ordered materials with emphasis on the short-range order in (Fe–B)-based crystalline and amorphous alloys

Y.D. Zhang [1], J.I. Budnick, J.C. Ford and W.A. Hines

Department of Physics and Institute of Material Science, University of Connecticut, Storrs, CT 06269, USA

In this review, we summarize recent developments in nuclear magnetic resonance (NMR) studies on (Fe–B)-based crystalline and amorphous alloys, focusing on the application of NMR in identifying the existence of short-range order (SRO), determining the types of SRO, characterizing the behavior of the SRO and exploring the effect of the SRO on the magnetic properties for the Fe–B system. NMR experiments reveal that certain local environments surrounding the B atoms exist in both crystalline and amorphous Fe–B alloys. The type of SRO existing in this rapidly quenched system can be either o-Fe_3B or bct-Fe_3B, or a mixture, depending on the composition and processing factors, especially the carbon content and quenching speed. The SRO originates from a strong covalent bonding between the B and Fe atoms. As this interaction plays the same role in both crystalline and amorphous Fe–B alloys, the SRO which occurs in the amorphous Fe–B alloys is similar to the SRO which exists in their crystalline counterparts. NMR, in combination with magnetization measurements, provides evidence indicating that the SRO existing in the amorphous Fe–B alloys has a significant effect on their soft magnetic properties and that different types of SRO may act differently, thus providing an opportunity to improve the magnetic properties by changing the SRO. In connection with reviewing the achievements of NMR studies in recent years, brief comments concerning the advantages and potential of NMR experiments in the investigation of other magnetically-ordered materials will also be presented.

1. Introduction

Today, magnetics and magnetic materials science have developed to such a stage that the understanding and control of materials on an atomic scale becomes more and more important. Nuclear magnetic resonance (NMR), because of its capability of providing both structural and magnetic information concerning very local atomic environments, appears to be a powerful tool in the fundamental study and exploration of magnetic materials more than ever before. In this review, we summarize and systematize recent achievements of NMR studies in rapidly quenched (Fe–B)-based crystalline and amorphous alloys. The reasons for focusing on the Fe–B system are twofold. First, The Fe–B system is a typical representative of transition metal–metalloid (TM–M) compounds, crystalline and amorphous alloys which have been extensively studied for several years owing to the diversity of their magnetic properties and variety of their technological applications [1]. Therefore, we hope that a detailed analysis of the Fe–B system can provide some insight for gaining a microscopic picture of similar TM–M systems. The relatively simple binary Fe–B system serves as a prototype for understanding the more complicated and commercially available systems which contain additional atomic constituents (one such system, which will be described here, is Metglas 2605 CO ($Fe_{67}Co_{18}B_{14}Si_1$)). Second, by concentrating completely on one system, it is intended to show how the various aspects of NMR can be integrated to provide a comprehensive description of the microscopic structure and its relationship to some magnetic properties with practical application. We hope

---

[1] On leave from Lanzhou University, Lanzhou, China.

that the typical examples shown below will encourage a more extensive utilization of NMR techniques in future studies.

Fe and B atoms can form a variety of compounds such as orthorhombic FeB (o-FeB) [2], body-centered-tetragonal Fe_2B (bct-Fe_2B) [2], orthorhombic Fe_3B (o-Fe_3B) [3,4], body-centered-tetragonal Fe_3B (bct-Fe_3B) [3] and face-centered-cubic $Fe_{23}B_6$ (fcc-$Fe_{23}B_6$) [3,4]. For the equilibrium alloy system, B has less than 1% solid solubility in Fe. By using rapid quenching, metastable crystalline $Fe_{100-x}B_x$ alloys can be made with compositions ranging from $x = 1$ to 12 [5] and amorphous alloys can be made with x ranging from 12 to over 35 [6]. The amorphous range can be expanded to approximately $x = 90$ if vapor quenching techniques, which provide a much higher cooling rate, are used [7]. In industrial applications, (Fe–B)-based amorphous alloys, with some additional elements such as Ni, Co, C, P and Si, form the most popularly used amorphous soft magnetic materials. On the other hand, $Fe_{80}B_{20}$ with a few percent of rare earth Nd substitution can develop very hard magnetic properties when the amorphous ribbons are annealed within a certain temperature range [8]. All of this wide variety of structures and properties have made Fe–B a very attractive system to study during the last two decades. Because instabilities and inhomogeneities of the phase constituents and short range order (SRO) exist in both crystalline and amorphous Fe–B based alloys, macroscopic magnetic measurements alone are not sufficient. Complementary studies on a microscopic scale are necessary for a comprehensive understanding of the magnetic behavior and the application properties of the materials. Hyperfine interaction studies, including the Mössbauer effect (ME) and NMR, have been very successful in exploring the microscopic structure of Fe–B systems. Panissod, Durand and Budnick [9] and Panissod [10] have summarized the early experimental studies on hyperfine fields (HFs) in amorphous alloys, including Fe–B alloys. In this review, we summarize recent advances in NMR studies on Fe–B and related alloys, including the commercially available Metglas 2605 CO ($Fe_{67}Co_{18}B_{14}Si$) alloy, and compare the NMR results with magnetic and structural analyses. Initially, more attention was paid to the HFs at the Fe nuclei. In this review, we summarize the role and advantages of NMR measurements on metalloid nuclei, e.g., ^{10}B and ^{11}B, for identifying SRO and/or phase precipitates. Here by the term "SRO" is implied a certain degree of atomic geometry surrounding the B atom. The results show that: (i) owing to a strong covalent bonding between the Fe and B atoms, a certain geometry surrounding the B atoms, namely SRO, always exists in Fe–B systems regardless of whether the material is metastable crystalline or amorphous in structure; (ii) there are two types of SRO, bct-Fe_3B and o-Fe_3B; (iii) the type and amount of SRO in a particular alloy are dependent on the composition of the alloy and processing factors, among which carbon content and quenching rate are crucial; (iv) the SRO may have considerable influence on the macroscopic properties of the material.

2. Characteristics of NMR in magnetically ordered materials

For an atom possessing a nuclear magnetic moment, there are magnetic hyperfine interactions, namely, the interactions between that nuclear moment and the unpaired (*magnetic*) electrons of both the atom itself and those of nearest neighboring atoms. These interactions are usually represented by an effective field, namely, the hyperfine field in that the hyperfine interaction energy is equivalent to the potential energy that the nuclear magnetic moment has in this HF. In the case of the paramagnetic state of a solid, the direction of the HF varies randomly due to thermal motion of the d electrons with frequencies much higher than an NMR frequency under normal conditions. Therefore, the hyperfine interaction has a much reduced effect and the NMR frequency of a paramagnetic substance is mainly determined by the externally applied magnetic field. However, in the case of a magnetically-ordered solid, the direction of the atomic magnetic moment is fixed due to the exchange interaction, and so is the HF direction. In this case, the HF produces a large change in the NMR behavior. For a discussion of the origin of the HFs in magnetically-ordered materials, the reader is referred to refs. [11–13]. Hyperfine fields at the nuclei of ferromagnetic materials can be as large as 10^5 Oe for 3d elements and 10^7 Oe for 4f elements. The hyperfine interaction becomes the dominant factor for producing the splitting of the nuclear energy levels. Thus, NMR can be observed without any external magnetic field and the NMR spectra of magnetically-ordered materials reflect the HF distributions. In the case of 3d transition metal compounds or transition metal alloys, the hyperfine field at a magnetic nucleus, H_{hf}, is assumed to arise from three contributions [14]: (1) a core polarization due to the exchange interaction between the on-site moment of the 3d electrons and the inner s electrons, H_{cp}, (2) a spin polarization of the conduction s electrons due to the on-site moment of the atom itself, H_s, and (3) an overall (transferred) conduction electron polarization due to the moments in neighboring shells, H_{sp}, which is herein called the transferred hyperfine field (THF). Hence, we can write the total HF at the nucleus of a 3d magnetic atom located at a certain crystalline site (i) as

$$H_{hf}(i) = H_{cp}(i) + H_s(i) + H_{sp}(i). \quad (1)$$

In general, the first two terms are proportional to the on-site magnetic moment, while the third term is proportional to the number of its nearest neighbor magnetic atoms and their magnetic moments. Thus

$$\begin{aligned} H_{cp}(i) + H_s(i) &= a\mu(i), \\ H_{sp}(i) &= c\sum_j n_j(i)\mu(j) \end{aligned} \quad (2)$$

and

$$H_{hf}(i) = a\mu(i) + c\sum_j n_j(i)\mu(j), \quad (3)$$

where $\mu(i)$ is the on-site magnetic moment of the ith atom, $\mu(j)$ is the moment of the neighboring j-site magnetic atom, a and c are the hyperfine coupling constants, and $n_j(i)$ is the number of the j-site magnetic atoms surrounding the ith atom.

In the case of non-magnetic atoms, the HF at its nucleus arises only from the THF and can be simply written as

$$H_{hf}(i) = H_{sp}(i) = c\sum_j n_j(i)\mu(j). \quad (4)$$

The magnetic properties of the transition metal materials originate from the magnetic moments of the 3d electrons. On the other hand, the 3d electrons play a key role in producing the hyperfine interactions in magnetic materials. Such a correlation between magnetism and hyperfine interactions provides a strong basis from which valuable information concerning magnetic properties can be obtained from hyperfine interaction studies. It is just because of this correlation that the importance of NMR studies increased rapidly soon after the first observation in a magnetically-ordered material [15].

The second significant characteristic of NMR behavior in magnetically-ordered materials is the radio frequency (rf) field enhancement effect [16]. For non-magnetic materials, the NMR is excited directly by an external rf field, h_e. In the case of magnetically-ordered materials, the rf field excites both the nuclear moment and the magnetic moments of the 3d electrons as well. The motion of these electron moments produces a corresponding variation in the HF, and thus results in an rf component, h_{hf}, of the hyperfine field. Therefore, the actual rf field at the nucleus is the sum of these two rf fields. Overall, the magnitude of h_{hf} is proportional to the susceptibility of the material studied. Because of the relatively large magnitude of the HF, $h_{hf} \gg h_e$, and their ratio is the rf field enhancement factor. For the domain rotation process in a magnetically-ordered material, the enhancement factor can be written as [16]

$$\eta_d = \frac{h_{hf}}{h_e} \propto \chi_d \frac{H_{hf}}{M_s} \quad (5)$$

and for the domain wall displacement process,

$$\eta_w = \frac{h_{hf}}{h_e} \propto \chi_w \frac{H_{hf}D}{M_s \delta}. \quad (6)$$

In the expressions above, χ_d and χ_w are the susceptibilities corresponding to domain rotation and domain wall displacement, respectively, H_{hf} is the hyperfine field, M_s is the saturation magnetization of the material, D is the size of domain and δ is the thickness of the domain wall.

In addition to the enhancement of the rf field, the hyperfine interaction also enhances the NMR signal observed. The mechanism of this enhancement lies in the fact that the resonance precession of a nucleus causes a corresponding motion of the electron moments which couple, via the hyperfine interaction, with this nucleus. Since a spin magnetic moment of an electron is much larger than that of a nucleus, the NMR signal observed in a ferromagnetic material comes mainly from the motion of the magnetic electrons and is much stronger than that induced by the nuclear precession alone. For multidomain particles, the signal enhancement arises from the motion of the domain walls, whereas for the single-domain case, the enhancement must come from the rotation of the magnetization within the domain itself. Consequently, for the latter, the enhancement is considerably smaller making single-domain NMR (nuclei within the domain) more difficult to observe than multidomain NMR (nuclei within the domain walls) due to the smaller signal intensity.

These characteristics make NMR in magnetically-ordered materials a powerful approach with diverse capabilities. Concerning the application of NMR, the reader is referred to some early reviews [9,10,17–20]. Here, we briefly summarize some areas where NMR can be employed directly.

(i) Since the NMR frequency of a magnetic atom is roughly linearly related to its local atomic magnetic moment (eq. (3)), this correlation provides a potential for studying basic magnetic properties, including the determination of the atomic moments. This is particularly useful when dealing with rather complex systems with inequivalent sites.

(ii) If we take the nuclei which exist naturally in the substance being studied as microprobes, NMR can *see* the atomic configuration on a very local atomic scale. This technique is very useful for exploring the structure of microcrystallites or SRO in magnetic materials.

(iii) As the NMR excitation for the nuclei in magnetic materials is related to the magnetization mechanism and ease of the magnetization process (eqs. (5) and (6)), NMR can be used to explore the macroscopic magnetic properties of magnetic materials. In the case of a multiphase material, NMR can provide information about the magnetization response of each individual phase constituent.

(iv) Experimentally, NMR experiments have multiple selectivities as it can be performed on various nuclei and at various frequencies. In the case of an alloy or compound containing more than one NMR species, NMR experiments can be performed on the different nuclei thus contributing to a more complete description. In addition, many elements contain more than one isotope and, in the case where the NMR spectra of different nuclei overlap each other, isotope substitution (i.e., utilizing an isotopically enriched starting material in the sample fabrication) can sometimes be used to resolve the spectra.

Solid state NMR experiments on non-magnetic substances usually involves the use of high pulsed rf power, high resolution, rather complicated pulse sequences and wide range frequency/magnetic field scanning. For magnetically-ordered materials, the above-mentioned characteristics of NMR eases the NMR experimental conditions:

(i) The rf field enhancement effect reduces, by a factor of 10 to 10^3, the rf power required to excite nuclear transitions.

(ii) The hyperfine fields at the nuclei of non-equivalent sites are different and, as a consequence, the NMR spectra of these nuclei are usually shifted far from one another; the distribution of the HF in a magnetically-ordered material is usually wide. These characteristics reduce the resolution requirement for the NMR spectrometer. For such broad spectra, the spin-echo pulse sequence is frequently advantageous. The reader can find a detailed description of NMR experimental methods for magnetically-ordered materials in ref. [20].

(iii) The NMR signal enhancement makes it possible to observe NMR signals from unfavorable isotopes. For instance, with an enhancement factor of about 10^3, the ^{57}Fe NMR spin-echo signal in pure Fe can be easily observed at room temperature, while it is extremely hard to find the ^{57}Fe NMR signal in a non-magnetic substance containing Fe.

As has been shown by the achievements obtained to date, NMR is indeed an approach which involves the use of inexpensive equipment, but from which a great deal of information can be obtained concerning atomic magnetic moments, crystal structures and structural distortion, defects, SRO, atomic configuration in surface or interface region, configuration, ferromagnetic–paramagnetic transitions (magnetic phase transitions), ferromagnetic–superparamagnetic transitions and magnetization processes, etc.

3. Structure and hyperfine interaction results for FeB, Fe_2B, bct-Fe_3B and o-Fe_3B

Several well defined intermetallic Fe–B compounds can be made: FeB, Fe_2B, bct-Fe_3B, o-Fe_3B and $Fe_{23}B_6$. We will omit the crystallographic notation on all of the phases except Fe_3B, since it is necessary to distinguish between the two forms of Fe_3B. All of these compounds are magnetic. Because of the relevance of the structure and atomic magnetic moments of these Fe-borides to the Fe–B alloys, there have been considerable theoretical and experimental investigations concerning the crystal structures, electronic states, charge transfer, atomic magnetic moments

Table 1
Crystal structure and HF information for FeB, Fe_2B, o-Fe_3B and bct-Fe_3B

Crystal structure	Lattice parameter (Å)	Z (mol./cell)	Site	Nearest neighbors		HF (4.2 or 77 K) (kOe)	NMR frequency (MHz)	Atomic moment (μ_B)	Hyperfine coupling constant (kOe/μ_B)	
									A	C
o-FeB	a = 5.508 b = 2.953 c = 4.063	4	Fe	10Fe 6 B		130 [31]	(17.8)	0.9 [33]	144	
			B	6 Fe 2 B	1 Fe [a]		12.4 [32]			2.3
bct-FeB	a = 5.110 b = 4.243	2	Fe	11 Fe 4 B		252 [25] 244	(34.5) (33.4)	1.6 [34]	15.5	
			B	8 Fe 2 B			41.2 [35] 40.9 [36]			3.23
o-Fe_3B	a = 4.450 b = 5.437 c = 6.657	4	Fe_I(8g)	11 Fe 3 B		(254)	35.0 [37]	(1.69) [b]	(150)	
			Fe_{II}(4c)	12 Fe 2 B		(291)	40.0 [37]	(1.94) [b]	(150)	
			B	6 Fe 3 Fe [a]			36.3 [38,39]			3.44
bct-Fe_3B	a = 8.635 c = 4.285	8	Fe_I(8g)	10 Fe 3 B		305 [40]	(41.8)	(2.03) [b]	(150)	
			Fe_{II}(8g)	10 Fe 3 B		284 [40]	(38.9)	(1.89) [b]	(150)	
			Fe_{III}(8g)	10 Fe 4 B		242 [40]		(1.61) [b]	(150)	
			B	6 Fe 3 Fe [a]			34.7 [39,41]			3.14
bcc α-Fe	2.866	2	Fe	8 Fe		341	46.7	2.20	154	

[a] These Fe atoms are farther than the nearest neighbor Fe atoms.
[b] Calculated atomic magnetic moment from the value of HF, assuming $A = 150$ kOe/μ_B.

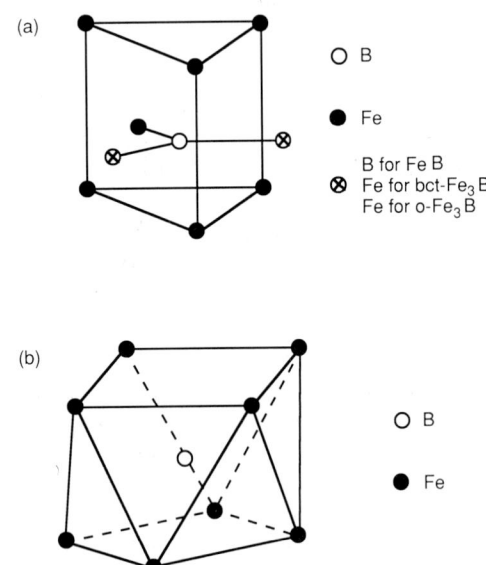

Fig. 1. Near neighbor atomic configuration surrounding a B atom: (a) trigonal prism coordination for FeB, bct-Fe$_3$B and o-Fe$_3$B phases; (b) Fe coordination to a boron atom in the Fe$_2$B phase.

and hyperfine interactions for these compounds [21–43]. Concerning the theoretical studies, we refer the reader to ref. [21]. Table 1 lists some of the structural and HF parameters for these compounds which are related to the discussion below.

In the Fe–B system, only FeB and Fe$_2$B are stable compounds. The FeB compound has an orthorhombic structure with 4 formula units per unit cell. In this phase, each B atom is located in the center of a trigonal prism formed by 6 Fe atoms, which are designated to be the nearest neighbor Fe atoms (see table 1). A seventh Fe atom is situated at a slightly longer distance outside the prism [2]. Each B atom has two nearest B neighbor atoms connected through a zigzag chain. Fig. 1a schematically illustrates the trigonal prismatic configuration surrounding a B atom. Fe$_2$B, has a body-centered-tetragonal structure with 2 formula units per unit cell. In this phase each B atom is surrounded by 8 Fe atoms at identical distances (2.1799 Å) as shown in fig. 1b. Crystallographically, all Fe atoms in the lattice are equivalent; however, because of the HF anisotropy, there exist two magnetically inequivalent Fe sites for this compound [25].

The bct-Fe$_3$B, o-Fe$_3$B and Fe$_{23}$B$_6$ phases are metastable compounds in the Fe–B system. The bct-Fe$_3$B phase can be made through the crystallization of an amorphous Fe$_{75}$B$_{25}$ (a-Fe$_{75}$B$_{25}$) alloy at about 420°C. The o-Fe$_3$B and Fe$_{23}$B$_6$ are very unstable and easily transform to Fe$_2$B plus bcc α-Fe. To date, no single phase o-Fe$_3$B or Fe$_{23}$B$_6$ specimen has been produced. They can be obtained by annealing amorphous Fe–B alloys at rather high temperatures, following a particular procedure which produces multiphase precipitates [3,4,6,35]. The o-Fe$_3$B phase is isostructural to Fe$_3$C [26–28], in which each unit cell contains 4 formula units. This phase consists of a single B site and two inequivalent Fe sites, Fe(8g) and Fe(4c), with a population ratio of 2 : 1 [3,28]. The bct-Fe$_3$B phase is isostructural to tetragonal Fe$_3$P, in which each unit cell consists of 8 formula units [29,30]. This phase consists of a single B site and three inequivalent Fe sites, Fe$_I$(8g), Fe$_{II}$(8g) and Fe$_{III}$(8g), with identical populations. In both the bct-Fe$_3$B and o-Fe$_3$B structures, each B atom is located in the center of a trigonal prism formed by 6 Fe atoms, 4 Fe(8g) and 2 Fe(4c) for the case of o-Fe$_3$B and 2 Fe$_I$(8g), 2 Fe$_{II}$(8g) and 2 Fe$_{III}$(8g) for the case of bct-Fe$_3$B, at the vertices; three other Fe atoms are situated further away as shown in fig. 1a. Although the B atoms in FeB, bct-Fe$_3$B and o-Fe$_3$B phases have the same 6 nearest Fe neighbors, the scales of these prisms and, consequently, the B–Fe and Fe–Fe distances are different. For the FeB phase, the B–B covalent bonding is stronger than the B–Fe bonding and, consequently, each B atom has two B atoms as nearest neighbors; for the other three iron borides, Fe$_2$B, o-Fe$_3$B and bct-Fe$_3$B, the interaction between the B and the Fe atoms is much stronger than the B–B interaction, and the B atoms have no other B atoms as nearest neighbors.

For the HF at the Fe nuclei in a majority of cases, the magnitude of H_{sp} (the second term in eq. (3)) is rather small in comparison with the contributions due to the on-site moment, particularly core polarization. Hence, eq. (3) can be approximately written as

$$H_{hf,Fe}(i) \approx A\mu_{Fe}(i). \qquad (7)$$

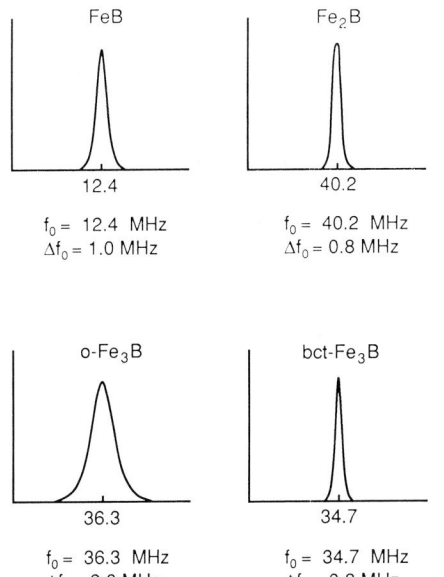

Fig. 2. ^{11}B NMR spectra in FeB, Fe$_2$B, o-Fe$_3$B and bct-Fe$_3$B compounds at 4.2 K. For each compound, the peak frequency and the line width are indicated. See table 1 and text. (Note: the ^{57}Fe spectra, which also exist in the frequency range, have been removed.)

For FeB and Fe$_2$B, whose atomic magnetic moments have been measured, the values of the hyperfine coupling constant A calculated from eq. (7) are 145 and 155 kOe/μ_B, respectively (see table 1 and refs. [33,34]). There are no directly measured Fe magnetic moment data available for the o-Fe$_3$B and bct-Fe$_3$B phases. Taking a representative value of 150 kOe/μ_B, these moments can be obtained from their corresponding hyperfine fields, which are also listed in table 1.

Fig. 2 illustrates the ^{11}B NMR spectra for the FeB, Fe$_2$B, o-Fe$_3$B and bct-Fe$_3$B compounds. These NMR frequencies are marked in the figure and also listed in table 1. According to eq. (4), the THF at the B nuclei in the Fe–B system can be written as

$$H_{sp}(B) = c \sum_j n_j \mu_{Fe}(j). \qquad (8)$$

From the experimental values of $H_{sp}(B)$ and $\mu_{Fe}(j)$, the hyperfine coupling constant c for each compound was calculated and the results are also listed in table 1. It can be seen from table 1, that for the o-Fe$_3$B, bct-Fe$_3$B and Fe$_2$B compounds, the values of c are close to each other (3.14–3.44 kOe/μ_B), implying a similarity of the local atomic neighboring configurations for the B atoms. However, the value for FeB, 2.3 kOe/μ_B, is smaller even though the B–Fe distances in FeB are a little smaller than those in the other Fe-borides [21].

In the study of the Fe–B system, the NMR measurements on B nuclei play a important role for the following reasons: (i) The B atoms are located at the centers of the structural units for all of these phases, (ii) the NMR signal intensity from the B nuclei is much greater than from the Fe nuclei, (iii) natural boron contains 18.83% ^{10}B and 81.17% ^{11}B isotopes, both of them are good NMR probes. The corresponding gyromagnetic ratios are $\gamma(^{10}B) = 0.4574$ MHz/kOe and $\gamma(^{11}B) = 1.3660$ MHz/kOe. The ^{10}B and ^{11}B have the same NMR spectrum profiles in a given substance (neglecting the quadruple interaction), but the ^{11}B NMR frequency is shifted by a factor of $\gamma(^{11}B)/\gamma(^{10}B) = 2.987$ to a higher frequency range. In the case where the ^{11}B spectrum overlaps that of ^{57}Fe, this enables an identification of the ^{11}B spectrum by using the ^{10}B spectrum, (iv) a common feature of these iron borides is that while their Fe sites have rather complex environments with more nearest neighbor Fe and B atoms spread over a large range of distances, the B atoms in these compounds have relatively simple near neighbor atomic configurations. As a consequence, the B NMR frequencies have reasonably well defined values even if the microphases are not completely developed. This enables the application of the B NMR spectra, to a certain extent, as fingerprints in identifying the type of SRO or phase precipitates present in the rapidly quenched Fe–B.

4. Short-range order in metastable crystalline Fe–B alloys

By using melt spinning techniques, metastable crystalline Fe$_{100-x}$B$_x$ alloys can be made with B concentrations up to $x = 12$ [5]. Information concerning the distribution of the B atoms in the

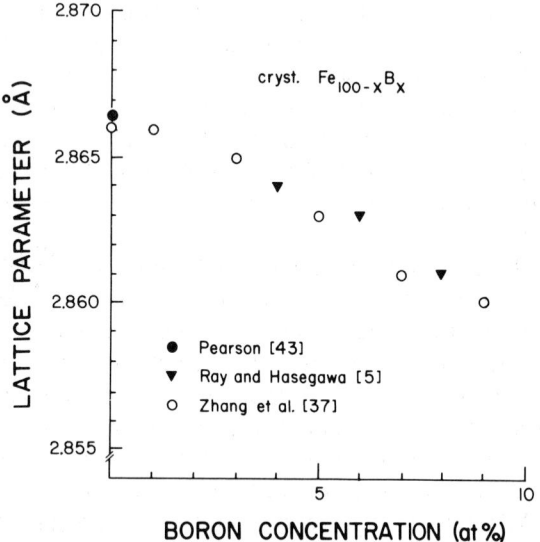

Fig. 3. Room temperature lattice constant (based on the bcc α-Fe structure), in ångström as a function of B concentration, x, in at% for the metastable crystalline $Fe_{100-x}B_x$ alloys: open circles [37]; solid triangles [5]; solid circles [43].

single bcc α-Fe phase [5]. Fig. 3 shows the lattice parameter obtained by XRD from metastable crystalline $Fe_{100-x}B_x$ ($x = 1, 3, 5, 7$ and 9) as a function of the B concentration [37]. With their small atomic radii, B, C and N are usually considered to be either interstitial or substitutional atoms in the lattice of transition metals or transition metal alloys. Based on the fact that the lattice parameter of α-Fe decreases with increasing B concentration, it was then concluded that the Fe–B crystalline alloys were essentially metastable bcc solid solutions in which the B atoms substituted for Fe atoms in a statistically random way [5].

XRD is a good method for structural analysis; however, in order to produce a resolved X-ray diffraction line, the number and lateral dimensions of the lattice planes responsible for the diffraction should contain sufficient repeat periods. For a normal crystalline powder, this prerequisite is easily satisfied. In the case of SRO, however, if the SRO involves an atomic configuration with a size much less than that required, then no resolved diffraction line can be formed. Consequently, XRD may not be a good means for discerning SRO on a nanometer scale unless high intensity synchrotron sources are used.

crystalline alloys, the electronic structure of B and Fe atoms and the character of B–Fe bonding is very important for understanding the properties of the Fe–B amorphous alloys. It was found that the X-ray diffraction (XRD) lines obtained from these crystalline ribbons can be indexed by a

Fig. 4 shows the spin-echo NMR spectrum

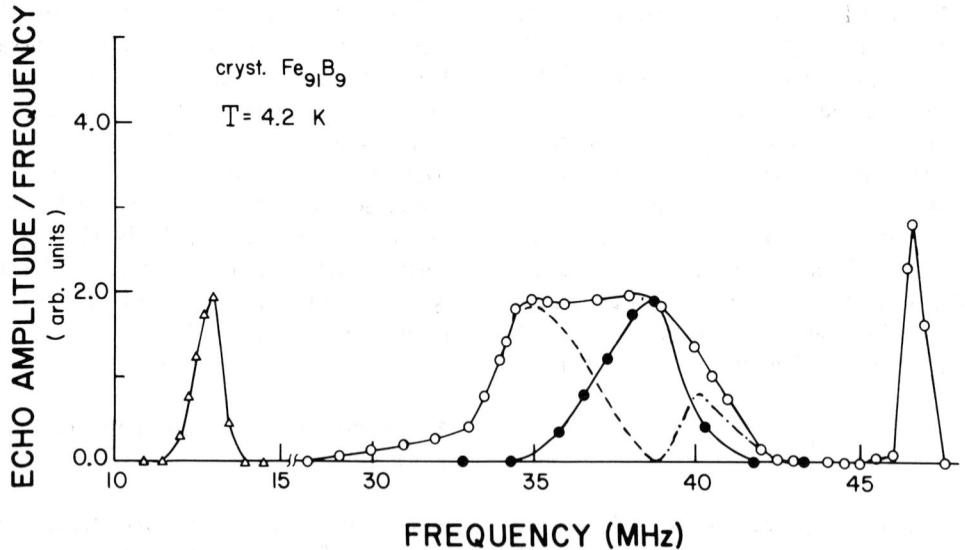

Fig. 4. Spin-echo NMR spectrum of the metastable crystalline $Fe_{91}B_9$ alloy at 4.2 K: open circles with solid line, total $^{11}B + {}^{57}Fe$ spectrum; open triangles with solid line, ^{10}B spectrum; closed circles with solid line, adjusted ^{10}B spectrum; dot–dashed line and dashed line, ^{57}Fe spectrum associated with Fe in the 4c and 8g sites of o-Fe_3B SRO, respectively [37].

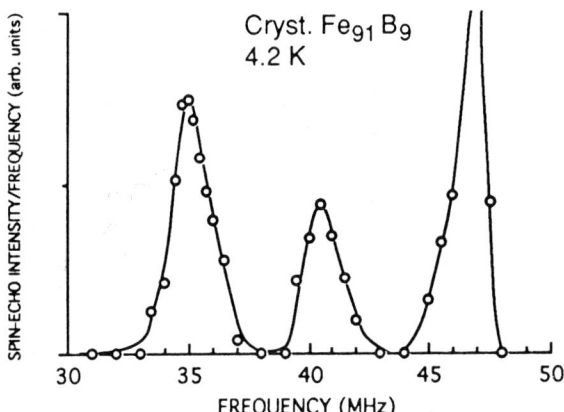

Fig. 5. ^{57}Fe spin-echo NMR spectrum obtained from a ^{10}B enriched metastable crystalline Fe$_{100-x}$B$_x$ alloy at 4.2 K. As the ^{11}B nuclei have been removed, the two ^{57}Fe resonances in the o-Fe$_3$B SRO become well resolved.

obtained from a melt spun Fe$_{91}$B$_9$ crystalline sample at 4.2 K over the frequency range from 10 to 50 MHz [37]. The figure shows that, in addition to two narrow resonance lines centered at 13.0 and 46.7 MHz, there is a very wide resonance over the frequency range from 33 to 42 MHz with less pronounced structure. The 46.7 MHz peak originates from the ^{57}Fe NMR in bcc α-Fe; the 13.0 MHz resonance comes from the ^{10}B NMR. Multiplying the frequency scale of the ^{10}B spectrum by a factor of $\gamma(^{11}B)/\gamma(^{10}B) = 2.987$ and adjusting its peak intensity, we obtained the ^{11}B NMR spectrum which is centered at 39.0 MHz as plotted in fig. 4. By subtracting the ^{11}B spectrum from the total intensity, two other resonance lines are obtained which are centered respectively at 35.0 and 40.0 MHz with an intensity ratio of about 2:1. In order to confirm this spectral decomposition, a Fe$_{91}$B$_9$ crystalline alloy was fabricated using 98% ^{10}B enriched B powder as a starting material. Fig. 5 shows its NMR spectrum at 4.2 K. In addition to the 13.0 and 46.7 MHz lines, the 35.0 and 40.0 MHz peaks became clearly resolved as the ^{11}B isotope has been removed. These two lines originate from ^{57}Fe nuclei. Fig. 6 shows the NMR spectra of crystalline Fe$_{100-x}$B$_x$ alloys with $x = 1, 3, 5, 7$ and 9. In the figure, the locations of the three ^{57}Fe peaks and a single ^{11}B peak are indicated by the arrows. It was found that the locations of these four resonances were independent of the B concentration over the range studied. With increasing B concentration, the 46.7 MHz resonance intensity decreased while the other lines increased. After the removal of the ^{11}B sub-spectrum, which could be done reasonably well for the $x = 5, 7$ and 9 compositions, it was found that the two ^{57}Fe peaks maintained an intensity ratio of 2:1. The fact that the intensity of the 35.0 MHz resonance is approximately twice that of the 40.0 MHz line rules out the possibility that the 40.0 and 35.0 MHz resonances come from the Fe atoms having one and two nearest neighbor B atoms, respectively, in bcc α-Fe. Therefore, the metastable crystalline Fe–B alloys are not an α-Fe(B) solid solution; instead, there is SRO in the alloys. Comparing the NMR data from the iron borides mentioned above with the present results, it is concluded that the SRO is similar to o-Fe$_3$B in nature. A complimentary ME experiment on these alloys revealed 2 distinct ^{57}Fe HFs, consistent with the ^{57}Fe HF values in the o-Fe$_3$B compound and having a site occupation ratio of 2:1 [46]. Also, similar conclusions based on NMR, ME and XRD experiments, have been reported in refs. [6,45].

Based on the NMR experiments described above, in concert with the Mössbauer measurements, it can be concluded that the metastable crystalline Fe$_{100-x}$B$_x$ with $1 \leq x \leq 9$ are not simply a bcc α-Fe(B) solid solution, instead, the system is a dispersion of very small regions with o-Fe$_3$B SRO in the bcc α-Fe matrix. A high resolution transmission electron microscopy (HRTEM) investigation which was carried out later confirmed the existence of this SRO and indicated that the regions were (5–15 Å) in diameter [44].

5. Short-range order in amorphous Fe–B alloys

Short-range order in amorphous alloys, especially in transition metal–metalloid TM$_{100-x}$M$_x$ amorphous alloys has been an extensively investigated topic during the last decade. In the expression above, TM can represent a combination of transition metals and M can represent a combination of high valence metalloid or glass formers. Several techniques, such as TEM [48,49], extended X-ray absorption fine structure (EXAFS) [50–52], neutron diffraction [53], electron scattering [54], ME and NMR have been used to explore and characterize their amorphous structures. ME has been used extensively and the reader can find a summary of the achievements of ME studies on amorphous magnetic materials in refs. [9,55]. With regard to the $Fe_{100-x}B_x$ amorphous alloys, ME experiments have found evidence showing that the bct-Fe_3B SRO exists in the high B concentration alloys [56–60]. Here we concentrate on the NMR studies of the SRO in Fe–B amorphous alloys [35,45,47,61–63].

In order to get more information concerning the SRO in the as-quenched amorphous alloys, it is useful to carefully anneal the alloys and investigate their crystallization processes. This is based on the idea that if a SRO has already exist in the as-quenched alloys, then, with the appropriate heat treatment, this type of SRO will nucleate and grow first. Therefore, annealing makes the SRO more evident if the thermal activation is just

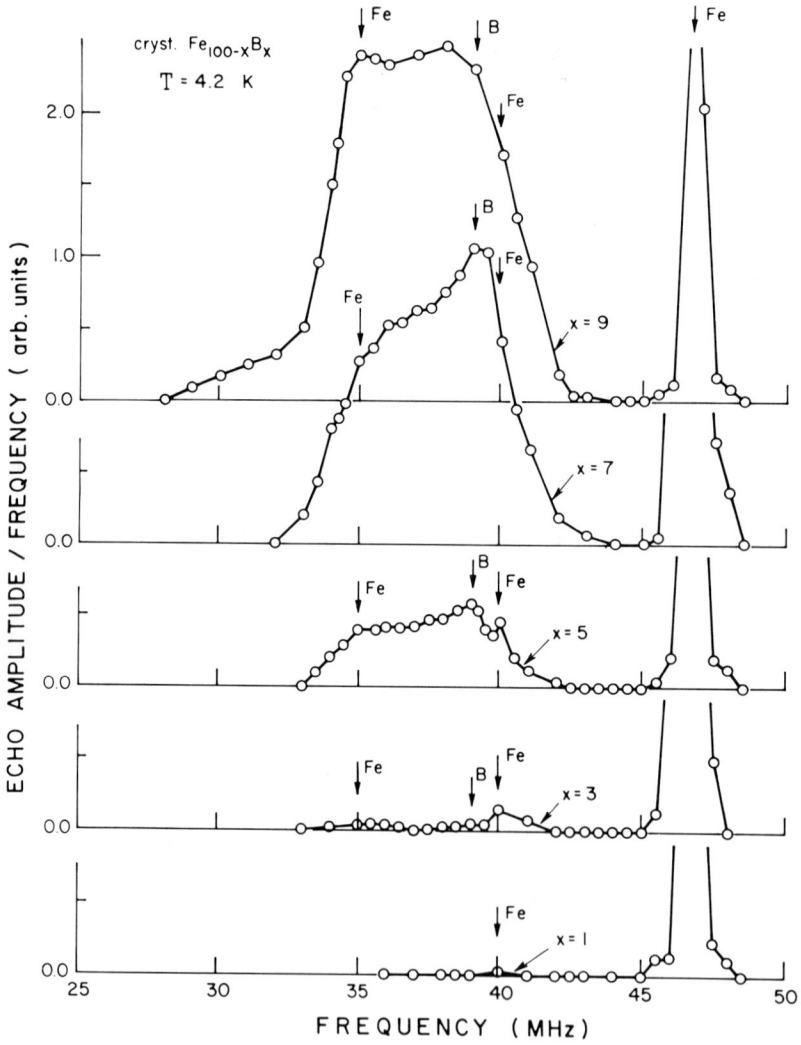

Fig. 6. Spin-echo NMR spectra of metastable crystalline $Fe_{100-x}B_x$ alloys ($x = 1, 3, 5, 7$ and 9) at 4.2 K. The emergence of the ^{11}B and three distinct ^{57}Fe resonances are indicated by the arrows.

beyond the critical level required for further nucleation of the SRO. We note that this concept has been used in a model calculation of crystallization processes of amorphous alloys [64].

Figs. 7 and 8 show the spin-echo NMR spectra obtained at 4.2 K for the $Fe_{75}B_{25}$ and $Fe_{86}B_{14}$ (as-quenched and annealed) alloys, which represent essentially the highest and lowest B concentrations for the amorphous regime, respectively [35]. It was found that the ^{11}B spectra for the various a-$Fe_{100-x}B_x$ amorphous alloys (a designates the amorphous state) are broad and similar to each other. For a-$Fe_{75}B_{25}$, the ^{11}B spectrum exists over a broad frequency range centered at 34.7 MHz. We note that this value is just the same as the ^{11}B NMR frequency in the bct-Fe_3B

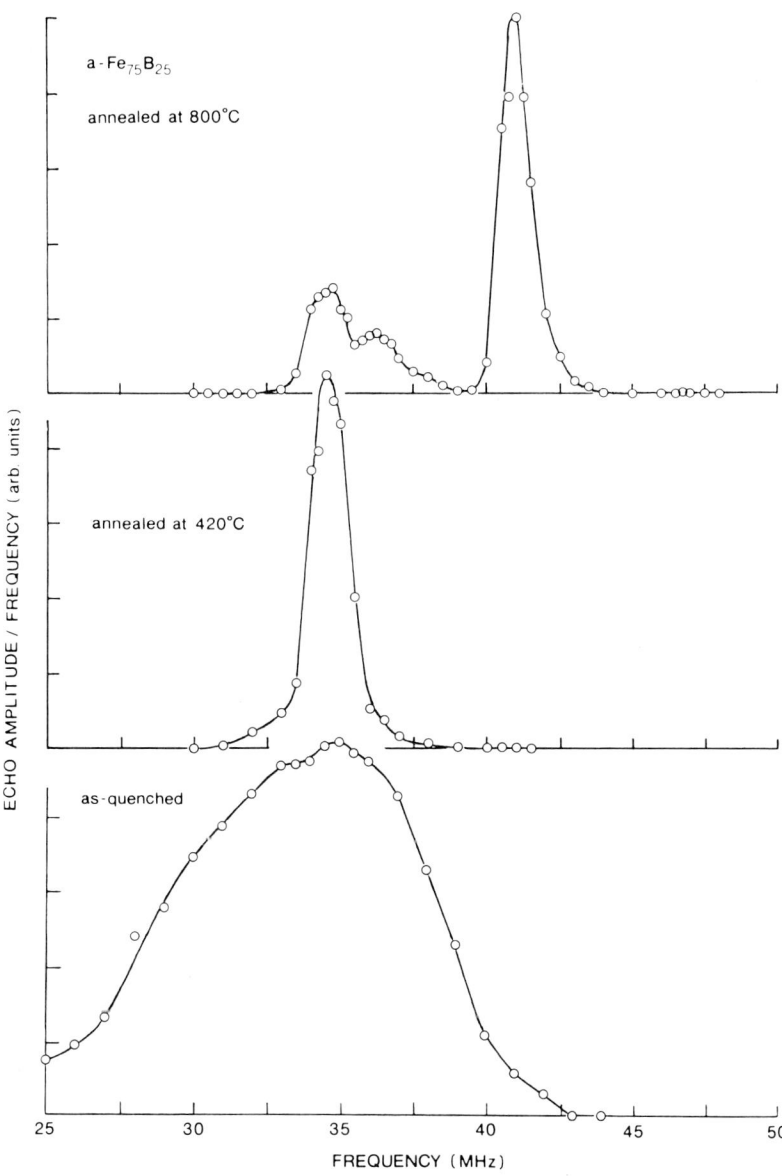

Fig. 7. ^{11}B spin-echo NMR spectra for $Fe_{75}B_{25}$ at 4.2 K: (a) as-quenched, (b) annealed at 420°C (40 min) and (c) annealed at 800°C (5 min) [35].

phase (see table 1). When the sample is annealed at 420°C, which is just slightly beyond the crystallization temperature, the XRD and ^{11}B NMR measurements (shown in fig. 7b) indicates a near perfect bct-Fe$_3$B phase structure. Fig. 7c shows the ^{11}B NMR spectrum obtained from an a-Fe$_{75}$B$_{25}$ sample annealed at 800°C. Comparing the peak frequencies in the spectrum with those obtained from the Fe-boride phases (table 1), we assign the peaks at 34.8, 36.3 and 41.3 MHz to be ^{11}B in the bct-Fe$_3$B, o-Fe$_3$B, Fe$_2$B phases, respectively, and the 46.7 MHz resonance to be the ^{57}Fe in bcc α-Fe. From the fact that the ^{11}B spectra obtained from as-quenched a-Fe$_{75}$B$_{25}$ is centered at approximately 34.7 MHz and that bct-Fe$_3$B precipitates are produced first, it be-

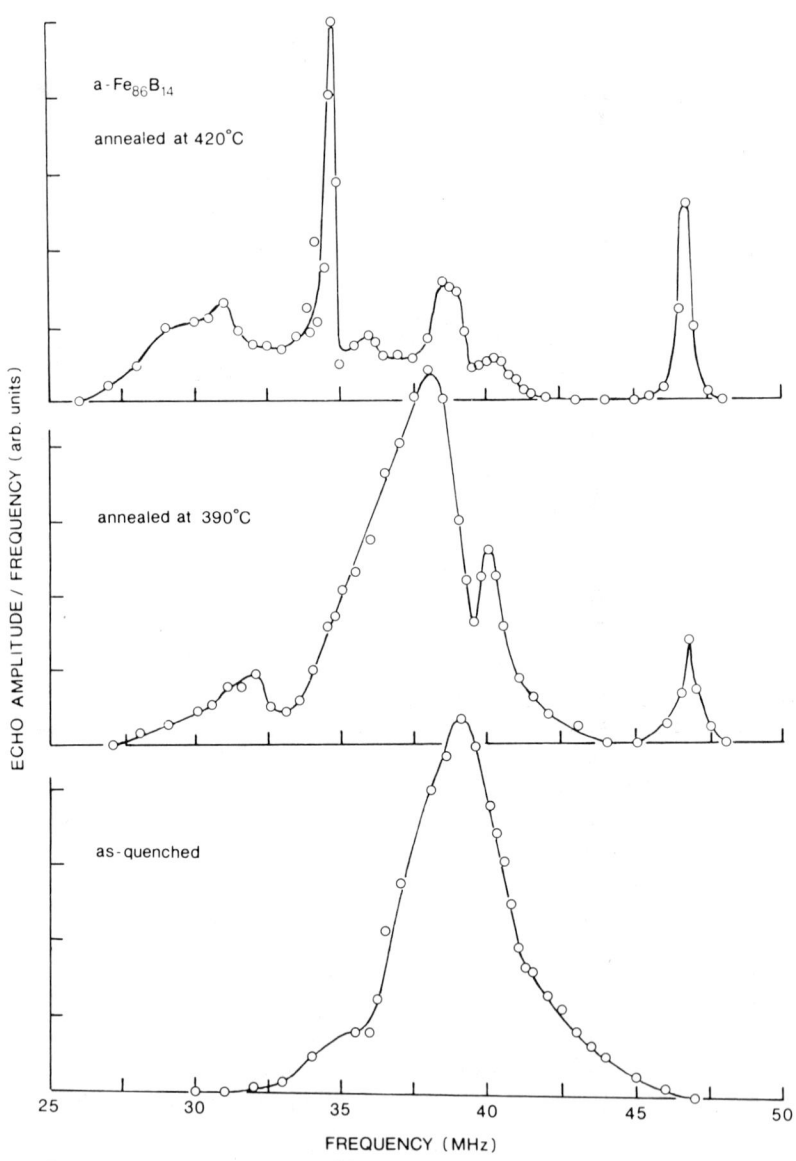

Fig. 8. ^{11}B spin-echo NMR spectra for Fe$_{86}$B$_{14}$ at 4.2 K: (a) as-quenched, (b) annealed at 390°C (40 min) and (c) annealed at 420°C (40 min) [35].

comes clear that as-quenched a-$Fe_{75}B_{25}$ contains mostly bct-Fe_3B SRO.

A systematic NMR and XRD experiment was performed for the a-$Fe_{86}B_{14}$ alloy annealed at various temperatures. It was found that the crystallization process of this alloy started at 390°C. Below 420°C, the XRD shows mainly the amorphous pattern plus bcc α-Fe peaks. Fig. 8 shows the spin-echo NMR spectra obtained at 4.2 K from three $Fe_{86}B_{14}$ samples: as-quenched, annealed at 390°C and annealed at 420°C. For the sample annealed at 390°C, the ^{11}B spectrum became narrower and a peak appeared at 40 MHz which was identified, via a ^{10}B resonance measurement, to be the ^{57}Fe resonance in the o-Fe_3B SRO (see table 1). The spectrum for the sample annealed at 420°C shows three peaks with centers at 34.8, 36.3 and 41.3 MHz, which are associated with the ^{11}B for bct-Fe_3B, o-Fe_3B and Fe_2B, respectively. In addition, a small contribution from the amorphous phase is still present, along with the peak for bcc α-Fe. It is interesting to note that while the XRD shows only the bcc α-Fe pattern, NMR measurements provide strong evidence for the existence of both bct-Fe_3B and o-Fe_3B SRO in the annealed samples, and that the o-Fe_3B microphase is produced first. We note that the o-Fe_3B SRO is never found when annealing high boron concentration amorphous alloys at low temperature. A recent NMR study has also confirmed the formation of o-Fe_3B SRO in low B concentration Fe–B amorphous alloys [45].

The ^{11}B NMR spectra obtained from a-$Fe_{100-x}B_x$ alloys have two significant features: (i) they are very broad in comparison with those from the Fe-borides shown in fig. 2 and (ii) the peak frequency decreases as the B concentration increases. Fig. 9 shows the ^{11}B peak frequency as a function of the B concentration. According to eq. (5), the value of the THF at a B nucleus depends on the magnetic moments of Fe atoms surrounding it. Therefore, the shift of the NMR frequency and broadening of the spectra reflect the variation of Fe magnetic moments in the alloys. Figs. 10a and b show the distributions of Fe magnetic moments for a-$Fe_{86}B_{14}$ and a-$Fe_{76}B_{24}$ obtained from ME experiments [40]. In both figures, the magnetic moments of the Fe

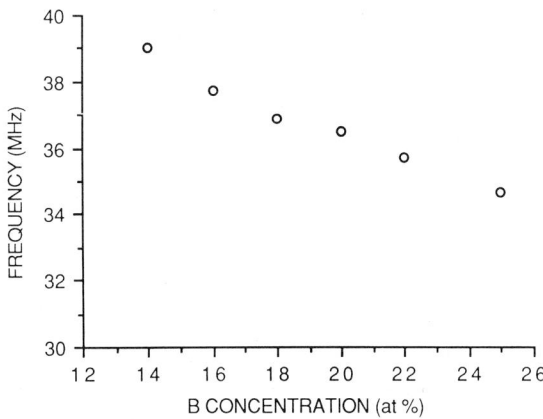

Fig. 9. ^{11}B NMR peak frequency in MHz as a function of B concentration x, in at% for amorphous $Fe_{100-x}B_x$ alloys.

atoms in the o-Fe_3B and bct-Fe_3B phases are marked. Based on the Fe moment distribution, the value of the average Fe magnetic moment can be obtained. For a-$Fe_{86}B_{14}$, this value is 2.05μ_B, which is greater than the average moment of the Fe atoms at the vertices for o-Fe_3B and bct-Fe_3B phases (1.77μ_B for o-Fe_3B, 1.84μ_B for bct-Fe_3B, see table 1). On the other hand, for a-$Fe_{75}B_{25}$,

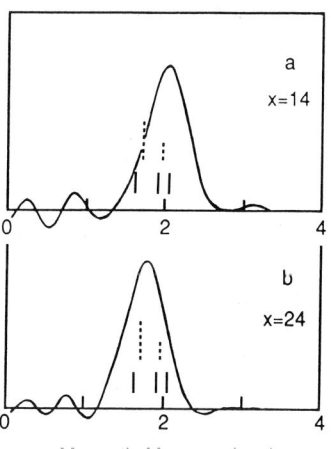

Fig. 10. Distribution of Fe magnetic moments in amorphous $Fe_{100-x}B_x$ alloys obtained from ME experiments at 4.2 K and assuming the hyperfine interaction coupling constant $A = 150$ kOe/μ_B [40]: (a) $x = 14$; (b) $x = 25$. In the figures, the magnetic moment values of of the Fe atoms in the bct-Fe_3B and o-Fe_3B phases are marked: solid lines for the bct-Fe_3B; dashed lines for the o-Fe_3B.

the average Fe moment obtained from fig. 10b is 1.83μ_B, which is approximately equal to the average Fe moment in bct-Fe$_3$B. This may explain why the ^{11}B NMR peak frequencies for the low B concentration alloys are higher than 36.3 MHz (^{11}B in o-Fe$_3$B); also, this may account for the monotonic change of the ^{11}B frequency with B concentration (shown in fig. 9). The spread of the Fe moment values in amorphous alloys and the coexistence of o-Fe$_3$B and bct-Fe$_3$B SRO may be the major reasons leading to the line broadening of the B NMR spectrum. In addition, the spread of the Fe moment values in amorphous alloys makes it difficult to discern the ^{57}Fe resonance.

It can be concluded that NMR studies, together with ME and other investigations, prove the existence of both o-Fe$_3$B and bct-Fe$_3$B SRO in Fe–B amorphous alloys, including a preference for o-Fe$_3$B SRO in the low boron concentration regime and bct-Fe$_3$B SRO in the high boron concentration regime. Perhaps the change in the type preference is not intrinsic, but dependent on the cooling speed of the alloy from the melt during quenching, which may vary with composition of the alloy.

6. Short-range order in amorphous Fe–B–C alloys

Another, perhaps more important, factor affecting the occurrence of o-Fe$_3$B SRO in amorphous Fe–B alloys is the addition of carbon atoms. It has been found that a small amount of C impurity leads to a significant increase of o-Fe$_3$B SRO [65]. This situation also exists in crystalline Fe–B compounds; e.g., stable Fe$_3$B$_{1-x}$C$_x$ compounds can be made to be orthorhombic by the addition of carbon with x greater than 0.1 [39]. As noted earlier, a pure single phase o-Fe$_3$B sample has never been made.

Fig. 11 shows the ^{11}B NMR spectra obtained at 8 K from Fe$_{80}$B$_{20-x}$C$_x$ amorphous alloys (x = 0, 2, 4, 7 and 9) [66]. For Fe$_{80}$B$_{20}$ (x = 0), the ^{11}B NMR shows a broad distribution with a peak at 34.9 MHz, which is assigned to be the ^{11}B in a bct-Fe$_3$B SRO. The spectrum is asymmetric in that the high frequency side has a larger intensity,

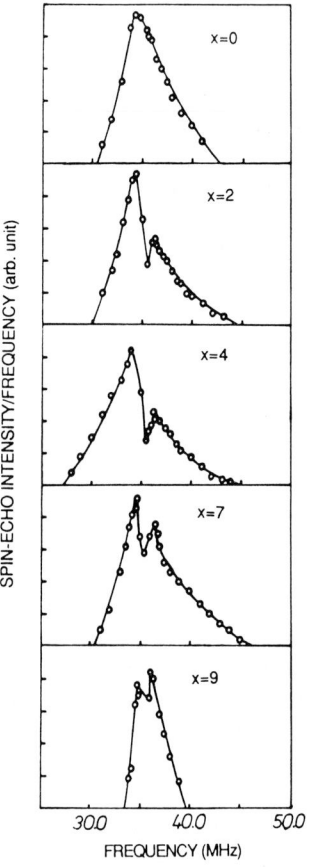

Fig. 11. ^{11}B spin-echo NMR spectra of Fe$_{80}$B$_{20-x}$C$_x$ amorphous alloys at 8 K [92].

implying that the material may contain some amount of o-Fe$_3$B SRO. Upon substitution of carbon atoms for boron atoms, another peak centered at 36.3 MHz appears, which must originate from the ^{11}B in an o-Fe$_3$B SRO. The figure clearly shows that the amount of the o-Fe$_3$B SRO increases significantly as the carbon concentration is increased.

It can be concluded that the appearance of o-Fe$_3$B SRO in Fe–B amorphous alloys depends on: (i) composition, (ii) technological conditions, especially quenching speed, (iii) carbon impurity. Among these factors, carbon is the most important factor. The parameters above may account for the different results which existed in the previous reports concerning the type of SRO in Fe–B amorphous alloys.

7. NMR study of Metglas 2605 CO ($Fe_{67}Co_{18}B_{14}Si_1$)

As described above, considerable research effort has been devoted to amorphous metallic alloys or "metallic glasses" of the general form $TM_{100-x}M_x$. This is due to both a desire to understand the fundamental nature of disordered metallic systems as well as the potential for as wide a variety of technological applications. With regard to the latter, the Fe–B-based amorphous alloys (known as the "Metglas 2605" alloys [67], where 26 and 05 represent the atomic numbers of Fe and B, respectively) have been available commercially for several years. Certainly, the relatively simple $Fe_{100-x}B_x$ binary amorphous alloys described above have served as a prototype system for understanding the more complicated systems containing additional atomic constituents. One system of particular significance is Metglas 2605 CO, which has a composition $Fe_{67}Co_{18}B_{14}Si_1$. This alloy, after the appropriate magnetic annealing treatment, exhibits an enhanced magnetomechanical coupling factor, k_{33}, a factor which is a measure of the efficiency of conversion of magnetic energy into mechanical energy [68]. Because of this property, Metglas 2605 CO is utilized in magnetic transducers. We will review here a series of NMR experiments [69,71], supplemented by magnetization [70,72], Mössbauer [71,73] and X-ray diffraction measurements [74], which contributed to an understanding of the nature of the SRO in this alloy and how it is influenced by magnetic annealing treatments. This work also demonstrates how NMR spectra can be unfolded through the use of more than one isotope, and thereby be used to identify the structure and composition of various phases.

Three Metglas 2605 CO samples were constructed with the following treatments: (i) no annealing treatment, (ii) annealing for 10 min at 369°C in a magnetic field of 6.1 kOe which lies in the plane of the ribbon and transverse to its length, and (iii) annealing for 15 min at 427°C in a magnetic field of 6.1 kOe which lies in the plane of the ribbon and transverse to its length. The sample prepared from the original material, (i) above, is designated "as-quenched" or "ASQ". The sample prepared by process (ii) is designated "annealed" or "ANN", and it is this treatment which yields the large value of k_{33} [68]. The sample prepared by process (iii) is designated "crystalline" or "XTL", and has a drastically reduced value of k_{33}. For all three samples, detailed spin-echo NMR spectra (and, consequently, the HF distributions) were measured for

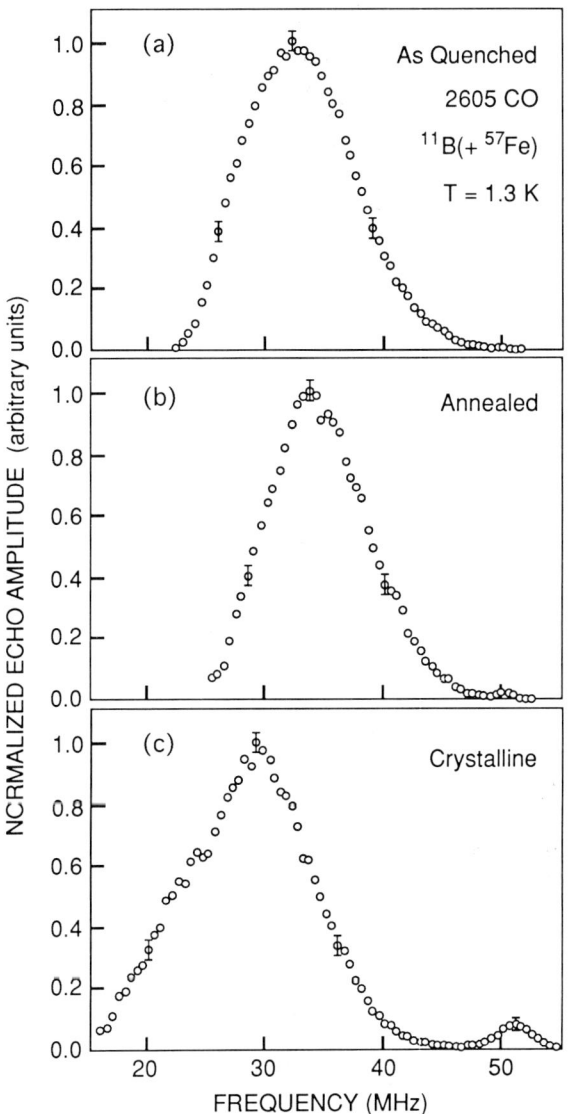

Fig. 12. $^{11}B(+^{57}Fe)$ corrected/normalized spin-echo spectra obtained at 1.3 K for the Metglas 2605 CO samples: (a) ASQ, (b) ANN and (c) XTL.

the ^{10}B and ^{11}B nuclei at 1.3 K (except only a small portion of the ^{10}B spectrum for the XTL sample was observed), and for ^{59}Co nuclei at 77 and 4.2 K. The ^{10}B spectra were observed over a frequency range from 10 to 15 MHz. The so-called ^{11}B spectra, which were found to exist over the frequency range from 20 to 50 MHz, while being due principally to the ^{11}B nuclei, also contain a contribution from the ^{57}Fe nuclei. For accuracy and clarity, the spectra obtained over this frequency range will be denoted ^{11}B ($+^{57}$Fe). The ^{59}Co spectra were observed over the frequency range from 120 to 320 MHz. For all of the Metglas 2605 CO NMR spectra presented here, the spin-echo amplitude has been normalized with a calibration signal and corrected for the frequency, thus providing a true representation of the HF distributions. For a complete discussion on the excitation conditions, see ref. [69].

Figs. 12a, b and c show the corrected/normalized ^{11}B($+^{57}$Fe) spin-echo NMR spectra obtained at 1.3 K for the (Metglas 2605 CO) ASQ, ANN and XTL samples, respectively. Figs. 13a and b show the corrected/normalized ^{11}B spin-echo NMR spectra obtained at 1.3 K for ASQ and ANN samples, respectively. In order to make a direct comparison with the corresponding spectra in fig. 12, the frequency scale for the spectra shown in fig. 13 have been multiplied by a correction factor reflecting the gyromagnetic ratios of ^{11}B and ^{10}B, i.e., gamma(^{11}B)/gamma(^{10}B) = 2.987. Only a small ^{10}B resonance was observed for the XTL sample at 10 MHz. From a consideration of the ^{59}Co spectrum (see below), it is believed that most of the ^{10}B spectrum for the XTL sample exists below 10 MHz, which is the lower limit of the NMR spectrometer. The ^{57}Fe spectra (shown in figs. 14a and b for the ASQ and ANN samples respectively) were obtained by subtracting the ^{10}B spectra from the corresponding ^{11}B($+^{57}$Fe) spectra, and represent the ^{57}Fe HF distribution (see ref. [69] for details). For a complex system such as Metglas 2605 CO, the subtraction process can be quite difficult and, as can be seen in fig. 14, there is considerable uncertainty in the ^{57}Fe HF distribution. Nevertheless, some peaks were clearly evident, while others can only be designated as possible. The corresponding HF peak and width values for ^{10}B (and ^{11}B) as well as ^{57}Fe are summarized in table 2. Finally, figs. 15a, b and c show the corrected/normalized ^{59}Co spin-echo NMR spectra obtained at 4.2 K for the ASQ, ANN and XTL samples, respectively. Although not shown here, spectra were also obtained at 77 K. For the ASQ and ANN samples, four peaks are clearly evident and the corresponding HF peak and width values for ^{59}Co are summarized in table 3. As discussed below, a consideration of the clearly resolved peaks in the ^{59}Co spectra obtained for these samples supported the identification of corresponding peaks in the somewhat noisy ^{57}Fe spectra.

The composition of Metglas 2605 CO ($Fe_{67}Co_{18}B_{14}Si_1$), is similar to the composition of the Fe–B amorphous alloys discussed above, but with

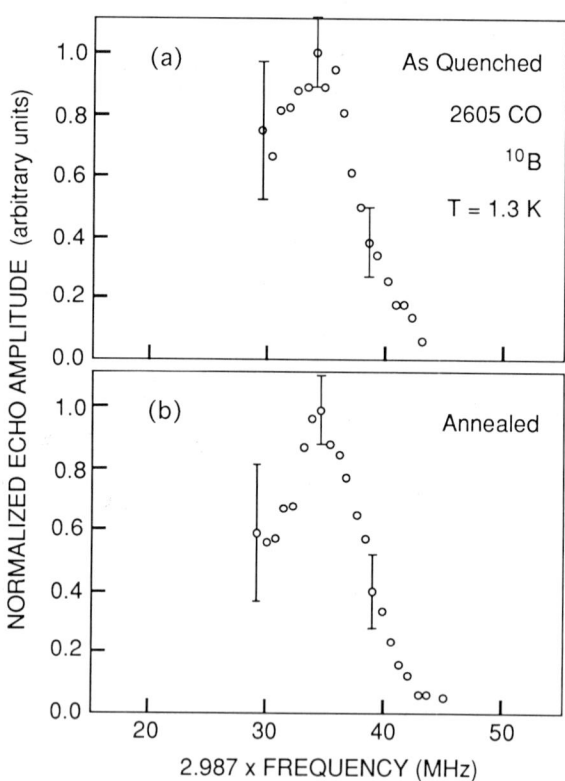

Fig. 13. ^{10}B corrected/normalized spin-echo NMR spectra obtained at 1.3 K for the Metglas 2605 CO samples: (a) ASQ and (b) ANN. In order to make a direct comparison with the spectra in fig. 12, the frequency scale has been multiplied by a correction factor reflecting the gyromagnetic ratio of ^{10}B and ^{10}B, i.e., gamma(^{11}B)/gamma(^{10}B) = 2.987.

about 21% of the Fe atoms replaced by Co and about 7% of the B atoms replaced by Si. It is well known that a replacement of Fe by another TM (such as Co or Ni) in many crystalline alloys (e.g., $Fe_{3-x}TM_xSi$, see ref. [75], results in a direct substitution of the TM atoms for the Fe atoms at the Fe sites. Similarly, in an alloy such as $Fe_3B_{1-x}C_x$, the second metalloid substitutes directly for the first metalloid at the metalloid sites without a drastic change in the structure [39]. Since there is very little Si content in Metglas 2605 CO, it is reasonable to approximate the composition as $(Fe_{0.8}Co_{0.2})_{85}B_{15}$ and assume that the local atomic structure can be compared to that of the Fe–B amorphous alloys of similar composition. If the Co atoms substitute randomly for the Fe atoms, and the transferred HF contribution, H_{sp}, is small compared to the on-site moment contributions, $H_{cp} + H_s$, then it might be

Table 2
Boron and iron hyperfine field peak and width values (in kOe) for Metglas 2605 CO

Nucleus		As-quenched	Annealed	Crystal-lized
^{10}B, ^{11}B	HF [a]	24.6 ± 0.2	25.2 ± 0.2	< 22
	ΔHF [a]	5.5 ± 0.4	6.0 ± 0.4	?
^{57}Fe	HF	217 s [b]	234 s	210 ± 3 s
		250 w	(255)	370 m
		(266)	283 w	
		277 m	299 m	
		(293)	(310)	
		315 m	321 w	

[a] HF = peak values; ΔHF = halfwidth values.
[b] Refers to the relative intensity of the peak; s = strong, m = medium, and w = weak; ( ) were questionable peaks.

expected that the ^{57}Fe HF values listed in table 2 are directly proportional to those obtained for ^{59}Co listed in table 3. Indeed, this is the case and the scale factors are 1.32 and 1.33 for the ASQ and ANN samples, respectively. The nearly constant scale factors were the first experimental indication that the Co atoms reside in an environment very similar to that of the Fe atoms. This result was later confirmed by XANES and EXAFS measurements [76]. It might be noted that the moment assignments of 2.1 and 1.7μ_B for the Fe and Co atoms, respectively, which were obtained from magnetization measurements on Metglas 2605 CO, have a ratio of 1.2. The HF at the non-magnetic ^{10}B (and ^{11}B) sites is entirely due to a transferred HF contribution, H_{sp} from the neighboring TM atoms. Assuming that Co is substituted randomly for Fe and using the moment assignments above, the B HF can be estimated. Using the HF coupling constant of 12.9 kOe/μ_B obtained for the transferred HF from Fe in the amorphous $Fe_{85}B_{15}$ alloy [69], and 7.32 kOe/μ_B obtained for the transferred HF from Co in the crystalline Co_3B alloy [77], and taking into account the relative Fe and Co compositions, a value of 24.2 kOe is obtained for the B HF. This is in good agreement with the measured values for the ASQ and ANN samples listed in table 2. Comparing the ANN and the ASQ samples, the increase in the B HF of 2.4% is in good agreement with, the increase of 2.6% which is

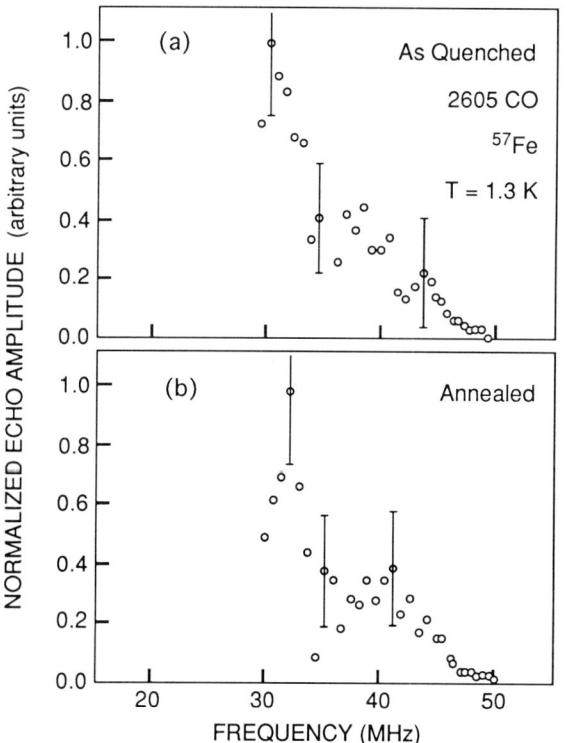

Fig. 14. ^{57}Fe hyperfine field distribution obtained at 1.3 K by subtracting the spectra in fig. 13 from the corresponding spectra in fig. 12 for the Metglass 2605 CO samples: (a) ASQ and (b) ANN.

observed for the saturation magnetization [70]. Finally, it is interesting to compare the ^{57}Fe HF values obtained for the ASQ and ANN samples (see table 2). For the ANN sample, the lower three values (234, 283 and 299 kOe) are within 2% of those for bct-Fe$_3$B. The three corresponding values for the ASQ sample also scale with those for bct-Fe$_3$B, although they are about 10% smaller than the ANN values. In view of the small changes in the B HF (2.4%) and the saturation magnetization (2.6%) that are caused by the magnetic annealing, this relatively large difference is not understood at the present time. It has been suggested that the ^{57}Fe HF results indicate that a bct-Fe$_3$B-like SRO exists in both the ASQ and ANN samples, and that the lower three values referred to above are due to the Fe atoms at the vertices of the tetragonal prism configurations [69]. The highest value, which is 321 kOe for the ANN samples, has been attributed to those Fe atoms in regions between the prisms which have no B atoms as first near neighbors [69].

Longworth and Budnick [73] have measured the ^{57}Fe HF distribution for the Metglas 2605 CO samples at 4.2 K using Mössbauer spectroscopy. They observed extremely broad HF distributions with average values of 306, 305 and 257 kOe for ASQ, ANN and XTL, respectively, and attributed these distributions to the amorphous phase in the bulk of the sample. (The NMR spectra presented above are characteristic of the bulk.) However, their results showed that there was another phase present to some degree in all of the samples. This other phase was characterized by a ^{57}Fe HF value of 371 kOe and was attributed to bcc α-Fe with about 20% Co substituted for Fe. This phase, denoted bcc-Fe(Co), was present in ASQ only on the surface of the sample that was in contact with the rapidly spinning wheel during the melt spinning process (denoted "wheel side"), making up about 7% of the surface composition. The bcc-Fe(Co) phase was present in ANN on the wheel side surface (24%). on the opposite surface (17%) and in the bulk (4%). The bcc-Fe(Co) phase was present in XTL on the wheel side surface (49%), on the opposite surface (46%), and in the bulk (45%). The ^{57}Fe HF peak at 370 kOe measured by NMR for the XTL sample is in excellent agreement with that of 371 kOe measured by the Mössbauer experiments, and is due to this crystalline phase, i.e., Fe$_{80}$Co$_{20}$. The corresponding ^{59}Co NMR peak at 292 kOe confirms this identification. This result demonstrates the usefulness of NMR, as well Mössbauer spectroscopy, in identifying the composition of a phase. It should be noted that the lattice d-spacings for bcc α-Fe which are measured by conventional X-ray diffraction techniques are insensitive to the substitution of up to 30% Co. However, the HF values are strongly dependent on the amount of Co which enabled a value of 20% to be obtained for the Co content of the bcc-Fe(Co) phase in the XTL sample. NMR has the additional advantage that both the ^{57}Fe HF and ^{59}Co HF can be easily used for this purpose. The other ^{57}Fe and ^{59}Co HF peaks for the XTL sample are attributed to the residual amorphous phase with altered composition.

Table 3
Cobalt hyperfine field peak and width values (in kOe) for Metglas 2605 CO

Temperature		As-quenched	Annealed	Crystallized		
77 K	HF [a]	213 ± 2.5	213 ± 2.5	139 ± 2.5,	292 ± 2.5	
	ΔHF [a]	104 ± 5	104 ± 5	55 ± 5,	39 ± 2.5	
4.2 K	HF	⎧163 ± 2.5 ⎪188 ± 2.5 ⎨213 ± 2.5 ⎩238 ± 2.5	⎧178 ± 2.5 ⎪213 ± 2.5 ⎨223 ± 2.5 ⎩238 ± 2.5	139 ± 2.5,	149 ± 2.5	292 ± 2.5
	ΔHF	104 ± 5	114 ± 5	72 ± 5,	33 ± 5	

[a] HF = peak values; ΔHF = halfwidth values.

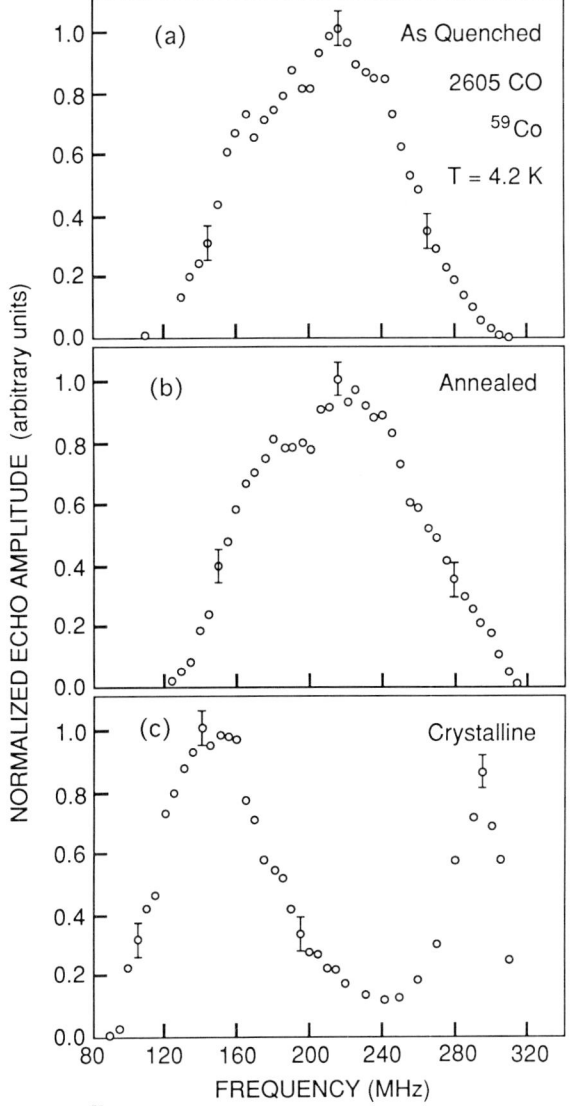

Fig. 15 ^{59}Co corrected/normalized spin-echo NMR spectra obtained at 4.2 K for the Metglas 2605 CO samples: (a) ASQ, (b) ANN and (c) XTL.

boride phase with a strong preferred orientation that approximately matched bct-Fe$_3$B. Finally, Hayes et al. [72] measured the magnetic anisotropy using a low field magnetization technique. A value of about 1800 erg/cm^3 was obtained for the magnetic anisotropy in the ASQ sample, with the easy axis parallel to the ribbon length, while the ANN sample had a value of 4300 erg/cm^3, with the easy axis perpendicular to the ribbon length. A third sample was annealed under similar thermal conditions as that of the ANN sample, but in zero magnetic field. It was found that this sample exhibited essentially the same magnetic anisotropy as the ASQ sample; however, its saturation magnetization was the same as that for the ANN sample. This result would suggest that the thermal anneal does result in some atomic rearrangement, while the presence of the magnetic field during annealing causes a further modification of the local atomic structure, as evidenced by the NMR results, and is responsible for the increase in magnetic anisotropy and the high value of the magnetomechanical coupling factor, k_{33}.

8. Hyperfine fields at the Fe nuclei in Fe–B amorphous alloys: a comparison between NMR and ME

As far as Fe–B amorphous alloys are concerned, the combination of the information obtained from ^{11}B NMR measurements and the information obtained from the ^{57}Fe ME measurements is particularly useful in order to obtain a complete description. For instance, figs. 16a and b show the results from studies of NMR and ME on the as-quenched, 390°C annealed and 420°C annealed Fe$_{86}$B$_{14}$ samples. The ^{11}B NMR spectra show that the alloy ribbons have already started to nucleate crystalline regions at 390°C with a growth of o-Fe$_3$B SRO. They partially crystallize after annealing at 420°C, decomposing into bct-Fe$_3$B, o-Fe$_3$B and bcc α-Fe crystallities. On the other hand, the ^{57}Fe ME spectra, even for the sample annealed at 420°C, consists of only two sextets. One of them is the typical amorphous pattern, corresponding to a mean HF of 254–258

Choi et al. [74], through analysis of their X-ray diffraction data, also found that a thin layer of surface crystallinity existed on the wheel side of the ASQ sample. This layer was identified as consisting of bcc α-Fe crystallites (with an undetermined Co content) oriented in a fiber texture configuration, and small amounts of an Fe–Co-

kOe at room temperature; while the other corresponds to bcc α-Fe [60,78]. A similar result occurred for the as-quenched $Fe_{80}B_{20-x}C_x$ amorphous alloys. As shown in fig. 11, the ^{11}B NMR spectra clearly show o-Fe_3B and bct-Fe_3B SRO, while the ME measurements on all five samples still give a amorphous pattern [79]. By combining the information provided by the ^{11}B NMR and by the ^{57}Fe ME experiments, a complete description of the SRO for the Fe–B amorphous alloys can be determined as follows. Since there are strong covalent interactions between the central B atom and adjacent Fe atoms, the B atom remains bonded with its nearest Fe neighbors in the amorphous state, very much like that in the compositionally related crystalline state. Therefore, the resolved ^{11}B peaks which are characteristic of the corresponding atomic configuration can still be observed. The similarity of the near neighbor atomic configurations of the boron atoms in the amorphous alloys and in the crystalline alloys makes B NMR useful in identifying SRO structure. On the other hand, since this configuration only involves a relatively small number of Fe atoms and each SRO region is independent from one another, a large majority of the Fe atoms are still involved in the amorphous state, therefore, the ME measurements result in an amorphous pattern.

It turns out from above comparison that the NMR technique is crucial for exploring very local atomic environments and, once again, the recent studies show the importance of a combined application of ME and NMR. By using ME alone, the existence of SRO for low B concentrations in the Fe–B system may not be revealed; whereas based on the NMR results alone, some amorphous alloys like a-$Fe_{80}B_{20-x}C_x$ might be mistakenly identified to be in a crystalline structure.

9. Discussion of the origin of SRO in (Fe–B)-based alloys

There is a variety of magnetic binary and ternary alloys, whose bulk magnetic properties depend not only on the composition, but on the distribution of the constituent atoms. Consider

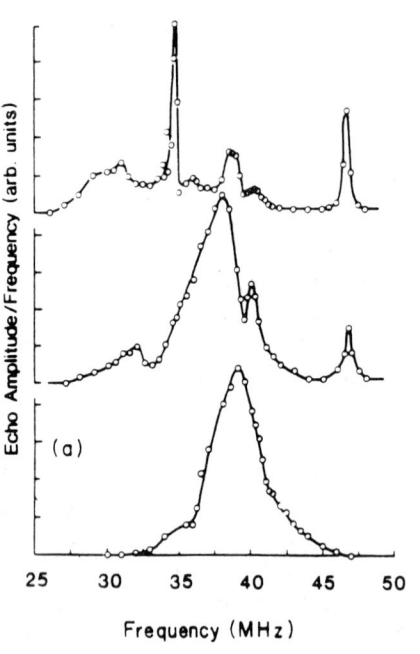
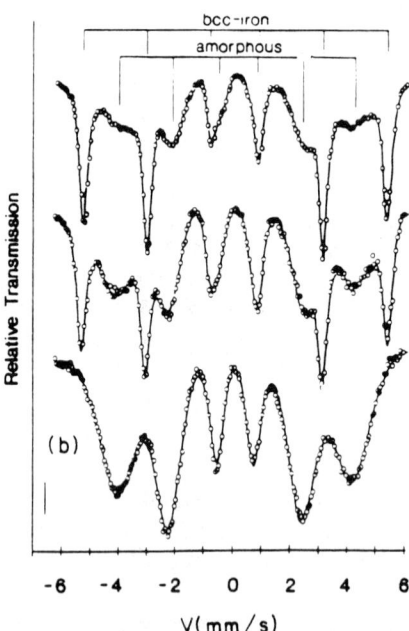

Fig. 16. Comparison of the spin-echo NMR spectra at 4.2 K (a) and the room temperature Mössbauer spectra (b) for the $Fe_{86}B_{14}$ amorphous alloy. From bottom to top: as-quenched, annealed at 390°C and annealed at 420°C.

the dilute binary alloy system $A_{100-x}B_x$, where A represents a solvent species and B a solute species. According to a statistical model, B atoms are located in the lattice formed by A atoms, either substitutionally or interstitially, in a statistically random way when the alloy is quenched from the melt. However, unless the B atoms are the same as the A atoms, both in nature and in size, there must exist either attractive or repulsive forces between the different species atoms which cause a preference in the arrangement, either to surround or avoid one another. When the interatomic interaction between the A and B atoms is small in comparison with the thermal vibration energy of the atoms in the liquid state, an approximately disordered distribution can be established; and, furthermore, if the quenching process is fast enough such that diffusion cannot occur, a disordered alloy can be obtained. However, it has been found that a large number of magnetic alloys are not actually disordered, this kind of ordering is called chemical short range order (CSRO). There have been extensive investigations on the types and features of CSRO and their effect on the magnetic properties of the alloys. NMR has been successfully employed in characterizing CSRO [42,80–82].

With regard to the local atomic configuration of an Fe–B alloy, or a TM–B alloy in general, the situation is quite different. Here, the most important factor is that the interaction between the B and TM atom is so strong that it cannot be destroyed even at temperatures much higher than its melting point. Concerning the amorphous structure, the dense random packing of hard spheres (DRPHS) proposed by Bernal [83–85] and the trigonal prism packing proposed by Gaskell [86,87] are the models used most often. In the DRPHS model, the relatively big TM atoms pack randomly, leaving interstitial voids for the small metalloid atoms. In the trigonal prism model, the basic structural unit is a somewhat distorted trigonal prism very much like that shown in fig. 1a, in which the TM atoms occupy the six vertices and the three further neighbors and the metalloid atom is located at the center of the trigonal prism, namely similar to the bct-Fe_3B or o-Fe_3B. There have been considerable discussions of these two models. As has been pointed out in ref. [10], NMR provides information concerning the geometry of the local atomic environment which is directly related to the structure of the amorphous state. As far as the Fe–B system is concerned, ref. [10] NMR experiments have provided evidence for the existence of the trigonal prism as the structural unit in this system. Therefore, the trigonal prism model seems to be a good pattern for the local atomic structure [86,87] for the Fe–B system, or for the TM–M systems in general. The covalent interaction between the B and Fe atoms is strong and anisotropic in nature; it produces a certain kind of ordered local atomic geometry (in accordance with the B–Fe bonding) during melting of the constituents. As soon as the bonding forms, it cannot be destroyed by further heating and it remains in the alloy. The type of SRO depends on the ratio of the TM to M atoms which, in the molten state, determines the symmetry. For a B concentration less than 25%, both o-Fe_3B and bct-Fe_3B SRO are possible; however, o-Fe_3B is very unstable and easy transforms to bct-Fe_3B. High cooling speed as well as C impurity can promote the o-Fe_3B SRO.

One of the most important problems concerning the SRO is whether the SRO existing in a material is similar to its crystalline counterpart. This is dependent on the interaction which results in the SRO. For Fe–B amorphous alloys, the SRO originates from the strong bonding between B and Fe, and this interaction plays the same role in both crystalline and amorphous alloys. This is why the SRO which occurs in amorphous Fe–B alloys is similar to that in their crystalline counterparts.

Here, we would like to emphasize that the trigonal prism structure is formed by the central B atom. A model has been proposed in which Fe–B amorphous alloys possess a local bct-Fe_3B SRO and, as the B concentration is decreased from 25%, some B atoms are replaced by Fe atoms with the bct-Fe_3B like SRO being retained. However, the trigonal prism cannot survive without a B atom sitting at the center so as to provide the bonding required.

The role that Fe–B bonding plays in forming a

structure and, hence, leading to certain magnetic properties can also be noted from the history of the discovery of $Nd_2Fe_{14}B$. A long time was spent searching for hard magnetic properties by making RE–iron alloys, where RE is a rare earth element. This search which was based on the idea of using an RE element for obtaining high anisotropy and using Fe for a high magnetic moment, did not succeeded. It was not until a small amount (about 8%) of B atoms were added to the Nd–Fe system that the desired hard magnetic properties resulted. In the permanent magnetic phase $Nd_2Fe_{14}B$, the B and Fe are again bonded in a trigonal prism configuration frame [88].

10. Effect of the short-range order of amorphous Fe–B alloys on magnetic properties

Perhaps SRO is not favorable for soft magnetic properties in magnetic materials as any kind of order in the atomic configuration may cause magneto-crystalline anisotropy. However, in many circumstances, the existence of SRO cannot be avoided, as is the case for Fe–B alloys. Therefore, a deep understanding of the effect of the SRO on the magnetic properties and thermal stability of the material becomes necessary. There has been considerable work concerning the effects of thermal treatment, alloy constituents, impurities and quenching stress on the magnetic properties of amorphous alloys. It is possible that these factors play a role by influencing the SRO in these alloys. However, it is very difficult to correlate the SRO with the magnetic properties of a material as it requires the application of specific techniques which can provide both microscopic and macroscopic information. The NMR technique has a good potential for solving this problem. In the case of spin-echo NMR, the optimum rf magnetic field required for maximum spin echo signal is determined by considering the following expression [20,89]:

$$\gamma \eta_{d,w} h_e \tau = k\pi, \quad (9)$$

where γ is the gyromagnetic ratio of the nucleus and the constant k is determined by the pulse sequence used. It turns out from this equation that the optimum value of rf magnetic field, h_e, required to excite the nuclei for a maximum spin echo signal is determined by the enhancement mechanism, $\eta_{d,w}$ which is directly related to the ease of magnetization for the material as shown in eqs. (5) and (6) [16]. Therefore, the relative intensities of the NMR sub-spectra can be used to provide a measure of relative amounts of the various SROs, and h_e provides information concerning the magnetic properties.

The relationship between h_e and the magnetic properties of a material has been observed in previous NMR experiments [90–92]. The NMR study of the Fe–B–C amorphous alloys further shows the role that NMR can play in studying the magnetic properties [93]. Magnetic measurements were performed on the as-quenched $Fe_{80}B_{20-x}C_x$ amorphous alloy ribbons, whose ^{11}B NMR spectra are shown in fig. 11. The results indicate that the coercivity of the alloy decreases with increasing C concentration. When these ribbons were annealed at 460 and 800°C, XRD, NMR and ME measurements identified the existence of o-Fe_3B and bct-Fe_3 phase precipitates. As mentioned above, the NMR frequencies for the bct-Fe_3B and o-Fe_3B phases are 34.7 and 36.3 MHz,

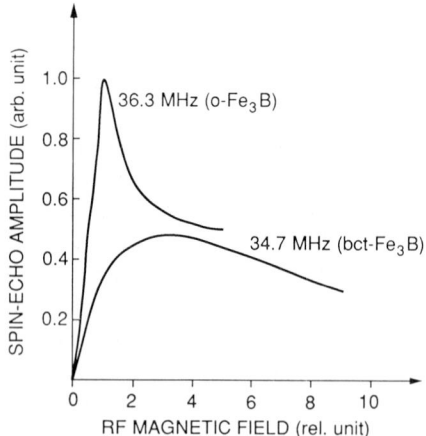

Fig. 17. The ^{11}B NMR signal intensities corresponding to bct-Fe_3B and o-Fe_3B peaks as a function of rf excitation field for $Fe_{80}B_{16}C_4$ annealed at 460°C for 40 min: solid circles, peak at 34.7 MHz (bct-Fe_3B); solid triangles, 36.3 MHz (o-Fe_3B) [92].

respectively. Fig. 17 shows the spin-echo intensity as a function of the rf field for the annealed sample measured at 34.7 and 36.3 MHz. The result indicates that the amplitude of rf field required to get the maximum ^{11}B spin echo signal from the o-Fe$_3$B phase at 36.3 MHz is only about half as much as required for exciting the ^{11}B NMR in the bct-Fe$_3$B phase at 34.7 MHz. This means that the rf field enhancement factor for the o-Fe$_3$B phase aggregate is twice as large as for the bct-Fe$_3$B phase aggregate. Consequently, this leads to a conclusion that the o-Fe$_3$B phase may have lower coercivity. Assuming that this argument is also true for the case of SRO in amorphous alloys, then it is reasonable to attribute the reduction of coercivity with C concentration to an increase in the amount of o-Fe$_3$B SRO (fig. 11).

11. Conclusions

1. SROs exist in both crystalline and amorphous Fe$_{100-x}$B$_x$ alloys. In the case of the (metastable) crystalline alloys ($0 < x < 12$), the type of SRO is o-Fe$_3$B, while in the case of the amorphous alloys ($12 < x < 25$), there exist either o-Fe$_3$B SRO or bct-Fe$_3$B SRO, or a mixture of these two types. The crucial factors influencing the type of SRO existing in the amorphous alloys are carbon content and quenching speed. The fact that low B containing alloys favor o-Fe$_3$B and high B concentration alloys favor bct-Fe$_3$B SRO may be actually related to quenching speed.

2. The SRO originates from the strong B–Fe bonding. As the interaction exists mostly in the nearest B–Fe pairs, the SRO involves only the nearest Fe atom shell surrounding the B atom. The fact that the B atom is located at the center of the SRO (trigonal prism) makes B NMR measurements particularly useful in exploring and characterizing the SRO.

3. NMR experiments, together with magnetization measurements, reveal that short-range order affects the magnetic properties of the Fe–B amorphous alloys. In general, SRO may have significant influences on the application properties, such as soft magnetic properties, mechanical properties and thermal stability. Much work still remains using combined microscopic and macroscopic experimental techniques.

4. Because of the similarity in nature of bondings between B and TM atoms and, generally speaking, between C and TM atoms, and between P and TM atoms, similar SRO may exist in other transition metal borides, carbides or phosphates and (TM–B)-, (TM–C)- and (TM–P)-based binary, ternary crystalline alloys as well as amorphous alloys. NMR studies on these systems are also useful. In fact, a great number of NMR and ME studies have obtained conclusions consistent with the present study on the (Fe–B)-based alloys (e.g., see refs. [10,94]).

In conclusion, we would like to point out that as a kind of microscopic probe technique, NMR has been effective in the following fields: (1) multiphase structures, microphase precipitates and SRO, (2) structures, magnetic moments and atomic distributions in thin films and multilayered films, (3) magnetic structures near the Curie temperature, (4) magnetism and atomic configurations of superfine particles (the authors have carried out some NMR studies on very small Co particles (less than 20 Å in size). The results show that information concerning the magnetic moments, atomic configurations, ferromagnetism–superparamagnetism transition, can be obtained [95]), and (5) the magnetization process.

Acknowledgements

The authors wish to thank Dr. M. Wojcik for many helpful discussions during the preparation of the manuscript. This paper describes the principal results which were obtained over a period of time from a research program that has profited enormously from collaborations with Drs. M. Choi, S.H. Ge, R. Hasegawa, G.H. Hayes, J.D. Livingston, G.E. Longworth, L.T. Kabacoff, D.M. Pease, F.H. Sanchez and D.P. Yang. We deeply appreciate the efforts of all of the co-workers mentioned above for their ideas, useful discussions and technical assistance. We especially thank Dr. S.H. Ge and her colleagues for allowing us to use some of their results and analyses prior to publication.

References

[1] F.E. Luborsky, in: Ferromagnetic Materials, vol. I, ed. E.P. Wohlfarth (North-Holland, Amsterdam, 1980) chap. 6.
[2] R. Kiessling, Acta Chem. Scand. 4 (1950) 209.
[3] U. Herold and U. Koster, Z. Metallkde 69 (1978) 326.
[4] H. Franke, U. Herold, U. Koster and R. Rosenberg, in: Rapidly Quenched Metals, vol. III, ed. B. Cantor (Cameleon, London, 1978) p. 155.
[5] R. Ray and R. Hasegawa, Solid State Commun. 27 (1978) 471.
[6] T. Nakajima, E. Kita and H. Ino, J. Mater. Sci. 23 (1988) 1279.
[7] W. Hoving, F. van der Woude, K.H.J. Buschow and I. Vincze, J. Non-Cryst. Solids 61 & 62 (1984) 421.
[8] K.H.J. Buschow, D.B. de Mooij and R. Coehoorn, J. Less-Common Met. 145 (1988) 601.
[9] P. Panissod, J. Durand and J.I. Budnick, Nucl. Instr. and Meth. 199 (1982) 99.
[10] P. Panissod, Hyperfine Interactions 25 (1985) 607.
[11] R.E. Watson and A.T. Freeman, Phys. Rev. 123 (1961) 2027.
[12] A.J. Freeman and R.E. Watson, in: Magnetism, vol. IIA, eds. G.T. Rado and H. Suhl (Academic Press, New York, 1965) p. 167.
[13] M.A.H. McCausland and I.S. Mackenzie, Adv. Phys. 28 (1979) 305.
[14] V. Niculescu, J.I. Budnick, W.A. Hines, K. Raj, S. Pickart and S. Skalski, Phys. Rev. B 19 (1979) 452.
[15] A.C. Gossard and A.M. Portis, Phys. Rev. Lett. 3 (1959) 163.
[16] A.V. Zalesskij and I.S. Zheludev, Atom. Energy Rev. 141 (1976) 133.
[17] A.M. Portis and R.H. Lindquist, in: Magnetism, vol. IIA, eds. G.T. Rado and H. Suhl (Academic Press, New York, 1965) p. 357.
[18] E.A. Turov and M.P. Petrov, Nuclear Magnetic Resonance in Ferro- and Antiferromagnets (Halstead Press–Wiley, New York, 1972).
[19] P.C. Riedi, Hyperfine Interactions 49 (1989) 335.
[20] I.D. Weisman et al., in: Techniques of Metals Research, eds. E. Passaglia et al., vol. IV, part II, p. 357.
[21] W.Y. Ching, Y.N. Xu, B.N. Harmon J. Ye and T.C. Leung, Phys. Rev. B 42 (1990) 4460.
[22] R.E. Watson and L.H. Bennett, Phys. Rev. B 43 (1991) 11642.
[23] D.J. Joyner, O. Johnson, D.M. Hercules, D.W. Bullett and J.H. Weaver, Phys. Rev. B 24 (1981) 3122.
[24] O. Johnson, D.J. Joyner and D.M. Hercules, J. Phys. Chem. 84 (1980) 542.
[25] I.D. Weisman, L.J. Swartzendruber and L.H. Bennett, Phys. Rev. 177 (1969) 465.
[26] R. Kiessling, J. Phys. Chem. Solids 26 (1967) 17.
[27] W.G. Wyckoff, Crystal Structures (Interscience, New York, 1965) p. 114.
[28] W.K. Choo and R. Kaplow, Metall. Trans. 8A (1977) 417.
[29] E.J. Lisher, C. Wilkinson, T. Ericsson, L. Häggström, L. Lundgren and R. Wappling, J. Phys. C 7 (1974) 1344.
[30] S. Rundqvist, Acta Chem. Scand. 16 (1962) 1.
[31] L. Takacs, M.C. Cadeville and I. Vincze, J. Phys. F 5 (1975) 800.
[32] B. Lemius and R. Kuentzler, J. Phys. F 10 (1980) 155.
[33] R.S. Perkins and P.J. Brown, J. Phys. F 4 (1974) 906.
[34] P.J. Brown and J.L. Cox, Philos. Mag. 23 (1970) 705.
[35] Y.D. Zhang, J.I. Budnick, J.C. Ford, W.A. Hines and F.H. Sanchez, J. Appl. Phys. 61 (1987) 3231.
[36] H. Abe, H. Yasuoka and A. Harai, J. Phys. Soc. Jpn. 21 (1966) 77.
[37] Y.D. Zhang, W.A. Hines, J.I. Budnick, M. Choi, F.H. Sanchez and R. Hasegawa, J. Magn. Magn. Mater. 61 (1986) 162.
[38] J.M. Dubois and G. Le Caer, Acta Metall. 32 (1984) 2101.
[39] Y.D. Zhang, J.I. Budnick, F.H. Sanchez, W.A. Hines, D.P. Yang and J.D. Livingston, J. Appl. Phys. 61 (1987) 4358.
[40] C.L. Chien, D. Musser, E.M. Gyorgy, R.C. Sherwood, H.S. Chen, F.E. Luborsky and J.L. Water, Phys. Rev. B 20 (1979) 283.
[41] V.S. Pokatilov, Sov Phys. Dokl. 29 (1984) 234.
[42] Y.D. Zhang, J.I. Budnick F.H. Sanchez and R. Hasegawa, J. Appl. Phys. 67 (1990) 5870.
[43] Pearson Handbook of Lattice Spacings and Structure of Metals and Alloys (Pergamon Press, New York, 1958); Phys. Met. Metall. 57 (1984) 88.
[44] J.I. Budnick, F.H. Sanchez, Y.D. Zhang, M. Choi, W.A. Hines, Z.Y. Zhang, S.H. Ge and R. Hasegawa, IEEE Trans. Magn. MAG-23 (1987) 1937.
[45] V. Pokatilov and N. Djakonova, Hyperfine Interactions 59 (1990) 525.
[46] F.H. Sanchez, J.I. Budnick, Y.D. Zhang, W.A. Hines and M. Choi, Phys. Rev. B 34 (1986) 4738.
[47] V.S. Pokatilov, Phys. Met. Metall. 57 (1984) 88.
[48] J.L. Walter, S.F. Bartram and R.R. Russell, Met. Trans. 9A (1978) 803.
[49] P. Tlomak, S.J. Pierz, L.J. Paulson and W.E. Brower, Jr., Mater. Sci. Eng. 97 (1988) 369.
[50] E. Zschech, V.K. Fjodorov and M.A. Sheromov, Nucl. Instr. and Meth. A 282 (1989) 586.
[51] G.S. Cargill III, J. Non-Cryst. Solids 61 & 62 (1984) 261.
[52] E. Zschech, W. Blau, K. Kleinstuck, H. Hermann, N. Mattern, M.A. Kozlov and M.A. Sheromov, J. Non-Cryst. Solids 86 (1986) 336.
[53] W. Matz, H. Hermann and N. Mattern, J. Non-Cryst. Solids 93 (1987) 217.
[54] A. Scherer and O.T. Inal, J. Mater. Sci. 22 (1987) 193.
[55] A.K. Bhatnagar, Hyperfine Interactions 24–26 (1985) 637.

[56] C.T. Limbach and U. Gonser, J. Non-Cryst. Solids 106 (1988) 399.
[57] I. Vincze, D.S. Boudreaux and M. Tegze, Phys. Rev. B 19 (1979) 4896.
[58] I. Vincze, T. Kemeny and S. Arajs, Phys. Rev. B 21 (1980) 237.
[59] J.M. Dubois and G. Le Caer, Nuc. Instr. and Meth. 199 (1982) 307.
[60] F.H. Sanchez, Y.D. Zhang, J.I. Budnick and R. Hasegawa, J. Appl. Phys. 66 (1989) 1671.
[61] J.G. Zhao, B.G. Shen and W.S. Zhan, J. Magn. Magn. Mater. 50 (1985) 119.
[62] J.C. Ford, J.I. Budnick, W.A. Hines and R. Hasegawa, J. Appl. Phys. 55 (1984) 2286.
[63] V.S. Pokatilov, Dokl. Akad. Nauk SSSR 290 (1986) 345.
[64] P. Svec and P. Duhaj, Mater. Sci. Eng. B 6 (1990) 265.
[65] S. Arajs, R. Caton, M.Z. El-Gamal, L. Granasy, J. Baloggh, A. Gziraki and I. Vincze, Phys. Rev. B 25 (1982) 127.
[66] M.X. Mao, S.H. Ge, Z.H. Cheng, G.L. Cheng, C.L. Zhang and Y.D. Zhang, to be published.
[67] Metglas is a registered trademark of Allied Corp.
[68] C.U. Modzelewski, H.T. Savage, L.T. Kabacoff and A.E. Clark, IEEE Trans. Magn. MAG-17 (1981) 2837.
[69] J.C. Ford, PdD thesis, Univ. of Connecticut (1988).
[70] J.C. Ford, W.A. Hines, J.I. Budnick, A. Paoluzi, D.M. Pease, L.T. Kabacoff and C.U. Modzelewski, J. Appl. Phys. 53 (1982) 2288.
[71] J.C. Ford, W.A. Hines, J.I. Budnick, M. Choi, H. Hayes, G.E. Longworth, D.M. Pease and D.P. Yang, J. Magn. Magn. Mater. 54-57 (1986) 245.
[72] G.H. Hayes, W.A. Hines, D.P. Yang and J.I. Budnick, J. Appl. Phys. 57 (1985) 3511.
[73] G.E. Longworth and J.I. Budnick, AERE Tech. Rep. R-11526 (1984).
[74] M. Choi, D.M. Pease, W.A. Hines, J.I. Budnick, G.H. Hayes and L.T. Kabacoff, J. Appl. Phys. 54 (1983) 4193.
[75] V.A. Niculescu, T.J. Burch and J.I. Budnick, J. Magn. Magn. Mater. 39 (1983) 223.
[76] G.H. Hayes, J.I. Budnick, M. Choi, S.M. Heald, D.M. Pease and D.E. Sayers, Bull. Am. Phys. Soc. 29 (1984) 515.
[77] M. Wojcik, H. Lerchner, P. Deppe, F.S. Li, M. Rosenberg and J.D. Livingston, J. Appl. Phys 55 (1984) 2288.
[78] F.H. Sanchez, Y.D. Zhang and J.I. Budnick, Phys. Rev. B 38 (1988) 8508.
[79] S.H. Ge, G.L. Cheng, F.S. Li, D.S. Xiua and Y.D. Zhang, to be published.
[80] A.I. Gusev, Phys. Stat. Sol. (b) 156 (1989) 11.
[81] V. Pierron-Bohnes, M.C. Cadeville and F. Gautier, J. Phys. F 13 (1983) 1689.
[82] V. Pierron-Bohnes, I. Mirebeau, E. Balanzat and M.C. Cadeville, J. Phys. F 14 (1984) 197.
[83] J.I. Finney, Proc. Roy. Soc. A 319 (1970) 497.
[84] J.D. Bernal, Proc. Roy. Soc. A 280 (1964) 299.
[85] R.D. Watts and I.J. McGee, Liquid State Chemical Physics (Wiley, New York, 1976).
[86] P.H. Gaskell, J. Non-Cryst. Solids 32 (1979) 207.
[87] P.H. Gaskell, Acta Metall. 29 (1981) 1203.
[88] J.F. Herbst, J.J. Croat, F.D. Pinkerton and M.B. Yellon, Phys. Rev. B 29 (1984) 4178.
[89] E.L. Hahn, Phys. Rev. 80 (1950) 580.
[90] P. Panissod, A. Qachaou, J. Durand and R. Hasegawa, Nucl. Instr. and Meth. 199 (1982) 231.
[91] M.B. Sterns, Phys. Rev. 162 (1967) 496.
[92] K.M.B. Alves, N. Alves, L.C. Sampaio, S.F. da Cunha and A.P. Guimaraes, J. Appl. Phys. 67 (1990) 5897.
[93] S.H. Ge, G.L. Cheng, M.X. Mao, Y.D. Zhang, W.A. Hines and J.I. Budnick, Phys. Rev. B (submitted).
[94] P. Panissod, D.A. Guerra, A. Amamou, J. Durand, W.L. Johnson, W.L. Carter and S.J. Poon, Phys. Rev. Lett. 44 (1980) 1465.
[95] To be published.

Rare earth–cobalt permanent magnets

Karl J. Strnat

KJS Associates, Dayton, OH 45403 and University of Dayton, Dayton, OH 45469, USA

and

Reinhold M.W. Strnat

KJS Associates, 1712 Springfield St., Dayton, OH 45403, USA

This paper reviews the historical background and the development of rare earth–cobalt-based permanent magnets from basic science studies on rare earth–transition metal alloys in the 1960's to today's broad spectrum of commercial magnet types and their applications. It puts the RE–Co magnets in perspective relative to older magnet types and also traces the path to the subsequent development of the related Nd–Fe–B magnets. The treatment is qualitative, with emphasis on the relationship between fundamental properties of the compounds and the interaction between microstructure and magnetic domain walls that makes high coercivity and the exceptional hard magnetic properties of the rare-earth magnets possible. The various kinds of RE–Co magnets in production and use today, some of their engineering properties, and economic aspects governing their applicability, cost and availability are also discussed. Many references provide a guide to the special literature regarding the physics, metallurgy, manufacture, product selection and properties of rare earth–cobalt magnets.

1. Introduction

The family of "rare-earth permanent magnets" (REPM) has evolved in the last 25 years. The REPM quickly surpassed all earlier magnets, with current best values of energy product and coercivity both being 5–10 times those of alnicos and ferrites. While they are too costly for universal use, the REPM – together with the hard-magnetic ferrites – are rapidly broadening applications for permanent magnets in general [1–4]. As a commercial product, rare-earth magnets had their start in the mid-1970s, about 20 years after the hexaferrites.

The REPM are metal magnets. Their magnetic components are alloys of 3d-transition metals (TM) with elements of the rare-earth group (RE or R), i.e. the 4f-elements La (57) through Lu (71) and Y (39). The RE in fig. 1 are used in magnets with Co as the main constituent. We

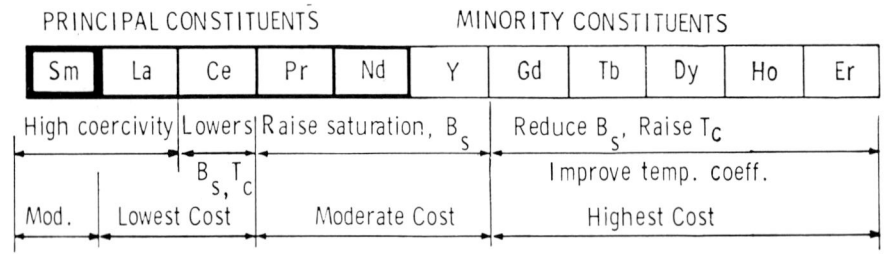

Fig. 1. Rare-earth elements used in RE–Co magnets. The influence of the RE on basic magnetic properties (relative to Sm) and comparative cost of the RE component are indicated [1].

presently distinguish two major subgroups: RE–Co-based magnets (especially Sm–Co) and RE–Fe-based magnets typified by Nd–Fe–B. The latter "third generation" REPM, in rapid development since 1983, use cheaper, more plentiful raw materials than Sm–Co and offer still higher energy density at room temperature (r.t.); but they also have greater stability problems at room and elevated temperatures. This paper is concerned with the RE–Co magnets. The RE–Fe–B magnets are the topic of a companion paper [5].

The technical and scientific literature about the REPM has become very extensive. We must rely heavily on review articles quoted here as guides to original work, also some conference proceedings and dedicated journal issues. A continuing series of "International Workshops on Rare-Earth Permanent Magnets and their Applications" and complementary "Symposia on Coercivity and Anisotropy" has produced eleven sets of books chronicling the REPM development [6–16].

The REPM are the latest fruits of about 60 years of efforts to identify new substances qualified for permanent magnet (PM) use. Candidate materials must have: (1) a high spontaneous magnetisation, M_s, at least around normal r.t.; (2) a Curie temperature, T_c, high enough for the contemplated device use. (The lowest T_c values now accepted by design engineers are near 300°C, for Nd–Fe–B and Mn–Al–C); (3) a mechanism for creating a sufficiently high intrinsic coercive force, H_{ci}. "Sufficient" is now usually defined as $H_{ci} \geq 0.5 B_{is}$, the minimum permitting a close approach to the theoretical energy product limit, $(2\pi M_s)^2 - (0.5 B_{is})^2$.

Fe–Co alloys with 30–50% Co have the best known combinations of B_{is} at 25°C and high T_c. The theoretical energy limit for fully dense Fe–Co with $B_{is} = 24.5$ kG is about 150 MGOe. But the highest H_{ci} attainable with single-domain Fe–Co in Alnico and elongated single-domain (ESD) magnets, using shape anisotropy as the energy barrier, was only ≈ 2000 Oe (160 kA/m). Approaching 150 MGOe would require a six times higher H_{ci}. Further PM development along these lines had to strive for magnet materials based on Fe–Co with $H_{ci} > 12$–13 kOe.

Magnetocrystalline anisotropy offers a better way than particle shape to generate high coercivity. Ferromagnetic crystals having a single easy axis of magnetisation, as small "single-domain particles", should have an H_{ci} about equal to their anisotropy field, $H_A = 2K/M_s$. The 1950s and 1960s brought a systematic search for candidate substances having strong crystal anisotropy. The hexagonal ferrites were the first practical materials of this kind and have found very broad commercial application; but they are oxides having low M_s. The RCo_5 group of intermetallic compounds was next and continuing research led to the broader family of rare-earth magnet alloys. Generally, RE additions were found to strongly enhance crystal anisotropy.

2. Early work on magnetic rare earth–transition metal alloys

Until the 1950s individual rare earths in metallic form were almost unavailable and of no industrial interest. The significance of the RE in nuclear fission caused the development of a technology for their separation and reduction which made the elemental metals commercially available and brought on a surge of scientific interest in RE alloys. During the 15-year period before 1965 many RE–TM compounds were prepared and systematic studies of their properties began [17]. Investigations of the magnetic ordering in RE–TM combinations yielded information important for the later magnet development. Initially, groups at Bell Telephone Laboratories [18], the University of Pittsburgh [19] and the US Naval Research Laboratories [20] were particularly active. Hubbard et al. reported that $GdCo_5$ has uniaxial anisotropy and that a powder had a coercive force of 8 kOe (0.64 MA/m). This was a first indication that RCo_5 compounds held promise for permanent magnets, but it was generally ignored because the low saturation of $GdCo_5$ disqualifies it as a practical magnet material.

Other laboratories joined the systematic study of RE–TM magnetism: CNRS Grenoble [21], the USAF Materials Laboratory [22] jointly with the University of Dayton [23], the Technical University of Vienna [24] and Philips Research Laboratories [25]. While the main interest was in basic magnetic properties, all had to study phase diagrams and the crystal structures of many intermetallic phases which were then unknown or poorly described.

A review of early results identified the RCo_5 in which R is a "light" RE (Ce, Pr, Nd, Sm) or Y as possible new PM materials [27,28]. Some R_2Co_{17} phases were soon added to the list [21,26] and so was $LaCo_5$ [25,32]. Studies of the anisotropy using single crystals [28–30] or oriented powders [31,32] showed that YCo_5 and most other RCo_5 compounds have an extremely large crystal anisotropy with a single easy axis of magnetisation. This led to the prediction that these could indeed become outstanding new permanent magnets [27,28,31]. Finding an unfavorable easy-basal-plane anisotropy for Y_2Co_{17} [28] initially discouraged consideration of 2–17 alloys for magnets until the R_2Co_{17} with Sm, Er and Tm, were also shown to be anisotropic with an easy axis [33–35].

Fine-particle theory suggested that micron-size powders of these RCo_5 and R_2Co_{17} compounds should have very high coercivity. Aligned compacts should be good magnets. But the practical realization of useful REPM was not so simple. In powders produced by mechanical grinding, H_{ci} reached only 5–10% of the anisotropy field. This was adequately high only for $SmCo_5$, so the initial magnet development concentrated on this compound [31]. Compacted powder magnets were prepared in several laboratories, but they were not very stable.

3. History of the RE–Co magnet development

The development of sintering techniques [36,37] finally made fully dense and stable $SmCo_5$ magnets possible and became the basis for the first commercial REPM. One can also use as the RE-component Ce, Pr, Nd and La – singly or in combinations including the natural RE blend known as mischmetal. But to consistently obtain high H_c, it is necessary to include some Sm in the mixture. It was further discovered that the partial substitution of Cu for Co in $SmCo_5$ and in $CeCo_5$ allowed formation of a precipitate in the RCo_5 matrix that causes usefully high coercive force without comminution [38,39]. It was first thought that the way was opened for processing RCo_5 into magnets by casting, similar to Alnico, but this proved too complex. Cu-containing, precipitation hardened magnets are now also produced by powder metallurgy. In the Cu-containing magnets it was possible to replace some Co by Fe and to increase the total amount of transition metals beyond 1:5, to $R(TM)_{5+x}$ where $x < 1$.

The realization of practical 2–17 magnets took another decade. Fabrication methods that had worked well for $SmCo_5$, when applied to Sm_2Co_{17}, yielded only coercivities in the 1–3 kOe range. Quasi-binary intermetallics $R_2(Co_{1-x}Fe_x)_{17}$ were shown to exist for most rare earths. Modest iron additions stabilize the easy-c-axis anisotropy in all light-RE systems (except with Nd) and increase M_s, while depressing T_c only slightly [33,34,40]. With this knowledge a "second generation" of RE–Co-based magnets was predicted with potential energy products up to ≈ 60 MGOe [41]. However, it remained difficult to obtain sufficient coercivity, even with R = Sm.

Various experiments produced only marginally useful H_c-values. Small additions of other 3d-metals proved beneficial, yet serious problems persisted [42–44]. The most pragmatic approach extended the range of precipitation-hardenable $R(TM)_z$ compositions with the help of Cu to $z \approx 7.2$, leading to sintered magnets in which the main phase has the 2–17 structure [45]. Coercivities of 4–10 kOe were achieved. By also substituting some Ce for Sm, and Fe and/or Mn in the TM position, a whole family of commercial magnets was developed, all characterized by general wall pinning at homogeneous precipitates.

Then, in 1976, energy products in excess of those for $SmCo_5$ (up to 30 MGOe at $H_c = 12$ kOe) were first achieved with sintered magnets of $Sm(Co, Fe, Mn, Cr)_{8.5}$ [46] and high coercivity was obtained in sintered $Sm(Co,TM)_8$ magnets containing some Fe and Cu [47]. These first true 2–17 magnets all exhibited "nucleation controlled" magnetization reversal like $SmCo_5$. They have not become commercial products.

A very important next step was the discovery that adding a little zirconium, in Sm(Co, Fe, Cu, Zr)$_z$, coupled with a relatively complex heat-treatment, yielded coercivities of ≈ 6.5 kOe (520 kA/m) by general wall pinning in alloys with more Fe, less Cu and higher z-values than previously possible [48]. The best energy product was also > 30 MGOe (240 kJ/m^3) at 1.5 wt% Zr and z = 7.4. These magnets have a much smaller (negative) temperature coefficient of H_{ci} than the Mn, Cr-containing 2–17, making them much more useful at elevated temperatures, where 2–17 magnets with their high Curie points were expected to excel. Adding titanium or hafnium to similar alloys with R = Ce or Sm has comparable beneficial effects on microstructure and coercivity [49,50]. Then it was found that much higher coercivity, 10 to > 25 kOe, was possible using more Zr (up to 3 wt%) and longer heat-treatments. The best energy product was 33 MGOe (263 kJ/m^3), H_{ci} = 13 kOe, z = 7.67 [51]. With Sm(Co, Cu, Fe, Zr)$_z$ alloys for bonded magnets it was even possible to raise H_{ci} up to 26 kOe at z = 8.35 and to 8 kOe for an alloy with z = 8.94 (TM-richer than the 2–17 composition!) [52].

All successful "2–17" magnets have a small-scale microstructure within the main grains, visible only in the electron microscope. It consists of 2–17 cells separated by a 1–5 boundary phase, the crystal lattices being coherent. The boundary phase pins the magnetic domain walls [53]. Zr, Hf or Ti promote the optimal formation of this precipitate [54,55].

The REPM offer a possibility – unique among PM materials – to alter the temperature dependence of B_r within wide limits by alloying, e.g. to achieve a low temperature coefficient around 25°C. Such temperature compensation requires partial substitution of Sm in a 1–5 or 2–17 alloy with a heavy RE (Gd, Tb, Dy, Ho or Er) [56–59].

The RE–Co magnet alloys were initially prepared by melting together the constituent metals. The later development of calciothermic reduction methods was economically important. They use RE oxides as starting materials, combine several production steps and yield the alloy in powder form. The "reduction–diffusion" (R–D) method, using CaH$_2$ as the reductant [60], and the "co-reduction process", which employs calcium metal vapor [61], significantly reduced alloy prices [62,63]. However, induction melting continues to be extensively used.

Bonded versions of the REPM are now becoming very important. The first SmCo$_5$ laboratory magnet samples were epoxy bonded, and it was clear that bonded RE–TM magnets had great technological and economic appeal. But early commercial polymer-matrix magnets of SmCo$_5$ were not stable enough for most uses [64]. Their ongoing slow development has by now resulted in a broad spectrum of products employing a variety of magnet alloys, binders and production methods. The use of 2–17 alloys, combined with better polymers or soft metals as matrix materials, yields magnets of good stability. Injection-molded, extruded and calendered products have been introduced. The availability of Nd–Fe–B then opened the prospect of much cheaper high-energy matrix magnets (with, however, additional stability problems).

4. Fundamentals of rare earth–cobalt magnets

4.1. Physical metallurgy and crystal structures

Designations like "RE–Co" magnets, "Sm–Co", "1–5" and "2–17" are simplistic. All REPM contain more than two elements and are multiphase, nonequilibrium metallurgical systems with complex microstructures. The distinction between RE–Co and RE–Fe-based magnets is also becoming blurred. But to understand the REPM we must first review selected binary RE–TM alloy systems, the compounds, their crystal structures and basic magnetic properties.

Technologically most important are Sm–Co alloys. There are four compounds in the Co-rich portion of the Sm–Co equilibrium phase diagram that are magnetic at room temperature (r.t.). The RE–Co systems with other rare earths are similar, with minor systematic variations.

Fig. 2 summarizes the binary RE–Co phases that are stable and magnetic at r.t. (or easily retained in a metastable state). For each idealized composition listed, the r.t. crystal symmetry is given and the bars indicate with which RE the phases form. 2–17 or 1–5 are the principal flux-producing phases in magnets. Others, down to 1–3, are present as secondary minor phases that influence the coercivity and its temperature dependence. Various metastable phases, such as RCo_4, R_7Co_{29} and R_4Co_{17} form easily with R = La, Ce, Pr and Nd in the composition range between 5–19 and 1–5. The crystalline structures of all phases from 1–3 to 2–17 are closely related. All exist in hexagonal form and all but the 1–5 also have a rhombohedral modification. The c-axis is always magnetically unique, an easy or hard magnetization direction.

4.2. Basic magnetism of rare earth–transition metal compounds

Spontaneous magnetization. Of the binary RE–TM intermetallics, only those rich in Co or Fe and particularly the ferromagnetic compounds of the low-moment light rare earths (LRE), Ce, Pr, Nd and Sm and of nonmagnetic La and Y, have a sufficiently high spontaneous magnetization to be of interest for magnets. Their r.t. saturation values are summarized in fig. 3. The general rule for the 4f–3d exchange interaction is evident: the LRE couple ferromagnetically with Co or Fe, yielding high saturation, while the heavy RE (HRE) couple antiparallel with resulting low M_s

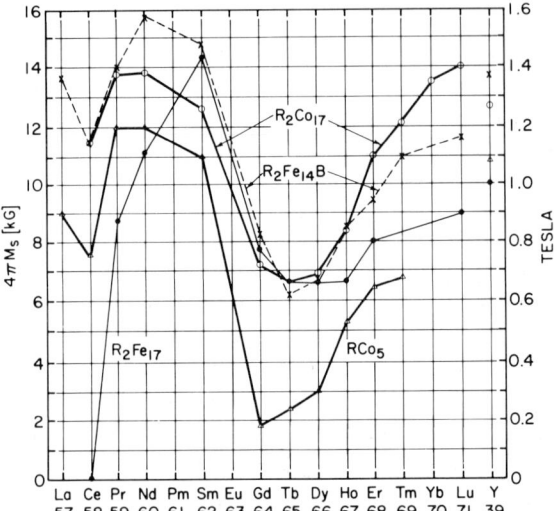

Fig. 3. Room-temperature spontaneous magnetization values ("saturation induction") of transition metal-rich RE-TM-based phases of interest for permanent magnets [103].

for compounds of the high-moment HRE, Gd through Er [19,21,26]. Included for comparison is a line for the ternary compounds $R_2Fe_{14}B$ which obey the same trend [65].

Curie temperature. Magnet alloys must remain magnetic up to reasonably high temperatures, say 300°C or more. Fig. 4 summarizes the Curie points of the same families of RE–TM compounds. The

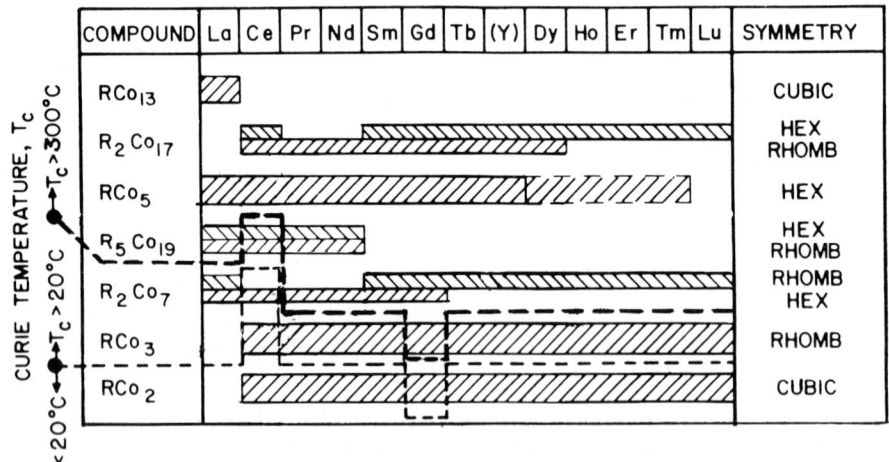

Fig. 2. Summary of magnetic binary R–Co intermetallics. The two dashed lines separate compounds having Curie points above and below 300°C, and room temperature, respectively.

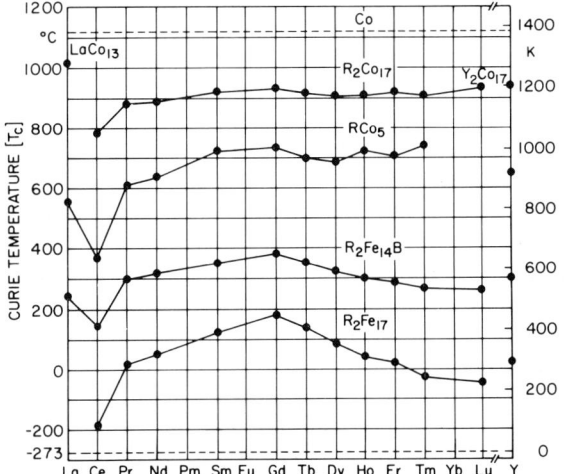

Fig. 4. Curie temperatures of transition metal-rich RE–TM-based phases of interest for permanent magnets [103].

R_2Co_{17}, with $T_c = 800-950°C$, are best. T_c for most RCo_5 is comfortably high, while the R_2Fe_{17} Curie points are too low for practical magnets. However, introducing some boron to form the ternary compounds $R_2Fe_{14}B$ raised T_c by 200–300°C over the corresponding R_2Fe_{17} values, which made these phases with R = Nd and Pr useful magnet materials [65]. Recently it was discovered that interstitial nitrogen also expands the lattice of R_2Fe_{17} and raises T_c and $4\pi M_s$, thus creating another related alloy series that may have practical potential for magnets [66].

Temperature variation of magnetization. In the RE–Co compound families 1–5 and 2–17 (and other RE–TM), two basic forms of the temperature dependence of M_s are found. The ferromagnetic compounds of Sm and the LRE show Brillouin behavior: M_s declines as T increases. Compounds of the high-moment HRE, however, are ferrimagnets with the TM and RE spin lattices opposed. If the magnetization of the HRE sublattice at 0 K is greater than that of the Co(TM)-sublattice, M_s vs. T-curves exhibit a compensation point above which there is a range in which M_s rises with increasing T. This has the important practical consequence that an internal temperature compensation of the useful flux is possible by mixing LRE and HRE in the alloy. Judicious choice of the HRE and its amount can make M_s fairly independent of T in the antici-

Table 1
Basic magnetic properties of interest for RE–Co magnets (room temperature values, except T_c) [2,3,41,44,88]

	B_r (kG)	T_c (°C)	K_1 (10^7 erg/cm^3)	H_A (kOe)	Theoret. $(BH)_m$ [a] (MGOe)
YCo_5	10.6	630	5.5	130	28.1
$LaCo_5$	9.1	567	6.3	175	20.7
$CeCo_5$	7.7	380	6.4	210	14.8
$PrCo_5$	12.0	620	8.1	170	36.0
$NdCo_5$	12.2	637	0.24	5	37.2
$SmCo_5$ [b]	11.4	727	11–20	250–440	32.5
$MMCo_5$ [c]	9.0	495	6.4	180	20.2
$MM_{0.8}Sm_{0.2}Co_5$ [c]	9.8	500	7.8	200	24.0
$Sm_{0.6}Gd_{0.4}Co_5$	7.3	727	7.7	264	13.3
Sm_2Co_{17}	12.5	920	3.2	65	39.0
$Sm_2(Co_{0.7}Fe_{0.3})_{17}$	14.5	840	3.0	52	52.6
$Sm_2(Co_{0.8}Fe_{0.1}Mn_{0.1})_{17}$	13.1		4.3	81	42.9
$Sm(Co_{0.87}Cu_{0.13})_{7.8}$	10.9	847	3.3	77	29.7

[a] The energy product limit is calculated as $(0.5 B_s)^2$.
[b] A wide range of values has been published for the anisotropy (H_A, K_1). Discrepancies may be due to composition variations, x, of $SmCo_{5+x}$, but also to insufficient measuring fields and the use of powders instead of single crystals.
[c] Mischmetal (MM) is produced with a broad compositional range. The properties reported for $MMCo_5$ vary widely depending on the RE mixture used, especially its Ce content, which is rarely reported.

pated use range of the magnet. If H_{ci} is high enough, the remanence will fairly closely follow M_s.

Magnetic anisotropy. In ferromagnets with strong easy-axis (EA) anisotropy, high H_c should be achievable in several ways, depending on the micromagnetic processes that dominate magnetization reversal – domain nucleation, localized or homogeneous wall pinning, or coherent spin reversal in single-domain particles. Theoretically H_{ci} could in either case be as high as the anisotropy field, making H_A an important quantity to know. It is the field needed to saturate a crystal in a "hard" basal plane direction perpendicular to c. The crystal anisotropy is thus the third important basic magnetic property for the RE–TM compounds. For good PM materials one wants to have EA anisotropy with a high H_A in the intended use temperature range. Almost all of the binary RCo_5 phases show this behavior near r.t. Only $TbCo_{5.1}$ has an easy basal plane, while $NdCo_5$ has a weak EA anisotropy and develops an easy cone with near-zero anisotropy on cooling below about 20°C. Among the binary R_2Co_{17} phases EA behavior is rather the exception, shown only by the compounds of Sm, Er and Tm. But moderate substitutions of certain third elements, especially iron, for some of the cobalt can induce EA anisotropy in most other R_2Co_{17} (except those where R = Nd or Dy).

Summary of basic magnetic properties. Table 1 lists relevant fundamental properties for phases on which practical R–Co magnets are based. The theoretical energy-product limit, calculated from the r.t. saturation, is fairly closely approached (75–80%) by the best laboratory magnets of $SmCo_5$, $PrCo_5$ and $(Sm, Gd)Co_5$, while there is no hope to realize the $(BH)_{max}$ potential of $NdCo_5$, given the low r.t. anisotropy (and therefore H_c). Some numbers are still uncertain, especially for mischmetal (MM) alloys since the composition of this RE blend varies from supplier to supplier.

5. Practical magnets from R–Co alloys

5.1. General aspects

Real magnets are more complex than the intermetallics identified as potential hard-magnetic materials. They are not stoichiometric compounds; there are always several phases present; the metallurgical microstructure is typically complex and not in thermodynamic equilibrium. Some magnets are even composites containing nonmagnetic binders.

The processes of magnetization change in the REPM are much more complicated than predicted by fine-particle theory. It is impractical to prepare single-crystal particles that show classical single-domain behavior (reversing magnetization in a single event at a high field strength near H_A) and it is certainly not possible to make magnets of a useful size of many such single-domain particles. In real particles (or grains of sintered magnets) the reversal always proceeds by movement of domain walls or the local formation of small reversed domains followed by wall motion. The variety of microstructures in the REPM provides several mechanisms for domain nucleation and the pinning of domain walls. An understanding of the structural features and their interaction with the walls is essential for controlling the engineering properties of the magnets, especially the coercivity.

The REPM are never just binary alloys (with the possible exception of "$SmCo_5$"). Certain elements are purposely added to achieve special magnetic properties (HRE for temperature compensation, Fe to raise M_s, etc.); or for economic reasons (Ce, La, Fe); or to form a desirable microstructure (Cu, Zr, etc.). Nonmetallic impurities, primarily oxygen, are inevitable and they may indeed play a crucial role as domain-wall pins in some REPM.

5.2. Powders, compacts and magnetic instability

In all early attempts to make REPM, R–Co alloys were prepared as coarse-grained ingots and comminuted into powders with a high percentage of single crystals. The particles were saturated and oriented by a magnetic field, and fixed with a resin binder or by compaction. The coercivity of powders prepared by mechanical grinding increases with decreasing particle size, reaches a maximum ($\approx$ 10–20 kOe for SmCo$_5$, but only between 1 and 4 kOe for YCo$_5$, PrCo$_5$, MMCo$_5$ and Sm$_2$Co$_{17}$) and declines on further grinding. It becomes increasingly difficult to align the finest particles due to a progressive disruption of the crystal lattice by plastic deformation near the particle surface. It is possible to increase H_{ci} several-fold by removing the disturbed outer layer with acids or by deactivating it with Zn [67]. Stress relief by annealing after grinding can also increase H_{ci}, as can grinding in liquid nitrogen. Magnets with good initial properties can be made from SmCo$_5$ powders (5–20 μm) compacted with or without a binder by uniaxial die pressing, isostatic compaction or a combination of the two. However, the properties of such compacts are unstable, especially at higher temperatures in air [68–70]. The primary effect of the aging is a loss of coercivity that is initially quite rapid and severe, then continues at a lower rate. It can be slowed but not prevented by compaction to high density or by various binders.

These observations suggest that small particles reverse their magnetization by the shifting of domain walls, initiated by "nucleation" events on their surface if the particles are first saturated. High fields can remove walls completely or compress residual reversed domains to tiny volumes in a disturbed lattice region where their walls are strongly pinned. The surface layer of cold-worked particles can effectively pin such wall fragments. Conversely, easy renucleation of domains would also most likely occur on the disturbed surface, in places where the anisotropy is lower, or where large local demagnetizing fields exist due to surface topography, or where RE(Sm) depletion by oxidation may form soft-magnetic second phases such as Co. So it is not surprising that the "nucleation" fields, and thus H_c, depend sensitively and in a complex way on the particle surface condition.

However, moderately fine powders of the copper-modified alloys, properly heat treated, do not suffer the described aging loss [69,71]. Their homogeneous intra-granular precipitate impedes the movement of domain walls wherever they are located, so H_c does not strongly depend on the surface condition of these particles.

5.3. Chemical stability problems

Surface corrosion of particles is one cause of the H_{ci}-loss during air aging, but it can cause stability problems in a more direct way as well. All RE have a high affinity for oxygen and in air (particularly at high humidity and/or elevated temperature) the RE near surfaces progressively corrodes and the alloy composition changes. This is most severe for powders, but compacts and even sintered bodies will also slowly oxidize. The oxidation of massive bodies is facilitated by open pores or microcracks. The affinity for oxygen of the different RE varies significantly, with the LRE being less stable than the HRE and Y. Among the LRE, Sm, Nd and Ce act relatively much better than La and Pr [71].

This has implications for magnet fabrication and use: Particles in polymer-matrix magnets and sometimes even massive sintered magnets need protective coatings; micron-size particles are often pyrophoric and there is a definite danger of accidental fires or explosions in REPM production. Even when this is minimized by precautions, it is necessary to protect fine powders by proper handling and storage and fast processing to avoid compositional shifts. For sintered magnets, high density, avoidance of microcracks and minimizing excess RE-rich phases in the grain boundaries are essential for good corrosion resistance and magnetic stability. This is even more important for alloys with a high content of La, Pr or mischmetal than for SmCo$_5$. (These problems are significantly more severe yet with Nd–Fe–B.)

5.4. Types of magnetization behavior

It is customary to distinguish two basic behavior patterns and accordingly classify magnets as "nucleation controlled" (type A) or "pinning controlled" (type B), identifiable by typical initial magnetization curves and the dependence of B_r and H_{ci} on the peak magnetizing field (fig. 5).

These are two limiting cases; the mechanisms are sometimes found mixed (type C), transitions between them can occur as the temperature changes and the same magnet can show different behavior after different heat-treatments.

In type A magnets, the movement of existing walls within a grain is easy, while the "nucleation" of a reversed domain after removal of the walls is difficult. This is characteristic of binary $SmCo_5$ (sintered or bonded) and its derivatives; also of sintered Nd–Fe–B. The virgin curve rises steeply; after thermal demagnetization there are many walls present in each grain that can be moved easily through the undisturbed crystal lattice by a small driving field. When the field is reduced again from a low peak value, H_m, and then increased in the negative direction, the shape of the demagnetization curve traced ("minor loop") still indicates mostly reversible wall motion with low remanence and coercivity. As H_m is increased, B_r rises slowly and H_{ci} very slowly. Increasingly grains are cleared of walls and reversed domains must be formed at the higher nucleation field in each such grain before its magnetization can be reversed again. (Most grains act independently.) Very high magnetizing fields are required to fully develop the best possible "major" demagnetization curve. In spite of extensive experimental and theoretical efforts, details of this so-called nucleation process are still unclear. One can argue whether (or when) a truly new nucleus is formed in a previously saturated grain, or whether the reversal starts from a residual domain left from the prior opposite magnetization state. The term "nucleation" must be broadened to include the anchoring/unpinning of wall fragments at highly localized sites in the grain boundary regions of sintered magnets.

In type B magnets, small reversed domains exist in all grains at all times or they form easily in low demagnetizing field, but obstacles present in most of the magnet volume strongly impede the movement of walls. This behavior is exhibited by "bulk hardened" Cu-containing magnets with compositions from 1–5 to 2–17. The virgin curve is almost horizontal at first, then rises to near saturation in a narrow range of H_m. When H is reduced and reversed, there is little magnetization change until the negative field range near H_{ci} is reached. The coercive field and remanence of minor loops at first remain near zero as H_m increases, then they rise steeply to their respective "saturation" levels. H_m only slightly greater than H_{ci} is sufficient to fully develop the second-quadrant curve. These magnets are thus easy to "charge". The homogeneous precipitate inside the main-phase grains pins walls wherever they are located. For the precipitates to be effective pins they must be crystallographically coherent with the matrix, have a magnetic saturation,

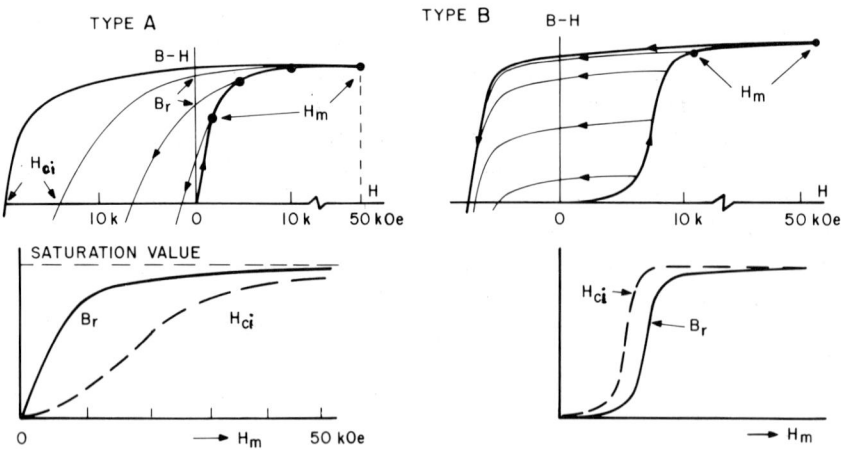

Fig. 5. Schematic magnetization curves for "nucleation controlled" (type A) and "pinning controlled" (type B) magnets [103].

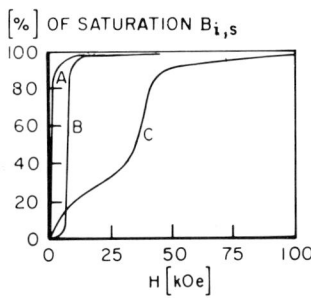

Fig. 6. Virgin magnetization curves for magnets of the three basic behavior types: A is nucleation controlled, B is pinning controlled, and C is mixed (or pinning with very non-uniform pin strength [3]).

anisotropy or exchange energy significantly different from the matrix phase, and have the right size and spacing [54,72–76].

The newer precipitation-hardened 2–17 magnets with very high H_c show still another form of the virgin curve. (Type C in fig. 6.) This behavior has been interpreted as a mixture of localized and homogeneous wall pinning [76]. Observation of domain patterns in variable fields [77] show that the pinning strength varies strongly from grain to grain in these magnets, while it is quite uniform in the low-H_c version.

5.5. Microstructures of R–Co magnets

The metallurgical microstructure controls the coercivity. Much effort has been invested in studying the details of metallurgical and magnetic-domain structures [54,78–81]. The structural features effective in wall-pinning or nucleation are so small that they can often be seen only with high-resolution transmission electron microscopes. Identifying compositions and crystal structures of the significant precipitates or grain boundary phases sometimes still defies the capabilities of today's best microprobe analyzers. Our knowledge of the interaction between the structure and moving domain walls is largely based on clever conjecture and simple theoretical models. There are many differences of opinion and even contradictions in published "explanations". But the characteristic microstructures associated with different magnets and behavior types are now fairly well established.

SmCo$_5$-type magnets. Sintered "SmCo$_5$" and other magnets with type-A behavior can have nearly single-phase, featureless microstructures. (Although there is usually an excess of Sm present as magnetic Sm$_2$Co$_7$ and nonmagnetic Sm$_2$O$_3$ grains.) Important are the matrix grain size (typically < 10 μm in SmCo$_5$), crystal texture, the nature of the grain boundaries, and the kind and distribution of any secondary phases. Parallel alignment of the c-axes is usually desired and can be made nearly perfect, so that the orientation of adjacent grains often differs only in the basal plane. Together with the near absence of voids, oxides and other secondary phases in well-prepared magnets, this makes the grain boundaries thin and clean looking. However, H_c can be very high, so these boundaries must still block the movement of domain walls from grain to grain. There must be imperfections at such boundaries not visible in the microscope yet quite effective in disrupting the exchange coupling between grains, or otherwise capable of strongly pinning domain walls there. Extremely thin coherent second-phase layers in the grain boundaries of these "single-phase" magnets could well have such an effect. Intragranular precipitates are absent in high-H_c SmCo$_5$. But the SmCo$_5$ phase itself is unstable below 750–800°C. Prolonged heating to ≈ 700–750°C causes dissociation of the metastable 1–5 producing small 2–7 and 2–17 crystallites, with at least the latter serving as easy nucleation sites. The magnetic effect is a drastic reduction of H_c. To achieve optimum coercivity, sintered SmCo$_5$ magnets must be annealed above the eutectoid temperature (around 900°C) and then rapidly cooled to below ≈ 400°C to avoid dissociation.

R(Co,Cu)$_5$-type magnets. A homogeneous precipitate is visible inside the grains of the 1–5 phase which impedes domain-wall motion. The magnetization behavior is type B. Addition of some Fe lowers the decomposition temperature, stabilizes the 1–5 phase and broadens the homogeneity range of the phase toward higher transition metal contents.

$R_2(Co,Fe,Cu,M)_{17}$-type magnets. Magnets in which the principal phase is 2–17 are also made by a bulk hardening process that leads to type-B or -C behavior. All "2–17" magnets in commercial production now contain some iron to increase B_s, copper to permit precipitation hardening and a small amount of Zr, Hf or Ti to aid in forming the required microstructure. Common varieties have compositions in the range 25–27 wt% Sm, 14–20% Fe, 5–10% Cu, 1.5–3% Zr, balance cobalt and oxygen. The typical microstructure differs from that of the Sm(Co, Fe, Cu)$_5$ magnets. Its main feature is a network of small cells of a 2–17 phase (within the much larger grains), separated by a thin boundary phase of 1–5 stoichiometry and structure. The cells are roughly rhombic, elongated in the c-direction, but irregular in the base plane. In a very high-H_c 2–17 magnet the cells are about 100–200 nm wide and the cell walls 5–20 nm. The heavily twinned cell interior has a rhombohedral structure (2–17R). There are also thin layers parallel to the basal plane that run across many cells. This "platelet" or "z-phase" contains most of the Zr (or Hf, Ti) and is hexagonal 2–17H. All three phases are fully coherent in good magnets. A thorough model analysis of the development of this microstructure by Ray [55,81] takes into account most earlier work. All features visible in micrographs develop during isothermal aging around 800°C. Yet, a sample quenched from this state has low coercivity! The high H_{ci} develops only during subsequent slow cooling to about 400°C with no further changes seen in the micrographs. Microstructures of the low-H_c versions of 2–17 magnets (with less Zr) are similar, but all dimensions are smaller. Typical cell sizes are about 50 nm and the boundary phase is only 5–10 nm thick. TEM observations of stationary domain walls in thermally demagnetized samples [74,82] show that the walls run roughly parallel to the c-axis but zigzag with the boundary phase. The 1–5 cell walls impede domain-wall motion and because of the small cell size, the effect is similar to homogeneous wall pinning. The high-H_c version exhibits strong inhomogeneities from grain to grain, magnetically and chemically, while the low-H_c magnets are uniform [77,78].

5.6. Magnet types by allow composition

The rare earth–cobalt-based magnets now commercially available may be classified by composition as follows. We shall review their properties, technological importance and some economic factors.

(1) "SmCo$_5$" magnets have a composition somewhat Sm-rich of 1:5. The RE is pure Sm or a less expensive RE mixture containing only ≈ 70% Sm which yields magnetic properties similar to those of SmCo$_5$ [83], Extremely high H_{ci}- and H_k-values are easily achieved with SmCo$_5$.

(2) (Sm, Pr)Co$_5$, a variation with the more abundant praseodymium substituted for a part of the Sm, can yield higher energy products. Too much Pr reduces the coercivity and can adversely affect the long-term stability. However, Pr-oxide dispersions yielded SmCo$_5$-like coercive force and suggest improved elevated-temperature stability for sintered magnets in which 80% of the RE component is Pr [84].

(3) "MMCo$_5$" magnets use as the RE component mainly a RE mixture – either conventional Ce-rich mischmetal (MM) or a modification with reduced cerium content. B_r, H_{ci}, and $(BH)_{max}$ are less than for SmCo$_5$; and because of the lower T_c (about 500°C) and the greater oxygen affinity, the temperature stability is poorer. To keep H_{ci} at a useful level, 15–25% of the total RE must be Sm. Fig. 7 shows demagnetization curves of sintered MM$_{1-x}$Sm$_x$Co$_5$ magnets with different Sm contents, $x = 0.0$ to 0.4 [85]. Additions of magnesium together with copper can increase H_c to SmCo$_5$ levels without using any Sm [86].

Ce-rich MM corresponds to the natural mixture of RE in the major ores. It is the cheapest and most abundant RE product. In the mid-1970s MM-based magnets were seen as a

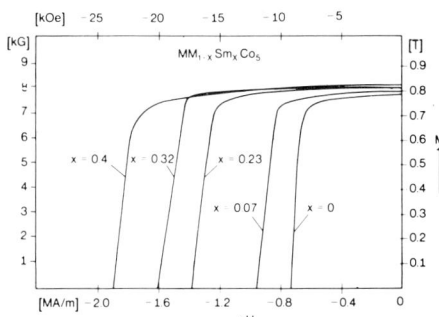

Fig. 7. Demagnetization curves of sintered $MM_{1-x}Sm_xCo_5$ magnets with different samarium contents [85].

promising REPM product for the future and were developed to production maturity [87,88]. But a supply crisis in 1978–80 made cobalt an expensive commodity for several years. Best advantage of the magnetic flux potential of Co is taken by combining it only with Sm. The cobalt price and supply have recovered, but now there exists Nd–Fe–B which needs neither Sm nor Co and is thus economically superior to MM–Co for many large-quantity consumer applications. These circumstances have so far kept $MMCo_5$ from finding its anticipated important place in the magnet market.

(4) $(Sm, HRE)Co_5$: A subgroup of 1–5 magnets in which different amounts of Gd or other heavy rare earths are used to reduce the temperature coefficient of B_r, even to near zero. Such magnets are needed in small volume for microwave tubes, measuring instruments, accelerometers and gyros, devices where cost is secondary to performance. They are very expensive compared with standard "$SmCo_5$" magnets. Stability is excellent.

(5) $Sm(Co, Fe, Cu)_{5-7}$ (pinning-type) magnets can have energy products comparable to sintered $SmCo_5$ while some of the Co is replaced by cheap Fe. Such magnets are also easier to magnetize. $Ce(Co,Fe,Cu)_5$ has properties poorer than $(MM, Sm)Co_5$, but is completely without Sm, using instead abundant Ce. Both types have found market niches, particularly in Japan. Combining Sm and Ce in any ratio has allowed the development of a whole family of $(Sm,Ce)(Co,Fe,Cu)_{5-7}$ magnets tailored to cost–performance compromises for different applications [89].

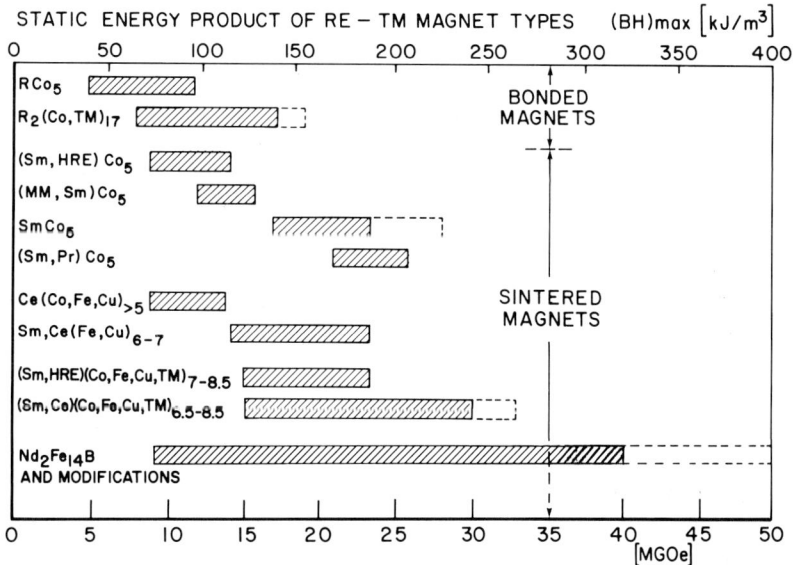

Fig. 8. Ranges of static energy product values for different types of RE–Co-based magnets. (Nd–Fe–B included for comparison.) Shaded bars are commercial products, dashed extensions are laboratory records [103].

(6) The precipitation hardened "2–17" alloys, particularly Sm(Co,Fe,Cu,Zr)$_x$, with $x = 7.2$–8.5, are now rapidly gaining technological importance. Low- and high-H_c versions are used in sintered and bonded magnets. The low-H_c type is an extension of the magnets discussed in (5), with similar applications. The high-H_c magnets are a cheaper – and generally better – replacement for SmCo$_5$ in all high-temperature applications requiring the best possible stability.

(7) (Sm,HRE)TM$_x$: "2–17" compositions can also be temperature-compensated with HRE elements. Magnets with HRE = Gd or Er are now in commercial production, used for the same special applications as compensated "1–5" magnets, see (4).

Fig. 8 is an overview of all these alloy types and the range of energy product values they offer. Bars indicate the range of production values, the dashed extensions laboratory records. Polymer-bonded products are included and – for comparison – generic Nd–Fe–B magnets.

6. Magnet manufacturing technology

6.1. Magnet fabrication methods – an overview

A number of techniques are used in factory or laboratory to prepare RE–TM magnets. Alloy preparation and magnet fabrication are typically sequential and discussed separately.

The elemental metals are fused together in an inert gas atmosphere (A, He) or in vacuum. Induction melting in alumina crucibles followed by chill casting is used in commercial production. Low-melting, near-eutectic RE–Co(Fe) master alloys can be produced by fused-salt electrolysis, then remelted into magnet alloys with TM additions. An economical commercial alternative to melting are the calciothermic methods of preparing alloys as a spongy powder from a mixture of oxide and elemental metal powders [62,63].

The production of magnets from the alloys on an industrial scale is generally by powder metallurgy: Pulverizing the alloys to a particle size of a few microns, compaction, sintering and a post-sintering heat-treatment [1–4,90,91,103]. Bonded magnets are made by crushing the ingot or a pre-sintered body to a somewhat larger particle size, then consolidating the powder into a magnet by mixing it with a polymeric binder and either die pressing, or injection molding, extruding or calendering the mixture into the desired shape [52,103].

Many laboratory methods have been used and some developed to a considerable degree. None are now used in commercial magnet production, although a few have potential for special applications. Casting magnets from bulk hardenable alloys was tried commercially around 1970 but did not succeed because too many parameters need to be controlled simultaneously. Sputtered films with useful coercivities and energy products have been prepared from many RE–TM alloys, potentially useful in some microwave device applications. Rapid quenching methods such as melt spinning and "liquid-dynamic compaction" of an atomized molten alloy, developed for the production of Nd–Fe–B in very fine-grained form, might also have some promise for preparing RE–Co magnets. All these techniques yield magnets with more or less isotropic properties. While this is desirable in some applications, the high cost of Sm–Co-based alloys generally dictates their use in grain-oriented form to take the best economic advantage of their high energy-product capability. A crystal texture with a magnetic easy axis or plane remains a development objective whose attainment could make some laboratory methods technologically useful. It may also be possible to produce useful deformation textures without a magnetic field by adapting techniques recently developed for rapidly quenched Nd–Fe–B [5] and for cast Pr–Fe–B [93].

6.2. Machining, handling and magnetizing

Sintered RE–Co magnets are hard and brittle. They must be sliced with diamond blades, surface ground or cut on electric spark-erosion machines in a demagnetized state.

The high H_{ci} and low recoil permeability of most REPM permits pre-charging and manipulation in open circuit; they will not self-demagnetize. But machining, shipping and handling in the magnetized state are difficult for larger high-energy magnets. It is often necessary to magnetize after partial assembly of a device. This can present problems. High-H_c 2–17 magnets and SmCo$_5$ first charged and then field-demagnetized, need up to 50 kOe (4 MA/m) for saturation. This requires expensive magnetizing equipment and may be impossible for larger assemblies. Some bulk hardened compositions require only 8–12 kOe to magnetize, like ferrites. Incompletely magnetized SmCo$_5$ magnets can exhibit severe thermal instabilities of the flux [94].

7. Bonded magnets

State of development. RE–TM-based bonded (or matrix) magnets were first introduced in the late 1960s, but their commercial development proceeded slowly. Difficulties relating to the chemical reactivity of the powders had to be overcome and inexpensive manufacturing methods developed. Early products using 5–20 μm particles of SmCo$_5$, bonded with epoxy, polyethylene chloride or ethylene vinyl acetate, deteriorated quickly above 50°C and were magnetically not very stable even at room temperature [95]. Coating the particles with metallic or organic diffusion barriers, careful selection of compatible polymeric binders, etc., brought much improvement. Most important was the introduction of alloys bulk-hardened with Cu. If used as coarser powders ($\approx$ 20–200 μm) these do not suffer the severe sensitivity of H_c to the particle surface condition, so they are much more stable. A soft metal matrix is another way of improving the elevated-temperature stability [70]. Alloys close to the 2–17 stoichiometry and with higher Fe content were developed specifically for bonded magnets [52], also ways to produce ingots with a coarse columnar grain structure, removing the need for using pre-sintered blocks. These developments, together with the adaptation of inexpensive molding methods such as extrusion and injection molding in an orienting field [96], have cleared the way for large-scale industrial applications of the bonded REPM.

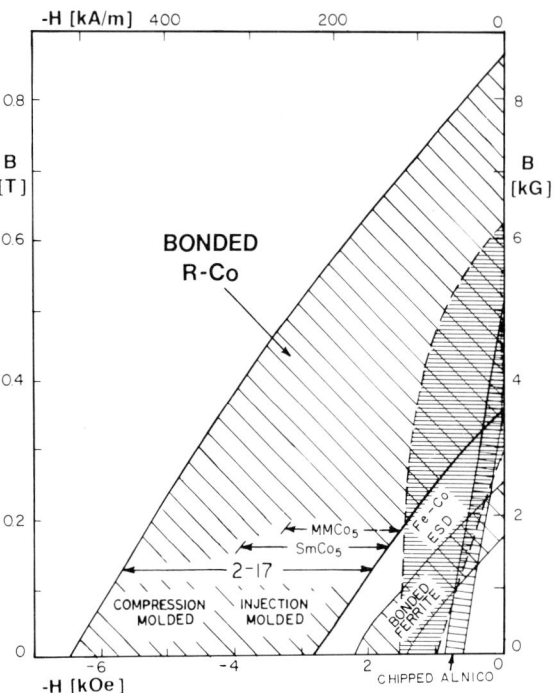

Fig. 9. Comparison of B,H-demagnetization curves for commercially produced bonded REPM and older matrix magnet types [103].

Subtypes and their properties. The best compression-molded 2–17 matrix magnets today offer r.t. properties equivalent to medium-grade sintered SmCo$_5$. They are far superior to any older, non-RE magnet type. Injection-molded magnets have significantly poorer B_r, H_{ci} and loop shape; their commercial attraction is low cost of processing. Plastic bonded REPM are now produced with energy products from 3 to >17 MGOe (of 2–17 alloys on the upper end, SmCo$_5$ in the middle, alloys with mischmetal substitution in the low-B_r region). Fig. 9 shows the range of B, H-demagnetization curves available. Metal-matrix magnets with 2–17 alloy, a Sn–Pb binder and Cu filler were prepared to obtain magnet sets with precisely controlled variable saturation [97].

Molding, machining, handling and magnetizing. The most important advantage of matrix bonding is the ability to mold magnets and even subassemblies to final shape with close tolerances, permitting inexpensive mass production. There is little cutting scrap or grinding loss, so the expensive and supply-limited magnet alloys are conserved. If machining is necessary, it can often be done with conventional machine tools. Bonded magnets cause fewer handling problems than sintered REPM. Their mechanical behavior ranges from somewhat ductile (metal matrix) and elastic or even flexible (plastic or rubber binders) to moderately brittle (at dense packing). They have a lower B_r than their sintered counterparts, so forces are less and the danger of breakage or injuries during assembly is reduced. Anisotropic bonded magnets come off the production line magnetized and it is often desirable to manipulate them that way to avoid additional demagnetizing and recharging.

8. Engineering properties of production magnets

General comments. Presenting engineering design data for magnets is beyond the scope of this paper. Readers needing information on salient magnetic and typical mechanical, electrical and thermal properties should consult the references in this chapter and magnet manufacturers' product literature. Note that a set of properties cannot simply be correlated to alloy composition alone; orientation and processing are also important. For reasons of market niches, materials supply, other economic considerations and patent protection a large number of different REPM types are in production, with little industry standardization as yet.

High and low-temperature magnetic properties. R-Co-based magnets (rather than the cheaper Nd-Fe-B) are often used in devices operating above $\approx 150°C$. Therefore, relatively much information has been generated about the magnetic properties and their long-term stability at elevated temperatures up to 200-400°C [3,70,92,98-101]. For special applications at cryogenic temperatures there is as yet little engineering data available. Basic studies of magnetization, anisotropy field and coercivity show that for Sm-Co-based 1-5 and 2-17 magnets all these quantities increase on cooling to near absolute zero. Such magnets, when charged at room temperature and then cooled, thus get only "better". Nd-Co compounds and $PrCo_5$ develop an easy-cone anisotropy on cooling which decreases the coercivity and distorts the demagnetization curve. Magnets with a high Nd content and perhaps those containing much Pr, are therefore not suitable for use in cryogenic devices. For special new applications in nuclear reactors and particle-beam devices the behavior of magnets in a radiation environment is of interest and now being studied. High doses of neutron and ionizing radiation were reported to severely degrade the open-circuit flux of 1-5 and 2-17 sintered Sm-Co and even more so of Nd-Fe-B magnets. Some of this degradation may be an indirect effect of internal heating by absorbed radiation and preventable by proper cooling.

Thermal cycling and aging mechanisms. Flux changes during short heating-cooling cycles and the long-term aging depend on the operating point of the magnet on its demagnetization curve. (The operating permeance, $p = B_d/H_d$.) At lower temperatures the irreversible loss is strictly magnetic in origin and recoverable by remagnetizing. It is due to thermally activated domain-wall motion under the influence of (self-)demagnetizing fields. This loss component increases rapidly with decreasing $|p|$ [95,102]. The reversible loss is almost independent of p and tracks the temperature variation of M_s. At higher temperatures "irrecoverable losses" occur due to slow changes in the metallurgical microstructure and oxidation, in bonded magnets also to a softening of the binder and a deterioration of the alloy-binder interface [70]. These effects become significant at 250-300°C for sintered $SmCo_5$, at 350-400°C for sintered 2-17, but at much lower temperatures (50-180°C) for the various types of bonded magnets.

Fig. 10 shows the reversible losses for five important sintered magnet types. This "loss" (or the temperature coefficient of B_d – the average slope of such curves) is primarily determined by the composition of the alloy. Note that the two 2-17 magnets have similar reversible losses despite their very different coercivity! This loss is closely related to the Curie temperature of the

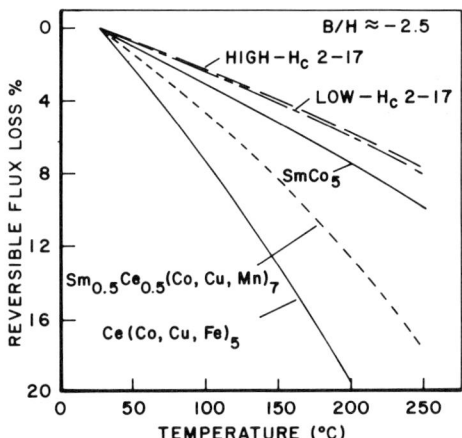

Fig. 10. Reversible temperature variation of the open-circuit flux (at a permeance $p \approx -2.5$) for five commercial sintered magnet types. Measured after prestabilization by heating for 5 h at 250°C; at 200°C for Ce(Co, Cu, Fe)$_5$ [101].

main phase. The more cerium an alloy contains the lower is its T_c and the greater the reversible loss.

"Irreversible losses" and "aging effects" are difficult to separate. When magnets are at elevated temperatures for long periods of time, additional time-dependent losses of the operating flux occur which are called aging losses. The domain-wall creeping responsible for the "irreversible" losses is a function not only of temperature but also of time. At 100 or 150°C, the magnetic domain structure may not reach an equilibrium state for several hours, near r.t. it may take hundreds of hours. The flux initially drops fast, then its rate declines and finally it levels off. But at higher temperatures the effects of metallurgical structure changes (such as the eutectoid decomposition of SmCo$_5$ or the coarsening of precipitates), of oxidation or of binder deterioration become noticeable. They show up as a flux reduction from the temporarily stabilized level and can eventually lead to a catastrophic loss of properties, definitely limiting a magnet's useful lifetime. Irreversible and aging losses are related to the intrinsic coercivity: they are more severe when the initial H_{ci} is low and worse for lower operating points. After heating a charged magnet above the subsequent use temperature for several hours the flux may remain constant for a long time period. Such pre-aging, properly done, is effective as a method of thermal stabilization [98].

Summary of physical properties. Engineers designing a magnetic device or machine also need information on the mechanical, electrical and thermal properties of the magnets. This can, e.g. be found in the refs., [1–4,70,92,98–103] and many commercial brochures. Published values for some mechanical strength properties cover a very wide range. The strength of sintered REPM – like that of all brittle materials – depends strongly on unavoidable minor flaws in the individual sample. Sintered magnets should never be made stress-bearing components of a machine.

9. Applications of the rare-earth permanent magnets

The high energy product (especially in dynamic applications) makes possible a dramatic miniaturization of devices in which the magnet was previously a major part of the volume and weight (e.g. magnetron tubes, magnetic couplings). Their high coercivity makes the REPM ideally suited for devices such as magnetic bearings which were barely feasible before the advent of the REPM. And they are ideal for many types of electric motors, generators and actuators, thus accelerating the trend toward a permanent-magnetic concept of electric machine design.

The high cost and relative scarcity of the RE, especially Sm – but also of cobalt – long discouraged widespread industrial use. However, imaginative circuit redesign to take best advantage of the unique properties of the REPM, and careful analysis of possible cost savings in manufacturing and system performance (including energy savings in operation) convinced designers that the REPM were the best choice in many cases after all. Their use has grown rapidly in the last decade. Fear of a Sm shortage fueled development of the 2–17 magnets, of a variety of Ce and MM–Co-based alloys and eventually of Nd–Fe–B magnets. As a consequence, application engineers now have a wide choice of commercially available REPM.

The Nd-Fe-based magnets have displaced Sm-Co in many applications where limited temperature capabilities and poorer chemical stability are of no concern. This has relieved the pressure on the Sm supply and price, making RE-Co magnets more available for those critical uses where they offer the best or only solution.

The first conventional devices to use $SmCo_5$ were microwave and mm-wave power devices (such as travelling wave tubes). These can also take advantage of the still better energy density and high-temperature capabilities of the 2-17 Sm-Co. The internally temperature-compensated magnets were also developed for such tubes, but they are finding additional uses (accelerometers, gyros). These are all applications in which Sm-Co will have no competition from Nd-Fe-B. Other novel devices have recently been developed in which the REPM focus or guide charged particle beams: magnetic lenses, deflection dipoles, multipolar undulators and wigglers for generating synchrotron radiation [104]; these, too, seem generally tied to Sm-Co. A family of highly original PM field sources is now also evolving [105]. These are possible only with REPM, Co- or Fe-based. Other devices that could only be built with REPM are certain magnetic bearings for ultra-centrifuges and turbomolecular pumps. An exciting new mechanical force application is in levitation and propulsion systems for "magnetic cushion railroads" – potentially a large volume use for REPM (Sm-Co, Nd-Fe-B or perhaps MM-Co). Another large quantity use is developing in magnetic resonance imaging devices, where PMs are competing with superconductors and the REPM with hard-magnetic ferrites, Sm-Co with Nd-Fe-B. REPM are used extensively in all modern electronic weighing systems, with applications from laboratory to supermarket. They also found extensive applications in consumer electronics: for loudspeakers, earphones, microphones, etc.; here Nd-Fe-B has largely replaced Sm-Co. In medicine, tiny (even implantable) hearing aids, catheters, stoma seals and artificial heart-assist devices utilize REPM.

But the largest and fastest growing market for the REPM is in small motors and actuators. Computer peripheral devices are important: disk-drive motors, recording head positioners, line and dot-matrix printers. Miniature REPM motors are used in tape or head drives of video cameras and players, in automated film cameras, miniature audio tape recorders, electronic typewriters and other business machines, with future applications seemingly unlimited. These small motors are the main application for injection-molded 2-17 and Nd-Fe-B magnets with polymer matrix. Larger REPM motors built to produce very high torques are used in machine tools, industrial robots and in aircraft for moving flight control surfaces. Large electric motors of several 100 kVA rating are also under development for a variety of uses. In some of these high-power applications, it is likely that RE-Co will be the magnets of choice because of high operating temperatures and long operating life requirements.

References

[1] K.J. Strnat, J. Magn. Magn. Mater. 7 (1978) 351.
[2] H. Nagel, in: Thyssen Edelstahl Techn. Ber., vol. 6, eds. H. Brandis et al. (Thyssen Edelstahlwerke AG, Krefeld, Germany, 1980) p. 5.
[3] W. Ervens, in: Tech. Mitteil. Krupp. Forschungsberichte, vol. 40 (Krupp Gemeinschaftsbetriebe, Essen, Germany, 1982) no. 3, p. 99.
[4] K.J. Strnat, Proc. IEEE 70 (1990) 923.
[5] J.F. Herbst and J.J. Croat, J. Magn. Magn. Mater. 100 (1991) paper 1123.
[6] Proc. 2nd Intern. Workshop on Rare-Earth Permanent Magnets and Appl. – Dayton, ed. K.J. Strnat (Univ. of Dayton, Ohio, 1976).
[7] Proc. 3rd Intern. Workshop RE Perm. Magnets and Appl. – San Diego, ed. K.J. Strnat (Univ. of Dayton, Ohio, 1978).
[8] Proc. 4th Intern. Workshop RE Perm. Magnets and Appl. – Hakone, eds. H. Kaneko and T. Kurino (Soc. Non-Trad. Technol., Tokyo 1979).
[9] Proc. 5th Intern. Workshop RE Perm. Magnets and Appl. – Roanoke, ed. K.J. Strnat (Univ. of Dayton, Ohio, 1981).
[10] Proc. 6th Intern. Workshop RE Perm. Magnets and Appl. – Baden, ed. J. Fidler (Tech. Univ. Vienna, Austria, 1982).
[11] Proc. 7th Intern. Workshop RE Perm. Magnets and Appl. – Beijing, eds. X. Pan. W. Ho and C. Yu (China Acad. Publish., Beijing, 1983).
[12] Proc. 8th Intern. Workshop RE Perm. Magnets and Appl. – Dayton, ed. K.J. Strnat (Univ. of Dayton, Ohio, 1985).

[13] Proc. 9th Intern. Workshop RE Perm. Magnets and Appl. – Bad Soden, eds. C. Herget and R. Poerschke (Deut. Phys. Ges., Bad Honnef, 1987).
[14] Proc. 10th Intern. Workshop RE Perm. Magnets and Appl. – Kyoto, eds. T. Shinjo and T. Kurino (Soc. Non-Trad. Technol., Tokyo, 1989).
[15] Proc. 11th Intern. Workshop RE Perm. Magnets and Appl. – Pittsburgh, ed. S.G. Sankar (Carnegie-Mellon Univ., Pittsburgh, PA, 1990).
[16] Proc. 2nd Intern. Symp. Magn. Anisotropy and Coercivity in RE–TM Alloys – San Diego, ed. K.J. Strnat (Univ. of Dayton, Ohio, 1978).
[17] W.E. Wallace, Rare Earth Intermetallics (Academic Press, New York, 1973) p. 112.
[18] E.A. Nesbitt, H.J. Williams, J.H. Wernick and R.C. Sherwood, J. Appl. Phys. 33 (1962) 1674.
[19] K. Nassau, L.V. Cherry and W.E. Wallace, J. Phys. Chem. Solids 16 (1960) 131.
[20] W.M. Hubbard, E. Adams and J. Gilfrich, J. Appl. Phys. 31 (1960) 3685.
[21] R. Lemaire, Cobalt 32 (1966) 133, 33 (1966) 201.
[22] K.J. Strnat, G.I. Hoffer and A.E. Ray, IEEE Trans. Magn. MAG-2 (1966) 489.
[23] A.E. Ray, K.J. Strnat and D. Feldmann, in: Proc. 3rd Rare-Earth Res. Conf., ed. K. Vorres (Gordon and Breach, New York, 1964) p. 443.
[24] H.R. Kirchmayr, IEEE Trans. Magn. MAG-2 (1966) 493.
[25] K.H.J. Buschow and W.A.J.J. Velge, J. Less-Common. Met. 14 (1967) 11.
[26] K.J. Strnat, G.I. Hoffer, J.C. Olson and W. Ostertag, J. Appl. Phys. 37 (1966) 1252.
[27] K.J. Strnat, G.I. Hoffer, J.C. Olson, W. Ostertag and J.J. Becker, J. Appl. Phys. 38 (1967) 1001.
[28] G.I. Hoffer and K.J. Strnat, IEEE Trans. Magn. MAG-2 (1966) 487.
[29] E. Tatsumoto, J. Okamoto, H. Fujii and C. Inoue, J. de Phys. 32 Suppl. (1971) C1-550.
[30] A.S. Ermolenko, in: Proc. Intern. Conf. on Magnetism, vol. 1, part 1, ICM-73 (Nauka, Moscow, 1973) p. 231.
[31] K.J. Strnat, Cobalt 36 (1967) 133.
[32] K.H.J. Buschow and W.A.J.J. Velge, Z. Angew. Phys. 26 (1969) 157.
[33] A.E. Ray and K.J. Strnat IEEE Trans. Magn. MAG-8 (1972) 861.
[34] H.G. Schaller, R.S. Craig and W.E. Wallace, J. Appl. Phys. 43 (1972) 3161.
[35] K.S.V.L. Narasimhan, W.E. Wallace, R.D. Hutchens and J.D. Greedan, in: AIP Conf. Proc., vol. 18 (Am. Inst. Physics, New York, 1974) p. 1212.
[36] D.K. Das, IEEE Trans. Magn. MAG-5 (1969) 214.
[37] M.G. Benz and D.L. Martin, Appl. Phys. Lett. 17 (1970) 176.
[38] E.A. Nesbitt, R.H. Willens, R.C. Sherwood, F. Buehler and J.H. Wernick, Appl. Phys. Lett. 12 (1968) 361.
[39] Y. Tawara and H. Senno, Japan. J. Appl. Phys. 7 (1968) 966.
[40] S. Yajima, M. Hamano and H. Umebayashi, J. Phys. Soc. Japan 32 (1972) 861.
[41] K.J. Strnat, IEEE Trans. Magn. MAG-8 (1972) 511.
[42] K.J. Strnat and A.E. Ray, in: Goldschmidt informiert, ed. F. Kornfeld (Th. Goldschmidt AG, Essen, Germany, 1975) 4/75, no. 35, p. 47.
[43] Y. Tawara and K.J. Strnat, IEEE Trans. Magn. MAG-12 (1976) 954.
[44] W. Ervens, in: Goldschmidt informiert, ed. F. Kornfeld (Th. Goldschmidt AG, Essen, Germany, 1979) 2/79, no. 48, p. 3.
[45] H. Senno and Y. Tawara IEEE Trans. Magn. MAG-10 (1974) 313.
[46] H. Nagel, in: AIP Conf. Proc. 29, eds. J.J. Becker et al. (Am. Inst. Physics, New York, 1976) p. 643.
[47] Y. Tawara and H. Senno, ref. [6], p. 340.
[48] T. Yoneyama, S. Tomizawa, T. Hori and T. Ojima, in: ref. [7], p. 406.
[49] K. Inomata, T. Oshima and N. Yamamiya, in ref. [7], p. 420.
[50] T. Nezu, M. Tokunaga and Z. Igarashi, in: ref. [8], p. 437.
[51] R.K. Mishra, G. Thomas, T. Yoneyama, A. Fukuno and T. Ojima, J. Appl. Phys. 52 (1981) 2517.
[52] T. Shimoda, K. Kasai and K. Teraishi, in: ref. [8], p. 335.
[53] J.D. Livingston and D.L. Martin, J. Appl. Phys. 40 (1977) 1350.
[54] J.D. Livingston, in: Symp. Proc. on Soft and Hard Magn. Materials and Applications, eds. J.S. Salsgiver et al. (Amer. Soc. Metals, Metals Park, Ohio, 1986) p. 71.
[55] A.E. Ray, ref. [54], p. 105.
[56] F.G. Jones and M. Tokunaga, IEEE Trans. Magn. MAG-12 (1976) 968.
[57] K.S.V.L. Narasimhan, J. Appl. Phys. 52 (1981) 2512.
[58] D. Li, E. Xu, J. Liu and Y. Du, IEEE Trans. Magn. MAG-16 (1988) 988.
[59] K.J. Strnat and A. Tauber, J. Less-Common. Met. 93 (1983) 269.
[60] R.E. Cech, J. Metals 26 (1974) 32.
[61] H.-G. Domazer, in: Proc. 3rd Eur. Conf. on Hard Magnetic Materials, ed. H. Zijlstra (Bond voor Materialenkennis, Den Haag, 1974) p. 140.
[62] C. Herget, ref. [42], p. 3.
[63] F.G. Jones, ref. [13], p. 737.
[64] K.J. Strnat, A.J. Kleman and H. Blaettner, ref. [6], p. 387.
[65] M. Sagawa, H. Fujimura, H. Yamamoto, Y. Matsuura and K. Hiraga, IEEE Trans. Magn. MAG-20 (1984) 1584.
[66] J.M.D. Coey and Hong Sun, J. Magn. Magn. Mater. 87 (1990) L251.
[67] J.J. Becker, IEEE Trans. Magn. MAG-5 (1969) 211.
[68] K.J. Strnat, IEEE Trans. Magn. MAG-6 (1970) 182.
[69] R.J. Cremer, G.W. Reppel and F. Demmel, ref. [10], p. 385.
[70] R.M.W. Strnat and S. Liu, ref. [10], p. 469.
[71] R.M.W. Strnat and H.-L. Luo, ref. [10], p. 457.

[72] A. Menth and H. Nagel, ref. [16], p. 56.
[73] B. Barbara, ref. [16], p. 137.
[74] J. Fidler and P. Skalicky, ref. [10], p. 585.
[75] H. Oesterreicher, J. Less-Common Met. 99 (1984) L-17.
[76] K.-D. Durst and H. Kronmüller, ref. [12], p. 725.
[77] D. Li and K.J. Strnat, J. Appl. Phys. 55 (1984) 2103.
[78] L. Rabenberg, R.K. Mishra and G.J. Thomas, J. Appl. Phys. 53 (1982) 2389.
[79] J. Fidler, P. Skalicky and F. Rothwarf, IEEE Trans. Magn. MAG-19 (1983) 2725.
[80] G.C. Hadjipanayis, ref. [54], p. 89.
[81] A.E. Ray, J. Appl. Phys. 67 (1990) 4972.
[82] F. Rothwarf et al., ref. [10], p. 567.
[83] H.-G. Domazer and K.J. Strnat, ref. [6], p. 348.
[84] M.H. Ghandehari and J. Fidler, IEEE Trans. Magn. MAG-12 (1985) 1973.
[85] H. Nagel and A. Menth, ref. [42], p. 42.
[86] J.W. Walkiewicz, E. Morrice and M.W. Wong, IEEE Trans. Magn. MAG-19 (1983) 2053.
[87] D.L. Martin, J.T. Geertsen, R.P. Laforce and A.C. Rockwood, in: Proc. 11th Rare Earth Res. Conf. (1974) p. 342.
[88] M.G.H. Wells and K.S.V.L. Narasimhan, ref. [44], p. 15.
[89] Y. Tawara and T. Chino, ref. [44], p. 10.
[90] M.A. Bohlmann, ref. [13], p. 233.
[91] J. Ormerod, J. Less-Common Met. 111 (1985) 49.
[92] H.F. Mildrum, G.A. Graves and Z.A. Abdelnour, ref. [9], p. 313.
[93] T. Shimoda, K. Akioka, O. Kobayashi, Y. Yamagami and A. Arai, ref. [15], vol. 1, p. 17.
[94] J.D. Livingston and D.L. Martin, IEEE Trans. Magn. MAG-20 (1983) 140.
[95] H.F. Mildrum and K.M.D. Wong, ref. [6], p. 35.
[96] M. Hamano, ref. [13], p. 683.
[97] R.M.W. Strnat, J.P. Clarke, H.A. Leupold and A. Tauber, J. Appl. Phys. 61 (1987) 3463.
[98] H.F. Mildrum, M.F. Hartings, K.M.D. Wong and K.J. Strnat, IEEE Trans. Magn. MAG-10 (1974) 723.
[99] S.C. Hanna and M.S. Walmer, ref. [12], p. 309.
[100] T. Shimoda, E. Natori and C. Tomita, ref. [12], p. 297.
[101] D. Li, H.F. Mildrum and K.J. Strnat, J. Appl. Phys. 63 (1988) 3984.
[102] D.L. Martin and M.G. Benz, IEEE Trans. Magn. MAG-8 (1972) 35.
[103] K.J. Strnat, in: Ferromagnetic Materials, vol. 4, eds. E.P. Wohlfarth and K.H.J. Buschow (North-Holland, Amsterdam, 1988) p. 131.
[104] K. Halbach, ref. [12], p. 123.
[105] H.A. Leupold, E. Potenziani III and J.P. Clarke, ref. [13], p. 109.

Neodymium–iron–boron permanent magnets

J.F. Herbst

Physics Department, General Motors Research Laboratories, Warren, MI 48090-9055, USA

and

J.J. Croat

Delco Remy Division - MAGNEQUENCH, General Motors Corporation, 6435 South Scatterfield Road, Anderson, IN 46013, USA

Permanent magnet research and technology have been propelled into a new era by the rare earth–iron–boron materials, $R_2Fe_{14}B$. Energy products surpassing all previous values have been attained in magnets based on $Nd_2Fe_{14}B$, the prototypical compound. In this review we place Nd–Fe–B in the historical context of permanent magnet evolution, summarize the intrinsic properties of the $R_2Fe_{14}B$ phases, and discuss the properties of practical Nd–Fe–B magnets produced by the two methods in present commercial use.

1. Introduction

A primary aspect of research in magnetism is the discovery and development of progressively more powerful permanent magnet materials. From a fundamental viewpoint this activity is stimulated by the challenge to understand and enhance intrinsic properties such as the magnetization and magnetocrystalline anisotropy. Strong motivation on the technological side is provided by continual widening of the scope of applications for hard magnets as their properties are improved. A new and exciting era in permanent magnet research and development has begun with the advent of the $R_2Fe_{14}B$ materials, in which R is a rare earth element. Energy products substantially exceeding all previous values have been achieved in magnets based on the prototypical compound, $Nd_2Fe_{14}B$. That fact coupled with economic advantages relative to predecessor samarium–cobalt materials has generated considerable technological interest, and the new compounds have also created a broad area for basic scientific inquiry.

This brief review is structured as follows. We first survey the historical evolution of permanent magnet materials and place neodymium–iron–boron in that context. Intrinsic characteristics of the $R_2Fe_{14}B$ compounds are then described, and we go on to discuss the properties of practical Nd–Fe–B magnets prepared by the two methods in current commercial use, namely, the rapid solidification technique of melt spinning and the traditional powder metallurgy, or sintering, approach.

2. Historical survey

Although the earliest permanent magnet, lodestone (comprised principally of magnetite, Fe_3O_4), was known to the ancients, the greatest strides in magnet development have occurred over the past hundred years. Each advance has been connected with the discovery of a new class of materials characterized by ever more desirable properties. This evolution has been monitored generally by one figure of merit for a permanent magnet, the maximum energy product $(BH)_{max}$, which provides a measure of the field that can be produced outside a unit volume of magnet material. In broad terms the larger $(BH)_{max}$, the greater the potential for reducing the size and weight of a device by replacing either electromagnets or permanent magnets having lower energy product. The so-called theoretical maximum energy product, the largest value realizable in principle, is an intrinsic quantity defined by

$$(BH)^*_{max} = (4\pi M_s)^2/4, \qquad (1)$$

where M_s is the saturation magnetization. This can be achieved only if the magnet retains M_s in a reverse field at least as large as $(4\pi M_s)/2$.

Fig. 1 is a semilogarithmic plot of the highest $(BH)_{max}$ values versus time during the past century for the five classes of technologically important permanent magnet materials. The straight line drawn between the earliest magnet steels and the Nd–Fe–B magnets underscores the exponential increase of $(BH)_{max}$ in this period. The steels, having additives such as C, Co, Cr or W, are characterized by rather high remanence, $B_r \sim 10$ kG, but their coercivities are no larger than a few hundred oersted and severely limit $(BH)_{max}$. These alloy systems essentially exploit the magnetic properties of elemental iron, for which $4\pi M_s = 21.5$ kG implies an enormous $(BH)^*_{max}$ of 115 MGOe. The anisotropy field H_a (the external field required to rotate the magnetization from the easy to a hard direction) is only ~ 500 Oe for iron, however, and $(BH)^*_{max}$ cannot be realized. The additives to the steels simply serve to translate the meager anisotropy field of iron into useful coercivity.

Alnicos, alloys of Al, Ni, Co and Fe with minor additions of other elements, were the next class of materials to advance the energy product. They were the first magnets to be aptly designated permanent because of their resistance to stray magnetic fields, mechanical shock and elevated temperatures. The shape anisotropy of finely dispersed, elongated, single domain particles is responsible for the coercivity, which can be more than twice that of the best magnet steels and which has enabled energy products as large as ~ 13 MGOe to be realized [1].

Based on the hexagonal oxides $BaFe_{12}O_{19}$ and $SrFe_{12}O_{19}$, the hard ferrites are included in fig. 1 because of their commercial importance even though they do not reside on the increasing $(BH)_{max}$ trend line. The coercivity of ferrite magnets derives from the uniaxial magnetocrystalline anisotropy, rather than the shape anisotropy, of

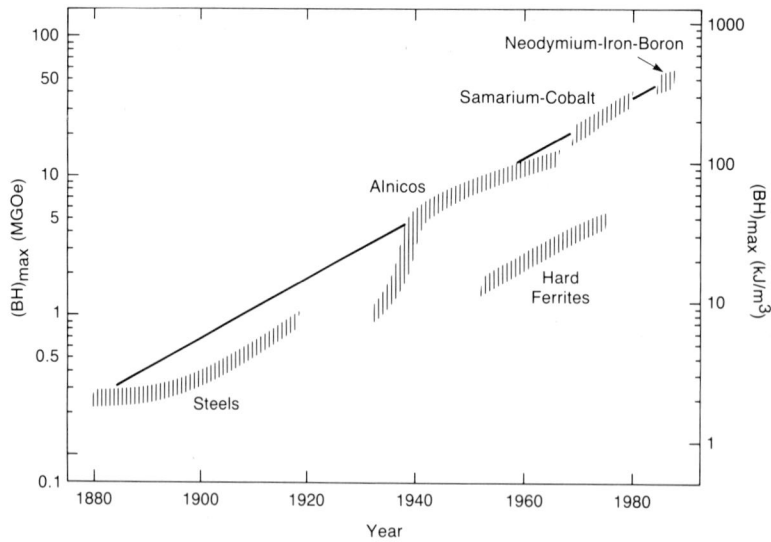

Fig. 1. Chronological trend of the maximum energy product $(BH)_{max}$ achieved in various permanent magnet materials.

single domain particles and can be substantially larger than that of the alnicos. The remanent magnetization is relatively low, however, and $(BH)_{max}$ is no larger than 5 MGOe. Per unit energy product the hexagonal ferrites are the most inexpensive hard magnets, and their high electrical resistivity has made them useful in a variety of high frequency applications [2].

Over the past three decades the energy product has been increased by rare earth–transition metal (R–TM) materials in which the the R component provides most of the magnetocrystalline anisotropy intrinsically responsible for the coercivity in a practical magnet while the magnetization arises principally from the TM sublattice. Investigation of such systems began with magnetization measurements on the RCo_5 compounds having the hexagonal $CaCu_5$ crystal structure, and determinations of the anisotropy indicated that the RCo_5 compounds were promising candidates for permanent magnet development. $SmCo_5$ became the focus of technological efforts, and energy products in the 16–28 MGOe range have been obtained by sintering [3]. For the compounds in which the rare earth 4f shell has a non-zero orbital magnetic moment, theoretical analyses demonstrated that the behavior of the R ion under the combined influence of the exchange and crystalline electric fields controls the easy magnetization direction and accounts for much of the large uniaxial anisotropy.

In view of their richer cobalt content and, hence, higher magnetization and larger $(BH)_{max}^*$ values, the R_2Co_{17} compounds were also heavily investigated as hard magnet materials. The cobalt sublattice in the R_2Co_{17} phases, in contrast to the RCo_5 series, is characterized by basal plane rather than uniaxial anisotropy, and the total anisotropy is unfortunately lower. Nevertheless, the R_2Co_{17} magnetization and the RCo_5 anisotropy have been exploited via the technique of precipitation hardening. Energy products as large as 33 MGOe have been obtained in samarium–cobalt-based alloys featuring a two-component cellular microstructure in which a Sm_2Co_{17}-type phase is surrounded by a $SmCo_5$-type boundary phase [4]. The development of both R_2TM_{17} and $R-TM_5$ permanent magnets has been reviewed extensively by Kumar [5] and by Strnat [6].

The samarium–cobalt materials are attractive replacements for electromagnets in many applications, but their use has been constrained by economic disadvantages. Samarium is the least abundant, and hence most expensive, of the light rare earths, and this unfavorable circumstance is compounded by the fact that the price and availability of cobalt are subject to large, unpredictable excursions. The bulk of the world's cobalt is mined in Zaire, and in particular the 1978 civil war there diminished supplies, radically increased prices, and, perhaps in retrospect most importantly, helped spur searches for alternative R–TM systems having economically more preferable elemental constituents.

Iron-based materials having the characteristics of samarium–cobalt had long been desired; however, no suitable compounds were known in 1978. R–Fe phases with the $CaCu_5$ structure do not exist, and the R_2Fe_{17} compounds in which R is a light rare earth have unacceptably low Curie temperatures. Despite the apparent absence of appropriate compounds, two lines of research led by the time of the 1983 Conference on Magnetism and Magnetic Materials in Pittsburgh to the preparation of iron-based rare earth hard magnets. One approach relied on the rapid solidification, by melt spinning, of rare earth–iron–metalloid alloys [7–11], and the other involved conventional powder metallurgy [12].

Croat et al. [7,13] reported energy products as large as 14 MGOe in melt-spun ribbons of $Nd_{0.135}Fe_{0.817}B_{0.048}$, work which evolved from earlier studies of rapidly solidified R–Fe binary alloys [14,15]. Extending previous investigations of crystallized amorphous $La_{0.05}Tb_{0.05}Fe_{0.74}B_{0.16}$ materials [16], Koon and Das [8] found $(BH)_{max} \sim 13$ MGOe in annealed melt-spun $La_{0.020}Nd_{0.130}Fe_{0.783}B_{0.067}$ and $La_{0.005}Pr_{0.145}Fe_{0.783}B_{0.067}$. Hadjipanayis and coworkers [9,17,18] explored a variety of rapidly quenched rare earth–iron–metalloid systems and observed an energy product of 13 MGOe in heat-treated $Pr_{0.16}Fe_{0.76}Si_{0.03}B_{0.05}$. These series of inquiries were encouraged in part by the seminal finding of Clark [19] that amorphous $Tb_{0.33}Fe_{0.67}$ developed a coercive force of ~3 kOe after annealing. Clark's result motivated examination of amorphous or metastable precursors, preparation of which necessitates rapid cooling methods, as a possible route toward a high energy product iron-based magnet. With the working assumption that binary alloys were not promising, Sagawa et al. [12] applied traditional sintering methods to ternary light rare earth–iron systems and obtained $(BH)_{max} \sim 36$ MGOe for $Nd_{0.15}Fe_{0.77}B_{0.08}$. All of these high $(BH)_{max}$ materials, rapidly solidified as well as sintered, contained a novel ternary crystalline compound. Although its chemical formula was tentatively and variously identified as $R_3Fe_{21}B$ [9,20], $R_3Fe_{20}B$ [17], $R_3Fe_{20}B_2$ [21], $R_5Fe_{25}B_3$ [22] and as $R_3Fe_{16}B$ in an earlier crystallographic study of R–Fe–B systems [23], the correct stoichiometry, $R_2Fe_{14}B$, and detailed crystal structure were soon determined [24–26]. The larger energy product reported by Sagawa et al. [12] arose from the crystallite orientation afforded by the powder metallurgy procedure.

Considerable technological interest has centered on $Nd_2Fe_{14}B$ because of its excellent intrinsic properties $[(BH)^*_{max} \sim 64$ MGOe, $H_a \sim 73$ kOe] and economic advantages over samarium–cobalt materials. Practical magnets with energy products in the 40–50 MGOe range, values significantly larger than any previously attained (cf. fig. 1), have been prepared from melt-spun [27,28] and sintered [29] alloys, and large-scale production programs employing both approaches have been implemented. The spectrum of applications for Nd–Fe–B magnets continues to expand. Scientifically, the existence of an entire $R_2Fe_{14}B$ series has stimulated a great deal of research on their properties and the physics underlying those properties.

3. Crystal structure

$Nd_2Fe_{14}B$ emerged as the prototypal $R_2Fe_{14}B$ compound since magnets having the largest energy products contain it as the principal constituent. Roughly comparable hard magnetic properties can be realized with $Pr_2Fe_{14}B$, but the Nd phase has received more attention as a magnet material because it has somewhat higher magnetization and neodymium is more abundant than praseodymium. The exact stoichiometry and crystal structure of $Nd_2Fe_{14}B$ were first established by neutron powder diffraction analysis [24] and confirmed by two independent single-crystal X-ray investigations [25,26].

Fig. 2 displays the $Nd_2Fe_{14}B$ unit cell. The lattice symmetry is tetragonal (space group $P4_2/mnm$), and each unit cell contains four formula units or 68 atoms. There are six crystallographically distinct iron sites, two different rare earth positions, and one boron site. Each $Nd_2Fe_{14}B$ unit cell consists of an eight-layer repeat structure perpendicular to the c-axis. Spatial relationships among several of the layers are illustrated in fig. 3. All the Nd and B atoms, but only the four Fe(c) atoms out of the total iron

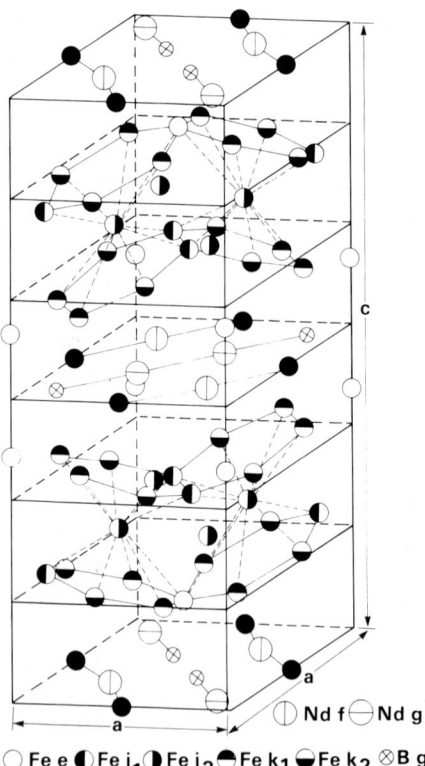

Fig. 2. Tetragonal unit cell of $Nd_2Fe_{14}B$, the prototypical structure of the $R_2Fe_{14}B$ compounds. The c/a ratio in the figure is exaggerated to emphasize the puckering of the hexagonal iron nets.

hard magnet materials. The iron layers in fig. 3(b) closely resemble the nets in the σ phase formed in the Fe–Cr and Fe–Mo systems. Among a number of features common to $Nd_2Fe_{14}B$ and the $CaCu_5$ structure of $SmCo_5$ is the centering of R atoms within hexagonal prisms. Additional analogies can be identified since many rare earth–transition metal (R–TM) structures are derivatives of the $CaCu_5$ type. For example, the Th_2Zn_{17} structure can be generated from $CaCu_5$ by appropriate atomic replacements; it characterizes Sm_2Co_{17} and Nd_2Fe_{17}, the binary compound nearest in stoichiometry to $Nd_2Fe_{14}B$. The $Fe(j_2)$ sites in $Nd_2Fe_{14}B$ and the Fe(c) sites in Nd_2Fe_{17} are crystallographic and magnetic cognates inasmuch as each has the largest number of near-neighbor Fe atoms and the largest magnetic moment in its structure.

A connection to TM–metalloid systems is evident from the boron coordination in $Nd_2Fe_{14}B$. Each boron occupies the center of a trigonal prism (fig. 4) formed by the three nearest iron atoms above and the three below the basal (or $z = 1/2$) plane. Fig. 3(a) shows that the triangular prism faces participate in completing the hexagonal Fe nets over the square basal plane units. The prisms pucker the Fe nets since the Fe(e) and $Fe(k_1)$ atoms in them are displaced significantly toward the B-containing planes as compared with

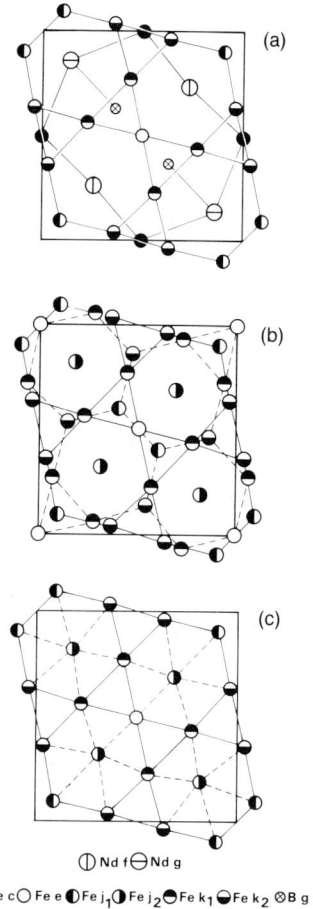

Fig. 3. (a) Projection of the basal plane and the first Fe layer ($z \approx 1/6$) in $Nd_2Fe_{14}B$. (b) Projection of the three Fe layers between the basal and $z = 1/2$ planes. (c) Projection of the first Fe layer and the neighboring layer of $Fe(j_2)$ atoms ($z \approx 1/4$).

number of 56 reside in the $z = 0$ and $z = 1/2$ mirror planes [fig. 2, fig. 3(a)]. Between these planes the other Fe atoms form three puckered nets. The $Fe(k_1)$, $Fe(k_2)$, $Fe(j_1)$ and Fe(e) sites comprise two slightly distorted hexagonal arrays rotated by $\sim 30°$ with respect to one another; they enclose a net of $Fe(j_2)$ sites located above or below the centers of the hexagons in the neighboring layers [fig. 3(b)]. Fig. 3(c) makes clear the essentially perfect triangular coordination of the $Fe(j_2)$ atoms and each of the hexagonal layers.

Many parallels exist between $Nd_2Fe_{14}B$ and other structures, including those of the Sm–Co

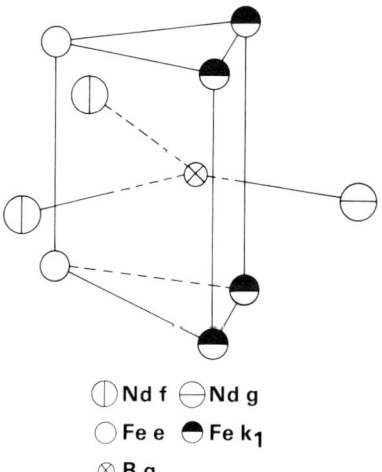

Fig. 4. Trigonal prism containing a boron atom in the $Nd_2Fe_{14}B$ structure.

the other Fe atoms in the nets. Linking the Fe layers above and below the planes containing Nd and B, the prisms evidently contribute to the stability of the structure. Such trigonal prisms are fundamental to the structure of many transition metal–metalloid systems, both crystalline (e.g., FeB, Fe_3C, Fe_3P) and amorphous.

The $R_2Fe_{14}B$ structure has been found to form with yttrium, thorium and all the rare earth elements except europium and radioactive promethium. Only two families of $Nd_2Fe_{14}B$-type compounds are known in which Fe or B is totally replaced by another element, namely, $R_2Co_{14}B$ and $R_2Fe_{14}C$. Partial substitution of R, Fe or B with maintenance of the $R_2Fe_{14}B$ structure is possible with many other elements, and $R_2Fe_{14}BH_x$ and $R_2Co_{14}BH_x$ interstitial hydride series exist.

4. Intrinsic properties of the $R_2Fe_{14}B$ compounds

4.1. Systematics

Lattice constants and summary magnetic data for the $R_2Fe_{14}B$ phases are listed in table 1. The decrease in the radii of the trivalent lanthanide ions with increasing atomic number Z, the lanthanide contraction, is apparent in the decrease of the c lattice parameter through the rare earth (La–Lu) series. As Z increases, the addition of another electron to the 4f shell does not completely screen the larger nuclear charge, and the radius of the rare earth ion contracts. $Ce_2Fe_{14}B$ deviates from the trend since Ce is essentially tetravalent in the $R_2Fe_{14}B$ structure, as is Th; all other R constituents are trivalent. The a lattice parameter declines much more slowly with Z, suggesting that the basal plane dimension is influenced more by the particularly stable trigonal Fe–B prisms rather than by the size of the R ions.

An estimate of the average iron moment in the series, $\mu_{Fe} \approx 2.1\mu_B$, is afforded by the 4 K magnetizations of the La, Ce, Lu, Y and Th compounds in table 1 since those materials can be regarded as magnetically blank with respect to the R component. The proximity of μ_{Fe} to the 2.2 μ_B moment of elemental α-Fe is a remarkable fact considering that α-Fe is body-centered cubic with but one symmetry site while $R_2Fe_{14}B$ is tetragonal with six crystallographically distinct Fe sites. The magnitude of μ_{Fe} suggests that the six iron sublattices in $R_2Fe_{14}B$ are ferromagnetically

Table 1
Lattice constants a, c, saturation magnetizations M_s (4 K) and $4\pi M_s$ (295 K), anisotropy fields H_a (295 K), spin reorientation temperatures T_s, and Curie temperatures T_c of the $R_2Fe_{14}B$ compounds [30]

Compound	a (Å)	c (Å)	4 K M_s (μ_B/f.u.)	295 K $4\pi M_s$ (kG)	295 K H_a (kOe)	T_s (K)	T_c (K)
$La_2Fe_{14}B$	8.82	12.34	30.6	13.8	20	–	530
$Ce_2Fe_{14}B$	8.76	12.11	29.4	11.7	26	–	424
$Pr_2Fe_{14}B$	8.80	12.23	37.6	15.6	75	–	565
$Nd_2Fe_{14}B$	8.80	12.20	37.7	16.0	73	135	585
$Sm_2Fe_{14}B$	8.80	12.15	33.3	15.2	150	–	616
$Gd_2Fe_{14}B$	8.79	12.09	17.9	8.9	24	–	661
$Tb_2Fe_{14}B$	8.77	12.05	13.2	7.0	220	–	620
$Dy_2Fe_{14}B$	8.76	12.01	11.3	7.1	150	–	598
$Ho_2Fe_{14}B$	8.75	11.99	11.2	8.1	75	58	573
$Er_2Fe_{14}B$	8.73	11.95	12.9	9.0	8	325	554
$Tm_2Fe_{14}B$	8.73	11.93	18.1	11.5	8	313	541
$Yb_2Fe_{14}B$	8.71	11.92	23	12	...	115	524
$Lu_2Fe_{14}B$	8.70	11.85	28.2	11.7	26	–	535
$Y_2Fe_{14}B$	8.76	12.00	31.4	14.1	26	–	565
$Th_2Fe_{14}B$	8.80	12.17	28.4	14.1	26	–	481

coupled and collinear; no experimental evidence to the contrary has emerged. For the compounds in which the rare earth supports a moment the average R moment μ_R can be estimated from table 1 by, for instance,

$$\mu_R = \tfrac{1}{2}\left[M_s^{4\,K}(R_2Fe_{14}B) - M_s^{4\,K}(Y_2Fe_{14}B)\right]. \quad (2)$$

Two inferences can be drawn from the resulting values.

First, the sign of μ_R indicates that the R and Fe moments are coupled ferromagnetically (antiferromagnetically) for the light (heavy) rare earths, an observation which holds for a broad range of crystalline as well as amorphous R–TM systems. Since Hund's rules apply to the R 4f shell, so that $J = L - S$ ($L + S$) for the light (heavy) lanthanides, the inference is equivalent to the statement that the rare earth 4f and iron 3d *spin* moments are always antiparallel, i.e., the 3d–4f exchange interaction is invariably antiferromagnetic. Although understanding of this ubiquitous characteristic is incomplete, it is likely that the rare earth 5d electrons mediate the 3d–4f interaction, as suggested by Campbell [31]. In this picture the 4f spins induce a local 5d moment via ferromagnetic, atomic 4f–5d exchange; 3d–5d exchange then generates the indirect 3d–4f interaction, which can be expected to be antiferromagnetic since the rare earths reside at the beginning of a transition metal series with respect to their 5d character while iron, cobalt and nickel are in the second half of the 3d series.

The second inference emerging from the estimates specified by eq. (2) is that each μ_R closely approaches the corresponding free ion R moment, gJ. This may appear surprising since an R ion having orbital angular momentum $L \neq 0$ can dominate the magnetocrystalline anisotropy (compare, for example, the H_a entries for $La_2Fe_{14}B$ and $Nd_2Fe_{14}B$ in table 1), pointing to significant crystal field splitting and a ground state R moment reduced from its free ion value. The fact that $\mu_R \approx gJ$ is rather to be interpreted as an indication that the influence of the exchange field arising from the Fe magnetization and acting on the 4f moments is stronger than that of the crystal field, even though the latter controls the direction of the 4f moments [32]. A corollary implication is that the total R and Fe sublattices can be considered collinear, at least in first approximation. This holds for most members of the $R_2Fe_{14}B$ series; however, the Nd, Ho, Er and Tm compounds change easy direction at a spin reorientation temperature T_s (included in table 1), and there is evidence that the total R moment or components of it deviate from collinearity with the Fe sublattice at low temperatures where higher order terms in the crystal field acting on the R ions become relatively more important.

Measured values of the Curie temperatures T_c for the $R_2Fe_{14}B$ materials are listed in the final column of table 1. Notwithstanding the essential identity of the average Fe moment with that of elemental Fe, the T_c entries are all substantially lower than $T_c(Fe) = 1043$ K, a property shared by all the known binary R–Fe compounds. In fig. 5 $T_c(R_2Fe_{14}B)$ is displayed as a function of $\sqrt{G}$, where $G \equiv (g-1)^2 J(J+1)$ is the De Gennes factor for the rare earth ion; $\sqrt{G}$ is the effective R spin. The T_c values for the La, Ce, Lu, Y and Th compounds, which have no R moment, show that the scale of T_c for the series is set by the exchange interactions among the iron spins. The substantial excursion from $T_c(Ce_2Fe_{14}B) = 424$ K to $T_c(Y_2Fe_{14}B) = 565$ K emphasizes the sensitiv-

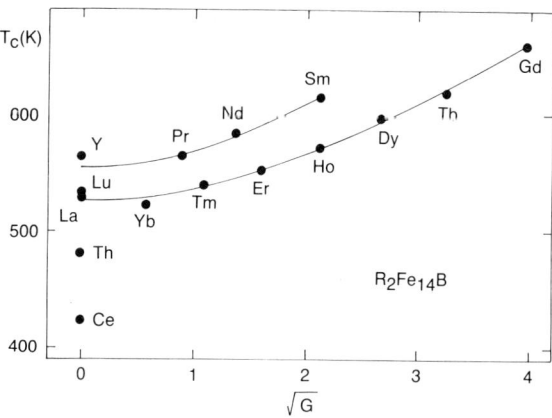

Fig. 5. Curie temperatures T_c of the $R_2Fe_{14}B$ compounds as a function of $\sqrt{G}$, where $G \equiv (g-1)^2 J(J+1)$ is the De Gennes factor of the rare earth ion.

ity of the electronic structure to the nature of the R ion and the lattice constants. For the compounds with an R moment the T_c data in fig. 5 divide naturally into two groups depending on the character of the R–Fe moment coupling. T_c is one smooth function of $\sqrt{G}$ for the ferromagnetic (light R) materials and a second smooth function of $\sqrt{G}$ for the ferrimagnetic (heavy R) phases. The solid lines in fig. 5 are fits to the T_c data from a simple mean field model in which Fe–Fe and R–Fe exchange parameters are allowed to differ for the light and heavy R compounds; the R–Fe exchange controls the T_c variation through the series and is significantly stronger for the light R materials [30].

Several distinct types of easy magnetization behavior occur in the $R_2Fe_{14}B$ series. The Fe sublattice anisotropy favors the tetragonal c-axis as the easy direction, which obtains for all temperatures below T_c in the La, Ce, Lu, Y and Th compounds having $\mu_R = 0$ as well as for the Pr, Gd, Tb and Dy members with $\mu_R \neq 0$. $Sm_2Fe_{14}B$ features basal plane $\langle 100 \rangle$ magnetization throughout the magnetically ordered regime. The Nd, Ho, Er, Tm and Yb phases change easy direction at the spin reorientation temperatures T_s listed in table 1. In $Er_2Fe_{14}B$, $Tm_2Fe_{14}B$ and $Yb_2Fe_{14}B$ the easy direction is a $\langle 100 \rangle$ basal plane direction below T_s and a $\langle 001 \rangle$ direction above T_s. $Nd_2Fe_{14}B$ and $Ho_2Fe_{14}B$ exhibit more complex behavior. Above T_s both compounds are uniaxial, but below T_s the magnetization cants away from the c-axis toward the [110] direction by an angle which increases with decreasing temperature to $\sim 30°$ at 4 K in $Nd_2Fe_{14}B$ and to $\sim 20°$ at 4 K in $Ho_2Fe_{14}B$.

4.2. Crystal field-exchange models

While most of the magnetization is provided by the Fe sublattice, the magnetocrystalline anisotropy of the $R_2Fe_{14}B$ compounds whose R component has non-zero orbital moment is dominated by the rare earth contribution. If $L \neq 0$ the 4f charge distribution is aspherical, interacts non-trivially with the crystalline electric field (CEF), and directional preferences emerge which influence the R moment alignment because the 4f spin–orbit coupling is strong. The large Fe–Fe exchange couples the six distinct Fe moments to form a single overall Fe magnetization which can be considered to generate an effective R–Fe exchange field H_{ex}. Excellent progress toward a detailed understanding of the anisotropy and magnetic structure of the $R_2Fe_{14}B$ systems has been made with models focussing on the interplay between the rare earth CEF and the R–Fe exchange interactions.

In such a description the Hamiltonian for an R ion within the lowest J multiplet can be written as [33]

$$\mathcal{H}_R(i) = \mathcal{H}_{CEF}(i) + 2(g-1)\boldsymbol{J} \cdot \boldsymbol{H}_{ex} + g\boldsymbol{J} \cdot \boldsymbol{H}, \quad (3)$$

where H is an applied field and

$$\begin{aligned}\mathcal{H}_{CEF}(i) &= \sum_{n,m} B_n^m(i) O_n^m \\ &= B_2^0(i)O_2^0 + B_2^{-2}(i)O_2^{-2} + B_4^0(i)O_4^0 \\ &\quad + B_4^{-2}(i)O_4^{-2} + B_4^4(i)O_4^4 + B_6^0(i)O_6^0 \\ &\quad + B_6^{-2}(i)O_6^{-2} + B_6^4(i)O_6^4 \\ &\quad + B_6^{-6}(i)O_6^{-6}. \end{aligned} \quad (4)$$

The CEF acting on each R ion has orthorhombic point symmetry and differs for ions residing on the basal or midway planes of fig. 2, so that four R sites (distinguished by the index i) must be considered in principle. The O_n^m are operator equivalents, and the B_n^m are specified by

$$B_n^m = \theta_J^n \langle r^n \rangle A_n^m, \quad (5)$$

where θ_J^n is the Stevens factor (α_J, β_J, γ_J for $n = 2$, 4, 6, respectively) dependent only the quantum numbers of the R 4f shell, $\langle r^n \rangle$ is the radial matrix element of r^n for the 4f shell, and the A_n^m are coefficients of the spherical harmonics of the CEF potential. If the $n = 2$ terms are the largest among the B_n^m, the sign of the principal anisotropy constant K_1, which often dictates anisotropy preference, is governed by the sign of α_J. There is in fact perfect correspondence between the sign of α_J and the low-temperature easy direction for the Pr, Sm, Tb, Dy, Er, Tm and

Yb members of the $R_2Fe_{14}B$ series: uniaxial if $\alpha_J < 0$, basal plane if $\alpha_J > 0$. The correlation is blurred by the moment canting below T_s in $Nd_2Fe_{14}B$ and $Ho_2Fe_{14}B$, but it correctly specifies the c-axis easy above T_s.

Various forms of the CEF-exchange model have been employed by many investigators. These efforts, phenomenological insofar as their goal is determination of the parameters in eq. (3) capable of explaining experimental results, demonstrate collectively that the model contains most of the basic physics responsible for the anisotropy and magnetic structure of the $R_2Fe_{14}B$ materials.

Boltich and Wallace [34] found early on that the existence of a spin reorientation in $Nd_2Fe_{14}B$ and the absence of such a transition in $Pr_2Fe_{14}B$ could be explained with reasonable choices of B_2^0 and B_4^0 in eq. (3). Their work showed that $n > 2$ CEF terms, rather than anisotropy competition between f and g sites, control the moment canting. The most comprehensive investigation of all nine $L \neq 0$ $R_2Fe_{14}B$ compounds within the CEF-plus-exchange framework has been that of Yamada et al. [33], who used the system Hamiltonian

$$\mathcal{H} = \sum_{i=1}^{4} \mathcal{H}_R(i) + 28 K_1^{Fe}(T) \sin^2\theta - 28 \mu_{Fe}(T) \cdot H \tag{6}$$

to calculate magnetization curves and derive magnetic structures. The Fe anisotropy constant $K_1^{Fe}(T)$ and moment $\mu_{Fe}(T)$ were taken to have the same temperature dependence as in $Y_2Fe_{14}B$. Yamada et al. [33] found good accommodation of experiment for the heavy R materials using eq. (3) for the ground state J multiplet, but excited multiplets were discovered to have appreciable impact on the results for the light R compounds, for which the appropriate generalization of eq. (3) was employed. Comparisons of theoretical and experimental magnetization curves at 4 and 290 K for $Pr_2Fe_{14}B$ and $Nd_2Fe_{14}B$ are shown in fig. 6. Solid lines represent calculations in which excited multiplets are included; dashed lines correspond to the ground multiplet alone. The effect of the excited state is particularly apparent in the 4 K $Pr_2Fe_{14}B$ results in fields where first-order

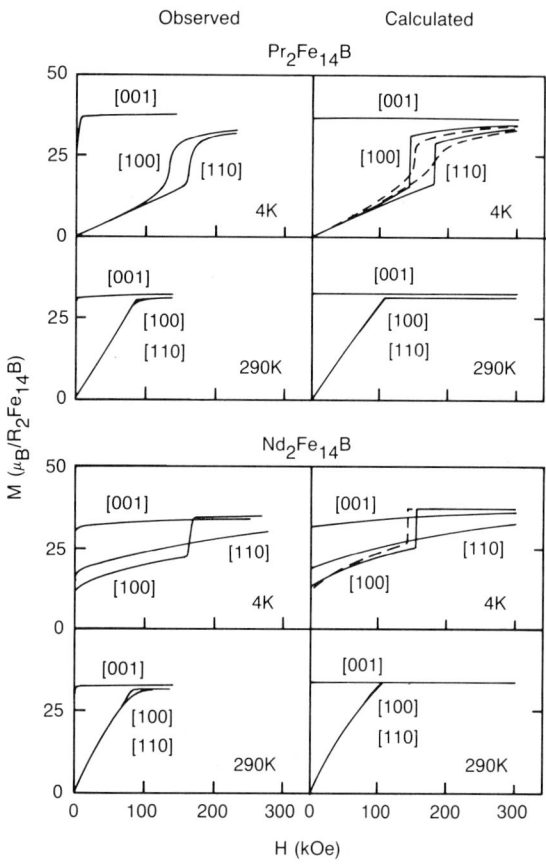

Fig. 6. Observed and calculated 4 and 290 K magnetization curves for $Pr_2Fe_{14}B$ and $Nd_2Fe_{14}B$ (adapted from ref. [33]).

magnetization processes (FOMPs) have been observed in the [100] and [110] directions. The calculated $Nd_2Fe_{14}B$ equilibrium spin structure below $T_s^{calc} = 130$ K is slightly non-collinear with all moments in the $(1\bar{1}0)$ plane.

Cadogan et al. [35] developed a new method for determining CEF and exchange parameters which was applied initially to $Nd_2Fe_{14}B$ and subsequently to other $R_2Fe_{14}B$ compounds [36]. Principally designed to reduce the amount of iterative computing, the technique relies on comparison of calculated anisotropy constants with those deduced from experimental magnetization curves. Overall the new approach successfully describes the compounds to which it has been applied. In the case of $Nd_2Fe_{14}B$, for example, Cadogan et al. [35] found the FOMP in the [110] direction at 4 K (see fig. 6) to be reproduced well and $T_s^{\text{calc}} = 138$ K.

Given the parameter freedom afforded by a crystal field-exchange model, it is not surprising that application under various simplifying conditions (neglect of f-g site dependences and/or excited multiplets, restriction of the number of B_n^m terms in the CEF, constraint of certain ratios to point charge results, etc.) leads to differences in the parameters emerging from the analyses despite, in many instances, comparable success in simulating magnetization data. More definitive parameter evaluation must await accurate interpretation of microscopic measurements, especially inelastic neutron diffraction. Loewenhaupt et al. [37] have made initial efforts in this direction.

5. Practical magnets

In order to serve as a base for a viable permanent magnet, a stoichiometric compound must satisfy the three principal requirements of (1) a Curie temperature T_c significantly exceeding any magnet operating temperature, (2) high spontaneous magnetization M_s and (3) large magnetocrystalline anisotropy as measured, for instance, by the anisotropy field H_a. The upper limit for the remanence B_r of a practical magnet is imposed by $4\pi M_s$, and that for the intrinsic coercivity H_{ci} is set by H_a. Furthermore, fabrication of a practical magnet from a compound possessing suitable intrinsic properties demands a specific preparation method and an associated starting ingot composition. The technical magnetic characteristics B_r, H_{ci} and energy product $(BH)_{\text{max}}$ are extrinsic inasmuch as they depend crucially on the microstructure of the material. The microstructure involves the size, shape and orientation of the crystallites of the compound and also the nature and distribution of secondary phases which usually control domain wall formation and motion and hence determine the magnetization and demagnetization behavior. Depending on the preparation technique and the secondary constituents, the starting composition yielding optimum magnetics can deviate appreciably from the stoichiometry of the major phase. Two quite distinct methods are in commercial use for producing Nd–Fe–B magnets: the rapid solidification technique of melt spinning and the traditional powder metallurgy (sintering) approach.

5.1. Melt-spun materials

Melt spinning has been employed for some time in the production of magnetically soft amorphous alloys for such applications as transformer cores, but its use in hard magnet technology represents a fundamental departure from conventional powder metallurgy methods. The melt spinning procedure involves ejection of molten starting alloy through a crucible orifice onto the surface of a substrate disc, usually copper because of its high thermal conductivity, rotating with surface velocity v_S. Since the rare earths are very reactive chemically, the process is carried out in an inert atmosphere, most often argon gas. The cooling or quench rate attainable by melt spinning can be as high as 10^6 K/s, and that rate can be varied by changing v_S, which in first approximation is directly proportional to the cooling rate. Ribbons typically 30–50 μm thick, 1–3 mm wide, and having lengths on the millimeter or centimeter scale depending on v_S are produced.

5.1.1. Ribbon properties

The microstructure and magnetic properties of melt-spun Nd–Fe–B ribbons are sensitively dependent on the quench rate (i.e., on v_S), with H_{ci} and $(BH)_{\text{max}}$ each exhibiting a maximum as a function of v_S. This point is illustrated in fig. 7, which displays demagnetization curves obtained by Croat et al. [7] for powdered and compacted $Nd_{0.135}Fe_{0.817}B_{0.048}$ ribbons melt spun at $v_S = 14$, 19 and 35 m/s together with corresponding scanning electron micrographs of ribbon fracture surfaces. High quench rates ($v_s \gtrsim 30$ m/s) produce essentially amorphous, "overquenched" materials with negligible H_{ci} and $(BH)_{\text{max}}$. Quench rates near $v_S = 19$ m/s in the example of fig. 7 yield ribbons having the highest coercivities ($H_{ci} \approx 14$ kOe) and energy products [$(BH)_{\text{max}} \sim 14$ MGOe]. The SEM micrograph for $v_S = 19$ m/s in fig. 7

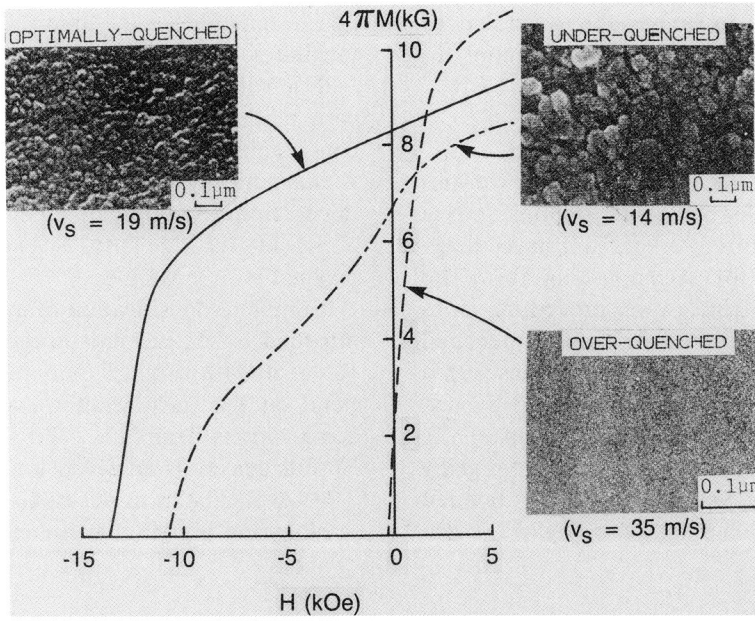

Fig. 7. Room temperature demagnetization curves and corresponding scanning electron micrographs of melt-spun $Nd_{0.135}Fe_{0.817}B_{0.048}$ ribbon fracture surfaces.

shows that these optimally-quenched ribbons consist of spheroidal $Nd_2Fe_{14}B$ grains with an average diameter of ~ 300 Å. Transmission electron microscopy (TEM) studies [38] have revealed that the distribution of c-axes is spatially, and hence magnetically, isotropic throughout most of the ribbon volume and that a very thin (~ 20 Å) Nd-rich, B-deficient amorphous phase of approximate composition $Nd_{0.7}Fe_{0.3}$ (close to the eutectic in the binary Nd–Fe phase diagram) is present in the intergranular regions (see fig. 8). $Nd_2Fe_{14}B$ comprises about 95% of the volume. Some preferential orientation of the easy c-axes normal to the ribbon plane has been produced by decreasing the quench rate [39] or by increasing the melt temperature [40], but at the expense of coercivity. As v_S is reduced below the value leading to optimum magnetics, ribbons comprised of progressively larger $Nd_2Fe_{14}B$ crystallites and characterized by decreasing H_{ci} and $(BH)_{max}$ are formed (cf. the $v_S = 14$ m/s results in fig. 7). From the technological perspective it is significant that largely amorphous, overquenched ribbons, melt spun at quench rates required only to exceed a minimum value rather than confined to

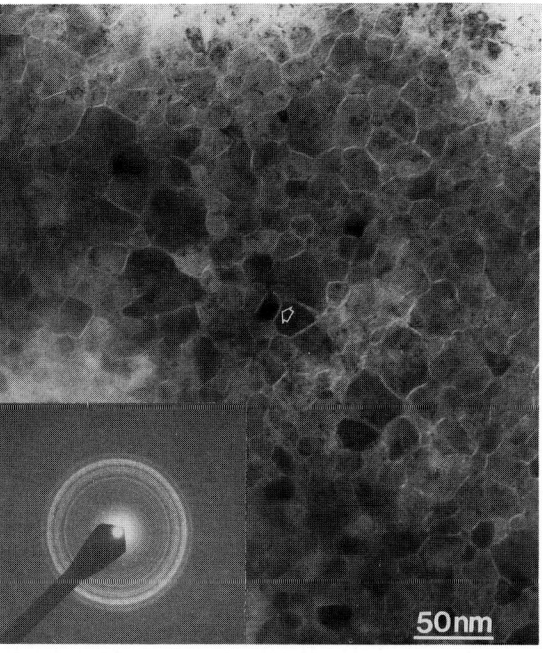

Fig. 8. Bright-field transmission electron micrograph of optimally-quenched Nd–Fe–B ribbon [38]. The inset is a selected area diffraction pattern whose rings indicate random distribution of $Nd_2Fe_{14}B$ grains. Note the very thin intergranular phase indicated by the arrow.

a narrow interval, can be suitably annealed to nearly duplicate the properties of optimum direct-quenched materials [13].

5.1.2. Consolidation methods

Production of a bulk magnet requires consolidation of the brittle melt-spun ribbons into a dense form. Lee and coworkers [41] have developed three means for accomplishing this. The first and most straightforward procedure is to cold press coarsely ground, optimally-quenched ribbons or annealed overquenched ribbons with a bonding medium such as epoxy. Due to the extremely fine particle size no degradation of the properties accompanies pulverization, making melt-spun materials very desirable for bonded magnet production. The flat geometry of the ribbon fragments (~ 0.5 mm $\times 0.5$ mm $\times 40$ μm) facilitates efficient packing so that densities of $\sim 85\%$ with respect to $Nd_2Fe_{14}B$ can be realized with modest pressures of 600–700 MPa. As its constituent ribbons, the resulting bonded magnet is magnetically isotropic. The remanence is reduced from the optimally-quenched ribbon value $B_r \sim 8$ kG (cf. fig. 7) to ~ 7 kG, and the energy product is ~ 9 MGOe.

Complete densification of melt-spun ribbons is afforded by the second procedure, hot pressing. Requisite pressures P and temperatures T depend on the starting alloy composition, but for compositions near $Nd_{0.135}Fe_{0.817}B_{0.048}$ (as in fig. 7), full density is achieved for $P \sim 100$ MPa and $T \sim 970$ K. The most desirable grain size is developed by hot pressing overquenched ribbons; vari-

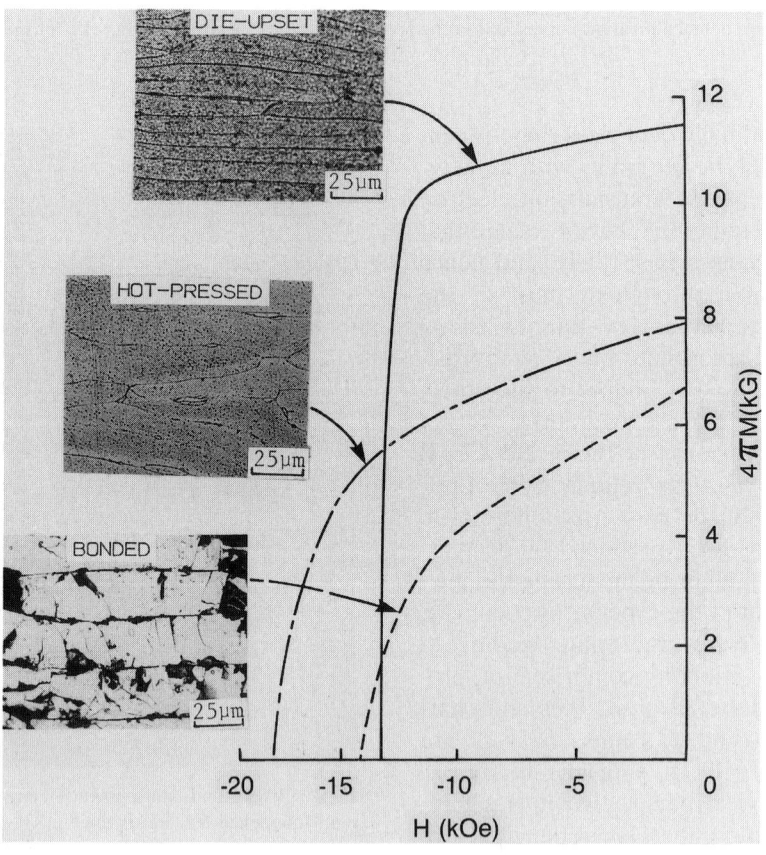

Fig. 9. Room temperature demagnetization curves and optical micrographs of bonded, hot-pressed and die-upset Nd–Fe–B magnets prepared from melt-spun ribbons.

ation of the magnetics with pressing conditions has been investigated by Gwan et al. [42] as well as by Lee and colleagues [41]. Energy products of the hot-pressed magnets are in the 10–20 MGOe range.

Uniaxial hot pressing introduces only a slight (~ 10%) crystallographic alignment of the easy c-axes parallel to the press direction. Much greater alignment (~ 75%), and consequently much larger energy products, can be obtained by the third method, in which an initial hot press is followed by another in a die cavity of greater diameter. This second hot press in a large die, designated die upsetting, produces bulk lateral plastic flow and an accompanying reduction in ribbon thickness. Nozawa et al. [43] and Tokunaga et al. [44] have examined the dependence of the magnetics on composition and on process parameters such as strain rate, working temperature, and the extent of deformation; greatest magnetic alignment was achieved only if the original sample height was reduced by at least a factor of four during die upsetting. As their hot-pressed precursors, die-upset magnets are fully dense, and energy products as large as 45 MGOe have been attained [28].

Fig. 9 shows optical micrographs of these three forms of rapidly solidified Nd–Fe–B magnets and also representative room temperature demagnetization curves measured with the applied field parallel to the press direction. It is clear that the hot-pressed ribbons deform plastically to fill the available volume without developing the cracks apparent in the bonded magnet. The diminished ribbon thickness resulting from the lateral plastic flow in the die-upset magnet is evident.

The optical micrographs of fig. 9 convey the overall packing features of the ribbons comprising the magnets, but it is the microstructure on the dimensions of grains and intergranular regions *within* the ribbons that controls the macroscopic magnetic properties. The grain texture of a hot-pressed magnet, prepared from overquenched ribbons, is similar to that of optimally-quenched ribbons in fig. 7, although somewhat coarser [45,46]. Die-upsetting, however, modifies the spheroidal grains in the hot-pressed ribbons to platelets ~ 3000 Å in diameter and ~ 600 Å

Fig. 10. Transmission electron micrograph of a die-upset Nd–Fe–B magnet (courtesy R.K. Mishra).

thick [45–47], as shown in the TEM micrograph of fig. 10. The platelets are stacked transverse to the press direction with the easy c-axis perpendicular to the face of each grain. A crystalline (fcc) phase, again near the $Nd_{0.7}Fe_{0.3}$ composition, forms in the intergranular regions [46]. The growth and stacking of these platelet grains is responsible for the magnetic alignment and the increased energy product. Mishra et al. [46,48] and Li and Graham [49] have demonstrated that the mechanism for grain alignment during die upsetting involves grain boundary sliding and anisotropic grain growth. These findings agree qualitatively with the results of Tenaud et al. [50], who demonstrated that the a-axis is the easy growth direction of $Nd_2Fe_{14}B$ single crystals and inferred that, under an applied stress, only grains with a-axes perpendicular to the press direction can be expected to grow. Powdered die-upset material has been employed to prepare anisotropic bonded magnets [43,44,51].

5.1.3. Cobalt substitution and elemental additives

Technological magnet materials offer a dual opportunity with regard to compositional modification: altering (a) the intrinsic properties of the major component and/or (b) the nature of the secondary phases and their influence on the technical characteristics, especially the coercivity.

An enhanced Curie temperature is desirable for some applications, making Co substitution of interest since T_c increases monotonically with x in the stoichiometric $R_2Fe_{14-x}Co_xB$ compounds. In $Nd_2Fe_{14-x}Co_xB$, T_c rises from 585 K for $x = 0$ to 700 K at the modest Co level of $x = 2$, while $T_c(Nd_2Co_{14}B) \approx 1000$ K is essentially identical to $T_c(SmCo_5)$. Melt-spun Nd–(Fe,Co)–B materials containing $Nd_2Fe_{14-x}Co_xB$ as the principal constituent have been prepared [52]; as is generally the case, optimization of a given technical property such as $(BH)_{max}$ or H_{ci} requires tailoring of the starting composition [53]. A uniaxial → basal plane spin reorientation occurs at a temperature $T_s < T_c$ for $x \gtrsim 9$ in the $Nd_2Fe_{14-x}Co_xB$ system, however, due to competing Fe and Co sublattice anisotropies. This intrinsic behavior limits the practical magnetics: B_r, H_{ci} and $(BH)_{max}$ all vanish as T approaches T_s rather than T_c in that Co concentration range.

Doping with other elements can increase significantly the coercivity of melt-spun Nd–Fe–B materials. The relevant concentrations are low, < 5 at%, making it likely that modifications to grain boundary composition and microstructure are involved more than changes to the $Nd_2Fe_{14}B$ matrix phase. Addition of niobium or zirconium improves H_{ci} of melt-spun ribbons [54], as does copper, which TEM work has shown to segregate preferentially to the intergranular regions [55]. Gallium addition to the starting ingots can enhance H_{ci} as well as the high temperature behavior of die-upset Nd–Fe–B magnets [43,44,56]. In an effort to explore diffusion alloying of dopants by combining small amounts of fine elemental powders with Nd–Fe–B ribbons prior to hot deformation, Fuerst and Brewer [57] found H_{ci} enhanced by 10–90% in hot-pressed and die-upset magnets containing Ag, Au, Cd, Cu, Ir, Mg, Ni, Pd, Pt, Ru or Zn.

5.2. Sintered magnets

Powder metallurgy processing involves a sequence of orient, press, sinter operations on finely divided alloy powders and has been long employed in the commercial production of ferrite and Sm–Co magnets. Magnets prepared in this general way are most often simply designated as sintered.

5.2.1. Neodymium–iron–boron

Sagawa et al. [12] were the first to demonstrate that such methods can be applied to Nd–Fe–B. Their canonical procedure can be summarized as follows. Induction-melted ingots having the composition $Nd_{0.15}Fe_{0.77}B_{0.08}$, substantially enriched in Nd and B relative to both stoichiometric $Nd_2Fe_{14}B$ ($\approx Nd_{0.12}Fe_{0.82}B_{0.06}$) as well as to the optimum melt-spinning composition $\approx Nd_{0.13}Fe_{0.82}B_{0.05}$, are crushed, milled and then pulverized to powder with a particle size of ~ 3 μm. The powder is aligned in a magnetic field (~ 10 kOe) and pressed ($P \sim 200$ MPa) perpendicular to the alignment direction. The resulting compact is sintered at ~ 1370 K in argon gas for one hour and cooled quickly. A post-sinter anneal at ~ 900 K, again under Ar, improves the coercivity to its maximum value. Extensive discussions of the methodology can be found in refs. [29,58–61]. The success of powder metallurgical fabrication of Nd–Fe–B magnets is due not only to the superior intrinsic properties of $Nd_2Fe_{14}B$ but also to several favorable metallurgical features, such as the fact that $Nd_2Fe_{14}B$, Nd and the tetragonal compound $Nd_{1+\epsilon}Fe_4B_4$ form a ternary eutectic which makes liquid phase sintering possible, enabling high densification without significant grain growth [29]. The demagnetization curve of a representative sintered magnet, having an energy product of 36 MGOe, is shown in fig. 11; it resembles that of the aligned, die-upset magnet in fig. 9. As for melt-spun Nd–Fe–B and other permanent magnet materials, improvements in the remanence at the expense of coercivity and vice versa are possible by altering the composition and processing parameters. Narasimhan [62] achieved an energy product of 45 MGOe using a composition near $Nd_{0.15}Fe_{0.80}B_{0.05}$, and Sagawa et al. [29] reported the record value $(BH)_{max} \approx 50.6$ MGOe for a sintered $Nd_{0.13}Fe_{0.81}B_{0.06}$ magnet; in both cases B_r and H_{ci} are respectively higher and lower than the corresponding values for the magnet of fig. 11.

Sintered Nd–Fe–B magnets are relatively complex multiphase systems containing, in addition to $Nd_2Fe_{14}B$, all or most of the following: $Nd_{1+\epsilon}Fe_4B_4$, Nd-rich phases, α-Fe, Nd oxides and pores. Fig. 12 is a composition image of a

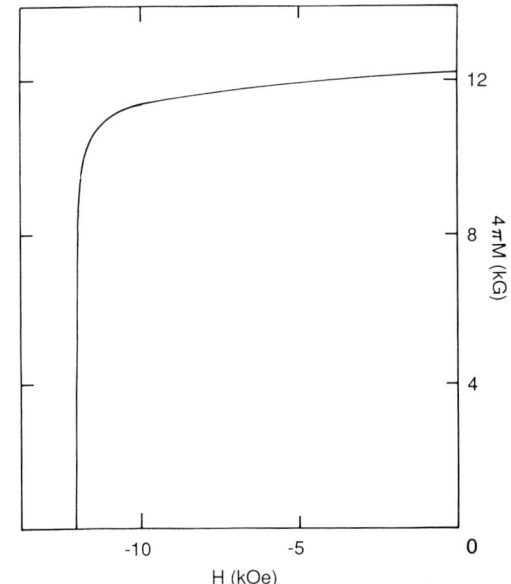

Fig. 11. Room temperature demagnetization curve of a sintered $Nd_{0.15}Fe_{0.77}B_{0.08}$ magnet characterized by $(BH)_{max} \approx 36$ MGOe (adapted from ref. [12]).

representative $Nd_{0.15}Fe_{0.77}B_{0.08}$ magnet; the contrast with the simpler microstructure of melt-spun materials in figs. 8 and 10 is clear. The major $Nd_2Fe_{14}B$ component occupies $\sim 85\%$ of the volume with an average grain size of ~ 10 μm, approximately two orders of magnitude larger than that in optimum melt-spun ribbons. Boron-richer $Nd_{1+\epsilon}Fe_4B_4$, frequently identified in early papers as $Nd_2Fe_7B_6$, forms irregularly distributed, heavily faulted grains of roughly the same size as those of $Nd_2Fe_{14}B$. A neodymium-rich, fcc phase of approximate composition $Nd_{0.95}Fe_{0.05}$ appears at grain boundary junctions and in thin intergranular layers (see fig. 12); it serves as a liquid-phase sintering aid in the powder metallurgy process. It now appears that smoothing of the intergranular layers is the principal microstructural effect accompanying the post-sintering heat treatment which enhances the coercivity; the reduction of irregularities likely impedes reverse domain formation by diminish-

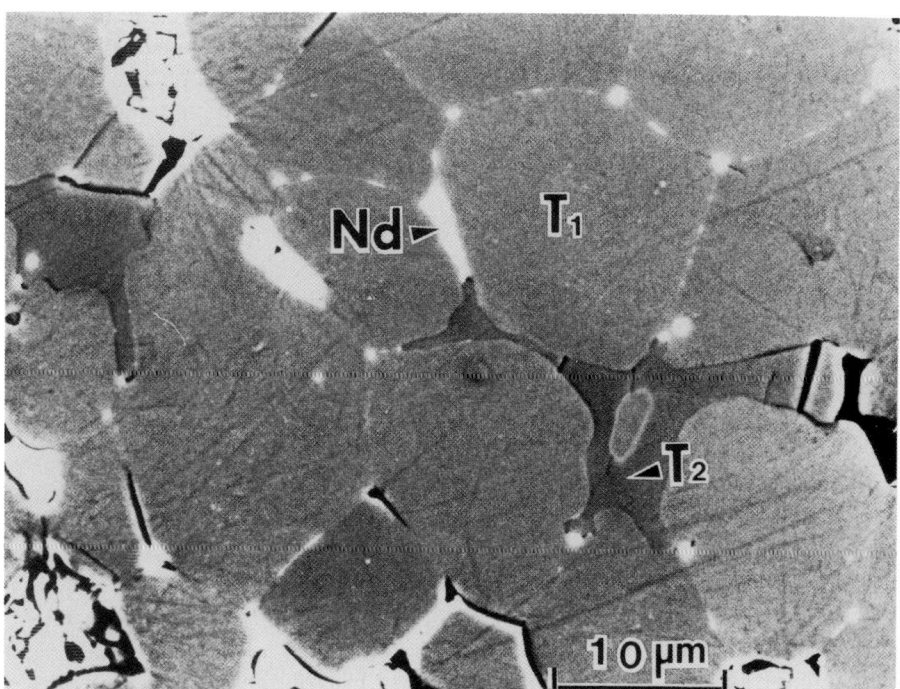

Fig. 12. X-ray composition micrograph of a sintered $Nd_{0.15}Fe_{0.77}B_{0.08}$ magnet; T_1, T_2 and Nd denote $Nd_2Fe_{14}B$, $Nd_{1+\epsilon}Fe_4B_4$ and a Nd-rich phase, respectively [29].

ing the local demagnetization fields produced by sharp projections [63].

5.2.2. Improvements via compositional modification

As is the case for melt-spun magnets, the primary impetus for the many investigations of substitutions and additions in sintered Nd–Fe–B magnets is extension of the useful operating temperature range. Routes to this goal include enhancement of T_c and H_{ci}, improvement (i.e., reduction) of the temperature coefficients of B_r and H_{ci} which can be defined by

$$\alpha(T) \equiv -d \ln B_r(T)/dT \qquad (7a)$$

and

$$\beta(T) \equiv -d \ln H_{ci}(T)/dT \qquad (7b)$$

or discrete versions of these, and diminution of the irreversible flux losses accompanying high temperature exposure. Sagawa et al. [64] initially demonstrated that Co substitution increases T_c and decreases α, observing $T_c = 740$ K and $\alpha = 0.07\%/$K for $x = 0.2$ in $Nd_{0.15}(Fe_{1-x}Co_x)_{0.77}B_{0.08}$ as compared with $T_c = 585$ K and $\alpha = 0.12\%/$K for $x = 0$. Arai and Shibata [65] achieved $(BH)_{max} = 42$ MGOe and $\alpha = 0.02\%/$K over the 300–400 K range in $Nd_{0.16}Fe_{0.66}Co_{0.11}B_{0.07}$. Cobalt substitution introduces microstructural changes, including the formation of the soft magnetic Laves phase $Nd(Fe,Co)_2$ [66], and it is also accompanied by loss of coercivity [67,68]. Partial replacement of Nd by Dy substantially enhances H_{ci} and reduces β, chiefly by increasing the anisotropy field of $Nd_2Fe_{14}B$, although B_r is lowered somewhat [64]. For the $(Nd_{0.53}Dy_{0.47})_{0.15}Fe_{0.77}B_{0.08}$ composition Sagawa et al. [68] achieved $(BH)_{max} \geq 20$ MGOe and $H_{ci} = 50$ kOe, the largest room temperature coercivity ever observed in a neodymium-based sintered magnet. Combined Dy and Co substitution is effective in raising T_c and H_{ci} and in decreasing α and the irreversible losses [69].

Aluminum substitution provides a fascinating example of opposing intrinsic and microstructural effects. Despite the fact that it *degrades* the anisotropy field of $Nd_2Fe_{14}B$, aluminum can *enhance* H_{ci} of sintered Nd–Fe–B magnets by as much as a factor of two [70]. Evidently the coercivity improvement is a consequence of changes in the microstructure. Knoch et al. [71] have shown that Al substitution decreases the wetting angle between $Nd_2Fe_{14}B$ and the surrounding liquid during sintering and have made the quite plausible suggestion that the improved wettability affords more complete separation of $Nd_2Fe_{14}B$ grains and enhances H_{ci} by reducing exchange interactions among them. Aside from the capability of greatly enhancing H_{ci}, aluminum substitution decreases the saturation magnetization and Curie temperature, but these effects can be overcome by partial replacement of Fe with Co. Mizoguchi et al. [72] achieved $B_r = 13.2$ kG, $H_{ci} = 11.0$ kOe, $(BH)_{max} = 41.0$ MGOe, $T_c = 770$ K, and $\alpha = 0.07\%/$K, half the value for an unsubstituted reference magnet, for the $Nd_{0.15}Fe_{0.625}Co_{0.16}B_{0.055}Al_{0.01}$ composition.

Low levels of gallium and niobium can also beneficiate the coercivity and thermal stability of sintered magnets. In a survey of elemental substituents in Nd–(Fe,Co)–B Endoh et al. [73] found Ga to be most effective in enhancing H_{ci} and reported that the Ga-substituted magnets feature better magnetic properties and lower irreversible losses than those containing Al or Dy. Tsutai et al. [74] recorded $B_r = 13.1$ kG, $H_{ci} = 12.2$ kOe, $(BH)_{max} = 40$ MGOe, $\alpha = 0.08\%/$K and $\beta = 0.5\%/$K over the 300–450 K interval for $Nd_{0.145}Fe_{0.63}Co_{0.16}Ga_{0.01}B_{0.055}$. Knoch et al. [71] have speculated that Ga acts similarly to Al by improving wettability of the matrix phase. Niobium addition has been shown to increase H_{ci} in Nd–Fe–B, (Nd,Dy)–Fe–B and (Nd,Dy)–(Fe,Co)–B and to reduce irreversible losses in such alloys; small (~ 500 Å), finely dispersed Nb-containing precipitates appear within the hard magnetic grains and may be responsible for improving H_{ci} by serving as domain wall pinning sites [75].

5.3. Coercivity mechanisms

Most of the numerous efforts to understand the coercivity of Nd–Fe–B magnets rely on interpretations based on either the nucleation of reverse domains or the pinning of domain walls by inhomogeneities, the mechanisms earlier found pertinent to Sm–Co materials. These possibilities have been distinguished by guidelines such as the initial susceptibility behavior, domain structure, microstructure and the dependence of H_{ci} on magnetizing field, temperature and the angle between the applied field and the alignment direction.

Three related parameters of relevance to the coercivity issue are the domain wall width δ, the domain wall energy density σ_W and the critical single-domain particle diameter D_c; for Nd–Fe–B at room temperature $\delta \sim 50$ Å, $\sigma_W \sim 30$ mJ/m^2 and $D_c \sim 0.3$ μm. D_c represents the diameter of an isolated sphere below which formation of a domain wall is energetically unfavorable and above which a multidomain configuration is stable. Its value provides a sharp qualitative distinction between sintered Nd–Fe–B magnets, whose average grain diameter $D_g \sim 10$ μm is much larger than $D_c \sim 0.3$ μm, and melt-spun ribbons and hot-pressed magnets, for which $D_g \sim 300$ Å $= 0.03$ μm $\ll D_c$. This contrast alone suggests that different coercivity mechanisms may be at work in the two classes. For the die-upset magnets there is no clear distinction since $D_c \sim D_g$.

5.3.1. Sintered magnets

It is the prevailing view that sintered Nd–Fe–B magnets at room temperature are nucleation-controlled systems with domain wall pinning relegated to a comparitively minor role. The results of a variety of experiments and analyses have been advanced as evidence for this interpretation. Domain observations show that in the thermally demagnetized state each Nd$_2$Fe$_{14}$B grain has a multidomain structure, consistent with the fact that $D_g \gg D_c$, and that domain walls inside each grain move quite easily in small applied fields. The latter characteristic is compatible with both the steep virgin magnetization curve and with TEM studies indicating that the Nd$_2$Fe$_{14}$B grains are mostly defect-free. Measurements of minor hysteresis loops show that the coercivity saturates in magnetizing fields $H \geq H_{sat}$ with $H_{sat} < H_{ci}$; it has been argued that H_{sat} may be considered the field associated with the pinning of domain walls at or near grain boundaries and that, since $H_{ci} > H_{sat}$, the coercivity must be controlled by the nucleation of reverse domains [76].

Most analyses of the temperature dependence of H_{ci} also favor a nucleation-dominated coercivity mechanism for sintered Nd–Fe–B magnets. In the work of Kronmüller et al. [77] H_{ci} is written as

$$H_{ci} = \frac{2K_1}{M_s}\alpha_K\alpha_\psi - N_{eff}(4\pi M_s), \qquad (8)$$

where K_1 is the first anisotropy constant, α_K describes inhomogeneity of the magnetocrystalline anisotropy in grain boundary regions whose anisotropy is lower than that of the matrix phase and in which reverse domain nucleation is supposed to occur, α_ψ accounts for misaligned grains and N_{eff} is an average effective local demagnetization factor. Kronmüller et al. [77] consider a planar inhomogeneity of thickness r_0 and compare predictions for both pinning and nucleation descriptions, for which different forms of α_K and α_ψ enter eq. (8). The pinning approximation is found to yield contradictory results, while in the nucleation case plots of H_{ci}/M_s vs. $(2K_1/M_s^2)\alpha_K^{nuc}\alpha_\psi^{nuc}(\min)$, where $\alpha_\psi^{nuc}(\min)$ is chosen to give a lower limit for H_{ci}, can be fit with a spectrum of r_0 values and are linear over an extended temperature range. Results for a Nd$_{0.15}$Fe$_{0.77}$B$_{0.08}$ magnet are shown in fig. 13; the two r_0 values are upper and lower bounds. According to Kronmüller et al. [77] the departures from linearity for $T \gtrsim 450$ K indicate that domain wall pinning governs the coercivity in that high temperature range.

A simpler version of eq. (8) has been employed frequently to analyze the temperature dependence of H_{ci}:

$$H_{ci} = cH_a - N_{eff}(4\pi M_s), \qquad (9)$$

where c and N_{eff} are temperature-independent constants. Accommodation of data by eq. (9) is usually assumed to support a nucleation-controlled coercivity mechanism, with the first term on the right interpreted as the nucleation field of a reverse domain. It has been found that eq. (9) adequately describes $H_{ci}(T)$ for Pr–Fe–B mag-

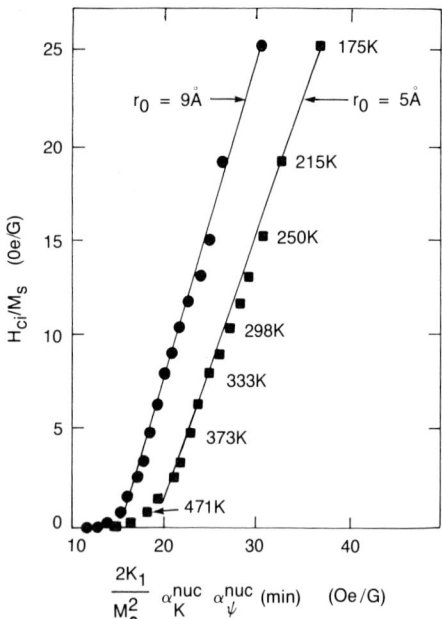

Fig. 13. H_{ci}/M_s vs. $(2K_1/M_s^2)\alpha_K^{nuc}\alpha_\psi^{nuc}(\min)$ for a sintered $Nd_{0.15}Fe_{0.77}B_{0.08}$ magnet (after ref. [77]).

nets of various compositions and Co-substituted Nd–Fe–B materials [68,78]. With $c = 1$ eq. (9) specifies the lower bound for the coercivity of an isolated, single domain particle, so that expressions such as eqs. (8) and (9) represent empirical extensions of the non-interacting single particle relation to bulk magnets.

An especially noteworthy aspect of the research on coercivity mechanisms is that it has produced valuable suggestions relevant to improving sintered magnets which can be summarized as follows. First, small $R_2Fe_{14}B$ grains are desirable for reducing the volume impacted by reverse domain nucleation at a defect, and the grains should be smooth to limit the deleterious effect of local demagnetizing (stray) fields at sharp edges and protrusions [77,78]. Second, a thin, smooth, defect-free grain boundary layer serves to magnetically isolate the $R_2Fe_{14}B$ crystallites and provides a barrier to demagnetization of neighboring grains [63]. Finally, it is helpful to minimize the amount of secondary phases such as $R_{1+\epsilon}Fe_4B_4$ and oxides, as well as pores, which can act as nucleation sites [79].

5.3.2. Melt-spun materials

Research on the origin of coercivity in melt-spun systems, especially hot-pressed and die-upset magnets, is less extensive than that for sintered materials, and the spectrum of opinion is not as nearly unilateral as it is in the case of sintered magnets. Lorentz microscopy investigations show that domain walls are pinned at grain boundaries in optimum ribbons [38]. The same statement applies to hot-pressed magnets, whose microstructure is essentially identical to that of the ribbons, and to die-upset magnets, in which the domain walls are parallel to the c-axis (or press direction) and can pass through many grains, unaffected if they intersect the faces of the platelets at right angles but pinned at grain edges [46,47]. Durst and Kronmüller [76] caution, however, that such observations, made on thin samples in the demagnetized state, may not provide definitive evidence for a pinning mechanism.

Virgin magnetization curves of melt-spun Nd–Fe–B ribbons exhibit much lower initial susceptibility than those of aligned sintered magnets. Furthermore, the coercivity develops more slowly with applied field, attaining its maximum, high-field value $H_{ci} = H_{ci}(\infty)$ only for $H > H_{ci}(\infty)$. These characteristics argue in favor of a pinning-dominated mechanism [80,81], as does the disparate behavior of the magnetic viscosity for ribbons and sintered magnets [82]. On the other hand, Durst et al. [83] assert that minor loop measurements on isotropic systems such as the ribbons are not meaningful in regard to the magnetic hardening mechanism. To demonstrate the point those authors prepared an isotropic $Nd_{0.15}Fe_{0.77}B_{0.08}$ sintered magnet and found that $H_{ci}(\infty)$ is reached only for $H \gg H_{ci}(\infty)$, similar to the situation for ribbons but the reverse of the case for aligned sintered magnets; if the latter are nucleation-controlled, the implicit logic is that $H_{ci}(H)$ data on isotropic systems, whether rapidly solidified or sintered, may not be inconsistent with a nucleation mechanism.

Conflicting inferences regarding the coercivity mechanism have been drawn from analyses of

$H_{ci}(T)$ for melt-spun Nd–Fe–B ribbons. Hadjipanayis et al. [84] reported good agreement of their measurements with the strong domain wall pinning model of Gaunt [85] in which H_{ci} is given by

$$(H_{ci}/H_0)^{1/2} = 1 - (75k_B T/4bf)^{2/3}, \quad (10)$$

where H_0 is the critical field in the absence of thermal activation, $4b$ is the range of interaction between domain wall and pin, and f is the maximum restoring force per pin. Pinkerton and Fuerst [86] found, somewhat surprisingly, that $H_{ci}(T)$ is very well described by eq. (9) with $c = 0.25$ and $N_{eff} = 0.26$ (for sintered Nd–Fe–B, $c \approx 0.37$ and $N_{eff} \approx 1$ [78]). Since the value of c is less than the $c \approx 0.3$ limit above which Kronmüller et al. [77] expect that eq. (9) cannot be valid for a pinning description (in which c is then interpreted as the pinning strength of an inhomogeneity such as a grain boundary), the results of Pinkerton and Fuerst [86] are at least not incompatible with pinning-dominated coercivity. In contrast, Durst and Kronmüller [76] obtained good accommodation of their $H_{ci}(T)$ data using eq. (8) with $\alpha_\psi = 1$, $\alpha_K = \delta/\pi r_0$ and inferred that melt-spun ribbons are nucleation-controlled systems for $T \lesssim 520$ K; their result for the width of the soft magnetic inhomogeneity, $r_0 \approx 21$ Å, is in excellent agreement with the grain boundary thickness measured directly by Mishra [38].

For (isotropic) hot-pressed and (anisotropic) die-upset magnets coercivity develops quickly with applied field but approaches saturation only in fields exceeding $H_{ci}(\infty)$. Pinkerton and Van Wingerden [81] ascribe the low-field behavior to domain propagation, with subsequent pinning in intergranular regions making further wall movement in larger magnetizing fields progressively more difficult. Magnetic viscosity measurements also indicate that the reversal mechanism for hot-pressed magnets is distinct from that of sintered materials [87]. In agreement with Pinkerton and Van Wingerden [81], Givord et al. [87] conclude that the high-field part of the virgin magnetization curve of hot-pressed magnets is controlled by the same irreversible, pinning-dominated processes governing demagnetization on the hysteresis loop. Pinkerton and Fuerst [86] analyzed $H_{ci}(T)$ for a die-upset Nd–Fe–B magnet and found that eq. (9) approximately describes the data with $c = 0.29$ and $N_{eff} = 0.72$; however, a quantitatively improved fit was obtained using eq. (10) for strong domain wall pinning [88]. Over a wide temperature range eq. (10) also quite successfully models $H_{ci}(T)$ for die-upset Pr–Fe–B [89].

6. Remarks

Neodymium–iron–boron magnets produced by either melt spinning or sintering can be found on many products ranging from dc automotive motors to computer disc drives and household appliances. The number of applications continues to expand steadily because of the economic advantages of Nd–Fe–B as well as the larger energy products of the new magnets relative to their Sm–Co predecessors.

Despite the swift progress from the laboratory discoveries to commercialization, there is broad opportunity for further research on $R_2Fe_{14}B$ materials. The largest energy products realized thus far (~ 50 MGOe) fall substantially short of $(BH)^*_{max} \sim 64$ MGOe, underscoring the necessity for continued investigation of alloy modifications and alternate preparation methods. Promising among the latter are mechanical alloying [90], hot deformation of ingots [91], and the hydrogenation, disproportionation, desorption process [92]. From a fundamental perspective our understanding of the intrinsic properties of these systems, and of R–TM materials in general, can be improved. Open problems include the following. First, the mechanisms underlying the TM anisotropy and its variation with temperature and composition require elucidation. Second, there is a need for accurate determination of crystal field and exchange energies, site-dependent magnetic moments, hyperfine fields and nuclear quadrupole splittings from both experiments and electronic structure calculations. The complexity of the crystal structure notwithstanding, semi-empirical [93] and first-principles [94] band structure work on $R_2Fe_{14}B$ compounds has been reported. Finally, it is desirable that concepts and models for coercivity be developed which supersede the single particle paradigm and more fully incorporate the features of real interacting systems.

References

[1] U. Enz, in: Ferromagnetic Materials, vol. 3, ed. E.P. Wohlfarth (North-Holland, Amsterdam, 1982) p. 1.
R.A. McCurrie, ibid., p. 107.
[2] H. Kojima, ibid., p. 305.
H. Stäblein, ibid., p. 441.
M. Sugimoto, ibid., p. 393.
[3] K.J. Strnat, IEEE Trans. Magn. MAG-23 (1987) 2094.
[4] R.K. Mishra, G. Thomas, T. Yoneyama, A. Fukuno and T. Ojima, J. Appl. Phys. 52 (1981) 2517.
[5] K. Kumar, J. Appl. Phys. 63 (1988) R13.
[6] K.J. Strnat, in: Ferromagnetic Materials, vol. 4, ed. E.P. Wohlfarth and K.H.J. Buschow (North-Holland, Amsterdam, 1988) p. 131.
K.J. Strnat and R.M.W. Strnat, J. Magn. Magn. Mater. 100 (1991) 38.
[7] J.J. Croat, J.F. Herbst, R.W. Lee and F.E. Pinkerton, J. Appl. Phys. 55 (1984) 2078.
[8] N.C. Koon and B.N. Das, J. Appl. Phys. 55 (1984) 2063.
[9] G.C. Hadjipanayis, R.C. Hazelton and K.R. Lawless, J. Appl. Phys. 55 (1984) 2073.
[10] D.J. Sellmyer, A. Ahmed, G. Muench and G. Hadjipanayis, J. Appl. Phys. 55 (1984) 2088.
[11] J.J. Becker, J. Appl. Phys. 55 (1984) 2067.
[12] M. Sagawa, S. Fujimura, N. Togawa, H. Yamamoto and Y. Matsuura, J. Appl. Phys. 55 (1984) 2083.
[13] J.J. Croat, J.F. Herbst, R.W. Lee and F.E. Pinkerton, Appl. Phys. Lett. 44 (1984) 148.
[14] J.J. Croat, Appl. Phys. Lett. 37 (1980) 1096; J. Appl. Phys. 52 (1981) 2509; J. Magn. Magn. Mater. 24 (1981) 125; Appl. Phys. Lett. 39 (1981) 357; J. Appl. Phys. 53 (1982) 6932; IEEE Trans. Magn. MAG-18 (1982) 1442; J. Appl. Phys. 53 (1982) 3161.
[15] J.J. Croat and J.F. Herbst, J. Appl. Phys. 53 (1982) 2294, 2404.
[16] N.C. Koon and B.N. Das, Appl. Phys. Lett. 39 (1981) 840.
N.C. Koon, B.N. Das and J.A. Geohegan, IEEE Trans. Magn. MAG-13 (1982) 1448.
B.N. Das and N.C. Koon, Metall. Trans. 14A (1983) 953.
[17] G.C. Hadjipanayis, R.C. Hazelton and K.R. Lawless, Appl. Phys. Lett. 43 (1983) 797.
[18] R.C. Hazelton, G.C. Hadjipanayis, K.R. Lawless and D.J. Sellmyer, J. Magn. Magn. Mater. 40 (1984) 278.
[19] A.E. Clark, Appl. Phys. Lett. 23 (1973) 642.
[20] H.H. Stadelmaier, N.A. ElMasry and S. Cheng, Mater. Lett. 2 (1983) 169.
[21] F. Spada, C. Abache and H. Oesterreicher, J. Less-Common Met. 99 (1984) L21.
[22] A.V. Deryagin, E.N. Tarasov, A.V. Andreev, V.N. Moskalev and A.I. Kozlov, JETP Lett. 39 (1984) 629.
[23] N.F. Chaban, Yu.B. Kuzma, N.S. Bilonizhko, O.O. Kachmar and N.V. Petriv, Dopv. Akad. Nauk Ukr. RSR, Ser. A, No. 10 (1979) 873.
[24] J.F. Herbst, J.J. Croat, F.E. Pinkerton and W.B. Yelon, Phys. Rev. B 29 (1984) 4176.
J.F. Herbst, J.J. Croat and W.B. Yelon, J. Appl. Phys. 57 (1985) 4086.
[25] C.B. Shoemaker, D.P. Shoemaker and R. Fruchart, Acta Cryst. C 40 (1984) 1665.
[26] D. Givord, H.S. Li and J.M. Moreau, Solid State Commun. 50 (1984) 497.
[27] R.W. Lee, Appl. Phys. Lett. 46 (1985) 790.
[28] J.J. Croat, J. Less-Common Met. 148 (1989) 7.
[29] M. Sagawa, S. Hirosawa, H. Yamamoto, S. Fujimura and Y. Matsuura, Jpn. J. Appl. Phys. 26 (1987) 785.
[30] J.F. Herbst, Rev. Mod. Phys. (1991) in press.
[31] I.A. Campbell, J. Phys. F 2 (1972) L47.
[32] S. Sinnema, R.J. Radwański, J.J.M. Franse, D.B. de Mooij and K.H.J. Buschow, J. Magn. Magn. Mater. 44 (1984) 333.
[33] M. Yamada, H. Kato, H. Yamamoto and Y. Nakagawa, Phys. Rev. B 38 (1988) 620.
[34] E.B. Boltich and W.E. Wallace, Solid State Commun. 55 (1985) 529.
[35] J.M. Cadogan, J.P. Gavigan, D. Givord and H.S. Li, J. Phys. F 18 (1988) 779.
[36] J.P. Gavigan, D. Givord, H.S. Li, O. Yamada, H. Maruyama, M. Sagawa and S. Hirosawa, J. Magn. Magn. Mater. 70 (1987) 416.
D. Givord, H.S. Li, J.M. Cadogan, J.M.D. Coey, J.P. Gavigan, O. Yamada, H. Maruyama, M. Sagawa and S. Hirosawa, J. Appl. Phys. 63 (1988) 3713.
[37] M. Loewenhaupt, M. Prager, A.P. Murani and H.E. Hoenig, J. Magn. Magn. Mater. 76&77 (1988) 408.
M. Loewenhaupt, I. Sosnowska and B. Frick, Phys. Rev. B 42 (1990) 3866.
[38] R.K. Mishra, J. Magn. Magn. Mater. 54–57 (1986) 450.
[39] D. Dadon, Y. Gefen and M.P. Dariel, IEEE Trans. Magn. MAG-23 (1987) 3605.
R. Coehoorn and J. Duchateau, Mater. Sci. Eng. 99 (1988) 131.
G-H. Tu, Z. Altounian, D.H. Ryan and J.O. Ström-Olsen, J. Appl. Phys. 63 (1988) 3330.
[40] J. Yamasaki, A. Furuta and Y. Hirokado, IEEE Trans. Magn. MAG-25 (1989) 4120.
[41] R.W. Lee, Appl. Phys. Lett. 46 (1985) 790.
R.W. Lee, E.G. Brewer and N.A. Schaffel, IEEE Trans. Magn. MAG-21 (1985) 1958.
[42] P.B. Gwan, J.P. Scully, D. Bingham, J.S. Cook, R.K. Day, J.B. Dunlop and R.G. Heydon, in: Proc. Ninth Intern. Workshop on Rare-Earth Magnets and Their Applications, eds. C. Herget and R. Poerschke (Deutsche Physikalische Gesellschaft, Bad Honnef, 1987) p. 295.
[43] Y. Nozawa, K. Iwasaki, S. Tanigawa, M. Tokunaga and H. Harada, J. Appl. Phys. 64 (1988) 5285.

[44] M. Tokunaga, Y. Nozawa, K. Iwasaki, S. Tanigawa and H. Harada, J. Magn. Magn. Mater. 80 (1989) 80.
[45] R.K. Mishra, in: High Performance Permanent Magnet Materials, Materials Research Soc. Symp. Proc., vol. 96, eds. S.G. Sankar, J.F. Herbst and N.C. Koon (Materials Research Society, Pittsburgh, 1987) p. 83.
[46] R.K. Mishra, J. Appl. Phys. 62 (1987) 967.
[47] R.K. Mishra and R.W. Lee, Appl. Phys. Lett. 48 (1986) 733.
[48] R.K. Mishra, E.G. Brewer and R.W. Lee, J. Appl. Phys. 63 (1988) 3528.
R.K. Mishra, T-Y. Chu and L.K. Rabenberg, J. Magn. Magn. Mater. 84 (1990) 88.
[49] L. Li and C.D. Graham, Jr., J. Appl. Phys. 67 (1990) 4756.
[50] P. Tenaud, A. Chamberod and F. Vanoni, Solid State Commun. 63 (1987) 303.
[51] L.J. Eshelman, K.A. Young, V. Panchanathan and J.J. Croat, J. Appl. Phys. 64 (1988) 5293.
J.J. Croat, IEEE Trans. Magn. MAG-25 (1989) 3550.
[52] J. Wecker and L. Schultz, Appl. Phys. Lett. 51 (1987) 697.
C.D. Fuerst and J.F. Herbst, J. Appl. Phys. 63 (1988) 3324.
[53] C.D. Fuerst and J.F. Herbst, Appl. Phys. Lett. 54 (1989) 1068; J. Appl. Phys. 66 (1989) 1782.
[54] O. Kohmoto, T. Yoneyama, and K. Yajima, Jpn. J. Appl. Phys. 26 (1987) 1804.
J. Wecker and L. Schultz, J. Magn. Magn. Mater. 83 (1990) 189.
[55] J.F. Herbst, C.D. Fuerst, R.K. Mishra, C.B. Murphy and D.J. van Wingerden, J. Appl. Phys. 69 (1991) 5823.
[56] M. Endoh, M. Tokunaga, E.b. Boltich and W.E. Wallace, IEEE Trans. Magn. MAG-25 (1989) 4114.
[57] C.D. Fuerst and E.G. Brewer, Appl. Phys. Lett. 56 (1990) 2252; J. Appl. Phys. 69 (1991) 5826.
[58] J. Ormerod, J. Less-Common Met. 111 (1985) 49.
[59] K.H.J. Buschow, in: Ferromagnetic Materials, vol. 4, eds. E.P. Wohlfarth and K.H.J. Buschow (North-Holland, Amsterdam, 1988) p. 1.
[60] E. Burzo and H.R. Kirchmayr, in: Handbook on the Physics and Chemistry of Rare Earths, vol. 12, eds. K.A. Gschneidner, Jr. and L. Eyring (North-Holland, Amsterdam, 1989) p. 71.
[61] J.S. Cook and P.L. Rossiter, in: CRC Critical Reviews in Solid State and Materials Sciences, vol. 15 (CRC Press, Boca Raton, 1989) p. 509.
[62] K.S.V.L. Narasimhan, J. Appl. Phys. 57 (1985) 4081.
[63] S. Hirosawa and Y. Tsubokawa, J. Magn. Magn. Mater. 84 (1990) 309.
[64] M. Sagawa, S. Fujimura, H. Yamamoto, Y. Matsuura and K. Hiraga, IEEE Trans. Magn. MAG-20 (1984) 1584.
[65] S. Arai and T. Shibata, IEEE Trans. Magn. MAG-21 (1985) 1952.
[66] J. Fidler, IEEE Trans. Magn. MAG-21 (1985) 1955.
[67] B.-M. Ma and K.S.V.L. Narasimhan, J. Magn. Magn. Mater. 54–57 (1986) 559.
[68] M. Sagawa, S. Hirosawa, K. Tokuhara, H. Yamamoto, S. Fujimura, Y. Tsubokawa and R. Shimizu, J. Appl. Phys. 61 (1987) 3559.
[69] W. Li, L. Jiang, D. Wang, T. Sun and J. Zhu, J. Less-Common Met. 126 (1986) 95.
[70] M. Zhang, D. Ma, X. Jiang and S. Liu, in: Proc. Eighth Intern. Workshop on Rare-Earth Magnets and Their Applications, ed. K.J. Strnat (University of Dayton, Dayton, 1985) p. 541.
[71] K.G. Knoch, B. Grieb, E.-Th. Henig, H. Kronmüller and G. Petzow, IEEE Trans. Magn. MAG-26 (1990) 1951.
K.G. Knoch, E.-Th. Henig and J. Fidler, J. Magn. Magn. Mater. 83 (1990) 209.
[72] T. Mizoguchi, I. Sakai and K. Inomata, Appl. Phys. Lett. 48 (1986) 1309.
[73] M. Endoh, M. Tokunaga and H. Harada, IEEE Trans. Magn. MAG-23 (1987) 2290.
[74] A. Tsutai, I. Sakai, T. Mizoguchi and K. Inomata, Appl. Phys. Lett. 51 (1987) 1043.
[75] J. Hu, Y. Wang, X. Li, L. Yin, M. Feng, D. Dai, T. Wang, J.G. Zhao and Z. Wang, J. de Phys. 49 (1988) C8-601.
S.F.H. Parker, P.J. Grundy and J. Fidler, J. Magn. Magn. Mater. 66 (1987) 74.
Y. Xiao, K.J. Strnat, H.F. Mildrum and A.E. Ray, in: Proc. Ninth Intern. Workshop on Rare-Earth Magnets and Their Applications, eds. C. Herget and R. Poerschke (Deutsche Physikalische Gesellschaft, Bad Honnef, 1987) p. 467.
M. Tokunaga, H. Harada and S.R. Trout, IEEE Trans. Magn. MAG-23 (1987) 2284.
[76] K.-D. Durst and H. Kronmüller, J. Magn. Magn. Mater. 68 (1987) 63.
[77] H. Kronmüller, K.-D. Durst and M. Sagawa, J. Magn. Magn. Mater. 74 (1988) 291.
[78] S. Hirosawa, K. Tokuhara, Y. Matsuura, H. Yamamoto, S. Fujimura and M. Sagawa, J. Magn. Magn. Mater. 61 (1986) 363.
M. Sagawa and S. Hirosawa, in: High Performance Permanent Magnet Materials, Materials Research Soc. Symp. Proc., vol. 96, eds. S.G. Sankar, J.F. Herbst and N.C. Koon (Materials Research Society, Pittsburgh, 1987) p. 161.
[79] G. Schneider, E.-T. Henig, H.H. Stadelmaier and G. Petzow, in: Proc. Fifth Intern. Symp. on Magnetic Anisotropy and Coercivity in Rare-Earth-Transition Metal Alloys, eds. C. Herget, H. Kronmüller and R. Poerschke (Deutsche Physikalische Gesellschaft, Bad Honnef, 1987) p. 347.
[80] G.C. Hadjipanayis, R.C. Dickenson and K.R. Lawless, J. Magn. Magn. Mater. 54–57 (1986) 557.
[81] F.E. Pinkerton and D.J. van Wingerden, J. Appl. Phys. 60 (1986) 3685.
[82] G.B. Ferguson, K. O'Grady, J. Popplewell and R.W. Chantrell, IEEE Trans. Magn. MAG-25 (1989) 3449.
[83] K.-D. Durst, H. Kronmüller, and G. Schneider, in: Proc.

Fifth Intern. Symp. on Magnetic Anisotropy and Coercivity in Rare-Earth–Transition Metal Alloys, eds. C. Herget, H. Kronmüller and R. Poerschke (Deutsche Physikalische Gesellschaft, Bad Honnef, 1987) p. 209.

[84] G.C. Hadjipanayis, K.R. Lawless and R.C. Dickerson, J. Appl. Phys. 57 (1985) 4097.

[85] P. Gaunt, Phil. Mag. B 48 (1983) 261; Can. J. Phys. 65 (1987) 1194.

[86] F.E. Pinkerton and C.D. Fuerst, J. Appl. Phys. 67 (1990) 4753.

[87] D. Givord, C. Heiden, A. Hoëhler, P. Tenaud, T. Viadieu and K. Zeibig, IEEE Trans. Magn. MAG-24 (1988) 1918.

[88] F.E. Pinkerton and C.D. Fuerst, J. Magn. Magn. Mater. 89 (1990) 139.

[89] F.E. Pinkerton and C.D. Fuerst, J. Appl. Phys. 69 (1991) 5817.

[90] L. Schultz, K. Schnitzke and J. Wecker, J. Magn. Magn. Mater. 80 (1989) 115.

[91] T. Shimoda, K. Akioka, O. Kobayashi and T. Yamagami, J. Appl. Phys. 64 (1988) 5290.

[92] T. Takeshita and R. Nakayama, in: Proc. Tenth Intern. Workshop on Rare-Earth Magnets and Their Applications (Society of Non-Traditional Technology, Tokyo, 1989) p. 551.
P.J. McGuiness, X.J. Zhang, H. Forsyth and I.R. Harris, J. Less-Common Met. 162 (1990) 379.

[93] J. Inoue and M. Shimizu, J. Phys. F 16 (1986) 1051.
B. Szpunar, W.E. Wallace and J. Szpunar, Phys. Rev. B 36 (1987) 3782.

[94] Z-Q. Gu and W.Y. Ching, Phys. Rev. B 36 (1987) 8530.
W.Y. Ching and Z-Q. Gu, J. Appl. Phys. 63 (1988) 3716.
D.J. Sellmyer, M.A. Engelhardt, S.S. Jaswal and A.J. Arko, Phys. Rev. Lett. 60 (1988) 2077.
S.S. Jaswal, Phys. Rev. B 41 (1990) 9697.
R. Coehoorn, in: Proc. NATO Advanced Study Institute on Supermagnets, Hard Magnetic Materials, eds. G.J. Long and F. Grandjean (Kluwer, Dordrecht, 1991) p. 133.

Permanent magnet materials based on tetragonal rare earth compounds of the type $RFe_{12-x}M_x$

K.H.J. Buschow

Philips Research Laboratories, NL 5600 JA Eindhoven, Netherlands

A review is given of the occurrence and composition of ternary rare earth compounds $RT_{12-x}M_x$ based on the tetragonal $ThMn_{12}$ structure. Most of the compounds form with T = Fe but there are also examples with T = Co, Ni or Mn. Particular attention is paid to the site preference of the nonmagnetic component M, (Al, Si, Ti, V, Cr, Mo, W, Re). A discussion is given of the magnetic properties of the $RT_{12-x}M_x$ compounds and the corresponding nitrides obtained after charging with nitrogen gas. The important role played by the intersublattice coupling on the magnetocrystalline anisotropy of these materials is stressed. Advantages and disadvantages of the $RFe_{12-x}M_x$ compounds as permanent magnet materials are discussed.

1. Introduction

Ternary compounds based on the tetragonal $ThMn_{12}$ structure in which rare earth elements (R) are combined with a 3d transition metal (T) and a nonmagnetic element (M) have already been known for many years [1,2]. These compounds have the composition RT_4Al_8, T representing Cr, Mn, Fe or Cu. Only recently it was discovered that ternary compounds of the $ThMn_{12}$ type are formed also for more 3d rich compositions [3–11]. Although ternary rare earth compounds of the type $RFe_{12-x}M_x$ can exist for high Fe concentrations ($x \geq 1$) when the M element is Ti, V, Cr, Mo, W or Si, the compound RFe_{12} itself does not exist.

All the compounds $RFe_{12-x}M_x$ are characterized by fairly high values of the Curie temperature and saturation magnetization. The magnetocrystalline anisotropy in these materials is of a sufficient size to consider them as candidates for permanent magnet applications. In fact, there are several reports describing that fairly large values of the coercivity can be attained under special conditions.

There are also numerous investigations of a more fundamental nature in which the magnetic properties were analysed in terms of the combined effect of crystal field interactions and exchange interactions. An excellent review article on this subject has been written by Li and Coey [12]. Attractive features of the $RFe_{12-x}M_x$ compounds are their relatively simple crystal structure and the possibility of varying the T component. Several of these compounds have the additional advantage that the M concentration can be varied over a fairly large range. In the present report a brief review will be given of the more salient properties of the $ThMn_{12}$ compounds. First the crystal structure and phase relationships will be discussed. This will be followed by a discussion of the magnetic properties. Finally, in the last section a review will be given of the hard magnetic properties obtained thusfar on several of the $ThMn_{12}$ type ternaries.

2. Crystal structure and phase relationships

A schematic representation of the crystal structure of the $ThMn_{12}$ type is given in fig. 1. The R atoms in the $RT_{12-x}M_x$ compounds occupy the corners and the centres of tetragonal prisms. The T and M atoms occupy one or more of the positions 8j, 8i, 8f shown in the figure. The

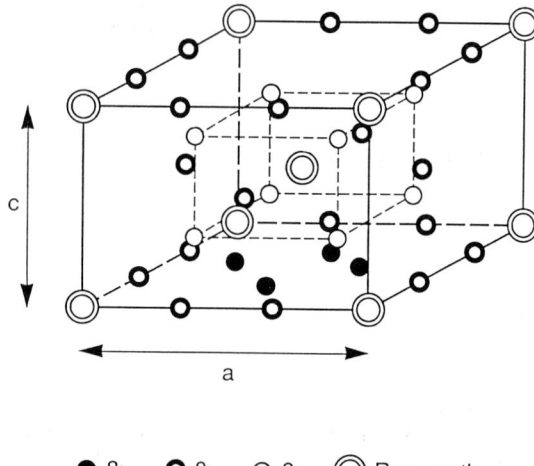

● 8j ◐ 8i ○ 8f ◎ Rare earth

Fig. 1. Schematic representation of the unit cell of compounds of the tetragonal $ThMn_{12}$ structure.

pounds and the corresponding preferred site occupation of the T and M atoms is given in table 1. For most of the compounds listed the preferred site occupations were derived from neutron diffraction data. In YFe_6Al_6 Mössbauer spectroscopy had led to the assumption that the Fe atoms occupy mainly the 8i and 8j sites [32] but recent high resolution powder diffraction shows that Fe is very reluctant to occupy the 8i site [30].

It may be inferred from the results listed in table 1 that the M component in T rich compounds of the type $RT_{12-x}M_x$ shows a preference for the 8i site when M equals another transition metal (Ti, V, Mo and possibly also W and Re). The situation changes, however, when M represents an s, p element (Al, Si). In these cases the M atoms avoid the 8i site and occupy preferentially the 8f and 8j sites. This latter occupation scheme is no longer valid in M-rich compounds. As may be seen from the table the s, p element has now a preference for the 8i site.

Attempts have been made to explain the various types of site occupations by means of size effects and by means of enthalphy effects [9,11,12]. It appears that neither of these two effects alone can give a satisfactory account of the site prefer-

occupation of the various sites by T and M atoms is by no means at random. Each of the atoms shows a strong preference for one of the three sites but this preference strongly depends on the nature of the other nonlanthanide component. A survey of the various types of $RT_{12-x}M_x$ com-

Table 1
Crystallographic and magnetic properties of $ThMn_{12}$ type ternaries. Approximate site perferences are given in columns 2–4, the corresponding references are those in front of the semicolon. The average moment per T atom in $YT_{12-x}M_x$ and the Curie temperatures of $RT_{12-x}M_x$ compounds with R = Y and Gd are given in columns 5, 6 and 7, respectively. The corresponding references are those behind the semicolon

$RT_{12-x}M_x$	8i	8j	8f	μ_T	T_c(R = Y)	T_c(R = Gd)	Refs.
$RFe_{11}Ti$	Ti,Fe	Fe	Fe	1.8	520	600	[13,6];[17,20]
$RFe_{10}V_2$	V,Fe	Fe	Fe	1.7	540	610	[14–16];[18–20]
$RFe_{10}Cr_2$	Cr,Fe	Cr,Fe	Cr,Fe	1.7	510	580	[21];[20–23]
$RFe_{10}Mo_2$	Mo,Fe	Fe	Fe	1.7	360	430	[11];[4,19,20]
$RFe_{10.8}W_{1.2}$	–	–	–	2.1	500	570	;[4,19,20]
$RFe_{10.8}Re_{1.2}$	–	–	–	> 1.8	460	518	;[24,25]
$RFe_{10}Si_2$	Fe	Si,Fe	Si,Fe		540	610	[11];[20,26]
$RCo_{11}Ti$	Ti,Co	Co	Co	1.6	1040	–	[10];[10]
$RCo_{10}V_2$	–	–	–	0.7	458	492	;[7,25]
$RCo_{10}Mo_2$	Mo,Co	Co	Co	1.3	730	–	[27];[27,28]
$RNi_{10}Si_2$	Ni	Ni	Si,Ni	≈ 0	–	< 4.2	[9];[25]
RMn_4Al_6	Al	Al	Mn	≈ 0	< 4.2	< 4.2	[29];[2,31]
RCr_4Al_6	Al	Al	Cr	≈ 0	–	8	[29];[2]
RFe_4Al_6	Al	Al	Fe	0.75	185	172	[30];[31,32]
RFe_6Al_6	Al	Al,Fe	Fe	–	330	345	[30];[12,32]
$RFe_{10}Al_2$ [a]	Fe	Al,Fe	Al,Fe	1.5	–	500	[8];[8]

[a] Data listed refer to a metastable phase $GdFe_{10}Al_2$.

ences observed and that at least a combination of both effects is required.

In view of the interest of the Fe-rich ThMn$_{12}$ type phases as possible permanent magnet materials several investigators have studied the phase relationships in the corresponding ternary systems [33–35]. Results of Neiva et al. [34] for the system Sm–Fe–Ti have been reproduced in fig. 2. The main difference of the ternary section obtained (at 1000°C) by Neiva et al. with those obtained by Kim et al. (1000°C) and Jang and Stadelmaier (900°C) is the occurrence of a phase Sm(Fe,Ti)$_9$ instead of a phase Sm(Fe,Ti)$_{11}$. It follows from all these investigations that it is not possible to obtain alloys in which SmFe$_{11}$Ti is in equilibrium with a phase that is nonmagnetic at room temperature. This can be seen also from the section shown in fig. 2 where the various two- or three-phase regions involve at least one of the phases α-Fe, Sm$_2$(Fe,Ti)$_{17}$, Sm(Fe,Ti)$_9$ or TiFe$_2$. All of these phases can be characterized as being magnetically soft so that their presence, even in small amounts, may have a detrimental influence on the generation of coercivities in these materials. This will be discussed in more detail in section 5, together with the occurrence of metastable phases.

3. Magnetic properties

Values of the Curie temperature of compounds of the type RT$_{12-x}$M$_x$ have been included in table 1 for the cases R = Y. Since Y has no magnetic moment the values listed can then be taken as representing the strength of the intrasublattice interaction of the T sublattice. Slightly higher values are used for magnetic R components owing to the effect of the R–T intersublattice interaction. A survey of the concentration dependence of T_c in various YFe$_{12-x}$M$_x$ compounds is given in fig. 3 [20].

The R–T intersublattice interaction is important for permanent magnet applications in at least two different aspects. The first is the Curie temperature enhancement mentioned already above. Of more importance is, however, its influence on the magnetocrystalline anisotropy and its temperature dependence. The crystalline magnetic anisotropy consists of a contribution of the T sublattice and a contribution of the R sublattice. Most permanent magnet materials owe their excellent properties to the high values of the latter contribution. It is well known that the R sublattice anisotropy originates from a crystal field induced single ion anisotropy. For this reason the role played by the R–T exchange interaction in determining the strength of the R sublattice anisotropy is not always fully realized when dealing with practical aspects. In fact it is this very interaction which couples the R sublattice anisotropy to the T sublattice. But at the same time is there an enhancement of the single-ion anisotropy of the R sublattice by means of the molecular field experienced by the R moments due to the T sublattice.

In practice it proves helpful to describe the rare earth sublattice anisotropy by a phenomenological expression of the type

$$E_{A,R} = K_1 \sin^2\theta + (K_2 + K_2 \cos 4\phi) \sin^4\theta + (K_3 + K_3 \cos 4\phi) \sin^6\theta, \quad (1)$$

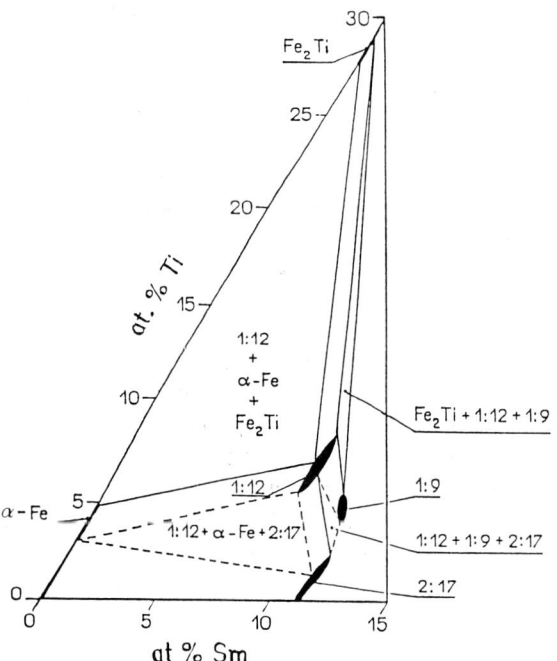

Fig. 2. Fe-rich part of the isothermal section at 1000°C of the Sm–Fe–Ti phase diagram. After Neiva et al. [34].

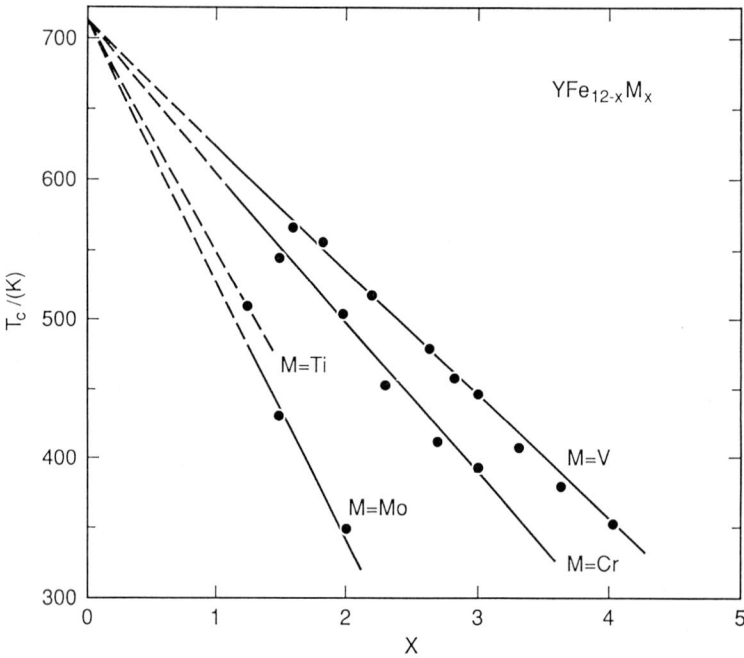

Fig. 3. Concentration dependence of the Curie temperature.

where θ and ϕ are the polar angles for the R sublattice magnetization relative to the crystallographic axes. From the transformation properties of the Stevens operator equivalents it is possible to derive relations between the anisotropy constants K_i and the crystal field parameters $B_{nm} = \theta_{nm}\langle r^n \rangle A_{nm}$ [36,37]. The lowest order term can be given as

$$K_1 = -\left[\tfrac{3}{2}B_{20}\langle O_{20}\rangle + 5B_{40}\langle O_{40}\rangle + \tfrac{21}{2}B_{60}\langle O_{60}\rangle\right]. \quad (2)$$

Similar expressions also hold for the other anisotropy constants appearing in eq. (1). A more detailed discussion of the effect of higher order terms on the magnetic properties of $RT_{12-x}M_x$ compounds has been given by Li and Coey [12]. Confining our attention to the lowest order term, one sees that the strength of the anisotropy is determined not only be the crystal field parameter B_{nm} (or A_{nm}) but also by the expectation value of the Stevens operators O_{nm}. For instance, at cryogenic temperatures the expectation value of O_{20} can be given by

$$\langle O_{20}\rangle_{T=0} = \langle 3J_z^2 - J(J+1)\rangle_{T=0} = 2J^2 - J \quad (3)$$

when there is a collinear ferrimagnetic ordering of the R and T moments and the magnetic moments of the R sublattice become equal to $\mu_R = -g\langle J_z\rangle \mu_B = gJ\mu_B$. The situation is, however, quite different at room temperature where, owing to the almost negligible R–R coupling, the expectation values $\langle O_{n0}\rangle$ and hence K_1 would become almost zero if the R–T coupling were absent. This example may illustrate that the intersublattice coupling strength is of equal importance for obtaining high room temperature anisotropies in rare earth based magnet materials as the crystal field parameters A_{nm}. For this reason special attention has been devoted to investigate the R–T coupling strength in various 4f–3d compounds in more detail [37,38].

A relatively simple method that can be used for obtaining experimental values of the intersub-

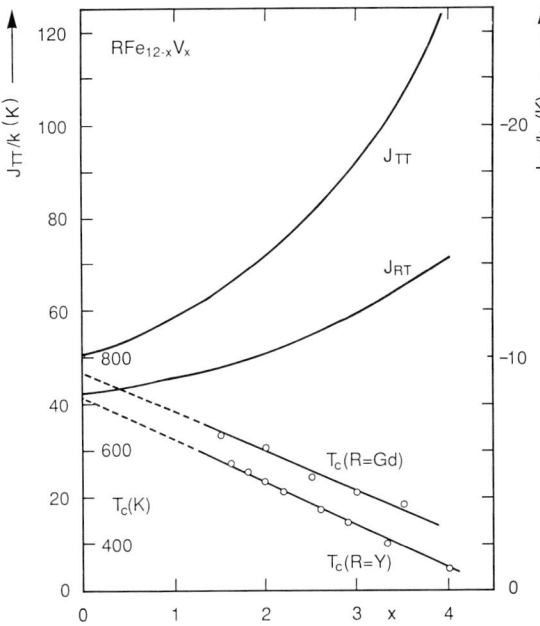

Fig. 4. Concentration dependence of the exchange coupling constants J_{GdFe} and J_{FeFe} as derived from the concentration dependence of T_c in GdFe$_{12-x}$V$_x$ and YFe$_{12-x}$V$_x$ shown in the lower part of the figure. After Zhong et al. [37].

lattice-coupling constant is based on the meanfield approach. By using the Curie temperatures of compounds in which $\mu_R \neq 0$ ($T_c = T_{c,R}$) in conjunction with Curie temperatures of compounds (R = Y, La, Lu) in which $\mu_R = 0$ ($T_c = T_{c,0}$) one easily derives for the intersublattice coupling constant J_{RT} (defined by the Hamiltonian $H = \Sigma_{RT} 2 J_{RT} S_R \cdot S_T$)

$$J_{RT}^2 = 9k^2 T_{c,R}(T_{c,R} - T_{c,0}) / 4 Z_{RT} Z_{TR} S_T(S_T + 1) G, \quad (4)$$

where G is the De Gennes factor $G = (g_R - 1)^2 J_R(J_R + 1)$, and k is Boltzmann's constant. The quantities Z_{RT} and Z_{TR} represent coordination numbers in the crystal structure considered.

Eq. (4) has been applied to various compounds of the series RFe$_{12-x}$V$_x$ with R = Y and Gd [37]. Results for T_c of both series and the corresponding values of J_{RFe} are shown in fig. 4. Included in this figure is also the concentration dependence of the intrasublattice constant J_{FeFe}, which can be obtained from the Curie temperatures of

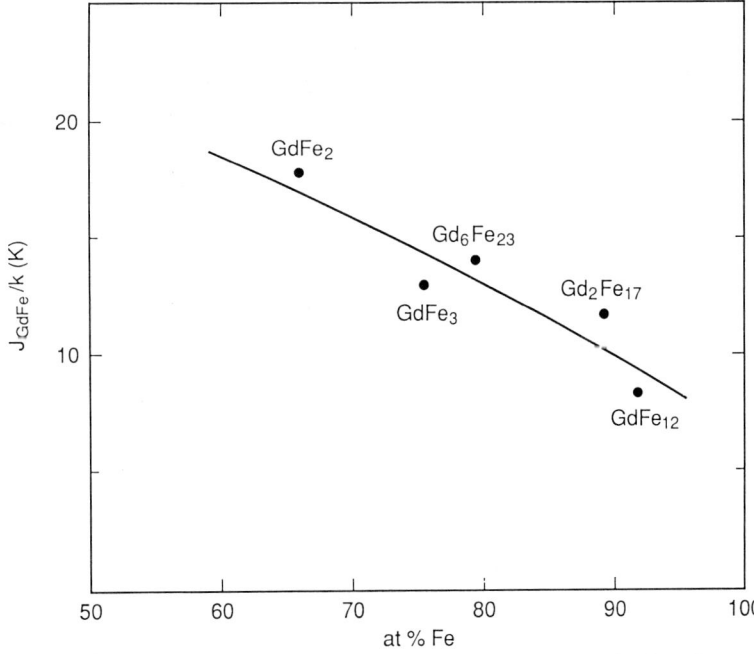

Fig. 5. Intersublattice magnetic coupling constant J_{GdFe} in binary rare earth compounds plotted versus Fe concentration. The values for GdFe$_2$, GdFe$_3$, Gd$_6$Fe$_{23}$ and Gd$_2$Fe$_{17}$ are from ref. [37]. The value for the fictitious compound GdFe$_6$ was obtained by extrapolation (see text). All values are absolute values.

$YFe_{12-x}V_2$ by the mean-field expression

$$J_{TT} = 3kT_c/2Z_{TT}S_T(S_T + 1). \quad (5)$$

Owing to the fact that the J_{RFe} values in $GdFe_{12-x}V_x$ show a fairly low variation with concentration in the low V content regime one may easily obtain the value J_{GdFe} applicable to $GdFe_{12}$ by extrapolation to $x = 0$. The latter value is compared in fig. 5 with J_{GdFe} values derived by the same method for other binary Gd–Fe compounds [38].

Inspection of the data plotted in fig. 5 shows that the intersublattice coupling constants have a tendency to decrease with Fe concentration. Brooks et al. [39] performed band structure calculations on RFe_2 compounds and found that the intersublattice coupling constant depends on the average values of the 5d electron spin of R ($\bar{S}_{5d}$) and the 3d electron spin of Fe ($\bar{S}_{Fe}$) via the relation

$$J_{RFe} \propto K_{4f5d}\bar{S}_{5d}/\bar{S}_{3d}, \quad (6)$$

where K_{4f5d} is the 4f–5d exchange integral. From magnetic measurements on the compounds YFe_2, YFe_3, Y_6Fe_{23}, Y_2Fe_{17} it is known that $\bar{S}_{3d}$ increases with Fe concentration which might partially explain the decrease of J_{RFe} with Fe concentration shown in the figure. Unfortunately there is no direct information on $\bar{S}_{5d}$. However, Coehoorn [41] made band structure calculations of the various Y–Fe compounds mentioned and these calculations show clearly that one may expect a decreasing trend of $\bar{S}_{4d}$. If one assumes that a similar behaviour is followed also by $\bar{S}_{5d}$ it will become clear that the ratio $\bar{S}_{5d}/\bar{S}_{3d}$ decreases strongly with Fe concentration. In terms of eq. (6) this might explain why one observes a strong decrease of the intersublattice coupling constant J_{RFe} with increasing Fe content (fig. 5).

As may be deduced already from the results shown in fig. 3, the $RFe_{12-x}M_x$ compounds with M = V have the largest range of solid solubility. This fact has been advantageously used by Verhoef et al. [19] to determine the saturation moment of the fictitious compound YFe_{12} by extrapolation of the high-field magnetization data of $YFe_{12-x}V_x$ ($1.6 \leq x \leq 3.5$). The value obtained in this way corresponds to an Fe moment of 2.07 μ_B/Fe. High-field magnetization measurements were made by the same authors also for several other $YFe_{12-x}M_x$ compounds and the corresponding average moment per atom $Fe_{1-z}M_z$ in $Y(Fe_{1-z}M_z)_{12}$ can be compared with the extrapolated Fe moment in YFe_{12}. It then follows that the moment reduction increases in the sequence W < Cr < V < Ti. The differences in moment reduction associated with the various M components were tentatively explained in terms of valence differences and the Slater–Pauling curve. More direct information regarding the moment formation in $YFe_{12-x}M_x$ compounds was obtained by Jaswal et al. [40] and Coehoorn [41] on the basis of electronic band structure calculations. Results of the calculations made for YFe_8M_4 compounds (in which the M atoms were assumed to be ordered at the 8i site) showed that the moment reduction increases in the sequence Cr < V < Ti, in agreement with experiment [41]. It furthermore follows from the band structure calculations that the magnetic moments of the Fe atoms occupying the 8f site in YFe_{12} are substantially lower than those residing at the 8i and 8j site. This is true also in the ternary compounds. Results of neutron diffraction experiments [14,15] and ^{57}Fe Mössbauer spectroscopy generally showed that the largest Fe moments are associated with the 8i sites, indeed. Apparently controversial results were obtained, however, in ^{57}Mössbauer investigations for the relative magnitude of the Fe moments residing at the 8j and 8f sites [42–52]. This is a consequence of the difficulty to make a distinction between the corresponding subspectra [47]. A subspectra assignment on the basis of band-structure calculations was made by Denissen et al. [48].

The magnetic structures and the corresponding easy moment directions in the various $RFe_{12-x}M_x$ compounds are governed by the competition the Fe sublattice anisotropy and the R sublattice anisotropy. The Fe sublattice anisotropy favours an easy magnetization direction parallel to the c-axis and the corresponding K_1 value has about the same magnitude as in $R_2Fe_{14}B$ and $R_2Fe_{14}C$ compounds. However, the crystal field induced rare earth sublattice anisotropy in $RFe_{14-x}M_x$ is quite different from that found in $R_2Fe_{14}B$ and $R_2Fe_{14}C$, both in sign and magnitude. This may be illustrated, for instance, by comparing the magnetic structures of $Dy_2Fe_{14}B$ and $DyFe_{11}Ti$. The former compound is a normal ferrimagnet with the magnetic moments oriented along the c direction at all temperatures. By contrast, $DyFe_{11}Ti$ is a ferrimagnet with $M \parallel c$ only in the limited temperature range $200\ K \leq T \leq T_c$. Below 200 K two spin reorientations occur, leading to magnetic structures in which the easy magnetization direction deviates from the c-axis and the magnetic moment is no longer collinear [52,55].

A similar complex temperature dependence of the spin structure was observed also for $TbFe_{11}Ti$ while in $NdFe_{11}Ti$ and $ErFe_{11}Ti$ there is only a single spin-orientation transition. A survey of the temperature dependence of the spin structures of the $RFe_{11}Ti$ compounds is presented in ref. [12].

The temperature dependence of the spin structures and the field dependence of the magnetization in various crystallographic directions (including three first order magnetization processes) observed by Hu et al. [52] on a $DyFe_{11}Ti$ single crystal was analysed by these authors in terms of a two-sublattice model based on exchange and crystal field effects. The set crystal field parameters A_{nm} obtained from this analysis were used by Hu et al. to explain the spin reorientations observed in the remainder of compounds.

When the crystal field parameters A_{nm} of $RFe_{12-x}M_x$ compounds are compared with those associated with the $R_2Fe_{14}B$ compounds (see for experimental values of A_{nm} derived from single crystals the review of Li and Coey [12]) one finds that the leading term A_{20} is of a different sign in the tetragonal $ThMn_{12}$ structure ($A_{20} < 0$) than in the tetragonal $Nd_2Fe_{14}B$ structure ($A_{20} > 0$). This means that the rare earth sublattice contributes to the uniaxial anisotropy only in those $RFe_{12-x}M_x$ compounds in which the second order Stevens factor $\theta_2 = \alpha_j$ is positive (Sm, Er, Tm). Since, furthermore, the Er and Tm sublattices couple antiparallel to the Fe sublattice the corresponding total magnetizations are low, leaving only the Sm compound as a candidate for permanent magnet purposes.

The comparison of the crystal field parameters of the $ThMn_{12}$ compounds with those in $Nd_2Fe_{14}B$ type compound also reveals that the leading term A_{20} appearing in the anisotropy expression (eqs. (1) and (2)) is much smaller in the former than in the latter. On the other hand, Hu et al. [52] showed that the magnetocrystalline anisotropy in $SmFe_{11}Ti$ is much greater than would be expected on the basis of the A_{20} value available. A possible explanation was offered by Li et al. [53] in terms of considerable crystal field and exchange induced mixing of higher multiplet levels into the J ground state multiplet level of Sm. In fact, a larger (negative) value of A_{20} has been found from measurements on a single crystal $SmFe_{11}Ti$ by Kaneko et al. [54]. These results provide a third example showing the influence of the intersublattice coupling on the anisotropy.

Spin structures of the $RFe_{11}Ti$ and $RFe_{10}V_2$ compounds have been reviewed by Li and Coey [12]. Generally it can be stated that $M \parallel c$ for those compounds in which the R sublattice does not contribute to the anisotropy (R = Y, Gd and Lu) or contributes only weakly (R = Ho and Tm). The R sublattice anisotropy is able to force the easy magnetization direction away from the c-axis at lower temperatures for R = Nd, Tb, Dy and Er but does so at substantially lower temperatures in $R_2Fe_{10}V_2$ than in $RFe_{11}Ti$. This change in easy magnetization direction may proceed by more than one magnetic phase transition. For instance, single crystal measurements [52,55] showed that there is an easy axis to easy cone transition in $DyFe_{11}Ti$ at about 220 K while a different cone structure is reached at still lower temperatures. A most surprising result is furthermore that the Er compound adopts a spin structure in $RFe_{10}V_2$ in which $M \perp c$. On the basis of $A_{20} < 0$ one would expect the R sublattice anisotropy to favour $M \parallel c$ if A_{20} could be regarded as leading term in the expression of the anisotropy energy (eqs. (1) and (2)). The general conclusion derived from the temperature dependence of the spin structures is that the anisotropy is determined largely by crystal field parameters of higher order than A_{20} in the $RFe_{10}V_2$ and $RFe_{11}Ti$ series. A similar situa-

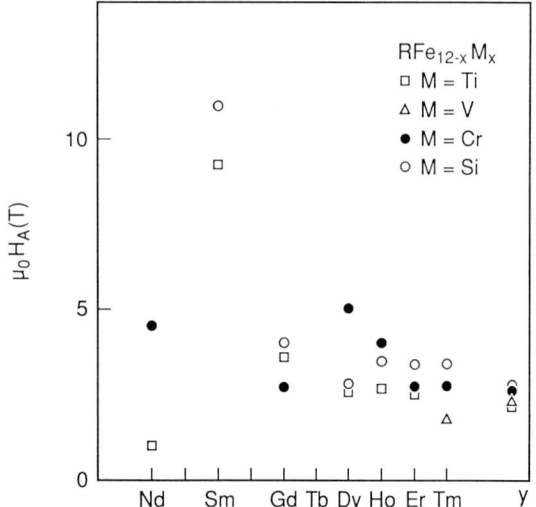

Fig. 6. Anisotropy field observed in various $RFe_{12-x}M_x$ compounds plotted versus the R component.

tion is found also in the $RFe_{10}Cr_2$ and $RFe_{10}Si_2$ series [50].

The experimental data available for the compounds $RFe_{12-x}M_x$ with M = Cr, Mo, W, Re and Si are still rather limited and do not make it possible to present diagrams of the temperature dependences of the magnetic structures. In the present report we will restrict ourselves therefore to giving a survey of the room temperature anisotropy field, since these quantities play a role in the possible application of these compounds as permanent magnets. These anisotropies have been collected in fig. 6 using data published by Stefanski et al. [56], Stefanski and Wrzeciono [57], Solzi et al. [16,58] and Hu et al. [17]. No data are given for the Tb compounds which appear to have an easy plane in most cases. For more details the reader is referred again to the review article of Li and Coey [12] or to the original papers.

The data presented in fig. 6 show that the Sm compounds have by far the largest anisotropy fields and hence are the most suitable candidates for permanent magnet purposes. Several methods used to generate coercivities in the $ThMn_{12}$ type materials will be discussed in section 5.

4. Interstitial nitrides of 1:12 phases

Strong changes in magnetic properties can be realized when intermetallic compounds are charged with nitrogen gas. In compounds of the type R_2Fe_{17} absorption of N_2 not only strongly increases the Curie temperature but it also leads to substantial changes of the rare earth sublattice anisotropy (see for instance ref. [12] and papers cited therein). Recently performed experiments on $RFe_{11}Ti$ compounds have shown that also these compounds are able to absorb large quantities of nitrogen gas when heated in an N_2 atmosphere at about 500°C, albeit their absorption capacity is lower than that of R_2Fe_{17} phases [59–64]. From the increase in the unit cell dimension after charging it may be derived that the nitrogen content in $RFe_{11}TiN_x$ is somewhat below $x = 1$.

The interstitial solution of nitrogen atoms leads to strong enhancements of the Curie temperatures. This may be seen from fig. 7 where T_c values in $RFe_{11}Ti$ are compared with those of the corresponding nitrides. Of equal importance is the change in rare earth sublattice anisotropy accompanying the N_2 uptake which is due to the closeness of the interstitial N atoms to the R atoms. In $RFe_{11}TiN_x$ it leads to a complete sign reversal of the R sublattice anisotropy (A_{20} changes from negative to positive), favouring an

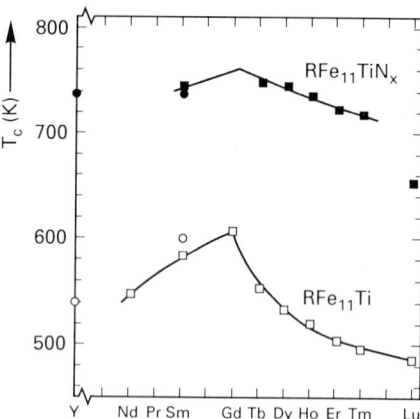

Fig. 7. Curie temperatures of various $RFe_{11}Ti$ compounds and their nitrides $RFe_{11}TiN_x$. Data were taken from Yang et al. [59] (open and filled circles) and from Coey et al. [61] (open and filled squares).

easy c-axis for R elements with $\alpha_J < 0$ and an easy magnetization direction perpendicular to the c-axis for R elements with $\alpha_J > 0$. The nitride $SmFe_{11}TiN_x$ is an example of the latter category. The easy magnetization direction, in contrast to $SmFe_{11}Ti$, is perpendicular to the c-direction at all temperatures below T_c. This shows that the Sm sublattice anisotropy is sufficiently strong to dominate the Fe sublattice anisotropy that favours an easy c-axis. The nitride $ErFe_{11}TiN_x$ is an example where the R sublattice anisotropy is able to dominate the Fe sublattice anisotropy only at cryogenic temperatures, leading to a spin reorientation at $T_{SR} = 45$ K [60]. Owing to the sign reversal of the rare earth sublattice anisotropy one now finds that $NdFe_{11}TiN_x$ rather than $SmFe_{11}TiN_x$ is a candidate for permanent magnet purposes.

5. Permanent magnets

Fe-rich ternary compounds derived from the tetragonal $ThMn_{12}$ structure have been proposed as inexpensive alternatives to permanent magnet materials. As described in section 3 their Curie temperatures and magnetizations have sufficiently high values for technical applications. In the previous section it was shown that generally only the Sm compounds qualify for materials with an adequately high uniaxial magnetic anisotropy.

There are numerous examples where sufficiently high coercive forces have been attained in these materials. In most of the investigations magnetic hardening was obtained by melt spinning [65–73]. Results of magnetic hardening by melt spinning obtained by Pinkerton and Van Wingerden [72] and Okada et al. [73] are shown in fig. 8. Concentrating on alloys of approximately the same Sm content, the results show that some improvement of the coercive field can be obtained by partially substituting Ti for V. Generally most authors reported fairly low values of the remanence. In order to be able to apply these materials as permanent magnets it will be necessary to improve their loop squareness and/or to generate texture in the ribbons.

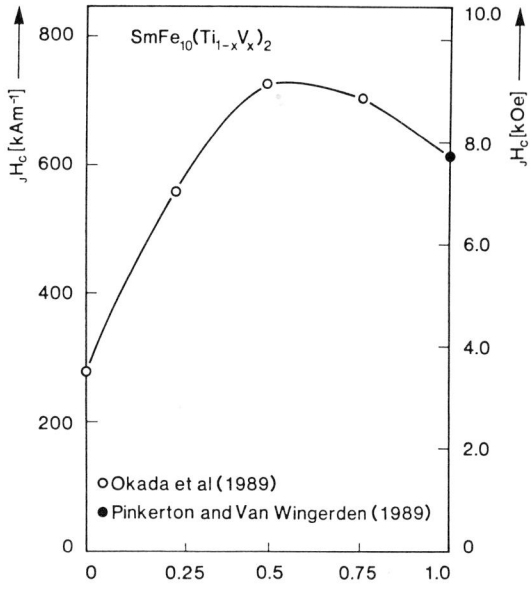

Fig. 8. Concentration dependence of the coercivity in $SmFe_{10}(Ti_{1-x}V_x)_2$ compounds.

An attractive process for manufacturing permanent magnets has been offered by Schultz and Wecker [67] who applied mechanical alloying. Magnets of various composition were prepared by milling the elemental powders with steel balls in a closed container followed by a diffusion heat treatment (600–900°C). The magnets are isotropic.

Schultz et al. [66] found evidence that a novel ternary compounds exists in the ternary Sm–Fe–Ti system close to the composition $Sm_{20}Ti_{10}Fe_{70}$. This compound seems to be metastable and can be obtained by vapour deposition [74], by melt spinning or via mechanical alloying [66]. Although a structure determination has not been made there are indications from X-ray diffraction that the structure of this compound is based on the structure of Nd_5Fe_{17} described by Moreau et al. [75]. Anyhow, Kamprath et al. [74] as well as Schultz et al. [65] reported giant coercivities for this material with coercivities of more than 64 kOe. The drawback of these magnets is that the remanence is still somewhat on the low side. On the other hand, the $ThMn_{12}$ type materials show a much better corrosion resistance than materials based on $Nd_2Fe_{14}B$ type compounds. The reason for this is probably the absence of an Nd-rich intergranular phase in the $ThMn_{12}$ type materials.

References

[1] O.S. Zarechnyuk and P.I. Kripyakevich, Sov. Phys. Cryst. 7 (1963) 436.
[2] K.H.J. Buschow, J.H.N. van Vucht and W.W. van den Hoogenhof, J. Less-Common Met. 50 (1976) 148.
[3] K.H.J. Buschow, D.B. de Mooij, M. Brouha, H.H.A. Smit and R.C. Thiel, IEEE Trans. Magn. MAG-24 (1988) 1611.
[4] A. Müller, J. Appl. Phys. 64 (1988) 249.
[5] K. Ohashi, Y. Tawara and R. Osugi, J. Less-Common Met. 139 (1988) L1.
[6] K. Ohashi, T. Yokoyama, R. Osugi and Y. Tawara, IEEE-Trans. Magn. MAG-23 (1987) 3101.
[7] M. Jurczyk and O.D. Chistyakov, Phys. Stat. Sol. 115 (1989) K229.
[8] X.Z. Wang, B. Chevalier, T. Berlureau, J. Etourneau, J.M.D. Coey and J.M. Cadogan, J. Less-Common Met. 138 (1988) 235.
[9] O. Moze, R.M. Ibberson and K.H.J. Buschow, Solid State Commun. 78 (1991) 473.
[10] Y.C. Yang, L.S. Kong, H. Sun, J.L. Yang, Y.F. Ding, B.S. Zhang, C.T. Ye and L. Jin, J. Appl. Phys. 67 (1990) 4632.
[11] K.H.J. Buschow, J. Appl. Phys. 63 (1988) 3130.
[12] H.S. Li and J.M.D. Coey, in: Magnetic Materials, vol. 6 (North-Holland, Amsterdam, 1991).
[13] O. Moze, L. Pareti, M. Solzi and W.I.F. David, Solid State Commun. (to be published).
[14] R.B. Helmholdt, J.J.M. Vleggaar and K.H.J. Buschow, J. Less-Common Met. 138 (1988) L11, 144 (1988) 209.
[15] W.G. Haije, J. Spijkerman, F.R. de Boer, K. Bakker and K.H.J. Buschow, J. Less-Common Met. 162 (1990) 285.
[16] M. Solzi, L. Pareti, O. Moze and W.I.F. David, J. Appl. Phys. 64 (1989) 5084.
[17] J. Hu, T. Wang, S. Zhang, Y. Wang and Z. Wang, J. Magn. Magn. Mater. 74 (1988) 22.
[18] F.R. de Boer, Huang Ying-kai, D.B. de Mooij and K.H.J. Buschow, J. Less-Common Met. 135 (1987) 199.
[19] R. Verhoef, F.R. de Boer, Zhang Zhi-dong and K.H.J. Buschow, J. Magn. Magn. Mater. 75 (1988) 319.
[20] K.H.J. Buschow and D.B. de Mooij; in: Concerted European Action on Magnets, eds. I.V. Mitchell, J.M.D. Coey, D. Givord, I.R. Harris and R. Hanitsch (Elsevier Applied Science, London, 1989) p. 63.
[21] P.C.M. Gubbens, G.J. Boender, A.M. van der Kraan and K.H.J. Buschow, Hyperfine Interactions (to be published).
[22] P. Stefanski, A. Kowalczyk and A. Wrzeciono, J. Magn. Magn. Mater. 81 (1989) 155.
[23] F.M. Yang, O.A. Li, R.W. Zhao, J.P. Kuang, F.R. de Boer, J.P. Liu, K.V. Rao, G. Nicolaides and K.H.J. Buschow, J. Less-Common Met. (1991) in press.
[24] M. Jurczyk, J. Less-Common Met. 166 (1990) 335.
[25] J.H.V.J. Brabers, T.H. Jacobs and K.H.J. Buschow, unpublished results.
[26] P. Stefanski and A. Wrzeciono, J. Magn. Magn. Mater. 82 (1989) 125.
[27] C. Lin, Y.X. Sun, Z.X. Liu and G. Li, J. Appl. Phys. 69 (1991) 5554.
[28] S.F. Cheng, V.K. Sinha, B.M. Ma, S.G. Sankar and W.E. Wallace, J. Appl. Phys. 69 (1991) 5605.
[29] O. Moze, R.M. Ibberson, R. Caciuffo and K.H.J. Buschow, J. Less-Common Met. 166 (1990) 329.
[30] O. Moze, R.M. Ibberson and K.H.J. Buschow, J. Phys.: Condens. Matter. 2 (1990) 1677.
[31] K.H.J. Buschow and A.M. van der Kraan, J. Phys. F 8 (1978) 921.
[32] I. Felner and I. Nowik, J. Magn. Magn. Mater. 54–57 (1986) 163; J. Phys. Chem. Solids 40 (1979) 1035.
[33] T.S. Jang and H.H. Stadelmaier, J. Appl. Phys. 67 (1990) 4957.
[34] A.C. Neiva, F.P. Missell, B. Grieb, E.Th. Henig and G. Petzow, J. Less-Common Met. 170 (1991) 293.
[35] Y.B. Kim, S. Sugimoto, M. Okada and M. Homma, J. Less-Common Met. 171 (1991) in press.
[36] P.A. Lindgård and O. Danielsen, Phys. Rev. B 11 (1975) 351.
C. Rudowics, J. Phys. C 18 (1985) 1415.
[37] X.P. Zhong, F.R. de Boer, D.B. de Mooij and K.H.J. Buschow, J. Less-Common Met. 163 (1990) 123.
[38] J.P. Liu, F.R. de Boer and K.H.J. Buschow, J. Magn. Magn. Mater. 98 (1991). 291.
[39] M.S.S. Brooks, L. Nordström and B. Johansson, Phys. B 172 (1991) 95.
[40] S.S. Jaswal, Y.G. Ren and D.J. Sellmyer, J. Appl. Phys. 67 (1990) 4564.
[41] R. Coehoorn, Phys. Rev. 39 (1989) 13072.
[42] C. Christides, A. Kostikas, D. Niarchos and A. Simopoulos, J. de Phys. 49 (1988) C8–539.
[43] B.P. Mu, H.S. Li, J.P. Gavigan and J.M.D. Coey, J. Phys.: Condens. Matter. 1 (1989) 755.
[44] P.C.M. Gubbens, A.M. van der Kraan and K.H.J. Buschow, Hyperfine Interactions 40 (1988) 389.
[45] Th. Sinnemann, M. Rosenberg and K.H.J. Buschow, J. Less-Common Met. 146 (1989) 223.
[46] Th. Sinneman, K. Erdmann, M. Rosenberg and K.H.J. Buschow, Hyperfine Interactions 50 (1989) 657.
[47] Th. Sinnemann, M.V. Wisniewski, M. Rosenberg and K.H.J. Buschow, J. Magn. Magn. Mater. 83 (1990) 259.
[48] C.J.M. Denissen, R. Coehoorn and K.H.J. Buschow, J. Magn. Magn. Mater. 87 (1990) 51.
[49] Z.W. Li, X.Z. Zhou, A.H. Morrish and Y.C. Yang, J. Phys.: Condens. Matter 2 (1990) 9621.
[50] P. Stefanski and A. Kowalczyk, Solid State Commun. 77 (1991) 397.
[51] Z.W. Li, X.Z. Zhou and A.H. Morrish, J. Appl. Phys. 69 (1991) 5602.

[52] B.P. Hu, M.S. Li, J.P. Gavigan and J.M.D. Coey, Phys. Rev. 41 (1990) 2221.
[53] H.S. Li, B.P. Hu, J.P. Gavigan, J.M.D. Coey, L. Pareti and O. Moze, J. de Phys. 49 (1988) C8-541.
[54] T. Kaneko, M. Yamada, K. Ohashi, Y. Tawara, R. Osugi, H. Yoshida, G. Kido and Y. Nakagawa, Proc. Intern. Workshop on Rare Earth Magnets, Kyoto (May 1989) p. 191.
[55] A.V. Andreev, M.I. Bartashevich, N.V. Kudrevatykh, S.M. Razgonyciev, S.S. Sigaev and E.N. Tarasov, Physica B 167 (1990) 139.
[56] P. Stefanski and A. Wrzeciono, J. Magn. Magn. Mater. 82 (1989) 125.
[57] P. Stefanski, A. Kowalczyk and A. Wrzeciono, J. Magn. Magn. Mater. 81 (1989) 155.
[58] M. Solzi, R.H. Xue and L. Pareti, J. Magn. Magn. Mater. 88 (1990) 44.
[59] Y.C. Yang, X.D. Zhang, L.S. Kong, Q. Pan, Y.T. Hou, S. Huang and L. Yang, J. Less-Common Met. 170 (1991) 37.
[60] Y.C. Yang, X.D. Zhang, L.S. Kong, Q. Pan and S.L. Ge, Appl. Phys. Lett. 58 (1991) 2042.
[61] J.M.D. Coey, Hong Sun and D.P.F. Hurley, J. Magn. Magn. Mater. 101 (1991) 310.
[62] L.X. Liao, Z. Altounian and D.H. Ryan, J. App. Phys. (MMM '91).
[63] Y.Z. Wang and G.C. Hadjipanayis, J. Appl. Phys. (MMM '91).
[64] M. Anagnoston, C. Christides and D. Niarchos, J. Appl. Phys. (MMM '91).
[65] J. Ding and M. Rosenberg, J. Magn. Magn. Mater. 83 (1990) 257; J. Less-Common Met. 161 (1990) 263, 369.
[66] L. Schultz, K. Schnitzke and J. Wecker, J. Magn. Magn. Mater. 83 (1990) 254.
[67] L. Schultz and J. Wecker, J. Appl. Phys. 64 (1988) 5711.
[68] D. Cochet-Muchy and S. Paidassi, J. Magn. Magn. Mater. 83 (1990) 249.
[69] Y.Z. Wang and G.C. Hadjipanayis, J. Magn. Magn. Mater. 87 (1990) 375.
[70] M.S. Anagnostou and D. Niarchos, J. Magn. Magn. Mater. 88 (1990) 100.
[71] H. Sun, Y. Otani, J.M.D. Coey, C.D. Meekison and J.P. Jakubovics, J. Appl. Phys. 67 (1990) 4659.
[72] F.E. Pinkerton and D.J. van Wingerden, IEEE Trans. Magn. MAG-25 (1989) 3306.
[73] M. Okada, K. Yamagishi and M. Homma, Mater. Trans. JIM 30 (1989) 374.
[74] N. Kamprath, N.C. Liu, H. Hedge and F.J. Cadieu, J. Appl. Phys. 64(1988) 5720.
[75] J.M. Moreau, L. Paccard, J.P. Nozières, F.P. Missel, G. Schneider and V. Villas-Boas, J. Less-Common Met. 163 (1990) 245.

Magnetic properties of the Kondo lattice

C. Lacroix

Laboratoire Louis Néel, CNRS, 166 X, 38042 Grenoble-Cedex, France

We review the magnetic properties of the Kondo lattice model using the functional integral technique: we discuss the magnetic and non-magnetic phases, and some properties of these phases. Comparison with other calculations is also made. Finally some open questions are discussed.

1. Introduction

In the last few years a new class of rare earth or actinide compounds has received much interest: these materials were termed first anomalous rare earth compounds or intermediate valence systems, then Kondo lattice compounds or heavy fermions systems [1–4]. In these compounds, (mainly Ce and U compounds) various types of ground states are observed: non-magnetic Fermi liquid, ordered magnetic compounds, and superconducting compounds. The existence of superconducting transitions has been discussed by many authors, in connection with the possibility of non-phononic pairing interaction and exotic superconductivity. This subject will not be discussed in this review where we concentrate on the magnetic properties of the normal phase. Some of these magnetic compounds exhibit a reduced magnetic moment at $T = 0$ K, which indicates a singlet ground-state; above the Néel temperature they usually behave as Kondo Ce^{3+} compounds. At high temperature magnetic and non-magnetic Kondo lattice compounds are similar, but at low temperature the non-magnetic ones show some typical behaviour: large effective masses m^*, metamagnetic transitions, magnetic correlations down to $T = 0$ K without ordering, etc.

All these anomalous magnetic properties are attributed to the competition between the RKKY interaction and the Kondo effect, which are of the same order of magnitude in these compounds, due to the position of the 4f or 5f level near the Fermi level.

Since the first theoretical study by Doniach [5], a large number of authors have tried to understand these features. The starting point for a theoretical investigation is the periodic Anderson model:

$$H = \sum_{k,\sigma} \epsilon_{k\sigma} C^+_{k\sigma} C_{k\sigma} + E_f \sum_{im} f^+_{im} f_{im}$$
$$+ V \sum_{im\sigma} (C^+_{i\sigma} f_{im} + H_c)$$
$$+ \frac{U}{2} \sum_{i,m \neq m'} n_{im} n_{im'}, \quad (1)$$

where two types of electrons are considered:

- conduction electrons with a wide conduction band ϵ_k,
- localized f electrons, with orbital degeneracy $N = 2j + 1$; the Coulomb repulsion U between f electrons is usually taken to be very large;
- V is the hybridization between the two types of electrons.

In the limit of nearly integral valence (i.e. $\langle \sum_m f^+_{im} f_{im} \rangle \to 1$) this Hamiltonian can be replaced by the Kondo lattice model through a Schrieffer–Wolff transformation; for an N-fold degenerate f level, the Kondo lattice model gives: [6]

$$H = \sum_{k,\sigma} \epsilon_k C^+_{k\sigma} C_{k\sigma} + \frac{J}{N} \sum_{i,mm'} C^+_{im} C_{im'} f^+_{im'} f_{im}. \quad (2)$$

In the case of a two-fold degenerate f-level the last term simply describes an antiferromagnetic interaction $J S_i \cdot \sigma_i$ between the f electron spin S_i and the conduction electron spin σ_i. In the following we generally restrict to this case.

There are two characteristic energies in this model: the Kondo temperature of a single impurity, $T_K = D \, e^{-1/\rho J}$ and the RKKY interaction $I \sim \rho J^2$. The competition between these two effects will be the main topic discussed in this paper. In the next section we discuss briefly the functional integral method, for the Kondo lattice model. In section 3 the magnetic phase is described within this method. In section 4 we discuss the properties of the non-magnetic phase.

2. The functional integral method

The study of the two Hamiltonians (1) and (2) has given rise to a large number of new methods, among them:

- variational methods [7–9],
- functional integral [10,11] and slave boson [12,13] techniques,
- self-consistent Greens functions calculations [14,15],
- perturbation in U [16] and V [17],
- numerical simulation [18–21].

In the present paper the functional integral method will be used; the advantages of this method are manifold:

- the Anderson and Kondo lattice Hamiltonians are replaced by similar effective Hamiltonians,
- the mean field approximation becomes exact in the large N limit [22] and $1/N$ expansions can be performed [6],
- the Kondo resonance and the heavy fermion behaviour are obtained in the mean field approximation,
- it provides a simple description of the competition between RKKY interactions and Kondo singlet formation.

In the functional integral formalism, the partition function of the Kondo lattice is written, through a Stratanovitch–Hubbard transformation, as an integral over scalar fields: the two body interaction (last term of eq. (2)) is then linearized. However there are several ways of making this Stratanovitch–Hubbard transformation; in the usual method two scalar Bose-like fields $\lambda_i(\tau)$ and $\phi_i(\tau)$ are introduced [6,10,11,22], which can be interpreted as time and site dependent chemical potential, and s–f hybridization. However this two fields method does not permit to study the magnetic properties at the mean field level; thus, in the following we introduce four auxiliary fields: $\lambda_i(\tau)$, $\phi_i(\tau)$ and two "magnetic fields", $\mu_i(\tau)$ and $\mu'_i(\tau)$ [10,23–25]. The partition function is then written as:

$$Z = \text{Tr} \int D\phi D\lambda D\mu D\mu' \exp(-\beta F), \quad (3)$$

where

$$F = \frac{1}{\beta} \int_0^\beta \left[H_{\text{eff}}(\tau) - \frac{J}{2} \sum_i \phi_i^2(\tau) \right.$$
$$\left. + \frac{J}{4} \mu_i(\tau) \mu'_i(\tau) \right] d\tau, \quad (4)$$

$$H_{\text{eff}}(\tau) = \sum_{k,\sigma} \epsilon_k C^+_{k\sigma} C_{k\sigma}$$
$$+ \sum_{i,\sigma} \frac{J\phi_i(\tau)}{2} (f^+_{i\sigma} C_{i\sigma} + C^+_{i\sigma} f_{i\sigma})$$
$$- \sum_{i,\sigma} \frac{J\mu_i(\tau)}{4} (C^+_{i\uparrow} C_{i\uparrow} - C^+_{i\downarrow} C_{i\downarrow})$$
$$- \sum_{i,\sigma} \frac{J\mu'_i(\tau)}{4} (f^+_{i\uparrow} f_{i\uparrow} - f^+_{i\downarrow} f_{i\downarrow})$$
$$+ \sum_i i\lambda_i(\tau) \left[\sum_\sigma f^+_{i\sigma} f_{i\sigma} - 1 \right]. \quad (5)$$

In the effective Hamiltonian (5) the two body interaction has been linearized but the effective fields are site and time dependent. The same type of expression is obtained using the slave boson technique for the Anderson lattice [12,26], but without the two "magnetic fields" μ and μ'. In the Anderson lattice model, magnetism is not obtained at the mean field level, but higher order terms must be introduced [6,27].

The simplest approximation to eqs. (3)–(5) is to neglect the time dependence (static approximation). By doing this, the quasiparticles are described by a "renormalized band model" similar to that obtained by other methods [28].

3. The magnetic phase

If the time dependence of the Bose fields is neglected, the mean field solutions are obtained. At low temperature the partition function (3) can then be evaluated in the saddle point approximation. The Bose fields are then determined self-consistently:

$$\phi_i^0 = \langle f_{i\sigma}^+ C_{i\sigma} \rangle, \quad \mu_i^0 = \langle f_{i\uparrow}^+ f_{i\uparrow} - f_{i\downarrow}^+ f_{i\downarrow} \rangle,$$
$$\mu_i'^0 = \langle C_{i\uparrow}^+ C_{i\uparrow} - C_{i\downarrow}^+ C_{i\downarrow} \rangle,$$

λ_i^0 is determined by the condition $\langle \Sigma_\sigma f_{i\sigma}^+ f_{i\sigma} \rangle = 1$.
(6)

Within these approximations, Hamiltonian (5) is then equivalent to an s–f band model, with renormalized parameters: the s–f hybridization is given by $\tilde{V}_i = J\phi_i^0/2$, the position of the f level by λ_i^0, and the splittings between up and down spin bands are related to μ_i^0 and $\mu_i'^0$. Usually the uniform approximation is also made, but it is also possible to study non-uniform solutions, for example antiferromagnetic states.

In the uniform approximation, a ferromagnetic phase is found below a critical value of the parameter ρJ [10,23,24,29]. This critical value is obtained when T_K is of the same order as the absolute value of the RKKY interactions I. This magnetic phase is characterized by a reduced magnetic moment for the f electrons: $\mu = \sqrt{1 - 4\phi^2}$ and a small conduction electron polarization: $\mu' = -(\rho J/2)\mu$.

The critical value $(\rho J)_c$ is different for ferro- or antiferromagnetic ordering, but this approximation always gives $(\rho J)_c$ of the order of 0.5 to 1. Experimentally this critical value has been observed in many systems, by applying pressure (or chemical pressure): a large number of cerium systems show this transition (see ref. [30] for example); the experimental critical value $(\rho J)_c$ is usually found to be of the order of 0.1. Several remarks can be made concerning this discrepancy:

- for $\rho J > 0.5$ the Kondo lattice model is no longer valid because charge fluctuations must be taken into account (intermediate valence regime),
- the crystallographic structure certainly plays a role because cubic systems are always magnetic ($CeAl_2$, $CePb_3$) and non-cubic systems are often non-magnetic ($CeAl_3$, $CeCu_2Si_2$),
- crystal field effects must be taken into account [31,32],
- the large N limit gives a smaller critical value [26],
- finally, two impurities results show a difference between ferro- and antiferromagnetic interactions: in the ferromagnetic case, Kondo effect always occurs, whereas in the antiferromagnetic case there is a critical value for ρJ [33,34,40]. These two impurities results show that a mean field treatment of the RKKY interactions is not satisfactory, specially for antiferromagnetic interactions. We come back to this point in the next sections.

Recent experiments indicate that some "non-magnetic" compounds such as UPt_3 [35], $CeAl_3$ [36] or URu_2Si_2 [37] may have a very small ordered moment. Models derived from the Kondo lattice have been developed to account for these results [38,39]. This weak antiferromagnetism can be interpreted as a spin density wave which occurs in the heavy fermion band.

In conclusion of this section, low temperature properties of the magnetic phase are qualitatively described by the mean field approximation: existence of a critical value of ρJ, reduction of magnetic moments at $T = 0$ K.

However a better treatment is needed to answer some open questions: the two impurities results give some insight for the different behaviour in case of ferro- or antiferromagnetic coupling. The same difference certainly exists in the lattice, but the mean field approximation does not reflect this difference. Also the RKKY interactions are modified by the Kondo effect when temperature decreases, and this is also beyond the mean field approximation.

Above the Néel temperature the magnetic Kondo compounds are similar to the non-magnetic ones, which are discussed in the next part.

4. The non-magnetic phase

4.1. Mean field solution

Within the mean field approximation, the quasiparticle band structure of the non-magnetic phase can be obtained easily [10]: the Kondo resonance is split by a coherence gap, the width of this gap is of the order of T_K. The Kondo temperature is the only energy scale. Similar pictures are obtained in the slave boson method [12], in Gutzwiller approximation [8] and in other variational methods [7]. The low temperature susceptibility and specific heat are well described in this approximation: in the metallic case, the Fermi level lies in a region of high density of states, the effective mass and static susceptibility scale as $1/T_K$, the Wilson ratio in the mean field approximation is equal to 1 [6], as in non-interacting systems.

If the Fermi level is exactly in the gap this model describes an insulating Kondo lattice: this occurs if the total number of electrons is exactly 2. Such an insulating behaviour has been observed in CeNiSn [41]. However if the degeneracy of f levels and conductions electrons is taken into account, some of the bands do not hybridize and the gap is replaced by a pseudo-gap.

As temperature increases this mean field description is no longer appropriate: it can be extended at non-zero temperature [42,43] if λ_0 and ϕ_0 are renormalized, but a non-physical second order phase transition arises at $T \geq T_K$ and the cross-over between high and low temperature regimes is not described correctly.

In this mean field solution the energy scale, T_K, is similar to the single impurity one (except for a renormalization factor n which is the number of conduction electrons per localized spin [10]). However the physics of the Kondo lattice is certainly different from the single impurity, because the screening cloud of a single impurity extends on several interatomic distances [74] and interference effects should be very large. Variational methods show that in the lattice the spins are compensated collectively [44].

At non-zero temperature, fluctuations become important. These fluctuations around the mean field solution (λ_0, ϕ_0) lead to effective interactions. Two types of fluctuations can be considered:

– Kondo type fluctuations, i.e. fluctuations of the parameters $\lambda_i(\tau)$ and $\phi_i(\tau)$,
– magnetic fluctuations, i.e. fluctuations of $\mu_i(\tau)$ and $\mu'_i(\tau)$.

To the second order these two types of fluctuations are decoupled and they will be considered separately.

4.2. Kondo fluctuations

The Gaussian fluctuations of the fields λ and ϕ have been considered by a large number of authors; they correspond to the lower order correction in a $1/N$ expansion [13,45–49]. Including these fluctuations, the non-physical transition at $T \geq T_K$ disappears and interactions between quasiparticles are created. These fluctuations are site and time dependent. Local fluctuations are dominant for the specific heat and for the resistivity: the correct value of the Wilson ratio is obtained [13]: $R = N/(N-1)$ which gives $R = 2$ for the spin $1/2$ case; the resistivity due to Kondo fluctuations is quadratic in T, as expected for a Fermi liquid [48,50]: $\rho \sim A(T/T_c)^2$, where T_c is the coherence temperature, below which the resistivity of the lattice differs from that of a single impurity.

Spatial fluctuations also destroy the coherent state: the hybridization gap disappears progressively when temperature increases [51], and at high temperature the single impurity regime takes place. The same behaviour with temperature has been obtained by other methods [52–54]. The coherence temperature T_c can also be defined as the temperature at which the pseudogap has disappeared [54].

For a quantitative calculation, the starting point is the eqs. (3), (4) and (5). Considering only the parameters λ and ϕ the partition function can be written as:

$$Z = \int D\phi D\lambda \, \exp\{-\beta[F(\lambda_0, \phi_0) + F_k]\}, \qquad (7)$$

where $F(\lambda_0, \phi_0)$ is the mean field free energy and F_k is the contribution of the 2nd order fluctuations [13,48,49]:

$$F_k = \sum_{q,\nu} |\phi(q,\nu)|^2 A(q,\nu) + \sum_{q,\nu} |\lambda(q,\nu)|^2 B(q,\nu)$$
$$+ 2\sum_{q,\nu} \phi(q,\nu)\lambda(-q,-\nu)C(q,\nu), \qquad (8)$$

where A, B and C depend on the mean field solution (λ_0, ϕ_0). At low temperature F_k renormalizes the mean field solution. At high temperature it becomes the dominant contribution and gives rise to the logarithmic behaviour, $\ln T/T_K$, observed for example in the transport properties [51,55]. The coefficients A, B and C do not depend on q at temperature $T \gg T_K$, indicating that only local Kondo fluctuations are important at high temperature.

Three different regimes can be defined when temperature increases:

(1) $T < T_c$: Fermi liquid, or coherent regime, T_c has been estimated to $\approx 0.1 T_K$ [52];
(2) $T_c < T < T_K$: incoherent Kondo regime;
(3) $T > T_K$: logarithmic regime.

Regimes 2 and 3 are also observed for a single Kondo impurity. In the lattice case, the coherent ground state appears only at very low temperature.

4.3. Magnetic fluctuations

Magnetic fluctuations can be observed directly by neutron experiments, which show that antiferromagnetic correlations are present at very low temperature in several non-magnetic Kondo lattice compounds [56,57]: calculations of the static and dynamical susceptibility ($\chi(q)$ and $\chi(q,\nu)$) have been done by several methods to interpret these experiments [26,58–65]. They can also be obtained within the same formalism as above: the contribution to the free energy due to magnetic fluctuations can be written as:

$$F_M = \frac{J^2}{4} \sum_{q,\nu} |\mu(q,\nu)|^2 X^0_{cc}(q,\nu)$$
$$+ \frac{J^2}{4} \sum_{q,\nu} |\mu'(q,\nu)|^2 X^0_{ff}(q,\nu)$$
$$+ 2\sum_{q,\nu} \left(\frac{J}{4} - \frac{J^2}{4} X^0_{cf}(q,\nu) \right)$$
$$\times \mu(q,\nu)\mu'(q,\nu), \qquad (9)$$

where X^0_{cc}, X^0_{ff} and X^0_{cf} are "non-interacting" susceptibilities calculated in the coherent Kondo phase. From expression (9) it is then possible to obtain the renormalized susceptibilities. Neglecting X^0_{cf}, one obtains for the f electrons susceptibility:

$$X_{ff}(q,\nu) = \frac{X^0_{ff}(q,\nu)}{1 - J^2 X^0_{ff}(q,\nu) X^0_{cc}(q,\nu)}. \qquad (10)$$

Large dynamic fluctuations are expected if the denominator of expression (10) is small. In X^0_{ff} and X^0_{cc} intra- and inter-band transitions contribute [58]. Generally, the q and ν dependence of X^0_{cc} can be neglected: $X^0_{cc} \sim \rho(E_F)$. For small q and ν one gets: $X^0_{ff} \sim (1/T_K)[1 - aq^2 + ic\nu]$ with $a \sim m/m^*$ and $c \sim 1/q$. Thus if ferromagnetic fluctuations are large, a paramagnon behaviour is obtained at low temperature with a $T^3 \ln T$ contribution to the specific heat [66,48].

However neutron experiments show that the magnetic fluctuations are large for a finite value of q, Q, corresponding to a maximum of X^0_{ff}. Close to this wave-vector Q, we can write: $X^0_{ff} \sim X^0_{ff}(Q)[1 - a(Q-q)^2 + ic\nu]$ as above, but c is now finite, giving a linear contribution to the specific heat, which can be quite large close to the magnetic instability [67]. The effective mass is then enhanced by the magnetic fluctuations.

At finite frequency $X_{ff}(q, \nu)$ has a maximum for $\nu \sim T_K$, due to interband transitions across the hybridization gap. This maximum has a weak q dependence: it corresponds to localized fluctuations and disappears at high temperature [64,65].

Antiferromagnetic fluctuations play an important role also in heavy fermion superconductors: UPt_3, $(U,Th)Be_{13}$, URu_2Si_2, $CeCu_2Si_2$,... [68]. All these superconductors are either close to an antiferromagnetic instability, or ordered with small moments indicating that the mechanism for superconductivity in these materials is probably connected with magnetic fluctuations.

4.4. The metamagnetic transition

Metamagnetic transitions have been observed in several non-magnetic Kondo lattices: $CeRu_2Si_2$ [69], UPt_3 [70], $CeCu_6$ [71], indicating that these compounds are very close to a magnetic instability. In $CeRu_2Si_2$ a strong volume anomaly is also observed at the critical field: this transition has been interpreted as a consequence of the Kondo-volume collapse mechanism [72] with ferromagnetic exchange interactions [73]: in the presence of a molecular field H, the Kondo temperature is renormalized as $T_K(H) = (T_K^2 + H^2)^{1/2}$; at $T = 0$ K the ground state energy is renormalized in the same way. On the other hand the Kondo temperature T_K has a strong volume dependence: the Grüneisen parameter $\Gamma = -\partial \ln T_K / \partial \ln V$ if of the order of 200 in $CeRu_2Si_2$. In the absence of magnetic field this large Γ value is the origin of the $\gamma-\alpha$ type transitions observed in several Ce systems [72]. With a magnetic field, T_K increases and the transition occurs in compounds in which it cannot occur without magnetic field. However a large ferromagnetic exchange interaction is necessary [73], whereas antiferromagnetic interactions are observed by neutron experiments.

A different model has been proposed in ref. [38] where the transition takes place inside the renormalized bands; in this model the volume anomaly would be only a consequence of the transition (usual magnetovolume effect). In section 4.5 another model will be presented, which also shows a metamagnetic transition, due to frustration effects.

4.5. Magnetic frustration effects in heavy fermions

Most of the non-magnetic Ce compounds show antiferromagnetic correlations. These correlations are induced by the RKKY interactions, which are long range interactions. In a lattice with such long range antiferromagnetic interactions, frustration generally occurs, because it is not possible to satisfy all antiferromagnetic interactions. This frustration will favour the Kondo effect, because, if a site is strongly frustrated, it can be more favourable to stabilize a Kondo singlet state on this site. This mechanism has been suggested recently for Ce compounds [75,76]. We have developed similar considerations for the RMn_2 compounds [77] which can be considered as "3d heavy fermions". The model developed in ref. [77] can be transposed for the Kondo lattice: using the development of the free energy to second order in μ, the magnetic contribution can be rewritten, in the static approximation, as:

$$F_M = D \sum_i \mu_i^2 - \tfrac{1}{2} \sum_{i \neq j} J_{ij} \mu_i \mu_j,$$

where

$$D = \sum_q \left[\frac{1}{X_{ff}^0(q)} - J^2 X_{cc}^0(q) \right],$$

$$J_{ij} = -\sum_q e^{iq(R_i - R_j)} \left[\frac{1}{X_{ff}^0(q)} - J^2 X_{cc}^0(q) \right]. \quad (11)$$

D can be interpreted as the energy necessary to induce a moment on site i ($D > 0$ for a non-magnetic ground state); D and J_{ij} are related to the susceptibilities X_{ff}^0 and X_{cc}^0 and they depend on temperature through the temperature dependence of λ_0 and ϕ_0 (section 3). Close to the magnetic non-magnetic instability, D is small ($D \sim T_K - I$ at $T = 0$ K); J_{ij} is the RKKY interaction, renormalized by the Kondo effect. The study of Hamiltonian (11) gives the following results:

- there exists a critical value D_c below which ordering occurs. This D_c is much larger for ferromagnetic than for antiferromagnetic interactions: thus, antiferromagnetic interactions strongly decrease the ordering temperature, or even suppress the ordered phase [75].
- application of a magnetic field in the non-magnetic phase induces metamagnetic transitions [76,77] which can be accompanied by magnetovolume effects, as in the RMn_2 compounds.
- for some values of the parameters, new ordered phases can be stabilized [76,77]: in these phases magnetic and non-magnetic sites coexist. An example of such a new phase could be CeSb [78] where magnetic planes alternate with non-magnetic planes.
- taking into account the volume and temperature dependence of the parameters D and J_{ij}, it is possible to have a coherent description of the static magnetic properties of heavy fermions.

5. Conclusion

In this review we have discussed the magnetic properties of the Kondo lattice within the functional integral treatment. The following points have been emphasized:

- there is a critical value of ρJ at which a magnetic non-magnetic transition occurs;
- in the magnetic phase, magnetic moments can be reduced by the Kondo effect;
- in the non-magnetic phase fluctuations must be taken into account. This can be done easily in the functional integral formalism and renormalization of the ground state parameters (e.g. the effective mass) can then be calculated, as well as interactions between quasiparticles;
- several energy scales have been defined: Kondo temperature, coherence temperature, magnetic transition temperature.

However I would stress again that several important points still need to be clarified:

- the most important one concerns the antiferromagnetic interactions: in the presence of antiferromagnetic RKKY interactions it is expected that the f spins partially compensate each other to form a nearly singlet state, even in the absence of Kondo effect; thus if it is the case, complete screening by the Kondo effect is obtained more easily, and a smaller number of conduction electrons are involved in this screening process. This can be an answer to Nozières argument [74] who pointed out that, in the lattice at low temperature, the number of conduction electrons available for the Kondo screening is too small. A theoretical treatment of this effect cannot start from the mean field description of section 3;
- close to the magnetic–non-magnetic instability, magnetic fluctuations are coupled with Kondo type fluctuations and it is thus necessary to make the development to the 4th order to take into account such coupling;
- a different picture of the magnetic instability has been developed [79,80], which is closer to an itinerant magnetism description: in this model the heavy quasiparticles form a narrow band of width $\approx T_K$. The residual interactions between quasiparticles are repulsive and also of the order of T_K. Thus heavy fermions would always be close to a magnetic instability because the interaction energy is of the same order as the kinetic energy. The connection between the two descriptions must be clarified;
- finally the study of the Kondo lattice is important not only for the comprehension of heavy fermion materials, but also for other types of systems: we have already pointed out that the RMn_2 compounds have some similarities with heavy fermions [77]. The magnetic properties of high T_c materials, look also like the heavy fermion ones, but with a different energy scale [81]. For example the Zhang–Rice singlet is analogous to the Kondo singlet [82]. In both RMn_2 and high T_c materials, the same problem of the competition between magnetic intersite interactions and local moment compensation is emphasized.

References

[1] Proc. Intern. Conf. on Anomalous Rare Earths and Actinides (Grenoble, 1986) J. Magn. Magn. Mater. 63&64 (1987).
[2] Proc. 6th Intern. Conf. on Crystal Field Effects and Heavy Fermion Physics (Frankfurt, 1988) J. Magn. Magn. Mater. 76&77 (1989).
[3] Proc. Intern. Conf. on the Physics of Highly Correlated Electron Systems (Santa Fe, 1989) Physica B 163 (1990).
[4] Proc. 6th Intern. Conf. on Valence Fluctuations (Rio de Janeiro, 1990) Physica B 171 (1991).
[5] S. Doniach, Physica B 91 (1977) 231.
[6] N. Read, D.M. Newns and S. Doniach, Phys. Rev. B 30 (1984) 3841.
[7] B.H. Brandow, Phys. Rev. B 33 (1986) 215.
[8] T.M. Rice and K. Ueda, Phys. Rev. Lett. 55 (1985) 995.
[9] P. Fazekas and H. Shiba, Intern. J. Mod. Phys. B 5 (1991) 289.
[10] C. Lacroix and M. Cyrot, Phys. Rev. B 20 (1979) 1969.
[11] N. Read, D.M. Newns and A.C. Hewson, J. Phys. C 16 (1983) L 1079.
[12] P. Coleman, Phys. Rev. B 28 (1983) 5255, B 29 (1984).
[13] N. Read and D.M. Newns, J. Phys. C 16 (1983) 3273.
[14] N. Grewe, Z. Phys. B 67 (1987) 323.
[15] C.I. Kim, Y. Kuramoto and T. Kasuya, Solid State Commun. 62 (1987) 627.
[16] K. Yamada, K. Okada, K. Yosida and H. Hanzawa, Progr. Theor. Phys. 77 (1987) 1097.
[17] H. Keiter and G. Morandi, Phys. Rep. 109 (1984) 227.
[18] R.M. Fye and D.J. Scalapino, Phys. Rev. Lett. 65 (1990) 25.
[19] R.M. Fye and J.E. Hirsch, Phys. Rev. B 40 (1989) 4780.
[20] K. Ueda, J. Phys. Soc. Japan 38 (1989) 3465.
[21] R. Jullien, J.N. Fields and S. Doniach, Phys. Rev. Lett. 38 (1977) 1500.
[22] N. Read and D.M. Newns, J. Phys. C 16 (1983) 3273.
[23] C. Lacroix and M. Cyrot, J. Magn. Magn. Mater. 15–18 (1980) 65.
[24] V.Yu. Irkin and M.I. Katsnelson, J. Phys.: Condens. Matter 2 (1990) 8715; Z. Phys. B 82 (1991) 77.
[25] P. Coleman and N. Andrei, J. Phys.: Condens. Matter 1 (1989) 4057.
[26] S. Doniach, Phys. Rev. B 35 (1987) 1814.
[27] T. Yamamoto and F. Ohkawa, J. Phys. Soc. Japan 57 (1988) 3562.
[28] P. Fulde, J. Phys. F 18 (1988) 601.
[29] M. Lavagna, C. Lacroix and M. Cyrot, J. Phys. F 13 (1983) 1007.
[30] F. Steglich, C. Geibel, S. Horn, U. Ahlheim, M. Lang, G. Sparn, A. Loidl, A. Krimmel and W. Assmus, J. Magn. Magn. Mater. 90&91 (1990) 383.
[31] P. Schlottman, Z. Phys. B 55 (1984) 293.
[32] S.M.M. Evans, J. Phys.: Condens. Matter 2 (1990) 9097.
[33] B. Jones and C.M. Varma, Phys. Rev. B 40 (1989) 324.
[34] P. Schlottmann and J.W. Rasul, Physica B 163 (1990) 544.
[35] G. Aeppli, E. Bucher, C. Broholm, J.K. Kjems, J. Baumann and J. Hufnagl, Phys. Rev. Lett. 60 (1988) 615.
[36] H. Nakamura, Y. Kitaoka, K. Asayama and J. Flouquet, J. Phys. Soc. Japan 57 (1988) 2644.
[37] T.M.M. Palstra, A.A. Menovsky, J. van den Berg, A.J. Dirkmaat, P.H. Kes, G.J. Nieuwenhuys and J.A. Mydosh, Phys. Rev. Lett. 58 (1987) 1467.
[38] K. Miyake and Y. Kuramoto, J. Magn. Magn. Mater. 90&91 (1990) 438.
[39] P. Coleman and J. Gan, in ref. [4], p. 3.
[40] T. Saso, Physica B 165&166 (1990) 405.
[41] T. Takabatake, Y. Nakazawa, M. Ishikawa, T. Sakakibara, K. Koga and I. Ogura, J. Magn. Magn. Mater. 76&77 (1988) 87.
[42] S.M. Evans, T. Chung and G.A. Gehring, J. Phys.: Condens. Matter 1 (1989) 10473.
[43] V.I. Belitskii and A.V. Goltsev, Sov. Phys. JETP 69 (1990) 1026.
[44] H. Shiba and P. Fazekas, Progr. Theor. Phys. 101 (1990) 403.
[45] N. Read, J. Phys. C 18 (1981) 2651.
[46] K. Harigaya, J. Phys.: Condens. Matter 2 (1990) 3259.
[47] Z. Tesanovic and T. Valls, Phys. Rev. B 34 (1986) 1918.
[48] A. Auerbach and K. Levin, Phys. Rev. Lett. 57 (1986) 877.
[49] P. Coleman, Theory of Heavy Fermions and Valence Fluctuations (Springer, Berlin, 1985) p. 163.
[50] P. Coleman, J. Magn. Magn. Mater. 63&64 (1987) 245.
[51] C. Lacroix, J. Magn. Magn. Mater. 63&64 (1987) 239.
[52] H. Kaga and H. Kubo, Solid State Commun. 65 (1988) 257.
[53] N. Grewe, Solid State Commun. 50 (1984) 19.
[54] H. Kaga, H. Kubo and T. Fujiwara, Phys. Rev. B 37 (1988) 341.
[55] M. Lavagna, C. Lacroix and M. Cyrot, J. Phys. F 12 (1982) 745.
[56] J. Rossat-Mignod, L.P. Regnault, J.L. Jacoud, C. Vettier, P. Lejay, J. Flouquet, E. Walker, D. Jaccard and A. Amato, J. Magn. Magn. Mater. 76&77 (1988) 376.
[57] G. Aeppli, A. Goldman, G. Shirane, E. Bucher and M.C. Lux-Steiner, Phys. Rev. Lett. 58 (1987) 808.
[58] H. Kaga, Phys. Rev. B 39 (1989) 9296.
[59] Y. Kuramoto, Physica B 156&157 (1989) 789.
[60] Y. Kuramoto and K. Miyake, J. Phys. Soc. Japan 59 (1990) 2831.
[61] B. Welslau and N. Grewe, Physica B 165&166 (1990) 387.
[62] N. Grewe, Solid State Commun. 66 (1988) 1053.
[63] A. Auerbach, Ju.H. Kim, K. Levin and M.R. Norman, Phys. Rev. Lett. 60 (1988) 623.
[64] D.L. Cox, N.E. Bickers and J.W. Wilkins, J. Appl. Phys. 57 (1985) 3166.
[65] Y. Kuramoto, Z. Phys. B 53 (1983) 37, B 54 (1984) 293, B 57 (1984) 95.
[66] M.T. Beal-Monod, J. de Phys. 41 (1980) 1109.
[67] T. Moriya, Phys. Rev. Lett. 24 (1970) 1433.
[68] M.R. Norman, J. Magn. Magn. Mater. 76&77 (1988) 513.
[69] J.M. Mignot, J. Flouquet, P. Haen, F. Lapierre, L. Puech and J. Voiron, J. Magn. Magn. Mater. 76&77 (1988) 97.

[70] J.J.M. Franse, A. de Visser, A. Menovsky and P.H. Frings, J. Low Temp. Phys. 52 (1985) 61.
[71] A. Sumiyama, Y. Oda, H. Nagano, Y. Onuki, K. Shibutani and T. Komatsubara, J. Phys. Soc. Japan 55 (1986) 1294.
[72] M. Lavagna, C. Lacroix and M. Cyrot, Phys. Lett. A 90 (1982) 210.
J.W. Allen and R.M. Martin, Phys. Rev. Lett. 49 (1982) 1106.
[73] F.J. Ohkawa, Solid State Commun. 71 (1989) 907.
[74] P. Nozières, Ann. Phys. 10 (1985) 19.
[75] B.R. Coles, S. Oseroff and Z. Fisk, J. Phys. F 17 (1987) L 169.
[76] K.I. Kugel and D.I. Khomskii, Proc. 17th Soviet Conf. on the Physics of Magnetic Phenomena (Donietsk, 1985) p. 107.
[77] R. Ballou, C. Lacroix and M.D. Nunez-Regueiro, Phys. Rev. Lett. 66 (1991) 1910; in RMn_2 systems frustration is due to the crystallographic structure. In heavy fermions it is due to the long range RKKY interactions.
[78] J. Rossat-Mignod, P. Burlet, J. Villain, H. Bartholin, W. Tscheng-si and D. Florence, Phys. Rev. B 16 (1977) 440.
[79] K. Miyake, Theory of Heavy Fermions and Valence Fluctuations (Springer, Berlin, 1985) p. 256.
[80] H. Jichu, T. Matsuura and Y. Kuroda, Progr. Theor. Phys. 72 (1984) 366.
[81] D. Pines, Physica. B 163 (1990) 78.
[82] F.C. Zhang and T.M. Rice, Phys. Rev. B 37 (1988) 3759.

Rare earth intermetallics

D. Gignoux and D. Schmitt

Laboratoire Louis Néel, C.N.R.S., 166 X, 38042 Grenoble-Cédex, France

A survey of the magnetic properties of rare earth intermetallic compounds investigated during the last fifteen years is presented. Four main topics concerning normal rare earths are emphasized. The first part deals with 3d magnetism, in particular its instabilities and magnetocrystalline anisotropy. Then the large diversity of the magnetic structures and their evolution with temperature are stressed. The third part proposes a catalogue of the quite different metamagnetic processes observed. The associated phase transitions are also discussed. Finally the last part is devoted to the evolution of the models used to interprete quantitatively the different experimental results.

1. Introduction

During the last 15 years, research in magnetism can be characterized by a boom in the field of rare earth (R) based materials, in particular the metallic ones. Before this period, magnetism was mainly studied on 3d transition elements in particular Fe, Co and Ni. Currently, rare earth intermetallics are in a prominent situation not only from a fundamental point of view but also for their large number of applications. Rare earth intermetallics play an important role in a large number of topics developed in this volume, in particular those devoted to permanent magnets, heavy fermions, valence fluctuations, Kondo lattices, magnetostrictive materials, spin glasses and random anisotropy systems. Taking into account that these aspects of magnetism in rare earth intermetallics are treated elsewhere, we do not develop them hereunder. We are mainly concerned by basic properties of intermetallic compounds with normal rare earths. The four main sections of the paper are devoted to the following topics: i) 3d magnetism, ii) magnetic structures, iii) metamagnetism and phase transitions and iv) quantitative analysis. Our purpose is to present the main results obtained in these fields over the last fifteen years. We also discuss the evolution of research in these areas.

2. 3d magnetism

The R–M systems, where M is a 3d transition metal, form an outstanding tool for the study of 3d band magnetism and in particular the interactions, instabilities and anisotropies of such magnetism. For a given M element, a series of compounds with different rare earths crystallize in the same crystallographic structure and thus have practically the same band structure. It is then possible to study the 3d magnetism under several conditions depending on the rare earth (non magnetic or magnetic, isotropic or not, sign of the anisotropy, ...). Many studies are devoted to these systems and it is not possible here to describe all the results obtained. Hereunder we focus on three original aspects of 3d magnetism discovered in R–M intermetallics.

2.1. Onset of magnetism in Co and Ni based alloys

These compounds are formed by the association of the 3d band with the 5d band (4d for Y) with higher energy. The electronegativity difference between the constituents gives rise to a transfer of 5d (4d) electrons towards the unfilled band. Since the screening of the nuclear potentials by the electrons is modified, the two bands close in on one another leading to 3d–5d (or

0304-8853/91/$03.50 © 1991 – Elsevier Science Publishers B.V. All rights reserved

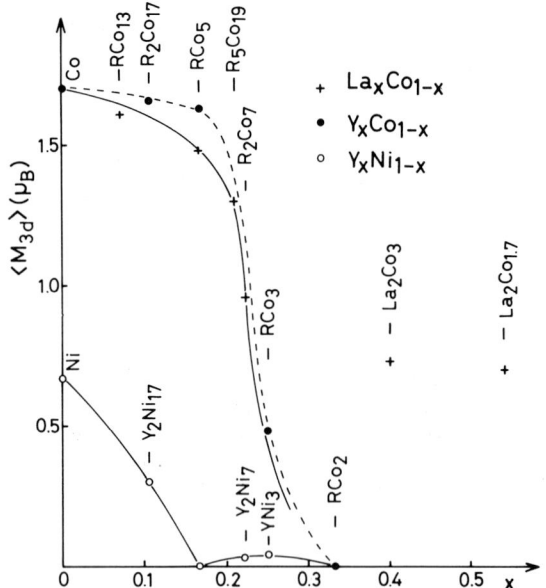

Fig. 1. Mean value of the 3d moment as a function of the rare earth amount in the compounds of the La–Co, Y–Co and Y–Ni systems.

3d–4d) hybridized states [1,2]. The Fermi level of the compounds often lies in this region. This itinerant description of 3d magnetism is the most appropriate for Co and Ni in which, due to the width of the 3d band, the U/W ratio is smaller than 1 (with Mn and Fe this ratio is closer to one and accordingly magnetism is more localized). Starting from pure Ni or Co, the progressive increase of the R percentage leads first to a decrease of the density of state at the Fermi level $n(\epsilon_F)$. For a critical concentration range (around RCo_2 for cobalt and RNi_5 for nickel) alloys are close to the conditions required for the onset of magnetism (Stoner criterion) and magnetic instabilities can be observed, each behaviour strongly depending on the fine structure of $n(\epsilon)$ near ϵ_F. However resurgence of 3d magnetism appears for a slightly larger R amount and then disappears definitively as shown in fig. 1. Three types of characteristic behaviours are mainly observed for R concentrations near or larger than the critical one, namely Collective Electron Metamagnetism (CEM), Very Weak Itinerant Ferromagnetism (VWIF) and Co antiferromagnetism.

CEM, predicted in 1962 by Wolhfarth and Rhodes [3], refers to the transition from a nonmagnetic to a magnetic state when the field acting on the band is larger than a critical value H_M. A maximum in the thermal variation of the susceptibility, having the same origin, is also predicted. Such behaviour is expected to occur in paramagnetic compounds close to the Stoner criterion when $n(\epsilon_F)$ has a strong positive curvature. During the period covered by this review paper such a behaviour has been observed and deeply studied in a large number of R-3d compounds, namely $ThCo_5$ [4], $Ce(CoCu)_5$ [5] and Y_2Ni_{17} [6]. But the best example of such behaviour concerns the RCo_2 series on which many experimental and theoretical studies have been devoted. As shown in fig. 1, these compounds are at the limit of the onset of Co magnetism. Whereas with magnetic rare earths Co is magnetic with a moment close to $1\mu_B$, in YCo_2 and $LuCo_2$ it is nonmagnetic [7]. These latter compounds are enhanced Pauli paramagnets but the field and thermal effects indicate the possibility of CEM. Indeed, in YCo_2 the susceptibility exhibits a broad maximum around 230 K and, at 4.2 K, the susceptibility increases by about 20% between 0 and 35 T [8]. In fact CEM was not observed because this maximum magnetic field was smaller than H_M. With magnetic rare earths the high magnetization state was reached thanks to the molecular field due to the rare earth. Moreover the first order transition observed at the Curie temperature in some of these compounds has been ascribed to the collapse of the Co magnetic moment at this temperature [7,9,10]. In 1977 [11] polarized neutron diffraction studies on $TmCo_2$ and $HoCo_2$ showed that H_M is smaller than 100 T (around 70 T with Tm). From a theoretical point of view, the first calculations [8] led to a value of H_M much larger (142 T). Later, more realistic band structure calculations led to H_M values around 80 T [12–14]. In order to directly observe the metamagnetic transition in this system, a large effort was undertaken to depress the critical field by substitution effects and in 1987 this was obtained by substituting Co by a small amount of Al in YCo_2 and $LuCo_2$ [15] where the transition occurs below 40 T. Finally, the most dramatic event in this fascinating story occurred quite recently thanks to the availability of magnetic measurement up to 94 T: as shown in fig. 2 the metamagnetic transition was directly observed in YCo_2 and $LuCo_2$ around 70 T [16] in agreement with the values calculated and those experimentally determined with magnetic rare earth. Finally it is worth noting that for the model to give account simultaneously of field and thermal effects in these compounds spin fluctuations have to be taken into account.

The resurgence observed in YNi_3 [17] and Y_2Ni_7 [18] must be associated with the fine structure of $n(\epsilon)$ in the region of 3d–4d hybridized states, such that $n(\epsilon_F)$ is large enough for the Stoner criterion to be fulfilled. The Ni moment is very small and VWIF behaviour similar to that of $ZrZn_2$ are observed on the bulk magnetic properties [17]. However, contrary to $ZrZn_2$, magnetization is not diffuse but localized on the Ni sites. This originates from the difference in the magnetic states in the band. Indeed, in $ZrZn_2$, magnetism is due to 4d electrons with a bonding character at the bottom of the 4d band while in YNi_3 the 3d electrons, which contribute to magnetism, lie at the top of the 3d band and have an antibonding character.

In the $LaCo_{1-\epsilon}$ and La_2Co_3 compounds, although the La amount is larger than the critical

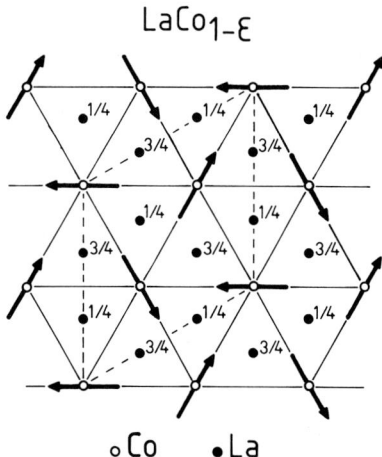

Fig. 3. Magnetic structure of $LaCo_{1-\epsilon}$. Moments of Co atoms of the same chain, parallel to the c axis, are parallel.

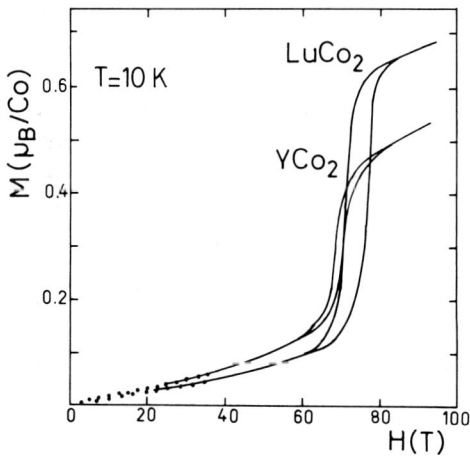

Fig. 2. Magnetization curves of YCo_2 and $LuCo_2$ at 10 K in pulsed ultra-high magnetic field up to 94 T (after ref. [16]). The magnetization data measured in a long pulse field are also plotted as closed circles.

concentration (RCo_2) for the onset of magnetism, one observes a resurgence of magnetism characterized by an antiferromagnetic ordering with a large Néel temperature (146 and 315 K respectively) [19]. This resurgence as well as its antiferromagnetic character (quite unusual in a Co based metallic system) can be understood in the light of the crystallographic structure of these compounds. In $LaCo_{1-\epsilon}$ La atoms form a ABAB-type hexagonal packing while Co atoms lie along chains parallel to the 6-fold axis. The Co–Co distance (2.36 Å) in a chain is much smaller than the distance between chains (4.89 Å). This leads to the exciting phenomenon of quasiunidimensionality of Co lattice. This compound has a triangular structure with Co atoms belonging to the same chain ferromagnetically ($M_{Co} = 0.7\mu_B$) coupled in a direction perpendicular to c. These chains are divided into three sublattices the magnetizations of which make an angle of 120° (fig. 3). Ferromagnetism inside a chain arises from the quasione dimensional character of Co. As shown by Weinert and Freeman [20], the large reduced dimensionality of linear chains gives rise to large moments and exchange values as compared to those of the bulk. The triangular magnetic structure between chains results from the frustration of negative interactions between neighbouring chains in a triangular (hexagonal) lattice. This unusual negative interaction in metallic Co systems has been ascribed to the indirect Co–La–Co interaction through the 3d–5d hybridization bearing strong analogy with superexchange in insulators. When the 3d magnetism is established the 3d–5d hybridization induces a polarization of the 5d electrons. The system takes then advantage of a spin distribution which minimizes the 5d electron kinetic energy and gives rise to an antiferromagnetic coupling between chains. For La_2Co_3 the same interpretation works as the crystallographic structure can be described as a packing of alternating layers of Co and La characterized by a Co–Co distance in the layer much smaller than that from layer to layer.

2.2. 3d magnetocrystalline anisotropy

Although crystalline electric field (CEF) coupling is much larger on 3d ions than on 4f ions, 3d metals and alloys are known to have a small magnetocrystalline anisotropy. This is particularly true in systems with cubic symmetry as can be observed in pure Fe and Ni, whereas anisotropy can be one order of magnitude larger in pure Co which is hexagonal. As the magnetocrystalline anisotropy of itinerant electron systems is difficult to describe with a theoretical model on account of the great complexity of the collective character of the electronic states, the weakness of the 3d anisotropy in metallic systems was commonly admitted and there was no clear-cut knowledge of the mechanisms which indeed rule this anisotropy. Actually it had been evidenced a large 3d contribution to the uniaxial anisotropy in the hexagonal RCo_5 compounds. In particular in YCo_5 the anisotropy field at 300 K reaches 140 kOe [21] whereas it is only 10 kOe in pure Co. In the last few years it has been shown that such 3d anisotropy, even larger, is not only present in the whole RCo_5 series but also in other rare earth–3d intermetallics such as the $RCo_{1-\epsilon}$ (this phase exists only with Y, Pr and Nd) [22] and RMn_2 [23] compounds. The understanding of this anisotropy led to a better knowledge of the 3d magnetism in metallic systems. This anisotropy arises precisely from the itinerant character of magnetic electrons (or holes) and its phenomenological approach [22], recently confirmed by band calculation [24], is the following. The anisotropy depends on the one hand on the orbital character of the electronic states responsible for magnetism and on the other hand on the spin–orbit coupling. The energy and the orbital character of these collective states has two origins. The first one is analogous to that which leads to the anisotropy in insulators or in 4f ions and corresponds to the diagonal elements of the band Hamiltonian in the individual state representation (these elements are one center overlap integrals usually named α integrals [25]). The second one, associated with the itinerant character, is due to the two center overlap integrals. These latter, named the β integrals [25], mix the individual states belonging to neighbouring atoms and lead to the electronic energy dispersion. This latter strongly depends on the atomic packing and will depend on the orbital character of the electronic states under consideration. Let us consider for instance the case of hexagonal YCo_5. The cobalt chemical bonds have a strong planar character (in the layers perpendicular to c), in particular Co–Co and Y–Co bonds for the Co atoms at the 2c site which hold the largest anisotropy and orbital character [26]. This leads to a much larger dispersion of the $|2,\pm 2\rangle$ states (linear combinations of the $x^2 - y^2$ and xy orbitals) than the other states. The Fermi level is expected to lie near the top of the band where the electronic states have a strongly $|2,\pm 2\rangle$ orbital character leading to an orbital moment and hence, through the spin–orbit coupling, to a spin moment along c. The anisotropy then favours the c axis and accounts for the large uniaxial anisotropy of this compound. In hexagonal $LaCo_{1-\epsilon}$ the situation is reversed: the Co–Co are shortest along the c axis and hence this latter is the hard magnetization direction.

2.3. Instability and frustration of Mn magnetism

It is well established that in metallic Mn based systems the Mn moment depends upon the interatomic distance. Moreover, contrary to the case of Co or Ni, in these systems magnetic interactions between nearest neighbours are negative and vary with distance. In this context the RMn_2 compounds, on account of their simple crystallographic structure, are particularly interesting for the study of the instability of the itinerant electron antiferromagnetism in a lattice where interactions are frustrated.

In these compounds, which crystallize either in the C14 hexagonal or the C15 cubic Laves phase structure, the Mn atoms occupy the corners of regular tetrahedra. As in the case of the two dimensional triangular lattice, this topology leads to highly frustrated lattices when magnetic interactions between magnetic atoms are negative. Moreover, Mn moments, in the RMn_2 series, are very close to the magnetic–nonmagnetic instability. This magnetic instability was made evident by the large magnetovolume effects especially near the Néel temperature [27]. Thermal expansion and nuclear magnetic resonance measurements were interpreted in terms of the dependence of the Mn moments as a function of distance. As shown in fig. 4, below a critical Mn–Mn distance Mn remains nonmagnetic. It is the case with heavy rare earths such as Ho, Er, Tm and Lu. The magnetic ordering is only characteristic of the rare earth and no volume anomaly occurs at the ordering temperature. Above the critical distance, i.e. with light rare earths like Pr, Nd, Sm and Gd, large Mn moments with antiferromagnetic interactions are found. Complex magnetic orderings [28] are then observed due to the lattice frustration and a sharp volume increase (about 1%) is observed at the Néel temperature. This first order transition, which arises from the volume dependence of exchange energy is associated with a discontinuity of the Mn moment. Near the critical Mn–Mn spacing are the compounds with Dy, Tb and Y. In these compounds magnetism is extremely sensitive to external parameters such as temperature, pressure, magnetic field and alloying. YMn_2 shows a first order transition at T_N accompanied by a giant volume change of about 5%, which is ascribed to a substantial reduction of the effective Mn moment at T_N, namely $2.6\mu_B$ below T_N and near $1.6\mu_B$ just above T_N. Moreover the large thermal expansion coefficient above T_N can be interpreted by a rapid recovery of the amplitude of the magnetic moment with increasing temperature because the magnetic volume change is proportional to $\langle \mu_{Mn}^2 \rangle$ (fig. 5). This has

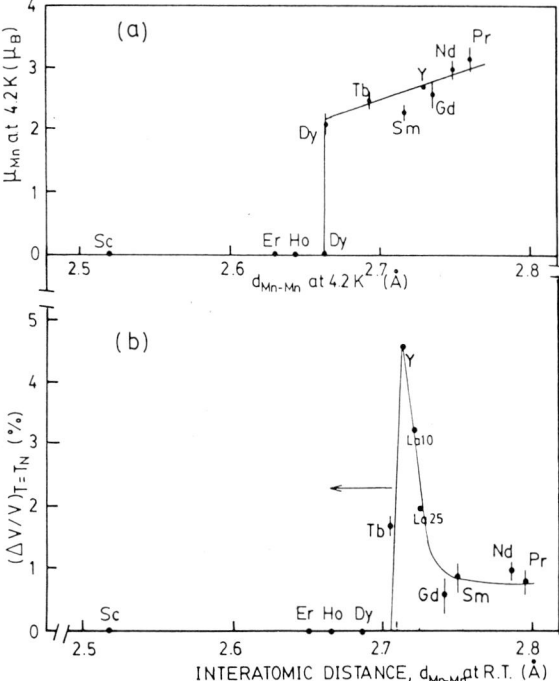

Fig. 4. Variations of Mn moment and volume change at T_N with the interatomic distance, d_{Mn-Mn} (after ref. [27]). (a) Mn moment in μ_B, at 4.2 K in RMn_2 as a function of d_{Mn-Mn} at 4.2 K; (b) volume change $\Delta V/V$ at T_N plotted against d_{Mn-Mn} at room temperature. La10 and La25 represent $Y_{0.9}La_{0.1}Mn_2$ and $Y_{0.75}La_{0.25}Mn_2$ respectively.

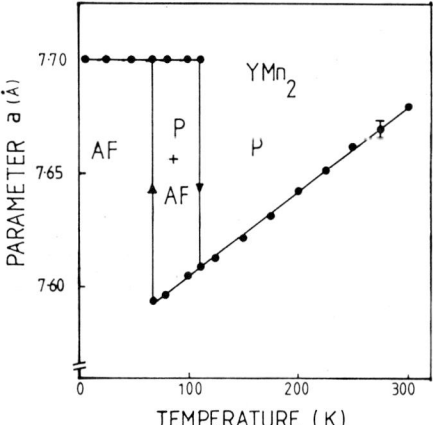

Fig. 5. Variation of the lattice parameter as a function of temperature for YMn_2, indicating a first order transition at approximately 100 K with a 20 K hystereris (after ref. [29]).

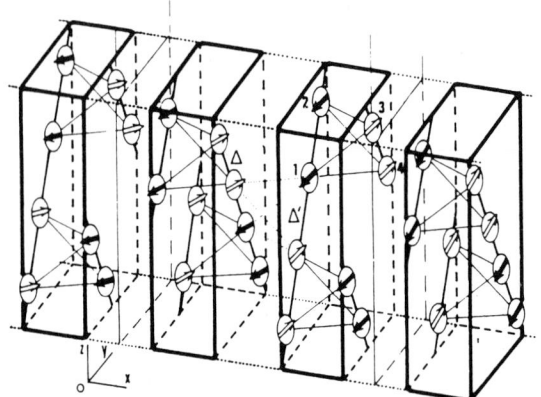

Fig. 6. Magnetic structure of YMn$_2$ (after ref. [23]). Only the Mn atoms are shown.

been directly confirmed by a polarization analysis performed at different temperatures in the paramagnetic phase [30]. As a result of frustration a complex magnetic structure, with a long wavelength distorted helical component, has been determined [23] (fig. 6). In the antiferromagnetic TbMn$_2$ compound, at low temperature, metamagnetic transitions with large field hysteresis are observed [31]. Application of pressure induces a dramatic decrease of the ordering temperature of the Mn ions, at a rate of 36 K/kbar [32]. The substitution of only 3% of Tb by nonmagnetic and smaller Sc atoms is enough to destroy the Mn moment [33]. The case of DyMn$_2$ is particularly interesting, as although all the Mn sites are chemically equivalent, NMR [34] and neutron diffraction experiment [35] have shown that only a fraction of them bears a magnetic moment (fig. 4). The theoretical approach of band magnetism instability in highly frustrated crystallographic structures has been carried out quite recently [36]. A model derived from the Hubbard Hamiltonian accounts for the complex ordered phases where magnetic and nonmagnetic sites coexist and for the unusual dependence on external parameters.

3. Magnetic structures

During the last fifteen years, the evolution of the knowledge of the magnetic structures in rare earth intermetallics can be characterized by the following aspects.

i) A large number of new compounds have been studied. In particular, as the first experiments were performed on pure rare earth and then on binary (or pseudo-binary) compounds the number of studies devoted on ternary compounds is always increasing. Among these latter it is worth notice the RM$_2$X$_2$ compounds which crystallize in the same tetragonal structure (ThCr$_2$Si$_2$-type) for a large number of M and X elements (M = Mn, Fe, Co, Ni, Cu, Al, Ru, Rh, ... and X = Si, Ge, Ga, ...).

ii) For a given compound quite extensive neutron diffraction experiments on powder can be carried out thanks to the improvement of the diffractometers (efficiency, accuracy, temperature control, ...). Whereas fifteen years ago magnetic structures were determined only at one temperature (generally 4.2 K), it now is readily possible to study in a short time the evolution of the magnetic structure in the whole temperature range below the ordering temperature.

iii) An increasing number of single crystals are available on which it is possible to measure the intensities with a better accuracy and to determine unambiguously the propagation vectors.

In rare earth based intermetallic compounds, magnetic structures mainly result from the compromise between different interactions and thermal effects. These interactions are three types. First, the bilinear exchange interaction of RKKY type is long range and oscillatory with distance. This leads to a Fourier transform $J(q)$ which is maximum for a value of q which has no reason to be commensurate with the crystallographic reciprocal lattice. The second interaction is the crystalline electric field (CEF). In uniaxial systems, i.e. mainly in tetragonal and hexagonal compounds, CEF often favours either Ising systems or $X-Y$ systems. At last the quadrupolar interactions generally are one order of magnitude smaller. However, they can play an important role, in particular in cubic systems where they can, for instance, stabilize multiaxial magnetic structures (see below).

It is impossible to systematically present the extremely wide variety of observed magnetic structures. So we present hereunder the results which seem for us particularly characteristic of the considered period. The most characteristic feature is that a majority of compounds exhibits several different magnetic orderings below the ordering temperature associated with complex magnetic phase diagrams.

3.1. Sine wave modulated structures

Due to the RKKY interactions many compounds, where only rare earth is magnetic, are antiferromagnetic and, at least just below T_N, the magnetic cell is incommensurate with the crystallographic one. In this temperature region, magnetic periodicity is mainly driven by exchange interactions and the associated propagation vector is that for which $J(q)$ is maximum. Generally these structure are sine wave modulated. Few helimagnetic structures are observed (the best examples remains those of pure rare earths such as Tb, Dy and Ho metals [37,38]) because they can be only stabilized in case of an easy plane in which the anisotropy is negligible. Gd compounds are good candidates for helimagnetic structures but to our knowledge, due to the small number of magnetic structures determined with Gd (see section 3.7), the only clear example is $GdBe_{13}$ [39]. Coming back to the modulated case, when temperature is lowered, magnetic structures result from the compromise between exchange interactions and magnetocrytalline anisotropy. Various situations are then observed.

3.1.1. Sine wave modulated structures stable down to 0 K

When the rare earth ground state, in the absence of a magnetic field (exchange and/or applied), is a nonmagnetic state (generally a singlet), at 0 K the magnetic moment is an increasing function of the exchange field and sine wave modulation can be stable down to very low temperatures [40,41]. For a normal rare earth this can occur in a limited number of cases. Indeed the R^{3+} ion must be a non-Kramers one and the local symmetry must be low enough so that CEF can lead to a singlet ground state. A good example of such a situation is $PrNi_2Si_2$ [42] where no clue of a squaring up of the modulated structure can be detected at low temperature. The existence of a nonmagnetic singlet ground state is confirmed by other measurements. Indeed a quantitative analysis of neutron spectroscopy, specific heat and single crystal magnetic measurements has shown that the CEF ground state is a singlet around 30 K below the first excited state [43]. It is worth noting the low temperature sine wave modulated of the $CeAl_2$ Kondo lattice where the singlet ground state results from the lifting of the doublet CEF ground state degeneracy by the Kondo interaction [44].

3.1.2. Progressive squaring up

In most cases the ground state is magnetic and at low temperature, all rare earth magnetic moments must be equal with a finite value (antiphase structure). This squaring up can be discontinuous or progressive. In this latter case, the propagation vector is temperature independent or varies weakly and continuously with temperature. The squaring manifests itself by the onset of higher order harmonics ($3Q, 5Q,...$) in the neutron diffraction pattern as observed for instance in $HoRu_2Ge_2$ [45]. As the associated intensities are much weaker than those associated with Q, they are not always possible to evidence on powder patterns but, as in HoAg and TmAg [46], they can be observed by neutron diffraction on single crystal. Although not detected, such squaring occurs in many compounds where the propagation vector remains incommensurate at low temperature. For sure, this squaring does occur in compounds with Kramers ions. Many tetragonal RM_2X_2 (i.e. $DyNi_2Si_2$, $HoNi_2Si_2$, $TbFe_2Si_2$, $HoFe_2Si_2$,...) compounds enter in this category [47].

One of the best observations of a progressive squaring up is the orthorhombic $TbCu_2$ compound [48]. At any temperature below $T_N = 55$ K, the propagation vector is $Q - (1/3, 0, 0)$ and, due to this long period commensurate value, the Fourier component of the Tb moment can be gathered into two terms: one of amplitude β associated with this vector which corresponds to the $Q, 5Q, 9Q,...$ harmonics; the other of amplitude α which can be describe with a $Q' = (1, 0, 0)$

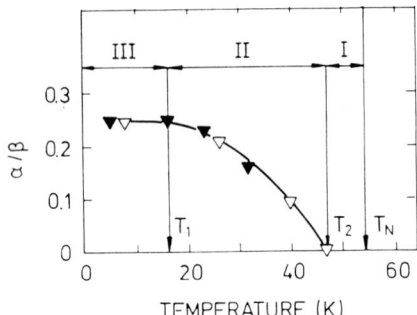

Fig. 7. Temperature dependence of the ratio α/β (after ref. [48]).

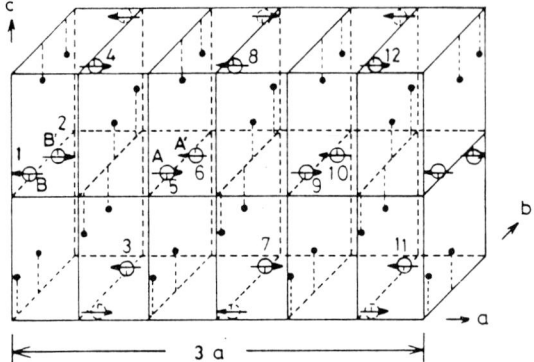

Fig. 8. Low temperature magnetic structure of TbCu$_2$ and DyCu$_2$. The crystallographic unit cell contains 4 atoms.

antiferromagnetic propagation vector and which corresponds to the $3Q$, $7Q$,... harmonics. Between T_N and $T_2 = 47$ K, α is zero and the structure is sine wave modulated. Between T_2 and $T_1 = 16$ K the α/β ratio increases (fig. 7) up to the value which corresponds to a collinear antiphase structure with equal moment M such that $\alpha = M/3$ and $\beta = 4M/3$ (fig. 8). Note that T_1 and T_2 do not correspond at any transition because there is no discontinuity of the Tb magnetic moment neither of the propagation vector. This behaviour probably occurs also in DyCu$_2$ [49]. For long period commensurate structures, the harmonics can be gathered into more than two terms. In some cases, such as Tm metal [50] and CeAl$_2$Ga$_2$ [51], one of these terms falls on nuclear peaks leading to a small ferromagnetic component.

3.1.3. First order transition towards an equal moment state

Quite frequent is the situation where the equal moment structure occurs through a first order transition associated with a discontinuity of the propagation vector which locks on a value corresponding to a higher symmetry point of the reciprocal space, as shown in table 1 where we have reported the characteristics of some compounds which enter in this category. Figs. 9 and 10 illustrate the properties of PrCo$_2$Si$_2$ and DyGa$_2$ respectively.

Some comments can be done at the light of these examples:

– One can observe, below T_N, more than one transition, each one being associated either with

Table 1
Characteristics of some compounds

Compound	Symmetry	T (K)	q	Moment direction	Low T structure	Ref.
PrCo$_2$Si$_2$	tetra.	$17 < T < T_N = 30$	(0, 0, 0.777)	[0 0 1]		
		$9 < T < 17$	(0, 0, 0.926)	[0 0 1]		
		$T < 9$	(0, 0, 1)	[0 0 1]	antiferro	[52]
DyGa$_2$	hex.	$8.7 < T < T_N = 11$	(0.433, 0.433, 0)	[1 1 0]		
		$6.1 < T < 8.7$	(0.433, 0.433, 0)	[1 −1 0]		
		$T < 6.1$	(1/2, 1/2, 0)	[1 −1 0]	antiferro	[53]
NdRu$_2$Si$_2$	tetra.	$10 < T < T_N = 24$	(0.12, 0.12, 0)	[0 0 1]		
		$T < 10$	(0, 0, 0)	[0 0 1]	ferro	[54]
PrGa$_2$	hex.	$3.2 < T < T_N = 7.3$	(0.149, 0.149, 0.022)	[1 1 0]		
		$T < 3.2$	(0.149, 0.149, 0)	[1 1 0]	unbalanced	
			$0.149 \approx 7/47$		antiphase	[55]

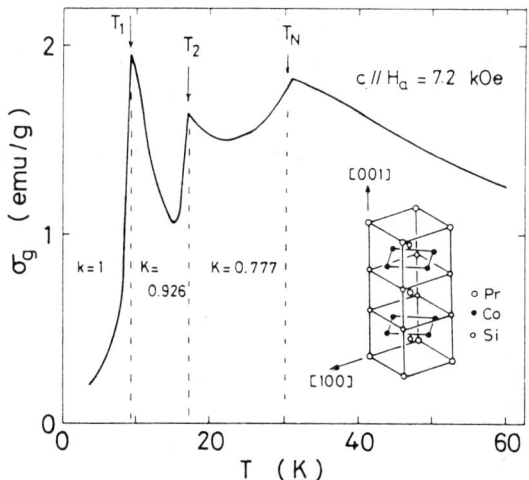

Fig. 9. PrCo$_2$Si$_2$: temperature dependence of magnetization along the c-axis at 7.2 kOe and the crystal structure (after ref. [52]).

a change of the propagation vector (PrCo$_2$Si$_2$) or with a change of the magnetization direction (DyGa$_2$). This latter property, observed in several compounds, originates from CEF effects and is discussed in section 3.6.

– First order transitions are not always visible from one type of measurement alone. For instance in DyGa$_2$ at T_2 a sharp maximum of heat capacity and a clear cut change of slope of resistivity are observed whereas a weak susceptibility anomaly is hardly visible. The reverse situation can also occur: as in TbNi$_2$Si$_2$ for instance [43], a first order transition is clearly visible in susceptibility and resistivity but is not observed in heat capacity, the latent heat being very weak. Different types of measurements are then necessary to evidence all the transitions.

– The low temperature magnetic structures belong to three categories, namely a simple antiferromagnetic structure (i.e. PrCo$_2$Si$_2$ and DyGa$_2$), a ferromagnetic structure (i.e. NdRu$_2$Si$_2$) and an antiphase structure associated with a long period magnetic cell (i.e. PrGa$_2$). In this latter situation, three cases have to be considered: a truly incommensurate value of Q, and a lock in on two types of commensurate cell (in PrGa$_2$ $Q = (\tau, \tau, 0)$ with $\tau = m/n$) with compensated or uncompensated antiphase structures. The only way to detect this latter is the observation of a ferromagnetic component by magnetization measurements. The comparison between this component and the magnetic moment on each atom allows the determination of m and n.

3.2. Complex phase diagram of CeSb

Although CeSb is a semimetal, particular attention has to be paid on this very well known cubic compound which exhibits a quite complex phase diagram (fig. 11) [56,57]. The 15 distinct phases, 7 of them being successively stabilized in zero field when temperature is decreased, correspond to long period commensurate structures. However the most unusual feature of CeSb comes from the coexistence, in the so-called AFP (antiferro–paramagnetic) and FP (ferro–paramagnetic) phases observed several regions, of magnetic and non(or para)magnetic Ce atoms. Recent magnetic excitation spectra, obtained from inelastic scattering with polarized neutron, have shown that the CEF ground state of magnetic Ce atoms is a Γ_8 state with large moment ($2.1\mu_B$) whereas that of paramagnetic Ce atoms is a Γ_7 state with a smaller moment ($0.7\mu_B$) [58]. The understanding of this complex diagram and the nature of nonmagnetic atoms is at the origin of the development of many theoretical studies. Some keywords, such as CEF effects, p–f mixing, devil's staircase and ANNNI (anisotropic next nearest neighbours Ising) model have been considered but no appreciable results have been obtained. More recently the complex phase diagram has been rather well accounted for by introducing an incommensurate mean field model [59]. In this model, paramagnetism on one part of Ce atoms arises because there is no exchange field on their site. Finally dramatic change of this phase diagram has been observed under pressure [60].

3.3. Nd metal: a long story

Although the investigation of the magnetic properties and structures of pure rare earth began more than 25 years ago, the phase diagram of Nd metal has been for long a puzzle [61,62] and it is only recently that the magnetic structures have been elucidated [63,64]. Nd exhibits a remarquable sequence of increasingly complex magnetic structures below $T_N = 19.9$ K. This complexity arises from the frustration of negative interactions in a double hexagonal close packed structure, comprising sites of locally hexagonal and cubic symmetry. Just below T_N, the magnetic moments of the hexagonal site order in a longitudinally polarized, incommensurately modulated structure with $Q_h = (0.147, 0, 0)$. This multi-domain single-Q structure exists only in a small temperature range. Below $T_2 = 19.1$ K, a series of

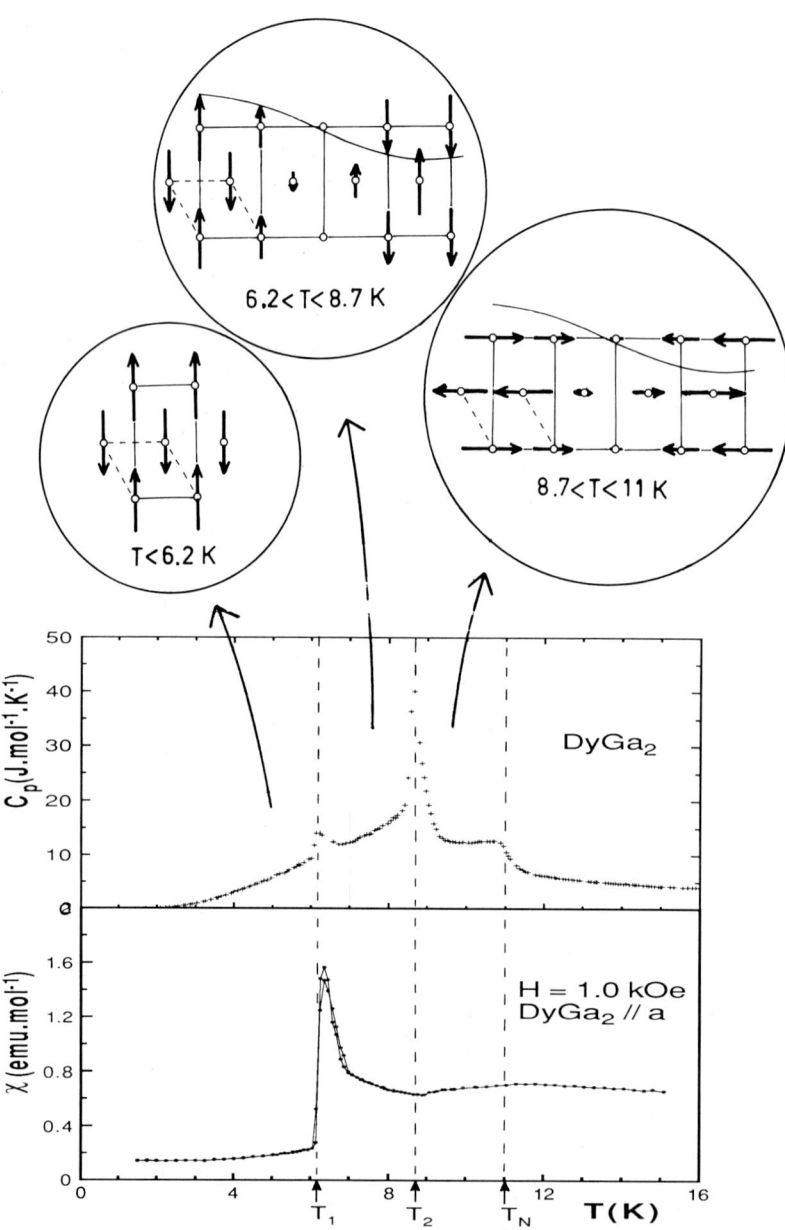

Fig. 10. DyGa$_2$: temperature dependences of the specific heat and of the susceptibility. Magnetic structures of the different phases are also sketched.

multiple-Q structures is observed. Thermal expansion measurements indicates the following transition temperatures $T_3 = 8.2$ K, $T_4 = 7.5$ K, $T_5 = 6.2$ K and $T_6 = 5.7$ K. Between T_2 and T_3, the moments display a multi-domain double-Q structure. Such a double-Q structure was demonstrated by neutron diffraction under applied field, as it is generally used to raise the undetermination between single and multi-Q structures. Below T_3, an additional array of neutron diffraction satellites appears with a modulation vector $\boldsymbol{Q}_c = (0.18, 0, 0)$: this transition corresponds to the ordering of the moments of the cubic site. The transition at T_4 is characterized by a change in the propagation vector. Below T_5 and, in particular at 4.5 K, a larger number of satellites is observed. It has been shown that they arise from a six domains 4-Q structure, the four propagation vectors being along or near $\langle 100 \rangle$ directions. This 4-Q quite complex structure can be visualised in terms of two coupled 2-Q structures formed from $\boldsymbol{Q}_1$ and $\boldsymbol{Q}_2$ on the one hand and from $\boldsymbol{Q}_3$ and $\boldsymbol{Q}_4$ on the other hand as illustrated in fig. 12. The moments associated with the first and second

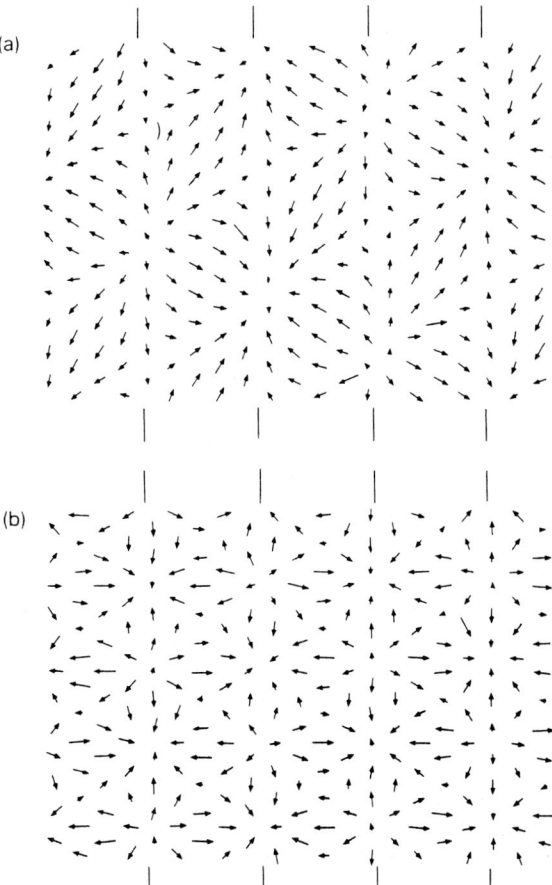

Fig. 12. Structure of Nd metal at 4.5 K calculated by the approximation that moments associated with $\boldsymbol{Q}_1$ and $\boldsymbol{Q}_2$ (a) and those with $\boldsymbol{Q}_3$ and $\boldsymbol{Q}_4$ (b) are on different layers ("hexagonal" and "cubic" respectively) of the lattice. However, the actual structure is more complicated and must have moments associated with all four $\boldsymbol{Q}$ vectors in both layers (after ref. [64]).

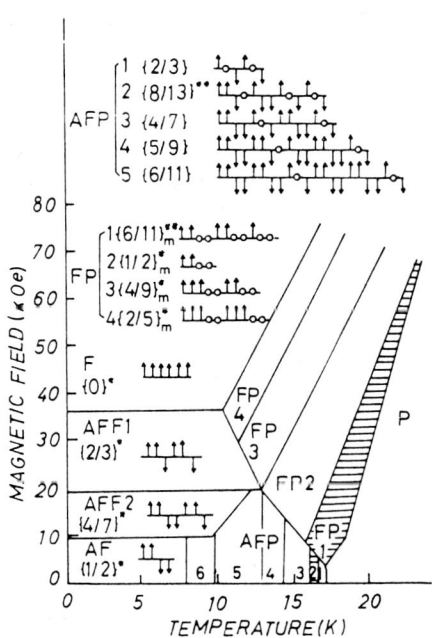

Fig. 11. Phase diagram and spin structures in CeSb determined by Rossat-Mignod et al. [57].

pairs are predominantly on the hexagonal and cubic sites respectively. However, the 4 propagation vectors are locked together by interactions between the moments on the two types of sites. Detailed analysis of the satellite intensities indicates small moments components on cubic sites associated with $\boldsymbol{Q}_1$ and $\boldsymbol{Q}_2$, and modulated components on the hexagonal sites associated with $\boldsymbol{Q}_3$ and $\boldsymbol{Q}_4$. Phases between T_3 and T_5 are still not quite clear. Moreover no change in the neutron diffraction pattern has been observed at T_6.

3.4. Quadrupolar interactions and magnetic ordering

In the recent past, multipolar interactions have been extensively studied and quantitatively analyzed in rare earth intermetallic compounds [65]. Quadrupolar couplings play a minor role in uniaxial compounds where quadrupoles spontaneously exist and are ordered by CEF effects. However these couplings are of particular importance in cubic compounds because quadrupoles appear only through a lowering of the symmetry, which occurs for instance at the magnetic ordering temperature. Quadrupolar interactions are often strong enough to compete with the Heisenberg interactions, as it is evidenced by the occurrence of quadrupolar ordering in the paramagnetic range of several compounds. Even when they are dominated by the bilinear interactions, they play a decisive role in the minimization of the free energy of the 4f ions systems and have important consequences on the magnetic properties [66], in particular in the determination of the magnetic structures. In antiferromagnetic compounds with cubic symmetry, the spin structures may have the same bilinear energy while they are described by one (collinear arrangements) or more (multiaxial arrangements) propagation vectors of the same star. However collinear and multiaxial arrangements are favoured by FQ (ferroquadrupolar) and AFQ (antiferroquadrupolar) interactions respectively. In $AuCu_3$ and CsCl-type cu-

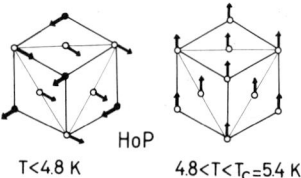

Fig. 14. Magnetic structures of HoP. At low temperature, the structure results from the competition of ferromagnetic and antiferroquadrupolar energies.

bic compounds, antiferromagnetic interactions very often coexist with AFQ ones. As soon as CEF favors $\langle 111 \rangle$ axes, a multiaxial structure is established as shown in fig. 13 for the DyAg compound where it is triple-Q with $Q = (1/2, 1/2, 0)$ [67]. Another interesting situation is that of HoP pnictide where CEF favors $\langle 100 \rangle$ and AFQ interactions compete with ferromagnetic ones. So HoP orders in a collinear ferromagnetic structure at 5.4 K. At 4.8 K, when the AFQ energy is strong enough, a first order transition toward a noncollinear ferromagnetic structure ("flopside") takes place (fig. 14) [68]. As one can expect complex magnetization processes are observed in such compounds (see section 4).

3.5. Spin reorientations due to competing interactions

In uniaxial rare earth–3d transition metal R–M compounds, where transition metal is magnetic, the magnetocrystalline anisotropy can be of opposite sign for rare earth and for 3d atom. These local anisotropies then are in competition with the R–M exchange interaction which favors the collinearity of magnetic moments of both elements. The difference of the thermal variations of both anisotropies can then lead to a spin reorientation. Such reorientations were known to occurs in the hexagonal RCo_5 compounds more than fifteen years ago [7]. From that times they have been observed and deeply studied not only in the RCo_5 compounds but also in a large number of compounds which are well known for their permanent magnet properties, namely some R_2M_{17}, $R_2F_{14}B$ and tetragonal $R(M_{1-x}M'_x)_{12}$ compounds with $ThMn_{12}$ type structure. The situation of $DyCo_5$ is illustrated in fig. 15 [69]. The c

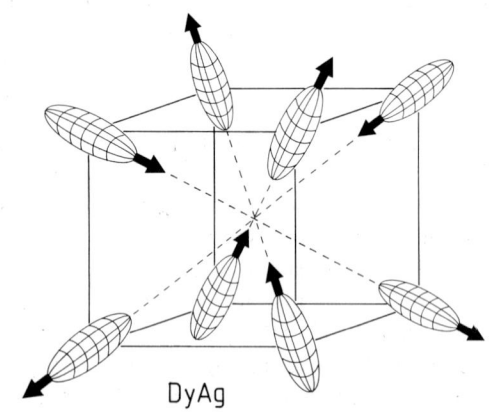

Fig. 13. Magnetic structure of DyAg. Quadrupoles are also sketched.

axis is the easy magnetization direction for Co whereas it is the hard direction for Dy. At low temperature the Dy anisotropy is the largest, so, because of the negative Dy–Co exchange interaction, Dy and Co moments are antiparallel and both perpendicular to c. As magnetic moment and hence anisotropy decreases faster for Dy than for Co, at high temperature the Co anisotropy overcomes the Dy one and the resulting magnetization is parallel to c. In the intermediate range of temperature (325 K < T < 367 K), the total magnetization continuously rotates. Note that, at high temperature, Dy moment is not parallel to c: its intermediate direction is a compromise between its anisotropy and exchange interaction. In $SmMn_2Ge_2$, where a competition between Sm and Mn anisotropies is also present, the more complex structures observed at different temperatures [70] come from the frustration of negative interaction between Mn atoms as it has been mentioned in section 2.3.

3.6. Spin reorientation due to CEF splitting on rare earth

Thermally induced spin reorientations have been commonly observed in cubic rare earth compounds which does not result from competing anisotropy either because only rare earth is magnetic either because only rare earth contributes significantly to the magnetocrystalline anisotropy.

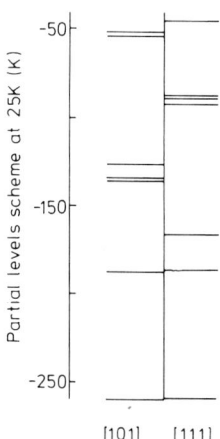

Fig. 16. Low lying CEF state of HoZn at 25 K according to whether the moment lies along the [101] or [111] direction.

Such transitions occurs for instance in NdZn, TbZn and HoZn [71], $NdAl_2$ and $HoAl_2$ [72], $NdCo_2$ and $HoCo_2$ [73], NdCd and TbCd [74]. This property originates from CEF effects on the rare earth ion. To illustrate what occurs let us consider the case of HoZn. Fig. 16 shows the splitting of the Ho^{3+} ground state multiplet by CEF and exchange field, at the rotation temperature ($T_R = 25$ K) according to whether the moment lies along the [101] or [111] direction. The ground state is slightly lower along [101] but the third excited state is much lower along [111]. So, at low temperature, the free energy is minimum when magnetization is parallel to [101]. However when temperature increases, because of the entropy term, the free energy becomes the lowest along the [111] direction leading to the spin reorientation. Such transition is first order as calculated and observed from specific heat which shows a sharp anomaly. More recently a spin reorientation of this kind has been observed in the hexagonal $DyGa_2$ compound (fig. 10) [53].

3.7. Magnetic structures of Gd based compounds

Because the neutron absorption cross section of natural gadolinium is extremely high (σ_a between 15 000 and 55 000 b) at the wavelengths usually available, magnetic structures of most of the Gd based compounds were for long unknown. Neutron diffraction experiments on Gd metal

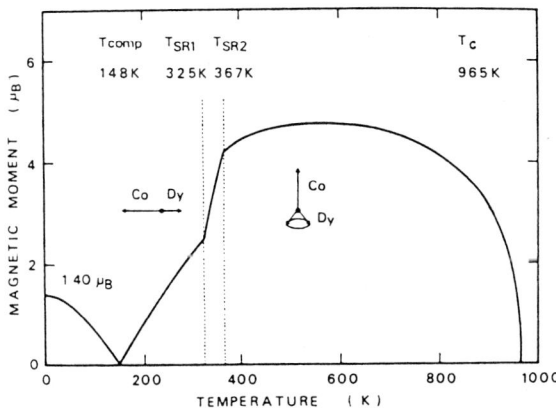

Fig. 15. Temperature dependence of the spontaneous magnetic moment in $DyCo_5$ (after ref. [70]).

were carried out in 1972 [75] on a single crystal specimen of the much less absorbing isotope ^{160}Gd. As a consequence, the study of this type of compounds was not studied as deeply as for the other rare earths. Gd based intermetallics though are of great interest on account of the S character of Gd^{3+} ion. As a result, contrary to the other rare earth which are anisotropic, magnetic structures, magnetization processes and thermodynamical properties are almost exclusively driven by bilinear exchange interactions, i.e. $J(Q)$. Absorption cross section of Gd is two orders of magnitude smaller at $\lambda = 0.5$ Å ($\sigma_a = 250$ b). This short wavelength now is available in some nuclear reactors. So during the last few years neutron diffraction experiments have been carried out on antiferromagnetic Gd compounds. Commensurate and incommensurate magnetic structures have been determined. However, neutron diffraction on polycrystalline samples does not always allow to choose between heli- or modulated structures. Recent experimental and theoretical studies of thermodynamical properties are quite promising in the lifting of this ambiguity. New and original results were obtained concerning the specific heat of amplitude modulated magnetic structures; especially the specific heat at the ordering temperature T_N is reduced by a factor of 2/3 with regard to the classical λ anomaly of equal moment structures of ferro-, antiferro- or helimagnetic type. In addition, under certain conditions, the maximum of specific heat at T_N is shifted toward lower temperatures. This is illustrated in fig. 17 which shows some calculated variations (a) and which compares those of $GdCu_2Si_2$ and $GdNi_2Si_2$ (b) which have magnetic stucture with equal and modulated moments respectively [76].

4. Metamagnetism and phase transitions

The concept of metamagnetism spread considerably over the last fifteen years. First limited to single-step phase transitions between a simple antiferromagnetic structure and a paramagnetic saturated state by the application of an external magnetic field, it has been progressively extended

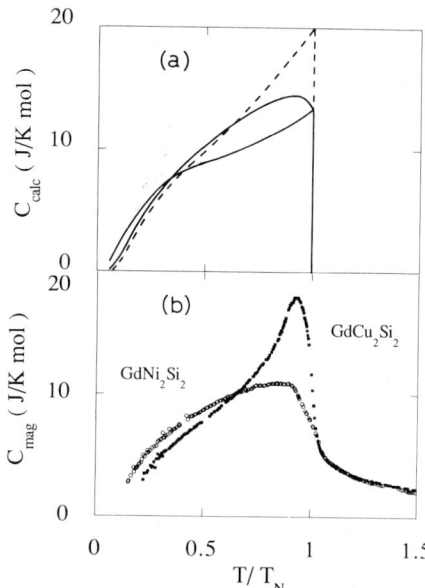

Fig. 17. Magnetic contribution to the specific heat in Gd compounds. (a) Theoretical variations: dashed and full lines correspond to equal moment and amplitude modulated structures respectively. (b) Experimental variations in $GdCu_2Si_2$ (equal moments) and $GdNi_2Si_2$ (sine wave modulated structure).

to all the processes breaking noncollinear magnetic structures in a more or less complex way, and more generally to all the types of field-induced magnetic phase transitions [77–79]. It is the purpose of this section to emphasize the specific role of the rare earth intermetallics in such an evolution by showing, through several examples, how each of the various types of metamagnetic processes is associated with a particular physics depending on the material under consideration, and how it involves the balance between the exchange interactions and various anisotropy terms.

Before 1975, most of the metamagnetic materials were nonmetallic systems, oxides, vanadates, garnets, linear chain systems,... and few rare earth intermetallics were concerned [77]. Only some isolated instances were thoroughly investigated, such as rare earth metals [38,80] and some rare earth pnictides like DySb and CeSb [81], while other systems were just evoked without many details [77]. Starting from 1975, the number of rare earth intermetallics exhibiting a metamagnetic behaviour began to grow up to finally explode after 1985. This evolution is connected with three main reasons: i) The first one is the important progress in the elaboration of large single crystals which are absolutely necessary to evidence and to understand the metamagnetic mechanisms; ii) The second reason is the strong increase of the field of investigation by considering ternary compounds which form a wealthy family; iii) The third reason is the increase of performance of the experimental facilities, in particular the high field measurements and the neutron diffraction experiments.

The early distinction which occurred among the metamagnetic systems was between those exhibiting spin–flip and spin–flop transitions according to the strength of the magnetocrystalline anisotropy [77]. More precisely, spin–flip (single-step) metamagnetic transitions are characterized by a simple *reversal* of all the spins antiparallel to the applied magnetic field, this situation arising in highly anisotropic materials. Oppositely, in the spin–flop systems, the metamagnetic transition proceeds via spin *rotations*, and that occurs in isotropic or weakly anisotropic materials.

4.1. Spin-flip systems

Typical examples of spin-flip compounds are $DyPO_4$ [82], $HoPO_4$ [83] and $DyVO_4$ [84]. It is worth noting that this class of compounds includes mainly insulators or semimetals, while few rare earth intermetallics are concerned, these latter exhibiting often more than one magnetization step (see below). A very nice example of single-step transition in a metallic system is provided by the orthorhombic $TbCu_2$ compound [85]. At 4.2 K, the magnetic structure of this compound is collinear antiferromagnetic and the magnetic unit cell is tripled along the a direction [49] (see fig. 8). The magnetic moments are parallel to the a direction, their stacking sequence being either $+ + -$ or $- - +$ along the propagation vector. When the field is applied along the easy magnetization direction, one single spin–flip transition occurs at 20 kOe (see fig. 18). Above this critical field, the magnetization reaches the saturated value, i.e. the ferromagnetic arrangement is re-

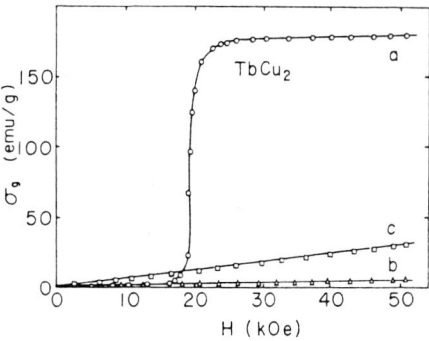

Fig. 18. Magnetization curves along each principal axis in the orthorhombic $TbCu_2$ compound at 4.2 K (after ref. [85]).

covered. It has been shown that a 12-sublattice model can account for such a behaviour in a limited region of the exchange couplings between the magnetic ions [86].

4.2. Spin-flop systems

The second class of metamagnetic materials, i.e. the spin-flop systems, includes more isotropic compounds where there is typically a rotation of the magnetic moments toward a direction perpendicular to the applied field, without changing drastically the position of the spins relatively to each other. Again, a well-known example is a nonmetallic system, namely $GdAlO_3$ [87,88]. Few thorough studies on single crystals are available in rare earth intermetallics. One example is the orthorhombic $GdCu_6$ compound [89], where the magnetization curves along the b and c axes show small discontinuous steps at around 3 T before varying linearly up to the saturated value (see fig. 19). Other good candidates are Gd compounds which are simple collinear antiferromagnets, such as $GdCu_2Si_2$ [90]. It can be noticed that Gd intermetallic compounds often exhibit in fact more complicated magnetic structures, in particular incommensurate ones, so that the metamagnetic processes are more complex than a simple spin-flop transition (see below). Other candidates are anisotropic rare earth systems where there are several equivalent easy magnetization directions, such as the fourfold axes in a cubic symmetry, or the twofold ones in a tetragonal symmetry. In that case, and despite the presence of a noticeable anisotropy, spin-flop transitions may occur when the magnetic field is applied along one of the easy magnetization directions: They correspond to the vanishing of the domains for which the magnetic moments are parallel to the applied field. This has been observed for example in the cubic PrB_6 compound where the anomalies on the magnetization curves however are very tiny [91].

4.3. Uniaxial multi-step systems

Beyond the above oversimplified distinction between spin-flip and spin-flop transitions, the strong growing of detailed magnetic studies over the last fifteen years led to widely extend the notion of metamagnetism to more complex magnetization processes. The strongest increase of the number of metamagnets found in the literature concerns for a large extent rare earth intermetallic antiferromagnets having a noticeable uniaxial anisotropy, mainly the tetragonal and hexagonal systems. In these compounds, much more than one metamagnetic transition are often observed before the saturated state, each of them

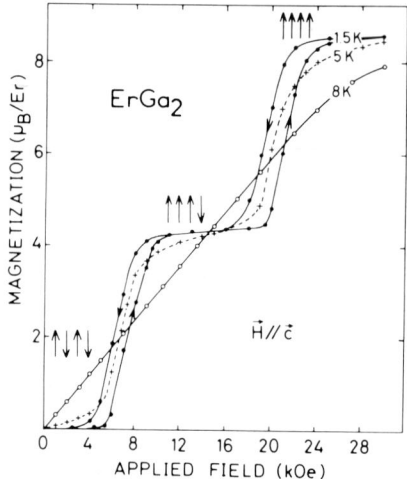

Fig. 20. Magnetization curves along the easy c axis in the hexagonal $ErGa_2$ compound at various temperatures (after ref. [93]).

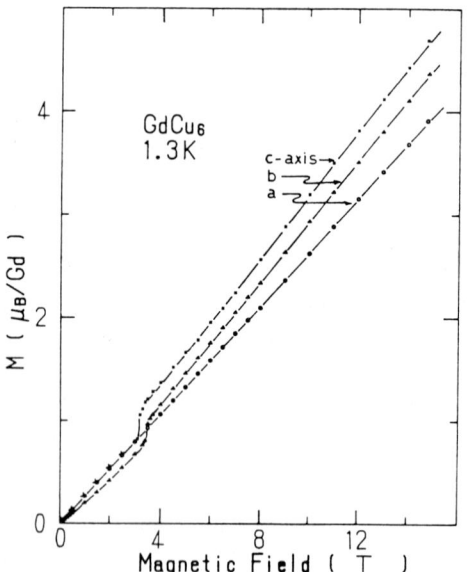

Fig. 19. Magnetization curves along each principal axis in the orthorhombic $GdCu_6$ compound at 1.3 K (after ref. [89]).

corresponding to an intermediate magnetic phase [79,92]. These multi-step magnetic processes can be very sharp at low temperature where the magnetic moments are frozen along the c direction, and a more or less pronounced hysteresis may be present for each step. The multi-step behaviour can be related to the long range character of the exchange interactions in these metallic systems, allowing a great number of spin configurations to be very close to each other with regard to their free energy. These latter varying not on the same way as a function of the applied magnetic field for the different configurations, several of them can cross at successive critical fields leading to the multi-step process.

Many examples are available to illustrate the multi-step metamagnetism in rare earth intermetallics. A first example is the hexagonal $ErGa_2$ compound where a double-step process occurs when the magnetic field is applied along the c easy magnetization direction [93] (see fig. 20). In zero field, the magnetic structure is characterized by the propagation vector $Q = (0, 1/2, 0)$. In the intermediate phase, as shown by neutron diffraction on a single crystal, one half of the magnetic moments antiparallel to the field have been flipped, leading to a stacking sequence $+ + + -$ instead of $+ - + -$ and to a net magnetization of one half of the saturated value. This process has been satisfactorily interpreted in a simple model involving a huge uniaxial anisotropy and exchange interactions between first and second nearest neighbours in the plane perpendicular to c [94].

A similar behaviour is observed in the body centered tetragonal $DyCo_2Si_2$ compound where the zero field sequence $+-+-$ for the moments is replaced by the sequence $+++-$ along the c easy magnetization direction, before reaching the saturated state [95] (see fig. 21). Several other examples can be found in this large family of ternary compounds having the $ThCr_2Si_2$-type structure, such as $PrCo_2Si_2$ [52], $NdCo_2Si_2$ [96], $NdRu_2Si_2$ [97], $CeAl_2Ga_2$ [98], $TbNi_2Si_2$ [92,99], ... Some compounds with an orthorhombic structure also exhibit multi-step metamagnetism, such as $DyCu_2$ [85,100,101], $CeZn_2$ [102,103] and $NdCu_6$ [104]. In $CeZn_2$, two intermediate magnetic phases have been determined under field, while in $NdCu_6$, up to five steps have been observed at 0.6 K in the magnetization process along the b easy direction. A surprisingly strong uniaxial behaviour has been also found in the cubic semimetallic pnictide CeSb, leading to a multi-step magnetization process associated with

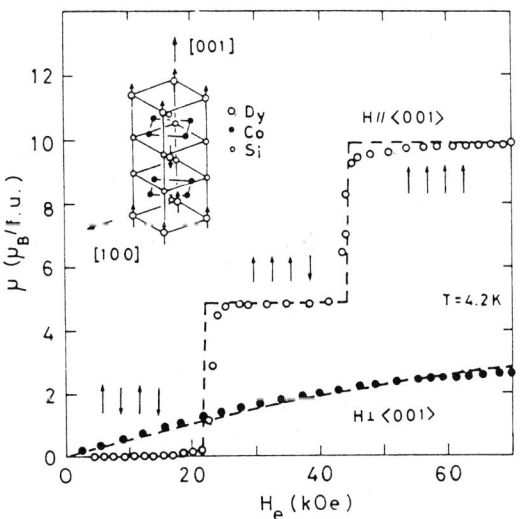

Fig. 21. Magnetization curves along and perpendicular to the c axis in the tetragonal $DyCo_2Si_2$ compound at 4.2 K (after ref. [95]).

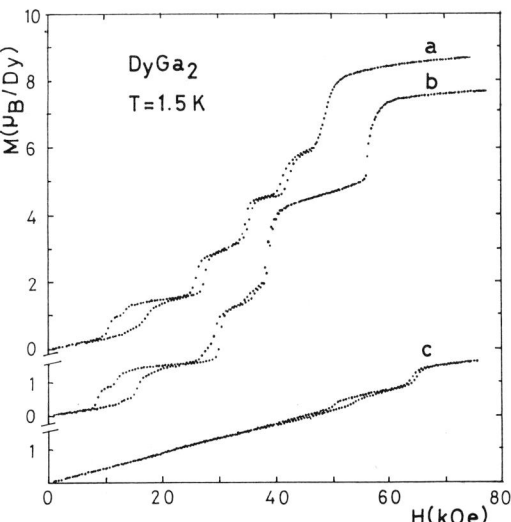

Fig. 22. Magnetization curves along each principal axis in the hexagonal $DyGa_2$ compound at 1.5 K (after ref. [53]).

its very complex phase diagram [105]. A great amount of work remains to be done on these systems in order to determine all the intermediate phases by neutron diffraction on single crystals.

4.4. Planar multi-step systems

A still more complex multi-step behaviour can occur when the magnetic moments are confined perpendicular to the principal axis of an uniaxial system, and when there is a noticeable anisotropy inside this basal plane. A typical example is provided by the hexagonal $DyGa_2$ compound where the magnetic moments are antiferromagnetically aligned along the [120] direction at low temperature [53]. Because of the three equivalent easy magnetization directions within the basal plane, a great number of intermediate phases have been evidenced for a magnetic field applied along the [120] axis and up to 7 successive steps have been detected (see fig. 22). The corresponding phase diagram is then very complex and the various steps are likely associated with a mixing of spin-flip and spin-flop transitions leading to more or less canted spin arrangements. It can be noticed that a phase diagram slightly different but as complex as the previous one has been also evidenced when the magnetic field is applied along the other high symmetry direction of the basal plane, namely the [100] direction.

4.5. Multi-axial multi-step systems

Another type of metamagnetic transition has been observed several years ago in cubic rare earth pnictides, these compounds exhibiting under an applied magnetic field canted intermediate structures similar to the "flopside" antiferromagnetic structure evidenced in HoP [81]. An early example is provided by DySb [77,106], but similar behaviours have been also observed in other pnictides such as DyBi [107] and HoSb [108]. Negative two-ion quadrupolar interactions have been found to be essential to explain these flopside phases [109]. Indeed, such negative couplings between 4f quadrupoles may favour a bi-axial (canted) arrangement for the magnetic moments, either in zero field, as in HoP, or under an applied field, as in DySb, DyBi and HoSb.

The same antiferroquadrupolar interactions are at the origin of the metamagnetic transitions observed more recently in other cubic rare earth intermetallics [110]. A very nice example is provided by $TmGa_3$ [111]. This compound orders at low temperature into an antiferromagnetic state described by the propagation vector $Q = (1/2, 1/2, 0)$. A thorough analysis of all the magnetic properties on a single crystal led to the conclusion that the antiferroquadrupolar pair interactions within the trigonal symmetry favours a multi-axial (triple-Q) spin arrangement, the magnetic moments pointing along the four threefold axes. A two-step metamagnetic behaviour has been observed along the three main symmetry directions (see fig. 23), leading to three magnetic phase diagrams similar to each other, except for the value of the critical fields, in particular the highest one along the [001] axis. Neutron diffraction on a single crystal allowed to determine the intermediate phase in each case, i.e. a bi-axial or a quadri-axial spin structure where only one fraction of the moments antiparallel to the field has rotated, either along its own direction or along another equivalent one. Such a behaviour has been explained by a strong competition between:

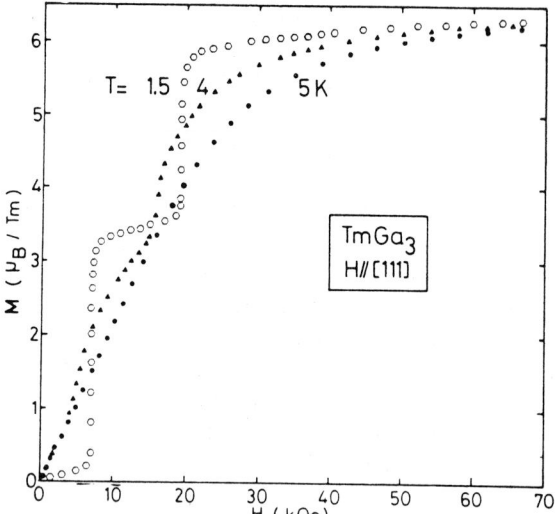

Fig. 23. Magnetization processes for a [111] magnetic field direction in $TmGa_3$ (after ref. [111]).

i) a magnetocrystalline anisotropy favouring the ⟨111⟩ directions, ii) a negative exchange interaction favouring an antiferromagnetic arrangement of the spins, and iii) a negative quadrupolar pair coupling favouring a mutually perpendicular arrangement of the quadrupoles.

A similar behaviour has been found in the cubic DyCu compound [112], although the critical fields are much higher in that case than in $TmGa_3$. In the isomorphous DyAg compound, the spontaneous triple-Q magnetic structure is identical to that of DyCu and $TmGa_3$. However, the breaking of this multi-axial structure occurs through a complex multi-step mechanism, namely through 2, 3 and 4 transitions along the [001], [101] and [111] directions, respectively [67,113] (see fig. 24). Again, antiferroquadrupolar pair interactions are necessary to explain this kind of metamagnetism and have been effectively determined in many of these compounds [65]. Multi-step metamagnetic transitions observed in other cubic rare earth intermetallic systems such as PrAg [114] and PrCd [115] have probably the same origin.

4.6. Incommensurate metamagnetic systems

As quoted above (see section 3), many rare earth intermetallic compounds exhibit an incom-

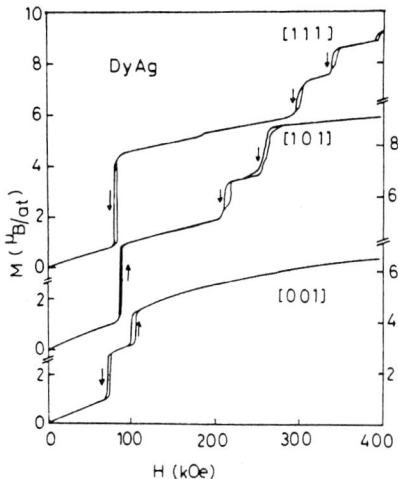

Fig. 24. Magnetization curves along the three main symmetry directions in DyAg at 4.2 K (after ref. [67]).

also be present, as in the cubic RBe_{13} compounds, in particular $TbBe_{13}$ [118,119].

In incommensurate amplitude modulated systems, the magnetic moments are frozen along one single direction. As a consequence, the metamagnetic behaviour occurs when the magnetic field is applied along this easy magnetization direction. This is the case of Er and Tm metals in a limited temperature range below T_N where the single-step metamagnetic transition is rather sharp [38]. It can be noticed that along the hard magnetization directions, multi-step behaviours have been also observed, but they are more complex to be analyzed. Other examples are available among the rare earth intermetallic compounds where the metamagnetic transition is often very smooth, i.e. there is no discontinuity in the magnetization processes, such as in the cubic $CeAl_2$ [120] and the hexagonal HoAlGa [121] compounds. In this latter compound, the magnetic structure is modulated between $T_N = 31$ K and $T_1 = 18.5$ K and antiphase below T_t, so that two different metamagnetic behaviours can be observed as a function

mensurate magnetic structure below their ordering temperature, at least in a limited temperature domain, owing to the long range character of the exchange interactions. These structures may be of three types, namely helimagnetic, amplitude modulated and antiphase. These latter two cases, where the moments remain collinear, are often associated with each other because of entropy considerations [65]. Each of these three configurations may lead to metamagnetic transitions. In helimagnetic systems, the magnetocrystalline anisotropy fixes the moments to be confined within a plane, and the anisotropy usually remains weak inside this plane. When a magnetic field is applied in the plane of the helical structure, this latter distorts and may undergo a metamagnetic transition toward a fan structure before reaching the ferromagnetic state. This situation has been observed in rare earth metals such as Dy [38] and more recently Tb [116]. In this latter compound, the fan phase is observed in a narrow temperature range between $T_c = 219.8$ K and $T = 222.8$ K, while between 222.8 K and $T_N = 228.1$ K, the ferromagnetic state is achieved immediately above the critical field (see fig. 25). In Ho, a multi-step metamagnetic behaviour is observed [38], suggesting the existence of complex intermediate "helifan" structures [117]. In nonuniaxial systems, similar metamagnetic behaviours may

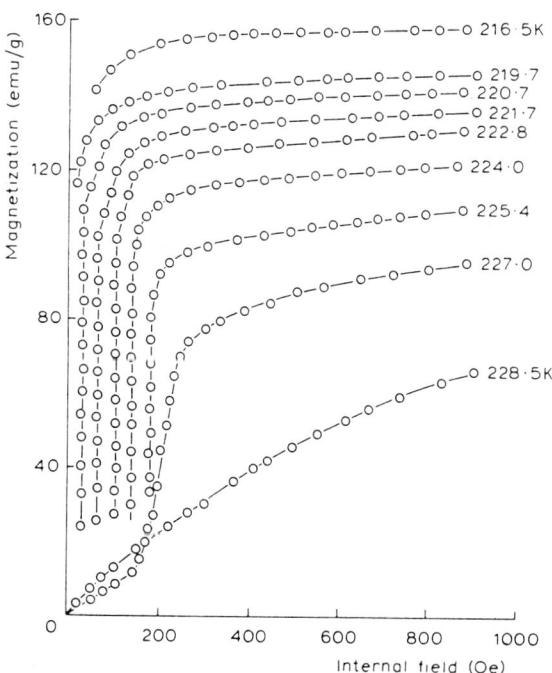

Fig. 25. Magnetization curves along the **b** axis of Tb at various temperatures in the helimagnetic phase (after ref. [116]).

of the temperature (see fig. 26). At low temperature, a triple-step magnetization process occurs along the *c* easy magnetization direction, while one single smooth transition is evidenced in the modulated phase.

Generally, amplitude modulated structures are not stable down to 0 K and may transform progressively, when the temperature decreases, to an antiphase structure where all the moments are equal. When a magnetic field is applied perpendicular to the easy magnetization direction, subtle metamagnetic behaviour may take place. The hexagonal $GdGa_2$ compound appears to be in this situation. Below $T_N = 22$ K, this compound exhibits an incommensurate magnetic structure with the propagation vector $Q = (0.39, 0.39, 0)$. Thermodynamical considerations have shown that this structure is not helimagnetic but amplitude modulated, therefore antiphase at low temperature [76]. The magnetization curve along the *c* axis does not exhibit the linear behaviour expected for a helimagnet having a weak anisotropy, but rather one smooth transition near 14 T [122] (see fig. 27). This kind of metamagnetic transition seems to be closely related to the antiphase character of the incommensurate magnetic structure. It is worth noting that another type of metamag-

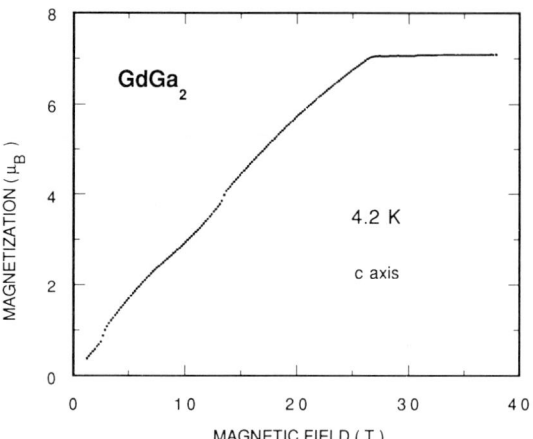

Fig. 27. High field magnetization curve along the *c* axis in $GdGa_2$ at 4.2 K (after ref. [122]).

netic process takes place in lower field (near 2.5 T), namely a spin-flop transition. A similar behaviour has been also evidenced on a single crystal of the orthorhombic $GdCu_2$ compound [123].

4.7. Ferromagnetic metamagnetic systems

The existence of metamagnetic transitions is not limited to pure antiferromagnets. Under certain conditions, ferromagnets may also exhibit metamagnetic processes. This is the case of the orthorhombic RNi compounds where this behaviour arises from the noncollinearity of the magnetic structure [124]. In these low symmetry systems, the local easy magnetization directions are not all parallel to each other so that the magnetic ions can be divided in two sublattices. Ferromagnetic interactions then result in a noncollinear (canted) arrangement for the moments, with both ferromagnetic and antiferromagnetic components. When the magnetic field is applied along this latter one, and if the magnetocrystalline anisotropy is large enough, a spin-flip mechanism may occur, corresponding to the reversal of the moments of one sublattice. This has been observed for example in DyNi [125] (see fig. 28).

Another case of metamagnetic ferromagnet is associated with a change of easy magnetization direction due to the magnetocrystalline aniso-

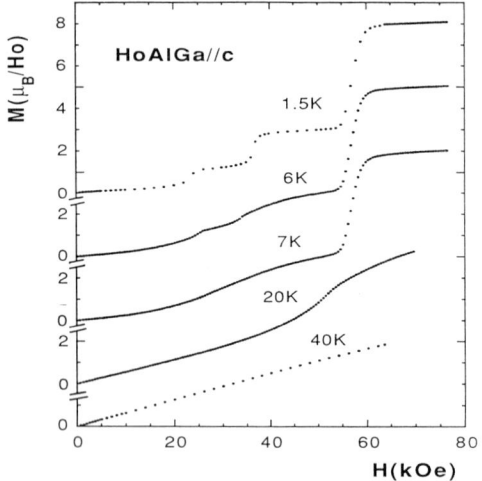

Fig. 26. Magnetization curves along the *c* axis in HoAlGa at various temperatures in the antiphase state (below 18.5 K) and in the modulated phase (between 18.5 and 31 K) (after ref. [121]).

tropy. More precisely, this latter usually induces an anisotropic raising of degeneracy of the crystal field states along two different crystallographic directions, leading to a temperature variation of the free energy quicker along one direction than along the other one. Both free energies may then cross each other at a given temperature where a spin reorientation occurs, as in cubic HoZn for example [71]. A metamagnetic behaviour is associated with this change of easy magnetization direction: At 4.2 K, when the magnetic field is applied along the [111] hard magnetization direction, the moments rotate progressively starting from the [101] easy axis, but at a threshold field, there is a strong increase of the magnetization when the moments suddenly align along the field. The same explanation is true in the hexagonal HoRu$_2$ compound where the jump of magnetization is still more spectacular because the moments rotate from near the a easy axis toward the c direction, i.e. over almost 90° [126] (see fig. 29).

This kind of metamagnetism has been also observed in the hexagonal PrCo$_5$ ferromagnet [127], in the rhomboedral ErFe$_3$ ferrimagnet [128] as in numerous other ferrimagnetic rare earth intermetallics containing 3d ions [129,130]. In these rare earth–3d compounds, the change of easy magnetization direction arises from the com-

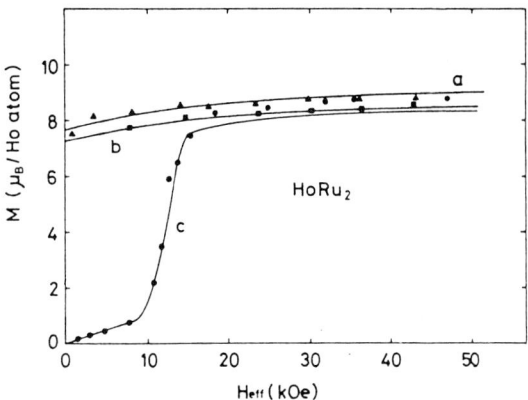

Fig. 29. Magnetization curves in HoRu$_2$ at 4.2 K along the three main symmetry directions of the hexagonal unit cell; Full lines are calculated curves including crystal field and exchange couplings (after ref. [126]).

petition between the magnetocrystalline anisotropy of both constituants, as soon as they are antagonistic (see section 3). More complex multi-step magnetization processes may also occur in easy-plane ferrimagnets as in hexagonal R$_2$T$_{17}$ (T = Co, Fe), where the sixfold symmetry of the basal plane increases the number of intermediate spin configurations [131,132].

4.8. Paramagnetic metamagnetic systems

Metamagnetic processes do not require a magnetically ordered material to be evidenced. Paramagnets may also exhibit field induced phase transitions under certain conditions of magnetocrystalline anisotropy, i.e. for particular CEF levels schemes. The most spectacular effect arises in case of levels crossing, as predicted many years ago about TmSb [133]. More recently, this effect has been observed in Pr metal at a field of about 32 T applied along the c axis [134] (see fig. 30). It manifests through a single-step transition which is as sharper as the temperature is lowered. This is explained by the crossing of two CEF levels on the hexagonal sites, namely a singlet which is the ground state in zero field and an excited level which is more magnetic than the singlet. The transition is of first-order only at 0 K where the ground state changes in a discontinuous way as a function of the magnetic field. It can be noticed

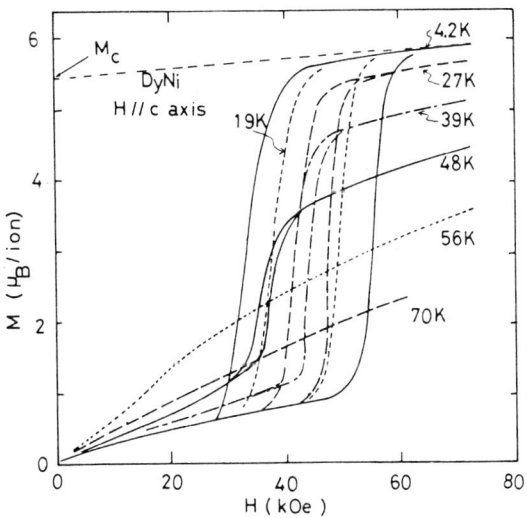

Fig. 28. Temperature dependence of the magnetization along the c axis in orthorhombic DyNi (after ref. [125]).

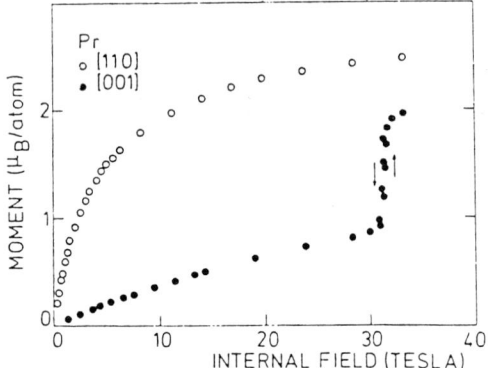

Fig. 30. High field magnetization data in Pr metal at 4.2 K along and perpendicular to the *c* axis; Note the weak hysteresis of the transition at 32 T (after ref. [134]).

that, this effect being much more pronounced at very low temperature, singlet or more generally nonmagnetic ground state systems are more appropriate to present this behaviour than other systems, because thay may remain paramagnetic down to 0 K even in the presence of magnetic interactions. Nevertheless, this condition is not absolutely necessary.

Another example where a crossing of CEF levels has been anticipated is the hexagonal $PrNi_5$ compound for a magnetic field applied along the [100] direction [135,136]. A weak transition indeed has been evidenced around $\approx$ 19 T at 1.5 K [122] (see fig. 31). Temperatures as low as 0.3 K

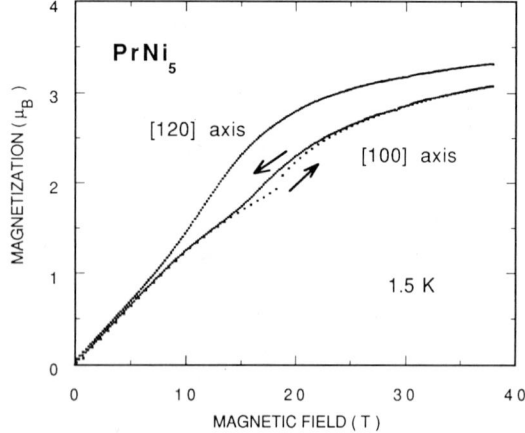

Fig. 31. High field magnetization curves in $PrNi_5$ along the two main symmetry directions of the basal plane at 1.5 K (after ref. [122]).

should be needed to observe a sharper stepwise process. This compound appears to be particularly interesting because, for a magnetic field applied along the [120] direction, an anticrossing of two CEF levels has been observed around 12 T [136], leading to another kind of metamagnetic transition which remains smooth even at 0 K (see fig. 31). In that case, the transition originates in Van Vleck matrix elements between the nonmagnetic ground state and excited levels states. In a general way, both types of metamagnetic process occur when a non- or weakly magnetic ground state evolves into a highly magnetic state as a function of an applied field, the continuous or discontinuous character of the transition depending on detailed composition of the ground state, i.e. if this latter includes or not the J_{max} component.

It is worth noting that a useful criterion of occurrence of a metamagnetic transition is provided by the sign of the third-order magnetic susceptibility [137]: This quantity indeed represents the initial curvature of the magnetic curves, so that a positive value is a sufficient (but not necessary) condition to observe an inflexion point at a certain field, i.e. a metamagnetic transition. However it does not allow to anticipate if this transition under field will be sharp or smooth, or what is its origin. At last, it has been shown that, more generally, a multi-step metamagnetic behaviour can be expected depending on the exact CEF level scheme, associated to several successive level crossings [138]. However, the corresponding critical fields are often predicted to be much higher than the experimentally available magnetic fields. This large magnitude explains why few examples are given in the literature. Nevertheless the increasing availability to have high magnetic fields and low temperatures should allow to change this situation in the near future.

4.9. Quadrupolar metamagnetic systems

If antiferroquadrupolar interactions may induce multi-step magnetization processes in a multiaxial antiferromagnetic structure (see section 4.5), ferroquadrupolar couplings may act on the magnetization in the paramagnetic state. This

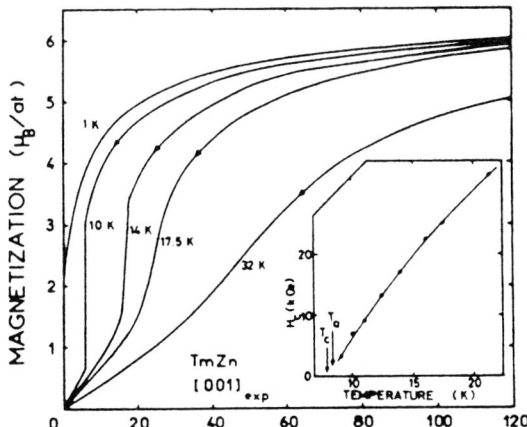

Fig. 32. Magnetization curves in TmZn along the [001] direction at various temperatures; the inset shows the temperature dependence of the critical field (after ref. [139]).

effect has been clearly evidenced in the quadrupolar compound TmZn [139] where a single-step transition occurs above the ordering temperature along the [001] easy magnetization direction (see fig. 32). The first-order transition arises when the internal field exceeds the critical field necessary to induce the ferromagnetic ordering. Again, this behaviour may be associated with a positive value for the third-order magnetic susceptibility [140]: It has been shown that this latter receives a positive contribution from the ferroquadrupolar coupling which is able to over-compensate the negative contribution arising from the CEF interactions. This effect has been also observed in several other rare earth compounds [65], such as TmCu [141], PrPb$_3$ [142], CeAg [143], TbP [144],... although in most cases this does not give rise to a first-order transition, but rather to a smooth transition.

5. Quantitative analysis

The knowledge of the magnetic properties in rare earth intermetallic compounds has been widely increased over the last fifteen years, not only in an experimental point of view, but also with regard to theoretical or phenomenological models used to account for the experimental results. Such a quantitative analysis started in the 70's by taking into consideration the Crystalline Electric Field (CEF) effects. This was motivated by the growing number of experimental evidences emphasizing the role of the CEF, in particular in the paramagnetic state, and led to several international conferences specifically dedicated to this problem. During this period, the main preoccupation has been to determine the CEF parameters, starting from the analysis of many magnetic properties which are strongly influenced by the CEF level spacing [145]. There have then been a wide variety of methods to achieve this goal, namely, among others, spectroscopic methods (optical, neutron, resonance,...), thermal effects (specific heat, magnetic susceptibility, elastic constants,...) and transport properties (electrical and thermal conductivity, thermoelectric power,...). Several systematics of CEF parameters have thus been obtained on various rare earth intermetallic series, mainly in cubic symmetry where only two parameters are involved [145–147].

The next step in the quantitative analysis of the magnetic properties has been to include the above CEF interaction together with exchange and Zeeman couplings in order to account for the anisotropic properties of ferromagnetically ordered materials, by a self-consistent diagonalization of the corresponding Hamiltonian. Three-dimensional magnetization processes in the ferromagnetic phase have then been investigated, mainly in cubic symmetry, as in HoAl$_2$ [72] or in HoCo$_2$ [73] compounds, but also in hexagonal ones, as for example in DyNi$_5$ [148].

In the early 80's, it appeared very soon that the above interactions (CEF + exchange + Zeeman) were insufficient to well describe all the magnetic properties in several intermetallic compounds. It was particularly the case in cubic symmetry, where only few parameters were needed and known with accuracy. Additional couplings were then included, namely the quadrupolar interactions, as in the RZn series [71] or in the cubic ErNi$_2$ compound [149] for example. A great amount of experimental and theoretical studies has been done in order to develop the tools needed to achieve a good knowledge of these quadrupolar interactions [65,66,150]. These latter are divided in two terms, namely the one-ion magnetoelastic coupling and the two-ion quadrupolar interaction. They have been found to deeply influence most of the magnetic properties, in the paramagnetic as well as in the ordered phase. Note that only few noncubic systems have been investigated in this point of view, such as the hexagonal $TmNi_5$ compound [151].

The semi-metallic pnictides have to be considered apart, since they seem to behave in a more complex way. In addition to quadrupolar interactions, two-ion magnetoelastic coupling as well as spin fluctuations have been found in DySb for example [152]. In cerium based pnictides, a strong p–f mixing has been claimed to explain the complicated phase diagram [153]. In rare earth–3d intermetallic compounds, a quantitative analysis of the magnetic properties is more difficult to achieve because the itinerant character of 3d magnetism prevents to describe its anisotropy in a simple manner. Only phenomenological models can then be used, assuming a constant magnitude for the 3d magnetic moments [132].

The quantitative investigation of the ordered state of antiferromagnetic compounds has been developed lately, because this requires to use many-sublattice models which are heavier to be handled. A 12-sublattice model has been investigated for example about $TbCu_2$ to explain the multi-step magnetization processes in orthorhombic systems [86]. The case of incommensurate (modulated) structures is still more difficult to be analyzed because it involves many-spin arrangements [59,154,155].

Considering the large increase of recent studies on single crystals of antiferromagnetic systems having more or less complex magnetic structures (simple antiferromagnetic, helimagnetic, modulated, antiphase,...), one may anticipate that a big effort will be done in the next decade to quantitatively understand the various and sometimes exciting magnetic properties. It is worth noting that the compounds exhibiting such properties are often of a lower symmetry than the cubic one. Therefore attention will have to be paid too in order to get a reliable estimation of the numerous crystal field parameters, by a coherent analysis of various physical properties.

6. Conclusion

The above survey shows the extreme wealth of rare earth intermetallics in magnetism. In the light of these results it is possible to state some predictions in further developments. At this moment, a large amount of studies are devoted to the determination of magnetic phase diagrams in a large number of compounds, in particular hexagonal and tetragonal. A large experimental effort will probably be invested to get an accurate knowledge of the different magnetic structures of the phase diagrams and on the determination of the field and temperature transitions. Single crystal of good quality are of fundamental importance. At the same time quantitative interpretations would lead to the understanding of the observed phenomenon and to the determination of the parameters describing the different interactions. This includes mainly exchange and CEF parameters. Determination of the parameters of smaller interactions such as the quadrupolar ones, the magnetoelastic coupling and eventually the anisotropy of bilinear interactions could also be obtained in some cases. As shown previously these latter can play a crucial role in magnetic properties. If theoretical models are available for the paramagnetic state and below T_c in ferromagnets, in antiferromagnets further theoretical models are needed to account for the field and temperature dependences of magnetization and structures below T_N. Furthermore an accurate analysis of the thermodynamical properties such as specific heat, resistivity or magnetic susceptibility taking into account the magnetic structure involved can be a useful tool. Finally Mn magnetism is not very well understood and further studies are anticipated in compounds with this 3d element not only in binary but also in ternary compounds such as the tetragonal RMn_2X_2 which seem quite promising.

References

[1] M. Cyrot and M. Lavagna, J. de Phys. 40 (1979) 763.
[2] M. Shimizu, J. Inoue and S. Nagawawa, J. Phys. F 14 (1984) 2673.

[3] E.P. Wohlfarth and P. Rhodes, Phil. Mag. 7 (1962) 1817.
[4] D. Givord, J. Laforest and R. Lemaire, J. Appl. Phys. 50 (1979) 7489.
[5] D. Givord, J. Laforest, R. Lemaire and Q. Lu, J. Magn. Magn. Mater. 31 (1983) 191.
[6] D. Gignoux, R. Lemaire and P. Molho, J. Magn. Magn. Mater. 21 (1980) 119.
[7] R. Lemaire, Cobalt 32 (1966) 132.
[8] D. Bloch, D.M. Edwards, M. Shimizu and J. Voiron, J. Phys. F 5 (1975) 1217.
[9] G. Petrich and R.L. Mössbauer, Phys. Lett. A 26 (1968) 403.
[10] F. Givord and J.S. Shah, Comptes Rendus 274 (1972) 923.
[11] D. Gignoux, F. Givord and W.C. Koehler, Physica B 86–88 (1977) 165.
[12] M. Cyrot, D. Gignoux, F. Givord and M. Lavagna, J. de Phys. 40 (1979) C5-171.
[13] H. Yamada and M. Shimizu, J. Phys. F 15 (1985) L175.
[14] H. Yamada, T. Tohyama and M. Shimizu, J. Phys. F 17 (1987) L163.
[15] T. Sakakibara, T. Goto, K. Yoshimura, M. Shiga, Y. Nakamura and K. Fukamichi, J. Magn. Magn. Mater. 70 (1987) 126.
[16] T. Goto, T. Sakakibara, K. Murata, H. Komatsu and K. Fukamichi, J. Magn. Magn. Mater. 90&91 (1990) 700.
[17] D. Gignoux, R. Lemaire and P. Molho, J. Magn. Magn. Mater. 21 (1980) 307.
[18] R. Ballou, B. Gorges, P. Molho and P. Rouault, J. Magn. Magn. Mater. 84 (1990) L1.
[19] D. Gignoux, R. Lemaire, R. Mendia-Monterroso, J.M. Moreau and J. Schweizer, Physica B 130 (1985) 376.
[20] M. Weinert and A.J. Freeman, J. Magn. Magn. Mater. 38 (1983) 23.
[21] D. Givord, J. Laforest, R. Lemaire and Q. Lu, J. Magn Magn. Mater. 31–34 (1983) 191.
[22] R. Ballou and R. Lemaire, J. de Phys. 49 (1988) C8-523.
[23] R. Ballou, J. Deportes, R. Lemaire, Y. Nakamura and B. Ouladdiaf, J. Magn. Magn. Mater. 70 (1987) 129.
[24] R. Ballou and H. Yamada, private communication.
[25] J. Friedel, in: The Physics of Metals, vol. 1, ed. Ziman (Cambridge Univ. Press, Cambridge, 1969).
[26] J. Schweizer and F. Tasset, J. Phys. F 10 (1980) 2799.
[27] H. Wada, H. Nakamura, K. Yoshimura, M. Shiga and Y. Nakamura, J. Magn. Magn. Mater. 70 (1987) 134.
[28] R. Ballou, J. Deportes, R. Lemaire and B. Ouladdiaf, J. Appl. Phys. 63 (1988) 3487.
[29] W. Hussen, K.U. Neumann, D.J. Parry, K.R.A. Ziebeck, R. Ballou, J. Deportes, R. Lemaire and B. Ouladdiaf, J. Magn. Magn. Mater. 84 (1990) 281.
[30] J. Deportes, B. Ouladdiaf and K.R.A. Ziebeck, J. Magn. Magn. Mater. 70 (1987) 14.
[31] R. Ballou, J. Deportes, R. Lemaire, P. Rouault and J.L. Soubeyroux, J. Magn. Magn. Mater. 90&91 (1990) 559.
[32] J. Voiron, R. Ballou, J. Deportes, R.M. Galera and E. Lelievre, J. Appl. Phys. 69 (1991) 5678.
[33] M. Shiga, J. Hirokawa, H. Wada and Y. Nakamura, J. Magn. Magn. Mater. (to appear).
[34] K. Yoshimura, M. Shiga and Y. Nakamura, J. Phys. Soc. Jpn. 55 (1986) 3585.
[35] C. Ritter, S.H. Kilcoyne and R. Cywinski, J. Phys.: Condens. Matter 3 (1991) 727.
[36] R. Ballou, C. Lacroix and M.D. Nunez Regueiro, Phys. Rev. Lett. 44 (1991) 1910.
[37] H.R. Child, W.C. Koehler, E.O. Wollan and J.W. Cable, Phys. Rev. 138 (1965) A1655.
[38] B. Coqblin, The Electronic Structure of Rare-Earth Metals and Alloys: The Magnetic Heavy Rare-Earths (Academic Press, New York, London, 1977) and references therein.
[39] F. Vigneron, M. Bonnet, A. Herr and J. Schweizer, J. Phys. F 12 (1982) 223.
[40] D. Gignoux, R. Lemaire and D. Paccard, Phys. Lett. A 41 (1972) 187.
[41] D. Gignoux, J.C. Gomez-Sal, R. Lemaire and A. de Combarieu, Solid State Commun. 21 (1977) 637.
[42] J.M. Barandiarán, D. Gignoux, D. Schmitt and J.C. Gomez-Sal, Solid State Commun. 57 (1986) 941.
[43] J.A. Blanco, D. Gignoux, J.C. Gomez-Sal and D. Schmitt, ICM'91, Edinburgh, J. Magn. Magn. Mater. 104–108 (1992) to be published.
[44] B. Barbara, M.F. Rossignol, J.X. Boucherle, J. Schweizer and J.L. Buevoz, J. Appl. Phys. 50 (1979) 2300.
[45] J.K. Yakinthos and E. Roudaut, J. Magn. Magn. Mater. 68 (1987) 90.
[46] P. Morin, D. Schmitt and C. Vettier, J. de Phys. 46 (1985) 39.
[47] A. Szytuła and J. Leciejewicz, Handbook on the Physics and Chemistry of Rare Earths, vol. 12, eds. K.A. Gschneidner, Jr. and L. Eyring (North-Holland, Amsterdam, 1989) p. 132.
[48] V. Sima, Z. Smetana, B. Lebech and E. Gratz, J. Magn. Magn. Mater. 54–57 (1986) 1357.
[49] B. Lebech, Z. Smetana and V. Sima, J. Magn. Magn. Mater. 70 (1987) 97.
[50] W.C. Koehler, J. Appl. Phys. 36 (1965) 1078.
[51] D. Gignoux, D. Schmitt, M. Zerguine, E. Bauer, H. Pillmayr, J.Y. Henry, V.N. Nguyen and J. Rossat-Mignod, J. Magn. Magn. Mater. 74 (1988) 1.
[52] T. Shigeoka, H. Iwata, H. Fujii, T. Okamoto and Y. Hashimoto, J. Magn. Magn. Mater. 70 (1987) 239.
[53] D. Gignoux, D. Schmitt, A Takeuchi and F.Y. Zhang, J. Magn. Magn. Mater. 97 (1991) 15.
[54] B. Chevalier, J. Etourneau, P. Hagenmuller, S. Quezel and J. Rossat-Mignod, J. Less-Common Met. 111 (1985) 161.
[55] A.R. Ball, D. Gignoux and D. Schmitt, ICNS'91, Oxford, to appear in Physica B.
[56] J. Rossat-Mignod, P. Burlet, S. Quezel, J.M. Effantin, D. Delacote, H. Bartholin, O. Vogt and D. Ravot, J. Magn. Magn. Mater. 31–34 (1983) 398.
[57] J. Rossat-Mignod, P. Burlet, L.P. Regnault and C. Vettier, J. Magn. Magn. Mater. 90&91 (1990) 5.
[58] L.P. Regnault, J.L. Jacoud, C. Vettier, T. Chattopadhyay, J. Rossat-Mignod, T. Suzuki, T. Kasuya and O. Vogt, Physica B 156&157 (1989) 798.

[59] M. Date, J. Phys. Soc. Jpn. 57 (1988) 3682.
[60] T. Chattopadhyay, P. Burlet, J. Rossat-Mignod, H. Bartholin, C. Vettier and O. Vogt, J. Magn. Magn. Mater. 63&64 (1987) 52.
[61] R.M. Moon, J.W. Cable and W.C. Koehler, J. Appl. Phys. 35 (1964) 1041.
[62] B. Lebech, J. Appl. Phys. 52 (1981) 2019.
[63] K.A. McEwen and S.W. Zochowski, J. Magn. Magn. Mater. 90&91 (1990) 94.
[64] E.M. Forgan, E.P. Gibbons, K.A. McEwen and D. Fort, Phys. Rev. Lett. 62 (1989) 470.
[65] P. Morin and D. Schmitt, in: Handbook on Ferromagnetic Materials, vol. 5, eds. E.P. Wohlfarth and K.H.J. Buschow (North-Holland, Amsterdam, 1990) p. 1.
[66] R. Aléonard and P. Morin, J. Magn. Magn. Mater. 84 (1990) 255.
[67] P. Morin, J. Rouchy, K. Yonenobu, A. Yamagishi and M. Date, J. Magn. Magn. Mater. 81 (1989) 247.
[68] P. Fischer, A. Furrer, E. Kaldis, J.K. Kjems and P.M. Levy, Phys. Rev. B 31 (1985) 456.
[69] T. Tsushima and M. Ohokoshi, J. Magn. Magn. Mater. 31–34 (1983) 197.
[70] H. Fujii, T. Okamoto, T. Shigeoka and H. Iwata, Solid State Commun. 53 (1985) 715.
[71] P. Morin and D. Schmitt, J. Phys. F 8 (1978) 951.
[72] B. Barbara, J.X. Boucherle, M.F. Rossignol and J. Schweizer, Physica B 86–88 (1977) 83.
[73] D. Gignoux, F. Givord and R. Lemaire, Phys. Rev. B 9 (1975) 3878.
[74] R. Aléonard and P. Morin, J. Magn. Magn. Mater. 50 (1985) 128.
[75] R.M. Moon, W.C. Koehler, J.W. Cable and H.R. Child, Phys. Rev. B 5 (1972) 991.
[76] J.A. Blanco, D. Gignoux, P. Morin and D. Schmitt, J. Magn. Magn. Mater. 90–91 (1990) 166.
[77] E. Stryjewski and H. Giordano, Advan. Phys. 26 (1977) 487.
[78] A. Kasten and H.G. Kahle, J. Magn. Magn. Mater. 54–57 (1986) 1325.
[79] M. Date, J. Magn. Magn. Mater. 90&91 (1990) 1.
[80] P.B. Fynbo, J. Phys. F 7 (1977) 2179.
[81] F. Hulliger, J. Magn. Magn. Mater. 8 (1978) 183.
[82] C.S. Koonce, B.W. Mangum and D.D. Thornton, Phys. Rev. B 4 (1971) 4054.
[83] A.H. Cooke, S.J. Swithenby and M.R. Wells, J. Phys. C 6 (1973) 2209.
[84] A.H. Cooke, C.J. Ellis, K.A. Gehring, M.J.M. Leask, D.M. Martin, B.M. Wanklyn, M.R. Wells and R.L. White, Solid State Commun. 8 (1970) 689.
[85] Y. Hashimoto, H. Fujii, H. Fujiwara and T. Okamoto, J. Phys. Soc. Jpn. 47 (1979) 67.
[86] I. Kimura, J. Magn. Magn. Mater. 52 (1985) 199.
[87] K.W. Blazey, H. Rohrer and R. Webster, Phys. Rev. B 4 (1971) 2287.
[88] K.W. Blazey and H. Rohrer, Phys. Rev. 173 (1968) 574.
[89] S. Takayanagi, Y. Onuki, K. Ina, T. Komatsubara, H. Wada, T. Watanabe, T. Sakakibara and T. Goto, J. Phys. Soc. Jpn. 58 (1989) 1031.
[90] J.M. Barandiarán, D. Gignoux, D. Schmitt, J.C. Gomez-Sal, J. Rodriguez Fernandez, P. Chieux and J. Schweizer, J. Magn. Magn. Mater. 73 (1988) 233.
[91] P. Morin, S. Kunii and T. Kasuya, J. Magn. Magn. Mater. 96 (1991) 145.
[92] H. Fujii and T. Shigeoka, J. Magn. Magn. Mater. 90&91 (1990) 115.
[93] M. Doukouré and D. Gignoux, J. Magn. Magn. Mater. 30 (1982) 111.
[94] D. Gignoux, D. Schmitt and M. Zerguine, J. Magn Magn. Mater. 66 (1987) 373.
[95] N. Iwata, K. Honda, T. Shigeoka, Y Hashimoto and H. Fujii, J. Magn. Magn. Mater. 90&91 (1990) 63.
[96] T. Shigeoka, N. Iwata, Y. Hashimoto Y. Andoh and H. Fujii, J. de Phys. 49 (1988) C8–431.
[97] T. Shigeoka, M. Saeki, H. Iwata, T. Takabatake and H. Fujii, J. Magn. Magn. Mater. 90&91 (1990) 557.
[98] D. Gignoux, D. Schmitt and M. Zerguine, J. de Phys. 49 (1988) C8–4334
[99] J.A. Blanco, D. Gignoux, D. Schmitt and C. Vettier, J. Magn. Magn. Mater. 97 (1991) 4.
[100] H. Iwata, Y. Hashimoto, T. Kimura and T. Shigeoka, J. Magn. Magn. Mater. 81 (1989) 354.
[101] Y. Hashimoto, A. Yamagishi, T. Takeuchi and M. Date, J. Magn. Magn. Mater. 90&91 (1990) 49.
[102] H. Fujii, M. Akayama, Y. Uwatoko, Y. Hashimoto and T. Kitai, J. de Phys. 49 (1988) C8–417.
[103] J. Voiron, P. Morin, D. Gignoux and R. Aléonard, J. de Phys. 49 (1988) C8–419.
[104] Y. Onuki, K. Ina, M. Nishihara, T. Komatsubara, S. Takayanagi, K. Kameda and N. Wada, J. Phys. Soc. Jpn. 55 (1986) 1818.
[105] J. Rossat-Mignod, P. Burlet, J. Villain, H. Bartholin, Wang Tcheng-Si, D. Florence and O. Vogt, Phys. Rev. B 16 (1977) 440.
[106] G.E. Everett and P. Streit, J. Magn. Magn. Mater. 12 (1979) 277.
[107] F. Hulliger, J. Magn. Magn. Mater. 15–18 (1980) 1243.
[108] T.O. Brun, F.W. Korty and J.S. Kouvel, J. Magn. Magn. Mater. 15–18 (1980) 298.
[109] D. Kim and P.M. Levy, J. Magn. Magn. Mater. 27 (1982) 257.
[110] P. Morin, J. Rouchy, D. Schmitt and E. du Trémolet de Lacheisserie, J. Magn. Magn. Mater. 90&91 (1990) 105.
[111] P. Morin, M. Giraud, P. Burlet and A. Czopnik, J. Magn. Magn. Mater. 68 (1987) 107.
[112] R. Aléonard, P. Morin and J. Rouchy, J. Magn. Magn. Mater. 46 (1984) 233.
[113] A. Yamagishi, K. Yonenobu, O. Kondo, P. Morin and M. Date, J. Magn. Magn. Mater. 90&91 (1990) 51.
[114] P. Morin and D. Schmitt, Phys. Rev. B 26 (1982) 3891.
[115] R. Aléonard and P. Morin, J. Magn. Magn. Mater. 42 (1984) 151.
[116] R.D. Greenough and N.F. Hettiarachchi, J. Magn. Magn. Mater. 31–34 (1983) 178.
[117] J. Jensen and A.R. Mackintosh, Phys. Rev. Lett. 64 (1990) 2699.

[118] E. Bucher, J.P. Maita, G.W. Hull, R.C. Fulton and A.S. Cooper, Phys. Rev. B 11 (1975) 440.
[119] J.M. Bouton, R. Clad, A. Herr, P. Mériel, A. Meyer and F. Vigneron, J. Magn. Magn. Mater. 15–18 (1980) 49.
[120] B. Barbara, J.X. Boucherle, J.L. Buevoz, M. Rossignol and J. Schweizer, Solid State Commun. 24 (1977) 481.
[121] D. Gignoux, D. Schmitt, A. Takeuchi and F.Y. Zhang, J. Magn. Magn. Mater. 98 (1991) 333.
[122] A. Ball, D. Gignoux, D. Schmitt and A. de Visser, unpublished.
[123] M.K. Borombaev, R.Z. Levitin, A.S. Markosyan, V.A. Reimer, E.V. Sinitsyn and Z. Smetana, Sov. Phys. JETP 66 (1987) 866.
[124] D. Gignoux and R. Lemaire, Solid State Commun. 14 (1974) 877.
[125] K. Sato, Y. Yosida, Y. Isikawa and K. Mori, J. Magn. Magn. Mater. 54–57 (1986) 467.
[126] T. Okamoto, H. Fujii, Y. Andoh and H. Fujiwara, J. Magn. Magn. Mater. 52 (1985) 208.
[127] G. Asti, F. Bolzoni, F. Leccabue, R. Panizzieri, L. Pareti and S. Rinaldi, J. Magn. Magn. Mater. 15–18 (1980) 561.
[128] R. Ballou, J. Deportes, B. Kebe and R. Lemaire, J. Magn. Magn. Mater. 54–57 (1986) 494.
[129] G. Asti, in: Ferromagnetic Materials, vol. 5, eds. K.H.J. Buschow and E.P. Wohlfarth (North-Holland, Amsterdam, 1990) p. 397.
[130] A.S. Lagutin and R.F. Druzhinina, J. Magn. Magn. Mater. 90&91 (1990) 85.
[131] S. Sinnema, J.J.M. Franse, A. Menovsky and R.J. Radwański, J. Magn. Magn. Mat. 54–57 (1986) 1639.
[132] J.J.M. Franse, R.J. Radwański and S. Sinnema, J. de Phys. 49 (1988) C8-505.
[133] B.R. Cooper, Phys. Lett. 22 (1966) 244.
[134] K.A. McEwen, G.J. Cock, L.W. Roeland and A.R. Mackintosh, Phys. Rev. Lett. 30 (1973) 287.
[135] E. Leyarovski, J. Mrachkov, A. Gilewski and T. Mydlarz, Phys. Rev. B 35 (1987) 8668.

[136] M. Reiffers, Y.G. Naidyuk, A.G.M. Jansen, P. Wyder, I.K. Yanson, D. Gignoux and D. Schmitt, Phys. Rev. Lett. 62 (1989) 1560.
[137] P. Morin, J. Rouchy and D. Schmitt, Phys. Rev. B 37 (1988) 5401.
[138] J. Mrachkov and E. Leyarovski, Physica B 150 (1988) 404.
[139] P. Morin, J. Rouchy and D. Schmitt, Phys. Rev. B 17 (1978) 3684.
[140] P. Morin and D. Schmitt, Phys. Rev. B 23 (1981) 5936.
[141] C. Jaussaud, P. Morin and D. Schmitt, J. Magn. Magn. Mater. 22 (1980) 98.
[142] P. Morin, D. Schmitt and E. du Trémolet de Lacheisserie, J. Magn. Magn. Mater. 30 (1982) 257.
[143] P. Morin, J. Magn. Magn. Mater. 71 (1988) 151.
[144] G. Raffius and J. Kötzler, Phys. Lett. A 93 (1983) 423.
[145] B. Lüthi, J. Magn. Magn. Mater. 15–18 (1980) 1.
[146] M. Loewenhaupt, B. Frick, U. Walter, E. Holland-Moritz and S. Horn, J. Magn. Magn. Mater. 31–34 (1983) 187.
[147] M. Giraud, P. Morin, J. Rouchy and D. Schmitt, J. Magn. Magn. Mater. 59 (1986) 255.
[148] G. Aubert, D. Gignoux, B. Hennion, B. Michelutti and A. Nait-Saada, Solid State Commun. 37 (1981) 741.
[149] D. Gignoux and F. Givord, J. Magn. Magn. Mater, 31–34 (1983) 217.
[150] M. Giraud, P. Morin and D. Schmitt, J. Magn. Magn. Mater. 52 (1985) 41.
[151] V.M.T.S. Barthem, D. Gignoux, D. Schmitt and G. Creuzet, J. Magn. Magn. Mater. 78 (1989) 56.
[152] R. Aléonard, P. Morin, D. Schmitt and F. Hulliger, J. Phys. F 14 (1984) 2689.
[153] T. Kasuya, Y.S. Kwon, T. Suzuki, K. Ishiyama K. Takegahara, J. Magn. Magn. Mater. 90–91 (1990) 389.
[154] P. Bak, Rep. Prog. Phys. 45 (1982) 587.
[155] J.A. Blanco, D. Gignoux and D. Schmitt, Phys. Rev. B 43 (1991) 13145.

Valence fluctuation and heavy fermion behaviour in rare earth and actinide based compounds

D.T. Adroja

Indian Institute of Technology, Powai, Bombay 400 076, India

and

S.K. Malik

Tata Institute of Fundamental Research, Bombay 400 005, India

A brief historical review of some of the experimental work reported on the mixed valent (MV) or valence fluctuating (VF) and heavy fermion (HF) systems is given. The characteristic physical properties of MV and HF systems are discussed. The salient features of the theoretical models are outlined. Results on some systems are presented.

1. Introduction

The elements starting from lanthanum (atomic number $Z = 57$) and ending at lutetium ($Z = 71$) are known as the lanthanides or the rare earths and are characterized by the progressive filling of their 4f shell. Scandium ($Z = 21$) and yttrium ($Z = 39$) are also included in the category of rare earths [1] because of their similar physical properties. Rare earths form a variety of intermetallic compounds and alloys with transition metals and metalloids [2–10]. Experimental investigations on these intermetallic compounds and alloys in the recent past have revealed that some of the compounds containing Ce, Eu and Yb (and also at times Pr, Sm and Tm) exhibit interesting physical properties/phenomena associated with the 4f electrons, such as magnetic ordering with anomalously high ordering temperature [11–13], Kondo effect [14–15], superconductivity [16–17], heavy fermion behavior [18–24] and mixed valence or valence fluctuation behaviour [25–37] (We shall use mixed valence, valence fluctuation or interconfiguration fluctuations synonymously to mean the same). Among these, the phenomena of heavy fermion (HF) and valence fluctuation (VF) have attracted considerable attention in recent years both experimentally (see refs. [18–37] and other references cited therein) and theoretically [38–48].

In spite of the availability of a large number of experimental data and a variety of theoretical models [38–48] on the VF and HF systems, there is still no single theoretical model which explains all the physical properties of these systems in a satisfactory way. Therefore, this has sustained interest in identifying new VF and HF systems and carrying out systematic studies using different experimental techniques to completely bring out various aspects of these systems. In this paper we review briefly some of the earlier work reported on MV and HF behaviour of rare earth intermetallic compounds. The characteristic physical properties of these systems are discussed.

The salient features of the theoretical models are outlined. Some of the recent results obtained in our laboratory are presented.

2. Historical development

2.1. Mixed valent systems

A valence change under pressure was first observed in cerium metal in the early 1940s [49–52]. Subsequently it was found that on application of pressure, Ce undergoes an isomorphic first order phase transition from γ-Ce to α-Ce accompanied by ~15% reduction in the cell volume [53]. The interpretation of the γ–α phase transition as a valence change was first suggested independently by Zachariasen [50] and Pauling [51]. The relationship between lattice constants and valence indicated a nonintegral valence of 3.67 in α-Ce [54,55]. Further, a valence of 3.63 in α-Ce was estimated from the Hall effect measurements [54].

A first order isostructural semiconductor–metal transition under pressure was observed in SmS by Jayaraman and co-workers [27,28,56–58]. It was found that the magnetic susceptibility in the collapsed high pressure metallic phase saturated to a constant value at low temperatures with no sign of magnetic order, and at high temperatures, it was intermediate between the values expected for Sm^{2+} and Sm^{3+} ions. The constant low temperature susceptibility was in sharp contrast to what one would have expected if the transition had proceeded directly to the trivalent state. The valence change in SmS was also studied by cationic substitution, i.e., $Sm_{1-x}R^{3+}_x S$ (R = rare earth) [27,28]. For a critical concentration of R, there was an abrupt change in the lattice parameter and a change in colour from black to golden yellow. This was attributed to a change in the valence state of Sm ions. Further experimental investigations on $SmS_{1-x}As_x$ showed that the valence transition in SmS could also be induced by anionic substitution [59].

The compound SmB_6 was another Sm compound which exhibited, at atmospheric pressure, the same magnetic anomaly as seen in the metallic phase of SmS. L_3-X-ray absorption measurements revealed a mixture of 30% $4f^6$ and 70% $4f^5$ Sm ions [60], whereas the Mössbauer measurements yielded a single line with isomer shift value in between the values expected for $4f^6$ and $4f^5$ configurations. Further, the isomer shift was found to be temperature independent between 1 and 1000 K [61].

The coexistence of VF and antiferromagnetic ordering was observed in TmSe [40,62], YbP [63] and Yb_3Pd_4 [64] compounds. The occurrence of a localized moment for TmSe in the entire temperature range was established [40] by magnetic susceptibility measurements. The coexistence of VF and superconductivity was also observed in $CeRu_3B_2$, $CeOs_3B_2$ [17,65], $CeRu_3Si_2$ [66,67], $CeRu_2$ [68,69] and $CeCo_2$ [30]. Evolution of magnetism from mixed valent system is observed in going from mixed valent $CeIr_3B_2$ to trivalent $CeIr_3Si_2$ [70]. Dehybridization of the 4f shell has been observed [71] in Si substituted $CeRh_3B_2$. The compound $CeRh_3B_2$ orders ferromagnetically with T_C of 115 K which is much larger than the T_C of 105 K of isostructural $GdRh_3B_2$ in spite of the fact that the saturation moment in Ce compound is only $0.4\mu_B$ compared to nearly $7\mu_B$ in the Gd compound. Hybridization of the Ce-4f electrons with the conduction electrons is thought to be responsible for the anomalously high T_C of the Ce compound. As boron is partially replaced by Si, the Ce moment in $CeRh_3B_{2-x}Si_x$ increases but the T_C decreases very rapidly. It has been suggested that dehybridization of the Ce-4f shell takes place on Si substitution which results in the decrease in T_C, inspite of an increase in the Ce moment.

[151]Eu Mössbauer measurements by Bauminger et al. [72] (1973) showed that Eu ion in $EuCu_2Si_2$ was in the intermediate valence state. This was perhaps the first Eu-based compound identified as an MV system. Later on, isostructural $YbCu_2Si_2$ was also found to be an MV system [73,74]. A sharp valence transition from 2.2 to 2.9 was observed around 15 K in the case of $EuPd_2Si_2$ [75]. From the studies on the $La_{1-x}Eu_xPd_2Si_2$ system, Croft et al. [76] showed that the sharp valence transition in $EuPd_2Si_2$ is due to the cooperative effect between Eu–Eu atoms. Stud-

ies on many dilute systems such as $Y_{0.97}Eu_{0.03}Cu_2Si_2$ [77] and $Sc_{0.97}Eu_{0.03}Al_2$ [78] showed that VF is a single ion property.

Studies of magnetic susceptibility, electrical resistivity (ρ), L_3-XANES and ^{151}Eu Mössbauer (in case of Eu compounds) showed that $CeIr_2Si_2$ [79] and $EuIr_2Si_2$ [80,81] exhibit VF behaviour. The low temperature resistivity (ρ) of $CeIr_2Si_2$ and $EuIr_2Si_2$ exhibits T^2 and T^3 dependence, respectively. The T^3 dependence of the resistivity has been also observed in $EuCu_2Si_2$ and $YbCu_2Si_2$ [73] compounds. From a detailed analysis of the $\rho(T)$ data of a large number of VF systems, Wohlleben and Wittershagen [82] introduced the dynamic alloy model. They have shown that the formalism of their model can reproduce, at least semi-quantitatively, a variety of resistivity anomalies.

A survey of the available data reveals that a large number of Ce, Sm, Eu and Yb based compounds exhibit MV behaviour. Some of these compounds are listed in table 1.

The MV systems show anomalies in their physical properties. Generally an indication of the MV behaviour is provided by the following characteristic features:

- Anomalies in the lattice constant or unit cell volume of the compound concerned are usually the first indication of the MV behaviour. The lattice constants in a series of isostructural rare earth (R) intermetallic compounds in which the R-ion is in the stable 3+ state, decrease smoothly with increasing atomic number Z of the R ion. This is known as the lanthanide contraction [83]. The compounds in which the R-ion is in the pure 2+ or 4+ state show deviation from the lanthanide contraction. The lattice parameters of the MV compounds deviate from lanthanide contraction and lie intermediate between integral valence states.

Table 1

List of Ce, Sm, Eu and Yb based compounds which exhibit MV behaviour at ambient conditions, on application of high pressure and on alloying

Ambient conditions	High pressure	Alloying
$CePd_3$, $CeSn_3$, Ce_3Al	Ce (8 kbar)	$Ce(Rh_{1-x}Pd_x)_3$
CeN, $CeBe_{13}$, CeB_4, $CeSi_2$	SmS (6.5 kbar)	$Ce_{1-x}Y_x$
CeM_2 (M = Fe, Co, Ru, Rh, Ir)	SmSe (30 kbar)	$CePt_{2-x}Ir_x$
CeNiSi, CeNiSn, CeNiIn	SmTe (50 kbar)	$CePt_{2-x}Rh_x$
CeRhIn, CeRhSb, CeIrGe	TmTe (20 kbar)	$CeLa_{1-x}Th_x$
$CeIr_2Si_2$, $CeCo_2Si_2$, $CeNi_2Si_2$	CeP (100 kbar)	$Ce_{1-x}Th_x$
$CeFe_2Si_2$, $CeOs_2Si_2$, $CePt_2Si_2$	EuO	$CePt_{1-x}Ni_xSi$
$CeNi_2Ge_2$, $CeCo_2Ge_2$	YbS, YBSe, YbTe	$Sm_{1-x}Th_y^{4+}S$
$Ce_{0.5}Rh_3B_2$, $CeIr_3B_2$		$Sm_{1-x}RE_xS$
$CeOs_3B_2$, $CeRu_3B_2$, $CeRu_3Si_2$		$SmS_{1-x}As_x$
$Ce_{1-x}Y_xIn_3$ ($x > 0.6$)		$EuPd_3Si_x$
SmB_6, Sm_4Bi_3		$EuPd_3B_x$
$EuIr_2$, $EuRh_2$, $EuNiSi_2$, EuPtP		$Eu_{1-x}La_xRh_2$
$EuCu_2Si_2$, $EuPd_2Si_2$, $EuIr_2Si_2$		$YbAu_{1-x}Ag_x$
$EuNi_2P_2$, $EuFe_4Al_8$, $Eu_2Ni_3Si_5$		
Eu_4As_3		
YbB_6, YbCu, YbAg, YbAu, YbZn		
YbB_4, YbC_2, $YbAl_3$, $YbAl_2$, YbSi		
$YbCu_{3.5}$, $YbCu_{4.5}$, YbB_{12}, Yb_4Bi_3		
Yb_4Sb_3, $YbIr_3$, YbCuAl, YbCuGa		
$YbAlB_4$, $YbInCu_4$, $YbAgCu_4$		
YbInAu, $YbInAu_2$, YbPdIn		
$YbNi_2Ge_2$, $YbPd_2Si_2$, $YbCu_2Si_2$		
$YbZn_2Si_2$, $YbZn_2As_2$		

- The presence of a prominent absorption peak (called white line) at the L_3-edge of R ion is of considerable interest in the study of valence state. The existence of a two-peak pattern (two white lines separated by ~ 7 to 10 eV) is an indication of MV behaviour of the R ion [84]. From the ratio of the intensities of the two peaks (deconvoluted), the average valence of R ion can be estimated. In the case of MV system, a replicate splitting for the 3d (or 4d) core levels is observed in X-ray photoelectron spectra (XPS). The values of replicate splittings for the 3d (~ 7 to 10 eV) and 4d (~ 8 eV) are used as guidelines for the identification of MV systems. Since, the VF characteristic time (10^{-13} s) is very small, the experimental response to this phenomenon is highly dependent on the probing time of the technique used. L_3-edge and XPS with characteristic probing times of ~ 10^{-16} s, which are much smaller than the VF time, provide simultaneously the signatures of the two different configurations, $4f^n$ and $4f^{n-1}$. These two techniques have, therefore, been extensively used for the study of MV compounds.
- The Mössbauer spectrum of an MV compound shows a single line with isomer shift intermediate between values for the two integral valence states. The observation of a single Mössbauer line suggests that the valence fluctuations occur on a time scale (10^{-13} s) faster than the Mössbauer probe time which is typically 10^{-11} s.
- Magnetic susceptibility (χ) of MV systems tends to a constant value at low temperatures. Most of the Ce- and Yb-based MV compounds show a maximum in $\chi(T)$ at some temperature. In the high temperature range, $\chi(T)$ of MV compounds follow Curie–Weiss behaviour with an effective magnetic moment (μ_{eff}) intermediate between the values of μ_{eff} for the two integral valence states.
- The electronic contribution to specific heat (γ_{ele}) of the MV systems is several times higher than those of normal metals.
- The electrical resistivity (ρ) of MV compounds shows a deviation from linear T behaviour at high temperatures. Ce-based MV compounds

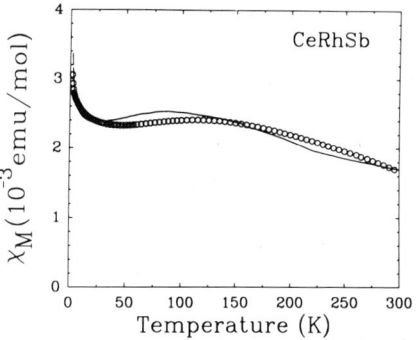

Fig. 1. Temperature dependence of the magnetic susceptibility of CeRhSb (from ref. [90]). Solid line is a fit to the data based on Coqblin–Schrieffer model.

show T^2 dependence of resistivity at low temperatures while $\rho(T)$ for some of the Eu- and Yb-based compounds exhibits T^3 dependence.

Recently, it has been shown by various experiments that the equiatomic ternary compounds CeNiSn [85,86], CeNiIn [87], CeRhIn [88], CeIrGe [89], CeRhSb [90] and YbCuGa [91] exhibit VF behaviour. A typical plot of susceptibility as a function of temperature is shown in fig. 1 for CeRhSb. The resistivity of CeNiSn and CeRhSb shows pseudo-gap formation in the electronic density of states at E_F while that of CeNiIn exhibits Kondo type behaviour. Fig. 2 shows the temperature dependence of the electrical resistivity of CeRhSb and LaRhSb between 4.2 and 300 K. The resistivity of CeRhSb increases as temperature is lowered from 300 K, shows a maximum at about 113 K and then decreases with decreasing

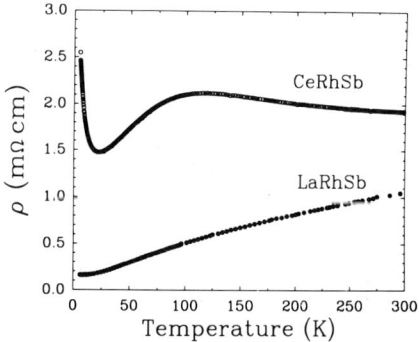

Fig. 2. Temperature dependence of the electrical resistivity (ρ) of CeRhSb and LaRhSb (from ref. [90]).

temperature. The resistivity shows a minimum at 21 K below which it rises rapidly. This rise continues down to very low temperatures. On the other hand, the resistivity of LaRhSb shows the usual metallic behaviour throughout the temperature range investigated. The rapid rise in the resistivity of CeRhSb may be interpreted to imply a gap formation in the electronic density of states. The gap energy is estimated from the plot of $\ln(\rho)$ vs. T^{-1}, shown in fig. 3. A linear relation is observed between the two in the temperature range of 5 K to about 17 K from which a gap energy of ~ 4 K is inferred.

The ternary compound CePdSb also exhibits unusual properties [92]. It orders ferromagnetically with T_M of 17 K while isostructural GdPdSb is ordered antiferromagnetically with T_N of 15.5 K. Therefore, not only is the ordering different in the two compounds, the ordering temperature of CePdSb is also anomalously high compared to that of GdPdSb when scaled by the De Gennes factor. Moreover, the resistivity of CePdSb goes through a broad maximum at about 150 K (fig. 4) which is a characteristic feature of dense Kondo systems. Thus, CePdSb may represent a ferromagnetic Kondo lattice system.

2.2. Heavy fermion systems

The phenomenon of HF was first observed in CeAl$_3$ by Andres et al. [93] in 1975. The discovery of HF superconductivity in CeCu$_2$Si$_2$ by Steglich et al. [16] in 1979 enhanced the activity in this field. This was followed by the discovery of

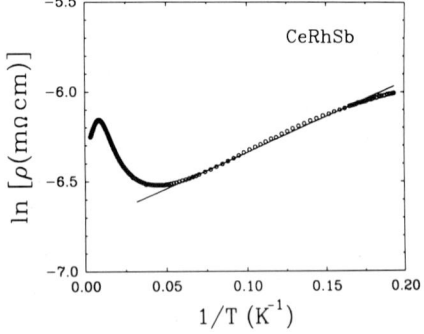

Fig. 3. Plot of log(ρ) versus inverse temperature for CeRhSb (from ref. [90]).

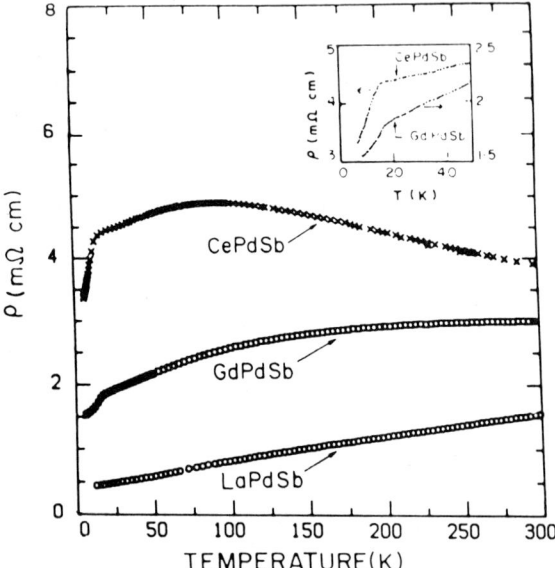

Fig. 4. Temperature dependence of the electrical resistivity (ρ) of CePdSb, GdPdSb and LaPdSb (from ref. [92]).

other HF superconductors, UBe$_{13}$ [94], UPt$_3$ [95] and URu$_2$Si$_2$ [96–98]. URu$_2$Si$_2$ was the first heavy fermion system in which a magnetically ordered state (T_N = 17.5 K) and superconductivity (T_c = 1.5 K) were found to coexist [96–98]. Later on, it was found that CePb$_3$, an HF antiferromagnet with $T_N \approx 1.1$ K, became an HF superconductor in an applied magnetic field of 15 T at 0.2 K [99]. The studies on the ternary equiatomic compounds of Ce showed that CePtSi [100,101] and CePtIn [88] are nonmagnetic HF systems, while CePdIn [102] is HF with antiferromagnetic ground state. The magnetic susceptibility and resistivity measurements on a polycrystalline CePt$_{1-x}$Ni$_x$Si (x = 0.01 to 1.0) showed the evolution of mixed-valence state from the heavy fermion state [103]. Recent NMR measurements at low magnetic fields have provided evidence for the existence of antiferromagnetic ordering in CeAl$_3$ below 1.2 K [104].

Like VF systems, the physical properties of the HF compounds are also anomalous. These are characterized by extremely large values of γ_{ele} (up to 10^3 times that of a normal metal) indicating the presence of electrons with very large effective mass at E_F. The susceptibility of HF compounds shows Curie–Weiss behaviour with μ_{eff} close to that of the 3+ R-ion at high temperatures, but exhibits an enhanced Pauli paramagnetic susceptibility at low temperatures. The low temperature resistivity shows T^2 behaviour which is similar to that observed in the case of MV system. The resistivity of some of the HF compounds increases as the temperature is lowered, passing through a broad maximum (Kondo-like behaviour), and eventually showing a sharp drop at very low temperature. This sharp drop in the resistivity is due to a transition from an incoherent to a coherent scattering of conduction electrons by the localized moments. The mechanism responsible for the coherent scattering remains still unclear, although both RKKY and quadrupolar exchange interactions have been suggested as possible sources [105].

Several HF systems have been discovered. These are collected in table 2. According to their ground state properties, HF systems are classified mainly into three categories; (i) nonmagnetic or

Table 2
Partial list of Ce, Yb and U based heavy fermion systems

Ground state	Compounds	$\gamma_{\mathrm{ele}}(0)$ (mJ/mol K^2)	T_{N} (K)	T_{c} (K)	Ref.
para-magnetic	CeCu$_6$	1600			[107]
	CePtSi	800			[100,101]
	CePtIn	700			[87]
	CeRuSn$_3$	1400 (0.6 K)			[108]
	CeInPt$_4$	2500			[109]
	CeRh$_2$Pt$_3$	149			[110]
	CePtSi$_2$	1700 (1.25 K)			[121]
	CeRu$_2$Si$_2$	350			[111,112]
	CeCu$_4$Al	2000 (<1 K)			[113]
	CeCu$_3$Al$_2$	540 (1.6 K)			[114]
	CeCu$_4$Ga	1900			[115–118]
	CeCu$_3$Ga$_2$	730 (1.45 K)			[119]
	YbCuAl	260			[120]
magnetic	CeAl$_3$	1620	<1.2 [a]		[94,104]
	CeB$_6$	300	2.3 (3)		[122,123]
	CePb$_3$	280, 1000 [b]	1.1		[99,124]
	CePdIn	330	1.8		[102]
	CeInCu$_2$	1200	1.6		[125]
			1.1		
	CePt$_2$Sn$_2$	3700	0.88		[106]
	YbPdCu$_4$	200	0.8		[126]
	U$_2$Zn$_{17}$ [d]	198, 535 [c]	9.7		[127]
	UCd$_{11}$	250, 840 [c]	5.0		[18]
	UAgCu$_4$	310	18.15		[128]
super-conducting	CeCu$_2$Si$_2$	830		0.53	[16,18]
		1270		0.64	
	UBe$_{13}$	1100		0.97	[94]
	UPt$_3$	450	5.0	0.54	[95]
	URu$_2$Si$_2$	65.5 [e]	17.5	1.5	[129,130]
	U$_2$PtC$_2$	75 [e]		1.47	[131,132]

[a] From NMR measurements.
[b] Value in an applied field of 11 T.
[c] Value above the magnetic ordering temperature.
[d] The heavy fermion nature in this compound has been questioned by other workers (see note added to ref. [127]).
[e] These values are somewhat small to qualify them as heavy fermion materials.

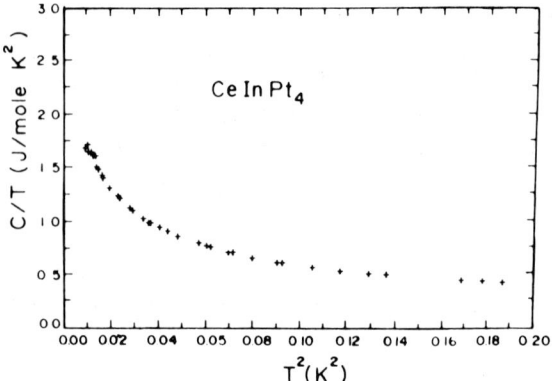

Fig. 5. Plot of C/T vs. T^2 for CeInPt$_4$ (from ref. [109]).

paramagnetic, (ii) magnetic and (iii) superconducting. Compounds belonging to the nonmagnetic category show pure Kondo lattice character without magnetic ordering down to low temperatures, as observed, e.g., in CePtSi and CePtIn. Some of the HF compounds exhibit magnetic ordering (mainly antiferromagnetic) at low temperatures as observed, e.g., in CePdIn [102] and CePt$_2$Sn$_2$ [106]. Since the magnetic ordering is established despite competition with the Kondo state, the observed ordering is anomalous (compared with the ordinary magnetic ordering). The compound CeCu$_2$Si$_2$ shows superconductivity ($T_c \approx 0.6$ K, sample dependent) and is the first HF superconductor.

Recently, magnetic susceptibility, electrical resistivity and low temperature specific heat measurements have been carried out on a new ternary cubic compound, CeInPt$_4$, which indicate that this represents a heavy fermion system [109] with effective mass among the highest reported. In the high temperature limit its susceptibility conforms to that of Ce^{3+} ions with effective magnetic moment of about $2.54\mu_B$ but with a rather large paramagnetic Curie temperature of -225 K. Considerable deviation from Curie–Weiss behavior is observed in the susceptibility below 100 K which is attributed to the effect of crystalline electric fields. The specific heat (C) shows an upturn in C/T vs. T^2 plot below 2 K (fig. 5). The electronic specific heat coefficient γ increases with decreasing temperatures and attains a value of 1750 mJ mol^{-1} K^2 at 0.1 K. As $T \to 0$ K, γ approaches a value of 2500 mJ mol^{-1} K^2.

3. Theoretical models on VF and HF phenomena

Several theoretical models have been proposed [38–48] to explain the VF and HF phenomena. Hirst [38,133] proposed a simple configuration-based approach to explain the VF behaviour. According to this model, each local shell has a discrete spectrum of many electron levels characterized by a different integral occupation number n. The energy, $E(n)$ of a 4f^n ionic configuration is given by

$$E(n) = \tfrac{1}{2} U_{ff} n(n-1) + Vn, \qquad (1)$$

where U_{ff} is the Coulomb repulsion between two localized 4f electrons, n is the integral number of localized electrons and V is the nuclear potential. The above expression can be rewritten as

$$E(n) = \tfrac{1}{2}(n - n_{\min})^2 + U_{ff} + \text{const}, \qquad (2)$$

where

$$n_{\min} = -V/U_{ff} + \tfrac{1}{2}.$$

Thus, we see that $E(n)$ has a parabolic dependence on n (fig. 6). The value of n at the minimum

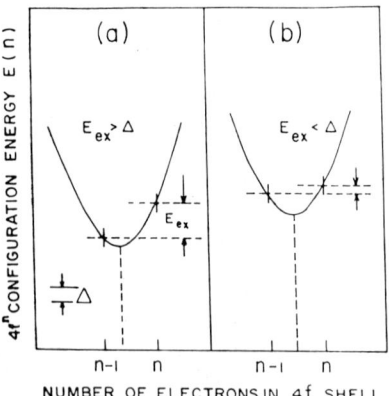

Fig. 6. Energy levels in the ionic model as a function of 4f occupation n: (a) The configurational stability of the 4f shell, as observed in normal rare earth intermetallics. (b) This represents the situation when interconfiguration fluctuations between 4f^n and 4f^{n-1} take place.

of the parabola is represented by n_{min}. The excitation energy, E_{ex}, is the energy difference between the ground state energy $E(n)$ and the first excited state energy $E(n-1)$; $E_{ex} = E(n) - E(n-1)$. The value of E_{ex} varies between zero for half integral n_{min} to $U_{ff}/2$ for integral value of n_{min}. In a metallic environment, the rare earth ions are surrounded by conduction electrons. The effect of mixing between the conduction electrons and the local 4f electrons is to broaden the 4f level. In the Friedel–Anderson theory [134,135], the width Δ of the 4f level is given by

$$\Delta = \pi \langle |V_{kf}|^2 \rangle \rho(E_f), \quad (3)$$

where V_{kf} is the matrix element of the mixing interaction between the conduction electrons and the local 4f electrons and $\rho(E_f)$ is the conduction electron density of states at the energy of the local level E_f. Under normal circumstances i.e., when $\Delta \ll E_{ex}$, the valence state of the rare earth is integral. If $j \neq 0$ for the rare earth ion, one can get magnetic ordering at low temperatures. It may so happen that for some value of U_{ff} and V, $E_{ex} \leq \Delta$. Under these conditions the mixing interaction can cause fluctuation between two 4f configurations $4f^n$ and $4f^{n-1}$ resulting in valence fluctuations even at $T \to 0$. The transitions from $4f^n$ to $4f^{n-1}$ give rise to an electron in the conduction band at E_F. This results in a very high density of states at E_F and accounts for the large electronic contribution to specific heat.

The mixed configuration in the ground state is understood in terms of the hybridization of the two configuration $4f^n(6s5d)^m$ and $4f^{n-1}(6s5d)^{m+1}$. The hybrid wave function is schematically given by

$$|\psi\rangle = a_n |4f^n(6s5d)^m\rangle$$
$$+ a_{n-1} |4f^{n-1}(6s5d)^{m+1}\rangle,$$
$$|a_n|^2 \neq 0 \quad \text{and} \quad |a_{n-1}|^2 \neq 0, \quad (4)$$

where $(6s5d)^m$ refers to the conduction band. Thus this process gives rise to a fluctuation between two valence states and the fluctuation time is of the order of 10^{-13} s. Each fluctuating valence system is associated with an intrinsic fluctuation temperature, $T_{sf} \sim \Delta/k_B$ and the fluctuation frequency is given by $\omega = k_B T_{sf}/\hbar$ where k_B is the Boltzman constant.

On the basis of the Hirst model, Sales and Wohlleben [39] gave a phenomenological Inter-Configurational Fluctuation (ICF) model to explain the susceptibility behaviour of the VF compounds. This model gives a good quantitative fit to the observed $\chi(T)$ results, for instance, on $CeIr_2Si_2$, $YbCu_2Si_2$, $YbAl_3$ and YbB_4 compounds [39,79]. Later, Sales and Vishawanathan [73] included the crystalline electric field (CEF) effect in the ICF model. This gives a qualitatively good fit to $\chi(T)$ of the HF compound, $CeCu_2Si_2$. In another approach, the Anderson's single impurity model [135] has been extended to VF phenomenon. Based on this model, Verma and Yafet [136] calculated $\chi(T)$ of VF systems using the variational method. This gives a good quantitative fit to the observed susceptibility of VF compounds, SmS, $YbAl_3$ and $CeAl_3$. Further work by Verma has shown that the ground state of most of the Ce and Yb based VF systems is the metallic Fermi liquid state [25]. However, some peculiar properties of anomalous rare earth compounds have not been explained within the single impurity model. The single impurity model has been generalized for rare earth atoms regularly spaced on a lattice site. The generalized model is called the periodic Anderson model. The results of alloy analog approximation for temperature dependent static susceptibility and specific heat show qualitatively all the features typical of VF systems [43]. Although this reproduces the overall resistivity behaviour (as well as thermo-electric power) in the case of a hybridization gap at E_F, it explains only the high temperature behaviour of ρ of the metallic VF systems.

Coqblin and Schrieffer [137] (C–S) have proposed a model to explain the Kondo effect in the case of $J \geq 1/2$. This model is derived from the degenerate Anderson model by means of a canonical transformation, known as the Schrieffer–Wolf transformation [138]. Following Coqblin and Schrieffer, Rajan has calculated the magnetic susceptibility, $\chi(T)$ and specific heat, $C(T)$ in zero magnetic field for angular momentum values form $J = 1/2$ to $7/2$ by solving the exact thermodynamic equation of the C–S model with the

Bethe ansatz [139]. The χ and C as $T \to 0$ are given by the following equations:

$$\chi(0) = \frac{\nu(\nu^2 - 1)(\mu_B g_J)^2}{24\pi k_B T_0}, \quad (5)$$

$$\frac{C(T \to 0)}{T} = \gamma_{\text{ele}} = \frac{(\nu - 1)\pi k_B}{6T_0}, \quad (6)$$

so that the following relationship is found for the low-temperature linear coefficient of specific heat, γ_{ele} and susceptibility, $\chi(0)$, at zero temperature;

$$\frac{\chi(0)}{\gamma_{\text{ele}}} = \frac{g_J^2 \mu_B^2 (J+1)}{\pi^2 k_B^2}, \quad (7)$$

where $\nu = 2J + 1$. The characteristic temperature T_0 is related to the Kondo temperature through the Wilson number, W, $T_K/T_0 = W$ [141]. These results are the same as those obtained from a Fermi liquid model with no interaction, except for the enhancement factor, $\nu/(\nu - 1)$.

Rajan has numerically solved a set of coupled nonlinear integral equations obtained from the Bethe ansatz solution of the C–S model to derive the free energy at an arbitrary temperature. The $\chi(T)$ and $C(T)$ at an arbitrary temperature were obtained by differentiating the free energy. This exact $\chi(T)$ shows that, at low temperatures, the effective impurity moment goes to zero, indicating the complete screening of the impurity spin by the electron gas. For $J > 1$, $\chi(T)$ exhibits a peak, which becomes more and more pronounced as J increases. Such peaks have been observed experimentally for Ce ($J = 5/2$) and Yb ($J = 7/2$) impurities in metals [30,140]. Fig. 7 shows the susceptibility of YbAgCu$_4$ which exhibits such a peak [140].

Ramakrishanan and Sur [141–143] have calculated the physical properties, such as valence, susceptibility, specific heat, etc., for an MV system using perturbative theory. Effect of alloying has also been explicitly taken into account. Their theory could give satisfactory explanation of, for the first time, most of the properties of MV systems such as the temperature dependence of valence, the positive T^2 slope of the low-temperature $\chi(T)$, a broad maximum in the $\chi(T)$, etc.

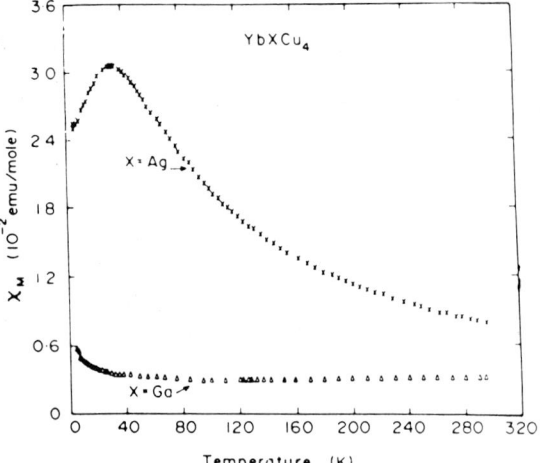

Fig. 7. Magnetic susceptibility of YBAgCu$_4$ as a function of temperature (from ref. [140]).

Kaplan and Mahanti [144] proposed an essentially localized d electron model to explain the various properties of Sm mixed valent compounds. This model is fundamentally different from that of the Anderson model in the sense that most of the d states involved in the model are localized. There are certain experimental observations to support the localization nature of d electrons. For example, in SmB$_6$ at sufficiently low temperature, the 5d electron of the $4f^5 5d$ configuration does not contribute to the conductivity [30]. Further, this model differs from others in the nature of the mixing interaction. In the ionic model, the mixing is a one electron process which removes one electron from the 4f shell and places it in the conduction band and vice versa. In the localized electron model, the localized f and d electronic states are allowed to mix via an inter-electronic dipole interaction. In the absence of the dipole interaction, the two electrons per site would combine in the localized orbitals to produce a singlet ground state in accordance with the Hund's rule. This model has been treated variationally by Kaplan et al. [145,146] and they find a mixed valence state for large dipole interactions. Later, they extended this model to include the effect of the lattice [147]. The localized d electron model is successful in predicting the phase diagram of Sm monochalcogenides and valence stability to an integral valence state at high pressure. Although, this model explains several observed features of MV compounds, it has been criticized on the question of two-electron hybridization.

The Fermi liquid model was first proposed by Varma [25] and later by Beal-Monod and Lawrence [148]. The support to the Fermi liquid theory comes from the experimental observation of (i) a finite susceptibility $\chi(0)$ at $T=0$ without any evidence for magnetic ordering and (ii) a large value of μ_{ele} of the VF systems. At low temperature $\chi(T)$ in this model is given by

$$\chi(T) = \chi(0)\left[1 + a\left(\frac{T}{T_{sf}}\right)^2\right]. \quad (8)$$

The parameters involved in the coefficient of T^2 in the above expression depend on the treatment of the Fermi liquid interaction [16]. In the paramagnon theory of Lawrence and Beal-Monod, which involves nearly ferromagnetic interaction, a depends on the derivative of the density of states at E_F and hence it can take either sign. This model predicts that in all the VF systems, one should, in principle, observe a maximum in χ at some temperature (as $\chi(T) \propto T^2$ at low temperatures) and $\chi(T) \propto 1/(T-\Theta)$ at high temperatures and a finite value of χ at $T=0$.

Another Fermi liquid approach to the VF has been given by Newns and Hewson [41]. It is based on the earlier observation of Hewson that the magnetic scattering leads to a loss of coherence in the excitation spectrum. Thus they were led to a description of the VF state as a local resonance and showed that the properties of many experimentally observed properties followed directly.

References

[1] J.A. Christiansen, Am. Chem. Soc. 82 (1960) 5525.
[2] K.N.R. Taylor, Adv. Phys. 20 (1971) 551.
[3] W.E. Wallace, Prog. Rare Earth Science and Technology, vol. III (1971) p. 1.
[4] W.E. Wallace, Rare Earth Intermetallics (Academic Press, New York, 1973).
[5] K.H.J. Buschow, Rep. Prog. Phys. 40 (1977) 1179.
[6] K.H.J. Buschow, Rep. Prog. Phys. 42 (1979) 1373.
[7] K.A. Gschneidner Jr., J. Less-Common Met. 100 (1984) 1.
[8] A. Iandelli, in: Handbook on the Physics and Chemistry of Rare Earths, vol. 2, eds. K.A. Gschneidner, Jr. and L. Eyring (North-Holland, Amsterdam, 1979) p. 1.
[9] H.C. Ku and G.P. Meisner, J. Less-Common Met. 78 (1981) 99.
[10] E. Hovestreydt, N. Engel, K. Kleep, B. Chabot and E. Parthe, J. Less-Common Met. 85 (1982) 247.
[11] S.K. Dhar, S.K. Malik and R. Vijayaraghavan, J. Phys. C 14 (1981) 321.
[12] V. Murgai, S. Raaen, L.C. Gupta and R.D. Parks, in: Valence Instabilities, eds. P. Wachter and H. Boppart (North-Holland, Amsterdam, 1982) p. 537.
[13] B.H. Grier, J.M. Lawrence, V. Murgai and R.D. Parks, Phys. Rev. B 29 (1984) 2644.
[14] M.B. Maple, L.E. Delong and B.C. Sales, in: Handbook on the Physics and Chemistry of Rare Earths, vol. 1, eds. K.A. Gschneidner Jr. and L. Eyring (North-Holland, Amsterdam, 1978) p. 797.
[15] H. Kadomatsu, H. Tanaka, M. Kurisu and H. Fujiwara, Phys. Rev. B 33 (1986) 4799.
[16] F. Steglich, J. Aarts, C.D. Bredl, W. Lieke, D. Meschede, W. Franz and J. Schafer, Phys. Rev. Lett. 43 (1979) 1892.
[17] K.S. Athreya, L.S. Hausermann-Berg, R.N. Shelton, S.K. Malik, A.M. Umarji and G.K. Shenoy, Phys. Lett. A 113 (1985) 330.
[18] G.R. Stewart, Rev. Mod. Phys. 56 (1984) 755.
[19] N.B. Brandt and V.V. Moshchalkov, Adv. Phys. 33 (1984) 373.
[20] F. Steglich, Springer Series in Solid-State Sciences 62 (1985) 23.
[21] H.R. Ott, Helv. Phys. Acta 60 (1987) 62.
[22] B.R. Coles, Contemp. Phys. 28 (1987) 143.
[23] U. Rauchschwalbe, Physica B 147 (1987) 1.
[24] F. Steglich, C.D. Bredl, F.R. de Boer, M. Lang, U. Rauchschwalbe, H. Rietschel, R. Schefzyk, G. Sparn and G.R. Stewart, Phys. Scripta T19 (1987) 253.
[25] C.M. Verma, Rev. Mod. Phys. 48 (1976) 219.
[26] Valence Instabilities and Related Narrow Band Phenomenon, ed. R.D. Parks (Plenum Press, New York, 1977).
[27] A. Jayaraman, in: Handbook on the Physics and Chemistry of Rare Earths, vol. 2, eds. K.A. Gschneidner, Jr. and L. Eyring (North-Holland, Amsterdam, 1979) p. 575.
[28] J.M. Robinson, Phys. Rep 51 (1979) 1.
[29] A.C. Hewson, J. Magn. Magn. Mater. 12 (1979) 83.
[30] J.M. Lawrence, P.S. Risenborough and R.D. Parks, Rep. Prog. Phys. 48 (1981) 1.
[31] Valence Fluctuations in Solids, eds. L.M. Falicov, W. Hanke and M.B. Maple (North-Holland, Amsterdam, 1981).
[32] Valence Instabilities, eds. P. Wachter and H. Boppart (North-Holland, Amsterdam, 1982).

[33] Proc. Intern. Conf. on Valence Fluctuations, Cologne, eds. E. Müller-Hartmann, B. Roden and D. Wohlleben, J. Magn. Magn. Mater. 47&48 (1985).
[34] B.D. Padalia and B. Darshan, Indian J. Phys. 60B (1986) 128.
[35] Theoretical and Experimental Aspects of Valence Fluctuations and Heavy Fermions, eds. L.C. Gupta and S.K. Malik (Plenum Press, New York and London, 1987).
[36] S.N. Vaidya, S.K. Sikka and W.B. Holzapfel, ref. [32], p. 617.
[37] M.N. Nyayage, S.H. Devare, S.K. Malik, D.T. Adroja and H.G. Devare, Phys. Lett. A 151 (1990) 547.
[38] L.L. Hirst, Phys. Kondens. Mater. 11 (1970) 255.
[39] B.C. Sales and D. Wohlleben, Phys. Rev. Lett. 35 (1975) 1240.
[40] B. Coqblin, A.K. Bhattacharjee, R. Jullien and J. Flouquet, J. de Phys. 6 (1980) C5-297.
[41] D.M. Newns and A.C. Hewson, J. Phys. F 10 (1980) 2429.
[42] A.M. Tsvelick and P.B. Wiegmann, Adv. Phys. 32 (1984) 453.
[43] G. Czycholl, Phys. Rep. 143 (1986) 277.
[44] D.M. Newns and N. Read, Adv. Phys. 36 (1987) 799.
[45] T.M. Rice, Phys. Scripta T19 (1987) 246.
[46] P. Schlottmann, Phys. Rep. 181 (1989) 1.
[47] S.M.M. Evans, T. Chung and G.A. Gehring, J. Phys. Condens. Matter 1 (1989) 10473.
[48] S.M.M. Evans and G.A. Gehring, J. Phys. Condens. Matter 1 (1989) 10487.
[49] P.W. Bridgman, Proc. Am. Acad. Arts Sci. 62 (1927) 207.
[50] W.H. Zacharisen, cited in A.W. Lawson and T. Tang, Phys. Rev. 76 (1949) 301.
[51] L. Pauling, cited in A.F. Schuck and J.H. Sturdivant, J. Chem. Phys. 18 (1950) 145.
[52] K.A. Gschneidner Jr., Rare Earth Alloys (Van Nostrand, Princeton, 1961).
[53] D.C. Koskenamaki and K.A. Gschneidner, Jr., in: Handbook on the Physics and Chemistry of Rare Earths, vol. 2, eds. K.A. Gschneidner, Jr. and L. Eyring (North-Holland, Amsterdam, 1979) p. 797.
[54] K.A. Gschneidner, Jr. and R. Smoluckowski, J. Less-Common. Met. 5 (1963) 374.
[55] E. Franceshi and G. Olscese, Phys. Rev. Lett. 22 (1969) 1299.
[56] A. Jayaraman, V. Narayanamurti, E. Bucher and R.G. Maines, Phys. Rev. Lett. 25 (1970) 368.
[57] E. Bucher, V. Narayanamurti and A. Jayaraman, J. Appl. Phys. 42 (1971) 1741.
[58] A. Jayaraman, A.K. Singh, A. Chatterjee and S. Usha Devi, Phys. Rev. B 9 (1974) 2513.
[59] F. Holtzberg, O. Pena, T. Penney and R. Tournier, ref. [26], p. 507.
[60] E.E. Vainshtein, S.M. Blokhin and Yu.B. Paderno, Sov. Phys. Solid State 6 (1965) 2318.
[61] R.L. Cohen, M. Eibschutz and K.W. West, Phys. Rev. Lett. 24 (1970) 383.

[62] H.R. Ott, K. Andres and E. Bucher, AIP Conf. Proc. 24 (1974) 40.
[63] R. Pot, W. Boksch, G. Leson, B. Politt, H. Schmidt, A. Freimuth, K. Keulerz, J. Langen, G. Neumann, F. Oster, J. Rohler, U. Walter, P. Weidner and D. Wohlleben, Phys. Rev. Lett. 54 (1985) 481.
[64] B. Politt, D. Durkop and P. Weidner, ref. [33], p. 583.
[65] S.K. Malik, A.M. Umerji, G.K. Shenoy and M.E. Reeves, J. Magn. Magn. Mater. 54–57 (1986) 439.
[66] H. Berz, Mater. Res. Bull. 15 (1980) 1489.
[67] U. Rauchschwalbe, W. Lieke, F. Steglich, C. Godart, L.C. Gupta and R.D. Parks, Phys. Rev. B 30 (1984) 444.
[68] W. Schmitt and G. Güntherodt, ref. [33], p. 542.
[69] D. Wohlleben and J. Rohler, J. Appl. Phys. 55 (1984) 1904.
[70] A.M. Umarji, S.K. Dhar, S.K. Malik and R. Vijayaraghavan, Phys. Rev. B 36 (1987) 8929.
[71] S.K. Malik, G.K. Shenoy, S.K. Dhar, P.L. Paulose and R. Vijayaraghavan, Phys. Rev. B 34 (1986) 8196.
[72] E.R. Bauminger, D. Froindlich, I. Nowick, S. Ofer, I. Felner and I. Mayer, Phys. Rev. Lett. 30 (1973) 1053.
[73] B.C. Sales and R. Viswanathan, J. Low Temp. Phys. 23 (1976) 449.
[74] T.K. Hatwar, Ph.D. Thesis, I.I.T. Bombay, India (1981) unpublished.
[75] E.V. Sampathkumaran, L.C. Gupta, R. Vijayaraghavan, K.V. Gopalkrishnan, R.G. Pillay and H.G. Devare, J. Phys. C 14 (1981) L237.
[76] M. Croft, J.A. Hodges, E. Kemly, A. Krishnan, V. Murgai, L.C. Gupta and R.D. Parks, Phys. Rev. Lett. 48 (1982) 826.
[77] B. Bittins, K. Keulerz, A. Scherzberg, J.P. Sanchez, W. Boksch, H.F. Bravn, J. Rohler, H. Schneider, P. Weidner and D. Wohlleben, Z. Phys. B 62 (1985) 21.
[78] W. Franz, F. Steglich, W. Zell, D. Wohlleben and F. Pobell, Phys. Rev. Lett. 45 (1980) 64.
[79] B. Buffat, B. Chevalier, M.H. Tuilier, B. Lioret and J. Etourneau, Solid State Commun. 59 (1986) 17.
[80] Sujata Patil, R. Nagarajan, L.C. Gupta, R. Vijayaraghavan and B.D. Padalia, Solid State Commun. 63 (1987) 955.
[81] B. Chaveliar, J.M.D. Coey, B. Lloret and J. Etourneau, J. Phys. C 19 (1986) 5421.
[82] D. Wohlleben and B. Wittershagen, Adv. Phys. 34 (1985) 403.
[83] V.M. Goldschmit, T. Brath and G. Lunde, Skr. Nor. Vidensk-Akad. Oslo, I, Mat-Naturvidensk kl 7 (1925) 59.
[84] B.D. Padalia and Sujata Patil, Crystal Properties Preparation 16 (1988) 111.
[85] T. Takabatake, F. Teshima, H. Fujii, S. Nishigori, T. Suzuki, T. Fujita, Y. Yamaguchi and J. Sukurai, Phys. Rev. B 31 (1990) 9607.
[86] T. Takabatake, Y. Nakazawa and M. Ishikawa, Jpn. J. Appl. Phys. 26 (1987) 547.
[87] H. Fujii, Y. Uwatoko, M. Akayama, K. Satoh, Y. Maeno,

T. Fujita, J. Sakurai, H. Kamimura and T. Okamoto, Jpn. J. Appl. Phys. 26 (1987) 549.

[88] D.T. Adroja, S.K. Malik, B.D. Padalia and R. Vijayaraghavan, Phys. Rev. B 39 (1989) 4831.

[89] P. Rogl, B. Chevalier, M.J. Besnus and J. Etourneau, J. Magn. Magn. Mater. 80 (1989) 305.

[90] S.K. Malik and D.T. Adroja, Phys. Rev. B 43 (1991) 6277.

[91] D.T. Adroja, S.K. Malik, B.D. Padalia, S.N. Bhatia, R. Walia and R. Vijayaraghavan, Phys. Rev. B 42 (1990) 2700.

[92] S.K. Malik and D.T. Adroja, Phys. Rev. B 43 (1991) 6295.

[93] K. Andres, J.E. Graebner and H.R. Ott, Phys. Rev. Lett. 35 (1975) 1979.

[94] H.R. Ott, H. Rudigier, Z. Fisk and J.L. Smith, Phys. Rev. Lett. 50 (1983) 1595.

[95] G.R. Stewart, Z. Fisk, J.O. Willis and J.L. Smith, Phys. Rev. Lett. 52 (1984) 679.

[96] T.T.M. Palstra, A.A. Menovsky, J. van den Berg, A.J. Dirkmaat, P.H. Kes, G.J. Nieuwenhuys and J.A. Mydosh, Phys. Rev. Lett. 55 (1985) 2727.

[97] W. Schlabitz, J. Baumann, B. Pollit, U. Rauchschwalbe, H.M. Mayer, U. Ahlheim and C. Bredl, Z. Phys. B 62 (1986) 171.

[98] M.B. Maple, J.W. Chen, Y. Dalichaouch, T. Kohara, C. Rossel, M.S. Torkachvili, M.W. Elfresh and J.D. Thompson, Phys. Rev. Lett. 56 (1986) 185.

[99] C.L. Lin, J.E. Crow, T. Mihalisin, J. Brooks, G. Stewart and A.I. Abou-Aly, J. Magn. Magn. Mater. 54–57 (1986) 379.

[100] W.H. Lee and R.N. Shelton, Phys. Rev. B 35 (1987) 5396.

[101] M. Sera, T. Satoh and T. Kasuya, Jpn. J. Appl. Phys. 26 (1987) 551.

[102] Y. Meno, M. Takahashi, T. Fujita, Y. Uwatoko, H. Fujii and T. Okamoto, Jpn. J. Appl. Phys. 26 (1987) 545.

[103] W.H. Lee, H.C. Ku and R.N. Shelton, Phys. Rev. B 36 (1987) 5739.

[104] H. Nakamura, Y. Kitaoka, K. Asayama and J. Flouquet, J. Phys. Soc. Japan 57 (1988) 2644.

[105] J.S. Schilling, Phys. Rev. B 33 (1986) 1667.

[106] J.D. Thompson, Z. Fisk, J.L. Smith, M. Selsane and G. Godart, Phys. Rev. Lett. (1990) submitted.

[107] G.R. Stewart, Z. Fisk and M.S. Wire, Phys. Rev. B 30 (1984) 482.

[108] T. Fukuhara, I. Sakamoto, H. Sato, S. Takayanagi and N. Wada, J. Phys. Condens. Matter 1 (1989) 7484.

[109] S.K. Malik, D.T. Adroja, M. Slaski, B.D. Dunlap and A. Umezawa, Phys. Rev. B 40 (1990) 9378.

[110] D.T. Adroja et al., unpublished.

[111] L.C. Gupta, D.E. MacLaughlin, C. Tien, C. Godart, M.A. Edwards and R.D. Parks, Phys. Rev. B 28 (1983) 3673.

[112] J.D. Thompson, J.O. Willis, C. Godart, D.E. MacLaughlin and L.C. Gupta, Solid State Commun. 56 (1985) 169.

[113] E. Bauer, E. Gratz and N. Pillmayr, Solid State Commun. 62 (1987) 271.

[114] E. Bauer, N. Pillmayr, E. Gratz, D. Gignoux and D. Schmitt, J. Magn. Magn. Mater. 67 (1987) L143.

[115] E. Bauer, N. Pillmayr, E. Gratz, D. Gignoux D. Schmitt, K. Winzer and J. Kohlmann, J. Magn. Magn. Mater. 71 (1988) 311.

[116] J. Kohlmann, K. Winzer and E. Bauer, Europhys. Lett. 5 (1988) 514.

[117] J.O. Willis, R.H. Aiken, Z. Fisk, E. Zirngibl, J.D. Thomson, H.R. Ott and B. Batlogg, ref. [34], p. 57.

[118] S.K. Dhar and K.A. Gschneidner Jr., J. Magn. Magn. Mater. 79 (1989) 151.

[119] E. Bauer, N. Pillmayr, E. Gratz, D. Gignoux and D. Schmitt, Phys. Lett. A 124 (1987) 445.

[120] W.C.M. Mattens, R.A. Elenbass and F.R. de Boer, Commun. Phys. 2 (1977) 147.

[121] W.H. Lee, K.S. Kwan, P. Klavins and R.N. Shelton, Phys. Rev. B 42 (1990) 6542.

[122] T. Kasuya, K. Taegahara, Y. Aoki, T. Suzuki, S. Kunii, M. Sera, N. Sato, T. Fujita, T. Goto, A. Tamaki and T. Komatsubara, Phys. Rev. Lett. 52 (1984) 359.

[123] J. Flouquet, P. Haen and C. Vettier, J. Magn. Magn. Mater. 29 (1982) 159.

[124] C.L. Lin, J. Teter, J.E. Crow, T. Mihlisin, J. Brooks, A.I. Abou-Aly and G.R. Stewart, Phys. Rev. Lett. 54 (1985) 2541.

[126] C. Rossel, K.N. Yang, M.B. Maple, Z. Fisk, E. Zirngiebl and J.D. Thompson, Phys. Rev. B 35 (1987) 1914.

[127] H.R. Ott, H. Rudigier, Z. Fisk and J.L. Smith, Phys. Rev. Lett. 50 (1983) 1595.
K.A. Gschneidner, Jr., J. Tang, S.K. Dhar and A. Goldman, Physica B 163 (1990) 507.

[128] J.D. Thompson, Z. Fisk and H.R. Ott, J. Magn. Magn. Mater. 54–57 (1986) 397.

[129] Y. Dalichaouch, M.B. Maple, J.W. Chen, T. Kohara, C. Rossel, M.S. Torikachivli and A.L. Giorgi, Phys. Rev. B 41 (1990) 1829.

[130] C. Boholm, J.K. Kjems, W.J.L. Buyers, P. Matthews, T.T.M. Palstra, A.A. Menovsky and J.A. Mydosh, Phys. Rev. Lett. 58 (1987) 1467.

[131] G.P. Meisner, A.L. Giorgi, A.C. Lawson, G.R. Stewart, J.O. Willis, M.S. Wire and J.L. Smith, Phys. Rev. Lett. 53 (1984) 1829.

[132] F. Marabelli and P. Wachter, Solid State Commun. 74 (1990) 1075.

[133] L.L. Hirst, AIP Conf. Proc. (1975) p. 11.

[134] J. Friedel, Nuovo Cimento Suppl. 7 (1958) 187.

[135] P.W. Anderson, Phys. Rev. 124 (1961) 41.

[136] C.M. Varma and Y. Yafet, Phys. Rev. 13 (1976) 2950.

[137] B. Coqblin and J.R. Schrieffer, Phys. Rev. 185 (1969) 847.

[138] J.R. Schrieffer and P.A. Wolff, Phys. Rev. 149 (1966) 491.

[139] V.T. Rajan, Phys. Rev. Lett. 51 (1983) 308.

[140] D.T. Adroja, S.K. Malik, B.D. Padalia and R. Vijayaraghavan, J. Phys. C 20 (1987) L307.
[141] T.V. Ramakrishnan and K. Sur, Phys. Rev. B 26 (1982) 1798.
[142] T.V. Ramakrishnan, ref. [31], p. 13; ref. [32], p. 351.
[143] T.V. Ramakrishnan, J. Magn. Magn. Mater. 63&64 (1987) 529.
[144] T.A. Kaplan and S.D. Mahanti, Phys. Lett. A 51 (1975) 265.
[145] T.A. Kaplan, S.d. Mahanti and M. Barma, ref. [26], p. 153.
[146] T.A. Kaplan, S.D. Mahanti and M. Barma, in: High Pressure and Low Temperature Physics, eds. C.W. Chu and J.A. Woollam (Plenum Press, New York, 1978) p. 141.
[147] S.D. Mahanti, T.A. Kaplan and M. Barma, Phys. Lett. A 58 (1976) 43.
[148] M.T. Beal-Monod and J.M. Lawrence, Phys. Rev. B 21 (1980) 5400.

Neutron scattering of lanthanide materials

R.M. Moon and R.M. Nicklow

Oak Ridge National Laboratory, P.O. Box 2008, Oak Ridge, TN 37831-6393, USA

A selected review of recent neutron scattering results on lanthanide materials is presented. Topics covered include the magnetic structures of the elemental metals, magnetic superlattices, enhanced nuclear magnetism, magnetic excitations and critical scattering.

1. Introduction

Neutron scattering has played a crucial role in advancing our understanding of the lanthanides. It has provided a detailed description of the rich array of exotic magnetic structures found in these materials; it has confirmed the accuracy of theoretical wavefunctions for the 4f electrons; and it has provided the fundamental data on the collective magnetic excitations and on the low-lying single-ion excitations. Without neutron scattering results, our microscopic image of the lanthanides would be dark and murky, full of doubtful issues.

Most of the neutron results are consistent with a few well-established theoretical concepts. The atomic magnetic moments are associated with 4f electrons, and these can be described by wavefunctions for the free +3 ions, modified by the crystalline electric field in which the iron is situated. The 4f electrons are well localized on a particular site so there is negligible direct exchange interaction between neighboring rare-earth ions. In the elemental metals and intermetallic compounds, there is an indirect exchange via the conduction electrons, the well-known Ruderman–Kittel–Kasuya–Yosida (RKKY) interaction. This interaction, which depends on details of the band structure of the conduction electrons, is a major factor in determining the periodicities of the magnetic structures. Finally, magnetostrictive strain can influence the crystal field, resulting in a modification of the magnetic structure. In the heavy rare-earth materials, the exchange interaction is dominant and the 4f electrons are separated in energy from the conduction band, resulting in a situation which is easier to deal with theoretically. In the light rare-earth materials, the 4f electrons are more extended in space, the crystal-field energies may be comparable to the exchange energy and the conduction electron band may overlap the 4f band, resulting in a difficult theoretical situation.

We will not attempt a comprehensive review of all the neutron scattering results on lanthanide materials. Rather, we will focus on a few selected topics in an attempt to demonstrate that this is still an active and fertile field of research. In selecting these topics, we give preference to new areas but also include some topics which are continuations of work begun in the early days of neutron scattering. Excellent reviews of this early work have been given by Koehler [1] and Mackintosh and Bjerrum Møller [2]. A somewhat later review by Sinha [3] covers both structural work and inelastic scattering results. A comprehensive treatment of both theory and experiment on the magnetic structures and excitations in rare-earth metals is soon to be published by Jensen and Mackintosh [4].

2. Magnetic structures

2.1. Elemental metals

Almost certainly, the magnetic structure of Nd is the longest-running problem in the history of

0304-8853/91/$03.50 © 1991 – Elsevier Science Publishers B.V. All rights reserved

neutron scattering. The first experiment was performed in 1963 [5], and there are still details of the structures to be determined. There are several different structures depending on the temperature, most of which have multiple propagation vectors and all of which involve multiple moment components, multiple atomic sites and multiple domains.

These modulated moment structures can be described analytically as

$$\mu_i(r) = \sum_{i,k} \mu_{ijk} \hat{x}_k \cos(q_j \cdot r + \phi_{ijk}), \quad (1)$$

where i distinguishes one of four successive basal planes needed to specify the structure in the dhcp unit cell, j runs over the number of distinct propagation vectors in a single domain, and k runs over a set of Cartesian axes. The position of magnetic neutron diffraction peaks can be used to determine the q_j, and the intensities of these peaks can be used to determine the μ_{ijk} and ϕ_{ijk}. For Nd, the observed q_j are all perpendicular to c^*, and therefore occur in multiples of six related by the hexagonal symmetry. A set of six equivalent satellites of allowed nuclear peaks may result from a single q structure (giving peaks at $\pm q$) and three domains with their qs rotated in 120° steps, or from a triple-q, single-domain structure. Indeed, the resolution of this ambiguity was a central theme in much of the magnetic structure work on Nd [6–8]. Two different types of experimental observations have been used to solve this problem. McEwen [9] and Forgan, Gibbons and McEwen [10] have used the judicious application of magnetic fields to create single-domain samples, thus revealing the fundamental q vectors with no ambiguity. These same authors [11] have used the occurrence of particular higher harmonics to get the same information. This technique is successful because there is a tendency at low temperature for the structure to "square up" so that third-order harmonics should be included in eq. (1). Not only terms of the type $3q_i$ should be included, but in a multiple q structure, terms like $2q_i \pm q_j$ and $q_i \pm q_j \pm q_k$ should be present. By observing the position of these third-order satellites, it is possible to determine which distinct q

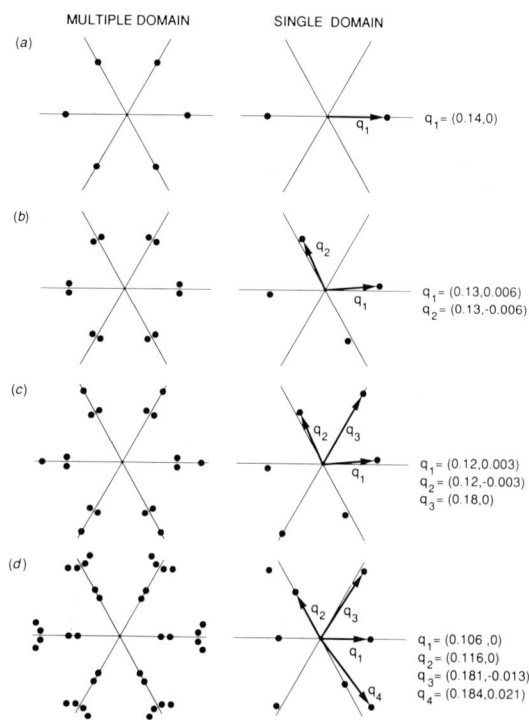

Fig. 1. Diffraction patterns of magnetic satellites observed at different temperatures around a typical nuclear peak in Nd. The q vectors are specified in orthogonal components parallel and perpendicular to the nearest {100} direction in units of a^*. Only the first-order peaks are shown. The multiple domain pattern is the one actually observed in zero applied field.

vectors are present in a single domain. These observations can be made in zero applied field.

Our current knowledge of the fundamental q vectors is summarized in fig. 1 for various temperature ranges. Note that single-q, double-q, triple-q and quadruple-q structures have been observed at different temperatures. The temperature range for the single-q structure (fig. 1a) is taken from the thermal expansion study of Zochowski and McEwen [12] who found first-order phase transitions at 19.9 and 19.1 K. This finding is consistent with the renormalization-group theory of Bak and Lebech [6] who found that a single-q structure required a first-order transition at T_N. The double-q structure (fig. 1b) is found between 19.1 and 8.2 K. The satellite spots show some motion in this temperature range [13], and the major component of the ordered moment is on the hexagonal sites and is oriented close to the propagation vector. The triple-q structure (fig. 1c) is thought [14] to exist between 8.2 and 6.2 K. The new satellites appearing at $q = 0.18\tau_{100}$ are primarily associated with moments on the cubic sites. Below 6.2 K the quadruple-q (fig. 1d) structure is found. There may be further complexities still to be revealed by neutron measurements because additional transitions at 7.5 and 5.7 K have been found in thermal expansion [12] and heat capacity [15] results. A complete summary of the various sets of q vectors, including a bewildering new array found in an applied magnetic field, is given by McEwen and Zochowski [14].

The magnetic structure of Pr is similar to the single-q structure of Nd. These materials both have the dhcp crystal structure and a similar propagation vector $(0.13\tau_{100})$ for the magnetic structure. A significant difference is that for Pr, the transition is driven by a mixed nuclear–electronic interaction. This will be discussed further in a later section.

The dhcp phase of Ce is difficult to study because there are typically two other phases present at low temperature. An early study by Wilkinson et al. [16] on a polycrystalline mixed-phase sample showed magnetic Bragg peaks associated with the dhcp phase below the Néel temperature of about 12.5 K. Only 3 weak peaks were observed, so it was not possible to do a definitive structure determination. More recently, Gibbons et al. [17] have measured a single-phase (dhcp) sample of $Ce_{0.75}Y_{0.25}$, and they suggest that the structure they deduced might also apply to pure Ce. In both studies the magnetic peaks indicate a propagation vector of $0.5\tau_{100}$. In the model of Gibbons et al., the moments are in the basal plane, perpendicular to the propagation vector. Assigning zero moment to the Y atoms, the Ce atoms have $0.91\mu_B$ on the hexagonal sites and $0.38\mu_B$ on the cubic sites.

There has been no recent work on the magnetic structures of Sm metal, but it is worth restating the structures reported by Koehler and Moon [18] in a more compact form. The crystal structure has equally populated twins related by a 180° rotation about the hexagonal c axis. In a type 1 region, the stacking sequence is ABABCB-

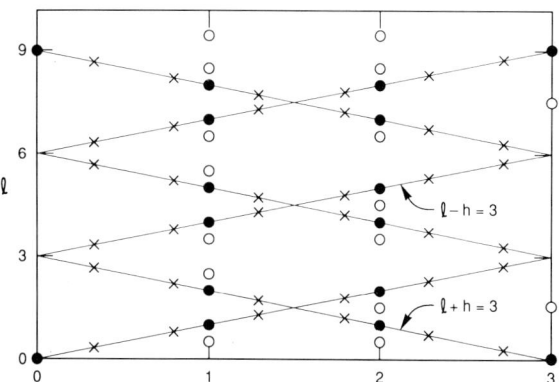

Fig. 2. A map showing the location of Bragg peaks in the $(h0l)$ plane of reciprocal space for Sm below 13.8 K.

CAC, and the type 2 sequence is obtained by interchanging B & C layers. The first, fourth and seventh layers have face-centered cubic near-neighbor symmetry, and the remainder have hexagonal symmetry. The hexagonal sites order at 106 K with a structure given by

$$\mu_h(r) = (2/\sqrt{3})\mu_h \hat{z} \sin(q_h \cdot r), \qquad (2)$$

where $q_h = \frac{3}{2}\tau_{001}$. This puts equal moments of magnitude μ_h on all hexagonal sites in a $0 + + 0 - -$ sequence, where the zeros fall on the cubic sites. Below 13.8 K additional magnetic reflections appear which are associated with ordering of the cubic sites. The hexagonal magnetic satellites do not change when the cubic sites order. A map of the $h0l$ reciprocal lattice plane is shown in fig. 2, giving the location of all the Bragg peaks observed below 13.8 K. The nuclear and magnetic peaks on the lines $l + h = 3n$ belong to type 1 Sm, and the peaks on the lines $l - h = 3n$ belong to type 2 Sm. The cubic site magnetic structure is

$$\mu_c(r) = \sqrt{2}\mu_c \hat{z} \cos(q_c \cdot r - \pi/4), \qquad (3)$$

where $q_c = \frac{1}{4}\tau_{10\bar{1}}$ for type 1 Sm and $q_c = \frac{1}{4}\tau_{101}$ for type 2 Sm, and where the constant phase factor has been selected to put equal moments of magnitude μ_c on all cubic sites in a $+ + - -$ sequence along the propagation vector. Note that cubic satellites appear around positions like (303) where there is no nuclear intensity. The cubic site structure factor for the (303) position is nonzero but is exactly canceled by the hexagonal site structure factor. The structures defined by eqs. (2) and (3) are the same as described in ref. [18], but in this format, can almost be written by inspection of fig. 2, given that modulated moments of the type $\cos(q \cdot r + \phi)$ will produce magnetic peaks displaced from the nuclear peaks by $\pm q$.

We will not review the magnetic structures of the heavy rare-earth metals which have been described in considerable detail in refs. [1,3]. However, the case of Ho is exceptional because there has been substantial activity in recent years on the low-temperature ($T < 50$ K) structures of this material. At about 20 K, the basal-plane spiral structure locks in to a propagation vector of $c^*/6$ and a ferromagnetic component develops along the c axis. Using X-ray scattering, Gibbs et al. [19] found evidence for lock-in behavior just above this temperature at other commensurable wave vectors. They explained this behavior in terms of spin discommensurations or spin slips. Cowley and Bates [20] obtained new neutron scattering data which they successfully analyzed on the basis of a spin-slip model. The underlying physics in a spin-slip model is that the basal-plane anisotropy favors spin alignment along one of the six-fold easy axes, while exchange may favor a different turn angle. A q vector of $c^*/6$ results when pairs of layers are aligned along successive easy directions. A spin slip occurs in this sequence when a particular easy axis has only one layer of spins. The modification of the Ho magnetic structure when a magnetic fold is applied in the basal plane has been studied theoretically by Jensen and Mackintosh [21] who found structures called helifans, which are consistent with experimental observations. These structures are intermediate between the helix and the fan. Cowley et al. [22] have introduced a different competing interaction by applying a magnetic field along the c axis. They found a series of commensurable spin-slip structures in the basal-plane moments with the distance between spin slips decreasing with increasing field.

2.2. Magnetic superlattices

Advances in materials processing techniques have been exploited to make an interesting new class of materials in which layers of magnetic materials are sandwiched between layers of nonmagnetic materials in a multilayer structure. In particular, rare-earth elements have been used in the magnetic layers and yttrium in the intervening layers. Neutron diffraction experiments have revealed the detailed magnetic structures of these materials with fascinating implications regarding the magnetic coupling of the rare-earth layers through the Y layers. This work has been reviewed by Majkrzak et al. [23] and Rhyne et al. [24].

For Dy–Y multilayers, with the c axis normal to the layers, Rhyne et al. [25,26] found helical magnetic structures with the propagation vector along c, as in pure Dy in the temperature range 85–178 K. Their samples had Dy layers about 16 atomic planes thick (46 Å) and Y layers ranging from 10–22 planes. From the width of the magnetic reflections, they found coherence lengths ranging from 580–245 Å, with the longer coherence length associated with the shortest Y thickness. These coherence lengths ranged from 8–2.4 bilayer spacings, indicating phase coherence and preservation of chirality across the Y layers. The suggested mechanism is a spin-density wave in the conduction bands of Y and Dy which couples to the 4f moments via the RKKY interaction. From the observed magnetic intensities, a phase change across the Y layers is deduced which indicates a temperature independent propagation vector in Y of 0.31 Å^{-1}. The propagation vector in the Dy layers is significantly different and is temperature dependent, as in pure Dy. However, the multilayers show no phase transition to ferromagnetism as found in pure Dy at 85 K. In general, the Ho–Y [23] case is similar to Dy–Y.

Different behavior has been reported by Majkrzak et al. [27] for Gd–Y multilayers. In this case, the Gd moments within a single layer are all coupled ferromagnetically and are oriented in the basal plane. In going from one Gd layer to the next, there is either a phase change of 0 or π, depending on the thickness of the intervening Y layer. If the number of Y atomic planes is 6 or 20, there is ferromagnetic coupling between Gd layers, but antiferromagnetic coupling is found for a Y thickness of 10 atomic planes.

In the case of Er–Y multilayers, Erwin et al. [28,29] found that the c-axis components order in a linear-spin-density-wave state below 78 K with a coherence length extending over 3 bilayers. Below 30 K the basal-plane components order with a different propagation vector than for the c-axis components. Ferromagnetism of the c-axis components, observed in pure Er, never develops in the Er–Y multilayers.

Very recent work includes studies of Dy/Y/Sc/Y multilayers [30], Dy/Lu multilayers [31] and Er/Lu multilayers [32].

In general, the magnetic structures observed in these multilayer materials are similar to those found in the pure metals, but there are differences caused by modifications of the exchange and magnetoelastic interactions. A more complete understanding of these materials may lead to the ability to tailor magnetic structures for specific device applications.

2.3. Enhanced nuclear magnetism

In singlet electronic-ground-state compounds, nuclear spins have an important role to play in the onset of long-range magnetic order. It is well known that the electronic exchange interaction may fail to produce spontaneous order in such systems if it is not strong enough. Mean-field theory reveals a quantitative criterion in terms of the parameter $\eta = 4K\alpha^2/\Delta$, where $2K$ is the ion–ion exchange interaction, $\alpha = \langle E | J_z | G \rangle$ and Δ is the energy difference between the first excited state $|E\rangle$ and the ground state $|G\rangle$. For $|\eta| > 1$, the electronic system will spontaneously order at finite temperature, and if $|\eta| < 1$, it will not. When the hyperfine and nuclear Zeeman terms are included in the Hamiltonian, Murao [33] has shown that spontaneous order can develop in both the nuclear and electronic systems even when $|\eta| < 1$. The nuclear spins on different sites are coupled via the hyperfine interaction on a single site and the electronic exchange interaction between sites. The temperature dependence of such a transition can be unusual.

The sensitivity of neutrons to the state of nuclear spin order arises from the spin dependence of the neutron–nucleus interaction. When the neutron, with spin $\frac{1}{2}$, interacts with a nucleus of spin I, it can do so in states of total spin $I + \frac{1}{2}$ or $I - \frac{1}{2}$, and associated with these two states are scattering lengths b^+ and b^-. Consequently, a nuclear polarization of magnitude $\langle I_z \rangle$ leads to a contribution to the neutron scattering length which is proportional to $\langle I_z \rangle (b^+ - b^-)$. By comparison, the scattering length due to an electronic magnetic moment μ is proportional to $f(Q)\mu_\perp$, where $f(Q)$ is the form factor and $\mu_\perp$ is the component of μ perpendicular to the scattering vector Q. Thus, in principle, it is possible to measure separately both the nuclear spin and electronic spin polarizations because the corresponding neutron scattering cross sections depend differently on $|Q|$ and on the relative orientations of Q and the polarization directions.

The case of PrCu$_2$ is one where η is slightly less than 1, and Kawarazaki et al. [34] have observed neutron diffraction peaks with a Néel temperature of 58 mK. These peaks appear as satellites of allowed nuclear reflections with a propagation vector given by $0.24a^* \pm 0.68c^*$, where a^* and c^* are reciprocal lattice basis vectors of the orthorhombic cell. The full details of the structure have not yet been determined, but there is a very interesting relationship in real space (fig. 3) between the crystal structure and the observed propagation vector.

Pure Pr is another case where η is slightly less than 1. McEwen and Sterling [35] observed several magnetic satellites with a propagation vector of $0.13a^*$, suggesting similarity to the Nd case. The temperature dependence of these peaks was reported by Bjerrum Møller et al. [36] and by Stirling and McEwen [37]. The satellites can be detected at 1 K, although they are broad and weak. The intensity remains nearly constant down to 100 mK and increases dramatically below 60 mK. It was presumed that this ordering was driven by the nuclear spins through the hyperfine-exchange pathway. Evidence showing that nuclear spin ordering contributes to the intensity below

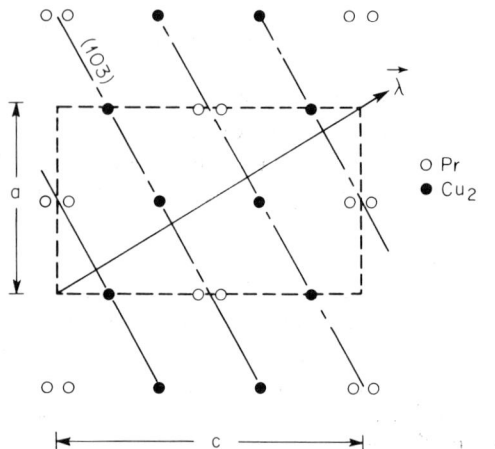

Fig. 3. Projection of the $PrCu_2$ structure onto the $a-c$ plane. The orthorhombic cell is indicated by the dashed lines. The vector λ gives the direction of propagation and wavelength of the modulated magnetic structure. The indicated diagonal planes are parallel to (103) planes and are very nearly planes of constant phase.

60 mK was reported by Kawarazaki et al. [38] and by McEwen and Stirling [39] using the neutron polarization analysis technique.

$HoVO_4$ is a case where $|\eta| \ll 1$ so that the electronic system is very far from spontaneous order. Susceptibility measurements by Suzuki et al. [40] showed an ordering temperature of 4.5 mK. Bleaney [41] predicted a magnetic structure which has now been confirmed by neutron diffraction measurements by Suzuki et al. [42]. Temperatures down to 2.7 mK were obtained in this neutron study by adiabatic demagnetization of the sample itself.

It is anticipated that more work in this interesting research area will follow as the coupling of dilution refrigerators and neutron spectrometers becomes routine. A more complete review has been given by Nicklow et al. [43].

3. Excitations in lanthanides

Neutron scattering is a particularly powerful technique for studies of excitations in magnetically ordered materials for two reasons: (1) the magnetic moment of the neutron provides a means, as in structure studies, of identifying the magnetic signal apart from all other scattering cross sections, providing even the identification of transverse and longitudinal excitations separately, and (2) the neutron energy and wave length are of similar magnitude to those of typical excitations, thereby allowing a very precise measurement of the excitation spectrum (e.g., spin-wave dispersion) to be made throughout the Brillouin zone. Measurements of the spin-wave dispersion have been indispensable for determining the form of the exchange interaction in magnetic materials.

3.1. Heavy metals

By the mid 1970s the principal features of the dispersion relations for the spin waves in most (Gd to Er) of the elemental heavy rare-earth metals had been measured by neutron scattering and analyzed to obtain detailed information about the isotropic exchange and anisotropic interactions [2,3]. Generally, the results confirmed the theoretical predictions about the form of the Fourier transform of the exchange interaction, $J(q)$. The maximum value of $J(q)$ occurs at the wave vector, q_0, that describes the spin ordering configuration of the metal; $q_0 = 0$ for ferromagnetic Gd, and q_0 is finite for the metals having helical or other types of oscillatory spin ordering. The neutron studies also revealed the importance of magneto-elastic and two-ion anisotropic interactions in the description of the excitations of these metals.

During the 1980s the study of excitations has included such topics as excitations in Tm, large crystals of which have only recently become available; commensurate phases of usually incommensurate structures like Ho; and the temperature dependence near to and well above the magnetic ordering temperature in Gd.

3.1.1. Thulium

The magnetic excitations in thulium have been investigated recently by two research groups [44,45]. The two sets of experimental results are essentially the same and show that the magnetic spectrum at low temperature is dominated by a single magnon like mode with an energy between 8.5 and 10 meV (fig. 4). The small dispersion and

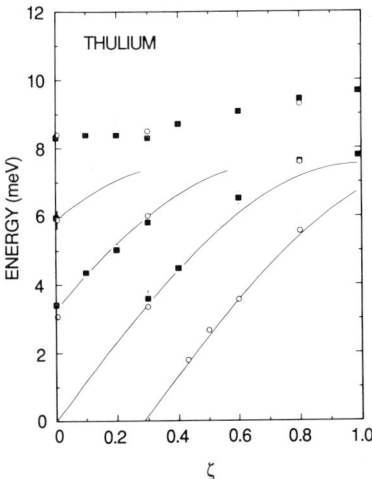

Fig. 4. Dispersion relations along the c direction of the excitations observed in Tm at 5 K. The lower energy points and curves correspond to phonons [44]. The closed symbols represent data obtained for the neutron scattering vector $\mathbf{Q} = (0, 0, 2+\zeta)$, in reduced units, while for the open symbols $\mathbf{Q} = (1, 1, \zeta)$.

contrast, for example, with that for Tm shown in fig. 4. However, the early neutron scattering study by Koehler et al. [46] and the recent X-ray study by Gibbs et al. [19] show that Ho may form a helical but commensurate structure. Recent measurements by Larsen et al. [47] and Patterson et al. [48] of the excitations in these commensurate structures of Ho have provided improved understanding of both the structure and excitations in Ho.

The results of Larsen et al. [47] are shown in fig. 5. A gap is clearly seen in the results obtained for the commensurable-helical structure. However, of particular interest is the fact that the gap in the periodic structure is much smaller than predicted theoretically. This leads to the suggestion that there exists an anisotropic contribution to the two-ion interactions. Furthermore, a dipolar contribution to the spin wave dispersion at $q = 0$ is clearly seen, and its magnitude, compared

large energy gap is consistent with the observed trend of the other heavy rare earths; the ratio of the exchange interaction to the magnitude of the crystal-field anisotropy decreases from Gd to Tm. In fact, the energy of the magnon in Tm is quite close to that estimated from the first dipolar transition in the crystal-field level scheme. The lower energy curves in fig. 4 are phonon branches that are also observed in the low-temperature measurements.

At elevated temperatures, additional neutron intensity is observed in the energy range well below that of the magnon mode. Furthermore, the neutron scattering peak at the magnon energy persists up to and well above the transition to the paramagnetic phase. McEwen et al. [45] have shown that these features are also consistent with transitions between crystal field energy levels.

3.1.2. Ho-commensurate periodic structure

The magnetic structures of Tb, Dy and Ho are generally considered to be helical with periodicities that are incommensurate with the crystal lattice periodicity. The standard theoretical treatment of this type of structure leads to a magnon dispersion that has no energy gap at $q = 0$, in

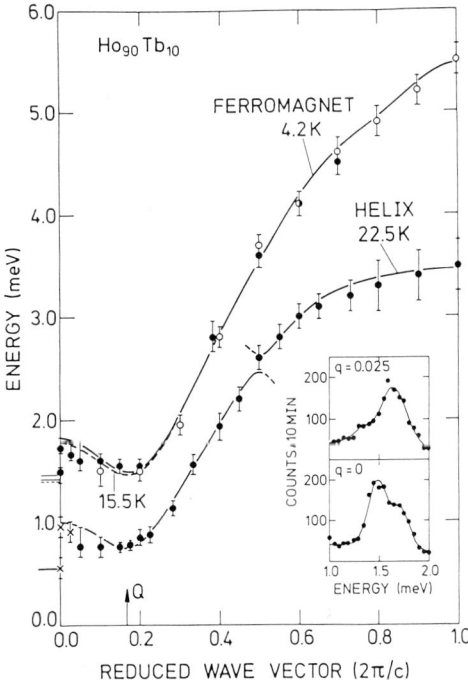

Fig. 5. Dispersion relations for the magnetic excitations in the c direction in $Ho_{90}Tb_{10}$. The inset shows that the dipolar discontinuity between the c direction and the basal plane directions is easily seen in the data. The curves are the calculated dispersion relations [47].

to that of the maximum in $J_c(Q) - J_c(0)$, is sufficiently large, that the absolute maximum in the two-ion interaction energy along the c direction is shifted from $q = q_0$ to $q = 0$, thereby leading to the stability of the conical structure in Ho below 19 K [47].

3.1.3. High-temperature excitations in Gd

During the 1980s there was considerable work carried out on the spin fluctuations of ferromagnets at finite temperatures. Much of the interest was centered on the existence (or not?) of propagating spin waves above the ordering temperature T_c. While results for itinerant electron systems remain controversial, the Heisenberg systems, EuO [49,50], EuS [51] and Gd [52,53] clearly exhibit a crossover from spin diffusion at small q to damped spin-wave behavior at "large q".

The results for Gd in fig. 6 show the contrasting behavior for small q [panels (a)–(c)] and large q [panels (d)–(f)] near T_c for q parallel to the c axis. The data are described excellently by a damped-harmonic-oscillator function [54]. Similar results are observed for q perpendicular to c,

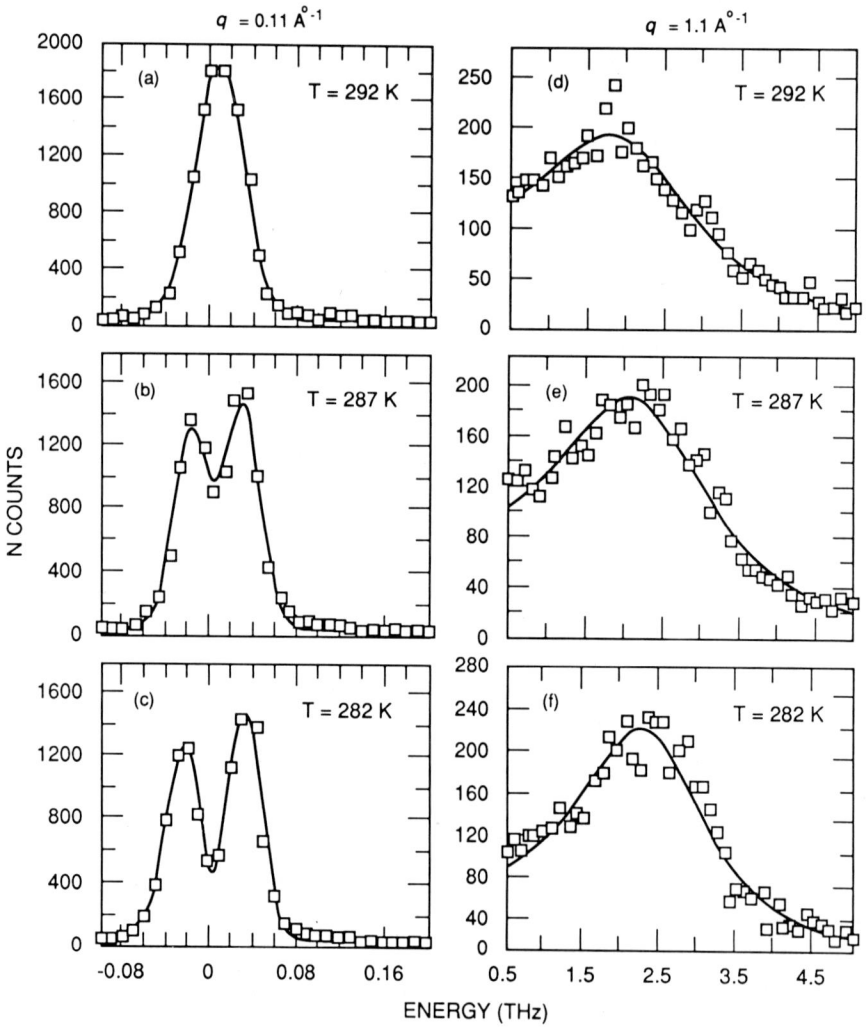

Fig. 6. Magnetic excitation spectra of Gd for temperatures just below T_c and for small and large values of the wave vector q in the c direction [52].

except that peaks in the observed spectra at large q persist to temperatures $T = 2.9T_c$. The results for Gd indicate that there is no simple relationship between the existence of spin waves above T_c, and the spin-pair correlations at high temperature. However, there was found to be a relationship with the spin-wave damping arising from mode–mode coupling. Consequently, the existence of spin waves at high temperature in Gd apparently is dependent on the spin dynamics, not on the spin statics.

3.2. Longitudinally modulated structures

Longitudinally modulated magnetic structures are found in both the light and heavy rare-earth elements: Pr, Nd, Er and Tm. One difference between the light and heavy metals is the temperature range in which the modulated structure is stable. For the light metals Pr and Nd, the modulated phases are stable at low temperature (e.g.,

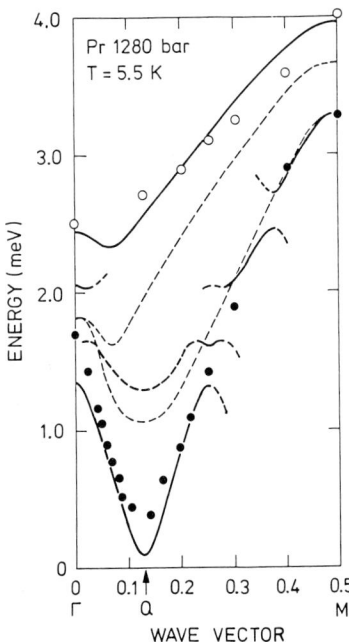

Fig. 8. The dispersion relation of the optical excitations in the longitudinally modulated phase of Pr at 5.5 K and a pressure of 1280 bar. The solid and open circles indicate the longitudinal and transverse branches. The lines are calculated dispersion curves [60].

$T < 2$ K). Whereas for the heavy metals Er and Tm, the modulated phase is stable only quite near to the Néel temperature, i.e., only at high temperature. The excitations in all these elements have been investigated during the past decade.

The results obtained for Er [55] show no well-defined peaks in measurements of the transverse excitations (fig. 7). This seems to be consistent with the original conclusions of Cooper [56], based on linear spin-wave theory for a system with weak anisotropy, that excitations in a sinusoidal magnet are not well-defined spin waves. However, subsequent work by Liu [57], Lingård, [58], Ziman and Lindgård [59], and Jensen et al. [60], showed that under certain assumptions (usually $T \to 0$ K and/or large anisotropy), peaks may appear in the spectrum of transverse excitations. While erbium does not seem to be described by these theories, the excitations in Tm [45] (large anisotropy) and Pr [60] and Nd [61] ($T \approx 0$ K and large anisotropy) can be understood, in part, in

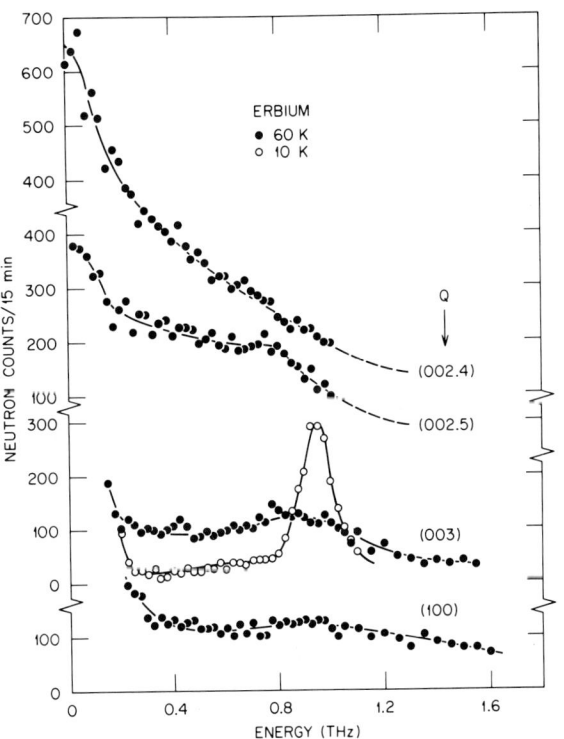

Fig. 7. Constant-Q measurements of the energy distribution of neutrons scattered by Er metal at 60 and 10 K [55].

terms of exchange-coupled single-ion crystal field excitations. Such excitations may exist in both paramagnetic and magnetically ordered phases and can have transverse or longitudinal character. Results obtained for Pr [60] are shown in fig. 8.

3.3. Critical scattering

Recent neutron-scattering studies of the critical properties of several of the heavy metals with oscillatory magnetic structures, Dy [62], Ho, [62] and Er [63] have been carried out to test various theoretical predications about the relationship of critical exponents and the number n of independent components of the order parameter. Theoretical work by Bak and Mukamel [64] gave $n = 4$ for Tb, Dy and Ho, and they derived values for β, ν and γ which can be measured by neutron scattering. These exponents characterize the sublattice magnetization in the ordered phase, the inverse correlation length and the staggered sus-

Table 1
The experimental values for the critical exponents ν and γ determined for Dy and Ho [62] are compared with the theoretical values

	ν	γ
	experiment	
Dy	0.57 ± 0.05	1.05 ± 0.07
Ho	0.57 ± 0.04	1.14 ± 0.10
	theory	
ref. [64]	0.70	1.39
ref. [65]	0.53	1.10

ceptibility. Recent work by Kawamura [65] considers that these metals correspond to $n = 2$, but that the twofold degeneracy of the chirality of the ordered phases leads to a new universality class with different exponents than those for $n = 2$ or 4.

Neutron determinations of β from measurements of the temperature dependence of the antiferromagnetic Bragg peak for Tb [66], Dy [67] and Ho [68] have given different, inconclusive results, presumably because of large effects of extinction. The recent work of Gaulin et al. [62] on Dy and Ho was carried out to measure instead ν and γ in the paramagnetic phases. The experimental data for κ, the inverse correlation length, are shown in fig. 9, and the resulting exponents are given in table 1. The measured exponents agree better with the theoretical results of Kawamura [65].

The exponent β for Er ($n = 2$) was determined recently by Du Plessis [63] through measurements of the temperature dependence of the antiferromagnetic Bragg peak. They find that β crosses over from a value of 0.39 near T_N to a mean-field value of 0.5 away from T_N. The value expected for β for $n = 2$ is 0.35.

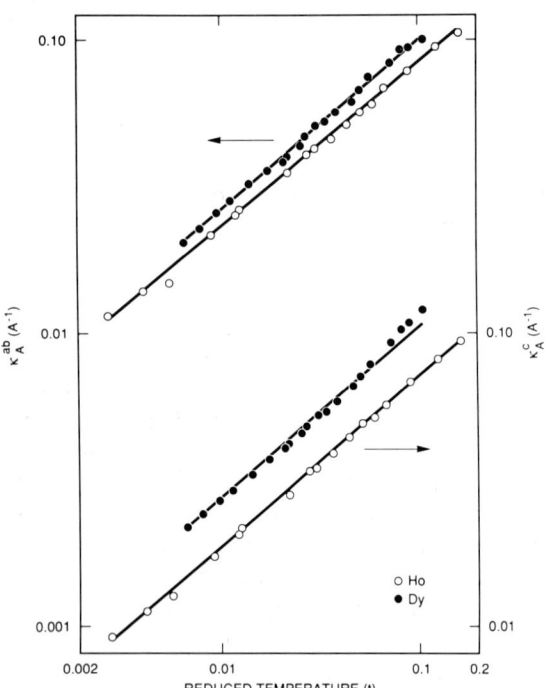

Fig. 9. The antiferromagnetic inverse correlation lengths along c and in the basal plane as a function of temperature for Dy (solid circles) and Ho (open circles). The lines are power law functions of $t = (T - T_N)/T_N$ [62].

4. Conclusion

Space does not allow a more comprehensive summary. Important topics not covered are magnetic form factors of the lanthanide materials, heavy fermion systems, high-temperature superconductors, mixed-valence materials, samarium-cobalt magnets and neodymium–iron–boron magnets. Most of these topics are reviewed in other articles in this volume.

Acknowledgement

Research supported by the Division of Materials Sciences, US Department of Energy under contract DE-AC05-84OR21400 with Martin Marietta Energy Systems, Inc.

References

[1] W.C. Koehler, in: Magnetic Properties of Rare Earth Metals, ed. R.J. Elliott (Plenum Press, London and New York, 1972) p. 81.
[2] A.R. Mackintosh and H. Bjerrum Møller, in: Magnetic Properties of Rare Earth Metals, ed. R.J. Elliott (Plenum Press, London and New York, 1972) p. 187.
[3] S.K. Sinha, in: Handbook on the Physics and Chemistry of Rare Earths, vol. 1, ed. K.A. Gschneidner and L. Eyring (North-Holland, Amsterdam, 1978) p. 489.
[4] J. Jensen and A.R. Mackintosh, Rare Earth Magnetism: Structures and Excitations (Oxford Univ. Press, Oxford, to be published).
[5] R.M. Moon, J.W. Cable and W.C. Koehler, J. Appl. Phys. 35 (1964) 1041.
[6] P. Bak and B. Lebech, Phys. Rev. Lett. 40 (1978) 800.
[7] R.M. Moon, W.C. Koehler, S.K. Sinha, C. Stassis and G.R. Kline, Phys. Rev. Lett. 43 (1979) 62.
[8] B. Lebech, J. Als-Nielsen and K.A. McEwen, Phys. Rev. Lett. 43 (1979) 65.
[9] K.A. McEwen, Physica B 136 (1986) 385.
[10] E.M. Forgan, E.P. Gibbons and K.A. McEwen, J. Phys. C 8 (1988) 337.
[11] E.M. Forgan, E.P. Gibbons, K.A. McEwen and D. Fort, Phys. Rev. Lett. 62 (1989) 470.
[12] S. Zochowski and K.A. McEwen, J. Magn. Magn. Mater. 54–57 (1986) 515.
[13] B. Lebech and J. Als-Nielsen, J. Magn. Magn. Mater. 15–18 (1980) 469.
[14] K.A. McEwen and S.W. Zochowski, J. Magn. Magn. Mater. 90 & 91 (1990) 94.
[15] E.M. Forgan, C.M. Muirhead, D.W. Jones and K.A. Gschneidner Jr., J. Phys. F 9 (1979) 651.
[16] M.K. Wilkinson, H.R. Child, C.J. McHargue, W.C. Koehler and E.O. Wollan, Phys. Rev. 122 (1961) 1409.
[17] E.P. Gibbons, E.M. Forgan and K.A. McEwen, J. Phys. F 17 (1987) L101.
[18] W.C. Koehler and R.M. Moon, Phys. Rev. Lett. 29 (1972) 1468.

[19] Doon Gibbs, D.E. Moncton, K.L. D'Amico, J. Bohr and B.H. Grier, Phys. Rev. Lett. 55 (1985) 234.
[20] R.A. Cowley and S. Bates, J. Phys. C 21 (1988) 4113.
[21] J. Jensen and A.R. Mackintosh, Phys. Rev. Lett. 64 (1990) 2699.
[22] R.A. Cowly, D.A. Jehan, D.F. McMorrow and G.J. McIntyre, Phys. Rev. Lett. 66 (1991) 1521.
[23] C.F. Majkrzak, Doon Gibbs, P. Böni, Alan I. Goldman, J. Kwo, M. Hong, T.C. Hsieh, R.M. Fleming, D.B. McWhan, Y. Yafet, J.W. Cable, J. Bohr, H. Grimm and C.L. Chien, J. Appl. Phys. 63 (1988) 3447.
[24] J.J. Rhyne, R.W. Erwin, J. Borchers, M.B. Salamon, R. Du and C.P. Flynn, Physica B 159 (1989) 111.
[25] J.J. Rhyne, R.W. Erwin, J. Borchers, S. Sinha, M.B. Salamon, R. Du and C.P. Flynn, J. Appl. Phys. 61 (1987) 4043.
[26] R.W. Erwin, J.J. Rhyne, M.B. Salamon, J. Borchers, S. Sinha, R. Du, J.E. Cunningham and C.P. Flynn, Phys. Rev. B 35 (1987) 6808.
[27] C.F. Majkrzak, J.W. Cable, J. Kwo, M. Hong, D.B. McWhan, Y. Yafet, J.V. Waszczak and C. Vettier, Phys. Rev. Lett. 56 (1986) 2700.
[28] R.W. Erwin, J.J. Rhyne, J. Borchers, M.B. Salamon, R. Du and C.P. Flynn, J. Appl. Phys. 63 (1988) 3461.
[29] J.A. Borchers, M.B. Salamon, R.W. Erwin, J.J. Rhyne, R.R. Du and C.P. Flynn, Phys. Rev. B 43 (1991) 3123.
[30] R.W. Erwin, J.A. Borchers, J.J. Rhyne, T.F. Tsui and C.P. Flynn, Bull. Am. Phys. Soc. 36 (1991) 785.
[31] R.S. Beach, A. Matheny, M.B. Salamon, C.P. Flynn, J.A. Borchers, R.W. Erwin and J.J. Rhyne, Bull. Am. Phys. Soc. 36 (1991) 785.
[32] J.A. Borchers, R.W. Erwin, J.J. Rhyne, R.S. Beach, A. Matheny, C.P. Flynn and M.B. Salamon, Bull. Am. Phys. Soc. (1991) 785.
[33] T. Murao, J. Phys. Soc. Japan 50 (1981) 3240.
[34] S. Kawarazaki, N. Kunitomi, Y. Morii, H. Suzuki, R.M. Moon and R.M. Nicklow, Solid State Commun. 49 (1984) 1153.
[35] K.A. McEwen and W.G. Stirling, J. Phys. C 14 (1981) 157.
[36] H. Bjerrum Møller, J.Z. Jensen, M. Wulff, A.R. Mackintosh, O.D. McMasters and K.A. Gschneidner, Jr., Phys. Rev. Lett. 49 (1982) 482.
[37] W.C. Stirling and K.A. McEwen, in: Multicritical Phenomena, eds. R. Pynn and A.T. Skejltorp (Plenum, New York, 1984) p. 213.
[38] S. Kawarazaki, N. Kumitomi, J.R. Arthur, R.M. Moon, W.G. Stirling and K.A. McEwen, Phys. Rev. B 37 (1988) 5336.
[39] K.A. McEwen and W.G. Stirling, Physica B 156 & 157 (1989) 754.
[40] H. Suzuki, N. Nambudriad, B. Bleaney, A.L. Allsop, G.J. Bowden, I.A. Campbell and N.J. Stone, J. de Phys. 39 (1978) C6-800.
[41] B. Bleaney, Proc. Roy. Soc. (London) A 370 (1980) 313.
[42] H. Suzuki, T. Ohtsuka. S. Kawarazaki, N. Kunitomi,

R.M. Moon and R.M. Nicklow, Solid State Commun. 49 (1984) 1157.
[43] R.M. Nicklow, R.M. Moon, S. Kawarazaki, N. Kunitomi, H. Suzuki, T. Ohtsuka and Y. Morii, J. Appl. Phys. 57 (1985) 3784.
[44] J.A. Fernandez-Baca, R.M. Nicklow, Z. Tun and J.J. Rhyne, Phys. Rev. B 43 (1991) 3188.
[45] K.A. McEwen, U. Steigenberger and J. Jensen, Phys. Rev. B 43 (1991) 3298.
[46] W.C. Koehler, J.W. Cable, H.R. Child, M.K. Wilkinson and E.O. Wollan, Phys. Rev. 151 (1966) 414.
[47] C.C. Larsen, J. Jensen and A.R. Mackintosh, Phys. Rev. Lett. 59 (1987) 712.
[48] C. Patterson, D.F. McMorrow, H. Godrin, K.N. Clausen and B. Lebech, J. Phys.: Condens. Matter 2 (1990) 3421.
[49] H.A. Mook, Phys. Rev. Lett. 46 (1981) 508.
[50] P. Boni and G. Shirane, Phys. Rev. B 33 (1986) 3012.
[51] H.G. Bohn, A. Kollmar and W. Zinn, Phys. Rev. B 30 (1984) 6504.
[52] J.W. Cable, R.M. Nicklow and N. Wakabayashi, Phys. Rev. B 32 (1985) 1710.
[53] J.W. Cable and R.M. Nicklow, Phys. Rev. B 39 (1989) 11732.
[54] P. Lindgård, Phys. Rev. B 27 (1983) 2980; J. Magn. Magn. Mater. 54–57 (1986) 981.
[55] R.M. Nicklow and N. Wakabayashi, Phys. Rev. B 26 (1982) 3994.
[56] B.R. Cooper, Solid State Phys. 21 (1968) 393.
[57] S.H. Liu, J. Magn. Magn. Mater. 22 (1980) 93.
[58] P.A. Lindgård, J. Magn. Magn. Mater. 31–34 (1983) 603.
[59] T. Ziman and P.A. Lindgård, Phys. Rev. B 33 (1986) 1976.
[60] J. Jensen, K.A. McEwen and W.G. Stirling, Phys. Rev. B 35 (1987) 3327.
[61] K.A. McEwen and W.G. Stirling, J. Magn. Magn. Mater. 30 (1982) 99.
[62] B.D. Gaulin, M. Hagen and H.R. Child, J. de Phys. 49 (1988) C8-327.
[63] P. de V. du Plessis, G.H.F. Brits and G.A. Eloff, J. de Phys. 49 (1988) C8-353.
[64] P. Bak and D. Mukamel, Phys. Rev. B 13 (1976) 5086.
[65] H. Kawamura, J. Appl. Phys. 63 (1988) 3086.
[66] O.W. Dietrich and J. Als-Nielsen, Phys. Rev. 162 (1967) 315.
[67] G.H.F. Brits and P. de V. du Plessis, J. Phys. F 18 (1988) 2659.
[68] J. Eckert and G. Shirane, Solid State Commun. 19 (1976) 911.

Neutron scattering studies of magnetic properties of actinide systems

G.H. Lander

Commission of the European Communities, Joint Research Centre, Institute for Transuranium Elements, Postfach 2340, 7500 Karlsruhe, Germany

and

G. Aeppli

AT&T Bell Laboratories, Murray Hill, NJ 07974, USA

In this article we review neutron scattering studies of the magnetic properties of actinide systems over the last 15 years, with particular emphasis on the work since 1984 and the reviews by Rossat-Mignod et al. and Buyers and Holden. In section 2 the results obtained on the actinide dioxides UO_2, NpO_2 and PuO_2 at spallation sources and the recent observation of intermultiplet transitions in UPd_3 and UPt_3 are presented. In section 3 a discussion is given of magnetization densities, starting with the almost localized system such as PuSb and continuing to more recent work on itinerant intermetallic systems such as UNi_2, UFe_2 and $PuFe_2$. In these latter systems the ratio μ_L/μ_S of the orbital and spin moments may be determined by neutrons and compared to theory. In section 4 we present work on compounds with the NaCl crystal structure. This includes complex magnetic phase diagrams, as found for example in UAs and NpAs, the discussion of the results of critical scattering experiments, and neutron inelastic experiments on ferromagnets such as PuSb and UTe. In section 5 the work on single crystals of U-containing heavy fermions is presented. In these systems neutron scattering has been able to characterize the small magnetic moments and the nature of the magnetic correlations, which are frequency dependent. A summary of the significant progress and problems remaining is given in section 6.

1. Introduction

The actinides form a series at the end of the periodic table that consist of a gradual filling of the 5f-electron shell. The elements that concern us here are Th, Pa, U, Np, Pu, Am, Cm and Bk. Heavier actinides are not available in sufficient quantities to envisage neutron scattering experiments in the forseeable future. Thorium is always tetravalent and the 5f state is too far above the Fermi level to be populated. In Pa^{2+} a single 5f electron state should be stable but the energy band is so wide that the 5f and 6d states are essentially indistinguishable, and these electrons form part of the conduction band. Thus it is only in U and the transuranium elements that 5f localization occurs and magnetic properties are found. We shall focus on this aspect in this review. In an earlier article on actinide research with neutrons 1955–1974 the focus was on magnetic structure and the magnitude of the magnetic moments [1]. More recently, the focus has shifted to studying more details of the magnetic behavior [2,3]. In particular, in the heavy fermion materials there is great interest in the magnetic correlations that appear at low temperature and may be responsible, or at least related to, the formation of superconductivity in some of these materials.

Because of limited space we have chosen to cover four topics in this review, which we believe illustrate the importance and unique role of neutron scattering in actinide research. In section 2 we discuss the work at spallation sources on oxides, which has made a major contribution in our understanding of these systems. Some new results on the observation of intermultiplet transitions are also reported. In section 3 we discuss work on magnetization densities; concentrating on the recent results on determining the ratio of orbital spin moments μ_L/μ_S and its relation with band-structure calculations. In section 4 we discuss systems with the NaCl structure; single crystals of which have been the subject of many neutron experiments [3]. In section 5 a brief review of work on heavy-fermion (uranium) systems is presented.

2. Single-ion effects

By "single-ion effects" we mean phenomena that are not dependent on the momentum transfer Q, apart from the usual magnetic form factor. The most obvious of these effects are crystal-field energy levels. A recent review [4] has discussed this in great detail with respect to the lanthanide (4f) series, and we shall not repeat the introduction here. For example, for an ion like U^{4+} with a $5f^2$ configuration the degeneracy of the $2J + 1 = 9$ states is lifted by the geometric arrangement and electrostatic interaction induced by the neighbouring ions. For materials with cubic symmetry, e.g. the oxides, the strength and type of this interaction may be defined by two parameters V_4 and V_6, which represent the strength of the 4th and 6th order crystal-field (CF) potentials, respectively. These two parameters define the ground-state configuration and thus the type of magnetism.

Determining such crystal-field energy levels has been a major preoccupation of light and neutron spectroscopy. Optical spectroscopy [5] is restricted to transparent materials (usually ions in dilute solution) but has excellent energy resolution (≈ 1 cm^{-1} = 0.12 meV) and a very large spectral range (often > 2 eV). Neutrons, on the other hand, can penetrate opaque materials but have resolution worse than optical spectroscopy and a spectral range limited until recently to ≈ 100 meV. The latter is sufficient for most 4f systems but the greater spatial extent of the 5f electrons as compared to those of the 4f shell means that V_4 and V_6 are greater in the 5f series.

The crystal-field energy levels for UO_2 were,

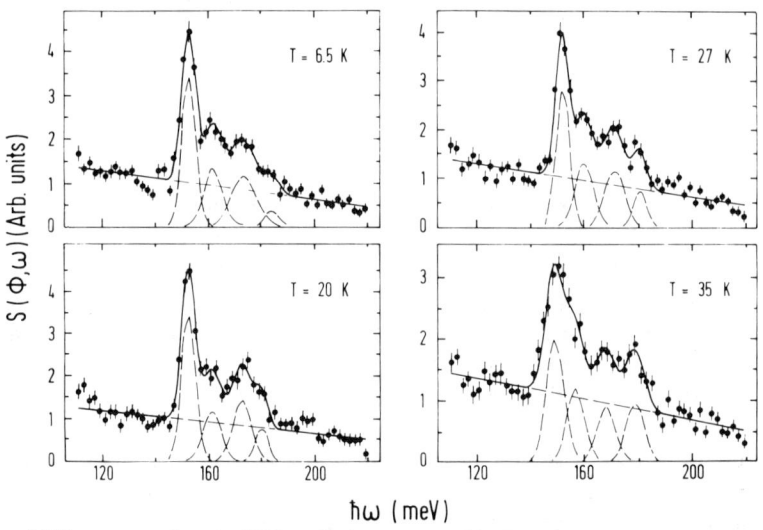

Fig. 1. Neutron spectra of UO_2 measured at the ISIS spallation source with $E_i = 290$ meV for different temperatures. The Néel temperature of UO_2 is 30.8 K. The smooth line is the fit to four Gaussian line shapes and a sloping background. These five components are shown by the dashed lines (from ref. [10]).

in fact, estimated in a classic paper by Rahman and Runciman [6] in 1965, but it was not until chopper spectrometers had been constructed at modern spallation neutron sources [7,8] that the CF states were seen directly. The first experiments indicating the positions of the CF energy levels at ≈ 140 meV were performed at the Intense Pulsed Neutron Source at Argonne National Laboratory and reported by Kern et al. [9] in 1985. The higher intensity and resolution of the HET chopper spectrometer at the ISIS spallation source (Rutherford–Appleton Laboratory) allowed a more complete study [10] to be made and one of these spectra taken at 5 K is shown in fig. 1. From this and other spectra the values of V_4 and V_6 were deduced as -125 and $+25$ meV, respectively. These are only ≈ 50% of the values deduced by Rahman and Runciman [6], indicating that screening of the effective charges at the oxygen sites is more substantial than first-principles calculations estimate. The value of V_4 is about twice that in the corresponding 4f compounds, which is approximately the same ratio as indicated by optical spectroscopy of actinide ions in dilute solution [5].

What was not expected in these studies was the extremely well resolved "hyperfine splitting" of the $\Gamma_5 \to \Gamma_4$ and $\Gamma_5 \to \Gamma_3$ transitions. It was known from diffraction work [11] that below T_N (30.8 K) in UO_2 the oxygen atoms are displaced from their equilibrium positions and the highly resolved inelastic lines reflect the destruction of the cubic symmetry around the uranium ions. To simulate these splittings it is necessary to introduce additional terms in the CF Hamiltonian. Interestingly, the analysis of the inelastic spectra has been able to decide between two possible magnetic structures [2] in UO_2 that both give identical diffraction patterns. The distortion of the oxygen cube around the uranium ion has been ascribed to the electrostatic forces induced by the highly asymmetric uranium magnetic quadrupoles [12].

The situation in NpO_2 is not yet completely clear despite a number of inelastic scattering experiments [13–15]. The ground state is a Γ_8 quartet, which is separated from the next nearest state (also a Γ_8) by ≈ 55 meV. The matrix elements for this transition are large so a clear sharp transition should be seen, but this is not the case. Probably an interaction occurs with an optic phonon involving motions of the oxygen atoms and this gives rise to a q dependence where q is a vector in the first Brillouin zone, of the CF transition. Since with a polycrystalline sample we measure the density-of-states across the Brillouin zone, q-dependent effects give rise to a broad peak. The inelastic experiments have recently [15] been able to shed further light on a very old riddle of what exactly happens at the '25 K transition' in NpO_2. Magnetic and Mössbauer experiments [16] showed that the material does not order magnetically and yet a sharp peak is seen in the specific heat. At low energy (see fig. 2) the use of polarization analysis has shown that the Γ_8 ground state is split into two doublets at 25 K with an energy gap of 7 meV. As in UO_2, this is a manifestation of the ordering of the magnetic quadrupoles. In this case, due to the special val-

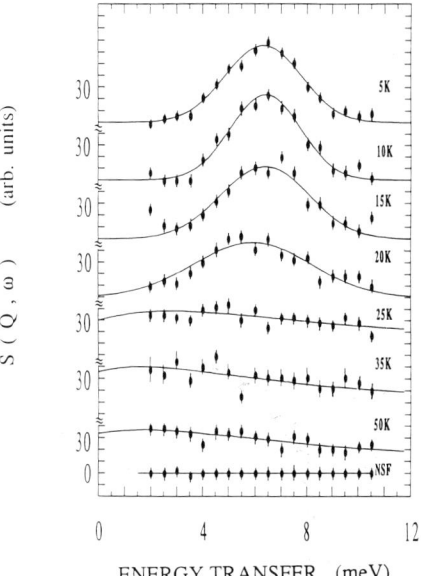

Fig. 2. Polarization analysis spectrum measured from polycrystalline NpO_2 at the IN20 three-axis spectrometer (ILL) as a function of temperature. The magnetic contribution is identified as the spin-flip cross-section. The non-spin-flip (NSF) is independent of temperature. The solid lines are best fits to the data. The transition temperature in NpO_2 is 25 K (from ref. [15]).

ues of V_4 and V_6 the dipole ordered moment in the quantization direction is essentially zero. Such an unusual situation was actually predicted for NpO_2 by Erdos et al. [17].

The experiments on PuO_2 have required the use of the special ^{242}Pu isotope, which was kindly supplied for both the Argonne [18] and Rutherford [19] experiments by Los Alamos National Laboratory. The initial experiments at Argonne discovered a broad CF peak at 110 meV, but the sample also contained some impurity with an OH radical which gave a peak at 90 meV. In PuO_2 the ground state is a nonmagnetic Γ_1 singlet, with the first excited state being a Γ_4 triplet. Although quadrupole ordering cannot effect the Γ_1 singlet, it lifts the degeneracy of the Γ_4 triplet and hence can split the Γ_1–Γ_4 transition.

To summarize the work on the oxides, two things are clear. First, the values of V_4 and V_6 are consistent for all three materials and smaller than previously predicted [6]. This gives us confidence in using Russell–Saunders LS coupling schemes as a good first approximation, although some caveates need to be attached to this statement – see ref. [18] on PuO_2. It is known from the lanthanides series that the CF potentials are larger for ionic-like materials (e.g. the oxides) than for those containing conduction electrons, because the latter contribute to the screening [4]. Thus, for conducting actinide systems, where the CF parameters remain something of an enigma (see section 3) we at least know from the oxides the upper limits of V_4 and V_6. Second, in all three materials there is clear evidence for important effects involving the ordering (either statically or dynamically) of the magnetic quadrupoles. Early work on this possibility in UO_2 by Sasaki and Obota [20] needs to be followed up theoretically now that considerable more experimental information is available. Of course the quadrupole ordering effects that occur in the oxides are, strictly speaking, cooperative effects. We should therefore expect to see dispersion in the energy levels; this is not, however, the case in UO_2 where the levels are extremely sharp in a polycrystalline sample (see fig. 1).

Finally, before leaving these experiments involving dispersionless excitations, we note the

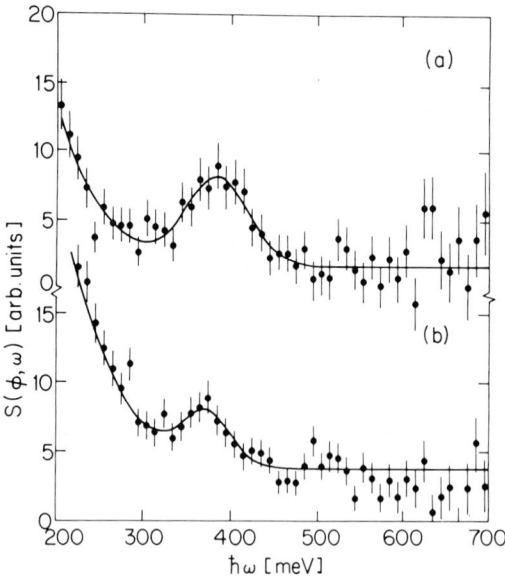

Fig. 3. Neutron inelastic scattering from (a) UPd_3, (b) UPt_3 using an incident energy of 800 meV at the ISIS spallation source. The peaks are assigned to the $^3H_4 \rightarrow {}^3F_2$ transition (from ref. [21]).

observation of intermultiplet transitions (i.e. involving a change in the J quantum number) on UPd_3 and UPt_3 as reported by Osborn et al. [21]. Earlier work by the same authors reported [22] on the rare-earth elements Pr and Nd, but the experiments on UPd_3 and UPt_3 are much more difficult as the transitions are at ≈ 400 meV (see fig. 3) and the kinematic conditions require that neutrons of incident energies of over 1 eV be used. This stretches present-day technology; in fact recent work on USb by Jones et al. [23] has shown that it is even more difficult to see the corresponding IMT in trivalent uranium, such as in the compound USb. The successful experiments show that IMT's exist not only in well-studied materials [3] such as UPd_3, but also in heavy-fermion systems such as UPt_3. They suggest also that the ground state of UPt_3 contains two 5f electrons.

3. Magnetization densities

The wavelength (λ) of thermal neutrons (≈ 1.5 Å) is comparable to the extent of the 5f electron densities $U_{5f^2}(r)$; it follows, therefore, that as the scattering factor $Q = 4\pi \sin \theta/\lambda$, where 2θ is the angle through which the neutrons are scattered, is increased, the scattering intensity decreases. This decrease is called a form factor and is exactly analogous to the well-known scattering factor in X-rays, which is also a decreasing function of Q. The correct definition of the form factor $f(Q)$ is

$$f(Q) = \int M(r) \exp(iQ \cdot r) \, dr,$$

where $M(r)$ is the magnetization density. A determination of $f(Q)$ can therefore give valuable information about the ground state wavefunction [24], and such studies have been applied widely in both the 3d and 4f series.

An early, pioneering experiment in the actinides was performed by Wedgwood [25] on single crystals of US. Unfortunately, US is probably best described as an itinerant system so that the $f(Q)$ could not be easily fit with wavefunctions derived with localized 5f concepts. Such concepts were, however, able to account for the form factors of USb [26] and PuSb [27]. The latter is shown in fig. 4. If the Pu ion is trivalent we should anticipate a $5f^5$ ground state, with Russell–Saunders coupling giving $J = \frac{5}{2}$, $L = 5$, $S = -\frac{5}{2}$. In addition to establishing a $J = \frac{5}{2}$ configuration the measurements on PuSb also showed that a Γ_8 crystal-field ground state was the ground state. This is oblate in shape and relates to the critical scattering effects discussed in the next section. In fact, the analogy with $Sm^{3+}:4f^5$ would predict a Γ_7 state but hybridization effects [28] drive the Γ_8 state lower in energy.

A knowledge of the magnetization [27] (value of μf at $Q = 0$) allows us to say that the form factor of PuSb (fig. 4) exhibits a "hump", or maximum near $\sin \theta/\lambda \approx 0.3$ Å^{-1}. What causes this? Most $f(Q)$ functions show a monotonic decrease. To examine this we need to expand $f(Q)$ in terms of Bessel transforms of the magnetization density [2],

$$f(Q) = \langle j_0 \rangle + \sum_{i=2,4,6} C_i \langle j_i \rangle,$$

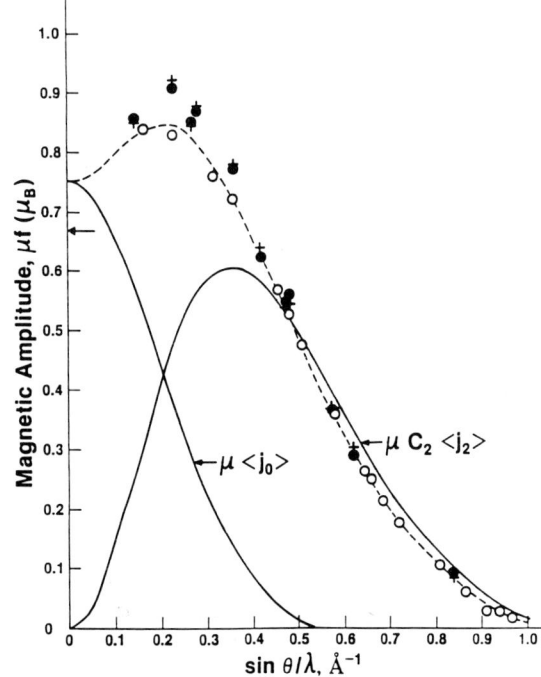

Fig. 4. The magnetic amplitude as a function of $\sin \theta/\lambda$ for PuSb. The individual values of $\mu \langle j_0 \rangle$ and $\mu C_2 \langle j_2 \rangle$, where $C_2 = 3.8$, are also shown for comparison. The open points are fit with the broken curve. The solid points are fit with the crosses, the difference arising from anisotropy in the ground-state wavefunction. The arrow on the μf axis shows the total moment derived from magnetization. The discrepancy between this and the neutron derived value is due to the conduction electrons (from ref. [27]).

where the C_i functions are $C_i(\theta, \phi)$, functions of the polar coordinates only, and the Q dependence is given by $\langle j_i(Q) \rangle$, where

$$\langle j_i(Q) \rangle = \int_0^\infty U_{5f}^2(r) j_i(Qr) \, dr,$$

where $U_{5f}^2(r)$ is the 5f electron density. Within the same approximation it may be shown that the spatial parts of the spin and orbital components, μ_S and μ_L, respectively may be written as

$$f_L(Q) \approx \langle j_0 \rangle + \langle j_2 \rangle,$$
$$f_S(Q) \approx \langle j_0 \rangle.$$

Since $\mu = \mu_L + \mu_S$ we can quickly deduce that

the total form factor in the dipole approximation is

$$f(Q) = \langle j_0 \rangle + C_2 \langle j_2 \rangle,$$

where $C_2 = \mu_L/\mu$. The "hump" in fig. 4 is caused by the near cancellation of μ_L and μ_S, which for less than a half-filled shell are opposite in sign, so that C_2 becomes large. A large $\langle j_2 \rangle$ contribution then appears with the $\langle j_0 \rangle$ terms in $f_L(Q)$ and $f_S(Q)$ cancelling. However, the value of C_2 in PuSb (= 3.8) is in good accord with single-ion theory with a $5f^5$ configuration.

We now turn to a series of intermetallic compounds in which the single-ion theory predictions of the ratio μ_L and μ_S and the C_2 factor fail. The first of these to be examined with enough precision to obtain a value of C_2 was UNi$_2$ by Fournier et al, in 1985 [30], they found $C_2 \approx 6$, whereas for U compounds we anticipate values less than 2. Band calculations more recently have been able to predict the values of C_2 and, in fact, predicted [31] a very large value in UFe$_2$ before it was observed [32]. The form factor measured for the U site in UFe$_2$ is shown in fig. 5, together with a schematic of the spin and orbital magnetization densities. Similar experiments and theory have been performed for NpCo$_2$ [33] and PuFe$_2$ [34].

A useful way of discussing these experiments and band-structure results is to plot the ratio $-\mu_L/\mu_S$ versus 5f count as done in fig. 6. Note that $C_2 = (2-g)/g$ and $\mu_L/\mu_S = C_2/(1-C_2)$. The crosses are determined from intermediate coupling and single-ion theory. The ions in many actinide compounds fall on this curve, as would *all* rare-earth materials. Transition metal materials with μ_L quenched and $g \approx 2$ would obviously

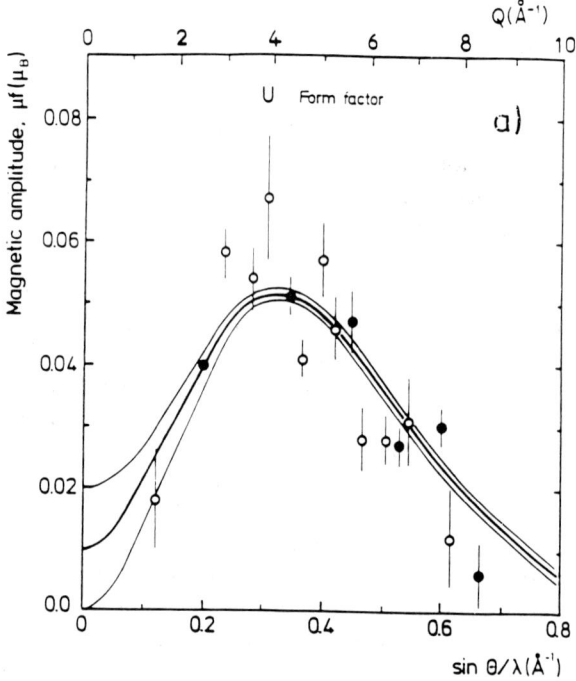

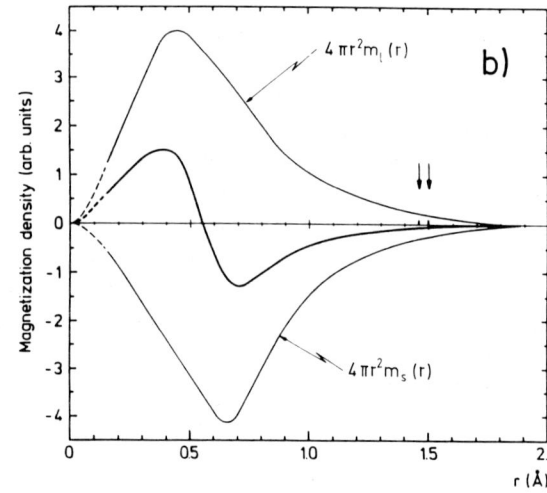

Fig. 5. (a) The magnitude of the magnetic scattering on the uranium sublattice in UFe$_2$ as a function of the scattering angle. The extrapolation to $\sin\theta/\lambda = 0$ gives the total moment, in this case almost zero. Note the maximum in the scattering cross section for $\sin\theta/\lambda \approx 0.3$ Å^{-1}. The different solid lines indicate the limits of the extrapolated moment on the uranium, $0 \le \mu \le 0.02\mu_B$. (b) Schematic representation of the orbital ($4\pi r^2 M_L$) and spin ($4\pi r^2 M_S$) components of the uranium magnetization density in UFe$_2$ as a function of the distance r from the nucleus (oscillations at small r have been omitted for clarity). The difference (bold line) gives the total magnetization density, which clearly has two peaks, one positive and one negative. The integrated area of the two is the same, so that in this case the total moment is zero. The form factor is the Fourier transform of the bold line; this has a maximum at $Q > 0$ because of the reversal of sign of the bold line (from ref. [32]).

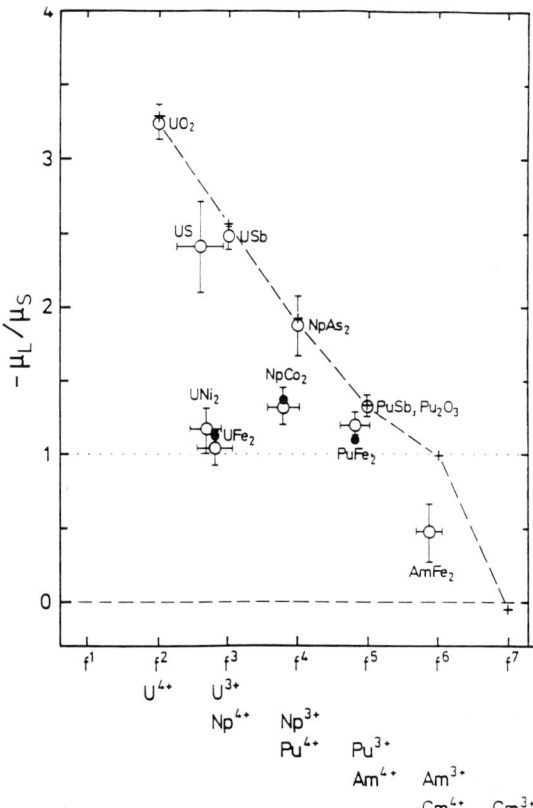

Fig. 6. Dependence of ratio between orbital μ_L and spin μ_S moments on f count in actinide materials. The crosses are derived from spectroscopic g values, which include intermediate coupling, and assume localized f electrons. Experimental values derived from neutron experiments are shown as open points. The values derived from the theory discussed in this paper are shown as solid points (from ref. [35]).

fall near the $\mu_L/\mu_S = 0$ line. The new aspect are the intermetallic compounds and they must surely be considered as itinerant systems. The theoretical values (solid circles) are in good agreement for the ratio μ_L/μ_s, although the individual values of μ_L and μ_s are too large, often by a factor of more than two [36]. The plot shows that the hybridization of 5f and 3d electrons results in a partial delocalization of the former, with a corresponding reduction in their orbital moment μ_L. The largest effects appear to occur in uranium material; this is to be expected because the 5f electrons are most extended in the uranium. The hybridization also modifies the magnetic behavior of the 3d atom; for example in UFe the iron moment [32] is reduced from the elemental value of 2.2 to $0.6\mu_B$.

In conclusion, studies of magnetization densities can give extremely important information on the ground-state wavefunction. In particular, the experiments on intermetallic systems as discussed in connection with fig. 6 show that certain properties (in this case orbital magnetism in itinerant systems) are unique to the actinides and have no parallel in other electron systems.

4. Studies of systems with the NaCl structure

The simple rocksalt (fcc) structure is formed with all actinide ions and elements of group VA (e.g. N, P, As etc. – the pnictide series) and group VI (e.g. S, Se, Te – the chalcogenide series). Because of their simple structure and variety of magnetic properties, a vast number of studies have been done not only on the pure compounds, but also on a series of pseudo-binary solid solutions. Some of this work, especially of the carbides and nitrides, has been motivated by the fact that these materials are potential fuels for advanced type of reactors. More importantly, however, in motivating the field was the development by Oscar Vogt at ETH, Zürich, in the early '70's of the mineralization method for the growth of single crystals. In the early '80's this technology was transferred to the European Institute for Transuranium Elements, Karlsruhe. A steady supply of large crystals for both uranium compounds (from Zürich) and transuranium (Np, Pu) compounds has allowed many workers to make contributions with such diverse techniques as magnetization, resistivity, specific heat, optical reflectivity, Mössbauer spectroscopy, μSR, neutron elastic and inelastic scattering, angle-resolved photoemission, and X-ray synchrotron scattering. In this short article we have no space to cover even a fraction of this work and refer to the many chapters in the Handbook of the Physics and Chemistry of the Actinides, vols. 1 and 2, see refs. [2,3] in particular.

We choose instead to cover three series of neutron experiments, which are representative of the type of studies that have been conducted over the last fifteen years.

4.1. Magnetic phase diagram of NpAs

The magnetic phase diagram of NpAs is shown in fig. 7. The general overview of the phase diagram with a field applied [37] is shown in fig. 7(a); more details of the high-temperature part [38] is shown in fig. 7(b). Phase diagrams of this level of complication are common in these materials, although NpAs is a fairly extreme case. The key aspect of the magnetic structures is that they consist of strongly correlated (001) planes of actinide moments with the planes weakly coupled together. The moments are directed perpendicular to the planes, i.e. in the [001] direction, so the spin structure is a longitudinal one. Such a coupling scheme closely resembles two-dimensional spin structures found, for example, in K_2NiF_4 [39] and the newly discovered doped La_2CuO_4 systems [40]. However, in the latter materials the large difference between the intra- and interplane exchange arises from the difference in the distances. But the actinide systems are cubic; this is the central point of the physics, a point to which we shall return below.

To return to fig. 7. There are a number of features worth noting. First, at high temperature, the structure is incommensurate with a wavevector somewhat less than $q = \frac{1}{4}$, therefore a repeat cell $(1/q)$ of slightly more than 4 unit cells. This phase can be easily destroyed by a magnetic field, converted into a paramagnetic state. At lower temperature, $T_0 < T < T_{IC}$ the material locks into a commensurate behavior, in this case $\frac{1}{4}$, corresponding to almost a square wave with a modulation $4+$, $4-$, $4+$, $4-$ etc. Application of a magnetic field to this structure results in first a ferromagnetic state, and then at the relatively low field of ≈ 30 kOe a paramagnetic state is induced.

At $T_0 \approx 140$ K a dramatic transition occurs in NpAs in which a first-order transition to a type I $+-+-$ $(q = 1)$ magnetic structure occurs. The structure is of the $3k$ type [2] in which the resultant moment is actually along $\langle 111 \rangle$ but the diffraction experiments pick out a particular component, m_x, m_y or m_z. The magnetic structure now has cubic structure so that the lattice symmetry also returns to cubic [41]. The $3k$ structure is a common feature of many of the actinide NaCl compounds, particularly for the heavier anions Sb

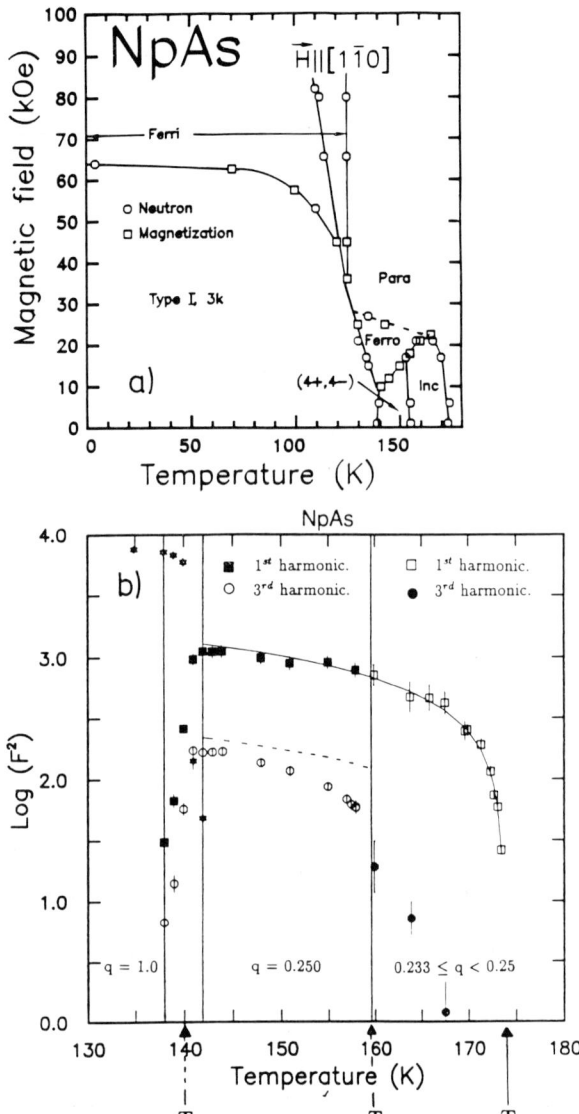

Fig. 7. Magnetic phase diagram of NpAs (a) as a function of magnetic field applied in a [110] direction, taken from ref. [37]. (b) More details of the high-temperature part taken from ref. [38]. Note in the latter the growth of the 3rd harmonic which in the $q = 0.250$ results in NpAs forming a pure square wave, $4+$, $4-$.

and Bi, and is extremely stable against an applied magnetic field. In fig. 7(a) we see that fields of over 60 kOe are required even before a single ferromagnetic component is induced.

The transition to the type I structure in NpAs is almost certainly an electronic one. On cooling into this phase the resistivity increases [41] by a factor of almost 100. Further work is needed to determine whether it is the 5$\underline{f}$ electrons that are partially localizing at T_0, and thus causing a reduction in the resistivity. Certainly the materials with the 3k structure exhibit properties resembling localized systems. One of the normal criteria of a localized–delocalized transition is a volume change; however, magneto-elastic effects [42] often prevent a clear identification of a purely electronic transition.

4.2. Magnetic critical scattering effects

The highly anisotropic nature of the magnetic correlations in NaCl-type actinide systems was first demonstrated [43] in USb in 1978, shortly after single crystals of these materials became available. More recently some aspects of these effects in USb have been remeasured with better energy analysis [44,45]. The most complete work is probably on UAs [46], which is discussed in detail in ref. [2]. Similar effects are found also in the Ce-monopnictides [47].

The anisotropic behavior is best seen just above the ordering temperature of antiferromagnets. Because the (001) planes are strongly correlated within the plane and only weakly between planes, the fluctuations within the plane have a mean correlation length $\xi_\perp$ (the $\perp$ sign to note that these correlations are perpendicular to the wave-vector direction q) that is longer than the mean interplanar correlation length $\xi_\parallel$. It is the ratio $R = \xi_\perp / \xi_\parallel$ that can be measured experimentally and is given for a variety of actinide and cerium materials in table 1, which is taken from ref. [38]. The parameters β and γ refer to the temperature dependence of the order parameter below T_N and the correlation lengths above T_N, respectively. In the few studies done with high precision R is found independent of T, as is expected on the basis of symmetry arguments.

The energy dependence of the correlation lengths is small and has not been examined in great detail as yet. The one study where this was addressed [45], on USb, found that scaling argu-

Table 1
Summary of critical parameters in Ce, U, Np and Pu monopnictides. R is the anisotropy ratio. Numbers in parenthesis refer to standard deviations of the least significant digit. See ref. [38] for the original references to the earlier work

	Lattice parameter (Å)	T_N (K)	β	ν	R
CeAs	6.078	8	–	–	0.6(1)
CeSb	6.412	16	first-order transition		1.8(2)
CeBi	6.487	25.4	0.317(5)	0.63(6)	2.5(2)
UN	4.890	≈ 54	0.31(3)	0.84(5)	2.8(3)
UAs	5.779	124	first-order transition		3.8(5)
USb	6.191	212.2	0.32(2)	0.68(4)	5.0(5)
NpAs	5.838	173.6	0.38(1)	0.73(2)	2.9(5)
NpSb	6.254	199.0	0.257(5)	–	4.5(10)
NpBi	6.438	192.5	0.31(2)		
PuSb	6.225	85.3	0.31(2)	0.58(5)	1.8(3)
classical mean field			0.5	0.5	
3D Heisenberg			0.345	≈ 0.7	
3D Ising			0.3125	0.64	
2D Ising			0.125	1.0	

ments could be used to relate parameters of the inelasticity to the static susceptibility. We should expect such a relationship.

In table 1 there is a rough correspondence of the value of β to the 3D Ising value of 0.3125, and ν should be $\approx 2\beta$. (There are difficulties in measuring ν in UN; see the original reference.) However, an exact calculation of these parameters is only possible within the framework of a model such as employed by Kaski and Selke [48] and called the ANNI model. Since some of these materials are near so-called Lifschitz instabilities [46] (first-order transitions, for example, occur in CeSb and UAs), the theoretical estimates for the critical exponents are difficult to perform.

The anisotropy ratio, R, has been considered for Ce and Pu systems by Kioussis and Cooper [49], Hu et al. [50] and Cooper et al. [51]. The R values for these systems do not exceed 3, and they obtain reasonable agreement with experiment. We should emphasize that the introduction of anisotropic interactions is central to the theories of Cooper [52] (1985) so the first-principles calculation of these R values represents an important test for these calculations. As noted by Burlet et al. [53], the R value is largest fo USb and NpSb both of which have triple-k structures. This suggests that these strong interactions, which give rise to a high ordering temperature, are also important in stabilising the triple-k structure, and have been discussed [46,54] in terms of direct mixing between the 5f electrons and the anion p-band. Evidence for such mixing can be found in Mössbauer experiments at the Sb site in USb and NpSb [55].

4.3. Inelastic response function for NaCl-type compounds

The measurements of the inelastic response function, $S(Q, \omega)$, represents the most difficult challenge in actinide systems – concomitantly, it gives the most microscopic information to the theorists. We confine our attention in this section to the materials with the NaCl structure, all of which order magnetically [2,3]. More generally, a large number of nonordering uranium-containing compounds have been studied in polycrystalline form. These measurements are usually confined to fitting $S(Q, \omega)$ to a single Lorentzian function with a half width of inelastic scattering Γ as the only parameter. The Q dependence cannot be determined usefully, except in a few special cases, from experiments on polycrystalline materials, and a usual U form factor is assumed. The interest is in the absolute value of Γ and its T dependence. This linewidth is often compared with the spin-fluctuation temperature T_{SF}, as determined from magnetic susceptibility and resistivity measurements. A comparison of work of this kind on cerium and uranium systems is given by Shapiro [56]. A recent study [57] of USn_3 also discusses the pros and cons of these type of experiments.

A complete review of the work on magnetically ordered NaCl-type compounds up to 1985 has been given by Buyers and Holden [2]. The most important point to emphasize in this field is the *absence* of well-defined magnetic excitations in those materials such as UN, UAs, US with the closest U–U separation and the smallest moments. At the time of these first measurements, especially on UN [58], this situation was unique in that well-defined excitations are always associated with ordered magnetism. We can now see that these experiments presaged those on heavy-fermion systems (see section 5 below), in which no conventional excitations have been found either. These broad features are especially difficult to study in neutron scattering; the intensity when integrated over any volume of (Q, ω) space is small, and very large (> 1 cm^3) samples are required. Careful experiments have been performed [58] on UN showing broad scattering at the X-point extending up to over 50 meV in energy transfer and with an apparent gap of ≈ 12 meV. It is intriguing to ask why no sharp excitations are seen in these systems. There are two schools of thought in this, although quantitative calculations have not been performed for either model. The first model (derived from the Anderson Hamiltonian) assumes that there is strong mixing between 5f electrons and the 6d and 7s conduction electrons. This mixing reduces the effective 5f moment and gives rise to ill-defined magnetic excitations. A direct consequence of this model is the anisotropic mixing discussed earlier in respect to the effects seen in critical scattering. The second qualitative idea concerning this loss of sharp excitations goes back to our observation of usual form factors discussed earlier, and the effective reduction, rather than a complete quenching as is the case for transition metals, of the orbital moment. The loss of a complete relationship between the orbital and spin throughout the Brillouin zone, as a consequence of the 5f electrons being itinerant, means that the Wigner–Eckert theorem does not apply. Under these circumstances it is not at all clear that sharp well-defined excitations should exist, even if there is an ordered moment at low temperature. Clearly, this aspect requires further theoretical work before a complete understanding will be available.

In this class of experiments on systems with the NaCl structure, two important experiments have been done since the review of Buyers and Holden [2]. The first we shall discuss, is the experiments [59] on single crystals of ^{242}PuSb. Both this and the experiment on UTe have the common feature that they have been performed on *single-domain* crystals. Since these materials are highly anisotropic, a point made earlier, and well known from magnetization experiments [60,61] and recent neutron work to measure the anisotropy constant directly [62], it is important to attempt to examine this at the microscopic level. For antiferromagnets with ordering wave vector q this is possible because the behavior from the components $[00q]$ appears at a magnetic zone center of $[000] + [00q]$, which is clearly at a different place from the zone center $[q00]$. However, for ferromagnets $q = 0$, so this separation in reciprocal space does not occur. Fortunately, if they are cooled in a modest field (≈ 5 kOe) a single-domain state is induced; the fields can then be removed with only a small loss of $\approx 5\%$ in the surface domains reverting to random.

We now discuss the experimental results for PuSb [59] with the aid of fig. 8, which has been taken from the theoretical work on this material [63]. The first point to notice is the almost complete lack of dispersion in the experimental curve. This is a little unexpected as PuSb orders at ≈ 90 K and becomes ferromagnetic at 65 K with an ordered moment of $0.75\mu_B$ per Pu. The earlier form factor work [27] (fig. 4) had shown that the ground state is a Γ_8 quartet and the predominant wave function is $M_J = |\frac{5}{2}\rangle$ as shown in level 1 on the right-hand side of fig. 8. The strongest transition is to level 4, i.e. L_{41} corresponding to a J_- operator, and the theory predicts this as the strongest with some dispersion, more in fact than observed experimentally. We assume that the observed scattering is from the L_{41} mode so that a slight renormalization of the theoretical L_{41} mode would bring better agreement with experiment. In the theory only the static properties, e.g. T_C etc., are used to predict the dynamic behavior.

The anisotropy gap at Γ (4 THz $\approx$ 16 meV $\approx$ 185 K) is large and consistent with very large anisotropies in these ferromagnet systems. For a ferromagnet the anisotropy gap is given by $E_0 = 2K_1/\mu$, where K_1 is the anisotropy constant and μ is the ordered moment. If $E_0 = 4$ THz in PuSb (fig. 8) the $K_1 \approx 1.6 \times 10^8$ erg/cm^3. This is comparable to the highest value of $\approx 5 \times 10^8$ erg/cm^3 ever observed [64] in a cubic compound containing rare-earth ions, viz. TbFe$_2$ and it is not surprising that magnetization experiments [61] on PuSb were not able to move the magnetic moment away from the easy $\langle 100 \rangle$ direction. The theoretical anisotropy is consistent with the experiment. (Brooks et al. [65] predicted similar large anisotropies with one-electron band theory including orbital polarization.)

However, possibly the most interesting aspect of the experimental results is that the lowest energy mode occurs at the X-point with q perpendicular to the moment direction μ (extreme left of fig. 8). We can be more precise than this, the experiments on a single-domain sample show that the mode that shows the lowest energy at X is that with the fluctuations parallel to the propagation direction, i.e. a *longitudinally polarized fluctuation*. As yet, this cannot be reproduced by the theory [63]. If we think qualitatively of this mode it is as if PuSb wants to adopt a longitudinal antiferromagnetic structure with $q = 1$.

The second example we shall present is the recent result [66] on a single-domain crystal of UTe. The dispersion curve for this is given in fig. 9. This material had been investigated previously

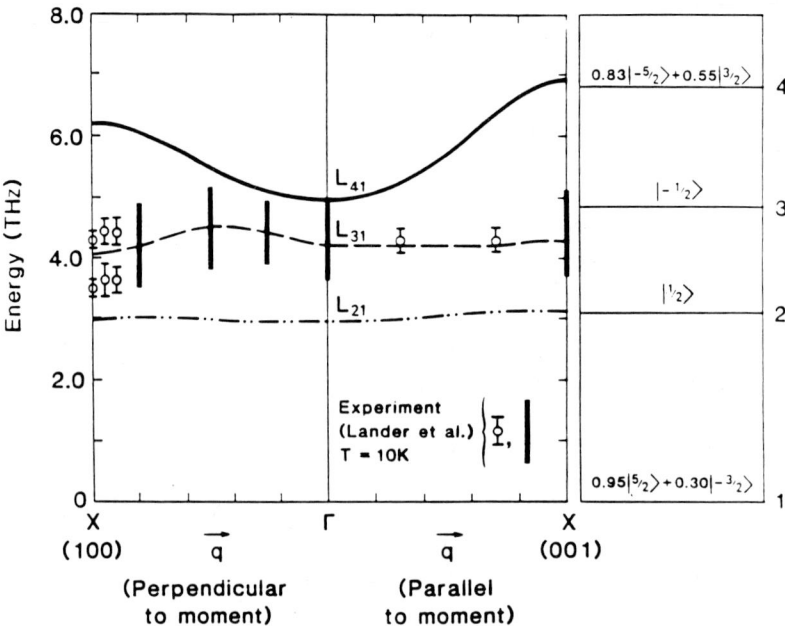

Fig. 8. Dispersion curves for PuSb (solid vertical bars and open points) as taken from ref. [59], together with the theoretical results derived by Banerjea and Cooper [53]. Molecular-field states are shown to the right taking the [100] axis of quantization. Both theory and experiment are at low temperature ($T = 10$ K).

[2], but not in a single-domain state. Although at low q the dispersion curve is isotropic, this is not true at larger q, so that measurements on a multidomain sample represent a complex average of the excitation spectra. Measuring the anisotropy gap, of course, does not depend on having a single domain so that fig. 9 is in agreement with the earlier work [2]. If the spin-wave propagation is parallel to the moment direction (all the uranium ferromagnets have $\langle 111 \rangle$ easy axes) then the dispersion curve is defined across the whole zone. The spin wave stiffness D in the familiar expression $E = E_0 + Dq^2$ is ≈ 10 THz Å^2, consistent with a ferromagnet ordering at 100 K with $2.2\mu_B$. The suprising feature is when the propagation is perpendicular to μ, left-hand side of fig. 9. In this case it becomes very difficult to see the excitation spectrum for $q > 0.3$ rlu, and this is indicated by the wide hatched area in fig. 9.

Other features of the UTe spectra are also puzzling. The excitations at $T = 5$ K are broader than the experimental resolution function at all q values, and as the temperature is raised the excitations become strongly damped. Experiments [67] on polycrystalline material at the Rutherford spallation source ISIS have shown that $S(Q, \omega)$ for UTe at low temperature also contains a broad inelastic component that extends out to 80 meV (≈ 20 THz) and which increases in amplitude at high temperature.

All features of the UTe neutron inelastic spectrum point to a system with strong hybridization between the 5f and conduction-electron states, and on the borderline of instability towards an itinerant system (such as UN or US). A similar strong interaction between the 5f and conduction states was deduced by Schoenes et al. [68] from transport measurements. Clearly, under these

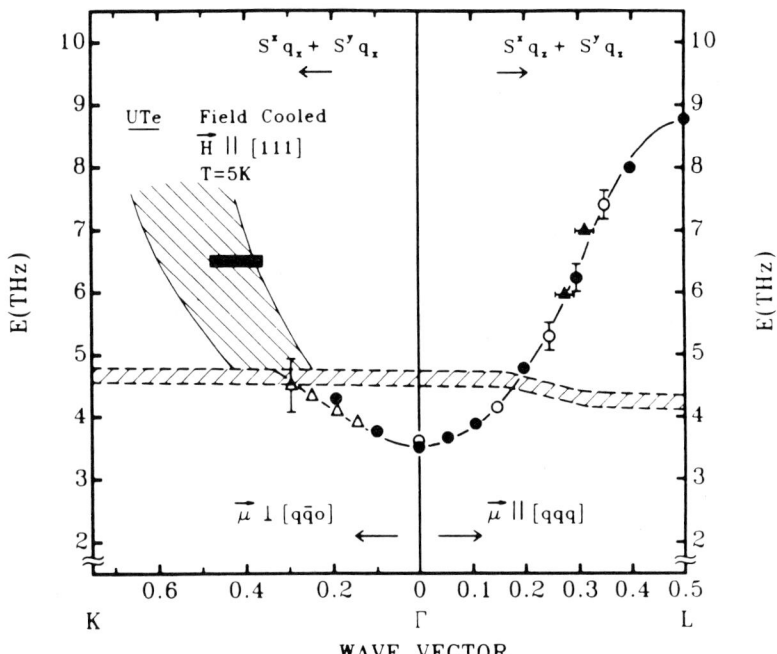

Fig. 9. Dispersion curve for UTe along the q_z (parallel to the moment) and q_x (perpendicular to the moment) directions at 5 K. The narrow hatched line shows the optic-phonon frequency. The different symbols correspond to different spectrometer conditions. The wide hatched area on the left-hand side corresponds to the region where we are unable to assign specific peak positions (from ref. [66]).

conditions, it is unreasonably to anticipated theory to be in perfect agreement with experiment. Hu et al. [69] have made an initial attempt to understand what happens with their model when the small level of hybridization for Ce and Pu compounds is increased to handle the more complex situation involving $5\underline{f}^2$ and $5\underline{f}^3$ configurations. They do find that the excitation spectra are damped, in particular in the case of UTe they find that all excitations are degenerate for $Q \parallel \mu$ but are not for $q \perp \mu$. In the latter case with more than one excitation at larger q values it may approximate the experimental situation in which a broad 'hump' of scattering is observed for $q > 0.3$ rlu.

To summarize the neutron inelastic scattering on NaCl structure systems. Although it may be convenient conceptually to start from a lanthanide localized $\underline{f}$ picture + crystal field + molecular field, the most important interactions are those between the $\underline{f}$ and conduction states. Thus UN is closer to a heavy-fermion system (see V below) than it is to UO_2 (see section 2), but both the relatively large ordered moments (0.7 to $2.8\mu_B$) of the NaCl systems and their strong anisotropy are, of course, more characteristic of predominantly localized $\underline{f}$ systems. This dichotomy remains the principle challenge to theory. The microscopic information provided by neutron inelastic scattering remains the most exacting from that point of view.

5. Heavy fermion systems

Heavy fermion systems [70] are intermetallic compounds where on cooling, the magnetic entropy is not lost through conventional phase transitions, but is converted into fermionic form. Because the conversion generally takes place at low temperatures ($T \lesssim 100$ K), the associated Sommerfeld constants and Pauli susceptibilities χ_0 can be up to three orders of magnitude larger than in conventional metals. Even more remarkable than heavy fermion paramagnetism is the usual superconductivity in four Ce- and U-intermetallics. Because magnetic fluctuations account for the heavy fermion phenomenon and are strongly favored as the source of heavy fermion superconductivity, there have been extensive neutron scattering investigations of these compounds [71–73]. Initially, single crystals large enough for such studies were unavailable, and experiments were restricted to polycrystalline and powder samples [71]. These early experiments demonstrated that the local moments responsible for the high-temperature Curie–Weiss susceptibility accounted at low T for a continuum of magnetic fluctuations with a bandwidth of the order of the renormalized Fermi temperature, and with weight, obtained from Kramers–Kronig analysis, comparable to χ_0. Thus, Occam's razor applied to the polycrystalline data indicated that the magnetic fluctuations in the heavy fermion compounds are analagous to those in almost ferromagnetic single component Fermi liquids such as ^{3}He. Subsequent neutron scattering measurements [72,73] on single crystals showed that this conclusion is very far from the truth. Indeed, *antiferromagnetic* local moment fluctuations dominate the magnetic response of heavy Fermion systems. Although a small (accounting for less than 20% of χ_0) ideal Fermi-liquid-like contribution to $\chi''(Q, \omega)$ must be present to account for de Haas–van Alphen oscillations [74], it still awaits definite experimental verification. Thus, the remainder of this section is devoted to the antiferromagnetic order and fluctuations found in the three U-based heavy Fermion compounds, U_2Zn_{17}, URu_2Si_2 and UPt_3, for which single crystal neutron scattering experiments have been reported.

5.1. U_2Zn_{17}

U_2Zn_{17} has two U moments per (rhombohedral) unit cell which, according to neutron diffraction [75], order antiparallel to each other at $T_N = 9.7$ K. Only roughly a third of the paramagnetic moment orders below T_N. The remainder accounts for a large linear term ($\gamma = 200$ mV/mol K^2) in the $T \to 0$ specific heat C, which also includes a cubic term of magnetic origin [76]. The latter are typically associated with spin waves in three-dimensional antiferromagnets. Consideration of the magnitudes of the ordered moment,

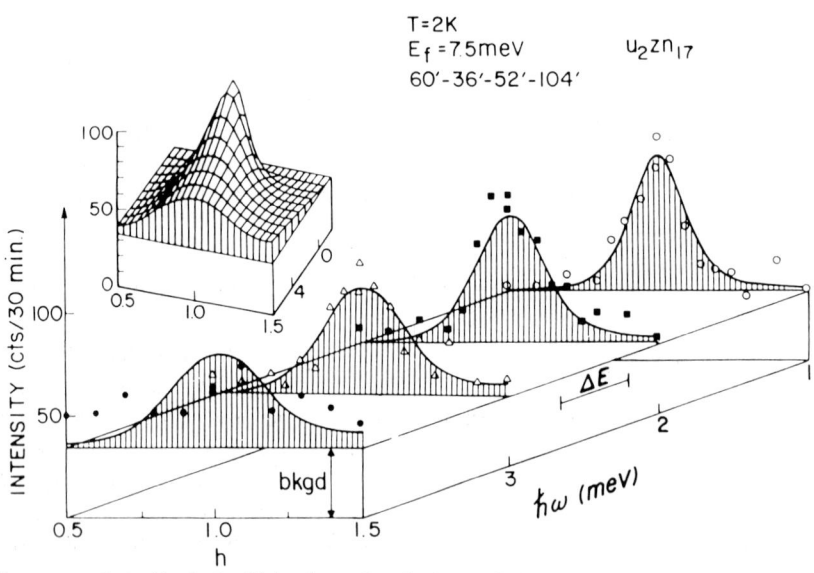

Fig. 10. Constant $\hbar\omega$ scans along $(h, 0, 3-h)$ in the ordered phase of U_2Zn_{17} at $T = 2$ K. Solid lines are from fit to the 3-parameter model described in text. Inset shows the corresponding model function in a perspective view. From ref. [77].

and the linear and cubic terms in the specific heat suggests that U_2Zn_{17} is a heavy electron analog of Cr, which is thought to have a partially gapped Fermi surface. Emboldened by this analogy, we set out to find the spin waves in U_2Zn_{17} [77]. Because U_2Zn_{17} is a heavy fermion system which by definition would have low Fermi velocity, one felt that the spin waves would be easily resolved. Fig. 10, which shows the outcome of the inelastic scattering experiments, demonstrates that this is not the case. Instead of appearing in two ridges converging at the (102) point, the scattering is concentrated in a single rod parallel to the energy ($\eta\omega$) axis and centered at (102). Thus, U_2Zn_{17} is not so much a model itinerant antiferromagnet as it is the heavy fermion partner of UN. Polarized beam measurements [78] on a single domain sample of U_2Zn_{17} show that the q-dependent scattering is primarily transverse, while a weakly q and ω-dependent background of roughly equal intensity is due to longitudinal fluctuations. The latter are of sufficient amplitude and occupy enough phase space that they could well be the end-product in the decay channel which prevents the spin waves from propagating.

On warming through T_N, it is no longer possible to distinguish between longitudinal and transverse modes. At any given T, the generalized susceptibility of a lattice of coupled Kondo impurities, whose ingredients are a two-spin (RKKY) coupling J_0' and an effective single-ion susceptibility $\chi_0(\omega) = \chi_0 \Gamma/(\Gamma - i\omega)$, corresponds well to the inelastic data. However, as fig. 11 shows, the parameters entering the model depend on temperature in a very interesting way. Most notably, the magnetic transition which occurs in T_N when $4\chi_0 J_0' \to 1$, is driven by an increase in the coupling J_0' rather than an increase in the single ion susceptibility χ_0 as is the case for ordinary magnets. Furthermore, an *increase*, rather than the conventionally anticipated decrease in the local fluctuation rate Γ coincides with the transition. This effect correlates with the decrease in the linear specific heat [76] and can be attributed to the rise in Kondo temperature which occurs when a smaller f-moment is screened by the conduction electrons. In contrast, the T-dependent J_0' is by its nature not a single ion effect. However, it

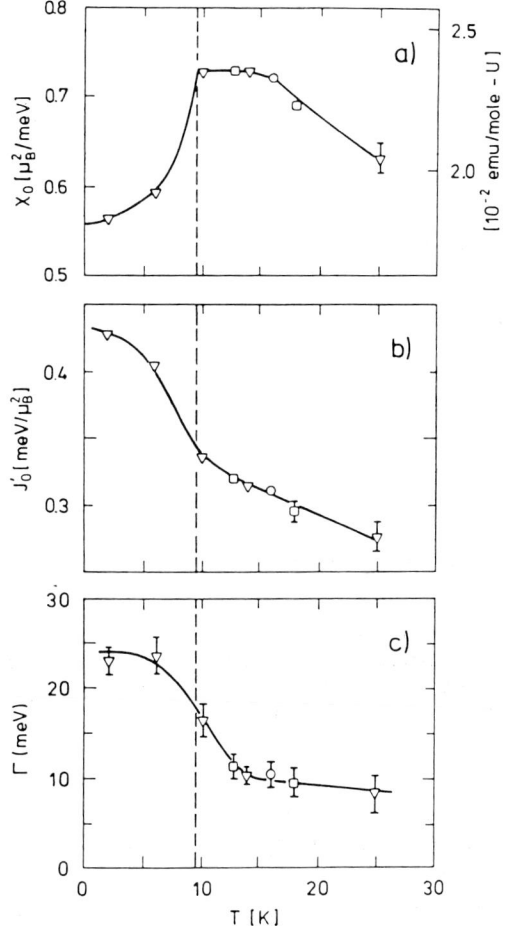

Fig. 11. Temperature dependence of model parameters obtained from fits to U_2Zn_{17} data. From ref. [77]. The different symbols correspond to different experimental runs.

represents the most direct experimental evidence for the renormalization, predicted for two Kondo impurities coupled to the same conduction sea [79], of the RKKY interaction.

5.2. URu_2Si_2

URu_2Si_2 has been a popular heavy fermion system because it is often superconducting below $T_c \approx 1$ K and can be easily prepared in single crystalline form [80]. Strictly speaking, it is only heavy (with modest $\gamma = 180$ mJ/mol K^2) for temperatures above $T_N = 17.5$ K, where a peculiar (see below) Néel transition occurs. For $T < T_N$,

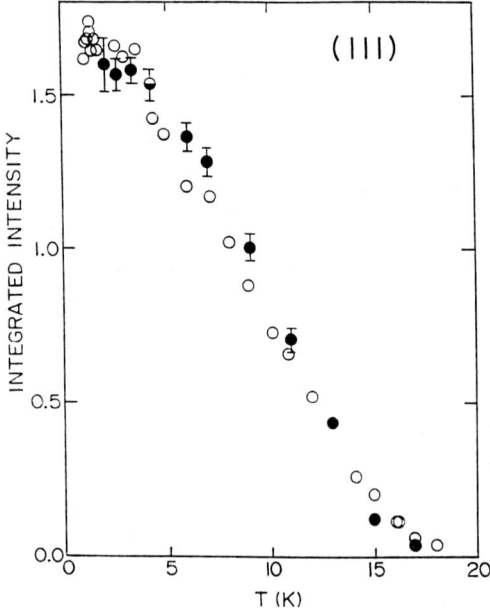

Fig. 12. Temperature dependence of integrated intensity of the (111) magnetic Bragg peak for URu_2Si_2. Solid and open circles are neutron and X-ray scattering results, respectively. From ref. [83].

$\gamma = 50$ mJ/mol K^2, comparable to the value for UN below its Néel temperature. Neutron scattering [78,81,82] has revealed that URu_2Si_2 is different from other heavy fermion systems in a sense much more profound than suggested by the small value of γ. One surprise [81] was that the ordered moment ($\mu_0 = 0.04\mu_B$ per U atom) appearing below T_N is much smaller than that which one would naïvely deduce from the specific heat data. Furthermore, its temperature dependence is very sample-sensitive, following (see fig. 12) in the best samples the form $|\mu_0(T)|^2 \sim (T_N - T)$ over a wide temperature range [83]. Both of these results are in sharp contrast to the behavior of U_2Zn_{17} and $U_{0.95}Th_{0.03}Pt_3$ [84], where the naïve thermodynamic estimates of μ_0 do apply and $\mu_0(T)$ evolves conventionally.

While not being associated with the establishment of an ordinary Néel state, the transition at T_N in URu_2Si_2 is marked by a dramatic change in magnetic dynamics [81,82]. Fig. 13 shows how on lowering T, propagating excitations with well-defined energy emerge from continuum scattering not unlike that in all other heavy Fermion systems. Frames (a)–(c) show energy scans with $Q = (1, 0.4, 0)$ taken at three temperatures, while frames (d)–(f) give the T-dependence of the amplitude A, damping Γ and peak frequency Δ, characterizing the magnetic response measured in scans like those in (a)–(c). Of special significance is that the growth of Δ is much more order-parameter-like than that of μ_0.

URu_2Si_2 is very reminiscent of the well-understood rare-earth singlet ground state magnets. Broholm and coworkers [78] attempted to model URu_2Si_2 in the same manner, but could not reconcile the behavior and size of the magnetic moment with the inelastic neutron scattering and thermodynamic data. They concluded that a primary order parameter different from a simple dipole moment must therefore account for the transition at T_N. The most likely candidate is quadrupolar, as also proposed for UPd_3 [85] and suggested by the λ-like anomaly in the thermal expansion [86] of URu_2Si_2 at T_N. Theorists [87], however, have suggested more exotic possibilities, among them order parameters consisting of products of dipole operators at different sites.

5.3. UPt_3

UPt_3 is doubtless the most studied heavy fermion system. Almost everything, ranging from de Haas–van Alphen oscillations [74] to muon spin relaxation [88], has been measured in this compound. Of special importance is the power law dependence on T of various bulk properties in the superconducting state. Two properties, namely absorption of transverse ultrasound [89] and magnetic penetration depth [90], have been shown to obey power laws T^x where x depends on crystallographic direction. Thus, among all superconductors, UPt_3 is the most likely to possess an anisotropic paired state. Such states are not generally thought to arise from electron–phonon coupling. Instead, many theories [91] involve antiferromagnetic fluctuations as the mediating bosons. It was thus very gratifying that around the time when these theories were developed, inelastic neutron scattering measurements [73] revealed antiferromagnetic correlations in UPt_3.

Fig. 14 shows representative low-T data in the form of constant-$\hbar\omega$ scans. Applying Fourier's theorem, these data together with others show that neighboring U moments in this hexagonal closed-packed material are antiferromagnetically correlated. In addition to potentially bearing responsibility for the pairing in UPt$_3$, the antiferromagnetic correlations develop below 20 K, thus also accounting for the previously unexplained bulk susceptibility maximum near this temperature and related "metamagnetic" phenomena in the applied field-T plane [92].

After the discovery of superconductivity in UPt$_3$, researchers immediately set out to study the effect of alloying on the superconducting and normal state properties. They found that superconductivity was destroyed by <1% substitutions of other elements on both the U and Pt sites [93]. In addition, when sufficiently large amounts of certain elements were introduced, the alloys underwent well-defined bulk phase transitions at temperatures of order 6 K. Neutron diffraction experiments [84] showed that for U$_{0.95}$Th$_{0.05}$Pt$_3$ and U(Pt$_{1-x}$Pd$_x$)$_3$ these transitions were to magnetic ground states with ordered moments $\mu_0 \approx 0.6\mu_B$/U ion, similar to those found in U$_2$Zn$_{17}$ and UN. The thermodynamic behavior as well as the T-dependent order parameter are also much

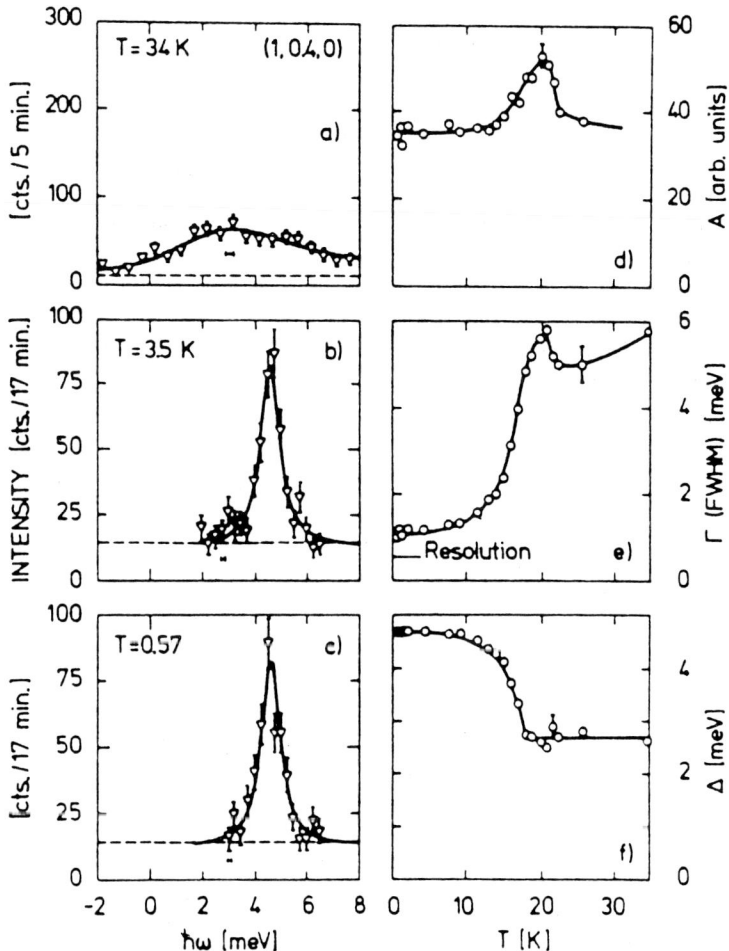

Fig. 13. Constant $q = (1, 0.4, 0)$ scans in the paramagnetic (a), antiferromagnetic (b) and superconducting (c) phases of URu$_2$Si$_2$. Frames (d)–(f) show the temperature dependence of amplitude A, damping Γ and frequency Δ of excitation found in scans such as (a)–(c). From ref. [78].

more similar to what is found for U_2Zn_{17} than for URu_2Si_2. The main surprise about the order in the alloys based on UPt_3 was that it occurred at a wave vector different from that at which the medium energy ($\hbar\omega \approx 5$ meV) fluctuations peaked in the pure compound. This motivated an examination [94] of the *low-frequency* fluctuations in pure UPt_3. Fig. 15 shows the essential result: in contrast to the single peak found at (0, 0, 1) in the constant $\hbar\omega = 8$ meV scan with $Q = (\xi, 0, 1)$, there are peaks at $(\pm\frac{1}{2}, 0, 1)$, which are ordering vectors [84] of (U, Th) (Pt, Pd)$_3$, for $\hbar\omega = 0.5$ meV. Therefore, the nature of the magnetic correlations in UPt_3 is frequency-dependent. More surprising still was the subsequent discovery of *elastic* magnetic scattering at $(\pm\frac{1}{2}, 0, 1)$. The corresponding magnetic "order", which sets in at a Néel temperature of 5 K, has much in common with that found for URu_2Si_2. Most prominent is the small magnetic moment ($\mu_0 \approx 0.02\mu_B$/U atom) and a wide range of validity for the linear relation $|\mu_0(T)|^2 \sim (T - T_N)$. The elastic peak is also not resolution-limited, its width corresponding to a correlation length of order 200 Å.

At present, it is not clear whether the small moment ordering in UPt_3 is intrinsic, but short-ranged because of impurities and/or defects, or is actually induced by impurities and/or defects. Nevertheless, many authors [95] have invoked it in schemes to explain the double superconducting

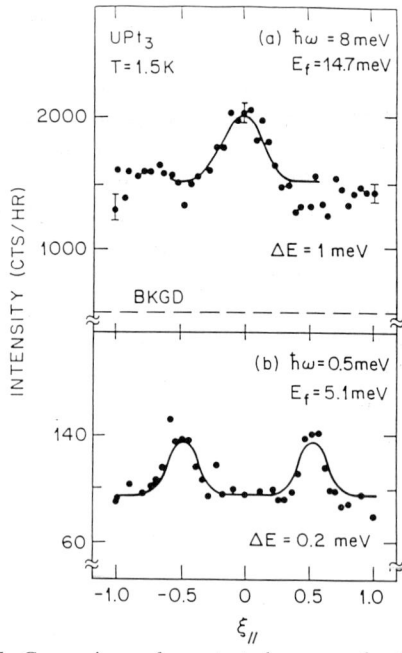

Fig. 15. Comparison of constant $\hbar\omega$ scans for UPt_3 with $Q = (\xi, 0, 1)$, obtained using (a) thermal neutrons ($\hbar\omega = 8$ meV) and (b) cold neutrons ($\hbar\omega = 0.5$ meV). From refs. [73,94].

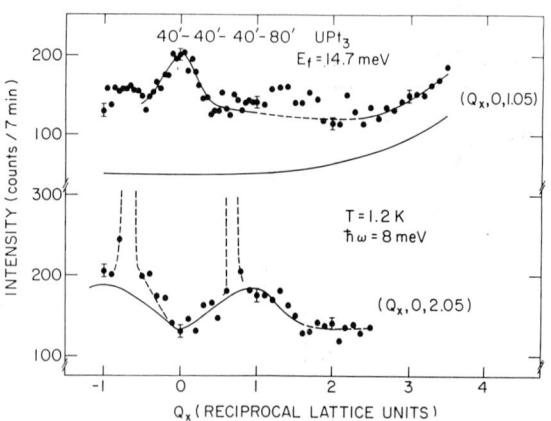

Fig. 14. Constant $\hbar\omega = 8$ meV scans in UPt_3. Solid curve below data in (a) shows Q^4 dependence of two-phonon scattering contribution. Dashed lines in (b) indicate positions of intense phonon peaks. From ref. [73].

transitions [96,97] in UPt_3. Furthermore, neutron diffraction studies [98] clearly demonstrate the coupling of μ_0 and the superconducting order parameter in UPt_3. Fig 16 shows how the Bragg-like intensity I corresponding to μ_0 depends on T and H in the superconducting state of UPt_3. The important result is that at the lowest fields and temperatures, I is actually reduced with respect to its value I_0 at T_c. At 0.1 K, an applied field of 1 T is sufficient to restore I_0, which also represents the field-independent value of I at T_c. An interesting point which we will not discuss in more detail here is that the Bragg intensity becomes field and T-dependent not when the samples cease to superconduct, as established by the ac susceptibility measurements, also shown in fig. 16, but near the locus of torsional oscillator and ultrasound anomalies found [97] within the superconducting state; the arrows in fig. 16 represent the reported portions of the torsional oscillator anomalies. These observations, together with the fact that H is parallel to c, and thus perpendicu-

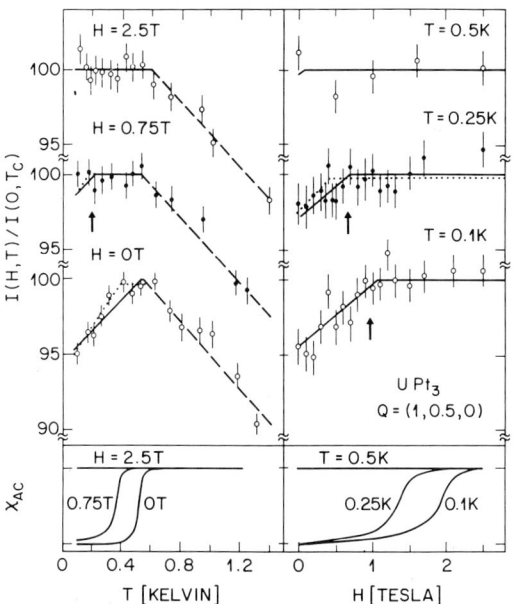

Fig. 16. Temperature dependent (left-hand column) and field-dependent (right-hand column) $(1, \frac{1}{2}, 0)$ magnetic Bragg intensities (upper frames) and bulk ac susceptibilities (lower frames) for UPt$_3$. From ref. [96].

lar to μ_0, clearly demonstrate that the anomalous behavior of I can only be due to the superconductivity of UPt$_3$. The data of fig. 16 therefore represent an explicit demonstration of the coupling between the superconducting order parameter and magnetic moments in a heavy fermion system. This conclusion does not depend on knowledge about the origin of the elastic magnetic scattering: should it be due to defects in a highly susceptible medium, I is actually a measure of the staggered susceptibility χ_{AF}, and hence the magnetic fluctuations, of superconducting UPt$_3$. The reduction in I in the superconducting state would then be due to a reduction in χ_{AF}. In this context, it is noteworthy that the dc $Q = 0$ susceptibility χ_0, as measured by polarized neutron diffraction in an external field [99], does not (to within statistical error) appear to change when UPt$_3$ becomes superconducting. The outcome here is different from what occurs for ordinary superconductors such as V$_3$Si [100] where the reduction in the induced moment corresponds to the spin susceptibility (superconductivity does not affect the orbital susceptibility).

6. Conclusions

Although pioneering work had been performed on magnetic excitations in UO$_2$ by Cowley and Dolling [101] in 1968, it was not until the production of large UX type crystals in Zürich in ≈ 1976 that neutron scattering began to address the questions of hybridization in actinide research. This work covered up to ≈ 1984 in refs. [2,3], has now been extended to a few transuranium materials. As expected, they confirm the unique hybridization effects that occur in the early 5f electron series.

Spallation sources have identified crystal-field levels at high energy in the oxides. Although neutron and optical spectroscopy agree on the parameters, these do not fit with those derived from susceptibility. These successes in the oxides further underline the contrast in attempting to start from a CF-like model in all metallic 5f systems. The statement by Buyers and Holden [3] that well-defined CF states appear to exist only in UPd$_3$ is as true today as it was in 1984, despite many experiments on other compounds.

On the other hand, the essential "f" nature of the electrons responsible for the magnetism in metallic systems results in a sizeable orbital moment being present. This was recognized theoretically by Brooks and Kelly [102] in 1983. Recent neutron [35] and theoretical work [36] has introduced the concept of a variable μ_L/μ_S ratio (see fig. 6), and the possibility that complete cancellation may occur. Certainly one of the important future points is to transfer the ideas from this work to the heavy fermion field; at present most theories in this field assume a spin-only ground state. The consequences of introducing a large orbital moment for the dynamic behavior and understanding $S(Q, \omega)$ are equally interesting.

The discovery of superconductivity in U-based heavy fermions in 1983 led to an enormous surge of activity in actinides. Of the microscopic probes available, neutrons and de Haas–van Alphen studies have given the most important results on which many of the current theories are based. Notable achievements have been in characterizing the small moment ordering in UPt$_3$ and URu$_2$Si$_2$, identifying (in UPt$_3$) that the nature of the correlations is frequency dependent (see fig. 15), and an explicit demonstration (again in UPt$_3$) that the magnetic and superconducting order parameters are indeed coupled (see fig. 16).

Much remains to be done. Lying as they do intermediate between 3d (band) and 4f (localized) magnetism, 5f systems provide a rich variety of behavior and an important testing ground for most theories of solid-state physics. With the production of new materials, e.g. transuranics, amorphous systems, multilayers containing actinides and new heavy fermions, we anticipate that neutron scattering applied to these materials will be an indispensible tool for the foreseeable future.

Acknowledgements

Our copious multi-authored references amply testify to the essential collaborative nature of this research. Space does not permit us to list all the collaborators; may we simply say that the long nights and weekends at various neutron sources are made bearable by the enthusiasm and friendship of numerous talented collegues.

References

[1] G.H. Lander and M.H. Mueller, in: The Actinides: Electronic Structure and Related Properties, vol. 1, eds. A.J. Freeman and J.B. Darby (Academic Press, New York, 1974) p. 303.
[2] J. Rossat-Mignod, G.H. Lander and P. Burlet, in: Handbook on the Physics and Chemistry of the Actinides, vol. 1, eds. A.J. Freeman and G.H. Lander (North-Holland, Amsterdam, 1984) p. 415.
[3] W.J.L. Buyers and T.M. Holden, ibid. vol. 2, p. 239.
[4] P. Fulde and M. Loewenhaupt, Advan. Phys. 34 (1986) 589.
[5] W.T. Carnall and B.G. Wybourne, J. Chem. Phys. 40 (1964) 3428;
W.T. Carnall and H.M. Crosswhite, in: The Chemistry of the Actinide Elements, vol. 2, eds. J.J. Katz, G.T. Seaborg and L.R. Morss (Chapman and Hall, London, 1986) p. 1235.
[6] H.U. Rahman and W.A. Runciman, J. Phys. Chem. Solids 27 (1966) 1833.
[7] C.G. Windsor, Pulsed Neutron Scattering (Taylor and Francis, London, 1981).
[8] J.M. Carpenter, G.H. Lander and C.G. Windsor, Rev. Sci. Instr. 55 (1984) 1019.
[9] S. Kern, C.K. Loong and G.H. Lander, Phys. Rev. B 32 (1985) 3051.
[10] G. Amoretti, A. Blaise, R. Caciuffo, J.M. Fournier, M.T. Hutchings, R. Osborn and A. Taylor, Phys. Rev. B 40 (1989) 1856.
[11] J. Faber and G.H. Lander, Phys. Rev. B 14 (1976) 1151.
[12] R. Siemann and B.R. Cooper, Phys. Rev. B 20 (1979) 2869.
[13] S. Kern, J. Morris, C.K. Loong, G. Goodman, G.H. Lander and B. Cort, J. Appl. Phys. 63 (1988) 3598.
[14] J.M. Fournier, A. Blaise, G. Amoretti, R. Caciuffo, J. Larroque, M.T. Hutchings, R. Osborn and A. Taylor, Phys. Rev. B 43 (1991) 1142.
[15] G. Amoretti, A. Blaise, R. Caciuffo, D. Di Cola, J.M. Fournier, M.T. Hutchings, G.H. Lander, R. Osborn, A. Severing and A. Taylor, Phys. Rev. B (in press).
[16] J.M. Friedt, F.J. Litterst and J. Rebizant, Phys. Rev. B 32 (1985) 257.
[17] P. Erdos, G. Solt, Z. Zołnierek, A. Blaise and J.M. Fournier, Physica B 102 (1980) 164.
[18] S. Kern, C.K. Loong, G.L. Goodman, B. Cort and G.H. Lander, J. Phys.: Condens. Matter 2 (1990) 1933.
[19] R. Caciuffo, G. Amoretti, A. Blaise, J.M. Fournier, M.T. Hutchings, G.H. Lander, R. Osborn, A. Severing and A.D. Taylor, Physica B (to appear).
[20] K. Sasaki and Y. Obota, J. Phys. Soc. Japan 28 (1970) 1157.
[21] R. Osborn, K.A. McEwen, E.A. Goremychkin and A.D. Taylor, Physica B 163 (1990) 37.
[22] A.D. Taylor, R. Osborn, K.A. McEwen, W.G. Stirling, Z.A. Bowden, W.G. Williams, E. Balcar and S.W. Lovesey, Phys. Rev. Lett. 61 (1988) 1309.
R. Osborn, E. Balcar, S.W. Lovesey and A.D. Taylor, Intermultiplet Transitions using Neutron Spectroscopy, Handbook on the Physics and Chemistry of Rare Earths, eds. K.A. Gschneider, Jr. and L. Eyring (North Holland, Amsterdam, to appear).
[23] D. Jones, W.G. Stirling, G.H. Lander, R. Osborn, A.D. Taylor, K. Mattenberger and O. Vogt, Physica B (to appear).
[24] W. Marshall and S.W. Lovesey, Theory of Thermal Neutron Scattering (Oxford Univ. Press, London, 1971) p. 152.
[25] F.A. Wedgwood, J. Phys. C 5 (1972) 2427.
[26] G.H. Lander, M.H. Mueller, D.M. Sparlin and O. Vogt, Phys. Rev. B 14 (1976) 5035.
[27] B.R. Cooper, P. Thayamballi, J.C. Spirlet, W. Muller and O. Vogt, Phys. Rev. Lett. 51 (1983) 2418.
G.H. Lander, A. Delapalme, P.J. Brown, J.C. Spirlet, J. Rebizant and O. Vogt, Phys. Rev. Lett. 53 (1984) 2262.
[28] G.J. Hu, N. Kioussis, A. Banerjea and B.R. Cooper, Phys. Rev. B 38 (1988) 2639.
[29] E. Balcar and S.W. Lovesey, Theory of Magnetic Neutron and Photon Scattering (Clarendon Press, Oxford, 1989) p. 110.

[30] J.M. Fournier, A. Boeuf, P. Frings, M. Bonnet, J.X. Boucherle, A. Delapalme and A. Menovsky, J. Less-Common Met. 121 (1986) 249.

[31] M.S.S. Brooks, O. Eriksson, B. Johansson, J.J.M. Franse and H.P. Frings, J. Phys. F. 18 (1988) L 33.

[32] M. Wulff, G.H. Lander, B. Lebech and A. Delapalme, Phys. Rev. B 39 (1989) 4719.
B. Lebech, M. Wulff, G.H. Lander, J. Rebizant, J.C. Spirlet and A. Delapalme, J. Phys.: Condens. Matter 1 (1989) 10229.

[33] M. Wulff, O. Eriksson, B. Johansson, B. Lebech, M.S.S. Brooks, G.H. Lander, J. Rebizant, J.C. Spirlet and P.J. Brown, Europhys. Lett. 11 (1990) 269.

[34] M. Wulff, G.H. Lander, J. Rebizant, J.C. Spirlet, B. Lebech and P.J. Brown, Phys. Rev. B 37 (1988) 5577.

[35] B. Lebech, M. Wulff and G.H. Lander, J. Appl. Phys. 69 (1991) 5891.
G.H. Lander, M.S.S. Books and B. Johansson, Phys. Rev. B 43 (1991) 13672.

[36] O. Eriksson, M.S.S. Brooks and B. Johansson, Phys. Rev. B 41 (1990) 9087.
O. Eriksson, M.S.S. Brooks, B. Johansson, R.C. Albers and A.M. Boring, J. Appl. Phys. 69 (1991) 5897.

[37] P. Burlet, D. Bonnisseau, S. Quezel, J. Rossat-Mignod, J.C. Spirlet, J. Rebizant and O. Vogt, J. Magn. Magn. Mater. 63 & 64 (1987) 151.

[38] D.L. Jones, W.G. Stirling, G.H. Lander, J. Rebizant, J.C. Spirlet, M. Alba and O. Vogt, J. Phys.: Condens. Matter 3 (1991) 3551.

[39] R.J. Birgeneau, J. Skalyo and G. Shirane, Phys. Rev. B 3 (1971) 1736.

[40] See, for example, T.R. Thurston, R.J. Birgeneau, M.A. Kastner, N.W. Preyer, G. Shirane, Y. Fujii, K. Yamada, Y. Endoh, K. Kakurai, M. Matsuda, Y. Hidaki and T. Murakami, Phys. Rev. B 40 (1989) 4585 and references therein.

[41] A.T. Aldred, B.D. Dunlap, A.R. Harvey, D.J. Lam, G.H. Lander and M.H. Mueller, Phys. Rev. B 9 (1974) 3766.

[42] H.W. Knott, G.H. Lander, M.H. Mueller and O. Vogt, Phys. Rev. B 21 (1980) 4159.

[43] G.H. Lander, S.K. Sinha, D.M. Sparlin and O. Vogt, Phys. Rev. Lett. 40 (1978) 523.

[44] B. Halg and A. Furrer, Phys. Rev. B 34 (1986) 6258.

[45] M. Hagen, W.G. Stirling and G.H. Lander, Phys. Rev. B 37 (1988) 1984.

[46] S.K. Sinha, G.H. Lander, S.M. Shapiro and O. Vogt, Phys. Rev. B 23 (1981) 4556.

[47] B. Halg and A. Furrer, J. Appl. Phys. 55 (1984) 1860.

[48] K. Kaski and W. Selke, Phys. Rev. B 31 (1985) 3128.

[49] N. Kioussis and B.R. Cooper, Phys. Rev. B 34 (1986) 3261.

[50] G.J. Hu, N. Kioussis, B.R. Cooper and A. Banerjea, J. Appl. Phys. 61 (1987) 3385.

[51] B.R. Cooper, G.J. Hu, N. Kioussis and J.M. Wills, J. Magn. Magn. Mater. 63&64 (1987) 121.

[52] B.R. Cooper, in ref. [1], vol. 2 (1985) p. 435.

[53] P. Burlet, J. Rossat-Mignod, G.H. Lander, J.C. Spirlet, J. Rebizant and O. Vogt, Phys. Rev. B 36 (1987) 5306.

[54] P. Burlet, S. Quezel, D. Bonnisseau, J. Rossat-Mignod, J.C. Spirlet and J. Rebizant, Solid State Commun. 67 (1988) 999.

[55] J.P. Sanchez, K. Tomala, J. Rebizant, J.C. Spirlet and O. Vogt, Hyperfine Interactions 54 (1990) 701.

[56] S.M. Shapiro, Physica B 136 (1986) 365.

[57] M. Loewenhaupt and C.K. Loong, Phys. Rev. B 41 (1990) 9294.

[58] T.M. Holden, W.J.L. Buyers, E.C. Svensson and G.H. Lander, Phys. Rev. B 30 (1984) 114.

[59] G.H. Lander, W.G. Stirling, J. Rossat-Mignod, J.C. Spirlet, J. Rebizant and O. Vogt, Physica B 136 (1986) 409.

[60] D.L. Tillwick and P. de V. du Plessis, J. Magn. Magn. Mater. 3 (1976) 329.

[61] B.R. Cooper, P. Thayamballi, J.C. Spirlet, W. Mueller and O. Vogt, Phys. Rev. Lett. 31 (1983) 2418.

[62] G.H. Lander, M.S.S. Brooks, B. Lebech, P.J. Brown, O. Vogt and K. Mattenberger, Appl. Phys. Lett. 57 (1980) 989; J. Appl. Phys. 69 (1991) 4803.

[63] A. Banerjea and B.R. Cooper, Phys. Rev. B 34 (1986) 1607.

[64] A.E. Clark, in: Ferromagnetic Materials, ed. E.P. Wohlfahrt (North-Holland, Amsterdam, 1980) p. 540.

[65] M.S.S. Brooks, B. Johansson, O. Eriksson and H.L. Skriver, Physica B 144 (1986) 1.

[66] G.H. Lander, W.G. Stirling, J. Rossat-Mignod, M. Hagen and O. Vogt, Phys. Rev. B 41 (1990) 6899.

[67] R. Osborn, M. Hagen, D.L. Jones, W.G. Stirling, G.H. Lander, K. Mattenberger and O. Vogt, J. Magn. Magn. Mater. 76&77 (1988) 429.

[68] J. Schoenes, B. Frick and O. Vogt, Phys. Rev. B 36 (1984) 6578.

[69] G.J. Hu, B.R. Cooper and G.H. Lander, Physica B 156&157 (1989) 822.
G.J. Hu and B.R. Cooper, unpublished.

[70] See reviews by G.R. Stewart, Rev. Mod. Phys. 56 (1984) 755.
P.A. Lee, T.M. Rice, J.W. Serene, J.L. Sham and J.W. Wilkins, Comm. Cond. Mat. Phys. 12 (1986) 99.
C.M. Varma, in: Trends in Theoretical Physics, eds. P.J. Ellis and Y.C. Tang (Addison Wesley, New York, Reading, 1990) p. 353.

[71] G. Aeppli, E. Bucher and G. Shirane, Phys. Rev. B 32 (1985) 7579.
A.I. Goldman, S.M. Shapiro, G. Shirane, J.L. Smith and Z. Fisk, Phys. Rev. B 33 (1986) 1627.
S. Horn, E. Holland-Moritz, M. Loewenhaupt, F. Steglich, H. Scheurer, A. Benoit and J. Flouquet, Phys. Rev. B 23 (1981) 3171.
Y.J. Uemura, C.F. Majkarzak, G. Shirane, C. Stassis, G. Aeppli, B. Batlogg and J.P. Remeika, Phys. Rev. B 33 (1986) 6508.
U. Walter, D. Wohlleben and Z. Fisk, Z. Phys. B 62 (1986) 235.

[72] G. Aeppli, H. Yoshizawa, Y. Endoh, E. Bucher, J. Hufnagl, Y. Onuki and T. Komatsuba, Phys. Rev. Lett. 57 (1986) 122.
L.P. Regnault, W.A.C. Erkelens, J. Rossat-Mignod, P. Lejay and J. Flouquet, Phys. Rev. B 38 (1988) 4481.

[73] G. Aeppli, A.I. Goldman, G. Shirane, E. Bucher and M.-Ch. Lux-Steiner, Phys. Rev. Lett. 58 (1987) 808.
A.I. Goldman, G. Shirane, G. Aeppli, E. Bucher and O. Hufnagl, Phys. Rev. B 36 (1987) 8523.

[74] P.H.P. Reinders, M. Springford, P.T. Coleridge, R. Boulet and D. Ravot, Phys. Rev. Lett. 57 (1986) 57.
L. Taillefer and G. Lonzarich, Phys. Rev. Lett. 60 (1988) 1570.

[75] D.E. Cox, G. Shirane, S.M. Shapiro, G. Aeppli, Z. Fisk, J.L. Smith, J. Kjems and H.R. Ott, Phys. Rev. B 33 (1986) 3614.

[76] H.R. Ott, H. Rudiger, P. Delsing and Z. Fisk, Phys. Rev. Lett. 52 (1984) 1551.

[77] C. Broholm, J. Kjems, G. Aeppli, Z. Fisk, J.L. Smith, S.M. Shapiro, G. Shirane and H.R. Ott, Phys. Rev. Lett. 58 (1987) 917.

[78] G. Broholm, thesis (University of Copenhagen, 1988).

[79] B. Jones and C.M. Varma, Phys. Rev. Lett. 58 (1987) 843.
E. Abrahams and C.M. Varma, unpublished.

[80] W. Schlabitz, J. Baumann, B. Pollit, U. Rauchschwalbe, H.M. Mayer, U. Ahlheim and C.C. Bredl, Z. Phys. B 62 (1986) 171.
M.B. Maple, J.W. Chen, Y. Dalichaouch, T. Kohara, C. Rossel, M.S. Torakachvili, M.W. McElfresh and J.D. Thompson, Phys. Rev. Lett. 56 (1986) 185.
T.T.M. Palstra, A.A. Menovsky, J. van den Berg, A.J. Dirkmaat, P.H. Kes, G. J. Nienwenhuys and J.A. Mydosh, Phys. Rev. Lett. 55 (1985) 2727.

[81] C. Broholm, J.K. Kjems, W.J.L. Buyers, P.T. Matthews, T.T.M. Palstra, A.A. Menovsky and J.A. Mydosh, Phys. Rev. Lett. 58 (1987) 1467.

[82] U. Walter, C.-K. Loong, M. Loewenhampt and W. Schlabitz, Phys. Rev. B 33 (1986) 7875.

[83] E.D. Isaacs, D.B. McWhan, R.N. Kleiman, D.J. Bishop, G.E. Ice, P. Zschak, B.D. Gaulin, T.E. Mason, J.D. Garrett and W.J.L. Buyers, Phys. Rev. Lett. 65 (1990) 3185.
T.E. Mason, B.D. Gaulin, J.D. Garrett, Z. Tun, W.J.L. Buyers and E.D. Isaacs, Phys. Rev. Lett. 65 (1990) 3189.

[84] A.I. Goldman, G. Shirane, G. Aeppli, B. Batlogg and E. Bucher, Phys. Rev. B 34 (1986) 6564.
P.H. Frings, B. Renker and C. Vettier, J. Magn. Magn. Mater. 63&64 (1987) 202.

[85] H.R. Ott, K. Andres and P.H. Schmidt, Physica B 102 (1980) 148.
W.J.L. Buyers, A.F. Murray, T.M. Holden, E.C. Svensson, P. de V. Du Plessis, G.H. Lander and O. Vogt, ibid. p. 291.

[86] A. de Visser, F.E. Kayzel, A.A. Menovsky, J.J.M. Franse, J. van den Berg and G.J. Niewenhuys, Phys. Rev. B 34 (1986) 8168.

[87] P. Coleman and J. Gan, Physica B (1991) to be published.
L. Gorkov, preprint (1991).

[88] R.H. Heffner et al., Phys. Rev. B 39 (1989) 11345.

[89] B.S. Shivaram, J.H. Jeong, T.F. Rosenbaum and D.G. Hinks, Phys. Rev. Lett. 56 (1986) 1078.

[90] C. Broholm, G. Aeppli, R.N. Kleiman, D.R. Harshman, D.J. Bishop, E. Bucher, D.L. Williams, E.J. Ansaldo and R.H. Heffner, Phys. Rev. Lett. 65 (1990) 2062.

[91] Early papers are by K. Mayake, S. Schmitt-Rink and C.M. Varma, Phys. Rev. B 34 (1986) 6554.
M.T. Béal-Monod, C. Bourbonnais and V.J. Emery, Phys. Rev. B 34 (1986) 7716.
J.E. Hirsch, Phys. Rev. Lett. 54 (1985) 1317.

[92] P.H. Frings and J.J.M. Franse, Phys. Rev. B 31 (1985) 4355.

[93] A.P. Ramirez, B. Batlogg, A.S. Cooper and E. Bucher, Phys. Rev. Lett. 57 (1986) 1072.
A. de Visser, S.C.P. Klaase, M. van Sprang, J.J.M. France, A. Menovsky and T.T.M. Palstra, J. Magn. Magn. Mater. 54–57 (1986) 375.
G.R. Stewart, A.L. Giorgi, J.O. Willis and J. O'Rourke, Phys. Rev. B 34 (1986) 4629.

[94] G. Aeppli, E. Bucher, C. Broholm, J.K. Kjems, J. Baumann and J. Hufnagl, Phys. Rev. Lett. 60 (1988) 615.
See also C. Broholm, J.K. Kjems, G. Aeppli, E. Bucher and W.J.L. Buyers, in: Magnetic Excitations and Flutuations I, eds. U. Balucani, S.W. Lovesy, M.G. Rosetti and V. Tognetti (Springer Verlag, Berlin, 1987) p. 162.
P. Frings, B. Renker and C. Vettier, Physica B 151 (1988) 499.

[95] See, E.G. R. Joynt, Supercond. Sci. Tech. 1 (1988) 210.
K. Machida, M. Ozaki and T. Ohmi, J. Phys. Soc. Jpn. 58 (1989) 4116.
E.I. Blount, C.M. Varma and G. Aeppli, Phys. Rev. Lett. 64 (1990) 3074.

[96] R.A. Fisher et al., Phys. Rev. Lett. 62 (1989) 1411.

[97] R.N. Kleiman, P.L. Gammel, E. Bucher and D.J. Bishop, Phys. Rev. Lett. 62 (1989) 328.
A. Schenstrom et al., Phys. Rev. Lett. 62 (1989) 332.
V. Müller et al., Phys. Rev. Lett. 58 (1987) 1224.

[98] G. Aeppli, D. Bishop, C. Broholm, E. Bucher, K. Siemensmeyer, M. Steiner and N. Stüsser, Phys. Rev. Lett. 63 (1989) 676.

[99] C. Stassis, J. Arthur, C.F. Mankrzak, J.D. Axe, B. Batlogg, J. Remcitra, Z. Fisk, J.L. Smith and A. Edelstein, Phys. Rev. B 34 (1986) 4382.

[100] F.A. Wedgwood and C.G. Shull, Phys. Rev. Lett. 16 (1966) 513.

[101] R.A. Cowley and G. Dolling, Phys. Rev. 167 (1968) 464.

[102] M.S.S. Brooks and P.J. Kelly, Phys. Rev. Lett. 51 (1983) 1708.

Giant magnetostriction materials

N.C. Koon, C.M. Williams and B.N. Das

Naval Research Laboratory, Washington, DC 20375-5000, USA

One of the significant technical developments in magnetism of the early 1970's was the discovery of a new class of rare earth intermetallic compounds, the RFe_2 Laves phases, which were found to exhibit room temperature magnetostrictive strains approaching 2×10^{-3}, an order of magnitude larger than any previously known. Since that time both the fundamental and technical properties of these materials have been of intense interest, and they remain the subject of active research even today. The large strains available are useful in such applications as production of high amplitude, low frequency sound waves in water, certain types of strain gages, vibration compensation and compensation for temperature induced strains in large laser mirrors. Because the performance of these materials depends critically on such fundamental properties as the magnetic anisotropy, magnetization and grain orientation of the material, there has been a very strong interplay between fundamental studies and applications. In this article we briefly review the fundamental magnetic and magnetostrictive properties of the RFe_2 Laves phases, focusing especially on the complex behavior of the anisotropy and the success of crystal field theory in explaining it. We also present neutron measurements of magnetic excitation spectra and explain how they provide an understanding of the remarkable success of mean field theory for these systems.

1. Background and introduction

The rare earth elements, whose magnetic properties stem from the unfilled 4f shell, differ in important respects from the magnetic transition metals, whose magnetism is based on the 3d electrons. As a general rule the 3d wave functions are much more extended than the 4f and can exhibit direct overlap between wave functions on neighboring sites. In most metals the 3d orbital moment is quenched and the magnetic anisotropy and ordinary magnetostriction arise from spin orbit coupling as essentially a perturbation. The 4f electrons, on the other hand, tend to be localized deep inside the atoms and thus well shielded form direct interactions with 4f electrons on other atomic sites. They also have very much larger spin–orbit coupling than the transition metal elements. As a consequence of these properties the total angular momentum $J = L + S$ is usually a good quantum number and interactions between the 4f magnetic electrons and the environment outside the atom are mostly indirect via the various conduction electrons.

Although the the 4f coupling with the lattice is indirect, it is not generally weak. The large spin–orbit coupling and the highly anisotropic localized nature of the 4f electronic charge distribution for non-S-state rare earth elements therefore routinely result in large magnetic anisotropies as well as large magnetostrictive strains. Coupling of the orbital (L) and spin (S) angular momenta also results in large magnetic moments per atom (up to $10\mu_B$ for Ho). Large magnetostrictive strains are useful in transducers and other devices which convert electrical to mechanical energy, while high magnetic anisotropy and large magnetization are key ingredients for good permanent magnet materials. Because of the relatively weak indirect 4f–4f exchange, however, it is almost always necessary to alloy rare earth elements with more strongly exchange coupled 3d transition metals in order to achieve high magnetostriction or magnetic anisotropy at useful (300 K and above) temperatures.

The extraordinary magnetostrictive properties of the rare earth elements were first recognized in the early 1960's when Legvold, Alstad and

Rhyne [1] and Clark, De Savage and Bozorth [2] reported a basal plane magnetostriction of single crystal Dy approach 1% at low temperatures. Unfortunately, because of weak indirect exchange, the magnetic ordering temperatures of the pure rare earth metals are below room temperature (except for Gd), so that these extraordinary strains are not available at room temperature where practical devices must operate. The solution to this problem was pointed out by Callen [3] in an unusually prescient paper at an ONR sponsored workshop on Metallic Magnetoacoustic Materials in 1969. He suggested that there was great promise in study of the magnetostrictive properties of rare earth–transition metal intermetallics, where the strong magnetism of the transition metals could increase the rare earth magnetic order at high temperatures. The first realization of this promise came in 1971, when groups at the Naval Research Laboratory [4] and the Naval Ordinance Laboratory (now the Navy Surface Weapons Center) [5] independently reported extraordinary magnetostrictive strains in RFe_2 Laves phase compounds at room temperature. As an interesting coincidence, early in 1972 Atzmony, Dariel and co-workers reported determination of the temperature and composition spin orientation diagram of $Ho_xTb_{1-x}Fe_2$ and other ternary alloys from Mössbauer effect measurements [6]. That work, together with the general concept of crystalline electric fields in rare earth systems, gave important guidance about how to reduce magnetic anisotropy of the alloys and thereby improve the low field magnetostrictive properties. Subsequent measurements of magnetization, magnetostriction, magnetic anisotropy energy and magnetic excitations on mixed rare earth single- and polycrystals significantly improved the scientific understanding of these materials and contributed to their technological development and application.

Because of space limitations it is not possible to give a complete treatment of all the subjects important to understanding of high magnetostriction materials, especially the theory. In the following sections, therefore, only enough of each topic is introduced to provide a conceptual framework in which to understand the results.

For a more complete discussion of some of the early work and greater detail concerning the phenomenological theory of magnetostriction, see the review article by Clark [7]. Section 2 introduces crystal field theory in the mean field approximation and its relationship to magnetic anisotropy and magnetostriction. Section 3 presents the phenomenological theory of anisotropy and magnetostriction as well as experimental results from magnetization, magnetostriction and torque magnetometry experiments. Section 4 contains a crystal field analysis of the anisotropy and spin reorientation data, while section 5 presents neutron inelastic scattering measurements of magnetic excitations which show clearly that the low-lying modes are primarily mean field in character, thus explaining why crystal field theory in the mean field approximation works as well as it does for magnetic anisotropy and other properties of the Laves phases.

2. Crystal fields and the rare earth elements

In rare earth–transition metal intermetallics there are generally contributions to the magnetic anisotropy from both the rare earths and from the transition metal. A detailed understanding of magnetic anisotropy in transition metals has proved to be elusive, primarily because it requires band structure calculations in which the magnetic anisotropy appears as very small differences (due to spin–orbit coupling) of very large energies. For the rare earth iron Laves phases the magnetic anisotropy of the iron sublattice appears to be quite small compared to that of the rare earth elements, so for the most part it may be either neglected or treated as a quantity to be determined experimentally. The magnetic anisotropy of the rare earths, as we shall show, seems to be very well described by crystalline electric fields acting on the highly anisotropic 4f electron distribution. While any simple connection between neighboring charges and the electric field acting on the 4f electrons, such as that provided by the point charge model, cannot be expected to be valid because of screening effects [8], crystal field theory does provide at least the correct form and symmetry of the interactions at a given site in a rare earth intermetallic. One can therefore regard experimentally determined crystal field parameters as quantities to be calculated from a more complete theory, such as has recently been developed by Coehoorn [9]. One of the most important aspects of crystal field theory for the rare earth elements, however, is that it does provide a method (though somewhat approximate) of scaling from one rare earth element to another. This dramatically simplifies the process of trying to develop materials with particular properties, as we will show for the Laves phases.

In general a Hamiltonion which describes the magnetic anisotropy of the rare earth–transition metal intermetallic compounds can be written as [10]

$$H = H_{\text{exch}} + H_{\text{cf}} + H_{\text{me}} + K(\text{TM}),$$

where

$$H_{\text{exch}} = -\sum_{i,l,m} g_i \mu_B (\boldsymbol{H}_{\text{ex}}(i) + \boldsymbol{H}_a) \boldsymbol{J}(i)$$

and

$$H_{\text{cf}} = \sum_{i,l,m} B_l^m(i) O_l^m(i) \quad \text{(general symmetry)}.$$

In these expressions $H_{\text{ex}}(i)$ is the exchange field experienced by the rare earth spin on site i due primarily to the transition metal polarization, H_a is the applied magnetic field, H_{cf} is the crystal field Hamiltonian, $K(\text{TM})$ is the magnetic anisotropy of the transition metal sublattice, and H_{me} is the magnetoelastic interaction. The B_l^m are crystal field parameters and O_l^m are the crystal field operators, of which the most commonly used are the Stevens operator equivalents [11,12]. The O_l^m are nothing more than operators in the $(2J+1) \times (2J+1)$ rare earth spin manifold which describe the interaction of the electric field with the anisotropic 4f charge distribution. The symmetry of an individual rare earth site determines the allowed values of the B_l^m parameters since the electric fields must be invariant under symmetry operations of the site. In general the l indices range from 2–6 and m from $-l$ to l. For cubic site symmetry, as in the Laves phases, only B_4^0 and B_6^0 are independent parameters, while in the general case there can be as many as 15 independent parameters just for the crystal field.

The Zeeman splitting of the rare earths is assumed to scale with the magnetization of the transition metal and the spin part of the rare earth moment as [10–12]

$$g_i \mu_B \boldsymbol{H}_{\text{ex}}(i) = \Delta_0 (g_i - 1) \boldsymbol{M}_{\text{TM}}(T) / M_{\text{TM}}(0),$$

where Δ_0 is independent of the rare earth element and $g_i - 1$ is proportional to the spin part of the rare earth moment. The crystal field parameters are expected to scale as

$$B_l^m = \theta_l \langle r^l \rangle (1 - \sigma_l) A_l^m,$$

where the A_l^m are independent of the rare earth element and characterize the host environment, σ_l are the screening parameters [13], $\langle r^l \rangle$ are the radial integrals and θ_l are the reduced matrix elements. The reduced matrix elements θ_l, which describe the shape of the 4f charge distribution, are often referred to as α_J, β_J and γ_J, respectively, for $l = 2, 4$ and 6. In principle, all of the parameters except A_l^m and Δ_0 can be calculated from atomic properties, so that once the crystal field and exchange parameters are determined for one rare earth element in a given atomic site, they can be estimated for any of the others which are substituted into the same site. In the Laves phases, therefore, even for mixed crystals with more than one rare earth element, there are only three significant parameters needed to describe the magnetic anisotropy: Δ_0, A_4^0, A_6^0. The first describes the exchange coupling to the spin part of the rare earth moment due to conduction electron spin polarization produced by the transition metal sublattice, and the two crystal field terms describe the symmetry and coupling of the electric field to the anisotropic 4f charge distribution of the rare earth. It is assumed (and shown by the neutron inelastic scattering data for the Laves phases) that the rare earth–rare earth exchange is negligible.

The role of symmetry and the nature of the 4f interactions with the lattice can be visualized using fig. 1, where we illustrate schematically a rare earth spin in a site of cubic symmetry. In the sense of first order perturbation theory the lowest order term ($\alpha_J = \theta_2$) of the 4f charge distribution does not couple to the lattice, although that component of the charge distribution *is* primarily responsible for the magnetostrictive distortion which occurs as the spin direction is rotated. It does couple in second order, as we shall discuss later, giving rise to the so-called ΔK_1 effect. The lowest l value which produces a first order contribution to the magnetic anisotropy in cubic symmetry is four ($\beta_J = \theta_4$).

From fig. 1 we can see why cubic symmetry is important for magnetostrictive materials. Because the $l = 2$ component of the charge does not contribute to the magnetic anisotropy in first order, we can choose rare earth elements whose $\alpha_J(\theta_2)$ coefficients have the same sign, and thus the same sign of magnetostriction, but with opposite

Table 1
Values of $g - 1$ and reduced matrix elements for Tb, Dy, Ho and Er (from Hutchings [12])

	$g-1$	θ_2 (10^{-3})	θ_4 (10^{-5})	θ_6 (10^{-6})
Tb	1.500	-10.010	12.244	-1.121
Dy	1.333	-6.439	-5.920	1.035
Ho	1.250	-2.222	-3.330	-1.294
Er	1.200	2.590	4.440	2.070

(and hopefully compensating) coefficients $\beta_J(\theta_4)$ and $\gamma_J(\theta_6)$, thus creating materials with both large magnetostrictive strains and low magnetic anisotropy. In a material with lower than cubic symmetry the $l = 2$ term contributes to both the magnetostriction and anisotropy, so that for such a material it is very difficult, if not impossible, to have both high magnetostriction and low anisotropy. Knowledge of crystal field theory and the reduced matrix elements therefore provide excellent guidance for constructing better magnetostrictive materials.

For the heavy rare earth elements Tb, Dy, Ho and Er we list the values of $g - 1$ and the reduced matrix elements (θ_2, θ_4 and θ_6) in table 1, taken from Hutchings [12]. From this table and the preceding figure it is possible to infer rather simply the possible combinations of these rare earth elements which might be used to construct high magnetostriction, low anisotropy materials. Tb, Dy and Ho all have large negative θ_2 and would be expected to have the same sign of the magnetostrictive strain. For both Dy and Ho θ_4 is negative, while for Tb it is positive, so that for some combination of Tb and Dy or Tb and Ho it should be possible to "cancel" the lowest order allowed anisotropy term κ_4. In fact, for combinations of Tb, Dy and Ho it is possible to minimize *both* κ_4 and κ_6, although for other reasons that turns out to be not particularly useful in practical materials. Tb and Dy have another advantage for practical applications in that among the non-S-state rare earth elements they have the largest spin component of the moment and thus the largest exchange coupling to the conduction electron spin polarization.

In the following section we present the phenomenological theory of magnetostriction and

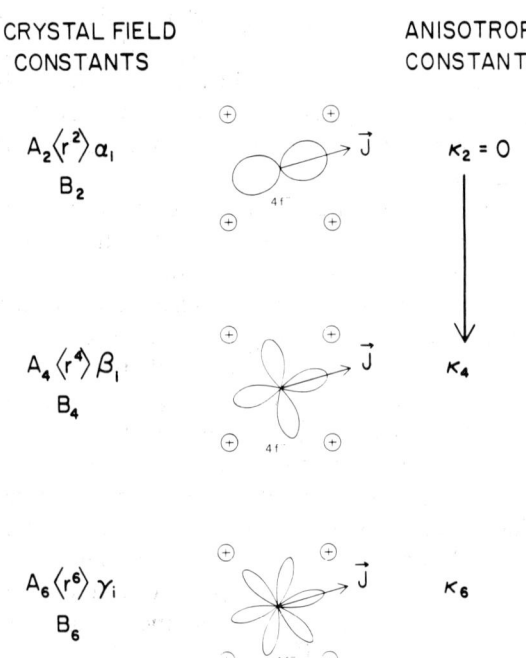

Fig. 1. Schematic diagram of a rare earth spin in a cubic environment showing the different symmetry components of the rare earth 4f charge distribution. On the left are the crystal field constants corresponding to the symmetry shown, and on the right are listed the main coefficients of the Kubic harmonic functions which result from the symmetry component shown. Note that κ_2 is forbidden for a lattice of cubic symmetry.

anisotropy in cubic materials along with selected experimental results. In section 4 we return to the task of using crystal field theory to calculate magnetic anisotropy and spin orientation diagrams.

3. Phenomenological theory and experimental results for magnetostriction and anisotropy

The phenomenological description of magnetic anisotropy and magnetostriction is usually given in terms of direction cosine expansions, with combinations of direction cosines chosen to reflect the underlying symmetry of the lattice. For the magnetic anisotropy of cubic materials the usual expansion was given by Akulov [14]:

$$E = K_0 + K_1(\alpha_1^2\alpha_2^2 + \alpha_2^2\alpha_3^2 + \alpha_1^2\alpha_3^2) + K_2(\alpha_1^2 + \alpha_2^2 + \alpha_3^2) + K_3(\cdots) + \cdots,$$

where the α's are direction cosines of the magnetization with respect to the cube axes. Unfortunately the terms of this expansion do not form an orthonormal set, so the experimental values obtained depend upon how many terms are retained in the expansion [15]. Because high order terms in the anisotropy energy are generally needed for rare earth elements, a more convenient expansion set to use are the Kubic harmonics, which have been tabulated by Mueller and Priestly [16]. Essentially the Kubic harmonics are a subset of the spherical harmonics in which the coefficients have been chosen to reflect cubic symmetry. In this expansion the anisotropy energy is given by

$$E = \sum_l \kappa_l H_l,$$

where each term in the expansion is orthonormal to every other and cubic symmetry is obeyed by every term separately. For cubic symmetry the lowest order term to appear is $l = 4$, while for uniaxial materials it would be $l = 2$.

For magnetostriction in cubic rare earth materials the higher order terms appear to be less important, so the usual direction cosine expansion first written down by Akulov [14] is adequate:

$$\delta l/l = \tfrac{3}{2}\lambda_{100}(\alpha_1^2\beta_1^2 + \alpha_2^2\beta_2^2 + \alpha_3^2\beta_3^2 - 1/3)$$
$$+ 3\lambda_{111}(\alpha_1\alpha_2\beta_1\beta_2 + \alpha_1\alpha_3\beta_1\beta_3$$
$$+ \alpha_3\alpha_1\beta_3\beta_1) + \cdots$$

In this expression the α's are direction cosines of δl and the β's are the direction cosines of the magnetization. A case of particular importance is when the magnetostriction is isotropic ($\lambda_{100} = \lambda_{111}$). When this occurs the strain depends only on the relative direction between the magnetization and the measuring direction and is independent of the crystal axes. If the material is a polycrystal this means that every grain will be strained identically as the magnetization is rotated uniformly and no magnetic energy will be stored in the elastic energy of interaction between grains. For more details concerning the phenomenological theory of magnetostriction see the review by Clark [7].

For an isotropic polycrystal the saturation magnetostriction at high fields is given in terms of the single crystal constants by

$$\lambda_s = 0.4\lambda_{100} + 0.6\lambda_{111}.$$

The field dependence of the magnetostriction of a polycrystalline material below saturation can be seen from the above relations to be potentially quite complicated, depending both on the magnetic anisotropy and on the relative and absolute values of λ_{100} and λ_{111}. For example, if λ_{100} is positive and the [100] direction is easy, but λ_{111} is negative and the same magnitude as λ_{100}, then the magnetostriction will be positive at low applied fields and become negative at saturation. This particular situation does occur, for example, in pure iron. If [111] is easy, on the other hand, everything else being the same, then the magnetostriction will be negative for all fields.

The importance of anisotropy in these high magnetostriction materials is evident in fig. 2, where we show the magnetostriction of polycrystalline TbFe$_2$ [17] as a function of applied magnetic field. While the strains are very large compared to a conventional magnetostrictive material such as Ni ($\lambda_s \approx 33 \times 10^{-6}$) driving fields on the order of many kOe are necessary to achieve most of the strain, and it has not reached its saturation value even at 25 kOe. Since most practical materials must operate at fields of a few tens or perhaps hundreds of oersteds, it is important to reduce the anisotropy so that large strains may be produced with low driving fields, as suggested by the dashed line in the figure.

In the early stages after the discovery of the interesting magnetostrictive properties of the Laves phases only polycrystals were available, so reduced anisotropy was sought primarily using crystal field theory and the spin orientation work of Atzmony, Dariel and co-workers as guidance [6], and success was evaluated from magnetostriction as a function of field on polycrystalline samples and through measurements of magnetomechanical coupling. Subsequently, single crystals became available, and both magnetization and single crystal torque measurements were used for much more precise measurements of anisotropy and spin reorientations. Since torque magnetometry provides much more precise and complete information on magnetic anisotropy and spin orientations, we focus on that method in the remainder of this section.

Torque magnetometry makes use of the fact that torque is the derivative of the magnetic free energy, so that a direct measure of the relative

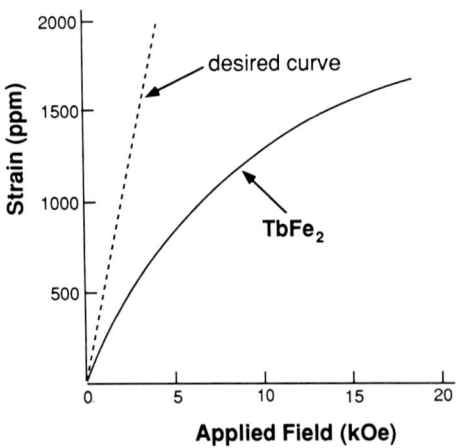

Fig. 2. Magnetostrictive strain of a polycrystalline $TbFe_2$ sample as a function of applied field at 300 K, showing the effects of large magnetic anisotropy.

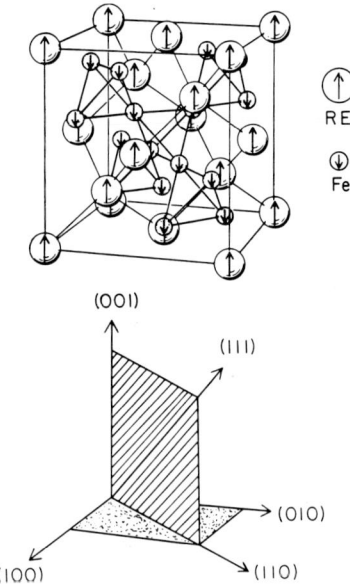

Fig. 3. Crystal structure of the cubic Laves phase compounds along with the principal planes in which the magnetization was rotated in making torque measurements.

magnetic free energy in a given plane can be obtained by numerical integration of torque curves. To obtain the largest amount of information requires that the applied field be large enough to ensure that the magnetic moment of the sample lies in the selected plane, as illustrated in fig. 3, although when the magnetic anisotropy is very large it is also possible to follow the easy axis directions by observing the zero torque angles. The resulting free energy curves are useful both for comparison with theory and because the easy axis direction can be determined simply by location of the minimum free energy point, as shown in fig. 4 for a typical compound of $(HoTb)Fe_2$. This compound, as do many of the ternary rare earth Laves phases, exhibits a complex set of spin reorientations as a function of both composition and temperature [6]. These can be easily followed from the free energy curves, as shown in fig. 5 for a crystal of $Ho_{0.846}Tb_{0.154}Fe_2$. The spin orientations which occur as a function of temperature for a closely related sample are presented in fig. 6. The data exhibit some rather remarkable features. The easy direction of magnetization is [110] at 300 K; it rotates continu-

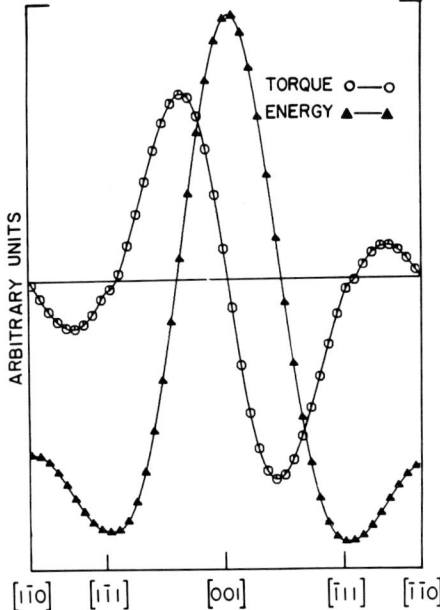

Fig. 4. Torque data and magnetic free energy of a typical (HoTb)Fe$_2$ compound. The easy axis direction is clearly seen [111]. The free energy was obtained by numerical integration of the torque.

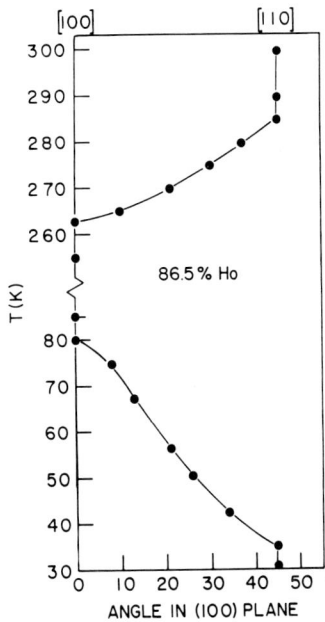

Fig. 6. The easy axis directions as a function of temperature for a single crystal of Ho$_{0.865}$Tb$_{0.135}$Fe$_2$. Data from refs. [18,19].

ously from [110] to [100] in the (001) plane as the temperature is lowered, then rotates back again at even lower temperatures. Other compositions exhibit similarly complex behavior. By combining data from several samples of different compositions, we obtain a spin orientation diagram of the easy axis directions in the Ho$_x$Tb$_{1-x}$Fe$_2$ alloy system as a function of composition and temperature as shown in fig. 7 [18]. In the light regions of the diagram the easy axis directions are along one of the principal axes. In the speckled region the easy axis is in the (001) plane between the [100] and [110].

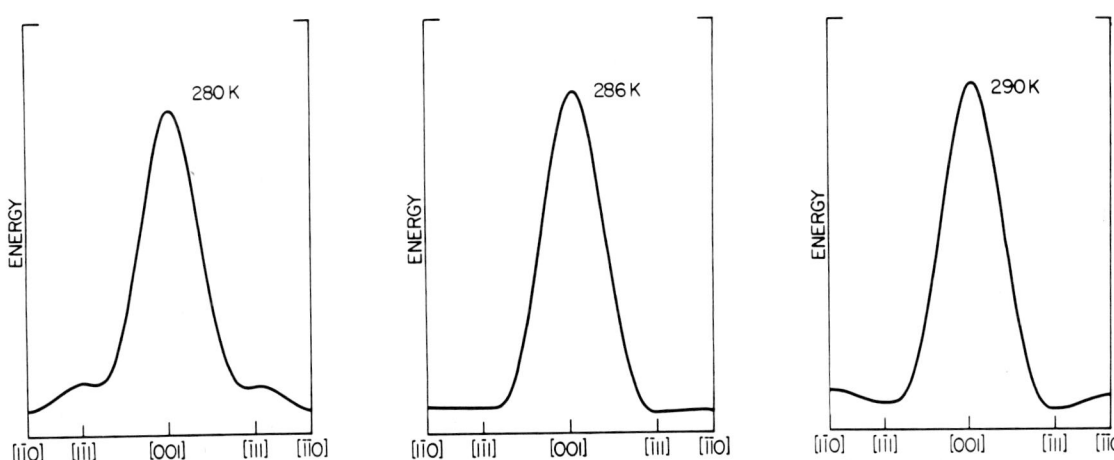

Fig. 5. Free energies as a function of angle in the (110) plane for a Ho$_{0.846}$Tb$_{0.154}$Fe$_2$ single crystal at various temperatures. A transition of the easy direction from [110] to [111] can easily be seen near 290 K. Data from ref. [18].

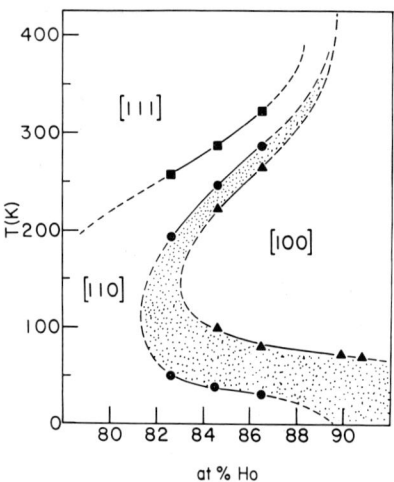

Fig. 7. Spin orientation diagram for $Ho_xTb_{1-x}Fe_2$. The solid squares mark spin reorientations between [110] and [111]. The solid dots and triangles, respectively, mark the boundaries between the regions where the [110] and [100] are easy and the dotted region where the easy axis direction is in the (100) plane between [100] and [110]. Data from ref. [18].

The spin orientation diagrams do not by themselves show that the anisotropy is low in the region of a spin reorientation. Indeed, the free energy plots in fig. 5 show that for the composition $Ho_{0.846}Tb_{0.154}Fe_2$ the free energy in the [001] direction is large at the reorientation point between [111] and [110]. This is reflected very clearly in the anisotropy constants as shown in figs. 8(a)–(c), where we plot the two lowest order coefficients κ_4 and κ_6 as a function of composition for the Ho–Tb, Dy–Tb and Ho–Tb–Dy Laves phase systems. Interestingly enough, in both the Ho–Tb and Dy–Tb case κ_4 changes sign near the spin reorientation, but for Ho–Tb the sign of κ_6 is positive at that point and for Dy–Tb it is negative. By combining the zero κ_4 compositions of the two pseudo-binaries, it is, in fact, possible to obtain a pseudo-ternary composition containing all three rare earths for which $\kappa_4 = \kappa_6 = 0$, as illustrated in fig. 8(c) [19]. For that composition the overall magnetic anisotropy is extremely low.

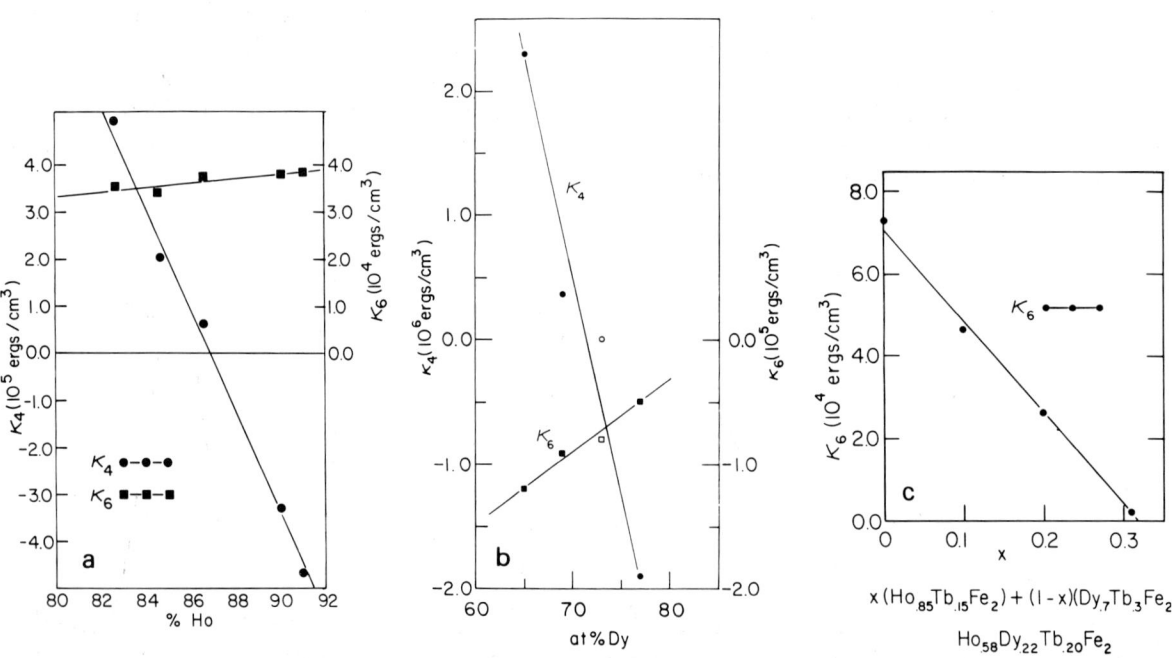

Fig. 8. Plots of the lowest order Kubic harmonic coefficients as a function of composition at room temperature for (a) $Ho_xTb_{1-x}Fe_2$, (b) $Dy_xTb_{1-x}Fe_2$ and (c) the sixth order coefficient κ_6 for $(Ho_{0.85}Tb_{0.15})_x(Dy_{0.7}Tb_{0.3})_{1-x}Fe_2$. Data from refs. [18,19].

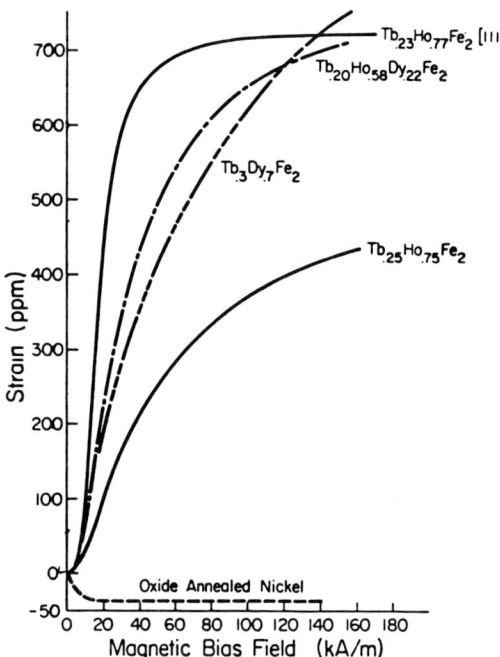

Fig. 9. Magnetostriction vs. applied field for several polycrystalline RFe$_2$ compounds and a [111] oriented single crystal as a function of applied magnetic field. Data from ref. [20].

In fact, it is so low that the dominant effect on the low field magnetostriction in an isotropic polycrystal of that composition does not come from the magnetocrystalline anisotropy, but from internal strains due to non-isotropic magnetostriction.

This can be seen qualitatively in fig. 9, which shows the magnetostrictive strain of several low anisotropy polycrystalline alloys as a function of magnetic field [20]. For comparison, the figure also shows data on a Tb–Ho [111] oriented single crystal. Except for the single crystal, the low anisotropy pseudo-ternary is clearly best at low fields. The main reason the single crystal is so much better is that it has been found for the cubic Fe Laves that quite generally $|\lambda_{111}| \gg |\lambda_{100}|$ [21], with much larger strains occurring along [111] than the [100]. This causes inhomogeneous strains in an isotropic polycrystal when a magnetic field is applied, so that even if a single crystal of such a material has no magnetocrystalline anisotropy, a polycrystal of the same material will have effective anisotropy because of the magnetoelastic interactions between grains. For this reason most high magnetostriction alloys today are made from a rapid zoned Tb–Dy material [22]. The rapid zoning produces a strong [211] texture along the zone direction. Because [211] is close to the high strain [111] direction, this texturing significantly reduces the magnetoelastic interaction between grains. In a sense, the strong [211] texture produces a pseudo-single crystal where the magnetoelastic interactions between grains is reduced and the effective crystalline anisotropy does not need to be quite so low.

4. Calculation of anisotropy energies and spin orientation diagrams

In order to construct a crystal field theory for these cubic systems we need to restrict our general set of equations to cubic symmetry and to include the effect of magnetostriction on the anisotropy [10]. For cubic site symmetry the crystal field Hamiltonian reduces to:

$$H_{cf}(i) = B_{4i}\left(O_{4i}^0 + 5O_{4i}^0\right) + B_{4i}\left(O_{6i}^0 - 21O_{6i}^4\right).$$

Technically one should include magnetoelastic effects by considering in detail the effects of strain on the crystal field through the magnetoelastic coupling. If we treat the problem empirically as a perturbation, however, we can use the fact that the magnetostriction at a fixed temperature in these alloys appears to scale linearly with composition [21], and that λ_{111} appears to obey the Callen and Callen [23] relation with temperature:

$$\lambda_{111}(T) = \lambda_{111}(0) I_{5/2}\left(I_{3/2}^{-1}(m)\right).$$

In this expression $I_{p/n}$ are normalized hyperbolic Bessel functions and m is the reduced magnetization of the rare earths. Under these conditions the magnetoelastic effect gives the usual ΔK_1 contribution to the magnetic free energy:

$$F_{me} = \Delta K_1\left(\alpha_1^2\alpha_2^2 + \alpha_1^2\alpha_3^2 + \alpha_2^2\alpha_3^2\right),$$

where α's are the direction cosines and

$$\Delta K_1 = \tfrac{9}{4}\left((C_{11} - C_{12})\lambda_{100}^2 - 2C_{44}\lambda_{111}^2\right).$$

The C_{ij}'s are the usual cubic elastic constants. Using Kubic harmonics this should probably be referred to as the $\Delta \kappa_4$ effect, but we remain with the customary terminology.

Because $|\lambda_{111}| \gg |\lambda_{100}|$ [21] for the Laves phase compounds, ΔK_1 is negative and the magnetostrictive anisotropy favors the [111] directions. All the elastic and magnetostriction constants appearing in these equations have been measured and are not adjustable [21]. The total magnetic free energy is then

$$F_{tot} = F_{me} + F_{cf} + K_1(\text{Fe}),$$

where F_{me} is the magnetoelastic free energy, $K_1(\text{Fe})$ is the free energy due to the iron sublattice, and the free energy due to the crystal field is

$$F_{cf} = -kT \sum_i \ln(Z_i)$$

and Z_i is the partition function for the ith ion and the sum extends over all rare earth ions in the crystal.

The only parameters undetermined are A_4^0, A_6^0, Δ_0 and $K_1(\text{Fe})$. In fig. 10 we show a fit to the spin reorientation data using $A_4^0 = 3.16$ meV/a_0^4, $A_6^0 = -0.12$ meV/a_0^6, $\Delta_0 = 28$ meV and $K_1(\text{Fe}) = -40$ μeV. The results are in quite good agree-

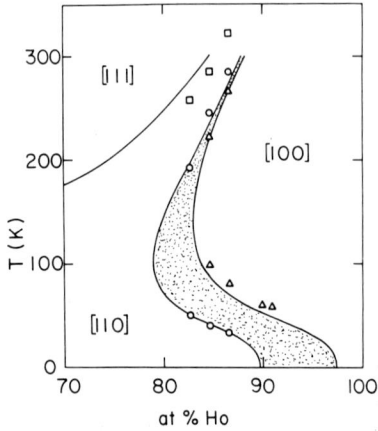

Fig. 10. The calculated and measured spin orientation diagram for $\text{Ho}_x\text{Tb}_{1-x}\text{Fe}_2$. The squares mark the experimental spin reorientations between [110] and [111]. The dots and triangles, respectively, mark the experimental boundaries between the regions where the [110] and [100] are easy and the dotted region where the easy axis direction is in the (100) plane between [100] and the [110]. Parameters used in the calculation are given in the text.

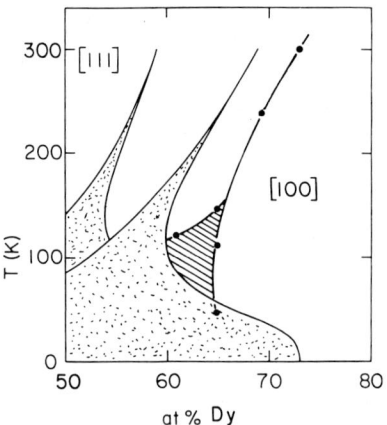

Fig. 11. Spin orientation diagram for $\text{Dy}_x\text{Tb}_{1-x}\text{Fe}_2$, the left curve is not including magnetostriction, while the right one does. The solid dots are experimental data marking spin orientations as measured on single crystals. The white regions are those in which the indicated principal axis directions are easy. The hatched (experimental) and dotted (theoretical) regions are those in which the easy axes are not along a principal axis direction. Results from ref. [18].

ment with the data, especially in view of the complexity of the spin orientations and with the fact that there is really no good reason to expect a mean field theory to work particularly well. The anisotropy of the Fe lattice is actually quite small and has very little effect on the diagram, so that it is effectively a three parameter fit. For the Ho–Tb system it is also true that the magnetostriction has a relatively small effect on the diagram. For Dy–Tb, however, which exhibits larger magnetostrictive strains, the effect is quite large, shifting the spin orientation diagram nearly 20% in composition at room temperature, as shown in fig. 11 [19].

Calculations for the easy axis directions as a function of temperature are compared with those actually observed for particular compositions in fig. 12. Although the agreement is not perfect, it is certainly quite good, given the limitations of a mean field theory and the limited number of parameters.

5. Neutron inelastic scattering results

The agreement between a simple mean field treatment of the crystalline electric field and the

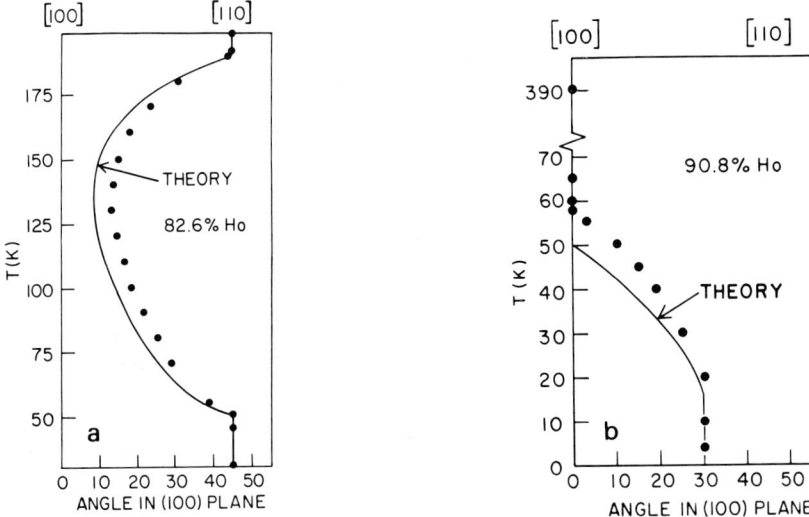

Fig. 12. The calculated and measured spin orientation curves of $Ho_xTb_{1-x}Fe_2$. (a) $x = 0.826$ and (b) $x = 0.908$.

spin orientation data raises the question of why the theory works as well as it does. Part of the answer can be obtained from neutron inelastic scattering experiments, which measure the normal modes of the magnetic system and which can also be used to determine the exchange and crystal field parameters. We have performed detailed studies of a number of the Laves phase com-

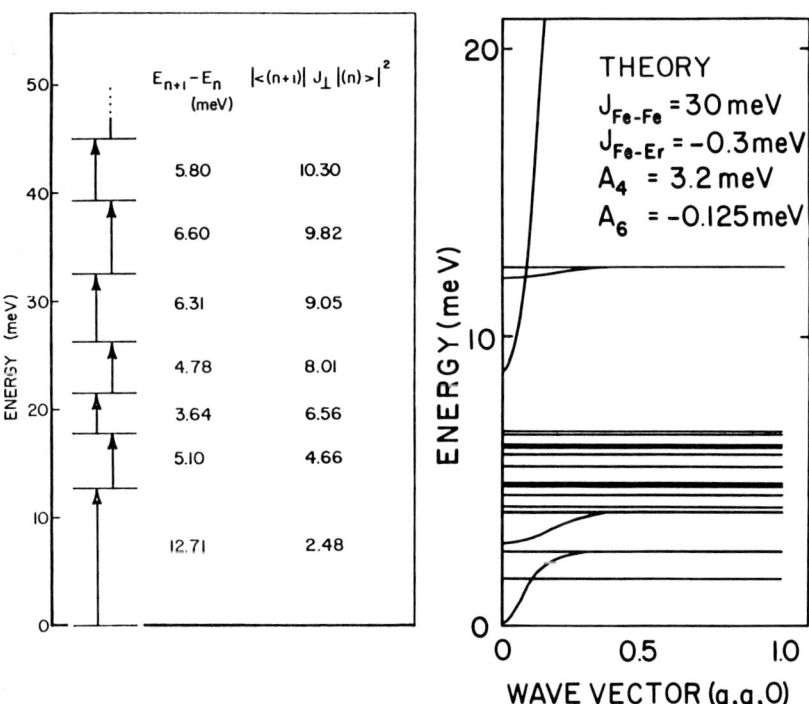

Fig. 13. Calculated magnetic excitation spectrum for $ErFe_2$ as determined from neutron inelastic scattering. On the left are the rare earth single ion crystal field levels. On the right is a calculated excitation spectrum for 300 K. From ref. [24].

pound, especially $ErFe_2$ [24]. In fig. 13 we present the single ion crystal field levels and the magnetic excitation spectrum of $ErFe_2$ at 300 K deduced from fits to the data. The theory used was a Green's function random phase approximation utilizing standard basis operators.

The theoretical fit presented in fig. 13 provides a great deal of useful information. The strongly dispersive mode intersecting $q = 0$ near 8.2 meV is a spin wave excitation of the iron sublattice, with the rare earth spins remaining rigid. The gap at $q = 0$ is due to the exchange field of the rare earths acting on the Fe spins. The flat modes correspond to out-of-phase excitations of the two rare earth spins in the primitive cell with the iron spins remaining rigid, and the low energy modes at small q are coupled excitations of the rare earth and iron sublattices. The spectra were calculated assuming no rare earth–rare earth exchange since that results in dispersion of the flat modes, in contradiction to the experiments. *The energies of the flat modes are exactly those calculated for the rare earth single ion crystal field levels, shown in the left part of the figure, which gives good physical insight into why mean field theory works so well for these particular compounds.*

The physical picture which emerges is that the very "stiff" iron sublattice provides the exchange interactions which cause the rare earth spins to order, yet the iron spin waves are largely at high energies over most of the Brillouin zone and therefore contribute little to the thermodynamics except at very high temperatures. Except for a very small "core region" of the zone, the low lying magnetic excitations are simply single ion excitations of the rare earth spins sitting in the combined exchange field from the iron plus the crystalline electric fields, which is precisely the definition of molecular field theory. If one were to use the calculated Green's functions to determine physical observables, then an integration over the Brillouin zone would lead to a large weighting of the low-lying single ion rare earth modes. This then explains why for these compounds crystal field theory in the mean field approximation can provide such a precise description of the observed phenomena.

Acknowledgements

One of us (NCK) would like to acknowledge fruitful collaborations with J.J. Rhyne and R.M. Nicklow in the neutron inelastic scattering experiments.

References

[1] S. Legvold, J. Alstad and J. Rhyne, Phys. Rev. Lett. 10 (1963) 509.
J. Rhyne and S. Legvold, Phys. Rev. 138 (1965) A507.
[2] A.E. Clark, B.D. De Savage and R. Bozorth, Phys. Rev. 138 (1965) A216.
A.E. Clark, R. Bozorth and B.F. De Savage, Phys. Lett. 5 (1963) 100.
[3] E. Callen, Proc. Metallic Magnetoacoustic Materials Workshop, Boston, MA, ed. F.S. Gardner (1969) p. 75.
[4] N.C. Koon, A. Schindler and F. Carter, Phys. Lett. A 37 (1971) 413.
[5] A.E. Clark and H. Belson, AIP Conf. Proc. no. 5 (1972) 1498; Phys. Rev. B 5 (1972) 3642.
[6] U. Atzmony, M.P. Dariel, E.R. Bauminger, D. Lebenbaum, I. Nowik and S. Ofer, Phys. Rev. Lett. 28 (1972) 244.
[7] A.E. Clark, Ferromagnetic Materials, vol. 1, eds. K.H.J. Buschow and E.P. Wohlfarth (North-Holland, Amsterdam, 1980) p. 531.
[8] P. Morin and D. Schmitt, Ferromagnetic Materials, vol. 5, eds. K.H.J. Buschow and E.P. Wohlfarth (North-Holland, Amsterdam, 1990) p. 1.
[9] R. Coehoorn, E.P. Wohlfarth lecture, J. Magn. Magn. Mater. 99 (1991) 55.
[10] N.C. Koon and C.M. Williams, J. Appl. Phys. 49 (1978) 1948.
U. Atzmony, M.P. Dariel, E.R. Bauminger, D. Lebenbaum, I. Nowik and S. Ofer, Phys. Rev. Lett. 28 (1972) 244.
[11] K.W.H. Stevens, Proc. Phys. Soc. (London) A 65 (1952) 209.
[12] M.T. Hutchings, Solid State Phys. 16 (1964) 227.
[13] A.J. Freeman and R.E. Watson, Phys. Rev. 139 (1965) A1606.
[14] N.S. Akulov, Z. Phys. 52 (1928) 389.
[15] C.M. Williams, N.C. Koon and J.B. Milstein, J. Phys. Chem. Solids 39 (1978) 823.
[16] F.M. Mueller and M.G. Priestly, Phys. Rev. 148 (1966) 638.
[17] N.C. Koon, unpublished data.
[18] C.M. Williams and N.C. Koon, Solid State Commun. 27 (1978) 81.
[19] N.C. Koon and C.M. Williams, in: Crystalline Electric Fields and Structural Effects in f-Electron Systems, eds. J.E. Crow, R.P. Guertin and T.W. Mihalisin (Plenum, New York, 1980) p. 75.

[20] R.W. Timme, private communication.
[21] A.E. Clark, J. Cullen, O. McMasters and H. Savage, Physica B 86–88 (1977) 73.
A.E. Clark, J.R. Cullen, O.D. McMasters and E.R. Callen, AIP Conf. Proc. 29 (1976) 192.
N.C. Koon and C.M. Williams, US Navy J. Underwater Acoustics 27 (1977) 127.
[22] H.T. Savage, R. Abbundi, A.E. Clark and O.D. McMasters, J. Magn. Magn. Mater. 15–18 (1980) 609.
[23] H.B. Callen and E.R. Callen, J. Phys. Chem. Solids 27 (1966) 1271.
[24] N.C. Koon and J.J. Rhyne, in: Crystalline Electric Fields and Structural Effects in f-Electron Systems, eds. J.E. Crow, R.P. Guertin and T.W. Mihalisin (Plenum, New York, 1980) p. 125.
See also R.M. Nicklow, N.C. Koon, C.M. Williams and J.B. Milstein, Phys. Rev. Lett. 36 (1976) 532.

Experimental study of Ce-based heavy-fermion compounds *

F. Steglich

Institut für Festkörperphysik, Technische Hochschule Darmstadt, W-6100 Darmstadt, Germany

In the first part of the paper, a brief survey is given on the development of heavy-fermion physics with special emphasis on Ce-based intermetallic compounds. Subsequently, selected topics of current interest are discussed, e.g., heavy-fermion band magnetism, heavy-fermion superconductivity and its relationship to a novel kind of lattice instability.

1. Early work

1.1. Dilute Kondo alloys and intermediate-valence compounds

Heavy-fermion physics, which is in the focus of current investigations of strongly correlated charge-carrier systems, has developed out of two roots, the Kondo-impurity [1] and the intermediate-valence (IV) [2,3] problems, respectively.

Originally, experimental studies of the Kondo-impurity problem were mainly devoted to 3d dopants in noble metals [1]. Because of the rather large spatial extent of the 3d wavefunction, however, inter-ionic correlations were not easy to separate from the more essential single-ion effects; these difficulties becoming extremely serious because of the relatively poor solubility of 3d ions in noble metals. The breakthrough in this field occurred at the end of the 1960s, when rare-earth (RE) elements of sufficient purity became available: Due to the strong localization of the 4f electrons inside the Xe core (with full $5s^2 5p^6$ shells), the direct 4f-wavefunction overlap between RE ions on adjacent lattice sites is negligible. In addition, RE generally can be substituted in a wide composition range for La, as well as for Lu and Y. Consequently, the single-ion effects can be studied reliably in systems like $La_{1-x}RE_xAl_2$ [4] at RE-concentration levels up to several at%.

Like in insulators, the diluted 4f shells in a metallic environment are characterized by the intra-shell Hund's rule correlations, yielding a $^{2S+1}L_J$ ($J = L \pm S$) ground state with a magnetic moment $\mu = -g_J \mu_B J$. Here S is the spin, L the orbital angular momentum, J the total angular momentum, g_J the Landé g-factor of the 4f shell and μ_B the Bohr magneton. Under the influence of the crystalline-electrical field (CF) (Δ_{CF}, the CF-excitation energy being smaller than the spin–orbit energy by typically one order of magnitude) the $(2J + 1)$-fold degeneracy of the ground state is partly removed.

The majority of the RE ions occur in the trivalent state only and are frequently labeled "normal" (or "stable-moment") RE. Some of the RE, i.e., Ce, Sm, Pr, Eu, Tm and Yb, with electronic configurations close to the stable $4f^0$, $4f^7$ and $4f^{14}$ configurations, however, show an ambivalent behavior: in insulators, they occur not only in the trivalent but also in an adjacent (tetra and divalent, respectively) state. If dissolved in a metallic environment, ambivalent RE ions can give rise to Kondo and IV phenomena.

Experiments performed on dilute alloys, particularly with Ce, but also with Yb, Sm and Pr dopants [5] revealed a richness of low-temperature anomalies, the so-called Kondo anomalies: upon cooling to below the Kondo temperature

---

* Dedicated to Professor Paul Kienle on the occasion of his 60th birthday.

T_K, for example, the resistivity increment increases, the "effective" local moment disappears and "heavy" quasiparticles are formed locally. The latter have 4f symmetry and can thus be considered weakly delocalized 4f-electrons [6]. It is reassuring that the aforementioned experimental efforts helped to develop a profound understanding of the Kondo-impurity problem [7]. As an especially exciting property of *superconducting* Kondo alloys we mention the possibility of more than one transition temperature in one sample. According to Müller-Hartmann and Zittartz [8], this can happen by reason of a strongly temperature- and energy-dependent pair-breaking rate of the Kondo ions in a superconducting matrix. The first example showing a re-entrant T_c vs. x phase boundary was the quasi-binary system $La_{1-x}Ce_x Al_2$ [9,10]. Later on, Winzer [11] was able to establish even three different T_c's, i.e., in $(La_{0.8}Y_{0.2})_{1-x}Ce_x$.

When Ce-impurities are exerted to sufficiently high pressure (external or internal), a transition into a "non-magnetic" state is induced, which essentially means a transition into an IV state [12–14]. Here the hybridization energy between 4f and conduction-band states, $\Delta = \pi V^2 N_F$ (V: average over hybridization matrix elements, N_F: conduction-band density of states at the Fermi energy, E_F), is at least comparable with, if not larger than, the valence-excitation energy ϵ. By contrast, $\epsilon/\Delta \approx 3-5$ for typical Kondo ions and $\epsilon/\Delta \gg 1$ for "stable-moment" RE. The strong 4f-conduction electron hybridization in the IV compounds containing ambivalent RE causes real charge fluctuations between the local 4f states and the conduction band ("valence fluctuations") which, well below the characteristic fluctuation temperature (100–1000 K), form a "homogeneous" IV state.

Heavy-fermion compounds are found at the concentrated limit of the Kondo alloys and are, therefore, frequently labeled Kondo-lattice systems. The characteristic Kondo-lattice temperature T^* replacing the single-ion T_K ranges between about a few tenths and a few tens of kelvin. Phenomenological differences between heavy-fermion and IV compounds have been established, e.g., for the magnetic neutron cross section [15,16], thermopower [17,18], thermal expansion [19,20] and de Haas–van Alphen (dHvA) effect [21].

1.2. Kondo-lattice systems

The experimental studies of Kondo-lattice systems began with resistivity measurements on Ce intermetallic compounds like $CeAl_2$ [22]. Typically, a negative temperature coefficient of the resistivity, $\rho(T)$, similar to that of dilute Kondo alloys is found for $T > T^*$. At low temperatures, however, the resistivity drops and reaches a residual value which is comparable to that of transition metals, of course depending on the disorder in the sample. The low-T resistivity behavior, therefore, manifests a major difference between the Kondo lattice and the Kondo alloy (fig. 1): Whereas the maximum resistivity (as $T \to 0$) in the latter reflects a resonance scattering off the single Kondo impurities at E_F, the coupling of the phases of the scattered carriers in the former is visible as a "freezing out" of incoherent scattering below the maximum temperature T_{max} ($\approx T^*$). Extended ("Bloch") states are formed by the weakly delocalized 4f electrons in the coherent low-temperature phase of the Kondo lattice [24], i.e., at $T \le T_{coh} \ll T^*$. In this low-T regime the resistivity is dominated by a contribution quadratic in temperature as first discovered by Andres et al. [25] for $CeAl_3$ (fig. 2). The gigantic coefficient of this electron–electron scattering law, AT^2, indicates a strong renormalization of the conduction-band states at E_F. A is proportional to $(m^*)^2$, where the effective carrier mass

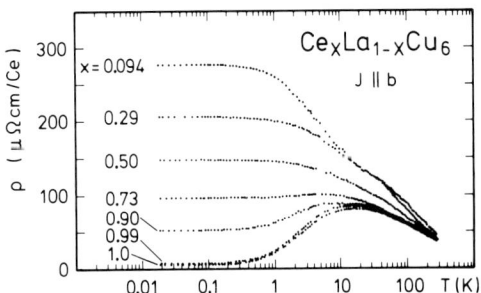

Fig. 1. Temperature dependence of electrical resistivity increment per mole of cerium for $Ce_xLa_{1-x}Cu_6$ [23].

is of the order of several hundred m_{el}, the free electron mass. The gigantic magnitude of the AT^2 term is the hallmark for a very slow propagation of the 4f electrons under the dominant influence of the intra-atomic correlation energies [26,27]. Correspondingly gigantic values were observed for the Sommerfeld coefficient in the electronic specific heat, $\gamma = 1.6$ J/K^2 mol, and the Pauli spin susceptibility, $\chi_0 \approx 4.5 \times 10^{-7}$ m^3/mol [25], both being proportional to m^* and exceeding the γ and χ_0 of simple metals like Na by about three orders of magnitude. At sufficiently high temperatures, where $\rho(T)$ shows a negative temperature coefficient, the susceptibility follows a Curie–Weiss law, characteristic of Ce-derived magnetic moments weakly coupled to the conduction electrons, and the Sommerfeld coefficient becomes as small as in ordinary d-band metals.

In the intermetallic compounds, the demagnetizing Kondo interaction (of dominant on-site character) is competing with the inter-site indirect exchange interaction, the Ruderman–Kittel–Kasuya–Yosida interaction [28]. Using the coupling constant for the exchange between the local spin and the conduction-electron spins, $g = N_F J$ ($J < 0$: exchange integral), $T_{RKKY} \sim g^2$, whereas $T^* \approx T_K \sim \exp(1/g)$. Because of their different dependences on g, $k_B T_{RKKY}$ exceeds $k_B T_K$ for sufficiently low values of $|g|$, but becomes dominated by it, if $|g|$ is sufficiently large. Doniach [29] has derived from the competition of the Kondo and RKKY interactions a generalized

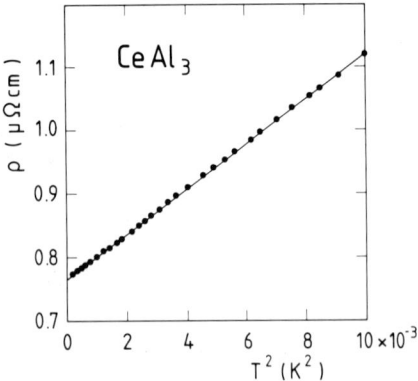

Fig. 2. Electrical resistivity of CeAl$_3$ below 100 mK as ρ vs. T^2 [25].

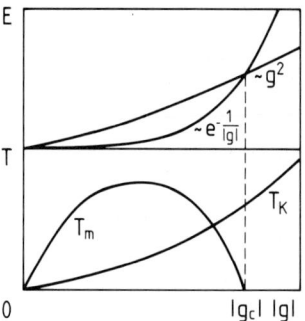

Fig. 3. Fundamental energies $k_B T_{RKKY} \sim g^2$ and $k_B T_K \sim \exp(-1/|g|)$ (upper part) as well as magnetic ordering temperature T_m and Kondo temperature T_K (lower part) as a function of microscopic coupling parameter $|g|$, according to Doniach [29].

"phase diagram" for a one-dimensional Kondo lattice as schematically reproduced in fig. 3. There exists a critical value $|g_c|$ of $|g|$, below which the systems show long-range magnetic order and above which they have a non-magnetic ground state. The maximum in the dependence of the magnetic ordering temperature T_m on $|g|$ reflects the competition between the increase in the RKKY interaction and the Kondo-derived reduction of the "effective" ordered moment μ_s, $T_m \sim \mu_s^2 T_{RKKY}$.

Most of the Ce-based heavy-fermion compounds show an antiferromagnetically ordered ground state, in which somewhat reduced 4f moments are coupled below the Néel temperature $T_N = T_m$ via the RKKY interaction. Such a state will be labeled "local-moment magnetism" (LMM) in the following. As was first demonstrated for CeAl$_2$, the moderate moment reduction [30] is accompanied by an also moderate carrier renormalization: The Sommerfeld coefficient $\gamma \approx 0.135$ J/K^2 mol of CeAl$_2$ [31] exceeds the γ value of the ferromagnet Fe by about a factor of twenty, but is smaller than the estimate for fictitious paramagnetic CeAl$_2$ by more than a factor of ten, see fig. 4. As another early example for LMM we mention CeB$_6$ [32,33]. The oscillatory nature of the RKKY interaction may result, under fortunate conditions, in ferro- rather than antiferromagnetic order. Examples are CeSi$_x$, $x < 1.85$ [34,35], CeCu$_2$ [36] and CeRu$_2$Ge$_2$ [37]. Disorder on the non-f ligand sites or on intersti-

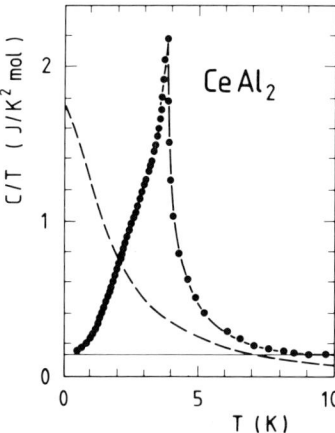

Fig. 4. Specific heat as C/T vs. T for CeAl$_2$ [31]. Solid line through data points is guide to the eye. Thin horizontal line indicates low-temperature value, $\gamma_0 = 135$ mJ/K^2 mol. Dashed line is Bethe-ansatz result for $S = 1/2$ Kondo impurity with $T_K = 3.5$ K [7].

tials surrounding Ce sites can even lead to a spin-glass type of ordering on a periodic Ce-lattice, as was established first for CeCu$_{6.5}$Al$_{6.5}$ [38] and subsequently confirmed for CePd$_3$B$_{0.3}$ [39] and CePtGa$_3$ [40].

Transitions from local-moment magnetism to a non-magnetic heavy-fermion state have been induced by application of pressure [14] or suitable alloying [5,12,13], i.e., by increasing $|g|$. Only a few Ce compounds behave as non-magnetic heavy Fermi liquids already at ambient pressure, e.g., CeCu$_6$ [41,42] and CeRu$_2$Si$_2$ [43], besides CeAl$_3$ [25].

1.3. Instabilities of the heavy Fermi liquid

Heavy-Fermi liquid phases, though being of much interest on their own right, have become a favored subject of modern condensed-matter physics when their inherent tendency towards instabilities against superconducting and (itinerant) magnetic phase transitions was recognized. The corresponding transition temperatures lie below T^*, i.e., the driving interactions between the heavy fermions must be limited by the characteristic (on-site) energy $k_B T^*$.

In fig. 5, specific-heat results, along with upper-critical-field, $B_{c2}(T)$, data are shown for the first heavy-fermion superconductor CeCu$_2$Si$_2$. Both the values of the specific-heat jump height [45] and the initial $B_{c2}(T)$ slope [46] scale with the huge Sommerfeld coefficient $\gamma \approx 0.7$ J/K^2 mol. This proves that the heavy fermions (with effective mass $m^* \approx 200$–$300 m_0$ [46]) form the Cooper pair states in this material. Consequently, no superconductivity is found in the non-f homolog LaCu$_2$Si$_2$ [45]. The observation that the typical heavy-fermion energy, $k_B T^*$, is much smaller than the typical phonon energy, $k_B \Theta_D$ (Θ_D: Debye temperature), led to the suggestion that in CeCu$_2$Si$_2$ the BCS-type deformation–potential coupling cannot be efficient for the Cooper-pair formation [45]. On the other hand, the seeming analogy [47] to the Fermi liquid ^{3}He and its superfluid transition stimulated suggestions of an unconventional order parameter, having a symmetry lower than the crystal [26], as well as of a magnetic pairing mechanism. For a recent survey on the superconducting properties of heavy-fermion compounds, see ref. [48].

The discovery of bulk superconductivity in CeCu$_2$Si$_2$ was received with much scepticism which, in fact, hindered a more rapid development of this field in the following years. One argument of the critics involved the complex ternary Ce–Cu–Si phase diagram [49] which made it difficult to disentangle intrinsic from extrinsic effects [50–53]. For example, single crystals grown from stoichiometric melts were not superconducting [51,54–56], whereas bulk-superconducting single crystals of small dimensions ($1 \times 1 \times 0.1$ mm^3)

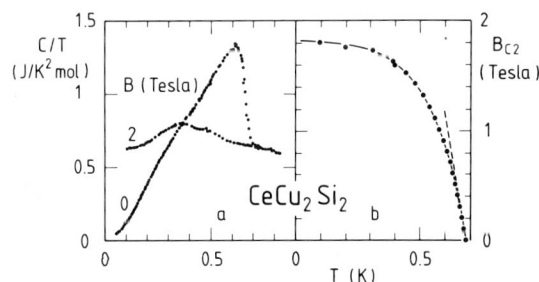

Fig. 5. Specific heat as C/T vs. T (a) and upper critical field, B_{c2} vs. T (b), of CeCu$_2$Si$_2$. Solid line in (b) is guide to the eye, and dashed line indicates $B_{c2}(T)$ slope at T_c (-13 T/K). Data in (a) and (b) taken from the same polycrystalline sample [44].

became available when melts with a Cu excess were used [51,52]. Apart from such concerns about sample quality, the strong pair-breaking capability of *dilute* Ce^{3+} ions in a BCS superconductor [9–11] made it difficult to believe in bulk superconductivity of a trivalent Ce-based *intermetallic compound*.

Heavy-fermion superconductivity attracted much greater interest, however, after it became confirmed in the three U-based compounds UBe_{13} [57], UPt_3 [58] and URu_2Si_2 [59]. The recent discovery [60] of superconductivity above 30 K in some cuprates, whose transport properties seem to be strongly influenced by the intra-atomic correlations on the Cu-3d shell, initiated additional motivation to understand superconductivity of the most highly correlated electrons, the heavy fermions. The present state of the art may be best characterized by a continuing lack of a unified microscopic theory [6,61,62,26,27], but an increasing manifold of experimental discoveries [63,64, 53,65,27], e.g., phase diagrams with more than one superconducting phase in $U_{1-x}Th_xBe_{13}$ [66,67] and UPt_3 [68], the coexistence between heavy-fermion superconductivity and antiferromagnetic ordering with $T_N \approx 10T_c$ in both URu_2Si_2 [59,69–71] and UPt_3 [72] as well as the new heavy-fermion superconductors UNi_2Al_3 [73] and UPd_2Al_3 [74].

A magnetic (Stoner-type) instability in a heavy-fermion compound was first reported for $NpSn_3$ [75]. Later examples are U_2Zn_{17} [76], $NpBe_{13}$ [77], URu_2Si_2 [70,71] and UPt_3 [72]. Though the classification as "heavy-fermion band magnets" (HFBM) is controversial for the first two compounds [78,27], the extremely low ordered moments $\mu_s = (2-3) \times 10^{-2}\mu_B/\text{U-atom}$ found in both URu_2Si_2 and UPt_3 appears to justify this assignment for the latter: According to the theoretical treatment of a Kondo lattice utilizing an idealized hybridized band model, such a low value for μ_s is expected in an itinerant magnetic state [79]. Among the Ce-based compounds, only $CePb_3$ has been identified so far as HFBM: Its Sommerfeld coefficient well below $T_N \approx 1$ K is as high as 1 J/K^2 mol [80], and its ordered moment is reduced by about a factor of ten compared to that of the free Ce^{3+} ion in the Γ_7 CF ground state [81].

In the remainder of this paper, we wish to discuss typical ground-state properties and recent developments in the study of Ce-based heavy-fermion compounds. We address in section 2 the Pauli paramagnets, before turning to local-moment magnets (LMM) (section 3) as well as to a system in which an alloying-induced transition between LMM and HFBM can be monitored (section 4). Recent results on $CeCu_2Si_2$, emphasizing an intimate relationship between heavy-fermion superconductivity, magnetism and lattice instability are the subject of section 5. The paper is concluded in section 6 by a short perspective.

2. Strongly enhanced Pauli paramagnetism

Before discussing in some detail the low-temperature properties of the prototypical systems $CeAl_3$ [25], $CeCu_6$ [41,42] and $CeRu_2Si_2$ [43], we wish to emphasize again the dominating role of on-site effects in determining the thermodynamic properties of a Kondo-lattice. This is demonstrated in figs. 6 and 7, showing the specific-heat and magnetic-susceptibility results [23] up to $T = 30$ and 300 K, respectively, for a series of

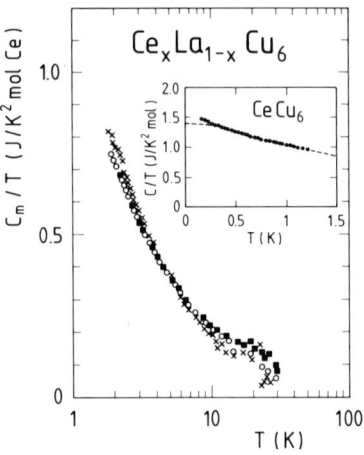

Fig. 6. Temperature dependence of the 4f-derived specific heat per mole of cerium as C_m/T vs. T for $Ce_xLa_{1-x}Cu_6$ with $x = 1$ (○), 0.8 (■) and 0.5 (×) [23]. Inset shows low-T data for $CeCu_6$ [82]. Dashed curve indicates result for $S = 1/2$ Kondo impurity with $T_K = 4.2$ K [7].

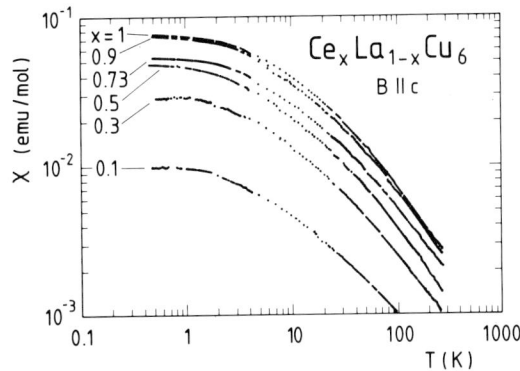

Fig. 7. Temperature dependence of the magnetic susceptibility per mole of formula unit for $Ce_xLa_{1-x}Cu_6$ [23].

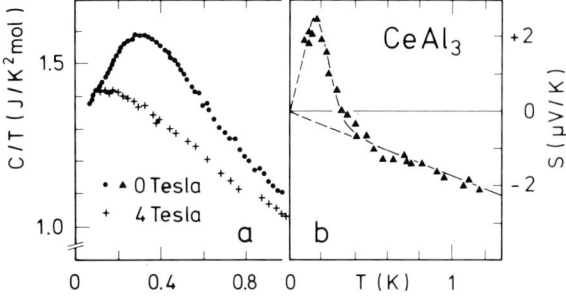

Fig. 8. Low-temperature specific-heat (a), as C/T vs. T at $B = 0$ and 4 T [83], and thermopower (b), as S vs. T [44], for $CeAl_3$.

$Ce_xLa_{1-x}Cu_6$ alloys. For both quantities, a rather uniform temperature dependence of the Ce-derived contribution is found, in wide ranges of the La–Ce composition. In the inset of fig. 6, the low-T specific-heat data [82] of the compound $CeCu_6$ are compared to the numerical result for an $S = 1/2$ Kondo impurity ($T_K = 4.2$ K) as obtained by the Bethe-ansatz calculation [7]. Apart from deviations seen at the lowest temperatures – which can be more pronounced in other compounds, as we shall see below – the Ce-derived part C_m/T vs. T is found to be in close agreement with single-ion theory. The same holds for the susceptibility results in fig. 7 (for the sake of clarity displayed per mole of the formula unit, rather than per mole of Ce). Obviously, the large effective carrier mass m^*, being proportional to both the Sommerfeld coefficient $\gamma = C_m/T$ and the Pauli spin susceptibility χ_0, (i) is of local origin and (ii) forms only upon cooling to way below the Kondo-lattice temperature T^*. Because of inter-site correlations, the latter is expected to slightly exceed T_K [26,27] as is found, e.g., for $Ce_xLa_{1-x}Cu_6$. On the other hand, in $Ce_xLa_{1-x}Al_2$ the reduction of the unit-cell volume on going from the dilute alloy to the $CeAl_2$ compound causes a ratio $T^*/T_K \approx 10$ [31].

Clear phenomenological differences between the dilute Kondo alloy and the Kondo lattice, which are so pronounced in the temperature dependence of the electrical resistivity (fig. 1), can, under favorable circumstances, also be observed in the low-T specific heat. A maximum in C/T vs. T is found, e.g., in $CeAl_3$ (fig. 8a) and normal-state $CeCu_2Si_2$ (fig. 5) near $T_{coh} \approx 0.4$ K [83], the temperature below which $\rho = AT^2$ (cf. fig. 2). The $C(T)/T$ maximum has been ascribed [83] to pseudogap formation in the resonant 4f density of states (DOS) at E_F due to 4f-conduction band hybridization [84–86]. A corresponding change of sign in the thermoelectric power [44], which probes the DOS slope at E_F, demonstrates the asymmetry of the pseudogap with respect to the Fermi energy [82] (fig. 8b). In order to get this kind of fine structure in the DOS, a coherent 4f band has to form, which is the case below T_{coh}. A direct manifestation for coherent heavy quasiparticles in the low-T phase of heavy-fermion compounds is provided by dHvA oscillations in the magnetization [87,88]. Effective masses up to

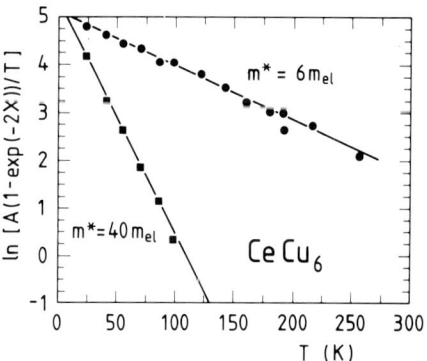

Fig. 9. De Haas–van Alphen amplitude as a function of temperature for two oscillatory components of the magnetization of $CeCu_6$, indicating two different effective masses [87]. $X = 2\pi^2 k_B T/\hbar\omega_c$ ($\omega_c/2\pi$: cyclotron frequency).

$40 m_{el}$ have been observed in CeCu$_6$ [87] (fig. 9) and up to $90 m_{el}$ in UPt$_3$ [88].

To build up a non-magnetic ground state in a dense Kondo system such as CeAl$_3$, antiferromagnetic short-range correlations are prerequisite: As was pointed out by Nozières [89], in order to screen the moments of only a small fraction of Kondo ions, one would already have to exhaust the whole system of conduction electrons. Thus, antiferromagnetic correlations develop upon decreasing temperature and support the quenching of the local moments, so that a macroscopic singlet state is formed. Fig. 10 shows that for CeCu$_6$ the correlation length for these antiferromagnetic couplings is anisotropic and increases upon cooling, until below T^* it saturates at a finite value of the order of the nearest-neighbor distance.

If the short-range correlations are sufficiently strong, an external magnetic field can cause a "metamagnetic" type of transition. This has been discovered for UPt$_3$ [91] and was confirmed later for the non-magnetic heavy-fermion compound CeRu$_2$Si$_2$ [92]: For the magnetic field applied along the tetragonal c-axis, an upturn occurs in the magnetization curve near 8 T (fig. 11). Mignot et al. [92] were able to demonstrate clearly, via pressure experiments, that the energy scale for the short-range antiferromagnetic correlations is

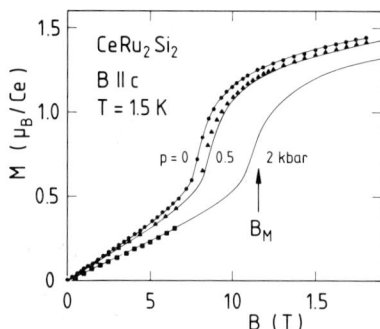

Fig. 11. Pressure dependence of the magnetization curve, M vs. B ($B \| c$), of CeRu$_2$Si$_2$ at 1.5 K. Dots (squares): experimental data at $p = 0$ (2 kbar). Closed triangles: deduced from the latter using an approximate linearized Maxwell equation for $\delta p = 0.5$ kbar. Solid lines: scaled curves for 0.5 and 2.0 kbar using $\partial \ln(k_B T^*)/\partial p = 171$ Mbar^{-1}. The value for the "metamagnetic field" B_M indicated corresponds to a pronounced peak in the magnetoresistance as measured at 1.2 K [92].

identical with the characteristic Kondo-lattice energy, $k_B T^*$.

3. Local-moment magnetism

LMM in heavy-fermion compounds has to be considered intermediate between magnetic ordering in stable-moment rare-earth systems, like Gd-metal, and itinerant magnetism, e.g., of the Fe-group metals. Intermediate values of the Sommerfeld coefficient γ (= a few 100 mJ/K^2 mol) above T_{RKKY} ($> T^*$) indicate moderate renormalizations due to the Kondo effect already at these elevated temperatures. Usually, the γ value in the magnetically ordered state ($T \ll T_m$) is somewhat reduced compared to the one in the paramagnetic state. This reflects a reduction of the heavy-fermion DOS at E_F caused by magnetic ordering.

As already mentioned in subsection 1.2, in realistic compounds the RKKY interaction can be strongly anisotropic. As a consequence, the magnetic structures in these materials are often rather complex. We wish to illustrate this point below by discussing two exemplary cases, CeAl$_2$ and CeCu$_2$Ge$_2$, the latter being doped with low Ni concentrations. In CeAl$_2$, a type-II antiferro-

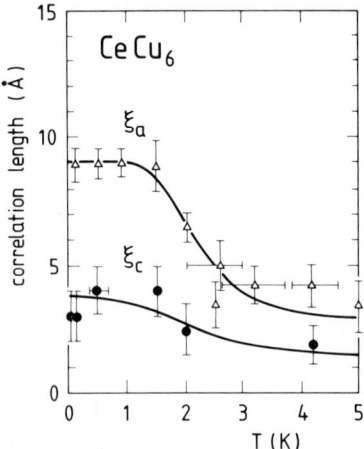

Fig. 10. Magnetic correlation lengths along a and c directions as a function of temperature for CeCu$_6$ [90]. Lines through data points are guides to the eye.

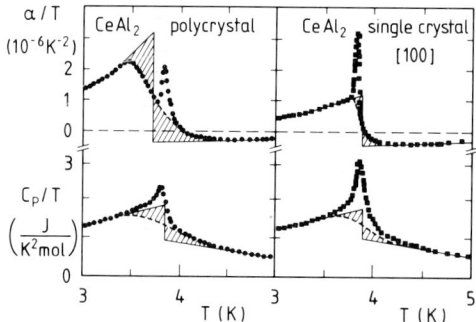

Fig. 12. Linear thermal-expansion coefficient as α/T vs. T (top) and specific heat as C_p/T vs. T (bottom) for polycrystalline (left) and single-crystal (right) CeAl$_2$. α measurement on single crystal was done along [100]. Dashed lines give interpolations between high- and low-temperature data points, and solid lines are used to replace broad transitions by idealized sharp ones at T_{N1} (with positive jumps $\Delta\alpha_1$ and ΔC_{p1}, respectively). A second sharper transition at T_{N2} is characterized by $\Delta\alpha_2 < 0$ and $\Delta C_{p2} > 0$ [93].

magnetic structure, involving a ferromagnetic coupling within and an antiferromagnetic one between adjacent (111) planes, exists which is modulated by an incommensurate wave vector $(\frac{1}{2}+\tau, \frac{1}{2}-\tau, \frac{1}{2})$, $\tau = 0.11$ [30]. The thermal-expansion results in fig. 12, especially on the polycrystalline sample, demonstrate two closely-spaced phase transition anomalies [93]: The broader transition at T_{N1} shows a positive jump $\Delta\alpha_1$, whereas the sharper one at T_{N2} is characterized by $\Delta\alpha_2 < 0$. In the specific heat the corresponding jumps ΔC_{p1} and ΔC_{p2} both are positive and, therefore, are not so easy to separate – though their superposition can be inferred from the change of slope seen in the $C_p(T)/T$ data of fig. 12. With the Ehrenfest relation

$$t_p = \frac{1}{T_N}\left(\frac{\partial T_N}{\partial p_h}\right)_{p_h \to 0} = 3V_{\text{mol}}\frac{\Delta\alpha}{\Delta C_p},$$

one gets a positive (hydrostatic) pressure coefficient for the ordering temperature T_{N1}, but a negative one for T_{N2}. The former has been observed through calorimetric [94] and the latter one through susceptibility [95,96] and neutron-diffraction [97] experiments, cf. figs. 13a and b. These two coexisting magnetic structures, which react quite differently to lattice defects [93], seem to realize the limiting cases of weak and strong local exchange coupling (fig. 3), a fact which could as yet not be explained.

Let us now turn to the compound CeCu$_2$Ge$_2$, a LMM with $T_{N1} = T_{Nh} = 4.1$ K [98]. $T_{RKKY} = 7$ K, the temperature below which short-range magnetic correlations show up [99], nearly coincides with $T^* \approx (8 \pm 2)$ K [99]. In fact, this compound is found near the critical value of the coupling constant $|g_c|$ in Doniach's schematic phase diagram (fig. 3), a modification of which is displayed for CeM$_2$Si$_2$ and CeM$_2$Ge$_2$ homologs (M = Cu, Ag, Au, Ni, Ru) in fig. 14. Here the microscopic

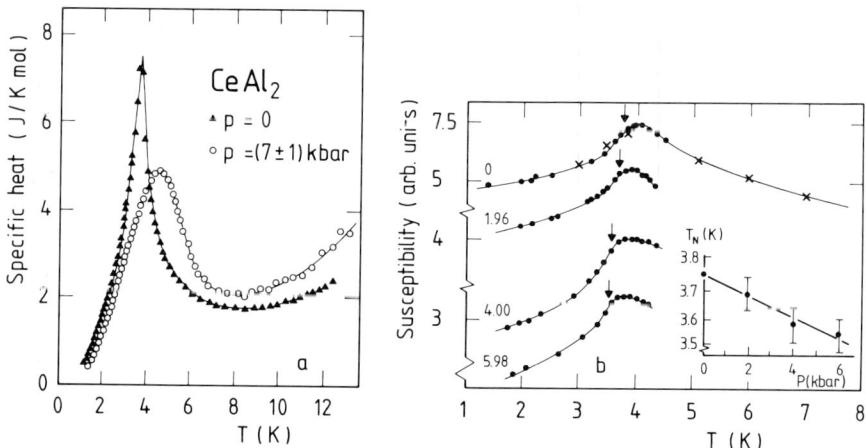

Fig. 13. Temperature dependences of the specific heat [94], at ambient pressure and $p = 7$ kbar (a), as well as of the magnetic susceptibility [95]; at $p = 0$, 1.96, 4.00 and 5.98 kbar (b), for CeAl$_2$.

coupling parameter $|g|$ is simulated by the inverse of the Ce–M distance r.

From its location close to the magnetic–nonmagnetic transition one would expect $CeCu_2Ge_2$ to exhibit a Néel temperature with negative pressure coefficient: compression of the volume should further destabilize the magnetic 4f-configuration. Most surprisingly, the thermal-expansion jump associated with the formation of antiferromagnetism is positive [101], as is the specific-heat jump (fig. 15). Ehrenfest's relation thus yields $t_{ph} = (1/T_{Nh})(\partial T_{Nh}/\partial p_h)_0 > 0$. Up to now pressure experiments have not been performed on this compound. Substituting Ni on Cu sites one can achieve a reduced average unit-cell volume as well. The results for a 2 at% Ni sample also shown in fig. 15 furnish a rather complex behavior: Compared to $CeCu_2Ge_2$, (i) T_{Nh} is reduced, although (ii) $\Delta\alpha_h$ remains positive, (iii) a second phase transition, established to be an intrinsic property of the $Ce(Cu_{1-x}Ni_x)_2Ge_2$ systems with $0.02 \leq x < 0.3$ [102], occurs at $T_{N\ell} < T_{Nh}$. The corresponding thermal-expansion jump (i.e. $t_{p\ell}$) is negative. According to Tachiki [102] the lower transition at $T_{N\ell}$ in the low-Ni doped systems may reflect a superposition of two incommensurate spin structures. These results, which are far from being understood, underline that the Ce–M separation is not the only parameter determining

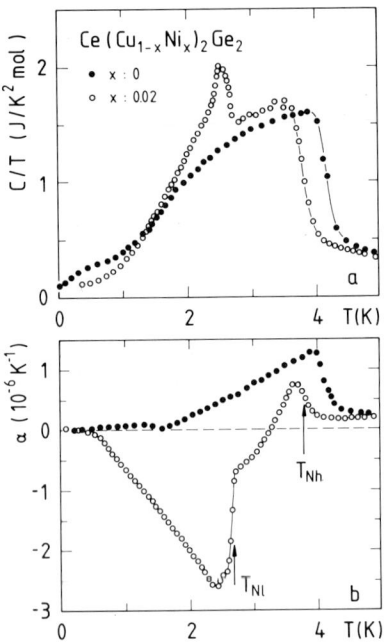

Fig. 15. Low-temperature phase transitions in $CeCu_2Ge_2$ and $Ce(Cu_{0.98}Ni_{0.02})_2Ge_2$ as revealed by the specific heat in a plot C/T vs. T (a) [102] and the linear thermal expansion, α vs. T (b) [101]. For the 2 at% Ni alloy, Néel temperatures T_{Nh} and $T_{N\ell}$ are indicated, see text.

the magnetic state of these systems. Band-structure effects will certainly play an important role, since for the dopant Ni the 3d states are close to E_F, whereas they lie way below E_F for Cu. In addition, two very different types of magnetic structure seem to develop in the pure $CeCu_2Ge_2$ and in dilute $Ce(Cu, Ni)_2Ge_2$ on the one hand and for Ni concentrations $x > 0.2$ on the other, as is discussed in the following section.

4. Transition from local-moment to heavy-fermion band magnetism

To study the transition between LMM and heavy-Fermi liquid in a Ce-based compound like $CeAl_2$, application of hydrostatic pressure is an appropriate tool [14]. Because of the obvious difficulties in determining various properties like specific heat, transport coefficients and thermal expansion in a pressure cell, alloying experiments have been frequently performed. Usually, the f-

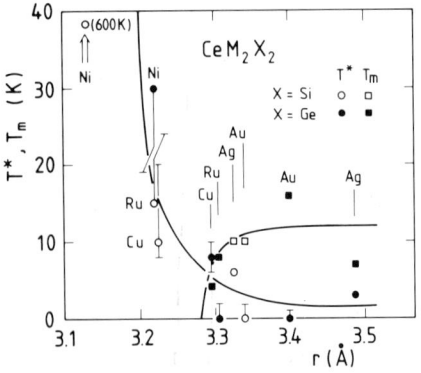

Fig. 14. Kondo-lattice temperature T^*, from low-T quasielastic neutron line width (circles) as well as entropy, thermal expansion and thermopower (bars), and magnetic ordering temperature T_m in CeM_2Si_2 and CeM_2Ge_2 vs. r, the Ce–M distance (M: Cu, Ag, Au, Ni, Ru). Lines are guides to the eye [100].

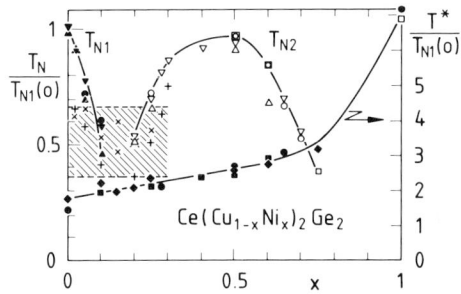

Fig. 16. Néel temperatures T_{N1}, T_{N2} (left) and Kondo-lattice temperature T^* (right), normalized by $T_{N1}(x=0)$, vs. Ni-concentration for $Ce(Cu_{1-x}Ni_x)_2Ge_2$. Results are shown from specific heat (▼, ▽, +), thermal expansion (▲, △, ×) and dc susceptibility (●, ○) (left) as well as quasielastic neutron scattering (●), thermal expansion (■) and resistivity (♦) measurements (right). Lines are guides to the eye. Hatched region marks systems with two phase transitions at T_{Nh} and $T_{N\ell}$, cf. fig. 15 [107].

ion Ce^{3+} was replaced by a non-magnetic trivalent dopant of smaller ionic radius, e.g., Y^{3+}. However, in such a situation disorder in the Ce-sublattice can lead to spin-glass-type ordering which complicates the magnetic phase diagram [13]. Therefore, in the last years magnetic–non-magnetic phase transitions have mostly been monitored for such compounds in which *non-f ligand atoms* were partially substituted by dopants [103].

As a recent example we address the system $Ce(Cu_{1-x}Ni_x)_2Ge_2$. Whereas $CeCu_2Ge_2$ is a LMM, $CeNi_2Ge_2$ [104] belongs to the class of non-magnetic heavy-Fermi-liquid systems, with $T^* = 30$ K and $\gamma = 0.4$ J/K^2 mol [104]. The quasi-binary alloys have been investigated in the whole concentration range by specific heat, thermal expansion, thermopower, resistivity, susceptibility [105] and magnetization [106] experiments. The dependence of the magnetic ordering temperatures T_{N1} and T_{N2} on the Ni concentration x is shown in fig. 16. In accord with Doniach's phase diagram (fig. 3), substitution of the smaller Ni for Cu results in a precipitous depression of $T_{N1}(x)$ for $x < 0.2$. Instead of the expected non-magnetic heavy-Fermi-liquid state, however, a second type of antiferromagnetic ordering below $T_{N2}(x)$ develops near $x = 0.15$. By contrast, the Kondo-lattice temperature T^* as determined by different experimental techniques exhibits a continuous increase, in agreement with expectation.

By neutron powder diffractometry utilizing the multidetector instrument D1b at the high-flux reactor of the Institut Laue–Langevin (Grenoble), Loidl et al. [107] could characterize the magnetic structures below $T_{N1}(x)$ and $T_{N2}(x)$, respectively. Like for the pure $CeCu_2Ge_2$ compound [99], an incommensurate spiral spin arrangement with relatively long ordering wave vector, $q_0 = (0.28, 0.28, 0.41)$, was obtained for the 10 at% Ni alloy. On the other hand, an also incommensurate spiral structure, but with a much shorter q_0, i.e. $(0, 0, 0.14)$, was inferred from these data for $Ce(Cu_{0.5}Ni_{0.5})_2Ge_2$. The ordered Ce-moment is $0.3\mu_B$ for $x = 0.5$. No magnetic Bragg reflections can be resolved from the powder spectra for the $x = 0.65$ alloy, although long-range antiferromagnetic order is clearly visible in the bulk measurements. This implies that $\mu_s < 0.2\mu_B$, since moments of this size cannot be extracted from powder diffraction.

The larger modulation vectors found for low Ni concentrations are of the typical order of Fermi-surface diameters and, thus, indicative of RKKY interactions. Kondo-reduced local Ce moments $\mu_s = (0.5–0.74)\mu_B$ characterize $CeCu_2Ge_2$ and the lightly Ni-doped systems as members of the class of LMM discussed in section 4. The short propagation vector and the smaller ordered moment determined for $x = 0.5$ describe a modulated spin arrangement which extends over almost ten unit cells. Correspondingly short q_0 vectors – along with small ordered moments μ_s – have to be considered hallmarks of HFBM [79]: By comparing the theoretical predictions by Grewe and Welslau [79] with the experimental $T^*(x)$ and $T_N(x)$ dependences of fig. 16, Loidl et al. [107] could establish a surprisingly good agreement for $x \geq 0.5$. Further support for a magnetically ordered state developing out of a heavy Fermi liquid derives from the magnetic neutron cross section which is dominated for $Ce(Cu_{1-x}Ni_x)_2Ge_2$ by single-site relaxation processes, while for the $CeCu_2Ge_2$ compound additional inter-site (RKKY) interactions between local Ce moments are found [99].

The magnetic phase diagram of fig. 16 suggests

that HFBM is suppressed near $x = 0.75$. Arguments for a non-magnetic Fermi-liquid state at higher Ni concentrations involve the lack of any phase-transition anomalies in the bulk properties, i.e., the specific heat and transport coefficients [105]. However, in such a situation HFBM associated with a *very* small μ_s value cannot be ruled out completely [100], as has been demonstrated in an exemplary way for UPt$_3$ [72]. Neutron-diffraction work on single crystals of Ni-rich Ce(Cu$_{1-x}$Ni$_x$)$_2$Ge$_2$ alloys are in progress to decide if long-range ordering with correspondingly tiny moments exists up to $x \approx 0.85$, the limiting value extrapolated from the $T_{N2}(x)$ dependence (fig. 16).

In ref. [100] it was noticed that, combining this extrapolation with the $T^*(x)$ curve for $x > 0.75$, one gets $T_{N2} = 0.6$–0.8 K for $T^* = 15$ K, the Kondo-lattice temperature for the heavy-fermion superconductor CeCu$_2$Si$_2$. Indeed, for the latter a number of experimental techniques, such as magneto-resistivity [108], Cu-NMR [109] and μSR [110] reveal – in just this temperature window – anomalies which have been tentatively ascribed [109,111,100] to the development of some long-range antiferromagnetic order. The ordered Ce-moment involved was estimated [110] to be quite small, $\mu_s < 0.1 \mu_B$. A survey on the low-temperature magnetic phase diagram of CeCu$_2$Si$_2$ is given in the following section.

5. Heavy-fermion superconductivity, magnetism and lattice instability in CeCu$_2$Si$_2$

We begin this section by discussing calorimetric and dilatometric results that were obtained recently [112,113] on a "new generation" of CeCu$_2$Si$_2$ single crystals grown by the cold-boat technique and subsequently annealed in a Cu atmosphere [114]. Their high crystalline perfection could be demonstrated convincingly by measurements of dHvA oscillations in the magnetization as a function of the applied field [115], as well as by sharp superconducting transitions at $T_c \approx 0.63$ K. On the other hand, before annealing the same crystals show only spurious superconductivity with low T_c (0.2–0.3 K) and a small Meissner volume (< 0.4%), which reflects inhomogeneities in the Cu concentration. For true bulk experiments, e.g., specific heat and thermal expansion, these "as grown" crystals can, therefore, serve as "non-superconducting" reference samples.

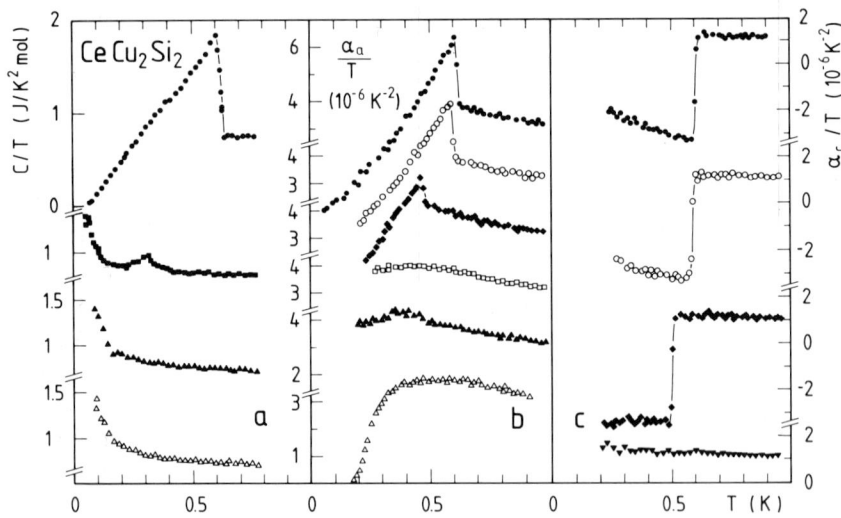

Fig. 17. Specific heat as C/T vs. T (a), linear-thermal-expansion coefficient along a-axis as α_a/T vs. T (b) and along c-axis as α_c/T vs. T (c) of CeCu$_2$Si$_2$ single crystal at $B = 0$ (●), 0.5 T (○), 1.5 T (♦), 2 T (■), 3 T (▼), 5 T (□), 6 T (▲) and 6.5 T (△) [113]. The anomaly at 0.35 K in the 2 T data of (a) indicates superconducting transition.

Figs. 17a–c display the specific-heat results in a plot C/T vs. T as well as the thermal-expansion results taken along both the a- and c-axis, as α_a/T vs. T and α_c/T vs. T, on bulk superconducting $CeCu_2Si_2$ single crystals for $B = 0$ and several values of the external magnetic field [113]. The data for $T > T_c$ are typical for a heavy Fermi liquid: In relation to the already very large Sommerfeld coefficient (as $T \to 0$), $\gamma \approx 770$ mJ/K^2mol, the coefficient α/T is enhanced by almost two more orders of magnitude, as is expressed by the "volume Grüneisen parameter", $\Gamma = -\partial \ln T^*/\partial \ln V = V_{mol}c_B(2\alpha_a + \alpha_c)/C = 63$ ($V_{mol} = 50.3$ cm^3: molar volume, $c_B = 1200$ kbar: bulk modulus [116]). This characterizes a very strong coupling of the heavy fermions to the breathing mode ("Grüneisen-parameter coupling"), which shows a 50% anisotropy: The "orientation-dependent Grüneisen parameters" Γ_a and Γ_c amount to 80 and 40, respectively [117].

The superconducting transition of the new high-quality crystals is characterized by pronounced jump anomalies in $C(T)$ as well as in $\alpha_a(T)$ and $\alpha_c(T)$. The idealized reduced specific-heat-jump height, $\Delta C/C_n(T_c) = 1.48$ (obtained by replacing, in the usual way [45], the broadened transition as measured by a sharp one), exceeds the BCS value (1.43). The specific-heat data points well below T_c ($T \geq 0.1$ K) lie above the exponential BCS law. Entropy conservation requires that a fictitious superconductor with the same T_c, but a "BCS-like" $C(T)$ dependence, would show an even larger ratio $\Delta C/C_n(T_c)$. Thus $CeCu_2Si_2$, like UBe_{13} [57], must be considered a strong-coupling superconductor.

The thermal-expansion jumps measured along the respective a- and c-axes are of opposite sign. Comparing them with ΔC, one gets the uniaxial pressure derivatives of T_c (as $p_i \to 0$; i: a, c) via Ehrenfest's relation

$$\frac{1}{T_c}\left(\frac{\partial T_c}{\partial p_i}\right)_0 = V_{mol}\frac{\Delta \alpha_i}{\Delta C}.$$

The hydrostatic pressure derivative $(\partial T_c/\partial p_h)_0 = 2(\partial T_c/\partial p_a)_0 + (\partial T_c/\partial p_c)_0 = 3.5$ mK/kbar. This value is nearly identical with the one directly measured calorimetrically on polycrystalline samples in a pressure cell [118]. Employing the appropriate elastic constants c_{ij} as determined on such new single crystals by Lüthi's group, one can calculate [113] the uniaxial strain dependencies (as $\epsilon_i \to 0$)

$$\left(\frac{\partial T_c}{\partial \epsilon_a}\right)_0 = -(c_{11} + c_{12})\left(\frac{\partial T_c}{\partial p_a}\right)_0 - c_{13}\left(\frac{\partial T_c}{\partial p_c}\right)_0$$
$$= -(20 \pm 9)\text{ K}$$

and

$$\left(\frac{\partial T_c}{\partial \epsilon_c}\right)_0 = -2c_{13}\left(\frac{\partial T_c}{\partial p_a}\right)_0 - c_{33}\left(\frac{\partial T_c}{\partial p_c}\right)_0$$
$$= +(10 \pm 8)\text{ K}.$$

The large uncertainties are caused by the uncertainty in the bulk modulus c_B [116] which enters the calculation of the transverse mode c_{13}.

We conclude that negative strain applied along the a-axis, namely a reduction of the Ce–Ce separation, gives rise to a T_c increase, whereas the opposite effect, though much less pronounced, is found along the c-axis. Since the intra-plane coupling between Ce ions is dominated by the 4f–5d hybridization, which tends to destabilize the magnetic 4f-configuration of Ce^{3+} via raising T^*, the observed strain dependence $(\partial T_c/\partial \epsilon_a)_0$ demonstrates clearly that a sufficient compensation of the local 4f moments is prerequisite for heavy-fermion superconductivity in $CeCu_2Si_2$.

The systematic investigation of high-quality single crystals revealed that the upper-critical-magnetic-field curves determined resistively [52] on Bridgman-grown single crystals for the field along the a- and c-axes is an intrinsic property of $CeCu_2Si_2$, as was also found above for the hydrostatic pressure derivative of T_c. The $B_{c2}(T)$ values read off the thermal-expansion results of figs. 17b and c coincide with the previous results by Assmus et al. [52], see fig. 18: The pronounced anisotropy in $B_{c2}(T)$ can well explain the anisotropy of the thermal-expansion jump $\Delta\alpha$ as a function of the magnetic field, applied along the same direction used for the length measurement, i.e.:

$$A = \frac{\Delta\alpha_c(B)}{\Delta\alpha_c(0)} \bigg/ \frac{\Delta\alpha_a(B)}{\Delta\alpha_a(0)} = 2.7 \text{ at } B = 1.5\text{ T}.$$

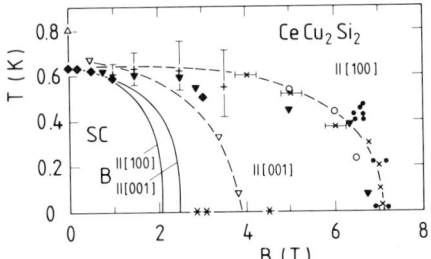

Fig. 18. $B-T$ diagram for $CeCu_2Si_2$, containing: B_{c2}-curves, defining the superconducting (SC) to normal transition, for fields parallel to the a- and c-axes (solid lines), temperatures at which Cu-NMR intensity on a polycrystalline sample drops from 90 to 10% upon cooling (+), magnetic fields where the isothermal longitudinal magnetoresistance $(B \| j \| a)$ shows a minimum (×), onset of magnetic correlations inferred from μSR (△), peak positions in dHvA oscillations for $B \| a$ (●) and $B \| c$ (★), step-like phase transition anomalies in the longitudinal elastic constants $C_{11}(T, B \| a)$ (▼) and $C_{33}(T, B \| c)$ (▽) and phase-transition-like anomalies in the coefficient of thermal expansion $\alpha_a(T, B \| a)$ (○). Also included are $\alpha(T, B \| a)$-derived transition temperatures $T_1(B)$ (◆) of a non-bulk-superconducting crystal [113].

Starting from the appropriate thermodynamic relation [113]

$$\Delta \alpha_i(B) = \frac{1}{V_{mol}\mu_0} \left(\frac{\partial B_{c2}}{\partial T}\right)^2_{p_i} \left(\frac{\partial T_c}{\partial p_i}\right)_B \left(\frac{\partial M}{\partial B}\right)_{p_i, T, B_{c2}}$$

$(i: a, c)$

and assuming that both the uniaxial pressure derivatives $(\partial T_c / \partial p_i)_B$ and the susceptibility $(\partial M / \partial B)_{p_i, T, B_{c2}} = [1.16(2\kappa^2 - 1)]^{-1}$ (κ: Ginzburg–Landau parameter) are independent of the orientation of the external field with respect to the symmetry axes of the crystal, one finds

$$A = \left[\left(\frac{\partial B_{c2}^c}{\partial T}\right)^2 \bigg/ \left(\frac{\partial B_{c2}^a}{\partial T}\right)^2\right]_{B_{c2}^c = B_{c2}^a = 1.5\,T} = 2.9 \pm 1,$$

in close agreement to the experimental value.

We now turn to the second phase transition which apparently exists, besides superconductivity, in $CeCu_2Si_2$ at low temperatures. Fig. 18 indicates those B-fields (applied along both the a- and c-axis) at which (i) isothermal magnetoresistivity measurements reveal a "kink" [108], (ii) dHvA oscillations show maxima [115] and (iii) the elastic constants $c_{11}(T)$ and $c_{33}(T)$ exhibit "steps"

[113]. In addition, we have marked in the figure those temperatures $T_1'(B)$ at which the intensity of the Cu-NMR of polycrystalline samples disappears [109] as well as the one below which internal magnetic fields develop in the $(B = 0)$ μSR measurement [110]. As is seen in fig. 17b, anomalies develop in the thermal expansion for $B \geq 5$ T, whose positions are close to the ones obtained by the techniques listed above. Interestingly enough, corresponding anomalies are neither found in $\alpha(T)$ for $B < 5$ T [113] nor in the specific heat up to $B = 6.5$ T (fig. 17a). Since $C_{11} \sim \Gamma^2 C$ and $\alpha \sim \Gamma C$, a very large and field-dependent Grüneisen parameter associated with the second phase transition may be concluded [113].

In order to shed more light on the nature of this transition documented for bulk-superconducting $CeCu_2Si_2$ samples by so different probes and tentatively ascribed [111] to long-range antiferromagnetic ordering, Lang et al. [113] have studied several of the "as-grown" single crystals, that are lacking bulk superconductivity. Most surprisingly, mean-field-type anomalies in $C(T)$ and $\alpha(T)$ are found at $T_1 = 625$ mK for zero-field (figs. 19a and b), although no related features can be resolved in the presence of bulk superconductivity. Clearly T_1 coincides with T_1', the critical temperature of the second phase transition in the superconducting samples. Upon increasing external B-field, T_1 is found to be depressed more rapidly than T_1', see fig. 18. Employing the negative jumps in $\alpha_a(T)$ and $\alpha_c(T)$ at T_1 (fig. 19b) one gets for the volume-expansion coefficient, $\beta = 2\alpha_a + \alpha_c$, a change of sign from plus to minus upon cooling through T_1. This goes along with a gigantic negative hydrostatic pressure derivative, $(\partial T_1 / \partial p_h)_0 = -200$ mK/kbar, as calculated from the Ehrenfest relation which links $\Delta \alpha_a$ and $\Delta \alpha_c$ with ΔC (fig. 19a).

Also, the strain dependences, $(\partial T_1 / \partial \epsilon_a)_0 = +(290 \pm 50)$ K and $(\partial T_1 / \partial \epsilon_c)_0 = (160 \pm 70)$ K are much larger compared to the analogous relations for T_c. This, along with the different signs of $(\partial T_1 / \partial \epsilon_a)_0$ and $(\partial T_c / \partial \epsilon_a)_0$, clearly distinguish the new transition from the superconducting one. An only minor change in the low-field susceptibility $\chi(T)$ at T_1 makes a magnetic transition unlikely as well [113]. Therefore, Lang et al. [113] have

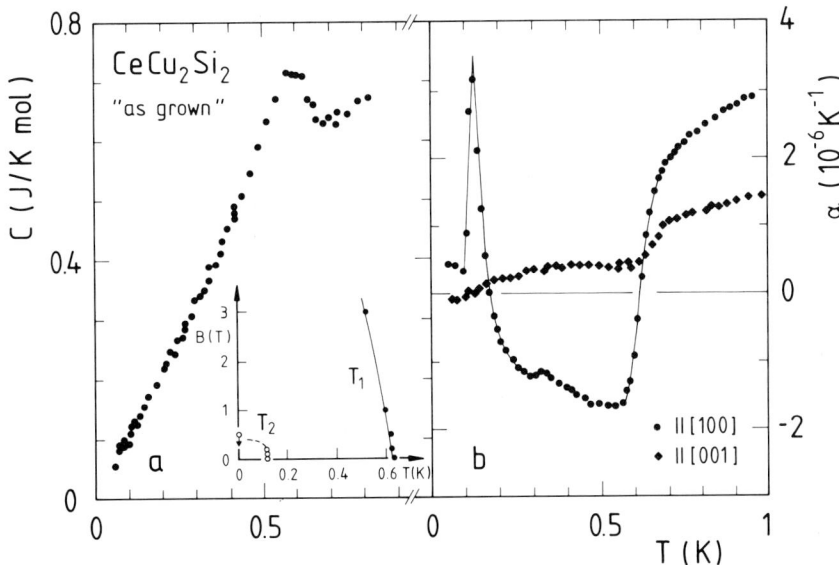

Fig. 19. Low-temperature properties of "as grown", i.e., non-bulk-superconducting, $CeCu_2Si_2$ single crystal. (a) specific heat, C vs. T; (b) linear thermal-expansion coefficients measured along [100] and [001] directions, α_a vs. T (●) and α_c vs. T (♦). Inset: field dependences of the phase transition anomalies as derived from α_a vs. T measured at different B-fields [113].

ascribed the phase-transition anomalies of fig. 19, related ones being observed in the elastic constants [113], to a "lattice instability". From the quite low value of T_1 and its dependence on magnetic field (inset of fig. 19) this structural transition was suggested to be driven by some mode (e.g., an anisotropic volume mode), which couples strongly to the heavy electronic degrees of freedom.

The microscopic origin of the new phase transition is presently not known. Quadrupolar ordering is most unlikely, in view of both the temperature dependence of the various elastic constants [113] and the CF-level scheme of Ce^{3+} in $CeCu_2Si_2$ [119]. However, the possibility of a charge-density wave in the heavy-Fermi-liquid phase should be seriously considered, since it has already been established theoretically under fortunate conditions [120]. Also a spin-density wave associated with a very small ordered moment can presently not be ruled out definitively, though being in seeming conflict with outward appearance. Irrespective of its underlying mechanism, the new phase transition is closely related to both magnetic and superconducting phenomena: In bulk superconducting samples, strong "dynamical magnetic correlations" [121] develop below T_1' – presumably due to changes in the 4f-ligand hybridization. The expansion of the volume observed upon cooling for "as grown" crystals is expected to enhance the stability of the magnetic 4f-configuration. Therefore, the anomaly found in $\alpha(T)$ at $T_2 = 115$ mK (fig. 19b) should be ascribed to an antiferromagnetic transition, which appears suppressed in the presence of bulk superconductivity.

Most importantly, the transition temperature T_1 coincides with the highest T_c that can be achieved in $CeCu_2Si_2$ by optimal treatment – similar to recent observations made for certain high-T_c superconductors [117,122]. Hopefully, a deeper insight in the nature of this new phase transition will help to better understand the pairing mechanism in the prototypical heavy-fermion superconductor $CeCu_2Si_2$.

6. Outlook

Ce-based heavy-fermion compounds have played an outstanding role to establish the behavior of the Kondo lattice and several of its novel, partly unexpected, ground states: the coherent heavy-Fermi-liquid phase, local (reduced) moment magnetism, heavy-fermion band magnetism and heavy-fermion superconductivity. Since their characteristic energies ($\leq k_B T^*$ and $\approx k_B T_{RKKY}$) are comparable, these different ground-state properties compete for stability and even can coexist. Recent investigations on $CeCu_2Si_2$ single crystals reveal the existence of an as yet unexplained lattice instability as an inherent property of the heavy-fermion system.

Apart from the "classical" heavy-fermion compounds with usually high carrier concentration discussed in this report, metallic Ce [21,123] and Yb [124] systems with very low carrier concentration and even semiconducting Sm-compounds [125], all of which showing heavy-fermion-type of phenomena, are presently receiving increasing attention. These investigations promise an improved understanding of the local processes underlying the fundamental Kondo interaction.

To conclude, it can be hoped that the recent progress made in the materials development on the one hand and the theoretical methods on the other will furnish a much more detailed knowledge of the Ce-based heavy-fermion compounds in the next few years. This will presumably have great impact on the whole field of the physics of strongly correlated electrons.

Acknowledgements

I would like to thank my colleagues P. Fulde, N. Grewe, B. Lüthi, A. Loidl, G. Weber, U. Ahlheim, W. Assmus, C.D. Bredl, R. Caspary, C. Geibel, M. Lang and G. Sparn for a fruitful cooperation over the years on the topics described in this article. Financial support by the Deutsche Forschungsgemeinschaft, partly under the auspices of the SFB 252, is gratefully acknowledged.

References

[1] For a review, see, e.g., G. Grüner and A. Zawadowski, Rep. Progr. Phys. 37 (1974) 1497.

[2] For an early review, see D.K. Wohlleben and B.R. Coles, in: Magnetism, vol. V, ed. H. Suhl (Academic Press, New York and London, 1973) p. 3.

[3] For a more recent review, see, e.g., J.M. Lawrence, P.S. Riseborough and R.D. Parks, Rep. Progr. Phys. 44 (1981) 1.

[4] See, e.g., F. Steglich, Festkörperprobleme (Adv. Solid State Phys.) XVII (1977) 319.

[5] For a review, see, e.g., M.B. Maple, L.E. DeLong and B.C. Sales, in: Handbook on the Physics and Chemistry of the Rare Earths, vol. 1, eds. K.A. Gschneidner, Jr. and L. Eyring (North-Holland, Amsterdam, 1978) p. 797.

[6] D.M. Newns, to be published.

[7] For a review, see N. Andrei, K. Furuya and J.K. Loewenstein, Rev. Mod. Phys. 55 (1983) 331.

[8] E. Müller-Hartmann and J. Zittartz, Phys. Rev. Lett. 26 (1971) 428.

[9] G. Riblet and K. Winzer, Solid State Commun. 9 (1971) 1663.

[10] M.B. Maple, W.A. Fertig, A.C. Mota, D. Wohlleben and R. Fitzgerald, Solid State Commun. 11 (1972) 829.

[11] K. Winzer, Solid State Commun. 24 (1977) 551.

[12] M.B. Maple and D.K. Wohlleben, AIP Conf. 18 (1973) 447.

[13] J. Aarts, F.R. de Boer, S. Horn, F. Steglich and D. Meschede, in: Valence Fluctuations in Solids, eds. L.M. Falicov, W. Hanke and M.B. Maple (North-Holland, Amsterdam, 1981) p. 301.

[14] For a review, see J.S. Schilling, Adv. Phys. 28 (1979) 657.

[15] S. Horn, F. Steglich, M. Loewenhaupt and E. Holland-Moritz, Physica B 107 (1981) 103.

[16] For a review, see P. Fulde and M. Loewenhaupt, Adv. Phys. 34 (1985) 589.

[17] D. Jaccard and J. Sierro, in: Valence Instabilities, eds. P. Wachter and H. Boppart (North-Holland, Amsterdam, 1982) p. 409.

[18] S. Horn, W. Klämke and F. Steglich, ibid., p. 459.

[19] J. Flouquet, J.C. Lasjaunias, J. Peyrard and M. Ribault, J. Appl. Phys. 53 (1982) 2127.

[20] E. Ruderer, R. Schefzyk, H. Biesenkamp, W. Klämke, S. Horn and F. Steglich, J. Magn. Magn. Mater. 45 (1984) 219.
R. Schefzyk, J. Heibel, F. Steglich, R. Felten and G. Weber, J. Magn. Magn. Mater. 47 & 48 (1985) 83.

[21] T. Kasuya, O. Sakai, H. Harima and M. Ikeda, J. Magn. Magn. Mater. 76 & 77 (1988) 46.

[22] H.J. van Daal and K.H.J. Buschow, Solid State Commun. 7 (1969) 217.

[23] Y. Onuki and T. Komatsubara, J. Magn. Magn. Mater. 63 & 64 (1987) 281.

[24] N.F. Mott, Phil. Mag. 30 (1974) 403.

[25] K. Andres, J.E. Graebner and H.R. Ott, Phys. Rev. Lett. 35 (1975) 1779.

[26] P. Fulde, J. Keller and G. Zwicknagl, Solid State Phys. 41 (1988) 1.

[27] N. Grewe and F. Steglich, in: Handbook on the Physics and Chemistry of the Rare Earths, vol. 14, eds. K.A. Gschneidner, Jr. and L. Eyring (North-Holland, Amsterdam, 1991) p. 343.
[28] M.A. Rudermann and C. Kittel, Phys. Rev. 96 (1954) 99.
T. Kasuya, Progr. Theor. Phys. 16 (1956) 45.
K. Yosida, Phys. Rev. 106 (1957) 895.
[29] S. Doniach, Physica 91 B (1977) 231.
[30] B. Barbara, J.X. Boucherle, J.L. Buevoz, M.F. Rossignol and J. Schweizer, Solid State Commun. 24 (1977) 481.
[31] C.D. Bredl, F. Steglich and K.D. Schotte, Z. Phys. B 29 (1978) 327.
[32] K. Winzer and W. Felsch, J. de Phys. 39 (1978) C6-832.
[33] M. Kawakami, S. Kunii, T. Komatsubara and T. Kasuya, Solid State Commun. 36 (1980) 435.
[34] H. Yashima, N. Sato, H. Mori and T. Satoh, Solid State Commun. 43 (1982) 595.
[35] W.H. Lee, R.N. Shelton, S.K. Dhar and K.A. Gschneidner, Jr., Phys. Rev. B 35 (1987) 8523.
[36] E. Gratz, E. Bauer, B. Barbara, S. Zemirli, F. Steglich, C.D. Bredl and W. Lieke, J. Phys. F 15 (1985) 1975.
[37] A. Böhm, R. Caspary, U. Habel, L. Pawlak, A. Zuber, F. Steglich and A. Loidl, J. Magn. Magn. Mater. 76 & 77 (1988) 150.
[38] U. Rauchschwalbe, U. Gottwick, U. Ahlheim, H.M. Mayer and F. Steglich, J. Less-Common Met. 111 (1985) 265.
[39] S.K. Dhar, K.A. Gschneidner, Jr., C.D. Bredl and F. Steglich, Phys. Rev. B 39 (1989) 2439.
[40] J. Tang, K.A. Gschneidner, Jr., R. Caspary and F. Steglich, Physica B 163 (1990) 201.
[41] Y. Onuki, Y. Shimizu and T. Komatsubara, J. Phys. Soc. Japan 53 (1984) 1210.
[42] G.R. Stewart, Z. Fisk and M.S. Wire, Phys. Rev. B 30 (1984) 482.
[43] L.C. Gupta, D.E. MacLaughlin, Cheng Tien, C. Godart, M.A. Edwards and R.D. Parks, Phys. Rev. B 28 (1983) 3673.
[44] F. Steglich, C.D. Bredl, W. Lieke, U. Rauchschwalbe and G. Sparn, Physica B 126 (1984) 82.
[45] F. Steglich, J. Aarts, C.D. Bredl, W. Lieke, D. Meschede, W. Franz and H. Schäfer, Phys. Rev. Lett. 43 (1979) 1892.
[46] U. Rauchschwalbe, W. Lieke, C.D. Bredl, F. Steglich, J. Aarts, K.M. Martini and A.C. Mota, Phys. Rev. Lett. 49 (1982) 1448.
[47] W. Lieke, U. Rauchschwalbe, C.D. Bredl, F. Steglich, J. Aarts and F.R. de Boer, J. Appl. Phys. 53 (1982) 2111.
[48] F. Steglich, Springer Series in Solid-State Sciences 90 (1990) 306.
[49] M. Ishikawa, H.F. Braun and J.L. Jorda, Phys. Rev. B 27 (1983) 3092.
[50] C.D. Bredl, H. Spille, U. Rauchschwalbe, W. Lieke, F. Steglich, G. Cordier, W. Assmus, M. Herrmann and J. Aarts, J. Magn. Magn. Mater. 31–34 (1983) 373.
[51] H. Spille, U. Rauchschwalbe and F. Steglich, Helv. Phys. Acta 56 (1983) 165.
[52] W. Assmus, M. Herrmann, U. Rauchschwalbe, S. Riegel, W. Lieke, H. Spille, S. Horn, G. Weber, F. Steglich and G. Cordier, Phys. Rev. Lett. 52 (1984) 469.
[53] For a review, see F. Steglich, Springer Series in Solid-State Sciences 62 (1985) 23.
[54] F.G. Aliev, N.B. Brandt, V.V. Moshchalkov and S.M. Chudinov, Solid State Commun. 45 (1983) 215; J. Low Temp. Phys. 57 (1984) 61.
[55] G.R. Stewart, Z. Fisk and J.O. Willis, Phys. Rev. B 28 (1983) 172.
[56] Z. Kletowski, J. Less-Common Met. 95 (1983) 127.
[57] H.R. Ott, H. Rudigier, Z. Fisk and J.L. Smith, Phys. Rev. Lett. 50 (1983) 1595.
[58] G.R. Stewart, Z. Fisk, J.O. Willis and J.L. Smith, Phys. Rev. Lett. 52 (1984) 679.
[59] W. Schlabitz, J. Baumann, B. Pollit, U. Rauchschwalbe, H.M. Mayer, U. Ahlheim and C.D. Bredl, Abstr. of the 4th Intern. Conf. on Valence Fluctuations (Cologne, 1984) (unpublished); Z. Phys. B 62 (1986) 71.
[60] J.G. Bednorz and K.A. Müller, Z. Phys. B 64 (1986) 189.
[61] C.M. Varma, Comments on Solid State Phys. 11 (1985) 221.
[62] P.A. Lee, T.M. Rice, J.W. Serene, L.J. Sham and J.W. Wilkins, Comments on Condensed Matter Phys. 12 (1986) 99.
[63] G.R. Stewart, Rev. Mod. Phys. 56 (1984) 755.
[64] N.B. Brandt and V.V. Moshchalkov, Adv. Phys. 33 (1984) 373.
[65] H.R. Ott, Progr. Low Temp. Phys. XI (1987) 217.
[66] H.R. Ott, H. Rudigier, Z. Fisk and J.L. Smith, Phys. Rev. B 31 (1985) 1651.
[67] U. Rauchschwalbe, F. Steglich, G.R. Stewart, A.L. Giorgi, P. Fulde and K. Maki, Europhys. Lett. 3 (1987) 751.
[68] R.A. Fisher, S. Kim, B.F. Woodfield, N.E. Phillips, L. Taillefer, K. Hasselbach, J. Flouquet, A.L. Giorgi and J.L. Smith, Phys. Rev. Lett. 62 (1989) 1411.
L. Taillefer, Physica B 163 (1990) 278.
[69] T.T.M. Palstra, A.A. Menovsky, J. van den Berg, A.J. Dirkmaat, J.G. Nieuwenhuys and J.A. Mydosh, Phys. Rev. Lett. 55 (1985) 2727.
[70] M.B. Maple, J.W. Chen, Y. Dalichaouch, Y. Kohara, C. Rossel, M.S. Torikachivili, H.W. McElfresh and J.D. Thompson, Phys. Rev. Lett. 56 (1986) 185.
[71] C. Broholm, J.K. Kjems, W.J.L. Buyers, P. Matthews, T.T.M. Palstra, A.A. Menovsky and J.A. Mydosh, Phys. Rev. Lett. 58 (1987) 1467.
[72] G. Aeppli, E. Bucher, A.I. Goldman, G. Shirane, C. Broholm and J.K. Kjems, J. Magn. Magn. Mater. 76 & 77 (1988) 385.
[73] C. Geibel, S. Thies, D. Kaczorowski, A. Mehner, A. Grauel, B. Seidel, U. Ahlheim, R. Helfrich, K. Petersen, C.D. Bredl and F. Steglich, Z. Phys. B 83 (1991) 305.
[74] C. Geibel, C. Schank, S. Thies, H. Kitazawa, C.D. Bredl, A. Böhm, M. Raü, A. Grauel, R. Caspary, R. Helfrich, U. Ahlheim, G. Weber and F. Steglich, Z. Phys. B 84 (1991) 1.

[75] R.J. Trainor, M.B. Brodsky, B.D. Dunlap and G.K. Shenoy, Phys. Rev. Lett. 37 (1976) 1511.
[76] H.R. Ott, H. Rudigier, P. Delsing and Z. Fisk, Phys. Rev. Lett. 52 (1984) 1551.
[77] G.R. Stewart, Z. Fisk, J.L. Smith, J.O. Willis and M.S. Wire, Phys. Rev. B 30 (1984) 1249.
[78] G.M. Kalvius, S. Zwirner, U. Potzel, J. Moser, W. Potzel, F.J. Litterst, J. Gal, S. Fredo, I. Yaar and J.C. Spirlet, Phys. Rev. Lett. 65 (1990) 2290.
[79] N. Grewe and B. Welslau, Solid State Commun. 65 (1988) 437.
[80] C.L. Lin, J. Teter, J.E. Crow, T. Mihalisin, J. Brooks, A.I. Abou-Aly and G.R. Stewart, Phys. Rev. Lett. 54 (1985) 2541.
[81] C. Vettier, P. Morin and J. Flouquet, Phys. Rev. Lett. 56 (1986) 1980.
[82] F. Steglich, U. Rauchschwalbe, U. Gottwick, H.M. Mayer, G. Sparn, N. Grewe, U. Poppe and J.J.M. Franse, J. Appl. Phys. 57 (1985) 3054.
[83] C.D. Bredl, S. Horn, F. Steglich, B. Lüthi and R.M. Martin, Phys. Rev. Lett. 52 (1984) 1982.
[84] R.M. Martin, Phys. Rev. Lett. 48 (1982) 362.
[85] N. Grewe, Solid State Commun. 50 (1984) 19.
[86] N. d'Ambrumenil and P. Fulde, Springer Series in Solid-State Sciences 62 (1985) 195.
[87] P.H.P. Reinders, M. Springford, P.T. Coleridge, R. Boulet and D. Ravot, Phys. Rev. Lett. 57 (1986) 1631.
[88] L. Taillefer and G.G. Lonzarich, Phys. Rev. Lett. 60 (1988) 1570.
[89] P. Nozières, Ann. de Phys. 10 (1985) 19.
[90] J. Rossat-Mignod, L.P. Regnault, J.L. Jacout, C. Vettier, P. Lejay, J. Flouquet, E. Walter, D. Jaccard and A. Amato, J. Magn. Magn. Mater. 76 & 77 (1988) 376.
[91] J.J.M. Franse, P.H. Frings, A. de Visser, A. Menovsky, T.T.M. Palstra, P.H. Kes and J.A. Mydosh, Physica B 126 (1984) 116.
[92] J.-M. Mignot, J. Flouquet, P. Haen, F. Lapierre, L. Puech and J. Voiron, J. Magn. Magn. Mater. 76 & 77 (1988) 97.
[93] R. Schefzyk, W. Lieke and F. Steglich, Solid State Commun. 54 (1985) 525.
[94] A. Berton, J. Chaussy, G. Chouteau, B. Cornut, J. Peyrard and R. Tournier, in: Valence Instabilities and Related Narrow-Band Phenomena, ed. R.D. Parks (Plenum Press, New York, 1977) p. 471.
[95] B. Barbara, M. Cyrot, C. Lacroix-Caen and M.F. Rossignol, J. de Phys. (Paris) 40 (1979) C5-340.
[96] M.C. Croft, R.P. Guertin, L.C. Kupferberg and R.D. Parks, Phys. Rev. B 20 (1979) 2073.
[97] B. Barbara, M.F. Rossignol, J.X. Boucherle and C. Vettier, Phys. Rev. Lett. 45 (1980) 938.
[98] F.R. de Boer, J.C.P. Klaasse, P.A. Veenhuizen, A. Böhm, C.D. Bredl, U. Gottwick, H.M. Mayer, L. Pawlak, U. Rauchschwalbe, H. Spille and F. Steglich, J. Magn. Magn. Mater. 63 & 64 (1987) 91.
[99] G. Knopp, A. Loidl, K. Knorr, L. Pawlak, M. Duczmal, R. Caspary, U. Gottwick, H. Spille, F. Steglich and A.P. Murani, Z. Phys. B 77 (1989) 95.

[100] F. Steglich, C. Geibel, S. Horn, U. Ahlheim, M. Lang, G. Sparn, A. Loidl, A. Krimmel and W. Assmus, J. Magn. Magn. Mater. 90 & 91 (1990) 383.
[101] A. Zahn, Diploma Thesis, TH Darmstadt (1990) unpublished.
[102] F. Steglich, G. Sparn, R. Moog, S. Horn, A. Grauel, M. Lang, M. Nowak, A. Loidl, A. Krimmel, K. Knorr, A.P. Murani and M. Tachiki, Physica B 163 (1990) 19.
[103] For an early investigation, see, e.g., J. Lawrence, Phys. Rev. B 20 (1979) 3770.
[104] G. Knopp, A. Loidl, R. Caspary, U. Gottwick, C.D. Bredl, H. Spille, F. Steglich and A.P. Murani, J. Magn. Magn. Mater. 74 (1988) 341.
[105] G. Sparn, R. Caspary, U. Gottwick, A. Grauel, U. Habel, M. Lang, M. Nowak, R. Schefzyk, W. Schiebeling, H. Spille, M. Winkelmann, A. Zuber, F. Steglich and A. Loidl, J. Magn. Magn. Mater. 76 & 77 (1988) 153.
[106] H. Nakotte and F.R. de Boer, unpublished results.
[107] A. Loidl, A. Krimmel, K. Knorr, G. Sparn, M. Lang, S. Horn, F. Steglich, B. Welslau, N. Grewe, H. Nakotte, F.R. de Boer and A.P. Murani, to be published.
[108] U. Rauchschwalbe, F. Steglich, A. de Visser and J.J.M. Franse, J. Magn. Magn. Mater. 63 & 64 (1987) 347.
[109] H. Nakamura, Y. Kitaoka, H. Yamada and K. Asayama, J. Magn. Magn. Mater. 76 & 77 (1988) 517.
[110] Y.J. Uemura, W.J. Kossler, X.H. Yu, H.E. Shone, J.R. Kempton, C.E. Stronach, S. Barth, F.N. Gygax, B. Hitti, A. Schenck, C. Baines, W.F. Langford, Y. Onuki and T. Komatsubara, Phys. Rev. B 39 (1989) 4726.
[111] F. Steglich, J. Phys. Chem. Solids 50 (1989) 225.
[112] W. Assmus, W. Sun, G. Bruls, D. Weber, B. Wolf, B. Lüthi, M. Lang, U. Ahlheim, A. Zahn and F. Steglich, Physica B 165 & 166 (1990) 379.
[113] M. Lang, R. Modler, U. Ahlheim, R. Helfrich, P.H.P. Reinders, F. Steglich, W. Assmus, W. Sun, G. Bruls, D. Weber and B. Lüthi, Phys. Scripta (forthcoming).
[114] W. Sun, M. Brand, G. Bruls and W. Assmus, Z. Phys. B 80 (1990) 249.
[115] M. Hunt, P. Meeson, P.-A. Probst, P.H.P. Reinders, M. Springford, W. Assmus and W. Sun, J. Magn. Magn. Matter 90 & 91 (1990) 374; J. Phys.: Condens. Matter 2 (1990) 6859.
[116] I.L. Spain, F. Steglich, U. Rauchschwalbe and H.D. Hochheimer, Physica B 139 & 140 (1986) 449.
[117] M. Lang, Dissertation, TH Darmstadt (1990) unpublished.
[118] A. Bleckwedel and A. Eichler, Solid State Commun. 56 (1985) 693.
[119] S. Horn, E. Holland-Moritz, M. Loewenhaupt, F. Steglich, H. Scheuer, A. Benoit and J. Flouquet, Phys. Rev. B 23 (1981) 3171.
[120] N. Grewe, private communication (1990).
H. Goebel, Diploma Thesis, TH Darmstadt (1989) unpublished.
[121] H. Nakamura, Y. Kitaoka, T. Iwai, H. Yamada and K. Asayama, J. Phys.: Condens. Matter (forthcoming).
Y. Kitaoka, H. Nakamura, T. Iwai, K. Asayama, U.

Ahlheim, C. Greibel, C. Schank and F. Steglich, J. Phys. Soc. Japan 60 (1991) 2122.
[122] M. Lang, R. Kürsch, A. Grauel, C. Geibel, F. Steglich, H. Rietschel, T. Wolf, Y. Hidaka, K. Kumagai, Y. Maeno and T. Fujita, to be published.
[123] T. Kasuya, Y.S. Kwon, T. Suzuki, K. Nakanishi and K. Takegahara, J. Magn. Magn. Mater. 90 & 91 (1990) 389.
[124] A. Ochiai, T. Suzuki and T. Kasuya, J. Phys. Soc. Japan 59 (1990) 4129.

[125] U. Ahlheim, K. Fraas, P.H.P. Reinders, F. Steglich, O. Nakamura, T. Suzuki and T. Kasuya, Proc. ICM '91, Edinburgh, J. Magn. Magn. Mater. (submitted), paper 3480A.
P.H.P. Reinders, K. Fraas, U. Ahlheim, C.D. Bredl, C. Schank, R. Caspary, F. Steglich, A. Ochiai, T. Suzuki and T. Kasuya, ibid. (submitted), paper 3481A.

Uranium-based heavy-fermion superconductors: an experimental survey

A. de Visser and J.J.M. Franse

Van der Waals–Zeeman Laboratorium, Universiteit van Amsterdam, Valckenierstraat 65, 1018 XE Amsterdam, Netherlands

The uranium-based heavy-fermion superconductors were discovered almost one decade ago. Here, we present an experimental survey of their interesting normal and superconducting state properties. It appears that most of the unusual normal-state properties can be attributed to the proximity of an antiferromagnetic instability and the presence of competing electronic interactions. The discovery of a superconducting instability in these strongly-correlated electron systems came totally unexpected. The parameters describing the superconducting state yield strong deviations from the standard BCS behaviour. Accumulating evidence has been gathered for a nontrivial superconducting pair function ($L \neq 0$). We illustrate recent developments by a number of prime studies, like high-field measurements and alloying experiments, and give special attention to multicomponent superconductivity in UPt_3 and $(U, Th)Be_{13}$.

1. Introduction

In the past decade a new research area in solid-state physics has received much attention: heavy-fermion physics. As a matter of fact, heavy-fermion systems have been around from the very beginning of the Journal of Magnetism and Magnetic Materials. In 1975, Andres, Graebner and Ott [1] reported the anomalous low-temperature specific heat of $CeAl_3$, with an extremely large electronic specific-heat coefficient: $\gamma = 1630$ mJ/mol K^2. In the case of nontransition metals, γ attains values of the order of 1 mJ/mol K^2, whereas in the transition metals this coefficient can reach values about ten times as high. The anomalously large γ-value observed for $CeAl_3$ is ascribed to the low-temperature ($T < 10$ K) formation of a highly-correlated electron band close to the Fermi level, due to the hybridization of the 4f (Ce) electrons with the s and p electrons of the Al atoms. The partial delocalization of the f electrons gives rise to a description of the low-temperature properties in the Fermi-liquid model [2], with quasiparticles with an effective mass, m_{eff}, of the order of 100 times the free electron mass, m_e. Several other systems with high γ-values have been discovered after 1975, primarily Ce (4f) and U (5f) intermetallic compounds. Among them: $CeCu_2Si_2$ [3], $CeCu_6$ [4], $CeRu_2Si_2$ [5], UBe_{13} [6], UPt_3 [7], UCd_{11} [8] and U_2Zn_{17} [9]. These compounds are nowadays known as the heavy-fermion compounds.

The discovery in 1979 of heavy-fermion superconductivity in $CeCu_2Si_2$ by Steglich and co-workers [5], later followed by the discovery of superconductivity in UBe_{13} [6] and UPt_3 [10], came as a big surprise. The hallmark of heavy-fermion superconductivity is that the very quasiparticles that form the heavy-electron bands take part in the superconducting condensate. The mass renormalization of a factor 100 implies that the Fermi-velocity of the superconducting quasiparticles is of the order of the sound velocity in the solid. This remarkable renormalization is not easily reconciled within the standard BCS theory. However, decisive proof for it is provided by the large anomaly observed in the electronic specific heat at the superconducting phase transition, indicating that virtually all the heavy-electrons become superconducting. The unexpected superconducting instability in a strongly correlated electron gas, where the interactions are princi-

0304-8853/91/$03.50 © 1991 – Elsevier Science Publishers B.V. All rights reserved

pally magnetic in nature, has attracted much attention to the heavy-fermion superconductors. It has been clear from the very beginning, that the superconducting properties of these compounds deviate from standard BCS behaviour, thereby giving rise to speculations upon nontrivial pairing ($L \neq 0$) [11].

Other heavy-fermion compounds remain in a Pauli paramagnetic ground state ($CeCu_6$ [12] and $CeRu_2Si_2$ [13]), as has been investigated for temperatures as low as 10 mK, or exhibit some type of antiferromagnetism ($CeAl_3$ [14], UCd_{11} [8] and U_2Zn_{17} [9]). The (normal-state) properties of most of the heavy-fermion systems indicate the proximity of an antiferromagnetic instability. The large effective mass is built up by a wealth of (competing) electronic interactions and excitations, among which the on-site Kondo (lattice) effect and the inter-site Ruderman–Kittel–Kasuya–Yosida (R-KKY) interaction are thought to play the major roles [15].

The heavy-fermion systems are not only of interest from the experimentalists point of view, but also form a challenge for theoretical physicists. An adequate description of strongly correlated electron systems, that focuses on the superconducting and antiferromagnetic instability, is still lacking (see refs. [15–17] for an introduction to the theoretical aspects of heavy-fermion systems). In this article we discuss the main issues that have emerged after one decade of uranium-based heavy-fermion superconductivity. We survey normal and superconducting state properties of UBe_{13}, UPt_3 and URu_2Si_2, concentrating on the experimental aspects. We illustrate the unusual properties by a number of prime experiments reported in the literature.

2. Heavy-fermion superconductivity in UBe_{13}

The discovery of UBe_{13} as the second heavy-fermion superconductor was reported in 1983 by Ott and co-workers [6]. Resistivity, susceptibility and specific heat data taken on unannealed single-crystalline samples revealed an enormous specific heat coefficient ($\gamma = 1100$ mJ/mol K^2) and a superconducting transition at $T_c = 0.85$ K. The

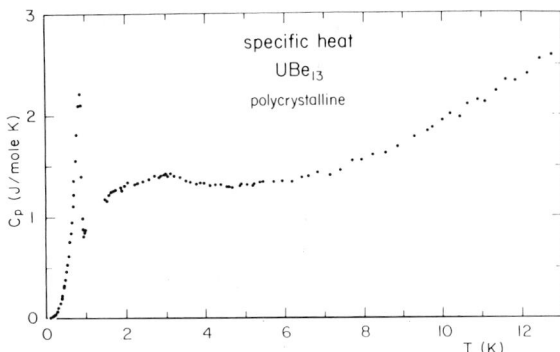

Fig. 1. Specific heat of polycrystalline UBe_{13} in a plot of C versus T (after Ott et al. [19]). The low-temperature contribution centered at 3 K is attributed to the Kondo-lattice effect. Superconductivity sets in at 0.9 K.

large discontinuity observed in the specific heat at T_c (fig. 1) showed convincingly that superconductivity in UBe_{13} is a bulk property. It is interesting to note that superconductivity in UBe_{13} had already been reported in 1973 [18], but at that time was ascribed to precipitated superconducting U filaments in the polycrystalline samples.

The anomalous low-temperature properties of UBe_{13} are clearly demonstrated by the specific heat [19] (fig. 1) and the electrical resistivity [19] (see fig. 2). The steady rise of $\rho(T)$ with decreasing temperature indicates the presence of Kondo-phenomena. Below ~ 20 K, $\rho(T)$ levels

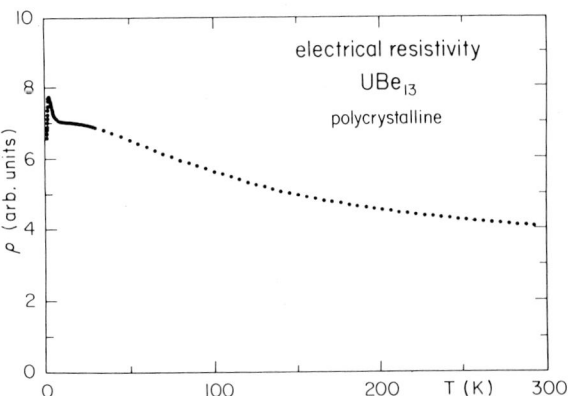

Fig. 2. Electrical resistivity of polycrystalline UBe_{13} (after Ott et al. [19]). The coherent Kondo-lattice state is formed below the low-temperature peak at 2.5 K ($\rho_{max} = 250$ $\mu\Omega$ cm).

off, but then starts to rise again and a maximum ($\rho_{max} \approx 250$ μΩ cm) appears at $T_{max} = 2.5$ K. The electrical resistivity of UBe_{13} is rather similar to $\rho(T)$ of $CeCu_2Si_2$ [20]. At high temperatures, contributions from the single-ion Kondo and crystalline electric field are present, while at very low temperatures a coherent Kondo-lattice is formed, as evidenced by the resistance drop below T_{max}. From the high-temperature susceptibility [19] an effective moment $p_{eff} = 3.08\mu_B$ is deduced, which falls in between the free-ion values for one f electron ($2.54\mu_B$) and two ($3.58\mu_B$) or three ($3.62\mu_B$) f electrons. Deviations from the Curie-Weiss behaviour appear in the high-temperature susceptibility below ~150 K. The negative Curie-Weiss constant ($\theta = -53$) points to the presence of antiferromagnetic correlations. The low-temperature susceptibility [19] is enhanced, $\chi(T = 1.5 \text{ K}) = 1.5 \times 10^{-2}$ emu/mole, indicating strong Pauli-paramagnetism, however, the ratio $\chi(T = 1.5 \text{ K})/\gamma$ is close to 1, as γ is enhanced accordingly. Inelastic neutron-scattering experiments performed at 10 K on polycrystalline samples [21], yield a broad quasi-elastic contribution with a width of 13 ± 2 meV. This energy scale is close to the temperature where deviations from the Curie-Weiss behaviour appear. The high-field magnetization [22] is nearly linear up to 24 T at 4.2 K, while a small increase in the differential susceptibility, $\chi_{diff} = \Delta\sigma/\Delta H$, is observed above 12 T at 1.25 K. As most of the heavy-fermion compounds exhibit some type of antiferromagnetic order, this might also be expected for UBe_{13}. However, sensitive μSR studies [23] and neutron-scattering experiments yield thus far negative results. Recently, anomalies observed in the low-temperature magnetostriction [24] were taken as evidence for antiferromagnetic order below $T_N = 8.8$ K. However, this was not confirmed by a more recent magneto-volume study [25]. An additional anomaly observed in the specific heat at high magnetic field [26] possibly indicates antiferromagnetic order below ~100 mK. Measurements of the thermoelectric power at very high pressures (67 kbar) [27] might indicate pressure induced long-range order at a temperature of a few kelvin. It is obvious that these claims need verification.

2.1. Unconventional superconducting properties of UBe_{13}

The superconducting properties of UBe_{13} are quite unusual [6]. First of all a remarkably large jump in the electronic specific heat is observed at T_c: $\Delta C/C \approx 2.5$ (the standard BCS-value is 1.43). Furthermore, the temperature dependence of the specific heat in the superconducting state deviates strongly from the usual exponential behaviour. Instead it was noticed that $C(T)$ varies approximately as T^3, suggesting (strong coupling) p-wave superconductivity (point nodes in the gap) [28]. However, subsequent specific-heat data revealed significant deviations from the T^3 dependence, indicating the importance of impurity scattering [29,30]. The London penetration depth [31], investigated for several single- and polycrystalline samples, shows also a power law temperature dependence, $\lambda_L \sim T^2$, which is primarily attributed to impurity scattering. NMR measurements [32] yield a spin-lattice relaxation rate $1/T_1 \sim T^3$, suggesting a gap that vanishes along lines on the Fermi surface. Obviously, the analyses of the different power-law temperature dependences are at variance with each other. Therefore, the power-law behaviour cannot be

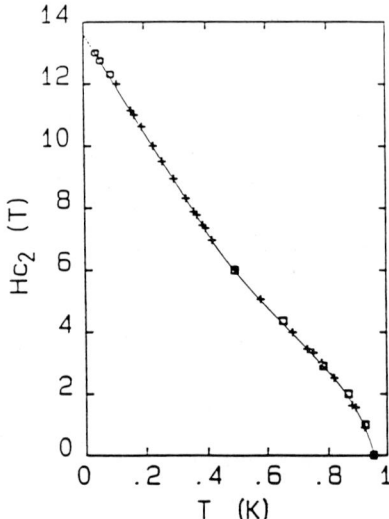

Fig. 3. The upper critical field of UBe_{13} determined resistively (after Brison et al. [33]). Note the unusual upwards curvature and the quasi-linear behaviour below 450 mK.

used at present for the definite determination of the symmetry of the order parameter. The upper-critical field for $T \to 0$, $B_{c2}(0)$, amounts to 13 T [33], an extremely large value for a low-T_c superconductor (see fig. 3). The initial slope, $dB_{c2}/dT|_{T \to T_c}$, is nearly vertical which hampers a correct experimental determination. Estimates vary from -26 [6] to -200 [29] T/K. B_{c2} has an unusual upwards curvature near $T/T_c = 0.5$, and a quasi-linear behaviour below this value, possibly indicating the presence of two different regimes in the superconducting phase.

It is undoubtedly clear that the superconducting properties of UBe_{13} are unconventional. Nevertheless, decisive proof for a nontrivial pairing state cannot be deduced from the experiments discussed above. However, impurity studies have revealed several other remarkable features, yielding further convincing evidence for unconventional superconductivity.

2.2. The superconducting phase diagram of $(U, Th)Be_{13}$

The influence of substitution on the uranium sites of a few percent of several impurity elements on the superconducting transition temperature of UBe_{13} has been studied by Smith and co-workers [34]. Most impurities tested, i.e. Sc, Ce, Lu, Y, Zr, Gd and La cause a monotonic suppression of T_c. Also a few percent of Cu or Ga, substituted on the Be sites, suppresses T_c, particularly rapidly in the case of Cu [35]. However, most remarkable effects occur for a few percent of Th substituted for U [36]. Specific-heat measurements by Ott and coworkers [37] revealed a new phase transition in the superconducting state for concentrations between ~ 2 and ~ 4 at% Th (see fig. 4). Tracing $T_c = T_{c1}$ and the temperature, T_{c2} where the second anomaly in $C(T)$ is observed, as function of Th content results in a remarkable phase diagram (fig. 5) [38]. Resistivity and susceptibility measurements proof that the upper transition in the concentration range $0.019 < x < 0.042$ is to the superconducting state. The second phase transition also turns up in the lower critical field B_{c1} [38]. Below T_{c2}, B_{c1} rises more rapidly, which can be explained as an

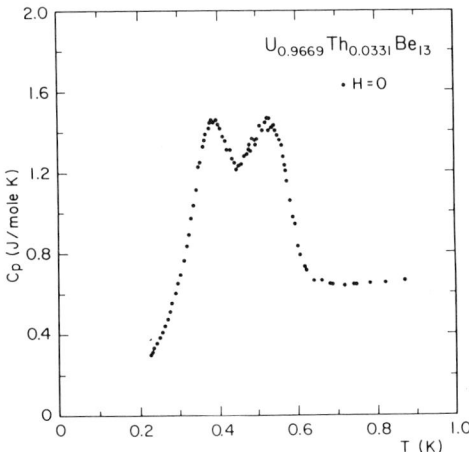

Fig. 4. Specific heat of $U_{0.9669}Th_{0.0331}Be_{13}$ in a plot of C versus T (after Ott et al. [37]). The upper transition at T_{c1} is to the superconducting state. The second transition at T_{c2} possesses some magnetic moment.

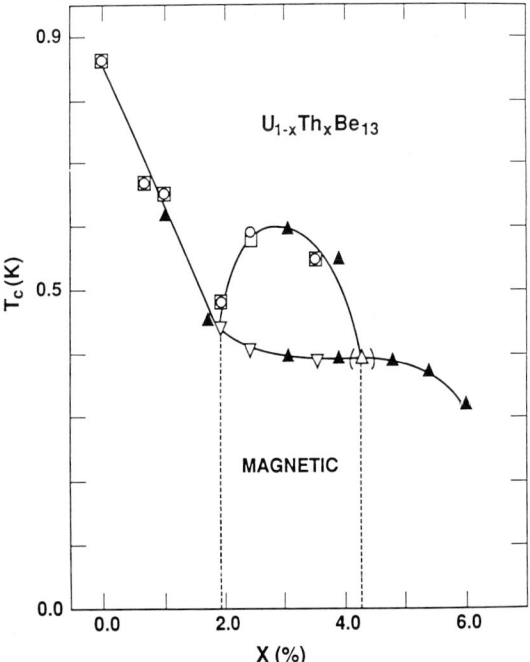

Fig. 5. Phase-diagram for $U_{1-x}Th_xBe_{13}$ (after Heffner et al. [38]): ($\square$) T_{c1} from χ_{ac}; ($\bigcirc$) T_{c1} from magnetization $M(H)$; ($\triangledown$) T_{c2} from kink in $H_{c1}(T^2)$; ($\blacktriangle$, $\triangle$) T_{c1} and T_{c2} from specific heat [37].

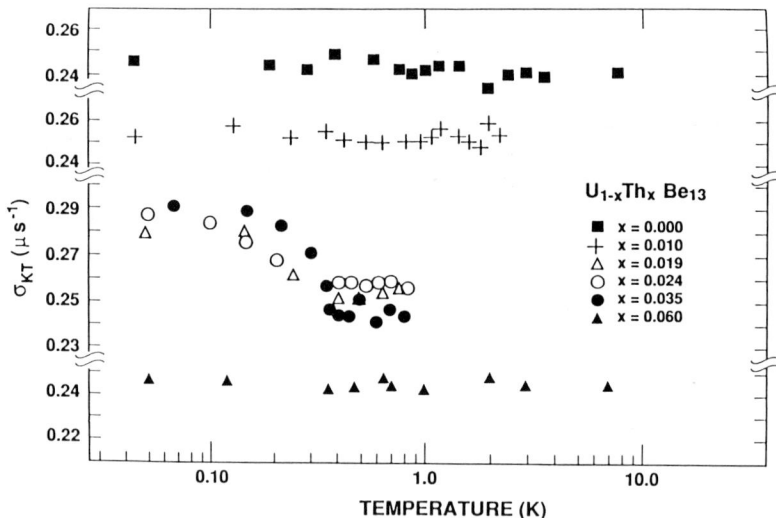

Fig. 6. Temperature dependence of the zero-field μSR rate $\sigma_{KT}(T)$ for various Th concentrations x in $U_{1-x}Th_xBe_{13}$ (after Heffner et al. [38]). The increase below T_{c2} for $x = 0.019$, 0.024 and 0.035 indicates the presence of magnetic correlations.

increase of the superfluid density (or a decrease of the effective mass). μSR measurements [38] have revealed that the phase below T_{c2} possesses some magnetic moment (fig. 6). The nature of the low-temperature state is rather puzzling. Several scenarios have been proposed in order to explain the transition at T_{c2} (see ref. [38] and references therein), among which: i) a combined antiferromagnetic and superconducting transition; ii) a transition to a magnetic (time-reversal-violating) superconducting phase; and iii) an antiferromagnetic spin-density wave transition. The first two proposals require a multicomponent (vector) order parameter. It has been argued that the large size of the jump in the specific heat at T_{c2} (see fig. 4) discards the spin-density wave scenario, as the Fermi-surface is already largely consumed by the superconducting transition. The complex superconducting phase diagram (fig. 5) would therefore provide strong evidence for a multicomponent order parameter in UBe_{13}. It is clear that the anomalous superconducting and normal state properties of UBe_{13} are still rather puzzling, and that further studies might yield more surprising results. In this respect, we mention the unexpected observation, made recently, for B doping, where for polycrystalline $UBe_{12.97}B_{0.03}$ a specific heat jump at T_c is measured, that is almost twice as large as for pure UBe_{13} [39].

3. Spin fluctuations and superconductivity in UPt_3

The first detailed investigation of the low-temperature properties of UPt_3 was reported by Frings et al. [7] in 1983. Specific-heat measurements revealed an anomalous upturn at low-temperatures, yielding an enhanced γ-value of 420 mJ/mol K^2. The upturn could be described with an additional $T^3 \ln(T/T^*)$-term suggesting the presence of pronounced spin-fluctuation phenomena [40,41]. Measurements on single-crystalline samples showed that the magnetic properties of UPt_3 are strongly anisotropic, with the (hexagonal) basal plane as the easy-plane for magnetization. The susceptibility, $\chi(T)$, has a pronounced maximum at $T_{max} = 18$ K, for a field direction in the basal plane, while no anomaly is observed for a field along the hexagonal (c) axis [7]. For temperatures $T < T_{max}$, the high-field magnetization [7] shows a metamagnetic transition at $B^* = 20$ T for a field in the basal plane, while a linear behaviour is observed along the c-axis (see fig. 7).

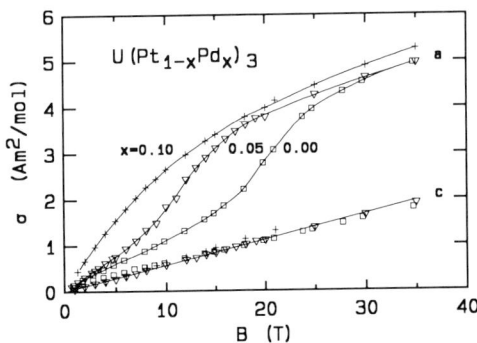

Fig. 7. High-field magnetization of single-crystalline U(Pt$_{1-x}$Pd$_x$)$_3$ at 4.2 K (after Franse et al. [53]). For $x = 0.00$ and 0.05 a metamagnetic-like transition appears for a field in the hexagonal plane.

The transition at 20 T is not a metamagnetic transition in the classical sense, because UPt$_3$ does not exhibit long-range order with large moments, as is evidenced by the absence of anomalies in the thermodynamic and transport properties. Therefore, it is often referred to as pseudo-metamagnetic or metamagnetic-like. The maximum in the susceptibility is likely related with the stabilization of intersite antiferromagnetic (Ruderman–Kittel–Kasuya–Yosida) interactions that give rise to deviations of the Curie–Weiss law below ~ 50 K. The metamagnetic-like transition is explained by a quenching of the antiferromagnetic spin fluctuations above B^* (the total increase in magnetization at the metamagnetic transition is roughly $0.5\mu_B$/U-atom). The high-field magnetoresistance $\rho(B)$ reveals a sharp maximum at B^* which is consistent with such a picture [42]. Solid evidence for antiferromagnetic spin-fluctuation phenomena comes from inelastic-neutron scattering experiments [43]. The fluctuation spectrum is quite complex, as different energy scales, are present. Experiments on polycrystalline samples yield a quasi-elastic contribution centered at 10 meV, that is related to the fluctuating local f-moment. The size of the fluctuating moment is of the order of $2\mu_B$/U-atom, which is of the same order as the effective moment deduced from the high-temperature Curie constant. Experiments on single-crystalline samples reveal a response centered at 6–8 meV, that evidences antiferromagnetic short-range order between nearest neighbour uranium atoms located in adjacent planes. The antiferromagnetic correlations disappear above T_{max}, whereas in-plane ferromagnetic correlations are present till about 150 K. At yet a lower energy (0.5 meV), a second type of antiferromagnetic (in-plane) correlations is found and, surprisingly, also a very weak long-range magnetic order appears with a Néel temperature $T_N = 5$ K [44]. The ordered moment equals $(0.02 \pm 0.01)\mu_B$/U-atom, and is directed along the b-axis. The order parameter has an unusual temperature dependence, it continues to grow linearly till far below T_N (see fig. 8). Note that the magnetic diffraction peaks are not resolution limited, indicating that some disorder must be present in the sample. Therefore, one cannot rule out completely that the appearance of the small-moment magnetism is related to crystallographic imperfections.

The discovery of superconductivity below 0.5 K by Stewart et al. [10] in 1984 brought UPt$_3$ into the select class of heavy-fermion superconductors. The unusual combination of superconductivity and strong spin-fluctuation phenomena immediately led to speculations upon a nontrivial pairing state, mediated by an electron–electron interaction. Since then an enormous experimental and theoretical effort had been made to unravel the intriguing aspects of heavy-fermion superconductivity in UPt$_3$ [45]. The most convincing evidence

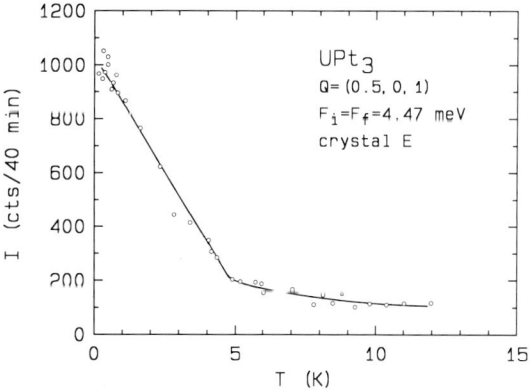

Fig. 8. Temperature dependence of the elastic peak intensity of $Q = (0.5, 0, 1)$ for UPt$_3$ (after Aeppli et al. [44]). Below $T_N = 5$ K, antiferromagnetic order is observed with a weak ordered moment (0.02 ± 0.01) μ_B along the b-axis.

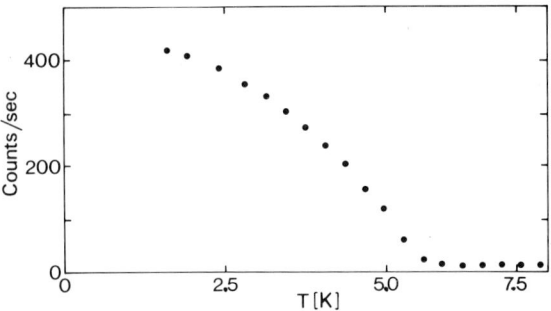

Fig. 9. Peak intensity of $Q = (0.5, 0.5, 1)$ as function of temperature for $U(Pt_{0.95}Pd_{0.05})_3$ (after Frings et al. [49]). Below $T_N = 5.8$ K, antiferromagnetic order is observed, with a fairly large ordered moment $(0.6 \pm 0.2)\mu_B$ along the b-axis.

alloying. The ordering vector is the same as for the small-moment magnetism observed in pure UPt_3. The antiferromagnetic order, with a maximum Néel temperature of 5.8 K, is only found in a rather small concentration range. The magnetic and superconducting phase diagram for the $U(Pt, Pd)_3$ [50] and $(U, Th)Pt_3$ [51] compounds is shown in fig. 10. The stabilisation of part of the U moment by alloying does, however, not remove the anomalous behaviour of the low-temperature specific heat, i.e. the contribution from the antiferromagnetic order is superimposed on the large heavy-fermion background. This is likely related to the competition of the RKKY and the Kondo effect, as will be discussed in the next section.

for unconventional superconductivity in UPt_3 stems from the observation of a multi-component superconducting phase diagram (see section 3.2). Superconductivity in UPt_3 is rapidly lost by doping with Pd [46] or Th [47]. Surprisingly, at increasing the Pd or Th concentration, long-range antiferromagnetic order appears [46,48]. The ordered moment amounts to $0.6\mu_B$ (see fig. 9) for 5 at% Pd [49] and Th [43] alloys, indicating that only part of the fluctuating moment orders. Note that the size of the ordered moment is roughly equal to the increase of the magnetization in pure UPt_3 at the metamagnetic-like transition, which suggests a common origin of the field induced moment and the moment stabilized by

3.1. Competition between RKKY and Kondo-effect in $U(Pt, Pd)_3$

Evidence for competing electronic interactions in heavy-fermion systems has for the greater part been gathered by alloying studies, i.e. by progressive replacements of one of the constituents. In this respect $U(Pt_{1-x}Pd_x)_3$ is an exemplary system. In pure UPt_3 the low-temperature properties are dominated by antiferromagnetic interactions, while by substituting small amounts of Pt by isoelectronic Pd, a crossover to a regime dominated by Kondo interactions is observed. This change in regime is most clearly demonstrated by the electrical resistivity, $\rho(T)$ (see fig. 11) [52]. For pure UPt_3, the gradual drop of $\rho(T)$ with decreasing temperature is ascribed to the stabilization of antiferromagnetic correlations, while for a Pd content of only 10 at%, a Kondo-like upturn is observed. The maximum in the susceptibility, $\chi(T)$, at $T_{max} = 18$ K, and the metamagnetic-like transition at a field of 20 T ($T \leq T_{max}$), characteristic of pure UPt_3, gradually decrease on alloying and are no longer observed for $x = 0.10$ (see fig. 7) [46,53], lending further support for a suppression of the antiferromagnetic correlations. The γ-value initially increases with Pd content and passes through a maximum near $x = 0.10$, evidencing that the heavy-fermion properties are preserved. However, $\partial\gamma/\partial B$ changes sign between $x = 0.07$ and $x = 0.10$, (as $\rho(B)$ does at low temperatures [54]), consistent with a crossover

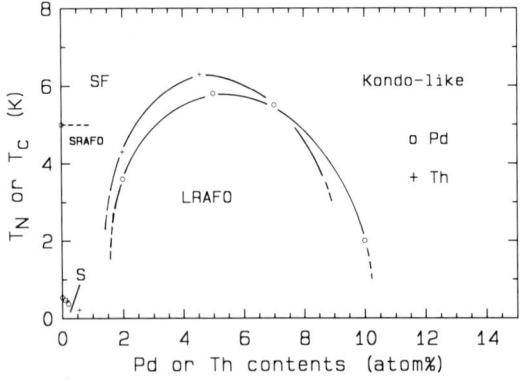

Fig. 10. Superconducting and magnetic phase diagram for $U(Pt, Pd)_3$ (○) and $(U, Th)Pt_3$ (+) (after Franse et al. [50]). S = superconductivity; LRAFO = long range antiferromagnetic order; SRAFO = short range (small moment) antiferromagnetic order; SF = spin fluctuations.

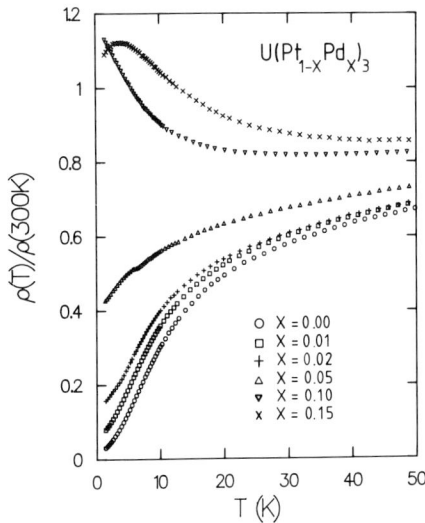

Fig. 11. Resistivity of $U(Pt_{1-x}Pd_x)_3$ as function of temperature (after Verhoef et al. [52]). The data are normalized to 1 at 300 K. Note the change to a Kondo behaviour between $x = 0.05$ and 0.10.

power law temperature dependence for the electronic excitation spectrum below T_c is strongly suggestive for an anisotropic gap function and unconventional $L \neq 0$ paring. In particular, detailed measurements of the acoustic attenuation [57] and the penetration depth [31,58] along different crystallographic directions, indicate a so-called hybrid gap function, i.e. a line of zeros at the equator and nodes at the poles. However, the temperature range where the validity of the power laws should be tested ($T \ll T_c$) has not been probed reliably yet. Furthermore, the contribution of impurity scattering might also give rise to deviations from the standard exponential BCS behaviour. Studies of the upper-critical field revealed several unusual features, in particular a kink for a field direction in the basal plane, first observed by Rauchschwalbe et al. [59], and a crossing of the $B_{c2}(T)$ curves for the different crystallographic directions at 200 mK [60]. It has been suggested that the latter feature provides strong evidence for odd-parity superconductivity [61]. Recently, another type of evidence for unconventional superconductivity in UPt_3 has come to light, namely the observation by Fisher et al. [62] of a double anomaly in the specific heat at the transition to the superconducting state (see fig. 12). The double transition has now been confirmed by several groups by specific-heat measurements on polycrystalline and single-crystalline samples prepared in different ways [63,64], and has further been observed in other thermodynamic properties, like the thermal expansion [65] and the sound velocity [66,67]. Studies of the effect of a magnetic field on the double transition reveal a phase diagram in the temperature-field plane with three (superconducting) phases, that meet at a tetra-critical point (see fig. 13). The phase diagrams recently determined from specific-heat [63,65], thermal expansion [65] and sound velocity [66,67] measurements, give rather consistent results. The thermodynamic studies essentially confirm the change in slope (at the tetra-critical point) of $B_{c2}(T)$ for fields in the hexagonal plane, as first reported by Rauchschwalbe et al. [59].

From group-theoretical work [68–71] it has been inferred that the double-peak structure in

to a Kondo regime. In the same Pd concentration range the coefficient of volume expansion, $\alpha_v(T)$, changes sign and the Grüneisen parameter, $\Gamma(T \to 0)$, shows a huge drop from 75 for pure UPt_3 to -300 for $x = 0.15$ [55]. This salient change in regime clearly indicates the presence of competing electronic interactions. However, it is not easy to determine the variation of the characteristic temperatures, T_{RKKY} and T_K, with Pd content, as both contributions have nearly equal energy scales and are both present over a wide concentration range [56]. Besides, as earlier mentioned, part of the fluctuating moment orders antiferromagnetically, giving rise to yet another superimposed contribution. In order to unravel the complex low-temperature properties of the $U(Pt, Pd)_3$ system, inelastic-neutron scattering experiments will certainly be very helpful.

3.2. Multicomponent superconductivity in UPt_3

The unusual combination of strong spin-fluctuation phenomena and superconductivity in UPt_3 attracts ample attention, in particular because of the speculations upon electron–electron mediated Cooper pairing. The observation of a

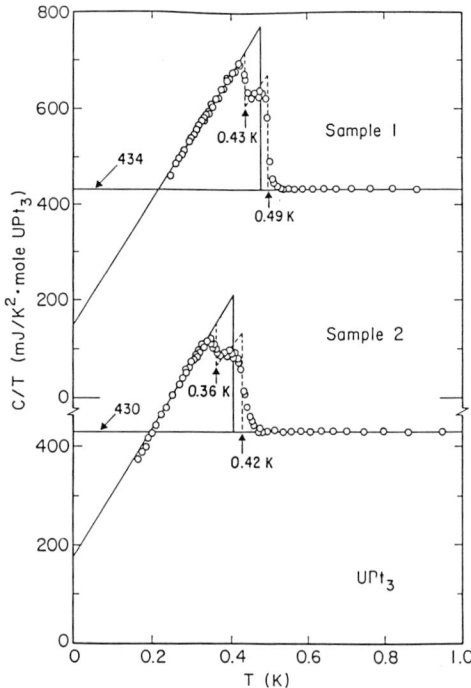

Fig. 12. The double superconducting transition in the specific heat for two polycrystalline UPt$_3$ samples, plotted as C/T versus T (after Fisher et al. [62]). The dashed (solid) lines represent two (one) ideally sharp transition(s), taking into account entropy conservation.

the multicomponent superconducting phase diagram, alloying experiments have been performed, with U substituted by Y and Pt substituted by Pd [74]. These two impurities behave differently as far as the effect on the distance in temperature between the two peaks in the C/T versus T curve is concerned. In the case of Y this distance remains about the same, whereas for Pd doping up to 0.2 at% this distance is increased. In view of the stabilisation of the uranium moment with

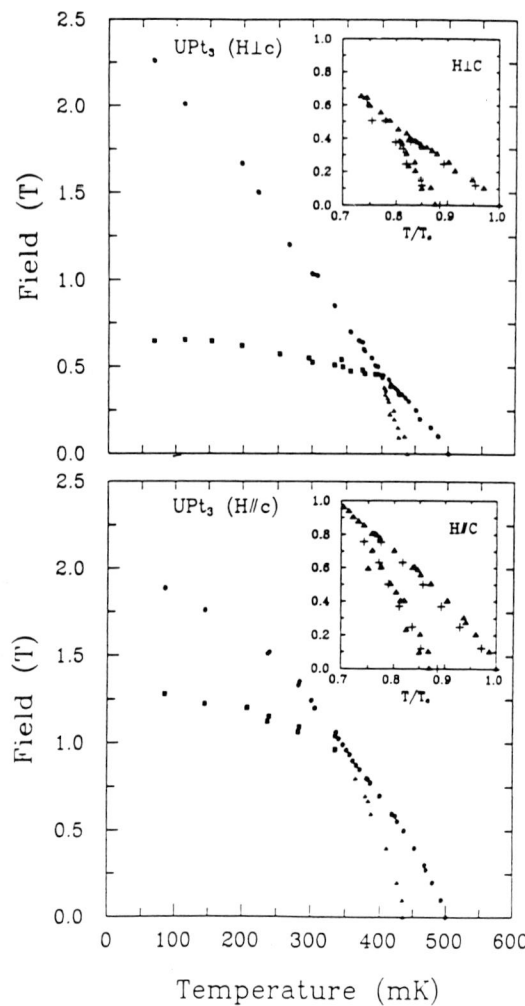

Fig. 13. Phase diagrams of superconducting UPt$_3$ for H/c and $H \perp c$ obtained from sound velocity measurements (after Adenwalla et al. [67]). The insets show a comparison of the phase boundaries determined by sound velocity (▲) and specific heat (+) measurements [63,65].

the specific heat at the superconducting transition might be ascribed to the lifting of the degeneracy of the superconducting vector order parameter by a symmetry breaking field. An appealing speculation is that the symmetry breaking field is provided by the weak antiferromagnetic moment that lies in the basal plane. Neutron-scattering experiments provide some evidence for a coupling of the superconducting and magnetic order parameter [72]. Applying phenomenological Ginzburg–Landau theory, the different order parameters that might describe the unusual superconducting state can be investigated. A comparison with the experimental results indicates that the order parameter belongs to the E_{1g} representation of the hexagonal group (d-wave pairing), however, this is still controversial. Also the existence of a tetra-critical point in the phase diagram is controversial according to thermodynamic constraints [73]. In order to further investigate

increasing Pd content (see section 3.1, note that for a 5 at% Y alloy no long-range ordered state has been observed [75]), the increased distance between the two specific-heat peaks could point to a strengthening of the symmetry-breaking field (the uranium magnetic moment), proving in that way the origin of the symmetry-breaking field. These suggestions have to be verified by neutron studies of the uranium moment in these doped UPt$_3$ compound. Also systematic metallurgical studies on the effect of heat treatment and/or defect structures have not been performed so far. Therefore, definite conclusions as to the interplay of superconductivity and weak antiferromagnetic order can not be drawn at present.

4. Antiferromagnetism and superconductivity in URu$_2$Si$_2$

In 1985, Schlabitz et al. [76], Palstra et al. [77] and Maple et al. [78] reported the existence of both an antiferromagnetic and a superconducting transition in the ternary compound URu$_2$Si$_2$. The anomalous properties of URu$_2$Si$_2$ are clearly illustrated by the electrical resistivity [79], as shown in fig. 14 for current directions along the tetragonal (c) axis and in the basal plane (a axis). On

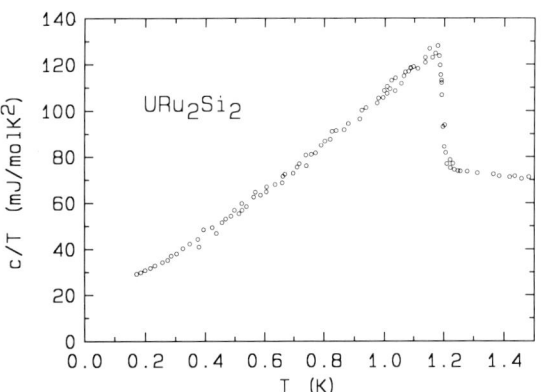

Fig. 15. Specific heat of annealed single-crystalline URu$_2$Si$_2$ in a plot of C/T versus T (after Hasselbach et al. [81]). One single sharp transition is observed.

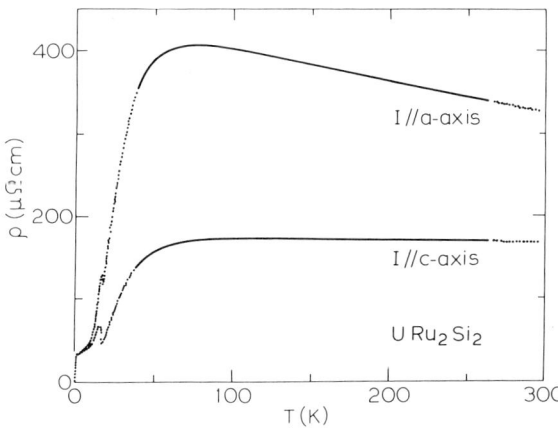

Fig. 14. Temperature dependence of the electrical resistivity of unannealed single-crystalline URu$_2$Si$_2$ for a current direction along the tetragonal (c) axis and in the plane (a) (after Palstra et al. [79]). The anomaly at 17.5 K indicates the transition to the antiferromagnetic state. Superconductivity occurs below 0.8 K.

cooling, $\rho(T)$ first increases, passes through a maximum near 70 K, and then drops rapidly, due to the formation of a coherent state. Such a high-temperature behaviour is normally ascribed to the Kondo(-lattice) effect. At 17.5 K a Cr-like anomaly evidences the antiferromagnetic transition. In the specific-heat data, the transition at 17.5 K turns up as a λ peak [76–78]. The exponential temperature dependence of $C(T)$ below T_N suggests that the magnetic order is accompanied by the opening of an energy gap over a part of the Fermi surface ($\Delta \approx 100$ K), indicating the presence of a spin-density wave. The Cr-like anomaly observed in the resistivity, and the strong increase in the Hall resistance below 17.5 K [80], are in accordance with such an interpretation. Superconductivity occurs below ~ 1.2 K. A large anomaly in the specific heat is observed at the superconducting transition (fig. 15) [76–78,81]. The upper-critical field, $B_{c2}(T)$, has an unusual shape (see fig. 16) [77]. Although B_{c2} is initially isotropic, a large anisotropy appears just below T_c. For a field directed along the c-axis $B_{c2}(0) = 1.8$ T, while for a field in the tetragonal plane $B_{c2}(0)$ has not been determined yet, but is probably of the order of 8 T. The γ-value of URu$_2$Si$_2$ is small when compared to UBe$_{13}$ and UPt$_3$, however, an analysis of the initial slope of the upper critical field, yields an effective mass in the order of $50 m_e$ [76], which justifies the classifica-

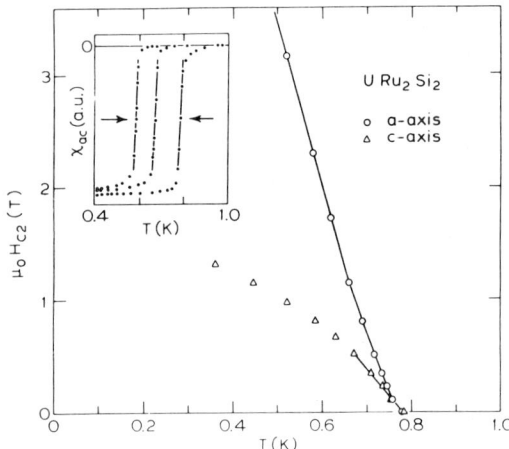

Fig. 16. Temperature dependence of the uppercritical field of unannealed single-crystalline URu_2Si_2 for a field direction along the tetragonal (c) axis and in the plane (a) (after Palstra et al. [77]). The inset shows the χ_{ac} signal for fields of 0, 0.52 and 0.81 T along the c axis.

tion of URu_2Si_2 as a moderate heavy-fermion superconductor.

Neutron-scattering measurements performed by Broholm et al. [82] confirm the antiferromagnetic order below $T_N = 17.5$ K. As in the case for UPt_3, the ordered moment is extremely small: $(0.03 \pm 0.01)\mu_B$/U-atom. A simple antiferromagnetic structure results, with the moments aligned along the tetragonal axis. However, because of the broad gradual increase of the magnetic moment, as follows from the unusual temperature dependence of the integrated elastic Bragg scattering, and the rather small magnetic coherence length, i.e. in the order of several hundredth Ångström, these results were rather controversial. Recently, neutron-scattering [83] and magnetic X-ray scattering [84] on samples grown with higher-purity uranium, have removed this controversy (see fig. 17). It is remarkable, that the anomalies in the thermodynamic properties at 17.5 K are rather pronounced, while the ordered moments that have been deduced from the neutron studies are not more than a few hundredths of a Bohr magneton. This has led to speculations that the transition at 17.5 K has primarily an electronic nature and that magnetic order is a second order effect.

In order to shed more light on the electronic properties, high-field magnetization and magnetoresistance measurements have been performed [85,86]. Surprisingly, three sharp transitions at 35.8, 37.3 and 39.4 T have been observed (at 1.5 K), for a field directed along the tetragonal axis (see fig. 18). For fields above the three-step magnetization process an uranium moment of the order of $1.5\mu_B$ results. The accompanying large jumps observed in the magnetoresistance point to complex electronic processes (fig. 19). Several mechanisms have been put forward in order to explain this intriguing high-field behaviour: (magnetic) crystal field levels that cross the Fermi level, a reconstruction of the Fermi surface and competing antiferromagnetic interactions. An analysis of the specific-heat and thermal-expansion data [87] above 17.5 K, and the high-temperature susceptibility data [77], point indeed to the presence of crystal-field effects [88]. The low-temperature specific heat, with the enhanced γ-value of 70 mJ/mol K^2, and the transition at 17.5 K, however, indicate that other processes must be taken into account.

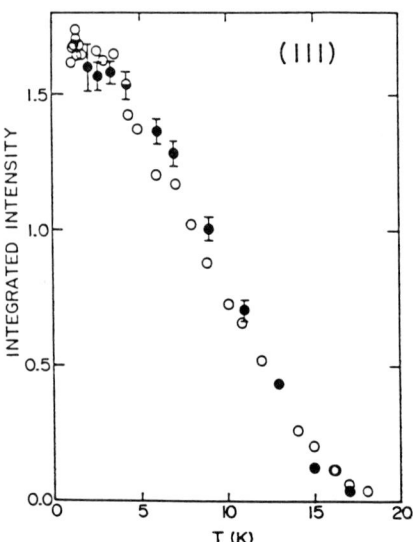

Fig. 17. The temperature dependence of the integrated intensitiy of the (111) magnetic Bragg peak, from neutron-diffraction experiments (●) and from X-ray magnetic scattering (○) (after Mason et al. [83]). Below $T_N = 17.5$ K, antiferromagnetic order is observed with a very weak ordered moment $(0.037 \pm 0.005)\mu_B$ along the tetragonal axis.

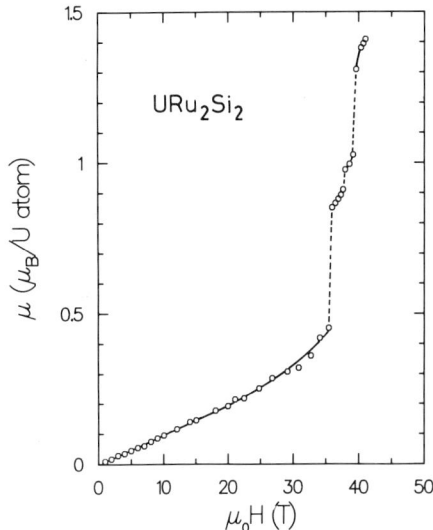

Fig. 18. High-field magnetisation of single-crystalline URu_2Si_2 at 1.5 K for a field along the tetragonal axis (after de Visser et al. [85]). A three-step magnetization process is observed as indicated by the dashed lines.

The superconducting transition is rather sample dependent, with values for T_c that vary between 0.8 and 1.4 K depending on the quality of

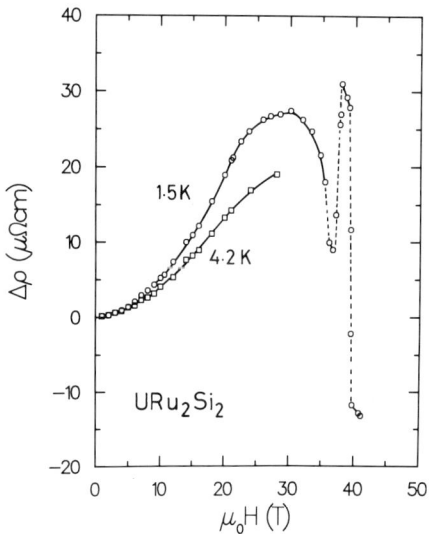

Fig. 19. High-field magnetoresistance of single-crystalline URu_2Si_2, for a field along the tetragonal axis (after de Visser et al. [85]). At 1.5 K three transitions take place as indicated by the dashed lines. The zero-field resistance equals 18.8 and 21.8 $\mu\Omega$ cm at 1.5 and 4.2 K, respectively.

starting materials and preparation methods. The earlier specific-heat data showed a rather broad transition or even in some cases, a double transition [89]. However, nowadays, samples prepared with high-purity uranium show one single sharp peak at the superconducting transition (see fig. 17) [81]. Below T_c, C/T follows a power law temperature dependence, like observed for UPt_3 and UBe_{13}, in contrast to the standard BCS exponential temperature dependence. The large anisotropy observed in the upper critical field (fig. 16) points to a strong coupling between the superconducting order parameter and the antiferromagnetic moment along the c-axis. The tetragonal symmetry, however, is not removed by this coupling and no splitting of the superconducting transition is expected, contrary to UPt_3. The power-law behaviour of the specific heat in the superconducting state with linear and quadratic terms in temperature points to a polar-like state with lines of zero-gap in the tetragonal plane.

5. Concluding remarks

The experimental results reviewed here clearly demonstrate the unequalled anomalous properties of the uranium-based heavy-fermion superconductors. Although many attempts to analyse the anomalous properties within current models have been undertaken, it appears that a satisfying description is in most cases still lacking. In the past years much insight has been gained from phenomenological models. However, we stress the need for microscopic models, that incorporate both superconductivity and antiferromagnetic interactions. We hope that this experimental review will further challenge theoretical physicists to develop such models. Most experiments have been performed on good-quality single-crystalline material. It appears, however, that metallurgical aspects are extremely important, in particular as to the small-moment antiferromagnetism and the (double) superconducting transition. Therefore, improved sample preparation techniques must be among the main goals, before a next series of detailing experiments is started. For experimentalists it will be a challenge to perform measurements at still higher-magnetic fields in order to investigate the suppression of the heavy-fermion state. Also neutron-scattering studies, in order to elucidate the competing electronic interactions will be very helpful. Only very few alloying studies have been performed thus far, often with remarkable results. New alloying studies will be necessary to further unravel the heavy-fermion properties. The unconventional superconducting properties of UBe_{13}, UPt_3 and URu_2Si_2 yield strong evidence for nontrivial ($L \neq 0$) pairing. Therefore, we believe that a wealth of exotic physics waits to be unveiled in the next decade.

Acknowledgement

The work of one of us (AdV) has been made possible by a fellowship of the Royal Netherlands Academy of Arts and Sciences.

References

[1] K. Andres, J.E. Graebner and H.R. Ott, Phys. Rev. Lett. 35 (1975) 1779.
[2] See e.g.: D. Pines and P. Nozières, in The Theory of Quantum Liquids, vol. I: Normal Fermi Liquids (Benjamin, New York, 1966).
[3] F. Steglich, J. Aarts, C.D. Bredl, W. Lieke, D. Meschede, W. Franz and H. Schäfer, Phys. Rev. Lett. 43 (1979) 1892.
[4] G.R. Stewart, Z. Fisk and M.S. Wire, Phys. Rev. B 30 (1984) 482.
T. Fujita, K. Satoh, Y. Onuki and T. Komatsubara, J. Magn. Magn. Mater. 47&48 (1985) 66.
[5] M.J. Besnus, J.P. Kappler, P. Lehmann and A. Meyer, Solid State Commun. 55 (1985) 779.
J.D. Thompson, J.O. Willis, C. Godart, D.E. Mac Laughlin and L.C. Gupta, Solid State Commun. 56 (1985) 169.
[6] H.R. Ott, H. Rudigier, Z. Fisk and J.L. Smith, Phys. Rev. Lett. 50 (1983) 1595.
[7] P.H. Frings, J.J.M. Franse, F.R. de Boer and A. Menovsky, J. Magn. Magn. Mater. 31–34 (1983) 240.
[8] Z. Fisk, G.R. Stewart, J.O. Willis, H.R. Ott and F. Hulliger, Phys. Rev. B30 (1984) 6360.
[9] H.R. Ott, H. Rudigier, P. Delsing and Z. Fisk, Phys. Rev. Lett. 52 (1984) 1551.
[10] G.R. Stewart, Z. Fisk, J.O. Willis and J.L. Smith, Phys. Rev. Lett. 52 (1984) 679.
[11] C.M. Varma, Bull. A. Phys. Soc. 29 (1984) 404.
P.W. Anderson, Phys. Rev. B 30 (1984) 1549.
[12] A. Sumiyama, Y. Oda, H. Nagano, Y. Onuki and T. Komatsubara, J. Phys. Soc. Jpn. 54 (1985) 877.
[13] J.L. Tholence, P. Haen, D. Jaccard, P. Lejay, J. Flouquet and H.F. Braun, J. Appl. Phys. 57 (1985) 3172.
[14] S. Barth, H.R. Ott, F.N. Gygax, B. Hitti, E. Lippelt, A. Schenck and C. Baines, Jpn. J. Appl. Phys. 26 Suppl. 3 (1987) 519.
[15] See e.g.: N.B. Brandt and V.V. Moschalkov, Adv. Phys. 33 (1984) 373.
[16] P.A. Lee, T.M. Rice, J.W. Serene, L.J. Sham and J.W. Wilkins, Comments Cond. Mat. Phys. 12 (1986) 99.
[17] P. Fulde, J. Phys. F 18 (1988) 601.
[18] E. Bucher, J.P. Maita, G.W. Hull, R.C. Fulton and A.S. Cooper, Phys. Rev. B 11 (1973) 440.
[19] H.R. Ott, H. Rudigier, Z. Fisk and J.L. Smith, Proc. NATO/CAP Inst. on Moment Formation in Solids, Vancouver Island, 1983, ed. W.J.L. Buyers (Plenum, New York, 1984) p. 305.
[20] W. Franz, A. Griessel, F. Steglich and D. Wohlleben, Z. Phys. B 31 (1978) 7.
[21] A.I. Goldman, S.M. Shapiro, G. Shirane, J.L. Smith and Z. Fisk, Phys. Rev. B 33 (1986) 1627.
[22] G. Remenyi, D. Jaccard, J. Flouquet, A. Briggs, Z. Fisk, J.L. Smith and H.R. Ott, J. Phys. 47 (1986) 367.
[23] R.H. Heffner et al., Phys. Rev. B 39 (1989) 11345.
[24] R.N. Kleiman, D.J. Bishop, H.R. Ott, Z. Fisk and J.L. Smith, Phys. Rev. Lett. 64 (1990) 1975.
[25] A. de Visser, N.H. van Dijk, K. Bakker, J.J.M. Franse, A. Lacerda, J. Flouquet, Z. Fisk and J.L. Smith, Phys. Rev. B (submitted).
[26] J.P. Brison, A. Ravex, J. Flouquet, Z. Fisk and J.L. Smith, J. Magn. Magn. Mater. 76&77 (1986) 525.
[27] S.Y. Mao, D. Jaccard, J. Sierro, Z. Fisk and J.L. Smith, J. Magn. Magn. Mater. 76&77 (1988) 241.
[28] H.R. Ott, H. Rudigier, T.M. Rice, K. Ueda, Z. Fisk and J.L. Smith, Phys. Rev. Lett. 52 (1984) 1915.
[29] U. Rauchschwalbe, Physica B 147 (1987) 1.
[30] J.P. Brison, Ph.D. Thesis, Université Joseph Fourier, Grenoble (1988) unpublished.
[31] F. Gross-Alltag, B.S. Chandrasekhar, D. Einzel, P.J. Hirschfeld and K. Andres, Z. Phys. B 82 (1991) 243.
[32] D.E. MacLaughlin, Cheng Tien, W.G. Clark, M.D. Lan, Z. Fisk, J.L. Smith and H.R. Ott, Phys. Rev. Lett. 53 (1984) 1833.
[33] J.P. Brison, J. Flouquet and G. Duetscher, J. Low Temp. Phys. 76 (1989) 453.
[34] J.L. Smith, Z. Fisk, J.O. Willis, B. Batlogg and H.R. Ott, J. Appl. Phys. 55 (1984) 1996.
[35] A.L. Giorgi, Z. Fisk, J.O. Willis, G.R. Stewart and J.L. Smith, Proc. LT17 Conf., eds. U. Ekern, A. Schmid, W. Weber and H. Wuhl (North-Holland, Amsterdam, 1984) p. 229.
[36] J.L. Smith, Z. Fisk, J.O. Willis, A.L. Giorgi, R.B. Roof, H.R. Ott, H. Rudigier and E. Felder, Physics B 135 (1985) 3.
[37] H.R. Ott, H. Rudigier, Z. Fisk and J.L. Smith, Phys. Rev. B 31 (1985) 1651.

[38] R.H. Heffner et al., Phys. Rev. Lett. 65 (1990) 2816.
[39] Z. Fisk and H.R. Ott, Intern. J. Mod. Phys B 3 (1989) 535.
[40] P.H. Frings and J.J.M. Franse, Phys. Rev. B 31 (1985) 4355.
[41] A. de Visser, J.J.M. Franse, A. Menovsky and T.T.M. Palstra, J. Phys. F 14 (1984) L191.
[42] A. de Visser, R. Gersdorf, J.J.M. Franse and A. Menovsky, J. Magn. Magn. Mater. 54–57 (1986) 383.
[43] G. Aeppli, E. Bucher, A.I. Goldman, G. Shirane, C. Broholm and J.K. Kjems, J. Magn. Magn. Mater. 76&77 (1988) 385.
[44] G. Aeppli, E. Bucher, C. Broholm, J.K. Kjems, J. Baumann and J. Hufnagl, Phys. Rev. Lett. 60 (1985) 615.
[45] For a review see: A. de Visser, A. Menovsky and J.J.M. Franse, Physica B 147 (1987) 81.
[46] A. de Visser, J.C.P. Klaasse, M. van Sprang, J.J.M. Franse, A. Menovsky, T.T.M. Palstra and A.J. Dirkmaat, Phys. Lett. A 113 (1986) 489.
[47] T. Vorenkamp, private communication.
[48] A.P. Ramirez, B. Batlogg, A.S. Cooper and E. Bucher, Phys. Rev. Lett. 57 (1986) 1072.
[49] P.H. Frings, B. Renker and C. Vettier, J. Magn. Magn. Mater. 63&64 (1986) 202.
[50] J.J.M. Franse, K. Kadowaki, A. Menovsky, M. van Sprang and A. de Visser, J. Appl. Phys. 61 (1987) 3380.
[51] K. Kadowaki, M. van Sprang, J.C.P. Klaasse, A.A. Menovsky, J.J.M. Franse and S.B. Woods, Physica B 148 (1987) 22.
[52] R. Verhoef, A. de Visser, A. Menovsky, A.J. Riemersma and J.J.M. Franse, Physica B 142 (1986) 11.
[53] J.J.M. Franse, M. van Sprang, A. de Visser and A.A. Menovsky, Physica B 163 (1990) 511.
[54] M. van Sprang, Ph.D. Thesis, Universiteit van Amsterdam, Amsterdam (1989) unpublished.
[55] A. de Visser, H.P. van der Meulen, B.J. Kors and J.J.M. Franse, Proc. ICM '91, Edinburgh, J. Magn. Magn. Mater. (1992) submitted.
[56] J.J.M. Franse, H.P. van der Meulen and A. de Visser, Physica B 165&166 (1990) 383.
[57] B.S. Shivaram, Y.H. Jeong, T.F. Rosenbaum, D.G. Hinks and S. Schmitt-Rink, Phys. Rev. B 35 (1987) 5372.
[58] C. Broholm, G Aeppli, R,N, Kleiman, D.R. Harshman, D.J. Bishop, E. Bucher, D.L. Williams, E.J. Ansalso and R.H. Heffner, Phys. Rev. Lett. 65 (1990) 2062.
[59] U. Rauchschwalbe, U. Ahlheim, F. Steglich, D. Rainer and J.J.M. Franse, Z. Phys. B 60 (1985) 379.
[60] B.S. Shivaram, T.F. Rosenbaum and D.G. Hinks, Phys. Rev. Lett. 57 (1986) 1259.
[61] C.H. Choi and J.A. Sauls, Phys. Rev. Lett. 66 (1991) 484.
[62] R.A. Fisher, S. Kim, B.F. Woodfield, N.E. Phillips, L. Taillefer, K. Hasselbach, J. Flouquet, A.L. Giorgi and J.L. Smith, Phys. Rev. Lett. 62 (1989) 1411.
[63] K. Hasselbach, L. Taillefer and J. Flouquet, Phys. Rev. Lett. 63 (1989) 93.
[64] T. Vorenkamp, Z. Tarnawski, H.P. van der Meulen, K. Kadowaki, V.J.M. Meulenbroek, A.A. Menovsky and J.J.M. Franse, Physica B 163 (1990) 564.
[65] K. Hasselbach, A. Lacerda, K. Behnia, L. Taillefer, J. Flouquet and A. de Visser, J. Low Temp. Phys. 81 (1990) 299.
[66] G. Bruls, D. Weber, B. Wolf, P. Thalmeier, B. Lüthi, A. de Visser and A. Menovsky, Phys. Rev. Lett. 65 (1990) 2294.
[67] S. Adenwalla, S.W. Lin, Q.Z. Ran, Z. Zhao, J.B. Ketterson, J.A. Sauls, L. Taillefer, D.G. Hinks, M. Levy and B.K. Sarma, Phys. Rev. Lett. 65 (1990) 2298.
[68] R. Joynt, Supercond. Sci. Technol. 1 (1988) 210.
[69] K. Machida, M. Ozaki and T. Ohmi, J. Phys. Soc. Jpn. 58 (1989) 4116.
[70] D.W. Hess, T. Tokuyasu and J.A. Sauls, J. Phys.: Condens. Matter 1 (1989) 8135.
[71] R. Joynt, V.P. Mineev, G.E. Volovik and M.E. Zhitomirsky, Phys. Rev. B 42 (1990) 2014.
[72] G. Aeppli, D. Bishop, C. Broholm, E. Bucher, K. Siemensmeyer, M. Steiner and N. Stüsser, Phys. Rev. Lett. 63 (1989) 676.
[73] S.K. Yip, T. Li and P. Kumar, Phys. Rev. B 43 (1991) 2742.
[74] M.C. Aronson, T. Vorenkamp, Z. Koziol, A. de Visser, K. Bakker, J.J.M. Franse and J.L. Smith, J. Appl. Phys. 69 (1991) 5487.
[75] K. Kadowaki, M. van Sprang, A.A. Menovsky and J.J.M. Franse, Jpn. J. Appl. Phys. 26 Suppl. 3 (1987) 1243.
[76] W. Schlabitz, J. Baumann, B. Pollit, U. Rauchschwalbe, H.M. Mayer, U. Ahlheim and C.D. Bredl, Z. Phys. B 62 (1986) 171.
[77] T.T.M. Palstra, A.A. Menovsky, J. van den Berg, A.J. Dirkmaat, P.H. Kes, G.J. Nieuwenhuys and J.A. Mydosh, Phys. Rev. Lett. 55 (1985) 2727.
[78] M.B. Maple, J.W. Chen, Y. Dalichaouch, T. Kohara, C. Rossel, M.S. Torikachvili, M.W. McElfresh and J.D. Thompson, Phys. Rev. Lett. 56 (1986) 185.
[79] T.T.M. Palstra, A.A. Menovsky and J.A. Mydosh, Phys. Rev. B 33 (1986) 6527.
[80] J. Schoenes, C. Schönenberger, J.J.M. Franse and A.A. Menovsky, Phys. Rev. B 35 (1987) 5375.
[81] K. Hasselbach, P. Lejay and J. Flouquet, Phys. Lett., in print.
[82] C. Broholm, J.K. Kjems, W.J.L. Buyers, P. Matthews, T.T.M. Palstra, A.A. Menovsky and J.A. Mydosh, Phys. Rev. Lett. 58 (1987) 1467.
[83] T.E. Mason, B.D. Gaulin, J.D. Garret, Z. Tun, W.J.L. Buyers and E.D. Isaacs, Phys. Rev. Lett. 65 (1990) 3189.
[84] E.D. Isaacs, D.B. McWhan, R.N. Kleiman, D.J. Bishop, G.E. Ice, P. Zschack, B.D. Gaulin, T.E. Mason, J.D. Garrett and W.J.L. Buyers, Phys. Rev. Lett. 65 (1990) 3185.
[85] A. de Visser, F.R. de Boer, A.A. Menovsky and J.J.M. Franse, Solid State Commun. 64 (1987) 527.
[86] K. Sugiyama and M. Date, J. Magn. Magn. Mater. 90&91 (1990) 461.
[87] A. de Visser, F.E. Kayzel, A.A. Menovsky, J.J.M. Franse, J. van den Berg and G.J. Nieuwenhuys, Phys. Rev. B 34 (1986) 8168.
[88] G.J. Nieuwenhuys, Phys. Rev. B 35 (1987) 5260.
[89] A.P. Ramirez, T. Siegrist, T.T.M. Palstra, J.D. Garret, E. Brück, A.A. Menovsky and J.A. Mydosh, to be published.

Normal state magnetism of the high T_c cuprate superconductors

David C. Johnston

Ames Laboratory–USDOE and Department of Physics and Astronomy, Iowa State University, Ames, IA 50011, USA

Magnetic measurements of various types have played an essential role in establishing the novel normal state characteristics of high transition temperature (T_c) superconductors with $T_c > 23$ K. Among these materials, the highest T_c's ($\leqslant 125$ K) are exhibited by the layered cuprates. In this paper, the normal state magnetic susceptibilities of the cuprates are reviewed and interpreted in the context of magnetic neutron scattering and other magnetic measurements, using the $La_{2-x}M_xCuO_4$-type and $YBa_2Cu_3O_{6+x}$-type materials as prototypical examples. The evolution of the magnetism upon doping the insulating antiferromagnetic "parent" compounds with $x = 0$ to form the high temperature superconductors is described. A recurrent property which differentiates these materials from conventional superconductors is the existence of strong antiferromagnetic correlations in the metallic state on the same sublattice of the structure in which the itinerant carriers reside.

1. Introduction

The remarkable discovery by Bednorz and Müller of superconductivity above 30 K in the La–Ba–Cu–O system in 1986 [1] soon led to a worldwide effort to explore these and related materials. Unanswered questions abounded. Was the superconductivity an interface effect or a bulk property? If a bulk property, what was the composition and crystal structure of the compound responsible for the indications of superconductivity? Could the superconducting transition temperature T_c be raised still further by chemical substitutions? If so, how high could T_c go? What was the microscopic mechanism for the superconductivity?

Progress on some of these questions was rapid. The composition of the Bednorz–Müller compound was found to be $La_{2-x}Ba_xCuO_4$, with $x \approx 0.1$–0.2, with the K_2NiF_4 structure [2]. The Sr-doped analogue was found to have a somewhat higher T_c, up to 40 K [3,4]. A large jump in T_c occurred with the discovery of superconductivity above 90 K in the compound $YBa_2Cu_3O_7$ [5]. The present record for verified bulk superconductivity is 125 K for $Tl_2Ba_2Ca_2Cu_3O_{10}$, discovered in early 1988 [6]. Thus, in two years beginning in 1986, the previous T_c record of 23 K was surpassed by more than 100 K. In the meantime, remarkable discoveries of superconductivity at temperatures up to ≈ 30 K in $Ba_{1-x}K_xBiO_3$ [7] and in $(K, Rb)_3C_{60}$ [8] were reported.

Measurements of the magnetic properties of the high T_c materials in both the superconducting and normal ($T > T_c$) states have played a central role in clarifying their nature and their similarities and differences with respect to the previously known conventional superconductors with T_c below 23 K, which are mostly intermetallic compounds and alloys (now termed "low T_c" superconductors). Of particular interest here are normal state magnetic measurements versus temperature, composition and structure, which have a bearing on the mechanism for the extraordinarily high T_c's. These measurements have shown that the Cu ions in the CuO_2 layers of the high T_c cuprates carry a (nearly) localized magnetic moment and that these moments are strongly coupled. Since the superconductivity is apparently associated with these CuO_2 layers, this feature strongly distinguishes the cuprates from all known low T_c materials. In the latter, the presence of a concentrated array of magnetic moments on the same sublattice of the structure in which the current carriers reside precluded superconductivity above 1 K. In the temperature range below T_c,

0304-8853/91/$03.50 © 1991 – Elsevier Science Publishers B.V. All rights reserved

magnetic measurements of various types have shed light on the lower (H_{c1}) and upper (H_{c2}) critical magnetic fields, on the magnetic field penetration depth, on flux pinning characteristics and on microscopic superconducting parameters such as the symmetry of the order parameter and the anisotropic superconducting coherence lengths.

The field of high T_c superconductivity has been interdisciplinary since its beginning five years ago. The cuprate materials, especially, are accessible to most research groups wishing to study them, and have been examined using the panoply of measurement techniques accessible to the physicist, chemist and materials scientist. The literature on just the magnetic properties is now very extensive, encompassing measurements such as electron spin (ESR), nuclear magnetic (NMR), nuclear quadrupole (NQR) and positive muon spin (μ^+SR) resonance spectroscopies, Mössbauer effect, magnetic neutron scattering and diffraction experiments, etc., in addition to magnetization and magnetic susceptibility $\chi(T)$ measurements.

The results of $\chi(T)$ measurements were a primary stimulus and guide to other experimental as well as theoretical studies of the high T_c compounds. Because of this and of space restrictions, and in order to provide a reasonably complete overview of the above novel magnetic character of the high T_c cuprates, this paper is limited mainly to a review of normal state $\chi(T)$ measurements of the cuprates and their interpretation. The subject matter is further limited mostly to results for the representative and most thoroughly studied La_2CuO_4- and $YBa_2Cu_3O_7$-type materials. Extensive measurements have been carried out on La_2CuO_4 itself and will be described in detail as an introduction to two-dimensional (2D) magnetism and to the magnetic properties of the other cuprates. The results and analyses of magnetic neutron scattering experiments are directly related to much of the interpretation of the $\chi(T)$ data, and will be briefly reviewed as appropriate. Only passing reference will be made to the important contributions of Raman spectroscopy and of NMR and NQR measurements to clarifying the magnetic properties of the normal state. The magnetism of rare earth ions and 3d ions other than Cu^{2+} in the cuprates will not be addressed. Recent reviews of many topics associated with high T_c superconductivity can be found elsewhere [9]. The understanding of the magnetic properties of the cuprates has historically progressed from the undoped insulating so-called "parent" compounds to the doped superconducting compositions, and this is the order in which these studies are presented below.

2. Undoped parent compounds

2.1. K_2NiF_4-type compounds

2.1.1. La_2CuO_4

The first known structure class of high T_c cuprate superconductors consisted of compounds of the type $La_{2-x}M_xCuO_4$ (M = Ca, Sr, Ba) with the body-centered-tetragonal (or orthorhombically distorted) K_2NiF_4 structure shown in fig. 1 [2,10]. The structure consists of (nearly) planar CuO_2 sheets formed from Cu-centered O_4 squares which are corner shared; two additional O atoms above and below each Cu atom give Cu

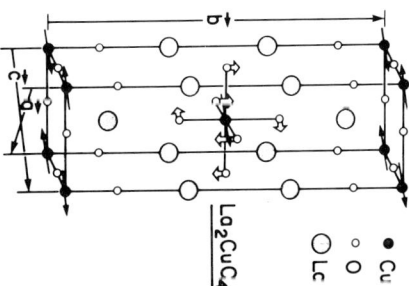

Fig. 1. Tetragonal crystal structure of La_2CuO_4, after ref. [10]. Below ≈ 530 K, the crystal distorts to an orthorhombic structure with lattice constants a, b and c as shown, accompanied by a rotation of the oxygen octahedra around the Cu atoms (shown). Below ≈ 300 K, La_2CuO_4 orders antiferromagnetically, with the spin arrangement shown by the arrows through the Cu atoms.

sixfold coordination by oxygen in which the oxygen atoms form an axially elongated octahedron. Between each adjacent pair of CuO_2 layers, the La and M atoms and axial oxygen atoms of the octahedra just described form two $La_{1-x/2}M_{x/2}O$ rocksalt layers. La_2CuO_4 undergoes a second-order structural distortion below $T_0 \approx 530$ K [11–14], from the high temperature tetragonal K_2NiF_4 structure to an orthorhombic structure in which the basal plane unit cell area increases by a factor of two. This distortion arises from a tilting ($\approx 5°$) of the CuO_6 octahedra about the tetragonal (110) direction as shown in fig. 1 [2,14].

The parent compound La_2CuO_4 has one formula unit per (tetragonal) primitive unit cell, containing an odd number (41) of valence electrons. From a conventional band picture, this compound would therefore be expected to be a metal and exhibit temperature-independent Pauli paramagnetism. State of the art band structure calculations confirmed these qualitative expectations [15], which were further confirmed with spin polarized calculations [16]. Substitution of trivalent La by the divalent M atoms to form the metallic, superconducting $La_{2-x}M_xCuO_4$ compositions ($x \approx 0.1$–0.2) would then be expected to reduce the itinerant electron carrier concentration slightly, but not result in qualitatively new physical properties.

In oxides, an alternative description is often useful in terms of an ionic model of the transition metal ion. For example, in La_2CuO_4, the oxidation states of the La and O ions are assumed to be $3+$ and $2-$, respectively, leaving Cu in an oxidation state of $2+$, with one hole in the d shell. Repulsion of two holes on the same Cu^{2+} ion could cause the compound to be an insulator, with the Cu^{2+} ions carrying a local magnetic moment with spin $S = 1/2$; magnetic ordering might then be expected. In the longitudinally elongated oxygen octahedron, the five orbital levels of the Cu ion in the presence of the crystalline electric fields are split in energy, leaving a hole in the highest energy $x^2 - y^2$ orbital. With this influence of the crystal fields, the orbital angular momentum and orbital magnetic moment of the Cu^{2+} ion are quenched to first order.

The early magnetic susceptibility, electrical resistivity and magnetic neutron diffraction measurements on the parent compound La_2CuO_4 [13,17] were strongly divergent from the band theoretical predictions, but in partial agreement with the ionic picture. The resistivity measurements showed that La_2CuO_4 is an insulator at low T, rather than a metal. The magnetic susceptibility $\chi(T)$ of a powder sample of La_2CuO_4 is plotted versus temperature T from 4 to 800 K in fig. 2(a), for several values of the oxygen content

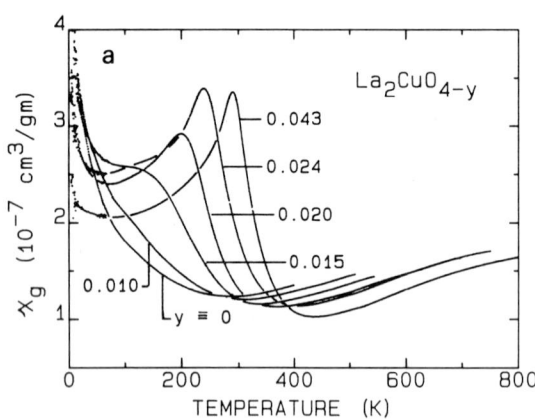 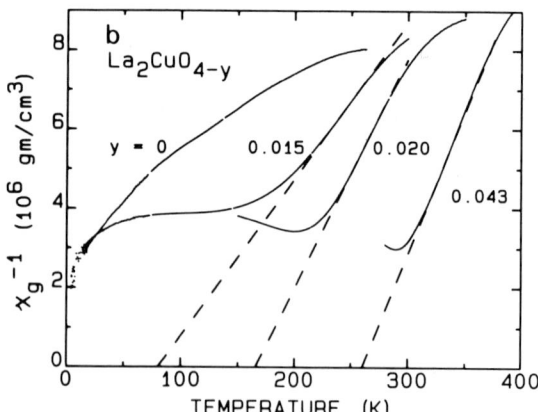

Fig. 2. (a) Magnetic susceptibility χ_g versus temperature for $La_2CuO_{\approx 4}$, for samples with slightly different oxygen contents [18]. The changes in y are accurate, but it is not known whether the total oxygen content is less than and/or greater than the precise value 4.00 for these samples. (b) Inverse magnetic susceptibility χ_g^{-1} versus temperature for some of the samples in (a).

differing by small amounts [18]. The $\chi(T)$ data show very little influence of the tetragonal to orthorhombic phase transition at, e.g., $T_0 \approx 530$ K for $T_N \approx 290$ K (but see below). $\chi(T)$ was found to be strongly temperature dependent rather than temperature independent as expected for Pauli paramagnetism. The pronounced peaks at temperatures up to ≈ 290 K suggest the occurrence of antiferromagnetic (AF) ordering of some type at those temperatures ($\equiv T_N$). However, $\chi(T)$ does not follow a Curie–Weiss type of temperature dependence [$\chi = C/(T - \Theta)$] above T_N, but rather decreases rapidly above T_N and then increases with T above about 300–400 K up to the experimental limit of 800 K. Just above T_N, plots of χ^{-1} versus T suggest the onset of ferromagnetic ordering, as shown in fig. 2(b). Thus, the nature of the magnetic ordering as revealed from powder susceptibility data was ambiguous. Neutron diffraction measurements of a powder sample with $T_N = 220$ K began to resolve the ambiguity, and indicated the occurrence of long-range antiferromagnetism below T_N, with a magnetic structure shown in fig. 1 [13]. The intensities of the magnetic neutron diffraction peaks at low temperature (10 K) were consistent with an ordered moment $\langle \mu \rangle = 0.48(15)\mu_B/$Cu lying in the CuO_2 plane along the tetragonal (110) direction; these peaks were subsequently confirmed to be magnetic in origin by polarized neutron diffraction studies [19]. The failure of conventional band theory to account for the observed properties of La_2CuO_4 is presumably due to the inadequacy of this theory in taking into account the strong Coulomb correlations present in this compound [16].

Several features of the above $\chi(T)$ and magnetic neutron diffraction data were puzzling: (i) The ordered moment was significantly less than the expected value $\langle \mu \rangle = gS\mu_B \approx 1.05\mu_B/$Cu, assuming a spectroscopic splitting factor $g \approx 2.1$ with $S = 1/2$; (ii) χ increased with T above 400 K, rather than decreased as expected from a Curie–Weiss behavior; and (iii) χ decreased much more rapidly with increasing T just above T_N than expected for antiferromagnetic ordering (fig. 2(b)). The first two features are understandable, at least semi-quantitatively, in an ionic picture, taking into account the two-dimensionality of the CuO_2 layers (feature (iii) will be addressed later). Because Cu^{2+} is a spin $1/2$ ion, the exchange interaction between adjacent Cu ions should be nearly isotropic, giving a nearest neighbor intralayer Heisenberg interaction

$$H_{ij} = JS_i \cdot S_j, \quad (1)$$

where J is the exchange coupling constant (J is positive for AF coupling in this notation). The $\langle \mu \rangle(0)$ expected from spin wave theory for the spin $1/2$ Heisenberg antiferromagnet on a 2D square lattice is $\beta g S \mu_B$, where the factor $\beta = 0.606$ is less than the classical Néel state value $\beta = 1$ due to quantum zero point spin deviations. The spin wave theory value of β has been closely confirmed by more recent calculations, which also indicate that the ground state of the model is antiferromagnetically ordered and that the system becomes so only at $T = 0$ [21–23]. Thus, assuming $g = 2.1$, theory predicts $\langle \mu \rangle(0) \approx 0.64 \mu_B/$Cu, in reasonable agreement with the above value of $0.48(15) \mu_B/$Cu found experimentally for La_2CuO_4 (see also below).

The theoretically predicted $\chi(T)$ for the 2D AF Heisenberg square lattice for $S = 1/2$ [20,24–28] is shown in fig. 3, which includes the Weiss mean field theory (MFT) prediction [24], the prediction from high temperature series expansion (HTSE) results [25], from Schwinger boson mean

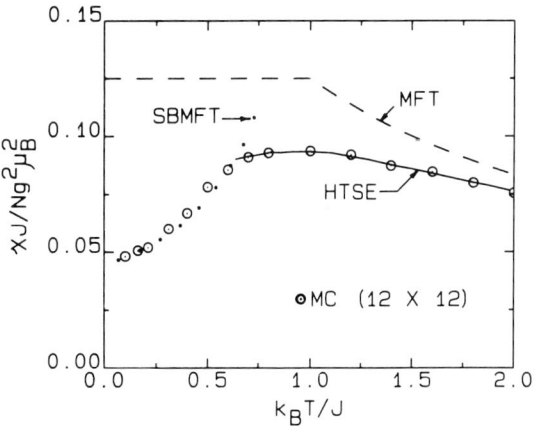

Fig. 3. Theoretical predictions for the spin susceptibility χ versus temperature T for the spin $1/2$ Heisenberg antiferromagnet on a square lattice.

field theory (SBMFT) and modified spin wave theory [26], and from quantum Monte Carlo (MC) calculations [27]. From the HTSE results, $\chi(T)$ shows a broad maximum at $T_{max} = 0.93(2)J/k_B$ [25] (close to the Weiss mean field prediction $T_N = 1.000J/k_B$ for antiferromagnetic ordering), signifying the onset with decreasing T of short-range dynamic AF ordering in the 2D spin system. The decrease in χ with decreasing T below 800 K in fig. 2(a) corresponds to the region of positive slope in fig. 3 at $T < 0.93 J/k_B$. Thus, a qualitative comparison of these two figures indicates that J is very large ($J/k_B \sim 1000$ K) in La_2CuO_4. High energy ($\hbar\omega \leqslant 140$ meV) inelastic neutron scattering measurements of the linear 2D spin wave dispersion relation (see also below) yield the accurate spin wave velocity (c) given by $\hbar c = 0.85(3)$ eV Å at 5 K and $0.75(3)$ eV Å at 300 K [29,30]. The spin wave velocity is related to the exchange coupling constant J via $\hbar c = Z8^{1/2}SJa$, where $Z = 1$ for classical spin waves, the spin $S = 1/2$ and $a = 3.81$ Å is the nearest neighbor Cu–Cu distance in the CuO_2 plane. Quantum corrections [31] yield $Z = 1.18(2)$. The value of $\hbar c$ at 5 K then gives $J = 134(5)$ meV, or $J/k_B = 1550(60)$ K, in good agreement with the value $J/k_B = 1480(70)$ K from 2-magnon Raman scattering experiments (including quantum renormalization effects) [32,33] and with the estimate of 1560 K from quantitative analysis of $\chi(T)$ above 530 K [34].

The existence of strong 2D short range dynamic ordering in La_2CuO_4 above T_N, implicit above from the discussion of the $\chi(T)$ data for $T > 400$ K [18], was proved explicitly via an extensive series of inelastic neutron scattering experiments on single crystals [29,30,35–37]. These experiments showed magnetic scattering in rods perpendicular to the CuO_2 planes above T_N as expected for 2D dynamic AF ordering. The energy-integrated intensity of this scattering is plotted versus temperature in fig. 4; also shown is the (elastic) 3D magnetic Bragg diffraction intensity below T_N [35,36]. The inelastic scattering intensity is seen to be nearly independent of T above $T_N = 195$ K up to 300 K and to die out only slowly below T_N. These and other extensive neutron scattering data [10,29,30,35–38] and theory

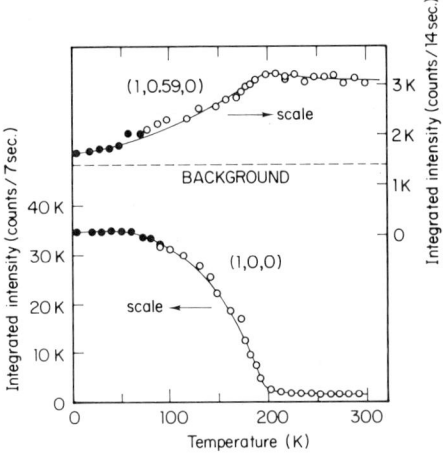

Fig. 4. Integrated intensity for the (100) Bragg peak (bottom) and inelastic scattering ridge (top) versus temperature for La_2CuO_4 with $T_N = 195$ K (after refs. [35,36].)

[39,40] show that La_2CuO_4 is a novel and nearly ideal example of a spin 1/2 Heisenberg antiferromagnet on a square lattice, with very strong exchange coupling between the nearest neighbor Cu^{2+} ions in the CuO_2 planes. 3D ordering at T_N can be induced by weak interlayer coupling when the magnetic correlation length within a CuO_2 plane becomes sufficiently large (≈ 100–1000 Å) [39]. Weak intralayer anisotropies also exist (see below).

In view of the above results and discussion, the sharp peak in $\chi(T)$ at the higher T_N in fig. 2(a) is anomalous [18]. The thermodynamic effects associated with long-range 3D AF ordering from well-developed short-range 2D AF ordering within the CuO_2 planes are expected to be very small; when the in-plane correlation length is already large at T_N, there is little entropy difference between the 2D disordered state at T_N and the 3D ordered state at $T = 0$. Indeed, high precision (0.1–1%) heat capacity measurements of La_2CuO_4 single crystal and powder failed to reveal any feature at T_N [41]. The magnetic heat capacity jump at T_N, $\Delta C_m(T_N)$, is expected [41] to be of order

$$\Delta C_m(T_N) \approx [S_m(T_N)/S_m(\infty)]\beta^2 \Delta C_m^{MF}(T_N),$$

(2)

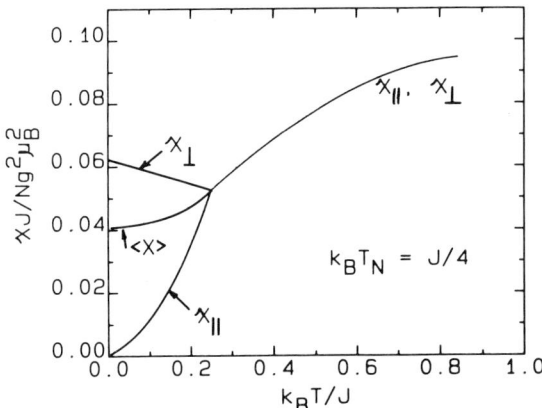

Fig. 5. Sketch of the expected anisotropic ($\chi_\parallel$, $\chi_\perp$) and powder average ($\langle \chi \rangle$) magnetic susceptibility versus temperature T for the spin 1/2 Heisenberg antiferromagnet on a square lattice which orders three-dimensionally at an assumed temperature $T_N = J/4k_B$ due to interplane coupling.

where $S_m(\infty)$ is the magnetic molar entropy in the high T limit, $S_m(\infty) = R \ln 2$ for $S = 1/2$, $\beta = 0.60$ is the above spin deviation factor, $\Delta C_m^{MF} = 1.5R$ is the mean field value, and $S_m(T_N)$ is the magnetic entropy at T_N, estimated using spin wave theory and the above value of J. Eq. (2) then yields $\Delta C_m(T_N) \approx 0.04$ J/mol K, which is only 0.3% of the mean field value and well below the resolution of the measurements of the magnetic plus lattice heat capacity at T_N.

In the case of powder $\chi(T)$ data, again only small effects are expected near T_N, as sketched in fig. 5. At the T_N of an antiferromagnet with a 3D spin lattice, one observes a sign discontinuity in the slope of $\chi(T)$. For the 2D to 3D ordering transition sketched in fig. 5 which is applicable to La_2CuO_4 from the above discussion, one expects to see only a small slope discontinuity at T_N for a power sample. The $\chi(T)$ data near T_N in fig. 2(a) are very different from this expected behavior. The reason for the anomalous peak in $\chi(T)$ at T_N was found to be due to hidden weak ferromagnetism in La_2CuO_4 below T_N [42–51]. This arises from the emergence of a Dzyaloshinsky–Moriya (DM) term in the spin Hamiltonian of the CuO_2 layers below the tetragonal–orthorhombic lattice transformation temperature $T_0 = 530$ K [42]. The DM interaction $\boldsymbol{D} \cdot (\boldsymbol{S}_i \times \boldsymbol{S}_j)$ tends to align adjacent Cu spins perpendicular to each other. In conjunction with the AF superexchange interaction, each CuO_2 layer develops a small ($\sim 10^{-3}\mu_B$/Cu) ferromagnetic canted moment component within a correlated area above T_N. As T approaches T_N from above, the total canted moment within a correlated area grows as the correlation length of the 2D AF ordering increases, leading to the incipient divergence. Below T_N, the canted ferromagnetic components in adjacent layers align antiferromagnetically, perpendicular to the CuO_2 layers, causing a decrease in χ associated with these components, and leading to the observed sharp cusp in $\chi(T)$ at T_N [42]. The $\chi(T)$ of La_2CuO_4 is anisotropic even above T_N, as shown in fig. 6 for a crystal with $T_N = 304(1)$ K [41]. The T dependence of the χ anisotropy above T_N is believed due to that of the ferromagnetic canted component [45].

From magnetoresistance, magnetization and magnetic neutron diffraction measurements, field-induced transitions are observed in single crystal La_2CuO_4 below T_N. When the applied magnetic field $\boldsymbol{H}$ is perpendicular to the CuO_2 layers (and to the ordering direction), a first order metamagnetic spin-flip transition is observed at a temperature-dependent critical field H_c [10,44–51], as shown in fig. 7(a) for the crystal of fig. 6 [45]. This transition occurs when the small canted (ferromagnetic) component of the ordered moment, initially antiparallel to the applied field in every other CuO_2 layer, flips to a

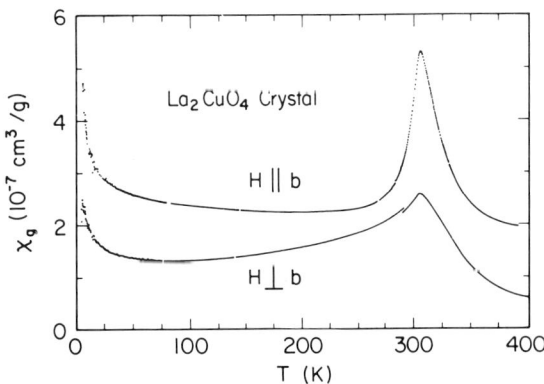

Fig. 6. Magnetic susceptibility χ_g versus temperature T for a single crystal of La_2CuO_4 with $T_N = 304$ K [41]. The b-axis is perpendicular to the CuO_2 planes.

parallel orientation. The field required for this transition to occur increases with decreasing T, as seen in fig. 7(a) and plotted in fig. 7(b) for the same crystal [45]. From fig. 7(b), there appears to be an appreciable $H_c(T_N)$; this was also observed previously and identified as a triple point [48].

Magnetic neutron diffraction experiments have verified the above interpretation of the nature of the field-induced transition [10]. A different field-induced transition occurs when H is in the plane of the CuO_2 layers and occurs at a higher field than the above spin flip transition [52]. This is a spin-flop transition in which the direction of the staggered moment, initially parallel to H, flops to a perpendicular orientation within the CuO_2 plane.

A self-consistent interpretation of the low energy inelastic neutron scattering results at low T, the field-induced transitions, the anomalous peak in $\chi(T_N)$, and inelastic light scattering experiments [53] was obtained [38,42,52] with the nearest-neighbor in-plane Hamiltonian

$$H_{ij} = S_i \cdot \hat{J} \cdot S_j, \qquad (3)$$

where

$$\hat{J} = \begin{pmatrix} J^{aa} & 0 & 0 \\ 0 & J^{bb} & J^{bc} \\ 0 & -J^{bc} & J^{cc} \end{pmatrix}$$

and $J^{aa} = J^{cc} \approx J^{bb}$. This H_{ij} includes an antisymmetric element J^{bc} in the tensor $\hat{J}$ arising from the DM interaction below T_0, as well as a weak out-of-plane spin exchange anisotropy $J^{cc} - J^{bb}$. These two anisotropies lead to energy gaps of 1.1(3) and 2.5(5) meV in the AF spin wave dispersion relations at 80 K for the in-plane and out-of-plane spin waves, respectively [38]. The values of the exchange constants were found to be $J = (J^{aa} + J^{bb} + J^{cc})/3 \approx 130$ meV as discussed above, $J^{bc} = 0.55$ meV and $J^{cc} - J^{bb} \approx 3$ μeV. The form $H_{ij} = JS_i \cdot S_j + D \cdot (S_i \times S_j)$ with $D = (D, 0, 0)$, used in ref. [48], is very similar to eq. (3), but does not include anisotropy in the diagonal exchange elements of $\hat{J}$. The effective exchange coupling constant $J_\perp$ between the Cu^{2+} ions in *adjacent* CuO_2 layers can be estimated using the relation [39]

$$J_\perp \beta^2 (\xi/a)^2 \approx k_B T_N, \qquad (4)$$

where $\beta = 0.60$ as above and $\xi \approx 200a$ is the in-plane correlation length at $T_N \approx 300$ K. This estimate yields $J_\perp \approx 2$ μeV $\sim 10^{-5} J$, showing that the coupling between layers is very weak compared with the intraplanar coupling.

2.1.2. $Sr_2CuO_2Cl_2$

In the absence of the orthorhombic distortion, the DM interaction should vanish and no cusp

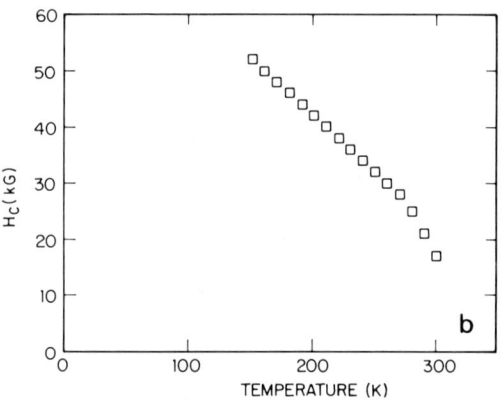

Fig. 7. (a) Magnetization M versus applied magnetic field H at several temperatures for the single crystal of La_2CuO_4 of fig. 6, with H parallel to the b-axis (perpendicular to the CuO_2 planes) [45]. The abrupt changes mark the critical field $H_c(T)$. (b) Critical field H_c versus temperature from plots as in (a).

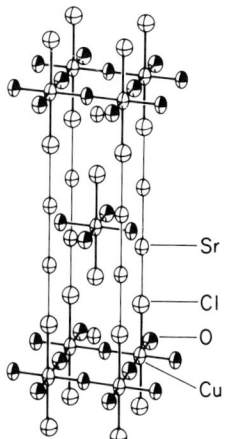

Fig. 8. Tetragonal crystal structure of $Sr_2CuO_2Cl_2$ [45,55].

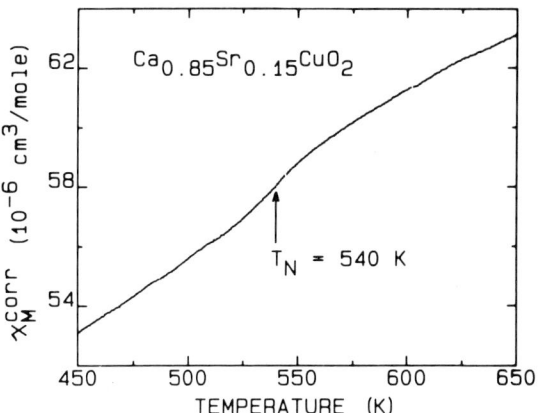

Fig. 10. Magnetic susceptibility χ_M^{corr} versus temperature for polycrystalline $Ca_{0.85}Sr_{0.15}CuO_2$ in the vicinity of the Néel temperature $T_N = 540(10)$ K; the data have been corrected for the contribution of Curie-term magnetic impurities [58].

should occur in $\chi(T)$ at T_N. This has been confirmed for the insulating tetragonal K_2NiF_4-type compound $Sr_2CuO_2Cl_2$, in which the apical oxygens of La_2CuO_4 are replaced by Cl and the La^{3+} by Sr^{2+}, as shown in fig. 8 [54,55]; the oxidation state of the Cu ions is still 2+. The compound does not exhibit any distortion from tetragonal symmetry down to at least 10 K [55,56]. The anisotropic $\chi(T)$ of a single crystal is shown in fig. 9 [56], where no cusp is in fact observed at $T_N \approx 300$ K. $\chi(T)$ increases with T above T_N, as expected for a 2D antiferromagnet with a large J. By fitting $\chi(T)$ above T_N to that of a spin 1/2 2D Heisenberg prediction, J/k_B was found to be ≈ 1800 K [56], similar to that in La_2CuO_4. A pronounced anisotropy is observed in $\chi(T)$, even above T_N (where the anisotropy is nearly independent of T up to 400 K). Magnetic neutron diffraction measurements on a single crystal with $T_N = 251(5)$ K yielded a magnetic structure identical to that in AF La_2CuO_4, with an ordered moment $\langle\mu\rangle(10\ K) = 0.34(4)\mu_B/Cu$ [56].

2.1.3. $Ca_{0.85}Sr_{0.15}CuO_2$

Powder $\chi(T)$ data for insulating tetragonal $Ca_{0.85}Sr_{0.15}CuO_2$ [57] (containing no out-of-CuO_2-plane anions), shown in fig. 10, also exhibit no cusp at $T_N = 540(10)$ K, but rather a weak inflection point in $\chi(T)$ at T_N [58]. Powder neutron diffraction measurements of the magnetic structure show that $T_N = 537(5)$ K, with $\langle\mu\rangle(10\ K) = 0.51(5)\mu_B/Cu$ lying within the basal plane [58], as in La_2CuO_4 and $Sr_2CuO_2Cl_2$.

2.2. Nd_2CuO_4 (T') type compounds

The T-type compounds typified by La_2CuO_4 have only been doped with holes (as opposed to electrons) to become metallic and superconducting, for example, by increasing the oxygen content above four oxygens per Cu atom or by substitut-

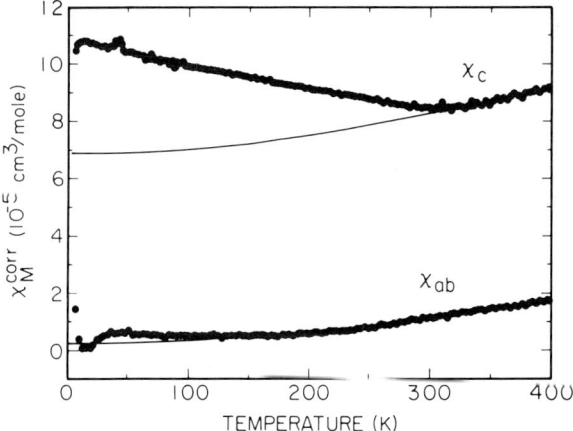

Fig. 9. Magnetic susceptibility χ_M^{corr} versus temperature for a single crystal of $Sr_2CuO_2Cl_2$, with the applied magnetic field parallel (χ_c) and perpendicular (χ_{ab}) to the tetragonal c-axis; the data have been corrected for the contribution of Curie-term magnetic impurities [56]. The solid curves are the behaviors expected in the absence of AF ordering at ≈ 310 K.

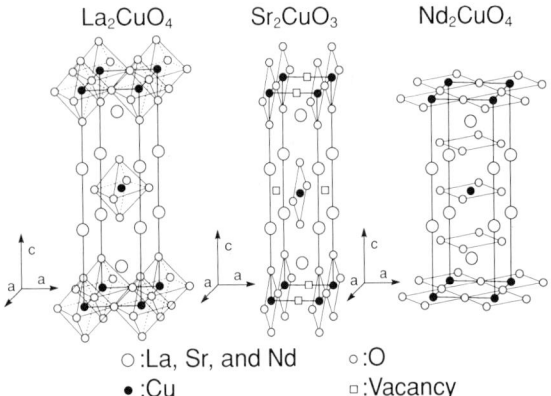

Fig. 11. Comparison of the crystal structures of (a) La_2CuO_4, (b) Sr_2CuO_3 and (c) Nd_2CuO_4.

ing a divalent alkaline earth element for the La. Therefore, the discovery of the electron-doped superconductors $L_{2-x}Ce_xCuO_4$ (L = Pr, Nd, Sm) [59,60] was of great interest; the Ce-doped Nd compound, for example, becomes superconducting below 25 K. These compounds have the tetragonal T' structure of Nd_2CuO_4 [61], shown in fig. 11. In this structure, the CuO_2 planes are the same as in the T structure, but the apical oxygens present in La_2CuO_4 are moved over to the unit cell faces, above the oxygens in the CuO_2 planes, leaving Cu in four-fold square-planar coordination. It is this seemingly innocuous change in the positions of two oxygen atoms per formula unit which apparently allows (and demands) the change from hole- to electron-doping. Indeed, it has not been reported up to now that the T-structure could be doped with electrons or the T'-structure with holes; there appears to be a basic chemical asymmetry here.

Not surprisingly, the magnetic properties of the Cu sublattice in the T'-type insulating parent L_2CuO_4 compounds are similar to those of La_2CuO_4, $Sr_2CuO_2Cl_2$ and $Ca_{0.85}Sr_{0.15}SuO_2$ above. The Cu sublattice in the parent compounds with L = Pr and Nd orders antiferromagnetically at $T_N \approx 250$ K [62–67]. The ordered Cu moment as $T \Rightarrow 0$ in Pr_2CuO_4 ($T_N = 255$ K) is $\approx 0.40(2)\mu_B/$Cu [66,67], aligned in the CuO_2 plane. Cu spin reorientation transitions are observed below T_N in Nd_2CuO_4 [62–65,67]. Inelastic neutron [66] and light [68–70] scattering experiments detect strong dynamic short range AF ordering in the CuO_2 layers above T_N in Pr_2CuO_4, Nd_2CuO_4 and Sm_2CuO_4, where the in-plane Cu–Cu exchange coupling constant J is found to be comparable to that (130 meV, see above) in La_2CuO_4. Values obtained for J were 110 [66] and 120 meV [70] for Pr_2CuO_4, 130 [66,70], 133 [68] and 108(6) meV [69] for Nd_2CuO_4, and 110(6) [69] and 130 meV [70] for Sm_2CuO_4. Thus, the intraplanar Cu–Cu exchange coupling is not significantly affected by the positions of the oxygens out of the CuO_2 plane.

2.3. $YBa_2Cu_3O_6$-type compounds

Shortly following the discovery of superconductivity at ≈ 92 K in $YBa_2Cu_3O_7$, T_c was found (see below) to be strongly affected by the oxygen content parameter x, which varies between $x = 0$ and 1 in the notation $YBa_2Cu_3O_{6+x}$. The evolution in the structure with oxygen content is shown in fig. 12 [71–80]. In contrast to all of the above-described compounds which contain Cu only in CuO_2 layers of the structures, $YBa_2Cu_3O_7$ contains Cu [denoted Cu(2)] in two CuO_2 layers per unit cell and, in addition, contains Cu [denoted Cu(1)] in a CuO_3 chain layer, with the Cu chains

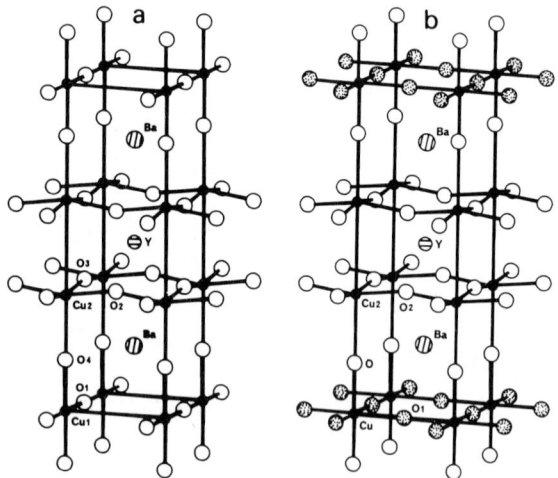

Fig. 12. (a) Orthorhombic crystal structure of $YBa_2Cu_3O_7$; (b) tetragonal structure of $YBa_2Cu_3O_{6+x}$ with $x \leqslant 0.4$. In $YBa_2Cu_3O_6$, the shaded oxygen atom positions in (b) are vacant. The figures were provided by A.J. Jacobson and J.M. Newsam.

running along the b axis of the orthorhombic unit cell. When the oxygen content is decreased, O atoms are removed from along the Cu(1) chain axis until, at $x \approx 0.4$, the structure becomes globally tetragonal and the oxygen atoms at the Cu(1) layer level distribute themselves statistically.

The insulating undoped parent compound of this system has the composition $YBa_2Cu_3O_6$ ($x = 0$), in which the Cu(2) in the CuO_2 layers are Cu^{2+} and the Cu(1) are Cu^{1+} in 2-fold coordination with O(4). Thus, the oxidation states of the ions are $Y^{3+}Ba_2^{2+}Cu(2)_2^{2+}Cu(1)^{1+}O_6^{2-}$. In view of the AF ordering observed for the insulating parent compounds containing CuO_2 layers described above, the occurrence of long-range AF order of the Cu(2) ions in $YBa_2Cu_3O_6$ was expected. Long-range AF ordering was first observed in μ^+SR experiments [81,82]. The magnetic structure determined by powder and single crystal neutron diffraction [83–88] is shown in fig. 13. The Cu(1) atoms were found to be nonmagnetic, consistent with the assignment of an oxidation state of 1+, and the ordered moment associated with the Cu(2) in the CuO_2 layers was found to be aligned in the basal a–b plane, with $\langle\mu\rangle(T \Rightarrow 0) = 0.66(7)\mu_B/Cu$ for a powder sample

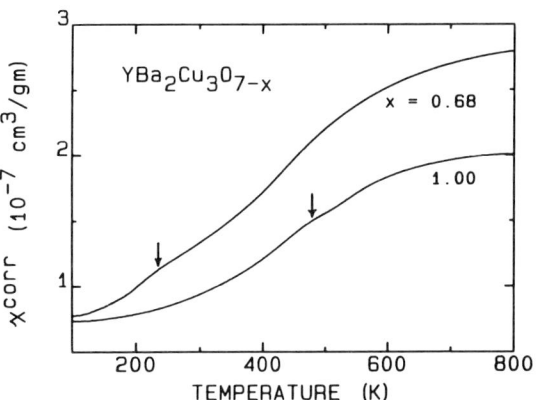

Fig. 14. Magnetic susceptibility χ^{corr} versus temperature for $YBa_2Cu_3O_6$ and $YBa_2Cu_3O_{6.32}$ powders [84]. The arrows indicate the respective Néel temperatures.

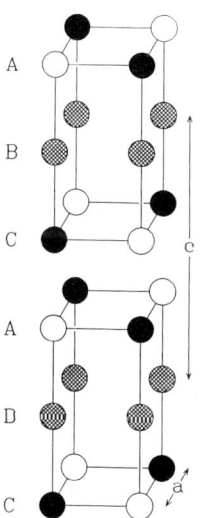

Fig. 13. Antiferromagnetic structure of $YBa_2Cu_3O_6$ [83,84]. The shaded circles represent nonmagnetic Cu(1) atoms. The filled and open circles represent Cu(2) atoms with oppositely aligned ordered moments in the a–b plane.

with $T_N \approx 500$ K [84] and $0.64(3)\mu_B/Cu$ for a single crystal with $T_N = 415(5)$ K [87]. These values are close to the theoretical value for the quantum spin 1/2 Heisenberg square lattice antiferromagnet ($g \approx 2.1$), as discussed above.

Powder $\chi(T)$ data for $YBa_2Cu_3O_6$ show only a slight inflection point at T_N [18,79,84], as shown in fig. 14 [18,84]. Anisotropic $\chi(T)$ data [89], shown in fig. 22 below, exhibit a discontinuity in slope from a negative to positive value for $H \parallel c$ upon increasing T through T_N, whereas the data with $H \perp c$ increase monotonically, with an inflection point at T_N. These data are consistent with the magnet structure below T_N in fig. 13 (cf. fig. 5).

The $\chi(T)$ data above T_N increase with increasing temperature, again indicating the presence of strong dynamic short-range AF order above T_N. This is consistent with the results of inelastic light scattering measurements which indicate that the in-plane Cu–Cu exchange coupling constant is $J = 100$ [90] to 120 meV [91], and with inelastic neutron scattering studies indicating $J > 80$ meV for a crystal with $T_N = 370$ K (x estimated to be 0.2) [92] and $J = (80^{+60}_{-30})$ meV for a crystal with $T_N = 260$ K ($x \approx 0.3$) [93]. Inelastic light scattering studies of the similar compounds $PrBa_2Cu_3O_6$ and $NdBa_2Cu_3O_6$ yielded J values of 114 and 116 meV, respectively [94].

3. Doped compounds

By doping the insulating AF parent compounds, superconducting materials are produced with remarkably high T_c. Doping is achieved by either changing the oxygen content as in $YBa_2Cu_3O_{6+x}$ ($0 \leqslant x \leqslant 1$) or La_2CuO_{4+x} (see below), by anion substitutions as in $Nd_2CuO_{4-x}F_x$ [95,96], or by cation substitutions as in $La_{2-x}Sr_xCuO_4$ or $Nd_{2-x}Ce_xCuO_4$; sometimes doping occurs by more than one mechanism simultaneously. Many studies have been carried out to establish how the transition from AF insulator to superconductor comes about. The phase diagrams obtained have many similarities which will become apparent below.

3.1. T-type compounds

3.1.1. La_2CuO_{4+x}

This system has an interesting history. In the early days of the high T_c field, samples of nominal composition La_2CuO_4 were synthesized which sometimes showed traces (~ 0.1 vol%) of superconductivity at ≈ 40 K [97,98]. By subjecting La_2CuO_4 to up to ~ 3 kbar of O_2 at 500–600°C, bulk superconductivity was obtained in both powders and single crystals [99,100]. It was found that below about 260–300 K, the system decomposes reversibly into an oxygen rich superconducting phase and oxygen poor, nearly stoichiometric La_2CuO_4 [100–102]. The ferromagnetic peak at $T_N \approx 250$ K [102] of the La_2CuO_4 phase, however, was greatly suppressed in magnitude, as is apparent in the anisotropic $\chi(T)$ of an oxygen loaded (3 kbar, 575°C) single crystal, shown in fig. 15 [103]; the reason may be associated with finely-divided granularity of phase separation. Above the phase separation temperature, χ is anisotropic and nearly temperature independent up to 300 K.

The above experiments would suggest that when the above Néel-ordered La_2CuO_4 phase contains the maximum amount of oxygen it can accomodate, the T_N has a minimum value of ≈ 250 K. On the other hand, with certain samples treated under modest (100 bar) pressure of O_2 at 500°C, T_N can be driven to $T \approx 0$ *without*

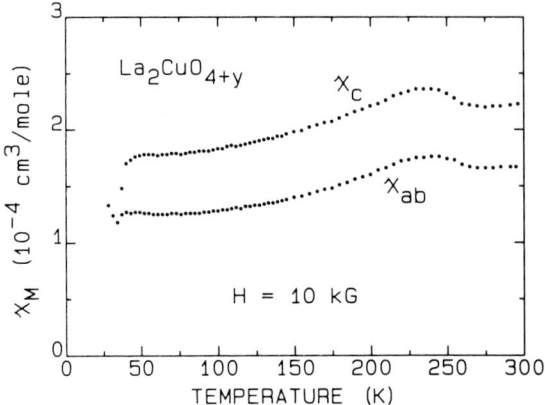

Fig. 15. Magnetic susceptibility χ_M versus temperature for a single crystal of La_2CuO_{4+x} for $H \parallel c$ (χ_c) and $H \perp c$ (χ_{ab}) [103].

the appearance of even traces of superconductivity (see fig. 2(a)) [18]. This contradiction demonstrates that there is a "hidden" variable that is not being accounted for, perhaps modest levels ($\sim 1\%$) of cation vacancies that can change form sample to sample. As a further example, for some powder samples of La_2CuO_4, T_N decreases only to ≈ 230 K with the same 100 bar O_2 treatment just described [18], in contrast to the behavior in fig. 2(a).

From fig. 2(a), the tetragonal to orthorhombic phase transition has only a small influence on $\chi(T)$. However, this influence is observable, as illustrated in expanded plots of the high T data in fig. 2(a) as shown in fig. 16(a) [18]. The slope discontinuities in $\chi(T)$ at T_0 are more clearly seen if the linear $\chi(T)$ behaviors above T_0 are subtracted from the data, as shown in fig. 16(b). T_0 is seen to decrease strongly with increasing oxygen content. The sign of the slope discontinuities is consistent with the onset of ferromagnetic correlations with decreasing T at T_0, which lead ultimately to the peak in $\chi(T)$ at T_N. From fig. 16(b), the magnitude of the slope discontinuity decreases with decreasing T_0. The phase diagram deduced from figs. 2(a) and 16 is shown in fig. 17 [18].

3.1.2. $La_{2-x}Sr_xCuO_4$

Of the T-structure class of compounds, the system $La_{2-x}Sr_xCuO_4'$ has been the most exten-

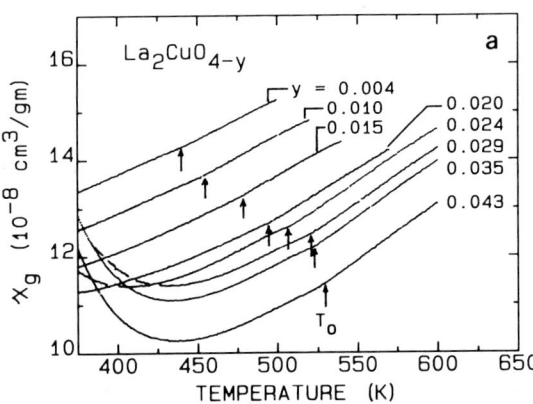
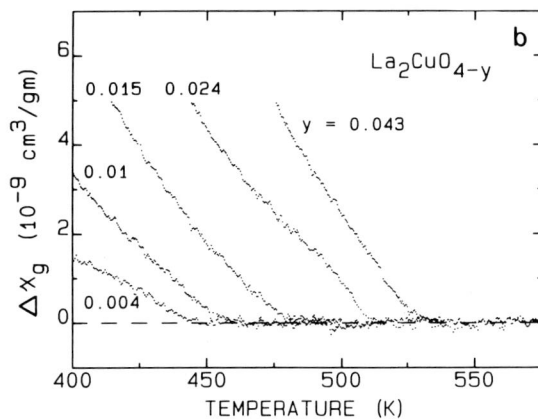

Fig. 16. (a) Expanded plots of the magnetic susceptibility χ_g versus temperature for La_2CuO_4 samples from fig. 2(a) [18]. The arrows indicate the respective tetragonal–orthorhombic transition temperatures T_0. (b) The data in (a) after subtraction of the linear background $\chi(T)$ above T_0.

sively studied. Electrical resistivity measurements indicate that the nonmetal to metal transition occurs at about $x = 0.06$ [104–106]. T_0 decreases approximately linearly with x, approaching zero at $x \approx 0.20$ [41]. The Néel temperature T_N decreases more rapidly with doping, reaching $T = 0$ by $x \approx 0.02$ [104–106]. As x increases further, $\chi(T)$ shows a smooth variation with x [18,106–

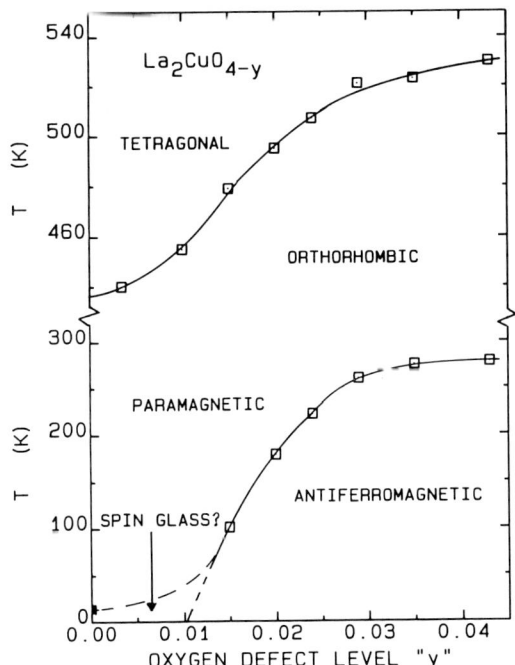

Fig. 17. Phase diagram for La_2CuO_{4-y}. Although the differences in y values are accurate, the absolute value of $4 - y$ is uncertain.

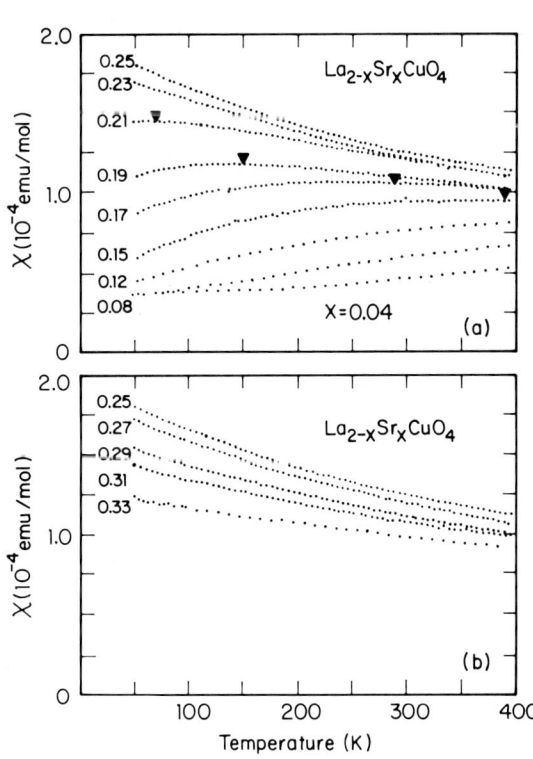

Fig. 18. Magnetic susceptibility χ versus temperature for polycrystalline $La_{2-x}Sr_xCuO_4$ (after ref. [108]). The arrows denote the maxima in $\chi(T)$.

112], as shown in fig. 18 [108]. The overall shape of $\chi(T)$ for each x is similar to that shown in fig. 3 above for the $S = 1/2$ Heisenberg AF on a square lattice, where, however, the scaling of the ordinate and abscissa depend on x. Above $x \approx 0.15$, the smooth maximum in $\chi(T)$ in fig. 3 apparently moves down into the temperature range of the measurements of fig. 18, and above $x \approx 0.21$, the maximum has apparently been suppressed to $T \approx 0$. The similarity in the shapes of the experimental curves and the theoretical prediction in fig. 3 has been exploited to derive how the effective exchange coupling constant J_{eff} and the effective moment μ_0 *per unit cell* (which includes the contributions both from the Cu^{2+} and the doped holes) vary with $x < 0.2$ [34], although this phenomenological treatment as yet has no theoretical justification; it was deduced that both μ_0 and J_{eff} are depressed rapidly with x. At present, there are no detailed theoretical predictions with which to compare the observed evolution in $\chi(T)$ with doping in the $La_{2-x}Sr_xCuO_4$ system or in any of the other high T_c cuprates.

The variation of T_c with x in the $La_{2-x}Sr_xCuO_4$ system is shown in fig. 19 [106,108,113]. T_c increases from $T = 0$ starting near $x = 0.06$, where the nonmetal to metal transition occurs, reaches a maximum of ≈ 38 K at $x \approx 0.15$ and then decreases to $T = 0$ again by $x \approx 0.26$ where the materials are still metallic [106,108,113].

From the similarity in the shapes of $\chi(T)$ of the parent AF insulators, of the insulating and of the superconducting compositions adjacent to the metal insulator transition and into the optimum superconducting composition region, it is clear that the AF correlations present above T_N in La_2CuO_4 persist in the doped metallic and superconducting compositions of the system $La_{2-x}Sr_xCuO_4$ [18]. Consequently, the superconductivity develops in the presence of dynamic AF short range order [18]. This interesting possibility has been addressed and confirmed by extensive inelastic neutron scattering measurements on single crystals for $x \leqslant 0.18$ [114–117]. These measurements have shown that the AF correlation length ξ decreases with doping as $\xi \approx 3.8/\sqrt{x}$ Å, which is the average separation between the doped holes [114]. It was also concluded that the doping does not change the Cu^{2+} moment itself appreciably [114], in apparent conflict with the above conclusion [34] from analysis of the $\chi(T)$ data.

In the intermediate composition region $0.02 \lesssim x \lesssim 0.06$ between the antiferromagnetic and superconducting metallic compositions, μ^+SR [118–120] and other experiments have indicated the presence of a spin glass phase below ≈ 10 K. Some experiments show the spin glass phase coexisting with the superconducting phase above $x = 0.06$ up to 0.15 [121–123]. However, more recent data [124] rule out this possibility for the optimum superconducting composition $x = 0.15$. Observation of the simultaneous presence of spin glass order and superconductivity may be due to chemical inhomogeneity in the samples [125–128].

3.1.3. $La_{2-x}Ba_xCuO_4$

Much less work has been done on this system than on the Sr-doped La_2CuO_4 system. Overall, the phase diagram is similar [129]. However, whereas $La_{2-x}Sr_xCuO_4$ shows a smooth variation of T_c with x with a maximum near $x = 0.15$ (fig. 19), $La_{2-x}Ba_xCuO_4$ shows a double-peaked structure in $T_c(x)$, with a central minimum at $T_c \approx 0$ at $x \approx 0.12$ [130,131]. This anomalous behavior was found to be correlated with a subtle structural transition below ≈ 80 K from the low temperature orthorhombic (LTO) structure encountered above for La_2CuO_4 to a new low temperature tetragonal (LTT) structure [131]. This LTO–LTT structural transformation was appar-

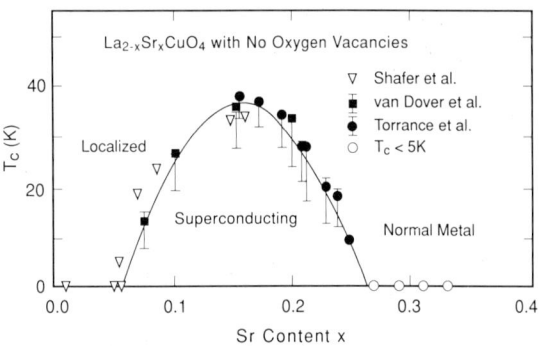

Fig. 19. Superconducting transition temperature T_c versus x for $La_{2-x}Sr_xCuO_4$ (after ref. [108]).

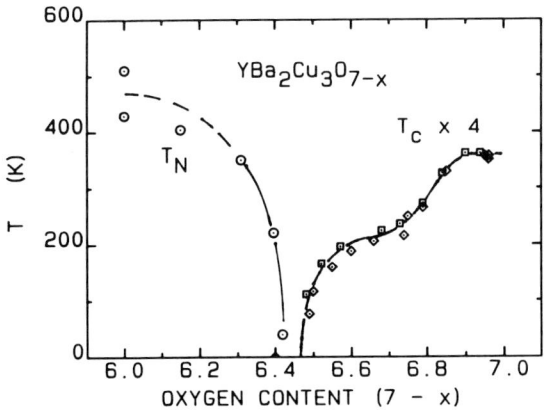

Fig. 20. Phase diagram of $YBa_2Cu_3O_y$ [84].

ently presaged by an anomalous variation in the orthorhombicity of the LTO phase in $La_{1.85}Ba_{0.15}CuO_4$ with decreasing temperature below ≈ 80 K [132]. The LTO–LTT transformation is accompanied by anomalous temperature dependences of the normal state resistivity and $\chi(T)$ [133], and by an extremely strong and anomalous variation with x in the oxygen isotope effect on T_c [134]. μ^+SR [135] and ^{139}La NQR [136] measurements indicate the onset of magnetic ordering of some type at low temperatures for $x = 0.125$, suggesting that the LTO to LTT transition may remove all or part of the Fermi surface; this would provide a natural explanation for the disappearance of T_c at $x = 0.12$.

3.2. $YBa_2Cu_3O_{6+x}$

The system usually remains nonmetallic (and nonsuperconducting) and tetragonal for $0 \leq x \leq 0.4$, although the precise x value at which the symmetry change to orthorhombic occurs and superconductivity first appears (and long-range AF ordering disappears, see below) depends on the details of sample preparation and annealing procedures [18,77–79,137–143]. The x doped oxygen atoms per formula unit are distributed on the Cu(1) plane in between the two CuO_2 planes, oxidizing Cu^{1+} ions in the former plane to Cu^{2+}. From magnetic neutron diffraction and μ^+SR measurements, T_N is nearly constant at 400–500 K for $x < 0.2$ and falls to 0 K at about $x = 0.4$ [18,84,86,87,144,145], as shown in fig. 20 [84]. Above $x \approx 0.4$, where the system becomes orthorhombic and metallic, the two well-known plateaus at $T_c \approx 60$ and 90 K are seen with increasing x, also shown in fig. 20 [18,79,139,141]. In single crystals of $YBa_2Cu_3O_{6+x}$ and $NdBa_2Cu_3O_{6+x}$, but not in powder samples, a second antiferromagnetic transition at $T_{N2} \lesssim 80$ K is sometimes observed by neutron diffraction for $x > 0$, associated with long-range magnetic ordering of the Cu^{2+} ions in the Cu(1) layer [146–150]. One neutron diffraction investigation indicates the coexistence of superconductivity and long-range AF order at $x = 0.55$ [151].

The variation in the powder $\chi(T)$ and anisotropic $\chi(T)$ with oxygen content [18,79,89] from

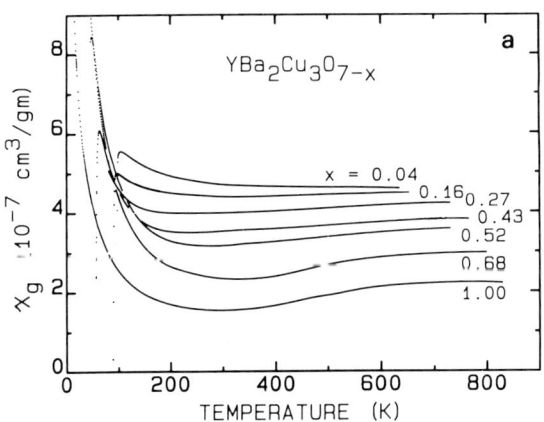
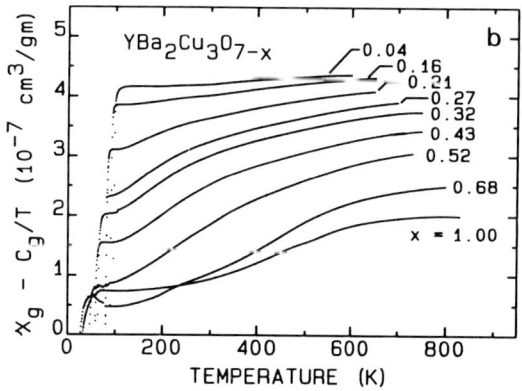

Fig. 21. (a) Measured powder magnetic susceptibility χ_g versus temperature for $YBa_2Cu_3O_y$ samples with $y = 6.00$ to 6.96 [79]. (b) The data in (a) corrected for Curie-type impurities/defects [18].

separate studies are shown in figs. 21 [18,79] and 22 [89], respectively. As in the $La_{2-x}Sr_xCuO_4$ system versus Sr doping, there is a smooth evolution in $\chi(T)$ with oxygen content from $x = 0$ to 1, with no discontinuities occurring at the insulator to metallic superconductor transition near $x = 0.4$. The strong temperature dependence of $\chi(T)$ for the insulating parent member $YBa_2Cu_3O_6$ is still present on both sides of the insulator–metal transition at $x = 0.4$. One may conclude from this that as in the $La_{2-x}Sr_xCuO_4$ system, the superconductivity in the metallic compositions with $T_c = 60$–80 K develops in the presence of strong short-range AF ordering [18]. Inelastic neutron scattering measurements on single crystals with $x \approx 0.5$–0.9 ($T_c = 50$–80 K) have confirmed and quantified this behavior [152–157]. On the other hand, $\chi(T)$ becomes nearly independent of temperature for the compositions ($x \approx 1.0$) with the highest T_c's (≈ 92 K). This has usually been interpreted in the literature to mean that the strong AF correlations and/or $Cu(2)^{2+}$ magnetic moments present at lower x are greatly weakened or absent in this composition. Again, however, it must be emphasized that theoretical predictions are not yet available to quantitatively interpret the evolution in $\chi(T)$ with doping level in the high T_c cuprates; hence, this last interpretation is questionable and appears [158,159] to be incorrect. Inelastic neutron scattering measurements on crystals with $x \approx 1.0$ with $T_c \approx 92$ K could help to settle the issue, but crystals large enough for these measurements are apparently not yet available.

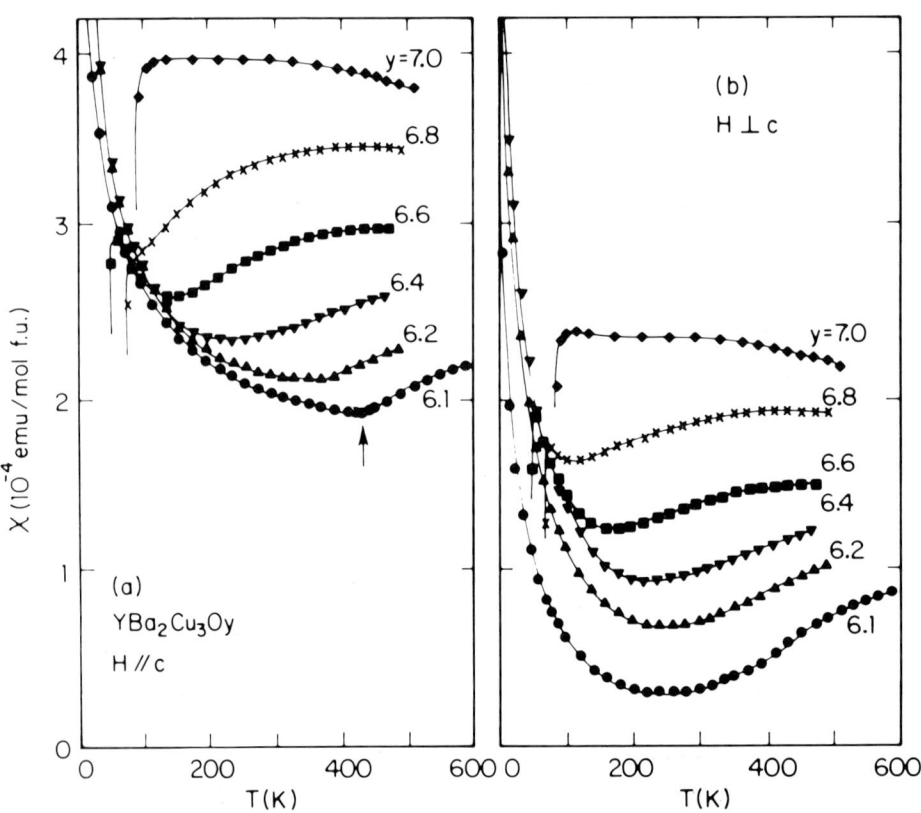

Fig. 22. Anisotropic magnetic susceptibility χ versus temperature T for $YBa_2Cu_3O_y$ samples with $y = 6.1$ to 7.0, for (a) $\boldsymbol{H} \parallel \boldsymbol{c}$ and (b) $\boldsymbol{H} \perp \boldsymbol{c}$ (after ref. [89]).

4. General magnetic properties

4.1. Ordered moment below T_N

The ordered moment $\langle\mu\rangle(T \Rightarrow 0)$ in the antiferromagnetically ordered cuprate compositions, as determined by magnetic neutron diffraction measurements, has been found to vary, depending on the value of T_N, as shown in fig. 23. T_N, in turn, depends on the structure of the compound and on the degree of doping. As seen in fig. 23, $\langle\mu\rangle$ decreases as $T_N \Rightarrow 0$ and has a maximum value of about $0.6-0.7\mu_B/Cu$ at the larger T_N values. This latter limit is equal within the errors to the theoretical prediction of $0.30g\mu_B/Cu$ for the quantum $S = 1/2$ 2D Heisenberg AF on a square lattice, if the g value is taken to be ≈ 2.1 as in most known insulators containing Cu^{2+}. The reason for the decrease in $\langle\mu\rangle$ at the lower T_N values has been addressed by μ^+SR measurements via the dependence of $\langle\mu\rangle(T)$ versus doping level in La_2CuO_{4+x} [161]. It was found that $\langle\mu\rangle(0)$ is nearly independent of T_N for $100 \text{ K} \leq T_N \leq 250 \text{ K}$. The difference between this result and fig. 23 indicates that the volume fraction of a sample which orders long-range increases with T_N, since the neutron diffraction results, on which fig. 23 is based, are computed as an average over the volume of the sample, whereas μ^+SR is a local probe. The reason that the volume fraction which orders long-range depends on T_N in the La_2CuO_{4+x} system remains obscure. The low value of $\langle\mu\rangle$ for $Sr_2CuO_2Cl_2$ ($0.34\mu_B/Cu$ [56]) may also be due in part to lack of long-range order throughout the volume of the crystal [162]. In the same context, it has been found that the temperature dependences of the ordered moment below T_N in both La_2CuO_4 and $Sr_2CuO_2Cl_2$ are different, depending on whether neutron diffraction or a local probe like μ^+SR, NQR or Mössbauer measurements are used to measure them [162].

Comparison of $\langle\mu\rangle(T)$ for $Sr_2CuO_2Cl_2$ and La_2CuO_4 as probed by μ^+SR with calculated curves from the 2D Heisenberg model with small intraplanar exchange coupling anisotropy indicates that the anisotropy is much less in the former compound, consistent with its higher lattice symmetry (tetragonal vs. orthorhombic) [163].

4.2. Superconducting fluctuation diamagnetism above T_c

Early in the development of the high T_c field, measurements of the electrical resistivity versus temperature for $YBa_2Cu_3O_7$ revealed negative deviations from the apparent higher temperature linear background resistivity upon cooling. These deviations progressively increased, starting at ~ 150 K, until the bulk T_c (92 K) was reached. These deviations arise from temporal thermodynamic fluctuations into the superconducting state, which become apparent far above T_c and decrease the normal state resistivity from that which otherwise would have been observed [164–168]. Analysis of these measurements indicate very short Ginzburg–Landau superconducting coherence lengths $\xi_c(0) \approx 14$ Å and $\xi_{ab}(0) \approx 1-2$ Å. Superconducting fluctuations have also been found to lead to a divergence in the heat capacity in the vicinity of T_c in highly homogeneous single crystals [169–171]; from these studies, estimates of the number of components to the superconducting order parameter have been made.

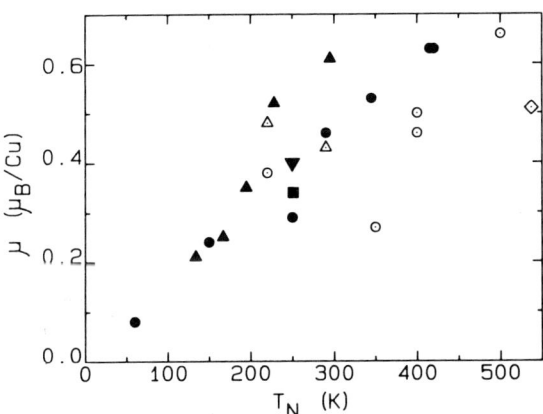

Fig. 23. Variation of the zero-temperature ordered copper magnetic moment μ with Néel temperature T_N for antiferromagnetically ordered cuprate compounds. The filled symbols are for single crystals of $La_2CuO_{\approx 4}$ (triangles pointing up) [35,160], $YBa_2Cu_3O_{6+x}$ (circles) [85–87], $Sr_2CuO_2Cl_2$ (square) [56] and Pr_2CuO_4 (triangle pointing down) [66]. The open symbols are for powder samples of $La_2CuO_{\approx 4}$ (triangles pointing up) [13,19], $YBa_2Cu_3O_{6+x}$ (circles) [84] and $Ca_{0.85}Sr_{0.15}CuO_2$ (diamond) [58].

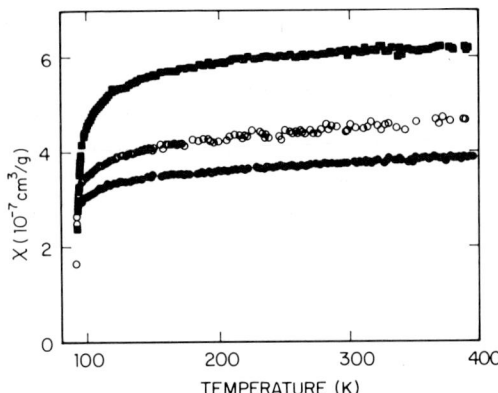

Fig. 24. Magnetic susceptibility χ versus temperature for $YBa_2Cu_3O_7$ samples with $H \parallel c$ (top), $H \perp c$ (bottom) and for random alignment (middle) [175].

The diamagnetic susceptibility associated with the superconducting fluctuations above T_c (superconducting fluctuation diamagnetism, or SFD) is observable in samples of the high T_c cuprates of sufficient purity with respect to magnetic impurities and/or isolated Cu^{2+} magnetic defects [18,164,172–176]. The highest magnetic purity single crystal or powder sample of $YBa_2Cu_3O_7$ studied to date appears to be a powder sample, for which the anisotropic $\chi(T)$ data are shown in fig. 24 for $H \parallel c$ (χ_c) and $H \perp c$ (χ_{ab}) [175,176]. These data were obtained after grain-aligning the free-flowing powder sample in situ in a SQUID magnetometer [176]. Negative curvature is observed in both sets of data below ≈ 200 K, or $\approx 2T_c$, attributed in part to SFD. The strongest SFD occurs with $H \parallel c$, as expected from theory [175]; within several K of T_c, its magnitude is of the order of the diamagnetic core susceptibility contribution. From theoretical fits to the data, microscopic parameters associated with the superconducting state are obtained, including the $T = 0$ Ginzburg–Landau coherence lengths $\xi_c(0) = 13.6$ Å and $\xi_{ab}(0) = 1.2$ Å; $\xi_c(0)$ is in agreement with the value calculated from the observed dH_{c2}/dT at T_c for $H \parallel c$ [177]. SFD has also been observed up to $\approx 2T_c$ in single crystal $Bi_2Sr_2CaCu_2O_8$, as shown in fig. 25 [178], and in polycrystalline samples of $La_{2-x}Sr_xCuO_4$ [175], $Bi_{2-x}Pb_xSr_2CaCu_2O_8$ [175,179] and $Bi_{2-x}Pb_xSr_2Ca_2Cu_3O_{10}$ [179].

The observation of SFD can be masked by the T-dependent Curie or Curie–Weiss paramagnetism of only a small amount ($< 1\%$) of paramagnetic impurities and/or defects, an amount usually undetectable by X-ray diffraction of the samples (see fig. 21(a)). Indeed, even in single crystals of $YBa_2Cu_3O_7$, this impurity/defect level has up to now been large enough to mask the SFD. Grinding the above high purity sample of this compound to a fine powder and the chemical reaction of the surface of the powder grains with epoxy have both been found to generate enough magnetic defects to obscure the SFD observed (fig. 24) for the free-flowing powder [176].

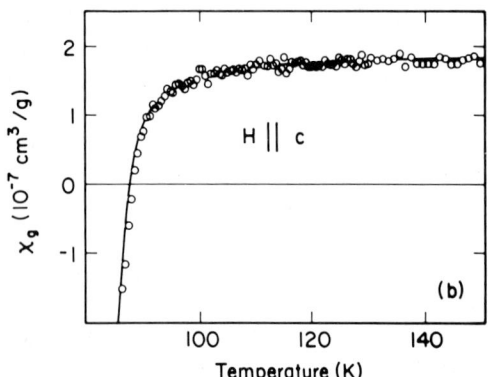
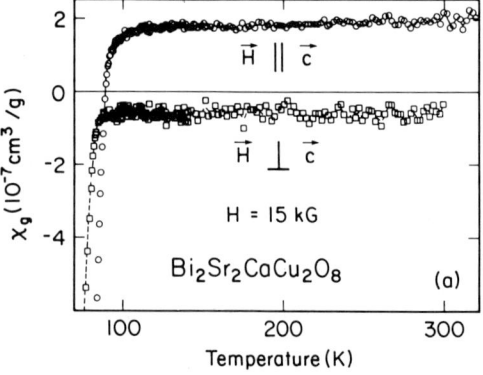

Fig. 25. (a) Magnetic susceptibility χ_g versus temperature for a single crystal of $Bi_2Sr_2CaCu_2O_8$ for $H \parallel c$ and $H \perp c$ [178]. (b) Expanded plot of the data for $H \parallel c$ in (a) below 150 K; the solid curve is a fit of superconducting fluctuation diamagnetism theory to the data [178].

4.3. Magnetic susceptibility anisotropy

Rather strong anisotropy in $\chi(T)$ has been observed for single crystals and aligned powders of the high T_c cuprates or antiferromagnetic parent compounds above T_c or T_N, respectively, as is apparent in the anisotropic $\chi(T)$ data for La_2CuO_4, $Sr_2CuO_2Cl_2$, La_2CuO_{4+x}, $YBa_2Cu_3O_{6+x}$, $YBa_2Cu_3O_7$ and $Bi_2Sr_2CaCu_2O_8$ shown above in figs. 6, 9, 15, 22, 24 and 25(a), respectively. Similar anisotropies are observed for single crystals of $La_{1.98}Ba_{0.02}CuO_4$ (above $T_N \approx 130$ K) and $La_{1.9}Sr_{0.1}CuO_4$ (above $T_c = 31$ K) [51].

Remarkably, for the above K_2NiF_4-type compounds and for $YBa_2Cu_3O_6$, the anisotropies $\Delta\chi \equiv \chi_c - \chi_{ab}$ (per mole of Cu^{2+} in the CuO_2 planes) at high temperatures are nearly the same and independent of T, irrespective of whether the compounds are antiferromagnetic insulators or high temperature superconductors at lower temperatures [175,178]. This suggests that the electronic properties in the vicinity of the Cu^{2+} ions in the CuO_2 planes are nearly the same and that in each compound $\Delta\chi$ has a common origin. This, in turn, is consistent with the above smooth variations in $\chi(T)$ in $La_{2-x}Sr_xCuO_4$ (fig. 19) and $YBa_2Cu_3O_{6+x}$ (figs. 21 and 22) as x changes across the respective insulator–metal transitions. The origin of $\Delta\chi$ is believed to be largely from the anisotropic T-independent Van Vleck paramagnetism of the Cu^{2+} ions, which retain their (nearly) localized character in the metallic state, and also partly from a weak anisotropy in the spin susceptibility arising from an anisotropic g factor of these ions [89,176,178,180–182].

5. Concluding remarks

The high T_c cuprates and parent compounds have provided a remarkably rich variety of normal state magnetic behaviors, from short-range static and dynamic antiferromagnetic ordering and long-range three-dimensional ordering to superconducting fluctuation diamagnetism. The observed two-dimensionality of the magnetic correlations in the CuO_2 planes motivated new theoretical understanding of the two-dimensional local moment quantum spin 1/2 Heisenberg antiferromagnet. Much current research is focused on developing a theoretical description and understanding of the transition from the insulating local moment antiferromagnet to the doped materials which exhibit high temperature superconductivity. The experiments demonstrate that the antiferromagnetic correlations survive the insulator-to-metal transition, and hence must be dealt with in realistic theories of the metallic state. The structure of the phase diagrams observed seems to suggest that these correlations are antagonistic to superconductivity, since the lowest T_c's in the metallic materials are associated with doping compositions adjacent to the insulator-to-metal phase boundaries where the correlations are most well-defined. The special role of copper and/or oxygen (if any) in these materials with regard to the high (30–125 K) T_c's remains to be seen. Thus, many basic questions remain unresolved, including the microscopic mechanism for the superconductivity and the related issue of the maximum realizable T_c. These pose a hope and a challenge for theorists and experimentalists alike.

Acknowledgements

The author is grateful to his collaborators and other colleagues for extensive interactions and discussions over the past five years to document and understand the novel magnetic properties of the cuprates and their possible relevance to high temperature superconductivity. Ames Laboratory is operated for the US Department of Energy by Iowa State University under Contract no. W-7405-Eng-82. This work was supported by the Director for Energy Research, Office of Basic Energy Sciences.

References

[1] J.G. Bednorz and K.A. Müller, Z. Phys. B 64 (1986) 189.
[2] J.D. Jorgensen, H.-B. Schüttler, D.G. Hinks, D.W. Capone, II, K. Zhang, M.B. Brodsky and D.J. Scalapino, Phys. Rev. Lett. 58 (1987) 1024.
[3] R.J. Cava, R.B. van Dover, B. Batlogg and E.A. Rietman, Phys. Rev. Lett. 58 (1987) 408.

[4] J.M. Tarascon, L.H. Greene, W.R. McKinnon, G.W. Hull and T.H. Geballe, Science 235 (1987) 1373.
[5] M.K. Wu, J.R. Ashburn, C.J. Torng, P.H. Hor, R.L. Meng, L. Gao, Z.J. Huang, Y.Q. Wang and C.W. Chu, Phys. Rev. Lett. 58 (1987) 908.
R.J. Cava, B. Batlogg, R.B. van Dover, D.W. Murphy, S. Sunshine, T. Siegrist, J.P. Remeika, E.A. Rietman, S. Zahurak and G.P. Espinosa, Phys. Rev. Lett. 58 (1987) 1676.
[6] Z.Z. Sheng and A.M. Hermann, Nature 332 (1988) 138.
S.S.P. Parkin, V.Y. Lee, E.M. Engler, A.I. Nazzal, T.C. Huang, G. Gorman, R. Savoy and R. Beyers, Phys. Rev. Lett. 60 (1988) 2539.
[7] L.F. Mattheiss, E.M. Gyorgy and D.W. Johnson, Jr., Phys. Rev. B 37 (1988) 3745.
R.J. Cava, B. Batlogg, J.J. Krajewski, R. Farrow, L.W. Rupp, Jr., A.E. White, K. Short, W.F. Peck and T. Kometani, Nature 332 (1988) 814.
[8] A.F. Hebard et al., Nature 350 (1991) 600.
K. Holczer, O. Klein, S-M. Huang, R.B. Kaner, K-J. Fu, R.L. Whetten and F. Diederich, Science 252 (1991) 1154.
[9] See, e.g., Physical Properties of High Temperature Superconductors, ed. D.M. Ginsberg (World Scientific, Singapore, 1989 (vol. I) and 1990 (vol II)).
S-W. Cheong, J.D. Thompson and Z. Fisk, Physica C 158 (1989) 109.
[10] M.A. Kastner, R.J. Birgeneau, T.R. Thurston, P.J. Picone, H.P. Jenssen, D.R. Gabbe, M. Sato, K. Fukuda, S. Shamoto, Y. Endoh, K. Yamada and G. Shirane, Phys. Rev. B 38 (1988) 6636.
[11] P. Lehuédé and M. Daire, Compt. Rend. Acad. Sci. (Paris) 276 (1973) C1011.
[12] J.M. Longo and P.M. Raccah, J. Solid State Chem. 6 (1973) 526.
[13] D. Vaknin, S.K. Sinha, D.E. Moncton, D.C. Johnston, J.M. Newsam, C.R. Safinya and H.E. King, Jr., Phys. Rev. Lett. 58 (1987) 2802.
[14] B. Grande, Hk. Müller-Buschbaum and M. Schweizer, Z. Anorg. Allg. Chem. 428 (1977) 120.
[15] L.F. Mattheiss, Phys. Rev. Lett. 58 (1977) 1028.
J. Yu, A.J. Freeman and J.-H. Xu, ibid. 58 (1977) 1035.
[16] T.C. Leung, X.W. Wang and B.N. Harmon, Phys. Rev. B 37 (1988) 384.
[17] D.C. Johnston, J.P. Stokes, D.P. Goshorn and J.T. Lewandowski, Phys. Rev. B 36 (1987) 4007.
S. Uchida, H. Takagi, H. Yanagisawa, K. Kishio, K. Kitazawa, K. Fueki and S. Tanaka, Japan. J. Appl. Phys. 26 (1987) L445.
R.L. Greene, H. Maletta, T.S. Plaskett, J.G. Bednorz and K.A. Müller, Solid State Commun. 63 (1987) 379.
[18] D.C. Johnston, S.K. Sinha, A.J. Jacobson and J.M. Newsam, Physica C 153–155 (1988) 572.
D.C. Johnston, (unpublished).
[19] S. Mitsuda, G. Shirane, S.K. Sinha, D.C. Johnston, M.S. Alvarez, D. Vaknin and D.E. Moncton, Phys. Rev. B 36 (1987) 822.

[20] P.W. Anderson, Phys. Rev. 86 (1952) 694.
T. Oguchi, ibid. 117 (1960) 117.
[21] J.D. Reger and A.P. Young, Phys. Rev. B 37 (1988) 5978.
[22] N. Trivedi and D.M. Ceperley, Phys. Rev. B 40 (1989) 2737.
[23] S. Liang, Phys. Rev. B 42 (1990) 6555, and references therein.
[24] See, e.g., C. Kittel, Introduction to Solid State Physics, 4th ed. (McGraw-Hill, New York).
[25] L.J. de Jongh, in: Magnetism and Magnetic Materials-1972, eds. C.D. Graham, Jr. and J.J. Rhyne, AIP Conf. Proc. no. 10 (American Institute of Physics, New York, 1973) p. 561.
L.J. de Jongh and A.R. Miedema, Adv. Phys. 23 (1974) 1.
[26] A. Auerbach and D.P. Arovas, Phys. Rev. Lett. 61 (1988) 617.
M. Takahashi, Phys. Rev. B 40 (1989) 2494.
[27] Y. Odabe, M. Kikuchi and A.D.S. Nagi, Phys. Rev. Lett. 61 (1988) 2971.
[28] G. Gomez-Santos, J.D. Joannopoulos and J.W. Negele, Phys. Rev. B 39 (1989) 4435.
[29] G. Aeppli, S.M. Hayden, H.A. Mook, Z. Fisk, S-W. Cheong, D. Rytz, J.P. Remeika, G.P. Espinosa and A.S. Cooper, Phys. Rev. Lett. 62 (1989) 2052.
[30] S.M. Hayden, G. Aeppli, H.A. Mook, S-W. Cheong and Z. Fisk, Phys. Rev. B 42 (1990) 10220.
[31] R.R.P. Singh, Phys. Rev. B 39 (1989) 9760.
[32] K.B. Lyons, P.A. Fleury, J.P. Remeika, A.S. Cooper and T.J. Negran, Phys. Rev. B 37 (1988) 2353.
K.B. Lyons, P.E. Sulewski, P.A. Fleury, H.L. Carter, A.S. Cooper, G.P. Espinosa, Z. Fisk and S.-W. Cheong, Phys. Rev. B 39 (1989) 9693.
[33] R.R.P. Singh, P.A. Fleury, K.B. Lyons and P.E. Sulewski, Phys. Rev. Lett. 62 (1989) 2736.
[34] D.C. Johnston, Phys. Rev. Lett. 62 (1989) 957.
[35] G. Shirane, Y. Endoh, R.J. Birgeneau, M.A. Kastner, Y. Hidaka, M. Oda, M. Suzuki and T. Murakami, Phys. Rev. Lett. 59 (1987) 1613.
[36] Y. Endoh, K. Yamada, R.J. Birgeneau, D.R. Gabbe, H.P. Jenssen, M.A. Kastner, C.J. Peters, P.J. Picone, T.R. Thurston, J.M. Tranquada, G. Shirane, Y. Hidaka, M. Oda, Y. Enomoto, M. Suzuki and T. Murakami, Phys. Rev. B 37 (1988) 7443.
[37] K. Yamada, K. Kakurai, Y. Endoh, T.R. Thurston, M.A. Kastner, R.J. Birgeneau, G. Shirane, Y. Hidaka and T. Murakami, Phys. Rev. B 40 (1989) 4557.
[38] C.J. Peters, R.J. Birgeneau, M.A. Kastner, H. Yoshizawa, Y. Endoh, J. Tranquada, G. Shirane, Y. Hidaka, M. Oda, M. Suzuki and T. Murakami, Phys. Rev. B 37 (1988) 9761.
[39] S. Chakravarty, B.I. Halperin and D.R. Nelson, Phys. Rev. Lett. 60 (1988) 1057; Phys. Rev. B 39 (1989) 2344.
[40] S. Tyc, B.I. Halperin and S. Chakravarty, Phys. Rev. Lett. 62 (1989) 835.
[41] K. Sun, J.H. Cho, F.C. Chou, W.C. Lee, L.L. Miller,

D.C. Johnston, Y. Hidaka and T. Murakami, Phys. Rev. B 43 (1991) 239, and references cited.

[42] T. Thio, T.R. Thurston, N.W. Preyer, P.J. Picone, M.A. Kastner, H.P. Jenssen, D.R. Gabbe, C.Y. Chen, R.J. Birgeneau and A. Aharony, Phys. Rev. B 38 (1988) 905.

[43] N.F. Oliveira, Jr., J.T. Nicholls, Y. Shapira, G. Dresselhaus, M.S. Dresselhaus, P.J. Picone, D.R. Gabbe and H.P. Jenssen, Phys. Rev. B 39 (1989) 2898.

[44] K. Fukuda, M. Sato, S. Shamoto, M. Onoda and S. Hosoya, Solid State Commun. 63 (1987) 811.

[45] W.C. Lee, D.C. Johnston, Y. Hidaka and T. Murakami, unpublished.

[46] S-W. Cheong, Z. Fisk, J.O. Willis, S.E. Brown, J.D. Thompson, J.P. Remeika, A.S. Cooper, R.M. Aikin, D. Schiferl and G. Gruner, Solid State Commun. 65 (1988) 111.

[47] S-W. Cheong, J.D. Thompson, Z. Fisk and G. Gruner, Solid State Commun. 66 (1988) 1019.

[48] S-W. Cheong, J.D. Thompson and Z. Fisk, Phys. Rev. B 39 (1989) 4395.

[49] B. Barbara, J. Beille and H. Dupendant, J. de Phys. 49 (1988) C8-2135.

[50] F. Zuo, X.D. Chen, J.R. Gaines and A.J. Epstein, Synthetic Metals 29 (1989) F729.

[51] K. Fukuda, S. Shamoto, M. Sato and K. Oka, Solid State Commun. 65 (1988) 1323.

[52] T. Thio, C.Y. Chen, B.S. Freer, D.R. Gabbe, H.P. Jenssen, M.A. Kastner, P.J. Picone, N.W. Preyer and R.J. Birgeneau, Phys. Rev. B 41 (1990) 231.

[53] R.T. Collins, Z. Schlesinger, M.W. Schafter and T.R. McGuire, Phys. Rev. B 37 (1988) 5817.

[54] B. Grand and Hk. Müller-Buschbaum, Z. Anorg. Allg. Chem. 417 (1975) 68.

[55] L.L. Miller, X.L. Wang, S.X. Wang, C. Stassis, D.C. Johnston, J. Faber, Jr. and C.-K. Loong, Phys. Rev. B 41 (1990) 1921.

[56] D. Vaknin, S.K. Sinha, C. Stassis, L.L. Miller and D.C. Johnston, Phys. Rev. B 41 (1990) 1926.

[57] T. Siegrist et al., Nature (London) 334 (1988) 231.

[58] D. Vaknin, E. Caignol, P.K. Davies, J.E. Fischer, D.C. Johnston and D.P. Goshorn, Phys. Rev. B 39 (1989) 9122.

[59] Y. Tokura, H. Takagi and S. Uchida, Nature 337 (1989) 345.

[60] H. Takagi, S. Uchida and Y. Tokura, Phys. Rev. Lett. 62 (1989) 1197.

[61] Hk. Müller-Buschbaum and W. Wollschläger, Z. Anorg. Allg. Chem. 414 (1975) 76.

[62] G.M. Luke, B.J. Sternlieb, Y.J. Uemura, J.H. Brewer, R. Kadono, R.F. Kiefl, S.R. Kreitzman, T.M. Riseman, J. Gopalakrishnan, A.W. Sleight, M.A. Subramanian, S. Uchida, H. Takagi and Y. Tokura, Nature 338 (1989) 49.

[63] G.M. Luke, L.P. Le, B.J. Sternlieb, Y.J. Uemura, J.H. Brewer, R. Kadono, R.F. Kiefl, S.R. Kreitzman, T.M. Riseman, C.E. Stronach, M.R. Davis, S. Uchida, H. Takagi, Y. Tokura, Y. Hidaka, T. Murakami, J. Gopalakrishnan, A.W. Sleight, M.A. Subramanian, E.A. Early, J.T. Markert, M.B. Maple and C.L. Seaman, Phys. Rev. B 42 (1990) 7981.

[64] S. Skanthakumar, H. Zhang, T.W. Clinton, W-H. Li, J.W. Lynn, Z. Fisk and S.-W. Cheong, Physica C 160 (1989) 124.

[65] S. Skanthakumar, H. Zhang, T.W. Clinton, I.W. Sumarlin, W-H. Li, J.W. Lynn, Z. Fisk and S-W. Cheong, J. Appl. Phys. 67 (1990) 4530.

[66] T.R. Thurston, M. Matsuda, K. Kakurai, K. Yamada, Y. Endoh, R.J. Birgeneau, P.M. Gehring, Y. Hidaka, M.A. Kastner, T. Murakami and G. Shirane, Phys. Rev. Lett. 65 (1990) 263.

[67] M. Matsuda, K. Yamada, K. Kakurai, H. Kadowaki, T.R. Thurston, Y. Endoh, Y. Hidaka, R.J. Birgeneau, M.A. Kastner, P.M. Gehring, A.H. Moudden and G. Shirane, Phys. Rev. B 42 (1990) 10098.

[68] S. Sugai, T. Kobayashi and J. Akimitsu, Phys. Rev. B 40 (1989) 2686.

[69] P.E. Sulewski, P.A. Fleury, K.B. Lyons, S-W. Cheong and Z. Fisk, Phys. Rev. B 41 (1990) 225.

[70] I. Tomeno, M. Yoshida, K. Ikeda, K. Tai, K. Takamuku, N. Koshizuka and S. Tanaka, Phys. Rev. B 43 (1991) 3009.

[71] T. Siegrist, S. Sunshine, D.W. Murphy, R.J. Cava and S.M. Zahurak, Phys. Rev. B 35 (1987) 7137.

[72] F. Beech, S. Miraglia, A. Santoro and R.S. Roth, Phys. Rev. B 35 (1987) 8778.

[73] M. François, E. Walker, J.-L. Jorda, K. Yvon and P. Fischer, Solid State Commun. 63 (1987) 1149.

[74] S. Katano, S. Funahashi, T. Hatano, A. Matsushita, K. Nakamura, T. Matsumoto and K. Ogawa, Japan. J. Appl. Phys. 26 (1987) L1049.

[75] J.S. Swinnea and H. Steinfink, J. Mater. Res. 2 (1987) 424.

[76] Y. Le Page, T. Siegrist, S.A. Sunshine, L.F. Schneemeyer, D.W. Murphy, S.M. Zahurak, J.V. Waszczak, W.R. McKinnon, J.M. Tarascon, G.W. Hull and L.H. Greene, Phys. Rev. B 36 (1987) 3617.

[77] J.D. Jorgensen, M.A. Beno, D.G. Hinks, L. Soderholm, K.J. Volin, R.L. Hitterman, J.D. Grace, I.K. Schuller, C.U. Segre, K. Zhang and M.S. Kleefisch, Phys. Rev. B 36 (1987) 3608.

[78] J.D. Jorgensen, B.W. Veal, W.K. Kwok, G.W. Crabtree, A. Umezawa, L.J. Nowicki and A.P. Paulikas, Phys. Rev. B 36 (1987) 5731.

[79] D.C. Johnston, A.J. Jacobson J.M. Newsam, J.T. Lewandowski, D.P. Goshorn, D. Xie and W.B. Yelon, Am. Chem. Soc. Symp. Ser. 351 (1987) 136.

[80] C.C. Torardi, E.M. McCarron, M.A. Subramanian, H.S. Horowitz, J.B. Michel, A.W. Sleight and D.E. Cox, Am. Chem. Soc. Symp. Ser. 351 (1987) 152.

[81] N. Nishida et al., Japan. J. Appl. Phys. 26 (1987) L1856.

[82] N. Nishida et al., J. Phys. Soc. Japan 57 (1988) 722.

[83] J.M. Tranquada, D.E. Cox, W. Kunnmann, H. Moud-

den, G. Shirane, M. Suenaga, P. Zolliker, D. Vaknin, S.K. Sinha, M.S. Alvarez, A.J. Jacobson and D.C. Johnston, Phys. Rev. Lett. 60 (1988) 156.
[84] J.M. Tranquada, A.H. Moudden, A.I. Goldman, P. Zolliker, D.E. Cox, G. Shirane, S.K. Sinha, D. Vaknin, D.C. Johnston, M.S. Alvarez, A.J. Jacobson, J.T. Lewandowski and J.M. Newsam, Phys. Rev. B 38 (1988) 2477.
[85] D. Petitgrand and G. Collin, Physica C 153–155 (1988) 192.
[86] P. Burlet, C. Vettier, M.J.G.M. Jurgens, J.Y. Henry, J. Rossat-Mignod, H. Noel, M. Potel, P. Gougeon and J.C. Levet, Physica C 153–155 (1988) 1115.
[87] J. Rossat-Mignod, P. Burlet, M.J. Jurgens, C. Vettier, L.P. Regnault, J.Y. Henry, C. Ayache, L. Forro, H. Noel, M. Potel, P. Gougeon and J.C. Levet, J. de Phys. 49 (1988) C8-2119.
[88] W-H. Li, J.W. Lynn, H.A. Mook, B.C. Sales and Z. Fisk, Phys. Rev. B 37 (1988) 9844.
[89] Y. Yamaguchi, M. Tokumoto, S. Waki, Y. Nakagawa and Y. Kimura, J. Phys. Soc. Japan 58 (1989) 2256; Proc. Tsukuba Seminar on High T_c Superconductivity, Tsukuba, Japan, 31 May–2 June, 1989 (Univ. Tsukuba, Tsukuba, 1989) p. 31.
[90] P. Knoll, C. Thomsen, M. Cardona and P. Murugaraj, Phys. Rev. B 42 (1990) 4842.
[91] K.B. Lyons, P.A. Fleury, L.F. Schneemeyer and J.V. Wasczak, Phys. Rev. Lett. 60 (1988) 732.
[92] M. Sato, S. Shamoto, J.M. Traunquada, G. Shirane and B. Keimer, Phys. Rev. Lett. 61 (1988) 1317.
[93] J.M. Tranquada, G. Shirane, B. Keimer, S. Shamoto and M. Sato, Phys. Rev. B 40 (1989) 4503.
[94] M. Yoshida, N. Koshizuka and S. Tanaka, Phys. Rev. B 42 (1990) 8760.
[95] A.C.W.P. James, S.M. Zahurak and D.W. Murphy, Nature 338 (1989) 240.
[96] J. Sugiyama, Y. Ojima, T. Takata, K. Sakuyama and H. Yamauchi, Physica C (in press).
[97] P.M. Grant, S.S.P. Parkin, V.Y. Lee, E.M. Engler, M.L. Ramirez, J.E. Vazquez, G. Lim, R.D. Jacowitz and R.L. Greene, Phys. Rev. Lett. 58 (1987) 2482.
[98] K. Sekizawa, Y. Takano, H. Takigami, S. Tasaki and T. Inaba, Japan. J. Appl. Phys. 26 (1987) L840.
[99] J. Beille, R. Cabanel, C. Chaillout, B. Chevalier, G. Demazeau, F. Deslandes, J. Etourneau, P. Lejay, C. Michel, J. Provost, B. Raveau, A. Sulpice, J.-L. Tholence and R. Tournier, Compt. Rend. Acad. Sci. Paris 304 (1987) 1097.
[100] J.D. Jorgensen, B. Dabrowski, S. Pei, D.G. Hinks, L. Soderholm, B. Morosin, J.E. Schirber, E.L. Venturini and D.S. Ginley, Phys. Rev. B 38 (1988) 11337, and references cited.
[101] P. Zolliker, D.E. Cox, J.B. Parise, E.M. McCarron III and W.E. Farneth, Phys. Rev. B 42 (1990) 6332.
[102] E.J. Ansaldo, J.H. Brewer, T.M. Riseman, J.E. Schirber, E.L. Venturini, B. Morosin, D.S. Ginley and B. Sternlieb, Phys. Rev. B 40 (1989) 2555.
[103] L.L. Miller, K. Sun, D.C. Johnston, J.E. Schirber and Z. Fisk, J. Less-Common Met. (to be published).
[104] D. Jerome, W. Kang and S.S.P. Parkin, J. Appl. Phys. 63 (1988) 4005.
[105] R. Yoshizaki, N. Ishikawa, M. Adajatsu, J. Fujikami, H. Kurahashi, Y. Saito, Y. Abe and H. Ikeda, Physica C 156 (1988) 297.
[106] H. Takagi, T. Ido, S. Ishibashi, M. Uota, S. Uchida and Y. Tokura, Phys. Rev. B 40 (1989) 2254.
[107] M. Oda, T. Ohguro, N. Yamada and M. Ido, J. Phys. Soc. Japan 58 (1989) 1137.
[108] J.B. Torrance, A. Bezinge, A.I. Nazzal, T.C. Huang, S.S.P. Parkin, D.T. Keane, S.J. LaPlaca, P.M. Horn and G.A. Held, Phys. Rev. B 40 (1989) 8872.
[109] K. Sreedhar and P. Ganguly, Phys. Rev. B 41 (1990) 371.
[110] M. Oda, T. Ohguro, H. Matsuki, N. Yamada and M. Ido, Phys. Rev. B 41 (1990) 2605.
[111] R. Yoshizaki, N. Ishikawa, H. Sawada, E. Kita and A. Tasaki, Physica C 166 (1990) 417.
[112] Y. Ando, M. Sera, S. Yamagata, S. Kondoh, M. Onoda and M. Sato, Solid State Commun. 70 (1989) 303.
[113] J.B. Torrance, Y. Tokura, A.I. Nazzal, A. Bezinge, T.C. Huang and S.S.P. Parkin, Phys. Rev. Lett. 61 (1988) 1127.
[114] R.J. Birgeneau, D.R. Gabbe, H.P. Jenssen, M.A. Kastner, P.J. Picone, T.R. Thurston, G. Shirane, Y. Endoh, M. Sato, K. Yamada, Y. Hidaka, M. Oda, Y. Enomoto, M. Suzuki and T. Murakami, Phys. Rev. B 38 (1988) 6614.
[115] G. Shirane, R.J. Birgeneau, Y. Endoh, P. Gehring, M.A. Kastner, K. Kitazawa, H. Kojima, I. Tanaka, T.R. Thurston and K. Yamada, Phys. Rev. Lett. 63 (1989) 330.
[116] R.J. Birgeneau, Y. Endoh, K. Kakurai, Y. Hidaka, T. Murakami, M.A. Kastner, T.R. Thurston, G. Shirane and K. Yamada, Phys. Rev. B 39 (1989) 2868.
[117] T.R. Thurston, R.J. Birgeneau, M.A. Kastner, N.W. Preyer, G. Shirane, Y. Fujii, K. Yamada, Y. Endoh, K. Kakurai, M. Matsuda, Y. Hidaka and T. Murakami, Phys. Rev. B 40 (1989) 4585.
[118] D.R. Harshman, G. Aeppli, G.P. Espinosa, A.S. Cooper, J.P. Remeika, E.J. Ansaldo, T.M. Riseman, D.Ll. Williams, D.R. Noakes, B. Ellman and T.F. Rosenbaum, Phys. Rev. B 38 (1988) 852.
[119] J.I. Budnick, B. Chamberland, D.P. Yang, Ch. Niedermayer, A. Golnik, E. Recknagel, M. Rossmanith and A. Weidinger, Europhys. Lett. 5 (1988) 651.
[120] B.J. Sternlieb, G.M. Luke, Y.J. Uemura, T.M. Riseman, J.H. Brewer, P.M. Gehring, K. Yamada, Y. Hidaka, T. Murakami, T.R. Thurston and R.J. Birgeneau, Phys. Rev. B 41 (1990) 8866.
[121] H. Kitazawa, K. Katsumata, E. Torikai and K. Nagamine, Solid State Commun. 67 (1988) 1191.
[122] T. Shinjo, T. Mizutani, N. Hosoito, T. Kusuda, T. Takabatake, K. Matsukuma and H. Fujii, Physica C 159 (1989) 869.
[123] A. Weidinger, Ch. Niedermayer, A. Golnik, R. Simon, E. Recknagel, J.I. Budnick, B. Chamberland and C. Baines, Phys. Rev. Lett. 62 (1989) 102.
[124] R.F. Kiefl, J.H. Brewer, J. Carolan, P. Dosanjh, W.N.

Hardy, R. Kadono, J.R. Kempton, R. Krahn, P. Schleger, B.X. Yang, H. Zhou, G.M. Luke, B. Sternlieb, Y.J. Uemura, W.J. Kossler, X.H. Yu, E.J. Ansaldo, H. Takagi, S. Uchida and C.L. Seaman, Phys. Rev. Lett. 63 (1989) 2136.

[125] D.R. Harshman, G. Aeppli, B. Batlogg, G.P. Espinosa, R.J. Cava, A.S. Cooper, L.W. Rupp, E.J. Ansaldo and D.Ll. Williams, Phys. Rev. Lett. 63 (1989) 1187.

[126] R.H. Heffner and D.L. Cox, Phys. Rev. Lett. 63 (1989) 2538.

[127] K. Yoshimura, T. Imai, T. Shimizu, Y. Ueda, K. Kosuge and H. Yasuoka, J. Phys. Soc. Japan 58 (1989) 3057.

[128] H. Yang, X. Zhan, C. Zhu, K. Wang, L. Cao and Z. Chen, Physica C 172 (1990) 71.

[129] T. Fujita, Y. Aoki, Y. Maeno, J. Sakurai, H. Fukuba and H. Fujii, Japan. J. Appl. Phys. 26 (1987) L368.

[130] A.R. Moodenbaugh, Y. Xu, M. Suenaga, T.J. Folkerts and R.N. Shelton, Phys. Rev. B 38 (1988) 4596.

[131] J.D. Axe, A.H. Moudden, D. Hohlwein, D.E. Cox, K.M. Mohanty, A.R. Moodenbaugh and Y. Xu, Phys. Rev. Lett. 62 (1989) 2751.

[132] D.McK. Paul, G. Balakrishnan, N.R. Bernhoeft, W.I.F. David and W.T.A. Harrison, Phys. Rev. Lett. 58 (1987) 1976.

[133] M. Sera, Y. Ando, S. Kondoh, K. Fukuda, M. Sato, I. Watanabe, S. Nakashima and K. Kumagai, Solid State Commun. 69 (1989) 851.

[134] M.K. Crawford, W.E. Farneth, E.M. McCarron, III, R.L. Harlow and A.H. Moudden, Science 250 (1990) 1390.

[135] Y.J. Uemura et al. (unpublished).

[136] J.H. Cho, F. Borsa and D.C. Johnston, unpublished.

[137] R.J. Cava, B. Batlogg, C.H. Chen, E.A. Rietman, S.M. Zahurak and D. Werder, Phys. Rev. B 36 (1987) 5719.

[138] R.J. Cava, B. Batlogg, C.H. Chen, E.A. Rietman, S.M. Zahurak and D. Werder, Nature 329 (1987) 423.

[139] A.J. Jacobson, J.M. Newsam, D.C. Johnston, J.P. Stokes, S. Bhattacharya, J.T. Lewandowski, D.P. Goshorn, M.J. Higgins and M.S. Alvarez, in: Chemistry of Oxide Superconductors, ed. C.N.R. Rao (Blackwell, London, 1988) p. 43.

[140] J.D. Jorgensen, H. Shaked, D.G. Hinks, D. Dabrowski, B.W. Veal, A.P. Paulikas, L.J. Nowicki, G.W. Crabtree, W.K. Kwok, L.H. Nunez and H. Claus, Physica C 153–155 (1988) 578.

[141] A.J. Jacobson, J.M. Newsam, D.C. Johnston, D.P. Goshorn, J.T. Lewandowski and M.S. Alvarez, Phys. Rev. B 39 (1989) 254.

[142] L. Rebelsky, J.M. Tranquada, G. Shirane, Y. Nakazawa and M. Ishikawa, Physica C 160 (1989) 197.

[143] R.J. Cava, A.W. Hewat, E.A. Hewat, B. Batlogg, M. Marezio, K.M. Rabe, J.J. Krajewski, W.F. Peck, Jr. and L.W. Rupp, Jr., Physica C 165 (1990) 419.

[144] J.H. Brewer, E.J. Ansaldo, J.F. Carolan, A.C.D. Chaklader, W.N. Hardy, D.R. Harshman, M.E. Hayden, M. Ishikawa, N. Kaplan, R. Keitel, J. Kempton, R.F. Kiefl, W.J. Kossler, S.R. Kreitzman, A. Kulpa, Y. Kuno, G.M. Luke, H. Miyatake, K. Nagamine, Y. Nakazawa, N. Nishida, K. Nishiyama, S. Ohkuma, T.M. Riseman, G. Roehmer, P. Schleger, D. Schimada, C.E. Stronach, T. Takabatake, Y.J. Uemura, Y. Watanabe, D.Ll. Williams, T. Yamazaki and B. Yang, Phys. Rev. Lett. 60 (1988) 1073.

[145] M.J. Jurgens, P. Burlet, C. Vettier, L.P. Regnault, J.Y. Henry, J. Rossat-Mignod, H. Noel, M. Potel, P. Gougeon and J.C. Levet, Physica B 156 & 157 (1989) 846.

[146] H. Kadowaki, M. Nishi, Y. Yamada, H. Takeya, H. Takei, S.M. Shapiro and G. Shirane, Phys. Rev. B 37 (1988) 7932.

[147] A.H. Moudden, G. Shirane, J.M. Tranquada, R.J. Birgeneau, Y. Endoh, K. Yamada, Y. Hidaka and T. Murakami, Phys. Rev. B 38 (1988) 8720.

[148] J.W. Lynn, W.-H. Li, H.A. Mook, B.C. Sales and Z. Fisk, Phys. Rev. Lett. 60 (1988) 2781.

[149] J.W. Lynn and W-H. Li, J. Appl. Phys. 64 (1988) 6065.

[150] W-H. Li, J.W. Lynn and Z. Fisk, Phys. Rev. B 41 (1990) 4098.

[151] D. Petitgrand and G. Collin, Physica B 156 & 157 (1989) 858.

[152] J.M. Tranquada, W.J.L. Buyers, H. Chou, T.E. Mason, M. Sato, S. Shamoto and G. Shirane, Phys. Rev. Lett. 64 (1990) 800.

[153] G. Shirane, J. Als-Nielsen, M. Nielsen, J.M. Tranquada, H. Chou, S. Shamoto and M. Sato, Phys. Rev. B 41 (1990) 6547.

[154] J. Rossat-Mignod, L.P. Regnault, M.J. Jurgens, C. Vettier, P. Burlet, J.Y. Henry and G. Lapertot, Physica B 163 (1990) 4.

[155] H. Chou, J.M. Tranquada, G. Shirane, T.E. Mason, W.J.L. Buyers, S. Shamoto and M. Sato, Phys. Rev. B 43 (1991) 5554.

[156] P. Bourges, P.M. Gehring, B. Hennion, A.H. Moudden, J.M. Tranquada, G. Shirane, S. Shamoto and M. Sato, Phys. Rev. B 43 (1991) 8690.

[157] P.M. Gehring, J.M. Tranquada, G. Shirane, J.R.D. Copley, R.W. Erwin, M. Sato and S. Shamoto, unpublished.

[158] C.H. Pennington, D.J. Durand, C.P. Slichter, J.P. Rice, E.D. Bukowski and D.M. Ginsberg, Phys. Rev. B 39 (1989) 274.

[159] M. Horvatic, P. Segransan, C. Berthier, Y. Berthier, P. Butaud, J.Y. Henry, M. Couach and J.P. Chaminade, Phys. Rev. B 39 (1989) 7332.

[160] K. Yamada, E. Kudo, Y. Endoh, Y. Hidaka, M. Oda, M. Suzuki and T. Murakami, Solid State Commun. 64 (1987) 753.

[161] Y.J. Uemura, W.J. Kossler, J.R. Kempton, X.H. Yu, H.E. Schone, D. Opie, C.E. Stronach, J.H. Brewer, R.F. Kiefl, S.R. Kreitzman, G.M. Luke, T. Riseman, D.Ll. Williams, E.J. Ansaldo, Y. Endoh, E. Kudo, K. Yamada, D.C. Johnston, M. Alvarez, D.P. Goshorn, Y. Hidaka, M. Oda, Y. Enomoto, M. Suzuki and T. Murakami, Physica C 153–155 (1988) 769.

[162] L.P. Le, G.M. Luke, B.J. Sternlieb, Y.J. Uemura, J.H. Brewer, T.M. Riseman, D.C. Johnston and L.L. Miller, Phys. Rev. B 42 (1990) 2182.

[163] L.P. Le, G.M. Luke, B.J. Sternlieb, Y.J. Uemura, J.H. Brewer, T.M. Riseman, D.C. Johnston, L.L. Miller, Y. Hidaka and H. Murakami, Hyperfine Interactions 63–65 (1991) (in press).

[164] P.P. Freitas, C.C. Tsuei and T.S. Plaskett, Phys. Rev. B 36 (1987) 833.

[165] B. Oh, K. Char, A.D. Kent, M. Naito, M.R. Beasley, T.H. Geballe, R.H. Hammond, A. Kapitulnik and J.M. Graybeal, Phys. Rev. B 37 (1988) 7861.

[166] T.A. Friedman, J.P. Rice, J. Giapintzakis and D.M. Ginsberg, Phys. Rev. B 39 (1989) 4258.

[167] M. Hikita and M. Suzuki, Phys. Rev. B 39 (1989) 4756.

[168] Y. Matsuda, T. Hirai, S. Komiyama, T. Terashima, Y. Bando, K. Iijima, K. Yamamoto and K. Hirata, Phys. Rev. B 40 (1989) 5176.

[169] S.E. Inderhees, M.B. Salamon, N. Goldenfeld, J.P. Rice, B.G. Pazol, D.M. Gisnberg, J.Z. Liu and G.W. Crabtree, Phys. Rev. Lett. 60 (1988) 1178, 2445(E).

[170] T. Taegreid, K. Fossheim, O. Traetteberg, E. Sandvold and S. Julsrud, Physica C 153–155 (1988) 1026.

[171] K. Fossheim, O.M. Nes, T. Laegreid, C.N.W. Darlington, D.A. O'Connor and C.E. Gough, Int. J. Mod. Phys. B 5 (1988) 635.

[172] T.R. McGuire, T.R. Dinger, P.J.P. Freitas, W.J. Gallagher, T.S. Plaskett, R.L. Sandstrom and T.M. Shaw, Phys. Rev. B 36 (1987) 4032.

[173] K. Kanoda, T. Kawagoe, M. Hasumi, T. Takahashi, S. Kagoshima and T. Mizoguchi, J. Phys. Soc. Japan 57 (1988) 1554.

[174] K. Kanoda, T. Takahashi, T. Kawagoe, T. Mizoguchi, M. Hasumi and S. Kagoshima, Physica C 153–155 (1988) 749.

[175] W.C. Lee, R.A. Klemm and D.C. Johnston, Phys. Rev. Lett. 63 (1989) 1012.

[176] W.C. Lee and D.C. Johnston, Phys. Rev. B 41 (1990) 1904.

[177] U. Welp et al., Phys. Rev. Lett. 62 (1989) 1908.

[178] D.C. Johnston and J.H. Cho, Phys. Rev. B 42 (1990) 8710.

[179] W.C. Lee, J.H. Cho and D.C. Johnston, Phys. Rev. B 43 (1991) 457.

[180] M. Takigawa, P.C. Hammel, R.H. Heffner, Z. Fisk, J.L. Smith and R.B. Schwarz, Phys. Rev. B 39 (1989) 300.

[181] S.E. Barrett, D.J. Durand, C.H. Pennington, C.P. Slichter, T.A. Friedmann, J.P. Rice and D.M. Ginsberg, Phys. Rev. B 41 (1990) 6283.

[182] R.E. Walstedt and W.W. Warren, Jr., Science 248 (1990) 1082.

Ab-initio calculations of the electronic structure of impurities and alloys of ferromagnetic transition metals

P.H. Dederichs, R. Zeller

Institut für Festkörperforschung, Forschungszentrum Jülich, Postfach 1913, W-5170 Jülich, Germany

H. Akai

Department of Physics, Nara Medical University, Kashihara, Nara 634, Japan

and

H. Ebert

Siemens AG, Research Laboratories, ZFE ME TPH11, W-8520 Erlangen, Germany

We review recent theoretical work on the electronic structure and the magnetic properties of ferromagnetic transition-metal alloys. All calculations are based on density-functional theory in the local-spin-density approximation. We report about calculations for dilute alloys using the KKR-Green's function method and for concentrated disordered alloys using the charge-self-consistent KKR-CPA method.

1. Introduction

Ferromagnetic alloys of iron, cobalt and nickel show a fascinating richness in their magnetic behavior. The magnetization of the metals can change drastically upon the addition of impurities. For transition-metal impurities the local moments can align parallel or antiparallel to the host moments. Concentrated alloys of cobalt and nickel show a very similar behavior as the corresponding dilute alloys, whereas the properties of Fe alloys are normally characterized by pronounced non-linear concentration dependence. Due to their magnetic properties there exist a huge amount of experimental data on these alloys, presumably more than on any other alloy system.

The early theoretical work of Friedel [1], Kanamori [2], Campbell and Gomes [3] and others has lead to a basic understanding of these alloys. Most important is the characterization of the alloys as either "strong" or "weak" ferromagnets. Strong ferromagnets have a completely filled majority d-band. Due to this their magnetization varies oppositely to the average valence of alloy. Thus by plotting the magnetization versus the electron-to-atom ratio these alloys all fall on the same straight line which represents the main branch of the Slater–Pauling curve [4,5]. For the weak ferromagnets the majority band is no longer completely filled, so that these alloys form the side branches of the Slater–Pauling curves. Whereas the above theoretical work was focussed on the dilute alloys, tight-binding CPA calculations e.g. by Hasegawa and Kanamori [6] have also given a basic understanding of the concentrated alloys.

The recent development of density-functional theory has lead to a breakthrough in our understanding of electronic structure and the magnetic properties of transition metals. We are now able to calculate the electronic and magnetic properties of these alloys fully self-consistent and without any adjustable parameters. The precision of the calculation is in many cases comparable or superior to the experiment and leads to an insight into the microscopic mechanism which cannot be obtained by any other method. In this paper we will give a review of the theoretical work on the electronic structure of the ferromagnetic alloys, especially the Fe- and Ni-alloys. Up to now very little work has been done for the Co-alloys, the behavior of which is known to be similar to the Ni alloys. We discuss both dilute and concentrated disordered alloys. Calculations for special ordered structures which can be obtained by usual band structure methods, will not be considered.

The organisation of the paper is as follows. In section 2 we shortly describe the theoretical methods used in the calculations, i.e. the KKR-Green's function method for impurity calculations and the KKR-CPA method for concentrated alloys. We discuss results obtained by the KKR-Green's function method for impurities in section 3 and for the host perturbations induced by the impurities in section 4. Using the 5d-impurities in Fe as an example we discuss in section 5 the importance of relativistic effects, leading e.g. to orbital moments and orbital hyperfine fields. In section 6 we present recent charge self-consistent KKR-CPA results for concentrated alloys, culminating in a very detailed Slater–Pauling curve as calculated from first principles. Also results for the lattice expansion of Ni alloys are given. In addition embedded cluster calculations studying local-environment effects are presented.

2. Calculational methods

For the theoretical description we apply density-functional theory (DFT) in the local-spin-density (LSD) approximation. For most calculations we use the LSD form given by Von Barth and Hedin [7] with the parameters as modified by Moruzzi, Janak and Williams [8]. We also checked the sensitivity of the results with respect to the more recent form given by Vosko, Wilk and Nusair [9] and found only minor changes except for critical cases like Mn impurities in Fe (see below). To solve the DFT one-particle equations we use multiple-scattering theory, i.e. the Korringa–Kohn–Rostocker Green's function (KKR-GF) method for the dilute impurity limit and the Korringa–Kohn–Rostocker coherent-potential approximation (KKR-CPA) for concentrated alloys.

2.1. KKR-GF method

The central quantities of spin-density-functional theory are the electronic charge density $n(r)$ and the magnetization density $m(r)$. They are connected to the spin densities for majority-spin ($+$) and minority-spin ($-$) electrons by $n(r) = n_+(r) + n_-(r)$ and $m(r) = n_+(r) - n_-(r)$. We determine the spin densities from the one-particle Green's function $G_\pm(r, r', E)$ as

$$n_\pm(r) = \sum_\nu |\phi_{\nu\pm}(r)|^2$$
$$-\frac{1}{\pi}\int_{E_{\text{bot}}}^{E_F} \text{Im}\, G_\pm(r, r, E)\, dE \quad (1)$$

by energy integration from the bottom E_{bot} of the valence band up to the Fermi level E_F. In (1) we treat the core states $\phi_{\nu\pm}(r)$ as an atomic problem. In a cell-centered representation the angular-momentum expansion for the Green's function can be written as [10]

$$G_\pm(r + R^n, r' + R^{n'}, E)$$
$$= \delta_{nn'} G^n_{S\pm}(r + R^n, r' + R^n, E)$$
$$+ \sum_{LL'} R^n_{L\pm}(r, E) G^{nn'}_{LL'\pm}(E) R^{n'}_{L'\pm}(r', E),$$
$$(2)$$

where the vectors r and r' are restricted to the Wigner–Seitz cells around the atomic positions R^n and $R^{n'}$. The first term $G^n_{S\pm}$ of (2) represents the Green's function for the spin-dependent single-cell potential $V^n_\pm(r)$ which vanishes outside the Wigner–Seitz cell Ω_n around the position R^n. This term is related to the free-space Green's function $g(r, r', E) = -(1/4\pi)\exp(i\sqrt{E}|r - r'|)/|r - r'|$ by an integral equation [10] with the integral restricted to cell Ω_n. The second term of

(2) represents the so-called back-scattering contribution. It contains the structural Green's function matrix $G_{LL'\pm}^{nn'}(E)$ which is related to its free-space counterpart $g_{LL'}^{nn'}(E)$ by an algebraic Dyson equation

$$G_{LL'\pm}^{nn'}(E) = g_{LL'}^{nn'}(E) + \sum_{n''L''L'''} g_{LL''}^{nn''}(E) t_{L''L'''\pm}^{n''}(E) G_{L'''L'\pm}^{n''n'}(E). \quad (3)$$

The t matrices are defined as

$$t_{LL'\pm}^{n}(E) = \int_{\Omega_n} J_L(r, E) V_\pm^n(r) R_{L'\pm}^n(r, E) \, dr, \quad (4)$$

where J_L denote products of spherical Bessel functions and spherical harmonics [10]. The single-cell wavefunction $R_{L\pm}^n(r, E)$ can be calculated from the integral equation

$$R_{L\pm}^n(r, E) = J_L(r, E) + \int_{\Omega_n} g(r, r', E) V_\pm^n(r') R_{L\pm}^n(r', E) \, dr'. \quad (5)$$

The Dyson equation (3) is solved in two steps by the intermediate use of the Green's function $\mathring{G}_\pm$ for the ideal periodic crystal. The Dyson equation which connects $\mathring{G}_\pm$ and g can be solved by Fourier transformation or alternatively by calculating the imaginary part of $\mathring{G}_\pm$ similarly to density-of-states integrals and the real part by a Kramers–Kronig integration [11]. The remaining Dyson equation, with $g_{LL'}^{nn'}(E)$ replaced by $\mathring{G}_{LL'\pm}^{nn'}(E)$ and with the t matrices $t_{LL'\pm}^{n}(E)$ replaced by their changes $\Delta t_{LL'\pm}^{n}(E)$ with respect to the ideal crystal,

$$G_{LL'\pm}^{nn'}(E) = \mathring{G}_{LL'\pm}^{nn'}(E) + \sum_{n''L''L'''} \mathring{G}_{LL''\pm}^{nn''}(E) \Delta t_{L''L'''\pm}^{n''}(E) G_{L'''L'\pm}^{n''n'}(E) \quad (6)$$

requires to solve a system of linear equations of finite dimensions determined by the number of perturbed atoms and angular momenta taken into account. Typically we use angular momenta up to $l = 3$ and several shells of perturbed host potentials around the impurity site. The efficiency of the KKR-GF method is greatly enhanced (i) by using group theory, which allows a decomposition of the structural Green's function matrices into irreducible submatrices, and (ii) by transforming the integral (1) along the real energy axis into a contour integral in the complex energy plane, where the Green's function becomes smooth and structureless. Therefore the integral can be integrated with a small number of energy mesh points. For more details we refer to ref. [11]. The formulation in this section is valid for potentials of arbitrary shape and not restricted to muffin-tin (MT) potentials. The previously applied MT approximation [11] or atomic-sphere approximation (ASA) make the t matrices diagonal in angular momentum and simplify the determination of the wavefunctions $R_L^n(r, E)$ which become products of solutions of the radial Schrödinger equation times spherical harmonics. Our full-potential KKR-GF programs are only just now developed so that all results given here have been obtained with the MT or ASA approximations.

2.2. KKR-CPA

The above description of the electronic structure by the corresponding Green's function within the KKR-GF formalism is an excellent starting point not only for impurity problems or dilute alloys, but also for concentrated ones. The philosophy behind more or less all alloy theories which has to be added to the KKR-GF formalism for that purpose, is to find a Green's function which represents the configuration average for the disordered alloy under consideration. If one ignores any correlation in the occupation of the various lattice sites, i.e. restricts oneself to random alloys, the best single-site theory at hand is the Coherent Potential Approximation (CPA) suggested by Soven [12]. The hypothetical CPA medium representing the binary alloy A_xB_{1-x} is found by the condition that the embedding of an A- or B-atom into this medium – weighted by the corresponding concentration x or $1 - x$, respectively – should not cause any additional scattering. This can be expressed very conveniently by means of the single site t-matrices t^α and the so-called scattering-path operators $\tau^\alpha (\alpha = A, B, CPA)$:

$$\tau^{CPA}(E) = x\tau^A(E) + (1-x)\tau^B(E), \quad (7)$$

$$= \frac{1}{\Omega_{BZ}} \times \int_{\Omega_{BZ}} d^3k \left[[t^{CPA}(E)]^{-1} - G(k, E) \right]^{-1}, \quad (8)$$

$$\tau^{A(B)}(E) = \left[(t^{A(B)})^{-1} - (t^{CPA})^{-1} + (\tau^{CPA})^{-1} \right]^{-1}. \quad (9)$$

Here $G(k, E)$ is the Fourier transform of the free-space Green's function. Since in eqs. (7)–(9) only site-diagonal scattering-path operators occur, the site indices n, n' – and also the angular-momentum indices L, L' as well as the spin indices – have been suppressed. The scattering operator $\tau^{nn'}_{LL'}(E)$ commonly used within the KKR-CPA formalism is linked to the KKR-Green's function $G^{nn'}_{LL'}(E)$ of the previous section by the relationship

$$\tau = tGt + t. \quad (10)$$

Because of the various numerical problems in solving the KKR-CPA equations (7)–(9), this formalism could be implemented only about 12 years ago by Stocks et al. [13]. However, in the meantime very efficient computer codes have been developed making charge self-consistent calculations possible [14]. In addition, corresponding relativistic [15] and spin-polarized relativistic [16] (see section 5) versions of the KKR-CPA formalism have been derived.

3. 3d and 4d impurities in Ni and Fe

In this section we discuss the local magnetic properties of transition-metal impurities in Ni and Fe. As results we present local densities of states (LDOS), local magnetic moments and hyperfine fields.

3.1. LDOS

Within the impurity cell Ω_n the LDOS are easily calculated from the Green's function as

$$n_\pm(E) = -\frac{1}{\pi} \int_{\Omega_n} \text{Im } G_\pm(r, r, E) \, dr. \quad (11)$$

As typical examples we show the LDOS of V and Fe impurities in Ni (fig. 1) and of V and Ni impurities in Fe (fig. 2), where for calculational convenience we have replaced the integration volume by a sphere of equal volume. Similar pictures for all 3d and 4d impurities in Ni and for 3d impurities in Fe are given in refs. [11,17].

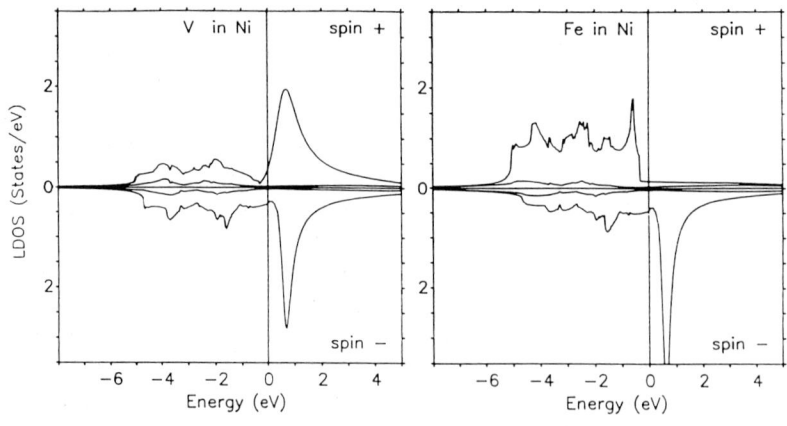

Fig. 1. Local density of states for V and Fe impurities in Ni. Two curves are shown for each spin direction. One gives the sum of s, p and f character, the other the total sum including the d character. The energies are given relative to the Fermi level.

The majority-spin LDOS for the early 3d impurities Sc to Cr in Ni and Fe are always characterized by Lorentzian-type virtual bound states (VBS) above the host d bands, as shown for the V impurities in figs. 1 and 2. These VBS arise from the hybridization of atomic-like 3d levels of the impurities with the host conduction electrons of s and p character. With increasing atomic number, from Sc to Cr, the VBS move towards the Fermi level and become narrower. Whereas the behavior of the minority-spin LDOS for the early impurities in Ni is similar to the one of the majority-spin LDOS, the situation for the minority-spin LDOS for these impurities in Fe is more complex. Since the energy position of the 3d levels of the impurities overlap with the minority-spin d band of the host Fe atoms, large hybridization effects occur. This leads to rather structured minority-spin LDOS. A typical example is shown in fig. 2 for the V impurity in Fe.

An important feature of the majority-spin LDOS of Fe and Co impurities in Ni is the small change, both in shape and magnitude, of the impurity LDOS compared to the host LDOS of pure Ni. The LDOS for Fe in Ni, shown in fig. 1, is almost identical to the one of pure Ni [8,11]. This means that the majority-spin electrons in Ni are almost unperturbed by Fe and Co impurities and leads, e.g., to the very low residual resistivities of these impurities in Ni. A similar, but not so pronounced trend for the majority-spin LDOS is also found for the late impurities Co and Ni in Fe.

The minority-spin LDOS for Fe and Co in Ni and for Co in Fe show more or less developed VBS above E_F, but also a considerable amount of states below E_F in the energy range of the host d bands. For Ni in Fe the minority-spin VBS just reaches the Fermi level which leads to a large positive contribution to the relative change $\Delta\gamma/\gamma$ of the electronic specific heat [17].

A very interesting feature of the transition-metal impurities is that the late transition-metal impurities have local moments being parallel to the host moment whereas the local moments of the early transition-metal impurities are antiparallel to the host moments. As firstly explained by Friedel [1–3], the transition from ferromagnetic to antiferromagnetic coupling occurs, when the virtual bound state in the majority band crosses the Fermi level, since then five majority d-states become unoccupied. Therefore locally more minority states are occupied to achieve charge neutrality and consequently the impurity moments are aligned oppositely to the host moments. The transition from this antiferromagnetic to ferromagnetic behavior occurs nearly at the atomic number $Z = 25$, i.e. for Mn impurities. This makes calculations for Mn impurities sensitive to numerical approximation and the results depend on whether angular momenta up to $l = 2$ or 3 are used and on the particular form of the LSD

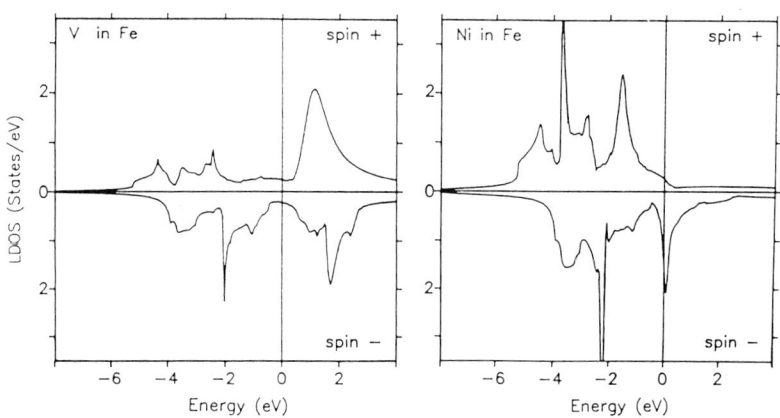

Fig. 2. Local density of states for V and Ni impurities in Fe. The energies are given relative to the Fermi level.

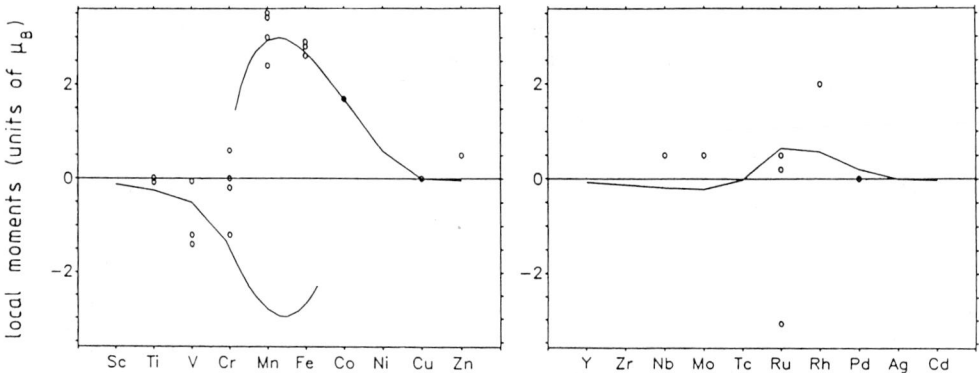

Fig. 3. Calculated local moments for 3d and 4d impurities in Ni (full curves) and experimental values (open circles). For experimental references see ref. [11].

approximation. For a detailed discussion we refer to ref. [17].

3.2. Local moments

By integrating the calculated spin densities over the impurity Wigner–Seitz spheres or by integrating the LDOS up to the Fermi level we obtain the local impurity moments as shown in figs. 3 and 4. The moments mainly arise from polarization of d states whereas s, p and f polarizations add only a small negative contribution (see table 1 of ref. [11]). The calculated moments for the 4d impurities are in general much smaller than the ones for the 3d impurities. The agreement with the experimental data, also shown in figs. 3 and 4, is mostly good. For a discussion of the discrepancies and the experimental problems we refer to refs. [11,17]. Particularly, the experimental Rh moment of $2\mu_B$ seems to be unreasonably large. A point worth studying is the behavior in the middle of the transition series where the moments switch from an antiparallel to a parallel alignment with the host magnetization. For this study we used non-integer atomic numbers and obtained two stable solutions for impurities in Ni between $Z = 24.2$ and 26.5 shown by the rounded curves in fig. 3. The transition range

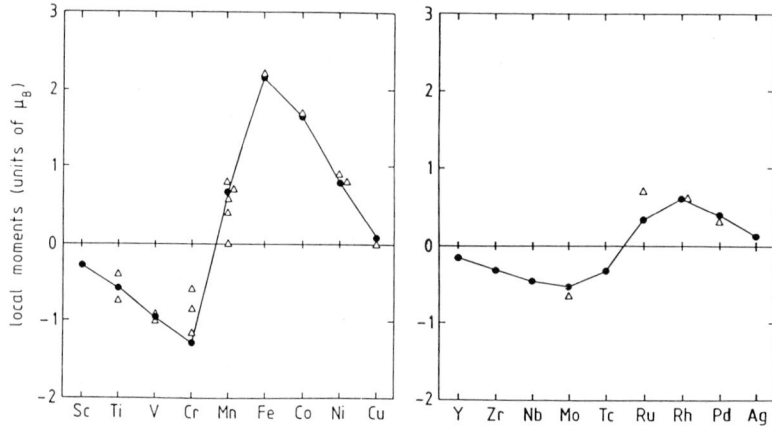

Fig. 4. Calculated local moments for 3d and 4d impurities in Fe (full curves) and experimental values (open triangles). For experimental references see ref. [17].

for 3d impurities in Fe is rather small and occurs near Z = 25. The best state of the art calculations show the transition to be of second order and to occur for nuclear charges slightly smaller than 25, so that for Mn a positive local moment of $0.69\mu_B$ is obtained in good agreement with experiment. For 4d impurities always only one stable solution was found. For more details see refs. [11,17].

3.3. Hyperfine fields of the impurities

A general discussion of the calculation of hyperfine interactions in metals is given in ref. [18]. Neglecting the spin–orbit interaction the hyperfine field for impurities with cubic symmetry (as the ones presented here) is dominantly given by the Fermi contact term

$$H_{hf} = \tfrac{8}{3}\pi\mu_B m(0), \qquad (12)$$

where $m(0)$ denotes the magnetization at the nuclear position. Formula (12) is only valid for a non-relativistic treatment. An often found misconception is to assume that it also holds in relativistic calculations, with the modification that $m(0)$ is replaced by the average of $m(r)$ over the nuclear volume. This is wrong and leads to a considerable overestimation of the relativistic contribution to the hyperfine fields. Instead one must replace Fermi's contact formula (12) by Breit's relativistic generalization. As it is demonstrated in ref. [19] the correct relativistic contact interaction field is given by

$$H_{hf} = \tfrac{8}{3}\pi\mu_B m_T, \qquad (13)$$

where m_T denotes the average of the relativistic magnetization over the Thomson radius $r_T = Ze^2/mc^2$ which is always much larger than the nuclear radius. This leads to much smaller relativistic enhancements of the hyperfine fields since the relativistic spin-density changes decrease with increasing distance from the nucleus. Fig. 5 shows calculated hyperfine fields for 3d and 4d impurities in Ni in comparison with the experimental data. The hyperfine fields are strongly negative for the ferromagnetically aligned impurities (Mn, Fe, Co and Ru, Rh, Pd) and moderately negative for the impurities with negative local moments (Sc, Ti, V, Cr and Y, Zr, Nb, Mo, Tc). The values for the antiferromagnetic impurities agree well with the experiments, however there are serious disagreements for the ferromagnetic impurities. Particularly, the calculated values for Mn and Fe, the impurities with the largest moments, are far too small. In the 4d series the values for Rh and Ru are also somewhat too small. The relativistic corrections are typically 10 and 30% for the ferromagnetic impurities and rather small for antiferromagnetic ones.

Figs. 6 and 7 show the corresponding results for the hyperfine fields of 3d and 4d impurities in

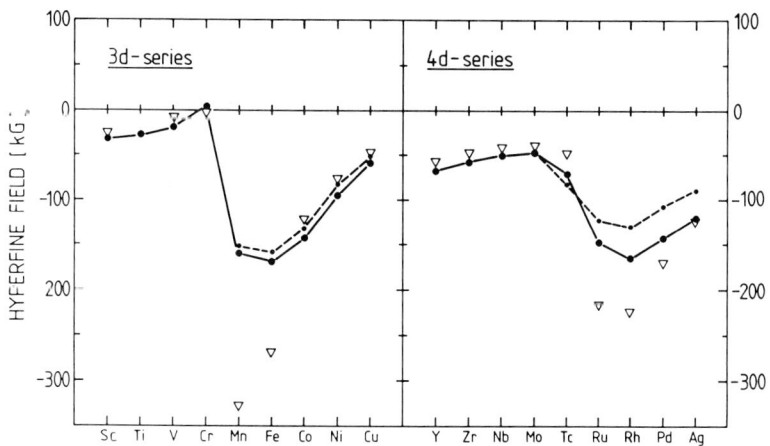

Fig. 5. Calculated hyperfine fields for 3d and 4d impurities in Ni. Non-relativistic (full curves) and semirelativistic (broken curves) values are given together with experimental data (open triangles). For experimental references see ref. [19].

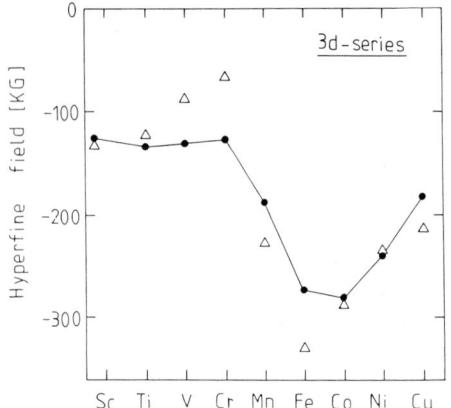

Fig. 6. Calculated hyperfine fields for 3d impurities in Fe (full curves) and experimental values (open triangles, for references see ref. [18]).

Fe. The general trend is similar as for impurities in Ni. The hyperfine fields are moderately negative for the antiferromagnetic 3d impurities Sc, Ti, V, Cr and strongly negative for pure Fe and the 3d impurities Co and Ni while the hyperfine field for Mn impurities is intermediate. There are again some serious deviations from experiment, particularly for pure Fe. The trend for the 3d impurities in Fe is similar as in Ni, good agreement for the antiferromagnetic impurities and some underestimation for the ferromagnetic ones. In our opinion the discrepancies are due to a failure of the local spin density approximation.

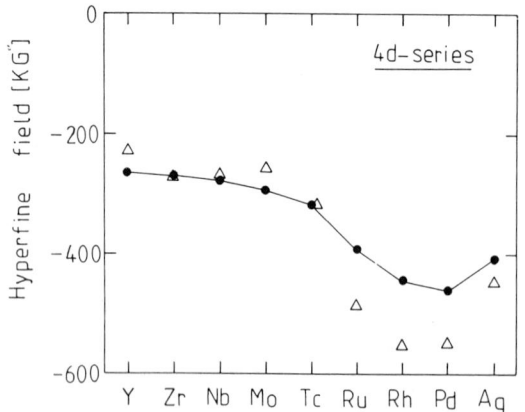

Fig. 7. Calculated hyperfine fields for 4d impurities in Fe (full curves) and experimental values (open triangles, for references see ref. [18]).

Especially the exchange interaction between the spin-polarized d-shell and the local core and valence s-electrons is apparently not described correctly. For more details we refer to refs. [18,19].

4. Host polarizations

For the understanding of the behavior of the ferromagnetic alloys, the changes of the host properties induced by the impurities are just as important as the properties of the impurities themselves. Therefore a calculation of the host disturbances around the impurities is essential for the understanding of the dilute alloys. 3d impurities in Ni are a very simple model case since they follow the behavior predicted by Friedel [1] and others [2,3] more or less exactly. The total change ΔM of the magnetization per impurity atom

$$\Delta M = \sum_n \delta M^n = \frac{dM}{dc}\bigg|_{c \to 0} \qquad (14)$$

can be easily measured from the concentration dependence of the alloy magnetization $M(c)$ in the dilute limit $c \to 0$. ΔM and the change ΔZ of the nuclear charge are directly related to the changes $\Delta N^\uparrow(E_F)$ and $\Delta N^\downarrow(E_F)$ of the band populations of the majority and minority bands.

$$\Delta Z = \Delta N^\uparrow(E_F) + \Delta N^\downarrow(E_F),$$
$$\Delta M = \Delta N^\uparrow(E_F) - \Delta N^\downarrow(E_F). \qquad (15)$$

In a simple tight-binding model with d-electrons only the majority d-band of Ni is completely filled. Therefore due to transition-metal impurities $\Delta N^\uparrow(E_F)$ can only change, if a virtual bound state of d-character splits off from the majority band and passes through E_F. Thus we have either a filled majority band with $\Delta N^\uparrow(E_F) = 0$ ("strong ferromagnet") or we have $\Delta N^\uparrow(E_F) = -5$ if the five-fold degenerate majority virtual bound state is located above E_F. For the magnetization changes we obtain from (15) either

$$\Delta M = -\Delta Z \quad \text{or} \quad \Delta M = -10 - \Delta Z. \qquad (16)$$

Clearly these simplifications arise due to the fact that we have only considered d-electrons. The

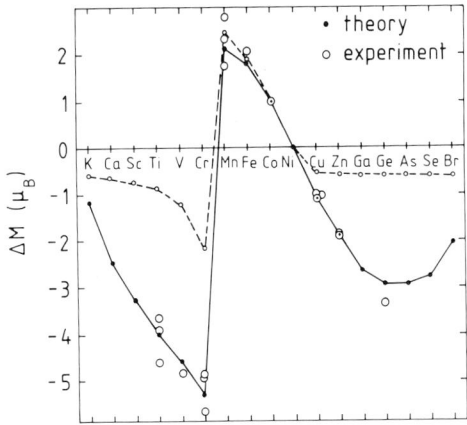

Fig. 8. Change ΔM of the magnetization per impurity atom for dilute Ni alloys with impurities of the third row. The experimental data are corrected for orbital contributions, so that only spin moments are compared. The broken line refers to the contribution to ΔM from the impurity site. For reference to the experimental data see ref. [20].

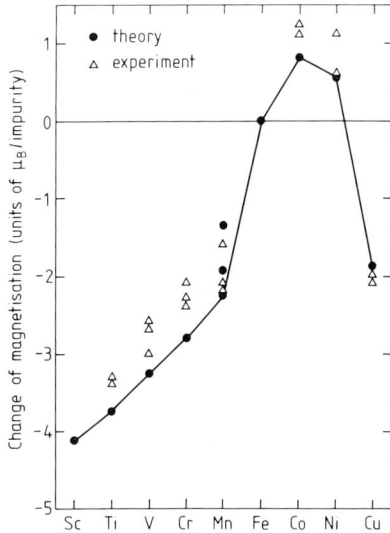

Fig. 10. Change ΔM of the magnetization per impurity atom for dilute Fe alloys with 3d-impurities. The triangles are results of magnetization measurements (for references see ref. [17]).

sp-hybridization leads to an important broadening of the virtual bound states, so that the transition from $\Delta N^\uparrow(E_F) = 0$ to $\Delta N^\downarrow(E_F) = -5$ should be more gradual. Figs. 8 and 9 show the results of the calculations for 3d- and 4sp-impurities in Ni [20] where the changes ΔM, $\Delta N^\uparrow(E_F)$ and $\Delta N^\downarrow(E_F)$ have been obtained by summing up the local changes up to the fourth shell. One indeed finds that the simple minded predictions of the tight binding model are well obeyed. The majority band is filled for Co, Fe and Mn impurities as well as for the early 4sp-impurities and the

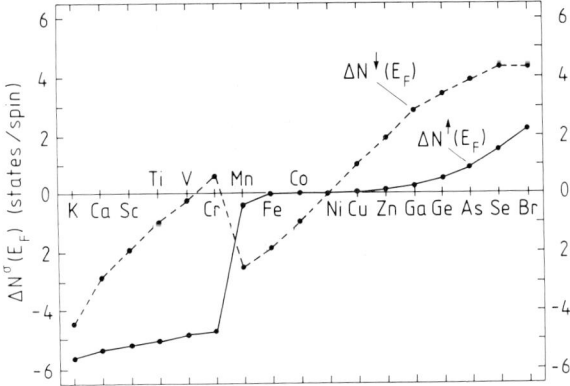

Fig. 9. Changes $\Delta N^\uparrow(E_F)$ and $\Delta N^\downarrow(E_F)$ of the majority and minority band populations for third row impurities in Ni.

minority band alone does all the screening. The change of the total magnetization follows nearly the straight line $\Delta M = -\Delta Z$. Contrary for the early transition-metal impurities Sc...Cr, the relation $\Delta N^\uparrow(E_F) = -5$ is reasonably well satisfied and the magnetization changes as $\Delta M \approx -10 - \Delta Z$. For a more detailed discussion we refer to ref. [20]. From fig. 8 it is seen that the agreement with the experimental data is excellent.

For the 4d- and 5d-impurities the above simple tight-binding model does not give a good description of the magnetic properties [20]. This is basically due to the much stronger hybridization resulting in very broad virtual bound states which are always partially occupied and partially empty.

In contrast to a similar behavior of the impurity moments in Fe and Ni (see figs. 3 and 4), the perturbations of the host moments show a very different trend, which is basically due to the fact that Ni is a strong ferromagnet with a completely filled majority band, whereas in the weak ferromagnet Fe about 0.2 to 0.3 majority d states are not occupied. Fig. 10 shows the changes ΔM of the magnetization per impurity atom calculated for 3d impurities in Fe together with experimental data [17]. Directly apparent is the different behavior for the early 3d impurities. Whereas in Ni the absolute value of ΔM is smallest at the beginning of the series and then increases up to Cr, in Fe the ΔM values are largest for Sc and then continuously decrease. Also the behavior of the late transition-metal impurities is very different. Whereas for Fe and Co impurities in Ni there is practically no host disturbance, in Fe one observes for Co and Ni impurities an enhancement of the neighboring moments being largest for the nearest neighbors but extending at least up to the 4th shell. The different behavior of dilute Fe and Ni alloys is even more clearly shown by the population changes $\Delta N^{\pm}(E_F)$ of both subbands which are given in fig. 11 for 3d impurities in Fe. It is seen that in Fe the majority band provides most of the screening whereas the changes ΔN^- in the minority band are rather small. Thus qualitatively speaking in Fe the role of both subbands is exchanged compared to Ni or Co. The reason for this behavior has been discussed by Malozemoff et al. [21]. The Fermi level of Fe falls into the minimum of the minority density of states, which acts like a pseudogap and tries to conserve the population of the minority band. For more details, including the results of 4d impurities in Fe, see ref. [17].

Since Fe is the most widely used Mössbauer atom, many experimental studies of host perturbations in dilute Fe alloys have been performed.

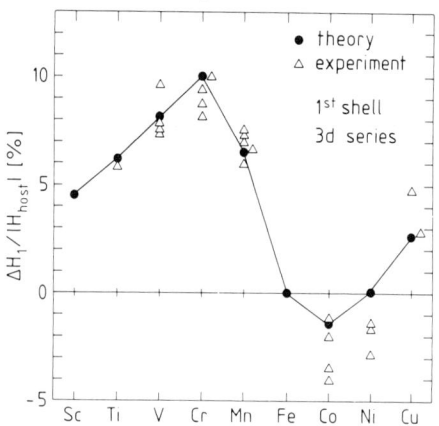

Fig. 12. Calculated change ΔH_1 of the hyperfine fields of Fe atoms in the first shell around 3d-impurities in Fe. The triangles denote experimental values.

The experiments are usually analysed in a two-shell model, i.e. by assuming that only the first two Fe neighbors around the impurity are strongly perturbed and yield resolvable satellites. Moreover, the contributions from different impurities are assumed to superimpose according to a random distribution.

In fig. 12 we give the calculated changes ΔH_1 for the hyperfine fields of the first Fe neighbors around 3d impurities. Given are the changes ΔH_1 in percent of the absolute values of the host hyperfine fields, i.e. $\Delta H_1 / |H_{\text{host}}| \times 10^2$. Positive (negative) values of ΔH_1 mean that the hyperfine field of the first neighbor is smaller (larger) than the unperturbed host field, which is negative. For discussion of the field of the second and further away neighbors we refer to ref. [22]. For the 3d-series the hyperfine field ΔH_1 is positive and strongly increasing in the series Sc to Cr. The field then slightly drops for Mn. For Co the first neighbor field is negative, for Ni it is practically zero and positive for Cu. The second neighbors show a similar trend, with ΔH_2 being rather large and positive for Sc, Ti, V and Cr, but rather small for Mn, Co and Ni. More details about these fields and similar results for 4d impurities in Fe are given in ref. [22]. As can be seen from fig. 12 the calculated fields ΔH_1 agree very well with experiment, at least for the early transition metals up to Mn. The behavior for Co and Ni

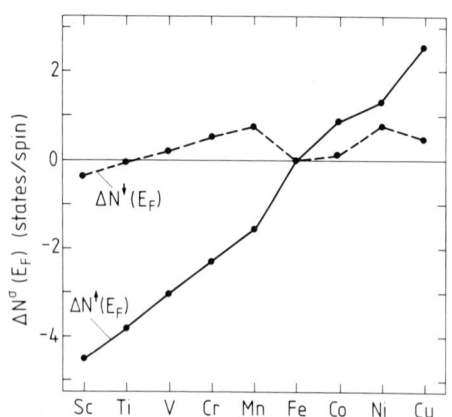

Fig. 11. Changes $\Delta N^+(E_F)$ and $\Delta N^-(E_F)$ of the majority and minority band populations for 3d-impurities in Fe.

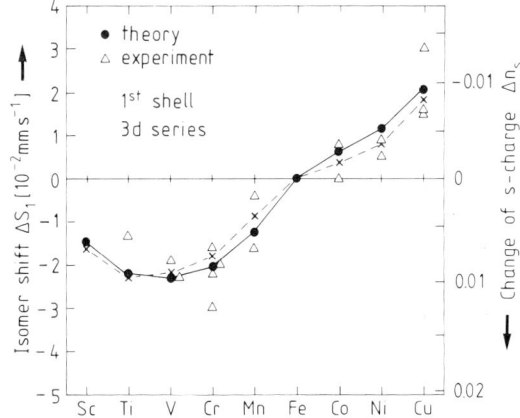

Fig. 13. Calculated isomer shifts of Fe atoms being nearest neighbors of 3d-impurities in Fe. The triangles are experimental values. The dashed line refers to the right inverted scale and gives the changes Δn_s of the number of s-electrons of the n.n. Fe atom.

impurities is more problematic and we refer to ref. [22] for a detailed discussion.

An interesting aspect of the Mössbauer effect is, that in addition to the hyperfine field as given by the magnetization density at the nucleus it allows also a determination of the isomer shift, i.e. the change of the charge density at the Fe nucleus. Extensive calculations of the isomer shifts of nearest-neighbor Fe atoms in dilute Fe alloys have been reported by Akai et al. [23]. By varying the nuclear charge of the impurity across the periodic table from $Z = 0$ (vacancy) to $Z = 52$ (Ba) a very systematic behavior is detected. The isomer shift of the n.n. Fe atom is small or negative at the beginning of each period, than continuously increases towards a maximum at the end of the period where it sharply decreases. More refined calculations for 3d impurities in Fe [22] are shown in fig. 13 together with experimental data. In general the agreement is good (see also refs. [18,23]). The calculations clearly show that all the trends of the isomer shifts in Fe alloys are essentially determined by the number of s-electrons of the n.n. Fe atom and that changes of the form of the s-wavefunctions are not very important. For a discussion of such polarization effects due to intra-atomic screening and their different significance for atoms and the considered alloys we refer to ref. [23]. An important conclusion of this work is that the isomer shift is in general not determined by the charge transfer. This is especially clear from the calculations for the sp-impurities in Fe [23] showing that the s-, p- and d-charges of the nearest neighbor Fe atoms exhibit a very different trend. For instance, for Ge, As and Se impurities the d-charge of the n.n. Fe atom strongly increases and essentially determines the charge transfer while the p- and s-changes are smaller and opposite in sign. It is only for transition-metal impurities, that all three charge contributions (s, p, d) exhibit similar trends, so that only for such systems the isomer shift can be related to the charge transfer.

5. Relativistic effects: 5d impurities in Fe

In the calculations presented in sections 3 and 4 relativistic effects have partly been incorporated in the scalar–relativistic approximation which only includes the Darwin and mass–velocity terms. Even for 3d- and 4d-transition metals the inclusion of these corrections has a noticeable influence on the magnetic properties and especially on the hyperfine interactions. However, this approach is leaving the spin as a good quantum number and therefore fails to describe phenomena that can be traced back to the symmetry-breaking property of the spin–orbit coupling. In order to be able to study such consequences of the spin–orbit coupling in magnetic materials, as for example, orbital contributions to the magnetic moments or hyperfine fields, the multiple-scattering formalism has recently been generalized accordingly within the framework of local-spin-density theory. This spin-polarized relativistic formalism (SPRKKR) is very similar to the treatment of non-spherical potentials mentioned briefly in section 2.1. The reason for this is that the necessarily spin-dependent potential, that enters the Dirac equation and accounts for the spin-polarization of the system, breaks the symmetry in spin-space. For this reason the resulting expression for the Green's function is completely analogous to eq. (2) (see e.g. ref. [24]).

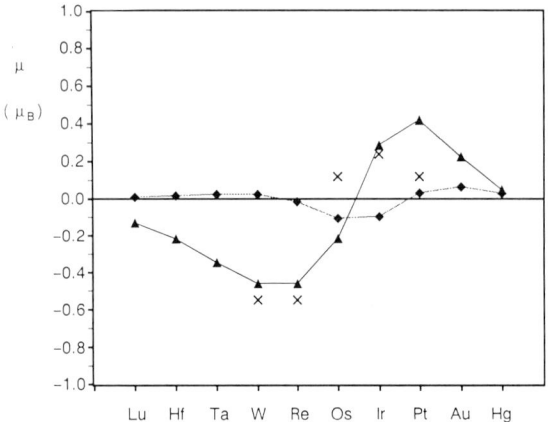

Fig. 14. Spin (▲) and orbital (♦) moments of 5d-transition metal impurities in Fe. The ×'s denote experimental values.

Impurity atoms of the 5d-transition metal row dissolved in the ferromagnets Fe, Co and Ni are especially interesting systems to study the various consequences of spin–orbit coupling in magnetic systems, since spin–orbit coupling is quite pronounced for these heavy atoms while the ferromagnetic host induces by hybridization a sizeable magnetic moment on the 5d-impurities which are normally non-magnetic.

Using the SPRKKR-formalism, we have recently performed corresponding investigations on the various magnetic properties of 5d-impurities in Fe [25–27]. The resulting magnetic moments are shown in fig. 14. Obviously the variation of the impurity moment along the 5d-series is very similar to the findings for the 3d- or 4d-impurities in Fe. This behavior can be traced back to the position of the majority virtual bound states with respect to the Fermi level, as discussed in sections 3 and 4. One important difference with respect to 3d and 4d impurities is that the transition from antiferromagnetic to ferromagnetic coupling is shifted to higher valences. Whereas in the 3d-series this transition occurs between Cr and Mn and in the 4d-series between Tc and Ru, in the 5d-series it occurs between Os and Ir. Thus Os, being isoelectronic to Ru and Fe, has a negative moment of $-0.21\mu_B$, as was recently confirmed experimentally. As was explained by Akai [27] the reason for this unusual behavior is the increase of the 5d-3d hybridization compared to the somewhat weaker 4d-3d and the much weaker 3d-3d hybridization. Therefore the majority virtual bound state is in the higher series systematically shifted to higher energies which explains the systematic shift of the transition towards the late transition-metal atoms. A second, somewhat less significant mechanism acting in the same direction is the relativistic shift of the d-level to higher energies which is more important for the higher series.

A direct consequence of spin–orbit coupling is the incomplete quenching of the angular momentum. The corresponding orbital magnetic moment could be calculated by our fully relativistic approach and is shown in fig. 14 again for the 5d-impurities in Fe. One of the most striking properties of these moments is that they are not always parallel to the spin magnetic moment as found before for the pure metals Fe, Co and Ni [28]. A rather simple explanation for this finding as well as for the variation of the orbital magnetic moment μ_{orb} in fig. 14 can be given if one considers spin–orbit coupling as a small perturbation. In this case the expectation value of the z-component of the angular-momentum operator l_z can be expressed – using some additional assumptions – by the spin–orbit-coupling parameter ξ and the spin-resolved local density of states at the Fermi level [28]

$$\langle l_z \rangle \approx \xi \left[n_{loc}^{(+)}(E_F) - n_{loc}^{(-)}(E_F) \right]. \qquad (17)$$

This expression makes clear that $\langle l_z \rangle$ and by that way also μ_{orb} may have a positive or a negative sign, depending on the spin-polarization $m(E) = n_{loc}^{(+)}(E) - n_{loc}^{(-)}(E)$ of the impurity at the Fermi level. For this reason one can expect for the 5d-impurities changes in sign of μ_{orb} along the row depending on the position of the impurity d-peaks, the virtual bound states respectively. As can be seen in fig. 15, where μ_{orb} is shown together with the spin-polarization $m^d(E_F)$ of the d-electrons, which dominate μ_{orb} completely, the simple picture based on eq. (17) describes the situation reasonably well.

The importance of accounting for relativistic effects in calculating hyperfine fields has recently be pointed out by Blügel et al. [19]. As has been

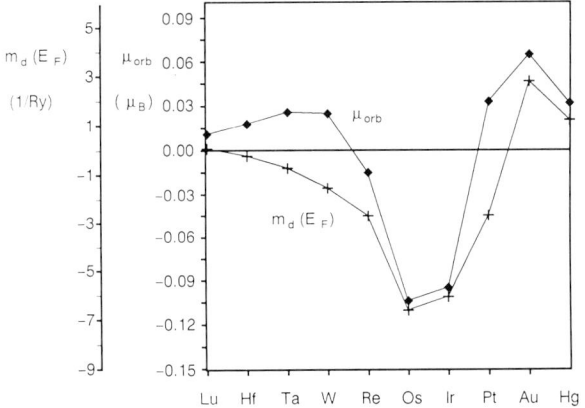

Fig. 15. Orbital moments μ_{orb} and the "magnetic" density of states $m_d(E_F) = n^{(+)}(E_F) - n^{(-)}(E_F)$ for 5d impurities in Fe.

shown by these and other authors before, relativistic effects in general enhance the hyperfine field due to s-electrons by around 5% for 3d-metals, around 15% for 4d-metals and even by a factor of 2–3 for 5d-metals. This enhancement can be treated properly within a scalar–relativistic approach giving results in very good agreement with fully relativistic calculations [19,29]. The orbital contributions B_{orb} to the hyperfine fields due to p- and d-electrons, on the other hand can be obtained only from spin-polarized, fully relativistic calculations. Corresponding results for the hyperfine fields are given in fig. 16. Although there are some deviations at the end of the row, the general trend of the experimental data is reproduced quite well. Because of the relatively small local impurity spin magnetic moments μ_{spin} (see fig. 14) the core polarization contribution to these fields, which is proportional to μ_{spin}, is rather small. The valence fields in turn are dominated by the s-electrons, which therefore are primarily responsible for the variation of the fields shown in fig. 16. Nevertheless, they are of the same order of magnitude as the orbital contribution, which closely follows the variation of μ_{orb}. For this reason B_{orb} is strongly negative for Os and Ir, -110 and -113 kG, respectively, while Pt and Au possess positive orbital fields: $+81$ and $+152$ kG, respectively.

These results clearly demonstrate that the orbital contributions to the hyperfine fields are very important to discuss the hyperfine properties of 5d-elements in a quantitative way.

6. Concentrated transition-metal alloys

6.1. KKR-CPA-LSD calculations

In this section we show some results of our KKR-CPA calculations for ferromagnetic alloys. The first example is the concentration dependence of Fe- and Ni-based 3d transition-metal alloys. Traditionally the saturation magnetizations of these alloys are plotted against the average electron number per atom, forming the so-called Slater–Pauling curve (see fig. 17). The experimental data align either along the main branch of the Slater–Pauling curve, having an angle of $-45°$ with the abscissa, or along several subbranches inclined by about $+45°$ and starting at pure Ni, Co or Fe. The overall trends of this curve were successfully explained many years ago by Hasegawa and Kanamori [6] on the basis of the tight-binding CPA combined with the Hartree–Fock approximation. Although their treatment was rather crude and, especially, the results depended on the parameters describing the electron–electron Coulomb and exchange interactions, it revealed the basic mechanism for the formation of the alloy magnetic moment. Two points were important in general. Firstly, though the average properties are well described by the

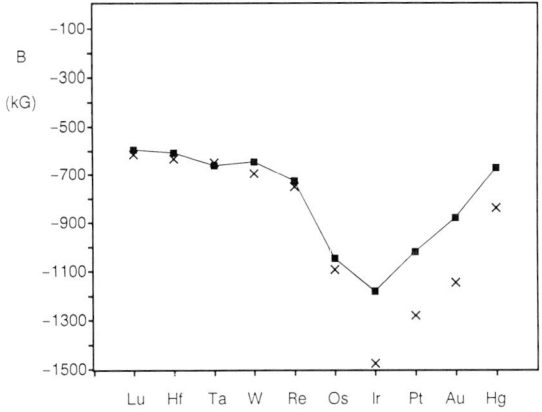

Fig. 16. Calculated hyperfine fields and experimental values ($\times$) for 5d impurities in Fe.

coherent medium, the constituent atoms still hold their intrinsic properties rather well. Thus Fe, Co and Ni have well defined local magnetic moments – as is consistent with the results of the neutron scattering experiments – not very much different from those of pure systems, namely around 2.2, 1.7 and $0.6\mu_B$, respectively. Secondly, in spite of the above, these local properties are affected by alloying, leading to a variety of bulk properties. A typical case is the Fe–Co alloy for which the Fe local moment is strongly enhanced by the existence of Co. This causes the well-known maximum in the saturation magnetization occurring near $Fe_{0.7}Co_{0.3}$.

All these general conclusions are still correct even when the model Hamiltonian is replaced by

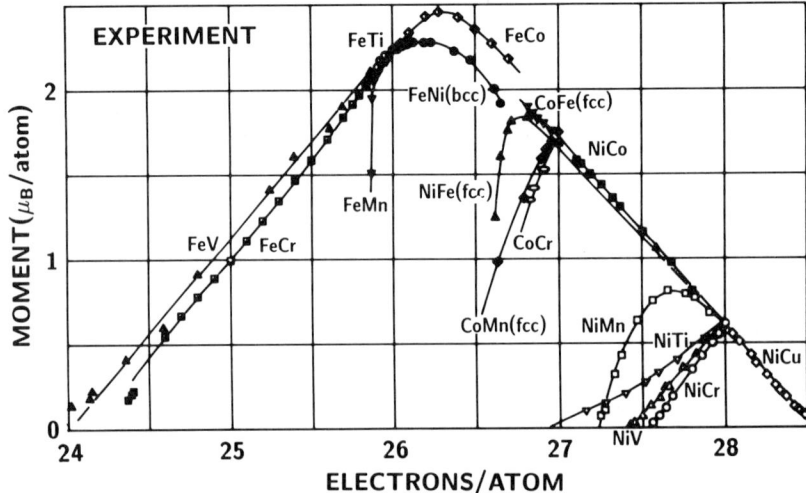

Fig. 17. Experimental values of the saturation magnetization of Fe-, Ni- and Co-based alloys vs. average number of electrons per atom, the so-called Slater–Pauling curve.

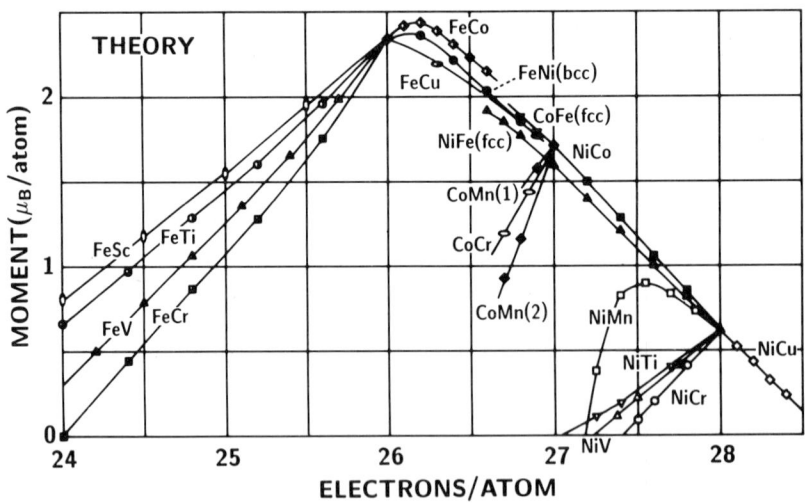

Fig. 18. Calculated magnetization of Fe-, Ni- and Co-based alloys vs. average number of electrons per atom. g factors are included in an approximate way, see text for details. The fcc, instead of hcp, structure is assumed for Co-based alloys. For Co–Mn, two solutions, CoMn(1) with a Mn local moment parallel to the bulk magnetisation and CoMn(2) with an antiparallel moment, are obtained.

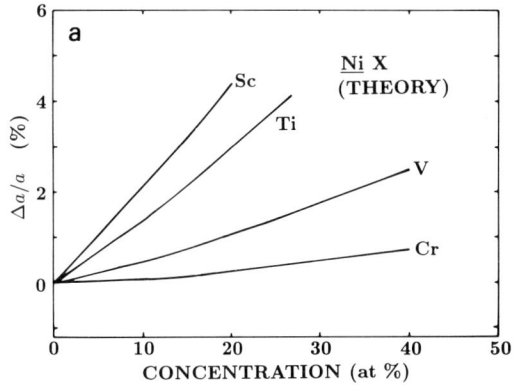
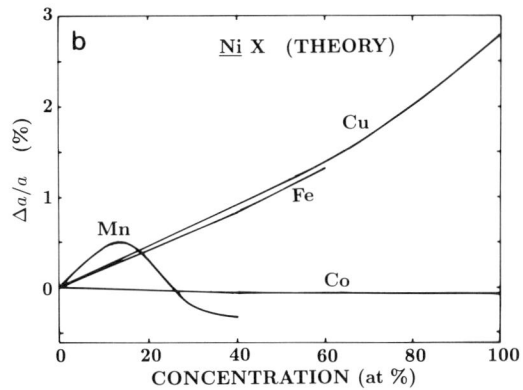

Fig. 19. Calculated ratios of the equilibrium lattice constant of Ni–X alloys to that of pure Ni vs. concentration of solute atom X. (a) X = Sc, Ti, V and Cr. (b) X = Mn, Fe, Co and Cu.

the Kohn–Sham equation of density-functional-theory. The new feature that is introduced by such an ab-initio calculation is its predictable power. Given the atomic numbers of the constituent atoms and their concentrations, the theory can predict all the ground state properties, including the equilibrium lattice constant and the saturation magnetization, without any adjustable parameters on the equally sound footing as the band-structure calculations for ordered systems. The calculations are characterized by two self-consistency cycles, a statistical one required by the CPA and the usual charge and magnetization self-consistency required by density-functional-theory in the LSD approximation. They enable us to calculate microscopic properties such as the spatial charge and spin distributions, which experimentally are observed by use of NMR, Mössbauer spectroscopy and other nuclear techniques.

Fig. 18 shows the calculated Slater–Pauling curve for which we used the lattice constants as determined by the minimum of the total energy. We include a g-factor correction by assuming $g = 2.078$, 2.17 and 2.19 respectively for Fe, Co and Ni, neglecting their concentration dependence and the contribution of other constituent atoms to the g-factor. The concentration depen-

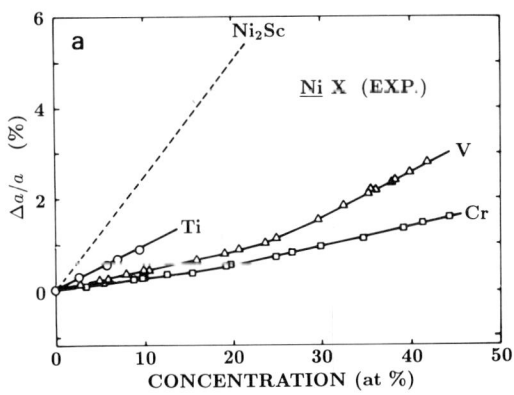
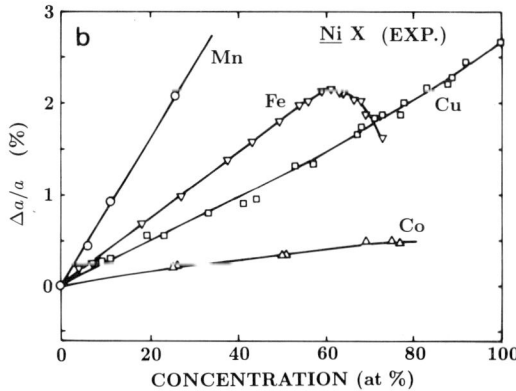

Fig. 20. Experimentally obtained ratios of the equilibrium lattice constant of Ni–X alloys to that of pure Ni vs. concentration of solute atom X. (a) X = Sc, Ti, V and Cr. (b) X = Mn, Fe, Co and Cu.

dence of the ratio of the equilibrium lattice constant of Ni-based alloys to that of pure Ni is shown in figs. 19a and b while figs. 20a and b show the corresponding experimental data (the reason why we prefer the ratios rather than the absolute values is that we can in this way offset the small discrepancy which exists even in pure Ni between the calculated and the experimental lattice constants.)

Though it is not difficult to point out some discrepancies between the theory and experiments, we conclude that the overall trends are well reproduced by the KKR-CPA-LSD calculation. We first discuss the general trends of our results briefly. The right half of the curves, which mainly consists of Ni-based fcc alloys, form a well-defined single line with $-45°$ slope, i.e. the main branch of the Slater–Pauling curve. The left half of the curve, which consists of Fe-based bcc alloys, on the other hand spreads rather widely though each series of the system forms a rather straight line. From this point of view we may say that this part of the Slater–Pauling curve resembles to the subbranches starting from Co and Ni.

The common feature seen in the alloys belonging to the subbranches is that the solute atoms (i.e. Sc, Ti, V and Cr in Fe, Co and Ni) have negative (antiparallel to the bulk magnetization) local magnetic moments. The negative local moment is associated with appearing of hole states above or at the top of the majority-spin d bands [1–3]. These hole states have a large amplitude at the solute atoms causing a negative local moment there, as well as the rapid decrease of the magnetization with the solute concentration as discussed in section 4 for the dilute alloys. The existence of the hole states in the majority-spin band also affects the concentration dependence of the magnetization. In the case that no holes exist in the majority-spin band, the main origin of the concentration dependence of the magnetization is the reduced number of available d electrons as determined by the average valence; the straight line of $-45°$ slope in the right half of the Slater–Pauling curve is explained in this way. Contrarily, in the case that holes exist in majority-spin band, it is mainly the missing number of majority d states that causes the concentration dependence. Thus the slope of the magnetization against the total number of electrons varies depending on the solute atoms; the smaller the difference in the number of the valence electrons, the steeper the slope.

The above discussion, however, is useful only for simple cases where the magnetic state is rather stable, typically at the region of the strong ferromagnetism of Ni. The question is whether we can understand more delicate cases such as the behavior of Ni–Fe, Ni–Mn and Fe–Mn alloys in the vicinity where ferromagnetism becomes instable. As for this point, it was pointed out [30] that the KKR-CPA with a single magnetic state for each constituent atom could not reproduce the experimental situation satisfactorily. Both Ni–Fe and Ni–Mn systems undergo a ferro- to non-magnetic transition which follows the initial increase of the magnetization. For Ni–Fe the calculation gives a first-order-like transition contrary to the experiments which show a smoother behavior. In this respect, the calculation seems to be satisfactory for the Ni–Mn cases. A detailed check, however, reveals that this agreement is only apparent. A magnetic transition in the present treatment inevitably accompanies a volume change through the magneto-volume coupling. In fact, as is shown in fig. 19b, the calculated lattice constant reaches a maximum at around 10 at% Mn, corresponding to the maximum of the magnetization, and then rapidly decreases with further increase of the Mn concentration. The experiments, fig. 20b, however show no specific behavior reconcilable with such a magnetic transition.

It is very plausible that we can improve the above situation, at least for the Ni–Mn case, by introducing more than one magnetic state for each constituent atom. The simplest extension may be to introduce two magnetic states for Mn, i.e., Mn↑ and Mn↓. Such a treatment was first proposed by Jo [31] in the framework of the tight-binding CPA. He explained in this way the existence of two NMR-signals of Mn as well as the magnetic transition of this system. A similar extension for the KKR-CPA is also possible and desirable in order to acquire a better understanding of this system. We expect this to be extremely important for the lattice expansion, which is determined by the strain derivative of the magnetization energy. Whereas the CPA describes average properties such as the average moment $\langle M \rangle$ very well it can fail for the magnetization energy depending essentially on $\langle M^2 \rangle$, especially in situations where $\langle M \rangle$ vanishes whereas $\langle M^2 \rangle$ is hardly unchanged.

As for the Ni–Fe (fcc) systems, the problem seems to be more serious [32]. In addition to the possible existence of more than one magnetic state, we have to consider the following possibility: Firstly, the LSD approximation may not be sufficient to take the renormalization caused by the electron–magnon scattering of the ground state into account [32]. Secondly since the magnetic instability in this case occurs near the boundary of fcc–bcc martensitic transformation, it is rather hard to simulate the experimental situation by the theory; inhomogeneities and the phase separation may also play a role.

On the other hand, we failed to obtain the magnetic transition of Mn–Fe systems which experimentally seems to take place in the region of low Mn concentrations. Since we are not convinced of the results, we omit this case from fig. 18. This is partly related to the fact that the magnetic state of Mn in Fe is very sensitive to the details of the calculation. The direction of the local magnetic moment, for example, is easily reversed by changing the charge density or the local potential slightly as discussed in sections 3 and 4. We may only say that the magnetic state looks stable even for a rather large Mn concentration. In such a situation, it is clearly necessary to introduce more than one magnetic state in addition to an improved numerics.

It is worthwhile to mention a few words about the reason of the initial increase of the magnetization of the Fe–Co system with Co concentration; the discussion follows that of Hasegawa and Kanamori [6]. Note that the local magnetic moment at the Co site cannot be much larger than that of pure Co which is a strong ferromagnet, and hence is smaller than that at the Fe site. This causes the difference in the exchange splitting between Fe and Co sites. The difference is nearly canceled in the majority-spin states by the potential difference caused by the nuclear charges, but is doubled in the minority-spin states. The potential difference in the minority-spin states thus produced now expels the amplitude of the Fe minority-spin states in the lower energy part of the d band towards the upper energy region. As a result the minority-spin states below the Fermi level lose some of their weight, which is followed by the charge accumulation at the Fermi level in the majority-spin states, leading to the enhancement of the Fe local moment. The situation for the Fe–Ni (bcc) system is similar. Here the enhancement in the magnetization, however, is not very pronounced because the local moment at the Ni site is not as big as the one of Co in Fe–Co. The replacement of Fe atoms by Ni atoms thereby easily degrades the enhancement.

Now we briefly discuss the calculated lattice constants given in fig. 19. As already pointed out, a serious discrepancy between the calculated and the experimental (fig. 20) concentration dependence is observed for Ni–Mn, for which we will not repeat the discussion. The discrepancy seen in Ni–Fe systems is also a case which is presumably caused by the problems in describing the magnetic transition. Except these two cases, the overall trends are well reproduced by the theory. Especially, it succeeded in correctly giving the deviation from Vegard's rule (i.e., linear dependence) experimentally observed for Ni–V, Ni–Cr and Ni–Cu.

The present results are still not satisfactory. Among various over-simplifications exploited in the calculation, the most serious one may be the angular-momentum truncation for which we tentatively choose $l_{max} = 2$. This truncation affects the result not only directly but also via the equilibrium lattice constant. Inclusion of the f state together with the extension to more magnetic states certainly will provide a more improved description of both the magnetic and the cohesive properties discussed here.

6.2. ECM-calculations

As outlined in section 2.2 the aim of the KKR-CPA is to give a configurational averaged description of random alloys. Even in systems where the assumption of randomness, i.e. the neglect of any short range order, is well justified, the dependence of local properties on the specific atomic surrounding may show up in experiments. An example for this is the inhomogenous line width of NMR-spectra or the occurrence of satellites. To study theoretically such phenomena Gonis et al. [34] proposed to embed clusters of given configurations into a CPA-medium, which represents the corresponding random alloy (ECM – embedded cluster method). This approach, which uses the CPA-medium as a reference medium, has been used recently by Banhart et al. [35] to study the nuclear spin lattice relaxation rate as a function of the occupation of the first two neighboring shells around a Cu- or Pt-atom, respectively, in Cu_xPt_{1-x}. The most important result of this work was that the average over all possible configurations, weighted according to their statistical probability, agreed perfectly with the CPA-results. This clearly ensures that the CPA-medium represents the configurationally averaged properties of a random alloy in a very reliable way. Because of the numerical effort in self-consistent calculations, corresponding studies of the configurational dependence of hyperfine fields in ferromagnetic alloys have been done until now only within the cluster-KKR-CPA formalism [36]. Using a six-shell cluster, Ebert et al. [36] have calculated the hyperfine fields of a central Ni- or Fe-atom which is surrounded by CPA-scatterers according to the alloy $Ni_{0.75}Fe_{0.25}$ and one neighboring shell completely occupied by either Fe- or Ni-atoms (the restriction of complete occupation of a neighboring shell was adopted to make use of the cubic symmetry of the resulting cluster). The hyperfine fields for Ni as central atom obtained in this way are shown in fig. 21. The numbers i label the fields which correspond to an occupation of shell i with only Ni- or Fe-atoms. In addition the CPA-result, i.e. the field with only CPA-scatterers as neighbors, is shown. As can be seen, all lines connecting the fields for the two extreme configurations on a shell – only Ni- or only Fe-atoms – pass through this point. This again demonstrates that the CPA in fact gives a very consistent description of the random alloy. The data shown in fig. 21 seem to indicate that the configuration dependence of the hyperfine fields can be described by the simple expression:

$$H_\alpha = H_\alpha^{CPA} + \sum_i \Delta H_\alpha^i (n^i - \overline{n}^i). \tag{18}$$

Here H_α^{CPA} is the CPA-result for the hyperfine field H_α of the atom α and $n^i(\overline{n}^i)$ is the actual (average) occupation number for the α-atoms on shell i. The hyperfine field coefficients ΔH_α^i are determined by the slope of the curves in fig. 21. Corresponding results for Fe in $Fe_{0.25}Ni_{0.75}$ are displayed in fig. 22 as a function of the distance of the neighboring shell from the central atom. Obviously the data show an oscillatory behavior and a great similarity with the shell dependence of the impurity atom induced changes in the Knight shift in Cu [37].

Expressions very similar to eq. (18) have been used in the past by many experimentalists to interpret the change of hyperfine fields due to ordering or local configurational changes in magnetic alloys. Unfortunately it seems quite difficult to deduce reliable hyperfine field coefficients from the experimental spectra. Nevertheless, reasonable good agreement between the theoretical results for ΔH_α^i in $Fe_{0.25}Ni_{0.75}$ and the corresponding experimental data has been found. More important, however, seems to be that the theoret-

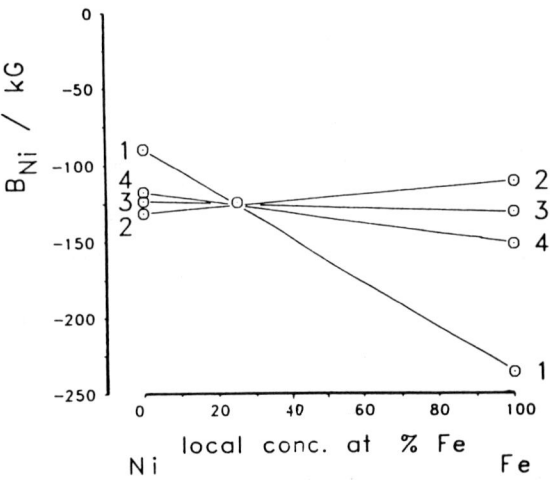

Fig. 21. Variation of the hyperfine fields of Ni with the configuration on one of the surrounding atomic shells. The numbers indicate the shell which is occupied by only Ni- (left) or Fe-atoms (right).

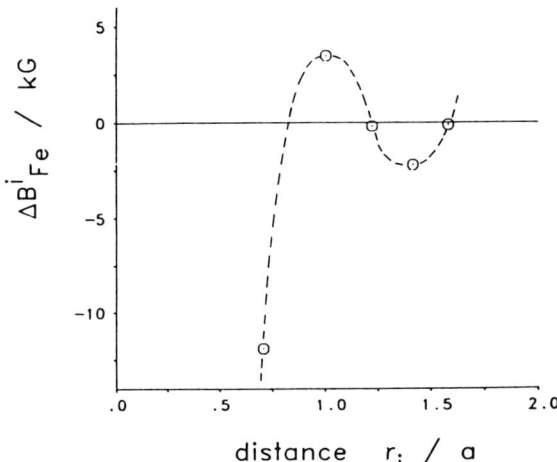

Fig. 22. Hyperfine field coefficients ΔB^i_{Fe} for Fe in $Fe_{0.25}Ni_{0.75}$ plotted against the distance of shell i from the central site – a is the lattice parameter.

ical coefficients are in accordance with the changes of the fields found directly by the experiments. With these coefficients and the average field H^{CPA}_α for $Fe_{0.25}Ni_{0.75}$ eq. (18) predicts a change by -10 and $+30$ kG for Ni and Fe atoms, respectively. The corresponding experimental data are -20 and $+12$ kG, respectively. The same set of data gives a change of B_{Fe} due to a positioning on a Ni-site instead on a Fe-site in ordered Ni_3Fe by -46 kG, while the corresponding experimental value is also -46 kG [36].

These results justify the use of eq. (18) although it has been derived only on the data for some few extreme configurations (see fig. 21). A further justification for the additivity expressed in eq. (18) could be given by calculations where two neighboring shells were occupied by Fe- or Ni-atoms. The results of these calculations deviated from the fields obtained using eq. (18) by less than 4% although the absolute value of the fields varied in a range of nearly 400% around the CPA-result.

In spite of the various limitations of the available cluster-KKR-CPA data for Fe_xNi_{1-x} [36] this work could demonstrate that the CPA provides an excellent starting point to investigate via the ECM the influence of atomic configurations on local magnetic properties in alloys.

References

[1] J. Friedel, Nuovo Cimento 10 suppl. no. 2 (1958) 287.
[2] J. Kanamori, J. Appl. Phys. 16 (1965) 929.
[3] I.A. Campbell and A.A. Gomes, Proc. Phys. Soc. London 91 (1967) 319.
[4] S. Chikazumi, Physics of Magnetism (Wiley, New York, 1964) p. 73.
[5] A.R. Williams, V.L. Moruzzi, A.P. Malozemoff and K. Terakura, IEEE Trans. Magn. MAG-19, (1983) 1983.
A.P. Malozemoff, A.R. Williams, K. Terakura, V.L. Moruzzi and K. Fukamichi, J. Magn. Magn. Mater. 35 (1983) 192.
[6] H. Hasegawa and J. Kanamori, J. Phys. Soc. Jpn. 31 (1971) 382, 33, (1972) 1599, 1606.
[7] U. von Barth and L. Hedin, J. Phys. C 5 (1972) 1629.
[8] V.L. Moruzzi, J.F. Janak and A.R. Williams, Calculated Electronic Properties of Metals (Pergamon Press, New York, 1978).
[9] S.H. Vosko, L. Wilk and M. Nusair, Can. J. Phys. 58 (1980) 1200.
[10] R. Zeller, J. Phys. C 20 (1987) 2347.
[11] R. Zeller, J. Phys. F 17 (1987) 2123.
[12] P. Soven, Phys. Rev. 156 (1967) 809.
[13] G.M. Stocks, W.M. Temmerman and B.L. Gyorffy, Phys. Rev. Lett. 41 (1978) 339.
[14] H. Akai, J. Phys. Soc. Jpn. 51 (1982) 1176.
H. Winter and G.M. Stocks, Phys. Rev. B 27 (1983) 882.
H. Akai, J. Phys.: Condens. Matter 1 (1989) 8045.
[15] J.B. Staunton, B.L. Gyorffy and P. Weinberger, J. Phys. F 10, (1980) 2665.
[16] H. Ebert, B. Drittler and H. Akai, submitted.
[17] B. Drittler, N. Stefanou, S. Blügel, R. Zeller and P.H. Dederichs, Phys. Rev. B 40 (1989) 8203.
[18] H. Akai, M. Akai, S. Blügel, B. Drittler, H. Ebert, K. Terakura, R. Zeller and P.H. Dederichs, Prog. Theor. Phys. Suppl. 101 (1990) 11.
[19] S. Blügel, H. Akai, R. Zeller and P.H. Dederichs, Phys. Rev. B 35 (1987) 3271.
[20] N. Stefanou, A. Oswald, R. Zeller and P.H. Dederichs, Phys. Rev. B 35 (1987) 6911.
[21] A.P. Malozemoff, A.R. Williams and V.L. Moruzzi, Phys. Rev. B 29 (1984) 1620.
[22] P.H. Dederichs, B. Drittler, R. Zeller, H. Ebert and M. Weinert, Hyperfine Interactions 60 (1990) 547.
[23] H. Akai, S. Blügel, R. Zeller and P.H. Dederichs, Phys. Rev. Lett. 56 (1986) 2407.
[24] P. Strange, H. Ebert, J.B. Staunton and B.L. Gyorffy, J. Phys.: Condens. Matter 1 (1989) 2959.
[25] H. Ebert, R. Wienke, G. Schütz and R. Zeller, J. Appl. Phys. 67 (1990) 4923.
[26] H. Ebert, R. Zeller, B. Drittler and P.H. Dederichs, J. Appl. Phys. 67 (1990) 4576.
[27] H. Akai, Hyperfine Interactions 43 (1988) 255.
[28] H. Ebert, P. Strange and B.L. Gyorffy, J. Phys. F 18, (1988) L135.
[29] H. Ebert, H. Winter, B.L. Gyorffy, D.D. Johnson and F.J. Pinsky, J. Phys. F 18 (1988) 719.

[30] H. Akai, P.H. Dederichs and J. Kanamori, J. de Phys. 49 (1988) C8-23.
[31] T. Jo, J. Phys. Soc. Jpn. 40 (1976) 715, 48 (1980) 1482.
[32] J. Kanamori and T. Jo, Electron Correlation and Magnetism in Narrow-Band Systems, ed. T. Moriya (Springer, Berlin, 1981) p. 109.
[33] J. Igarashi, Electron Correlation and Magnetism in Narrow-Band Systems, ed. T. Moriya (Springer, Berlin, 1981) p. 115.
[34] A. Gonis, G.M. Stocks, W.H. Butler and H. Winter, Phys. Rev. B 29 (1984) 555.
[35] J. Banhart, H. Ebert, J. Voitländer and P. Weinberger, Solid State Commun. 65 (1988) 693.
[36] H. Ebert, H. Winter, B.L. Gyorffy, D.D. Johnson and F.J. Pinski, Solid State Commun. 64, (1987) 1011.
[37] B. Drittler, H. Ebert, R. Zeller and P.H. Dederichs, Phys. Rev. B 39 (1989) 6334.

Theory of itinerant electron magnetism

Tôru Moriya

Department of Physics, Faculty of Science and Technology, Science University of Tokyo, Yamazaki, Noda 278, Japan

The present status of the theory of itinerant electron magnetism is outlined with particular emphasis on the developments in the recent two decades, characterized by the substantial advances in the theory of spin fluctuations. The possible applicability of the itinerant electron model to the high temperature superconductor oxides and the possible importance of the effects of antiferromagnetic spin fluctuations are also discussed.

1. Introduction

Magnetism of itinerant electron systems, in particular, those including d-electrons and sometimes even f-electrons, has been one of the central issues in condensed matter physics in this century after the advent of quantum mechanics. This problem has been a difficult one because of the effects of strong electron–electron correlations in narrow energy bands which enforce us to go beyond traditional Hartree–Fock (HF) or mean field theory. A number of approximation methods have been developed for this purpose.

The problem may be divided into two parts. One is to describe the magnetic ground state and the other is to describe the thermal excitations and the finite temperature properties of interacting itinerant electron systems.

The first part, i.e., the ground state properties are now believed to be described fairly well by the band theory which has advanced steadily and substantially in recent decades. In particular, the ground state with a strong magnetic long range order seems to be well described by the band theory. This may be more or less expected from the following simple physical arguments; two electrons with opposite spins are spatially well separated in the systems with strong magnetic polarizations, and thus the effects of electron–electron correlations are no longer very important there. As a matter of fact, the Fermi surfaces and the effective masses of the ferromagnetic transition metals obtained by recent state-of-the-art band calculations agree well with the observations.

In the systems with weak magnetic long range order, or those near the critical boundary for the appearance of magnetic long range order, the effects of electron correlations to the ground state properties are more significant; the effective masses are enhanced significantly beyond the values of the band calculations. However, the calculated Fermi surfaces seem to agree fairly well with the observations. In any case the band theory is believed to provide at least a good starting point of the theory. In this article we will not discuss this first part of the problem, referring the readers to refs. [1,2].

The second part of the problem, i.e. the thermal excitations and the finite temperature properties has been the more serious part of the problem, since the effects of electron–electron correlations are essentially important in this problem. In other words, the simple Stoner theory based on the mean field approximation at finite temperature turned out almost always to fail. The Stoner theory takes account of the thermal excitations of electron–hole pairs with opposite spins, each of which moves independently of each other in the mean potential field. This theory fails to explain the Curie–Weiss magnetic susceptibility observed in almost all ferromagnets far below their Fermi temperatures. Also, the calculated Curie temperature T_C is at least an order of magnitude too high compared with experiment. This disagreement is considered to be natural since the thermal excitations in the Stoner theory are unrealistic, neglecting the vitally important effect of the interaction between the excited electron and the hole. When this effect is taken into account the excitations become collective modes with quite different excitation energy spectra from the Stoner excitations of free electron–hole pairs. When an electron–hole pair makes a bound state it corresponds to a spin wave excitation and when the bound state is not formed the excitation becomes a collective mode of dissipative nature, for example, like paramagnons. Above the Curie temperature all the thermal excitations are of dissipative nature. Thus it is essential to deal with the exchange-enhanced collective modes of magnetic excitations of spin density fluctuations in the theory of finite temperature properties of itinerant electron magnetism.

Historically, the possible existence of local moments in metals was the earlier subject of investigations in 1950's and 1960's. This is due to the fact that the Curie–Weiss susceptibilities observed in almost all ferromagnets were long believed to be explained solely by the local moment theory by Langevin and Weiss. In 1960's the dynamical Hartree–Fock theory or the random phase approximation (hereafter abbreviated to RPA) was also developed to describe the collective excitations in the k-space. However this theory did not at all improve the Stoner theory for finite temperature properties; the RPA theory calculates the spin fluctuations around the Stoner equilibrium state only.

In 1970's the self-consistent renormalization (hereafter abbreviated to SCR) theory of spin fluctuations was developed [3–5]. This theory goes one step beyond the HF-RPA theory and deals with the coupled modes of spatially extended spin density fluctuations in a self-consistent fashion. This theory gave a new mechanism for the Curie–Weiss magnetic susceptibility associated with the spatially extended modes of spin fluctuations instead of the local moments as spatially localized spin fluctuations. Furthermore, the theory was successful quantitatively in explaining various physical properties of weak itinerant ferro- and antiferromagnets. In other words, the theory contains a limited number of parameters (defined in the k-space) which can be determined experimentally. When the number of independent experimental methods to determine the parameter values exceeds the number of parameters parameter-free tests of the theory become possible. This type of tests were carried out in these two decades and were so successful as to establish the SCR theory as the first fully satisfactory theory of itinerant electron magnetism for this class of materials. In other words, concerning the weak itinerant ferro- and antiferromagnets we are in the same position as concerned with the magnetic insulator compounds where the local moment theory was established with an innumerable number of tests by using the parameters defined in real space.

Thus in 1970's the theory of magnetism was advanced to the stage of having two established regimes in the opposite extremes, the local moment systems and weak itinerant ferro- and antiferromagnets. In the former regime the spin fluctuations are localized in real space while in the latter they are localized in k-space.

Natural direction of research after these advances seemed to develop a theory to fill in the gap between the two extremes, or to develop an interpolation theory for the intermediate regime. Such theories were advanced in late 1970's and 1980's by using the functional integral method within the static approximation. These theories were successful not only in explaining a number of so far unexplained experimental results qualitatively or semiquantitatively but also in providing a general perspective for the unified picture of magnetism of the d-electron systems. As a matter of fact this interpolation theory also enabled us to calculate the phase diagram for the Mott transition in magnetic compounds for the first time on the basis of the Hubbard model. The dynamical spin fluctuation theory, however, is still to be developed for the intermediate regime.

Since 1986, various high temperature superconductor cuprates were discovered [6,7]. They are doped Mott insulators including charge transfer insulators, and the relevant electrons are considered to form strongly correlated two-dimensional systems in the CuO_2 plane. Although various new

Table 1
Anomalous temperature dependences of physical quantities around the critical phase boundary for itinerant ferro- and antiferromagnetism predicted by the SCR theory. χ_Q: magnetic susceptibility, R: electrical resistivity, T_1: nuclear spin–lattice relaxation time, C_m: magnetic contribution to the specific heat, C.W. (Curie–Weiss): $T + \Theta$

	3-dimensional		2-dimensional		Normal
	ferro ($Q = 0$)	a.f.	ferro ($Q = 0$)	a.f.	Fermi liquid
χ_Q^{-1}	$T^{4/3} \to$ C.W.	$T^{3/2} \to$ C.W.	$T^{4/5} \to$ C.W.	$T + \theta$	$a + bT^2$
R	$T^{5/3}$	$T^{3/2}$	$T^{4/3}$	T	T^2
T_1^{-1}	$T\chi$	$T\chi_Q^{1/2}$	$T\chi^{3/2}$	$T\chi_Q$	T
C_m/T	$-\ln|1 - \alpha_u|$	$|1 - \alpha_u|^{-1/2}$		$-\ln|1 - \alpha_s|$	

approaches from the strong correlation limit have been developed for this problem, in particular, to explain various anomalous physical properties above the critical temperature T_c, it seems that these anomalous properties are well explained in terms of the antiferromagnetic spin fluctuations in two-dimensional itinerant electron systems as described by the SCR theory [8]. This fact seems to indicate the importance of studying magnetism and superconductivity simultaneously in strongly correlated itinerant electron systems.

As was mentioned briefly in the above, there have been quite interesting developments in these two decades in the field of itinerant electron magnetism and to make a brief survey of these developments is the purpose of the present article. In view of the fact that we already have several review articles which cover substantial part of these development [3–5,9], we will confine ourselves in this short article to explain the outline of the main streams of these developments without going into details for which we refer the readers to the previous review articles and recent papers. Thus no intention is made to cover all the important research activities or to give a balanced list of references.

In section 2 we discuss the present status of the self-consistent renormalization theory of spin fluctuations for weak itinerant ferro- and antiferromagnetism which established a new regime of magnetism in the opposite limit to the local moment regime. Recent application of the theory to high T_c cuprates will also be discussed. Before discussing on a unified picture of magnetism based on the interpolation theory in section 4, we briefly mention in section 3 about early theories of metallic magnetism based on the local moment picture, in particular their applicability and limitations. In section 5 we discuss, in view of the unified picture of magnetism, on the possible physical reasons why the itinerant electron model with antiferromagnetic fluctuations can be applied to high T_c oxides as doped Mott insulators. Finally a brief summary is given in section 6.

2. Self-consistent renormalization (SCR) theory of spin fluctuations and weak itinerant ferro- and antiferromagnetism

The discovery of weak itinerant ferromagnetism in various intermetallic compounds, such as $ZrZn_2$, Sc_3In, Ni_3Al, $MnSi$, etc. in late 1950's and 1960's and successful explanations of their physical properties in terms of the SCR theory of spin fluctuations in 1970's have opened a new horizon in the theory of itinerant electron magnetism. These investigations established a new category of magnetism in the opposite extreme to the long familiar local moment systems.

First of all, the discovery by the SCR theory of the new mechanism for the Curie–Weiss (CW) susceptibility due to the interacting extended modes of spin fluctuations has released us from a long-dominant prejudice of connecting the observed CW susceptibility directly with the existence of local moments. Another important point of this development is the quantitative success of the SCR theory just in the same sense as the successful establishment of the local moment theory in magnetic insulator compounds [10]. Thus the famous controversy over itinerant vs. localized models, started with the advent of the quantum mechanical theories of magnetism, was believed to be resolved, at least partly, into a well-defined problem of characterizing the nature of spin density fluctuations in each substance.

For weakly and nearly ferro- and antiferromagnetic metals the SCR theory predicted that the spin fluctuations give rise to various anomalous temperature dependences of various physical quantities different from the prediction of the normal Fermi liquid theory. These anomalies occur above a certain crossover temperature which decreases and tends to zero as the phase boundary is approached. Below the crossover temperature physical properties show the Fermi liquid-like behaviors. We summarize some of these anomalous properties in the left hand side of table 1. All of these predictions were borne out by subsequent experimental investigations in 1970's [3,4].

As for the quantitative test of the theory, the SCR theory contains a small number of parameters in the weakly ferromagnetic limit. The parameters may be chosen as follows: $m(0)$, the ordered moment per atom or $\chi(0,0)$, the magnetic susceptibility at $T=0$ K for substances with paramagnetic ground state, F_1, the mode–mode coupling constant for the long wavelength modes of spin fluctuations which may be deduced from the Arrott plot of the magnetization data, A, the exchange or transverse stiffness constant for the spin density, and Γ_0, the parameter characterizing the energy spreads of the spin fluctuation modes. The dynamical susceptibility above T_C, for example, is given in terms of these parameters as follows:

$$\chi(q,\omega) = \chi(q)/(1 - i\omega/\Gamma_q), \quad (2.1)$$

with

$$\chi(q) \approx \chi(0)/(1 + q^2/\kappa^2), \quad \Gamma_q = \Gamma_0 q(\kappa^2 + q^2),$$

$$\kappa^2 = \chi_0(0)/A\chi(0), \quad (2.2)$$

where $\chi(0)$ is obtained by solving the following equation which takes account of the mode–mode couplings in a self-consistent fashion, χ_0 being the susceptibility of non-interacting electrons:

$$\frac{1}{\chi(0)} = \frac{1}{\chi_{HF}(0)} + \tfrac{5}{3} F_1 S_L^2(T), \quad (2.3)$$

$$S_L^2(T) = \frac{3}{2\pi} \int d\omega \, \coth\!\left(\frac{\omega}{2T}\right) N_0^{-2} {\sum_q}' \operatorname{Im} \chi(q,\omega). \quad (2.4)$$

Here $\chi_{HF}(0)$ is the Hartree–Fock result for the uniform susceptibility given by $\chi_{HF} = \chi_0/(1-\alpha)$, $(1-\alpha)^{-1}$ being the exchange enhancement factor, and the second term in eq. (2.3) represents the effect of the mode–mode coupling or renormalization. Since eqs. (2.1) and (2.2) for $\chi(q,\omega)$ contain $\chi(0)$, the eqs. (2.1)–(2.4) must be solved consistently for $\chi(0)$. This is the key point of the SCR theory and the second term in eq. (2.3) is essential for the Curie–Weiss behavior.

One may notice that the structure of the dynamical susceptibility in eqs. (2.1)–(2.2) is the same as the RPA theory. However, the RPA uses $\chi_{HF}(0)$ for $\chi(0)$, neglecting the second term in eq. (2.3) or the effect of the mode–mode coupling in contrast with the SCR theory and thus fails to improve the Stoner theory. Furthermore, in the SCR theory the parameters α, A and Γ_0 are also renormalized in general, although the effect is not so dramatic as for $\chi(0)$. In actual comparisons between the theory and experiments we use the experimentally measured values for the parameters. $m(0)$ or $\chi(0,0)$ and F_1 are obtained directly from the static magnetization measurements, while Γ_0 and A can be obtained directly from the neutron scattering experiment and indirectly from various other measurements. Thus the measured values for the parameters may be regarded as the renormalized ones.

Various physical quantities such as the Curie temperature T_C, the magnetic susceptibility above T_C, the specific heat, nuclear spin relaxation rate and its magnetic field dependence, electrical and thermal resistivities, magneto-resistance, are now expressed practically in terms of these four parameters. Intensive and extensive experimental investigations of these properties were carried out in 1970's and early 1980's with overwhelming success.

We will not show details here referring to refs. [3–5,9] for interested readers.

One apparent difficulty with the SCR theory has been associated with the reported discovery of temperature-induced ferromagnetism in Y_2Ni_7 about ten years ago. Although the Stoner theory reproduced this result with the use of a certain model for the density of states, the SCR theory showed from a general point of view that the effect of spin fluctuations make such a phenomenon hardly possible [11]. This serious experimental challenge to the SCR theory, however, disappeared recently; the previous result of measurements was reported to be an artifact caused by gadolinium impurities after careful recent investigations [12].

As an important recent topic we next discuss the possible relevance of the SCR theory of antiferromagnetic spin fluctuations to the physical properties of high temperature superconductor cuprates. Intensive investigations after the discovery of the high T_c oxides have shown that their physical properties such as electrical resistivity, Hall effect, optical conductivity, nuclear spin-lattice relaxation rates, etc. are quite anomalous above the critical temperature T_c. These anomalous properties led many people to suspect the possible breakdown of the Fermi liquid picture. Since these high T_c cuprates are doped Mott insulators where the electron–electron correlations are expected to be quite strong and the conducting electrons or holes are believed to form two-dimensional systems in the CuO_2 plane, novel approaches from the strong correlation limit in two-dimensional models were developed and extended intensively [6,7].

From a rather different point of view, the antiferromagnetic spin fluctuations in two-dimensional itinerant electron systems were recently investigated by using the SCR theory [8]. As was seen in table 1, the SCR theory for the three-dimensional magnets predicted quite anomalous temperature dependences for various physical quantities near the critical boundary for the magnetic long range order. In this case the behaviors in the low temperature limit are just normal Fermi liquid-like and there is a certain crossover temperature above which we have a wide range of anomalous behaviors characteristic to spin fluctuations. The anomalous properties observed in high T_c oxides reminded us of these previous results in weak itinerant ferro- and antiferromagnets. Thus the SCR theory was extended to the two-dimensional itinerant electron systems.

Such an extension of the theory is straightforward and the results contain many interesting features. We discuss here only the antiferromagnetic case. The first of all it is shown that there is no antiferromagnetic long range order at finite temperatures even if the ground state is antiferromagnetic. When the ground state is antiferromagnetic, the staggered magnetic susceptibility diverges toward $T = 0$ exponentially; we have $\chi \propto \exp[KM_Q^2(0)/T]/T$ at low temperatures, where $M_Q(0)$ is the staggered magnetization at $T = 0$ and K is a constant. This behavior is similar to that in the two-dimensional Heisenberg antiferromagnets. When the ground state is paramagnetic (nearly antiferromagnetic) we have $\chi_Q(0) - \chi_Q(T) \propto T^2$ at low temperatures. At elevated temperatures $\chi_Q(T)$ obeys the CW law in both of the above cases. We will not go further into the details of the theory, but show in the right-hand side of table 1 the SCR predictions for various anomalous behaviors in the two-dimensional system around the critical phase boundaries. We find that these results for antiferromagnetic case just explain the reported anomalous behaviors of high T_c cuprates above T_c.

Furthermore, the theory was successful even quantitatively. Electrical resistivity, specific heat, nuclear spin–lattice relaxation rates of ^{63}Cu and infrared conductivity were consistently explained with the use of only four parameters. In particular, good parameter-free comparisons were reported for the infrared conductivity, after fixing all the parameter values from the other experimental results. Thus we expect that the SCR theory applies well to two-dimensional itinerant electron systems and the nearly antiferromagnetic two-dimensional itinerant electron system provides a good model for the high T_c cuprates. For further details we refer to the original papers [8]. After further extended discussions on the itinerant electron magnetism and on a unified picture of magnetism in sections 3 and 4, we will come back to this problem in section 5 and discuss about possible reasons why this model can apply to high T_c cuprates as doped Mott insulators.

3. Early theories based on the local moment picture

It has been well known for many years that the local moment theory of magnetism was well established for magnetic insulator compounds and in most of the rare earth metals. As was mentioned in the previous section we now have the other established theories for weak itinerant ferro- and antiferromagnets, which constitute the opposite limit to the local moment systems. It is thus important to fill in the gap between these two extremes. Before going into this subject in the next section let us discuss here some earlier attempts at the theory of metallic magnetism based on the local moment picture and their limitations when applied to d-metals. Such a review of rather old but initial theories may be worthwhile to be given here briefly, since more detailed calculations along this line, based on the results of band calculations, have been pursued around this decade.

The importance of electron–electron correlations in d-band metals was emphasized by Van Vleck who seemed to conceive a possible local moment picture in d-metals [13]. The theory of local moment in metals was first developed for dilute alloys [14,15]. The condition for the appearance of the local moment was explicitly discussed by Anderson with the use of the Anderson model within the Hartree–Fock approximation [15]. Anderson correctly connected this result to the Curie–Weiss susceptibility. This corresponds, as a matter of fact, to use a saddle point approximation in the functional integral formalism as was discussed later by Schrieffer et al. [16]. Alexander and Anderson extended the theory to two impurity problems and discussed the effective exchange coupling constants [17]. The present author generalized this theory to include all kind of pairs of neighboring moments and obtained simple rules relating the sign of the effective exchange constant and the occupied fraction of the d-shell. Then switching to the effective Heisenberg model, he discussed the magnetism of d-metals and their alloys with qualitative and semiquantitative success [18]. The same type of arguments were extended also to the Wolff tight-binding model with essentially the same results [19]. The functional integral theory was later applied to the two impurities problem and the local saddle point approximation turned out to correspond to the above local Hartree–Fock theory [16]. The remarkable success, though qualitative, of these early theories were then considered to be an encouraging signal for further extension of this line of approach. As a matter of fact Hubbard [20], for example, used a similar local moment idea much later in his calculation of the Curie temperature of iron by using the result of band calculations. In order to estimate the effective Heisenberg exchange constant he calculated the energy cost of turning a local moment from the direction of the ferromagnetic moment in the ground state of iron as a function of turning angle. Incidentally, the functional integral theory for the Hubbard model was first discussed by Cyrot by using the coherent potential approximation and the local saddle point approximation [21]. It seems important to notice, however, that these theories are of qualitative or semiquantitative significance and should by no means be regarded as establishing the local moment picture in metallic magnets.

Exactly speaking, the local moment theory or the Heisenberg model, for d-electrons are established only for the Mott insulators, where the energies of spin slip excitations or bound electron–hole pair excitation are well separated from the excitation energies of unbound electron–hole pairs. In magnetic d-metals these two types of excitation spectra overlap in general. Thus the concept of local moment cannot be well-defined and have only approximate significance at best. As will be discussed in the following section, the amplitude variation of the local spin density should generally be taken into consideration when we deal with metals. This fact should be kept in mind when we discuss theories for the intermediate regime.

To be more explicit, we expect that most of the magnetic d-metals are in the intermediate regime. Some of them are close to the local moment limit while there are some others close to the weakly ferro- and antiferromagnetic limit. For example, Heusler alloys such as Pd_2MnSn, where magnetic Mn atoms are well separated from each other, are practically in the local moment regime. $FePt_3$ are expected to be close to the local moment regime while Fe_3Pt are not quite so, according to the neutron scattering experiments. Neutron scattering experiments on Fe and Ni show that their spin wave energies extend to much higher energies than $k_B T_C$, indicating that they can hardly be described consistently by the local moment theory. For further discussions on various magnetic metals see for example ref. [4]. Thus, the remarkable qualitative success of the early theories based on the local moment picture should mean that in describing some of the physical properties of certain metallic magnets the spin fluctuations may fairly well be approximated by those arising from a set of local moments. The limitation of the local moment picture as discussed in the above should be kept in mind in extending more realistic theories based on the calculated band structures.

4. A unified picture of magnetism – an interpolation theory

Let us now compare the characteristic features of spin fluctuations in the two opposite extremes.

The first point is concerned with the local amplitude of the spin density which should be kept constant in the local moment systems but not in itinerant electron systems in general. As a matter of fact the mean square local amplitude of the spin density S_L^2 varies significantly with temperature in weak itinerant ferromagnets and its increase with temperature above T_C is considered to give the origin of the new mechanism of the Curie–Weiss susceptibility. Thus the variable nature of the amplitude of the spin density must be taken into account in the unified theory.

The second point is concerned with the spatial spin correlations or the short range magnetic order above the transition temperature T_C. It is well known that in the usual Heisenberg local moment ferromagnets the short range order decays rapidly with increasing temperature above T_C. When substantial short range order remains above T_C the magnetic susceptibility should deviate from the Curie–Weiss law significantly [22]. Since almost all the metallic ferromagnets exhibit good Curie–Weiss susceptibilities the local moment picture (the constant amplitude of the local spin density) and the short range magnetic order are not compatible in them.

On the other hand the long wavelength components of spin fluctuations are predominant in weak itinerant ferromagnets even far above the Curie temperature, i.e., strong short range order remains even far above T_C. This is the result of the SCR theory which gives the new mechanism for the Curie–Weiss susceptibility compatible with the short range magnetic order, and was confirmed in MnSi by neutron scattering experiment [23]. Thus the consideration of the short range magnetic order should be another important factor to be taken into account in the unified theory.

The third point of contrast is the energy spread of the spin fluctuations compared with $k_B T_C$. Now Γ_q denotes the energy spread of the Fourier q-component of the spin fluctuation above T_C. In the local moment systems the q-dependence of Γ_q is relatively weak except for the range of small q and we have $\langle \Gamma_q \rangle \approx k_B T_C$. While in weak itinerant ferromagnets Γ_q increases significantly with q and we have $\langle \Gamma_q \rangle \gg k_B T_C$.

It is now desired to make a theory of general spin fluctuations which gives the already known correct results in the opposite two extremes and smoothly interpolate between them. It seems to be a very hard task to make such a unified theory for dynamic spin fluctuations which should give a final resolution of the controversy over localized vs. itinerant pictures in metallic magnetism. As a step toward this goal various spin fluctuation theories were developed within the adiabatic or static approximation by using the functional integral formalism.

Since earlier functional integral approaches to this problem employed the local saddle point approximation which resulted in a constant amplitude of the local moment or the local spin density above T_C, we had clearly to proceed beyond this approximation, so that the variation in amplitude of the local spin density was taken into account.

Such a theory was first presented in late 1970's [24]. The spirit of this theory is to consider the following form of the free energy as a function of spin variables:

$$\psi = -\sum_q V_q(S_L^2)|S_q|^2 + N_0 L(S_L^2)$$

$$= -\sum_{j,l} V_{jl}(S_L^2)(S_j \cdot S_l) + N_0 L(S_L^2),$$

with

$$S_L^2 = N_0^{-2}\sum_q |S_q|^2, \quad \sum_q V_q(S_L^2) = 0, \quad (4.1)$$

where N_0 is the number of magnetic atoms and the functional forms of $V_q(S_L^2)$ or $V_{jl}(S_L^2)$ and $L(S_L^2)$ are to be calculated from a given band structure. The partition function is then calculated by taking an average of $e^{-\psi/k_B T}$ with respect to both direction and amplitude of the spin variables. This model may be regarded as a generalization of the Heisenberg model to include the variation in amplitude of the local moments. However, the above discussion, using classical spin variables, shows only the spirit of the theory and in actuality the Stratonovich–Hubbard technique was employed to deal with the quantum mechanical systems. Confining ourselves to the static approximation, we calculate $\psi[\xi]$, a free energy of the non-interacting system under the influence of the magnetic field $(\pi U k_B T)^{1/2}\xi_j$ at the site j, etc., and then take an average of $\exp(-\pi\sum_j \xi_j^2 - \psi[\xi]/k_B T)$ over the field variables $\{\xi_i\}$, leading to the partition function $\exp(-F/k_B T)$. The charge density fluctuations can also be taken into account in this scheme introducing a charge field variable η_j.

Various methods of calculating $\psi[\xi, \eta]$ from a given band structure and several methods of evaluating the functional integrals were presented. We refer to ref. [4] for these developments and just mention here that a closed form expression for $\psi[\xi, \eta]$, corresponding to eq. (4.1) was given [3,25]. This expression takes care of the arbitrary variation of the amplitude of local spin density and the pairwise spatial spin correlations and reduces to the known correct results in the local moment limit of the half-filled Hubbard model

and in the weakly ferromagnetic limit, though within the adiabatic approximation.

As a direct application of this formalism; the phase diagram in the (U/W)–T plane of the Hubbard model with a half-filled band, showing the Mott transition, was actually calculated for the first time [3,25]. There are a number of numerical calculations performed so far based on the band structures of magnetic transition metals. They are mostly successful semiquantitatively, although some of them assume a constant local amplitude of the spin density and the others either neglect or empirically parametrize the short range order [26].

In view of the fact that the established theories in the both extremes, in practice, are those including parameters, it is also desirable to make a parametrized theory for the intermediate regime. Although it is difficult to find any valid expansion parameters in the intermediate regime, it should be useful if the two extremes are continuously connected in terms of certain parameters even qualitatively.

For this purpose the longitudinal stiffness constant for the spin density was introduced. When this parameter is large enough we are in the local moment regime with fixed S_L^2, while small values for the parameter allow S_L^2 to change with temperature. Introduction of this parameter and the transverse stiffness constant (corresponding to the familiar exchange stiffness constant) makes it possible to discuss general features of various physical quantities in the intermediate regime interpolating between the extremes. For example, a general expression for the magnetic susceptibility was obtained, interpolating between the two extremes characterized by different mechanisms of the Curie–Weiss law.

One of the most important features of the unified picture is the temperature variation of S_L^2 which depends primarily on the band structure and the electron occupation. This new degree of freedom enabled us to explain various old enigmas or so far unexplained phenomena. For example various kinds of anomalous temperature dependences of the magnetic susceptibilities in FeSi, CoS_2, $CoSe_2$, $MnAs_{1-x}P_x$, etc. were explained primarily in terms of the specific temperature variations of S_L^2 in them [27]. The Invar phenomenon was associated with the band structures which give rise to strong temperature variations of S_L^2.

The phenomena of coexistence of ferro- and antiferromagnetism and the transition between them are reasonably described with the use of this degree of freedom or the variable nature of S_L^2 [4]. Thus the interpolation theory, though within the adiabatic approximation, was rather successful in giving a unified picture for a broad range of magnets and in elucidating various old problems that remained without explanation.

5. High T_c oxides as doped Mott insulators

As was mentioned in section 1 the magnetic ground states with strong spin polarizations are expected to be described primarily by the Hartree–Fock or band theory. This applies to the antiferromagnetic ground state of the Mott insulator, as represented by the Hubbard model with a half-filled band, where even the spin waves as elementary excitations are correctly described by the RPA, i.e, the RPA results for large U/t ratio agree precisely with those obtained from the corresponding Heisenberg model with the Anderson kinetic superexchange interaction [4].

These situations are the same as in the itinerant electron magnets where the ferromagnetic ground states and the spin waves are fairly well described by the HF-RPA scheme with certain correlation corrections.

Since the Hartree–Fock-RPA scheme is known to be correct in the weak coupling regime, it is natural to expect that the ground state and the elementary excitations from the ground state are fairly well described by this scheme in practically the whole range of interaction. At least this scheme should give a good starting point of the theory and the many body corrections, which are expected to be most significant around the boundary of magnetic phase transition, are expected to give rise to no serious qualitative change of the physical picture.

At finite temperatures, however, the HF-RPA scheme fails almost completely; it no longer gives any consistent description of the physical properties of the system, not to speak of its poor comparison with experiment.

In order to make any consistent descriptions of the physical properties at finite temperatures we must use more advanced theories of spin fluctuations in each regime; the SCR theory for weak itinerant ferro- and antiferromagnets, the local moment theories for the Mott insulators, and the interpolation theory (yet to be worked out beyond the adiabatic approximation) for the itinerant magnets in the intermediate regime.

In the above discussions we are dealing with the itinerant magnets and the Mott insulators on the same footing. This point of view made it possible to describe the phase diagram including the Mott transition in section 4. In some previous strong correlation theories of the Mott transition [28,29], the antiferromagnetic nature of the ground state is neglected and thus the theories practically deal with the relatively high temperature phase although the temperature does not explicitly come into the theories. The meanings of these theories should thus be taken with care; they may be useful for very qualitatively understanding of the nature of this problem but by no means indicate the breakdown of the HF or band theory in the antiferromagnetic ground state.

Now let us consider the high T_c oxides as doped Mott insulators. In most of the doped cuprates of our interest, antiferromagnetic long range order is reduced rapidly with doping and superconductivity seems to appear only in the range of small or vanishing antiferromagnetic long range order and a substantial number of charge carriers whose effective masses do not seem to be much enhanced.

It may be pertinent to note here about the effect of random impurity potential in this problem. This effect is expected to be important when the number of carriers is small and the Anderson localization takes place. When the number of carriers is substantial, as in the high T_c regime, the impurity potential is expected to be screened so that we may neglect it in the first place. The screening effect is also expected on the electron–electron interactions and the effective interaction may be reduced from the value in the insulator phase.

Now in view of the general arguments in the earlier part of this section and in the preceding section, the high T_c oxides seem to be well regarded as weakly or nearly antiferromagnetic itinerant electron systems and thus their normal states may be treated in terms of the SCR theory.

One should note here that the above arguments may apply certainly to three-dimensional systems but not quite evidently to two-dimensional systems of the present interest. However, the successful application of the SCR theory of two-dimensional itinerant antiferromagnets to interpret the anomalous physical properties of high T_c oxides, as mentioned in section 2, seems to indicate that the SCR theory is applicable to two-dimensional itinerant electron systems and the two-dimensional nearly antiferromagnetic itinerant electron system is a good model for the relevant electrons in high T_c oxides.

Since antiferromagnetic spin fluctuations are expected to be responsible for the anomalous physical properties, including the electrical resistivity, of high T_c cuprates above T_c, it may be natural to expect the same spin fluctuations to play an important role in the mechanism of superconductivity.

Antiferromagnetic spin fluctuations are known to be a possible origin of the anisotropic superconductivity of d-wave character [30]. This mechanism is believed to be realized in some of the heavy electron superconductors. The antiferromagnetic spin fluctuations in high T_c cuprates as evaluated from the SCR analyses discussed in section 2, were employed to study this mechanism of superconductivity in high T_c oxides within the weak coupling theory [8]. The result showed that the superconducting order was of B_{1g} type and T_c was estimated to be of the right order of magnitude. From experimental point of view it seems that, there is no generally accepted conclusion at present as to the type of superconductivity in high T_c oxides. The above theoretical analysis is still too simple to be convincing. However, in view of the success of the above mentioned analysis above T_c we expect the antiferromagnetic spin fluctuations to play an important role in the mechanism of superconductivity.

6. Conclusions

As was outlined in the above sections our understanding of itinerant electron magnetism has advanced significantly in these two decades, as far as the d-metals and their compounds are concerned. It has turned out that the ground state properties of the d-metals are fairly well described by the band theory with some many body corrections, while the thermal excitations and finite temperature properties should be treated by the appropriate theories of spin fluctuations beyond the HF-RPA scheme.

The old controversy over the itinerant and the localized pictures was concerned primarily with the nature of magnetic excitations rather than the ground state properties and was resolved into the problem of the nature of spin density fluctuations in each substance which varies from material to material, ranging between the local moment limit and the extended moment limit (localized in k-space).

We now have established theories of spin fluctuations in the two extremes, the local moment regime and the weakly ferro- and antiferromagnetic regime. As for the intermediate regime, where most of the d-metals and their compounds are considered to belong the theories at present are based on the adiabatic approximation for the spin fluctuations. Although the interpolation theory not only gave a general picture of this regime but also succeeded in explaining many old enigmas or unexplained phenomena, the dynamical theory of spin fluctuations is still to be developed in the intermediate regime.

Aside from the above mentioned traditional problems of itinerant electron magnetism, recent discovery of high temperature superconductor oxides seems to have introduced a challenging problem in this field. We have strongly correlated itinerant d-electrons in these systems, doped Mott insulators, where the antiferromagnetic spin fluctuations seem to play an important role. This fact makes us to expect that magnetism and superconductivity should somehow be related strongly. This problem will provide us with one of the most challenging area of investigation in coming years.

References

[1] C. Herring, Magnetism IV, eds. G.T. Rado and H. Suhl (Academic Press, New York, 1964).

[2] V.L. Moruzzi, J.F. Janak and A.R. Williams, Calculated Properties of Metals (Pergamon, New York, 1978).
O.K. Anderson, O. Jepsen and D. Glötzel, in: Highlights of Condensed-Matter Theory, eds. F. Bassani, F. Fumi and M.R. Tosi (North-Holland, Amsterdam, 1985) p. 59.

[3] For a review see: T. Moriya, J. Magn. Magn. Mater. 14 (1979) 1. Also see the following two references.

[4] T. Moriya, Spin Fluctuations in Itinerant Electron Magnetism, Springer Series in Solid State Sciences, vol. 56 (Springer, Berlin, 1985).

[5] T. Moriya, Unified Picture of Magnetism, in: Metallic Magnetism, Topics in Current Physics, vol. 42, ed. H. Capellmann (Springer, Berlin, 1987) p. 15.

[6] For a recent review see for example: High Temperature Superconductivity, eds. K.S. Bedell, D. Coffey, D.E. Meltzer, D. Pines and J.R. Schrieffer (Addison Wesley, Reading, 1990). See also the following reference.

[7] Physics and Chemistry of Oxide Superconductors, eds. H. Yasuoka and Y. Iye (Springer, Berlin, 1991).

[8] T. Moriya, Y. Takahashi and K. Ueda, J. Phys. Soc. Jpn. 59 (1990) 2905; in ref. [7].
T. Moriya and Y. Takahashi, J. Phys. Soc. Jpn. 60 (1991) 776.

[9] G.G. Lonzarich, J. Magn. Magn. Mater. 54-57 (1986) 612.

[10] P.W. Anderson; in: Solid State Physics, vol. 14, eds. F. Seitz and D. Turnbull (Academic, Press, New York, 1963) p. 99.

[11] T. Moriya, J. Phys. Soc. Jpn. 55 (1986) 357.

[12] R. Ballou, B. Gorges, P. Mollo and P. Rouault, J. Magn. Magn. Mater. 84 (1990) L1.

[13] J.H. Van Vleck, Rev. Mod. Phys. 25 (1953) 220.

[14] J. Friedel, Nuovo Cimento suppl. 7 (1958) 287.

[15] P.W. Anderson, Phys. Rev. 124 (1961) 41.

[16] J.R. Schrieffer, W.E. Evenson and S.Q. Wang, J. de Phys. 31 (1971) C1.

[17] S. Alexander and P.W. Anderson, Phys. Rev. 133 (1964) A1594.

[18] T. Moriya, Prog. Theor. Phys. 33 (1965) 157.

[19] M. Inoue and T. Moriya, Prog. Theor. Phys. 38 (1967) 41.

[20] J. Hubbard, Phys. Rev. B 19 (1979) 2626.

[21] M. Cyrot, J. de Phys. 33 (1972) 125; Phil. Mag. 25 (1972) 1031.

[22] For a current discussion see: B.S. Shastry, D.M. Edwards and A.P. Young, J. Phys. C 14 (1981) L665.

[23] Y. Ishikawa, Y. Uemura, C.F. Majzak, G. Shirane and Y. Noda, Phys. Rev. B 31 (1985) 5884.

[24] T. Moriya and Y. Takahashi, J. Phys. Soc. Jpn. 45 (1978) 397.

[25] T. Moriya and H. Hasegawa, J. Phys. Soc. Jpn. 48 (1980) 1490.

[26] For a review see ref. [4] and the following references: Electron Correlation and Magnetism and Narrow Band Systems, Springer Series in Solid State Sciences, vol. 29, ed. T. Moriya (Springer, Berlin, 1980); Metallic Magnetism, Topics in Current Physics, vol. 42, ed. H. Capellmann (Springer, Berlin, 1987).

[27] References may be found in ref. [4] except for $MnAs_{1-x}P_x$ for which see: K. Motizuki and K. Kato, J. Phys. Soc. Jpn. 53 (1984) 735.

[28] J. Hubbard, Proc. Roy. Soc. A281 (1964) 401.

[29] W.F. Brinkman and T.M. Rice, Phys. Rev. B 2 (1970) 4302.

[30] K. Miyake, S. Schmitt-Rink and C.M. Varma, Phys. Rev. B 34 (1986) 6554.
D.J. Scalapino, E. Loh, Jr. and J.E. Hirsch, Phys. Rev. B 34 (1986) 8190.

The random field Ising model

D.P. Belanger and A.P. Young

Physics Department, University of California Santa Cruz, Santa Cruz, CA 95064, USA

We discuss the current status of random field systems, particularly those with Ising symmetry. Both theory and experiment agree that, in the equilibrium state, there is a transition to an ordered state in three dimensions and no such transition in two dimensions. The critical behavior in three dimensions is, however, not very well understood. More work remains to be done to understand the dynamics, both in the critical region and the low temperature phase.

1. Introduction

All solid materials have some quenched randomness, either in the form of defects and impurities or as a general structural property. In some cases the randomness is an integral part of the characteristics. Alloys and glasses, for example, can be designed to have properties not achievable with pure elements or periodic structures. Even surprisingly small amounts of quenched randomness can greatly influence the phase transitions by which materials form. It is impossible to eliminate all impurities or defects from materials. If we are to fully understand the formation of materials, we must characterize the drastic effects quenched impurities and defects can have on phase transitions. Among the models describing the effects of quenched randomness, are the spin glass model [1], which applies when the exchange interactions are both random and frustrated, the random exchange model [2] where the exchange interactions are random but not frustrated, and the random field (RF) model [3–16], where random ordering fields compete with the long range order. The topic of this review is primarily the last one. Considerable progress on the random field Ising model (RFIM) has been made over the past decade and much work is still in progress. The systems most commonly used to study the random field model are magnetically dilute crystals with applied uniform fields. Since these become random exchange systems once the random field is removed, we will need to include some essential properties of random exchange systems as well. Most of the definitive work on the random field model has been done on the Ising model, in which the ordering involves alignment, of magnetic spins for example, along one unique direction in the system. Although this condition is never strictly satisfied in experimental systems, materials with uniaxial anisotropy will normally exhibit Ising behavior for temperatures close enough to the transition temperature T_c (T_N for antiferromagnets).

The Hamiltonian of the random field Ising model is

$$\mathcal{H} = -\sum_{\langle i,j \rangle} J_{i,j} S_i S_j - \sum_i h_i S_i. \qquad (1)$$

The interactions $J_{i,j}$ are positive, so the ground state is ferromagnetic in the absence of the fields, and the fields, h_i, are independent random variables with a symmetric distribution, $P(h)$: i.e.

$$[h_i]_{av} = 0, \quad [h_i^2]_{av} = h_r^2, \qquad (2)$$

where $[\cdots]_{av}$ denotes an average over the disorder, i.e. over $P(h)$. Typical distributions chosen in the theory are the Gaussian distribution, and the bimodal distribution in which $h_i = \pm h_r$ with equal probability. This article will discuss the theory for this model and also experiments on systems which should be in the same universality class, as far as critical phenomena are concerned. One can generalize the model to vector spins, in which case both the spins and the fields are replaced by vectors. However, the vector case has been less thoroughly studied because, as we shall see, it is not expected to have a transition in three dimensions.

Much of the experimental progress on the random field problem has been made using insulating magnetic systems, which constitute nearly ideal physical realizations of the model. The magnetic interactions are typically short ranged and therefore often well modeled using the interactions between the first one or two nearest neighbors only. If there is easy-axis spin anisotropy, the system has Ising critical behavior close to T_c. Of course, one cannot actually apply a field which varies randomly from site to site in real materials. However, a major breakthrough, from the experimental perspective, came with Fishman and Aharony's realization [17] that effective random fields can be generated in dilute or mixed Ising antiferromagnets upon the application of a *uniform* field along the uniaxial direction. The randomness in the local spin concentration on each sublattice tends to favor local alignment of the more populated sublattice with the field direction. This local effect competes with the long range antiferromagnetic order in the same manner that the random field in a ferromagnet competes with the long range ferromagnetic order. It was shown [18] that the static universal critical behavior is identical for ferromagnets with random fields and dilute antiferromagnets in uniform fields. This equivalence does not necessarily hold for dynamics. Of great experimental importance in the dilute antiferromagnetic systems is the fact that the easily controlled uniform field strength translates directly into precise control of the effective random field strength. The Hamiltonian of the site diluted antiferromagnet in a uniform field is

$$\mathcal{H} = \sum_{\langle i,j \rangle} J_{i,j} \epsilon_i \epsilon_j S_i S_j - \sum_i H \epsilon_i S_i, \quad (3)$$

in which the interactions give an antiferromagnetic ground state, ϵ_i is 1 if site i is occupied and 0 if empty, and H is uniform field. The uniform and random field strengths are related by [18]

$$h_r^2 = \frac{x(1-x)[T^{MF}/T]^2 (g\mu_B SH/k_B T)^2}{[1 + \Theta^{MF}(x)/T]^2}, \quad (4)$$

where x is the concentration, T^{MF} is the pure system mean field transition temperature and Θ^{MF} is the Curie–Weiss susceptibility parameter. A great deal of the progress on experimental RFIM systems has been made on the dilute antiferromagnets partly because the random field can be varied continuously and even turned off completely. This aids discriminating RFIM effects from random exchange effects and allows crossover effects to be characterized.

Random field behavior has also been observed, for example, in structural phase transitions [19], binary fluids in gels [20], binary fluids in porous media [21], and commensurate–incommensurate transitions with frozen impurities [22].

The first paper to discuss whether ordering could occur in random field systems was by Imry and Ma [23]. They gave a simple, and by now famous, argument that even weak random fields would destroy long range order (LRO) in Ising systems in dimension d, if $d < 2$. For isotropic vector systems the corresponding argument gives no order for $d < 4$. This is the reason that the random field problem with vector spins has been less thoroughly studied, as mentioned above. In critical phenomena jargon one says that the lower critical dimension, d_l, of the RFIM is equal to 2, whereas for the pure Ising model it is equal to 1.

The Imry–Ma work was followed by more sophisticated calculations, based on diagrammatical perturbation theory, which showed that, taking the most divergent diagrams at each order, the critical behavior of the RF system is the same as that of the corresponding pure system in two fewer dimensions [24–26]. If one assumes that the perturbation theory determines the critical behavior in all dimensions (which is, of course, open to doubt) then one would conclude that d_l should be 2 greater than that of the pure system, i.e. $d_l = 3$ for the RFIM, which would imply no LRO in $d = 3$. There is however, a rigorous proof [27], see also ref. [28], which shows that there *is*

LRO in $d = 3$ at low temperatures and weak random fields. This is consistent with the Imry–Ma argument but inconsistent with dimensional reduction. As we shall discuss, dimensional reduction fails [22] because frustration gives rise to a "many valley" structure in the configuration space, similar to the situation in spin glasses [1].

There was also some controversy as to whether the experiments indicated LRO in three dimensions. This arose because RF systems have very slow dynamics close to the transition, and especially below it. This gave rise to difficulties of deciding what was the equilibrium state. There is now, however, general agreement, both among experimentalists and theoreticians, the LRO occurs for weak random fields in 3-dimensional Ising systems. The critical behavior at this transition, while clearly quite different from that of the pure Ising model, is not, however, very well understood. In two dimensions, there is no transition at all in the RFIM, so there the difference with the pure system is even more marked than in $d = 3$.

In the next section we summarize the current state of the theory, followed by a review of experiments in section 3. The last section discusses current problems in the field.

2. Theory

We start by first discussing briefly the predictions of mean field theory [29]. As the random field is increased from zero, the transition temperature decreases until it goes to $T = 0$ at a critical field, h_r^c. For $h_r > h_r^c$ the system is disordered at all temperatures. For distributions of the random field, $P(h)$, which monotonically decrease with increasing $|h|$, such as a Gaussian distribution, the transition is always continuous. For distributions which have a minimum at zero field, such as the bimodal distribution, there is a tricritical point [29] on the critical line such that the transition becomes first order for larger critical fields. Experimentally obtainable fields generally only depress T_c by a few percent, and are therefore in the region where mean field theory predicts a second order transition.

Next we review the basic definitions of critical exponents. The specific heat has a critical exponent α so

$$C_p = A^{\pm} |t|^{-\alpha} + B, \tag{5}$$

where we have also included a constant background term. This can be either an asymmetric cusp ($\alpha < 0$), an asymmetric divergence ($\alpha > 0$) or, in the case $\alpha = 0$, a symmetric divergence

$$C_p = A \ln(|t|) + B. \tag{6}$$

We also need to define exponents for the correlation length,

$$\xi = \xi_0^{\pm} |t|^{-\nu}, \tag{7}$$

the susceptibility (staggered susceptibility for antiferromagnets)

$$\chi = \chi_0^{\pm} |t|^{-\gamma}, \tag{8}$$

and the magnetization (staggered for antiferromagnets),

$$M = M_0 |t|^{\beta}, \tag{9}$$

which is only nonzero for $t < 0$. In these equations $t = T/T_c - 1$ and the $+$ and $-$ symbols indicate temperatures above and below T_c, respectively. It is also necessary to describe the behavior of the correlations precisely at the transition, and one defines

$$\chi(\mathbf{q}) \equiv \left[\langle S_q S_{-q} \rangle - \langle S_{-q} \rangle \langle S_q \rangle \right]_{\mathrm{av}} \sim q^{-2+\eta}. \tag{10}$$

For systems without random fields, $\langle S_q \rangle = 0$ at and above T_c, but for RF systems, it turns out that each of the two terms in eq. (10) are more divergent than χ itself, so one needs to define another exponent $\bar{\eta}$ for the disconnected part, i.e.

$$\chi^{\mathrm{dis}}(\mathbf{q}) \equiv \left[\langle S_{-q} \rangle \langle S_q \rangle \right]_{\mathrm{av}} \sim q^{-4+\bar{\eta}}. \tag{11}$$

Away from T_c one generalizes eqs. (10) and (11) with the usual scaling ansatz,

$$\chi(\mathbf{q}) = \frac{1}{q^{2-\eta}} \tilde{\chi}(q\xi), \tag{12}$$

$$\chi^{\mathrm{dis}}(\mathbf{q}) = \frac{1}{q^{4-\bar{\eta}}} \tilde{\chi}^{\mathrm{dis}}(q\xi). \tag{13}$$

In mean field theory, $\eta = \bar{\eta} = 0$, and for $T > T_c$, $\tilde{\chi}(x) = 1/(1 + x^2)$, and $\tilde{\chi}^{\text{dis}}(x) = 1/(1 + x^2)^2$. Below T_c, χ^{dis} also has a delta-function piece at $q = 0$ coming from long range order. The elastic neutron scattering cross section is proportional to the structure factor,

$$S(q) = \left[\langle S_q S_{-q} \rangle\right]_{\text{av}} \equiv \chi^{\text{dis}}(q) + \chi(q). \quad (14)$$

Hence, the mean field structure factor is a sum of Lorentzian and Lorentzian-squared terms, plus a delta-function part below T_c, i.e.

$$S(q) = \frac{A}{q^2 + \kappa^2} + \frac{B}{(q^2 + \kappa^2)^2} + M^2\delta(0), \quad (15)$$

where $\kappa = \xi^{-1}$. This form has often been used to analyze experimental data. We should emphasize, however, that eq. (15) is only a mean-field approximation. It cannot be very accurate in $d = 3$ for the RFIM because, since it neglects the exponents η and $\bar{\eta}$, it incorrectly predicts a lower critical dimension of 4. (To see this, integrate the Lorentzian-squared term at T_c over q and note that the answer must be finite since it is a local quantity.) For antiferromagnets, the delta function is at the wave-vector of the magnetic ordering rather than at $q = 0$. Note that the Lorentzian-squared term vanishes in the absence of the random fields, at least above T_c (a random bond magnet is, however, expected to have such a term below T_c).

For transitions with zero random field, all critical exponents can be determined from two of them, ν and η by scaling laws. For RF systems the natural generalization [30–32] is that there are three independent exponents, ν, η and $\bar{\eta}$. Furthermore, the hyperscaling relations, which involve the dimension, d, have d replaced by $d - \theta$, where [30–32]

$$\theta = 2 - \bar{\eta} + \eta. \quad (16)$$

For example, we have

$$(d - \theta)\nu = 2 - \alpha, \quad (17)$$
$$\beta = \tfrac{1}{2}(d - \theta - 2 + \eta)\nu = \tfrac{1}{2}(d - 4 + \bar{\eta})\nu. \quad (18)$$

Schwartz and Soffer [33] have given a simple but elegant proof of the inequality

$$\bar{\eta} \leq 2\eta. \quad (19)$$

It has been suggested [34–36] that the inequality in eq. (19) may actually occur as an equality. If so, there would only be two independent exponents, as in zero random field systems. We will come back to this later.

As the transition is approached, it is common at larger $|t|$ to observe behavior resembling one universality class, only to see a crossover to another class upon closer approach. An example of crossover phenomena relevant to our discussions is from nonrandom field behavior to random field behavior as T_c is approached in weak fields. The manner in which this crossover occurs is integral to the understanding of the random field problem. We need to study the crossover behavior of the free energy scaling function

$$F = |t|^{2-\alpha} f(t h_r^{-2/\phi}), \quad (20)$$

where ϕ is the crossover exponent, t is the reduced temperature given by

$$t = \left(T - T_c(0) + bh_r^2\right)/T_c(0), \quad (21)$$

$T_c(0)$ is the transition temperature in zero field, and the term involving b is a mean field correction. Eq. (20) is valid in the limit that h and t tend to zero. For the RFIM $\phi = \gamma_{\text{pure}}$, where γ_{pure} is the susceptibility exponent of the pure system, but for the dilute antiferromagnet in a field, it turns out [37] that ϕ is somewhat larger than the susceptibility of the random-exchange system in zero field. For weak random fields, *all relevant reduced temperatures* scale with h_r in the combination $t/h_r^{2/\phi}$. For example, in three dimensions, the scaling function, f, in eq. (20), will have a singularity at some (negative) value of its argument, so the transition occurs at temperature $T_c(H)$, where the corresponding reduced temperature $t_c(H)$, related to $T_c(H)$ as in eq. (21), is given by

$$t_c(H) = -cT_c(0)h_r^{2/\phi} \quad (d = 3), \quad (22)$$

where c is a nonuniversal constant. Note that the relation between the applied field, H, and the effective random field, h_r, is given by eq. (4). In $d = 2$, there is no transition in a field, so the scaling function, f, does not have a singularity. It is, however, possible to define a temperature

$t_{\text{round}}(H)$, at which, say, the specific heat has a rounded maximum, where

$$t_{\text{round}}(H) = -cT_{\text{c}}(0)h_{\text{r}}^{2/\phi} \quad (d=2). \tag{23}$$

In addition, crossover to RFIM behavior will only be observed for

$$|t| < t_{\text{cross}}(H) = ah_{\text{r}}^{2/\phi}, \tag{24}$$

where a is another nonuniversal constant. Far from $T_{\text{c}}(0)$, and with small random fields the system shows critical-like behavior characteristic of the zero field random exchange Ising model (REIM). Only when $T_{\text{c}}(0)$ is closely approached, with eq. (24) satisfied, does the system cross over to random field behavior.

Having defined the exponents, we now discuss the static critical behavior of the RFIM. The first question to ask is whether or not even weak random fields can destroy LRO. (Note that sufficiently strong random fields will always destroy LRO, as predicted by mean field theory.) This question was first discussed by Imry and Ma [23]. They considered the possibility that the ordered state will break up into domains of typical size L, where, in each domain, the spins follow the net direction of the random field. The gain in random field energy in doing this is of order the square root of the volume of the domain, i.e. $\sim h_{\text{r}} L^{d/2}$. This will, however, cost exchange energy because bonds are broken on the surface of the domains. This energy is $\sim JL^{d-1}$, where J is the exchange interaction. One therefore has to minimize

$$-h_{\text{r}} L^{d/2} + JL^{d-1} \tag{25}$$

with respect to L. For $d > 2$ one finds that the exchange term dominates so domains do not form and one has LRO. By contrast, for $d < 2$, domains do form, with a size

$$L_{\text{c}} \sim (h_{\text{r}}/J)^{-2/(2-d)}. \tag{26}$$

This indicates that LRO is destroyed at $T = 0$ for weak random fields in $d < 2$, and $d_1 = 2$. A somewhat more sophisticated argument [38] shows that LRO is also destroyed in $d = 2$ and in this case the domain size is given by

$$\ln(L_{\text{c}}) \sim (J/h_{\text{r}})^2. \tag{27}$$

Applying similar arguments for isotropic vector spins one sees that the domain wall can be "spread out" over the size of the domain, which leads to an exchange energy varying like L^{d-2} and a lower critical dimension of $d_1 = 4$. Hence no LRO is expected in $d = 3$, as noted earlier. The Imry–Ma argument has two great virtues: (i) it is very simple and (ii) it seems to give the right value for the lower critical dimension.

The Imry–Ma argument does not, however, give the values of the critical exponents. The first attempts to calculate them used the continuum formulation of the model and applied perturbative renormalization group (RG) techniques. One finds that the upper critical dimension, d_{u}, above which one has mean field exponents, turns out to be $d_{\text{u}} = 6$, compared with $d_{\text{u}} = 4$ for systems without random fields. Furthermore, one finds that to all orders in perturbation theory [24–26] the critical exponents are those of the corresponding pure system in two fewer space dimensions, a phenomenon known as "dimensional reduction"; more precisely one gets:

$$\eta(d) = \bar{\eta}(d) = \eta_{\text{pure}}(d-2), \tag{28}$$

$$\nu(d) = \nu_{\text{pure}}(d-2). \tag{29}$$

Hyperscaling relations also go through with d replaced by $d - 2$ because θ in eq. (16) is equal to two. This is a beautiful and remarkable result. Unfortunately, it must be incorrect because, taken literally, it would predict a lower critical dimension of three, which disagrees with the Imry–Ma argument and, more seriously, with the rigorous proof [27,28] that there is LRO in $d = 3$ for weak disorder and low temperatures.

Insight into the difficulty can be obtained from the Parisi and Sourlas [26] derivation of dimensional reduction. They noted that the terms included in the perturbation theory correspond to averaging the local mean field equations. These are equations which relate the site magnetizations, m_i, to the random fields for a given configuration of the disorder. If one assumes that there is only one solution, which was implicitly done by Parisi and Sourlas, then dimensional reduction follows. However, it is not difficult to see [30] that there must be many solutions below the transition temperature of the pure system, i.e. everywhere along the random field phase diagram. Presumably, if one could incorporate the existence of many solutions and give them the correct statistical weight, one would obtain a correct theory with $d_l = 2$.

An analogous problem occurs in the mean field theory of spin glasses [1], where the energy landscape has many minima and it is necessary to include contributions to the statistical sum from many of them. This has been accomplished in the Parisi [39] theory, which used the technique of "replica symmetry breaking". It would obviously be useful to apply similar ideas for the RFIM. Although this has not been done for the bulk critical properties of the RFIM, it has been carried out for the related problem of an interface in the RFIM [40]. In this problem one assumes that LRO has been established and that there is an interface separating a region of up spins from a region of down spins. One then asks how far the interface wanders because of the random fields and thermal fluctuations. More precisely, let ω be the height of the ($d-1$ dimensional) interface above some reference height, and let us define an exponent ζ by

$$\left[\langle (\omega(x+1) - \omega(x))^2 \rangle\right]_{av} \sim l^{2\zeta}. \quad (30)$$

An Imry–Ma type argument then gives [40]

$$\zeta = (5-d)/3. \quad (31)$$

This is consistent with $d_l = 2$ because, in the interface approach, one argues that the ordering has disappeared when the wandering of the interface is as large as its width, i.e. $\zeta = 1$. One can also estimate ζ by field theory techniques, and one obtains a different result, namely [41]

$$\zeta = (5-d)/2, \quad (32)$$

to all orders in perturbation theory, which consistent with the incorrect value, $d_l = 3$. By setting up the problem in the replica formulation, Mézard and Parisi [40] have shown that result (32) corresponds to the replica symmetric solution. They then demonstrate that this solution is unstable and construct another solution which breaks replica symmetry, and which is exact in the limit

where the height of the interface is generalized to an m-component vector with $m \to \infty$. The replica symmetry breaking solution recovers the Imry–Ma result, (31), i.e. it is consistent with the apparently correct prediction that $d_l = 2$. It would clearly be desirable to carry out a similar calculation for the bulk critical behavior.

Because of these difficulties with perturbative renormalization group calculations, the best set of estimates for exponents have come from numerical calculations. Large scale Monte Carlo simulations have been carried out on the RFIM by Young and Nauenberg [42] who suggested that the transition may be first order, and by Ogielski and Huse [43] on the dilute antiferromagnet in a field who found a second order transition. Probably the most accurate estimates of exponents are by Ogielski [44] who used a sophisticated algorithm to obtain exact ground states for rather large random field systems. He found a second order transition with

$$\nu \approx 1.0, \quad \overline{\eta} \approx 1.1. \quad (33)$$

From the scaling relation (18) one sees that the order parameter exponent β is very small, so, though the transition seems to be continuous according to Ogielski's calculations, it is very difficult to distinguish it from a first order transition, which would correspond to $\beta = 0$. The best estimate of η is from Monte Carlo [43] and gives

$$\eta = 0.5 \pm 0.1, \quad (34)$$

so the Schwartz–Soffer inequality (19) may be satisfied as an equality. From eqs. (16), (17) one finds $\alpha \approx 0.5$, which, as we will see in section 3 below, is inconsistent with the experimental value $\alpha \approx 0$.

Finally, in this section, we discuss critical dynamics. Whereas at most transitions, the relaxation time, τ, diverges according to the conventional dynamical scaling form [45], $\tau \sim \xi^z$, where z is the dynamical exponent, an exponential divergence

$$\tau \sim \exp\left[C/(\xi^\theta)\right] \quad (35)$$

is predicted [30,32] for the RFIM. Some qualitative evidence in favor of eq. (35) has been found in simulations [43]. Eq. (35) does seem to have been confirmed by experiment [46], as discussed in the next section.

3. Experiments

We first discuss the case of $d = 2$, for which it is well established that there is no transition in the equilibrium state, as discussed in the previous section. For weak random fields, the system will appear to be approaching the transition with exponents of the zero random field system as the temperature is reduced. When t, the reduced temperature defined by eq. (21), reaches the crossover temperature, eq. (24), the behavior changes and the specific heat then has a rounded maximum at the reduced temperature, t_{round} in eq. (23). With a small enough random field, the rounded RFIM region will, in practice, be too small to be observed. However, even for arbitrarily small random fields, there can be no long range order in the two-dimensional system at low T, and, in principal, domain structure should be observable. In critical behavior studies, however, commonly used techniques such as neutron scattering are limited by instrumental resolution to the length scales of the order of 1000 Å, so a state with much larger domains is indistinguishable from one with long range order. The domain size, L_c, for $d = 2$, is given by eq. (27) and is enormous for small fields. Hence, for small enough fields, the system will appear to develop long range order and a phase transition characterized by eqs. (5)–(9) will appear to take place.

Random field specific heat results for $d = 2$ were obtained [47] using the layered magnetic system $Rb_2Co_xMg_{1-x}F_4$ with $x = 0.85$. This system is an extremely good two-dimensional Ising system. The pure Rb_2CoF_4 and the closely related K_2CoF_4 systems have been studied [48,49] extensively and all of the critical behavior is in excellent accord with the exact results for the pure two-dimensional Ising model [50]. The two-dimensional character of the ordering is a result of a very large [51] intraplanar exchange interaction relative to the interaction between planes. In the dilute system $Rb_2Co_{0.85}Mg_{0.15}F_4$, Ferreira et

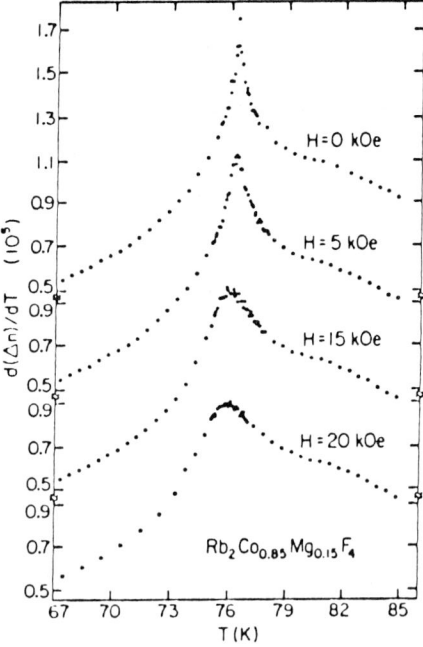

Fig. 1. The temperature derivative of the optical birefringence, $d(\Delta n)/dT$, vs. T for the $d = 2$ Ising antiferromagnet $Rb_2Co_{0.85}Mg_{0.15}F_4$ in applied fields $H = 0$, 0.5, 1.5 and 2.0 T. In dilute antiferromagnets $d(\Delta n)/dT$ is proportional to the magnetic specific heat. The data sets are offset for clarity. The largest field, $H = 2.0$ T, corresponds to a random field $h_r \approx 10^{-2}$ given by eq. (4). The rounding of the peak increases with H and occurs in equilibrium [47].

al. [47] used optical birefringence techniques to show that, whereas the zero field transition is sharp to the extent of the experimental resolution, a small random field $h \approx 10^{-2}$ at $H = 2$ T, severely rounds the transition, as shown in fig. 1. In zero field, the specific heat shows a nearly symmetric, logarithmic divergence. With the application of the field, the rounding of the transition is characterized by the scaling function [47]

$$C_m = g(th_r^{-2/\phi}) - A \ln(h) + D(t, h_r), \quad (36)$$

where $D(t, h_r)$ is a nonsingular background. Eq. (36) is obtained from eq. (20) by differentiating twice with respect to t and noting that $\alpha = 0$ for the pure 2-d Ising model corresponds to a logarithmic singularity. The data do indeed scale according to eq. (36), as shown in fig. 2. The crossover exponent is found to be $\phi = 1.58 \pm 0.22$, which agrees with the staggered susceptibility exponent $\gamma = 1.75$. All of the effects observed in the $d = 2$ specific heat experiments are consistent with theoretical predictions and computer simulation results [38].

In neutron scattering, no long range order is observed, independent of the field and temperature cycling [52], in proximity to the rounded specific heat peak at temperature $T_{\text{round}}(H)$. At temperatures much below $T_{\text{round}}(H)$, however, freezing does occur. If the sample is cooled in zero field to low temperatures, the long range order is preserved as the field is applied. The long range order is metastable in this case. Then, as the temperature is increased toward $T_{\text{round}}(H)$, the long range order collapses with time [52,53] and is completely gone before the temperature region of the rounded specific heat peak is entered, as shown in fig. 3. The temperature at which the metastable long range order is observed to decay depends on the times scales of the experiment, as shown by comparison of neutron scattering [52] ($t \approx 10^3$ s) and high field pulsed experiments [54] ($t \approx 10^{-5}$ s). A particularly important point to make about the $d = 2$ experiments is that the transition rounding, in the specific heat experiments for example, occurs in a region where no irreversibility is seen in any

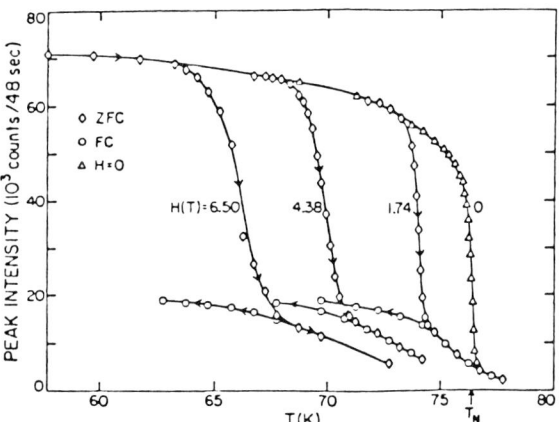

Fig. 3. The neutron scattering intensity vs. T at the (100) antiferromagnetic Bragg point in $Rb_2Co_{0.85}Mg_{0.15}F_4$ for $H = 0$, 1.74, 4.38 and 6.5 T. For $H > 0$, two procedures, cooling in the field (FC) and cooling in zero field and subsequently applying the field (ZFC), give different results at low T. However, in the region corresponding to the rounded birefringence peaks, both procedures yield identical results, indicating equilibrium. The curves are simply to aid the eye and the arrows indicate the direction in which data were taken. The large ZFC intensities, essentially identical with the $H = 0$ case, show that long range antiferromagnetic order is metastable at low T [52].

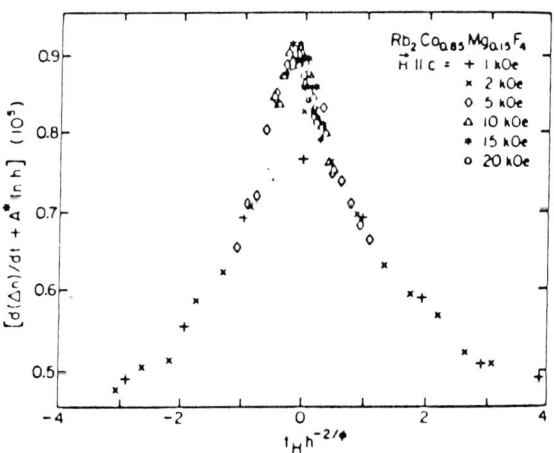

Fig. 2. $d(\Delta n)/dT + \bar{A} \ln(h)$ vs. $t_H h^{-2/\phi}$ in $Rb_2Co_{0.85}Mg_{0.15}F_4$ for applied fields $H = 0.1, 0.2, 0.5, 1.0, 1.5$ and 2.0 T, with $10^{-3} \le |t| \le 10^{-2}$, $\bar{A} = 1.64 \times 10^{-6}$, $\phi = 1.75$ and $h = g\mu_B SH/k_B T$, showing the $d = 2$ scaling behavior in eq. (36) [47].

experiment, i.e. the rounding of the transition occurs *in equilibrium*. This is in stark contrast with the situation for $d = 3$ discussed below, where rounding can be attributed to dynamics and nonequilibrium effects associated with the $d = 3$ RFIM transition.

Early neutron scattering experiments [55,56], in which the absence of long range order was first observed after cooling the sample in a field (FC) below $T_{\text{round}}(H)$ in $Rb_2Co_xMg_{1-x}F_4$, led to the conclusion that the transition was destroyed by random fields, in agreement with conclusions from the optical experiments done somewhat later.

An important observation [55] in $d = 2$ neutron scattering measurements is that well below the rounded transition for $H > 0$, the scattering lineshapes are very non-Lorentzian if the sample is FC, in contrast to the usual $H = 0$ lineshapes. For both $d = 2$ and 3, once the instrumental resolution corrections are made, the $H = 0$ scattering lineshapes are adequately described by a sum of a Lorentzian and a delta function Bragg scattering term, i.e. eq. (15) with $B = 0$.

We next discuss the situation for the $d = 3$ RFIM which has proven to be much more complex and difficult to study in the dilute antiferromagnets since dynamic effects play a major role in any characterization of the phase transition. Nevertheless, considerable progress has been made. For $d = 3$, $Fe_xZn_{1-x}F_2$ is the system most extensively studied and characterized. The anisotropy in FeF_2 is single ion [57] in character and since it is large compared with the exchange interaction, asymptotic $d = 3$ Ising critical behavior has been observed [58,59] for $|t| < 0.02$. The FeF_2 antiferromagnet forms extremely high quality single crystals when mixed with the isomorphic ZnF_2 diamagnet. Crystals with virtually any magnetic concentration can be grown and, with considerable effort, several very high quality crystals [60] have been obtained and characterized for uniformity in concentration.

Of paramount importance in any critical behavior measurement in a system with quenched randomness is the uniformity of the randomness on macroscopic scales. Since the transition temperature in general depends strongly on concentration, gradients in concentration produce variations in T_c throughout the sample. Gradients can obscure otherwise sharp transitions. Incorrect values for exponents can be obtained when fitting data to power law behavior if the gradients are not properly considered. The measurement of concentration gradients and their effect on critical behavior measurements have been described in detail [61,62]. Some of the discrepancies between the interpretations of experimental data from different experimental groups are attributable to the degree to which the gradients are considered. Typically, concentration gradients must be small enough so that the variation of $T_c(H)$ through the sample is on the order of 10^{-3} of $T_c(H)$ or less if one wants to observe RIFM critical behavior.

Since a transition must occur [27,28] in a $d = 3$ RFIM system, it will occur at a reduced temperature $t_c(H)$, given by eq. (22), where the reduced temperature is related to the actual temperature by eq. (21). For random antiferromagnets in a field, the crossover exponent is predicted [37] to be somewhat larger than the exponent γ for the zero random field system. This has been observed experimentally in the $Fe_xZn_{1-x}F_2$ system, for which the value $\gamma = 1.31 \pm 0.03$, measured [63] using neutron scattering techniques, is several percent less than the crossover exponent $\phi = 1.42 \pm 0.02$ evaluated from the scaling behavior of $t_c(H)$, in excellent agreement with the theoretical prediction [37] $\phi/\gamma \approx 1.1$. In the Ising system $Fe_xZn_{1-x}F_2$ and the weakly Ising system $Mn_xZn_{1-x}F_2$, the concentration dependence of the parameter c in eq. (22) has been characterized [64] by the empirical expression $c(x) = c'(x)T_N(1)/T_N(x)$, where $c'(x)$ has only a weak concentration dependence. Just as in the $d = 2$ case, the RFIM region becomes larger as H increases.

For the measurement of the specific heat in $d = 3$ systems, optical techniques have proven invaluable. The resolution of the optical techniques rivals [58] the best specific heat results in the pure Ising system FeF_2, with an advantage of being insensitive to phonon specific heat contributions [65]. In dilute systems, optical techniques have an added advantage over more conventional techniques in that the effects of concentration gradients can be greatly minimized by orienting the laser beam perpendicular to the concentration gradient. The first convincing experimental [66] indications of a phase transition in a $d = 3$ random field antiferromagnet were obtained using an optical linear birefringence technique. Similar results [67], shown in fig. 4, are obtained using a Faraday rotation optical technique [68]. Note that the RFIM critical behavior region increases with H as expected from eq. (24). The specific heat has also been measured directly with a heat pulse technique [69] in one of the most uniform magnetically dilute crystals, $Fe_{0.46}Zn_{0.54}F_2$, and the results are in excellent agreement with those of the optical techniques [64].

With the birefringence technique, the magnetic specific heat critical behavior of $Fe_{0.6}Zn_{0.4}F_2$ in zero field was shown to be strikingly different from that of pure FeF_2. In the pure case, the critical parameters $\alpha = 0.11 \pm 0.005$ and $A^+/A^- = 0.53 \pm 0.02$ are in excellent agreement with the

$d = 3$ pure Ising model. The dilute crystal in zero field yields [70] the parameters $\alpha = -0.09 \pm 0.03$ and $A^+/A^- = 1.6 \pm 0.3$, in agreement with theory [71,72] and satisfying the Harris criterion [2,73] that $\alpha < 0$ for random exchange systems.

For $d = 3$, the behavior observed near $T_c(H)$ depends on the field and temperature cycling procedure used. If the sample is FC (cooled in the field), the transition will appear rounded and at low T the system will be in a metastable domain state. If, on the other hand, the sample is cooled in zero field before the uniform field is applied (ZFC), the transition will appear much sharper, being rounded only by dynamic effects. The data shown in fig. 4 were obtained after ZFC. The temperature above which FC and ZFC procedures result in identical behavior is designated as the equilibrium boundary $T_{eq}(H)$. The field scaling of $T_{eq}(H)$ is identical to that of $T_c(H)$, see eq. (22), except that $T_c(H) < T_{eq}(H)$. It is not known at this time what time dependence $T_{eq}(H)$ may have. It may be determined consistently in many different experiments, indicating a rather weak time dependence at most. Upon ZFC, the application of a field causes the $Fe_xZn_{1-x}F_2$ specific heat critical behavior to change [66] from the the zero field asymmetric cusp, described earlier, to a peak approximating a symmetric, logarithmic divergence, implying $\alpha \approx 0$ and $A^+/A^- \approx 1$. The applied uniform fields used to observe the crossover to the new behavior produce effect random fields in this system on the order of 10^{-1} for $H = 2$ T – much larger than those created in the $d = 2$ system described previously where the transition was clearly rounded. As noted in section 2, the best theoretical estimate of α, obtained using scaling relations to relate α to other exponents determined numerically, is $\alpha \approx 0.5$, which disagrees with experiment.

The ZFC peaks in fig. 4 not only exhibit symmetric, logarithmic behavior over a range of $|t|$

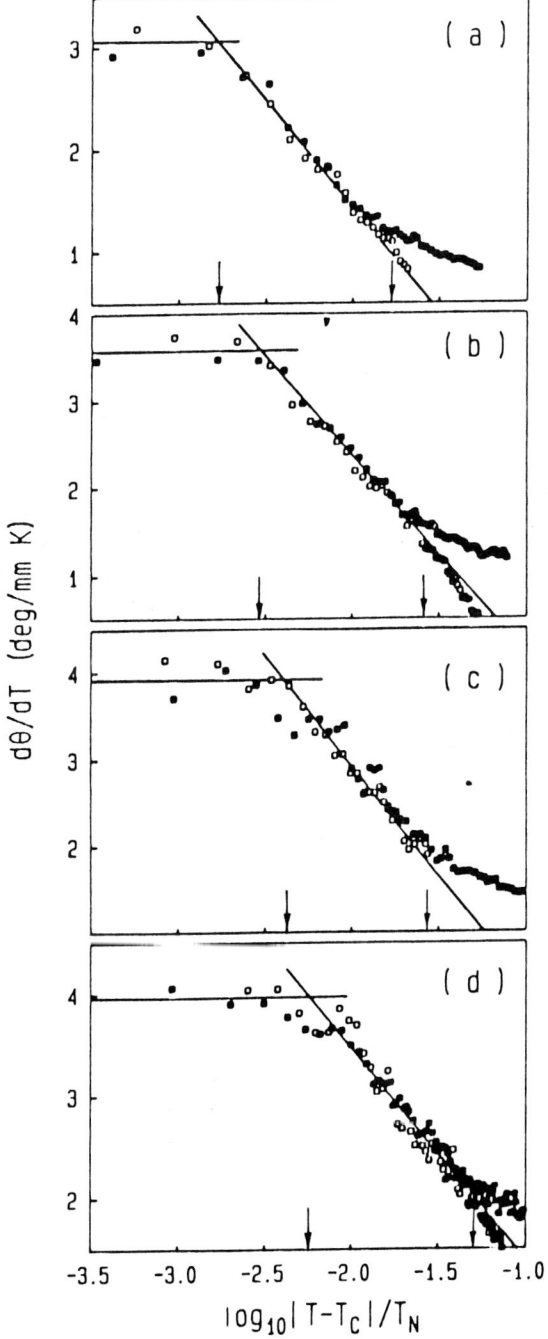

Fig. 4. $d\theta/dT$, which is proportional to the magnetic specific heat, vs. $\log_{10}(|t|)$ in the $d = 3$ Ising antiferromagnet $Fe_{0.47}Zn_{0.53}F_2$ for $H = 2$, 3, 4 and 5 T in (a), (b), (c) and (d), respectively. In each figure, the right arrow indicates $t_{cross}(H)$ at which the random exchange to random field crossover occurs, in accordance with eq. (24). The left arrow in each case indicates the onset of rounding from RF dynamics. Data between the arrows in each case are seen to approximate a symmetric, logarithmic divergence [67].

which increases with the applied field, but also rounding at smaller $|t|$. This rounding, which also increases with the strength of the random field, is due to the very slow dynamics of the RFIM near T_c. The relaxation time diverges exponentially according to eq. (35) and, when the time of the measurement is less than the relaxation time, the system drops out of equilibrium and rounding occurs. Nevertheless, in specific heat measurements the region of $|t|$ over which rounding takes place is observed to be much smaller than the RFIM critical behavior region. The nature of the H–T phase diagrams, including dynamic effects has been reviewed [7] in detail from an experimental perspective for both $d = 2$ and 3 RFIM systems.

The random field dynamics in $Fe_{0.46}Zn_{0.54}F_2$ were investigated by King, Mydosh and Jaccarino using ac susceptibility techniques [74], which, incidentally, is one of the first techniques used, by Rohrer and Scheel [75], to study random field effects experimentally. It was shown that activated dynamics could describe the data with the modified Vogel–Fulcher law [30,32] in eq. (35). If conventional dynamic scaling [45], $\tau \sim \xi^z$, were used to describe the data, an unusually large exponent, $z \approx 14$, is required [74]. Nash, King and Jaccarino [46] extended the measurements to the frequency range 5×10^{-3} Hz $\leq f \leq 10^5$ Hz, as shown in fig. 5, and were able to rule out conventional power law dynamics in favor of activated dynamics. Using the neutron scattering result [76] $\nu = 1.0 \pm 0.15$ (which is in good agreement with the theoretical estimate, eq. (33)) the value $\theta = 1.05 \pm 0.22$ is obtained, in agreement with the modified hyperscaling relation, eq. (17), using the birefringence [66] result $\alpha \approx 0$. As already noted, this value of α is in disagreement with the best theoretical estimates.

The nature of the critical slowing, as observed in the ac susceptibility measurements, is not very sensitive to the field and temperature cycling procedure used to prepare the sample. This implies that the fluctuations probed on these time scales are not sensitive to length scales as large as the metastable domains formed upon FC.

Neutron scattering techniques have also been extensively used to elucidate the nature of the

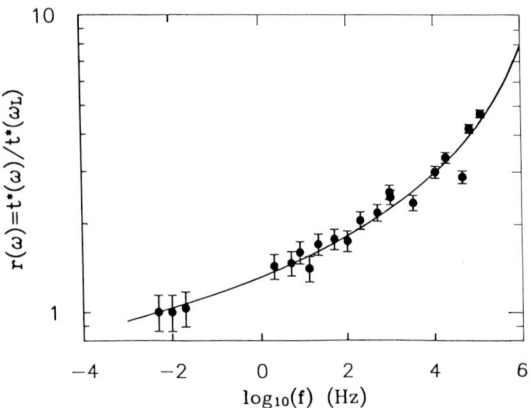

Fig. 5. Normalized dynamic rounding temperature ratio, $r(\omega) = t^*(\omega)/t(\omega_L)$, vs. $\log(\omega/2\pi)$, with $\omega/2\pi = 0.005$ Hz. The solid curve is a fit to activated dynamics scaling $r(\omega) = [\log(\omega/\omega_0)/\log(\omega_L/\omega_0)]^{-1/y}$ with $y = 1.05$ and $\omega_0 = 5.5 \times 10^7$/s. Conventional scaling would lead to a linear relation between $r(\omega)$ and $\log(\omega/2\pi)$ [46].

random field Ising transition in $d = 3$ dilute antiferromagnets. The application of neutron scattering to dilute antiferromagnets has been reviewed in detail [8]. The scattering lineshapes for the dilute Ising antiferromagnet at $H = 0$ are rather well described by Lorentzian functions of the form given in eq. (15), with $B = 0$, where the delta-function is really at the antiferromagnetic Bragg scattering point. The behavior is much like that of the pure FeF_2 system, but with slightly different exponents and amplitude ratios. The pure Ising exponents measured [59] in FeF_2 are $\nu = 0.64 \pm 0.01$ and $\gamma = 1.25 \pm 0.02$, whereas in $Fe_{0.46}Zn_{0.54}F_2$ the values obtained [63] are $\nu = 0.69 \pm 0.01$ and $\gamma = 1.31 \pm 0.03$. A non-Lorentzian component of the lineshape is predicted [77] for dilute antiferromagnets in zero field, but has not yet been detected.

Random fields alter the scattering lineshapes significantly. The first evidence of the non-Lorentzian nature of the $d = 3$ scattering profiles for $H > 0$ were obtained by Yoshizawa et al. [55,56] in the FC anisotropic $Co_xZn_{1-x}F_2$ system at low T. The scattering profile is dramatically altered in a similar way by the application of a field in the $Fe_xZn_{1-x}F_2$ system. However, in the latter case, the non-Lorentzian nature was observed [76] even above $T_{eq}(H)$, where the system

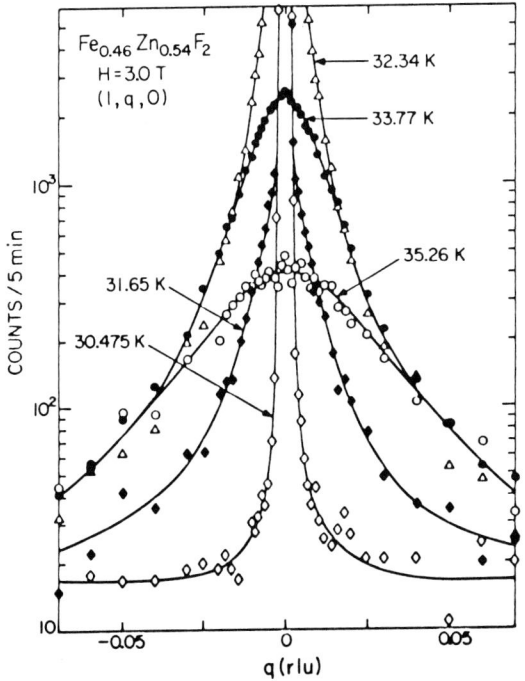

Fig. 6. Neutron scattering intensity vs. q in $Fe_{0.46}Zn_{0.54}F_2$ for $H = 3$ T at $T = 30.475, 31.65, 32.34, 33.77$ and 35.26 K. All of the data were obtained after ZFC. Data at temperatures $T = 32.34, 33.77$ and 35.26 K are above $T_c(H)$ and are well described by eq. (15). The lineshapes for $T < T_c(H)$ are not well understood. The solid curves are guides for the eye [90].

is in equilibrium. Empirically, a sum of Lorentzian and squared Lorentzian terms, i.e. eq. (15) with $M = 0$, has proven suitable [76] for characterizing the experimentally obtained scattering line shape above $T_c(H)$, as shown for $Fe_{0.46}Zn_{0.54}F_2$ in fig. 6. Mean-field arguments which give eq. (15) are presented in refs. [78,79]. Since the exact form of $S(q)$ is unknown, the sum of a Lorentzian and squared Lorentzian is often used for analysis, though, as argued in section 2, it is rather unsatisfactory.

The most extensive critical behavior analysis has been done [76] on the system $Fe_{0.6}Zn_{0.4}F_2$ for $T > T_{eq}(H)$. The correlation length–temperature dependence is found to be consistent with the scaling form in eq. (7), with an exponent $\nu = 1.0 \pm 0.15$. This is in agreement with the best theoretical result, see eq. (33). The exponent ν is not very sensitive to the precise form of the scattering function used to fit the data. The staggered susceptibility is more sensitive. If one takes χ to be the $q = 0$ intensity of the Lorentzian part of eq. (15), the exponent $\gamma = 1.75 \pm 0.2$ is found for $Fe_{0.6}Zn_{0.4}F_2$ above $T_{eq}(H)$. Note that one has $\gamma = (2 - \eta)\nu$, so the experiments suggest a positive value of η, though with a large relative error. However, $\eta = 0$ according to the Lorentzian form, so the method of analysis is not entirely consistent. Unfortunately, in the absence of a theoretical prediction for the structure function, it is difficult to do better. If the intensity of the squared Lorentzian term at $q = 0$ is evaluated as a power law in $|t|$, the exponent is 3.5 ± 0.3, about twice that of the Lorentzian. Other than the unusual line shape needed to analyze the data, the neutron scattering critical behavior results above $T_{eq}(H)$ are reasonably in accord with the usual behavior associated with an approach to a second order phase transition as T decreases.

In a study of another system, weakly anisotropic $Mn_{0.5}Zn_{0.5}F_2$, the correlation length was observed [80] to reach a finite size at which point it was assumed that further progress towards a second order transition was pre-empted by the occurrence of a first order transition. Another plausible description of the observed behavior is that the transition is simply rounded by concentration gradients, which were not independently measured in the sample. The gradients, along with the extreme random field dynamics and the fact that β is very small even if the transition is second order, preclude any definitive statements concerning the possibility of a first order nature of the random field transition at $T_c(H)$ in the dilute, anisotropic antiferromagnets at this time.

Some of the random field behavior seen in the $Fe_xZn_{1-x}F_2$ system has also been observed [81–83] in the Ising system $Fe_xMg_{1-x}Cl_2$ in which the layers are ferromagnetic and antiferromagnetic interactions exist between adjacent layers. For example, the crossover exponent $\phi = 1.44$ measured in this system is in agreement with the results from the $Fe_xZn_{1-x}F_2$ one. An additional effect measured [84] in $Fe_xMg_{1-x}Cl_2$ is the *non-linear* susceptibility which is in excellent agreement with theory [37]. Interestingly, the weakly Ising-like system $Mn_xZn_{1-x}F_2$, which does not

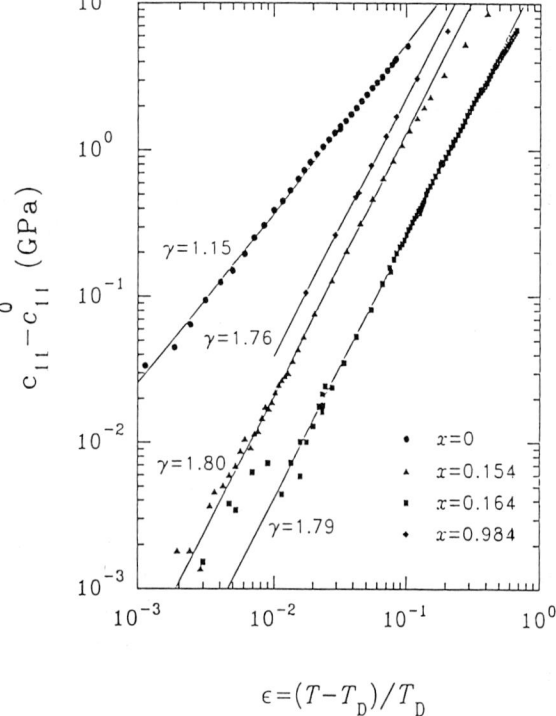

Fig. 7. Log–log plot of the soft-mode elastic constant, $c_{11} - c_{11}^0$, vs. reduced temperature in Dy(As$_x$V$_{1-x}$)O$_4$, with $x = 0, 0.514, 0.164$ and 0.984. For $x \neq 0$, the exponent values obtained from power law fits to the data, shown as straight lines in the figure, are in agreement with values from Monte Carlo and from dilute antiferromagnets [88].

show [58] asymptotic Ising-like specific heat behavior for $|t| > 10^{-3}$, qualitatively shows [85,86] much of the same RFIM behavior as the Ising-like Fe$_x$Zn$_{1-x}$F$_2$ system. The Mn$_x$Zn$_{1-x}$F$_2$ system yields a crossover exponent $\phi = 1.43 \pm 0.03$ in agreement with the other random field systems [87].

Experiments in systems other than dilute antiferromagnets have provided important independent tests of the random field Ising model problem. Ultrasonic measurements [88] in the Jahn–Teller structural system DyAs$_x$V$_{1-x}$O$_4$ have yielded the susceptibility exponent $\gamma = 1.79 \pm 0.07$, as shown in fig. 7. Since the structure of the spin correlation function is not utilized in the analysis of the acoustic data, the result for γ is independent of the uncertainties involved in the analysis of the neutron scattering data with regard to the precise spin–spin correlation function. The exponent from the acoustic measurements does agree with the less precisely determined neutron scattering value $\gamma = 1.75 \pm 0.2$. The agreement between the results of the very different techniques lends support to the procedure discussed earlier of using only the Lorentzian part of the neutron scattering profile to extract the staggered susceptibility in the dilute antiferromagnets. Another interesting effect which has been observed in the structural transition but not in the dilute antiferromagnets is the roughening of twinned domain walls by random fields [89].

Binary liquids in cross-linked gels [20] have been studied with small angle neutron scattering techniques using a Lorentzian plus squared Lorentzian lineshape. The correlation length exponent was found to be $\nu \approx 0.94$ for the Lorentzian term, in excellent agreement with the neutron scattering results for Fe$_{0.6}$Zn$_{0.4}$F$_2$. However, in this case the correlation length of the squared Lorentzian is apparently associated with the gel matrix and is not equal to the correlation length used in the Lorentzian term.

The analysis of the neutron scattering data in the nonequilibrium region below $T_{eq}(H)$ is tremendously more difficult since the lineshapes are definitely not well described [90] by eq. (15) even when the sample is ZFC. If the sample is cooled in zero field, long range order forms as T is reduced below $T_c(H)$. When the field is then applied well below the transition, a Bragg contribution is clearly observed, indicating long range antiferromagnetic order. The scattering intensity outside the Bragg region, set by instrumental resolution appears to be much more peaked near $q = 0$ then can be described even by a squared Lorentzian term [90]. Moreover, the scattering intensity near $q = 0$ shows strong evidence of dynamics near $T_c(H)$. Obtaining information about the critical behavior of χ and κ below $T_c(H)$ will be difficult until the RFIM dynamics and $S(q)$ are better understood. If the sample is cooled in a field, metastable domains persist below the phase transition. These domains contribute to the scattering at low temperatures in a manner approximated by a pure squared Lorentzian shape [91]

$$S(q) = B/(q^2 + \tilde{\kappa}^2)^2, \tag{37}$$

where $\tilde{\kappa}$ is some characteristic inverse length associated with the metastable domains and, except perhaps near $T_{eq}(H)$, is not the same as κ given in eq. (15). No Bragg peak forms at low T upon FC, indicating the absence of long range order.

Perhaps the most elusive part of the RFIM problem experimentally is the measurement of the static critical behavior of the staggered magnetization for the $d = 3$ transition. One might expect this to be easily obtained in dilute antiferromagnets from measurements of the Bragg intensity as a function of temperature. However, the high quality $Fe_xZn_{1-x}F_2$ crystals exhibit very significant extinction effects which preclude determination of the staggered magnetization critical behavior in bulk single crystals. Other techniques have been used to attempt a determination of the exponent β. Magnetic X-ray scattering has been done on the weakly Ising-like system $Mn_{0.5}Zn_{0.5}F_2$ using synchrotron radiation [92]. Although some indication of a reduction in the value of β was seen upon application of a field, no definitive results were obtained since the small applied field did not provide for a sufficiently wide crossover region to allow examination of the random field staggered magnetization. Capacitance and dilation measurements [93] in $Fe_{0.46}Zn_{0.54}F_2$ have indicated a value $\beta \approx 1/8$ or less from analysis of an inverse piezomagnetic effect. Recent studies of the structural random field system $Dy(As_xV_{1-x})O_4$ have provided another interesting result which shows how experimentally unsettled is the problem of the magnetization exponent value. Although γ is consistent with the results obtained [94] in dilute antiferromagnets, the value of β from the optical measurements is consistent with the *pure* Ising value. The determination of the critical behavior for the staggered magnetization is still an open question and an important one in light of the theoretical questions still to be answered.

Attempts have been made to study the dynam-

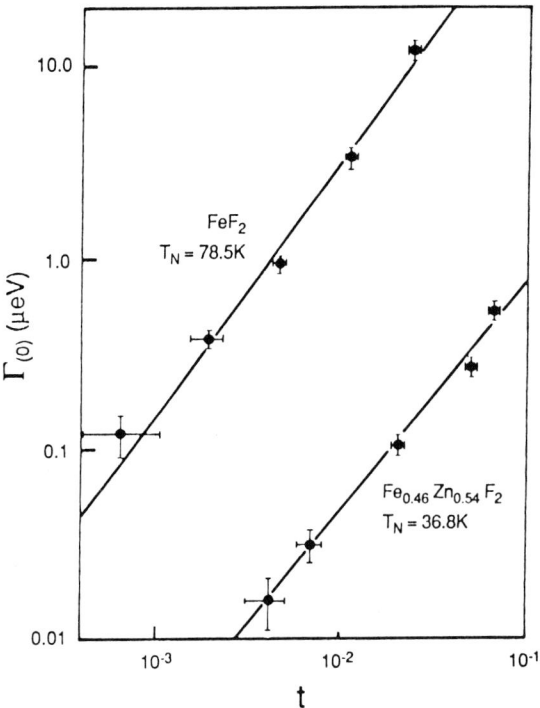

Fig. 8. The logarithm of the relaxation rate Γ vs. $\log(t)$ at $q = 0$ for FeF_2 and $Fe_{0.46}Zn_{0.54}F_2$. The dilute system has relaxation rates two orders of magnitude slower than the pure system over this range of $|t|$ [95].

ics of the antiferromagnetic ordering with neutron scattering. If successful, one could use such experiments to extract information about the q dependence of the dynamic critical behavior. The neutron spin echo technique proved [95] very useful with respect to the zero field $q = 0$ dynamics of FeF_2 and $Fe_{0.46}Zn_{0.54}F_2$. In $Fe_{0.46}Zn_{0.54}F_2$ at $H = 0$, the relaxation rates

$$\Gamma = 1/\tau \sim \xi^{-z}, \tag{38}$$

shown in fig. 8, were measured and the exponent $z = 1.7 \pm 0.2$ is shown to be smaller than the corresponding pure FeF_2 value $z = 2.1 \pm 0.1$. More strikingly, the relaxation rates are two orders of magnitude slower than those of FeF_2 for the range $0.005 < |t| < 0.1$. For typical random field experiments, such as specific heat or neutron scattering, a relatively large field $H > 1$ T is required to see crossover to "static" random field behavior for $|t| > 0.001$. The random field dynamics observed with the spin echo technique proved to be quite different. Relaxation is observed for $|t| \approx 0.1$ only for $h < 0.1$ T. Larger fields or smaller $|t|$ result in no observable relaxation. Hence, even when the static critical behavior appears to be that of the random exchange Ising model, random field dynamics dominate processes at the much shorter time scales of nanoseconds. Crossover to random field behavior is clearly a function of time as well as temperature and field [96].

There has also been much interest in *nonequilibrium* dynamics well below $T_c(H)$ in RFIM systems. Such dynamics have been studied in the $d = 3$ system $Fe_xZn_{1-x}F_2$ with $x = 0.46$ and 0.47 and in $Fe_xMg_{1-x}Cl_2$. As discussed earlier, it is now understood that domains do not signify the destruction of the phase transition for $d = 3$, but are metastable manifestations of extremely slow dynamics. The metastable domains formed upon FC can be utilized in the study of the dynamical processes of the approach to equilibrium well below $T_c(H)$. Neutron scattering studies have so far shown no change in the size of FC domains well below $T_c(H)$ as long as the field and temperature are not changed [97]. This result may seem to disagree with the calculations of domain dynamics in ferromagnets with random fields. The characteristic domain size is expected [98,99] to vary with time as

$$R(t) \sim H^{-\nu_H} \ln(t/\tau), \tag{39}$$

where $\tau = h/J$. The size of FC domains has been shown [100] to vary with H with the exponent $\nu_H = 3.3 \pm 0.2$. However, the time dependence in eq. (39) has not been observed. This may signify a difference between the ferromagnet in a random field and a dilute antiferromagnet in a uniform field. The vacancies in the dilute system act as pinning sites and must be explicitly included in a theory for domain wall evolution in these systems. Scaling arguments by Nattermann and Vilfan and others [101–103] do include vacancy pinning and predict extraordinarily long relaxation times for Ising systems at low T, consistent with neutron scattering observations. By removing the applied field after FC, random field pinning is removed.

Nevertheless, at low temperatures, $T < \frac{1}{2}T_c(0)$, neutron scattering observations [97] indicate that the domain size does not change significantly with time in the very Ising-like system $Fe_xZn_{1-x}F_2$ at low T. Thus, the vacancies by themselves act as extremely effective pinning sites. The effectiveness of the vacancy pinning comes about because narrow domain walls can significantly lower their energy by passing through numerous vacancies. Wall movement on the scale of many lattice spacings requires that the walls pass through well ordered, magnetically dense domains which act as large energy barriers.

While neutron scattering measurements indicate no time evolution on the macroscopic scale, relaxation does occur on the scale of a lattice spacing at $H = 0$. When a domain wall forms upon FC, the spins at the wall are aligned predominately along the field. Hence, the domain walls are a source of uniform magnetization. This can be observed using optical Faraday rotation [68], which is proportional to the uniform magnetization, or with squid magnetometers. It has been shown that the uniform magnetization varies inversely with the domain length scale, confirming that domain walls and not volume effects are the source of the uniform magnetization. With time, a wall will fluctuate, eventually leaving as many pairs of spins down as up with no net wall contribution to the Faraday rotation. An expression has been developed by Nattermann and Vilfan [101] for the time decay of the uniform magnetization (Faraday rotation)

$$\Delta M = AH_0^2 [T \ln(t/\tau_0)]^{-\psi} + B, \tag{40}$$

where H_0 is the field in which the sample is cooled, τ_0 is a characteristic time, $\psi \approx 0.4$ and B is a small constant volume term. At low T, the Nattermann–Vilfan equation was used to analyze data in $Fe_{0.47}Zn_{0.53}F_2$ [67] and $Fe_xMg_{1-x}Cl_2$ [81,82], but some systematic deviations were observed. Monte Carlo simulations [104], on the other hand, suggest that the power law dependence

$$\Delta M \sim t^{-x} \tag{41}$$

is a better description of the decay of the uniform

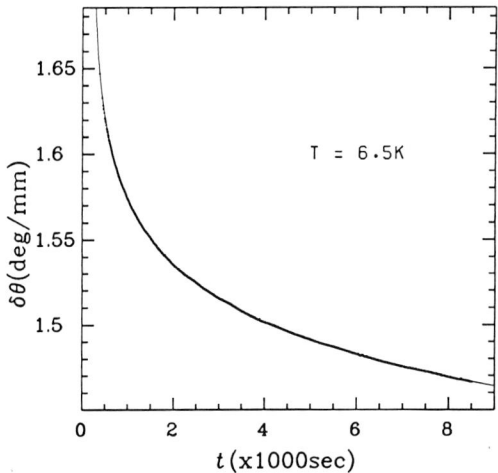

Fig. 9. Faraday rotation, θ, vs. time t in $Fe_{0.47}Zn_{0.53}F_2$ measured from the time the field $H = 7$ T was turned off after FC to $T = 6.5$ K. The solid curve represents a fit of the data to eq. (42) [105].

magnetization. Recently, more precise Faraday rotation measurements [105] on $Fe_{0.47}Zn_{0.53}F_2$ show that the slightly more general form

$$\Delta M \sim \exp\left[-A(\ln(t))^y\right] \quad (42)$$

provides even better fits of the data with y slightly greater than unity as shown in fig. 9. Equations of the same general form as eq. (42) have been suggested for the dynamics of the random field critical region equilibrium fluctuations [106] and for temperature in the region of Griffiths singularities above $T_c(x)$ for the dilute antiferromagnet and below the pure Néel temperature $T_c(1)$ [107,108]. There is no suitable theory predicting the behavior in eq. (42) for $H = 0$ in the experimental temperature region well below $T_c(x)$. It is still open whether the Nattermann–Vilfan theory holds for a range of T lower than has been investigated experimentally.

The extremely slow dynamics, observed at $H = 0$ after FC, suggest that vacancies may play an important role in the dynamics of random field dynamics at $H > 0$. Studies of the dynamics in high fields not too far below $T_c(H)$ have recently been done [109] using a squid magnetometer in $Fe_{0.47}Zn_{0.53}F_2$ to examine random field dynamics, as opposed to the random exchange dynamics at

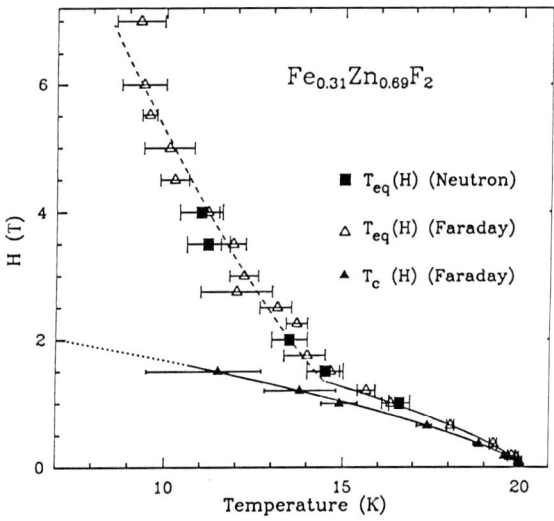

Fig. 10. H vs. T phase diagram for $Fe_{0.31}Zn_{0.69}F_2$. For $H < 1.5$ T, $T_{eq}(H)$ and $T_c(H)$ scale with $\phi = 1.4$, the RFIM crossover exponent. For $H > 2$ T, $T_{eq}(H)$ changes curvature and scales with $\phi = 3.4$. The region below $T_c(H)$ exhibits antiferromagnetic long range order. The region below $T_{eq}(H)$ and above $T_c(H)$ appears not to have antiferromagnetic long range order [115].

$H = 0$. In part of the H vs. T phase diagram at large H and not too small T, the behavior is described in the context of random field dynam-

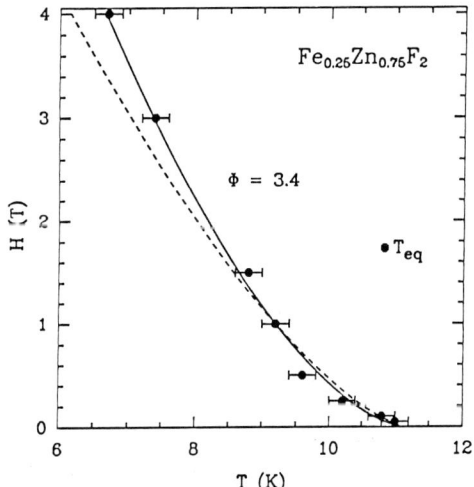

Fig. 11. H vs. T phase diagram for $Fe_{0.25}Zn_{0.75}F_2$, which has a concentration near the percolation threshold. $T_{eq}(H)$ scales with $\phi = 3.4$ and in no region of the diagram is long range order observed [110].

ics associated with the pinning of domain walls by random fields [99]. The interpretation of the dynamics as large scale wall motion, as opposed to wall reorientation on lattice spacing scales, has so far not been supported by neutron scattering measurements [97] which indicate little time dependence in the FC intensities at $q = 0$, though further scattering experiments are surely called for. At smaller H, dynamics more like that of the $H = 0$ REIM are observed.

Interesting new random field phenomena occur [110–116] in the dilute antiferromagnetic Ising system $Fe_xZn_{1-x}F_2$ when the concentration approaches the percolation threshold which, because the second nearest neighbor exchange is so dominant, is near $x = 0.24$. In this system, high field measurements [117] indicate that the exchange interactions are not significantly altered upon dilution and there is little evidence for frustration. At $x = 0.31$, REIM behavior is seen [115,116] near the transition, though some disorder remains at low temperatures. For small applied fields $H < 2$ T, the RFIM behavior is very much like that of the larger concentration samples. At these fields, both $T_c(H)$ and $T_{eq}(H)$ scale with H as shown in a manner similar to eq. (22), with ϕ consistent with the RFIM crossover exponent value 1.42. At higher fields, $H > 2$ T, the $T_{eq}(H)$ boundary changes curvature, scaling now with $\phi \approx 3.4$. From neutron scattering measurements [115,116], it has been determined that the region of the H vs. T phase diagram below $T_c(H)$ at low fields is antiferromagnetic. However, the large region below $T_{eq}(H)$ at large H and above $T_c(H)$ appears not to be associated with long range antiferromagnetic order. Very close to the percolation threshold $x_p = 0.25$, $T_{eq}(H)$ shows [112,114] the scaling behavior with $\phi = 3.4$ for all H. The low field behavior observed at larger x is completely absent. Neutron scattering measurements [118] show that no long range order occurs at this concentration even for $H = 0$. The high field behavior at magnetic concentrations just above percolation and the behavior seen at all fields very close to percolation are very suggestive of spin glass-like behavior. The $T_{eq}(H)$ boundary, with $\phi = 3.4$, resembles the equilibrium boundaries observed in spin-glass systems and is consistent with the De Almeida–Thouless [119] boundary predicted in mean field theory. Although some theoretical indications of such behavior exist for large random fields and large dimensions [120], the details are not understood in detail in the dilute antiferromagnets. It is not known to what extent the new behavior may reflect the dynamics associated with vacancies, represents a natural breakdown of the weak-field formulation of the RFIM, or perhaps originates from some other source.

4. Conclusions

While the existence of a long range ordered state in the RFIM in $d = 3$ (but not $d = 2$) is now well established by theory and experiment, there appear to be inconsistencies between theoretical predictions for critical exponents and the experimental values. Unfortunately, the theory and experiment do not both give accurate estimates for the *same* exponents. The best theoretical estimates are from numerical calculations for ν and $\bar{\eta}$, which are rather difficult to obtain accurately from experiment, whereas the most accurate experimentally-determined exponent is α, which has not been obtained directly from theory. To compare these exponents one therefore has to assume the validity of the scaling relations, eqs. (16), (17), which have not been directly verified. The experimental values, $\alpha \approx 0$ and $\nu \approx 1.0$, are similar to those of the pure 2-dimensional Ising model, which would suggest a dimensional reduction by 1. However, the early theoretical calculations which gave the dimensional reduction by 2 now appear to be incorrect, and there is no particular reason to believe that the exponents should be exactly equal to those of a pure system in *any* dimension. The best theoretical estimates are $\alpha \approx 0.5$, and $\nu \approx 1.0$, so we see that the values for ν agree, but there is a major discrepancy with α.

Concerning critical dynamics, there is experimental evidence for the ideas of "activated dynamic scaling". Unfortunately, this means that relaxation times increase so fast as T approaches T_c that the system is not in full equilibrium very close to T_c which makes accurate estimates of exponents very difficult. As with the much studied spin glass problem, a clear picture of dynamics in the low temperature phase is missing.

Acknowledgements

The work of APY was supported in part by the NSF grant no. DMR 87-21673 and the work of DPB was supported in part by the DOE grant no. DE-FG03-87ER45324.

References

[1] K. Binder and A.P. Young, Rev. Mod. Phys. 58 (1986) 801.
[2] A.B. Harris, J. Phys. C 7 (1974) 1671.
[3] V. Jaccarino, in: Condensed Matter Physics, the Theodore D. Holstein Symp., ed. R.L. Orbach (Springer, New York, 1987).
[4] V. Jaccarino and A.R. King, Hyperfine Interactions 17–19 (1984) 403.
[5] A.R. King and D.P. Belanger, J. Magn. Magn. Mater. 54–57 (1986) 19.
[6] V. Jaccarino and A.R. King, J. de Phys. 49 (1988) C8-1209.
[7] V. Jaccarino and A.R. King, Physica A 163 (1990) 291.
[8] D.P. Belanger, Phase Transitions 11 (88) 53.
[9] R.J. Birgeneau, Y. Shapira, G. Shirane, R.A. Cowley and H. Yoshizawa, Physica B 137 (1986) 83.
[10] R.A. Cowley, R.J. Birgeneau and G. Shirane, Physica A 140 (1986) 285.
[11] R.J. Birgeneau, R.A. Cowley, G. Shirane and H. Yoshizawa, J. Stat. Phys. 34 (1984) 817.
[12] P.-z. Wong, J.W. Cable and P. Dimon, J. Appl. Phys. 55 (1984) 2377.
[13] D.S. Fisher, G.M. Grinstein and A. Khurana, Phys. Today (December 1988) 56.
[14] T. Nattermann and J. Villain, Phase Transitions 11 (1988) 817.
[15] G. Grinstein, Fundamental Problems in Statistical Mechanics VI, ed. E.G.D. Cohen (North-Holland, Amsterdam, 1985).
[16] T. Nattermann, Ferroelectrics 104 (1990) 171.
[17] S. Fishman and A. Aharony, J. Phys. C 12 (1979) L729.
[18] J. Cardy, Phys. Rev. B 29 (1984) 505.
[19] J.T. Graham, M. Maliepaard, J.H. Page, S.R.P. Smith and D.R. Taylor, Phys. Rev. B 35 (1987) 2098.
[20] S.K. Sinha, J. Huang and S.K. Satija, in: Scaling Phenomena in Disordered Systems, eds. R. Pynn and A. Skjeltorp (Plenum, New York, 1985).
[21] S.B. Dierker and P. Wiltzius, Phys. Rev. Lett. 66 (1991) 1185.
[22] J. Villain, J. de Phys. Lett. 43 (1982) L551.
[23] Y. Imry and S.-K. Ma, Phys. Rev. Lett. 35 (1975) 1399.
[24] A. Aharony, Y. Imry and S.K. Ma, Phys. Rev. Lett. 37 (1976) 1364.
[25] A.P. Young, J. Phys. A 10 (1977) L257.
[26] G. Parisi and N. Sourlas, Phys. Rev. Lett. 43 (1979) 744.
[27] J. Bricmont and A. Kupiainen, Phys. Rev. Lett. 59 (1987) 1829.
[28] J.Z. Imbrie, Phys. Rev. Lett. 53 (1984) 1747.
[29] A. Aharony, Phys. Rev. B 18 (1978) 3318.
[30] J. Villain, J. de Phys. 46 (1985) 1843.
[31] A.J. Bray and M.A. Moore, J. Phys. C 18 (1985) L927.
[32] D.S. Fisher, Phys. Rev. Lett. 56 (1986) 416.
[33] M. Schwartz and A. Soffer, Phys. Rev. Lett. 55 (1985) 2499.
[34] M. Schwartz, J. Phys. C 18 (1985) 1843.
[35] M. Schwartz and A. Soffer, Phys. Rev. B 33 (1986) 2059.
[36] M. Schwartz, preprint.
[37] A. Aharony, Europhys. Lett. 1 (1986) 617.
[38] K. Binder, Z. Phys. B 50 (1983) 343; Phys. Rev. B 29 (1984) 5184.
[39] G. Parisi, J. Phys. A 13 (1980) 1887.
[40] M. Mézard and G. Parisi, J. Phys. A 23 (1990) L1229 and preprint.
[41] E. Brézin and H. Orland, unpublished; referenced in ref. [40].
[42] A.P. Young and M. Nauenberg, Phys. Rev. Lett. 54 (1985) 2429.
[43] A.T. Ogielski and D.A. Huse, Phys. Rev. Lett. 56 (1986) 1298.
[44] A.T. Ogielski, Phys. Rev. Lett 57 (1986) 1251.
[45] P.C. Hohenberg and B.I. Halperin, Rev. Mod. Phys. 49 (1977) 435.
[46] A.E. Nash, A.R. King and V. Jaccarino, Phys. Rev. B 43 (1991) 1272.
[47] I.B. Ferreira, A.R. King, V. Jaccarino, J.L. Cardy and H.J. Guggenheim, Phys. Rev. B 28 (1983) 5192.
[48] P. Nordblad, D.P. Belanger, A.R. King, V. Jaccarino and H. Ikeda, Phys. Rev. B 38 (1983) 278.
[49] R.A. Cowley, M. Hagen and D.P. Belanger, J. Phys. C 17 (1984) 3763.
[50] L. Onsager, Phys. Rev. 65 (1944) 117.
[51] H. Ikeda and M.T. Hutchings, J. Phys. C 11 (1978) L529.
[52] D.P. Belanger, A.R. King and V. Jaccarino, Phys. Rev. Lett. 54 (1985) 577.
[53] H. Ikeda, J. Phys. C 16 (1983) L1033.
[54] A.R. King, V. Jaccarino, M. Motokawa, K. Sugiyama and M. Date, J. Appl. Phys. 57 (1985) 3297.
[55] H. Yoshizawa, R.A. Cowley, G. Shirane, R.J. Birgeneau, H.J. Guggenheim and H. Ikeda, Phys. Rev. Lett. 48 (1982) 438.
[56] R.J. Birgeneau, H. Yoshizawa, R.A. Cowley, G. Shirane and H. Ikeda, Phys. Rev. B 28 (1983) 1438.
[57] M.T. Hutchings, B.D. Rainford and H.J. Guggenheim, J. Phys. C 3 (1970) 307.
[58] D.P. Belanger, P. Nordblad, A.R. King, V. Jaccarino, L.

Lundgren and O. Beckman, J. Magn. Magn. Mater. 31–34 (1983) 1095.
[59] D.P. Belanger and H. Yoshizawa, Phys. Rev. B 35 (1987) 4823.
[60] Many of the highest quality crystals used in the $d = 3$ experiments have been grown at the UCSB Material Preparation Laboratory by N. Nighman.
[61] A.R. King, I.B. Ferreira, V. Jaccarino and D.P. Belanger, Phys. Rev. B 37 (1988) 219.
[62] D.P. Belanger, A.R. King, I.B. Ferreira and V. Jaccarino, Phys. Rev. B 37 (1988) 226.
[63] D.P. Belanger, A.R. King and V. Jaccarino, Phys. Rev. B 34 (1986) 452.
[64] I.B. Ferreira, A.R. King and V. Jaccarino, Phys. Rev. B (1991) in press.
[65] J. Ferré and G.A. Gehring, Rep. Prog. Phys. 47 (1984) 513.
[66] D.P. Belanger, A.R. King, V. Jaccarino and J.L. Cardy, Phys. Rev. B 28 (1983) 2522.
[67] P. Pollak, W. Kleemann and D.P. Belanger, Phys. Rev. B 38 (1988) 4773.
[68] W. Kleemann, A.R. King and V. Jaccarino, Phys. Rev. B 34 (1986) 479.
[69] K.E. Dow and D.P. Belanger, Phys. Rev. B 39 (1989) 4418.
[70] R.J. Birgeneau, R.A. Cowley, G. Shirane, H. Yoshizawa, D.P. Belanger, A.R. King and V. Jaccarino, Phys. Rev. B 27 (1983) 6747.
[71] K.E. Newman and E.K. Riedel, Phys. Rev. B 25 (1982) 264.
[72] G. Jug, Phys. Rev. B 27 (1983) 609.
[73] L. Chayes, J. Chayes, D.S. Fisher and T. Spencer, Phys. Rev. Lett. 57 (1986) 2999.
[74] A.R. King, J.A. Mydosh and V. Jaccarino, Phys. Rev. Lett. 56 (1986) 2525.
[75] H. Rohrer and H.J. Scheel, Phys. Rev. Lett. 44 (1980) 876.
[76] D.P. Belanger, A.R. King and V. Jaccarino, Phys. Rev. B 31 (1985) 4538.
[77] R.A. Pelcovits and A. Aharony, Phys. Rev. B 31 (1985) 350.
[78] S.W. Lovesey, J. Phys. C 17 (1984) L213.
[79] P.M. Richards, Phys. Rev. B 30 (1984) 2955.
[80] R.J. Birgeneau, R.A. Cowley, G. Shirane and H. Yoshizawa, Phys. Rev. Lett. 54 (1985) 2147.
[81] U.A. Leitão, W. Kleemann and I.B. Ferreira, Phys. Rev. B 38 (1988) 4765.
[82] Y.A. Leitão and W. Kleemann, Phys. Rev. B 35 (1987) 8696.
[83] P-z. Wong and J.W. Cable, Phys. Rev. B 28 (1983) 5361.
[84] U.A. Leitão and W. Kleemann, Europhys. Lett. 5 (1988) 529.
[85] H. Ikeda, J. Phys. C 19 (1986) L811.
[86] Y. Shapira, N.F. Oliveira, Jr. and S. Foner, Phys. Rev. B 30 (1984) 6639.
[87] C.A. Ramos, A.R. King and V. Jaccarino, preprint.
[88] J.T. Graham, J.H. Page and D.R. Taylor, Phys Rev. B (1991) accepted.
[89] J.T. Graham, D.R. Taylor, D.R. Noakes and W.J.L. Buyers, Phys. Rev. B 43 (1991) 3778.
[90] D.P. Belanger, A.R. King, V. Jaccarino and R.M. Nicklow, Phys. Rev. Lett. 59 (1987) 930.
[91] P. Debye, H.R. Anderson, H.J. Brumberger, J. Appl. Phys. 28 (1957) 679.
[92] T.R. Thurston, C.J. Peters, R.J. Birgeneau and P.M. Horn, Phys. Rev. B 37 (1988) 9559.
[93] C.A. Ramos, A.R. King, V. Jaccarino and S.M. Rezende, J. de Phys. 49 (1988) C8-1241.
[94] D.R. Taylor and K.A. Reza, ICM '91 proceedings (1991) submitted.
[95] D.P. Belanger, B. Farago, V. Jaccarino, A.R. King, C. Lartigue and F. Mezei, J. de Phys. 49 (1988) C8-1229.
[96] Y. Shapir, Phys. Rev. B 35 (1987) 62; Phys. Rev. Lett. 54 (1985) 154; J. Phys. C 17 (1984) L809.
[97] D.P. Belanger, A.R. King and V. Jaccarino, Solid State Commun. 54 (1985) 79.
[98] J. Villain, Phys. Rev. Lett. 52 (1984) 1543.
[99] R. Bruinsma and G. Aeppli, Phys. Rev. Lett. 52 (1984) 1547.
[100] M. Hagen, R.A. Cowley, S.K. Satija, H. Yoshizawa, G. Shirane, R.J. Birgeneau and H.J. Guggenheim, Phys. Rev. B 28 (1983) 2602.
[101] T. Nattermann and I. Vilfan, Phys. Rev. Lett. 64 (1988) 223.
[102] T. Nattermann, Y. Shapir and J. Vilfan, Phys. Rev. B 42 (1989) 8577.
[103] D.A. Huse and C.L. Henley, Phys. Rev. Lett. 54 (1985) 2708.
[104] U. Nowak and K.D. Usadel, Phys. Rev. B 43 (1991) 851; Physica B 165 (1990) 211.
[105] S-J. Han, D.P. Belanger, W. Kleemann and U. Nowak, in preparation.
[106] D.A. Huse and D.S. Fisher, Phys. Rev. B 39 (1987) 6841.
[107] M. Randeria, J.P. Sethna and R.G. Palmer, Phys. Rev. Lett. 54 (1985) 1321.
[108] A.J. Bray and G.J. Rodgers, Phys. Rev. B 38 (1988) 9252.
[109] M. Lederman, J. Selinger, R. Bruinsma, J.M. Hammann and R. Orbach, in preparation.
[110] S.M. Rezende, F.C. Montenegro, U.A. Leitão and M.C. Coutinho-Filho, New Trends in Magnetism, eds. M.D. Coutinho-Filho and S.M. Rezende (World Scientific, Singapore, 1989).
[111] F.C. Montenegro, S.M. Rezende and M.D. Coutinho-Filho, J. Appl. Phys. 63 (1988) 3755.
[112] F.C. Montenegro, M.D. Coutinho-Filho and S.M. Rezende, Europhys. Lett. 8 (1989) 382.
[113] S.M. Rezende, F.C. Montenegro, M.D. Coutinho-Filho, C.C. Becerra and A. Paduan-Filho, J. de Phys. 49 (1988) C8-1267.

[114] F.C. Montenegro, U.A. Leitão, M.D. Coutinho-Filho and S.M. Rezende, J. Appl. Phys. 67 (1990) 5243.
[115] F.C. Montenegro, A.R. King, V. Jaccarino, S.-J. Han and D.P. Belanger, Phys. Rev. B (1991) to be published.
[116] D.P. Belanger, Wm. E. Murray, Jr., A.R. King, V. Jaccarino and R.W. Erwin, Phys. Rev. B (1991) to be published.
[117] A.R. King, V. Jaccarino, T. Sakakibara, M. Motokawa and M. Date, Phys. Rev. Lett. 47 (1981) 117.
[118] D.P. Belanger and H. Yoshizawa, unpublished.
[119] J.R.L. de Almeida and D.J. Thouless, J. Phys. A 11 (1978) 983.
[120] J.R.L. de Almeida and R. Bruinsma, Phys. Rev. B 35 (1987) 7267.

Quenching of spin fluctuations by high magnetic fields

K. Ikeda [a,†], S.K. Dhar [b], M. Yoshizawa [a] and K.A. Gschneidner, Jr. [c]

[a] *Department of Metallurgy, Iwate University, Morioka 020, Japan*
[b] *Tata Institute of Fundamental Research, Homi Bhabha Road, Colaba, Bombay 400 005, India*
[c] *Ames Laboratory * and Department of Materials Science and Engineering, Iowa State University, Ames, IA 50011, USA*

The quenching of spin fluctuations by magnetic fields has been observed in heat capacity and electrical resistivity measurements at low temperatures for a series of highly exchange enhanced magnetic materials. These include: the weak itinerant electron ferromagnets Sc_3In, $Zr_{1-x}Hf_xZn_2$ ($0 \leq x \leq 0.2$) and Ni_3Al; the strong Pauli paramagnets RCo_2 (R = Sc, Y and Lu), $TiBe_2$ and $Pd_{1-x}Ni_x$ ($0 \leq x \leq 0.01$); and the heavy fermion systems $CeSn_3$, $CeSi_x$ ($x \approx 1.85$) and UAl_2. The reported quenching of spin fluctuations in scandium and palladium by magnetic fields is reviewed, and it appears that the initial observations and conclusions are incorrect, and that fields greater than 10 and 40 T, respectively, will be necessary to quench spin fluctuations in these metals. The behaviors of these spin fluctuators have been grouped into six classes.

1. Introduction

The study of spin fluctuations in nearly and weakly ferromagnetic materials has been of considerable interest theoretically and experimentally for more than two decades. The exchange enhanced paramagnets or incipient ferromagnets are characterized by a large but finite magnetic susceptibility at $T = 0$ K. The ratio of the experimentally observed susceptibility to that of the free spins is called the Stoner enhancement factor, S. Values of S for some exchange enhanced paramagnets are: Pd 10, $TiBe_2$ 65, $LuCo_2$ 17, $CeSi_{1.90}$ 135 [1]. In analogy with the behavior of a ferromagnetic spin system just above the Curie temperature, one expects intuitively that for large S the dominant spin excitation modes at low temperatures would occur at low frequency, ω, and low q (large wavelengths). Based on a phenomenological model in which there is a short range repulsive Coulomb interaction between the electrons, it can be shown using the random phase approximation that the spectral weight function Im $\chi(q, \omega)$, where Im means the imaginary part of the generalized dynamic susceptibility, $\chi(q, \omega)$, develops a peak which is considerably enhanced and becomes narrower at low ω and q as S increases [2]. If the width of the peak is taken as a measure of the inverse lifetime of the corresponding spin fluctuations then they persist for very long times with increasing S. Although the spin fluctuations or paramagnons are strongly damped, their characteristic excitation energy $k_B T_{sf}$ (k_B is the Boltzman constant and T_{sf} the spin fluctuation temperature) is reduced by a factor S, i.e. $T_{sf} = T_F/S$ where T_F is the Fermi temperature. Using neutron elastic scattering, Low had already observed in 1964 that the non-local susceptibility of Pd had a range of about 6–10 Å [3]. Berk and Schrieffer invoked ferromagnetic spin correlation to explain the absence of singlet spin paired superconductivity in Pd [4].

The interaction of electrons with spin fluctuations contributes to their self-energy and gives rise to an effective mass [4], and therefore to an enhancement of the linear term of the electronic heat capacity, in a fashion similar to that arising from the more well known electron–phonon in-

---

[†] Deceased, see Dedication at the end of this review.
[*] Operated for the US Department of Energy by Iowa State University under contract no. W-7405-ENG-82. This work was supported by the Office of Basic Energy Sciences.

teraction. In addition a contribution of the form $T^3 \ln(T/T_{sf})$ to the electronic heat capacity is also predicted by the theory [5]. Calculations made by Rice [6] showed that the spin fluctuations contribute an extra T^2 term to resistivity at low temperatures. It was pointed out on theoretical grounds that paramagnons while being detrimental to singlet spin paired superconductivity could give rise to p wave or triplet spin paired superconductivity [7,8].

It was first pointed out by Brinkman and Engelsberg [9] and Béal-Monod et al. [10] that application of high magnetic fields offers a way of testing the spin fluctuation theory. If the magnetic field is sufficiently large so that the Zeeman splitting energy of opposite spin states is comparable to or larger than the characteristic spin fluctuation energy $k_B T_{sf}$ then the paramagnons no longer have enough energy to flip the spins and, therefore, the inelastic spin flip scattering is quenched leading to a decrease of the electronic heat capacity. Brinkman and Engelsberg explained that magnetic fields of the order of $H_{eff} = k_B T_{sf}/\mu_B$ are required to quench the spin fluctuation enhancement, but more recently Béal-Monod and Daniel [11] have shown that $H_{eff} = k_B T_{sf}/\mu_B S^{1/2}$.

In late 1980 Ikeda and Gschneidner [12] were the first to unequivocally observe the quenching of spin fluctuations in the exchange enhanced Pauli paramagnet $LuCo_2$ in applied magnetic fields. Although an earlier claim had been made in the case of UAl_2, the scatter of the experimental data at their highest field of 4.3 T made such a conclusion quite tenuous [13]; indeed subsequent studies on UAl_2 showed that fields of at least 8 T are required to make a measurable change in the heat capacity [14]. Since 1980 the quenching of spin fluctuations in applied magnetic fields in exchange enhanced paramagnetic and weakly ferromagnetic materials has been observed in a number of d and f based materials. Most of the evidence is a result of heat capacity and/or electrical resistivity measurements made in fields up to 10 T, and occasionally up to 17 T. For example, the coefficient of the linear term of the electronic heat capacity decreases by 10% for $LuCo_2$ [12,15], 13% for 0.47 at% Ni in Pd [16] and 27% for $CeSn_3$ [17] in an applied field of 9.8 T, while the heat capacity decreases by 10% at 2.1 K in UAl_2 [14] and by 12% at 5 K in $TiBe_2$ [18] in an applied field of 12 T.

Prior to the mid-1980's the T_{sf} values of these exchanged enhanced paramagnets were usually chosen as the maximum in the χ vs. T plot, but as pointed out by Ikeda et al. [15] these values are too large by at least one order of magnitude. They noted that a new definition of T_{sf} for these strong Pauli paramagnets, which contains more detailed characteristics of the band structure in addition to T_F and S, is required, and it should be based on experimental studies dealing with the effect of magnetic fields on spin fluctuations.

A few years later Konno and Moriya [19] in a theoretical analysis of the long wavelength paramagnon contribution to the heat capacity noted that the usual terms associated with spin fluctuations, the most prominent one being $T^3 \ln T$, would be valid only at $T < 0.1 T^*$, where they called T^* the characteristic temperature, and stated that T^* is different from T_{sf}. This is consistent with the ideas of Ikeda et al. [15] and the experimental results for the exchange enhanced paramagnets if one associates T^* with T_{max}, the maximum in the χ vs. T plot.

On the other hand, for the itinerant electron ferromagnet Sc_3In heat capacity [20] and electrical resistivity [21] measurements as a function of magnetic field suggest that the T_{sf} is equal to the Curie temperature at $H = 0$ T.

2. Theoretical background

2.1. Heat capacity in magnetic fields

For the case of uniform enhancement, i.e. S is the same at every site of the lattice, the full expression for heat capacity at temperatures below T_{sf} is given by [4,5,9,22,23]:

$$C = \gamma_0 T \left[\frac{m^*}{m} + \frac{3}{2} \left\{ \left(\frac{6\pi^2}{5} \right) \frac{(S-1)^2}{S} \left(\frac{T}{\overline{T}_F} \right)^2 \right. \right.$$

$$\left. \left. \times \left(\ln \frac{T}{\overline{p}_1 \overline{T}_F} + 1.78 \right) \right\} \right] + \beta_L T^3, \quad (1)$$

where

$$\bar{T}_F = \frac{4}{\pi}\left(\frac{S}{S-1}\right)T_{sf} \approx T_{sf} \quad (2)$$

and $m^*/m = 1 + \lambda_{ep} + \lambda_{spin}$ is the zero temperature many body mass enhancement, which includes the electron–phonon (λ_{ep}) and spin fluctuation (λ_{spin}) contributions, γ_0 is the electronic specific heat constant determined from the band structure density of states, $\bar{p}_1$ is a cut-off momentum and β_L is the coefficient of the lattice heat capacity and is related to the Debye temperature, Θ_D, at 0 K by the following expression:

$$\beta_L = (1.9437 \times 10^6)/\Theta_D^3, \quad (3)$$

where β is in the units of mJ/g at K^4. The above expression (eq. (1)) assumes a spherical Fermi surface and a contact exchange interaction, I. It is shown for Pd that the inclusion of band structure effects and the finite range of exchange interaction leads to a decrease in the predicted many body mass enhancement m^*/m [24]. An expression similar to eq. (1) has been obtained for alloys, i.e. Ni in Pd where the exchange enhancement is not uniform throughout the lattice [22,25]. For weakly ferromagnetic systems, Brinkman and Engelsberg [9] showed that

$$m^*/m \approx \tfrac{9}{2}\ln|1-\bar{I}|, \quad (4)$$

where $\bar{I} = IN(E_F)$ and $N(E_F)$ is the density of states at the Fermi level. More recently Makoshi and Moriya [23] and Lonzarich [26] have derived expressions for heat capacity in exchange enhanced paramagnets and weak ferromagnets which are qualitatively similar to the earlier results. That is, the simplified form of eq. (1) is commonly given as

$$C = \gamma_0 T\left[m^*/m + \alpha_0(T/T_{sf})^2\ln(T/T_{sf})\right] + \beta_L T^3, \quad (5)$$

where α_0 is equal to $(6\pi^2/5)(S-1)^2/S$ and $S = [1 - IN(E_F)]^{-1}$.

Fig. 1 shows a variation of the second term of eq. (5), the electronic heat capacity, $C_E^{II}/T \equiv \gamma_0\alpha_0(T/T_{sf})^2\ln(T/T_{sf}) \equiv C/T - \gamma_0(m^*/m) - \beta_L T^2$, as a function of T/T_{sf} for the three cases

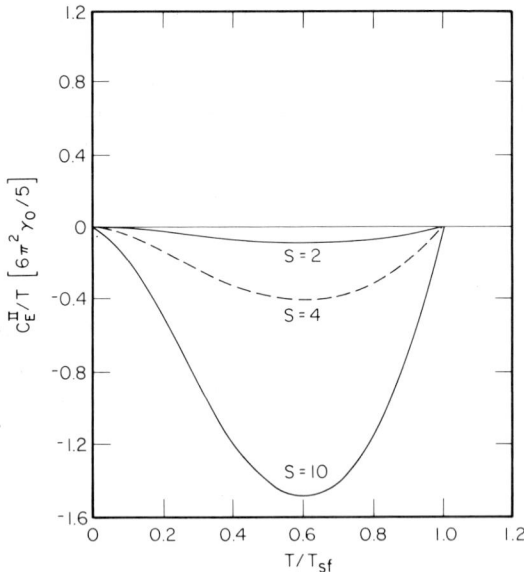

Fig. 1. Second term of electronic heat capacity [see eq. (5)] due to spin fluctuations as a function of T/T_{sf}, where T_{sf} is the characteristic temperature of spin fluctuations.

of $S = 2$, 4 and 10. As seen from this figure, the $T^3\ln(T/T_{sf})$ term of the heat capacity is negligibly small when S has a small value, whereas it shows a remarkable variation for the case of $S \geq 10$. Furthermore, this is a negative contribution to the heat capacity as long as $T < T_{sf}$. The significance of this is pointed out below.

Experimental studies of the magnetic field dependence of the low temperature heat capacity of spin fluctuators have shown that eqs. (1) and (5) do not correctly describe the influence of the magnetic field on the heat capacity [1,17,27], but they might not be expected to do so, since they were derived for $H = 0$. One major modification that would need to be made is that the coefficient β_L of the last term ($\beta_L T^3$) must be replaced by the coefficient β_T which embodies both a lattice (β_L) and a magnetic (β_M) component, i.e.

$$\beta_T = \beta_L + \beta_M(H). \quad (6)$$

Here β_L is the normal lattice contribution; and $\beta_M(H)$ is due to an induced moment on the magnetically active atom(s), which is equal to zero at $H = 0$, but increases with increasing field and seems to saturate for several materials at $H = 7$ T. It should be noted that a field dependence for β_T is not observed in all spin fluctuators. The first term of eqs. (1) and (5) exhibits a magnetic field dependence which is given by

$$\frac{m^*}{m} = \frac{\gamma}{\gamma_0} = 1 + \lambda_{ep} + \lambda_{spin}(H, T), \qquad (7)$$

where the paramagnon enhancement factor is reduced by a sufficiently high field, generally $H > 5$ T. Also if the temperature rises much above 15 K this enhancement factor vanishes. Finally we note that high magnetic fields will also reduce C_E^{II} contribution. Thus, the observed heat capacity change with magnetic field can be complex, because the two of the contributions (C_E^{II} and $\beta_T T^3$) will cause it to increase with increasing magnetic field, while the many body enhancement term causes it to decrease. Although the C_E^{II} contribution decreases with increasing field (e.g. it disappears at $H = 5$ T in the case of CeSn$_3$ [17]) the heat capacity increases because C_E^{II} is a negative quantity. Because the interplay of the field dependence of these three quantities is mixed, it is possible that a seemingly "dead" material (with respect to the field dependence of the observed heat capacity) is really quite active and that all of the quantities are varying with field. This is only evident when the experimental data are subject to a least squares fit to eq. (5) or its modified form (see eq. (18)). This effect has been observed in CeSn$_3$ samples (see section 5.1). Furthermore, the magnetic field dependences of the three terms do not vary in the same manner, i.e. the applied field may have no effect on one term and a substantial effect on one or both of the others. These variations have led to a classification of the magnetic field dependences of the heat capacity into six types [1], which will be discussed in more detail at the end of this review (section 6).

The magnetic field dependence of the electronic specific heat constant has been examined in some detail by several theorists. Béal-Monod et al. [10] using the Maxwell relation $\partial M/\partial T = \partial S_M/\partial H$ (where M is the magnetization and S_M is the entropy), have shown that the shift of the electronic specific heat constant at 0 K caused by an applied field H is given by the expression

$$\frac{\delta\gamma}{\gamma(0)} \approx 0.1 \left[\frac{S}{\ln S}\right] h^2, \qquad (8)$$

where $\delta\gamma = \gamma(0) - \gamma(H)$, $h = \mu_B H/k_B T_{sf}$, and μ_B is the Bohr magneton. On the other hand, Hertel et al. [28] have made a more detailed mathematical analysis of the influence of magnetic fields on the spin fluctuation contribution to the electronic specific heat and the electrical resistivity using a uniform enhancement model. In their paper they present plots of the effective mass enhancement and of the high and low temperature resistivity coefficients as a function of the reduced field, h, for metals which had Stoner enhancement factors of 50, 10 and 4.

2.2. Magnetoresistance

The electrical resistivity of highly exchange enhanced materials at low temperatures is given by

$$\rho(H, T) = \rho_0(H) + \rho_{ep} + A(H)T^2. \qquad (9)$$

The first term in eq. (9), the residual resistivity, includes the spin-fluctuation contribution (see eq. (17), section 3.3) along with the usual contributions due to impurities and lattice faults; the second term is the contribution due to the phonon scattering; and the third term is due to both the interband electron–electron scattering [29,30] and spin fluctuations [31,32]. In the materials containing transition metals ρ_{ep} usually consists of two terms, one due to interband s–d scattering (a T^3 term) and the other due to the intraband s–s scattering (T^5 term) [33], but this contribution is smaller than the third one of eq. (9) (i.e. the T^2 term) in highly enhanced materials.

Ueda [34] has calculated the effect of magnetic fields on spin fluctuations in the electrical resistivity and the nuclear spin–lattice relaxation rate of weakly ferromagnetic metals by using the self-consistent renormalized spin fluctuation theory. Ueda's conclusions on the spin fluctuation contribution to the magnetoresistance are summarized as follows: (a) there is a negative magnetoresistance for all cases; (b) when the magnetic field is small, the negative magnetoresistance is proportional to H and H^2 for the weakly and nearly ferromagnetic materials, respectively, whereas in strong fields, $\rho(H)/\rho(0) \propto H^{-1/3}$ holds for both materials; (c) at low temperatures the coefficient of T^2 term in the resistivity becomes smaller with increasing fields; and (d) the negative magnetoresistance becomes smaller as the temperature increases and has a cusp at the Curie temperature.

The magnetoresistance is defined by $\Delta\rho(H, T) = \rho(H, T) - \rho(0, T)$. According to Kohler's rule [35], the dependence of $\Delta\rho(H, T)$ of metals on the temperature, magnetic field and sample purity should be a function of τH only, where τ is the relaxation time due to the scattering of the conduction electrons. In the weak magnetic field limit ($\omega\tau \ll 1$, where ω is the cyclotron frequency) the magnetoresistance is proportional to H^2. This positive magnetoresistance with an H^2 dependence is due to the effect of a cyclotron orbital Lorentz force acting on the conduction electrons. This has been observed in many metals and semimetals [36]. The total magnetoresistance of itinerant electron magnets is therefore given by,

$$\Delta\rho(H, T) = \Delta\rho_c(H, T) + \Delta\rho_{sf}(H, T), \quad (10)$$

where $\Delta\rho_c(H, T)$ is the positive contribution due to the cyclotron motion of conduction electrons and $\Delta\rho_{sf}(H, T)$ is the negative contribution due to the spin fluctuations. At low magnetic fields, $\Delta\rho_c(H, T)$ is given by,

$$\Delta\rho_c(H, T) = B(T)H^2, \quad (11)$$

where the parameter $B(T)$ is proportional to $\rho(0, T)^{-1}$. When $B(T)$ decreases moderately with increasing temperature, the sign of $\Delta\rho(H, T)$ may change from positive to negative if $|\Delta\rho_c(H, T)| > |\Delta\rho_{sf}(H, T)|$ at low temperatures. Actually, a positive magnetoresistance has been observed only in the low temperature range, i.e. below 10 K for both $ZrZn_2$ [37] and Ni_3Al [38], and below 1 K for MnSi [39].

3. Weak ferromagnets

Experimental studies of the quenching of spin fluctuations in the weak itinerant electron ferromagnets Sc_3In, $Zr_{1-x}Hf_xZn_2$ and Ni_3Al have been made using heat capacity and electrical resistivity measurements. The ordering temperatures are ≈ 6 K for Sc_3In, 20 K for $ZrZn_2$ (which decreases with increasing Hf additions) and 0 K for Ni_3Al at the 3:1 ratio but increases rapidly when there is excess Ni. The ordering temperature of Sc_3In occurs at an awkward value, which complicates the interpretation and analysis of the low temperature properties. But for the other two compounds the ordering temperature is either zero, or sufficiently high that it poses no serious problem in this regard.

3.1. Heat capacity of Sc_3In

The low-temperature heat capacity of three DO_{19}-type Sc_3In alloys containing 24.1, 24.3 and 24.4 at% In with Curie temperatures at $H = 0$ of 5.5, 6.0 and 6.3 K, respectively, was measured in magnetic fields up to 9.98 T [20]. Fig. 2 shows the temperature variation of the heat capacity at five magnetic fields for the 24.3 at% In alloy which is typical of the other two alloys. The C/T vs. T^2 curve at zero field has a maximum around T_c, which was determined from magnetization measurements at $H < 0.1$ T and is indicated by the arrow in fig. 2. An upturn in the zero field heat capacity at the lowest temperatures is also evident. With increasing magnetic fields one sees that this maximum becomes significantly smaller. The 2.50 T curve crosses the zero field curve at 7.7 K, but at magnetic fields greater than 2.50 T, one does not find such an enhancement of heat capacity for $T > T_c$. The disappearance of the heat capacity peak of Sc_3In around the Curie temperature in high magnetic fields is thought to be due to the quenching of spin fluctuations. Since Sc_3In orders ferromagnetically, $T_{sf} \approx T_c$ [40] is assumed. Making use of the relationship

$$H_{eff} = k_B T_{sf}/\mu_B, \quad (12)$$

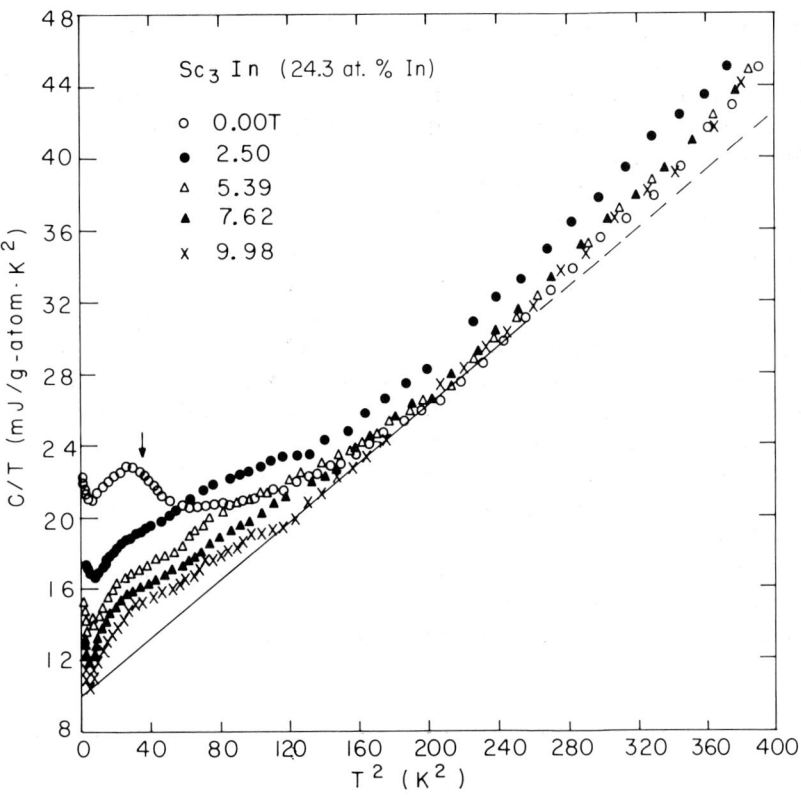

Fig. 2. Heat capacity of the Sc$_3$In alloy containing 24.3 at% In at five magnetic fields. The arrow shows T_c at $H = 0$. After Ikeda and Gschneidner [20].

the magnetic fields required to quench the spin fluctuations in Sc$_3$In, designated as H_{eff}(a), are given in table 1 for the three In–Sc alloys.

The quenching of the magnetic contribution (due to spin fluctuations) to the heat capacity by magnetic fields > 2.5 T enables one to better estimate the electronic specific heat constant, γ, and a "Debye temperature" at 0 K, Θ_D, see table 1. These were obtained from a least squares fit of the data around 1.5 K and between 10.5 and 14.0 K at 9.98 T to the equation:

$$C/T = \gamma + \beta T^2. \quad (13)$$

The "Θ_D" values for Sc$_3$In measured by Ikeda

Table 1
Curie temperature, electronic specific heat constants, "Debye temperature", H_{eff} values, magnetic entropy and ratio of enhancement factors of some Sc$_3$In alloys

Composition (at% In)	Curie temp. (K)	H_{eff} (a) (T)	H_{eff} (b) (T)	γ [a] $\left(\dfrac{mJ}{g\,at\,K^2}\right)$	γ_{total} $\left(\dfrac{mJ}{g\,at\,K^2}\right)$	Θ_D [b] (K)	S_M $\left(\dfrac{mJ}{g\,at\,K}\right)$	$\dfrac{\lambda_{spin}}{1+\lambda_{ep}}$
24.1	5.5	8.2	11.0	10.8	23.7	298	69	1.19
24.3	6.0	8.9	12.0	10.0	22.3	288	69	1.23
24.4	6.3	9.4	12.3	10.7	22.4	290	67	1.09

[a] Measured at 9.98 T, unenhanced (spin) γ.
[b] Fit parameter (eq. (13)) "Debye temperature", see text for further discussion.

and Gschneidner [20] are smaller than that determined from sound velocity measurements, $\Theta_D = 343$ K [41]. This implies an additional contribution to the T^3 term of heat capacity (β_M) over and above the lattice contribution (β_L), see eq. (6), similar to that observed in Pd–Ni [16] and CeSn$_3$ [17]. This extra contribution (β_M) is thought to be due to an induced moment on the magnetically active atom (Sc, Pd or Ce) by the magnetic field.

Fig. 3 shows the excess heat capacity, plotted as $\Delta C/T$, where $\Delta C = C - \gamma T - \beta T^3$, for the 24.3 at% In alloy at five magnetic fields. As seen in this figure, the heat capacity associated with the magnetic transition decreases with increasing magnetic field, and at 9.98 T the height of peak is $\approx 1/4$ of the zero field value. Moreover, another peak around 9 K is found in the non-zero magnetic fields curves, and is thought to be due to an induced magnetization on the Sc sites by the magnetic field. The magnetic entropies were estimated from this type of plot using the following relationship and neglecting the $\Delta C/T$ upturn at temperatures below 4 K

$$S_M = \int_0^{14 \text{ K}} (\Delta C/T) \, dT. \quad (14)$$

The zero field S_M values, which are given in table 1, are $\approx 1/10$ of the expected value from the local moment theory. Ikeda and Gschneidner [20] found that the magnetic entropy decreased linearly as a function of the magnetic field. The extrapolated values for $S_M = 0$ are tabulated in table 1 as $H_{\text{eff}}(b)$. The difference between $H_{\text{eff}}(a)$ and $H_{\text{eff}}(b)$ is thought to be due to the induced moment on the Sc atom by the magnetic field. As noted above, the induced moment also accounts for the second heat capacity peak and the β_M coefficient to the T^3 term of the heat capacity equation.

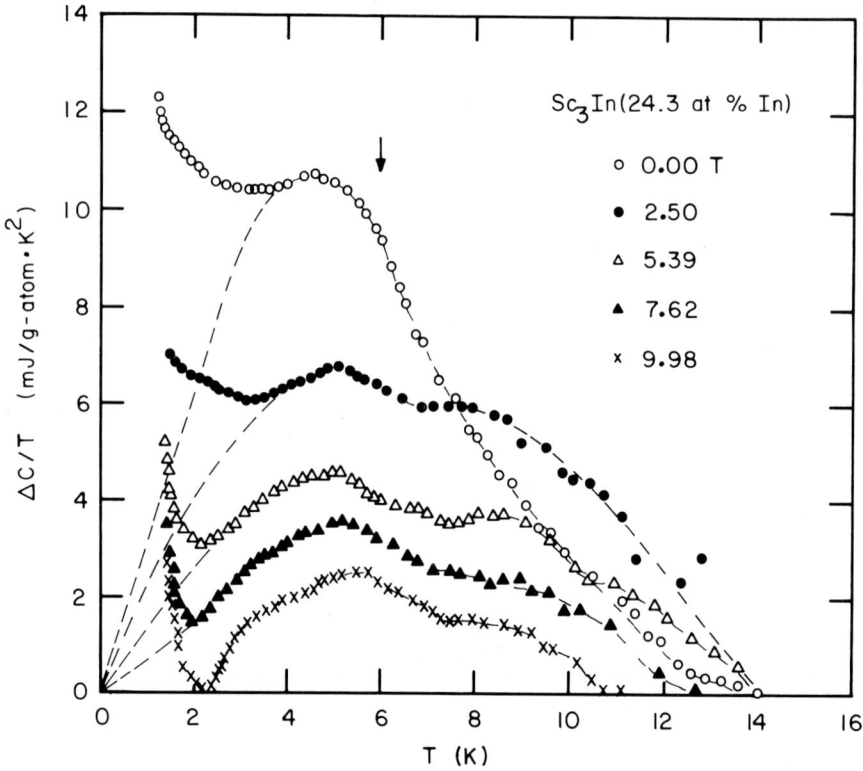

Fig. 3. $\Delta C/T$ vs. T for the Sc$_3$In alloy containing 24.3 at% In. The magnetic entropy was graphically estimated by neglecting the upturn at lower temperatures. After Ikeda and Gschneidner [20].

From a careful analysis of their zero field data and assuming that the spin fluctuations in Sc_3In are *completely* quenched at 9.98 T, the authors [20] were able to determine the total electronic specific heat constant at zero field (γ_{total}) and the ratio $\gamma_{spin}/(1+\lambda_{ep})$. These values are also listed in table 1.

3.2. Magnetoresistance of Sc_3In

The electrical resistivity and longitudinal magnetoresistance of Sc_3In were measured at magnetic fields up to 14 T [21] on the same specimens as used in the above-mentioned heat capacity studies [20]. Fig. 4 shows the temperature varia-

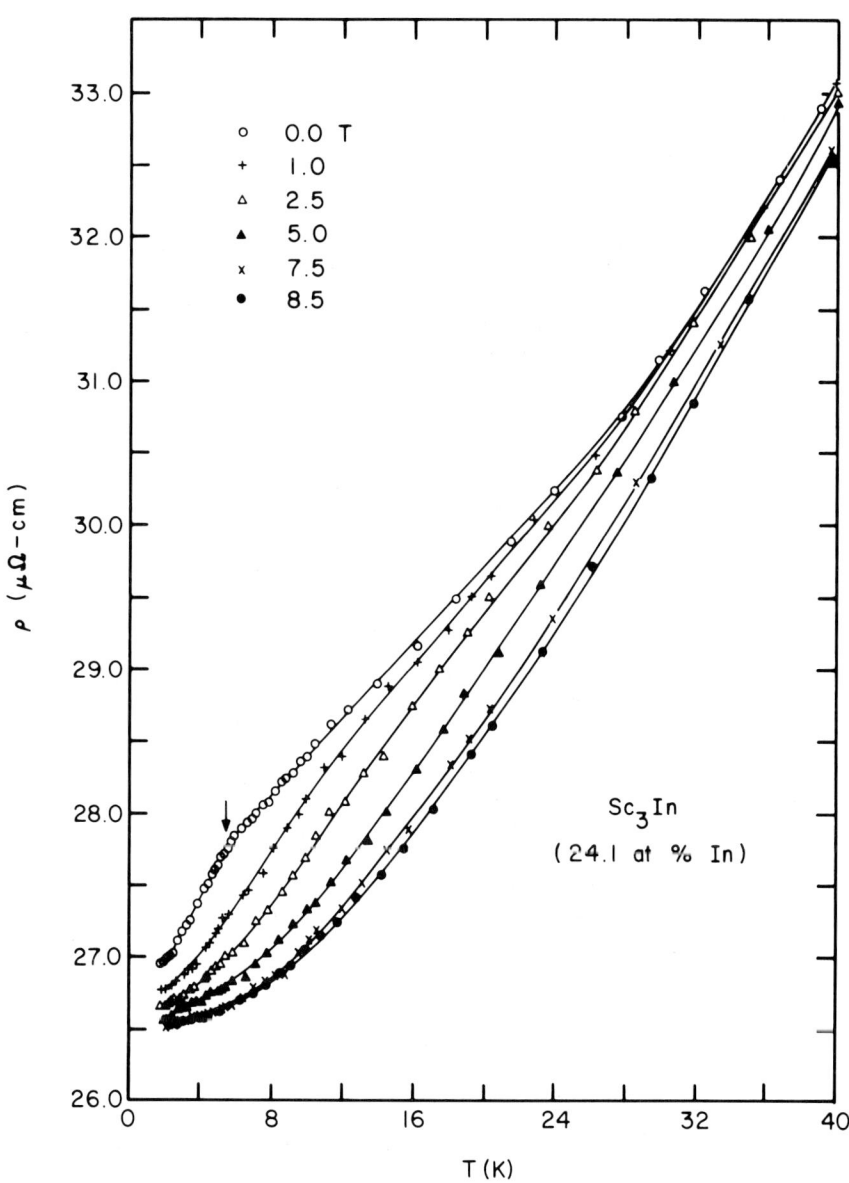

Fig. 4. Electrical resistivity at six magnetic fields for the Sc_3In alloy containing 24.1 at% In. The arrow shows T_c at $H = 0$ T. After Ikeda et al. [21].

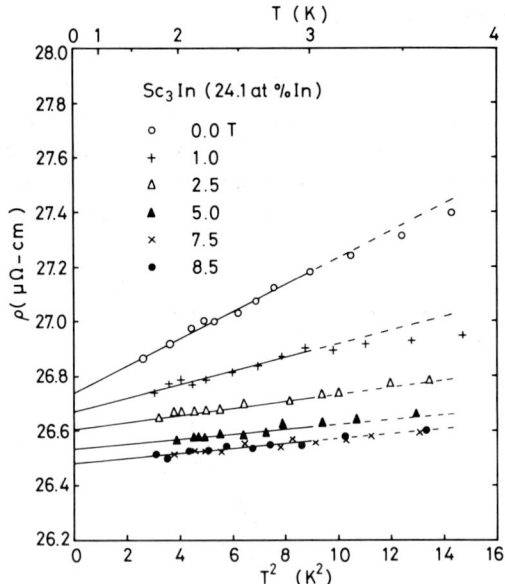

Fig. 5. T^2 plots of the electrical resistivity at six magnetic fields for the Sc_3In alloy containing 24.1 at% In. After Ikeda et al. [21].

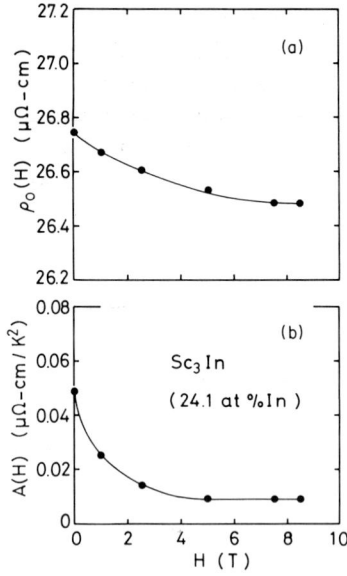

Fig. 6. (a) Residual resistivity, $\rho_0(H)$, and (b) the coefficient of T^2 term in the low temperature resistivity, $A(H)$, as a function of the applied magnetic fields for the 24.1 at% In–Sc alloy. After Ikeda et al. [21].

tion of the electrical resistivity at six magnetic fields up to 8.5 T for the 24.1 at% In alloy. A change in slope is observed near T_c in the ρ vs. T curve at $H = 0$. With increasing magnetic fields the anomaly around T_c disappears and at $H > 5$ T the ρ vs. T curves are similar to those of a normal paramagnetic metal. These magnetoresistance results are similar to those reported by Knapp et al. [42] and by Masuda et al. [43], and they are also consistent with the heat capacity measurements up to 10 T on the same specimens [20].

Fig. 5 shows the variation of ρ vs. T^2 for $T < 4$ K at six magnetic fields for the 24.1 at% In alloy. As is seen, the data fit the expression:

$$\rho(H, T) = \rho_o(H) + A(H)T^2, \qquad (15)$$

quite well below 3 K. Fig. 6 shows the field dependence of $\rho_0(H)$ and $A(H)$, which were obtained by a least squares fit of the data. An applied magnetic field slowly lowers $\rho_0(H)$ and causes an initial rapid drop in $A(H)$ which becomes constant at $H > 5$ T. The coefficient $A(H)$ may be expressed as a sum of two terms,

$$A(H) = A_0 + \Delta A(H), \qquad (16)$$

where A_0 is due to the electron–electron interaction [30] and $\Delta A(H)$ to spin fluctuations [31,32]. A_0 is independent of the applied magnetic field and was estimated from the data taken at $H > 7.5$ T. Fig. 7 shows the variation of $\Delta A(H)$ as a

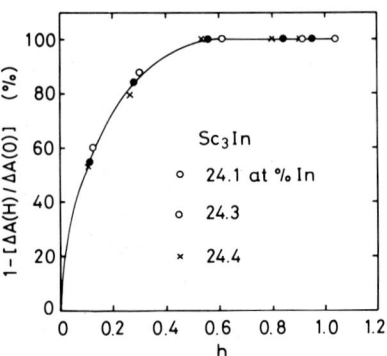

Fig. 7. Variation of the spin fluctuation contribution to the low temperature resistivity $[\Delta A(H)]$ as a function of the reduced magnetic field for three Sc_3In alloys. After Ikeda et al. [21].

function of the reduced magnetic field, $h = \mu_B H / k_B T_c$. All of the data points for the three Sc_3In alloys lie on the same curve, which tends to saturate at a lower field ($h \approx 0.6$) than was found in the heat capacity study ($h = 1.2$) [20]. The results of the heat capacity and electrical resistivity measurements show that the application of a magnetic field of ≈ 10 T completely quenches the spin fluctuations in Sc_3In and the assumption of $T_{sf} = T_c$ is appropriate for this weak ferromagnet.

Several years ago Kadowaki and Woods [44] pointed out that there is a linear relationship between the coefficient of T^2 term in the electrical resistivity, A, and the square of the electronic specific heat constant (γ^2) of heavy fermion compounds, see fig. 8. This relationship is also valid for some of the group VII and VIII elements and some A15 superconductors [45]. As is seen in fig. 8 this correlation holds as well for the weak itinerant ferromagnet Sc_3In (and for Ni_3Al and

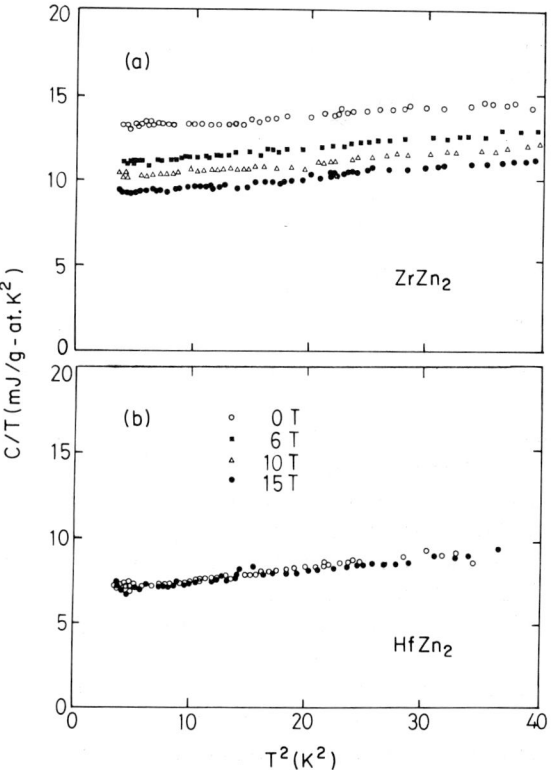

Fig. 9. Heat capacity of (a) $ZrZn_2$ and (b) $HfZn_2$ in four magnetic fields. After Ikeda et al. [47].

possibly $CeSn_3$ – two other compounds discussed in this review). This proportional relation between A and γ^2 is a natural development arising out of Fermi liquid theory [46].

3.3. $Zr_{1-x}Hf_xZn_2$

The cubic Laves phase ($MgCu_2$-type) compound $ZrZn_2$ is a weak ferromagnet with the Curie temperature $T_c = 20$ K, whereas the isostructural $HfZn_2$ compound is a strong Pauli paramagnet with the Stoner enhancement factor $S = 3.6$. Both magnetization and Curie temperature of the $Zr_{1-x}Hf_xZn_2$ alloys decrease with increasing Hf content (x) and become zero around $x = 0.16$. Ikeda et al. [47,48] have measured the low temperature heat capacity of $Zr_{1-x}Hf_xZn_2$ compounds in magnetic fields up to 15 T. As shown in fig. 9, the C/T vs. T^2 curves at $H = 0$, 6, 10 and 15 T for $ZrZn_2$ are

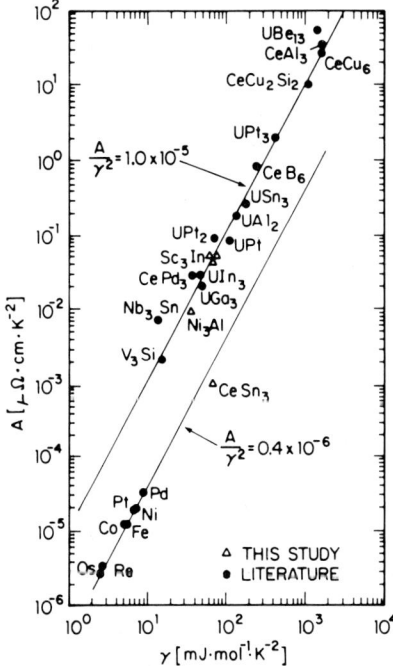

Fig. 8. Relationship between the coefficient of T^2 term in the low temperature resistivity, A [see eq. (15)], and the square of the electronic specific heat constant, γ^2, for weak ferromagnets, strong Pauli paramagnets, heavy fermions, superconductors and ferromagnets (see refs. [44,45]).

Table 2
Curie temperature, electronic specific heat constant, Debye temperature and spin enhancement factor of some $Zr_{1-x}Hf_xZn_2$ alloys

Composition x	Curie temp. (K)	$\gamma \left(\dfrac{mJ}{g\,at\,K^2} \right)$	Θ_D (K)	λ_{sf}
0	20	12.7	327	1.9
0.10	10	14.1	333	2.3
0.13	2.9	12.8	375	2.0
0.16	0	12.7	323	1.9
1.0	–	6.6	289	0.3

parallel to one another and shift to lower values with increasing magnetic fields. However, for $HfZn_2$ there is no difference between the zero field data and the 15 T data.

The decrease of γ with increasing fields in $ZrZn_2$, where the decrement ratio is 16, 23 and 31% at 6, 10 and 16 T, respectively, is probably due to the suppression and quenching of spin fluctuations. However, this phenomenon disappears at temperatures higher than ≈ 20 K as is expected because of renormalization of spin fluctuations [23]. The magnetic field dependence of γ in the $Zr_{1-x}Hf_xZn_2$ alloys for $0 \leq x \leq 0.2$ was most pronounced (33% at 15 T) in the $Zr_{0.9}Hf_{0.1}Zn_2$ composition, which also has the maximum γ value in this system (see table 2). The γ decrease under magnetic fields in $Zr_{1-x}Hf_xZn_2$ alloys is semiquantitatively explained by the numerical calculation on the quenching of spin fluctuations by Hertel et al. [28], if T_{sf} is assumed to be 53–65 K for these materials.

The spin enhancement factors for the $Zr_{1-x}Hf_xZn_2$ alloys were calculated assuming that (1) the density of states at the Fermi level obtained from the band structure calculations for $ZrZn_2$ [$N(E_F) = 1.62$ states/(eV atom spin)] [49] is independent of composition x, and (2) the mass enhancement factor due to electron–phonon interaction is given as $\lambda_{ep} = 0.42(1-x) + 0.38x$, i.e. we assume that λ_{ep} values of Zr, Hf and Zn metals (0.49, 0.38 and 0.38, respectively) remain the same in the alloy. These spin enhancement values are listed in table 2. The maximum decrease in γ due to the complete quenching of spin fluctuations by high magnetic fields may be expected to reach $\approx 60\%$ in weakly and nearly ferromagnetic $Zr_{1-x}Hf_xZn_2$ ($0 \leq x \leq 0.2$) alloys, because $\lambda_{sf}/(1 + \lambda_{total}) = 58$ and 62% for $ZrZn_2$ and $Zr_{0.9}Hf_{0.1}Zn_2$, respectively.

The electrical resistivity and longitudinal magnetoresistance measurements [48] were made in magnetic fields up to 8 T on the same $Zr_{1-x}Hf_xZn_2$ specimens as those used in the heat capacity measurements [47]. The ρ vs. T curves have a T^2 dependence at temperatures below ≈ 20 K and the data at each magnetic field followed eq. (15). Not only does the coefficient of T^2 term decreases with increasing magnetic fields, but the residual resistivity also decreases. This is also the case for Sc_3In. These behaviors are probably due to the quenching of spin fluctuations by magnetic fields. This can be shown from the following arguments. The residual resistivity is expressed by the equation:

$$\rho_0 = m^*/ne^2\tau_0, \qquad (17)$$

where m^*, n and e, respectively, are the effective mass, concentration and charge of the carrier, and τ_0 is the relaxation time of the scattering due to impurities, lattice faults and other imperfections. Since the magnetic field dependence of the effective mass $m^*(H) \propto \lambda_{spin}(H)$ (eq. (7)) and since λ_{spin} decreases with increasing field then we expect ρ_0 of these weak itinerant electron ferromagnets to decrease as the magnetic field is increased.

3.4. Ni_3Al system

Ni_3Al compound crystallizes in the simple Cu_3Au structure-type and exists over a homogeneity range from 72.5 to 77 at% Ni. In the composition range 73.5 to ≈ 75 at% Ni the alloys are strongly paramagnetic; for higher Ni concentrations itinerant electron ferromagnetism sets in [50]. The band structure calculated density of states [51] is found to be much lower than that obtained from zero field heat capacity data, which implies a considerable paramagnon enhancement. De Haas–van Alphen effect measurements on stoichiometric Ni_3Al in fields up to 10 T have

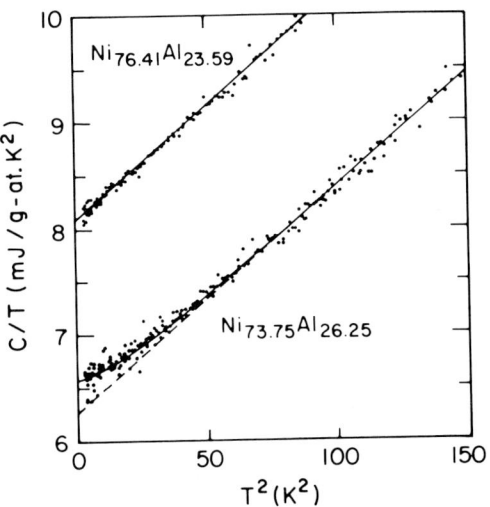

Fig. 10. The C/T vs. T^2 plots for $Ni_{73.75}Al_{26.25}$ and $Ni_{76.41}Al_{23.59}$ alloys. The solid line for the former is the fit to eq. (18) and for the latter to eq. (13). After Dhar et al. [54].

revealed appreciable electronic mass enhancements ranging from 2 to 3.4 depending on the character of band states [52]. A large body of experimental data based on the temperature dependence of magnetization, magnetoresistance, nuclear spin relaxation time, thermal expansion and the dependence of the Curie temperature on Ni concentration has been interpreted successfully invoking theoretical models of spin fluctuations.

Recently heat capacity has been measured in the 1.5 to 20 K range in applied fields up to 9.8 T on $Ni_{74.75}Al_{25.25}$, $Ni_{75}Al_{25}$ and $Ni_{75.6}Al_{24.4}$ [53] (these compositions differ slightly from the nominal values reported in ref. [53] but are the values reported by ref. [54] for the chemically analyzed samples) and in zero field only on $Ni_{73.75}Al_{26.25}$ and $Ni_{76.41}Al_{23.59}$ [54]. With the exception of $Ni_{76.41}Al_{23.59}$ alloy which shows a linear C/T vs.

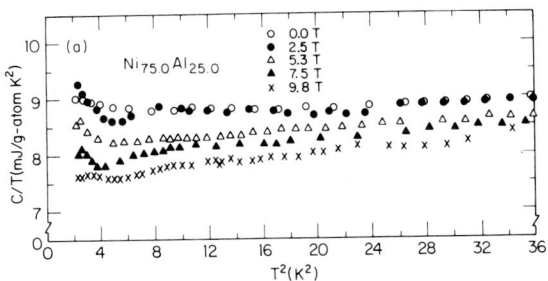 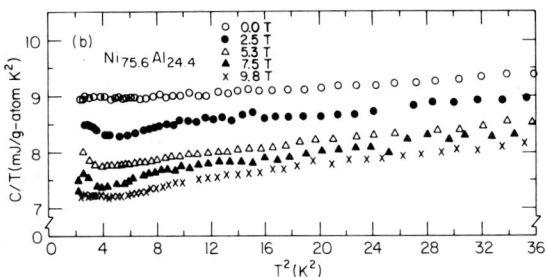

Fig. 11. The magnetic field dependence of the heat capacity of (a) the paramagnetic alloy $Ni_{75.0}Al_{25.0}$ and (b) the ferromagnetic alloy $Ni_{75.6}Al_{24.4}$. After Dhar and Gschneidner [53].

Table 3
Curie and spin fluctuation temperatures, electronic specific heat constant, Debye temperature and the fit parameters a) to the zero field heat capacity of some Ni_3Al alloys

Composition (at% Al)	T_c (K)	T_{sf} (K)	$A = \gamma$ $\left(\dfrac{mJ}{g\ at\ K^2}\right)$	B $\left(\dfrac{mJ}{g\ at\ K^4}\right)$	Θ_D (K)	D $\left(\dfrac{mJ}{g\ at\ K^4\ \ln K}\right)$
26.25	–	12.7	6.59	0.0035	465 b)	0.0062
25.25	–	17.7	8.32	−0.0114	462 c)	0.0107
25.00	–	18.1	8.93	−0.0274	467 c)	0.0161
24.40	38.2	–	8.95	−0.0035	465 b)	0.0083
23.59	68.9	–	8.07	–	465 b)	0

a) See eq. (18).
b) Assumed value, which was obtained from elastic constant and neutron scattering measurements.
c) Calculated from the β value obtained by fitting the data to eq. (13) in the range of 10 to 15 K.

T^2 in the 1.5 to 12 K range, the other four alloys show an upturn (concave upwards) in their C/T plots at temperatures below ≈ 12 K (see fig. 10). An applied field of 2.5 T has practically no effect on the heat capacity of the paramagnetic alloys (fig. 11a) but lowers the heat capacity over the entire temperature range of the ferromagnetic $Ni_{75.6}Al_{24.4}$ alloy (fig. 11b). A depression of the heat capacity of the paramagnetic alloys is observed at the next higher applied field of 5.3 T at temperatures below 12 K and heat capacity decreases further as the field is increased to 9.8 T (fig. 11a).

The zero field C/T plots can be fitted well by the expression

$$(C/T) = A + BT^2 + DT^2 \ln T \qquad (18)$$

and the values of the parameters are given in table 3. The field independence of the heat capacity in paramagnetic alloys at least up to 2.5 T (and presumably higher but lower than 5.3 T) appears to rule out the presence of magnetic clusters in these well prepared alloys. For large S the coefficient of the ln T term goes as $\approx S$ (section 2.1). Therefore as one approaches the ferromagnetic transition from the paramagnetic side the magnitude of D increases as actually observed (see fig. 12). On the ferromagnetic side as the spontaneous magnetization at $T = 0$ K increases and the spin fluctuations give way to spin waves, the upturn in C/T should decrease as observed experimentally (see table 3 and fig. 12, the D coefficient) and predicted theoretically by Makoshi and Moriya [23]. It is likely that in $Ni_{76.41}Al_{23.59}$ ($M_s = 0.114\mu_B/Ni$, $T_c = 68.9$ K) this situation prevails and accordingly the C/T vs. T^2 plot is linear (fig. 10). A linear behavior has also been reported by Ho et al. [55] in their alloys which were Ni-rich in content. In the maximum applied field of 9.8 T we find $\delta\gamma/\gamma(0) = -10.2\%$ and -15.2% for the paramagnetic $Ni_{74.75}Al_{25.25}$ and $Ni_{75}Al_{25}$ alloys, respectively.

The spin fluctuations in Ni_3Al have been investigated by using resistivity measurements [56]. The effect of spin fluctuations is reflected in the coefficient A of T^2 term of the resistivity. Fluitman et al. [56] investigated the Ni concentration

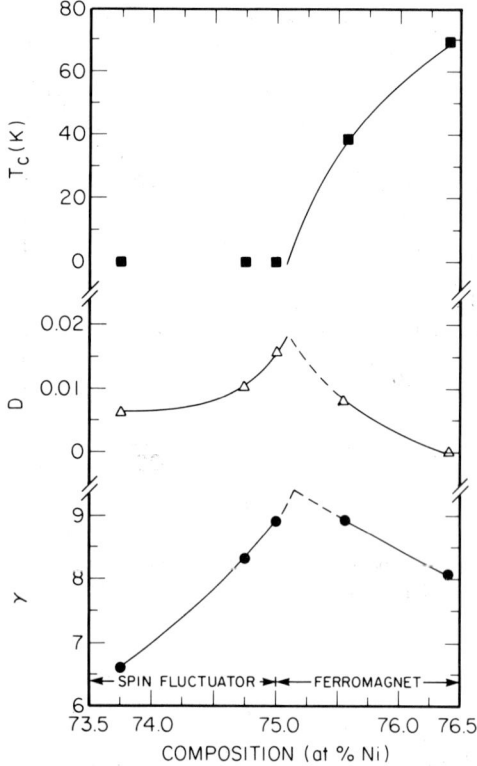

Fig. 12. Variation of the Curie temperature, the coefficient D (see eq. (18)), and electronic specific heat constant as a function of nickel concentration for the Ni_3Al alloys. Note the compositions are based on chemically analyzed alloys [54]. After Dhar et al. [53,54].

dependence around the stoichiometric composition, and found that A had a peak structure at $Ni_{75.5}Al_{25.5}$, i.e. reportedly the concentration at which ferromagnetism vanishes [50]. A similar behavior for γ and the coefficient D has been found in the heat capacity measurements (see fig. 12), except the critical concentration was reported to be (75.1 ± 0.2) at% Ni [54]. The difference is due to the fact that aluminum is lost by evaporation which means that the nickel concentration for nominal alloys is always low. The results reported by Dhar et al. [54] are based on chemically analyzed alloys. Recently Seki et al. [57] measured the magnetoresistance of $Ni_{75.07}Al_{24.93}$ alloy and they found that the coefficient A decreased by about 40% in 8 T.

4. Strong Pauli paramagnets

4.1. RCo_2 (R = Sc, Y and Lu)

The strong Pauli paramagnetic RCo_2 (R denotes Sc, Y and Lu) compounds, which have the cubic $MgCu_2$-type structure, exhibit a broad maximum in the magnetic susceptibility vs. temperature plot at a fairly high temperature (T_{max}) of ≈ 250, ≈ 370 and ≈ 600 K for YCo_2, $LuCo_2$ and $ScCo_2$, respectively. The heat capacity measurements on $ScCo_2$ (35 at% Sc), YCo_2 (35 and 36 at% Y) and $LuCo_2$ (35 at% Lu) were made between 1.5 and 20 K at five magnetic fields up to 9.98 T [12,15]. Fig. 13 shows the experimental results for $LuCo_2$. As can be seen, (1) all the C/T vs. T^2 curves are linear and there is no evidence for a $T^3 \ln T$ term in C, (2) the curves are parallel to one another, and (3) they fall with increasing magnetic field, especially for $H > 2.5$ T. For $ScCo_2$ a similar behavior was found, but for YCo_2 the straight lines for the various fields tend to cross one another with the intercept decreasing and the slopes increasing with increasing fields.

Fig. 14 shows the electronic specific heat constant and the Debye temperature at 0 K for $ScCo_2$ and $LuCo_2$ as a function of the magnetic field. As shown in this figure, Θ_D remains con-

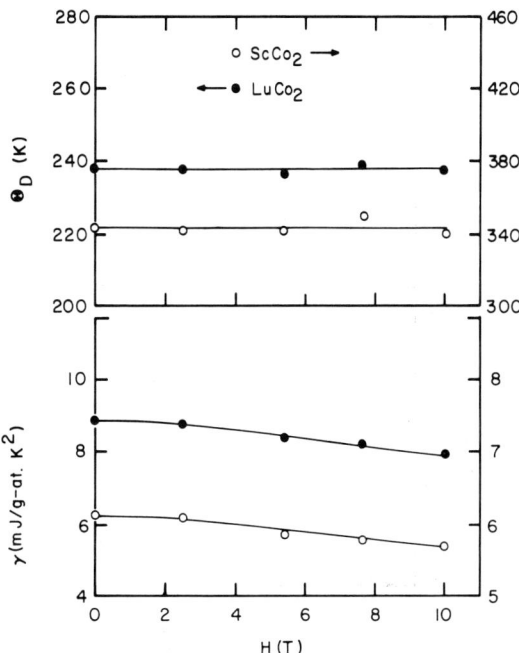

Fig. 14. Electronic specific heat constant γ and the Debye temperature Θ_D of $ScCo_2$ and $LuCo_2$ as a function of the magnetic field. After Ikeda et al. [15]. Reproduced by permission of Physical Review B.

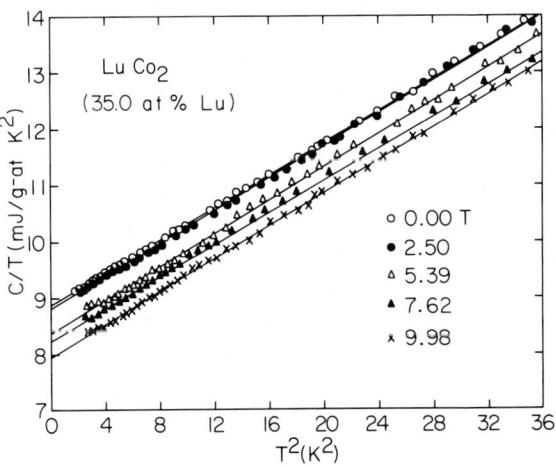

Fig. 13. Heat capacity of $LuCo_2$ below 6 K (36 K^2) at five magnetic fields. After Ikeda and Gschneidner [12], and Ikeda et al. [15]. Reproduced by permission of Physical Review B.

stant within the accuracy of measurements of ±2%, but γ decreases with increasing field as predicted by Béal-Monod et al. [10] and by Hertel et al. [28]. Barnea [58] estimated the characteristic spin fluctuation temperatures of YCo_2 and $LuCo_2$ to be 330 and 630 K, respectively, from the temperature dependence of the magnetic susceptibility. Barnea's T_{sf} values, and for that matter the T_{max} values if one equates T_{max} with T_{sf}, are unreasonably large, by about two orders of magnitude, as estimated from the depression of the electronic specific heat constant by the applied magnetic field. By using eq. (8) and $S = 16.5$, $H = 10$ T and $T_{sf} = 630$ K we obtain a value of $\delta\gamma/\gamma(0)$ of 0.05% for $LuCo_2$ which is more than two orders of magnitude smaller than the observed value of 10.3% at 9.98 T [12,15]. This paradox was noted earlier in the Introduction.

The characteristic spin fluctuation temperature T_{sf} for RCo_2 compounds can be estimated by comparing our experimental results with the theoretical models. Fig. 15 shows the shift of the

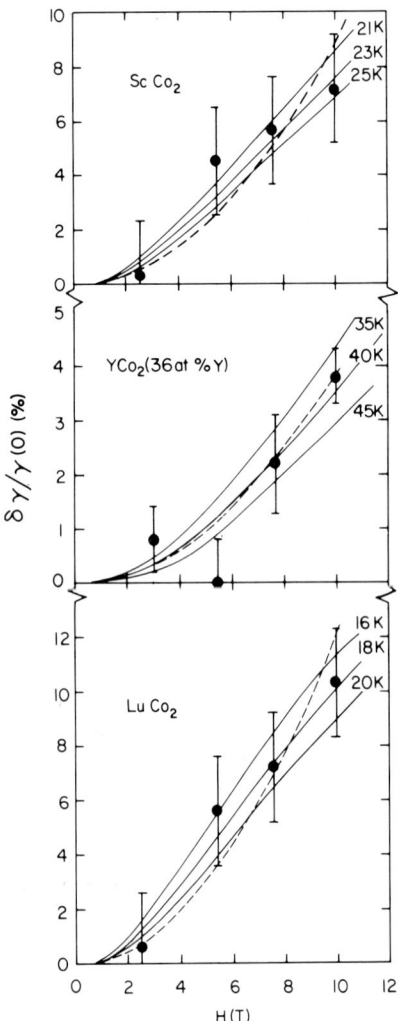

Fig. 15. The percentage shift of the electronic specific heat constant of $ScCo_2$, YCo_2 and $LuCo_2$ caused by an applied magnetic field. The solid and dashed lines are the theoretical results based on the models of Hertel et al. [28] and of Béal-Monod et al. [10], respectively, while the points are the observed values. The values listed by the solid curves are the chosen T_{sf} values. After Ikeda et al. [15]. Reproduced by permission of Physical Review B.

electronic specific heat constant at 0 K, $\delta\gamma/\gamma(0)$, of $ScCo_2$, YCo_2 and $LuCo_2$ caused by an applied field. The solid curves are taken from Hertel et al. [28] ($S = 10$, $\lambda_{spin} = 0.37$ and the three T_{sf} values shown in the figure). The dashed lines are the results of fitting the model of Béal-Monod et al. [10], i.e. eq. (8), by assuming $S = 15.5$ and $T_{sf} = 17$ K for $ScCo_2$, $S = 22.5$ and $T_{sf} = 29$ K for YCo_2 (36 at% Y), and $S = 16.7$ and $T_{sf} = 15$ K for $LuCo_2$. As can be seen from fig. 15, the experimental results are in good agreement with that of Hertel et al. [28] for $T_{sf} = (23 \pm 2)$ K for $ScCo_2$, T_{sf} (40 ± 5) K for YCo_2 and $T_{sf} = (18 \pm 2)$ K for $LuCo_2$.

By using the calculated density of states at the Fermi surface of YCo_2 [59], $N(E_F) = 0.367$ states/(eV atom spin), and the $N(E_F)$ values calculated from the zero field γ values, we determined $\gamma_{total} = 2.54$, 5.38, 4.89 and 4.12 for $ScCo_2$, YCo_2 (35 and 36 at% Y) and $LuCo_2$, respectively. The λ_{ep} values for RCo_2 compounds were calculated as the average value from the λ_{ep} values of the component elements, i.e. 0.20, 0.14, 0.35 and 0.42, for Sc, Y, Lu and Co, respectively. The resultant λ_{spin} values for these compounds are quite large (2 to 5), see table 4. This is consistent with the observations of the reduction of γ with increasing magnetic field.

Wohlfarth and Rhodes [60] were the first to predict that the itinerant electron metamagnetic transition in YCo_2 would occur at an applied field of ≈ 140 T. The metamagnetic transition was found in the magnetization measurements on the paramagnetic compounds YCo_2 and $LuCo_2$ [61], $Y(Co_{1-x}Al_x)_2$ [62,63], $Lu(Co_{1-x}Al_x)_2$ [64] and $Lu(Co_{1-x}Al_x)_2$ [65]. Recently, the magnetization measurements in magnetic fields up to ≈ 95 T have revealed the metamagnetic transitions at about 70 and 75 T on YCo_2 and $LuCo_2$, respectively, [66,67] which are significantly higher than the earlier values of ≈ 15 T and > 30 T, respectively, at 4.2 K, reported by Schinkel [61]. The metamagnetic behavior in these RCo_2 compounds is thought to originate with the anomalous band structure, which has a sharp and high peak in the density of states near the Fermi level [68]. The quenching of spin fluctuations by magnetic fields observed in the heat capacity of RCo_2 is surely related to this unusual band structure.

4.2. Scandium and palladium

Scandium and palladium are well known as exchange enhanced metals. Reported low temperature heat capacity measurements in high magnetic fields of high purity scandium and palladium (resistance ratios of 350 and 10 000, respectively) showed that the electronic specific heat constant decreased with increasing fields for both metals: 11% in scandium at 10 T [69], 8% in palladium at 11 T [70]. However, unpublished measurements by Stewart and Brandt [71] on two of the scandium single crystals used in the earlier study [69], and by two of the authors of this review [72] on two new electrotransport purified scandium samples showed that the heat capacity remained unchanged in fields up to 18 T in scandium. Similarly Stewart and Brandt [73] and our unpublished results [74] show that the heat capacity of palladium also did not change in magnetic fields up to 20 T. In the case of palladium, this has also been confirmed by de Haas–van Alphen measurements in fields up to 40 T, in which no evidence was obtained for a field dependence of either the mass enhancement factor or the Stoner factor [75]. The latest evidence strongly suggests that spin fluctuations in scandium or palladium are not quenched by magnetic fields as high as ≈ 20 T.

4.3. Palladium alloys doped with nickel

The Stoner enhancement factor S of palladium is thought to be ≈ 10. If one adds about 2.3 at% Ni, the system becomes ferromagnetic, i.e. $S \to \infty$. Low temperature heat capacity measurements of Ni–Pd alloys [76,77] showed a striking increase in the electronic specific heat constant, γ, as small amounts of Ni are alloyed with Pd, exhibiting a maximum near the critical concentration (2.3 at% Ni). Chouteau et al. [78] have also found that the concentration dependence of the coefficient of the T^3 term, β, decreases and at 5 at% Ni, β has regained the pure Pd value (see fig. 16). A strong magnetic field dependence of the electrical resistivity at low temperatures has been observed in the dilute paramagnetic Ni–Pd alloys by Schindler and LaRoy [79]. They measured the electrical resistivity of alloys containing 0–2.0 at% Ni at temperatures between 1.5 and 4.2 K and in magnetic fields up to 9.3 T, and have found that the coefficient of the T^2 term de-

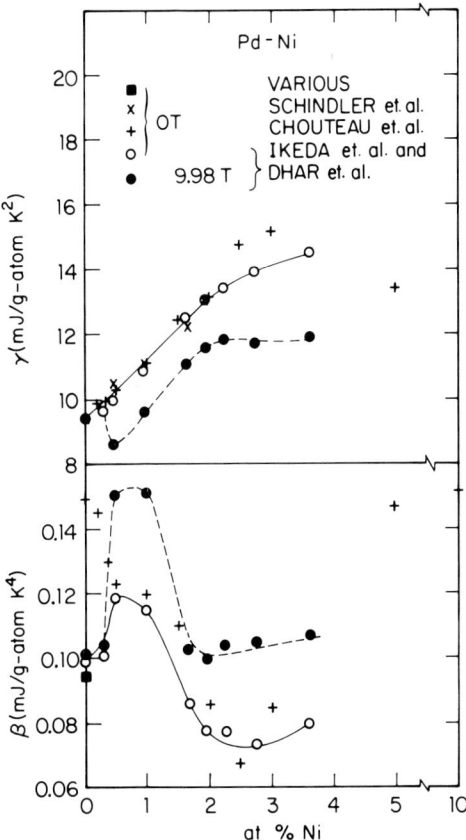

Fig. 16. Low temperature heat capacity results of several Ni–Pd alloys at 0 and 9.98 T. The data were taken from the publications by Schindler et al. [76], Chouteau et al. [77,78], Ikeda et al. [16], and Dhar et al. [74]. The zero field values of Cheung and Kouvel [80] for γ and β for three Pd–Ni alloys containing 1.85, 2.35 and 2.8 at% Ni differ considerably from the zero field values shown in the above plot.

creases with increasing fields, the tendency being stronger with increasing Ni concentrations.

Fig. 17 shows the results of the heat capacity measurements for the Pauli paramagnetic Pd–Ni alloys containing 0.97 at% Ni at zero and four magnetic fields [16]. All C/T vs. T^2 curves are linear and there is no evidence for a $T^3 \ln T$ term. With increasing magnetic fields, one can distinctly see that the C/T vs. T^2 curves fall and their slopes become steeper. This is also quite evident in fig. 18, which shows the field dependence of the two fit parameters of γ and β of eq. (13) of two Ni–Pd alloys. As shown the γ values

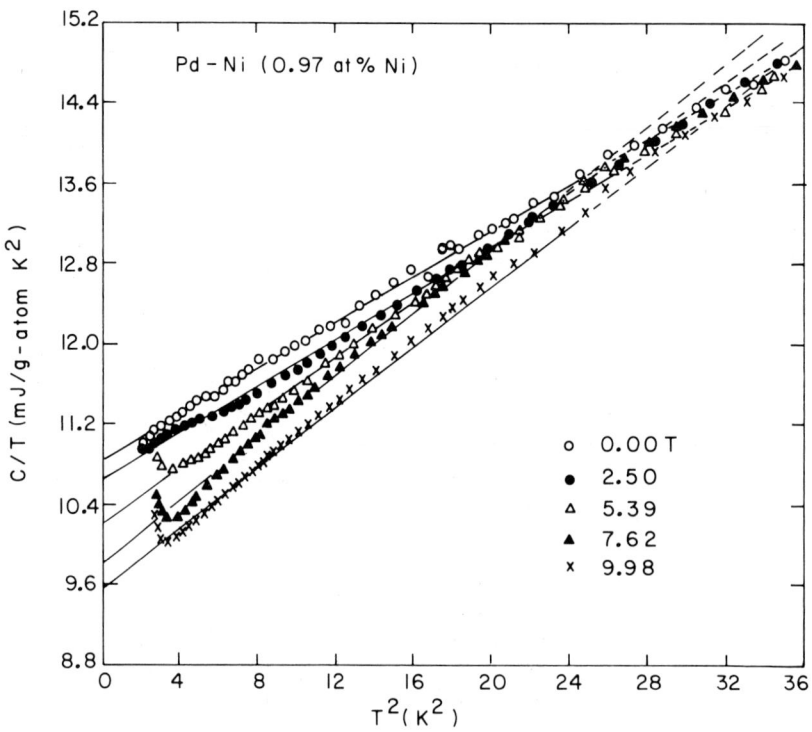

Fig. 17. Heat capacity of the Ni–Pd alloy containing 0.97 at% Ni at five magnetic fields. After Ikeda et al. [16]. Reproduced by permission of Physical Review B.

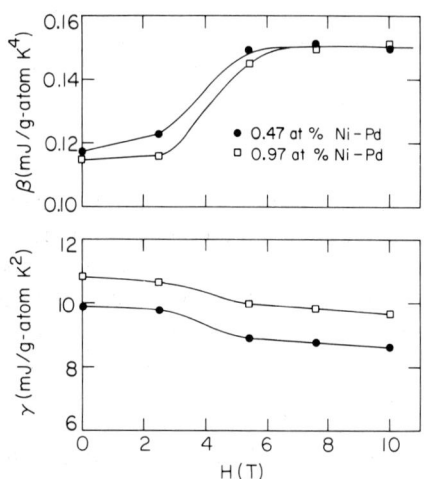

Fig. 18. Electronic specific heat constant γ and the coefficient β of the T^3 term for two dilute Ni–Pd alloys as a function of the magnetic field. After Ikeda et al. [16]. Reproduced by permission of Physical Review B.

for both samples decrease slowly: falling more rapidly between 2.5 and 5.39 T than at higher fields. A similar but opposite effect is found for β. Unpublished results by some of the authors [74] on alloys containing between 1.65 and 3.60 at% Ni show a similar behavior as in the 0.47 and 0.97 at% Ni–Pd alloys described above, except that the zero field data have a different temperature dependence at low temperature. Alloys containing 0.30 at% Ni or less behave exactly like pure palladium, i.e. the heat capacity is unaffected by magnetic fields up to 9.8 T [74]. The field dependence of γ for dilute Pd–Ni alloys is interpreted as the quenching of spin fluctuations in the exchange enhanced Pauli paramagnets with $T_{sf} = 20$ K. On the other hand, two possible mechanisms have been proposed to account for the increase in β. One, the increase is due to the magnetic quenching of localized spin fluctuations in the vicinity of the nickel atoms; or two, the increase is due to a magnetic contribution to the heat capacity which results from an induced magnetic moment on the nickel atoms by the applied magnetic field [16].

Cheung and Kouvel [80] carried out low temperature heat capacity measurements in magnetic fields of 0, 2 and 4 T on three Ni–Pd (1.85, 2.35 and 2.8 at% Ni) alloys near the critical composition for ferromagnetism, and found that γ decreased and β increased with increasing magnetic fields which are similar to the results shown in fig. 17. They concluded that the observed changes in the heat capacity in a magnetic field were due to magnetic clusters, and not due to the quenching of spin fluctuations. In view of the difference in the heat capacity vs. temperature curves for our alloys containing <1 at% Ni relative to those containing >1 at% Ni, it is possible that magnetic clustering (at least partially so) may be occurring in the higher concentration alloys, which would be consistent with Cheung's and Kouvel's conclusion. However, only measurements below 1 K in applied fields of alloys containing more than 1 at% Ni would give an unambiguous answer.

4.4. $TiBe_2$

$TiBe_2$ was first reported by Matthias et al. to be an itinerant antiferromagnet [81] but later work on well characterized single phase samples has shown it to be an exchange enhanced paramagnet with $S = 65$ and $\gamma = 56$ mJ/mol K^2. Stewart et al. were the first to find $T^3 \ln T$ temperature dependence in the heat capacity of this material and they measured the effect of field up to 7 T [82]. In a later work Stewart et al. have applied fields up to 17 T [83]. They found that quenching of spin fluctuations begins at 5.2 T, the field at which there is a measurable change in the heat capacity. The heat capacity decreased by 19% at 5.05 K in a field of 17 T, see fig. 19. The depression became less at higher temperatures and for temperatures greater than 16 K the change was practically negligible in the highest applied field of 17 T. These authors [83] extrapolated the value of the field for the full quenching

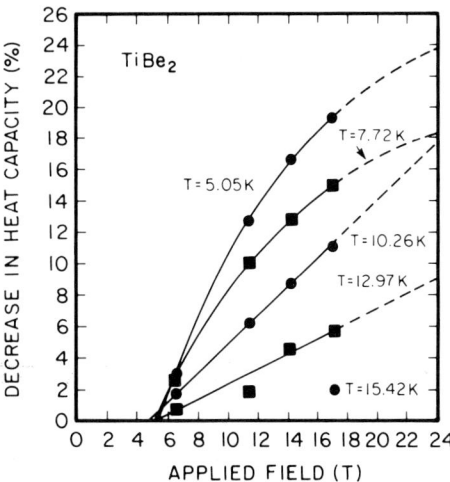

Fig. 19. The lowering of the heat capacity of $TiBe_2$ from the zero field values at five temperatures as a function of the applied magnetic field. After Stewart el al. [83].

of spin fluctuations to be 25 T, which they claim is consistent with the estimate of T_{sf} of (22 ± 4) and (50 ± 25) K [82]. High field (up to 20 T) magnetoresistivity measurements on $TiBe_2$ also give strong evidence for scattering due to spin fluctuations [84]. Some peculiarities in the magnetoresistance are observed around 5 T and anomalies in the susceptibility and differential susceptibility [85] are also observed at the same field.

5. Heavy fermion systems

5.1. $CeSn_3$

$CeSn_3$ has the fcc Cu_3Au-type structure. Its magnetic susceptibility exhibits a Curie–Weiss behavior between 250 and 800 K and a broad maximum at ≈ 135 K. The low temperature heat capacity of $CeSn_3$ [86] shows an increase of C/T with decreasing temperatures below ≈ 3 K and a large γ value, and its electrical resistivity follows a T^2 law for $T \leq 17$ K [87]. These experimental results show that $CeSn_3$ exhibits mixed valent properties, and that a change in the nature of the f states occurs from the band-like states of Ce^{4+} ($T < 40$ K) to the localized states of Ce^{3+} ($T \geq 40$

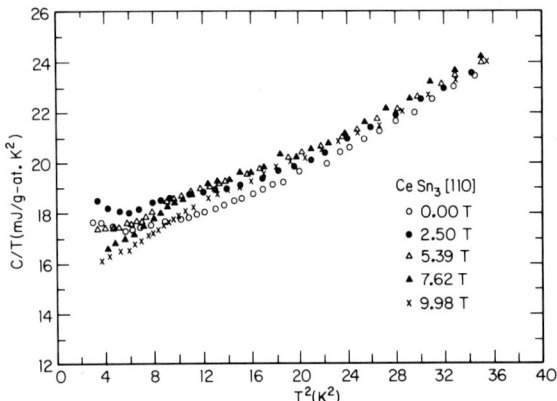

Fig. 20. Heat capacity of a single crystal of CeSn$_3$ at five magnetic fields with the [110] direction parallel to the field. After Gschneidner et al. [88].

K). Because of these interesting properties this compound is one of the typical heavy fermion materials which has been examined to see if the persistent spin fluctuations in such systems can be quenched by large magnetic fields.

Fig. 20 shows the results of the heat capacity measurements on CeSn$_3$ at five magnetic fields up to 9.98 T [17,88,89]. Because of the unusual shape of the C/T vs. T^2 curves at 0 and 2.50 T below ≈ 5 K, the heat capacity data for these two fields were fitted to eq. (18). Comparing eqs. (5) and (18) we find that $A = (m^*/m)\gamma_0 = \gamma$, $B = \beta_L - D \ln T_{sf}$ and $D = \alpha_0 \gamma_0 / T_{sf}^2$. On the other hand, the data at 5.39, 7.62 and 9.98 T were fitted to eq. (13). The magnetic field dependence of both γ and β changes rapidly at low applied fields, then more slowly with increasing magnetic fields for $H \geq 5$ T, similar to the behavior shown in fig. 18 for the Ni–Pd alloys. At 9.98 T the electronic specific heat constant had dropped by 27% from the zero field value for CeSn$_3$ [17] which is about $2\frac{1}{2}$ times the amount for the Ni–Pd alloys. In later experiments on single crystals [88] and several compositions ranging from CeSn$_{2.99}$ to CeSn$_{3.05}$ [89] the depression of γ by the magnetic field was significantly less, ranging from 8 to 21%. The larger shift in the first CeSn$_{3.0}$ sample [17] may be due to the large iron impurity concentration in that sample (175 ppm at.) compared to much lower values (2 to 20 ppm at.) in later samples [88,89].

For some of the later samples [89], in particular CeSn$_{3.05}$, from a cursory inspection of their C/T vs. T^2 plots one might conclude that the heat capacity was independent of the applied magnetic field, when all the curves were superimposed over one another. However, when the data were fitted to eqs. (18) and (13), as noted above, a definite variation in γ and β with field was observed. But since the field dependence of γ, β and the $T^3 \ln T$ terms have opposing signs ($-$, $+$, $+$ respectively), see section 2.1, the total heat capacity does not seem to change with field. Only when one makes a careful analysis of the data at each field do the trends become apparent.

Although not too evident in fig. 20, the coefficient of the "lattice" or T^3 contribution to the heat capacity, β, increases with increasing field. The neutron scattering results [90] have shown that an applied magnetic field of 4.23 T induces a magnetic moment on the Ce sites in CeSn$_3$. This induced moment gives rise to a magnetic contribution to the heat capacity at $H \geq 0$ T, C_M, much like a material which orders magnetically. The

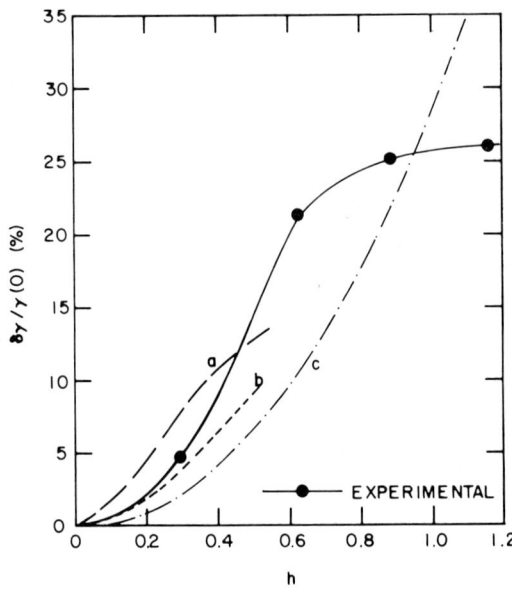

Fig. 21. Shift of the electronic specific heat constant of CeSn$_3$ caused by an applied field as a function of the reduced field. Three curves a, b and c are the theoretical results. After Ikeda and Gschneidner [17]. Reproduced by permission of Physical Review B.

C_M value of CeSn$_3$ in magnetic fields of 5 T, which was estimated by using the β_L value determined in the neutron scattering measurements [91], indicates that the magnitude of magnetic entropy due to these induced moments is quite small, $\approx 1/1000$ of the theoretical value for a spin of $s = \frac{1}{2}$.

The fit parameters B and D [to eq. (18)] may be used to determine the spin fluctuation temperature, i.e. $B = \beta_L - D \ln T_{sf}$, provided the β_L value is known. By using the Debye temperature at 0 K given by Stassis et al. [91] and the zero field B and D fit parameters, a value of 5.8 K is obtained for T_{sf} [17]. Fig. 21 shows the shift of the electronic specific heat constant at 0 K of CeSn$_3$ caused by an applied field as a function of the reduced field, $h = \mu_B H / k_B T_{sf}$. By using $T_{sf} = 5.8$ K, an $H_{eff} = 8.6$ T is obtained. This effective field for the quenching of spin fluctuations is consistent with the experimental result that γ is essentially constant above 7.6 T, see fig. 21. This figure also shows the theoretical curves a and b taken from Hertel et al. [28] (case a: $S = 10$ and $\gamma_{spin} = 0.37$, and case b: $S = 4$ and $\gamma_{spin} = 0.345$), and from Béal-Monod et al. [10] (curve c: $S = 4.23$ and $T_{sf} = 5.8$ K). The agreement is fair in the former cases (a and b) and poor in the latter case (c).

Tsang et al. [88] measured the field dependence of the low temperature heat capacity of a CeSn$_3$ single crystal with the magnetic field parallel to the [100] and [110] directions. A measurable anisotropy developed for magnetic fields > 2.5 T. The quenching of spin fluctuations was larger

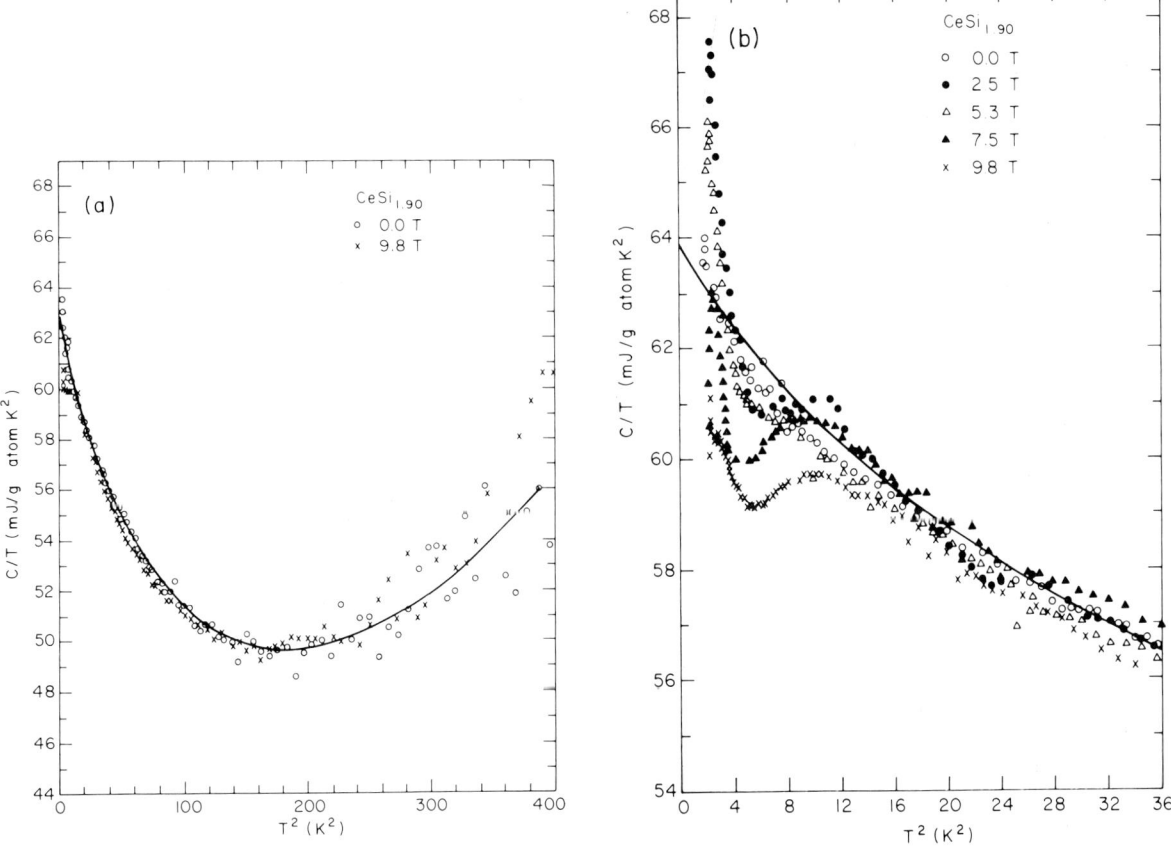

Fig. 22. Heat capacity of CeSi$_{1.90}$ in zero and applied magnetic fields up to 9.8 T: (a) from 1.5 to 20 K (2 to 400 K^2) and (b) from 1.5 to 6 K (2 to 36 K^2). After Dhar et al. [95]. Reproduced by permission of Physical Review B.

when the magnetic field was parallel to the [110] direction. Furthermore, a distinct difference was found in the magnetic susceptibilities below 5 K, i.e., $\chi_{110} > \chi_{100}$ by $\approx 5\%$ at 1.8 K. This was the first evidence of an anisotropicity in the quenching of spin fluctuations.

5.2. $CeSi_x$ system

Another Ce based system in which an unambiguous contribution of the form $T^3 \ln T$ to the heat capacity has been observed is the $CeSi_x$ system. $CeSi_x$ alloys exist over a solid solution range $1.55 < x < 1.95$. At room temperature below a silicon concentration of $x = 1.75$, a crystallographic distortion occurs and the crystal structure changes from tetragonal to orthorhombic [92]. There is a smooth progression from a magnetically ordered ferromagnetic dense Kondo system ($x < 1.85$) to compositions where the conventional single impurity Kondo effect dominates the exchange interaction, resulting in a ground state which is non-magnetic [93,94].

Dhar et al. measured the heat capacity in zero and applied fields up to 9.8 T on $CeSi_{1.83}$, $CeSi_{1.85}$ and $CeSi_{1.90}$ [95]. A contribution of the form $T^3 \ln T$ occurs for all the three alloys, see fig. 22a. In $CeSi_{1.83}$ and $CeSi_{1.85}$ the deviation from $T^3 \ln T$ behavior occurs approximately below 8 and 2.4 K, respectively, due to the onset of magnetic order at 5.5 and 1 K, respectively. Using paramagnon expression for the heat capacity, eq. (18), γ values of 184.6 and 269.6 mJ/mol Ce K^2, are obtained for $x = 1.90$ and 1.85, respectively. For $CeSi_{1.85}$ the γ value is at best a high temperature extrapolation value. The 0 K value for this alloy could be different as it orders magnetically at 1 K. For the paramagnetic $CeSi_{1.90}$ the spin fluctuation temperature and the Stoner enhancement factor are found to be 28 K and 135, respectively. Spin fluctuations in $CeSi_{1.90}$ are not quenched in fields up to 7.5 T but the data suggest that the quenching may be starting at 9.8 T in which the heat capacity is marginally lower than that in zero field (see fig. 22a).

Dhar et al. [95] also found that a bump develops in the heat capacity at ≈ 3 K at magnetic fields > 2 T in both $CeSi_{1.85}$ and $CeSi_{1.90}$ (see fig. 22b). The magnitude of the bump varies with the field: it is observed at 2.5 T, disappears at 5.3 T and reappears at 7.5 and 9.8 T, with the magnitude being the largest at 7.5 T (fig. 22b). The authors thought this bump was due to an induced magnetic moment on the cerium atoms by the applied magnetic field. They also explained the disappearance and reappearance of the heat capacity bump with changing field as being due to the moments on a small piece of Fermi surface, which is destroyed when the field is changed, and reappears at higher field when another part of the Fermi surface is favorable for an alignment of the induced moments. As noted in our discussion on Sc_3In (section 3.1) a bump was also found to develop in the heat capacity of this material at high magnetic field (see fig. 3) but for Sc_3In the peak occurs at ≈ 8 K. Dhar et al. [95] also suggested that a similar bump may occur in $TiBe_2$, but more experimental data are needed to confirm this.

5.3. UAl_2

UAl_2 was the first intermetallic compound in which a $T^3 \ln T$ contribution to the heat capacity was observed at low temperatures [13]. The heat capacity of an annealed single crystal of UAl_2, $\gamma = 142.3$ mJ/mol K^2, has been measured from 2 to 23 K in fields up to 12.5 T and from 4 to 16 K in fields to 17 T [14], see fig. 23. The heat

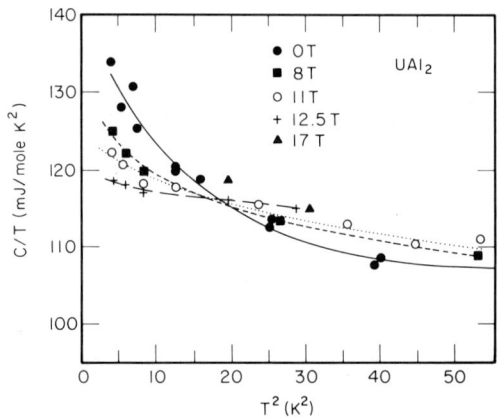

Fig. 23. Heat capacity of UAl_2 in zero and applied magnetic fields up to 17 T. After Stewart et al. [14].

Table 4
Summary of properties of exchange enhanced paramagnets

Material	Type	$\gamma \left(\dfrac{mJ}{mol\ M\ K^2}\right)$ [a]	$\dfrac{-\delta\gamma}{\gamma(0)}$ at 10 T (%)	$\dfrac{\delta\beta}{\beta(0)}$ at 10 T (%)	λ_{spin}	$\chi \times 10^4$ (at 0 K) (emu/g at)	S	T_{sf} (K)	T_c (K)
Sc	0	10.33	0	0	1.60	3.93	5.4	30	–
Pd	0	9.45	0	0	0.37	7.36	10	250	–
Pd (0.05%Ni)	0	9.48 [b]	0	0	0.38	7.7	10.5	–	–
Pd (0.10%Ni)	0	9.58 [b]	0	0	0.39	8.1	11	–	–
Pd (0.30%Ni)	0	9.95 [b]	0	0	0.49	9.3	12	–	–
ScCo$_2$	1	18.4	7.2	0	2.36	3.67	15.5	20	–
LuCo$_2$	1	26.6	10.3	0	3.85	3.96	16.7	16	–
YCo$_2$ (36%Y)	2	30.6	3.8	18	4.64	5.33	22.5	35	–
YCo$_2$ (35%Y)	2	33.1	–	–	5.13	6.00	25.3	35	–
Pd (0.47%Ni)	2	9.90 [b]	13.4	27	0.56	11.76	15.3	20	–
Pd (0.97%Ni)	2	10.85 [b]	11.9	31	0.65	18.36	23.9	20	–
Pd (1.65%Ni) [c]	2	12.53 [b]	11.1	20	0.98	71.8	93.0	20	–
Pd (1.95%Ni) [c]	2	13.10 [b]	11.8	28	1.08	–	–	20	–
Pd (2.25%Ni) [c]	2	13.41 [b]	11.9	35	1.13	–	–	–	–
Pd (2.75%Ni) [c]	–	13.98 [b]	16.0	47	1.23	–	∞	–	7
Pd (3.60%Ni) [c]	–	14.5 [b]	15.6	36	1.32	–	∞	–	24
CeSn$_{2.99}$	3	69.8	14.1	77	4.82	–	–	5.1	–
CeSn$_{3.00}$	3	65.0	8.2	75	4.36	6.0	17	3.9	–
CeSn$_{3.02}$	3	65.5	8.3	70	4.38	–	–	4.1	–
CeSn$_{3.05}$	3	66.5	10.3	74	4.42	5.4	15	4.4	–
CeSn$_{3.0}$[H$_{100}$]	3	75.0	15.4	65	5.67	7.2	20	5.6	–
CeSn$_{3.0}$[H$_{110}$]	3	75.0	21.1	120	5.67	8.0	22	5.6	–
Ni$_{73.75}$Al$_{26.25}$	3	9.38	–	–	0.04 [d]	–	10	12.7	–
Ni$_{74.75}$Al$_{25.25}$	3	11.24	10.2	2	0.29 [d]	–	21	17.7	–
Ni$_{75.00}$Al$_{25.00}$	3	11.91	15.2	8	1.40 [e]	–	41	18.1	–
Ni$_{75.60}$Al$_{24.40}$	–	11.05	15.8	25	1.21 [e]	–	∞	–	38.2
Ni$_{76.41}$Al$_{23.59}$	–	9.97	–	–	0.98 [e]	–	∞	–	68.9
UAl$_2$	4	142.2	5.7	–	15.6 [f]	15.1	38	45	–
U$_{0.85}$La$_{0.15}$Al$_2$	4	280	5.5	–	27 [f]	30.3	76	60	–
CeSi$_{1.90}$	5	184.6	1.8	–	46	71.5	135	29	–
CeSi$_{1.85}$	5	269.6	9.7	–	70	194 [g]	373 [g]	28	–
TiBe$_2$	5	56.3	9.7	–	15 [f]	30.7	65	50	–
Sc$_3$In(24.4In)	6	13.8 [h]	–	–	1.09	[i]	∞	6.3 [j]	6.3

[a] M = magnetic atom (M = Sc, Ti, Co, Ni, Ce and U).
[b] In units of mJ/g at K.
[c] Magnetic clustering may be occurring in this alloy.
[d] Calculated by reviewers assuming $\lambda_{ep} = 0.2$ and the paramagnetic density of states [99].
[e] Calculated by reviewers assuming $\lambda_{ep} = 0.2$ and the ferromagnetic density of states [99].
[f] Calculated by reviewers assuming $\lambda_{ep} = 0.3$.
[g] At 2 K.
[h] At 10 T.
[i] Ferromagnetic.
[j] Authors [20] assumed T_c (Curie temperature) = T_{sf}.

capacity is found to decrease only at temperatures below 4 K. In an applied field of 12.5 T, the heat capacity at 2.1 K is lowered by about 10%. Using $\gamma_0(1 + \lambda_{ep}) = 96$ as obtained from the high temperature extrapolation of C/T vs. T^2 plot of UAl_2, $\lambda_{ep} = 0.3$, $\chi = 15.1 \times 10^{-6}$ emu/g and the fit coefficients to the paramagnon equation for the heat capacity (eq. (18), however, the lattice term was taken as $BT^3 + CT^5$) these authors estimate $S = 4.35$ and $T_{sf} = 30$ K. On the other hand, if one uses the value of γ_0 of 8.4 mJ/mol K^2 as obtained from band structure calculation [96] then S turns out to be 38.4 and $T_{sf} \approx 45$ K. These authors [96] give another procedure for estimating S, γ_0 and T_{sf} which involves setting up an equation that contains fit coefficients to eq. (18), the experimentally measured susceptibility and S. The equation is then solved self-consistently. The values of S, γ_0 and T_{sf} obtained this way are 10.5 (13.6), 30.6 (23.7) and 39 (41) for $\lambda_{ep} = 0.3$ and (1.0), respectively. These latter values are more reasonable than the initial values suggested by Stewart et al. [14] for S, γ_0 and T_{sf} and they are also more consistent with values reported for other spin fluctuators, see table 4. The differential susceptibility of UAl_2 has been measured at low temperature as a function of magnetic field up to 40 T [97]. The results support the view that the low temperature anomalies in UAl_2 are due to spin fluctuations and not to the fine structure near the Fermi level. A characteristic field H_{sf} and a characteristic temperature T_{sf}, defined by the inflection points in $\chi(H)$ and $\chi(T)$, respectively, turn out to be nearly equivalent, i.e. $T_{sf} \approx 14$ K.

The effect of La substitution in UAl_2 has been reported for the alloy $U_{0.85}La_{0.15}La2$ [98]. Wire et al. found that the low temperature susceptibility increases by more than a factor of 2. The heat capacity has the $T^3 \ln T$ temperature dependence as in the parent UAl_2 and γ is found to be 238.3 mJ/mol K^2. The authors conclude that the La substitution caused an enhancement of S by 30% and that of T_{sf} by 50%. The latter observation appears to be inconsistent with the larger susceptibility observed in the La substituted alloy. Such apparently contradictory relation is observed in Ni_3Al alloys also [53].

6. Concluding remarks

The quenching of spin fluctuations can be categorized into six types of behaviors based on the variation of low temperature heat capacity as a function of temperature as the applied magnetic field is increased from 0 to at least 10 T. These are summarized in figs. 24–29 (parts a and b) going from the simplest behavior to the most complex and finally to the stage where the material actually orders magnetically near 0 K. The important physical properties of these materials and their type behavior are given in table 4.

Also listed in table 4 are a few materials which were examined at high magnetic fields and their heat capacities were found to be the same as that measured at zero field. These materials are considered to be "dead" with respect to an applied magnetic field at least to the maximum field used in the experiment. These are: scandium and palladium (see section 4.2) and palladium doped alloys containing 0.05 to 0.30 at% nickel (see section 4.3), and they are listed in table 4 as being type "0" materials.

There also appears to a be a rough correlation between the field dependence of the heat capacity and the temperature variation of the magnetic susceptibility. The χ vs. T dependences are shown in the same respective figs. 24–29 as part c. The relationship between the heat capacity and susceptibility behaviors are described in the following discussions of six types of behaviors of the field dependence of the low temperature heat capacities. The grouping of the various materials into the six categories is based primarily on the field dependence of the low temperature heat capacity.

6.1. Type 1 behavior

The temperature dependence of the heat capacity of a type 1 material at zero field is the same as that of a normal metallic material and has no $T^3 \ln T$ dependence (fig. 24a) but λ_{spin} is usually quite large (≈ 2). The only effect of high magnetic fields is to suppress γ (and λ_{spin}) – as is shown in fig. 24b. The β coefficient of the T^3 term in the heat capacity (eq. (13)) is constant

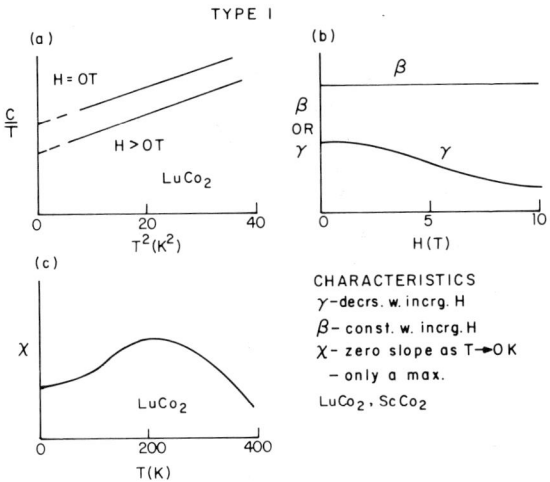

Fig. 24. Schematic representation of the low temperature heat capacity as a function of magnetic field and the magnetic susceptibility of a type 1 spin fluctuator.

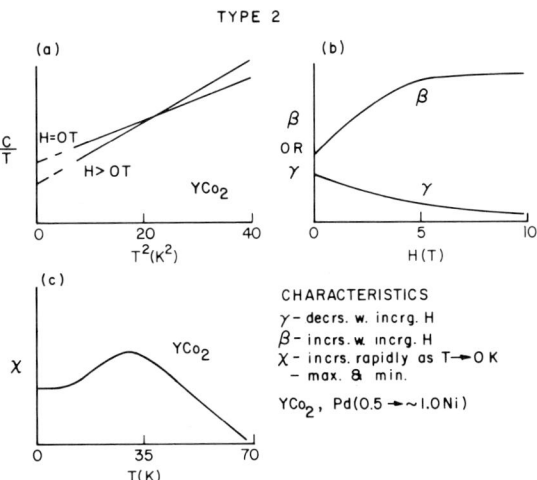

Fig. 25. Schematic representation of the low temperature heat capacity as a function of magnetic field and the magnetic susceptibility of a type 2 spin fluctuator.

and independent of the magnetic field. The known type 1 materials are $LuCo_2$ and $ScCo_2$.

The magnetic susceptibilities for these materials are shown in fig. 24c, where it is seen that the susceptibility of a type 1 material is characterized by a high temperature peak, and a nearly zero slope as $T \to 0$ K.

6.2. Type 2 behavior

For type 2 materials the temperature dependence of the heat capacity at zero field is the same as that for a normal metal and type 1 substances. But the application of high magnetic field has two effects: the γ is lowered due to the suppression of λ_{spin} and the slope β is increased due to an induced moment of the fluctuating electrons which are aligned by the field (see fig. 25a and b). The known type 2 materials include pure YCo_2 and Pd-alloys doped with nickel (0.47 to ≈ 1.0 at%). The magnetic susceptibility of these materials is similar to that of type 1 spin fluctuators (compare figs. 24c and 25c).

6.3. Type 3 behavior

All the remaining highly exchanged enhanced spin fluctuators (types 3, 4 and 5) exhibit a $T^3 \ln T$ behavior in their zero field heat capacities, however, the high field behaviors are different. Type 3 behavior is exhibited by $CeSn_3$, and possibly by Ni_3Al (see section 6.5), and is exemplified by the facts that as a high magnetic field is applied the $T^3 \ln T$ term disappears, γ is lowered, and the slope β increases, see fig. 26a and b. The weakness of the $T^3 \ln T$ term is probably due to the low T_{sf} (5.7 K). Because the $T^3 \ln T$ term disappears between 2.5 and 5 T the field

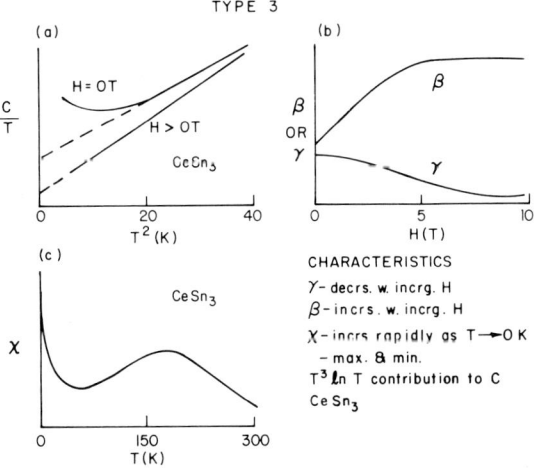

Fig. 26. Schematic representation of the low temperature heat capacity as a function of magnetic field and the magnetic susceptibility of a type 3 spin fluctuator.

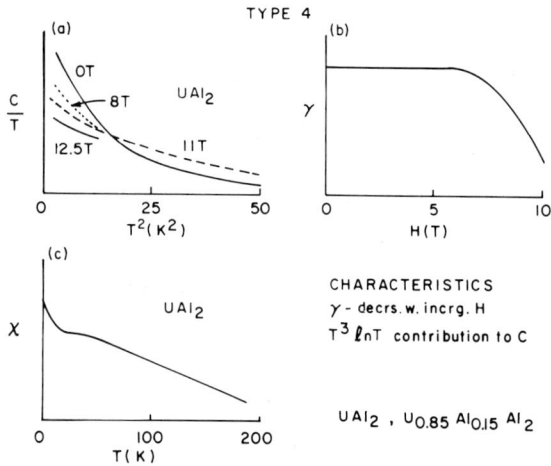

Fig. 27. Schematic representation of the low temperature heat capacity as a function of magnetic field and the magnetic susceptibility of a type 4 spin fluctuator.

dependences of γ and β are relatively easy to establish.

The magnetic susceptibility for CeSn$_3$ shows a significant departure from type and 1 and type 2 materials, with a minimum and fairly sharp upturn in χ as $T \to 0$ K (fig. 26c).

6.4. Type 4 behavior

UAl$_2$ based compounds UAl$_2$ and U$_{0.85}$Al$_{0.15}$Al$_2$ are the only known type 4 behavior materials, however, some of the properties of Ni$_3$Al fit the characteristics of type 4 behavior (see section 6.5). The $T^3 \ln T$ term is quite strong and is diminished by magnetic fields (fig. 27a), but does not disappear even at 17 T (fig. 23). The electronic specific heat constant is lowered, but it has not been established whether or not β increases with field (fig. 27b). It is possible that behaviors of types 3 and 4 are the same and differ only in degree. The crucial point being whether β varies or remains constant. If β remains constant then UAl$_2$ (type 4) remains a distinct and different type. We believe that β remains constant for UAl$_2$ because the heat capacity curve is smooth and does not exhibit any bumps which are introduced with an increasing magnetic field as is found in type 5 and 6 behaviors, see section 6.6 and 6.7.

The magnetic susceptibility is shown in fig. 27c where it is seen to have an inflection point at low temperature due to the merger of the characteristic maximum and minimum of type 3 behavior (fig. 26c). It is quite obvious that this χ vs. T behavior is intermediate between those of types 3 and 5, see figs. 26c and 28c, respectively.

6.5. Ni$_3$Al

The field dependence of electronic contribution and the $T^3 \ln T$ behavior of Ni$_3$Al are similar to CeSn$_3$. The γ values decrease $\approx 12\%$ for both phases (see table 4). The $T^3 \ln T$ term disappears between 7.5 and 10 T for Ni$_3$Al, and between 2.5 and 5 T for CeSn$_3$. But the field dependence of β (hardly changes) and the temperature dependence of the magnetic susceptibility of Ni$_3$Al are much more like those exhibited by UAl$_2$ and are quite different from those of CeSn$_3$. The behaviors would place Ni$_3$Al intermediate between CeSn$_3$ and UAl$_2$. This is also consistent with the spin fluctuation temperatures for these three materials, which are: ≈ 5 K for CeSn$_3$, ≈ 18 K for Ni$_3$Al and 45 K for UAl$_2$.

6.6. Type 5 behavior

The $T^3 \ln T$ upturn is quite strong in these materials, and high fields lower γ and the $T^3 \ln T$ upturn (fig. 28a and b). This behavior is quite similar to the behavior exhibited by type 4 materials, i.e. UAl$_2$ (fig. 27a and b), but at high fields a distinct bump develops in the temperature dependence of the heat capacity (fig. 28a). We believe this bump in the heat capacity develops because of an induced moment due to the alignment of the fluctuating electrons by the high magnetic field. It is noted that perhaps types 3 and 5 are really the same, but for the former the disappearance of the $T^3 \ln T$ term at low fields makes it appear that they are different. Furthermore, no bump is observed in the heat capacity in high fields of the type 3 material. These points will need further thought and consideration. The characteristic rapid rise in χ as $T \to 0$ K, see fig. 28c, suggests that this material is about ready to order, as a matter of fact CeSi$_{1.85}$ does order at 1

K. The materials that are classified type five materials are CeSi$_x$ ($x = 1.9$ and 1.85) and TiBe$_2$.

6.7. Type 6 behavior

The last type of behavior comes about when the interactions of the fluctuating electrons are strong enough so that at low temperatures the spins align spontaneously giving rise to "ferromagnetism". The effect of magnetic fields on the temperature dependence of the heat capacity is illustrated in fig. 29a and b, where it is seen that the application of field destroys the magnetic ordering in Sc$_3$In (the peak due to "ferromagnetic" ordering vanishes as $H \to \approx 12$ T). Not evident in fig. 29a is the development of a bump with increasing field at a temperature slightly larger than the ferromagnetic transition temperature, but it is clearly evident in the original C/T vs. T plots for three different Sc$_3$In samples. The bump is shown schematically in the $\Delta C/T$ vs. T plot of fig. 29b. The typical χ vs. T curve for a "ferromagnetic" material is shown in fig. 29c. The magnetization curve for Sc$_3$In is exactly like that expected for a ferromagnet (fig. 29d) and this is why the material is called an itinerant "ferromagnet". However, the quenching of "ferromagnetism" by high magnetic fields is unusual

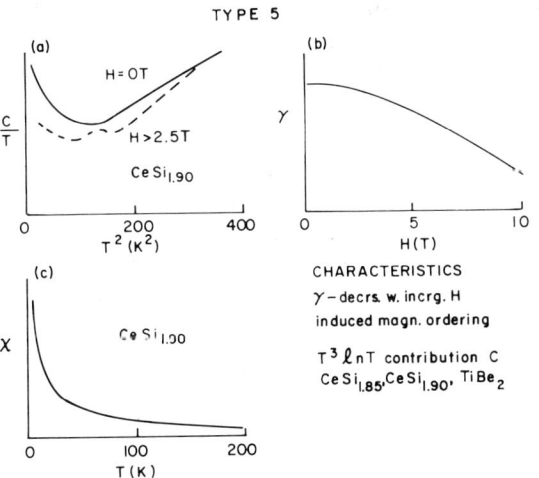

Fig. 28. Schematic representation of the low temperature heat capacity as a function of magnetic field and the magnetic susceptibility of a type 5 spin fluctuator.

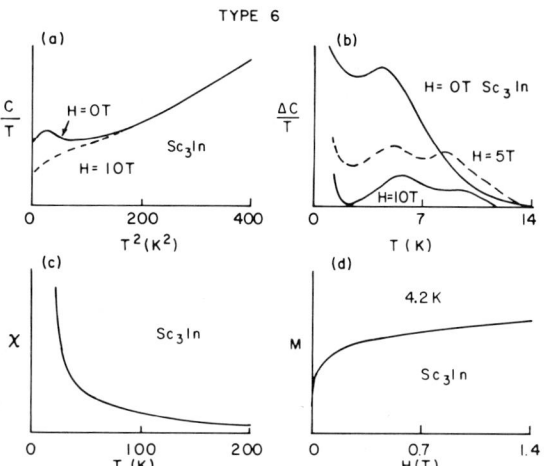

Fig. 29. Schematic representation of the low temperature heat capacity as a function of magnetic field, the magnetic susceptibility and the magnetization of a type 6 spin fluctuator.

and unexpected and that is why we have used quotation marks around the words ferromagnetism and ferromagnet for this material.

6.8. Ferromagnetic Ni$_3$Al and CeSi$_x$

In addition to "ferromagnetic" Sc$_3$In both Ni$_3$Al and CeSi$_x$ become ferromagnetic at sufficiently high Ni or Ce concentrations. In the case of Ni$_3$Al, no one has studied the magnetic field dependence of a ferromagnetic sample around the magnetic ordering temperature, so we are unable to compare its behavior relative to that of Sc$_3$In.

The heat capacity of CeSi$_{1.83}$, which orders at 5.5 K, has been studied as a function of magnetic field [45] and it behaves as a typical ferromagnet – the peak becomes rounded, and both the peak and the magnetic entropy are shifted to higher temperatures as the magnetic field is increased. The behavior is distinctly different from that exhibited by Sc$_3$In, see figs. 2, 3 and 29. Since the magnetic ordering behaviors of Sc$_3$In and CeSi$_{1.83}$ are quite different in nature, ferromagnetic CeSi$_{1.83}$ cannot be classified as exhibiting type 6 behavior. From an analysis of the magnetic entropy Dhar et al. [95] concluded that 17% of the entropy is associated with ferromagnetism and 83% with spin fluctuations. Thus the ferromagnetic CeSi$_x$ alloys (or at least CeSi$_{1.83}$) can be best classified as a ferromagnetic variant of type 5 behavior. It is for this reason that CeSi$_{1.85}$ is considered to be a type 5 spin fluctuator, even though it orders ferromagnetically at 1 K.

6.9. Epilogue

The quenching of spin fluctuations in highly exchanged enhanced Pauli paramagnets has been observed in approximately ten different materials. The quenching behaviors have been divided into six types according to the effect the applied magnetic field has on low temperature heat capacities. In type 1 the interaction between the fluctuating spins seems to be the weakest and it appears to increase successively until it is sufficiently strong enough to give rise to a spontaneous ordering of the spins at low temperature in type 6 materials. The temperature dependence of the magnetic susceptibility for these six types seems to support the general trend of increasing strength of the interactions as the type number increases. The variation of λ_{spin} and S for the six types generally (but not always) increase on going from type 1 materials to type 6 materials (see table 4). For example, λ_{spin} increases from values of ≈ 1 for type 1 and 2 materials, to ≈ 5 for type 3, to ≈ 10 for type 5 before dropping to ≈ 1 for the magnetically ordered type 6 substance. The Stoner enhancement values increase from ≈ 10 for type 1 to ≈ 50 or more for type 5, to ∞ for type 6 and ferromagnetically ordering type 5 materials (see section 6.8).

Clearly such simple parameters as λ_{spin}, the characteristic spin fluctuation temperature (T_{sf}) in themselves or together cannot explain the observed trends, but they do serve as useful guides. Another important parameter may be the number of electrons per atom that are involved, which varies from a low of ≈ 0.001 for CeSn$_3$ to a high of ≈ 0.1 for Pd containing Ni alloys. Unfortunately, this number is difficult to obtain for some of the materials, especially those that exhibit type 1 and 4 behaviors.

Although the observation of the quenching of spin fluctuations in these highly enhanced systems is slightly more than ten years old, some great strides have been made experimentally. The theory, which is much older (≈ 20 years) is inadequate and needs to be re-examined in light of the recent experimental studies, however, some recent work has helped to improve our knowledge on the subject. It is hoped that this review of the known materials and the systematization of behaviors will help us achieve a much better understanding of these unique class of materials.

Dedication

This review is dedicated to Prof. Kōki Ikeda who died on 5 April, 1991 after a brief illness due to a bacterial infection following an operation for an ulcer at the end of January. Before entering the hospital he completed the first draft for a major part of this review paper – his last contribution to the field of magnetic phenomena and magnetic materials.

Prof. Ikeda was born in Sakata, Japan on 23 February, 1946. He obtained his bachelor and doctor degrees from Tohoku University, Sendai in 1968 and 1973, respectively. He spent seven years as a doctoral and post doctoral fellow carrying out research at the Research Institute for Iron, Steel and other Metals (now called Institute for Materials Research), Tohoku University, before he joined the Faculty of Engineering, Iwate University, Morioka, Japan in 1975. Except for a two year visit at the Ames Laboratory, Iowa State University, Ames, Iowa, USA from 1978 to 1980, and a one year stay at the Laboratoire de Physique des Solides, Université de Paris-Sud, Orsay, France in 1982–1983, he has been a faculty member of the Department of Metallurgy, Iwate University. He was named a full Professor in 1989.

Prof. Kōki Ikeda studied the electrical and magnetic properties of materials containing the transition and rare earth elements by using electrical resistivity, low temperature heat capacity and magnetization measurements. His studies include: (1) the effect of magnetic field on the spin fluctuations, (2) the correlation between the electrical resistivity and magnetism, (3) the influence of the lattice faults on the magnetic properties, (4) the origin of the high upper-critical field in the ceramic superconductors, and (5) the correlation between superconductivity and lattice instability in lanthanum chalcogenides. He published 71 papers on these and related topics. During his studies at the Ames Laboratory, he and K.A. Gschneidner, Jr. discovered the unusual decrease of the electronic heat capacity and electrical resistivity in the weakly and nearly ferromagnetic materials, Sc_3In, $LuCo_2$, and other materials in applied magnetic fields. This decrease is caused by the quenching of spin fluctuations by high magnetic fields (up to 10 T). This discovery was recognized by the US Department of Energy as the Outstanding Scientific Accomplishment in Metallurgy and Ceramics in 1982. In 1990 Prof. Ikeda received the 48th Meritorious Honor Award of the Japan Institute of Metals for his work in the field of materials physics.

We are deeply saddened by his sudden and premature death. We will miss his insight into the physics of magnetic phenomena, his council and advice, and above all his warm and congenial friendship. He will be truly missed by all of us, especially those who have had the privilege and pleasure to work with him and to know him as a friend and companion.

K.A. Gschneidner, Jr.
S.K. Dhar
M. Yoshizawa

References

[1] K.A. Gschneidner, Jr. and S.K. Dhar, in: Magnetic Excitations and Fluctuations, eds. S.W. Lovesey, U. Balucani, F. Borsa and V. Tognetti (Springer-Verlag, Berlin, 1984) p. 177.
[2] S. Doniach, Proc. Phys. Soc. 91 (1967) 86.
[3] G.G. Low, Proc. Intern. Conf. on Magnetism, Nottingham, September 1964 (Institute of Physics, London, 1965) p. 133.
[4] N.F. Berk and J.R. Schrieffer, Phys. Rev. Lett. 17 (1966) 433.
[5] S. Doniach and S. Engelsberg, Phys. Rev. Lett. 17 (1966) 750.
[6] M.J. Rice, Phys. Rev. 159 (1967) 153.
[7] A. Layzer and D. Fay, Intern. J. Magn. 1 (1971) 135.
[8] D. Fay and J. Appel, Phys. Rev. B 16 (1977) 2325.
[9] W.F. Brinkman and S. Engelsberg, Phys. Rev. 169 (1968) 417.
[10] M.T. Béal-Monod, Shang-Keng Ma and D.R. Fredkin, Phys. Rev. Lett. 20 (1968) 929.
[11] M.T. Béal-Monod and E. Daniel, Phys. Rev. B 27 (1983) 4467.
[12] K. Ikeda and K.A. Gschneidner, Jr., Phys. Rev. Lett. 45 (1980) 1341.
[13] R.J. Trainor, M.B. Brodsky and H.V. Culbert, Phys. Rev. Lett. 34 (1975) 1019.
[14] G.R. Stewart, A.L. Giorgi, B.L. Brandt, S. Foner and A.J. Arko, Phys. Rev. B 28 (1983) 1524.
[15] K. Ikeda, K.A. Gschneidner, Jr., R.J. Stierman, T.-W.E. Tsang and O.D. McMasters, Phys. Rev. B 29 (1984) 5039.
[16] K. Ikeda, K.A. Gschneidner, Jr. and A.I. Schindler, Phys. Rev. B 28 (1983) 1457.
[17] K. Ikeda and K.A. Gschneidner, Jr., Phys. Rev. B 25 (1982) 4623.
[18] G.R. Stewart, J.L. Smith and B.L. Brandt, Phys. Rev. B 26 (1982) 3783.
[19] R. Konno and T. Moriya, Japan. J. Appl. Phys. 26, suppl. 26-3, (1987) 491 [Proc. 18th Intern. Conf. Low Temp. Phys., Kyoto].
[20] K. Ikeda and K.A. Gschneidner, Jr., J. Magn. Magn. Mater. 22 (1981) 207, 30 (1983) 273.
[21] K. Ikeda, K.A. Gschneidner, Jr., N. Kobayashi and K. Noto, J. Magn. Magn. Mater. 42 (1984) 1.
[22] S. Engelsberg, W.F. Brinkman and S. Doniach, Phys. Rev. Lett. 20 (1968) 1040.
[23] K. Makoshi and T. Moriya, J. Phys. Soc. Japan 38 (1975) 10.
[24] J.R. Schrieffer, J. Appl. Phys. 39 (1968) 642.
[25] P. Lederer and D.L. Mills, Phys. Rev. Lett. 20 (1968) 1036.
[26] G.G. Lonzarich, J. Magn. Magn. Mater. 54-57 (1986) 612.
[27] K.A. Gschneidner, Jr. and K. Ikeda, J. Magn. Magn. Mater. 31-34 (1983) 265, erratum 39 (1983) 320.
[28] P. Hertel, J. Appel and D. Fay, Phys. Rev. B 22 (1980) 534.
[29] W.G. Baber, Proc. Roy. Soc. (London) A158 (1937) 383.
[30] J. Appel, Phil. Mag. 8 (1963) 1071.
[31] D.L. Mills and P. Lederer, J. Phys. Chem. Solids 27 (1966) 1805.
[32] K. Ueda and T. Moriya, J. Phys. Soc. Japan 39 (1975) 605.
[33] A.H. Wilson, Proc. Roy. Soc. (London) A167 (1938) 580.
[34] K. Ueda, Solid State Commun. 19 (1976) 965.
[35] M. Kohler, Ann. Phys. 32 (1938) 211.
[36] J.-P. Jan, Solid State Physics, vol. 5, eds. F. Seitz and D. Turnbull (Academic Press, New York, 1957) p. 1.
[37] S. Ogawa, Physica B 91 (1977) 82.
[38] H. Sasakura, K. Suzuki and Y. Masuda, J. Phys. Soc. Japan 53 (1984) 352.

[39] K. Kadowaki, K. Okuda and M. Date, J. Phys. Soc. Japan 51 (1982) 2433.
T. Sakakibara, M. Mollymoto and M. Date, J. Phys. Soc. Japan 51 (1982) 2439.
[40] K.H.J. Buschow and H.J. van Daal, in: Magnetism and Magnetic Materials, AIP Conf. Proc. No. 5, eds. C.D. Graham and J.J. Rhyne (American Institute of Physics, New York, 1972) p. 1464.
[41] L.R. Testardi, L.M. Holmes, W.A. Reed and F.S.L. Hsu, Phys. Rev B 6 (1972) 3365.
[42] G.S. Knapp, L'.. Isaacs, H.V. Culbert and R.A. Conner, in: Magnetism and Magnetic Materials, AIP Conf. Proc. No. 5, eds. C.D. Graham and J.J. Rhyne (American Institute of Physics, New York, 1972) p. 467.
[43] Y. Masuda, T. Hioki and A. Oota, Physica B 91 (1977) 291.
[44] K. Kadowaki and S.B. Woods, Solid State Commun. 58 (1986) 507.
[45] The elemental values were taken from M.J. Rice, Phys. Rev. Lett. 20 (1968) 1439.
For the A15 superconductors from K. Miyake, private communication.
[46] K. Yamada and K. Yoshida, Prog. Theor. Phys. 76 (1986) 621.
[47] K. Ikeda, M. Yoshizawa, K. Kai and T. Nomoto, Physica B 165 & 166 (1990) 203.
[48] K. Ikeda, M. Yoshizawa and K. Okuno, to be published.
[49] R.A. de Groet, D.D. Koelling and F.A. Mueller, J. Phys. F 10 (1980) L235.
[50] F.R. de Boer, C.J. Schinkel, J. Biesterbos and S. Proost, J. Appl. Phys. 40 (1969) 1049.
[51] J.M. Buiting, J. Kubler and F.M. Mueller, J. Phys. F 13 (1983) L179.
[52] T.I. Sigfusson, N.R. Bernhoeft and G.G. Lonzarich, J. Phys. F 14 (1986) 2141.
[53] S.K. Dhar and K.A. Gschneidner, Jr., Phys. Rev. B 39 (1989) 7453.
[54] S.K. Dhar and K.A. Gschneidner, Jr., L.L. Miller and D.C. Johnston, Phys. Rev. B 40 (1989) 11488.
[55] J.C. Ho, R.C. Ling and D.P. Dandekar, J. Appl. Phys. 59 (1986) 1397.
[56] J.H.J. Fluitman, R. Boom, P.F. de Châtel, C.J. Schinkel, J.L.L. Tilanus and B.R. de Vries, J. Phys. F 3 (1973) 109.
[57] H. Seki, M. Yoshizawa, K. Ikeda, K. Shigematsu, to be published.
[58] G. Barnea, J. Phys. F 7 (1977) 315.
[59] M. Cyrot and M. Lavagna, J. de Phys. 40 (1979) 763.
[60] E.P. Wohlfarth and P. Rhodes, Phil. Mag. 7 (1962) 1817.
[61] C.J. Schinkel, J. Phys. F 8 (1978) L87.
[62] V.V. Aleksandryan, A.S. Lagutin, R.Z. Levitin, A.S. Markosyan and V.V. Snegirev, Sov. Phys. JETP 62 (1985) 153.
[63] T. Sakakibara, T. Goto, K. Yoshimura, M. Shiga and Y. Nakamura, Phys. Lett. A 117 (1986) 243.
[64] I.L. Gabelko, R.Z. Levitin, A.S. Markosyan and V.V. Snegirev, Sov. Phys. JETP Lett. 45 (1987) 458.

[65] K. Ishiyama, K. Endo, T. Sakakibara, T. Goto, K. Sugiyama and M. Date, J. Phys. Soc. Japan 56 (1987) 29.
[66] T. Goto, K. Fukamichi, T. Sakakibara and H. Komatsu, Solid State Commun. 72 (1989) 945.
[67] T. Sakakibara, T. Goto, K. Yoshimura, K. Murata and K. Fukamichi, J. Magn. Magn. Mater. 90 & 91 (1990) 131.
[68] M. Aoki and H. Yamada, Solid State Commun. 72 (1989) 21; J. Magn. Magn. Mater. 78 (1989) 377.
[69] K. Ikeda, K.A. Gschneidner, Jr., T.-W.E. Tsang and F.A. Schmidt, Solid State Commun. 41 (1982) 889.
[70] T.Y. Hsiang, J.W. Reister, H. Weinstock, G.W. Crabtree and J.J. Vuillemin, Phys. Rev. Lett. 47 (1981) 523.
[71] G.R. Stewart and B.L. Brandt, unpublished results on the heat capacity of scandium single crystals up to 18 T (c-axis) and 11 T (a-axis).
[72] S.K. Dhar and K.A. Gschneidner, Jr., unpublished results on the heat capacity of scandium metal up to 9.8 T.
[73] G.R. Stewart and B.L. Brandt, Phys. Rev. B 28 (1983) 2266.
[74] S.K. Dhar, K.A. Gschneidner, Jr., A.I. Schindler and O.D. McMasters, unpublished results on the heat capacity of palladium and several dilute Ni in Pd alloys.
[75] L.W. Roeland, J.C. Wolfrat, D.K. Mak and M. Springford, J. Phys. F 12 (1982) L267.
[76] A.I. Schindler and C.A. Mackliet, Phys. Rev. Lett. 20 (1968) 15.
[77] G. Chouteau, R. Fourneaux, K. Gobrecht and R. Tournier, Phys. Rev. Lett. 20 (1968) 193.
[78] G. Chouteau, R. Fourneaux, R. Tournier and P. Lederer, Phys. Rev. Lett. 21 (1968) 1082.
[79] A.I. Schindler and B.C. LaRoy, Solid State Commun. 9 (1971) 1817.
[80] T.D. Cheung and J.S. Kouvel, Phys. Rev. B 28 (1983) 3831.
[81] B.T. Matthias, A.L. Giorgi, V.O. Struebing and J.L. Smith, J. de Phys. Lett. 39 (1978) L441.
[82] G.R. Stewart, J.L. Smith, A.L. Giorgi and Z. Fisk, Phys. Rev. B 25 (1982) 5907.
[83] G.R. Stewart, J.L. Smith and B.L. Brandt, Phys. Rev. B 26 (1982) 3783.
[84] J.M. van Ruitenbeek, A.P.J. van Deursen, H.W. Myron, A.J. Arko and J.L. Smith, Phys. Rev. B 34 (1986) 8507.
[85] F. Acker, Z. Fisk, J.L. Smith and C.Y. Huang, J. Magn. Magn. Mater. 22 (1981) 250.
[86] J.R. Cooper, C. Rizzuto and G. Olcese, J. de Phys. 32 (1971) C1-1136.
[87] B. Stalinski, Z. Kletowski and Z. Henkie, Phys. Stat. Sol. (a) 19 (1973) K165.
[88] T.-W.E. Tsang, K.A. Gschneidner, Jr., O.D. McMasters, R.J. Stierman and S.K. Dhar, Phys. Rev. B 29 (1984) 4185.
[89] K.A. Gschneidner, Jr., S.K. Dhar, R.J. Stierman, T.-W.E. Tsang and O.D. McMasters, J. Magn. Magn. Mater. 47 & 48 (1985) 51.
[90] C. Stassis, C.-K. Loong, B.N. Harmon, S.H. Liu and R.M. Moon, J. Appl. Phys. 50 (1979) 7567.

[91] C. Stassis, C.-K. Loong, J. Zarestky, O.D. McMasters and R.M. Nicklow, Phys. Rev. B 23 (1981) 5128.
[92] W.H. Lee, R.N. Shelton, S.K. Dhar and K.A. Gschneidner, Jr., Phys. Rev. B 35 (1987) 8523.
[93] W.H. Dijkman, A.C. Moleman, E. Kesseler, F.R. de Boer and P.F. de Châtel, in: Valence Instabilities, eds. P. Wachter and H. Boppart (North-Holland, Amsterdam, 1982) p. 515.
[94] H. Yashima, H. Mori, T. Satoh and K. Kohn, Solid State Commun. 43 (1982) 193.
[95] S.K. Dhar, K.A. Gschneidner, Jr., W.H. Lee, P. Klavins and R.N. Shelton, Phys. Rev. B 36 (1987) 341.
[96] A.M. Boring, R.C. Albers, G.R. Stewart and D.D. Koelling, Phys. Rev. B 31 (1985) 3251.
[97] P.H. Frings and J.J.M. Franse, Phys. Rev. B 31 (1985) 4355.
[98] M.S. Wire, G.R. Stewart, W.R. Johanson, Z. Fisk and J.L. Smith, Phys. Rev. B 27 (1983) 6518.
[99] J. Xu, B.I. Min, A.J. Freeman and T. Oguchi, Phys. Rev. B 41 (1990) 5010.

Magnetic properties of diluted magnetic semiconductors

W.J.M. de Jonge and H.J.M. Swagten

Department of Physics, Eindhoven University of Technology (EUT), P.O. Box 513, 5600 MB Eindhoven, Netherlands

A review will be given of the magnetic characteristics of diluted magnetic semiconductors and the relation with the driving exchange mechanisms. II–VI as well as IV–VI compounds will be considered. The relevance of the long-range interaction and the role of the carrier concentration will be emphasized.

1. Introduction

Compound semiconductors in which a (small) fraction of cations is replaced by magnetic ions have been the subject of study for a long time [1]. It was Gałązka [2] who draw the attention to these compounds and first emphasized the crucial role of the interaction between the mobile carriers and the local moments as the driving mechanism behind the interesting semiconducting and magnetic properties. To distinguish this class of materials the name Semimagnetic Semiconductors (SMSC) was introduced in the late seventies [2,3]; nowadays, however, the class of materials is more commonly referred to as Diluted Magnetic Semiconductors (DMS).

To be more specific, most of the investigations so far deal with II–VI compounds diluted with Mn, such as $Cd_{1-x}Mn_xTe$, $Zn_{1-x}Mn_xTe$, $Hg_{1-x}Mn_xTe$ and the corresponding selenides and sulfides. Recently also results have been reported on IV–VI DMS and DMS containing other transition metal magnetic ions, such as Fe and Co. One of the interesting properties of this class of materials is the broad range of compositions which in general can be realized and the possibility to vary the band structure and lattice parameters with composition. This unique flexibility or tunability make these alloys particularly attractive for scientific research and potential applications, for instance in heterostructures (see the contribution of Prinz in this issue).

The characteristic features of DMS, which have attracted such a considerable attention during the past years, are directly related to the fact that DMS contain two interacting subsystems: the electronic system of the carriers and the (diluted) magnetic system of moments. The semiconducting properties all stem from the interaction between d electrons (the localized Mn^{2+} spins) and the s- and p-like orbitals of the band electrons, the so-called sp-d exchange (J_{sp-d}). This interaction amplifies so to say the Zeeman splitting of electronic levels and gives rise to effects such as a giant Faraday rotation [4], an extremely large negative magneto-resistance [5] and field-dependent metal–insulator transitions [6].

The magnetic properties of DMS are intimately related to the interactions between the magnetic ions (J_{d-d}), which are mediated by the carriers and therefore strongly related to J_{sp-d}. It has been found that in II–VI compounds J_{d-d} is antiferromagnetic (AF) and long ranged (with significant and systematic variations in range depending on covalency or bandgap). A spin-glass phase is found at low magnetic-ion concentrations (Mn^{2+} as well as Fe^{2+} and Co^{2+}) extending to the very dilute limit. At higher concentrations sometimes AF ordered regimes are reported. Since the carrier concentration is in general relatively low ($n, p < 10^{19}$ cm^{-3}), the well known Ruderman–Kittel–Kasuya–Yoshida (RKKY) interaction [7], which is proportional to the carrier density, is too small to account for the interaction between the magnetic ions. A number of alternative physical mechanisms, involving carriers in different bands, have been suggested [8]. Recent theoretical results, however, indicate that, at least for the interactions between nearest neighbors, superexchange might be the dominant mechanism [9]. These nearest-neighbor interactions cannot, however, be responsible for the observed spin-glass transition in the diluted limit, which is claimed to be triggered primarily by the long-range part of the exchange interaction [10].

In contrast to the II–VI DMS, several IV–VI DMS such as $Sn_{1-x}Mn_xTe$ and $(PbGe)_{1-x}Mn_xTe$, show ferromagnetic ordering phenomena, implicating a rather different exchange mechanism. Furthermore Story et al. [11] reported the observation of a carrier-concentration-induced ferromagnetic transition in $Pb_{0.25}Sn_{0.72}Mn_{0.03}Te$ (PSMT) above a critical concentration of carriers p_{crit}. Analogous effects were reported for other compositions including $Sn_{1-x}Mn_xTe$ [12]. Moreover, a carrier-induced transition from the ferromagnetic state to a spin-glass state was observed for these IV–VI DMS [13]. This unique demonstration of the influence of the carrier concentration on the magnetic properties is facilitated by the significant range of carrier concentrations (10^{19} to 10^{21} cm^{-3}) which can be realized in these systems by slight deviations from stochiometry. The relatively high carrier concentration favors the RKKY interaction as the dominant exchange mechanism between localized moments, which in combination with the particular band structure of these IV–VI DMS gives rise to the observed phenomena [13,14].

This paper will be focused on the magnetic behavior of DMS. Since excellent review papers have appeared on the II–VI Mn- [15] and Fe-diluted compounds [16], we will not try to surpass these comprehensive reviews, but instead emphasize a few recent topics which have not yet been included, i.e. the relevance of long-range interactions and the carrier-induced phenomena in the IV–VI DMS which we have both introduced above.

The organization of this paper is as follows. In section 2 we will briefly review the crystal structure, composition, band structure and sp–d exchange in DMS, which should serve as a background for the magnetic properties presented in sections 3 and 4, since they strongly depend on composition and electronic properties. The first part of section 3 will be devoted to the Mn-based II–VI group DMS. We will show how the experimentally observed thermodynamic behavior of the random diluted magnetic system is correlated with the d–d exchange (strength, range) and the underlying physical mechanisms. Subsequently, in section 3.2 the substitution of magnetic ions other than Mn^{2+}, such as Fe^{2+} or Co^{2+}, will be shortly discussed. In section 4, finally, we will focus on recent data of the carrier-induced magnetic phenomena in IV–VI DMS.

2. Crystal structure, composition, band structure and sp–d exchange

In table 1 the crystal structure and the composition range of all known Mn-, Fe- and Co-containing DMS are tabulated. In most cases crystals of high quality and large dimensions can be synthesized by using the Bridgman(–Stockbarger) method or, in some cases, by chemical transport or sintering. Dilution with Fe, and also Co, is not so straightforward as in the Mn case, which, tentatively, has been ascribed to the resemblance between the electronic configuration of Mn and the cations (both half or completely filled in contrast to Fe or Co).

The available structural investigations for the systems in table 1 reveal in almost all cases a Vegard-type linear dependence of the lattice constant with the concentration of magnetic ions. This linear behavior has been used for the determination of the concentration of magnetic ions, besides other more direct methods like electron probe micro analysis (EPMA). For II–VI compounds diluted with Mn^{2+}, the lattice parameters are shown in fig. 1. Other examples are illustrated in fig. 2a and b, for DMS containing FeSe and $Pb_{1-x}Mn_xTe$, respectively. The Vegard-type departure of the lattice parameter from that of the semiconducting host is, for all substances, clearly illustrated. One should note, however, that standard X-ray diffraction only probes the aver-

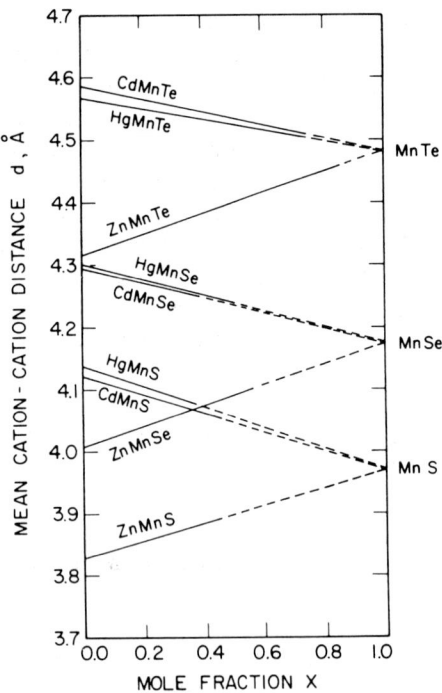

Fig. 1. Mean cation–cation distances d as a function of the Mn mole fraction x for $A_{1-x}Mn_xB$ alloys originating from the II–VI group (after ref. [20]). The lattice parameters can be obtained from d as follows, zinc blende: $a = d\sqrt{2}$, wurtzite: $a = d$ and $c = (8/3)^{1/2}d$; analytic expressions corresponding to the figure are given in refs. [15,20].

age bond lengths and not the microscopic features of the structure. Recently [23], extended X-ray absorption fine structures (EXAFS) studies on $Zn_{1-x}Mn_xSe$ convincingly demonstrated that on microscopic scale strong distortions from the zinc blende phase are visible. Moreover, both the Zn–Se and Mn–Se bond length remain constant over the entire concentration regime, which might be of great importance for the exchange constants in DMS.

As we quoted before, one of the attractive features in DMS is also the tunability of band parameters upon substitution. The most well-known example for this is the proportionality of the energy gap with composition, which has been established for almost all DMS. For Te-containing II–VI compounds the change of the gap by composition is plotted in fig. 3, illustrating the "opening" of the gap in $Hg_{1-x}Mn_xTe$.

The sp–d exchange, which as we pointed out before, is the driving mechanism behind the anomalous field-dependent physical behavior of DMS, involves the electronic band structure as well as the local magnetic ion 3d levels, which are superimposed on the band structure. As an example, fig. 4 shows a schematic illustration of Mn-$3d^5$ levels in an idealistic zinc blende band structure, which is a representative example for II–VI DMS having the direct gap at the Γ point. The majority (spin up) and minority (spin down) Mn levels are in reality narrow bands due to mixing or hybridization with s and p orbitals of the semiconductor. According to Larson et al. [9], the location and degree of hybridization of the Mn levels are extremely important for the strength of the exchange integrals in DMS, which we will discuss in some detail further on. U_{eff} in fig. 4 is the energy necessary to add one electron to a Mn^{2+} ($3d^5 \rightarrow 3d^6$) and amounts to ≈ 7 eV in $Cd_{1-x}Mn_xTe$. For Fe and Co compounds it is not yet well understood how the 3d levels are embedded within the band structure and what is their significance with respect to the exchange processes. In the following treatment of sp–d exchange we will confine ourselves to the Mn case.

For the description of the effect of the interaction on the electronic system one exploits the extension of the electron wave functions – the mobile electron interacts simultaneously with a large number of spins – in order to transform the Heisenberg-exchange Hamiltonian as follows

$$\sum_{R_i} J_{sp-d}(r - R_i) S_i \cdot \sigma \rightarrow \sigma_z \langle S_z \rangle x \sum_{R} J_{sp-d}(r - R), \quad (1)$$

where σ and S are the spin operators for band electron and $3d^5$ electron, respectively, J_{sp-d} is the electron–ion exchange constant, and r and R_i the coordinates of band electron and magnetic ion, respectively. Two approximations are made in this transformation, (i) the spin operator S has been replaced by its thermal average in the z direction $\langle S_z \rangle$, using the molecular-field approximation for paramagnetic ions experiencing a field B along the z axis and (ii) the summation no longer runs over the exact (random) position of the spins (R_i) but over all cation sites (R) weighted by the concentration magnetic ions x. Now, the exchange integral follows the periodicity of the semiconducting lattice (the virtual crystal approximation) and provides the possibility to solve the Landau-energy levels of the DMS system. Just as an illustration for a parabolic conduction band, the following effective g-factor can be derived [24]:

$$g_{\text{eff}} = g^* - N_0 \alpha x \langle S_z \rangle / \mu_B H$$
$$= g^* + \alpha M / g_{\text{Mn}} \mu_B^2 H, \quad (2)$$

where g^* is the band g-factor, N_0 the number of cations, $\alpha = \langle S | J_{\text{sp-d}} | S \rangle / \Omega_0$ (Ω_0 the volume of an elementary cell) the exchange integral for s-like conduction electrons, and g_{Mn} the Landé-factor for Mn^{2+}. Similar expressions can be found for the valence bands, the exchange integral now denoted as β instead of α. The principal difference between α and β stems from the symmetry of the electron wave functions within the band structure. In II–VI group semiconductors s-like electrons (e.g., of Γ_6 symmetry in $Cd_{1-x}Mn_xTe$) are associated with the conduction band and determine α. On the other hand, β is the exchange integral for p-like holes in the valence bands of these alloys (e.g., of Γ_8 symmetry in $Cd_{1-x}Mn_xTe$).

The right-hand side of the effective g-factor in eq. (2) is only present in nonzero external magnetic fields and includes the macroscopic magnetization of the system, and, consequently, the magnetic field, temperature and composition enter in the electronic properties of a DMS. Due to this, the electronic g-factors in DMS can become extremely large, in some cases exceeding 100, which is unique for semiconductors and leads to the so-called amplification of Zeeman splitting as mentioned in the Introduction. The splitting of the conduction band is schematically shown in fig. 5 for several cases contained in eq. (2). Some examples of the physical effects arising from the large spin splitting of the electronic levels are (i) excessive exchange splitting of exciton transitions in open-gap semiconductors [26]; (ii) a giant Faraday rotation [4], containing promising features for applications in, for example, optical devices or very precise magnetic field sensors; (iii) anomalous temperature and field dependence of Shubnikov–De Haas oscillations in narrow-gap semiconductors (such as $Hg_{1-x}Mn_xSe$ and $(Cd_{1-x}Mn_x)_3As_2$ [27,28]); (iv) bound magnetic polarons due to the polarization of Mn^{2+} spins around an electron trapped in an impurity potential [29], as well as giant negative magnetoresistance in $Hg_{1-x}Mn_xTe$ caused by the field-dependent mixing of exchange-enhanced splitted valence bands

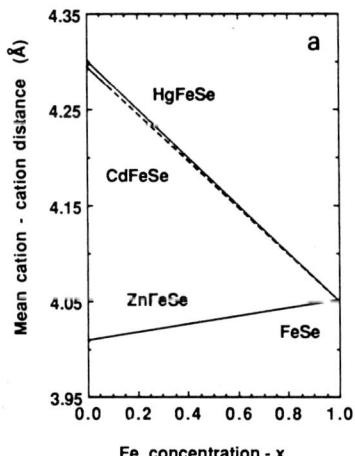
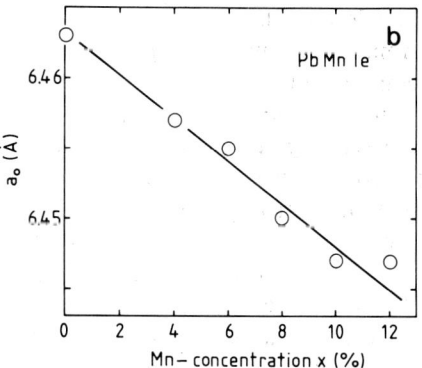

Fig. 2. (a) Mean cation–cation distances d as a function of the Mn mole fraction x for $A_{1-x}Fe_xSe$ (after ref. [21]); (b) lattice parameter a_0 for $Pb_{1-x}Mn_xTe$ as a function of the Mn mole fraction x (after ref. [22]).

Table 1
Crystal structure and composition range for Mn-, Fe- and Co-based diluted magnetic semiconductors [17–19]

Material	Type	Crystal structure	Composition range
$Zn_{1-x}Mn_xS$	II–VI	zinc blende	$0 < x \leq 0.10$
		wurtzite	$0.10 < x \leq 0.45$
$Zn_{1-x}Mn_xSe$		zinc blende	$0 < x \leq 0.30$
		wurtzite	$0.30 < x \leq 0.57$
$Zn_{1-x}Mn_xTe$		zinc blende	$0 < x \leq 0.86$
$Cd_{1-x}Mn_xS$		wurtzite	$0 < x \leq 0.45$
$Cd_{1-x}Mn_xSe$		wurtzite	$0 < x \leq 0.50$
$Cd_{1-x}Mn_xTe$		zinc blende	$0 < x \leq 0.77$
$Hg_{1-x}Mn_xS$		zinc blende	$0 < x \leq 0.37$
$Hg_{1-x}Mn_xSe$		zinc blende	$0 < x \leq 0.38$
$Hg_{1-x}Mn_xTe$		zinc blende	$0 < x \leq 0.75$
$(Cd_{1-x}Mn_x)_3As_2$	II–V	ref. [18]	$0 < x \leq 0.12$
$(Zn_{1-x}Mn_x)_3As_2$		ref. [18]	$0 < x \leq 0.15$
$Pb_{1-x}Mn_xS$	IV–VI	rocksalt	$0 < x \leq 0.05$
$Pb_{1-x}Mn_xSe$		rocksalt	$0 < x \leq 0.17$
$Pb_{1-x}Mn_xTe$		rocksalt	$0 < x \leq 0.12$
$Sn_{1-x}Mn_xTe$		rocksalt	$0 < x \leq 0.40$
$Ge_{1-x}Mn_xTe$		rhombohedral	$0 < x \leq 0.18$
		rocksalt	$0.18 < x \leq 0.50$
$Zn_{1-x}Fe_xS$	II–VI	zinc blende	$0 < x \leq 0.26$
$Zn_{1-x}Fe_xSe$		zinc blende	$0 < x \leq 1.00$
$Zn_{1-x}Fe_xTe$		zinc blende	$0 < x \leq 0.01$
$Cd_{1-x}Fe_xSe$		wurtzite	$0 < x \leq 0.20$
$Cd_{1-x}Fe_xTe$		zinc blende	$0 < x \leq 0.03$
$Hg_{1-x}Fe_xSe$		zinc blende	$0 < x \leq 0.20$
$Zn_{1-x}Co_xSe$	II–VI	zinc blende	$0 < x \leq 0.10$
$Zn_{1-x}Co_xS$		zinc blende	$0 < x \leq 0.14$
$Cd_{1-x}Co_xS$		wurtzite	$0 < x \leq 0.002$
$Cd_{1-x}Co_xSe$		wurtzite	$0 < x \leq 0.09$

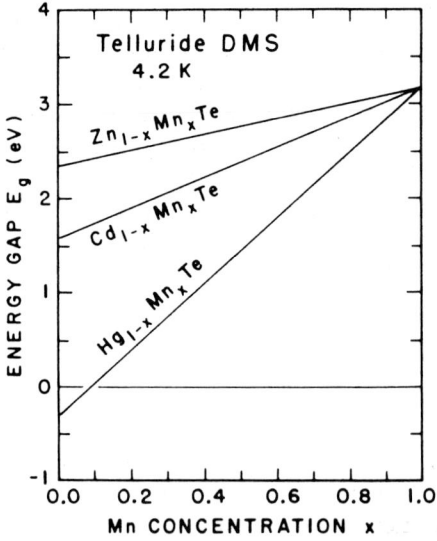

Fig. 3. Energy gap E_g vs. Mn concentration x for telluride DMS at low temperatures. A linear extrapolation of E_g to $x = 1$ gives an energy gap of 3.18 eV for zinc blende MnTe (after ref. [15]).

with the ferromagnetic $1/r$ Coulomb exchange, which tends to align the d-electron spin with the electron spin of conduction band states. In the case of valence band states, however, the integral

in acceptor levels [5]. All the available quantitative information on the exchange integrals α and β has been derived from the study of these physical phenomena, since in the magnitude of the effects the exchange constants (or a combination of them) are comprised. It is also worthwhile mentioning that through the study of magneto-optical properties (Faraday rotation, exciton splitting) it is possible to investigate the magnetization of the system; see eq. (2).

Table 2 contains a list of experimentally available α and β values, including recent investigations on Fe and Co compounds. The α exchange parameter is positive and generally associated

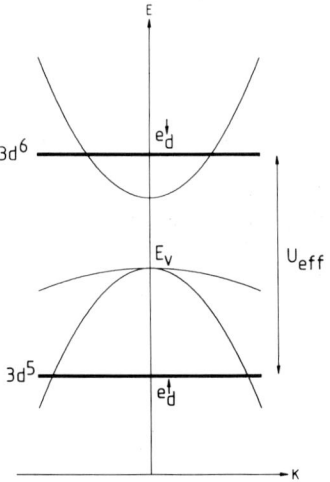

Fig. 4. Schematic band structure for a DMS with the direct gap at the Γ point. The Mn levels $e_d^\uparrow$ (occupied) and $e_d^\downarrow$ (unoccupied) are split by the energy U_{eff} (≈ 7 eV). The effect of p–d hybridization and crystal field on the Mn levels is neglected in this schematic representation, since the effect is relatively small.

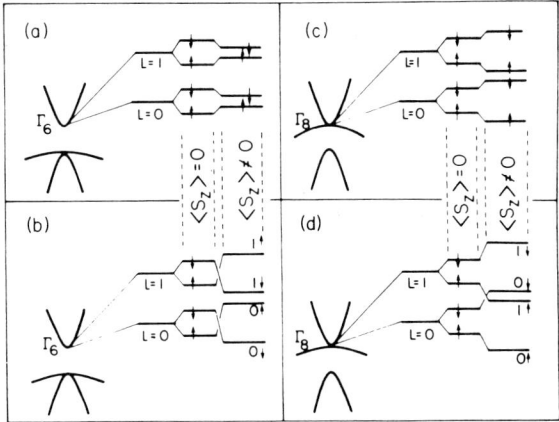

Fig. 5. Schematic illustration of the spin splitting of the first two Landau levels of the conduction band in the parabolic limit, with and without exchange contributions. For Γ_6-band electrons, (a) $\langle S_z \rangle$ reduces the size of the total spin splitting; (b) for increasing $\langle S_z \rangle$ the sign of the splitting is reversed due to the fact that $\alpha > 0$. In (c) and (d) the enhancement of the total spin splitting for Γ_8-band electrons is shown (after ref. [25]).

β is not only composed of the Coulomb term, but also of a much stronger antiferromagnetic sp–d hybridization term, which is absent for conduction band states due to symmetry reasons. This explains why β is larger then α and of opposite sign.

Table 2
Experimental sp–d exchange constants $N_0\alpha$ and $N_0\beta$ for Mn-, Fe- and Co-based diluted magnetic semiconductors [15,16,28,30]

Material	$N_0\alpha$ (eV)	$N_0\beta$ (eV)
$Zn_{1-x}Mn_xSe$	0.26	−1.31
$Zn_{1-x}Mn_xTe$	0.18	−1.05
$Cd_{1-x}Mn_xS$	0.22	−1.80
$Cd_{1-x}Mn_xSe$	0.23	−1.26
$Cd_{1-x}Mn_xTe$	0.22	−0.88
$Hg_{1-x}Mn_xSe$	0.4	−0.7
$Hg_{1-x}Mn_xTe$	0.4	−0.6
$(Cd_{1-x}Mn_x)_3As_2$	3.8	−2.3
$Zn_{1-x}Fe_xSe$	0.22	−1.74
$Cd_{1-x}Fe_xSe$	0.23	−1.9
$Cd_{1-x}Co_xSe$	0.28	−1.87

Special attention should be paid on the exchange parameter β, since it is directly correlated to the d–d interaction (section 3). According to the superexchange model of Larson et al. [9] developed for Mn-based DMS, one may roughly speaking state that a larger β ($\approx J_{p-d}$) leads to a larger d–d interaction between moments. The increase of β from Mn-, via Fe- to Co-containing II–VI DMS (see table 2) is indeed also recovered in the d–d interaction ($J_{d-d,Co} > J_{d-d,Fe} > J_{d-d,Mn}$: see table 3 in the next section). One should admit, however, that the exchange mechanisms in Co and Fe DMS are still not completely understood, although the dominance of sp–d hybridization (superexchange) has been suggested [16,30].

3. d–d exchange and magnetic properties of II–VI group DMS

In the following subsections we will concentrate on two different subjects. First of all, we will consider the Mn-containing DMS originating from the II–VI group. Subsequently, we will briefly introduce relatively new classes of DMS materials on the basis of Fe and Co.

3.1. Mn

The magnetic behavior of DMS is in principle determined by the exchange interactions J_{d-d} between the localized Mn ions in the diluted magnetic subsystem. Information about the exchange is contained in experimental data on thermodynamic properties like specific heat C_m, magnetization M, susceptibility χ and phase transitions. Data have been reported on a large number of II–VI Mn DMS and a rather characteristic behavior is observed. For low Mn concentration ($x < 0.2$) it can be summarized as:

- The high-temperature susceptibility χ shows a Curie–Weiss behavior, $\chi = C/(T - \Theta)$, with a negative Θ ($\Theta \equiv \Sigma_{ij} J_{d-d,ij}$) which indicates dominant antiferromagnetic Mn–Mn interactions.
- At a temperature $T = T_F$ a cusp is observed in the low-temperature susceptibility χ. No

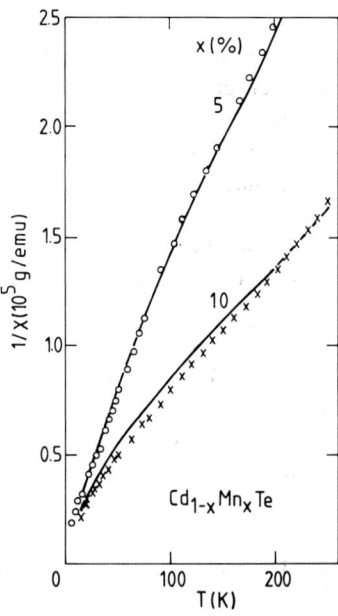

Fig. 6. Inverse high-temperature susceptibility [31] of $Cd_{1-x}Mn_xTe$; solid lines are ENNPA calculations with $J(R) = -10R^{-6.8}$ K.

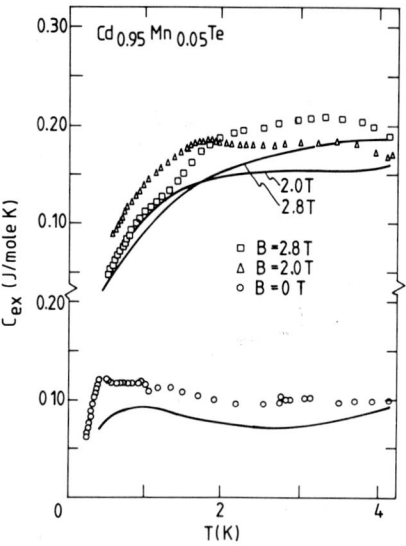

Fig. 7. Magnetic specific heat of $Cd_{0.95}Mn_{0.05}Te$ [31]; solid lines represent ENNPA calculations using $J(R) = -10R^{6.8}$ K.

anomaly is detected in C_m and strong irreversible behavior is seen in the magnetization. One may therefore conclude that at $T = T_F$, the freezing temperature, a transition to a spin-glass state takes place. T_F decreases continuously with x and seems to persist even in the limit $x \to 0$.

- The magnetic contribution to the zero-field specific heat, C_m, shows a broad maximum in the ^{4}He region and shifts to higher T with x, indicating a range of exchange-splitted Mn energy levels.
- The field dependence of the magnetization indicates AF coupling between Mn ions. In various cases apparent saturation is reached (far below the full saturation value of Mn^{2+}) followed by a succession of discrete magnetization steps which can be attributed to transitions between successive levels in strongly AF coupled Mn–Mn pairs.

We have illustrated this behavior in figs. 6–11 for $Cd_{1-x}Mn_xTe$ and $Zn_{1-x}Mn_xSe$.

The earliest attempts [31] to describe this magnetic behavior were based on a model in which the interaction was restricted to Mn ions on neighboring *lattice* sites. The interaction was taken to be AF in accordance to the evidence listed above. The use of this model was supported by the initial observations that the spin-glass transition did only exist for Mn concentrations x above the percolation limit x_p, which amounts to ≈ 15–20% in cubic systems (see fig. 11). This strongly pointed to nearest-neighbor interactions (J_{NN}) as the driving mechanism. However, no

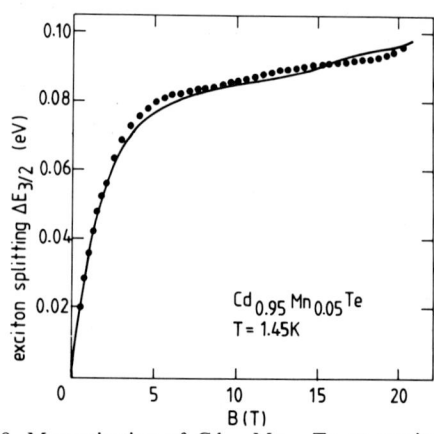

Fig. 8. Magnetization of $Cd_{0.95}Mn_{0.05}Te$ as monitorized by exciton splitting [32]; the solid line represents the ENNPA with $J(R) = -10R^{-6.8}$ K.

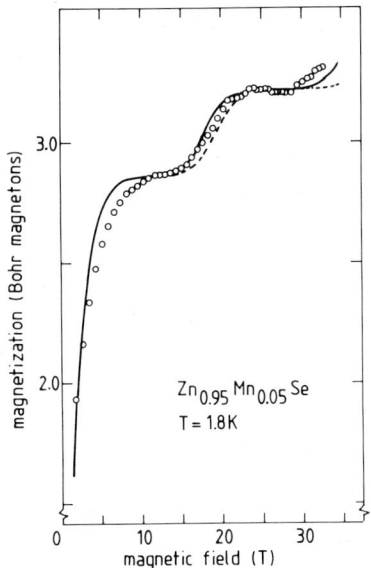

Fig. 9. Magnetization of $Zn_{0.95}Mn_{0.05}Se$ [33]; the solid line represents the ENNPA with $J_{NN} = -13$ K and $J(R) = -7R^{-6.8}$ K; see ref. [48].

consistent description of the data could be obtained with this nearest-neighbor model. Various exchange parameters were obtained, more than an order of magnitude apart, depending on the thermodynamic property under study and the random distribution of ions was questioned. The

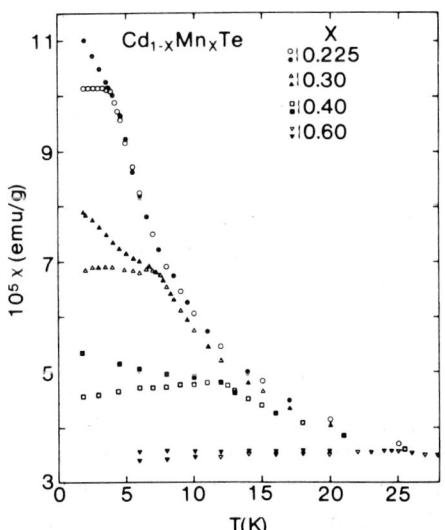

Fig. 10. ZFC and FC susceptibility of $Cd_{1-x}Mn_xTe$ as a function of T near T_F observed in a 15 Oe field [34].

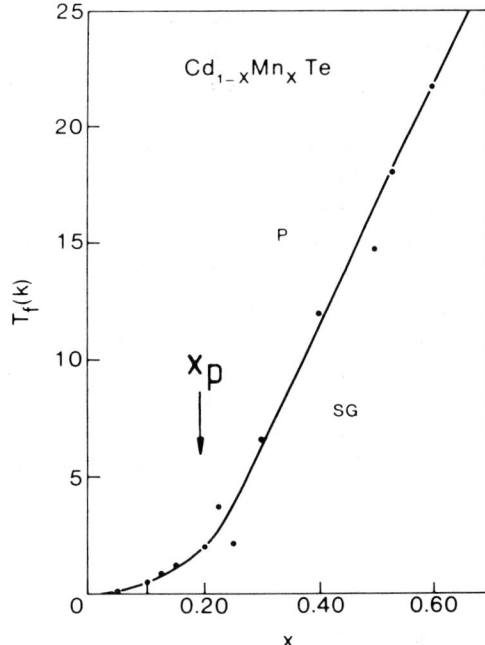

Fig. 11. Paramagnetic (P) to spin-glass (SG) phase diagram, T_F versus x, for $Cd_{1-x}Mn_xTe$ [31,34,35]. The solid line is a guide to the eye only. The arrow indicates the percolation threshold for NN interactions, x_p.

first clue to resolve this confusing situation was obtained when it became clear that the spin-glass transition was also observed at a Mn concentration far below the percolation limit for cubic lattices, as quoted above (see also fig. 11). This behavior was first noted in $(Cd_{1-x}Mn_x)_3As_2$, but has been observed since then for almost all DMS [10,15,18,36]. Existence of a spin-glass state down to the very dilute limit cannot be understood on the basis of nearest-neighbor interactions alone, but strongly indicates the existence of a long-ranged interaction as the driving mechanism behind this transition.

Before we proceed, let us summarize the actual situation as it emerges from the data. The magnetic subsystem in a DMS consists of a random array of magnetic moments statistically distributed over the available cation lattice sites and coupled to nearest neighbors as well as further neighbors by (long-ranged) exchange interaction. The various experimental data summarized above probe so to say different parts of the exchange distribution. The Curie–Weiss temperature Θ probes the sum of all interactions, which appears to be AF, while the low-temperature C_m and T_F probe mainly the (smaller) long-ranged tail of the interaction. Evidence about the (strong) NN interactions can be obtained for instance from the step-like behavior of the magnetization. This is only possible in those systems where the NN interaction is appreciable stronger than the coupling to further neighbors (we will return to this later in relation to the spatial extension of the long-range interaction) and the Mn concentration low enough to avoid contributions from more complicated statistical manifold NN clusters, in which case only NN pairs dominate. The energy level of such a pair is shown in fig. 12 together with the field dependence of the magnetization showing characteristic steps at $B = 2r|J_{NN}|/g\mu_B$, with $r = 1, 2, \ldots, 5$ [37]. The energy gaps between the singlet ground state and excited state have also been probed by neutron scattering [38,15] and Raman scattering [15]. The values for J_{NN} (and in some cases also J_{NNN}) obtained this way are tabulated in table 3. The table also contains data obtained by other methods, such as the high-temperature susceptibility, specific heat and electron paramagnetic resonance (EPR). However, in these cases J_{NN} or J_{NNN} are probed rather indirectly and subject to serious errors.

Information about the long-ranged part of the interaction can be obtained from the spin-glass freezing transition. For such a transition a scaling analysis should be applicable. Such a scaling analysis generally exploits the fact that for a continuous random distribution it is assumed that $R_{ij}^3 x$ = const, where R_{ij} denotes a typical distance between the ions. Implementation of this expression in a model for spin-glass freezing, given a known functional form for the radial dependence of the exchange interaction, then yields a theoretical prediction for $T_F(x)$ which can be compared with experimental data. Details of this procedure can be found elsewhere for a continuous as well as a discrete distribution of ions [48]. In the spirit of earlier analyses [78] the spin-glass-freezing condition is based on the conjecture that T_F is related to the interaction energy at the average distance ($\overline{R}$) between the magnetic ions,

$$k_B T_F \approx J(\overline{R}) S(S+1). \quad (3)$$

Given a power-law dependence of the interaction, the concentration dependence of T_F can be expressed as

$$\ln T_F \sim \tfrac{1}{3} n \ln x \quad \text{for } J(R) \sim R^{-n}. \quad (4)$$

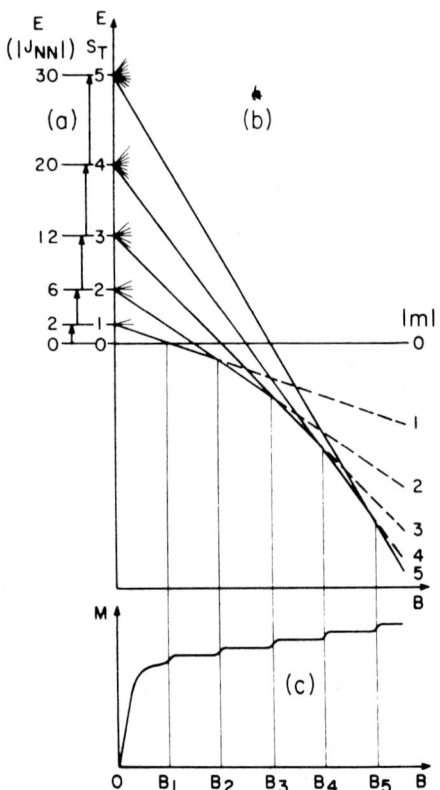

Fig. 12. (a) Energy level diagram, in units $|J_{NN}|$, for a pair of NN Mn^{2+} ions at $B = 0$; (b) Zeeman splitting of the energy levels in a magnetic field B; note the level crossings at B_n; (c) illustration of the magnetization curve at low temperatures, showing the apparent saturation ($M = M_s$), followed by the magnetization steps due to pairs. The magnetization steps due to triples or other clusters, as well as long-ranged coupled ions are not shown (after ref. [37]).

As the power-law behavior of $J(R)$ is chosen somewhat arbitrarily (although inspired by earlier theoretical predictions [8]) it may be replaced by more adequate functional forms. However, it describes the data remarkably well in the whole concentration range as is shown in fig. 13, where the freezing temperature for a number of Mn

Table 3
Nearest-neighbor and next-nearest-neighbor exchange constants (J_{NN}, J_{NNN}) for Mn-, Fe- and Co-based diluted magnetic semiconductors; EPR = electron paramagnetic resonance; EXCI = high-field exciton splitting; MAGN = steps in the high-field magnetization; MAGN* = high-field magnetization, fitted with pair approximations; NEUT = inelastic neutron scattering; RAM = Raman scattering; SUSC = high-temperature susceptibility; $T_F(x)$ = spin-glass freezing analysis, combined with thermodynamic properties

Material	$-J_{NN}$ (K)	Techn.	Ref.	$-J_{NNN}$ (K)	Techn.	Ref.
$Zn_{1-x}Mn_xS$	≈ 10 16.9 ± 0.6	MAGN	[39,40]	0.6	MAGN	[40]
	16.1 ± 0.2	NEUT	[41]	4.8 ± 0.1	SUSC	[42]
	13.0 ± 1.5	EPR	[43]	≈ 0.7	$T_F(x)$	[44]
$Zn_{1-x}Mn_xSe$	≈ 13 9.9 ± 0.9	MAGN	[39,33]	2.4 3 ± 2.4	SUSC	[46,47]
	≈ 12.6 12.2 ± 0.3		[45,37]	≈ 0.7	$T_F(x)$	[48]
	12.3 ± 0.2	NEUT	[41]			
	18 13.5 ± 0.95	SUSC	[39,46]			
$Zn_{1-x}Mn_xTe$	10.0 ± 0.8 10.1 ± 0.4	MAGN	[49,50]	0.6	MAGN	[52]
	9.25 ± 0.3 8.8 ± 0.1		[51,33]	4.6 ± 0.9 3.6 ± 2	SUSC	[47]
	9.0 ± 0.2		[37]			
	9.52 ± 0.05 7.9 ± 0.2	NEUT	[53,59]			
	12 11.85 ± 0.25	SUSC	[49,54]			
$Cd_{1-x}Mn_xS$	> 4 8.6 ± 0.9	MAGN	[39,33]	5.2 ± 0.3	RAM	[55]
	10.5 ± 0.3		[45]			
	9.65 ± 0.2 11.0 ± 0.2		[40]			
	10.6 ± 0.2	RAM	[55]			
$Cd_{1-x}Mn_xSe$	8.3 ± 0.7 7.9 ± 0.5	MAGN	[39,56]	1.6 ± 1.5	MAGN	[56]
	9 10.6 ± 0.2	SUSC	[39,54]	4.6 ± 2 5.4 ± 1.5	SUSC	[47]
	7.7 ± 0.3 8.1 ± 0.2	RAM	[57,55]			
$Cd_{1-x}Mn_xTe$	≈ 10 6.3 ± 0.3	MAGN	[39,56]	1.9 ± 1.1 1.1 ± 0.2	MAGN	[56,58]
	6.1 ± 0.3 6.2 ± 0.2		[45,37]	0.67	NEUT	[59]
	≈ 7.5 6.7	NEUT	[60,59]	1.2 ± 1	SUSC	[47]
	6.9 ± 0.15	SUSC	[54]			
	7.7 ± 0.3	EXCI	[32]			
	6.1 ± 0.2	RAM	[57]			
$Hg_{1-x}Mn_xSe$	6 ± 0.5 5.3 ± 0.5	MAGN	[61,52]			
	10.9 ± 0.7	SUSC	[54]			
	0.1	EPR	[62]			
$Hg_{1-x}Mn_xTe$	5.1 ± 0.5 4.3 ± 0.5	MAGN	[61,52]	1	SUSC	[63]
	15 15.7 7.15 ± 0.25	SUSC	[64,63,54]			
$(Zn_{1-x}Mn_x)_3As_2$	≈ 100	$T_F(x)$	[18,65]	≈ 2	$T_F(x)$	[18,65]
$(Cd_{1-x}Mn_x)_3As_2$	≈ 30	$T_F(x)$	[18,66]	≈ 5	$T_F(x)$	[18,66]
$Pb_{1-x}Mn_xS$	0.537	MAGN*	[67]			
	1.28	SUSC	[68]			
$Pb_{1-x}Mn_xSe$	≈ 1	MAGN*	[69]			
	1.67	SUSC	[68]			
$Pb_{1-x}Mn_xTe$	≈ 1	MAGN*	[69]			
	0.84	SUSC	[68]			
$Zn_{1-x}Fe_xSe$	22 ± 2	SUSC	[70]			
$Cd_{1-x}Fe_xSe$	11.25 ± 1.5 18.8 ± 2	SUSC	[71,70]			
$Hg_{1-x}Fe_xSe$	15. ± 1 18 ± 2	SUSC	[71,72]			
$Zn_{1-x}Co_xS$	47.5 ± 0.6	NEUT	[73]	2.25 ± 0.2	MAGN	[75]
	47 ± 6	SUSC	[74]			
$Zn_{1-x}Co_xSe$	50 ± 1	NEUT	[76]	3.04 ± 0.1	MAGN	[75]
	54 ± 8	SUSC	[74]			
$Cd_{1-x}Co_xSe$	37 ± 5	SUSC	[77]			

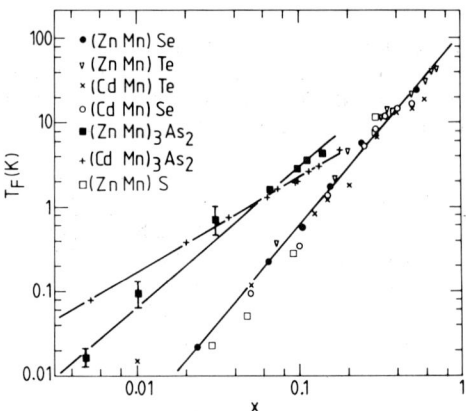

Fig. 13. Freezing temperatures as a function of the Mn-concentration x for several DMS on a logarithmic scale; the drawn lines yield the power dependence $J(R) \sim R^{-n}$ (see table 4); see refs. [10,18,36,44,48].

DMS is plotted versus concentration and yields a characteristic value for the radial decay of the interaction strength $J(R) \sim R^{-n}$ through the fitting parameter n. This analysis therefore provides a probe for the spatial extension of the d–d interactions strength in the various systems, which can hardly be obtained by any other way. In table 4 the range of interactions ($\sim 1/n$) available so far from spin-glass freezing data are listed.

Tables 3 and 4 contain all the characteristics about the exchange interaction as can be deduced from an analysis of the magnetic data. Consequently one should be able to describe the various data using these entities. The calculation of the thermodynamic properties of an infinite, random ensemble of spins, all coupled to each other by a long-ranged AF interaction $J \sim R^{-n}$, requires an approximate calculation method. It was shown that the so-called Extended Nearest-Neighbor Pair Approximation (ENNPA) is particular useful for this purpose [18,66,48]. The essential ingredient of this approximation, first suggested by Matho [79], is the assumption that the partition function of the ensemble of random spins can be factorized into the contribution of pairs and triples. In this case a pair (triple) consists of a Mn spin coupled to its neighboring Mn spin(s), which can be located anywhere as prescribed by the statistics in a random array.

Since the size of this paper is rather limited we will not report on the results for each DMS system in detail. Instead we exemplify the results by the figs. 6–9, which all refer to $Cd_{1-x}Mn_xTe$ and $Zn_{1-x}Mn_xSe$. Results for other systems show basically the same features. It should be emphasized that no efforts have been undertaken to adjust the parameters in order to obtain a better overall fit to the data.

From the comparison shown in figs. 6–9 one would like to conclude that, on the whole, the agreement between calculated results, using reported values for the interaction, and the data is fair and shows that a model which includes the long-range nature of the interaction is in principle capable of explaining the overall magnetic behavior in DMS. A more detailed comparison also for other systems can be found in refs. [18,10,80].

In addition it is interesting to note that the steplike behavior of the magnetization as shown in figs. 8, 9 and 12 is only observed for those DMS in which the spatial decay of the interaction is fast ($J \sim R^{-n}$, with $n > $). Indeed, for these systems one would expect that the basic assumption behind this effect, the existence of NN-coupled pairs which can be treated as isolated, is best satisfied. A slowly decaying interaction will yield a coupling of the pair to other moments in the magnetic subsystem which diffuses the transitions.

Given the results for the J_{d-d} for various DMS in tables 3 and 4 one may now ask how the observed systematics in strength, sign and range

Table 4
Type, NN distance, band gap and exponent n of various DMS [10,18,44,48]

Material	d_{NN} (Å)	E_g (eV)	n ($J \sim R^{-n}$)
$Zn_{1-x}Mn_xS$	3.83	3.8–3.9	7.6
$Zn_{1-x}Mn_xSe$	4.00	2.8–3.0	6.8
$Zn_{1-x}Mn_xTe$	4.31	2.3–2.8	6.8
$Cd_{1-x}Mn_xS$	4.12	≈ 2.5	≈ 6.8
$Cd_{1-x}Mn_xSe$	4.28	1.5–2.5	≈ 6.8
$Cd_{1-x}Mn_xTe$	4.58	1.5–2.7	≈ 6.8
$Hg_{1-x}Mn_xSe$	4.30	≈ 0	5
$Hg_{1-x}Mn_xTe$	4.55	≈ 0–1.1	5
$(Zn_{1-x}Mn_x)_3As_2$	2.94	≈ 1	4.5
$(Cd_{1-x}Mn_x)_3As_2$	3.17	≈ 0–0.2	4.0

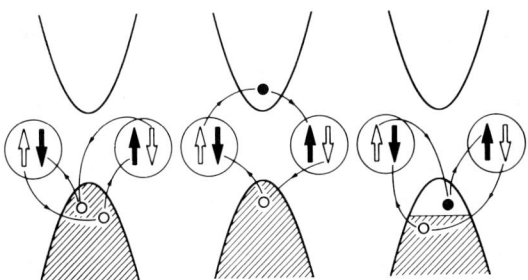

Fig. 14. Schematic representation of the exchange processes superexchange, Bloembergen–Rowland (BR) and Rudermann–Kittel–Kasuya–Yosida (RKKY). The filled valence bands, empty conduction bands, Mn spin states and transitions are shown (based on ref. [81]).

compare with potential mechanisms and theoretical predictions. Basically, for the spin–spin interaction in semiconductors three mechanisms can be defined, i.e., superexchange, Bloembergen–Rowland (BR) and Rudermann–Kasuya–Kittel–Yosida (RKKY) [7,8]. All mechanisms can be expressed in terms of virtual electron transitions (induced by exchange and/or hybridization) between band electrons or holes, and the d^5 electrons. In fig. 14 these processes are illustrated. Superexchange involves 2-electron (or hole) processes, the so-called intraband transitions. In BR on the other hand, only interband transitions are allowed, and finally, RKKY involves an intraband process at the Fermi level. As a consequence, one may expect that, roughly speaking, superexchange applies for open-gap materials (isolators, semiconductors), BR for small-gap materials (semiconductors) and RKKY for partially filled bands (metals, semimetals).

By a quantitative analysis Larson et al. [9] were able to demonstrate that in open-gap DMS [(Zn, Cd)$_{1-x}$Mn$_x$(S, Se, Te)] superexchange (involving only two-hole processes in the valence band, see fig. 14) is the dominant Mn–Mn exchange mechanism, while the one hole–one electron process (BR) accounts for only a few percent, and the contributions of two electron processes, involving only carriers at the Fermi-level (RKKY), is negligible. (For IV–VI DMS, however, the latter mechanism is the dominant mechanism, at least for higher carrier concentrations, as we shall see later on.) They were able to calculate J_{NN} and J_{NNN} in Cd$_{1-x}$Mn$_x$Te, and the detailed results were used to construct a simplified so-called three-level model, in which only the most relevant characteristics of the electronic structure are contained and which permits the extension of their results to other DMS. The relevant levels are schematically indicated in fig. 4 and determine together with the hopping parameter V_{pd} (characterizing the hybridisation between Mn d-orbitals and anion p-orbitals contained in the valence band), the strength of the NN exchange. Introducing a simplified k dependence of the levels yields an R dependence of the exchange $f(R_{ij})$ which is claimed to be "material insensitive", that is independent of electronic structure details within a class of materials closely related by symmetry:

$$J(R_{ij}) = -2V_{pd}^4 \big[(E_d + U - E_v)^{-2} U^{-1}$$
$$+ (E_d + U - E_v)^{-3} \big] f(R_{ij}), \quad (5)$$

where for $f(R_{ij})$ in Cd$_{1-x}$Mn$_x$Te-like symmetry is suggested

$$f(R_{ij}) = 51.2 \exp(-4.89 R_{ij}^2). \quad (6)$$

It has been shown [9] that the chemical trends, contained in the effects of anion and cation substitution on the energy levels and V_{pd}, compare very well with the experimental data as tabulated in table 3, demonstrating convincingly the applicability of the model.

Of particular interest to us in the present context are the predictions related to the spatial extension of the exchange; i.e. the "universality" of $f(R_{ij})$ within a certain class of materials and the actual quantitative decay of $f(R_{ij})$.

From the results shown in fig. 13 and tabulated in table 4, indeed a remarkable universal behavior of the radial dependence (characterized by $1/n$) can be observed for certain classes of materials, thus corroborating the predictions. In particular the universal behavior of the large class of wide-gap materials A$_{1-x}$Mn$_x$B where A = Cd, Zn and B = S, Se, Te is impressive, in view of the rather wide variety of bandgap, hybridization, lattice constants and NN-exchange constants found in this class (tables 3, 4). We note parenthetically

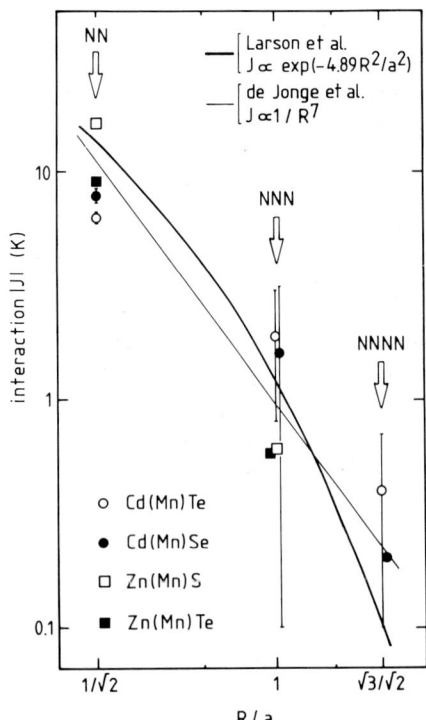

Fig. 15. Radial dependence of the exchange interaction $|J|$ based on (i) Larson's three-level superexchange model [9] (eq. (6)), and (ii) based on a scaling analysis of the spin-glass freezing temperature $T_F(x)$ (eq. (4), see refs. [18,48,10] and table 4), compared with experimental data (refs. [37,40,56] and table 3).

that the *coalescence* of the freezing temperatures as function of x for all these materials, as is apparent in fig. 13, represents an additional remarkable feature. This behavior seems at variance with the common wisdom that the freezing transition temperature should be proportional to the strength of the driving interaction.

To compare the quantitative prediction of $J(R_{ij})$ with the actual observations, we have plotted the results of the analytical radial dependence of $J(R)$ obtained from the analysis of $T_F(x)$ in fig. 15, together with the prediction of Larson [9] (eq. (6)) and a selection of reported values for J_{NN}, J_{NNN} and J_{NNNN} (see table 3). Obviously, the Gaussian decay suggested by eq. (6) and the power-law dependence employed in the analysis of $T_F(x)$ do not agree. One has to recall, however, that the Gaussian decay is only claimed to be valid up to roughly the fourth neighbor while the power law behavior probes mainly the tail of the interaction. The comparison in fig. 15 shows that the deviations between the experimental exchange interaction probed by $T_F(x)$ and the prediction based on superexchange are irrelevant in the plotted range, certainly when compared with the inaccuracy of the data obtained by other methods. One may therefore conclude that these data yield additional support for the superexchange mechanism and demonstrate the usefulness of the analysis of $T_F(x)$.

Finally, we would like to mention recent investigations dealing with the anisotropic part of the exchange. Larson et al. [9] were able to relate the observed tendencies of the EPR linewidth to their superexchange model, yielding evidence for a Dzyaloshinsky–Moriya anisotropic exchange $\mathcal{H}_{DM,i} = -D(R_{ij}) \cdot S_i \times S_j$, $|D_{NN}/J_{NN}|$ being roughly between 10^{-3} and 10^{-1} depending on the specific system. Though relatively small compared with the isotropic part of the interaction, this anisotropy can be of vital significance for the magnetic properties at low temperatures. According to the present understanding of spin-glasses, for instance, randomness combined with either competition or frustration seems to be essential ingredients for the existence of a spin-glass. It has been shown theoretically [82] that for a diluted magnetic array in an fcc host lattice coupled by long-range interactions, competing AF and F interactions seem essential. The additional anisotropy might be the clue in the open fundamental question about the spin-glass formation in these dilute arrays, which appear to be coupled by long-range AF interactions only. Moreover, the characteristic upturn of the magnetic specific heat at temperatures below 1 K strongly suggest the existence of DM anisotropy (see the discussion in ref. [48]), which therefore should be included in future calculations.

In this section we did not pay much attention to small gap DMS materials such as $(Cd_{1-x}Mn_x)_3As_2$ and $Hg_{1-x}Mn_xSe$. Existing data on these compounds [10,18,61,65,66,83] pointed to a much slower decay of the interaction $J(R)$ (between R^{-3} and R^{-5} instead of R^{-7}, see also fig. 13 and table 4), which, tentatively, might be ascribed to an increasing role of the BR mechanism and can be described in terms of a small or vanishing fundamental energy gap. There is, however, actually no conclusive theoretical support to explain the observed tendencies, neither for the strength of J_{NN} (being very large for II–V DMS and not accurately known for small gap II–VI DMS, see table 3) nor for the more extended tail of the interaction.

3.2. Fe and Co

As we quoted earlier, the majority of experimental studies so far have been devoted on the semiconductors containing Mn^{2+}. Recently, however, new systems based on Fe^{2+} and Co^{2+} have been investigated; see table 1 and refs. [84,85]. Due to the nonzero orbital momentum of the free Fe or Co ion, the ground state of these ions is split by the crystal field of the surrounding anions. This is schematically shown in fig. 16 and might greatly affect the low-temperature magnetic properties with respect to Mn.

In the Co case, however, the cubic crystal-field and spin–orbit coupling creates a fourfold ($S = 3/2$) degenerate $^4A_2$ ground state widely separated from higher lying states (fig. 16). Consequently, a strong analogy with Mn has been established from several investigations, mostly performed on $Zn_{1-x}Co_xSe$ and $Zn_{1-x}Co_xS$:

- From inelastic neutron scattering J_{NN} has been determined [73,76], yielding $J_{NN} = -47.5$ and -50 K for $Zn_{1-x}Co_xS$ and $Zn_{1-x}Co_xSe$, respectively, which is considerably higher than in the Mn case.
- The susceptibility at high temperatures [74,77] follows the Curie–Weiss law with negative Θ, consistent with the results from neutron scattering. For low temperatures spin-glass behavior is indicated [86]; the extension to small x provides evidence for a long-range interaction.
- A reduced magnetization indicates AF interaction between Co ions [85,75]; steps due to NNN and NNNN interaction are visible at low temperatures [75].
- The low-temperature specific heat clearly demonstrates Co–Co interaction beyond the NN [87,88].

In addition to these phenomena, Lewicki et al. [77,88] clearly demonstrated the existence of a trigonal splitting of the $^4A_2$ ground state in hexagonal $Cd_{1-x}Co_xS$ and $Cd_{1-x}Co_xSe$. A quantitative analysis of specific heat and susceptibility yielded $D \approx 1$ K. From the concentration depen-

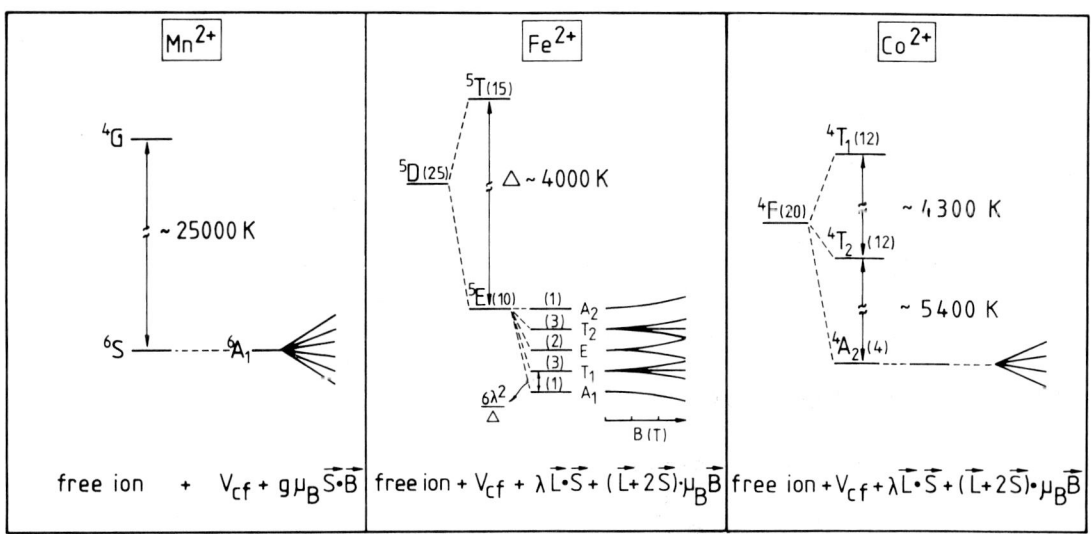

Fig. 16. Energy level diagram for isolated Mn^{2+}, Fe^{2+} and Co^{2+} ions. Effects of crystal field, spin–orbit interaction and magnetic field are shown (not to scale). The energy distances marked in the figure correspond to a ZnSe host lattice.

dence of the freezing temperature in $Zn_{1-x}Co_xS$ no power-law behavior similar to eq. (4) could be deduced. This was tentatively ascribed to additional dipolar interactions [86]. Nevertheless, in both $Cd_{1-x}Co_xS$ and $Zn_{1-x}Co_xSe$ the ENNPA calculations yielded a satisfactory description of the specific heat by using a power-law behavior for J_{d-d} [87,88], while the results for J_{NN}, J_{NNN} and J_{NNNN} deduced from magnetization steps seem to corroborate this. It is therefore conceivable that the interaction mechanisms in Co DMS are, on the whole, comparable to Mn and should be associated primarily with long-range superexchange. However, some basic questions such as the origin of the large value for J_{d-d} ($J_{d-d,Co} \approx 4J_{d-d,Mn}$) and the chemical trend in J_{NN} (increasing from S to Se, in contrast with Mn; see table 3), remains unsolved so far.

On the other hand, the magnetic properties of Fe compounds display rather deviant characteristics compared to Mn, which is intrinsically contained in the effect of crystal field and spin–orbit coupling. This is also sketched in fig. 16, showing the singlet ground state of Fe^{2+} which is responsible for the Van Vleck type of paramagnetism with field-induced magnetic moments only. This unique situation for DMS induces some particular magnetic properties usually not encountered for Mn, such as the abscence of a substantial low-temperature magnetic specific heat and a temperature-independent magnetization for low T. An extensive summary of these phenomena has been given by Twardowski [16].

Magnetic d–d interaction between Fe ions is evident from most of the magnetic properties. However, the specific energy structure of both single and interacting Fe ions, due to the influence of crystal field and spin–orbit coupling, complicates a quantitative analysis of the Fe interaction seriously. Since weakly coupled Fe ions are behaving quite similar to isolated Fe, it is even hard to collect proof for the *existence* of a long-range part of J_{d-d}. The only reliable information on J_{d-d} is therefore restricted to J_{NN} and stems from the high-temperature susceptibility [70], see table 3. In this respect, the analysis of the exponential onset of the specific heat [72] and the prediction of magnetization steps [89] may be relevant for future research. From the data available so far it seems that the exchange mechanism is governed primarily by the AF superexchange, analogously to the Mn case. No drastic influence from additional occupied Fe states located above the valence bands (and absent in Mn) could be inferred [16].

Ordering phenomena in Fe DMS are still far from understood. Some indications for spin-glass transitions are observed in $Hg_{1-x}Fe_xSe$ [71], $Cd_{1-x}Fe_xSe$ [71], $Zn_{1-x}Fe_xSe$ [90] and $Zn_{1-x}Fe_xS$ [91], but a analysis of these data with respect to the d–d interaction (strength, range) is not possible up to now. The restricted stable composition range for bulk Fe DMS is in this respect a limiting factor. The preparation of Fe-containing compounds by MBE, which has been possible in the case of $Zn_{1-x}Fe_xSe$ for all x, is of great value for future research [21].

4. Carrier-concentration-dependent magnetic properties of IV–VI group DMS

Compared to the II–VI DMS the scientific interest for IV–VI DMS has been rather modest so far, although increasing rapidly. Among the IV–VI DMS most of the attention has been focused on the Pb-containing compounds [19,22, 67–69,92–94], e.g. $Pb_{1-x}Mn_xTe$. The magnetic properties of these alloys closely resemble the characteristic behavior of the II–VI Mn-containing DMS described in section 3.1, although typically the direct gap is an order of magnitude smaller. The strength of the AF interaction is, however, much weaker than in the II–VI DMS (see table 3), leading to very small Curie–Weiss temperatures. Anderson, Gorska and co-workers [68,69] ascribed this strong reduction compared to the II–VI DMS primarily to the separation (d_{Mn-a}) between Mn and an adjacent anion. Within existing models on superexchange [8,9] the d–d interaction is very sensitive for this separation through the d_{Mn-a}^{-4} or $d_{Mn-a}^{-7/2}$ dependence of the hybridization parameter V_{pd}, yielding

$$J_{d-d} \sim d_{Mn-a}^{-14} \quad \text{or} \quad J_{d-d} \sim d_{Mn-a}^{-16}. \tag{7}$$

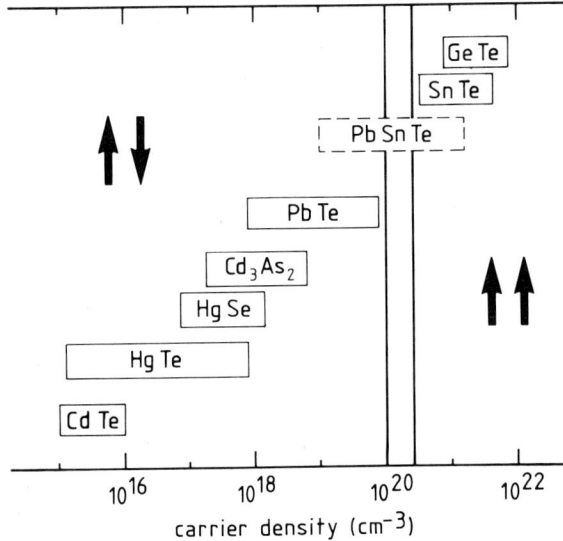

Fig. 17. Schematic representation of carrier concentrations in DMS based on Mn^{2+}. The antiparallel and parallel arrows indicate the observed antiferromagnetic and ferromagnetic interactions in those regimes, respectively. In $Pb_{1-x-y}Sn_y Mn_x Te$, carrier concentrations in both regimes are covered.

We should, however, not exclude that the observed differences are induced by the Mn-anion-Mn bond angle. According to a quantum-mechanical treatment of an idealized three-site molecule [8,54], the NN interaction varies with $\cos^2\phi$ which is minimal for the $\pi/2$ angle in the rocksalt structure of $Pb_{1-x}Mn_x Te$. On the other hand, Ginter et al. [95] indicated that exchange constants at the L point, where the direct gap is located in IV-VI DMS, are substantially diminished compared to the Γ point in the II-VI DMS.

One of the most interesting features in IV-VI DMS in the relation between magnetic properties and carrier concentration. In fig. 17 various DMS are arranged according to the carrier density. We already argued in the Introduction that for the II-VI compounds the small concentration of charge carriers ($< 10^{19}$ cm^{-3}) rules out the indirect RKKY mechanism between the Mn^{2+} ions and results in the antiferromagnetic superexchange as the dominant exchange process. This seems to be valid also for $Pb_{1-x}Mn_x Te$ and the corresponding selenides and sulfides with carrier densities never exceeding 10^{20} cm^{-3}. In striking contrast to this, the IV-VI DMS $Sn_{1-x}Mn_x Te$ [96], $Pb_{1-x-y}Ge_y Mn_x Te$ [97] and $Ge_{1-x}Mn_x Te$ [98] have generally (as-grown) higher carrier densities ($> 10^{20}$ cm^{-3}) and exhibit ferromagnetic interactions (positive Θ) and ferromagnetic ordering. The latter systems are gathered on the right-hand side of fig. 17. The RKKY interaction seems to be the most probable candidate to induce these ferromagnetic interactions, since its strength is roughly proportional to the number of charge carriers.

New experimental results on the role of the carrier concentration were obtained by Story et al. [11]. They observed that in $Pb_{1-x-y}Sn_y Mn_x Te$ an increasing carrier concentration transforms the system from essentially paramagnetic to ferromagnetic. Viewed in a wider perspective, this was the first observation of the fact that the carrier concentration can be a relevant parameter in the magnetic phase diagram [99]. It should be noted that the possibility to vary the carrier concentration reversibly over a relatively wide range (10^{19}–10^{21} cm^{-3}), by introducing a slightly nonstoichiometric composition by thermal annealing in a Te-rich or Sn-rich atmosphere (as shown in fig. 18), is a basic condition for this type of research. In order to illustrate the carrier-induced phenomena in IV-VI DMS we will concentrate on two different effects which will be discussed below.

4.1. Carrier-induced ferromagnetism

The first demonstration of the effect of the carrier subsystem on the magnetic subsystem was

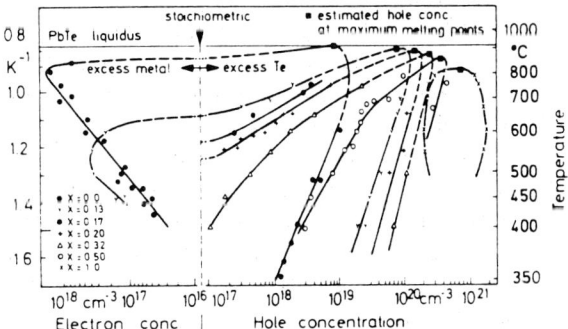

Fig. 18. Apparent carrier concentrations at 77 K (300 K for SnTe) versus isothermal annealing temperatures for $Pb_{1-x}Sn_x Te$; see ref. [101].

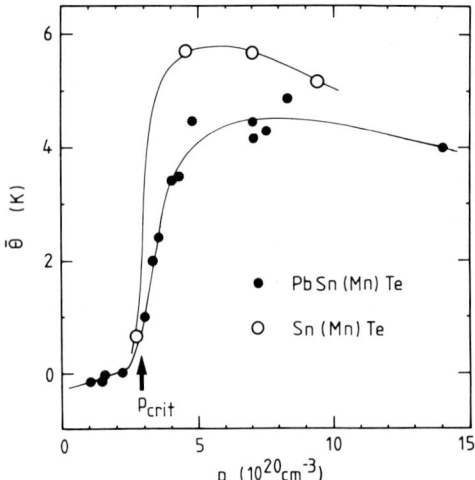

Fig. 19. Curie–Weiss temperature $\bar{\Theta}$ ($\equiv 0.03\Theta/x$ used for small variations in Mn-concentration) as a function of the carrier concentration p, for $Pb_{0.25}Sn_{0.72}Mn_{0.03}Te$ (full circles) and $Sn_{0.97}Mn_{0.03}Te$ (open circles); the lines are guides to the eye only; see refs [11,12,14].

found, as we quoted earlier, by Story et al. in $Pb_{0.25}Sn_{0.72}Mn_{0.03}Te$. The most convincing evidence appeared from the Curie–Weiss temperature Θ, depicted in fig. 19 and supplemented with more recent results on $Sn_{1-x}Mn_xTe$ [12] to actualize this compilation. Below a critical carrier concentration (p_{crit}) the total d–d interaction strength, as monitored by Θ, is negligible, whereas above p_{crit} a strong ferromagnetic interaction is evident. At p_{crit} the ferromagnetic interaction is switched on, so to say, and above p_{crit} a ferromagnetic ordering has been observed in the bulk magnetization, susceptibility and specific heat below a critical ordering temperature T_c which is proportional to Θ, as expected in a mean-field approximation. Recent neutron scattering data [102] provided decisive evidence for the genuine (long-range) ferromagnetic nature of this phase. When the carrier density is reduced below p_{crit}, the ferromagnetic transition disappears and the system shows the essentially paramagnetic or weakly AF characteristics of $Pb_{1-x}Mn_xTe$ discussed earlier.

It was suggested that the strong ferromagnetic interaction observed above p_{crit} should be attributed to a RKKY interaction, which seems the only mechanism strong enough and sufficiently long ranged to drive the ferromagnetic state. The RKKY interaction is given by [7]:

$$J_{RKKY}(R_{ij}) = \frac{m^* J_{s-d}^2 a_0^6 k_F^4}{32\pi^3 \hbar^2} \times \left[\frac{\sin(2k_F R_{ij}) - 2k_F R_{ij} \cos(2k_F R_{ij})}{(2k_F R_{ij})^4} \right], \quad (8)$$

where m^* is the effective mass of carriers, J_{s-d} the exchange integral between localized moment and free carriers, k_F the Fermi vector, R_{ij} the distance between spins and a_0 the lattice constant. However, calculations of Θ on the basis of this standard RKKY expression given by eq. (8) will yield a gradually increasing and continuous contribution to Θ in the carrier range plotted in fig. 19 and do not display a threshold-type of behavior at a critical carrier concentration as observed.

Triggered by observations from Hall effect studies [103] and the pressure dependence [104] of T_c, indicating the importance of contributions of carriers from other hands, Swagten et al. [14] presented a first interpretation of the threshold dependence of T_c at p_{crit}, based on a realistic two-band model for these compounds. The two bands denoted by VB1 and VB2 are located at the L point and along the Σ direction. The top of VB1 is higher than the top of VB2 and m_1^*, the effective hole mass of VB1 is much smaller than m_2^* of VB2 (see fig. 20). The total d–d exchange

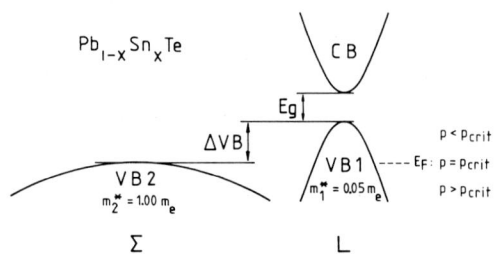

Fig. 20. Schematic illustration of the band structure of $Pb_{1-x}Sn_xTe$ used in the model calculations as described in the text [14].

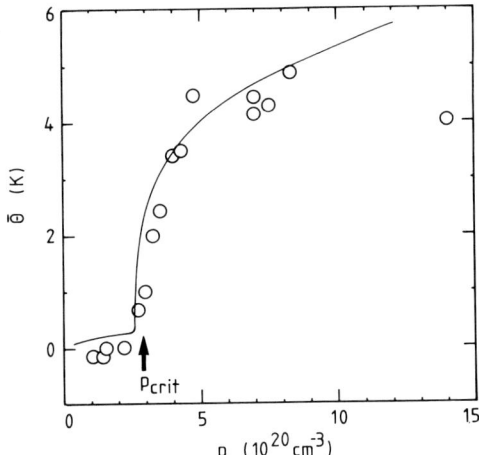

Fig. 21. Curie–Weiss temperature [scaled to $0.03\Theta/x$, used for small variations in Mn concentration] as a function of the carrier concentration p of $Pb_{0.25}Sn_{0.72}Mn_{0.03}Te$; data are taken from refs. [11,14]. The calculations are performed within the twoband RKKY model [14], supplemented with the multi-valley character of the relevant band structure [105] (parameters $m_1^* = 0.04$, $m_2^* = 1.00$, $J_{sp-d} = 0.09$ eV).

or Curie–Weiss temperature Θ is now assumed to be the sum of the contributions from the carriers in each band, each with its own effective mass:

$$\Theta_{RKKY} = \tfrac{2}{3}S(S+1) \times x \sum_i \left[(J_{i\text{-RKKY},VB1} + J_{i\text{-RKKY},VB2}) \right], \quad (9)$$

with i running over all (Mn −) lattice sites. For low carrier concentrations $p < p_{crit}$, E_F is located in the top of VB1, at p_{crit} the Fermi level enters also the top of VB2. The increase of Θ at p_{crit} is due to the large effective mass of VB2 ($m_2^*/m_1^* \approx 20!$) as follows from eq. (8). Calculations based on this simple model (supplemented with an exponential decay of J_{RKKY} to account for the finite mean free path), and using estimated values for the parameters, yielded a fair agreement with the experimental data as shown in fig. 21. Recent more detailed calculations of Story et al. [105] based on actual band structure calculations confirmed this result.

It should be noted, however, that although this rather crude and simple two-band RKKY model is capable of explaining the observed carrier-induced ferromagnetism, the quantitative agreement of the absolute magnitude of Θ, specifically in the ferromagnetic regime, may be somewhat fortuitous because it depends directly on the value for the s–d exchange constants for each valence band, $J_{s-d,1}$ and $J_{s-d,2}$, for which no direct experimental data are available.

Since $Pb_{1-x}Mn_xTe$ and $Sn_{1-x}Mn_xTe$ have comparable band structures, the results reported above might contain a clue with respect to the contrasting magnetic behavior of these compounds as was illustrated in fig. 17. It was therefore conjectured that, if the model holds, a reduction of p in $Sn_{1-x}Mn_xTe$ would yield a transformation of this well-known ferromagnet to a paramagnet. Such a transition was indeed observed recently [12] as shown in fig. 19. The same mechanism might also be responsible for the substantial decrease of the (ferromagnetic) Θ in $(Ge_{1-x-y}Pb_y)_{0.9}Mn_{0.1}Te$ with increasing Pb content, particularly since it was reported to be accompanied by a simultaneous decrease of the carrier density [97].

4.2. Carrier-induced ferromagnetic to spin-glass transition

The carrier-induced ferromagnetism discussed in the previous subsection was induced by the specific band structure. In contrast to this, we will now report some rather new data [13] which exploit the influence of the carrier concentration on exchange mechanisms itself. More specifically, we will report on a carrier-induced transition between a ferromagnetic and spin-glass state, which, as we will show, is intrinsically contained in the RKKY interaction mechanism.

It has been documented [106] that the aforementioned RKKY interaction is also responsible for the spin-glass behavior of canonical metallic spin-glasses like Cu(Mn). The period of the RKKY oscillations ($\sim p^{-1/3}$; see eq. (8)) is rather short compared with the average Mn–Mn distance due to the metallic carrier densities (10^{23} cm^{-3}), and therefore the competition between F and AF interactions induces the spin-glass formation [106]. There have been speculations about a phase transition which should be observed in metallic spin-glasses if the carrier concentration could be reduced. This reduction would increase the period of the spatial oscillations [see eq. (8)] till, ultimately, competition vanishes. Experimentally, however, such a transition has never been observed, mainly because of the restricted range of carrier concentrations which can be realized in metallic systems.

In this respect the IV–VI materials, introduced above, offer a unique opportunity to study the influence of the carrier concentration on the magnetic interaction, which is, as was shown, dominated by the RKKY mechanism.

Fig. 22 shows representative results of the magnetization, susceptibility and specific heat for $Sn_{1-x}Mn_xTe$ for two carrier concentrations. The data below $p \approx 10^{21}$ cm^{-3} show the well-known characteristics of a ferromagnetic transition, which have been treated previously. For carrier concentrations above $p \approx 10^{21}$ cm^{-3}, however, a completely different picture emerges from the data. The most obvious observation is that the ferromagnetic transition in M, χ and C_m at $T_c \approx \Theta$ has disappeared. The magnetization is strongly suppressed, while the lambda-anomaly in C_m is transformed into a broad rounded maximum and the sharp peak in the ac susceptibility is replaced in some cases by a cusp or an onset to a cusp at much lower temperatures. Moreover, it is important to note that, although the transition to the ferromagnetic state is suppressed when p is increased, the Curie–Weiss temperature Θ ($\sim \Sigma J_{ij}$) as a function of p displays no discontinuity, in accordance with the expectation for the RKKY interaction [12,14,105] (see inset fig. 22). Therefore, the breakdown of the ferromagnetic state with increasing carrier concentration is apparently not caused by a collapse of the interaction strength, but must be induced by a different ordering mechanism.

Although the present data have been obtained for temperatures only above $T \approx 1.5$ K and additional data are required, the behavior of M, χ and C_m of the system with high carrier concentration strongly resembles that of a spin-glass [106,82]. The observed cusplike behavior of χ can then be considered as an indication for the freez-

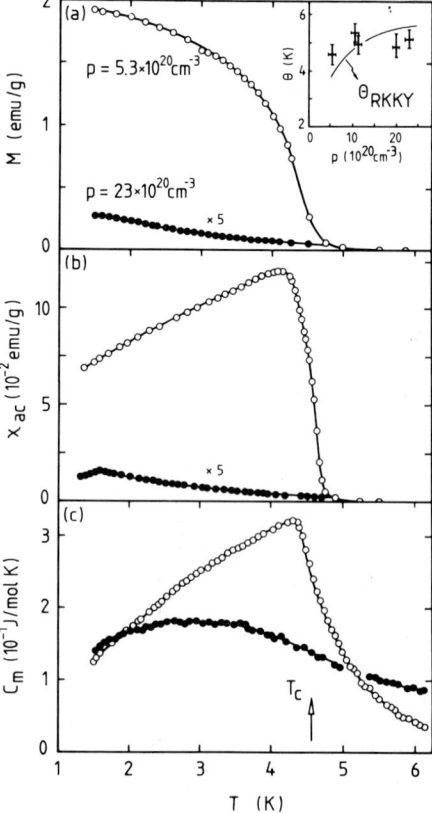

Fig. 22. (a) Magnetization, (b) ac susceptibility and (c) magnetic specific heat of $Sn_{0.97}Mn_{0.03}Te$, for $p = 5.3 \times 10^{20}$ cm^{-3} (open circles) and $p = 23 \times 10^{20}$ cm^{-3} (closed circles). The arrow indicates the ferromagnetic transition temperature T_c for $p = 5.3 \times 10^{20}$ cm^{-3}. The inset shows the Curie–Weiss temperature as a function of the carrier density; the crosses represent experimental data; the solid line shows calculations based on RKKY interaction [14,105].

ing temperature. Further analysis of C_m shows that above this temperature approximately 70% of the magnetic entropy [$xR \ln(2S + 1)$] is recovered, which is also observed in canonical spin-glasses.

In a diluted system, coupled by RKKY interactions, such a carrier-concentration-induced transition between a ferromagnetic state and a spin-glass state is conceivable, as we argued earlier. A more quantitative approach will be made in the following simple model. The RKKY-oscillations, as sketched in fig. 23 and given by eq. (8), have a periodicity of k_F^{-1} (or $p^{-1/3}$), and the first switch from F to AF occurs at a distance $R_{0,RKKY} \sim p^{-1/3}$. A second distance which plays a role is the mean distance between the Mn ions $\bar{R}_{Mn}$. When $\bar{R}_{Mn} \ll R_{0,RKKY}$ the ferromagnetic interactions between the Mn ions outweigh the oscillatory behavior, whereas for $\bar{R}_{Mn} \gg R_{0,RKKY}$ competition might be expected resulting in spin-glass behavior. In the spirit of an earlier suggestion of Mauger and Escorne [107] one might intuitively expect a transition between these regimes when $\bar{R}_{Mn} = R_{0,RKKY}$, and since $\bar{R}_{Mn} \sim x^{-1/3}$ and $R_{0,RKKY} \sim p^{-1/3}$, it follows that for this transition

$$x/p = \text{constant}. \quad (10)$$

The resulting phase boundary is shown in fig. 24, together with the existing data. The results show that, indeed, on the basis of this simple model a transition can be expected in the range of Mn concentrations (x) and carrier concentrations (p) studied in the present experiment, which is actually far below the range of p generally covered by

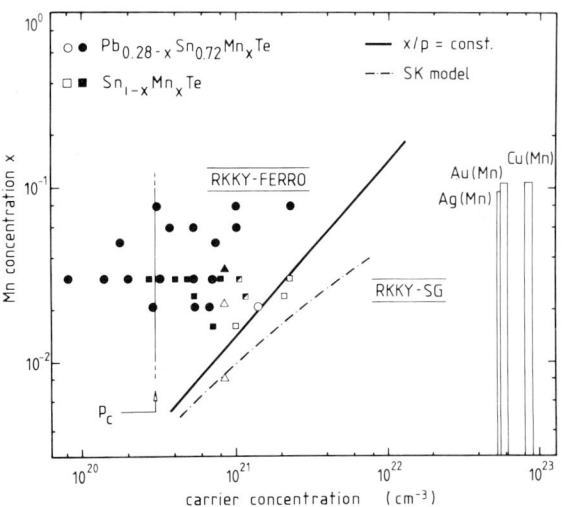

Fig. 24. Magnetic x–p phase diagram for RKKY systems. The closed symbols represent ferromagnetic $Pb_{1-x-y}Sn_y Mn_x Te$ or $Sn_{1-x}Mn_x Te$ (below p_{crit} the ferromagnetism is switched off), whereas the open symbols are spin-glass-like, including canonical spin-glasses such as Cu(Mn), Au(Mn) and Ag(Mn). The calculated phase boundaries between the ferromagnetic and spin-glass state are also shown (see the text). The triangular symbols denote data obtained for $Sn_{1-x}Mn_x Te$ from ref. [107].

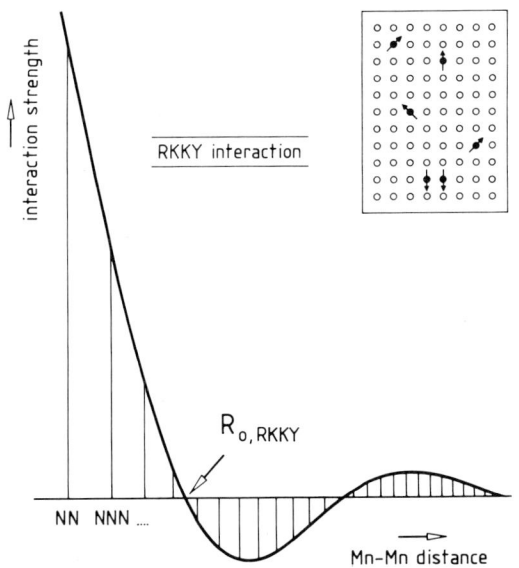

Fig. 23. Schematic representation of the RKKY interaction strength as a function of the distance between magnetic ions. The vertical bars indicate the available lattice positions for randomly diluted magnetic ions, starting with nearest-neighbor (NN), next-nearest neighbor (NNN), etc. The distance $R_{0,RKKY}$ ($\sim p^{-1/3}$) represents the first switch from F to AF interactions. The inset shows a schematic sketch of magnetic moments randomly diluted in the host matrix.

canonical metallic spin-glasses [106], which is also shown in fig. 24.

A more sophisticated approximation to this intuitive (geometrical) approach can be found in the well-known spin-glass model of Sherrington and Kirkpatrick (SK) [108]. The most important parameter in the SK model is $\eta \equiv J_0/\Delta J$, which relates the mean (J_0) and the rms deviation (ΔJ) of a Gaussian distribution of magnetic interactions to ordering phenomena of the spin system. More specifically, a pure spin-glass phase is predicted for η smaller than unity. Strictly speaking, the condition given by $\eta = 1$ does not apply in the present case, since one is dealing with a non-Gaussian distribution of the interaction strengths. Nevertheless, this phase boundary given by $\eta = 1$ has been calculated [109] for various values of x and p, and is also included in fig. 24. Again, the transition between a ferromagnetic and a spin-glass state is predicted by this model to occur in the range of carrier and manganese concentrations covered by our experiments, in agreement with our earlier geometrical model.

We would like to emphasize again that the ferromagnetic to spin-glass transition was traced by a change of the carrier concentration and the resulting change of the RKKY interaction *only*. In principle the phase transition can, of course, also be observed by changing the concentration x of magnetic ions [107]. In the latter case, however, additional contributions to the *direct* exchange between nearest-neighbor magnetic ions, if there are any, obscure the issue since their effectiveness increases with x. Such transitions have been observed [110] for instance in Ag(Fe), Au(Fe) and Pd(Mn).

5. Concluding remarks

In the present paper, we have focused on some selected topics related to the magnetic behavior of diluted magnetic semiconductors. We hope that, even within these limitations, the richness of the phenomena originating from the interaction between the electronic and magnetic subsystem may have become evident.

We would like to add that the flexibility offered by the DMS alloys in terms of composition, band structure and carrier concentration, as we have shown, facilitates tuning and manipulation of the physical properties, which will continue to be of great importance both for testing fundamental concepts as well as for device applications and artificial structures of these "group of solids on the interface between semiconductors and magnetic materials" [111].

Acknowledgements

The authors would like to acknowledge the valuable discussions and cooperation with R.R. Gałązka, T. Story, A. Twardowski, C.J.M. Denissen, K. Kopinga and J.A. Mydosh. Support by the Foundation for Fundamental Research on Matter (FOM) is also acknowledged. The research of Dr. Swagten has been made possible by a fellowship of the Royal Netherlands Academy of Arts and Sciences.

References

[1] See e.g., R.T. Delves, J. Phys. Chem. Solids 24 (1963) 885.
U. Sondermann, J. Magn. Magn. Mater. 2 (1976) 216.
A.V. Komarov, S.M. Rjabchenko, O.V. Terletski, I.I. Zheru and R.D. Ivanchuck, Zh. Eksp. Teor. Fiz. 73 (1977) 608 [Sov. Phys. JETP 46 (1977) 318].
G. Bastard, C. Rigaux, Y. Guldner, A. Mycielski, J.K. Furdyna and D.P. Mullin, J. de Phys. 39 (1978) C8-87.
[2] R.R. Gałązka, in: Proc. 14th Intern. Conf. on the Physics of Semiconductors, Edinburgh, 1978, IOP Conference Series, ed. B.L.H. Wilson (IOP, London, 1978).
[3] J.A. Gaj, J. Phys. Soc. Japan 49 (1980) 797.
[4] See e.g., J.A. Gaj, in: Semiconductors and Semimetals, Vol. 25, treatise eds. R.K. Willardson and A.C. Beer, volume eds. J.K. Furdyna and J. Kossut (Academic Press, Boston, 1988) p. 275.
D.V. Bartholomew, J.K. Furdyna and A.K. Ramdas, Phys. Rev. B 34 (1986) 6943.
[5] See e.g., A. Mycielski and J. Mycielski, J. Phys. Soc. Japan 49 Suppl. A (1980) 807.
Y. Shapira, in: Diluted Magnetic Semiconductors, vol. 89, eds. R.L. Aggarwal, J.K. Furdyna and S. von Molnar (Materials Research Society, Pittsburgh, 1987) p. 209.
[6] See e.g., M. Sawacki, T. Dietl, J. Kossut, J. Igalson, T. Wojtowicz and W. Plesiewicz, Phys. Rev. Lett. 56 (1986) 508.
[7] M. Rudermann and C. Kittel, Phys. Rev. 96 (1954) 99.
T. Kasuya, Prog. Theor. Phys. 16 (1956) 45.
K. Yoshida, Phys. Rev. 106 (1957) 893.
[8] N. Bloembergen and T.J. Rowland, Phys. Rev. 97 (1955) 1679.
P.W. Anderson, Phys. Rev. 155 (1959) 1, in: Solid State Physics, vol. 14, eds. F. Seitz and F. Turnbull (Academic Press, New York, 1963) p. 99.
C.E.T. Gonçalves Da Silva and L.M. Falikov, J. Phys. C 5 (1972) 63.
W. Geertsma, C. Hass, G.A. Sawatzky and G. Vertogen, Physica B 86–88 (1977) 1039.
G.A. Sawatzky, W. Geertsma and C. Hass, J. Magn. Magn. Mater. 3 (1976) 37.
[9] B.E. Larson, K.C. Hass, H. Ehrenreich and A.E. Carlsson, Phys. Rev. B 37 (1988) 4137.
B.E. Larson and H. Ehrenreich, Phys. Rev. B 39 (1989) 1747.
[10] W.J.M. de Jonge, A. Twardowski and C.J.M. Denissen, in: Diluted Magnetic Semiconductors, vol. 89, eds. R.L. Aggarwal, J.K. Furdyna and S. von Molnar (Materials Research Soc., Pittsburgh, 1987) p. 153.
[11] T. Story, R.R. Gałązka, R.B. Frankel and P.A. Wolff, Phys. Rev. Lett. 56 (1986) 777.
[12] H.J.M. Swagten, S.J.E.A. Eltink and W.J.M. de Jonge, in: Growth, Characterization and Properties of Ultrathin Magnetic Films and Multilayers, vol. 151, eds. B.T. Jonker, J.P. Heremans and E.L. Marinaro (Materials Research Soc., Pittsburgh, 1989) p. 171.

W.J.M. de Jonge, H.J.M. Swagten, S.J.E.A. Eltink and N.M.J. Stoffels, Semicond. Sci. Technol. 5 (1990) S131.
[13] W.J.M. de Jonge, T. Story, H.J.M. Swagten and P.J.T. Eggenkamp, to be published.
[14] H.J.M. Swagten, W.J.M. de Jonge, R.R. Gałązka, P. Warmenbol and J.T. Devreese, Phys. Rev. B 37 (1988) 9907.
[15] J.K. Furdyna, J. Appl. Phys. 64 (1988) R29; Semiconductors and Semimetals: Diluted Magnetic Semiconductors, vol. 25, treatise eds. R.K. Willardson and A.C. Beer, volume eds. J.K. Furdyna and J. Kossut (Academic Press, Boston, 1988).
[16] A. Mycielski, J. Appl. Phys. 63 (1988) 3279.
A. Twardowski, J. Appl. Phys. 67 (1990) 5108.
[17] For II–VI DMS see e.g., refs. [15,16] and A. Pajaczkowska, Prog. Cryst. Growth Charact. 1 (1978) 289; Landolt–Bornstein: Technology of III–V, II–VI and Non-tetrahedrally Bound Compounds, vol. 17, M. Schulktz and H. Weiss (Springer, Berlin, 1984) Pt. d; Landolt–Bornstein: Physics of II–VI and I–VII Compounds, Semimagnetic Semiconductors, vol. 17, ed. O. Madelung (Springer, Berlin, 1982) Pt. b; for Co DMS see section 3.2.
[18] For II–V DMS see e.g., C.J.M. Denissen, Ph.D. thesis, Eindhoven (1986) unpublished.
[19] For IV–VI DMS see section 4 and e.g., G. Bauer, in: Diluted Magnetic Semiconductors, vol. 89, eds. R.L. Aggarwal, J.K. Furdyna and S. von Molnar (Materials Research Soc., Pittsburgh, 1987) p. 107.
[20] D.R. Yoder-Short, U. Debska and J.K. Furdyna, J. Appl. Phys. 58 (1985) 4056.
[21] B.T. Jonker, J.J. Krebs, G.A. Prinz, X. Liu, A. Petrou and Salamanca-Young, in: Growth, Characterization and Properties of Ultrathin Magnetic Films and Multilayers, vol. 151, eds. B.T. Jonker, J.P. Heremans and E.L. Marinero (Materials Research Soc., Pittsburgh, 1989) p. 151.
[22] Z. Korczak and M. Subotowicz, Phys. Stat. Sol. (a) 77 (1983) 497.
[23] W.F. Pong, R.A. Mayanovic, B.A. Bunker, J.K. Furdyna and U. Debska, Phys. Rev. B 41 (1990) 8440.
[24] See e.g., G. Bastard, C. Rigaux, Y. Guldner, A. Mycielski, J.K. Furdyna and D.P. Mullin, Phys. Rev. B 24 (1982) 1961.
[25] J.K. Furdyna, J. Vac. Sci. Technol. 21 (1982) 220.
[26] See e.g., J.A. Gaj, J. Ginter and R.R. Gałązka, Phys. Stat. Sol. (b) 89 (1978) 655.
A. Twardowski, M. Nawrocki and J. Ginter, Phys. Stat. Sol. (b) 96 (1979) 497.
G. Rebman, C. Rigaux, G. Bastard, M. Menant, R. Triboulet and W. Giriat, Physica B 117 & 118 (1983) 452.
[27] See e.g., M. Jaczynski, J. Kossut and R.R. Gałązka, Phys. Stat. Sol. (b) 88 (1987) 73.
L.M. Roth and P.N. Argyres, in: Semiconductors and Semimetals: Diluted Magnetic Semiconductors, vol. 1, eds. R.K. Willardson and A.C. Beer (Academic Press, New York, 1966) p. 159.
[28] H.M.A. Schleijpen, Ph.D. thesis, Eindhoven (1987) unpublished.
[29] See e.g., P.W. Wolff, in: Semiconductors and Semimetals: Diluted Magnetic Semiconductors, vol. 25, treatise eds. R.K. Willardson and A.C. Beer, volume eds. J.K. Furdyna and J. Kossut (Academic Press, Boston, 1988) p. 413.
M. Nawrocki, R. Planel, G. Fishman and R.R. Gałązka, Phys. Rev. Lett. 46 (1981) 735.
[30] M. Nawrocki, F. Hamdani, J.P. Lascaray, Z. Golacki and J. Deportes, Solid State Commun. 77 (1991) 111.
[31] R.R. Gałązka, S. Nagata and P.H. Keesom, Phys. Rev. B 22 (1980) 3344.
[32] R.L. Aggarwal, S.N. Jasperson, P. Becla and R.R. Gałązka, Phys. Rev. B 32 (1985) 5132.
[33] J.P. Lascaray, M. Nawrocki, J.M. Bruto, M. Rakoto and M. Demianiuk, Solid State Commun. 61 (1987) 401.
[34] S.B. Oseroff and F. Gandra, J. Appl. Phys. 57 (1985) 3421.
[35] M.A. Novak, O.G. Symko, D.J. Zheng and S. Oseroff, Physica B 126 (1984) 469.
S.B. Oseroff, Phys. Rev. 25 (1982) 6584.
[36] See e.g., W.J.M. de Jonge, M. Otto, C.J.M. Denissen, F.A.P. Blom, C. van der Steen and K. Kopinga, J. Magn. Magn. Mater. 31–34 (1983) 1373.
M.A. Novak, O.G. Symko, D.J. Zheng and S. Oseroff, J. Appl. Phys. 57 (1985) 3419.
A. Mycielski, C. Rigaux, M. Menant, T. Dietl, M. Otto, Solid State Commun. 50 (1984) 257.
[37] S. Foner, Y. Shapira, D. Heiman, P. Becla, R. Kershaw, K. Dwight and A. Wold, Phys. Rev. B 39 (1989) 11793.
Y. Shapira, J. Appl. Phys. 67 (1990) 5090.
[38] T.M. Giebultowicz, J.J. Rhyne and J.K. Furdyna, J. Appl. Phys. 67 (1990) 5096.
[39] Y. Shapira, S. Foner, D.H. Ridgley, K. Dwight and A. Wold, Phys. Rev. B 30 (1984) 4021.
[40] Y. Shapira, S. Foner, D. Heiman, P.A. Wolff and C.R. McIntyre, Solid State Commun. 70 (1989) 355.
[41] T.M. Giebultowicz, J.J. Rhyne and J.K. Furdyna, J. Appl. Phys. 61 (1987) 3537.
[42] V. Spasojevic, A. Bajorek, A. Szytuła and W. Giriat, J. Magn. Magn. Mater. 80 (1989) 183.
[43] W.H. Brumage, C.R. Yarger and C.C. Lin, Phys. Rev. 133 (1964) A765.
[44] H.J.M. Swagten, A. Twardowski, W.J.M. de Jonge, M. Demianiuk and J.K. Furdyna, Solid State Commun. 66 (1988) 791.
[45] Y. Shapira and N.F. Oliveira Jr., Phys. Rev. B 35 (1987) 6888.
[46] J.K. Furdyna, N. Samarth, R.B. Frankel and J. Spałek, Phys. Rev. 37 (1988) 3707.
[47] A. Lewicki, J. Spałek, J.K. Furdyna and R.R. Gałązka, Phys. Rev. B 37 (1988) 1806.

[48] A. Twardowski, H.J.M. Swagten, W.J.M. de Jonge and M. Demianiuk, Phys. Rev. B 36 (1987) 7013.
[49] Y. Shapira, S. Foner, P. Becla, D.N. Domingues, M.J. Naughton and J.S. Brooks, Phys. Rev. B 33 (1986) 356.
[50] R.L. Aggarwal, S.N. Jasperson, P. Becla and J.K. Furdyna, Phys. Rev. B 34 (1986) 5894.
[51] G. Barilero, C. Rigaux and N.H. Hau, J.C. Picoche and W. Giriat, Solid State Commun. 62 (1987) 345.
[52] J.P. Lascaray, A. Bruno, J.C. Ousset, H. Rakoto, J.M. Broto and S. Askenazy, Physica B 155 (1989) 353.
[53] L.M. Corliss, J.M. Hastings, S.M. Shapiro, Y. Shapira and P. Becla, Phys. Rev. B 33 (1986) 608.
[54] J. Spałek, A. Lewicki, Z. Tarnawski, J.K. Furdyna, R.R. Gałązka and Z. Obuszko, Phys. Rev. B 33 (1986) 3407.
[55] D.U. Bartholomew, E.K. Suh, S. Rodriguez, A.K. Ramdas and R.L. Aggarwal, Solid State Commun. 62 (1987) 235.
[56] B.E. Larson, K.C. Hass and R.L. Aggarwal, Phys. Rev. 33 (1986) 1789.
[57] E.D. Isaacs, D. Heiman, P. Becla, Y. Shapira, R. Kershaw, K. Dwight and A. Wold, Phys. Rev. B 38 (1988) 8412.
[58] X. Wang, D. Heiman, S. Foner and P. Becla, Phys. Rev. B 41 (1990) 1135.
[59] T.M. Giebultowicz, J.J. Rhyne, W.Y. Ching, D.L. Huber, J.K. Furdyna, B. Lebech and R.R. Gałązka, Phys. Rev. B 39 (1987) 6857.
[60] T. Giebultowicz, B. Lebech, B. Buras, W. Minor, H. Kepa and R.R. Gałązka, J. Appl. Phys. 55 (1984) 2305.
[61] R.R. Gałązka, W. Dobrowolski, J.P. Lascaray, M. Nawrocki, A. Bruno, J.M. Broto and J.C. Ousset, J. Magn. Magn. Mater. 72 (1988) 174.
[62] B.B. Stojic, M. Stojic and B. Stosic, J. Phys.: Condens. Matter 1 (1989) 7651.
[63] A.B. Davydov, L.M. Noskova, B.B. Ponikarov and L.A. Ugodnikova, Sov. Phys. Semicond. 14 (1980) 869.
[64] H. Savage, J.J. Rhyne, R. Holm, J.R. Cullen, C.E. Carroll and E.P. Wohlfarth, Phys. Stat. Sol. (b) 58 (1973) 685.
[65] C.J.M. Denissen, S. Dakun, K. Kopinga, W.J.M. de Jonge, H. Nishihara, T. Sakakibara and T. Goto, Phys. Rev. B 36 (1987) 5316.
[66] C.J.M. Denissen, H. Nishihara, J.C. van Gool and W.J.M. de Jonge, Phys. Rev. B 33 (1986) 7637.
[67] G. Karczewski, M. von Ortenberg, Z. Wilamowski, W. Dobrowolski and J. Niewodnizanska-Zawadzka, Solid State Commun. 55 (1985) 249.
[68] J.R. Anderson, G. Kido, Y. Nishina, M. Gorska, L. Kowalczyk and Z. Golacki, Phys. Rev. B 41 (1990) 1014.
[69] M. Gorska and J.R. Anderson, Phys. Rev. B 38 (1988) 9120.
[70] A. Twardowski, A. Lewicki, M. Arciszewski, W.J.M. de Jonge, H.J.M. Swagten and M. Demianiuk, Phys. Rev. B 38 (1988) 10749.
[71] A. Lewicki, J. Spałek and A. Mycielski, J. Phys. C 20 (1987) 2005.

[72] A. Twardowski, H.J.M. Swagten and W.J.M. de Jonge, Phys. Rev. B 42 (1990) 2455.
[73] T.M. Giebultowicz, P. Klosowski, J.J. Rhyne, T.J. Udovic, J.K. Furdyna and W. Giriat, Phys. Rev. B 41 (1990) 504.
[74] A. Lewicki, A.I. Schindler, J.K. Furdyna and W. Giriat, Phys. Rev. B 40 (1989) 2379.
[75] Y. Shapira, T.W. Vu, B.K. Lau, S. Foner, E.J. McNiff Jr., D. Heiman, C.L.H. Thieme, C-M. Niu, R. Kershaw, K. Dwight, A. Wold and V. Bindilatti, Solid State Commun. 75 (1990) 201.
[76] T.M. Giebultowicz, F. Klosowski, J.J. Rhyne, T.J. Udovic, U. Debska, J.K. Furdyna and W. Giriat, Bul. Am. Phys. Soc. 34 (1989) 592.
[77] A. Lewicki, A.U. Schindler, I. Miotkowski and J.K. Furdyna, Phys. Rev. B 41 (1990) 4653.
[78] J.L. Tholence and R. Tournier, J. de Phys. 35 (1974) C4-229.
[79] K. Matho, J. Low Temp. Phys. 35 (1979) 165.
[80] C.J.M. Denissen and W.J.M. de Jonge, Solid State Commun. 59 (1986) 503.
[81] B.E. Larson, K.C. Hass, H. Ehrenreich and A.E. Carlsson, Solid State Commun. 56 (1985) 347.
[82] J.A. Mydosh, in: Proc. 3rd Intern. Conf. on Physics of Magnetic Materials, eds. W. Gorzkowski, H.K. Lachowicz and H. Szymczak (World Scientific, Singapore, 1986).
[83] R.R. Gałązka, W.J.M. de Jonge, A.T.A.M. de Waele and J. Zeegers, Solid State Commun. 68 (1988) 1047.
[84] For some early papers on Fe DMS see, A. Twardowski, M. von Ortenberg and M. Demianiuk, J. Crystal Growth 72 (1985) 401.
B.T. Jonker, J.J. Krebs, S.B. Qadri and G.A. Prinz, Appl. Phys. Lett. 50 (1986) 848.
[85] For some early papers on Co DMS see, J.J. Krebs, B.T. Jonker and G.A. Prinz, IEEE Trans. Magn. MAG-24 (1988) 2548.
B.T. Jonker, J.J. Krebs and G.A. Prinz, Appl. Phys. Lett. 53 (1988) 450.
[86] P.M. Shand, A. Lewicki, B.C. Crooker, W. Giriat and J.K. Furdyna, J. Appl. Phys. 67 (1990) 5246.
[87] A. Twardowski, H.J.M. Swagten, W.J.M. de Jonge and M. Demianiuk, Acta Phys. Polon. A 77 (1990) 167.
[88] A. Lewicki, A.I. Schindler, I. Miotkowski, B.C. Crooker and J.K. Furdyna, Phys. Rev. B 43 (1991) 5713.
[89] H.J.M. Swagten, C.E.P. Gerrits, A. Twardowski and W.J.M. de Jonge, Phys. Rev. B 41 (1990) 7330.
[90] H.J.M. Swagten, A. Twardowski, W.J.M. de Jonge and M. Demianiuk, Phys. Rev. B 39 (1989) 2568.
[91] H.J.M. Swagten, A. Twardowski, E.H.H. Brentjens, W.J.M. de Jonge, M. Demianiuk and J.A. Gaj, in: Proc. 20th Intern. Conf. on the Physics of Semiconductors, Thessaloniki, Greece, 1990, eds. E.M. Anastassakis and J.D. Joannopoulos (World Scientific, Singapore, 1990) p. 742.
[92] M. Escorne, A. Mauger, J.L. Tholence and R. Triboulet, Phys. Rev. B 29 (1984) 6306.

[93] V.C. Lee, Phys. Rev. B 34 (1986) 5430.
[94] H. Pascher, P. Rothlein, G. Bauer and M. von Ortenberg, Phys. Rev. B 40 (1989) 10469.
[95] J. Ginter, J.A. Gay and Le Si Dang, Solid State Commun. 48 (1983) 849.
[96] See e.g., R.W. Cochrane, F.T. Hedgcock and J.O. Strom-Olsen, Phys. Rev. B 8 (1973) 4262.
M. Escorne, A. Ghazalli and P. Leroux-Hugon, in: Proc. 12th Intern. Conf. on Physics of Semiconductors, ed. M.H. Pilkuhn (Teubner, Stuttgart, 1974) p. 915.
M. Mathur, D. Deis, C. Jones, A. Patterson, W. Carr and R. Miller, J. Appl. Phys. 41 (1970) 1005.
M. Inoue, H.K. Fun and H. Yagi, J. Phys. Soc. Jpn. 49 (1980) 835.
[97] T. Hamasaki, Solid State Commun. 32 (1979) 1069.
[98] R.W. Cochrane, F.T. Hedgcock and J.O. Strom-Olsen, Phys. Rev. B 9 (1974) 3013.
[99] In fact, the investigations reported in refs. [97,100] already demonstrated, though somewhat obscured, the carrier-induced ferromagnetism; however, these authors did not pay attention to the decisive role of the carrier density.
[100] F.T. Hedgcock, P.C. Sullivan, K. Kadowaki and S.B. Woods, J. Magn. Magn. Mater. 54–57 (1986) 1293.
A.V. Brodovoi, G.V. Lashkarev, M.V. Radchanko, E.I. Slyn'ko and K.D. Tovstyuk, Sov. Phys. Semicond. 18 (1984) 970.
[101] See e.g., Physics of IV–VI Compounds and Alloys, ed. S. Rabii (Gordon and Breach, London, 1974);
Narrow-gap Semiconductors and Related Materials, eds. D.G. Seiler and C.L. Littler (Hilger, Bristol, 1990).
[102] C.W.H.M. Vennix, E. Frikkee, H.J.M. Swagten, K. Kopinga and W.J.M. de Jonge, J. Appl. Phys. 69 (1991) 6025.
[103] M. Ocio, Phys. Rev. B 10 (1974) 4274.
[104] T. Suski, J. Igalson and T. Story, J. Magn. Magn. Mater. 66 (1987) 325.
[105] T. Story, L. Swierkowski, G. Karczewski, W. Staguhn and R.R. Gałązka, in: Proc. 19th Intern. Conf. on Physics of Semiconductors, ed. W. Zawadski (Warsaw, 1988) p. 1567.
T. Story, G. Karczewski, L. Swierkowski and R.R. Gałązka, Phys. Rev. B 42 (1990) 10477.
[106] For some reviews on spin-glasses, see: K. Binder and A.P. Young, Rev. Modern Phys. 4 (1986) 801.
K.H. Fischer, Phys. Stat. Sol. (b) 116 (1983) 357, 130 (1985) 13.
C.Y. Huang, J. Magn. Magn. Mater. 51 (1985) 1.
[107] A. Mauger and M. Escorne, Phys. Rev. B 35 (1987) 1902.
[108] D. Sherrington and S. Kirkpatrick, Phys. Rev. Lett. 35 (1975) 1792.
[109] The calculation of η requires a statistical treatment of the long-range RKKY interactions given by eq. (8), and will be published elsewhere.
[110] See e.g., J.A. Mydosh and G.J. Nieuwenhuys, in: Ferromagnetic Materials, vol. 1, ed. E.P. Wohlfarth (North-Holland, Amsterdam, 1980) p. 71.
[111] R.R. Gałązka, J. Crystal Growth 72 (1985) 364.

The Invar problem

E.F. Wassermann

Tieftemperaturphysik, Universität Duisburg, 4100 Duisburg, Germany

After briefly reviewing the history and practical applications of Invar, we present results of recent total energy calculations as a function of the magnetic moment and atomic volume $E(M, V)$. Based on these results, we analyze Invar-relevant physical properties for a broad variety of fcc ferromagnetic and antiferromagnetic 3d-alloy systems as a function of the electron concentration. The observed systematics can principally be understood within the framework of the band calculations mentioned. We also give an answer to the question, why close to the Invar range martensitic fcc–bcc transitions occur. Finally we show that a deeper understanding of all the observed anomalies results from an analysis of available data as a function of the lattice constant. Almost quantitative agreement between experiment and theory allows the conclusion that the moment–volume instabilities are undoubtedly responsible for the magnetic, magnetovolume and structural properties exhibited by 3d-alloy systems.

1. Introduction

The explanation of the "Invar-effect" has intrigued solid state physicists, since Ch.E. Guillaume found in 1897 that ferromagnetic (FM) face centred cubic (fcc) FeNi alloys at concentrations around $Fe_{65}Ni_{35}$ show almost constant "invariant" thermal expansion in a wide range around room temperature. For this discovery and his finding of the temperature-independent elastic behavior of FeNiCr alloys ("Elinvar") he was honoured with the Nobel Prize in physics in 1920. Inspite of numerous attempts, the origin of Invar was never fully understood and to date is still a challenging problem of solid state magnetism.

From their early finding, Invar and Elinvar type alloys have experienced widespread practical application. Invar materials like the classical FeNi alloys, ternary alloys of FeNiCo ("Super-Invar") or FeCoCr ("Stainless Invar"), which meet the requirement of dimensional stability with temperature, are used in bi-metals, precision tools and instruments, laser sources, lead frames and seismographic devices, to name a few. Very recently a new field for technical application of FeNi and FeNiCo Invar promising large scale consumption, was opened by using these materials for shadow masks in high resolution TV tubes. Elinvar materials like FeNiCr, or quarternary alloys of the system FeNiCoCr ("Super-Elinvar"; "Co-Elinvar"), which meet the requirement of elastic stability with temperature, find their main application in time recording instruments, in reeds, reed relays and delay lines. Best known and most important in the past, experiencing a certain revival todate is the use of antiferromagnetic (AF) FeCrMn Elinvar as hair spring and construction material for "antimagnetic" wrist watches.

In the field of basic research on magnetism, the widespread and continuous interest in Invar has mainly two reasons. Firstly, the observation of Invar-like anomalies remained by no means bound to ferromagnetic (FM) alloys with fcc structure. There are AF Invar alloys, and Invar properties are also found in materials with bcc, hexagonal or even amorphous structure, as well as in rare earth (RE)–transition metal compounds with Laves phase structure (e.g. $RECo_2$, $REMn_2$) or compounds like the hard ferromagnet $Fe_{14}Nd_2B$. Today we know that it is a key point that the systems are rich in at least one (but a certain, as we shall see) 3d-transition element component. There are neither purely 4f, 5f nor insulating Invar alloys. Seemingly, Invar is a problem of

0304-8853/91/$03.50 © 1991 – Elsevier Science Publishers B.V. All rights reserved

itinerant 3d-magnetism. Secondly, the anomaly in the thermal expansion is only one of many. The list of anomalous physical properties of Invar materials includes the temperature dependence of the lattice constant, heat capacity, magnetization, pressure dependence of the magnetization and of Curie-(Néel)-temperatures, spontaneous volume magnetostriction for both low and high T, thermal variation of high field susceptibility and of elastic constants, as well as of Young's and bulk moduli. The listings demonstrate the richness of the Invar-effect and at the same time the complexity incorporated in it. We have recently reviewed and outlined this spectrum [1], and the proceedings of an international symposium [2], which took place in Duisburg, Germany in 1989 present a comprehensive overview.

Much effort has been invested to explain the Invar phenomenon theoretically. Within the years more than 20 different models for the understanding of Invar have been published. We have reviewed these developments in detail [1] too. The early local models stressed the metallurgical and/or magnetic inhomogeneity as Invar relevant, since in the archetypical Invar system FeNi the magnetovolume effects reach a maximum in a concentration range, where a strong deviation of the average magnetic moment (or magnetization) from the Slater-Pauling curve is observed, and simultaneously a structural transition from the fcc γ-phase to the bcc α-phase occurs. With the detection of the Invar-effect on FM ordered Fe_3Pt, a system which neither shows mixed magnetic behavior nor a deviation of the moment from the Slater-Pauling curve (but lies close to a γ-α transition) all the local models came principally into doubt. We have also shown recently [3] through magnetization and thermal expansion measurements on fcc FeNi films with Fe-concentrations around 65 at% that inspite of the fact that the deviation of the moment from the Slater-Pauling curve is *not* observed in these films they exhibit Invar behavior in their thermal expansion.

Within the years, homgeneous models, based on ideas of weak itinerant ferromagnetism gave some progress in the understanding of Invar, but also failed in a close description of all the properties in the broad variety of systems listed above. The problem of providing correctly for electronic and phononic properties, occurring simultaneously in Invar, remained unsolved.

2. Moment–volume instabilities

Progress in a new theoretical understanding of Invar was made in self-consistent band structure investigations within the local density approximation, using the fixed-spin-moment (FSM) method [4–6]. Within this method the total energy of an alloy is calculated as a function of both, magnetic moment and atomic volume, resulting in so-called binding energy surfaces $E(M, V)$. So far, the calculations have been carried out only for ordered structures, but principally give correct answers, since order–disorder (though of influence) is not of crucial importance for the understanding of the Invar-effect. This is further supported by KKR-CPA calculations on disordered $Fe_{65}Ni_{35}$ [7], where the total energy as a function of the atomic volume shows similar behavior as in ordered Fe_3Ni.

In fig. 1 we show as an example recent results of FSM-calculations for ordered phases in the FeNi system [8]. Constant energy contours with a difference of 1 mRy ($\approx$ 150 K) are projected into the moment–volume plane. For Fe_3Ni, which is close in concentration to the archetype $Fe_{65}Ni_{35}$ Invar, additionally 0.5 mRy contours (dashed) are given. Pure fcc Fe at the equilibrium lattice constant is non-magnetic (NM), because the minimum in $E(V, M)$ lies on the abcissa. At expanded atomic volume fcc Fe shows metastable magnetic states, as other authors [9,10] also predicted. The results depend on details of the theoretical assumptions, which will not be discussed further here. Moreover, comparison with experiment is difficult, since pure fcc Fe can only be stabilized in form of small coherent, AF precipitations in e.g. a Cu matrix. As Ni is added to Fe, we see from fig. 1 that a FM state evolves (starting near Fe_3Ni), which becomes progressively more stable towards FeNi. Pure Ni is a stable ferromagnet as expected. In fig. 1 we also show the zero field solutions ($H = dE/dM = 0$; full lines) and zero pressure solutions ($p = -dE/dV = 0$; dashed lines) for Fe_3Ni, FeNi and Ni. According to these results the moments in FeNi and pure Ni are stable and the NM-state could only be reached in experiments with technically unrealistically high pressures.

The behavior is different in Fe_3Ni, where the moment versus volume curve (full line in fig. 1) drops discontinuously, parallel to the decrease of the zero pressure curve. This is the key result towards a new undestanding of Invar, because it shows that in Fe_3Ni the moment is unstable with respect to changes in volume, and relatively small pressures (≈ 100 kbar) are sufficient to induce a FM–NM transition. A similar moment–volume instability has also been found in total energy calculations within the FSM-method for Fe_3Pt [6].

We present these results in fig. 2a–d, because detailed information about the Invar behavior can be deduced from these calculations. Fig. 2a shows – analogous to fig. 1 – constant energy contours projected into the moment–volume plane, clearly exhibiting the high-spin (HS) state with large moment ($\approx 1.8\mu_B$) at large volume ($V_{HS} = 84$ bohr^3 $\hat{=}$ $r_{WS} = 2.72$ au $\hat{=}$ $a = 3.68$ Å) and the no moment (NM) state ($0\mu_B$) at a volume of about 79 au^3 ($r_{WS} = 2.66$ au $= a \hat{=} 3.6$ Å). Fig. 2b shows a detailed calculation around the saddle point of fig. 2a and reveals that there is also a low spin (LS) state in Fe_3Pt with $\mu = 0.2\mu_B$ and similar volume as the NM state. The NM- and the LS state energetically lie close together as shown in fig. 2c. There is an energy difference of $\Delta E = E_{NM} - E_{HS} = 1.3$ mRy ($= \approx 200$ K). Fig. 2d

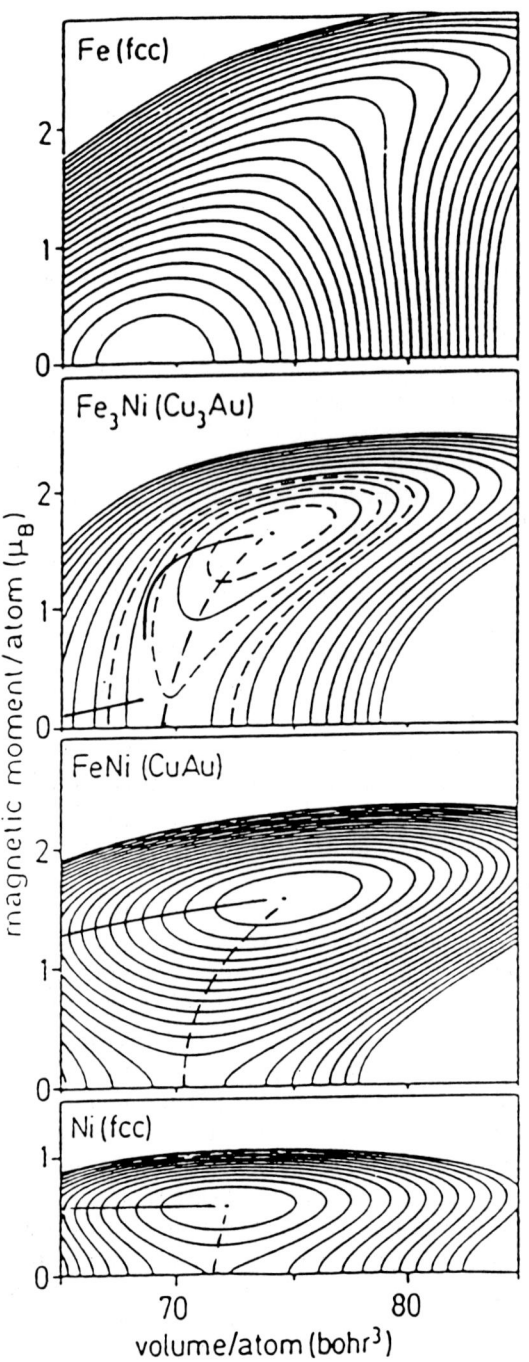

Fig. 1. Constant energy contours projected into the magnetic moment–volume plane as calculated with the Fixed Spin Moment (FSM) method by Mohn et al. [8] for the FeNi alloy series. All plots are on the same scale and all quantities are normalized to one atom per unit cell. The distance between contour lines is 1 mRy (≈ 150 K) with additional dashed contours at 0.5 mRy shown for ordered Fe_3Ni. The zero external field solutions ($H = dE/dM = 0$; full lines) and zero pressure solutions ($p = -dE/dV = 0$; dashed lines) are also shown. Note that pure fcc Fe, which is not stable in nature, is non-magnetic, while Ni and ordered FeNi are ferromagnets, as expected. Ordered Fe_3Ni, which is also not stable in nature, but lies close in concentration to the $Fe_{65}Ni_{35}$-Invar, shows an instability of the moment with respect to small changes in the volume (moment–volume instability; discontinuous full line).

finally exhibits that the zero pressure and zero field solutions cross twice in Fe_3Pt (full dots) and therefore the HS- and the LS- (as well as the NM) state are stable states of the system. In fact, two stable solutions, i.e. a HS- and a LS-state have also been found for Fe_3Ni by Moruzzi [4]. The results of the modern band calculations are therefore in some sense reminiscent of the Weiss 2γ-states model [11].

The correctness of these theoretical findings has been recently confirmed by low temperature Mössbauer measurements under pressure on $Fe_{68.5}Ni_{31.5}$ and ordered as well as disordered $Fe_{72}Pt_{28}$ [12]. Though the absolute values found experimentally differ somewhat from the theoretical predictions, it is beyond doubt that the presence of moment–volume instabilities forms the basis for the occurrence of the Invar-effect.

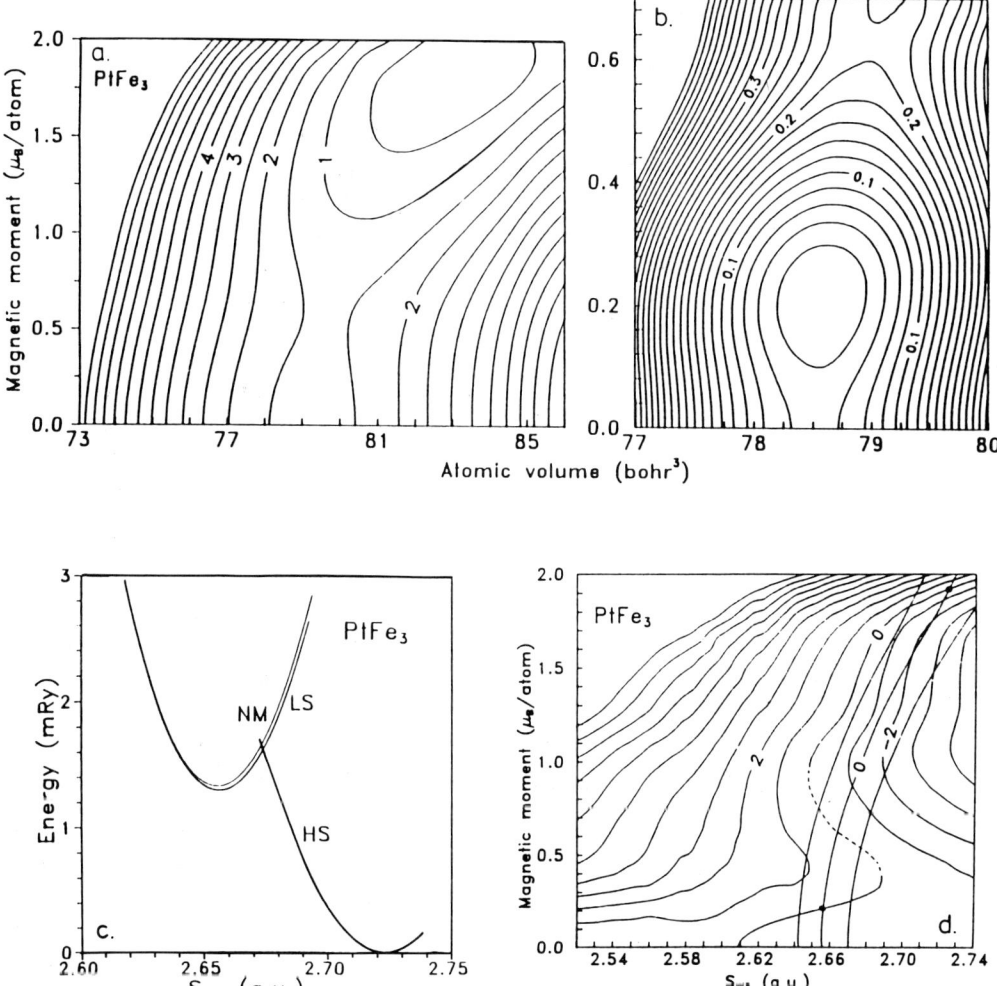

Fig. 2. (a) Total energy surface for ordered Fe_3Pt in the moment–volume plane; (b) blow up of the region around the low spin (LS) state in (a) with a distance of the contour lines of $\Delta E = 0.02$ mRy; (c) total energy of the high spin (HS), low spin (LS) and non-magnetic (NM) phases of Fe_3Pt as a function of the radius of the Wigner–Seitz cell S_{WS}; (d) contour plot of the magnetic field in the moment–S_{WS} plane for Fe_3Pt. Contours are plotted every 1 mRy/μ_B. Dashed line indicates the unphysical region of the $M(S_{WS})$-curve. Superimposed are the pressure contours for −50, 0 and 50 kbar. Note the two crossing points of zero pressure contour lines at the HS and LS state, respectively (all data from ref. [6]).

3. Instabilities as a function of electron concentration

3.1. Theoretical results

In fig. 3 we have plotted the energy differences between the NM- and HS-states, $\Delta E = E_{NM} - E_{HS}$ (left hand scale), versus the electron concentration e/a (counting 3d and 4s electrons) as taken from fig. 1 and reported in the literature for the FM FeNi-series [4,5,8,14] by various authors using different calculational methods. For pure FM fcc Fe the NM-state forms the ground state so that $\Delta E < 0$ (plus sign from ref. [13]; triangle from ref. [9]). The plot also contains a result for disordered $Fe_{65}Ni_{35}$ (full dot) from ref. [7] as well as for ordered Fe_3Pt from fig. 2 ($\Delta E = 1.3$ mRy [6]). Though for FeNi and γ-Fe there are some discrepancies in the absolute theoretical values for ΔE, the data in fig. 3, when interpolated linearly, show a general trend of $\Delta E(e/a)$, which can be used as the basis for an overall understanding of the itinerant magnetic behavior, magnetovolume effects and structural phase transitions not only for FeNi but – as comparison with experiments will show (see section 3.2 below) – for all fcc 3d-alloy systems. We make the following observations:

1) All alloys and compounds with e/a values within the hatched area (fig. 3) are stable fcc ferromagnets. Particularly, for e/a around 9 the energy difference ΔE between the FM (or HS) and NM states is large and positive (8–12 mRy). The critical pressure for transformation ($p_c = -2\Delta E/\Delta V$ [8]) from the FM to the NM state is also large. The systems thus show no or small magnetovolume effects.

2) In the range $8.5 \leq e/a \leq 8.7$ the energy difference ΔE is still positive, i.e. the systems are ferromagnets, but absolute values of ΔE and p_c are small. The systems therefore show large moment–volume instabilities (large spontaneous volume magnetostriction), equivalent to the occurrence of the Invar-effect.

3) On further reduction of the electron concentration into the range $e/a \leq 8.5$ two cases are to be distinguished:

i) Both ΔE and p_c approach zero, which would lead to even larger magnetovolume effects than in the Invar range. However, the fcc lattice cannot "bare" a volume increase larger than $\approx 2\%$ as observed in Invar. Consequently, when reducing e/a below 8.5, the systems undergo a structural phase transition from the dense packed fcc phase into the more open bcc structure (martensitic γ–α transition; see section 3.3 below).

ii) Reducing e/a by alloying Mn ($e/a = 7$) to e.g. FeNi leads to an increase in the lattice constants of the alloys, because of the large atomic volume of Mn. As a result, the magnetovolume effects are reduced and the γ–α transitions in the range around $e/a \approx 8.5$ are prevented. Instead, since Mn atoms tend to AF coupling, mixed exchange interactions oc-

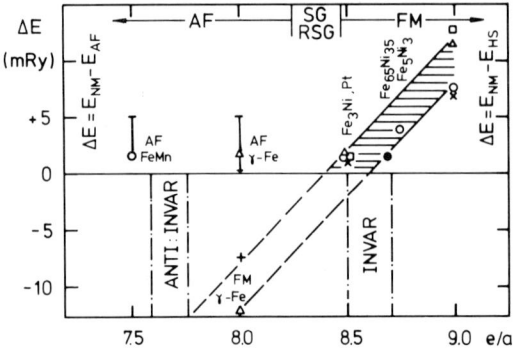

Fig. 3. Energy differences between non-magnetic (NM) and high spin (HS) instability states, $\Delta E = E_{NM} - E_{HS}$, for FM order (right hand scale) and between NM and antiferromagnetic ground states for AF order, $\Delta E = E_{NM} - E_{AF}$ (left hand scale) as a function of the electron concentration as calculated by different groups using various theoretical methods for pure FM γ-Fe (open triangle [9], plus sign [13]), the FeNi series (open triangles [4], open dots [5], open squares [8]), cross signs [14] and disordered $Fe_{65}Ni_{35}$ (full dot [7]) and Fe_3Pt (same symbol as for Fe_3Ni [6])). The FM data are interpolated linearly. The alloys are structurally stable within the hatched area. For $8.5 \leq e/a \leq 8.7$, where ΔE is of the order of 1 mRy we observe theoretically the moment–volume instabilities (cf. figs. 1 and 2) and experimentally (cf. fig. 4) the Invar effect. In the AF-range theoretical results are scarce. In AF γ-Fe (open triangle [9] and AF FeMn with CuAu-structure (open dot [16]) ΔE-values vary as a function of the lattice constant as indicated by the respective vertical bars in the figure. Experimentally, "Anti-Invar" is found in the range $7.55 \leq e/a \leq 7.75$ (cf. fig. 4) as indicated by the vertical dashed lines.

cur and one observes reentrant spin glass (RSG) or pure spin glass (SG) order for $e/a \approx 8.3$–8.5.

4) For alloys with $e/a < 8.3$ antiferromagnetic (AF) order is observed experimentally (cf. fig. 4a). However, it is difficult to provide correctly for AF within the FSM calculations, so that for $e/a < 8.3$ data are scarce. Moruzzi et al. [9] give results for pure AF γ-Fe ($e/a = 8$) and AF γ-Mn ($e/a = 7$). According to their work for fcc Fe, the AF-state forms the stable ground state for 2.54 au $\leq r_{WS} \leq 2.71$ au. Within this range, the energy difference $\Delta E = E_{NM} - E_{AF}$, which is plotted in fig. 3 (see left hand scale) varies from $\Delta E = 0$ for the equilibrium (zero pressure) lattice constant at $r_{WS} = 2.54$ au $= 3.44$ Å to $\Delta E \approx 5$ mRy at $r_{WS} = 2.71$ au $= 3.67$ Å. At the experimental lattice constant for fcc Fe $a = 3.58$ Å $= 2.64$ au, there is $\Delta E = E_{NM} - E_{AF} \approx 2$ mRy (see triangle in fig. 3). For γ-Mn it was shown in ref. [9] that the AF-state forms the stable ground state for all lattice constants. At the (extrapolated) experimental lattice constant of $a = 3.67$ Å, one would read $\Delta E = E_{NM} - E_{AF} \approx 5$ mRy from the data in ref. [9].

Recently, Jepsen and Herman [15] have performed linear muffin-tin-orbital band-structure calculations as a function of the lattice constant for a set of 15 ordered and disordered FeNiMn compounds with FM and AF order and fcc structure. Their results show that in the AF-range, depending on the atomistic environment, both Fe- and Mn-atoms can experience a moment–volume instability. A rigid band picture as applied above in the FM-range is thus not applicable for the AF alloys.

A first preliminary result of a total energy calculation with the FSM-method for fcc AF Fe_2Mn_2 ($e/a = 7.5$) was very recently accomplished by Podgorny [16]. He shows that for this ordered compound the Fe- and Mn-moments as a function of the lattice constant indeed behave very complex. A minimum of $\Delta E = E_{NM} - E_{AF} \approx 1$ mRy is found where, however, only the Fe-moments vanish while the Mn-atoms still have their full moment. The moment–volume instability observed in Fe_2Mn_2 is thus due mainly to the weakening of the AF interaction of the Fe-atoms.

This result can be corroborated experimentally (see below, section 4). In fact, $Fe_{50}Mn_{50}$ is an antiferromagnetic Invar alloy ("Anti-Invar") with a spontaneous volume magnetostriction of $\mu_{so} = 0.5 \times 10^{-2}$, calling for a small ΔE between the instability states. Moreover, $Fe_{50}Mn_{50}$ undergoes a structural γ-ϵ martensitic transition at low temperatures (≈ 80 K) [17], supporting our idea that when ΔE is small the systems are also close to a structural instability.

In total, the behavior of the AF alloys though more complex in detail, is analogous to the $\Delta E(e/a)$ behavior of the FM alloys. However, more theoretical data are necessary for AF alloys and compounds, and the discussion has shown that the rigid band picture, i.e. the dependence of the physical properties on e/a alone, is not sufficient to describe the situation in the AF range. We will see later that this difficulty can be removed when the physical properties of the AF-alloys along with those of the FM-alloys are analysed as a function of the lattice constant or atomic volume. This will allow much deeper insight and further going general conclusions (see section 4, below).

Finally, we have so far omitted the fact that Invar is a finite temperature effect, but the band structure calculations are done for $T = 0$. Though finite temperature calculations are at their beginning [8], it is beyond dispute that the experimentally observed broad spectrum of physical anomalies in 3d-transition metal alloys at finite temperatures also finds its origin in the existence of moment–volume instabilities. The physical nature of the excitations from the ground state in these systems is not yet fully understood. There are transverse spin fluctuations (spin waves as revealed by neutron scattering [1]) *and* longitudinal spin fluctuations coupled to fluctuations of the volume. The existence of these magnetoacoustic ("forbidden") modes has recently been demonstrated in neutron scattering experiments [18].

3.2. Comparison with experiments

The correctness of the above observations (1)–(4) is readily proven from a comparison with experimental data. For this reason we have plot-

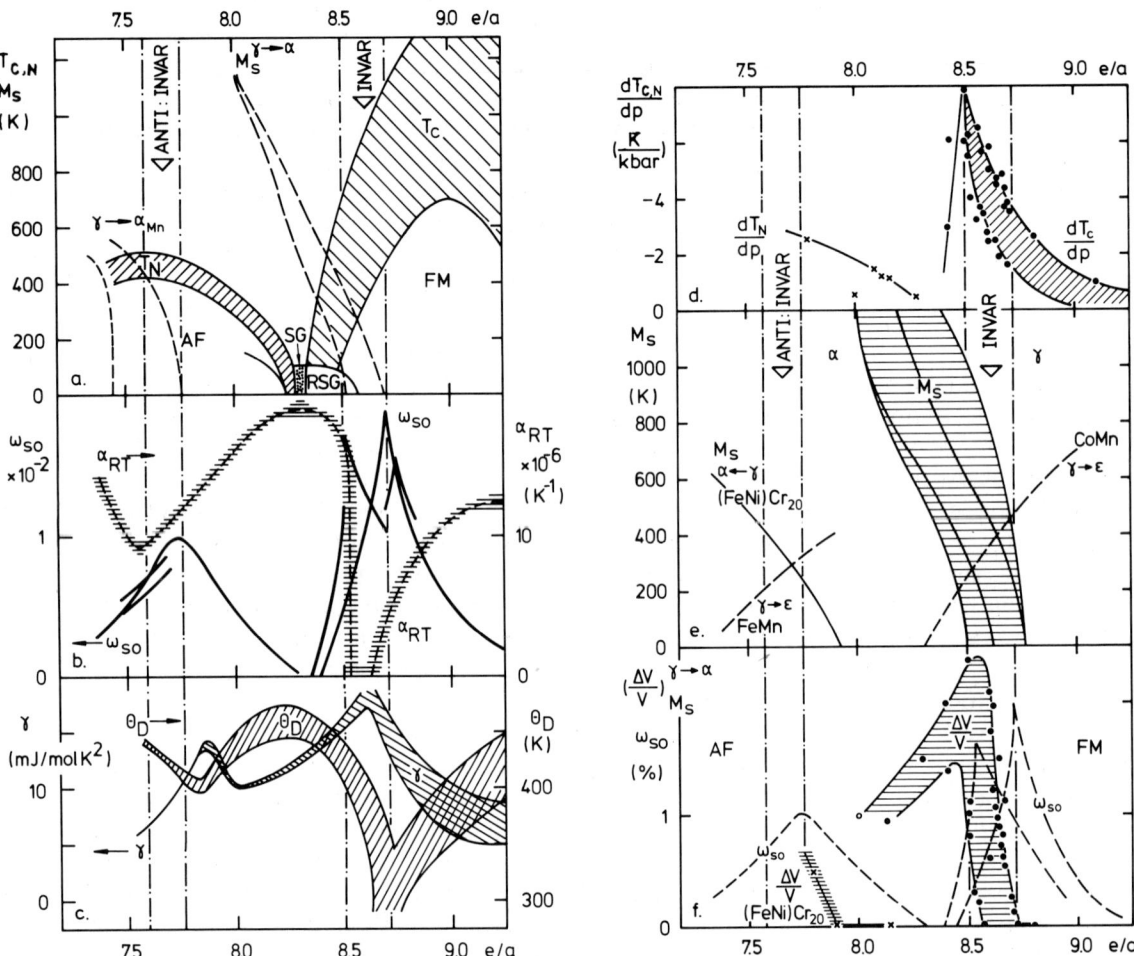

Fig. 4. Various experimentally determined physical quantities as a function of the electron concentration per atom (e/a) for fcc alloys in the 3d-series (binary systems FeMn, FeCo, FeNi, FePt, FePd; ternary systems FeNiCr, FeNiMn, FeNiCo). Data points are omitted for clearness (for details see ref. [1]) and only hatched areas are shown in which the experimental points lie. (a) Curie-temperatures (T_c) for ferromagnets, Néel-temperatures for antiferromagnets, reentrant-spin-glass range and pure spin-glass range (dotted) versus e/a. Validity ranges of the martensitic start temperatures for structural transition from the fcc γ-phase into the bcc α-phase, $M_S^{\gamma-\alpha}(e/a)$, are also given. In the AF range, structural transitions from the γ-phase to α_{Mn} are indicated. (b) Spontaneous volume magnetostriction extrapolated to zero temperature, $\omega_{so} = (\Delta V/V)_{T \to 0}$ and thermal expansion coefficient at room temperature $\alpha_{RT} = [d(\Delta l/l)/dT]_{T=300K}$ versus e/a. ω_{so} is maximum ($\approx 2\%$) in the FM-Invar range around $e/a \approx 8.6$–8.7, and simultaneously α_{RT} very small or even zero. Note that a similar behaviour of ω_{so} and α_{RT} is observed in the AF-Invar range around $e/a = 7.7$. (c) Debye-temperatures, Θ_D, and electronic γ-terms, both from specific heat measurements versus e/a. In analogy to the behaviour of $\omega_{so}(e/a)$ and $\alpha_{RT}(e/a)$ in (b), note the softening of the lattice (decrease of Θ_D) and increase in the electronic γ-term (proportional to the density of states) in both the FM- and AF-Invar ranges. (d) Pressure dependence of the Curie- and Néel-temperatures, $dT_{c,N}/dP$, versus e/a for FM and AF Invar systems. (e) Martensite starting temperatures M_S at which the structural fcc–bcc or $\gamma-\alpha$ transitions occur in FM alloys (hatched area, from ref. [19]) and AF (FeNi)$_{80}$Cr$_{20}$ [21] versus e/a. Note that the fcc–hex or $\gamma-\epsilon$ martensite transition lines in FM CoMn [21] and AF FeMn [20] run in an opposite sense as compared to the $\gamma-\alpha$ transitions. (f) Spontaneous volume magnetostriction ω_{so} (dashed curves, repeated from (b)) and relative volume change between the γ- and α-phase at the martensic start temperature M_S for FM alloys [19] (full dots and hatched area), pure Fe at the A_3-point [28] (open dot) and AF (FeNi)$_{80}$Cr$_{20}$ [19] versus e/a. Note the counterbalancing dependence between the decrease in ($\Delta V/V$) and increase in ω_{so} and vice versa.

ted in figs. 4a–f various experimentally determined quantities as a function of e/a for the systems FeNi, FeCo, FeMn, NiMn, FePt, FePd, FeNiCr, FeNiMn and FeNiCo. Data points are omitted for clearness, but have been given earlier in detail [1]. We therefore show only hatched regions in which the experimental points lie.

In fig. 4a we present the magnetic ordering temperatures (Curie-temperatures T_c for the ferromagnets, Néel-temperatures T_N for the antiferromagnets, glass-temperatures T_g for the reentrant- and pure spin glasses) as well as the martensitic start temperatures $M_s^{\gamma-\alpha}$ for the onset of structural $\gamma-\alpha$ transitions (dashed). The latter will be discussed below. In accordance with the above statements, there is pure FM order present for $e/a \geq 8.6$, pure itinerant SG order (dotted region) around $e/a \approx 8.3$ and AF order for $e/a \leq 8.3$. The spontaneous volume magnetostriction extrapolated to zero temperature, ω_{so} in fig. 4b peaks in the FM Invar range around $e/a \approx 8.7$ and in the AF Invar range at $e/a \approx 7.7$. Oppositely, the thermal expansion coefficient at room temperature, α_{RT}, in fig. 4b has minima, when ω_{so} peaks, and vice versa. Note further that around $e/a = 8.3$ in the SG-range ω_{so} vanishes and α_{RT} reaches values of almost 20×10^{-6} K^{-1}. As shown in fig. 4d, the maxima in ω_{so} (minima in α_{RT}) as a function of e/a are accompanied by maxima in the pressure dependence of the Curie-(Néel)-temperatures $dT_{C,N}/dp$.

As can be seen from fig. 4c, the extreme magnetovolume effects in the 3d-alloy series are accompanied by a softening of the lattice. The Debye-temperatures Θ_D as a function of e/a decrease around $e/a = 8.7$ and 7.7, respectively. Simultaneously, fig. 4c reveals that the γ-terms of the electronic specific heat determined at low temperatures are only slightly enhanced with respect to "normal" behavior, but clearly increase in the critical e/a-regions.

3.3. Martensitic transitions

As mentioned in the above statement (3i), the FM alloys not containing Mn undergo a martensitic phase transition from the γ (fcc) to the α (bcc) structure as shown by the martensitic start temperatures M_s versus e/a in fig. 4e (hatched region, from ref. [19]). $\gamma \to \alpha$ M_s-transitions are also observed for some AF-alloys. Data are scarce, however, in the literature and we show as an example in fig. 4e results for (FeNi)Cr$_{20}$ [20]. FeMn alloys undergo a martensitic transition from the γ-phase to the hexagonal ϵ-phase [17]. The same type of transition is observed in FM CoMn [21]. Note that the $\gamma-\epsilon$ M_s-curves in fig. 4e run in an opposite sense as compared to the $\gamma-\alpha$ transitions. The reason for that is not understood at present.

In the FM-alloys, the martensitic $\gamma-\alpha$ transitions occur in a well defined, narrow range of the electron concentration ($8.5 \leq e/a \leq 8.7$ for $T = 0$). This reminds of the Hume–Rothery rules, well established for non-magnetic alloys. While, however, non-magnetic alloys (e.g. CuZn) undergo an fcc–bcc structural transition with *increasing* electron concentration, the magnetic alloys undergo the same transition with *decreasing* electron concentration (the γ-phase is stable to the right hand side of the $M_s(e/a)$-curves in fig. 4e). We have recently explained this observation within the rigid band picture [19].

Technically, M_s-transitions at finite temperatures are of substantial importance in the so-called "shape memory alloys (SMA)" [22]. With respect to a change in temperature for constant e/a, one observes in non-magnetic SMA (CuZn, CuZnAl, NiTi, NiTiAl) a martensitic transition from a high temperature bcc-phase (which is mostly ordered) into a low temperature fcc-phase at the respective M_s. In magnetic SMA (FePt, FePd, FeNiAl, FeNiCoTi, FeMnSi), which show the Invar-effect in certain concentration ranges, just the opposite behavior is found. From fig. 4e it is apparent that the M_s-transition takes place from a high temperature (mostly disordered) fcc-phase into a low temperature bcc-phase. The bcc–fcc transition in the non-magnetic SMA is accompanied by a decrease in the volume of the lattice. In magnetic SMA the fcc–bcc transitions are accompanied by a volume increase, which depends on the concentration in such a way that in the Invar-range the volume change at the martensite temperature, $(\Delta V/V)_{M_s}^{\gamma-\alpha}$ vanishes. This is seen from fig. 4f, where $(\Delta V/V)$ at M_s is

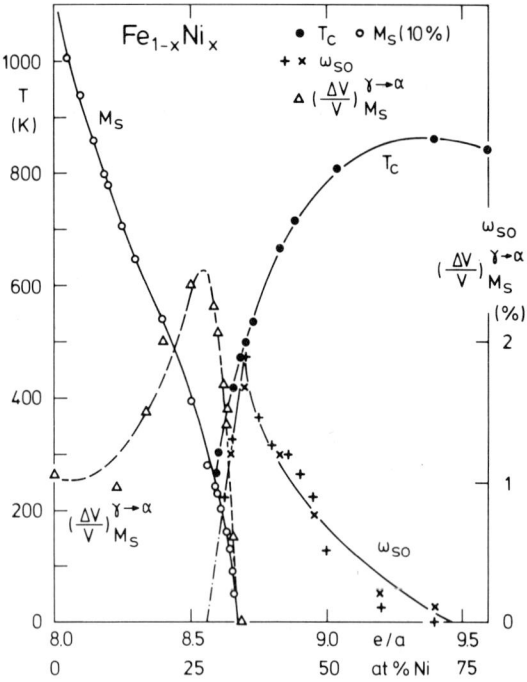

Fig. 5. Concentration dependence of the Curie-temperatures T_c [23], spontaneous volume magnetostriction ω_{so} [24,25], martensite starting temperature for 10% volume fraction of α-martensite M_S [26] and relative volume change at M_S $(\Delta V/V)_{M_s}^{\gamma-\alpha}$ [27] in the Fe–Ni system and in pure Fe at the A_3-point (1178 K) [28]. For details see text.

plotted versus e/a (dots and hatched region, for references see ref. [19]). Also shown again in fig. 4f are the respective, Invar typical spontaneous volume magnetostrictions $\omega_{so}(e/a)$ (dashed lines). Note the counterbalancing effect, namely that when ω_{so} is maximum ($\approx 2\%$), $(\Delta V/V)_{s}^{\gamma-\alpha}$ is zero, and when $\omega_{so} \to 0$, $(\Delta V/V)_{M_s}^{\gamma-\alpha}$ rises to about the same absolute value ($\approx 2.4\%$). In the AF-range, possibly similar effects occur, but again due to lack of data the situation is not as clear as in the FM-range.

While the presentation in fig. 4f is more meant to exhibit the general trends, we show in fig. 5 the results for one system, $Fe_{1-x}Ni_x$, to demonstrate the importance of the Invar-effect (or better moment–volume fluctuations) on the occurrence of the structural transitions in magnetic 3d-systems. Starting from the right hand side in fig. 5 at $e/a = 9.5$ (FeNi$_3$), we can see that the Curie-temperatures (full dots, from ref. [23]) start to decrease with decreasing e/a and simultaneously the spontaneous volume magnetostriction ω_{so} (crosses from ref. [24]; plus signs from ref. [25]) increases until it reaches a maximum of 1.95% in the Invar alloy $Fe_{65}Ni_{35}$ ($e/a = 8.7$). On further increase of the Fe concentration, ω_{so} starts to drop steeply along with a further decrease of T_c. Simultaneously, for $e/a \leq 8.68$ ($Fe_{66}Ni_{34}$) a martensitic transition starts to occur at low temperatures (open dots, from ref. [26]). M_s-values rise steeply with decreasing e/a, and for $e/a = 8.6$ ($Fe_{70}Ni_{30}$) we reach $M_s = T_c = 250$ K. Note the counterbalancing effect between the steep drop in ω_{so} and the sharp rise in the volume change (open triangles in fig. 5; from ref. [27]) $(\Delta V/V)$ at M_s accompanying the martensitic transition. Extrapolating ω_{so} to zero (dashed dotted line in fig. 5), i.e. into a range, where the γ-phase is not stable any more, we observe that a maximum in $(\Delta V/V)_{M_s}$ is reached for $e/a = 8.56$ ($Fe_{72}Ni_{28}$) when $\omega_{so} = 0$. For $e/a < 8.56$, $(\Delta V/V)_{M_s}$ drops again. In pure Fe ($M_s = 1178$ K), the volume change accompanying the γ–α transition at high temperatures is about 1%. Analogous behavior as in fig. 5 is found, if the respective data for e.g. FePt, FePd, FeNiCo$_{10}$ and FeNiCo$_{30}$ i.e. the magnetic SMA are analyzed in the same fashion.

The long known observation that the Invar-effect in any FM system is always found close to the γ–α stability limit has thus good reasons. The moment–volume fluctuations, causing the Invar-effect, also drive the system into the less dense-packed bcc structure. Sometimes a structural sequence fcc–fct–bcc is observed (e.g. FePd), but principally the behavior of all systems is similar.

A first theoretical reasoning to understand these structural transitions has been given through total energy calculations for Fe_3Ni with fcc and bct structure, respectively [29], using the Bain distortion [30] as the underlying microscopic transformation mechanism from fcc to bct. From these (preliminary) results one can see that for all volumes smaller than the equilibrium value of 74.25 au^3 (equivalent to $a = 3.52$ Å), always $E_{fcc} < E_{bct}$ as expected, since the fcc-phase is a stable phase for Fe_3Ni theoretically (not experimentally; cf. fig. 5). For large atomic volumes one can extrapolate that $E_{bct} < E_{fcc}$, since the $E_{bct}(V)$ drops much faster with increasing atomic volume V than $E_{fcc}(V)$. Fe$_3$Ni therefore not only shows Invar behavior, because the energy difference $\Delta E = E_{LS} - E_{HS}$ (cf. fig. 3) is small, but lies also close to a structural instability, because the energy difference $\Delta E = E_{bct} - E_{fcc}$ is small too.

In total, undoubtedly the Invar-effect and the martensitic $\gamma - \alpha$ transitions in fcc magnetic SM-alloys find their origin in the presence of the moment–volume instability inherent to the Fe-atoms. This is contrasted by the behavior of the non-magnetic SMA, in which it is apparent from high temperature neutron scattering [31] that damped phonons are responsible for the martensitic transformations.

4. Instabilities as a function of the lattice constant

4.1. Magnetic ordering temperatures versus $a_{4.2}$

In fig. 6 we have plotted Curie-temperatures T_c (positive ordinate) and Néel-temperatures T_N (negative ordinate) as taken from the magnetic phase diagrams published in ref. [1] (where all references are given) versus the lattice constants at 4.2 K for a broad variety of binary and ternary systems with fcc structure. Lattice constants are mostly taken from Pearson [32] and corrected for thermal expansion. The general trend as depicted in fig. 6 does, however, not change very much, if the ordering temperatures $T_{c,N}$ are plotted versus the respective lattice constants taken at room temperature.

The overall behavior of $T_{c,N}(a_{4.2})$ in fig. 6 on first glance is surprising. Almost all data (besides for some compositions in the ternaries FeNiPd, FeNiPd) form a continuous curve with oscillating character, reminding us of the Rudermann–Kittel–Kasuya–Yoshida (RKKY) type dependence. Starting on the left hand side, we can see that the ferromagnetic (FM) interactions, i.e. $T_c(a_{4.2})$ in the binaries FeNi, NiMn, CoMn and the ternary systems FeNiMn, FeNiCr and FeCoMn decrease in a narrow range of lattice constants and drop to zero at a value of $a_{01} = (3.57 \pm 0.01)$ Å. Note that for $T_c < 500$ K the curve $T_c(a_{4.2})$ for the Invar-systems FeNi, FeNiMn, FeNiCo and FeNiCr has positive slope. This reflects the negative pressure dependence of the Curie-temperatures of Invar (cf. fig. 4d), since T_c decreases when the lattice constant is reduced. Of importance with respect to the above discussions (section 3.3) is also the fact that at certain compositions and therefore certain values of the lattice constants the systems undergo a martensitic $\gamma - \alpha$ transition. Those compositions are marked with an arrow in fig. 6.

A further interesting fact observed from the plot in fig. 6 is that for $T_c < 100$ K in a very narrow range of lattice constants around the critical value $a_{01} = 3.57$ Å only spin glass (SG) ordering is observed in all systems which are structurally stable there. Transition temperatures given are thus actually SG freezing temperatures T_f.

In the AF-range the $T_N(a_{4.2})$ data in fig. 6, which are plotted "downwards" to provide for the dominance of negative exchange interactions, also form a continuous curve, reflecting the itineracy of the magnetic behavior of these systems. T_N-values stem here from the AF-ranges of the ternary systems FeNiMn, FeCoMn, FeNiCr and the binary systems NiMn, CoMn and FeMn. The latter system orders solely antiferromagnetically and data are given within its fcc stability range from 20 to 60 at% Mn (see crosses in fig. 6). This stability range can be enlarged on the Fe-rich side by alloying carbon to FeMn (see T_N-values indicated by crosses with perdendicular bars in fig. 6 for $a_{4.2} < 3.62$ Å), and on the Mn-rich side by alloying with (≈ 5 at%)Cu (same symbols as for FeMnC), resulting in the decrease of T_N (increase in fig. 6) above the maximum (minimum in fig. 6) observed for Fe$_{50}$Mn$_{50}$ ($T_N = 515$ K).

The "Anti-Invar" alloys are found in the AF-branch with negative slope in fig. 6. This means that T_N decreases with decreasing lattice constant, or in other words, a negative pressure dependence of the Néel-temperatures, a salient feature of "Anti-Invar" as already quoted (cf. fig. 4d).

The T_N-value of pure antiferromagnetic γ-Fe lies close to the critical instability value a_{01}. One can therefore conclude (and theoretical results, see below section 4.2 will support this) that it is

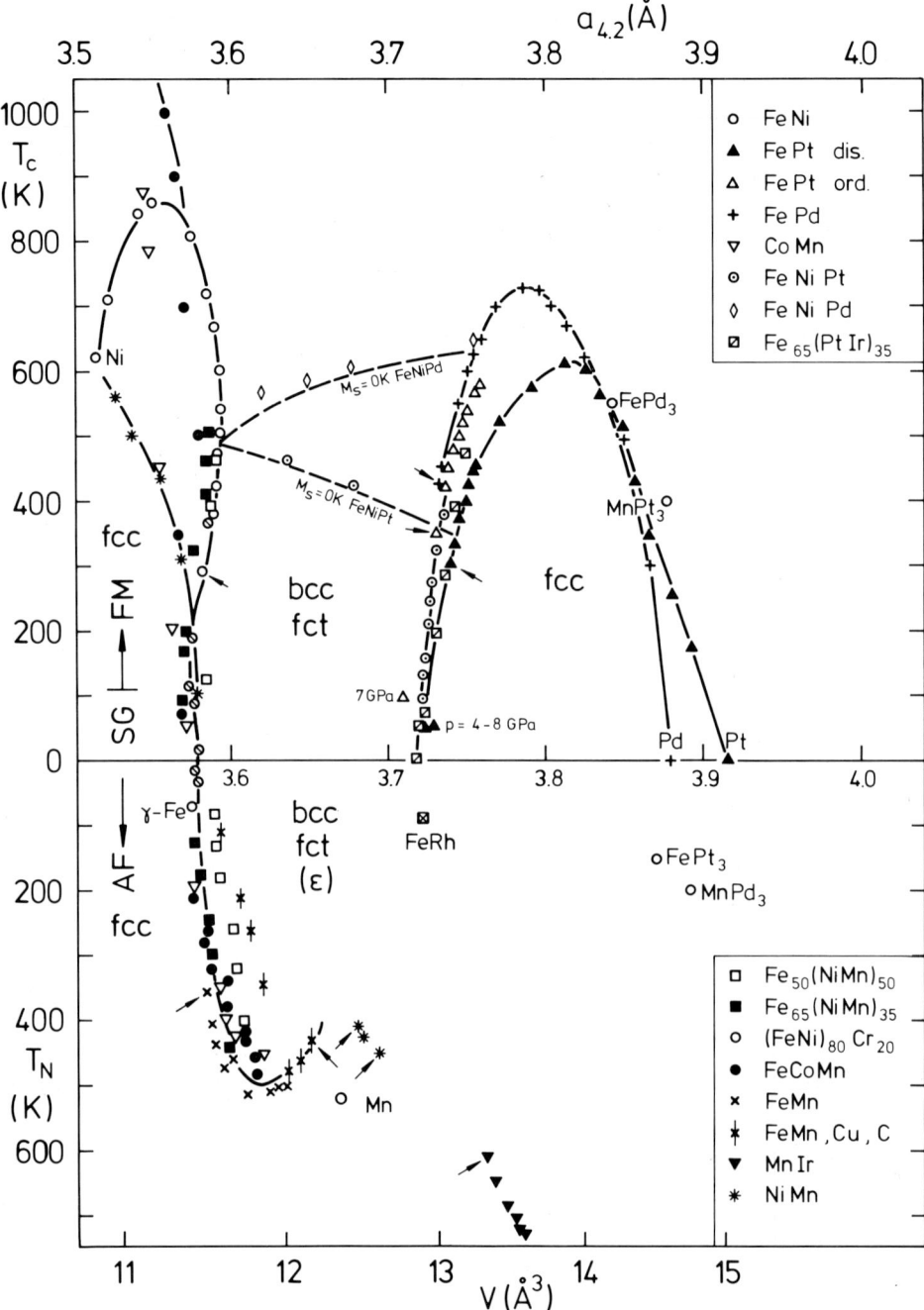

Fig. 6. Curie-temperatures (positive ordinate) and Néel-temperatures (negative ordinate) versus lattice constants at 4.2 K, $a_{4.2}$, for different alloy systems, characterized by the respective symbols given in the figure (all data from magnetic phase diagrams as published in ref. [1], where all the references are given). Note the oscillating RKKY-type behavior of all the data, with T_c and T_N reaching zero at the values $a_{01} = 3.57$ Å and $a_{02} = 3.72$ Å, respectively, where the γ-Fe and the γ-Mn moments vanish (cf. fig. 7a, b). At the respective compositions marked by an arrow, the systems undergo a martensitic fcc–bcc (or γ–α) transition, i.e. $T_c = M_s$ or $T_N = M_s$. Additionally the $M_s = 0$ K lines are given for the ternary systems FeNiPd and FeNiPt (dashed lines). In the region indicated by bcc, fct and ϵ, the fcc structure of all systems is instable. For details see text.

the instability of γ-Fe moment which dominates the magnetovolume effects of all FM and AF alloys in the left hand branch of the RKKY-type curve in fig. 6 which contain Fe. Surprisingly, the two FM systems, NiMn (see stars in fig. 6) and CoMn (downward open triangles in fig. 6) not containing Fe also have vanishing T_c for $a_{4.2} \to a_{01}$. This observation is not understood at present.

On the other hand, the T_N-values of AF NiMn and AF MnIr (downward closed triangles in fig. 6) obviously lie outside the general trend of the AF alloys as depicted in the figure. It must be therefore the AF γ-Mn moment which governs their magnetic behavior, but when extrapolated, T_N-values of both systems do not point to fcc γ-Mn. The Néel-temperature ($T_N = 530$ K) and lattice constant ($a_{4.2} = 3.67$ Å) of pure γ-Mn are extrapolated from investigations on MnCu alloys [33]. In nature, pure fcc Mn is only stable in the narrow range 1360–1445 K below the melting point.

For reasons of structural instability the rising branch of the $T_N(a_{4.2})$ curve between the minimum at around $T_N = 500$ K and antiferromagnetic fcc $Fe_{50}Rh_{50}$ ($a_{4.2} = 3.722$ Å, estimated value from $a_{RT} = 3.738$ Å [34], $T_N = 82$ K [35]), a metastable (quenched) phase of the preferably bcc ordering system $Fe_{1-x}Rh_x$, is missing. However, as one of the most important features of the plot in fig. 6, we observe that for lattice constants $a \geq a_{02} = (3.72 \pm 0.02)$ Å a new, continuous FM-branch of the RKKY-type curve sets in. Curie-temperatures in this branch stem from alloys of the systems FePt (ordered and disordered), FePd and ternaries like (FeNi)Pt and $Fe_{65}(PtIr)_{35}$. They rise to $T_c \approx 750$ K in FePd and then drop to zero for pure Pt and Pd at around $a_{4.2} = 3.88-3.92$ Å.

The well investigated Invar alloys $Fe_{72}Pt_{28}$ (ordered and disordered) and $Fe_{65}Pd_{35}$ can be found in the range of positive slope of the right hand FM-branch, close to the $\gamma - \alpha$ instability limits indicated by the arrows in fig. 6. When pressure is applied, $Fe_{72}Pt_{28}$ alloys (ordered and disordered), however, remain in the fcc-structure, but their Curie-temperatures drop drastically [12]. If we use the value $da/dp = 0.0064$ Å/GPa found experimentally for the pressure dependence of the lattice constant of $Fe_{72}Pt_{28}$ [36], which compares well with the theoretical value published for Fe_3Pt ($da/dp \approx 0.0067$ Å/GPa [6]), we arrive at the values for T_c given in fig. 6 for ordered $Fe_{72}Pt_{28}$ at 7 GPa: $T_c \approx 100$ K; $a_{4.2} \approx 3.71$ Å and for disordered $Fe_{72}Pt_{28}$ in the range 4–8 GPa: $T_c \approx 50$ K; $a_{4.2} = 3.72-3.73$ Å. These data fit well into the $T_c(a_{4.2})$ dependence found in (FeNi)Pt [37] and $(FePt)_{65}Ir_{35}$ [38], all together forming a continuous curve leading to a_{02} for zero moment or zero T_c, respectively. It is interesting to note that in $Fe_{65}(Pt_xIr_{1-x})_{35}$ for $x < 0.6$, SG-behavior is observed [38], which means that analogous to the behavior around the instability a_{01}, mixed exchange interactions also appear close to the instability around a_{02}. For $x = 0$ the alloy $Fe_{65}Ir_{35}$ shows almost zero susceptibility as a function of temperature [38], which means that the Fe-moment has vanished.

Alloys of FePt and FePd which are rich in the non-magnetic components Pt and Pd as well as the ordered FM compounds $FePd_3$ and $MnPt_3$ are found in the range between the maximum and $T_c = 0$ for pure Pt, Pd in fig. 6. They follow a $T_c(a_{4.2})$ dependence with negative slope, a behavior also observed in Co-rich ferromagnetic CoMn and FeCoMn, in Ni-rich FeNi and NiMn (see respective values in fig. 6) and in CoPt,Pd and NiPt,Pd (data not shown in fig. 6). The behavior is equivalent to a positive pressure dependence of the Curie-temperatures and reflects the local character of the magnetic moments of these systems in the concentration regions rich of the 4d(5d)-components.

Following through the total RKKY-type curve in fig. 6, we have thus a continuously changing magnetic behavior, with local moment character on either right and left hand side and itinerant behavior in the middle, where, however, the AF local branch is missing. In the itinerant range the T_c, T_N values reach zero, i.e. the moments vanish for a_{01} and a_{02}.

4.2. Magnetic moments as a function of the lattice constant

The $T_{c,N}(a_{4.2})$ behavior as depicted in fig. 6 and discussed in the preceding section can be

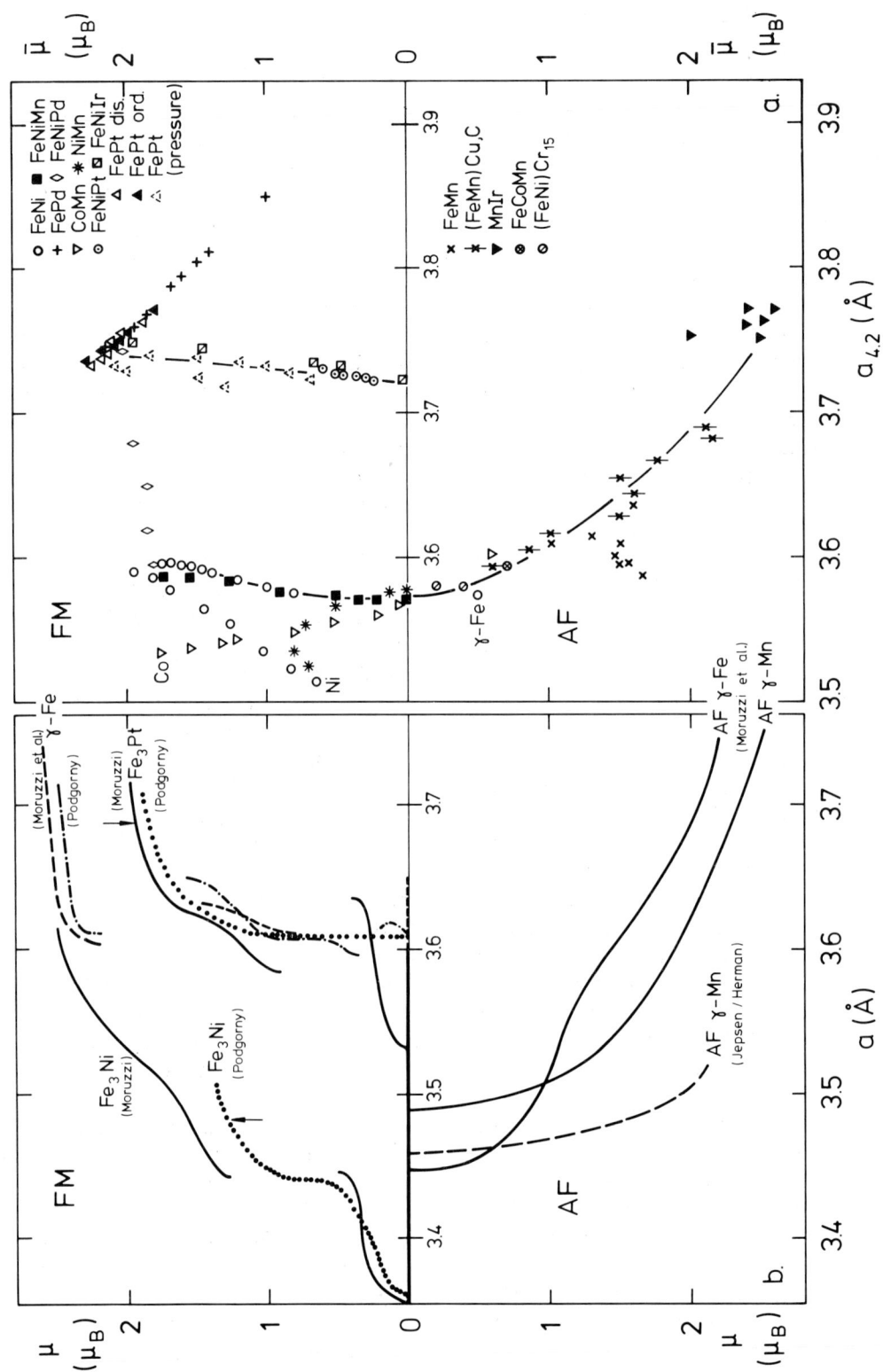

understood if we analyze in a similar fashion the available experimental data for the average magnetic moments of the respective alloy systems and compare them to the results of the theoretical FSM-calculations. The respective presentations are given in fig. 7.

Fig. 7a (on the right hand side) shows average moments $\bar{\mu}$ versus $a_{4.2}$ for ferromagnetically ordered alloys (upper half) and antiferromagnetic systems (results from neutron scattering) in the lower half. Same symbols as in fig. 6 are used for the different systems and references in detail have been given in ref. [1]. Grosso modo the plot $\bar{\mu}(a_{4.2})$ in fig. 7a is analogous to the plot $T_{c,N}(a_{4.2})$ in fig. 6. We see in fig. 7a that the average magnetic moments vanish at exactly the same critical value of $a_{01} = (3.57 \pm 0.01)$ Å for the FM alloys of the systems FeNi, NiMn, CoMn, FeNiMn as well as in the AF-ranges of FeMn, FeMnC, FeMnCu, FeCoMn and FeNiCr. A zero moment at $a_{02} = (3.72 \pm 0.01)$ Å is found in FePt (under pressure), (FeNi)Pt and $Fe_{65}(PtIr)_{35}$. Note that when comparing figs. 6 and 7a, for FePt (ordered and disordered) as well as FePd the moment still rises in a range of lattice constant 3.73 Å $\leq a_{4.2} \leq$ 3.8 Å, i.e. in the Invar-range, where the ordering temperatures T_c already decrease. This constant rise of $\bar{\mu}$ with decreasing $a_{4.2}$ expresses in a different fashion the absence of the deviation of the moment from the Slater–Pauling curve, well known for FePt and FePd in the Invar concentration range. It contrasts the behavior of FeNi (and other) Invar systems in the left hand branch of fig. 7a, where deviations of the moment from the Slater–Pauling curve are found and thus $\bar{\mu}(a_{4.2})$ and $T_c(a_{4.2})$ (c.f. fig. 6) decrease simultaneously on approach of the instability range.

Explanation for all these experimental observations can now be achieved when we look at fig. 7b, where the theoretical data for the moments versus lattice constants (these are in fact $a_0(T=0) = a_0$ values) are presented. Results are given for the behavior of the Fe-moment in FM Fe_3Ni [4,5], pure FM γ-Fe [5,9], Fe_3Pt [6,39] and AF γ-Fe [9] as well as AF γ-Mn [9,15]. The accordance between these theoretical data as depicted in fig. 7b and the experimental results in fig. 7a is indeed striking. For the first time a closed description of the moment–volume behavior of purely 3d and 3d–4d,5d alloys can be given. We note that in Fe_3Ni both theoretical results in fig. 7b predict that the FM Fe-moment is instable and drops discontinuously to a LS state at a lattice constant of $a_0 = (3.45 \pm 0.05)$ Å. At about the same value a_0, the AF γ-Fe and γ-Mn moments [15] vanish. This value of a_0 has to be compared to $a_{01} = 3.57$ Å as determined from the experiments and – though similiar behavior is observed – leads to a difference in the absolute value for the critical lattice constant for zero moment of $\Delta a_0 \approx 0.12$ Å, equivalent to $\approx 3.4\%$. Note that the second instability theoretically found for AF γ-Mn at $a_0 = 3.49$ Å [9] as well as the calculated $\mu(a)$ curve for AF γ-Mn could possibly explain the $T_N(a_{4.2})$ dependence of NiMn and MnIr, which lie outside the general RKKY-trend in fig. 6 (but not in fig. 7a!).

As shown in fig. 7b the moment of pure FM γ-Fe becomes instable in the same range of lattice constant as in Fe_3Pt, at around $a_0 = (3.61 \pm 0.01)$ Å. Undoubtedly, this instability is equivalent to the one found in the experimental alloy systems at $a_{02} = 3.72$ Å, though again there is a difference in absolute values between the theoretical and the experimental result of $\Delta a_0 = 0.11$

Fig. 7. (a) Average magnetic moments $\bar{\mu}$ for ferromagnetic systems (upper half) and antiferromagnetic systems (lower half) versus the lattice constant at 4.2 K (all data from ref. [1]). The different symbols for the respective alloy systems are the same as in fig. 6. Results for FePt alloys under pressure (from ref. [12]) are indicated by dashed triangles. Note that the moments vanish at the same lattice constants $a_{01} = 3.57$ Å and $a_{02} = 3.72$ Å, where the Curie- and Néel-temperatures in fig. 6 reach zero values. (b) Magnetic moments versus lattice constants for Fe_3Ni, Fe_3Pt, pure ferromagnetic γ-Fe and antiferromagnetic pure γ-Mn as calculated by Moruzzi [4,39], Moruzzi et al. [9], Podgorny [5,6] and Jepsen and Herman [15]. Note that the moments are instable with respect to small changes in the lattice constant in two regions, like in (a). However, the absolute values of the lattice constants, where the moment–volume instabilities occur, are theoretically about 3% smaller than the experimentally determined values.

Å or ≈ 3%. Theoretical curves in fig. 7b and experimental curves in fig. 7a when superposed thus perfectly match, but there is a difference in absolute values of the lattice constant of ≈ 3%. The values of the theoretically calculated lattice constant are therefore generally ≈ 3% too small. This does not sound very much but indeed means "water in the wine" with respect to the otherwise observed coherence between theory and experiment. The reason is probably that the exchange correlations and consequently the adhesive energy in the theoretical models are too strong. More theoretical data especially for AF systems are also necessary to achieve a better understanding.

We finally turn to a question for which a theoretical answer is also still lacking. It concerns the difference in the finite temperature behavior (or ground state excitations) in weakly FM $Fe_{65}Ni_{35}$ Invar on the one hand and strongly FM $Fe_{72}Pt_{28}$ on the other. In the e/a-plots (fig. 4) both systems show magnetovolume anomalies at almost equal electron concentration. When analyzed, however, as a function of the atomic volume (figs. 6 and 7) the differences between the two systems become obvious.

Of special interest within this context are alloy series of the ternary systems FeNiPd and FeNiPt. The Curie-temperatures of $Fe_{65}(Ni_xPd_{1-x})_{35}$ with $e/a = 8.7 = $ const, for $0 \le x \le 1$ almost follow the transition curve of $M_s(e/a) = 0$ K, which is found for $e/a \approx 8.68 = $ const (see dashed line in fig. 6) in this system. A similar situation is observed for $Fe_{65}(Ni_xPt_{1-x})_{35}$ ($e/a = 8.7$), where for $0 \le x \le 1$ the Curie-temperatures remain almost constant (≈ 500 K). The $M_s = 0$ K line, however, does not follow $e/a = $ const here, but starts at $Fe_{66}Ni_{34}$ ($e/a = 8.68$) and is found at $e/a = 8.5$ in $Fe_{75}Pt_{25}$ (disordered; $T_c = 350$ K; see respective dashed line in fig. 6). Both systems thus form a "bridge" between the two moment-instability ranges as discussed above, and one can study continuously the changes in magnetic and magneto-volume properties from FeNi to the FePt Invar. This leads to interesting results concerning e.g. the temperature dependence of the thermal expansion as investigated for $Fe_{65}(Ni_xPd_{1-x})_{35}$ [40] and a series $Fe_{65}Ni_{35}$–$Fe_{75}Pt_{25}$ [41]. Without dis-

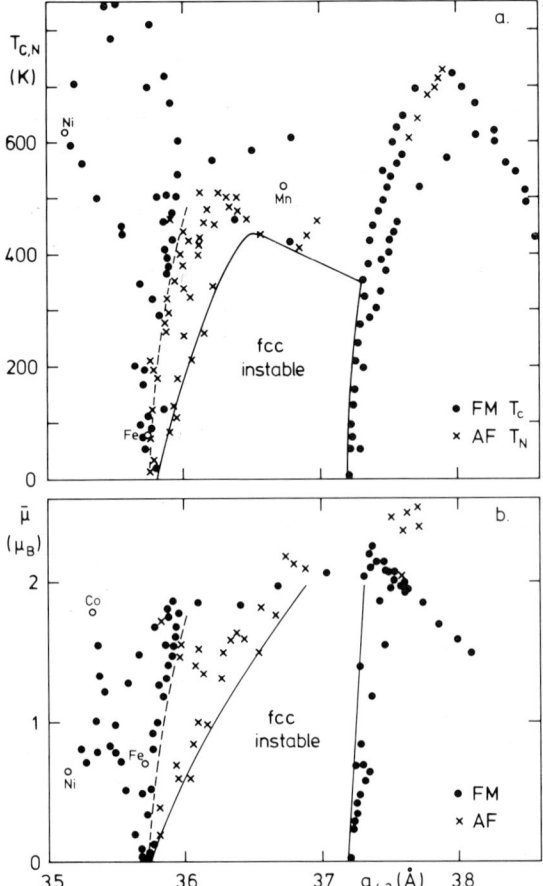

Fig. 8. (a) Experimentally determined Curie-temperatures T_c (full dots) and Néel-temperatures T_N (crosses) versus the lattice constant $a_{4.2}$. Actual data are the same as in fig. 6, however, without differentiating between all the alloy systems involved. Note the nearness of FM and AF order in the range around the instability $a_{01} = 3.57$ Å, while around $a_{02} = 3.72$ Å only FM order is observed. In the range closed in by the full lines the fcc structure is instable. (b) Experimentally determined average magnetic moments $\bar{\mu}$ versus the lattice constant $a_{4.2}$. Actual data are the same as in fig. 7a, again without differentiation between the respective alloy systems. FM ordering alloys are shown by the full dots, AF alloys by crosses. Note the analogy to the behavior as observed in (a).

cussing these results in detail here, within the framework of the present analysis at least some basic features responsible for the differences between $Fe_{65}Ni_{35}$ and $Fe_{75}Pt_{28}$ can be made plausible.

For this we show in figs. 8a, b the same data as

in figs. 6 and 7a, i.e. $T_{c,N}(e/a)$ and $\bar{\mu}(e/a)$, but now plotted in a different fashion. We have so to speak "folded up" in fig. 8 the AF-branches of figs. 6 and 7a and neglected the differences in symbols for the different alloy systems. In figs. 8a, b FM alloys are just indicated by full dots, AF alloys by crosses. One can now see that in the left hand instability range in figs. 8a, b where $Fe_{65}Ni_{35}$ (and other) Invar is found, ferromagnetism and antiferromagnetism lie closely together, while in the right hand instability range, where $Fe_{72}Pt_{28}$ (and $Fe_{65}Pd_{35}$) Invar is found, preferably ferromagnetism dominates at finite temperatures. We therefore are bound to say that the differences in magnetic excitations observed with increasing temperature in FeNi on the one hand and FePt on the other find their origin in the fact, that in FeNi we excite from a ferromagnetic high spin ground state to an antiferromagnetic low spin state, while in FePt the excitation is from a ferromagnetic high spin state to a ferrromagnetic low spin state. This difference, long debated from experimental side [42], has to be provided for in future finite temperature calculations. This type of calculations could also possibly give an answer why the fcc-phase is structurally instable in the respective range indicated in figs. 8a, b between the two instabilities at a_{01} and a_{02} and a temperature of roughly 400 K ($\approx$ 3 mRy). We think that a deeper theoretical understanding of Invar will very likely result from combining the band models mentioned and a molecular-orbital cluster type of calculation [43], in which the Invar-effect was explained to originate from the presence of strongly antibonding majority-spin orbitals and non-bonding minority-spin orbitals at the Fermi level in FeNi-clusters. Progress in this direction is under way [44].

Acknowledgements

The author is grateful for numerous valuable discussions with M. Acet, P. Entel, M. Podgorny and W. Pepperhoff. Work was supported by Deutsche Forschungsgemeinschaft within Sonderforschungsbereich (SFB) 166.

References

[1] E.F. Wassermann, Phys. Scripta T25 (1989) 209; in: Ferromagnetic Materials, vol. V, eds. K.H.J. Buschow and E.P. Wohlfarth † (North-Holland, Amsterdam, 1990) p. 240.
[2] ISOMES'89, Proc. Intern. Symp. on Magnetoelasticity and Electronic Structure of Transition Metals, Alloys and Films, Duisburg, Germany, March 1989, eds. E.F. Wassermann, K. Usadel and D. Wagner (North-Holland, Amsterdam, 1989) in Physica B 161 (1989).
[3] G. Dumpich, U. Kirschbaum, H. Mühlbauer, L. Vollbrandt and E.F. Wassermann, to be published.
[4] V.L. Moruzzi, Physica B 161 (1989) 99.
[5] M. Podgorny, Acta Phys. Polon. A 78 (1990) 941.
[6] M. Podgorny, Phys. Rev. B 43 (1991) 11300.
[7] D.D. Johson, F.J. Pinski, J.B. Staunton, B.L. Gyorffy and G.M. Stocks, Scandia Nat. Lab. Report SAND 89-8504 (1989).
[8] P. Mohn, K. Schwarz and D. Wagner, Phys. Rev. B 43 (1991) 3318.
[9] V.L. Moruzzi, P.M. Marcus and J. Kübler, Phys. Rev. B 39 (1989) 6957.
[10] M. Podgorny, Physica B 161 (1989) 105.
[11] R.J. Weiss, Proc. Roy. Phys. Soc. (London) 82 (1963) 281.
[12] M.M. Abd-Elmeguid and H. Micklitz, Phys. Rev. B 40 (1989) 7395.
[13] G.L. Krasko, Phys. Rev. B 36 (1987) 8565.
[14] E.G. Moroni and T. Jarlborg, Phys. Rev. B 41 (1990) 9600.
[15] O. Jepsen and F. Hermann, Phys. Rev. B 41 (1990) 6801.
[16] M. Podgorny, private communication.
[17] W. Stamm, H. Zähres, M. Acet and E.F. Wasserman, J. de Phys. 49 (1988) C8-315.
[18] P.J. Brown, I.K. Jassim, K.U. Neumann and K.R.A. Ziebeck, Physica B 161 (1989) 9.
[19] E.F. Wassermann, M. Acet and W. Pepperhoff, J. Magn. Magn. Mater. 90 & 91 (1990) 126.
[20] M. Acet, W. Stamm, H. Zähres and E.F. Wassermann, J. Magn. Magn. Mater. 68 (1987) 233.
[21] C. John, H. Zähres, M. Acet, W. Stamm, E.F. Wassermann and W. Pepperhoff, J. Appl. Phys. 67 (1990) 5268.
[22] E. Hornbogen, Metall 41 (1987) 488.
[23] J. Crangle and G.C. Hallam, Proc. Roy. Soc. London, Ser. A 272 (1963) 119.
[24] Y. Tanji and Y. Shirakawa, J. Japan. Inst. Met. (1970) 228.
[25] M. Hayase, M. Shiga and Y. Nakamura, J. Phys. Soc. Japan 34 (1973) 925.
[26] L. Kaufmann and M. Cohen, Trans. AIME 206 (1956) 1393.
[27] C.L. Magee and R.G. Davies, Acta Met. 20 (1972) 1031.
[28] S. Kachi and H. Asano, J. Phys. Soc. Japan 27 (1969) 536.
[29] P. Mohn and K. Schwarz, private communication.
[30] E.C. Bain, Trans. AIME 70 (1924) 25.
[31] A. Heiming, W. Petry, J. Trampenau, M. Alba, C. Herzig and G. Vogl, Phys. Rev. B 40 (1989) 11425.

[32] W.B. Pearson, Handbook of Lattice Spacings and Structures of Metals and Alloys, vol. 2 (Pergamon Press, Oxford, 1967).
[33] M. Acet, H. Zähres, W. Stamm, E.F. Wassermann and W. Pepperhoff, Physica B 161 (1989) 67.
[34] C.C. Chao, P. Duwez and C.C. Tsuei, J. Appl. Phys. 42 (1971) 4282.
[35] K. Sumiyama, M. Shiga and Y. Nakamura, Phys. Stat. Sol. (a) 13 (1972) K 75.
[36] G. Oomi and N. Mori, J. Phys. Soc. Japan 50 (1981) 2924.
[37] N. Kawamiya and K. Adachi, Trans. JIM 16 (1975) 327.
[38] M. Shiga, M. Kosaka and Y. Nakamura, Phys. Stat. Sol. (b) 50 (1972) 351.
[39] V.L. Moruzzi, private communication.
[40] A.K. Zatoplyaev, A.Z. Menshikov and S.M. Podgornych, Physica B 161 (1989) 25.
[41] A. Kussmann and K. Jessen, Archiv. Eisenhüttenwesen 29 (1958) 585.
[42] E.F. Wassermann, Adv. Solid State Phys. 27 (1987) 85.
[43] J. Kaspar and D.R. Salahub, Phys. Rev. Lett. 47 (1981) 54.
[44] P. Entel, private communication.

High energy spectroscopies and magnetism

R.J.H. Kappert, H.R. Borsje [1] and J.C. Fuggle

Spectroscopies of Solids and Surfaces, Research Institute for Materials, University of Nijmegen, Toernooiveld, NL-6525 ED Nijmegen, Netherlands

We present a review of high energy spectroscopy (HES) techniques in the field of magnetism, and show that the use of HES techniques can give valuable information on magnetic systems. There is a wide range of HES techniques, so that a large number of parameters relevant to magnetism can be probed. This review will present some illustrations thereof.

1. Introduction

The purpose of the present short review is to give an overview of the role played by HES in studies of magnetism. A list of the spectroscopies and their acronyms is given in table 1. We start by defining the term "high energy", which we will take to be the energy range for spectroscopies beyond the optical range (usually 10–200 eV). The range of significant in HES is very large, e.g., in dichroism experiments on the $M_{4,5}$-absorption edges of rare earths (RE) (see below) excitation energies of ≈ 1 keV are used, with a solution of ≈ 0.5 eV, but, as we will demonstrate, ground state effects in the meV range can be detected. The energy scales are high compared to the energy scales relevant to many magnetic phenomena, and this has led many people to doubt the relevance of HES to ground state properties. The reason for these doubts is not hard to see; the spectroscopies themselves and the theoretical methods required to understand the results are both complex and highly specialized. Nevertheless, the spectroscopies have begun to contribute significantly to our understanding of magnetic properties. Much of this understanding has not yet found its way to standard textbooks, which contributes to the apparent complexity and specialization of HES. In this paper we present an overview of the role HES can play in the field of magnetism, and show that methods not involving spin-polarized electrons can also contribute significantly to our understanding of magnetic materials. However, in many areas the reader will have to undertake further study of the literature for complete understanding of the issues involved.

Consistent with our aim of giving illustrations of the role of HES, we do not give an exhaustive account of the experimental methods, but instead use section 2 to give a brief indication of the nature of the measurements and the role they can

---

[1] Present address: DIMES, P.O. Box 5046, NL-2600 GA Delft, Netherlands.

Table 1
Acronyms of various HES spectroscopies

Acronym	Definition
AES [b]	Auger Electron Spectroscopy
PES [a,b]	Photoelectron Spectroscopy
UPS [a,b]	Ultraviolet Photoelectron Spectroscopy
IPES [a,b]	Inverse Photoemission Spectroscopy
BIS [b]	Bremsstrahlung Isochromat Spectroscopy
EELS [b]	Electron Energy Loss Spectroscopy
MXD	Magnetic X-ray Dichroism
MXS	Magnetic X-ray Scattering
SES [b]	Secondary Electron Spectroscopy
XAS	X-ray Absorption Spectroscopy
XPS	X-ray Photoelectron Spectroscopy

[a] When performed angle-resolved, place AR for acronym.
[b] When performed spin-polarized, place SP from acronym; spin-polarized BIS is abbreviated BISCEPS.

play. Then, in section 3, we give an account of the role played by the different techniques and spectroscopies in the study of some important questions. For convenience we will give our own work more prominence in this review than it deserves, because where it is possible to illustrate a point in this way it saves us a great deal of work. Most issues will be illustrated with examples of metallic magnets, but the material presented can generally be extended to include semiconductor and insulator systems as well.

1.1. Conceptual framework

A conceptual framework for many of the experiments we describe here is provided by figs. 1 and 2. First, fig. 1 gives the simplest possible band structure representation of a magnetic material. It is the picture of the Stoner model for itinerant electron magnets (e.g., the 3d-transition metals Fe, Co and Ni) [1,2]. In the Stoner model, the individual magnetic moments self-consistently constitute a mean magnetic field, causing the electron subband with electrons antiparallel to the mean field (spin-down or minority-spin electrons) to shift with respect to the subband with electrons parallel to the field (spin-up or majority-spin electrons), resulting in the so-called exchange splitting Δ. The shifting of the bands

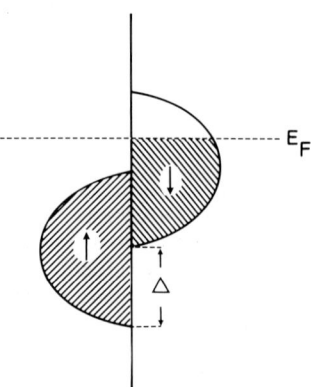

Fig. 1. Simple DOS of an itinerant electron ferromagnet. On the left, the spin-up d-band is depicted, which in this example is completely filled (strong ferromagnet, e.g., Ni; if the spin-up level is partially filled, as for Fe, the magnetic is called a weak ferromagnet). On the right the spin-down d-band is depicted. In this picture, the sp-band is neglected.

result in a redistribution of electrons among the spin-up (majority) and spin-down (minority) states. In order to preserve charge neutrality, the Fermi level has to shift accordingly. From the Stoner theory a criterion was derived, which can be used to calculate whether a material will exhibit spontaneous magnetization:

$$IN_0(E_F) > 1. \qquad (1)$$

Here, I is the intra-atomic exchange interaction and $N_0(E_F)$ is the paramagnetic density of states (DOS) at the Fermi level. Eq. (1) is known as the Stoner criterion, and describes the situation in which the magnetic state is energetically favourable. The shift of the bands results in a net spin-moment, because the occupation of spin-up and spin-down bands is no longer equal, and one may speak of an effective magnetic moment per atom μ.

This simple picture allows one to determine exchange splitting and effective magnetic moment per atom, but the picture is really to naïve, as we see when we compare with the results of a spin-dependent band structure calculation in fig. 2(a). This is a single-particle band structure in which the energies of the calculated bands are plotted as a function of the crystal momentum $\boldsymbol{k}$ for various directions in reciprocal space, with the spin-up bands as full lines and the spin-down bands as dashed lines. The associated DOS is shown in fig. 2(b), where the spin-down distribution is distinguished by the shaded area. Spin-up and spin-down contributions generally have different shapes, and it is clear that the exchange interaction does not lead to a "rigid shift". The exchange splitting is seen to be a function of $\boldsymbol{k}$, as for example in the case of Γ_1 band between Γ and H. The reason in this particular case is quite trivial. Near the Γ-point, the band has s-symmetry but near the H-point the band has mainly d-wave character and here the exchange splitting is larger. In other regions of the band structure the reasons for the variation in the apparent value of Δ may be more complex. A band structure, such as fig. 2(a) also contains other information relevant to the microscopic origin of magnetism: the dispersion of the bands relates to the

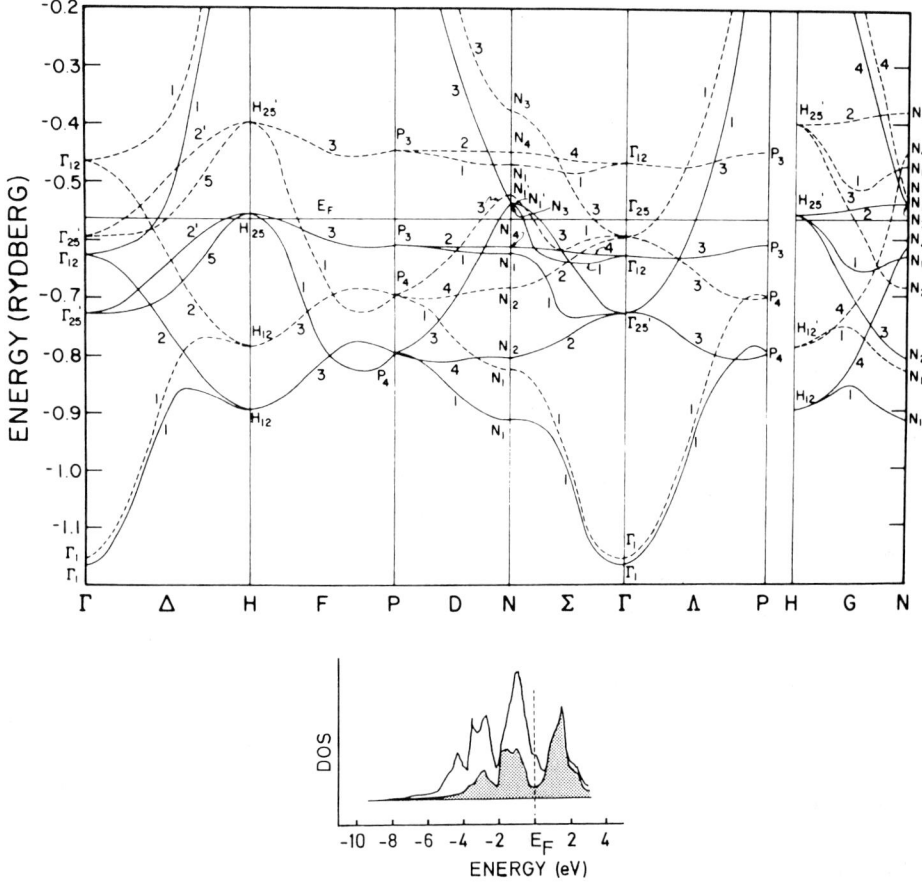

Fig. 2. (a) Energy bands of ferromagnetic iron calculated by Callaway and Wang [3]. Solid lines, majority-spin bands; dashed lines, minority-spin bands; (b) calculated DOS for Fe (solid line), adapted from ref. [4]. The shaded area corresponds to the spin-down DOS.

degree of localization; flat bands imply localized, atomic-like wave functions, whilst highly dispersed bands are a characteristic of delocalized wave functions. Experimental characterization of exchange splittings and band structure of the valence band is primarily a task for HES (sections 2.1 and 3.2).

Before proceeding further it is important to note a fundamental caveat in diagrams like fig. 2(a) which arises because, strictly speaking, the eigenenergies used to construct the diagram have no physical meaning and are not observables. The single electron generally interacts in a complex way with the other electrons. The Coulomb repulsion of the electron and the Pauli principle lead to a (spin-dependent) lower average electron density in the vicinity of the electron. One speaks of the exchange correlation hole of the electron, and the electron, together with its exchange-correlation hole, is described as a "quasiparticle". Quasiparticle energies do have physical meaning, and they are directly probed by HES. One often discusses the differences between single particle band structures and quasiparticle band structures in terms of self-energies, see, e.g., ref. [5]. In general, the self-energy is spin- and energy-dependent (and hence k-dependent for bands with significant dispersion). The self-energy has a real (shift) and imaginary (broadening) part, which means that one cannot simply relate energy differences of spectral peaks to those derived from figs. 2(a) or (b). We must stress that this is not a failure of HES, but simply because the single-particle eigenvalues have, strictly speaking, no physical meaning.

The concepts illustrated by figs. 1 and 2 do not exhaust the questions to which HES can be applied. For instance, nothing has been said about antiferromagnets and ferrimagnets, or the angular momentum contributions to the magnetic moment, the directions of local magnetic moments and the influence of crystal fields. Nevertheless we may now list some of the problems to which HES can contribute, i.e.,

1) What does the DOS look like and which states contribute near E_F?
2) How large is the effective Coulomb correlation?
(3) How large is the dispersion of the valence bands and are they more likely to be successfully described by localized or itinerant models?
4) How large are the exchange splittings and how do they vary as a function of energy?
5) How do the exchange effects vary as a function of temperature (particularly above the Curie and Néel temperatures)?
6) How important are the effects of self-energy?
7) What is the orientation of local moments?

2. Short description of experimental methods

2.1. (Inverse) photoemission of valence states

The occupied electronic states can be studied with photoemission. The unoccupied states, which are just as important for the physical properties of a solid [6], can be probed by inverse photoemission. Most of the principles needed to understand the spectra are common to both techniques.

In photoelectron spectroscopy (PES, in the ultraviolet (UV) region called UPS, in the X-ray region called XPS) [8–12] a sample is exposed to a beam of monoenergetic photons of sufficient energy $h\nu$ to ionize electrons from the solid (fig. 3(a)). The photoemitted electrons are subsequently energy analyzed. The binding energy BE of an electron, relative to the Fermi level, is related to the measured kinetic energy KE by:

$$KE = h\nu - BE. \qquad (2)$$

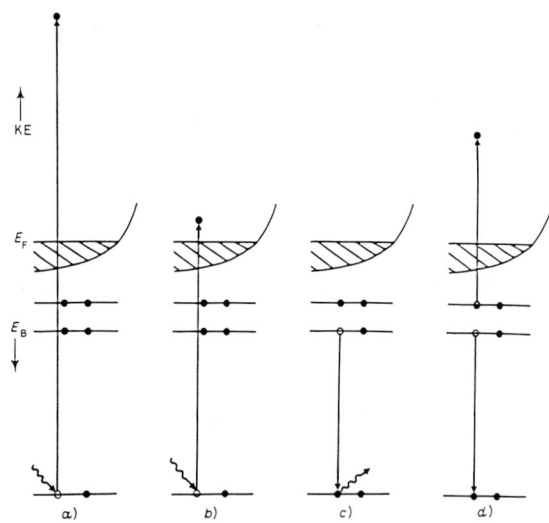

Fig. 3. Schematic diagram of the transitions involved in various spectroscopies in a single-particle approach. (a) Photoemission (here: XPS); (b) X-ray absorption spectroscopy (XAS); (c) X-ray emission (XES); (d) Auger electron emission (AES). Figure taken from ref. [7].

The observed intensities as a function of binding energy carry information on the occupied DOS, but should not be seen as the DOS itself.

The principle of inverse photoemission [13–16] is illustrated in fig. 4. A beam of electrons with energy E is incident on the sample surface. The electron entering the system will decay into an unoccupied level and the photon emitted by the radiative decay process can be detected. In the isochromat mode, the energy of detection is kept constant, while the energy of the electrons is swept. The spectrum thus obtained carries information on the unoccupied DOS. In the UV region, inverse photoemission is the counterpart of UPS, and is abbreviated IPES. The counterpart of XPS is for historical reasons called Bremsstrahlung Isochromat Spectroscopy (BIS). In both the low-energy techniques, IPES and UPS, k-conservation is important as the crystal momentum parallel to the surface is conserved in the process. Determination of k makes it possible to map the

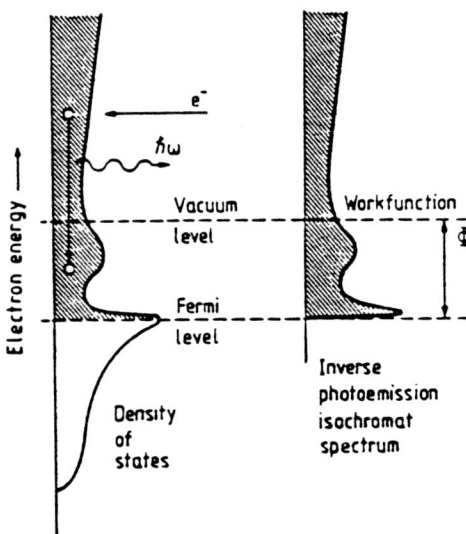

Fig. 4. Schematic representation of inverse photoemission spectroscopy (taken from ref. [15]). If the incident electron energy E is swept while holding $\hbar\omega$ constant (isochromat mode), the measured photon counting rate will replicate the unoccupied DOS, as indicated on the right.

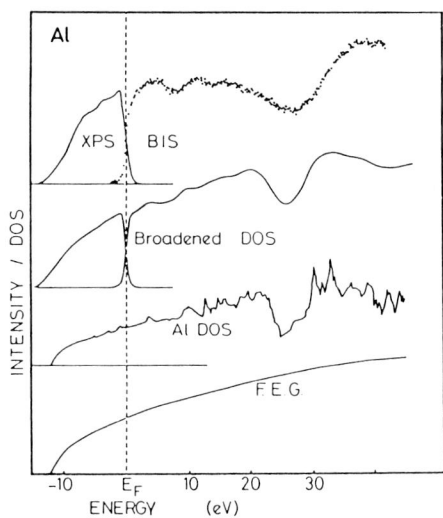

Fig. 5. Al, using data from refs. [17,18]. Bottom curve: Free electron gas (F.E.G.). Next: computed DOS. Next: DOS folded with a lifetime broadening. Top curves: XPS/BIS.

band structure in an ordered solid, by measuring the dispersion $E(k)$. This is done by angle-resolved (I)PES (AR(I)PES), see, e.g., refs. [8,13]. At the higher energies, used in XPS and BIS, k-conservation is less relevant, for reasons including the momentum associated with the photon, phonon effects and the extremely small angular resolution required to detect these effects.

The information to be obtained from (inverse) photoemission studies of the valence bands depends on the degree of localization of the valence electrons. For materials with strongly delocalized electrons, like aluminium in fig. 5, the observed XPS-BIS spectra closely resemble the DOS. Note that even here the spectra do not show the sharp peaks and valleys of the single-particle DOS. This is not a fault of the experiment, but reflects the lifetime broadening introduced by rapid decay of a state with an extra electron or hole far from the Fermi level. This is also referred to as the imaginary part of the self-energy and is not included in single-particle DOS calculations.

The other extreme is represented by materials with highly localized electrons where atomic correlations dominate, as for the 4f electrons of most RE. The Hubbard model [19] is often used for such materials, in which the Hubbard correlation energy, U_{eff}, is defined as the difference between the first ionization potential and the first electron affinity of the system. For instance, if we have an array of f^7 atoms, then U_{eff} is the energy required to change one atom to f^6 and another to f^8. The symbol U_{eff} is used to imply that an *effective* Coulomb correlation energy is being discussed, in which all the valence electrons have been allowed to adjust their distribution in response to removal or addition of a localized electron.

The ionization potential and electron affinity can be read directly from spectra like those of Gd in fig. 6. Such studies [20,21] have had a large impact. However, of even more interest are the

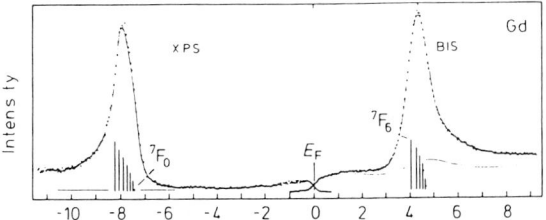

Fig. 6. Combined XPS and BIS spectra of Gd. Vertical bars indicate the positions and calculated intensities of the 4f final states. (From ref. [21].)

so-called "narrow band systems" [22], in which both band structure *and* atomic correlations must be taken into account to understand the electronic and magnetic properties of the material. In these cases the XPS/BIS spectra show characteristics intermediate between those of Al and Gd, and new features may appear, as we will see in section 3.

2.2. Core level photoemission

Core levels are not directly involved in magnetic phenomena because the shells are filled in the ground state and have negligible (diamagnetic) interaction with both external and internal magnetic fields. However, when a core electron is removed:

1) The spin and angular moment of the core hole will interact with those of the valence electrons, giving rise to multiplet structure in the core level XPS spectrum [23], which can be used to diagnose the local spin and magnetic moment in the ground state of the core-ionized atom. For example, in the case of an s core electron and a spin-only moment, one would expect two s core level peaks due to final states in which the core and valence electron moment are aligned parallel and antiparallel. The splitting of the two peaks would then be proportional to the size of the magnetic moment on the ionized atom. At one time this method was thought to have great potential as a local, element-specific probe of magnetic moments, but it is now clear that for transition metal (TM) compounds the initial and final state valence electron moments are not identical [24]. Thus, as a probe of the size of local moments, core level splittings are only useful for the case of highly localized moments (e.g., the RE 4f moments), where the screening effects described as point 3 below, are weak. However, spin-polarized studies of core level photoemission have shown that the final states do contain some memory of the ground state moments, see, e.g., ref. [25]. This is being used in attempts to probe the degree of magnetic ordering in photoelectron diffraction experiments [26], which are not described in detail here.

2) A second piece of information lies in the asymmetry of core level XPS peaks, which was shown by Mahan [27] and Doniach and Šunjić [28] to be related to the DOS at the Fermi level. It arises because creation of a core hole is accompanied by creation of many low energy, electron–hole pairs, which costs energy and thus reduces the kinetic energy of the photoelectron. Although there are differences in the asymmetry of peaks from levels of atoms in different alloys [29] related to the differences in the local, atom-selected DOS, this phenomenon has, so far, proved to be too complex to be used to extract quantitative information.

3) Finally, the screening of a core hole and its influence on the shape of the core level XPS lines provides insight into magnetic interactions. This, too, is a complex phenomenon but its analysis has led to very considerable and quantitative insight into magnetism and low temperature thermal properties of materials with very strongly correlated electrons, like Ce and many of its alloys and compounds. It is the partially filled valence states which are responsible for magnetism, but methods have been developed to relate the *core* level line shapes in XPS to the hybridization between valence states which give a surprising degree of insight [30,31]. This is discussed in more detail in section 3.5 for the hybridization of Ce 4f levels with other valence states.

2.3. Spin-polarized electrons

2.3.1. Detection of spin-polarized electrons

We will describe here techniques of detection of the spin-polarization of a beam of electrons. The spin-polarization P is usually defined as:

$$P = \frac{I_\uparrow - I_\downarrow}{I_\uparrow + I_\downarrow}, \qquad (3)$$

where $I_\uparrow$ and $I_\downarrow$ are the measured intensities of the majority and minority spins, respectively.

The oldest form of spin-detection technique for electrons, and probably the technique which is still used most, is based on the Mott scattering geometry [32–34]. The angular distribution of high energy electrons (typically 100–120 keV) scattered off heavy nuclei, e.g., Au, shows a spin asymmetry due to the spin–orbit interaction. In conventional Mott detectors, the incoming electrons are accelerated to the desired energy, scattered off a Au foil, and the scattering intensities are measured with two detectors positioned around scattering angles of $\pm 120°$. If we denote the ratio of detected intensities as Q then P is given by:

$$P = \frac{1}{S_{\text{eff}}} \frac{1-Q}{1+Q}, \qquad (4)$$

where S_{eff} is the effective Sherman function of the detector, which describes the spin-asymmetry in the scattering process, taking multiple scattering into account. Due to the high kinetic energy of the electrons, the scattering process is not very sensitive to contamination of the Au foil. There is therefore no need to keep the foil in ultra-high vacuum (UHV). Due to shielding of high voltage parts, conventional Mott detectors are rather large (several m^3), although smaller versions of the Mott detector have been designed (see, e.g., refs. [35,36]).

Low energy electron diffraction (LEED) provides an alternative to the Mott scattering geometry. A single crystal surface is used instead of a polycrystalline or amorphous target [37] and spin-polarization is reflected by intensity-asymmetry in diffraction spots [38], due to spin–orbit interaction in the scattering process. The LEED detector has a similar figure of merit but is much smaller than a Mott detector. However, the detector surface needs to be clean, i.e., the crystal has to be positioned in UHV, and it has to be cleaned before use. Spin–orbit and exchange interactions simultaneously cause the spin-selectivity in diffraction off a magnetized Fe(001) single crystal surface, which was recently proposed since its figure of merit is some 20 times higher than other techniques [39].

2.3.2. Spin-polarized electron sources

The electron source is an essential component in any experiment requiring beams of spin-polarized electrons, e.g., SPIPES or SPEELS. The most widely used sources are based on binary (e.g., GaAs) or tertiary (e.g., AlGaAs or GaInP) semiconductor compounds. Spin–orbit coupling in the valence band of the semiconductor can lead to an energy splitting of the valence bands. For GaAs, threshold excitation, i.e., excitation from the top of the valence band, can then lead to a 50% spin-polarized occupation of the conduction band if one used the dipole section rules associated with circularly polarized light [40]. The sign of the spin-polarization can be reversed by reversing the helicity of the exciting light.

To overcome the work function at the surface, the semiconductor is heavily p-doped, so that the Fermi level (and with it the vacuum level) is pulled down towards the valence band, and downward band-bending occurs at the surface. Deposition of Cs and oxygen on the surface will lower the vacuum level below the bulk conduction band. (This is called negative electron affinity (NEA).) Effective polarizations of NEA photocathodes are usually 30–35%. This is much less than the theoretical value of 50%, and is due to various depolarizing effects [41].

2.4. Spin-polarized (inverse) photoemission

Spin-polarized (angle-resolved) photoemission (SP(AR)PES, usually SP(AR)UPS) is performed by analyzing the spin of the photoelectron after energy analysis. Spin-polarized electrons are emitted if the spin degeneracy of the initial state is lifted by the exchange interaction, which allows the determination of the exchange splitting in the valence band (see section 3.2). However, spin–orbit splittings in the initial or final states of the photoemission process can, if the incident photons are circularly [42] polarized, also yield spin-polarized electrons [43,44]. The first spin-polarized photoelectrons [45] were observed around 1970 [46,47]. A breakthrough in the field came when it was realized it was not necessary to perform measurements in an externally applied field [48]. A year after this, the first spin-polarized photoemission experiment was reported [49], soon followed by spin-polarized angle-resolved measurements [50–52].

IPES performed with a spin-polarized electron source is called spin-polarized inverse photoemission, or SPIPES [53,54], allowing spin-selective probing of the unoccupied states. The earliest work includes SPIPES on Ni [55], spin-polarized ARIPES (SPARIPES) on Fe [56] and, recently, spin-polarized BIS (BISCEPS) on Ni [57].

2.5. Auger electrons

The Auger process is depicted in fig. 3(d). Once a core hole is created by photon or electron bombardment, it can undergo fluorescence or radiationless processes, and the latter is known as Auger decay. Auger decay is more probable than fluorescence decay for all but deepest core holes [58]. It is itself a single step process, controlled by a Coulomb matrix element, in which one electron drops into the core hole and one is emitted with kinetic energy KE given by

$$KE = BE_1 - BE_2 - BE_3 - U_A, \quad (5)$$

where BE_1 is the binding energy of the initial core hole and BE_2 and BE_3 are the binding energies of the final state holes. U_A represents the interaction between the two final state holes. U_A is the key to one of the main uses of Auger electron spectroscopy (AES) in studies of magnetic materials because for some transitions it is closely related to the Hubbard correlation energy, U_{eff}, as shown by Cini and Sawatzky [59,60].

A second use of Auger spectroscopy in magnetism concerns spin-polarized studies. If a material is magnetic it is clear that Auger transitions producing final state holes in the valence band may lead to strong spin-polarization of the Auger peaks. In fact multiplet splittings often lead to spin-polarized Auger peaks even when only core electrons are involved. In both cases the spin-polarization of the Auger peak carries information on the local magnetization of the atom involved because the Auger transition matrix element is very short ranged. The result from this short-ranged matrix element is to make spin-polarized AES (SPAES) [61-64] a good, element-sensitive probe of local magnetization.

2.6. Secondary electrons

Electrons incident on a solid (primary electrons) lose energy through electron–electron scattering processes. Valence band electrons that are emitted from a solid through a cascade of electron–electron scattering processes are called secondary electrons. Pioneering experiments found a large spin-polarization for electrons emitted at low kinetic energies [65,66]. The large spin-polarization was shown to be larger than the average band polarization, but to scale with the magnetization [67]. Low-energy secondary electrons have since been used for many applications, among which magnetic domain microscopy, magnetization measurements and magnetic depth profiling [68-70].

2.7. Electron energy loss

In electron energy loss spectroscopy (EELS), a beam of monoenergetic electrons is incident on a sample surface. In the scattering process various excitations, such as phonons, plasmons, electron–hole pairs, etc., may be created. If the excitations involve core electrons, the process is similar to XAS (see below) [71]. The scattered electrons are energy analyzed, and characteristic energy losses associated with the excitations are mapped. For the study of magnetic systems, excitations involving spin-flip processes are important (see sections 3.2 and 3.3). Because both incoming and outgoing particles are electrons, spin-polarized EELS (SPEELS) can be performed using a spin-polarized source, or by spin-analysis of the outgoing electrons, or both. The latter, i.e., SPEELS with both spin-polarized source *and* detector, allows the separation of EELS spectra in spin-flip and non-spin-flip transitions [72,73].

2.8. X-ray absorption

In XAS [74-76] (see fig. 3(b)) an electron is excited into an unoccupied state, on the same site. This excitation is then subject to dipole selection rules, of which the $\Delta l = \pm 1$ selection rule allows monitoring of the l-projected DOS. The absorption process is site- and element-sensitive. The XAS spectra are usually divided in a near-edge region, usually studied for information on electronic structure, and a region more than ≈ 50 eV away of the absorption edge, where oscillations can be observed in the absorption cross section, that can be related to inter-atomic distances. The latter process is called "extended X-ray absorption fine structure" (EXAFS).

The relevance of XAS to magnetism lies mainly in two applications. The first is magnetic X-ray dichroism (MXD). Magnetic dichroism is already a well-established technique in the optical region. Theoretical considerations on magneto-optical Kerr effects (MOKE) [77] in the $M_{2,3}$ ($3p \rightarrow 3d$) absorption edge of Ni [78] led to the search for MOKE effects in the X-ray region (see, e.g., ref. [79]), but the observed effects were small. The real breakthrough came with theoretical and experimental studies during the last ≈ 5 years in which very strong magnetic dichroism in the X-ray region was predicted and observed [80–86].

With circularly polarized light, electrons excited from a core level are spin-polarized [87]. The transitions of such electrons into exchange-split unoccupied states then allows the study of magnetic systems, because the absorption of circularly polarized light by a magnetic medium then depends on the polarization direction of the incident light with respect to the direction of the magnetization.

A second application concerns magnetic X-ray scattering (MXS). The magnetic cross section is considerably enhanced with respect to the charge scattering cross section at an absorption edge (see. e.g., ref. [88]). For instance, a fifty-fold enhancement was observed for scattering off Ho at the Ho L_3 ($2p_{3/2} \rightarrow 3d$ absorption) edge [89–91].

3. Illustrations of the uses of HES

In this section we analyze some of the contributions of HES to various problems in magnetism. We do not try to analyze whether "HES did it first", nor do we discuss the relative merits of HES to other techniques. These questions we defer to professional historians.

3.1. General trends in electronic structure

Here we give a short review of the electronic structure of metallic materials, discussing general trends in band widths and U_{eff} that have emerged from two decades of experimental and theoretical studies, before considering some elements in detail in section 3.2.

For many elements ARUPS/ARIPES allow one to follow the band structure in some detail and XPS/BIS studies give good data in the band widths. For highly correlated metals the on-site Coulomb interaction U_{eff} may be obtained directly from XPS/BIS studies [20,21] and for a few elements Auger spectroscopy also gives useful data. For intermediate cases more detailed modelling may be needed to interpret spectroscopic data, in terms of band widths and U_{eff}, but still an impressive amount of data is available. The spectra from free-electron-like metals like Al (fig. 5) basically yield a free-electron-like parabola for the DOS, overlayed with structure due to the crystal potential. Of more interest to use are the TM and RE. Tables 2 and 3 summarize some of the literature data on band widths, hybridization potentials and effective Coulomb interactions. Work is still going on in this field to refine the models used to interpret the data. However, in general, we have not given the latest values here, but have used data which is at a consistent level of sophistication in order to establish trends. These data are of sufficient accuracy to establish the major differences between the RE and TM.

In the pure elements the 3d and 4d band widths range from 5.4–8.5 and 6.3–9.6 eV, respectively. Contrary to many text book statements, there is not a major difference between the band widths at the beginning and end of the series, and the band widths only start to decrease dramatically at the end of the series, in the noble metals. There is also not a major difference between the band widths of the 3d and 4d elements.

Most of our available data on Coulomb correlation and U_{eff} in TM comes from Auger spectroscopy, using the Cini–Sawatzky methodology. This is only justified for nearly filled bands, so that it is only good for Ni, Pd, Cu and some compounds. Nevertheless the information it gives is important. At the right hand side of the series, U_{eff} is comparable to the 3d and 4d band widths, which explains very readily why both band dispersions and local Coulomb correlation effects have to be taken into account to explain the electronic and magnetic properties of these materials. We know of no values for U_{eff} at the left of the series whose reliability is similar to that at the right. It should also be noted that atomic multiplet effects detected in various spectroscopies [94,100,101], can make up a significant fraction of the total value of U_{eff}, and indeed they may even dominate it.

When we compare the values of U_{eff} for Ni with Pd, or Cu with Ag, we note that the 4d

Table 2
Selected values for occupied (W^-), unoccupied (W^+) and total (W) d-band widths and effective Coulomb correlation energies (U_{eff}). Data obtained from photoemission (W^-), inverse photoemission (W^+) and Auger spectroscopy (U_{eff}). All values in eV. The dual values are given where there is ambiguity in interpretation of experiment but they are also a good guide to the accuracy. Theoretical values are denoted (th)

Element	W^-	W^+	W	U_{eff}	Refs.
Sc	1.7 (th)	4.4–4.7	6.1–6.4		[92]
Ti	3.1 (th)	3.2–3.8	6.3–6.9		[92]
V	3.2 (th)	3.4–3.8	6.6–7.0		[92]
Cr	4.5 (th)	1.8–2.3	6.3–6.8		[92]
Mn	≈ 5	≈ 3.5	≈ 8.5		[92,93]
Fe	5.3 (th)	2.9–3.5	8.2–8.8		[92]
Co	5.5 (th)	1.4	6.9		[92]
Ni	5.0 (th)	0.4	5.4	3.65	[92,94]
Cu		≈ 2.6		7.1	[95]
MnO			2.5 (th)	7.8 (+th)	[96,97]
FeO			2.3 (th)	3.5 (+th)	[96,97]
CoO			2.0 (th)	4.9 (+th)	[96,97]
NiO			1.8 (th)	7.3	[96,97]
CuO				5.1 (+th)	[97]
Al$_3$Ni (Ni 3d)				3.9	[94]
MgCu$_2$ (Cu 3d)				7.8	[94]
Y	2.0	6.1	8.1		[92]
Zr	3.1	5.2–5.7	8.3–8.8		[92]
Nb	3.5	5.2–5.9	8.7–9.4		[92]
Mo	5.4	3.9–4.5	9.3–9.9		[92]
Ru	6.2	2.0	8.2		[92]
Rh	5.8	1.1	6.9		[92]
Pd	5.9	0.4	6.3		[92,98]
Ag			3.5	5.1	[95,99]

Table 3
The first ionization potentials (Δ_-), electron affinities (Δ_+) and effective Coulomb correlation energies (U_{eff}) for the 4f levels of the RE. All values in eV, obtained from ref. [21]. Also given are hybridization potentials Δ of the 4f levels derived from core level spectroscopies, in meV, with the references in brackets. Δ is expressed as $2\pi\rho_{max}V^2$, where ρ_{max} is the maximum density of valence states and V is the potential mixing the 4f and valence states

Element	$\Delta_+ \pm 0.2$	$\Delta_- \pm 0.1$	$U_{eff} \pm 0.3$	Δ
La	5.31			70 [102]
γCe	3.46	0.27–1.92	3.74–5.39	25 [103]
αCe				40 [104]
CeAl$_2$				30 [102]
CeNi$_2$				100 [102]
Ce$_7$Pd$_3$				30 [102]
CePd				90 [102]
CePd$_3$				150 [105]
CeAu$_2$				15 [102]
Pr	2.14	3.33	5.47	
Nd	1.72	4.65	6.37	
Sm	0.46	5.07	5.33	
Eu	8.63	1.50	10.14	
Gd	4.04	7.44	11.48	
Tb	2.76	2.23	4.99	
Dy	1.81	3.86	5.68	
Ho	1.93	4.89	6.82	
Er	2.15	4.70	6.86	
Tm	1.10	4.57	5.67	
Yb		1.27		
Lu		7.02		

element values are much smaller than those for the 3d elements. This is one of the major factors behind the difference in properties of the 3d and 4d elements (and apparently not the band widths, which are similar). Table 2 also gives some values of band widths and U_{eff} for a small range of compounds. For the TM, the band widths are usually smaller in compounds, partly because the number of nearest neighbour contacts decreases, and partly because inter-atomic distances tend to increase. The changes in U_{eff} are not as large, but can be significant in some cases. Really major differences appear when we compare the data for TM d-bands with that for RE 4f levels in table 3. The values of U_{eff} are not dramatically different but the band widths really are. RE 4f band widths must be measured in meV and are typically a hundred times smaller than the TM d band widths. Furthermore the width usually appears to

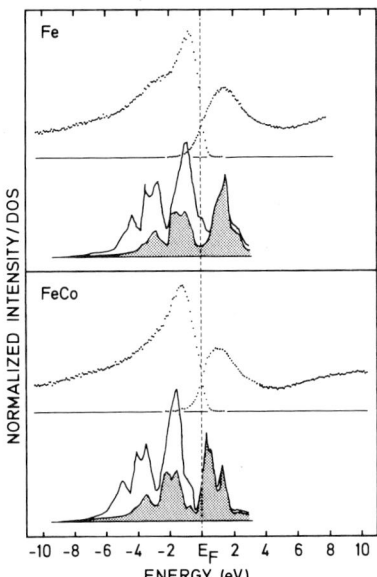

Fig. 7. XPS/BIS spectra of Fe and FeCo, with calculated DOS from ref. [4]. The shaded area in the DOS corresponds to the minority DOS. Figure obtained from ref. [107].

be due to hybridization with the more itinerant RE 5d and 6s states, rather than in direct f–f overlap [106].

3.2. The valence bands of magnetic materials

Magnetism originates from unequal amounts of spin-up and spin-down electrons in the valence band of magnetic materials. Therefore, the study of the valence band is clearly important. The study of the electronic structure, in particular the detection of the exchange splitting, allows the monitoring of magnetic properties. Here we give examples of the determination of the (k-dependent) exchange splitting, as probed by (inverse) photoemission, and the determination of the spin- and symmetry-dependent electronic structure by MXD. We will show that the accumulated data points to deficiencies in the Stoner model.

The valence band XPS/BIS spectra of bcc Fe are shown in the top panel of fig. 7, together with a calculated DOS. The shaded area denotes the minority-spin density. The calculated DOS states and the XPS/BIS spectra are in reasonable agreement with each other. This is by no means trivial, since the eigenvalues obtained from single-particle calculations have, strictly speaking, no physical meaning, as discussed in section 1.1. As can be seen from fig. 7, assignment of spin character to a specific peak in the XPS/BIS spectrum is, with the exception of the unoccupied minority peak, not possible. Moreover, spin-up and spin-down densities of state have, in general, different shapes, and a determination of the exchange splitting as in fig. 1 does not seem to be possible. However, it is still possible to determine the exchange splitting as a function of k. Single transitions in k-space can be studied with angle-resolved (inverse) photoemission. Assisted by detailed band-structure calculations, the obtained AR(I)PES spectra can be described in terms of transitions involving spin-up or spin-down bands. The spectra then can be deconvoluted to obtain the exchange splitting. This procedure has been applied in the past, see, e.g., ref. [108]. It has the disadvantage, that the deconvolution becomes increasingly more difficult as the exchange splitting becomes smaller.

Exchange splittings can be detected directly, without trouble in peak assignment, by spin-polarized techniques. In fig. 8 the normal emission ($k_\parallel = 0$) spin-resolved photoemission spectrum of Fe(001) is shown. Spin-up and spin-down photoemission intensities show peaks at different energies, reflecting the exchange-split band structure. The spin-up part is dominated by the bands $\Gamma_{25}'^{\uparrow}$ and $\Gamma_{12}'^{\uparrow}$, while in the spin-down part, the band $\Gamma_{25}'^{\downarrow}$ is visible (see fig. 2(a)). The energy difference between $\Gamma_{25}'^{\uparrow}$ and $\Gamma_{25}'^{\downarrow}$ (2.2 eV) is the exchange splitting Δ. The spin-split band structure of a magnetic material can be mapped, and the exchange splittings of the material can be determined at various points in the Brillouin zone.

Of course, not all bands are occupied and the exchange splitting between two *unoccupied* bands can be determined by spin-polarized inverse photoemission. An example hereof, again for Fe(001), is shown in fig. 9. In the top spectrum, the spin-integrated IPES spectrum is shown, while the lower spectrum is the spin-resolved spectrum, allowing the decomposition of the upper spectrum in majority-spin and minority-spin parts.

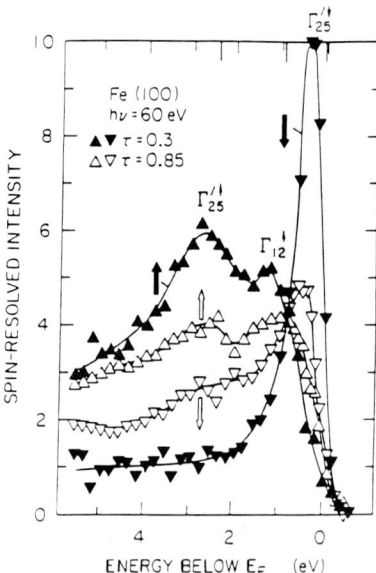

Fig. 8. Spin-polarized photoemission spectra of Fe(001) taken near the Γ-point [52]. Solid triangles represent data taken at $0.3T_c$. Open triangles represent data taken at $0.85T_c$.

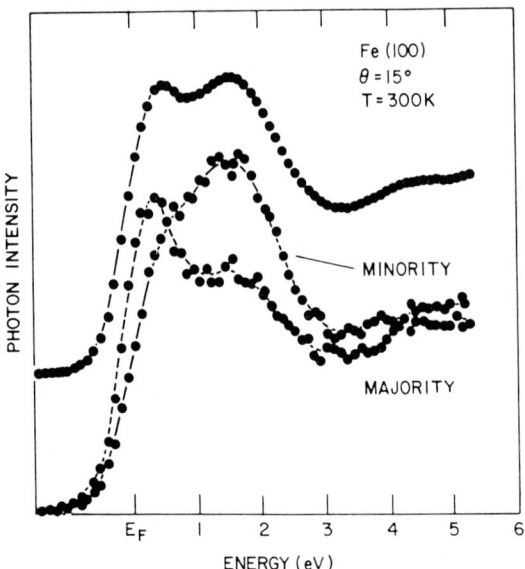

Fig. 9. Inverse photoemission data adapted from ref. [109]. The upper curve represents the total intensity spectrum for Fe(001), taken at $\theta = 15°$ angle of incidence and at room temperature. In the lower curves the spin character of the two-peak structure is identified by the spin-resolved spectra.

The transitions here are near the H-point and the exchange splitting at this point is found to be (1.6 ± 0.2) eV. If only the majority-spin band is occupied, the energy difference between the PES peak and the IPES peak is the exchange splitting, but only if there is no influence of the difference in charge state. While one may expect a distortion in a localized system due to different Coulomb interactions (in PES an electron is missing, while in IPES an electron is added), for delocalized magnets like Fe there appear to be little effects [110].

Another way of observing the exchange splitting in magnetic materials is the detection of Stoner excitations through (SP)EELS. The elementary transition in Stoner theory are single-particle spin-flip excitations, the so-called Stoner excitations: an electron with spin σ is transferred to an empty state with spin $-\sigma$. If there is no net momentum transfer involved in the transition, the energy needed for this transition equals the exchange splitting Δ. This allows the exchange splitting to be determined through the measurement of Stoner excitations with EELS. An incoming electron with spin σ can be "exchange scattered", i.e., the incoming electron fills an unoccupied orbital, and an electron with spin $-\sigma$ is emitted. This implies also a spin-asymmetry: there are less spin-up empty states than spin-down empty states, so the probability of a down-up exchange is larger than an up-down exchange, at least if the material has long-range magnetic order.

Glazer and Tosatti argued that Stoner excitations should be observable in both ordinary and spin-polarized EELS (SPEELS) [111]. This was confirmed, using unpolarized electron beams, by Hopster et al. [112] with, and by Modesti et al. [113] without spin-analysis. Although it has been argued that these effects are due to quantum interference between different scattering processes [114], the maximum was found at the energy loss corresponding to the average exchange splitting [112,113,115,116], albeit with a rather large energy spread [117].

The evolution of MXD in the last years allows the determination of the symmetry-projected conduction band, which for a metallic ferromagnet

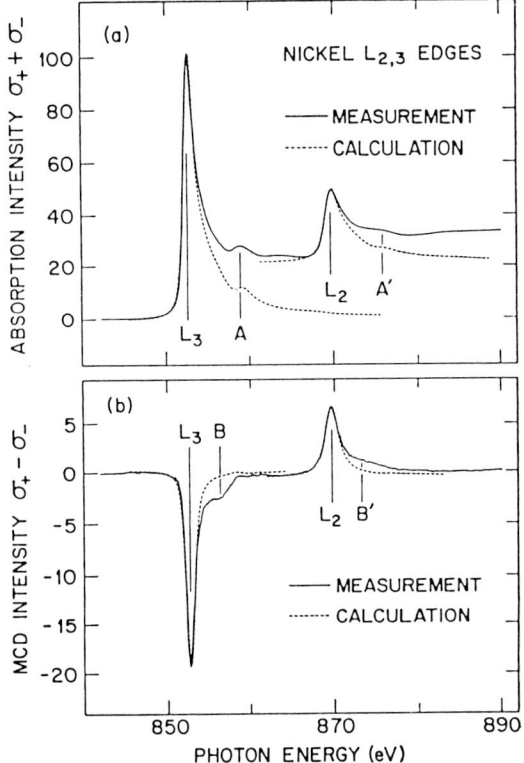

Fig. 10. Comparison between experimental soft X-ray spectra (full curves) and a tight-binding model calculation (dashed curves) of the L_3 and L_2 white lines in Ni [118]. (a) the total absorption $(\sigma^+ + \sigma^-)$; (b) the magnetic circular-dichroism (MCD) $(\sigma^+ - \sigma^-)$. The raw experimental MCD spectrum shown here has been magnified by a factor of 1.85 to account for incomplete photon polarization and sample magnetization [118].

consists of the d-electrons responsible for the magnetism, and of sp-electrons that are generally also exchange-split. The unoccupied d-levels of Ni can be probed by excitation of a p-electron, e.g., the $L_{2,3}$ absorption edge consist mainly of excitations of electrons from the 2p core level to the unoccupied 3d states. In fig. 10 the $L_{2,3}$ edges of Ni(111) are depicted, together with the asymmetry between left- and right-circularly polarized light. Analysis of these spectra in a one-electron model [118], fitting the asymmetry and total spectra not only yields a value for the local spin-moment, but also gives the contribution of the orbital moment to the total moment. For Ni, the spin and orbital moments were found to be 0.52 and $0.05\mu_B$, respectively, which are in good agreement with theory, see, e.g., refs. [119,120].

The feature B (B') cannot accurately be assigned by Chen et al. in their single-particle analysis (see the dotted line in fig. 10(b)), and is attributed by them to a many body-effect. We note, that for Ni many-body effects are important, as can be judged from the ratio of U_{eff} to W (table 2). Many-body effects in Ni show, e.g., as a satellite structure in XPS [98]. The feature B (B') was reproduced by a separate analysis of Jo and Sawatzky [121], taking correlation-effects into account and yielding values of spin and orbital moments that are nearly the same as those of Chen et al.

The p-projected unoccupied states can be detected from the absorption by an s-level. The difference in the absorption of left- and right-circularly polarized light permits the separation of the absorption into spin-dependent contributions. The 1s → 4p absorption of Fe, obtained with circularly polarized light, is shown in fig. 11 [82]. In the top panel, the total absorption spectrum is shown. In the lower panel, the difference between the absorption spectra obtained with the two different helicities is shown. Analysis shows

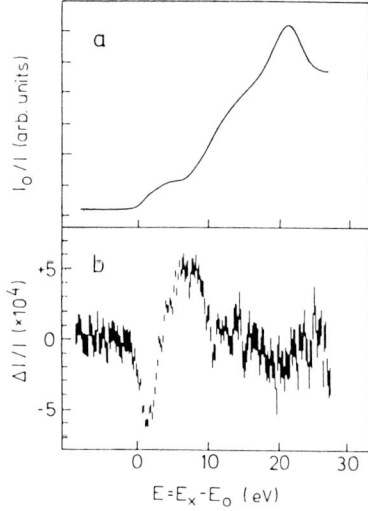

Fig. 11. (a) Absorption I_0/I of X-rays as a function of the energy E above the K-edge of iron and (b) the difference of the transmission $\Delta I/I$ of X-rays circularly polarized in and opposite to the direction of the spin of the magnetized d electrons. (Figure taken from ref. [82].)

Table 4
Exchange splittings Δ determined by (inverse) photoemission at various points in the k-space of Fe (see fig. 2). The listed calculated values stem from ref. [3]. All values (in eV) taken from table I of ref. [110]

k-point	Δ (exp.)	Δ (theor.)
Γ'_{25}	2.1	1.82
Γ_{12}	2.4	2.18
H_{12}	1.3	1.51
P_3	2.1	2.10
N_1	0.9	1.65

the empty p-like band at ≈ 2 eV is preferentially occupied by electrons with spin parallel to the spin of the d-electrons, whereas the band at ≈ 7 eV is a minority-spin band. Comparison of the results with a band-structure calculation confirm the 2 eV peak, but agreement at higher energies is poor [82]. This could be due to many-body effects, since a relativistic multiple-scattering analysis [86] seems to be consistent with the data, but quadrupole terms in the absorption process (and hence excitation to s and d final states) might also be important for these transitions [122].

In conclusion, we note that HES now provides various alternatives for measurements of exchange splittings Δ. It has been shown that Δ can be determined as a function of the crystal momentum by (inverse) photoemission. For Fe, experimentally determined values of $\Delta(k)$ were compiled by Santoni and Himpsel [110], and some of their values of Δ are listed in table 4. It is clear that Δ is not a constant over the valence band, and that exchange splittings from band structure calculations are not consistent with experimental values. We can identify at least three effects leading to variations in the exchange splitting over the valence band. First, it was already mentioned in section 1.1 that bands with different symmetry have different exchange splittings. Second, the self-energy is generally E- and k-dependent. This can have profound consequences for the electronic structure, as shown in the studies of Liebsch [123]. For Ni, it results in a strongly spin-polarized satellite structure [49]. Thirdly, in contrast to bonding wavefunctions, antibonding wavefunctions are zero between atoms, and renormalization of the wavefunctions leads to a higher density of antibonding states in the core of the atom [124–126], where the electron-density is larger. Thus, bonding and antibonding states generally feel different exchange interactions.

3.3. Finite temperature magnetism

In this section the question: "How does the exchange splitting change with temperature?" is addressed. For itinerant magnets the original Stoner theory predicts that the single-particle excitations reduce the mean field which, in turn, reduces the exchange splitting between the spin-up and spin-down bands. Hence, the macroscopic magnetization reduces in proportion to the (local) exchange splitting, until they both become zero at T_c. However, calculated transition temperatures from Stoner theory are far too high compared to experiment (usually by a factor 5–10) [127], due to the neglect of low-energy collective modes (e.g., spin waves) in Stoner theory. In modern models on magnetism [128–134] the magnetic moment slowly changes with temperature, and vanishes at temperatures much higher than T_c [133]. Determination of (local) exchange splittings is more difficult at elevated temperatures than it is at low temperatures, but still possible, and we will show that published data indicate that the local exchange splitting persists almost unchanged through the critical temperature T_c.

The study of elementary excitations, which according to Stoner theory must be spin-flip transitions (see section 3.2), offers a first possibility to monitor the temperature dependent exchange splitting: according to Stoner theory the energy at which the Stoner excitations are to be found will tend toward zero as T_c is approached due to the decrease of Δ. The number of unoccupied spin-up states increases, and therefore the ratio of up-down to down-up spin-flip transitions should approach unity in the limit of $T \to T_c$. This can be observed in SPEELS studies as the vanishing of the symmetry in the loss peak (fig. 12), which can also be explained by the disappearance of long-range magnetic order. The spin-flip energy, however, was found to be constant up to the Curie temperature in nickel (see fig. 12), indicating an

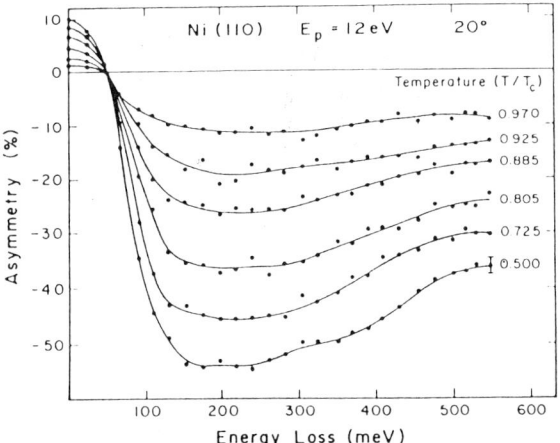

Fig. 12. Temperature dependent asymmetry spectra from the SPEELS experiments of ref. [136], where the asymmetry is defined as: $A = (1/|P_0|)(I^\uparrow - I^\downarrow)/(I^\uparrow + I^\downarrow)$, with P_0 the initial polarization of the electron beam. The temperature-dependent spectra were obtained at 12 eV primary energy, 20° off specular.

exchange splitting that has not noticeably changed [135,136].

Photoemission data on the exchange splitting of Fe as a function of temperature show bands with vanishing exchange splitting ("collapsing bands") and bands where the exchange splitting remains constant ("non-collapsing bands"). For Fe(001), spin-resolved photoemission [52,137] (see fig. 8) indicated that the emission from $\Gamma_{25'}^\downarrow$ decreases much more strongly than the photoemission from $\Gamma_{25'}^\uparrow$. The peak positions, however, do not change, indicating a constant exchange splitting. The results from inverse photoemission experiments show also both "collapsing" and "non-collapsing" behaviour [109,138]. The behaviour of the exchange splitting with temperature is thus k-dependent, and therefore the notion of a rigid band shift with temperature is false.

Note the broadening of the photoemission peaks in fig. 8 at higher temperatures. The broadening of the (inverse) photoemission spectra at higher temperatures is a consequence of the increased probability of non-vertical transitions in k-space. Due to dispersion, this causes broadening of the observed peaks and, hence, increased uncertainty in the determination of the exchange splitting. Accurate determination of the exchange splitting at higher temperatures with SPEELS is also hampered by the reduction of the asymmetry peak (see fig. 12). Moreover, a vanishing exchange splitting obtained from, e.g., (inverse) photoemission does not necessarily mean that the local moment (or local exchange splitting) has vanished. Remember that one measures *quasiparticle* energies, which are sensitive to detailed changes in the many-body state of the system. For instance, for Ni metal the exchange splitting is found to vanish at many points in the Brillouin zone, if the temperature goes through T_c. Recent calculations by Nolting and Borgieł [139–141], considering explicitly the 3d-electron correlations, showed that the exchange splitting indeed vanishes all over the *quasiparticle* band structure. However, these calculations also showed that the Ni local moment remains essentially unchanged as the temperature goes through T_c!

The determination of changes in the exchange splitting is therefore not an indication of changes in the local magnetic moment. Determination of the local moment with temperature is feasible with HES, as was shown by a recent photoemission study on magnetic impurities dissolved in Pd or Pt [142]. These materials form random alloys, and crystal momentum and k-conservation do not have the same relevance as in ordered systems. Photoemission in the Cooper minimum [143,144] suppresses the contribution of the host to the photoemission signal and permits inspection of the impurity local density of states (LDOS). The data for 10 at% Fe in Pd (T_c = 225 K) is shown in fig. 13. The valence band closely resembles the Pd valence band features because of the strong hybridization of the Fe impurity with the Pd host [144,145]. Due to the Cooper minimum effect, ≈ 50% of the photoemission signal stems from the Fe impurities. The solid line in the spectrum shows the room temperature signal where it differs from the liquid-nitrogen temperature signal (dots). The difference spectrum is also shown, magnified 5 times. The differences are minor and only visible in the valence band region.

The small changes observed are not compatible with the expected strong redistribution of the impurity LDOS with even small changes in the magnetic moment. To show the sensitivity of the photoemission signal to changes in the local electronic structure, in fig. 14 a calculation is presented for an Fe impurity in Pd. Changing the Fe moment 1% already results in an observable change in the LDOS. The weight of the majority-spin part of the LDOS can be observed to shift towards the Fermi level. Spurious effects, such as lifetime broadening and energy-dependence of the photoemission matrix elements were simulated in the top panel of fig. 14. Comparing this with the observed changes in the valence band spectrum it was concluded that the local moment does not change appreciably over the temperature range studied, and the results are incompatible with a collapse of the local magnetic moment at the impurity site at T_c when the long range magnetic order goes to zero. This further implies that the local exchange splitting is essentially invariant through T_c.

To conclude this section, we note that Stoner theory fails in the description of finite-tempera-

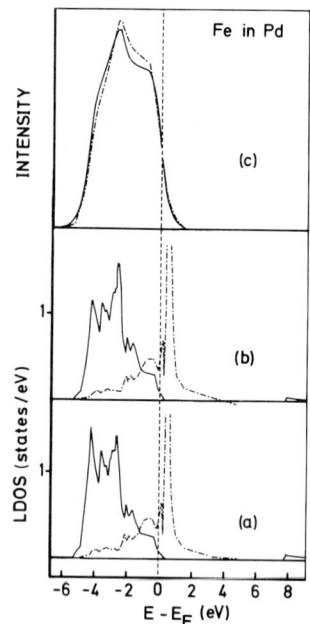

Fig. 14. Fe impurity in Pd [142]. (a) LDOS on Fe site obtained from Clogston–Wolff (C–W) model calculations. (b) LDOS on Fe site obtained from C–W calculations with the local magnetic moment reduced to 99% of the value of (a). (c) The occupied part of the LDOS on Fe broadened to simulate instrument (Gaussian, fwhm 0.6 eV) and lifetime (Lorentzian, fwhm 0.1 $(E - E_F)$ eV) effects. Single electron matrix elements (see ref. [126]) are taken into account by multiplying a sloping line of 15%/eV. Full curve: broadened DOS obtained from 9a). Dash–dotted curve: broadened DOS obtained from (b).

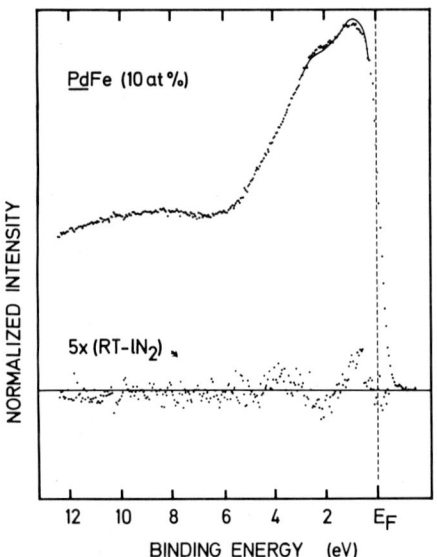

Fig. 13. Valence band photoemission spectra of PdFe, taken at 130 eV photon energy. The upper dotted curve represents the spectrum taken at liquid nitrogen temperature. The solid lines represent the room temperature spectrum, where it substantially differs from the liquid nitrogen temperature spectrum. The difference between these spectra, magnified 5 times, is presented by the lower dotted curve. (From ref. [142].)

ture magnetism of itinerant electron magnets, which is indicated by the invariance of the local magnetic moment, and the k-dependent behaviour of the exchange splitting, as a function of temperature. Controversy still exists, however, about the amount of short range magnetic order above T_c.

3.4. Surface magnetism and chemisorption

The loss of translational symmetry in the direction perpendicular to the surface can induce modifications in the surface electronic structure and result in different magnetic properties with respect to the bulk. The magnetic properties of an overlayer can differ even more from the bulk values, approaching the two-dimensional (2D) regime in the limit of vanishing overlayer–substrate interaction. However, the Mermin–Wagner theorem states that no long-range magnetic order exists in a 2D isotropic Heisenberg magnet with short-range magnetic interactions [146]. Overlayer magnetism exists, however, and deviations from the conditions leading to the Mermin–Wagner theorem (e.g., anisotropies or long-range magnetic interactions) are subject of investigation.

The study of surface and overlayer systems can be performed with a number of surface-sensitive HES techniques (see table 5). These techniques are applied for the study of the electronic structure of surfaces and overlayers. In this section we will show some examples thereof.

3.4.1. Surface magnetism

The magnetic properties of the surface can be investigated with surface sensitive techniques much like the bulk valence bands can be studied [147]. The first question to ask of the surface of a magnetic material is whether it is itself magnetic. That this is by no means a trivial question was shown by the report of magnetically "dead" Fe, Co and Ni surface layers [148,149]. At the moment, the consensus is that the magnetic moments of surface atoms are generally enhanced with respect to the bulk value. For Ni, the increased magnetic moment on the clean surface can be viewed [150] as a consequence of the band narrowing at the surface, accompanied by an upward electrostatic shift to maintain charge neutrality, resulting in a lower occupation of the minority spin band. An example of the detection of a magnetic surface through study of magnetic surface states will be shown. Furthermore, an example of the use of secondary electrons in surface magnetism will be given.

An elegant method to prove that the surface layer is magnetic is to study the magnetic properties of surface states. This can be done with (inverse) photoemission. Since the wave function of surface states is predominantly localized in the surface layer, detection of an exchange split surface state is a rather straightforward indication of surface magnetism. As an example, consider fig. 15. Here the IPES spectra of the S_1 surface state on Ni(110), situated around 6 eV above the Fermi level can be seen, spin-averaged in the left panel

Table 5
Sensitivities of HES spectroscopies to the spin moment S, the orbital moment L (total moment J), the surface electronic structure, the magnetic structure (magnitude and orientation of the magnetic moment), the site (or, equivalently, the element) and the symmetry (l). Neutron scattering is included to allow comparison. lim. denotes limited

Method	Magnetic moment		Surface	Bulk	Magnetic structure	Site	Symmetry
	spin S	orbit L (J)					
neutron	yes	yes	no	yes	yes	lim.	lim.
UPS	no	no	yes	yes	no	lim.	lim.
SPUPS	yes	no	yes	yes	lim.	lim.	lim.
XPS	no	no	lim.	yes	no	yes	lim.
IPES	no	no	yes	lim.	no	lim.	lim.
SPIPES	yes	no	yes	lim.	lim.	lim.	lim.
BIS	no	no	lim.	yes	no	lim.	lim.
BISCEPS	yes	no	lim.	yes	lim.	lim.	lim.
AES	no	no	yes	lim.	no	yes	lim.
SPAES	yes	no	yes	lim.	lim.	yes	lim.
SES	no	no	yes	lim.	no	no	no
SPSES	yes	no	yes	lim.	lim.	no	no
EELS	no	no	yes	lim.	no	lim.	no
SPEELS	yes	no	yes	lim.	lim.	lim.	no
XAS	no	no	yes	yes	no	yes	yes
MXD	yes	yes	yes	yes	yes	yes	yes
MXS	yes	yes	lim.	yes	yes	yes	yes

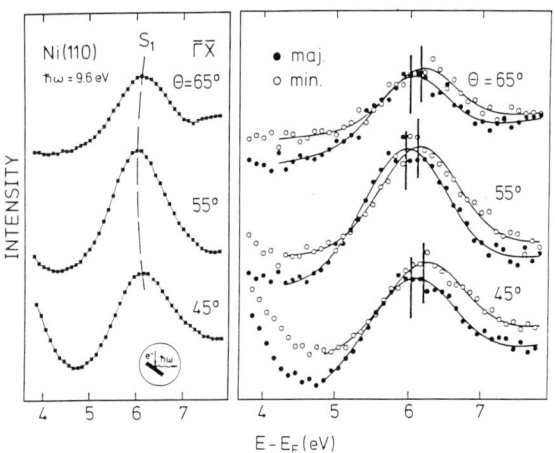

Fig. 15. Inverse photoemission spectra of the surface state S_1 around the $\overline{X}$ point of the surface Brillouin zone, spin-averaged (left) and spin-resolved (right), from ref. [53].

and spin-resolved in the right panel, for three angles of incidence θ. At least-squares fit yields an exchange splitting of ≈ 170 meV. The surface of Fe(001) was shown to be magnetic by Brookes et al., whose photoemission investigation yielded a minority surface state on the clean surface. These results prove that the surface layer of a clean ferromagnet is not magnetically "dead".

The temperature dependence of the surface magnetization, which can be determined by e.g., the secondary electron spin-polarization can yield additional information on the exchange coupling of the surface layer to the bulk. The magnetization at low temperatures at the surface, $M_s(T)$, should follow a $T^{3/2}$ law according to [69]:

$$M_s(T)/M_s(0) = 1 - \kappa C T^{3/2}, \qquad (6)$$

where C is a constant describing the decrease of bulk magnetization due to spin waves. According to Mathon and Ahmad [151], κ depends on $J_\perp/J$, where $J_\perp$ is the (interatomic) exchange interaction between surface and bulk, and J is the exchange interaction in the bulk. If this is a valid prediction, the measurement of $M_s(T)$ will yield information on the exchange interaction at surfaces. Mauri et al. have investigated the amorphous ferromagnet FeNiB$_{0.5}$ measuring the spin-polarization of low-energy secondary electrons. Their findings are shown in fig. 16. The solid line for $\kappa = 2$ seems to fit the experimental curve for the clean surface. However, Mauri et al. corrected for the probing depth and found an effective $\kappa = 1.3$ [69]. This means that for the clean surface of FeNiB$_{0.5}$, the exchange interaction in the direction perpendicular to the surface is reduced.

In summary, we have illustrated how HES techniques can be used to establish whether the true surface layer of a material exhibits magnetism, as demonstrated on a surface state with (inverse) photoemission. The electronic structure differs from that of the bulk, and HES techniques can be used to map those differences, e.g., the changed inter-atomic exchange interaction perpendicular to the surface.

3.4.2. Metallic overlayers

Thin magnetic films have attracted considerable interest in recent years (see, e.g., ref. [152] and the collection of papers edited by Pescia [153]). As an illustration of the possible uses, we

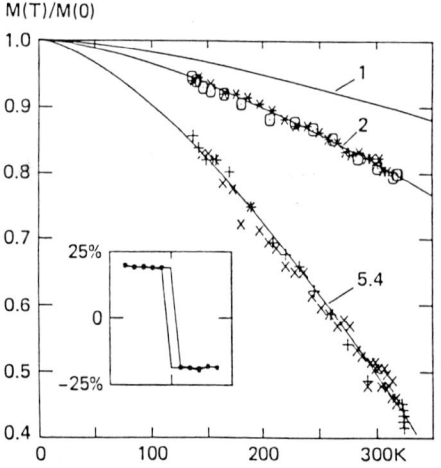

Fig. 16. Temperature dependence of the low-energy cascade relative spin polarization $P(T)/P(0)$ [69], for a clean surface of FeNiB$_{0.5}$ (cooling, circles; warming, asterisks) divided by $P(0) = 21\%$, and for the same surface covered with ≈ 0.5 monolayer of Ta (cooling, plusses; warming, crosses) divided by $P(0) = 12\%$. Solid lines are temperature dependencies calculated from eq. (6), with effective κ-values of $\kappa = 1$, 2 and 5.4. Inset: hysteresis loop measured with the clean surface at $T = 170$ K. The five points in saturation were averaged and the difference between the averages in positive and negative saturation yielded one datum point of $P(T)$.

will show that HES techniques are capable of detecting surface anisotropies, and yield the orientation of the overlayer magnetic moments with respect to the substrate moments.

The existence of anisotropies in an overlayer system with little interaction with the substrate, i.e., a quasi-2D system, allows the existence of an ordered magnetic system in accordance with the Mermin–Wagner theorem. If the overlayer–substrate hybridization is small, the overlayer system can be thought of as an approximation of the free monolayer (ML). Consider as an example, Fe on Ag(001), where the hybridization of the overlayer with the substrate is low, because the sp-density in Ag is not high and the d–d overlap will be small, because the substrate d-bands lie relatively far below the Fermi level. Also the spacing of the Fe atoms is larger than in bulk Fe, thus reducing d–d interaction in the overlayer. The calculated moment per Fe atom for a ML Fe on Ag(001) is $3.0\mu_B$ [154,155], which is much larger than the bulk value of $2.2\mu_B$ [3,156].

The orientation of spin moments in an Fe ML on Ag(001) was found from a theoretical treatment of Gay and Richter [157] to be oriented perpendicular to the film plane. Moreover, the orientation is thickness dependent, and the easy magnetization axis is directed in the plane of the film for larger thickness. The result of a spin-polarized PES experiment on epitaxial Fe films on Ag(001) by Jonker et al. [158] are consistent with the results of Gay and Richter in that they found the Fe ML on Ag to yield little spin polarization in the plane of the film. The spin-polarization increased for thicknesses from 2.5 ML upward, indicating a spin-polarization, and hence, magnetization, in the plane of the film. Similar results for fcc-Fe on Cu and Cu_3Au have been reported [159,160].

The detection of the magnitude of surface magnetic moments is a complicated matter (see section 3.6). However, the (relative) orientation of spin moments is much easier studied than the magnitude of the moments, and research has tended to probe these experimentally more accessible quantities. HES studies can provide a means for the detection of (relative) orientations. The systems in the studies thus far encountered in this

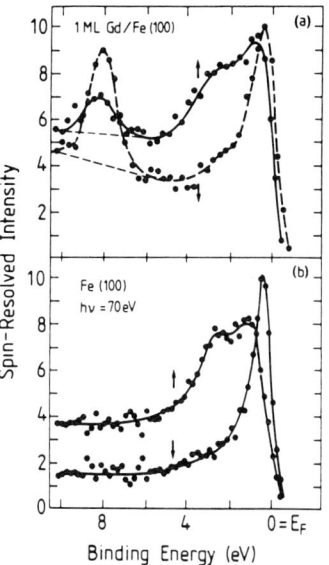

Fig. 17. (a) Spin-resolved energy distribution curves from 1 ML Gd on Fe(100) for normal emission and normal incident light at 70 eV photon energy, $T = 175$ K; (b) spin-resolved energy distribution curves from clean Fe(001) before Gd evaporation. (From ref. [161].)

paper show a ferromagnetic ordering. In the ferromagnetic state the magnetic moments are aligned parallel to each other, but configurations where the moments are aligned antiparallel (ferrimagnetic or antiferromagnetic structures) are also possible. Examples are the antiparallel configurations built up by 4f and 3d spins in a RE–TM system, as in fig. 17. The structure ≈ 8 eV below the Fermi level is formed by the Gd 4f levels (see fig. 6), while the Fe 3d majority-spin electrons are situated ≈ 3 eV below E_F, and the Fe 3d minority-spin electrons near the Fermi level, see figs. 7 and 8. The main intensity of the Gd 4f photoemission is concentrated in the minority-spin spectrum, indicating an antiparallel configuration with respect to the Fe majority spins. Note that near E_F the Fe majority-spin intensity increases while the minority-spin intensity decreases on deposition of Gd. At the same time, intensity is gained in the spin-down spectrum at the position of the spin-up peak. These are indications of a overall decrease of spin-polarization across the valence-band region. This might be the result of a decrease in magnetization in the interface region, or because of spin-flip scattering of the photoelectron in the Gd layer [161].

In conclusion, the orientation of the magnetic moments in an overlayer is an important parameter, either to describe the anisotropies in a system such as Fe/Ag(001), or the alignment of moments in an overlayer, to detect the exchange coupling to the substrate. Both systems have been studied with HES, and these studies have yielded the required information. Again, here HES yields unique information on ultra-thin layers, so that it is expected to play an increasingly important role if the present interest in metallic layers continues.

3.4.3. Chemisorption

"Is a surface still magnetic after chemisorption?" We have seen in the previous section that deposition of a metal on a surface might result in a slightly reduced magnetization of the substrate. The question is whether non-metallic matter will quench the magnetism. We will focus on the magnetic properties of substrate and adsorbate and show that the presence of magnetic moments for both is indicated by HES investigations.

Using SPUPS, Schmitt et al. [162] found a reduction of the splitting of majority and minority spin peaks after chemisorption of oxygen on a Ni(110) surface. This is shown in fig. 18. The total intensity of the photoemission peak after exposure to oxygen diminishes, and the two peaks end up closer to each other than before the chemisorption. This was interpreted as a quenching of the exchange splitting in the surface region. To explain the data it was necessary to assume up to four magnetically dead layers [163].

However, as discussed before, the exchange splitting is k-dependent. Therefore, in 1987, Schmitt et al. revisited their original experiment, extending their measurements with angular resolution [165]. Their findings are depicted in fig. 19. Observe that on deposition of oxygen or sulphur on the clean Ni(110) surface, the exchange splitting hardly changes. This in contrast to the results of the earlier study, depicted in fig. 18. The spectrum in the first study consisted of contributions of the $S_3^\uparrow$ and $S_4^\downarrow$ bands, that could not be

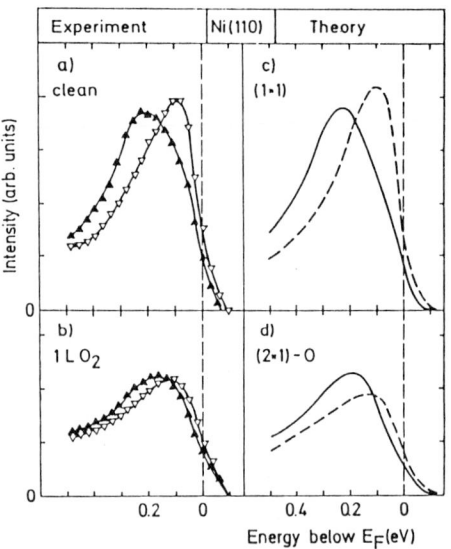

Fig. 18. Normal photoemission spectra from Ni(110) excited by unpolarized light ($h\nu = 16.85$ eV) at normal incidence. Experimental majority (solid triangles) and minority (open triangles) data for clean surface (panel (a)) and after exposure to 1L O_2 (panel (b)) [162]. Calculated results [163] for relaxed (1×1) surface (c), and for (2×1) saw-tooth reconstruction model with oxygen and four magnetically dead Ni layers (d). Figure taken from ref. [164].

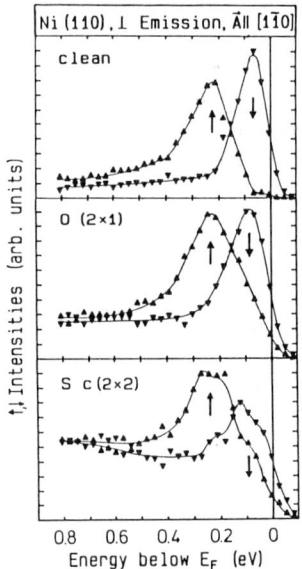

Fig. 19. Spin-resolved spectra for emission from Ni bands with S_4 symmetry for the clean, oxygen- and sulphur-covered (110) surfaces [165]. Triangles pointing up, majority-spin electrons; triangles pointing down, minority-spin electrons.

separated in that study. The almost unchanged intensity of the $S_3^\uparrow$ and the decrease of intensity from the $S_4^\downarrow$ bands led to the observed shift of the majority-spin spectrum towards lower binding energy, suggesting a decreased exchange splitting [165]. In the last study, emission from these two bands could be separated, and no change in exchange splitting is observed. This finding seems to be inconsistent with a dead surface layer, but that cannot be excluded completely, because of the reconstruction of the Ni(110) surface induced by oxygen [165]. The conclusion that the exchange splitting does not change much, if at all, is supported further by similar indication from AR-SPIPES measurements [166].

For oxygen, the influence on the substrate electronic structure seems small, and is mainly manifested in the reduction of the minority peak with respect to the majority peak. On adsorption of sulphur, however, the peaks develop shoulders at the position of the peak of opposite spin, and the relative weight of the peaks changes. This could be due to decrease of substrate magnetization. However, Schmitt et al. discuss the spectra in terms of a spin-dependent scattering in the adsorbate layer [165], which would also explain decrease of the minority-peak in fig. 19 for O/Ni(110).

Recent attention in research on chemisorption has focused on the magnetic behaviour of the adsorbates themselves. From theory it became clear that due to the hybridization of the adsorbate p-states with the metal d-states, the adsorbate might carry some magnetic moment (values of $\approx 0.2\mu_B$ for O/Fe(001) have been predicted (see, e.g., ref. [167]), with an *increase* of the surface magnetic moment of Fe). The evidence from HES supports this picture for oxygen or sulphur chemisorbed on Fe(001): using spin-polarized AES, polarization effects have been observed in an oxygen Auger peaks [168], and a 0.2 eV exchange splitting was found in an sulphur Auger peak [169]. Spin-polarized photoemission experiments detected a small magnetic moment on adsorbate oxygen and sulphur [170], and $k_\parallel$-dependent exchange splittings of adsorbate induced features on Fe(001) were recently reported [171,172]. An 80 meV exchange splitting has been observed for an O-2p derived empty band on Ni(110) (ref. [173], see also the discussion in ref. [53]).

In summary, HES has shown that chemisorption might reduce the substrate magnetization slightly, but does not kill the exchange splitting at the surface. Moreover, the adsorbate might aquire a magnetic moment of its own. It is true, that we have focused mainly on oxygen and sulphur chemisorbed on Ni and Fe, and these statements need not be true for every adsorbate and substrate. The chemisorption of, e.g., CO has been reported to reduce the magnetism in the Ni substrate [174].

3.5. Very narrow bands; electronic, magnetic and spectroscopic properties of Ce

In this section we will briefly explain some of the anomalous properties of Ce and then show how HES have contributed to our understanding of these properties. Ce is an element with many remarkable properties which are normally attributed to the 4f states. One example is the γ–α transition, which is a transition involving a nonmagnetic state. By 1950 it had been proposed [175] from chemical arguments that this transition involved promotion of the Ce 4f electron into the 5d bands, with consequent loss of the magnetic moment that would be expected of the Ce 4f electron, and the notion of mixed valence, with partial occupancy of the 4f level, emerged to explain the properties of Ce and its compounds. In the late 1960's model calculations [176] predicted a very sharp (10 meV) 4f state that was situated near (approximately 100 meV) the Fermi level. This picture, however, failed to take into account the strength of hybridization between the 4f levels and the Ce valence bands and has since been revised, mainly as a result of spectroscopic evidence.

The strength of the Ce 4f-valence band hybridization was considered by Kotani–Toyozawa [177] and Gunnarsson–Schönhammer (GS) [178]. When a core hole is created, as in XPS, all the wave functions of the other N-1 electrons of the material relax to adjust to the new potential. In the core-ionized state of lowest energy the valence electrons build up density around the atom where a core hole was created. The distribution of this extra charge may look like that in an atomic orbital, so that one can speak of "screening orbitals", and for RE the most effective screening orbital bears strong resemblance to a 4f orbital. The GS model is suitable for much solid state work because it incorporates partial occupation of the screening orbitals and hybridization effects, as well as the dynamics of photoemission. This is done by giving the screening orbital a width and energy in both the initial and final states which is proportional to the hybridization between the "localized" orbital and the other valence states, very much in the philosophy of the Anderson Hamiltonian [179]. This is illustrated in fig. 20, which also gives the results of numerical calculations [30,178].

In the initial state of photoemission the screening orbital is at an energy Δ_+ above the Fermi level, which can be several eV, as seen in XPS/BIS studies like those in fig. 6 and as represented in the inset of fig. 20. For γ–Ce it corresponds to the 4f^2 level which is actually at ≈ 4 eV [21]. In these models, the width W represents the coupling to the other levels of the material, but the experimental resolution, multiplet, and many-body effects make it impossible to take the width directly from the spectra. In the final state of the core level XPS experiment, the screening level is pulled down U_{ac} by the core hole potential. For the case of Ce this can be understood as if the core hole made the ionized Ce act like Pr, in which the 4f^2 state is occupied. The level is now Δ_- below E_F. The probability that the screening orbital will be occupied after core level photoemission is related to the energy of the orbital and its coupling to the other orbitals. If the screening orbital is narrow and far above E_F in the initial state, as shown in (a) of the inset of fig. 20, then its occupancy will be quite low. There is then very little overlap between the initial state and final states with the screening level occupied. This means that the probability of a transition to the "well screened" 4f^2 final state is small. Most of the XPS intensity will be found in the "poorly screened" state lying approximately Δ_- to higher BE and corresponding to the transition to a final state with the screening level almost unoccupied, as shown in curve (a). If, on the other hand, the width W of the screening level is of the same order as its position Δ_+, the level is strongly coupled to the system and its large tail below E_F represents considerable occupancy, even in the ground state. Conversely, in the final state W is also comparable to Δ_- and there is considerable unoccupied character in the well screened final state. The probability that the screening orbital becomes occupied in the final state of core-level photoemission is then high and most of the intensity is found in the so-called well screened peak, as shown in curve (d).

The right hand side of fig. 20 illustrates that there is a reasonable correlation between trends in core level line shapes and our expectations of the character of the screening orbital. In Ce, Th and Ti the screening orbitals have, respectively 4f, 5f and 3d character. In this series the 4f orbitals are the most strongly localized and decoupled from the rest of the system, whilst the 3d

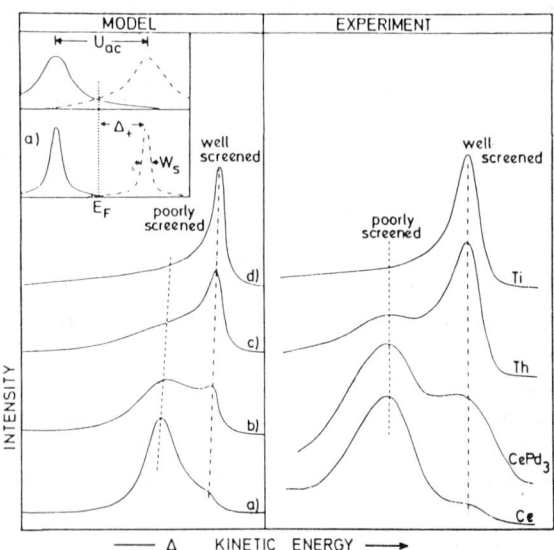

Fig. 20. Left: illustration of the Gunnarsson–Schönhammer model of screening and its implications for the core-level line-shapes. (From ref. [180].) The values of Δ_+/W are (d) 0.94, (c) 0.75, (b) 0.56, (a) 0.38 and U_{ac} was $1.5\Delta_+$ [30]. Right: Ce 3d$_{5/2}$, Th 4f$_{7/2}$ and Ti 2p$_{3/2}$ XPS peaks from Ce, CePd$_3$, Th and Ti. The peak binding energies are 883, 333 and 454 eV for the Ce, Th and Ti levels, respectively.

orbitals are the most strongly coupled. It is thus reasonable that Ce should have least spectral weight in the well-screened peak and Ti should have most. The values of the hybridization thus found from core level XPS studies range from ≈ 10–200 meV. High hybridization is found in materials which were traditionally regarded as "mixed valent", but in practice the Ce 4f counts are much higher than previously thought, and f-counts below 0.8 are rare in Ce intermetallic compounds [106]. It is now believed that the anomalous lattice contraction and the loss of magnetism in some Ce materials, such as α-Ce, is better explained by the hybridization of the Ce 4f levels than by promotion of a Ce 4f electron [31,181].

Another area involving magnetic interactions of the 4f levels of Ce is the low temperature thermal and transport properties. The idea of mixed valence, in its simplest form, implies that there must be a partially filled f band or level at, or very close to, the Fermi level in order to allow transfer of electrons between the 4f levels and the delocalized bands. Of course, such a narrow band at or near the Fermi levels has profound consequences for the ground state properties of a solid, such as specific heat, electrical resistivity and magnetic moments. However, it must be recognized that anomalously high low-temperature specific heats and electrical resistivities may be indicative of more than just a narrow band of one-electron states at the Fermi level, as in the case of magnetic impurities and the Kondo effect.

Impurities act as scattering centers, which at low temperatures will be the dominant source of electrical resistance, for instance. For non-magnetic scatterers, the resistivity decreases monotonically with decreasing temperature toward the so-called residual resistivity [182]. For dilute magnetic alloys, however, it has been known for some time that a resistivity minimum occurs at low (10–50 K) temperatures [183]. This is the so-called Kondo effect [184]. Kondo [185,186] showed that this arises only when the scattering center has a magnetic moment. At high temperatures, the impurity spin is free, which is manifested by the Curie–Weiss susceptibility. Lowering the temperature, the spin binds to a spin-polarization cloud in the electron gas, forming a single state. The temperature below which this occurs is called the Kondo temperature [187]. The Kondo effect is dependent on a sharp cut-off of the conduction wave vector distribution so that the divergence of the magnetic scattering distribution disappears at higher temperatures where the thermal distribution at the Fermi edge is rounded. The Kondo effect is a subtle and complex effect. It appears only when the magnetic interaction is treated to higher orders in perturbation theory, and it is far from trivial to extract the relevant physics even when one is given an adequate Hamiltonian to describe the magnetic interaction. Nevertheless, there is a widespread feeling [188] that, through the work of people like Gunnarsson, Schönhammer, Allen and Martin [178,189] we are slowly beginning to get more insight into the Kondo problem and in particular its relationship to the problem of mixed valence.

A complicated situation arises when the energy of a change in number of f electrons becomes comparable to the integrals mixing the localized f-levels with the delocalized conduction bands. For a Ce compound in which hybridization is not existent, with a 4f ionization energy of E_I eV (with respect E_F) and with a first electron affinity E_A above E_F, the XPS/BIS spectrum will show two peaks at $-E_I$ and $+E_A$ respectively. As the hybridization of the 4f level is increased new peaks appear at E_F in the XPS/BIS spectra. If the experimental resolution were high enough one would find that the *height* of the peaks at E_F would be inversely proportional to the strength of hybridization and the *weight* would increase with strength of hybridization. As an example of this effect we show in fig. 21 the PES and BIS spectrum of $CeNi_2$ and CeAl. The Kondo temperature of $CeNi_2$ is ≈ 900 K and a large peak appearing in both BIS and PES, centered slightly above E_F, is the Kondo resonance in this material. The Kondo temperature of CeAl is ≈ 1–10 K, and the weight at E_F due to the Kondo peak is experimentally insignificant here, although it plays a large role in transport properties.

This is also true of the core level spectra and it is probably safe to say that in combination these two methods have contributed greatly to our un-

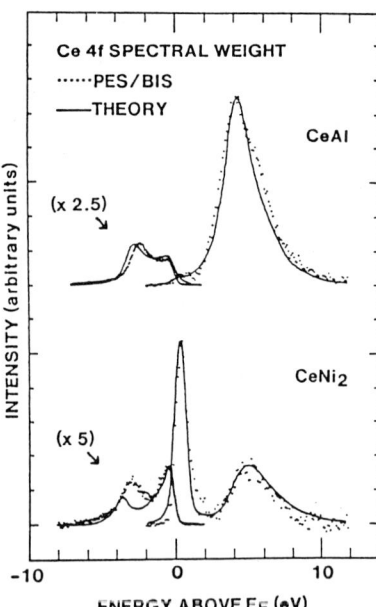

Fig. 21. Combined theoretical and experimental 4f PES and BIS spectra for CeAl and CeNi$_2$, as described in ref. [31]. The PES portion is increased by a scale factor for clarity. The Kondo temperature for CeAl and CeNi$_2$ are ≈ 10 and ≈ 900 K, respectively. The large peak appearing in both BIS and PES of CeNi$_2$, centered slightly above E_F, is the Kondo resonance in this material.

derstanding of the electronic structure of Ce and materials containing Ce. For the future we expect similar methods to be applied to TM and actinide compounds, where, in some cases, our understanding is probably now less complete than for Ce. It is now possible to obtain quite detailed information about the Ce 4f levels by modelling spectra such as those shown in fig. 21 and many materials have been studied [31].

3.6. Orientation of magnetic moments

The (relative) orientation of magnetic moments is of fundamental importance to the description of the magnetic properties in these systems. The determination of magnitude and orientation of both the spin moment S and the orbital moment L parts of the total magnetic moment (J) is the subject of this section.

In fig. 17 the photoemission spectrum of Gd on Fe(001) was depicted. The relative orientations of Gd and Fe spin moments could be determined because the Gd 4f and Fe 3d peaks are well separated in energy. However, photoemission is not an element-sensitive technique (table 5) and if the relevant peaks (partially) coincide in the PES spectrum, determination of the separate spin-moments will be very difficult. Much more suitable are techniques that are element-sensitive, such as (SP)AES. The spin-polarization of the Auger signal from the Gd/Fe(001) system is depicted in fig. 22. The fact that Auger electrons have characteristic energies allows the separation of overlayer and substrate contributions, and makes a comparison between the two possible. As can be seen from fig. 22, the Fe Auger electrons have a positive polarization, whereas the spin polarization of the Gd Auger electrons is largely negative. It is clear we are dealing with a magnetically ordered overlayer, and that the coupling of the Gd 4f spins to the Fe 3d spins is antiparallel. This seems to be the general behaviour of the RE spin moments on Fe or Ni [161,190,191].

Spin-polarized analysis yields information on the magnetization of a sample along one axis only. If the magnetization is directed perpendicular to that axis, the spin-polarization will be zero. Full determination of the orientation of the spin

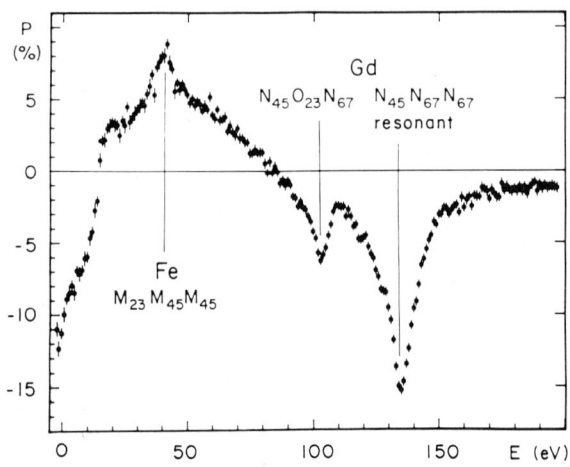

Fig. 22. Spin polarization versus kinetic energy of secondary electrons, including the labelled Auger transitions, from Gd ML film on Fe(100) at $T = 175$ K, excited with primary electrons of 2500 eV. (From ref. [63].)

means determination of the spin-polarization in three separate directions. The orientation of the spin-moment is therefore possible with the appropriate spin-polarized techniques. For the RE, however, the angular momentum of the 4f electrons generally dominates the magnetic moment. The angular momentum L cannot be detected by spin-polarized techniques (see table 5). For the description of RE and its alloys and compounds, that are known from, e.g., neutron scattering, to exhibit complex magnetic structures [192,193], it is also necessary to obtain information on the orientation of the magnetic moments.

In table 5 we present an overview of selected characteristics of several methods. Table 5 gives an indication of the sensitivities of these methods to properties of interest. Note that neutron scattering, magnetic X-ray scattering and magnetic dichroism in XAS are the only techniques giving direct information on the magnitude of L (and J). In the analysis of a magnetic alloy that consists of a variety of magnetic ions, the site selectivity of a spectroscopy is crucial. Moreover, in the presence of a complex magnetic structure, knowledge of the orientation of the total magnetic moment is a very important parameter, which must be known.

MXD on the $M_{4,5}$ (3d → 4f) absorption edges [80,84,85] can fulfil these conditions, because of the localized nature of the 4f final state. MXD spectra can even be obtained with linearly polarized light [80,84], as will be discussed here. The advantage of MXD with respect to, e.g., neutron scattering, lies in its surface sensitivity.

For RE, the transitions from the Hund's rule ground state $4f^n(J)$ to the numerous dipole-allowed final states $3d^94f^{n+1}(J')$ are well described by atomic Hartree–Fock calculations. The transitions can be divided in the three groups of spectral lines with $\Delta J = J' - J = -1, 0, -1$. In fig. 23, the absorption of a free Yb^{3+} ion is presented. In the final state of the absorption process the 4f shell is filled, allowing only the transition $J = 7/2 \rightarrow J' = 5/2$. In the presence of a magnetic field, the degenerate ground state will split into $2J + 1$ sublevels $M_J = -J, \ldots, +J$, where the relative population of the sublevels is determined by a Boltzmann distribution [84]. At low temperatures,

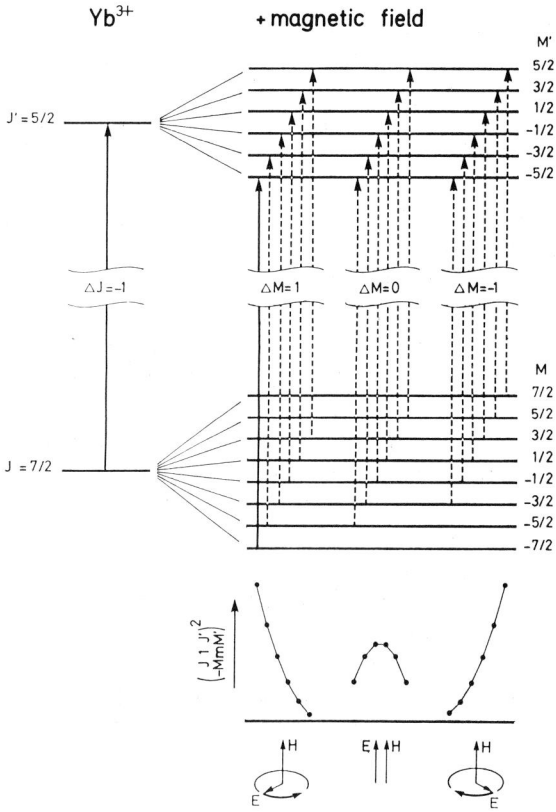

Fig. 23. Energy-level diagram of the $3d^{10}4f^{13} \rightarrow 3d^94f^{14}$ ($M_{4,5}$) transition of Yb^{3+} without (left) and with (right) a magnetic field [84]. The transitions $|JM\rangle \rightarrow |J'M'\rangle$ that are allowed by the optical dipole selection rules are indicated by the arrows. Their relative intensities are indicated by the dots. The polarization of the light with respect to the magnetic field H required for the different ΔM channels is indicated at the bottom.

with the thermal energy smaller than the magnetic splitting, only the lowest lying Zeeman level ($M_J = +J$) is occupied. Also, the final states are split by the magnetic field. However, the splittings of initial and final states, which are are of the order of meV, cannot be resolved due to the core-hole lifetime and instrumental broadenings, that are of the order of 1 eV.

The polarization dependence of the XAS process can now be used to overcome this, see fig. 23. One can visualize this as follows: if the polarization vector of the incident X-rays (E) is parallel to the magnetic field (H), only transitions with $\Delta M_J = 0$ are allowed, while for E perpendicular to H only transitions with $\Delta M_J = \pm 1$ are allowed. (Using circularly polarized light, the latter transitions can be resolved into $\Delta M_J = +1$ and $\Delta M_J = -1$ transitions, depending on the helicity of the light.) Because the ΔJ groups consist mainly of $-\Delta M_J$ transitions [84,85], the total absorption strength of each of the ΔJ groups changes with respect to the unpolarized absorption spectrum, as a function of the angle between polarization vector and magnetization vector. The ΔJ groups are well separated in energy, leading to immediately visible changes in the absorption spectrum as the angle of the polarization vector is changed with respect to the (internal) magnetic field H. For the case of Yb^{3+} at $T = 0$ K this would be most dramatic. Since only the lowest M_J level is occupied at this temperature, the only possible transition is the $\Delta M_J = +1$, i.e., if E is parallel to H, the absorption will vanish!

This mechanism will be most effective if only the lowest lying sublevel is occupied. At high temperatures, when all sublevels are equally populated, there is no longer any polarization dependence. We will refer to this spectrum as "unpolarized', since it is identical to the XAS spectrum obtained with unpolarized X-rays.

As an illustration, in fig. 24 the M_5 edge of Tb is shown. In the left panel, a calculation of the MXD effect is presented, whereas in the right panel, an experimental M_5 edge is shown, obtained from an Fe–Tb amorphous alloy. The alloy, which is a film of ≈ 1000 Å thick, exhibits a remanent magnetic field perpendicular to the film plane, as was shown by MOKE measurements. As depicted in the left panel of fig. 24, the Tb M_5 edge consists mainly of two peaks, flanked by shoulders. The left major peak will be dominant for $E \parallel H$, whereas the right major peak will dominate when $E \perp H$. A dominant left major peak is observed for the case where E is perpendicular to the surface, whereas for E (almost) parallel to the surface, the right major peak is found to be have the largest intensity. Consequently, the magnetic field vector is pointing out of the surface. From the temperature dependence of the MXD the magnitude of the local

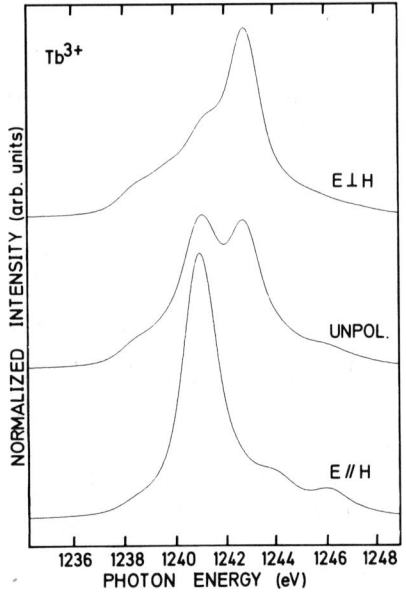
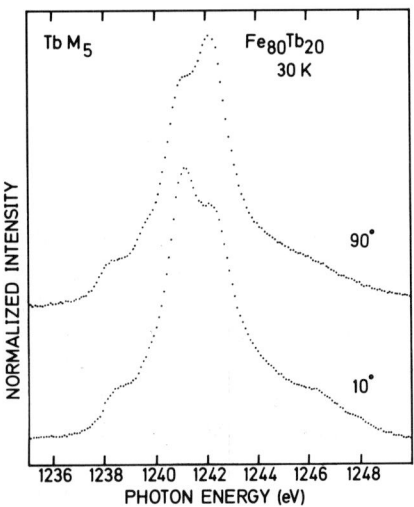

Fig. 24. MXD spectra of the Tb M_5 edge, for linearly polarized light. Left panel: Theoretical spectra for a Tb^{3+} free ion in a homogeneous magnetic field H (adapted from ref. [84]). Right panel: Experimental spectra of Tb ions in an $Fe_{80}Tb_{20}$ alloy, obtained with the polarization vector of the incident light perpendicular to the surface normal (upper curve: 90°) and almost parallel to the surface normal (lower curve: 10°).

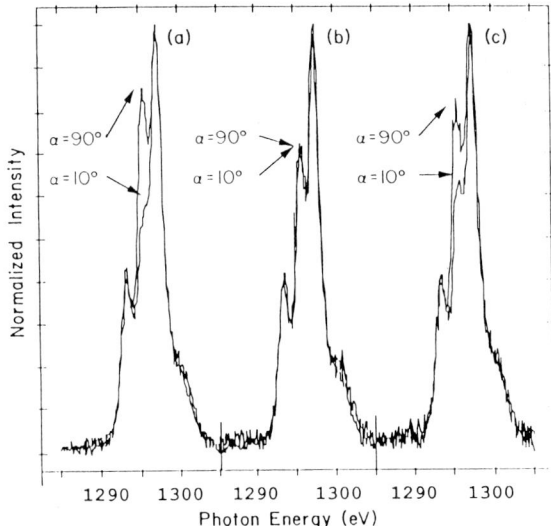

Fig. 25. Dy M_5 absorption edge measured at $\alpha = 10°$ and $\alpha = 90°$, where α is the angle between the polarization vector E of the linearly polarized light with respect to the surface normal [194]. (a) 0.2 ML Dy deposited at room temperature; (b) 0.1 ML deposited at 50 K; (c) same as (b), plus annealing at 900 °C.

magnetic field at the Tb site was determined to be ≈ 14 T, corresponding to a magnetic moment of $6.6\mu_B$.

The MXD technique thus yields size and orientation of local magnetic properties, as does neutron scattering. It is moreover also surface sensitive (see table 5), which is illustrated by fig. 25. Here the Dy M_5 edge is shown for submonolayer coverages of Dy on Si(111)7 × 7. The dichroism is clearly visible, even for 0.2 ML Dy.

One may wonder, however, whether a description in terms of magnetic fields is appropriate in this case. Surely, the substrate does not generate magnetic fields, and transition temperatures for Dy alloys lie generally far below room temperature, see, e.g., refs. [193,195]. A splitting of the atomic energy levels can be induced by a crystal electric field. The wavefunctions of these states can be written as linear combinations of atomic wavefunctions $|JM_J\rangle$. In the absence of a magnetic field, the atomic wavefunctions $|JM_J\rangle$ and $|J - M_J\rangle$ will have equal weight in the crystal-field wavefunction. MXD with linearly polarized light is sensitive to $\langle M_J^2 \rangle$, whereas if circularly polarized is used, MXD is sensitive to $\langle M_J \rangle$, where $\langle \rangle$ denotes a thermodynamic average. Thus, with linearly polarized light, the crystal field levels can be probed. In general, both Zeeman and crystal field splittings will be present, and can be resolved in MXD experiments using both linearly and circularly polarized light.

In summary of this section, we note that many high-energy spectroscopies yield information on the orientation of spin moments, particularly at surfaces. This source of information is particularly useful if it is site-specific. In addition to this, there are dichroism effects in XAS, as a result of magnetic and crystal-field effects. Use of these effects (and of magnetic X-ray scattering), is developing rapidly and will certainly yield unique information on the size of the orbital or total moments as well as the size of magnetic and crystal field splittings.

4. Concluding remarks and perspective

In this review we have presented some selected illustrations of the use of HES for research into magnetic materials. We have shown, that they can give detailed information on band widths, band dispersions, exchange interactions, Coulomb correlations, many-body effects, and the size and orientation of spin and orbital (or total) moments. Much of this information can only be obtainend from HES, and this is particularly true for surface effects.

It is common for reviewers to claim that their technique will be of growing importance in future research, and in this we are no different. However, as evidence we present the growing capital investment being made for these studies especially in synchrotron beamlines, and point out that this will inevitably lead to an increased number of published papers. Of particular interest is the area of X-ray dichroism, which will profit from new undulators producing helical magnetic fields, and hence, circularly polarized X-rays. These X-rays will be used primarily for dichroism studies in XAS, and resonant scattering.

Acknowledgements

We thank M. Altarelli, V. Dose, M. Donath, H. Ebert, H. Hopster, P.D. Johnson, M. Landolt, J. Kirschner, A. Kotani, G. Krill, D.T. Pierce, M. Sacchi and G.A. Sawatzky, who have helped us writing this paper by answerering questions by post or in discussions, or by reading the manuscript.

This work has been supported in part by the Stichting voor Fundamenteel Onderzoek der Materie (FOM) and the Stichting Scheikundig Onderzoek in Nederland (SON) with financial support from the Nederlandse Stichting voor Wetenschappelijk Onderzoek (NWO), and by the Committee for the European Development of Science and Technology (CODEST) program.

References

[1] E.C. Stoner, Proc. R. Soc. London A154 (1936) 656.
[2] E.C. Stoner, Rep. Prog. Phys. 11 (1947) 43.
[3] J. Callaway and C.S. Wang, Phys. Rev. B 16 (1977) 2095.
[4] K. Schwarz, P. Mohn, P. Blaha and J. Kübler, J. Phys. F 14 (1984) 2659.
[5] R.W. Godby, chap. 3 of ref. [6].
[6] J.C. Fuggle and J.E. Inglesfield, eds., Unoccupied Electronic States: Fundamentals for XANES, EELS, IPS and BIS. Topics in Applied Physics (Springer Verlag, Heidelberg, 1991) in press.
[7] J.C. Fuggle, in: Proc. Intern. School of Physics "Enrico Fermi", Photoemission and Absorption Spectr. of Solids and Interfaces with Synchrotron Radiation, ed. R. Rosei (North-Holland, Amsterdam, 1990).
[8] F.J. Himpsel, Adv. Phys. 32 (1983) 1.
[9] C.R. Brundle and A.D. Baker, eds., Electron Spectroscopy, vols. 1 and 2 (Academic Press, London, 1977).
[10] D. Briggs, ed., Handbook of X-ray and Ultraviolet photoelectron Spectroscopy (Heyden, London, 1977).
[11] M. Cardona and L. Ley, eds., Photoemission in Solids, vols. 26 and 27 of Topics in Applied Physics (Springer Verlag, Berlin, 1978).
[12] B. Feuerbacher, B. Fitton and R.F. Willis, eds., Photoemission and the Electronic Properties of Surfaces (Wiley, New York, 1978).
[13] V. Dose, Surface Sci. Rep. 5 (1985) 377.
[14] F.J. Himpsel, Comments Cond. Mat. Phys. 12 (1986) 199.
[15] N.V. Smith, Rep. Prog. Phys. 51 (1988) 1227.
[16] R. Schneider and V. Dose, chap. 9 of ref. [6].
[17] H. Hoekstra, W. Speier, R. Zeller and J.C. Fuggle, Phys. Rev. B 34 (1986) 5177.
[18] Y. Baer and G. Busch, Phys. Rev. Lett. 30 (1973) 289.
[19] J. Hubbard, Proc. R. Soc. London A276 (1963) 238, A277 (1964) 237.
[20] P.A. Cox, J.K. Lang and Y. Baer, J. Phys. F 11 (1981) 113.
[21] J.K. Lang, Y. Baer and P.A. Cox, J. Phys. F 11 (1981) 121.
[22] J.C. Fuggle, G.A. Sawatzky and J.W. Allen, eds., Narrow-Band Phenomena – Influence of Electrons with both Band and Localized Character, vol. 184 of NATO ASI Series B (Plenum Press, New York, 1988).
[23] C.S. Fadley, in: Electron Spectroscopy, ed. D.A. Shirley (North-Holland, Amsterdam, 1972).
[24] J.F. van Acker, Z.M. Stadnik, J.C. Fuggle, H.J. Hoekstra, K.H.J. Buschow and G. Stroink, Phys. Rev. B 37 (1988) 6827.
[25] F.U. Hillebrecht, R. Jungblut and E. Kisker, Phys. Rev. Lett. 65 (1990) 2450.
[26] B. Sinković, B. Hermsmeier and C.S. Fadley, Phys. Rev. Lett. 55 (1985) 1227.
[27] G.D. Mahan, Phys. Rev. 163 (1967) 612; Phys. Rev. B 11 (1979) 4814, and references therein.
[28] S. Doniach and M. Šunjić, J. Phys. C 3 (1970) 285.
[29] Z.M. Stadnik, J.C. Fuggle, T. Miyazaki and G. Stroink, Phys. Rev. B 35 (1987) 7400.
[30] J.C. Fuggle, M. Campagna, R. Zołnierek, R. Lässer and A. Platau, Phys. Rev. Lett. 45 (1980) 1597.
[31] J.W. Allen, S.J. Oh, O. Gunnarsson, K. Schönhammer, M.B. Maple, M.S. Torikachvili and I. Lindau, Adv. Phys. 35 (1986) 275.
[32] N.F. Mott, Proc. R. Soc. London A124 (1929) 425.
[33] C.G. Shull, C.T. Chase and F.E. Myers, Phys. Rev. 63 (1943) 29.
[34] J. Kessler, Polarized Electrons, vol. 1 of Series on Atoms and Plasmas (Springer Verlag, Berlin, second ed., 1985).
[35] F.B. Dunning, L.G. Gray, J.M. Ratliff, F.-C. Tang and G.K. Walters, Rev. Sci. Instrum. 58 (1987) 1706.
[36] M.R. Scheinfein, D.T. Pierce, J. Unguris, J.J. McClelland, R.J. Celotta and M.H. Kelley, Rev. Sci. Instrum. 60 (1989) 1, and although the authors probably feel this design should not be called a Mott detector, we have classified it as such since its basic principles are similar. It should be noted, however, that the design allows operation in the 100–200 V range.
[37] M.R. O'Neill, M. Kalisvaart, F.B. Dunning and G.K. Walters, Phys. Rev. Lett. 34 (1975) 1167.
[38] J. Kirschner and R. Feder, Phys. Rev. Lett. 42 (1979) 1008.
[39] D. Tillmann, R. Thiel and E. Kisker, Z. Phys. B 77 (1989) 1.
[40] D.T. Pierce and F. Meier, Phys. Rev. B 13 (1976) 5484.
[41] J. Kirschner, chap. 5 of ref. [147].
[42] Linearly polarized or even unpolarized light can also yield spin-polarized photoelectrons, either due to final

state effects, spin-dependent photoelectron diffraction or polarization effects in the transmission of the photoelectron through the surface (see, e.g., refs. [196,197]).

[43] A. Eyers, F. Schäfers, G. Schönhense, U. Heinzmann, H.P. Oepen, K. Hünlich and J. Kirschner, Phys. Rev. Lett. 52 (1984) 1559.

[44] J. Garbe, D. Venus, S. Suga, C. Schneider and J. Kirschner, Surface Sci. 178 (1986) 342.

[45] The measurement of the spin-polarization of the photoelectrons without analyzing their energies is called spin-polarized total photoyield. This was the technique used in the first spin-resolved photoemission experiments.

[46] G. Busch, M. Campagna, P. Cotti and H.C. Siegmann, Phys. Rev. Lett. 22 (1969) 597.

[47] U. Bänninger, G. Busch, M. Campagna and H.C. Siegmann, Phys. Rev. Lett. 25 (1970) 585.

[48] E. Kisker, W. Gudat, E. Kuhlmann, R. Clauberg and M. Campagna, Phys. Rev. Lett. 45 (1980) 2053.

[49] R. Clauberg, W. Gudat, E. Kisker, E. Kuhlmann and G.M. Rothberg, Phys. Rev. Lett. 47 (1981) 1314.

[50] R. Raue, H. Hopster and R. Clauberg, Phys. Rev. Lett. 50 (1983) 1623.

[51] H. Hopster, R. Raue, G. Güntherodt, E. Kisker, R. Clauberg and M. Campagna, Phys. Rev. Lett. 51 (1983) 829.

[52] E. Kisker, K. Schröder, M. Campagna and W. Gudat, Phys. Rev. Lett. 52 (1984) 2285.

[53] M. Donath, Appl. Phys. A 49 (1989) 351.

[54] M. Donath, V. Dose, K. Ertl and U. Kolac, Phys. Rev. B 41 (1990) 5509.

[55] J. Unguris, A. Seiler, R.J. Celotta, D.T. Pierce, P.D. Johnson and N.V. Smith, Phys. Rev. Lett., 49 (1982) 1047.

[56] H. Scheidt, M. Glöbl, V. Dose and J. Kirschner, Phys. Rev. Lett. 51 (1983) 1688.

[57] H.R. Borsje, H.W.H.M. Jongbloets, R.J.H. Kappert, J.C. Fuggle, S.F. Alvarado, R. Rochow and M. Campagna, Rev. Sci. Instrum. 61 (1990) 765.

[58] J.C. Fuggle, in: Electron Spectroscopy, eds. C.R. Brundle and A.D. Baker, vol. 4 (Academic Press, London, 1981).

[59] M. Cini, Solid State Commun. 24 (1977) 681; Phys. Rev. B 15 (1978) 2788.

[60] G.A. Sawatzky, Phys. Rev. Lett. 34 (1977) 504.

[61] M. Landolt and D. Mauri, Phys. Rev. Lett. 49 (1982) 1783.

[62] H. Mizuta and A. Kotani, J. Phys. Soc. Jpn. 54 (1985) 4452.

[63] M. Taborelli, R. Allenspach, G. Boffa and M. Landolt, Phys. Rev. Lett. 56 (1986) 2869.

[64] R. Allenspach, D. Mauri, M. Taborelli and M. Landolt, Phys. Rev. B 35 (1987) 4801.

[65] G. Chobrok and M. Hofmann, Phys. Lett. A 57 (1976) 257.

[66] J. Unguris, D.T. Pierce, A. Galejs and R.J. Celotta, Phys. Rev. Lett. 49 (1982) 72.

[67] M. Landolt, Appl. Phys. A 41 (1986) 83.

[68] K. Koike and H. Hayakawa, J. Appl. Phys. 57 (1985) 4244.

[69] D. Mauri, D. Scholl, H.C. Siegmann and E. Kay, Phys. Rev. Lett. 61 (1988) 758.

[70] D.L. Abraham and H. Hopster, Phys. Rev. Lett. 58 (1987) 1352.

[71] J. Fink, chap. 7 of ref. [6].

[72] J. Kirschner, Phys. Rev. Lett. 55 (1985) 973.

[73] D. Venus and J. Kirschner, Phys. Rev. B 37 (1988) 2199.

[74] R.D. Cowan, The Theory of Atomic Structure and Spectra (University of California Press, Berkeley, 1981).

[75] D.C. Koningsberger and R. Prins, X-Ray Absorption: Principles, Applications, Techniques of EXAFS, SEXAFS and XANES (Wiley, New York, 1988).

[76] D.D. Vvedensky, chap. 5 of ref. [6].

[77] J. Kerr, Phil. Mag. 3 (1877) 339.

[78] J.L. Erskine and E.A. Stern, Phys. Rev. B 12 (1975) 5016.

[79] E. Keller and E.A. Stern, EXAFS and Near Edge Structure III (Springer, Berlin, 1984).

[80] B.T. Thole, G. van der Laan and G.A. Sawatzky, Phys. Rev. Lett. 55 (1985) 2086.

[81] G. van der Laan, B.T. Thole, G.A. Sawatzky, J.B. Goedkoop, J.C. Fuggle, J.-M. Esteva, R. Karnatak, J.P. Remeika and H.A. Dabkowska, Phys. Rev. B 34 (1986) 6529.

[82] G. Schütz, W. Wagner, W. Wilhelm, P. Kienle, R. Zeller, R. Frahm and G. Materlik, Phys. Rev. Lett. 58 (1987) 737.

[83] G. Schütz, M. Knülle, R. Wienke, W. Wilhelm, W. Wagner, P. Kienle and R. Frahm, Z. Phys. B 73 (1988) 67.

[84] J.B. Goedkoop, B.T. Thole, G. van der Laan, G.A. Sawatzky, F.M.F. de Groot and J.C. Fuggle, Phys. Rev. B 37 (1988) 2086.

[85] J.B. Goedkoop, Ph.D. thesis, University of Nijmegen (1989).

[86] H. Ebert, P. Strange and B.L. Gyorffy, Z. Phys. B 73 (1988) 77.

[87] U. Fano, Phys. Rev. 178 (1969) 131, 184 (1969) 250.

[88] M. Blume, J. Appl. Phys. 57 (1985) 3615.

[89] D. Gibbs, D.R. Harshman, E.D. Isaacs, D.B. McWhan, D. Mills and C. Vettier, Phys. Rev. Lett. 61 (1988) 1241.

[90] J.P. Hannon and G.T. Trammell, Phys. Rev. Lett. 61 (1988) 1245.

[91] P. Carra, M. Altarelli and F. de Bergevin, Phys. Rev. B 40 (1989) 7324.

[92] W. Speier, J.C. Fuggle, R. Zeller, B. Ackermann, K. Szot, F.U. Hillebrecht and M. Campagna, Phys. Rev. B 30 (1984) 6921.

[93] J.C. Fuggle and D.J. Fabian, in: Proc. 1972 Intern. Symp. on X-ray Spectroscopy and Electronic Structure of Matter, eds. A. Fässler and G. Wiech (Fotodruck Frank, Munich, 1973).

[94] P.A. Bennett, J.C. Fuggle, F.U. Hillebrecht, A.

Lenselink and G.A. Sawatzky, Phys. Rev. B 27 (1983) 2194.
[95] J.C. Fuggle, L.M. Watson, D.J. Fabian and P.R. Norris, Solid State Commun. 13 (1973) 507.
[96] B. Kolita and L.M. Falicov, J. Phys. C 7 (1974) 299.
[97] G.A. Sawatzky, in ref. [22], p. 121.
[98] J.C. Fuggle, F.U. Hillebrecht, R. Zeller, Z. Zołnierek, P.A. Bennet and Ch. Freiburg, Phys. Rev. B 27 (1983) 2145, and references therein.
[99] J.A.D. Matthew, J. Phys. C 11 (1978) L47.
[100] D. van der Marel, Ph.D. thesis, University of Groningen (1985).
[101] D. van der Marel and G.A. Sawatzky, Phys. Rev. B 37 (1988) 10674.
[102] J.C. Fuggle, F.U. Hillebrecht, Z. Zołnierek, R. Lässer, Ch. Freiburg, O. Gunnarsson and K. Schönhammer, Phys. Rev. B 27 (1983) 7330.
[103] E. Wuilloud, H.R. Moser, W.-D. Schneider and Y. Baer, Phys. Rev. B 28 (1983) 7354.
[104] J.C. Fuggle, N.B. Mårtensson, J.W. Allen, S.-J. Oh, O. Gunnarsson and K. Schönhammer, unpublished.
[105] E. Wuilloud, W.-D. Schneider, B. Delley, Y. Baer and F. Hulliger, J. Phys. C 17 (1984) 4799.
[106] J.C. Fuggle, Physica B 130 (1985) 56, and references therein.
[107] J.F. van Acker, E.W. Lindeijer, J.C. Fuggle, Z.M. Stadnik and K.H.J. Buschow, unpublished.
[108] D.E. Eastman, F.J. Himpsel and J.A. Knapp, Phys. Rev. Lett. 44 (1980) 95.
[109] J. Kirschner, M. Glöbl, V. Dose and H. Scheidt, Phys. Rev. Lett. 53 (1984) 612.
[110] A. Santoni and F.J. Himpsel, Phys. Rev. B 43 (1991) 1305.
[111] J. Glazer and E. Tosatti, Solid State Commun. 52 (1984) 905.
[112] H. Hopster, R. Raue and R. Clauberg, Phys. Rev. Lett. 53 (1984) 695.
[113] S. Modesti, F. Della Valle, R. Rosei, E. Tosatti and J. Glazer, Phys. Rev. B 31 (1985) 5471.
[114] D.L. Mills, Phys. Rev. B 34 (1986) 6099.
[115] J. Kirschner, D. Rebenstorff and H. Ibach, Phys. Rev. Lett. 53 (1984) 698.
[116] D.L. Abraham and H. Hopster, Phys. Rev. Lett. 62 (1989) 1157.
[117] J. Kirschner and S. Suga, Surface Sci. 178 (1986) 327.
[118] C.T. Chen, N.V. Smith and F. Sette, Phys. Rev. B 43 (1991) 6785.
[119] H. Ebert, P. Strange and B.L. Gyorffy, J. Phys. F. 18 (1988) L135.
[120] O. Eriksson, B. Johansson, R.C. Albers, A.M. Boring and M.S.S. Brooks, Phys. Rev. B 42 (1990) 2707.
[121] T. Jo and G.A. Sawatzky, Phys. Rev. B 43 (1991) 8771.
[122] P. Carra and M. Altarelli, Phys. Rev. Lett. 64 (1990) 1286.
[123] A. Liebsch, Phys. Rev. Lett. 43 (1979) 1431; Phys. Rev. B 23 (1981) 5203.
[124] V.V. Nemoshkalenko and V.G. Aleshin, Teoreticheskie Osnovy Rentgenovky Emissionnoy Spektroskopii (Naukova Dumke, Kiev, 1974).
[125] A. Neckel, K. Schwarz, R. Eibler, P. Rastland and P. Weinberger, Mikrochim. Acta Suppl. 6 (1975) 257.
[126] W. Speier, J.C. Fuggle, P. Durham, R. Zeller, R.J. Blake and P. Sterne, J. Phys. C 21 (1988) 2621.
[127] O. Gunnarsson, J. Phys. F 6 (1976) 587; Physica B 91 (1977) 329.
[128] V. Korenman, J.L. Murray and R.E. Prange, Phys. Rev. B 16 (1977) 4032, 4048, 4058.
[129] R.E. Prange and V. Korenman, Phys. Rev. B 19 (1979) 4691.
[130] H. Capellmann, J. Phys. F 4 (1974) 1466; Z. Phys. B 34 (1979) 29.
[131] J. Hubbard, Phys. Rev. B 20 (1979) 4584.
[132] H. Hasegawa, J. Phys. Soc. Jpn. 49 (1980) 178.
[133] A.J. Pindor, J. Staunton, G.M. Stocks and H. Winter, J. Phys. F 13 (1983) 979.
[134] J. Staunton, B.L. Gyorffy, A.J. Pindor, G.M. Stocks and H. Winter, J. Magn. Magn. Mater. 45 (1984) 15.
[135] J. Kirschner and E. Langenbach, Solid State Commun. 66 (1988) 761.
[136] H. Hopster and D.L. Abraham, Phys. Rev. B 40 (1989) 7054.
[137] E. Kisker, K. Schröder, W. Gudat and M. Campagna, Phys. Rev. B 31 (1985) 329.
[138] J. Kirschner, Surface Sci. 162 (1985) 83.
[139] W. Nolting, W. Borgieł, V. Dose and Th. Fauster, Phys. Rev. B 40 (1989) 5015.
[140] W. Borgieł, W. Nolting and M. Donath, Solid State Commun. 72 (1989) 825.
[141] W. Borgiel and W. Nolting, Z. Phys. B 78 (1990) 241.
[142] R.J.H. Kappert, H.R. Borsje, J.F. van Acker, K. Horn, H. Haak, K.H.J. Buschow and J.C. Fuggle, Phys. Rev. B 43 (1991) 3259.
[143] J.W. Cooper, Phys. Rev. 128 (1962) 681.
[144] J.F. van Acker, P.W.J. Weijs, J.C. Fuggle, K. Horn, W. Wilke, H. Haak, H. Saalfield, H. Kuhlenbeck, W. Braun, G.P. Williams, D. Wesner, M. Strongin, S. Krummacher and K.H.J. Buschow, Phys. Rev. B 38 (1988) 10463.
[145] J.F. van Acker, Ph.D. thesis, University of Nijmegen (1990).
[146] N.D. Mermin and H. Wagner, Phys. Rev. Lett. 17 (1966) 1133.
[147] R. Feder, ed., Polarized electrons in surface physics (World Scientific, Singapore, 1985).
[148] L.N. Liebermann, D.R. Fredkin and H. Shore, Phys. Rev. Lett. 22 (1969) 539.
[149] L.N. Liebermann, J. Clinton, D.M. Edwards and J. Mathon, Phys. Rev. Lett. 25 (1970) 232.
[150] M. Weinert and J.W. Davenport, Phys. Rev. Lett. 54 (1985) 1547.
[151] J. Mathon and S.B. Ahmad, Phys. Rev. B 37 (1988) 660.
[152] L.M. Falicov, D.T. Pierce, S.D. Bader, R. Gronsky, K.B. Hathaway, H. Hopster, D.N. Lambeth, S.S.P. Parkin, G.

Prinz, M. Salomon, I.K. Schuller and R.H. Victora, J. Mater. Res. 5 (1990) 1299.
[153] D. Pescia, ed., Appl. Phys. A 49 (1989) 437.
[154] R. Richter, J.G. Gay and J.R. Smith, Phys. Rev. Lett. 54 (1985) 2704.
[155] C.L. Fu, A.J. Freeman and T. Oguchi, Phys. Rev. Lett. 54 (1985) 2700.
[156] S. Ohnishi, A.J. Freeman and M. Weinert, Phys. Rev. B 28 (1983) 6741.
[157] J.G. Gay and R. Richter, Phys. Rev. Lett. 56 (1986) 2728.
[158] B.T. Jonker, K.H. Walker, E. Kisker, G.A. Prinz and C. Carbone, Phys. Rev. Lett. 57 (1986) 142.
[159] C. Carbone, G.S. Sohal, E. Kisker and E.F. Wassermann, J. Appl. Phys. 63 (1988) 3499.
[160] R. Rochow, C. Carbone, Th. Dodt, F.P. Johnen and E. Kisker, Phys. Rev. B 41 (1990) 3426.
[161] C. Carbone and E. Kisker, Phys. Rev. B 36 (1987) 1280.
[162] W. Schmitt, H. Hopster and G. Güntherodt, Phys. Rev. B 31 (1985) 4035.
[163] R. Feder and H. Hopster, Solid State Commun. 55 (1985) 1043.
[164] R. Clauberg and R. Feder, chap. 14 of ref. [147].
[165] W. Schmitt, K.-P. Kämper and G. Güntherodt, Phys. Rev. B 36 (1987) 3763.
[166] L.E. Klebanoff, R.K. Jones, D.T. Pierce and R.J. Celotta, Phys. Rev. B 36 (1987) 7849.
[167] H. Huang and J. Hermanson, Phys. Rev. B 32 (1985) 6312.
[168] R. Allenspach, M. Taborelli and M. Landolt, Phys. Rev. Lett. 55 (1985) 2599.
[169] B. Sinković, P.D. Johnson, N.B. Brookes, A. Clarke and N.V. Smith, Phys. Rev. Lett. 62 (1989) 2740.
[170] P.D. Johnson, A. Clarke, N.B. Brookes, S.L. Hulbert, B. Sinković and N.V. Smith, Phys. Rev. Lett. 61 (1988) 2257.
[171] A. Clarke, N.B. Brookes, P.D. Johnson, M. Weinert, B. Sinković and N.V. Smith, Phys. Rev. B 41 (1990) 9659.
[172] P.D. Johnson, J. Electron Spectr. and Rel. Phen. 51 (1990) 249.
[173] G. Schönhense, M. Donath, U. Kolac and V. Dose, Surface Sci. Lett. 206 (1988) L888.
[174] C.S. Feigerle, A. Seiler, J.L. Peña, R.J. Celotta and D.T. Pierce, Phys. Rev. Lett. 56 (1986) 2207.
[175] W.H. Zachariassen, quoted by A.W. Lawson and T.Y. Tang, Phys. Rev. 76 (1949) 301.

L. Pauling, quoted by A.F. Schuck and J.H. Sturdivant, J. Chem. Phys. 18 (1950) 145.
[176] R. Ramirez and L.M. Falicov, Phys. Rev. B 3 (1971) 2425.
[177] A. Kotani and Y. Toyozawa, J. Phys. Soc. Jpn. 35 (1973) 1073.
[178] O. Gunnarsson and K. Schönhammer, Phys. Rev. Lett. 50 (1983) 604; Phys. Rev. B 28 (1983) 4315.
[179] P.W. Anderson, Phys. Rev. 124 (1961) 41.
[180] J.C. Fuggle, in ref. [22], p. 166.
[181] F. Patthey, B. Delley, W.-D. Schneider and Y. Baer, Phys. Rev. Lett. 55 (1985) 1518, 57 (1986) 270, 58 (1987) 1283 (E).
[182] N.W. Ashcroft and N.D. Mermin, Solid State Physics (Holt-Saunders, Philadelphia, 1976).
[183] W. Meissner and B. Voigt, Ann. Phys. (Leipzig) 7 (1930) 761, 892.
[184] In general, the problem of a correlated impurity in a metal is called a "valence fluctuation" or "mixed valency" problem [187]. If the impurity can be described by its spin only, it is named a Kondo system.
[185] J. Kondo, Prog. Theor. Phys. 32 (1964) 37.
[186] J. Kondo, in: Solid State Physics, vol. 23, eds. H. Ehrenreich, D. Turnbull and F. Seitz (Academic Press, New York, 1969).
[187] J. Zaanen, chap. 4 of ref. [6].
[188] J.W. Wilkins, Phys. Today 39 (1986) S22.
[189] R.M. Martin and J.W. Allen, J. Magn. Magn. Mater. 47 & 48 (1985) 257.
[190] O. Paul, S. Toscano, W. Hürsch and M. Landolt, J. Magn. Magn. Mater. 84 (1990) L7.
[191] C. Carbone, R. Rochow, L. Braicovich, R. Jungblut, T. Kachel, D. Tillmann and E. Kisker, Phys. Rev. B 41 (1990) 3866.
[192] D.H. Martin, Magnetism in Solids (MIT Press, Cambridge, 1967).
[193] K.A. Gschneidner, Jr. and L. Eyring, eds., Handbook on the Physics and Chemistry of Rare Earths, vol. 2 (North-Holland, Amsterdam, 1979).
[194] M. Sacchi, O. Sakho and G. Rossi, Phys. Rev. B 43 (1991) 1276.
[195] K.H.J. Buschow, Rep. Prog. Phys. 40 (1977) 1179.
[196] R. Feder, chap. 4 of ref. [147].
[197] B. Schmiedeskamp, B. Vogt and U. Heinzmann, Phys. Rev. Lett. 60 (1988) 651.

Spontaneous nuclear magnetic ordering in copper and silver at nano- and picokelvin temperatures

Pertti Hakonen, Olli V. Lounasmaa and Aarne Oja

Low Temperature Laboratory, Helsinki University of Technology, 02150 Espoo, Finland

Owing to the weak mutual interactions, spontaneous nuclear magnetic ordering in metallic copper and silver occurs at 60 nK and 560 pK, respectively. These extremely low spin temperatures can be reached by two-stage adiabatic nuclear demagnetization. Spin ordering has been investigated by employing magnetic susceptibility measurements on copper and silver and by using neutron diffraction techniques on copper. Three antiferromagnetic phases in the field–entropy plane have been discovered in copper, caused by competition between the dipolar and Ruderman–Kittel exchange interactions; only one ordered state has been found in silver. Negative spin temperatures have been produced in silver as well, and a clear ferromagnetic tendency was observed when $T < 0$. The theoretically calculated spin–spin interactions, ordering temperatures, magnetic phase diagrams and ordered spin structures are in good overall agreement with experimental data for these two metals.

1. Introduction

Nuclear spins in metals provide good models to investigate magnetism. The nuclei are well localized, their spins are isolated from the electronic and lattice degrees of freedom at low temperatures, and the interactions between nuclear spins can be calculated from first principles. Therefore, nuclear magnets are particularly suitable for testing theory against experiments. Because the nuclear magneton is small, the critical temperatures for spontaneous magnetic ordering are in the submicrokelvin range. Much higher nuclear ordering temperatures, around 1 mK, have been observed in solid ^{3}He [1], owing to the strong quantum mechanical exchange force enhanced by the large zero-point motion, and in Van Vleck paramagnets [2], like $PrNi_5$ [3], in which considerable hyperfine enhancement of the magnetic field occurs.

Experiments on nuclear magnetic ordering in metals are based on the pioneering studies of Kurti and his coworkers [4]. They established, in 1956, the feasibility of the nuclear demagnetization method. In spite of the limitations imposed by cryogenic techniques available at that time, the Oxford group succeeded in reaching 1 μK in the nuclear spin system of copper. The first studies of nuclear cooperative phenomena, on insulators like CaF_2 and LiH, were made by Abragam and Goldman and their coworkers at Saclay 15 years later [5,6].

Investigations of nuclear magnetic ordering in copper were started in Helsinki in the mid 1970's [7–9] by constructing a two-stage nuclear demagnetization cryostat, precooled by a dilution refrigerator; the apparatus, the first of its kind, was specially designed for nuclear ordering experiments. The first important results were obtained, however, only in 1982 [10,11] when magnetic susceptibility measurements showed that copper orders antiferromagnetically below its critical temperature of about 60 nK. Two years later, three antiferromagnetic phases were discovered [12,13] in a single crystal specimen of copper below the critical field $B_c = 0.25$ mT.

In order to determine the spin structures of antiferromagnetically ordered copper, neutron diffraction experiments were initiated in 1985, after a careful feasibility study [14], in a collaboration between the Risø National Laboratory in Denmark, the Hahn–Meitner Institute of Berlin, and our laboratory. A new cascade nuclear demagnetization cryostat was constructed in Helsinki and installed at Risø where the neutron diffraction experiments were carried out. These measurements extended, by an order of magnitude, the temperature regime at which neutron diffraction had been employed previously. In the fall of 1987, a clear antiferromagnetic (1 0 0) Bragg peak, characteristic of type-I order in an fcc lattice, was observed in two field regions: below 0.08 mT and between 0.12 and 0.25 mT [15]. The spin alignments at intermediate fields, however, remained unknown until experiments in 1989, when four new but equivalent antiferromagnetic Bragg peaks, $(1 \frac{1}{3} \frac{1}{3})$, $(1 -\frac{1}{3} -\frac{1}{3})$ and $\pm(0 \frac{2}{3} \frac{2}{3})$ were discovered [16]. Previously, a spin structure characterized by these reflections had not been found in any fcc antiferromagnet.

The work on silver started in Helsinki in 1986. In the first set of measurements [17], spin entropies were reduced to $0.5R\ln 2$, and an indication of antiferromagnetic ordering was observed. In more recent experiments, the entropy and susceptibility were measured, as a function of the spin temperature, down to 800 pK at $T > 0$ and up to -4.3 nK at $T < 0$ [18]. In 1990, conclusive proof of a transition into an antiferromagnetic state was obtained at $T_c = 560$ pK [19]; this is the lowest temperature ever produced and measured. The recent work on copper and silver has been reviewed briefly in refs. [20,21].

The Hamiltonian of the nuclear spin assembly $\{I_i\}$ in copper and silver consists of the dipole–dipole interaction and the indirect Ruderman–Kittel (RK) coupling [22], in addition to the Zeeman energy, viz.,

$$\mathscr{H} = \mathscr{H}_D + \mathscr{H}_{RK} - \hbar \boldsymbol{B} \cdot \sum_i \gamma_i \boldsymbol{I}_i. \quad (1)$$

Here $\boldsymbol{B}$ is the external field and γ is the gyromagnetic ratio. Natural copper and silver both have two stable isotopes; their γ's differ by 7 and 14%, respectively, but the spins are equal. The quadrupolar interaction vanishes strictly in silver, owing to the spin $I = \frac{1}{2}$, and it does so in an undistorted copper lattice ($I = \frac{3}{2}$) as well because of cubic symmetry. The dipolar forces depend on the orientations of the spins with respect to their positions in the crystalline lattice, viz.,

$$\mathscr{H}_D = (\mu_0 \hbar^2/4\pi) \sum_{i<j} \gamma_i \gamma_j r_{ij}^{-3}$$
$$\times \left[\boldsymbol{I}_i \cdot \boldsymbol{I}_j - 3(\hat{\boldsymbol{r}}_{ij} \cdot \boldsymbol{I}_i)(\hat{\boldsymbol{r}}_{ij} \cdot \boldsymbol{I}_j) \right], \quad (2)$$

where $\hat{\boldsymbol{r}}_{ij}$ is a vector between lattice sites i and j, while the RK interaction depends only on the relative alignment of the spins,

$$\mathscr{H}_{RK} = -\sum_{i<j} J_{ij} \boldsymbol{I}_i \cdot \boldsymbol{I}_j. \quad (3)$$

The strength of the RK interaction, with respect to the dipolar force, is usually described by a dimensionless parameter

$$\mathscr{R} = \sum_j J_{ij}/\mu_0 \rho \hbar^2 \gamma^2, \quad (4)$$

where ρ is the number density of nuclei. NMR experiments [23] have yielded $\mathscr{R} = -0.42 \pm 0.05$ for copper and $\mathscr{R} = -2.5 \pm 0.5$ for silver [18,24,25]. The free electron model can be employed to approximate the range function J_{ij} of the RK interaction [22]. More reliable values for J_{ij} have been obtained in recent calculations using realistic electronic band structures and wave functions [26–28]. The calculated parameters $\mathscr{R}$ are in good agreement with measurements.

The ordered spin structures are sensitive to the strength of the RK force because of the two competing coupling mechanisms: the dipolar force favors ferromagnetism in an fcc lattice while the RK interaction is antiferromagnetic [29]. The balance of this competition is delicate especially in copper as shown by the complicated theoretical interpretation of the observed three phases (see section 5).

In silver the spin–spin interactions are exchange dominated because of the relatively strong RK force. This makes silver a good model for the nearest-neighbor Heisenberg antiferromagnet in an fcc lattice. The ground state of such a system is an interesting problem because of geometric frustration [30] and prominent quantum fluctuations owing to the small spin, $I = \frac{1}{2}$, of silver nuclei.

2. Experimental techniques

The "brute force" nuclear cooling method, applicable to metallic samples, was employed in these experiments. In this straightforward technique [31], particularly suitable for copper and silver owing to their good thermal conductivity, the nuclear spins are first polarized to $p > 90\%$, using a high initial magnetic field B_i and a low precooling temperature T_i, and then adiabatically demagnetized to a small field B_f, whereby a very low final temperature, approximately given by

$$T_f = T_i \left(B_f^2 + b_{int}^2 \right)^{1/2} / B_i \quad (5)$$

results. Spin–spin interactions will then produce spontaneous alignment in the assembly of nuclei. To reach the ordered state, B_f has to be on the order of the internal field b_{int}, 0.36 mT in copper and 0.035 mT in silver.

During adiabatic demagnetization only the nuclei are cooled [31]; the spin temperature T of the nuclei and the common lattice and conduction electron temperature T_e can differ by several orders of magnitude. It is thus reasonable to speak about two distinct temperatures in the same specimen since the nuclei reach local thermal equilibrium among themselves in a time characterized by $\tau_2 \approx 10$ ms, the spin–spin relaxation time, whereas the approach to equilibrium between nuclear spins and conduction electrons is governed by the spin–lattice relaxation time $\tau_1 \gg \tau_2$. According to Korringa's law [32],

$$\tau_1 = \kappa / T_e, \quad (6)$$

where κ is Korringa's constant. At low fields, $\kappa = 0.5$ s K in copper and 3 s K in silver. For example, a conduction electron temperature of 50 µK yields a relaxation time of 3 h in copper. In silver the conduction electrons were typically cooled to 200 µK; the relaxation time is thus 4 h. Thermal decoupling of the nuclei and the conduction electrons at very low temperatures is a crucial factor since it guarantees a sufficiently slow warm-up of the spin system, giving enough time for the measurements. In high magnetic fields, κ is 2–3 times larger.

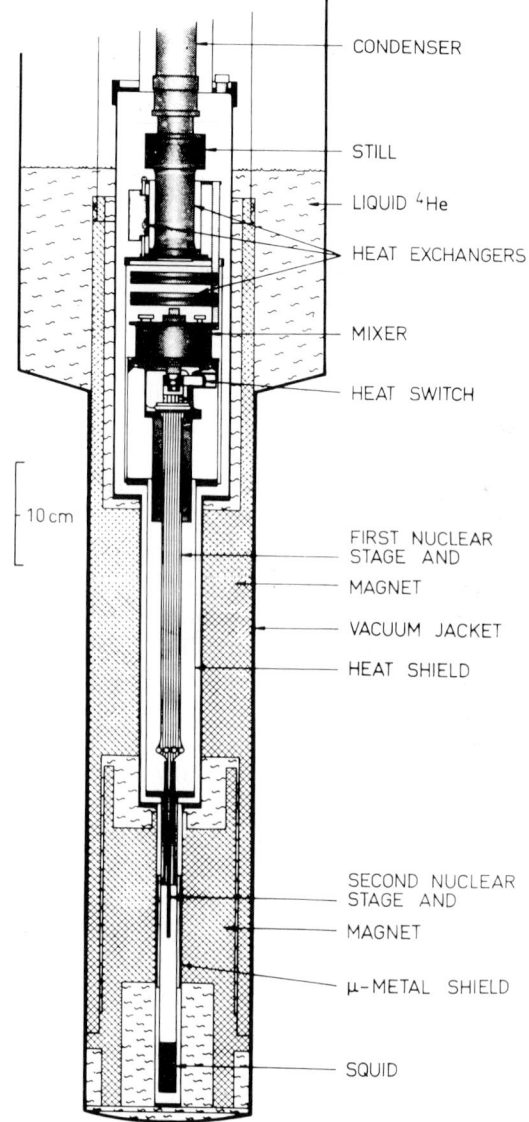

Fig. 1. Low temperature parts of the cascade nuclear demagnetization cryostat in Helsinki [8].

To obtain nuclear spin temperatures in the nano- or picokelvin range, a sophisticated cooling apparatus is needed. Fig. 1 shows the cascade nuclear refrigerator in Helsinki [8]. This apparatus has been operating successfully for almost 15 years. Precooling is done by a home-made dilution refrigerator, whose mixing chamber reaches a temperature of about 7 mK without a heat load. The early version [8] of the first nuclear stage was made of 10 mol of insulated copper wire. This corresponds to an effective size of 2.5 mol of the metal in the operating magnetic field of 8 T. Later versions [13,17,33] of this cooling stage were constructed from a bulk piece of copper. Slits were cut along its length to reduce eddy current heating during demagnetization.

The second nuclear stage is also the sample. The copper or silver specimen was connected to the precooling first nuclear stage by welding, without a heat switch. This means that the conduction electron temperature is the same in both nuclear stages and that, for thermal isolation of the nuclear spin system in the sample, the slow spin–lattice relaxation is sufficient. The effective size of the specimen is about 0.03 mol or 2–3 g. The external fields on the two nuclear stages could be changed independently by using the upper and lower superconducting magnets, which can produce maximum fields of 8 and 7.4 T, respectively.

A typical cascade nuclear cooling experiment is performed as follows (see fig. 1). The large upper nuclear stage is first magnetized to 8 T, and then both stages are precooled by the dilution refrigerator to 15 mK. The heat switch, constructed of tin and located below the mixer of the dilution machine, is then made superconducting and hence thermally isolating. Next, the first nuclear stage is demagnetized to 0.1 T, whereby it cools to 200 μK. During this process, the specimen itself is magnetized to 7 T. After a sufficient waiting time, typically 1.5 h, demagnetization of the first stage is continued to 20 mT, which produces a conduction electron temperature $T_e = 50$ μK in the upper nuclear stage. The nuclei are now at about $T = 200$ μK.

The 7 T field of the lower magnet is then reduced to zero rather quickly, in about 20 min; this keeps polarization losses low during demagnetization. The specimen is now in a field of 10 mT, generated by a small superconducting solenoid inside a μ-metal shield. After waiting for about 1 min, the noise in the SQUID detection equipment had decreased sufficiently for demagnetization to the low final field, usually between 0 and 0.3 mT, in which nuclear ordering is being investigated. The spin system is thereby cooled from 200 μK to below 50 nK in copper, while the conduction electrons reach 50 μK. In silver, demagnetization from 100 to 20 mT was omitted; the corresponding numbers are $T < 600$ pK and $T_e = 200$ μK. Immediately afterwards, the spin temperature starts to relax towards T_e. The magnetic susceptibility was then measured during the warmup.

The static nuclear susceptibility signal was too noisy to be recorded accurately with a SQUID. Therefore, the absorptive part $\chi''(f)$ of the complex susceptibility $\chi(f) = \chi'(f) - i\chi''(f)$, was measured at low frequencies, at which the skin effect would not prevent the magnetic field from penetrating the metallic specimen. From $\chi''(f)$, by using the Kramers–Krönig relation

$$\chi'(0) = (2/\pi)\int_0^\infty [\chi''(f)/f]\, df, \qquad (7)$$

one can calculate the static susceptibility $\chi'(0)$. Furthermore, from $\chi''(f)$, measured at fields $B \gg b_{int}$, one may obtain the nuclear polarization as well, viz.,

$$p = A\int \chi''(f)\, df, \qquad (8)$$

where A is a constant to be determined using the equations of the paramagnetic phase. In the case of silver

$$p = \tanh(\mu B/k_B T), \qquad (9)$$

where μ is the magnetic moment $\hbar\gamma I$.

As soon as polarization is known, one can compute the entropy because, at these extremely low temperatures, the only contribution to S is the nuclear spin entropy. For $I = \frac{1}{2}$ (silver),

$$S/R = \ln 2 - \tfrac{1}{2}[(1+p)\ln(1+p) + (1-p)\ln(1-p)]. \qquad (10)$$

Eqs. (9) and (10) are similar but more complex for $I = \frac{3}{2}$ (copper).

One of the difficult problems in these experiments was how to measure the absolute thermodynamic temperature T of the nuclear spin system. One cannot go into detail here, but let us just mention that we directly used a basic law of thermodynamics, $T = \Delta Q/\Delta S$. That is, the nuclear spin system was supplied with a small amount of heat ΔQ and the ensuing entropy increase ΔS was measured. The process, simple in principle, is complicated in practice, and several different procedures have been used [13,34]. In the ordered states, the spin temperature could be determined reliably only at $B = 0$; this is why the susceptibility and neutron diffraction data are usually presented as a function of time, instead of temperature; of course, T increases monotonously with time.

3. Susceptibility results on copper

Fig. 2 illustrates the most important of the early Helsinki results [11,13]. The entropy of the nuclear spin system of copper, in units of the maximum spin entropy $R \ln 4$, was measured against the temperature. There is a clear jump in S, which signifies a first order transition to an antiferromagnetic phase below $T_c \approx 60$ nK. S_{c1} is the lower and S_{c2} the higher critical entropy; the latent heat of the transition $L = T_c(S_{c2} - S_{c1}) = 0.09$ μJ/mol.

In 1984, experiments were made on a single crystal copper specimen [12]. The fcc lattice axes were aligned approximately along the sample

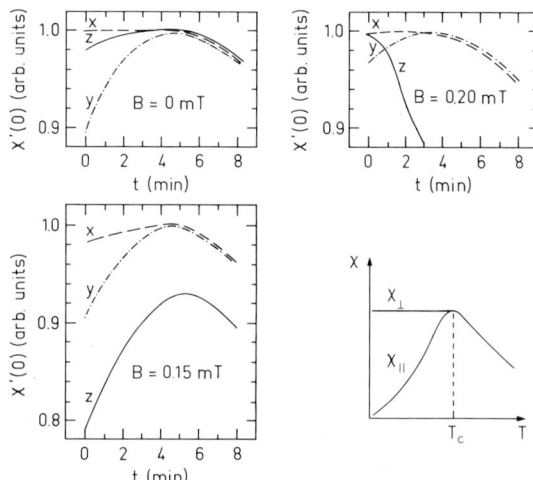

Fig. 3. Static nuclear susceptibility $\chi'(0)$ of copper, measured in the three Cartesian directions (x, y, z) as a function of time after demagnetization to $B = 0$, 0.15 and 0.20 mT [12,13]. These characteristically different behaviors (see schematic illustration, in the lower-right corner, for the temperature dependence of susceptibility perpendicular ($\chi_\perp$) and parallel ($\chi_\parallel$) to sublattice magnetization) indicate three antiferromagnetically ordered phases of copper nuclei. The suggested spin arrangements are shown in fig. 4.

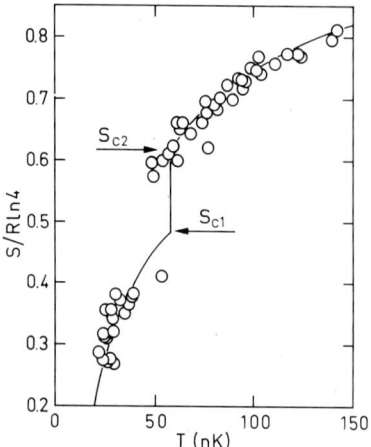

Fig. 2. Nuclear spin entropy of copper as a function of temperature in zero external magnetic field [11,13]. The vertical line indicates a first-order phase change at $T_c \approx 60$ nK.

edges. By an elaborate coil system, the susceptibility could be measured separately in the x-, y- and z-directions. The analysis of the data was based on the fact that, in antiferromagnets below the Néel temperature, $\chi'(0)_\perp$, the susceptibility transverse to sublattice magnetization, is constant while $\chi'(0)_\parallel$, the susceptibility parallel to sublattice magnetization, approaches zero as $T \to 0$; see schematic illustration in the lower right corner of fig. 3.

Consequently, the data in fig. 3 at $B = 0$ show that magnetization is mainly along the y-axis because the susceptibility changes are largest in this direction. At $B = 0.15$ mT, the sublattice magnetization has its biggest component in the z-direction but it also displays a smaller component in the y-direction. At $B = 0.20$ mT, the spins are leaning towards the external magnetic field because there is no longer antiferromagnetic behavior of the susceptibility in the z-direction. Furthermore, in contrast to the paramagnetic time dependence of $\chi'(0)_z$, the small initial increase in

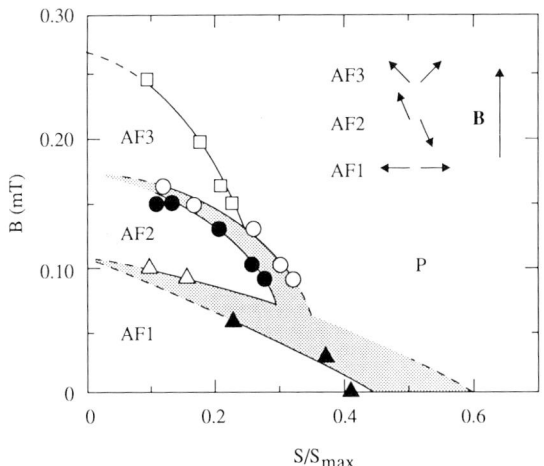

Fig. 4. Phase diagram of nuclear spins in copper in the coordinates of entropy and external magnetic field, with B applied approximately along the [0 0 1] crystalline axis [12]. The ordered antiferromagnetic states are marked by AF1, AF2 and AF3, and the shaded areas indicate regions in which first-order transitions take place. P stands for the paramagnetic phase. The different symbols refer to the experimental points which mark the various phase boundaries. The schematic spin configurations are shown in the upper-right corner, as deduced from the susceptibility data of fig. 3. For copper, the maximum spin entropy $S_{max} = R \ln 4$.

$\chi'(0)_y$ indicates an antiferromagnetic y-component of sublattice magnetization.

The postulated spin structures at 0, 0.15 and 0.20 mT are shown in the upper right corner of fig. 4 [12], which depicts the B-S phase diagram of copper; it was constructed by demagnetizing from different initial values of entropy, between 10 and 35% of $R \ln 4$, thus coming down on the diagram, and then by letting the specimen to warm up, thus moving horizontally to the right while the susceptibility was being measured. The low field phase is marked by AF1, the middle phase by AF2, and the high field phase by AF3; P stands for the paramagnetic state. The shadowed areas indicate regions where one characteristic behavior changes to the next and a latent heat is being supplied.

The data in figs. 2–4 show that quite a bit was found from these simple but technically difficult susceptibility measurements. The phase diagram of copper is surprisingly complex and it triggered several theoretical studies [26,27,35–39].

4. Neutron diffraction experiments

In order to obtain more details on the spin structure of antiferromagnetically ordered copper, neutron diffraction experiments were initiated by a Danish–Finnish–German group, including Michael Steiner of the Hahn–Meitner Institute and Kurt Clausen of Risø. Before these measurements, the method had been used to determine the magnetic structures in a few other nuclear spin systems, at much higher lattice and nuclear spin temperatures. These included LiH [40], solid ^{3}He [41], several Pr compounds [42], and Pr metal [43].

The structure determination is based on the scattering amplitude [44]

$$\mathscr{A} = b_0 + b\mathbf{I} \cdot \boldsymbol{\sigma}, \quad (11)$$

which arises from the strong interaction between a neutron and a nucleus. $\mathbf{I}$ and $\boldsymbol{\sigma}$ ($\sigma = \frac{1}{2}$) are the nuclear and neutron spins, respectively, and b_0 and b are the coefficients of the spin-independent and spin-dependent parts of the scattering length, respectively. The second term in eq. (11) carries information about $\mathbf{I}$ and gives rise to additional Bragg reflections in the magnetically ordered state. The scattered neutron intensity is proportional to the squared structure factor

$$F^2(\mathbf{k}) = \sum_{ij} \left\{ \left[b_0^2 + \tfrac{1}{2} b_0 b \langle \boldsymbol{\sigma} \rangle \cdot (\langle \mathbf{I}_i \rangle + \langle \mathbf{I}_j \rangle) \right. \right.$$
$$\left. \left. + \tfrac{1}{4} b^2 \langle \mathbf{I}_i \rangle \cdot \langle \mathbf{I}_j \rangle \right] \exp\left[-i \mathbf{k} \cdot (\mathbf{r}_i - \mathbf{r}_j)\right] \right\}, \quad (12)$$

where $\mathbf{k}$ is the scattering vector, $\langle \boldsymbol{\sigma} \rangle$ is the neutron beam polarization, and $\langle \mathbf{I}_i \rangle$ denotes the thermal average of the spin at site $\mathbf{r}_i$. In the ordered state at zero field, $\langle \mathbf{I} \rangle$ decreases with temperature and vanishes at $T = T_c$.

The first term in eq. (12) is the usual structure factor. Copper has an fcc lattice for which only reflections with Miller indices (hkl) all even or all odd are allowed. The second term, which adds onto the fcc peaks, depends on the angle between $\langle \boldsymbol{\sigma} \rangle$ and $\langle \mathbf{I}_i \rangle$ and yields the total nuclear magnetization along $\langle \boldsymbol{\sigma} \rangle$. This term has recently been measured in various magnetic systems to obtain the nuclear spin polarization and temperature in a strong magnetic field [45].

The third term in eq. (12) is the most important for investigations of the antiferromagnetic ordering. Depending on the spin structures, i.e. on $\langle I_i \rangle$, new Bragg peaks will appear, in addition to the ordinary lattice reflections. In zero external magnetic field, theoretical calculations [29] predicted antiferromagnetic modulation, exemplified by the propagation ($=$ scattering) vector $\boldsymbol{k} =$

Fig. 5. Experimental setup for neutron scattering measurements at Risø [15]. The cascade nuclear cooling cryostat is in the left side of the picture on the spectrometer turntable. Cold neutrons emerge from the lower right corner, then pass through the monochromator and the BeO filter. Next, they enter the cryostat and are scattered by the sample to the counter in the left foreground.

$(2\pi/a)(1, 0, 0)$, which yields a (1 0 0) Bragg reflection. In copper, the lattice constant $a = 3.61$ Å.

It is important to note that because of the isotropic neutron–nucleus scattering amplitude (see eq. (11)), the absolute directions of the spins cannot be determined. This is unlike in studies of electronic magnets, in which anisotropic magnetic interactions between a neutron and the magnetic moments of the electrons gives access to the spin directions; with nuclei, this interaction is vanishingly small.

The neutron experiments were carried out in the guide hall next to the DR-3 reactor at Risø. A new two-stage nuclear demagnetization cryostat, especially designed for studies of nuclear magnets by neutron diffraction, was constructed in Helsinki for these experiments. Much of the counting hardware was provided by the Hahn–Meitner Institute. The 4.7 Å neutron beam was first reflected by a graphite monochromator crystal. It then passed a $\lambda/2$ filter and hit the sample in the cryostat; the scattered neutrons were detected by a NaI counter. The cryostat and the detector could be moved independently in the scattering plane. Fig. 5 is a photograph from Risø.

The sample was made of a ^{65}Cu single crystal because the spin-dependent part of the scattering length for this isotope yields a 7-fold gain in the scattered intensity, as compared with natural copper. The problem that the experimentalists had initially been most afraid of was the additional warming caused by neutrons. The beam heating was measured to be 1–2 nW which made the spin–lattice relaxation faster by a factor of two [46]. The relatively high level of impurities in the sample, tens of ppm's, further shortened the relaxation time. The resulting $\tau_1 \approx 20$ min at low fields was, however, sufficiently long for detecting magnetic order.

After some struggling and frustration [47], the experiment succeeded. The neutron intensity at the (1 0 0) Bragg position, as a function of time (i.e., of temperature) after the external magnetic field had been reduced to zero, is shown in fig. 6; plenty of scattered neutrons were discovered! Long-range antiferromagnetism in copper had

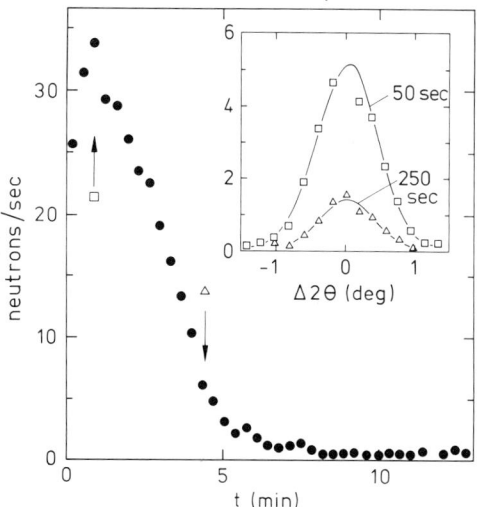

Fig. 6. Integrated neutron intensity of the (1 0 0) antiferromagnetic Bragg reflection for copper as a function of time after the zero field was reached [15]. The inset shows the shape of the Bragg peak as a function of the scattering angle $\Delta 2\theta$, obtained by a linear position-sensitive detector; neutron counts per channel are averaged over two 75 s intervals, centered at 50 s and 250 s, respectively. The solid lines are the best Gaussian fits to the experimental data.

thus been proven unambiguously [15,46]. The simultaneously measured longitudinal susceptibility displayed a maximum, which was employed in the earlier Helsinki measurements [13] as the signature of the antiferromagnetic state. The shape of the Bragg peak is illustrated in the inset; the amplitude decreases with time while the width, limited by instrumental resolution, remains almost constant [48].

An interesting dependence of the (1 0 0) Bragg peak on the external magnetic field was found as well [15]. A clear neutron signal was observed in two regions: below 0.08 mT (see fig. 6) and between 0.12 and 0.25 mT (see fig. 7, upper curve). In intermediate fields, 0.08–0.12 mT, the (1 0 0) reflection was very small (see fig. 7, lower curve), although the susceptibility clearly indicated antiferromagnetic order (see inset in fig. 7). This was a puzzle during the first set of neutron diffraction measurements at Risø.

The missing neutron intensity at intermediate fields was searched for in more recent experiments [16]. At first, it was thought that scanning

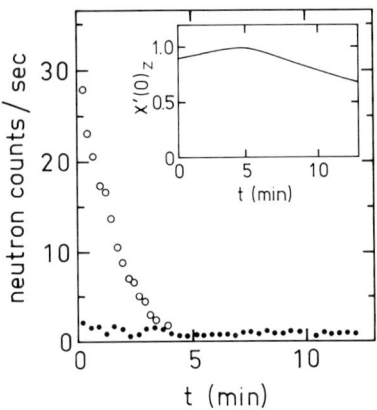

Fig. 7. Integrated neutron intensity as a function of time after final demagnetization to $B = 0.10$ mT (●) and to $B = 0.16$ mT (○), respectively. The inset shows the simultaneously measured longitudinal susceptibility $\chi'(0)_z$ at $B = 0.10$ mT [15].

for possible new Bragg reflections, which would have been a routine operation with electronic magnets, is hopeless. The ordered state could be maintained only for 5 min every 36 h and, furthermore, movements of the cryostat and the detector during a scan were expected to heat up the sample rapidly. Scanning, however, turned out to be possible, although it did speed up the warming of the sample, but not catastrophically.

Based on theoretical calculations [26,27], the scans were concentrated in the $[0\ \eta\ \eta]$ and $[\xi\ 0\ 0]$ directions of the k space. A most gratifying result was obtained in 1989 when a clear $(1\ \frac{1}{3}\ \frac{1}{3})$ reflection was observed at $B = 0.07$ mT. Later, the $(1\ -\frac{1}{3}\ -\frac{1}{3})$ and the $\pm(0\ \frac{2}{3}\ \frac{2}{3})$ reflections were observed as well; they are equivalent to the $(1\ \frac{1}{3}\ \frac{1}{3})$ peak under fcc symmetry. It is notable that the spin structure characterized by these signals has not been observed in any other fcc system.

The field dependence of the $(1\ \frac{1}{3}\ \frac{1}{3})$ reflection was investigated in separate experiments, as illustrated in fig. 8. At 0.08 mT, the $(1\ \frac{1}{3}\ \frac{1}{3})$ peak is very strong in comparison with the $(1\ 0\ 0)$ reflection. At 0.12 mT, the $(1\ \frac{1}{3}\ \frac{1}{3})$ signal increases during the first minute while the $(1\ 0\ 0)$ peak decreases; this can be interpreted as an ongoing first-order phase transition from a $(1\ 0\ 0)$ state to a $(1\ \frac{1}{3}\ \frac{1}{3})$ structure. At 0.04 mT, the behavior is different: both reflections decay at the same rate.

A contour diagram of the $(1\ \frac{1}{3}\ \frac{1}{3})$ and $(1\ 0\ 0)$ neutron intensities was constructed from 17 measurements at different magnetic fields [16]; it is illustrated in fig. 9. Note that the intensities are again presented as functions of the external magnetic field and of the warmup time, rather than field and temperature. The $(1\ 0\ 0)$ contours display the above-mentioned two maxima, at low and high fields. The $(1\ 0\ 0)$ signal is weakest when the $(1\ \frac{1}{3}\ \frac{1}{3})$ reflection is strongest, and vice

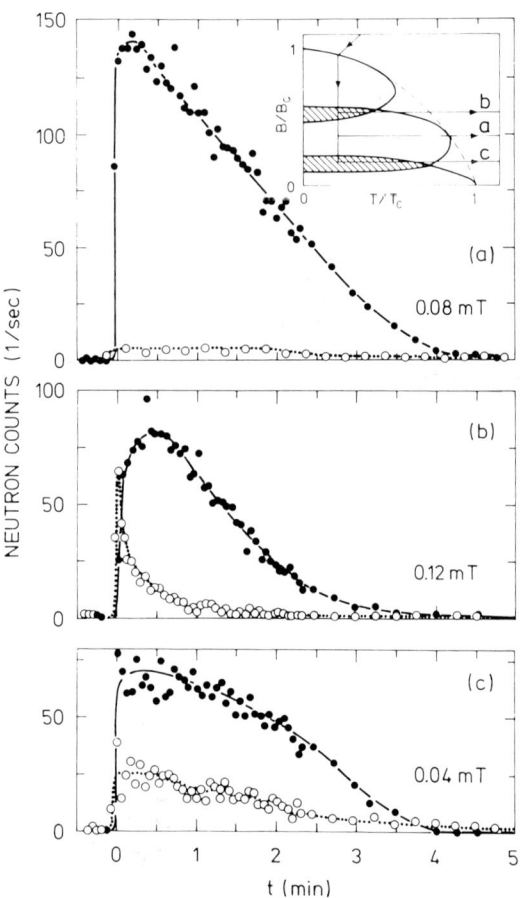

Fig. 8. (a) Time dependence of the $(1\ \frac{1}{3}\ \frac{1}{3})$ Bragg reflection for copper at $B = 0.08$ mT [16]. Inset: Schematic phase diagram in the $B-T$ plane. The lines with arrows indicate entrance, along an isentrope ($S = 0.15R \ln 4$), into the ordered phase and subsequent measurements in a constant field; a, b, and c correspond to curves in frames (a), (b) and (c), respectively. (b) Time dependencies of the $(1\ \frac{1}{3}\ \frac{1}{3})$ (●) and $(1\ 0\ 0)$ (○) reflections in the vicinity of the upper phase boundary. (c) Same at 0.04 mT field.

versa. The neutron intensity plot and the previously obtained B–S phase diagram, shown in fig. 4, appear similar, even though the directions of the external fields with respect to the crystalline axes of the specimens were [0 0 1] in Helsinki and [0 −1 1] in Risø. It looks that the neutron contour diagram reveals three phases as well, although the boundary between the low-field and the intermediate field phases is not well resolved.

An obvious extension to the neutron diffraction experiments described so far was to examine the phase diagram with the external magnetic field aligned along the other main crystallographic axes, besides the [0 −1 1] direction. A lot of successful work was done by Annila et al. [49] along these lines during 1990. In fig. 10 we present the main result of these studies. The antiferromagnetic states are bounded by the second order critical field line; the B_c curve was determined from the neutron diffraction and susceptibility data. In fields between 0.10 and 0.20 mT, the (1 0 0) order, shown with different degrees of shading, is strong in a wide span of directions around $\boldsymbol{B}\|[1\,0\,0]$ and over a narrower angular region about $\boldsymbol{B}\|[0\,1\,1]$. There is also pure (1 0 0) order near the origin below 0.01 mT.

The $(0\,\tfrac{2}{3}\,\tfrac{2}{3})$ phase, shown by double lines in fig. 10, has strong maxima around 0.07 mT, both in the [1 0 0] (equivalent to [0 0 1]) and in the [0 1 1] field directions, but in between the inten-

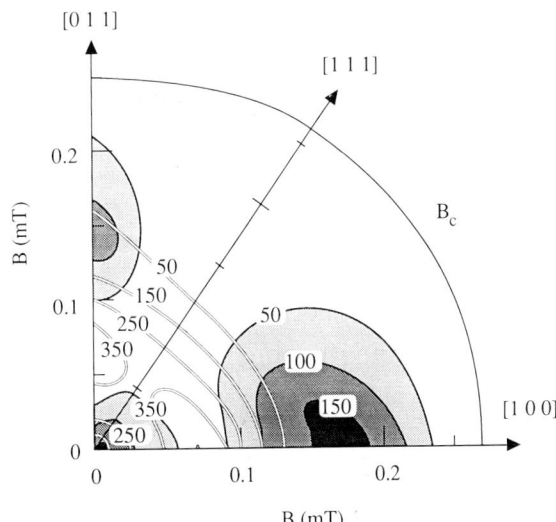

Fig. 10. Neutron intensity diagram for copper as a function of the magnetic field in the plane of the principal crystalline axes [49]. Counts/s for field directions equivalent under fcc symmetry were summed together. The region of the (1 0 0) intensity is marked with different degrees of shading (50, 100 and 150 counts/s, respectively), and the $(0\,\tfrac{2}{3}\,\tfrac{2}{3})$ intensity is shown by double contours (the curve nearest to origin corresponds to 150 counts/s). The antiferromagnetic phase is bordered by the second order B_c curve.

sity is somewhat less. We see that the three phases, AF1, AF2 and AF3, found in the susceptibility measurements with $\boldsymbol{B}\|[0\,0\,1]$ (see fig. 4) [12], are reproduced remarkably well by the neutron diffraction results.

The ordering vector of the high-field phase when $\boldsymbol{B}\|[1\,1\,1]$ is the main remaining puzzle; there is a large field region, which is antiferromagnetic according to susceptibility data, with no neutron intensity. In spite of much effort [49] to find a new Bragg reflection at these fields, no neutrons were discovered. Unfortunately, it was not clear where to look for because theory predicts (1 0 0) order [35,37,39]. There is thus some future work to do, both for experimentalists and theorists.

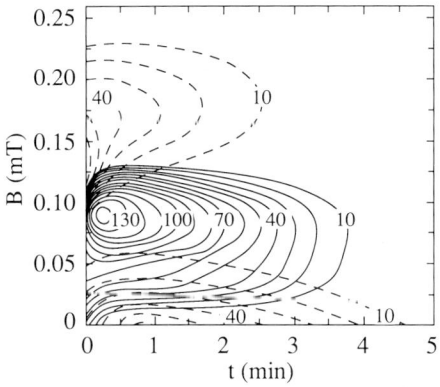

Fig. 9. Neutron intensity contour diagram of the $(1\,\tfrac{1}{3}\,\tfrac{1}{3})$ (solid lines) and the (1 0 0) (dashed lines) Bragg peaks for copper as functions of time and the external magnetic field [16]. The outermost contours, 10 counts/s, show approximately when long-range order disappears.

5. Theoretical spin structures

The ordered spin configurations of copper were studied theoretically first in 1979 [29]. Type-I antiferromagnetism [50] was predicted, characterized by the neutron reflections (1 0 0), (0 1 0) and (0 0 1). Because this structure displays, in the mean field (MF) theory [29], a continuous degeneracy, even in an arbitrary magnetic field [35], subsequent theoretical efforts concentrated on identifying the ground state among the degenerate type-I spin configurations [36–39].

After the observation of the (1 0 0) Bragg reflections in the low- and high-field phases, Lingård [36] proposed that quantum fluctuations would stabilize an incommensurate structure with $(0\ \eta\ \eta)$ order in intermediate fields. Apart from the commensurability, the subsequently observed $(0\ \frac{2}{3}\ \frac{2}{3})$ peak was consistent with this prediction.

The actual spin structures with the $(0\ \frac{2}{3}\ \frac{2}{3})$ order have been investigated by two theoretical approaches. Using the mean-field theory, Viertiö and Oja [52] succeeded in finding an analytical expression for the ground-state structure. The Hamiltonian, eqs. (2) and (3), can be expressed in matrix form as $\mathcal{H} = \sum_{i<j} \boldsymbol{I}_i \cdot \boldsymbol{A}_{ij} \cdot \boldsymbol{I}_j$. The characteristic energy of an antiferromagnetic state, with an ordering vector $\boldsymbol{k}$, can be found from the eigenvalue equation

$$\boldsymbol{A}(\boldsymbol{k}) \cdot \boldsymbol{e}_n(\boldsymbol{k}) = \lambda_n(\boldsymbol{k}) \boldsymbol{e}_n(\boldsymbol{k}), \quad n = 1, 2, 3, \quad (13)$$

where $\lambda_n(\boldsymbol{k})$ are the eigenvalues and $\boldsymbol{e}_n(\boldsymbol{k})$ the eigenvectors of the 3×3 Fourier-transformed interaction matrix $\boldsymbol{A}(\boldsymbol{k}) = \sum_j \boldsymbol{A}_{ij} \exp(-\mathrm{i}\boldsymbol{k} \cdot \boldsymbol{r}_{ij})$. Usually, the $\boldsymbol{k}$ vector giving the minimum eigenvalue determines the spin modulation, and the corresponding eigenvector gives the spin directions. In copper, the minimum eigenvalue, and hence the energy of the spin structure, displays a "double-well" character at the irreducible part of the first Brillouin zone: there is a local minimum at $2\pi/a\ (1, 0, 0)$ and another near $2\pi/a\ (0, \frac{2}{3}, \frac{2}{3})$. According to first-principles calculations [26,27], the global minimum of $\lambda_n(\boldsymbol{k})$ is at $2\pi/a\ (1, 0, 0)$, but $\lambda(0, \frac{2}{3}, \frac{2}{3})$ is higher only by 10%.

The assumption made in the theory [52] was that $\lambda(0, \frac{2}{3}, \frac{2}{3})$ is, in fact, slightly smaller than $\lambda(1, 0, 0)$. To understand why the $(0\ \frac{2}{3}\ \frac{2}{3})$ order is not, then, stable in zero field but the (1 0 0) state is, one must examine the corresponding eigenvalues and eigenvectors. The eigenvalue for the $2\pi/a\ (1, 0, 0)$ modulation is doubly degenerate and the spins are free to rotate in the plane perpendicular to $2\pi/a\ (1, 0, 0)$. In contrast, the eigenvalue for the $2\pi/a\ (0, \frac{2}{3}, \frac{2}{3})$ modulation is unique, which singles out a unique eigenvector as well. The easy-axis anisotropy results in an intrinsic ferromagnetic component associated with the $(0\ \frac{2}{3}\ \frac{2}{3})$ order as is evident from the spin pattern ↑↑↓ (up–up–down). Since the eigenvalue $\lambda(0)$ associated with ferromagnetic order is energetically unfavorable, an external field is required to stabilize ↑↑↓ antiferromagnetism but it is not needed for the (1 0 0) order.

The ground-state spin structures at $B \le B_c/3$ are illustrated in fig. 11 [52] for the [0 1 −1] field direction. Apart from some degeneracy, the MF equation for the ground state is

$$\langle \boldsymbol{I}_i \rangle = (0, d_0, -d_0) + (0, d_1, -d_1) \cos(\boldsymbol{k}_1 \cdot \boldsymbol{r}_i)$$
$$+ (0, d_2, d_2) \cos(\boldsymbol{k}_2 \cdot \boldsymbol{r}_i), \quad (14)$$

where $\boldsymbol{k}_1 = 2\pi/a\ (0, \frac{2}{3}, \frac{2}{3})$ and $\boldsymbol{k}_2 = 2\pi/a\ (1, 0, 0)$. Both the magnetization d_0 and the $(0\ \frac{2}{3}\ \frac{2}{3})$ order are proportional to the field, viz., $d_0 = -d_1/4 = IB/B_c\sqrt{2}$ with $I = \frac{3}{2}$, while the (1 0 0) order satisfies $2d_2^2 = I^2 - 18d_0^2$. In zero field, the structure is a pure (1 0 0) state, see fig. 11a. When the field is increased, the $(0\ \frac{2}{3}\ \frac{2}{3})$ order grows continuously at the expense of the (1 0 0) order but the two structures are present simultaneously, see fig. 11b. Exactly in the field $B_c/3$, antiferromagnetism is of the $(0\ \frac{2}{3}\ \frac{2}{3})$ type only. As illustrated in fig. 11c, this structure looks like ↑↑↓: two sublattices are oriented parallel to the field and one antiparallel to it. At fields above $B_c/3$, the MF theory predicts a 3-$\boldsymbol{k}$ state which is characterized by the simultaneous Bragg reflections (1 0 0), (0 1 0) and (0 0 1).

In order to compare the theory with experiments, the magnetic structure factors (see eq. (12)) were calculated in the B–T plane by following the warmup of the ground state in a Monte Carlo simulation [52]. The result well resembles the experimental phase diagram of fig. 9, giving strong support to the theory. In the [0 0 1] direction of the external magnetic field, used in the susceptibility measurements at Helsinki [12], three antiferromagnetic phases, separated by first-order

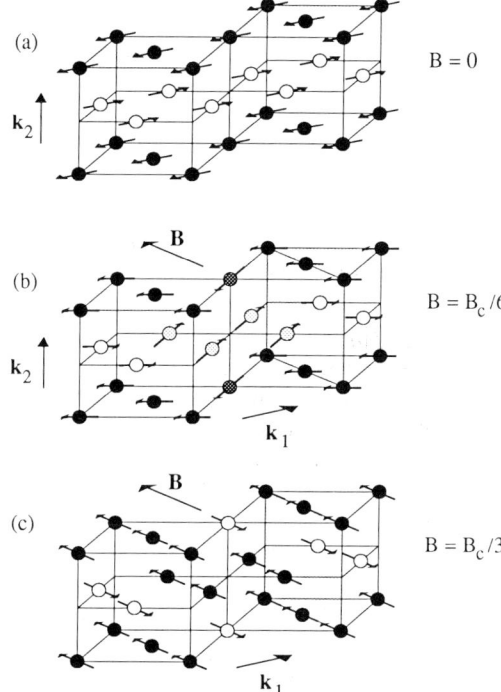

Fig. 11. Ordered spin structures $\langle I \rangle$ of copper for the [0 1 −1] alignment of the external magnetic field, as given by eq. (14). (a) $B = 0$; antiferromagnetic $k_2 = 2\pi/a$ (1, 0, 0) structure, made up of alternating ferromagnetic planes. (b) $0 < B < B_c/3$: mixed structure with ordering vectors k_2 and $k_1 = 2\pi/a$ $(0, \frac{2}{3}, \frac{2}{3})$, illustrated at $B = B_c/6$. (c) $B = B_c/3$: structure with $(0\ \frac{2}{3}\ \frac{2}{3})$ order [52]. $\langle I \rangle$ is dependent on temperature and approaches, at $B = 0$, zero when $T \to T_c$.

transitions, are reproduced by the theory (see fig. 4). The phases AF1 and AF2 both display simultaneous (1 0 0) and $(0\ \frac{2}{3}\ \frac{2}{3})$ order, while AF3 is a pure (1 0 0) state. Several theoretical predictions for other field directions were made as well [53]. Most recent neutron diffraction measurements [49] have shown that at least the regions where the $(0\ \frac{2}{3}\ \frac{2}{3})$ Bragg peak exists, when the field is applied in the various high-symmetry crystalline directions, are in agreement with the predictions.

Spin structures with $(0\ \frac{2}{3}\ \frac{2}{3})$ order have been investigated theoretically also by Lindgård [36]. By continuing his second-order perturbation theory approach on a cluster of spins, he also showed [51] that a phase with $\eta = \frac{2}{3}$ might, indeed, be stable around $B_c/3$ when $\boldsymbol{B} \| [0\ -1\ 1]$. The spin arrangement is a three-sublattice structure where one spin is along the field and two spins, opposite to each other, are almost perpendicular to the field. The phase diagram would consist of three antiferromagnetic states as a function of the field. The experimental observation of simultaneous (1 0 0) and $(0\ \frac{2}{3}\ \frac{2}{3})$ neutron signals at low fields would then result from smearing of the phase boundary.

6. Nuclear magnetism in silver

Experiments on nuclear magnetism in silver are more difficult than those on copper because of the factor-of-twenty times smaller magnetic moment: $-0.113\mu_N$ and $-0.130\mu_N$ for the two stable isotopes ^{107}Ag (51.8%) and ^{109}Ag (48.2%), respectively. Saturation magnetization of the sample, typically made from 80 pieces of 25 μm polycrystalline foil, is experimentally very difficult to reach; the equilibrium polarization at $B = 7$ T and $T = 200$ μK is 91.4% (the corresponding entropy is 0.26 R ln 2), while in copper it would be 99.9%. Achieving a high polarization in silver is a slow process: the spin–lattice relaxation time in a high magnetic field is 14 h at $T_e = 200$ μK.

Owing to coupling of the two silver isotopes during precession, the strength of the indirect RK interaction can be determined by measuring the relative intensities of the ^{107}Ag and ^{109}Ag NMR absorption signals as a function of spin polarization. An antiferromagnetic coupling, characterized by the parameter (see eq. (4)) $\mathscr{R} = -2.5 \pm 0.5$, has been found [25], in agreement with recent electronic band structure calculations [28,54]. This strong RK force is the origin of the large enhancement in the ^{109}Ag resonance line as observed in the spectrum of fig. 12a.

Interesting features of spin dynamics were investigated during the first set of the Helsinki NMR experiments on silver [24]. In high magnetic fields, where the isotopic lines are clearly separated, the systems of both spin species may be characterized by their own nuclear temperatures. As the isotopic lines start to coalesce with decreasing magnetic field, T_{107} and T_{109} begin to equalize with increasing speed.

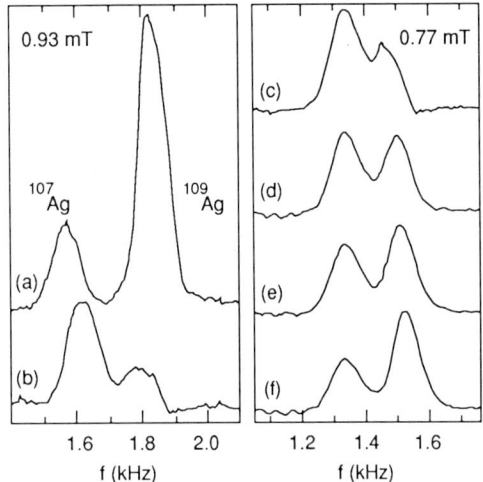

Fig. 12. NMR spectra demonstrating cross relaxation in silver [24]. (a → b) A difference, $T_{109} - T_{107}$, in the two Zeeman temperatures is created by nearly saturating the ^{109}Ag signal at $B = 0.93$ mT. (c) The initial state before cross relaxation; the field has been lowered to 0.77 mT for experimental convenience. (d) The first snapshot. Between (c) and (d) the field was reduced to 0.22 mT for 0.20 s; note the changes in the absorption signals. (e, f) Further snapshots; the last spectrum is already close to equilibrium, compare with (a). The cumulative time spent at 0.22 mT is 0.40 s for (e) and 1.20 s for (f).

A typical measurement is illustrated in fig. 12 [24]. The ^{109}Ag NMR signal at the 0.93 mT field was first saturated by applying a strong alternating field at the resonance frequency, which selectively heated up the ^{109}Ag spins, while T_{107} did not change. This temperature difference between the two spin species did not vanish in several hours. However, if the field was lowered to 0.22 mT for 0.20 s and then brought back to 0.77 mT, some cross relaxation had taken place. This procedure was repeated several times, which made it possible to obtain "snapshots" of the cross-relaxation process.

The characteristic time constant for equilibrium between the two spin species increased steeply with the external field. The surprising finding was the slowness of the cross relaxation, even though the NMR lines of ^{107}Ag and ^{109}Ag were almost completely overlapping at 0.22 mT. This verifies the somewhat counterintuitive theoretical prediction [55] that mutual flips of unlike spins, which become effective once the NMR lines overlap, do not lead to equalizing of temperature. Equilibrium between the two Zeeman reservoirs, i.e. $T_{107} = T_{109}$, is achieved only through the energy reservoir of spin interactions via single spin flips, which become frequent at low fields.

Studies of nuclear magnetism in silver have been extended to negative temperatures as well; the feasibility of such experiments was demonstrated in 1987 [17]. At positive temperatures the number of spins in an upper energy level is always smaller than in a lower level. The spin distribution is given by the Boltzmann factor, $\exp(-\mu B/k_B T)$. If one then quickly, in a time $t \leq \tau_2 = 10$ ms, reverses the magnetic field, the spins have no time to redistribute themselves among the energy levels. Consequently, there are more spins in an upper than in a lower level and, according to the Boltzmann factor, this corresponds to $T < 0$. At the absolute zero on the negative side all spins are in the highest energy level so that this state corresponds to an energy maximum. At $T < 0$, an isolated, isentropic system tries to maximize its energy instead of minimizing it. Because of antiferromagnetic interactions in silver, antiparallel alignment of spins minimizes the energy while parallel alignment corresponds to the energy maximum. Therefore, at $T < 0$, the ground state is ferromagnetic.

The Helsinki group [18] has made measurements of $\chi''(f)$ of silver, at $T > 0$ down to 1.0 nK and at $T < 0$ up to -4.3 nK. Fig. 13 shows the data; the experimental points follow quite nicely the Lorenzian lineshapes. Instead of absorbing, as at $T > 0$, the system is emitting energy when $T < 0$. From these measurements, again by using eq. (7), one can calculate $\chi'(0)$. Unfortunately, polarization losses during the field reversal were about 50%, which is the reason why the ordered state has not been reached yet when $T < 0$.

The ferromagnetic tendency at negative temperatures upon cooling is clearly demonstrated in fig. 14 [18] which shows the absolute value of the inverse static susceptibility of silver. The data display an antiferromagnetic Curie–Weiss law, $\chi'(0) = C/(T - \Theta_A)$ with $\Theta_A = -4.8$ nK, at positive temperatures; $C = 2$ nK is Curie's constant. At $T < 0$, a ferromagnetic law, $|\chi'(0)| =$

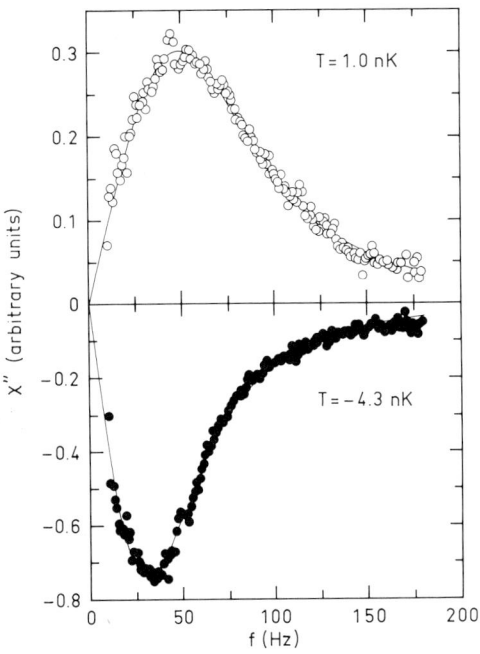

Fig. 13. NMR absorption $\chi''(f)$ for silver at $T = 1.0$ nK (upper picture) and at $T = -4.3$ nK (lower picture) [18]. Note that the vertical scales are different for $T > 0$ and $T < 0$.

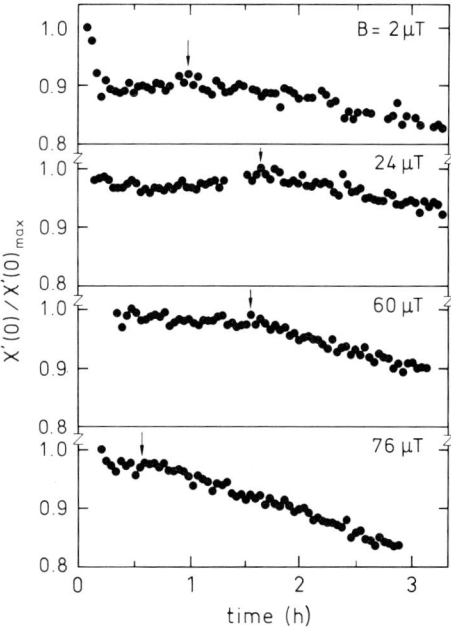

Fig. 15. Static susceptibility of silver nuclei as a function of time for four values of the external magnetic field [19]. Each set of data was separately scaled by $\chi'(0)_{max}$. The arrows indicate the transition points from the ordered to the paramagnetic state.

$C/(|T| - \Theta_F)$ with $\Theta_F = 2.8$ nK was obtained. According to the MF theory [50], $|\Theta_A| = |\Theta_F|$. By averaging the two measured values and taking into account a correction for the shape of the specimen [18], one finds $\mathcal{R} = \Theta/C = -2.5$ in accordance with values obtained using spin dynamics [24,25]. It is worth noticing that both sets of data in fig. 14 follow the Curie–Weiss law to the lowest temperatures, even though this is not the case in magnetic materials usually. There may be slightly more deviation from this law at negative than at positive temperatures.

Fig. 15 illustrates the measured static susceptibility for silver as a function of time, after demagnetization to various low fields [19]. We note, as in the case of copper (see fig. 3), that $\chi'(0)$ has a small maximum or a sharp kink before it starts to decrease; this is a sign of antiferromagnetic order.

However, the clearest indication of antiferromagnetism in silver comes from changes in the NMR spectra upon ordering as illustrated in fig. 16 [19,34]. The lineshape in fig. 16a displays an upward frequency shift of 20 Hz initially after demagnetization to a 2 μT field. The shift decreases with time and vanishes at T_c. However, if the signal in the paramagnetic state is subtracted from the lineshape, then a peak with constant frequency shift is resolved; the constancy of this

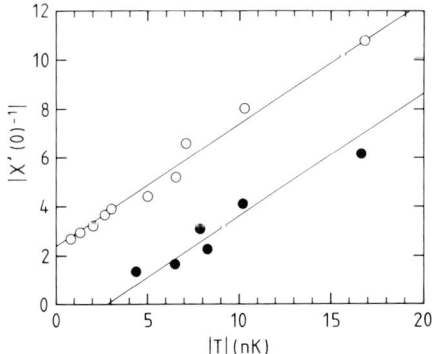

Fig. 14. Absolute value of the inverse static susceptibility $|\chi'(0)^{-1}|$ vs. absolute value of the temperature for silver, measured at $T > 0$ (○) and at $T < 0$ (●), respectively [18].

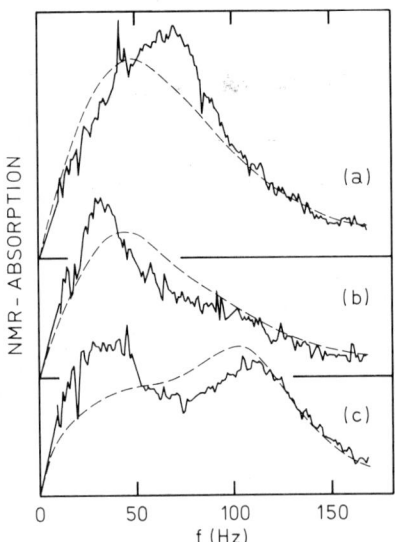

Fig. 16. NMR absorption vs. frequency in silver, measured 5 min after demagnetization was completed to the final field: (a) +2 μT, (b) −5 μT and (c) +24 μT, after a one minute stay at −5 μT [19]. The dashed curves depict the lineshapes in the paramagnetic phase about 1 h later.

frequency shift, characteristic to the antiferromagnetic state, points towards a first order transition in silver.

The measured NMR shifts depend on the demagnetization sequence employed. By applying a small inverted field, −5 μT, a frequency shift downwards was observed, as shown by fig. 16b. When the field was inverted back and increased to 24 μT (fig. 16c) two maxima were seen; these can be identified as two antiferromagnetic resonance modes in the ordered state [35]. Figs. 16a and b separately display a high and a low frequency mode, respectively, whereas fig. 16c illustrates both modes simultaneously.

Using the primary experimental data on the NMR frequency shifts (see fig. 16) and the static susceptibility, the field vs. entropy diagram of silver was constructed at $T > 0$ [19]; it is displayed in fig. 17. Only one ordered state could be singled out, although some features of the data indicate the presence of two phases [34]; the polycrystalline nature of the sample complicated definite conclusions. The presence of just one phase would be an indication of strong dominance by the isotropic RK interaction in silver.

The shape of the phase diagram of fig. 17 is interesting. Owing to the "bulge", silver appears to order antiferromagnetically more easily in a small than in zero field. The measured critical temperature at $B = 0$ is (560 ± 60) pK, whereas $T_c \approx (700 \pm 120)$ pK at 32 μT. This suggests that the shape of the phase diagram is the same in the B–T plane as in the B–S plane. This would indicate, according to Clausius–Clapeyron equation for first order magnetic transitions, that the magnetization is larger in the ordered state than in the paramagnetic state, pointing to ferromagnetism. All the susceptibility data, on the other hand, suggest antiferromagnetism. The bulge could be straightened, however, if $(\partial B/\partial T)_S < 0$ in low fields around T_c; this more likely possibility is, in fact, consistent with the measured $S(T)$-curves at 0 and 32 μT [34]. The presence of two ordered phases could produce the observed phase diagram as well; Ising-model calculations [56], for example, give a higher T_c for the high field phase than for the low field phase.

The observed general features of nuclear ordering in silver seem to agree with theoretical calculations [28,39,57,58]. In zero field, the measured transition temperature, 560 pK, is close to the value, 500 pK, predicted by Monte Carlo simulations; the first order nature of the transi-

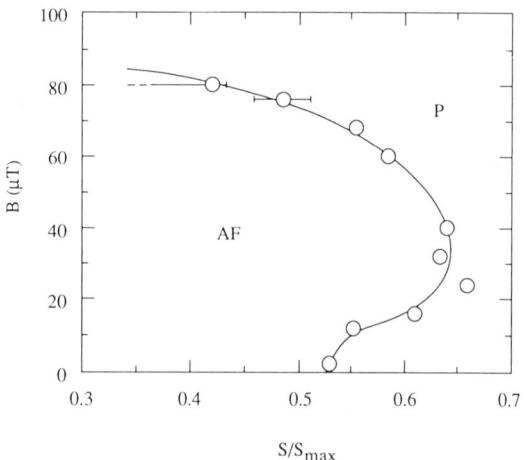

Fig. 17. Measured phase diagram of magnetically ordered silver [19] in the entropy vs. magnetic field plane: (AF) antiferromagnetic and (P) paramagnetic phase. The maximum molar entropy $S_{max} = R \ln 2$.

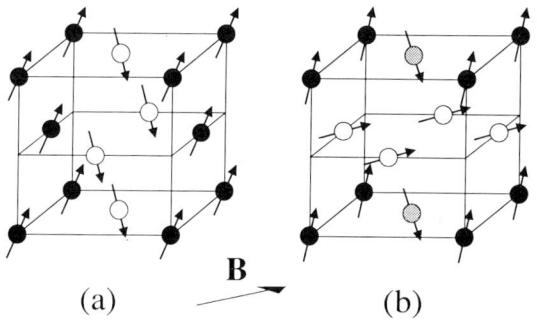

Fig. 18. Equilibrium configuration $\langle I \rangle$ of the nuclear spin system in silver at $T = 0.17$ nK, determined by Monte Carlo simulations [57] for the magnetic field applied in the [1 1 0] direction: (a) $B/B_c < 0.4$, (b) $B/B_c > 0.4$.

tion at small fields is also reproduced by these calculations [39,57]. Similar results have been obtained in computations with exchange interaction only [59]. Monte Carlo calculations have not determined the critical entropy of the transition; this would be a more accurate quantity to compare with experiments. Ising-model results [60] yield $S_c/S_{max} = 0.852$, which exceeds significantly the largest experimental value, 0.64, measured between 30 and 40 μT. The theoretically calculated critical field, $B_c = 140$ μT, is somewhat larger than the extrapolated value of 100 μT, obtained from the measured phase diagram.

Theoretical results [39,57] indicate the presence of a 1-k state at small magnetic fields ($B < 0.4B_c$, see fig. 18a) and a 3-k state, i.e., three propagation vectors, at large fields in the [1 1 0] direction (see fig. 18b); only a 1-k phase is present according to the spin-wave calculation [37] when the field is pointing in the [0 0 1] direction. Thus, the critical field for a transition between these two structures depends on the alignment of B, which would lead to stable side-by-side coexistence of the two ordered states in a polycrystalline sample. In fact, the experimental data on the NMR modes and their weights in silver (see fig. 16) may be interpreted this way [34], but the problem why only one antiferromagnetic resonance mode of both phases is excited in the experiments remains open. To settle the question of the number of ordered phases, and to verify the true long-range order in silver, neutron diffraction experiments are needed. Hopefully, in the near future, it also will be possible to reach the ferromagnetic phase when $T < 0$.

7. Conclusions and future prospects

Significant progress has been made during the last decade in studies of nuclear magnetic ordering in copper and silver. The intriguing magnetic phase diagram of copper nuclei, found first in the susceptibility measurements [12], was later investigated by using neutron diffraction as well [15,16]. The three phases were found to display two kinds of antiferromagnetic order, characterized by the Bragg reflections (1 0 0) and (0 $\frac{2}{3}$ $\frac{2}{3}$). Their variation with the external field has been explained by recent theoretical work. The calculated [52] spin structure at low fields is particularly interesting because it displays both (1 0 0) and (0 $\frac{2}{3}$ $\frac{2}{3}$) orders in a single magnetic domain. However, the ordered structure at high fields oriented along the [111] direction remains a puzzle and calls for further experimental and theoretical work.

Silver provides a good realization of the $I = \frac{1}{2}$ antiferromagnetic Heisenberg model in an fcc lattice, which is an interesting case because of geometrical frustration in this quantum system. A record-low temperature, somewhat below $T_c = 560$ pK has been obtained in silver, and the antiferromagnetically ordered phase has been reached unambiguously in this metal [19]. Susceptibility measurements at negative temperatures, extending to -4.3 nK, have shown a clear ferromagnetic tendency. The most recent experiments on silver have focused on ferromagnetic ordering at $T < 0$ [61]. Saturation of susceptibility has been observed in these measurements extending beyond -2 nK, but the identification of the actual T_c has turned out to be more difficult than at positive temperatures.

Although experimental data [17,19,34] on silver strongly suggest antiferromagnetic long range order, unambiguous proof awaits neutron diffraction measurements. Such experiments may be under way soon [62]. According to present plans, nuclear order in silver will be investigated by a Helsinki–Mainz–Risø collaboration, using the facilities of the Hahn–Meitner Institute.

What other experiments have been done on nuclear ordering in simple metals? Thallium with ferromagnetic exchange interaction has been investigated [63,64]. Since this metal is superconducting below $B_c = 17$ mT, these experiments have been performed at fairly high magnetic fields, $B > B_c$, which makes it difficult to resolve between a ferromagnetically ordered state and a polarized paramagnetic state. The measured NMR lines display a splitting, sometimes even multifold, at $p > 0.4$. This may be a precursor of ferromagnetic ordering [63,64], but it is more likely that its origin lies in the presence of two interacting spin species. It has been proposed that the splitting may originate either from bound spin wave modes localized on impurities, ^{203}Tl (29.5%) embedded in the ^{205}Tl matrix (70.5%) [65], or from ordinary energy level repulsion, competing with a merging term due to thermal fluctuations [66].

Some preliminary work has also been done on scandium [67,68] in which spin ordering takes place in the presence of a quadrupolar interaction; indications of this have been observed [69]. One complicating experimental problem in scandium is the presence of iron impurities which, however, should not prevent the observation of nuclear ordering since they tend to freeze into a spin-glass state around 1 mK [67].

Gold would be an obvious continuation of the susceptibility studies done in Helsinki. In fact, this material has already been picked up by a group in Florida [70]. In this metal there is the additional bonus that it might become superconducting [71]. One of the problems is to obtain a sufficiently pure specimen so that electronic magnetic impurities would not destroy superconductivity.

To study the interplay between superconductivity and nuclear ordering, experiments on rhodium are planned in Helsinki. This metal has the lowest critical temperature, 325 μK, and smallest critical field, 4.9 μT, of any superconductor [72]; B_c for nuclear ordering is expected to be higher.

Various assemblies of nuclear magnets are ideal for testing many interesting systems, such as spin glasses. By substitution of the magnetic ^{195}Pt with nonmagnetic isotopes of platinum, one can cover the whole "impurity" range from 0 to 100%, without otherwise affecting the properties of the metal. Because of the short spin-lattice relaxation time, $\tau_1 = 30$ ms K in platinum, one might need a three-stage nuclear refrigerator for these experiments.

Acknowledgements

We would like to thank A.J. Annila, B.N. Harmon, M.T. Huiku, T.A. Jyrkkiö, K. Kakurai, J.K. Kjems, P. Kumar, J. Kurkijärvi, J.M. Kyynäräinen, K.K. Nummila, K. Siemensmeyer, Y. Takano, J.T. Tuoriniemi, X.W. Wang, H. Weinfurter, H.E. Viertiö and S. Yin for collaboration. We are, in particular, indebted to Kurt Clausen, Per-Anker Lindgård and Michael Steiner, who were our senior colleagues during the neutron diffraction measurements at Risø. Our work was supported by the Academy of Finland.

Note added in proof

Ferromagnetic ordering has recently been found in silver at $T < 0$, with a critical initial polarization of about -50% [61]. The observed state seems to be characterized by an exceptionally small hysteresis. This can be understood by a theoretical analysis of the ferromagnetic domain configuration at negative temperatures [73]. Recent theoretical work has also yielded predictions for the antiferromagnetically ordered spin structure of copper when the external field is applied close to the [111] crystalline direction [74,75].

References

[1] W.P. Halperin, C.N. Archie, F.B. Rasmussen, R.A. Buhrman and R.C. Richardson, Phys. Rev. Lett. 32 (1974) 927.
[2] For examples of work on Van Vleck paramagnets, see K. Andres, E. Bucher, J.P. Maita and A.S. Cooper, Phys. Rev. Lett. 28 (1972) 1652.
J. Babcock, J. Kiely, T. Manley and W. Weyhmann, Phys. Rev. Lett. 43 (1979) 380.

[3] M. Kubota, H.R. Folle, Ch. Buchal, R.M. Mueller and F. Pobell, Phys. Rev. Lett. 45 (1980) 1812.
[4] N. Kurti, F.N. Robinson, F.E. Simon and D.A. Spohr, Nature 178 (1956) 450.
N. Kurti, Physica B 109 & 110 (1982) 1737.
[5] M. Chapellier, M. Goldman, V.H. Chau and A. Abragam, CR Acad. Sci. 268 (1969) 1530.
[6] A. Abragam and M. Goldman, Nuclear Magnetism: Order and Disorder (Clarendon Press, Oxford, 1982).
[7] G.J. Ehnholm, J.P. Ekström, J.F. Jacquinot, M.T. Loponen, O.V. Lounasmaa and J.K. Soini, Phys. Rev. Lett. 42 (1979) 1702.
[8] G.J. Ehnholm, J.P. Ekström, J.F. Jacquinot, M.T. Loponen, O.V. Lounasmaa and J.K. Soini, J. Low Temp. Phys. 39 (1980) 417.
[9] For a review of the early work on copper, see O.V. Lounasmaa, Phys. Today 32 (1979) 32.
[10] M.T. Huiku and M.T. Loponen, Phys. Rev. Lett. 49 (1982) 1288.
[11] M.T. Huiku, T.A. Jyrkkiö and M.T. Loponen, Phys. Rev. Lett. 50 (1983) 1516.
M.T. Huiku, T.A. Jyrkkiö, M.T. Loponen and O.V. Lounasmaa, AIP Conf. Proc. No. 103 (1983) 441.
[12] M.T. Huiku, T.A. Jyrkkiö, J.M. Kyynäräinen, A.S. Oja and O.V. Lounasmaa, Phys. Rev. Lett. 53 (1984) 1692.
[13] M.T. Huiku, T.A. Jyrkkiö, J.M. Kyynäräinen, M.T. Loponen, O.V. Lounasmaa and A.S. Oja, J. Low Temp. Phys. 62 (1986) 433.
[14] K. Clausen, M.T. Huiku, T.A. Jyrkkiö, M.T. Loponen, O.V. Lounasmaa, P.R. Roach and K. Sköld, Helsinki University of Technology Report TKK-F-A529 (1983).
[15] T.A. Jyrkkiö, M.T. Huiku, O.V. Lounasmaa, K. Siemensmeyer, K. Kakurai, M. Steiner, K.N. Clausen and J.K. Kjems, Phys. Rev. Lett. 60 (1988) 2418.
[16] A.J. Annila, K.N. Clausen, P.-A. Lindgård, O.V. Lounasmaa, A.S. Oja, K. Siemensmeyer, M. Steiner, J.T. Tuoriniemi and H. Weinfurter, Phys. Rev. Lett. 64 (1990) 1421.
[17] A.S. Oja, A.J. Annila and Y. Takano, J. Low Temp. Phys. 85 (1991) no. 112.
[18] P.J. Hakonen, S. Yin and O.V. Lounasmaa, Phys. Rev. Lett. 64 (1990) 2707.
[19] P.J. Hakonen, S. Yin and K.K. Nummila, Europhys. Lett. 15 (1991) 677.
[20] O.V. Lounasmaa, Phys. Today 42 (1989) 26.
[21] A.S. Oja, Physica B 169 (1991) 306.
[22] M.A. Ruderman and C. Kittel, Phys. Rev. 96 (1954) 99.
[23] J.P. Ekström, J.F. Jacquinot, M.T. Loponen, J.K. Soini and P. Kumar, Physica B 98 (1979) 45.
[24] A.S. Oja, A.J. Annila and Y. Takano, Phys. Rev. Lett. 65 (1990) 1921.
[25] A.S. Oja, A.J. Annila and Y. Takano, to be published.
[26] P.-A. Lindgård, X.-W. Wang and B.N. Harmon, J. Magn. Magn. Mater. 54–57 (1986) 1052.
[27] A.S. Oja, X.-W. Wang and B.N. Harmon, Phys. Rev. B 39 (1989) 4009.

[28] B.N. Harmon, X.-W. Wang and P.-A. Lindgård, J. Magn. Magn. Mater. (1991) to be published.
[29] L.H. Kjäldman and J. Kurkijärvi, Phys. Lett. A 71 (1979) 454.
[30] K. Binder and A.P. Young, Rev. Mod. Phys. 58 (1986) 801, see especially pp. 950–953 and references therein.
[31] O.V. Lounasmaa, Experimental Principles and Methods below 1 K (Academic Press, London 1974).
K. Andres and O.V. Lounasmaa, Progress in Low Temperature Physics, vol. 8, ed. D.F. Brewer (North-Holland, Amsterdam, 1982) p. 222.
[32] J. Korringa, Physica 16 (1950) 601.
[33] The original design for the solid block nuclear stage with slits is due to P. Roubeau.
[34] P.J. Hakonen and S. Yin, J. Low Temp. Phys. 85 (1991) no. 112.
[35] P. Kumar, J. Kurkijärvi and A.S. Oja, Phys. Rev. B 33 (1986) 444.
[36] P.-A. Lindgård, Phys. Rev. Lett. 61 (1988) 629.
[37] H.E. Viertiö and A.S. Oja, Phys. Rev. B 36 (1987) 3805.
[38] S.J. Frisken and D.J. Miller, Phys. Rev. Lett. 61 (1988) 1017.
[39] H.E. Viertiö and A.S. Oja, Quantum Fluids and Solids, AIP Conf. Proc. No. 194 (1989) 305.
[40] Y. Roinel, V. Bouffard, G.L. Bacchella, M. Pinot, P. Mériel, P. Roubeau, O. Avenel, M. Goldman and A. Abragam, Phys. Rev. Lett. 41 (1978) 1572.
[41] A. Benoit, J. Bossy, J. Flouquet and J. Schweizer, J. de Phys. Lett. 46 (1985) L923.
[42] R.M. Nicklow, R.M. Moon, S. Kawarazaki, N. Kunitomi, H. Suzuki, T. Ohtsuka and Y. Morii, J. Appl. Phys. 57 (1985) 3784.
A. Benoit, J. Flouquet, J.L. Genicon and J. Palleau, Physica B 108 (1981) 1103.
[43] S. Kawarazaki, N. Kunitomi, J.R. Arthur, R.M. Moon, W.G. Stirling and K.A. McEwen, Phys. Rev. B 37 (1988) 5336.
[44] H. Glättli and M. Goldman, in: Neutron Scattering, eds. K. Sköld and D.L. Price, Methods of Experimental Physics 23 (Academic Press, New York, 1987) part C, p. 241.
[45] M. Steiner, Phys. Scripta 42 (1990) 367.
[46] T.A. Jyrkkiö, M.T. Huiku, K. Siemensmeyer and K.N. Clausen, J. Low Temp. Phys. 74 (1989) 435.
[47] K.N. Clausen, Neutron News (1991) to be published.
[48] Some field-dependent decrease in the width of the (1 0 0) Bragg peak has been observed during warm-up. For more details, see K. Siemensmeyer, Ph.D. Thesis, Technischen Universität Berlin (1989).
[49] A.J. Annila, K.N. Clausen, A.S. Oja, J.T. Tuoriniemi and H. Weinfurter, Phys. Rev. B (1991) to be published.
[50] See, e.g., J.S. Smart, Effective Field Theories of Magnetism (Saunders, Philadelphia, 1966).
[51] P.-A. Lindgård, J. Magn. Magn. Mater. 90 & 91 (1990) 138.
[52] H.E. Viertiö and A.S. Oja, Phys. Rev. B 42 (1990) 6857.

[53] H.E. Viertiö and A.S. Oja, Physica B 165 & 166 (1990) 797, 799.
[54] D.J. Miller and S.J. Frisken, J. Appl. Phys. 64 (1988) 5630.
[55] M. Goldman, Spin Temperature and Nuclear Magnetic Resonance in Solids (Clarendon Press, Oxford, 1970), see particularly chap. 6 and references therein.
[56] J.M. Sanchez, D. de Fontaine and W. Teitler, Phys. Rev. B 26 (1982) 1465.
[57] H.E. Viertiö, Phys. Scripta T33 (1990) 168.
[58] S.J. Frisken and D.J. Miller, unpublished.
[59] T.M. Giebultowitz and J.K. Furdyna, J. Appl. Phys. 57 (1985) 3312.
[60] See, e.g., D.C. Mattis, The Theory of Magnetism II (Springer Verlag, Berlin, 1985).
[61] P.J. Hakonen, K.K. Nummila, R. Vuorinen and O.V. Lounasmaa, to be published.
[62] A.J. Annila, K.N. Clausen, P.J. Hakonen, P.-A. Lindgård, O.V. Lounasmaa, K.K. Nummila, A.S. Oja, K. Siemensmeyer, M. Steiner, J.T. Tuoriniemi, H. Weinfurter and H.E. Viertiö, Risø National Laboratory Report No. M-2874 (1990).
[63] G. Eska and E. Schuberth, Japan. J. Appl. Phys. Suppl. 26-3 (1987) 435.
[64] G. Eska, in: Quantum Fluids and Solids – 1989, eds. G.G. Ihas and Y. Takano, AIP Conf. Proc. No. 194 (New York, 1989) p. 316.
[65] D. Rainer, I. Fomin and G. Eska, to be published.
[66] A.S. Oja, A.J. Annila and Y. Takano, Phys. Rev. B 38 (1988) 8602.
[67] H. Suzuki, T. Sakon and N. Mizutani, Physica B 165 & 166 (1990) 795.
[68] L. Pollack, E.N. Smith, R.E. Mihailovich, J.H. Ross Jr., P.J. Hakonen, E. Varoquaux, J.M. Parpia and R.C. Richardson, Physica B 165 & 166 (1990) 793.
[69] H. Suzuki, private communication.
[70] Y. Takano, private communication.
[71] Ch. Buchal, R.M. Mueller, F. Pobell, M. Kubota and H.R. Folle, Solid State Commun. 42 (1982) 43.
[72] Ch. Buchal, F. Pobell, R.M. Mueller, M. Kubota and J.R. Owers-Bradley, Phys. Rev. Lett. 50 (1983) 64.
[73] H.E. Viertiö and A.S. Oja, J. Magn. Magn. Mater. (1991) submitted.
[74] P.-A. Lingård, J. Magn. Magn. Mater. (1991) submitted.
[75] A.S. Oja and H.E. Viertiö, J. Magn. Magn. Mater. (1991) submitted.

Magnetic recording materials since 1975

Geoffrey Bate

School of Engineering, Santa Clara University, Santa Clara, CA 95053, USA

When information is to be recorded and stored on a re-usable medium, magnetic recording, in one form or another, has been and is today the dominant technology. Magnetic particles or thin films having a coercivity of several hundred oersteds, are easily capable of retaining a magnetic pattern of recorded information (at densities of tens of thousands of bits/inch and track densities of thousands of tracks/inch) for hundreds of years and yet when desired, the pattern can be changed by simply writing the new information over the old. Since the recording process requires only a change in direction of the electron spins, the process is infinitely reversible and the new information may be read immediately with no development process being required. This paper deals with developments of the magnetic (and some important non-magnetic) properties of recording media that have occurred since 1975. It describes their current embodiment in particles of cobalt-modified iron oxides, chromium dioxide, barium ferrite and metals and in thin films of metals, alloys and oxides. The curious and currently unexplained aspects of the behavior of these important materials is noted.

1. Introduction

The earliest (and only partially successful) magnetic recording materials were wires of stainless steels (12% nickel, 12% chromium) which had been annealed so that single-domain particles of the ferritic phase precipitated in an austenitic matrix. Coercivities of 200–300 Oe are easily obtained in this way. Two practical problems limited the usefulness of the wires. They twisted so that the region of the wire which contacted the head during writing was not necessarily the same region that was close to the head during the reading operation. Secondly, the wires broke easily and could only be repaired by tying knots. For these reasons the wires were replaced in the 1940s and 1950s by splice-able tapes coated with synthetic particles of $\gamma\text{-Fe}_2\text{O}_3$ in a polymeric matrix. In 1956 the particles were also used to coat the 24″ diameter disks in the IBM Ramac (<u>Ran</u>dom <u>ac</u>cess) 305 disk machine. From then until the mid-1970s tape and disk coatings were principally of $\gamma\text{-Fe}_2\text{O}_3$ single-domain, particles. Magnetic disks used these particles until ca. 1990.

The mechanism of magnetization reversal in the acicular single-domain particles (length, typically, 0.3 μm, diameter ≈ 0.06 μm) is believed to be incoherent rotation (e.g., "curling") of the spins [1–4].

2. Magnetic particles

That $\gamma\text{-Fe}_2\text{O}_3$ should have been preferred over Fe_3O_4 is, at first, surprising since σ_s is higher in Fe_3O_4 particles (by 20%). Thus the coercivity is also higher (by 15%) in these acicular particles in which shape anisotropy is more important than magnetocrystalline anisotropy. Furthermore, Fe_3O_4 is the penultimate step in the preparation of $\gamma\text{-Fe}_2\text{O}_3$ particles and thus Fe_3O_4 particles should be marginally cheaper.

Particles of magnetite have acquired the reputation of being prone to oxidation to $\alpha\text{-Fe}_2\text{O}_3$ (non-ferrimagnetic) rather than to $\gamma\text{-Fe}_2\text{O}_3$. The long-term stability of information stored on media made of magnetite particles is also somewhat suspect because they are believed to suffer from magnetic "accommodation" effects. Accommodation is the phenomenon in which the material does not achieve a constant hysteresis loop at the first cycling of the applied field. Only after sev-

0304-8853/91/$03.50 © 1991 – Elsevier Science Publishers B.V. All rights reserved

eral cycles does the hysteretic behavior become constant [1]. Clearly this would be a highly inappropriate behavior in a magnetic recording material. A related (and equally undesirable property) is "print-through". This is found in tapes and occurs because two layers of magnetic coating are separated only by the thickness of the plastic substrate. The pattern of fields emanating from the recorded signal on one layer, passes through the substrate and may be strong enough to modify the magnetic state of the particles in adjacent layers. The effect is most serious for long wavelength patterns. It thus tends to be more of a problem in audio recording at low frequencies, where one occasionally hears echoes or "pre-echoes" of music recorded on adjacent layers of the tape. The solution is to control the shape and size distributions of the magnetic particles.

One of the principal advantages of magnetic recording is that the recorded information may be changed at any time, by simply recording the new information over the old. The old pattern of magnetization is, in principle, replaced by the new pattern, as long as the recording fields exceed those required to saturate the magnetic coatings. Usually this requires recording fields equal to or greater than four or five times the coercivity, or, in the case of particles of γ-Fe_2O_3, fields of 1000–1200 Oe.

In flexible disks for data recording the overwrite specification is that, after writing and reading a lowest density pattern of frequency 1F and then overwriting with the highest frequency 2F (at the same current), on re-reading the disk the remaining 1F signal must be ≤ 5% of its original level; i.e., the unerased 1F signal must be reduced by 26 dB. Achieving more complete erasure requires the use of overwriting signals at higher currents than those used for the recording of the 1F signal. It appears to be practically impossible to achieve total erasure even though the current used for overwriting is more than five times the amplitude of that used for writing the original 1F pattern. The 1F signal may be decreased by 90 dB but is still detectable [5]. This curiously non-reversible behavior of magnetic recording media has, not surprisingly, been a cause for concern among security agencies.

3. Magnetic properties and recording performance

Since the recorded information is stored as a pattern of magnetization along the track followed by the writing and reading heads, we require first that the magnetic material have high remanence, M_r. This, in turn, calls for high saturation magnetization, M_s. When the stored information is digital data, one (or more) reversals of magnetization correspond to each recorded bit. If the bits are to be stored at high density along the track (to reduce the cost of storage and to reduce the time taken to reach the desired record), the magnetization reversals must be sharply defined. This calls for a narrow distribution of switching fields ($\Delta H / H_c$) as defined in fig. 1.

The recorded pattern (fig. 2) of opposed magnetic dipoles is subject to demagnetization and hence, the coercivity, H_c (or more exactly, the remanence coercivity H_r) must be high. The driving force of demagnetization is the remanent magnetization M_r and the ability to resist demagnetization is described by H_r, thus some optimal value of the ratio M_r/H_r must be found and used in the recording medium.

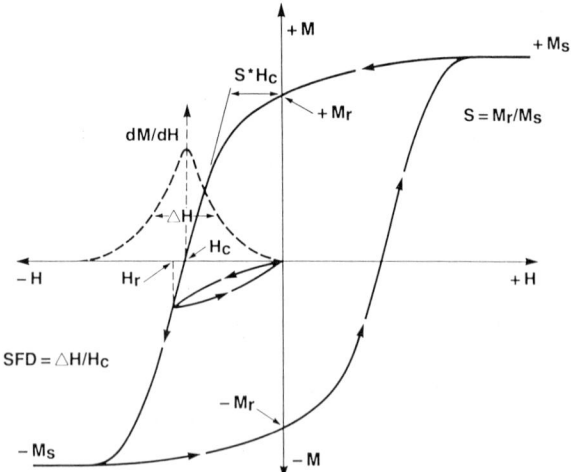

Fig. 1. Magnetic properties important in magnetic recording: coercivity, H_c, remanence coercivity, H_r, remanence, M_r/M_s. Switching Field Distribution (SFD) = $\Delta H/H_c$ (alternate form of switching field distribution = $1 - S^*$).

4. High-coercivity particles

Particles of unmodified iron oxide are now used only for low recording density applications such as telephone answering recorders. For high fidelity audio recording, video recording and high density data recording, particles having higher coercivity than 300 Oe are needed.

4.1. Cobalt-modified iron oxide

Cobalt, when substituted in small amounts (2–3%) in magnetite and γ-Fe_2O_3, has been known for many years to cause increases of hundreds of oersteds in the coercivity (by increasing the magnetocrystalline anisotropy). Unfortunately, the temperature sensitivity of H_c is also increased to unacceptable levels by the substitution of cobalt. The coercivity decreases with increasing temperature. Since the magnitude of the magnetostriction constant, λ_s, also increases with increasing cobalt content, magnetization of the cobalt-doped iron oxide particles is also sensitive to stress [6].

A solution to the problem of using cobalt to increase the coercivity of iron oxide particles without also increasing their temperature dependence, was reported by Umeki et al. in 1974 [7]. It consisted of immersing the iron oxide particles in a solution of cobaltous sulphate to which sodium hydroxide was then added. Cobaltous hydroxide $Co(OH)_2$ precipitates on the surface of the particles. On warming to 45°C for 8 h, cobalt ions diffuse into the surface of the iron oxide particles and bring about an increase in coercivity. In particles whose coercivity has been increased by the diffusion of Co ions into a surface layer of the particle, the temperature dependence of coercivity is much less than is found when cobalt is uniformly distributed throughout the volume of the particle. This was clearly demonstrated in an experiment of Kishimoto et al. in 1984 [8]. Starting with particles of γ-Fe_2O_3 suspended in an alkaline solution containing cobaltous and ferrous ions in a molar ratio (Co^{2+}/Fe^{2+}) of 0.5, they heated the suspension at 45°C for 8 h "to crystallize cobalt-ferrite on the γ-Fe_2O_3 particles. The resultant particles (cobalt-epitaxial iron oxide particles) had the uniaxial magnetic anisotropy along the acicular axis". After measuring the magnetic properties, the particles were heated at 450°C in air for 2 h after which "cobaltous ions were uniformly substituted into the γ-Fe_2O_3 particles". By changing the cobalt content from 3.6 to 7 wt%, the coercivity at 20°C was increased from 800 to 1300 Oe and the "squareness" (M_r/M_s) was increased from 0.70 to 0.72.

On measuring the effect of temperatures on H_c, it was found that (as shown in fig. 3) the temperature coefficient of coercivity was much higher when the cobalt was uniformly distributed throughout the particles than when it was confined to the surface of the particles. The second result was that the value of coercivity at room temperature for a given percent of cobalt was approximately the same whether the cobalt was on the surface or distributed throughout the volume of the particle.

The effect of cobalt on the magnetocrystalline anisotropy constant of magnetite (Fe_3O_4) was measured by Bickford et al. [9] in 1957 as a function of temperature. Slonczewski [10,11] explained the large change in magnetocrystalline anisotropy constant, K_1, that they observed in

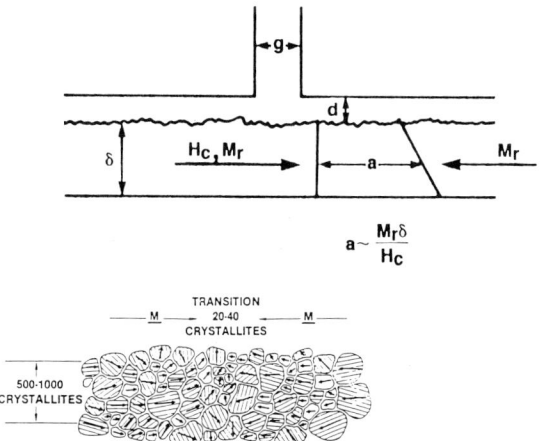

Fig. 2. (a) The recorded transition in digital recording. The magnetic medium, moving from left to right is initially magnetized in the negative direction. The recording field is presently in the positive direction. The transition region width a is inversely related to the maximum recording density. (b) When the recording medium consists of particles or crystallites, the transition region has irregular boundaries which are an important source of media noise.

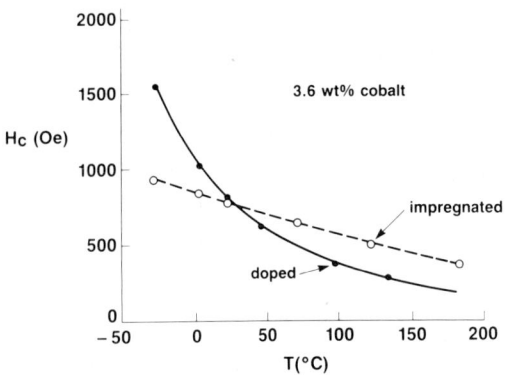

Fig. 3. Coercivity in 3.6 wt% Co-modified iron oxide particles as a function of temperature. "Doped" cobalt is distributed throughout the volume of each particle. "Impregnated" cobalt is confined to a surface layer of each particle.

terms of "one-ion" model involving the spin–orbit coupling of Co^{2+} ions placed in an electric field of trigonal symmetry in the lattice. There is at present no completely satisfactory explanation of the origin of the coercivity and its temperature dependence of cobalt surface-modified iron oxide particles and this is one of the most intriguing of the unsolved problems in the field of recording materials. The absence of understanding of the origin of coercivity in these particles has not prevented their use in high recording density applications such as video and high-performance audio tape, instrumentation tapes and flexible disks.

4.2. Other iron oxides

Even without the complications caused by the addition of cobalt the magnetic behavior of the iron oxides is difficult to understand. The final step in the preparation of particles of $\gamma\text{-}Fe_2O_3$ is the oxidation of Fe_3O_4 particles at low temperatures in the presence of water. In some commercial processes this is not the final step. The $\gamma\text{-}Fe_2O_3$ particles are reduced back to Fe_3O_4 and then re-oxidized and re-reduced, etc., so that the final commercial product may have a composition somewhere between Fe_3O_4 and Fe_2O_3. Horiishi and Takedoi [12] treated particles of $\gamma\text{-}Fe_2O_3$ with an aqueous solution of either ferrous salts or ferrous hydroxide at 50–100 °C and pH of 8.0–14.0 in a reducing atmosphere and then re-oxidized to $\gamma\text{-}Fe_2O_3$. The first cycle produced an increase in the particle coercivity from 370 to 420 Oe. The cycle could be repeated several times. The end product had an atomic ratio of $Fe^{3+} : Fe^{2+} = 1 : 0.23$; i.e., the material was midway between Fe_2O_3 and Fe_3O_4. These are often referred to as "Berthollide oxides".

Haneda and Kojima [13] investigated the effect of varying the proportion of magnetite in $(Fe_3O_4)_x(\gamma\text{-}Fe_2O_3)_{1-x}$ and found that the maximum coercivity was achieved at a higher value of $x = 0.75$. They suggested that their particles were composed of finely divided crystallites of $\gamma\text{-}Fe_2O_3$ and pure Fe_3O_4. Knowles [14] explained the magnetic properties of particles whose composition was intermediate between Fe_3O_4 and Fe_2O_3 as resulting from "an inner core of Fe_3O_4 with an outer layer of $\gamma\text{-}Fe_2O_3$. Owing to the large change in volume which occurs on oxidation, these ... stress each other. Over a certain range of composition, the stress field interacts with the magnetization to increase H_c, and it is also responsible for H_c increasing with time after the oxidation process. A similar situation applies to partially reduced $\gamma\text{-}Fe_2O_3$".

4.3. Chromium dioxide

The existence of a ferromagnetic oxide of chromium, CrO_2, was reported as early as 1935, but the material was not made in single-domain form and used a magnetic recording material until 1961 [12]. The process involved the decomposition of CrO_3 under pressure and usually in the presence of water and results in highly acicular particles (aspect ratios of 20 or 30 : 1 are not unusual) with a very regular prismatic shape. The appearance in the electron microscope is in marked contrast to that of $\gamma\text{-}Fe_2O_3$ and Co-γ particles. The two principal problems were the cost of preparation (largely due to batch processing) and the low Curie temperature ($\Theta_c \approx 120$ °C). Since 1961, much effort has been expended in attempts to reduce the former and increase the latter. Chromium dioxide particles are used commercially in high-quality audio tapes (an application now dominated by particles of cobalt-impregnated iron oxide) and in the tape cartridges for the IBM 3480 computer tape drive.

The low Curie temperature has been used to advantage in the thermoremanent copying of tapes. The recorded tape (made of magnetic materials with Θ_c considerably higher than that of CrO_2) is placed in (stationary) contact with an unrecorded tape of CrO_2 particles at temperatures between 25 and 120°C. The field pattern emanating from the master tape then records on the CrO_2 slave tape as the temperature is reduced to ambient. The advantage of thermoremanent transfer is that it can be carried out at much higher speeds than are possible by conventional copying; (i.e., by taking the amplified signal, reading from the master and using it to write conventionally on the slave tape).

Chromium dioxide tapes have been suspected of suffering from short-term and long-term loss of remanence. It is well established [16] that if a field sufficient to bring a sample of CrO_2 tape to saturation remanence is applied and is followed by a field in the reverse direction slightly less than the coercivity, the magnetization level will decrease with time even though the applied field is held constant (fig. 4). However, tapes made of particles of cobalt-treated γ-Fe_2O_3 change their level of magnetization in much the same way and so, to a lesser extent, do tapes made of untreated γ-Fe_2O_3 particles, as seen in fig. 4.

The long-term effect is, perhaps, more serious, particularly when CrO_2 particles are used in a medium like 1/2″ computer tape which is intended for archival storage of data. The concern arises because of the possible reactions between CrO_2 particles and the polymers used to isolate the particles and to bind them to the substrate which results in a loss of magnetization over time.

4.4. Metal particles

Single-domain particles of the ferromagnetic metals, principally iron may be prepared by three methods: 1) reduction in hydrogen of the oxides or oxyhydroxides, 2) decomposition of the borohydride by heat and 3) condensation of the vaporized metal. The advantages of using the metal, rather than the oxide, in the form of particles for

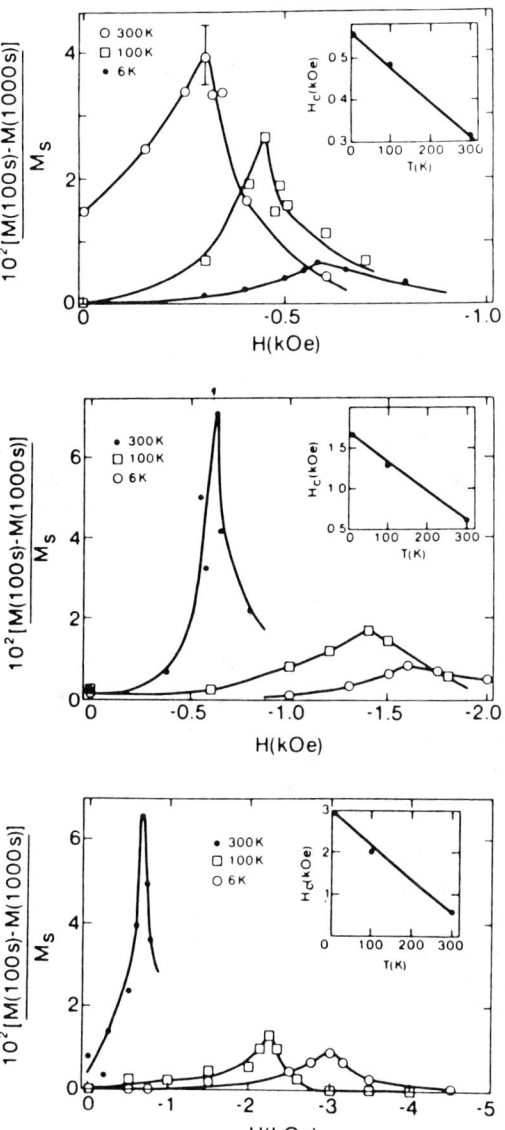

Fig. 4. Time decay of magnetization at 300, 100 and 6 K, for particles of: (a) γ-Fe_2O_3; (b) cobalt-surface modified γ-Fe_2O_3; (c) CrO_2. After applying a saturating field in the positive direction fields plotted along the abscissa are applied in the inverse direction and the magnetization is measured as a function of time that the plotted field is applied. $M(100\ s)$ is the magnetization after the reverse field is applied for 100 s. $M(1000\ s)$ is the magnetization after the reverse field is applied for 1000 s [16b].

recording are the higher saturation and remanent magnetizations and thus, in particles whose anisotropy is dominated by shape, higher coercivities. The disadvantages are higher cost of preparation and greater chemical reactivity with the surrounding organic binder materials and with the atmosphere.

The particles currently in use in high-quality audio cassettes (analog and digital) have coercivities up to 2100 Oe [17]. This is made up of the contribution of shape anisotropy (≈ 1900 Oe) and magnetocrystalline anisotropy (≈ 200 Oe). Since the outer shell of the particles is deliberately oxidized to provide a barrier against further corrosion, the saturation magnetization of the particles is somewhat less (984 to 1338 emu/cm^3) than the 1714 emu/cm^3 for pure iron. Similar values for the saturation magnetization are obtained if the iron particles are converted to iron nitride, Fe_4N.

4.5. Barium ferrite particles

Barium ferrite $BaO \cdot 6Fe_2O_3$ has the distinction of being the most widely used permanent magnet material. It contains no expensive ingredients and has, in fine particle form, coercivities of 2000–3000 Oe. The saturation magnetization of 60 emu/g at room temperature not surprisingly is less than that of the iron oxides. Before the material can be used in fine particle form for magnetic recording purposes, it is necessary to reduce the coercivity to ≥ 1000 Oe without reducing still further a magnetization level which is already low. A method of accomplishing both goals was achieved by Kubo et al. [17]. They melted a mixture of BaO, Fe_2O_3 and B_2O_3 and rapidly cooled the liquid by running it between steel rollers. The boron oxide provided a glass-like matrix within which single-domain platelets of barium ferrite formed and were then crystallized by reheating. The B_2O_3 was removed with hot acetic acid from the $BaO \cdot 6Fe_2O_3$ particles which had an average diameter of ≈ 0.08 μm and a thickness of ≈ 0.03 μm; i.e., the particles had the form of hexagonal platelets. To reduce the coercivity to levels suitable for a recording medium, cobalt and titanium in the form of CoO and TiO_2 were added to the starting material and incorporated in the crystal lattice, $BaFe_{12-2x}Co_xTi_xO_{19}$. When $x = 0.75$ the coercivity is reduced to ≈ 900 Oe, while the value of magnetic moment density $\sigma = 60$ emu/g (almost the same as its value at $x = 0$).

The addition of cobalt and titanium is believed to support the hexagonal axis as the easy axis of magnetization. Thus crystalline anisotropy encourages magnetization along the hexagonal axis while shape anisotropy supports magnetization in the basal plane. The coercivity is then given by the difference between the crystalline term $2K/M_s$ and the shape term NM_s or

$$\langle H_c \rangle = 0.48 \left(\frac{2K}{M_s} - NM_s \right),$$

where $K = +9.45 \times 10^5$ is the magnetocrystalline anisotropy constant (erg/cm^3), [0.48] is a "randomizing" factor which is applied to particle samples in which the c-axis of the platelets are not aligned. M_s is the saturation magnetization (emu/cm^3); $N \approx 2\pi$ is the demagnetizing factor along the c-axis. For a temperature of 20°C and $x = 0.8$; $H = 960$ Oe. As the temperature increases M_s becomes smaller and the first term ($2K/M_s$), grows larger while the second term ($-NM_s$) becomes smaller in magnitude and remains negative in sign. Thus both terms encourage H_c to grow larger with increasing temperature; an unusual behavior for a magnetic recording material.

Dispersion of the particles within the solvent and binder materials is difficult since, in addition to surface tension effects, the particles experience a magnetic force of attraction to other particles, thus forming a stack which is difficult to separate into individual particles.

Corradi et al. [18] and Kishimoto and Kitahata [19] showed that the increase in H_c with increasing temperature obtained by Kubo et al. [17] becomes more complicated as the range of temperatures investigated is increased. Below ≈ 250 K and above ≈ 450 K the coercivity of Co, Ti doped barium ferrite particles *decreases* with increasing temperature. No simple explanation of these results has been found.

Suzuki et al. [20] described a potentially important, new application of barium ferrite particles. They are used as the magnetic component of

Table 1
Table of properties of magnetic particles

Property	Particle				
	γ-Fe_2O_3	Co–Fe_2O_3	CrO_2	BaO·6 Fe_2O_3 + Co, Ti	Fe
Saturation magnetization, M_s [kA/m]	336–340	350–355	340–390	380	870–1100
M_r/M_s	0.5	0.5–0.8	0.5	0.6–0.7	0.23 ± 0.52
Curie temperature, T_c [K]	[863]	[863]	390–400	590	1040
Magnetocrystalline anisotropy constant, K_1 [J/m³]	-4.64×10^3	-5 to $+100 \times 10^3$	$+2.5 \times 10^4$	$+3.3 \times 10^5$	$+4.4 \times 10^4$
Saturation magnetostriction, λ_s	-5×10^{-6}	-5 to -15×10^{-6}	$+1 \times 10^{-6}$	$+10 \times 10^{-6}$	$+4 \times 10^{-6}$
Coercivity, H_c [kA/m]	20–28	43.8–60	36–53	26–157	30–131
Switching field distribution, $\Delta H/H_c$	0.26–0.61	0.30–0.6	0.35–0.6	0.16–0.6	0.5–0.75
Temperature coeff. of coercivity, $\Delta H_c/H_c$ °C (20–70°C)	-1×10^{-3}	impreg.^{d} -2.4×10^{-3} doped -10×10^{-3}	-5×10^{-3}	$+3.1 \times 10^{-3}$	-0.6×10^{-3}
Specific surface area (BET) [m²/g]	20–40	20–40	25–37	15–31	26
Density ρ	4.60	4.80	4.88–4.95	5.28	5.8
Crystal structure	cubic $a_0 = 0.25$ nm, m3m	cubic	tetragonal $a = 0.4218$ nm $c = 0.9182$ nm	hexagonal $c = 2.32$ nm $a = 0.589$ nm	bcc $a = 2861$ nm
Ferro- or ferrimagnetic	ferri	ferri	ferro	ferri	ferro
Particle size [μm]	needles $l = 0.3$ $d = 0.06$	needles $l = 0.3$ $d = 0.06$ equi ax. $l = 0.2$	needles $l = 0.5$ $d = 0.05$	hex. platelets dia = 0.08–0.1 thick = 0.001–0.025	needles $l = 0.3$ $d = 0.06$

the "slave" tapes in the contact duplication of DAT (Digital Audio Tape) masters. The coercivity of the barium ferrite tape was 600–700 Oe well below the coercivity, 2100 Oe, of the metal-particle master tape.

Table 1 summarizes the properties of magnetic particles for recording applications.

5. Thin film recording media

The particulate recording media discussed thus far have undoubtedly been conspicuously successful as storage media for audio, video and data over the past quarter century, but they are still far from ideal. Their magnetic constituent, in the form of fine particles of metals, alloys or oxides, occupies 40% of the coating volume in tapes and flexible disks and only 20% in hard disks. Since the 1960s many attempts have been made to displace particulate recording media with metallic thin films, but it is only in the last five years that those efforts have been successful. Particulate recording materials still dominate flexible media; e.g., computer tapes and flexible disks, but hard disks now increasingly are made of thin films deposited by vacuum deposition or, more commonly, by sputtering. It is appropriate, at this point to review briefly the salient characteristics of each type of medium.

5.1. Particulate coatings

The magnetic properties of the finished coating are determined almost entirely by the magnetic properties of the particles which can be measured in the "dry" state before the ink is mixed and coated on the substrate. Changes in magnetic properties from the dry powder to the finished coating are relatively slight and generally quite predictable.

The mechanical properties of the coating are determined overwhelmingly by the binder polymers (which occupy 60% of the coating volume), while the magnetic properties are determined by the particles. Thus, to a considerable extent it is possible to vary and control the mechanical properties independently of the magnetic properties.

In the manufacture of tapes or flexible disks, high coating speeds (hundreds of feet/minute) and wide rolls (≤ 2 m) can be used to achieve high production rates. Uniform properties within a roll of coated material (and from roll to roll) are achievable.

The materials and processes are generally well-understood and thus yields are high and coating costs are low.

5.2. Thin-film coatings

It is relatively easy to make coatings thinner than 0.25 μm by sputtering or vacuum evaporation but relatively difficult to make such thin coatings from particles. A thin coating is required to achieve high recording densities, particularly on disks where there is no erase head and a single head must achieve a compromise between the differing requirements of writing and reading. Thin coatings allow a greater length of tape to be packed into a reel of a given diameter and thus enable high volumetric densities to be achieved.

The intensity of magnetization is much higher in thin films since the magnetic constituent is not diluted by a binder. Furthermore the coercivity and other relevant magnetic properties are continuously variable (within limits) by changing the deposition conditions; e.g., in a system in which the film is deposited by sputtering, varying the target voltage, the bias voltage and the partial pressure of argon allows fine tuning of the coercivity and remanence of the deposited film.

In contrast, changing the magnetic properties of a particulate coating usually involves a major development effort. Changing the magnetic properties almost always involves changing the chemical behavior of the surface particles which then react differently with the solvent and, much more importantly, with the binder polymers, dispersants, etc. affecting the durability and coefficients of friction of the coating as well as its recording performance.

The saturation magnetization of thin films can be high, almost equal to that of the pure metal. The coercivity and other magnetic properties are continuously variable, within limits, by changing the deposition conditions and thus fine-tuning of the magnetic properties is possible to do in hours rather than the months required for particulate coatings.

Four methods of deposition have been used to prepare thin metallic films for recording purposes. They are: 1) electroplating, 2) autocatalytic plating, 3) evaporation and 4) sputtering.

5.3. Electroplated films

In this process the metallic ions are reduced from an aqueous solution at the cathode using electrons from an external source of current. The process can be continuous, inexpensive and can be used to deposit two metals at the same time if, as with cobalt and nickel, their deposition potentials are similar. It was found experimentally that, to obtain high-coercivity films, a high cobalt content was required. Alloy films of Co-Ni rather than Co-Fe are used because the deposition potentials of Co and Ni are very similar and films of Co-Ni were more likely to show good corrosion resistance than would films of Co-Fe. The basic plating path was introduced by Bonn and Wendell [21] and a practical form of it was described by Koretzky [22]. The electrolyte was made up of cobalt and nickel chlorides, ammonium chloride (found to enhance the compositional uniformity of the film) and sodium hypophosphite. The latter ingredient was used to add 1-3% phosphorous to the plated films where it segregated at the crystal boundaries and served to control the size of the crystallites and hence, the coercivity. The other parameters found to influence grain size and grain separation (and hence, noise in the finished films) were pH, temperature, substrate smoothness and, most importantly, plating current density.

5.4. Autocatalytically deposited films

It is possible to deposit from solution, films of cobalt and nickel alloys by a deposition process in which the potential required for the reduction of the metal ions is provided by a reducing agent, such as a hypophosphite or a borohydride, rather than by an external power supply. Each layer of atoms of Co and Hi deposited on the substrate serves to catalyze the deposition of a further layer. Films having a coercivity of ≤ 1800 Oe and saturation magnetization up to 80% of that of the bulk alloy of Co–Ni have been made autocatalytically. It is also possible to use the autocatalytic process to deposit films of Co–Ni–P continuously on polyethylene terephthalate films on which, to start the autocatalysis, a thin film of Ni–P has first been chemically deposited.

5.5. Films prepared by vacuum evaporation

Films of the ferromagnetic elements and their alloys deposited by vacuum evaporation are continuous, highly conducting but of coercivity < 50 Oe, much too low for recording purposes. Some way must be found to give the films a single-domain structure. A method of overcoming this problem was described by Speliotis, Bate, Alstad and Morrison [23]. They controlled the coercivity of the deposited film by varying the angle of incidence of the arriving metal atoms so that needle-shaped crystallites grow roughly along the direction of deposition. In this way coercivities < 1000 Oe could be achieved. Substrate temperatures were in the range 20–50°C and so the method could be used to make tapes on polyethylene terephthalate substrates. The metals were evaporated by electron bombardment and films of thickness in the range 0.01–0.4 μm were deposited.

This approach was used by Rossi et al. [24,25] shown in fig. 5. A magnetic film of Fe (42.5 at%), Co (42.5 at%), Cr (15 at%) was deposited at an angle of incidence of 60° while the substrate was rotated to ensure circumferential uniformity of the deposited film. It was found necessary to control to close tolerances the rate of deposition since the vapor pressure (and hence the rate of

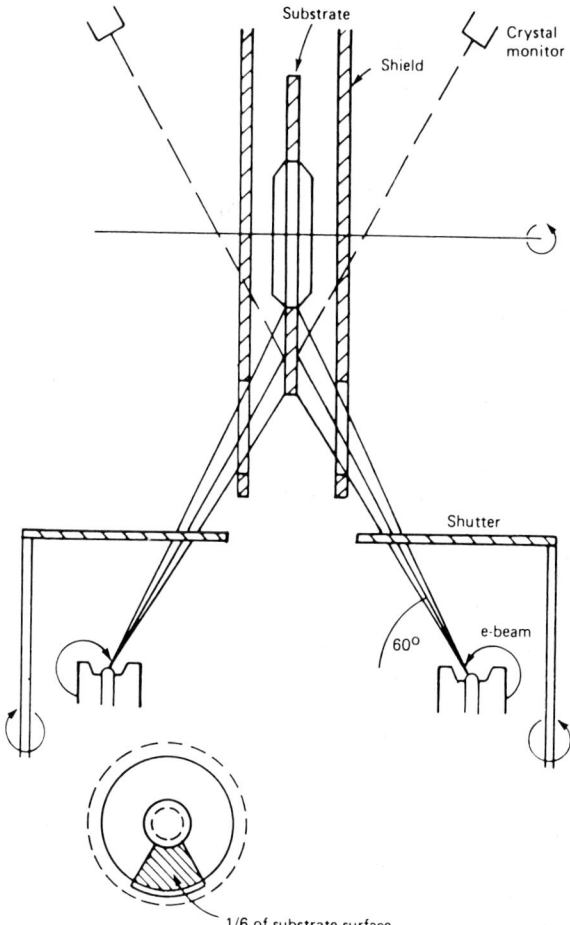

Fig. 5. Vacuum evaporation (by e-beam gun) at an angle of 60° to the surface of a rotating disk substrate. The angle and the thickness distribution are controlled by the apertures in the shields [24].

evolution) of chromium is appreciably higher than that of iron or cobalt. An underlayer of Haynes 188 stainless alloy was sputtered onto the conventional rigid-disk substrate of Al–Mg. This layer served to act as a corrosion barrier between the substrate and the metal film and also could be polished to a smooth but textured underlayer for the magnetic film. The purpose of texturing (which has become an important feature of thin-film rigid disks) is to prevent adhesive contact between disk and head. It is a critical step in the manufacturing process since if the texturing is too light the head and magnetic film may wring together, while texturing that is too heavy will introduce an unacceptable separation between head and magnetic coating. It should be remembered that, at current recording densities, a head-to-recording film distance of 4 μin. (1000 Å) may be intolerably large. In addition to eliminating head–disk sticking, the texturing can also be used to enhance the circumferential easy axis of magnetization of the FeCoCr layer.

5.6. Sputtered films

Sputtering is done in a vacuum chamber pumped down to a pressure of 0.75×10^{-6} to 0.75×10^{-7} Torr and filled with a gas, usually argon to a pressure of 10^{-3} to 7.5×10^{-2} Torr [2]. A potential difference of several thousand volts is applied between two parallel plates. These are the (positive) substrates (the surface on which the film is to be deposited) and the (negative) target (made of the material to be deposited). Argon ions are attracted to the target and knock out atoms of the target material. By virtue of their acquired momentum the dislodged atoms migrate toward and some are deposited on the substrate. One of the key advantages to sputtering over vacuum evaporation is that the composition of the deposited film is the same as that of the target and does not depend on the relative vapor pressure of the target's component elements. By changing the target composition, the composition of the deposited film can be changed in a predictable way and thus the task of examin-

Table 2
Typical properties of magnetic film media

Material	Deposition process [a]	Orientation [b]	Dominant crystal structure [c]	M_s (kA/m)	H_c (kA/m)	S, S^*	K_u (10^5 m^{-3})
Co	OIE	IPA	hcp	1100–1400	61–120		4 (bulk)
	NIE, SP	IPI	hcp	1100–1400	30– 60 [d]		
Fe	OIE	IPA	bcc	1600	60– 90		0.3–3.0
	SP	IPI	bcc	1600	10		
Ni	OIE	IPA	fcc	400	20– 28		
	SP	IPI	fcc	400			
Co–Ni	OIE	IPA	hcp/fcc	800–1200	30– 70		
Co–Fe	OIE	IPA	bcc	1400–1600	60–120	0.9, 0.9	
Co–Sm	NIE (e)	IPA	amorphous	500–1000	33– 55	1.1	
Co–P	EL, EP	IPI	hcp	800–1100	36– 96	0.9, 0.9	
Co–Re	SP	IPI	hcp/fcc	500– 750	18– 58	0.9, 0.9	
Co–Pt	SP	IPI	hcp/fcc	800–1400	→ 140		
Co–NiP	EI, EP	IPI	hcp/fcc	600–1000	40–120	0.8, 0.8	
Co–(30 at%) Ni:N$_2$	SP	IPI	hcp/fcc	650	80	0.96	
Co–Ni:O$_2$	OIE	IPA	hcp/fcc	300– 400	80	0.7–0.8	
Co–Ni–Pt	SP	IP*I	hcp/fcc	800– 900	60– 70	0.9, 0.97	
Co–Ni–W	SP	IPI	hcp/fcc	450	30– 50	0.8, 0.8	
Fe$_3$O$_4$	NIE, SP	IPI	is	400	17– 32		
γ-Fe$_2$O$_3$:Co	SP	IPI	is	200– 250	40–100	0.8, 0.8	
γ-Fe$_2$O$_3$:Os	SP	IPI	is	240	160	0.8, 0.8	
Co–(18 at%) Cr	SP	⊥	hcp	300– 550	80–100 (⊥)		−1.0
Co–(20 at%) Cr	SP	⊥	hcp	400	65– 95 (⊥)		0.15
Co–(22 at%) Cr	SP	⊥	hcp	300– 340	80–105 (⊥)		0.4

[a] OIE = Oblique Incidence Evaporation; NIE = Normal Incidence Evaporation; SP = SPuttered; EL = ElectroLess plated; EP = ElectroPlated.
[b] IPA = In-Plane Anisotropic; IPI = In-Plane Isotropic; ⊥ = perpendicular.
[c] hcp = hexagonal close-packed; fcc = face-centered cubic; bcc = body-centered cubic; is = inverse spinel.
[d] Coercivities for Co films deposited on Cr or W underlayers.
[e] Co–Sm values for films deposited in the presence of an orienting field.

ing the effect of varying the composition of binary, ternary or quaternary alloys is greatly simplified and accelerated. The principal independent variables (other than target composition) are the argon pressure and the target and substrate voltages.

Belk, George and Mowry [26] studied experimentally and theoretically the noise in thin-film longitudinally oriented magnetic films of CoP, CoNiP, CoRe, CoSm and found in each case that the disk noise in nV^2/Hz increased markedly with increasing bit density. Only with particulate disks of γ-Fe$_2$O$_3$ and a perpendicularly magnetized film of CoCr was there no increase (γ-Fe$_2$O$_3$) or decreasing noise (CoCr).

They showed experimentally and theoretically that the physical mechanism for the noise was the fluctuation that occurs from bit to bit in the geometry of the zigzag transitions separating the bit cells. The effect of changes in argon pressure, substrate bias and substrate temperature on noise level in similar, sputtered films of CoPtCr on a chromium underlayer, were measured by Yogi et al. [27]. Noise decreased as argon pressure increased and as substrate temperature and bias voltage decreased. These changes in the condition under which the films were made are ones that might be expected to produce grains that are increasingly more isolated from neighboring grains. Thus the current explanation of the results of Belk et al. [26] and Yogi et al. [27] appears to be that the noise arises principally because, in the case of Belk's films, the grains are sufficiently close together to enable exchange forces to correlate the spin directions of crystallites on opposite side of a recorded transition. By increasing the argon pressure and reducing substrate temperature and bias; as in Yogi's films, non-magnetic layers form at the surface of the grains, thus separating the grains and strongly reducing the extent of the exchange coupling. The modulation noise is thus reduced. The effects of argon pressure on magnetization, S^* (coercivity squareness), coercivity and noise level in Yogi's films are shown in fig. 6.

The structure, orientation and magnetic properties of thin films for recording media are summarized in table 2 (Arnoldussen [30]).

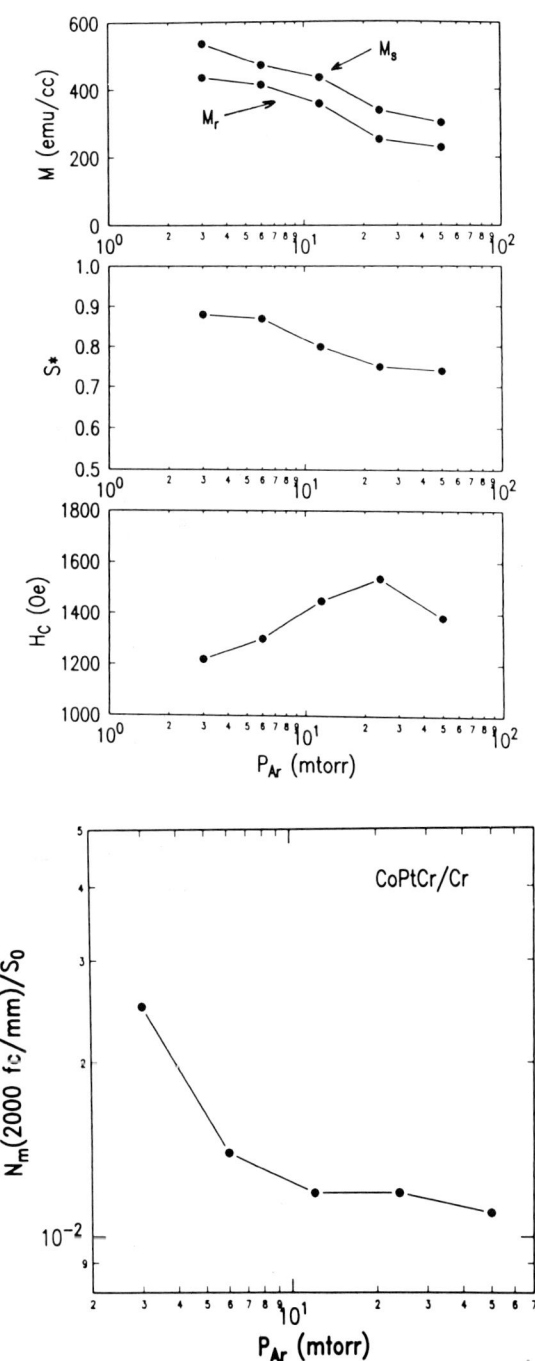

Fig. 6. The magnetic properties and normalized media noise of CoPtCr/Cr media as a function of argon sputtering pressure [27].

6. Conclusion

During the past 15 years some very difficult problems in magnetic recording materials and drives have been tackled and solved. As a result the performance (i.e., bit density, track density, access time to the data, data rate, reliability, etc.) of recording systems have been greatly improved. The information storage density (the product of bits/inch and tracks/inch) allows us to quantify one aspect of recording performance. In 1975 the IBM 3350 disk drives stored 3×10^6 bits/in.^{2} by flying the head 0.4 μm over the disk. In 1990 IBM published the results of successful laboratory demonstrations of disks operating at 1×10^9 bits/in.^{2} [27,28]. This involved "flying" the head 0.04 μm (i.e., 0.4 mean free paths in air) above the surface Of the disk in order to record and read at a linear density of 158 000 bits/in. The track density was 7470 tracks/in. In 1991 the areal density was increased in the Hitachi laborators to 2×10^9 bits/in.^{2} [29]. The IBM recording medium was built of five layers in which the sputtered, CoPrCr magnetic layer (12–40 nm thick) was isolated from corrosive interaction with the disk substrate by a layer of chromium 50–200 nm thick. A protective layer of hard carbon (12.5 nm thick) was deposited and was itself covered by a thin film of lubricant. It is significant that the magnetic part of the disk consisted of only one layer out of the five. This strongly suggests that, as we look forward to further improvements in magnetic recording media, over the next 15 years, the most challenging problems will increasingly be the non-magnetic ones.

References

[1] G. Bate, Recording Materials, in: Ferromagnetic Materials, vol. 2, ed. E.P. Wohlfarth (North-Holland, Amsterdam, 1980) chap. 7, p. 381.
[2] E. Koster and T.C. Arnoldussen, Recording Media, in: Magnetic Recording, vol. 1, Technology, eds. C.D. Mee and E.D. Daniel (McGraw-Hill, New York, 1987) chap. 3.
[3] W. Steck, J. de Phys. 46 (1985) 33.
[4] M.P. Sharrock, IEEE Trans. Magn. MAG-25 (1989) 4374.
[5] W.A. Manly, Jr., IEEE Trans. Magn. MAG-12 (1976) 758.
[6] P.J. Flanders, IEEE Trans. Magn. MAG-12 (1976) 770.
[7] S. Umeki, T. Uebori and Y. Imaoka, IEEE Trans. Magn. MAG-10 (1974) 655.
[8] M. Kishimoto, M. Amemiya and F. Hayama, J. Appl. Phys. 55 (1984) 2272.
[9] L.R. Bickford, J.M. Brownlow and R.F. Penoyer, Elec. Eng. 104B suppl. 5 (1957) 238.
[10] J.C. Slonczewski, J. Appl. Phys. 29 (1958) 448.
[11] J.C. Slonczewski, J. Appl. Phys. 32 (1961) 253S.
[12] N. Horiishi and A. Takedoi, Japan. Patent 75 101 299 (1975).
[13] K. Haneda and H. Kojima, J. de Phys. 40 (1979) C2583.
[14] J.E. Knowles, J. Magn. Magn. Mater. 22 (1975) 263.
[15] T.J. Swoboda, P. Arthur, N.L. Cox, J.N. Ingraham, A.L. Oppegard and M.S. Sadler, J. Appl. Phys. 32 (1961) 3745.
[16] (a) R.M. Kloepper, B. Finkelstein and D. Braunstein, IEEE Trans. Magn. MAG-20 (1984) 757.
(b) S.B. Oseroff, D. Clark, S. Schultz and S. Shtrikman, IEEE Trans. Magn. MAG-21 (1985) 1495.
[17] O. Kubo, T. Ido and H. Yokoyama, IEEE Trans. Magn. MAG-18 (1982) 1122.
[18] A.R. Corradi, D.E. Speliotis, G. Bottoni, D. Candolfo, A. Cechetti and F. Masoli, IEEE Trans. Magn. MAG-25 (1989) 4066.
[19] M. Kishimoto and S.-I. Kitahata, IEEE Trans. Magn. MAG-25 (1989) 4063.
[20] T. Suzuki, T. Ito, M. Isshiki and H. Saito, IEEE Trans. Magn. MAG-25 (1989) 4060.
[21] T.H. Bonn and D.C. Wendell, Jr., US Patent 2 644 787 (1953).
[22] H. Koretzky, Proc. 1st Austr. Conf. on Electrochemistry (1963) 417.
[23] D.E. Speliotis, G. Bate, J.K. Alstad and J.R. Morrison, J. Appl. Phys. 36 (1965) 972.
[24] E.M. Rossi, G. McDonnough, A. Tietze, T.C. Arnoldussen, A. Brunsch, S. Doss, M. Henneberg, F. Lin, R. Lyn, A. Ting and G. Trippel, J. Appl. Phys. 15 (1984) 2254.
[25] T.C. Arnoldussen, E.M. Rossi, A. Ting, A. Brunsch, J. Scheider and G. Trippel, IEEE Trans. Magn. MAG-20 (1984) 821.
[26] H.R. Belk, P.K. George and C.S. Mowry, IEEE Trans. Magn. MAG-21 (1985) 1350.
[27] T. Yogi, T.A. Hguyen, S.E. Lambert, G.L. Gorman and G. Castillo, IEEE Trans. Magn. MAG-26 (1990) 1578.
[28] T. Yogi, C. Tsang, T.A. Hguyen, K. Ju, G.L. Gorman and G. Castillo, IEEE Trans Magn. MAG-26 (1990) 2271.
[29] M. Futamoto, F. Kugiya, M. Suzuki, N. Takano, H. Fukuoka, Y. Matsuda, Y. Inaba, T. Takagaki, Y. Miyamura, K. Akagi, T. Nakao, H. Sawaguchi and T. Munemoto, 5th Joint MMM–Intermag Conf., Pittsburgh (1991) submitted.
[30] T.C. Arnoldussen, Proc. IEEE 74 (1986) 1526.

Magnetooptics

J.F. Dillon, Jr. [1]

AT&T Bell Laboratories, Murray Hill, NJ 07974, USA

The field of magnetooptics since 1975 encompasses a vast amount of fundamental and applied work. This review deals briefly with a few particularly interesting developments. Under the topic of new effects are a magnetic linear birefringence linear in *H* and a birefringence arising from the gradient in *M*. Research applications include the visualization of antiferromagnetic domains and the current search for evidence of anyons in high temperature superconductors. Several examples are given of recent applied work on both bulk and film waveguide magnetooptical isolators as well as the materials used in them.

1. Introduction

Since 1975 the field of magnetooptics has been active and diverse, so much so that for this review we can only touch a few topics. The choice is difficult, and in the end quite personal. The author's intent is to report briefly on several new effects, to examine some research applications of magnetooptics as well as technological applications. An example of a recent advance in magnetooptical materials is included. Brief mention of new measurement techniques are included in several of these segments. It will be apparent that the fundamental optical properties of magnetically ordered materials continue to receive careful study. The tools of magnetooptics continue to be of profound value in diverse areas of research and technology, and there have been dramatic developments on the materials side. We note here and below that three other articles in this issue "Imaging of Magnetic Domains and Walls" by Argyle [1], "SMOKE" by Bader [2] and "Magnetooptical Recording and Materials" by Bate [3] cover exceedingly significant areas of magnetooptics.

2. New and less familiar magnetooptical effects

In recent years several new magnetooptical (MO) effects (MOE) have been explored. They give us new fundamental understanding and provide valuable tools for visualizing magnetic and antiferromagnetic domain structure. In section 2.2 we discuss a linear magnetooptic effect which in some symmetries yields contrast between antiferromagnetic domains. Section 2.3 deals with a new birefringence related to the magnetization gradient. It provides an entirely new contrast mechanism in the examination of magnetic domain structure using the magnetooptic Kerr effect. However, first there are some general remarks to be made about these phenomena.

2.1. Overview

Far and away the most familiar MO effects are manifestations of magnetic circular birefringence (MCB). This can be readily observed as MO

---

[1] Present affiliation: Department of Applied Physics, Yale University, New Haven, CT 06520-2157, USA.

rotation (or the Faraday rotation) of the axis of linearly polarized light traveling parallel to **M**. Closely related is magnetic circular dichroism (MCD) which typically produces a small ellipticity in light which entered the medium as linearly polarized, and travels along **M**. Light traveling perpendicular to **M** encounters a magnetic linear birefringence and dichroism (MLB and MLD). These fundamental MO effects of a uniformly magnetized material are to be viewed as arising from the dielectric response of the material to the electric vector of the propagating light wave. This response is contained in the second rank dielectric tensors $\vec{\epsilon}$.

In fact, it turns out that a whole family of MOE may appear in $\vec{\epsilon}$. Eremenko and Kharchenko [4] have recently given a symmetry analysis of the optical properties of transparent magnetically ordered crystals. Nonmagnetic materials in an applied field show MCB and MCD for fields along the propagation vector **k** and MLB and MLD for fields perpendicular to **k**. The circular effects are proportional to **H** and the linear effects to H^2. Magnetically ordered materials may show analogous birefringences and dichroisms spontaneously. Under time reversal, i.e. reversing all the elementary magnetizations, the spontaneous circular effects reverse sign, but the linear effects do not. But magnetically ordered materials may have a lower space–time symmetry than the corresponding nonmagnetic material in a field, and they may have MO properties that have no analog in the material without order. Ref. [4] contains a table giving the connection between symmetry and optical properties of transparent magnetically ordered materials. Unfortunately it contains too much detail and requires too much explanation to quote here. Included in the table are properties proportional to applied field **H**, to wave vector **k**, to the product **H · k** and quadratic in **H**. The table is limited to transparent materials, and thus covers birefringences but not dichroisms. We know that in many cases dispersive and absorptive effects are of comparable importance.

The theory of Eremenko and Kharchenko is a symmetry theory. It specifies the tensors underlying possible effects, and makes it possible to determine relationships among tensor elements for each magnetic point group. Though a symmetry theory tells nothing about the size of the effects or their physical origin, ref. [4] does cite several papers dealing with microscopic theories.

2.2. Linear magnetooptic effect

According to ref. [4] contributions to the dielectric tensor by the linear magnetooptic effect (LMOE) are given by

$$\epsilon_{ij}(\text{LMOE}) = q_{ij\alpha}H_\alpha. \tag{1}$$

The tensor $q_{ij\alpha}$ is symmetric, and describes a linear birefringence proportional to magnetic field intensity. This birefringence changes sign on reversal of the field or when the directions of all the elementary magnetic moments in the crystal are reversed. It is not a possible effect in nonmagnetic crystals, but is allowed for magnetically ordered crystals whose magnetic point group does not contain the anti-inversion operation $\underline{\bar{1}} = \bar{1} \cdot \underline{1}$ ($\bar{1}$ is inversion of spatial coordinates; $\underline{1}$ is time reversal).

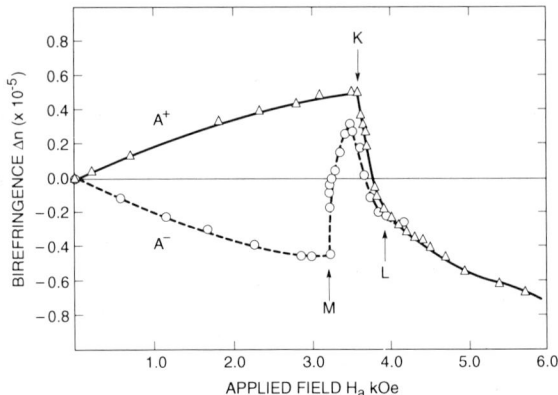

Fig. 1. Magnetic linear birefringence $\Delta n = n_\parallel - n_\perp$ measured in $Dy_3Al_5O_{12}$ in fields along [111] at 1.375 K. Light travels along [1$\bar{1}$0]. The solid line pertains to one antiferromagnetic state A^+ and the dashed line to the time reverse conjugate state A^-. Above L the crystal is in the paramagnetic state after having gone through the metamagnetic transition. Between M and K the spin system is in a mixed state. Note that the birefringence is nearly linear in H_a, and that it reverses sign when the signs of all the elementary moments are reversed. (From ref. [5].)

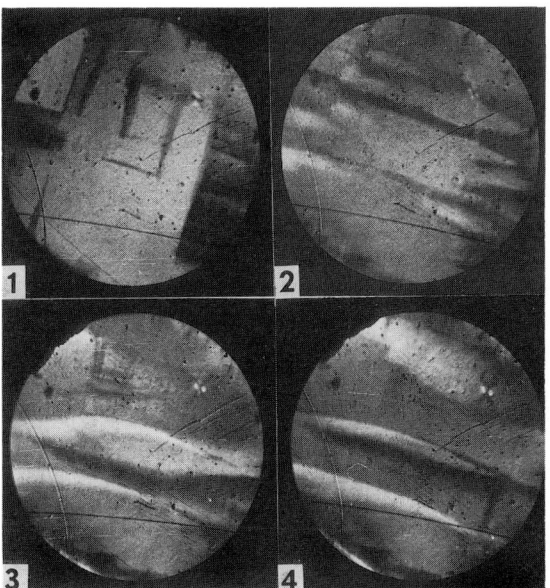

Fig. 2. Antiferromagnetic time reversal domains observed by Eremenko and Kharchenko in CoF_2 at temperatures 3–5 K below T_N. The light is along the 4-fold axis. The underlying contrast arises from the LMOE and a $\lambda/4$ plate was used to reveal it. (From ref. [4].)

The LMOE has been studied in such crystals as the weak ferromagnet $CoCO_3$; the antiferromagnets CoF_2 and $DyFeO_4$ and the metamagnet $Dy_3Al_5O_{12}$. In general the linear birefringence is accompanied by magnetic circular birefringence and the two must be disentangled. Fig. 1 from the author's work [5], shows the birefringence linear in field, and the fact that it changes sign when the elementary moments are reversed. At least in the crystals mentioned, the effects are quite large and can be used to visualize antiferromagnetic domains. Eremenko and Kharchenko [4] have explored the microscopy using LMOE contrast and published beautiful photographs showing the character and behavior of antiferromagnetic domains in plates of $DyFeO_3$ and CoF_2. Fig. 2 shows a variety of domain patterns observed in CoF_2 at 3–5 K below the T_N of ≈ 37.5 K. It is truly remarkable to have an experimental tool, the LMOE, which makes it possible to distinguish vividly the subtle difference between time reverse antiferromagnetic domains.

2.3. Gradient-related magnetic birefringence

Hubert and his collaborators [6–8] at Erlangen–Nürnberg have recently studied a new birefringence related to the spatial gradient of magnetization. This was first observed in reflection on normal incidence from metal samples in which M lies along various directions in the plane. The effect may be seen in the four part fig. 3. This shows domain patterns on the (100) face of an Fe(3% Si) crystal. These images were made by a digital subtraction technique in which the reference image is either of a saturated state or a dynamically blurred image. This reference image is subtracted from one with a domain pattern to obtain clear images in which nonmagnetic contrast has been eliminated.

In fig. 3(a) the magnetization in all domains lies parallel to either the polarizer or analyzer axis so there is no contrast between domains. Note that the 90° walls appear as bright and dark lines, but there is no distinctive contrast marking the 180° walls. In (b) we see the same pattern but with the polarizer turned through 90° relative to the sample. Again the 90° walls appear, but the contrast is reversed. In (c) the crystal has been rotated by 45° relative to the polarizer and we are viewing a similar but different domain pattern. The MLB in reflection (the Voigt effect) produces a strong contrast between domains in which M is parallel to $+45°$ and those in which it is parallel to $-45°$. Note that in this case the 90° domain walls are not seen, but the 180° walls appear as distinct light or dark lines. On rotating the polarizer 90° relative to the sample, the contrast of both domains and walls is reversed. The walls are seen by virtue of a birefringence associated with the gradient in the magnetization. The contrast is reversed when the sign of the gradient is changed. Clearly the contrast cannot be due to any normal component of M at the wall since that would not change when the crystal is rotated between the polarizer–analyzer system. Note that the Voigt effect contrast is the same when M is reversed indicating that it is quadratic in the magnetization. On the other hand, the gradient birefringence contrast reverses when the sign of the gradient is reversed

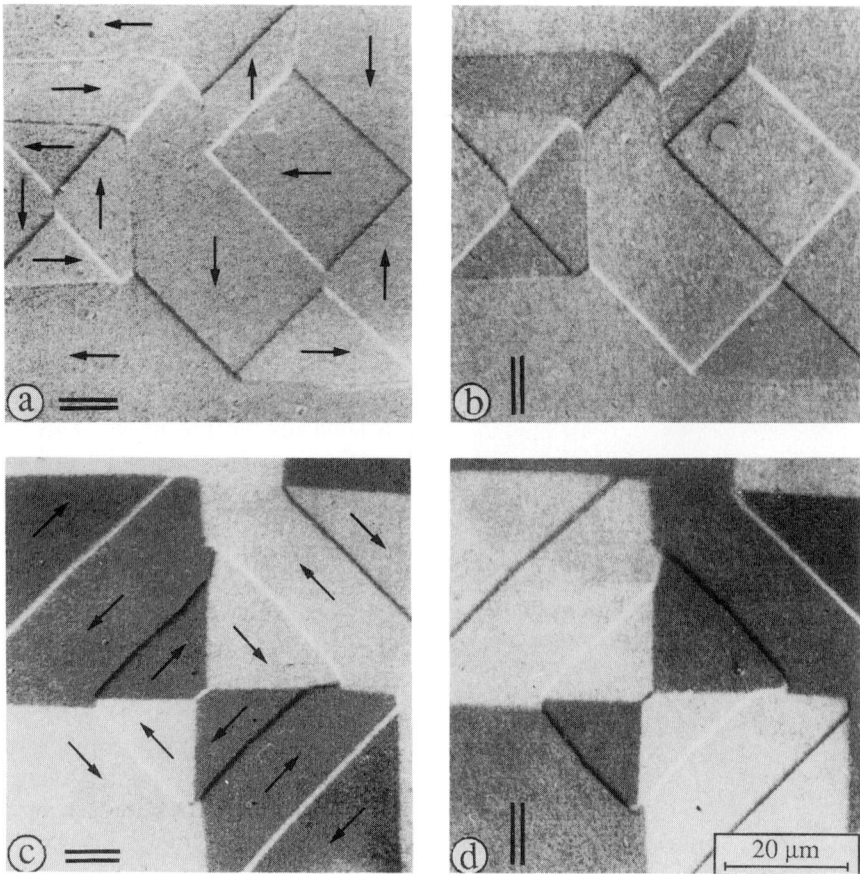

Fig. 3. Domain patterns illustrating the birefringence arising from the spatial gradient in the magnetization. The polarizer orientations in each case are indicated by the double line at lower left. The analyzer is crossed with respect to the polarizer. As noted in the text these photographs were made with a digital subtraction technique. (From ref. [8].)

and thus is linear in the components of M. Since M lies in the plane, the effect observed here would more accurately be called a magnetic planar gradient birefringence.

Thiaville, Hubert and Schäfer [8] formulated a highly symmetric representation of the gradient-related magnetic birefringence. They began by noting that a formal description of ordinary MCB can be given by

$$D^F = i\epsilon Q m \times E \quad \text{or} \quad D_j^R = i\epsilon Q \eta_{jkl} m_k E_l. \quad (2)$$

In this E is the electric vector of the incident light and D^F is the rotated component of the dielectric displacement vector, m is a unit vector parallel to the magnetization, i is the imaginary unit, ϵ is the ordinary dielectric constant. Though Q is a complex material constant, it is taken to be predominantly real. The quantity η_{jkl} is the totally antisymmetric third rank tensor. The material is taken to be isotropic except for the magnetization. Thiaville et al. arrived at an analogous expression for the gradient effect, i.e. an explicit form for the material tensor underlying the gradient related contribution to ϵ_{ij}. Initially they hypothesize that the gradient effect may make a contribution to D of the form

$$D_i^{gr} = U_{ijklm} E_j k_k \frac{\partial m_l}{\partial x_m}, \quad (3)$$

where k is the propagation vector of the light. The inclusion of k is important in that it makes U

an odd rank tensor and several mathematical theorems apply. In particular, it is possible for U to be isotropic or "spherical", i.e. to be independent of crystal direction. By examining the symmetry properties of the material tensor U and the experimental results, the authors were able to arrive at much simpler expressions for the dielectric component arising from the gradient effects. The reader is referred to the original paper.

One implication of the theory of ref. [8] is that at oblique incidence, gradients in the normal component of magnetization could be observed, a polar gradient effect. Though complicated by diffraction, observation of the polar effect on single crystals of Co_5Sm was reported in ref. [8] in qualitative agreement with the theory.

These authors have shown the gradient related birefringence in a series of beautiful experiments and they have given a symmetry theory tying together the observations. However, a theory of the physical origin is not yet in hand. The symmetry theory is immediately useful because of the need to understand all of the possible sources of contrast in domain observations. The planar effect constitutes a new valuable tool in unraveling the magnetization distributions underlying images obtained with perpendicular illumination. Finally, it is conceivable that some aspect of these effects may have technological application.

3. Research applications

The uses of MOE in magnetic research continues to expand. Surely most important is the ability to see domains, how the magnetization is actually disposed in a tremendous range of specimens gives a basis in reality to our understanding of magnetic behavior. It has simply made possible the understanding on which magnetic bubble technology and MO recording technology have been built. This visualization now extends to antiferromagnetic domain structure. The study of magnetic phase transitions is an important research application though we do not touch upon it. Other current topics are the MO study of magnetic layers only a few monolayers in thickness, and the recent explosion of interest in the MO properties of high T_c superconductors.

3.1. Ferro- and ferrimagnetic domain visualization

Argyle's [1] paper deals extensively with the imaging of domains and walls, and its current level of sophistication. Section 2.3 above and the underlying refs. [1–3] are concerned with a contrast mechanism which seems most useful in the polar Kerr configuration.

3.2. Antiferromagnetic domain visualization

It is also possible to see antiferromagnetic domains and domain walls. In the author's [9] early work on $Dy_3Al_5O_{12}$ visual contrast was observed between the two time reverse conjugate states designated A^+ and A^-. In the case of $FeCl_2$ near the metamagnetic transition it was possible [10] to see the antiferromagnetic domain walls by decorating them with a thin ribbon of nearly saturated paramagnet with its large magnetooptical rotation. In section 2.2 above we have glimpsed a little of the antiferromagnetic domain studies made possible by the use of the new LMOE.

3.3. SMOKE

Bader's papers [2] treats the use of SMOKE, surface magnetooptical Kerr effect, to characterize the magnetic behavior of films of transition metals in the monolayer range.

3.4. Search for anyons in high T_c superconductors

The search for a mechanism to explain high T_c superconductivity has led to a search for MO effects in these materials. This followed a suggestion by Laughlin [11] that there is possibly a spontaneous breaking of time reversal symmetry in some layer structure superconductors and that this might be observable in magnetooptical experiments. In these "anyon" theories the charge carriers are confined to two dimensions and obey fractional statistics. The ground states of such systems must simultaneously break two symmetries: time reversal (T) and two dimensional reflection (P). In general, the product symmetry (TP) is preserved. The theoretical idea [12] is that below some spontaneous symmetry breaking temperature T_{TP} there appears a correlated state with broken symmetry among the copper spins in a layer and that T_{TP} could be equal to or greater than the superconducting critical temperature T_c. The symmetry breaking can be of either sign and there are magnetic fields associated with it. There might be a coupling between layers such that the sign of the ordering in adjacent layers would be the same, thus "ferromagnetic". Alternatively, antiferromagnetic coupling is possible as is the absence of significant coupling. In any event, on cooling through T_{TP}, the ordering could be distributed in domains of indeterminate size. The presence of magnetic fields on cooling might drastically effect the overall ordering.

These are provocative theoretical suggestions concerning an important current problem, the mechanism of high temperature superconductivity. Central to the theories is the absence of time reversal symmetry. Wen and Zee [13] pointed out that independent of any model, the symmetry properties of these materials could be studied by using magnetooptical effects. After all, effects arising from magnetic circular birefringence and dichroism are of tremendous use in the study of magnetic materials, and these explicitly probe time reversal symmetry. The experiments undertaken by Lyons et al. [14] were designed to explore the time reversal symmetry of high T_c superconductors.

If light linearly polarized along x impinges normally on the xy surface of a magnetic medium (one lacking time reversal symmetry), the polarization of the reflected light may be somewhat different from that of the incoming light. Typically it will be elliptical with the major axis rotated slightly from the input orientation and there will be a small ellipticity. Conventionally this can be expressed as a complex polar Kerr rotation:

$$\hat{\Phi} = \frac{E_y^r}{E_x^r} = \phi_K' + i\phi_K''. \qquad (4)$$

Solution of Maxwell's equations shows that the normal modes of propagation in the material are the two circular polarizations (+CP and −CP). The ϕ_K' is the phase difference between the two CP after reflection; ϕ_K'' corresponds to the difference in amplitudes after reflection. We can write explicit expressions for ϕ_K' and ϕ_K'' in terms of of the real and imaginary parts of the off diagonal terms in the dielectric tensor ϵ_{12}' and ϵ_{12}'' and the complex refractive index $N = n - ik$. The quantity measured by Lyons et al. was ϕ_K'', and this contains both ϵ_{12}' and ϵ_{12}''.

One way to represent the effect of optical elements is by the use of 2×2 Jones matrices. Here we are interested in CB and CD in a normal incidence experiment, and it is natural to use *circular* Jones vectors and matrices. In this representation $e_+ = \begin{bmatrix} 1 \\ 0 \end{bmatrix}$ is plus-polarized light and $e_- = \begin{bmatrix} 0 \\ 1 \end{bmatrix}$ is minus-polarized light. The vector $\begin{bmatrix} e^{i\theta} \\ 1 \end{bmatrix}$ is linear polarization with a phase difference of θ between the two CP. Note that we adopt a convention in which polarization states are referred to the same set of axes before and after reflection. Thus the sign of circular polarization does not change on reflection from a perfect mirror. The most general sample may be represented by the matrix

$$M_s = \begin{bmatrix} A_+ & B_- \\ B_+ & A_- \end{bmatrix}, \qquad (5)$$

where the elements may be complex. When the input is polarized along x, the reflected amplitude is given by

$$M_s \begin{bmatrix} 1 \\ 1 \end{bmatrix} = \begin{bmatrix} A_+ + B_- \\ A_- + B_+ \end{bmatrix}. \qquad (6)$$

To the extent that M_s differs from a scalar multiple of the unit matrix, the polarization state will be changed.

The matrix M_s is a statement of the most general changes in polarization state. Ordinary circular birefringence and circular dichroism are contained in the real and imaginary differences between A_+ and A_-. For a material invariant under time reversal $A_+ = A_-$. On the other band, if $|B_+| \neq |B_-|$ the resultant change in polarization will be indistinguishable from circular dichroism unless some further polarization selection is employed.

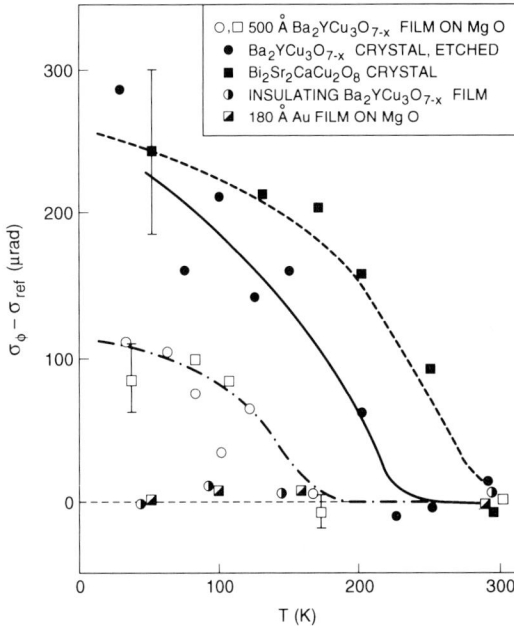

Fig. 4. Standard deviation (less a background value) of the measured CD on reflection for cuprate sample indicated. Note the difference in the onset temperatures for the two different single crystal compositions and that both are higher than that for the films. The lines are merely guides to the eye. The Au film on an MgO substrate was one of several controls. (From ref. [14].)

Using a novel technique which employed a rotating $\lambda/2$ plate, Lyons et al. measured the 515 nm ϕ_K'' at 25 μm diameter spots on the surface of superconducting samples as these were cooled from 300 K down to well below T_c. The rotating plate helps to eliminate polarization changes from linear birefringences and dichroisms. The procedure was to make measurements at a point, then move the spot about 100 μm and repeat this process perhaps 25 times. At high temperature, a set of 25 readings would average to zero with a standard deviation σ_ϕ of about 20 μrad. As the superconducting samples were cooled, $\bar{\phi}$ remained close to zero, but σ_ϕ increased dramatically below some characteristic temperature. Results for films and crystals of several layer structure cuprate compounds are shown in fig. 4. In these samples there was no anomaly at T_c though in another compound sharp structure was seen.

Consider three possible experimental subjects. For a sample which showed no dichroism, we would expect $\bar{\phi} = 0$ and σ_ϕ to be a measure of instrumental noise and that these would not vary with temperature. A one domain sample with CD would have a temperature dependent $\bar{\phi} = \pm \phi_0$ and σ_0 would be the instrumental value. Finally, a sample with a dichroism of ϕ_0 broken up into a number of positive and negative domains smaller than the spot size would have $\bar{\phi} = 0$ and an additional contribution to σ_ϕ of $\phi_0/\sqrt{N}$. If r is the radius of the domains and d is the diameter of the laser spot, σ_ϕ would be $\approx \phi_0 r/d$. This last case seems to correspond to the experimental results.

Weber et al. [15] at Dortmund used a complex technique to measure quantities proportional to CD on reflection from a single orthorhombic domain crystal of $YBa_2Cu_3O_7$. The crystal had $T_c = 92.5$ K. Light of $\lambda = 633$ nm was used. They found a CD which deviated from zero below ≈ 120 K and there was no anomaly at T_c. They also had similar results in a thin section of $Bi_2Sr_2CaCu_2O_8$ single crystal. The results were qualitatively similar to those of fig. 4 above, but the onset temperatures were markedly lower.

Particularly interesting were the effects of cooling in a magnetic field. In both reflection and transmission experiments it was found that the CD or CB after cooling in the absence of a magnetic field varied statistically in sign and magnitude. In field cooling experiments, the samples were cooled from 180 K to a temperature below the onset temperature but above T_c in a field of 200–500 Oe. It was found that the sign of the field on cooling had determined the sign of CD or CB. Weber and his colleagues interpreted this as indicating that there is a built in magnetic "moment" associated with the broken symmetry state and that a "Zeeman" interaction between this and the applied field determines the low temperature state.

Spielman et al. [16] at Stanford University used a fiber optic Sagnac interferometer to look for nonreciprocal circular birefringence in high quality films of $YBa_2Cu_3O_7$. The technique is of great sensitivity and is free of ambiguity in that it is specifically sensitive to nonreciprocal effects.

They found no measurable CB. In their interferometer counterpropagating beams travel around a (1 km) loop of polarization maintaining fiber and interfere at the output. The sample is introduced into the beam between quarter wave plates arranged so the two beams go through as the two circular polarizations. If the system rotates, the path length is different for the 2 beams and there is a phase shift when the beams are recombined. Thus it is a "fiber optic gyroscope" delicately sensitive to rate of rotation. The earth's rotation is in fact used for calibration. The instrument is useful here because nonreciprocal CB in the optical path also causes a phase shift, whereas all reciprocal effects cancel out. The Sagnac interferometer promises to be exceedingly useful in measuring MO effects in the presence of linear birefringences. Initially using 1.06 μm light and films of thicknesses 50, 200, 500 and 800 Å, the Stanford authors were unable to find any circular birefringence down to an accuracy of 2 μrad.

The central result of the Spielman et al. experiment clearly contradicts those of Lyons et al. and of the reflection experiments reported by Weber et al. Furthermore, the Stanford transmission experiment apparently contradicts the results of the Dortmund transmission experiments. These simple statements of conflicting results ignore the fact that the experiments were done at different frequencies and on different samples and that they measure somewhat different physical properties. These are exceedingly sensitive experiments. Lyons et al. worked with an instrumental error of about 20 μrad and saw effects of about 200 μrad. Spielman et al. reported an error level of 2 μrad. In contrast, the polar Kerr rotation of the amorphous films used in MO memory technology is perhaps 5000 μrad. Though techniques for preparing and characterizing high quality samples of $YBa_2Cu_3O_7$ and related compositions are now widely known, there has been no exchange of samples by the research groups involved. It is conceivable, for instance, that there are radical differences in domain size among the various samples which might be traceable to nucleation conditions and to some aspect of film microstructure not probed by present sample characterization. Even more interesting is the very real possibility that an understanding of the apparent contradictions lies in some subtle aspect of the physics of these materials. Further experiments as well as deeper theoretical understanding are needed.

Dzyaloshinskii [17] has carried out a symmetry analysis of a hypothetical anyonic state in which the magnetoelectric effect enters. The planes are taken to be ordered antiferromagnetically. In structures where there are two planes per unit cell (123 and 2212), the model predicts that ϕ''_K, can be nonzero as in the Lyons et al. experiments, and further that $\phi'_F = 0$ as in the Stanford experiment. The rotation seen in the Dortmund transmission experiment might be a reciprocal rotation which is linear in the ordering and in the propagation vector k. The model clearly predicts that in structures with one layer per unit cell, ϕ''_K would vanish. Further experiments on such a material La_2CuO_4 : Sr and using new techniques to distinguish the A^+, A^- from B^+ and B^- will be reported shortly. There is widespread interest in the unfolding story of the magnetooptical properties of the high temperature superconductors.

4. Technical applications

The two most prominent technological applications for magnetooptical effects are surely transmission devices such as isolators and magnetooptical memories. Another important use is in Faraday effect sensors [18], an optical method of measuring electric current or magnetic field. Ring laser gyroscopes depend on magnetooptical mirrors [19]. Both the memories and the materials used for them are treated in the article by Bate [3]. Here we concentrate on the current state of isolator development and examples of recent advances.

4.1. Bulk isolators

The development of very large MO isolators has been vigorously pursued in connection with laser fusion experiments. An excellent review has been given by Weber [20]. In contrast, the near

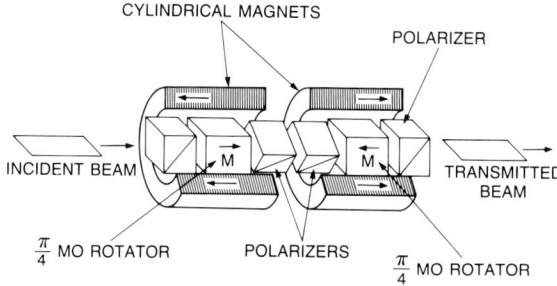

Fig. 5. Double isolator using YIG single crystals produced by Hitachi Metals, Ltd. (After ref. [21].)

IR isolators for use in communications are small, and the pressure to make them smaller is great. In the rotators of those available today, light propagates through a bulk crystal of ferrimagnetic garnet. The MO elements are made either from bulk flux grown crystals or from thick films grown by liquid phase epitaxy. The current state of isolator design may be seen in fig. 5. This is a sketch of a compact double isolator produced by Hitachi Metals [21]. The beam is about 1.5 mm in diameter. The polarizers are glass beam splitter cubes. The rotators are YIG single crystals which are magnetically saturated by cylindrical Sm–Co permanent magnets. The senses of rotation in the two halves are different so the output linear polarization is parallel to the input. To mitigate the temperature and wavelength variation of rotation in YIG, the lengths of the two rods are slightly different [22]. The design wavelength is (1.55 ± 0.02) µm. All of the many optical surfaces are AR coated and the insertion loss is only 0.9 dB. Over the temperature range $-20°C \leq T \leq 60°C$ and the wavelength range 1.53 µm $\leq \lambda \leq$ 1.57 µ, the isolation ratio exceeds 60 dB. Such devices as this are needed in high bit rate communications systems to protect semiconductor laser diodes from reflected light which greatly degrades their performance. A number of manufacturers produce high quality isolators using bulk crystals. The extinction ratio is high, the overall isolator is rather large and expensive and it is not easy to couple into integrated optics.

4.2. Film isolators

Isolators in which the light propagates in film waveguide offer the possibility of smaller, cheaper devices into which it is easier to couple than such bulk isolators as that in fig. 5. The sketch of fig. 6 illustrates one concept. Linearly polarized light with azimuth 45° from the laser at left enters a magnetized ridge waveguide film which is so thin as to only support two modes, and in transmission is rotated 45° At this point E lies in the plane, and the propagation mode is TE. A short metallized section of the ridge constitutes an analyzer since it strongly attenuates TM modes, and has no effect on TE modes. The outgoing signal from the laser couples into the fiber. Horizontally polarized radiation coming back from the optical system passes through the polarizer, but is rotated to $-45°$ on passage down the waveguide, a polarization to which the laser is not sensitive. Thus no input polarizer is needed. Note again that this is merely a conceptual sketch, many variants can be devised. The essence is propagation in film magnetized along k, and there are some complications in that.

Fig. 7(a) shows the Poincaré sphere representation of polarized light. Each point on this unit sphere [24] represents a possible polarization state of propagation in an infinite medium. The north pole, point C_+, is positive CP; the south pole, point C_-, is negative CP. Linear polarizations, represented by points L_ψ, lie around the equator at 2ψ where ψ is the azimuth. Points off the equator correspond to elliptical polarizations; the ellipticity increasing with the latitude 2χ. If light

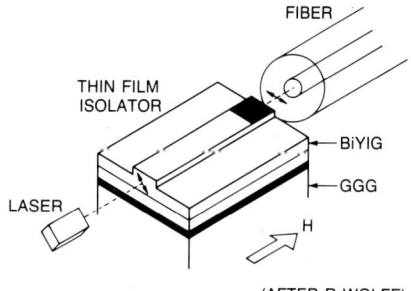

Fig. 6. Conceptual sketch of a thin film isolator as envisioned by R. Wolfe.

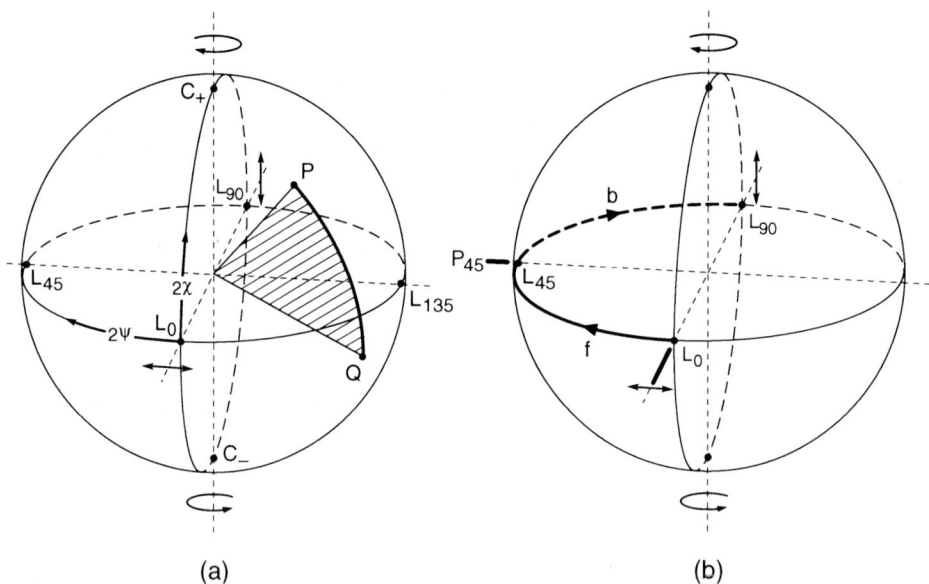

Fig. 7. (a) Poincaré sphere representation of polarized light; (b) repesentation of an ideal isolator in which incoming light only encounters MCB.

of unit intensity and general polarization Q falls on a general polarizer P, the transmitted intensity is $\cos^2(\overline{PQ}/2)$. In this $\overline{PQ}$ is the angle of the great circle arc between the points P and Q. This tells us immediately how an arbitrary polarization state is resolved onto a single polarizer. Furthermore, we see that orthogonal modes which have no projection on each other are 180° apart on the sphere, i.e. are antipodal points. If P′ is a polarizer orthogonal to P, the intensity transmitted through P′ is $\cos^2(\overline{P'Q}/2)$ or $\sin^2(\overline{PQ}/2)$.

There is a sense of rotation convention to be specified. As in the discussion of MO effects in superconductors in section 3.4, we use a single system of axes. The propagation is along $+z$ or $-z$. The sense of CP is determined by the sense in which E in an xy plane rotates regardless of which way the light is traveling. Thus, if $+CP$ were to encounter a perfect mirror in an xy plane, the reflected beam would also be $+CP$. Note particularly we are not using the concept of helicity.

On solving Maxwell's equations for propagation in a uniaxial medium we find that the normal modes are given by two orthogonal linear polarizations, L_0 and L_{90}, and these are antipodal points on the equator. In a magnetic medium which is otherwise isotropic, the normal modes are positive and negative CP, the two poles of the sphere. Linearly polarized light on entering the magnetic medium is resolved into the two normal CP's with equal amplitude each with its characteristic propagation constant. On exit these recombine with an altered phase, i.e. with a change in azimuth. This is circular birefringence. Quite generally, birefringence corresponds to a rotation on the sphere about the axis of the normal modes. The convention is that the trajectory moves in a right handed sense about the fast axis. If one CP had been attenuated more than the other, the emergent light would have been off the equator and thus elliptically polarized. This is circular dichroism.

This compact review of the Poincaré sphere has been given in order to discuss isolators in an insightful way. In the isolator of fig. 7(b), light traveling along z enters a plate in the xy plane, is resolved into C_+ and C_- which travel with different propagation constants. On exit they recombine with an altered phase relation to give linear polarization at some azimuth. The change in azimuth is, in fact, one half the phase difference. As the thickness of the plate increases from zero, the resultant state on exit follows a trajectory around the equator of the sphere. In this ideal case, the rotation is 45°. The input polarizer and output polarizers are P_0 and P_{45}, so light enters the MO material through P_0 at point L_0, its polarization is rotated to point L_{45}, and is then completely transmitted by P_{45}. Light traveling in the opposite direction enters through P_{45}, is rotated by the MCB to point L_{90}, and is then completely rejected by P_0.

The idealization of fig. 7 in general is not valid for film isolators because the normal modes are not circular. The solution of the propagation problem [25] in thin nonmagnetic films leads to two sorts of allowed modes, TM and TE. For sufficiently thin films only the lowest order mode of each type is possible, and these have different propagation constants β_{TM} and β_{TE} For such thin films we can use the Poincaré sphere representation. If we define the intrinsic birefringence of a film as $\Delta n_i = n_{TM} - n_{TE}$, we see that $\Delta \beta_i = 2\pi \Delta n / \lambda_0$. The phase difference between the two modes after a distance x is given by $\Delta \beta_i x$. On the Poincaré sphere the axis of normal modes for this nonmagnetic case would be the axis passing through the horizontal and vertical polarizations, i.e. the modes TE and TM or L_0 and L_{90}.

The intrinsic $\Delta \beta_i$ is but one of several contributions to a propagation constant difference between the TM and TE modes. Others may arise from misfit strain between the lattice constants of substrate and film and from a growth anisotropy. Since these are all along the same axes we may simply add them, $\Delta \beta_{LB} = \Delta \beta_i + \Delta \beta_{strain} + \Delta \beta_{anis}$, where the subscript LB designates linear birefringence. If we now endow the film with a magnetization along z, it carries with it an MCB, and thus a $\Delta \beta_{MCB}$. These linear and circular components add as vectors, $\Delta \beta = \sqrt{\Delta \beta_{LB}^2 + \Delta \beta_{MCB}^2}$, and the axis of normal modes on the Poincaré sphere is at angle $\tan^{-1}(\Delta n_{LB}/\Delta n_{MCB})$ from the north pole toward the TE–TM axis.

The fundamental difficulty with film isolators is now clear. In general, if a linear polarization enters the film at say L_0, and then exits after $\Delta \beta z = 45°$, the endpoint of the forward trajectory on the sphere E_{ef} is elliptical, and there will be some loss on encountering P_{45}. Even worse, the backward going beam will enter through P_{45}, be rotated to point E_{eb}, and fail to be extinguished by the input polarizer P_0. This is illustrated in fig. 8(a). For many years the film isolator problem was seen to be that of cancelling or circumventing the deleterious effects of the linear birefringences, and a number of solutions were reported. In a 1990 paper Dammann, Pross, Rabe and Tolksdorf [26] outlined an elegantly simple technique to surmount the problem of the unwanted linear birefringences, the "Dammann trick".

4.2.1. The "Dammann trick"

First note that TE mode defines the zero meridian on the Poincaré sphere, $2\psi_{TE} = 0$. Linear birefringences bring the axis of normal modes away from the poles, but always along this meridian. The Dammann trick consists in having the output polarizer at $-22.5°$ and the path length just long enough so the backward wave exits at $+22.5$. Under these conditions, the film waveguide with its overall elliptical birefringence acts exactly as a simple 45° rotator would. The Dammann trick depends on the realization that a linear polarization with azimuth $-\psi$ will, on transmission through the proper length of material, be rotated to a linear polarization with azimuth $+\psi$. Fig. 8(b) shows the trajectory connecting $-22.5°$ with $+22.5°$. It is an arc of a small circle in a plane normal to the normal mode axis. The isolator using this trick is also shown in the figure. The input polarizer for the forward beam $P_{67.5}$ is at 67.5°. With increasing length in the film, the polarization for light in the forward direction follows trajectory $\overline{L_{67.5}E_{ef}}$ to where it meets $P_{-22.5}$ a linear polarizer set at $-22.5°$. For unit intensity at the input and no absorption in the film, the intensity through the output polarizer would be $\cos^2(\overline{L_{22.2}E_{ef}}/2)$, which falls off quite slowly at first. This constitutes an unavoidable insertion loss. Light traveling in the backward direction enters the film at $L_{-22.5}$, and is rotated to $L_{+22.5}$ as if the film were a simple rotator. This polarization is now completely blocked by the input polarizer $P_{67.5}$. Dammann et al. give a complete analysis along with a descrip-

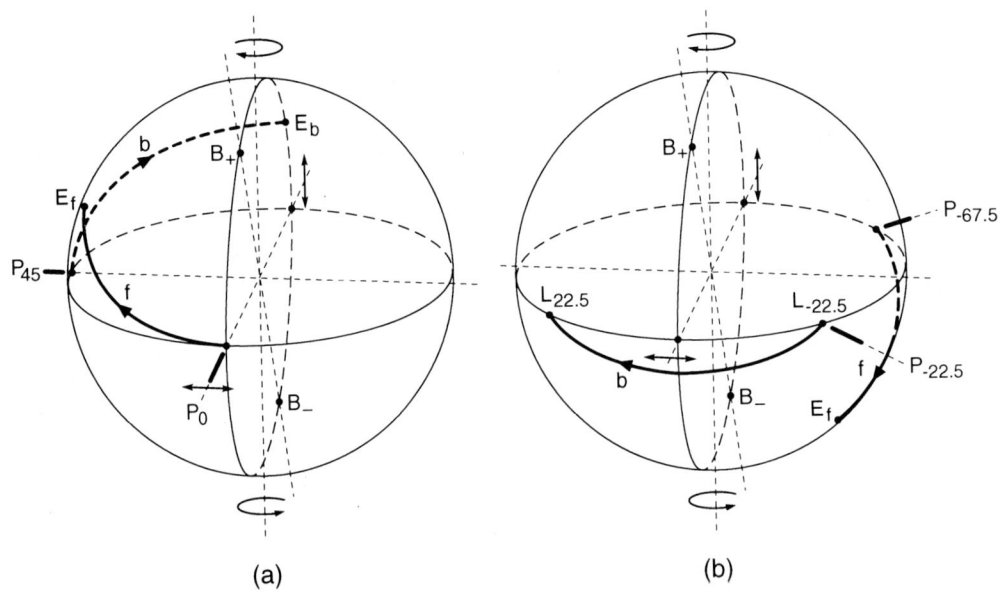

Fig. 8. (a) Poincaré sphere for propagation in a film showing the difficulties imposed by linear birefringences; (b) illustration of the "Dammann trick" to make a film isolator in spite of linear birefringences.

tion of experimental realizations which are completely in accord with the theory. They achieved an isolation ratio of 32 dB in preliminary experiments. Wolfe et al. [27] constructed an isolator of this type and explored how it could be made to operate over a substantial wavelength range (1.43 μm $\leq \lambda \leq$ 1.58 μm) by adjusting the input polarization angle.

This discussion of recent progress in thin film isolators should not be construed to mean that the problems are all solved. Thin film isolators are ultimately attractive because it should be possible to incorporate them in integrated optical structures. The Dammann trick requires polarization axes at 22.5° to the plane, and that is a major stumbling block in terms of circuit simplicity and manufacturability.

4.2.2. Mach–Zehnder interferometer

For propagation in thin films, nonreciprocal devices are possible that have no analog in bulk devices. Assume an axis system with x normal to a MO film, y in the plane and z along the propagation direction. Then suppose that there is a magnetization along y. On examination it turns out rather surprisingly that the TM modes couple to the magnetization. The coupling can occur in film because for TM modes there are finite components E_z. At a position off the center in the film, E will appear to rotate in the xz plane in one sense if the wave is traveling toward $+z$, and in the opposite sense if the wave is traveling toward $-z$. If the field distribution in the film is made to be asymmetric, say by loading one side with a thin layer of high dielectric constant material, the propagation constants for waves going in opposite directions will be different. Therefore the phase shift corresponding to a given propagation length will be different for the two waves. Note that this is a nonreciprocal phase shift of the TM modes; the TE mode see no effect; there is no coupling between TM and TE.

The archetypical device based on the coupling of TM modes with transverse M is the Mach–Zehnder interferometer first suggested by Witte and Auracher [28]. It has been studied primarily by two groups, one at Tokyo Institute of Technology [29], the other at Osaka University [30]. Fig. 9, redrawn from ref. [29] is a schematic of a circulator. It shows two 3 dB directional couplers and nonreciprocal phase shifter in two arms. Not shown is the fact that one arm is a reciprocal 90°

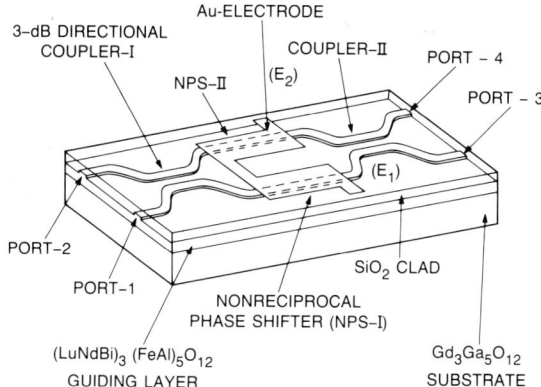

Fig. 9. Schematic diagram of a Mach–Zehnder circulator as fabricated by Mizumoto et al. Magnetic garnet waveguides, 3×0.05 μm^2, are close together to form 3 dB directional couplers. Between the directional couplers two arms in variable transverse fields of opposite sign constitute nonreciprocal phase shifters. (From ref. [29].)

longer than the other. The structure is grown on a $Gd_3Ga_5O_{12}$ (111) substrate. The composition of the guides is given as $(LuNdBi)_3(FeAl)_5O_{12}$, and they are 0.05 μm thick and 3 μm wide. The film has an in-plane anisotropy, a rotation of 900°/cm and a loss of about 10 dB/cm at $\lambda = 1.152$ μm. The overall length of the device shown is 8.5 mm. The magnetic structure is covered by a 0.5 μm cladding layer of SiO_2, and there is an Au electrode to provide magnetizing fields of opposite sense in the two arms. No data were given relative to the stray magnetic fields operating on the device, on the applied field at the phase shifter, or of the actual magnetic behavior of the active material.

Under test, light is input at port-1 or port-2, and the intensities out port-3 and port-4 are measured as the current through the electrode swept from −70 to +70 mA. Note that reversing the current is completely equivalent to reversing the direction of propagation. The data for this preliminary device show that the transmission from port-1 to port-4 changes by 3 dB on going from one current extreme to the other. The authors attribute this low value to fabrication errors. Nevertheless, they have demonstrated an exceedingly interesting device, an MO isolator/circulator which is inherently compatable with integrated optics, and does not require the inclusion of polarizers at awkward azimuths.

5. Magnetooptical materials

5.1. Materials for magnetooptical recording

There is continuing wide ranging quest for MO materials for use in applications and for materials with high MO oefficients. The complex material requirements for MO recording are discussed by Bate [3]. At present the materials of choice are rare earth–transition metal amorphous metals. The polar Kerr rotation is an important factor, but so also are such questions as corrosion and grain size. A major industry has been built on Kerr rotations of perhaps 0.4°!

5.2. Materials for transmission devices

In near infrared transmission devices for communications use, the magnetic garnets are invariably used. The garnet system with all its substitution possibilities can be tailored to enhance seemingly endless sets of requirements. Crystal growth techniques, particularly LPE, are advanced. Isolator material can now be grown as thick epifilms. With care compositions can be grown which have very low absorptions but high MO coefficients at the wavelength range of interest. In the next section we describe an example of the kind of sophisticated materials engineering which is now possible.

5.3. Fratello temperature compensated isolator material

When used in a high bit rate lightwave system, bulk isolators such as that illustrated in fig. 10 must operate over a specific temperature range, perhaps $0 \leq T \leq 85°$ C. This constitutes a significant difficulty because the specific rotation varies substantially over this range. Though a rotator may give exactly 45° rotation and thus a very high extinction ratio at some design temperature, significant deviation from this temperature will quickly degrade the isolator performance. Fratello et al. [31] have recently given an elegant solution to this problem which takes advantage the magnetization compensation in many ferrimagnets at some temperature T_{comp}.

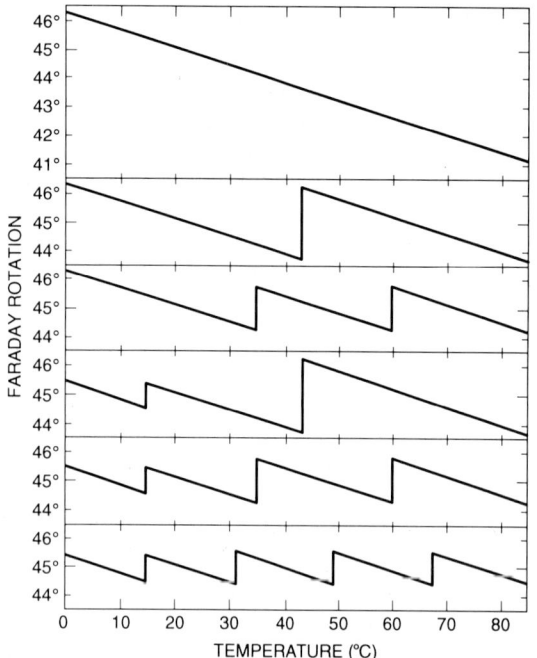

Fig. 10. Theoretical curves of MO rotation against temperature for a 45° rotator incorporating 0, 1, 2, 2, 3 and 4 compensation point layers. These are idealized in that rotation vs. temperature is taken to be linear over the whole range and the compensation point transition is assumed to be sharp. The rotation in each case is set to be 45° at 22°C. Note the way in which the maximum deviation from 45° is reduced. (From ref. [31].)

Our idealization assumes perfect polarizers and a lossless rotator. A deviation θ of 1° implies ER = 35 dB, 2° gives 29 dB and 3° gives 25 dB.

For compact isolators to be used in communications applications, the ferrimagnetic iron garnets are the materials of choice. A 45° rotator of flux grown yttrium iron garnet ($Y_3Fe_5O_{12}$ or YIG) at $\lambda = 1.3$ μm would be about 2.5 mm long, and have a temperature coefficient of rotation of about $-0.04°/°C$. The corresponding rotator of a bismuth-containing composition would be much shorter, cheaper and could be grown much more efficiently. The rotation of the 45° length would vary about $-0.07°/°C$. That means that a rotator with 45° rotation at 42.5°C would give 42° at 85°C and 48° at 0°C. The ER would fall from over 60 dB at the mid temperature to about 25 dB at the extremes. In practice isolators are designed to operate at room temperature, 22°C, and so the degradation at 85°C is even worse.

In a ferrimagnetic garnet the moments of the octahedral (a) and tetrahedral (d) sublattices are antiparallel to each other. That of the dodecahedral (c) lattice is parallel to that of the (a) sublattice. The net moment is the sum of these three contributions:

$$M(T) = M_a(T) - M_d(T) + M_c(T). \tag{8}$$

On the other hand, the contribution of each sublattice magnetization to the magnetooptical rotation is given by a different coefficient. We write

$$\Theta = A(\lambda)M_a(T) + D(\lambda)M_d(T) + C(\lambda)M_c(T). \tag{9}$$

In a rare earth–iron garnet, the temperature dependence of the magnetization of the rare earth ions on the dodecahedral lattice is much steeper than that of the iron sublattices, (a) and (b). When the rare earth magnetization M_c dominates at low temperature, it falls off more rapidly than $M_a - M_b$, and above a compensation temperature T_{comp}, the iron sublattice magnetization dominates. Below T_{comp} M_c lies parallel to a saturating magnetic field, and above T_{comp} M_c lies antiparallel. Thus on passing through the

In the discussion of fig. 7(b) we saw that an ideal isolator consists of an input linear polarizer at azimuth 0°, a 45° ± θ magnetooptic rotator and an output linear polarizer at 45°. In the forward direction, linearly polarized light with unit intensity enters the rotator at 0°, and leaves at 45° ± θ to pass through the output polarizer with an intensity proportional to $\cos^2\theta$. In the reverse direction, light would enter the rotator at 45°, leave it at 90° ± θ, and the intensity on the far side of the input polarizer would be $\sin^2\theta$. Usually one quotes an extinction ratio in dB of the transmitted power P_2 in the reverse direction divided by P_1, that in the forward direction,

$$\text{ER} = -10 \log \frac{P_2}{P_1} = -10 \log \frac{\sin^2\theta}{\cos^2\theta}. \tag{7}$$

compensation temperature, the MO rotation of a sample in a field abruptly changes sign.

The idea of Fratello et al. is to include in the isolator one or more layers of materials with values of T_{comp} within the operating range. Fig. 10 shows how the incorporation of compensation point layers can be used to dramatically reduce the overall deviation from 45°. An appropriate material for the compensation point layer is $(Bi_{0.8}Gd_{1.1}Tb_{1.1})(Fe_{4.6}Ga_{0.4})O_{12}$.

Ref. [31] lists the considerations entering into the composition selection. The compensation point layers would be grown on the substrate, if necessary polished to the correct thickness, then the principle thickness of the rotator would be grown on top. The authors even considered the possibility that a single layer could be grown in which T_{comp} is a continuous function of depth. The reference reports on the performance of a rotator incorporating a single T_{comp} layer.

In the design of isolators, the adjustment of MO rotation to give a constant value over a temperature range is a difficult and important problem. By introducing one or more compensation point layers, Fratello et al. have given us an elegant solution which draws on the most fundamental properties of ferrimagnets, and on the wonderful crystal chemical flexibility of the garnets.

Acknowledgements

The author particularly wishes to thank E.I. Blount, V. Fratello, K.B. Lyons and R. Wolfe for insightful discussions.

References

[1] B.E. Argyle, J. Magn. Magn. Mater. (1991), to be published.
[2] S.D. Bader, J. Magn. Magn. Mater. 100 (1991) 440.
[3] G. Bate, J. Magn. Magn. Mater. 100 (1991) 413.
[4] V.V. Eremenko and N.F. Kharchenko, Sov. Sci. Rev. A 5 (1984) 1.
[5] J.F. Dillon, Jr., L.D. Talley and E. Yi Chen, AIP Conf. Proc. MMM 34 (1976) 388.
[6] R. Schäfer and A. Hubert, Phys. Stat. Sol. (a) 118 (1990) 271.
[7] R. Schäfer, M. Rührig and A. Hubert, IEEE Trans. Magn. MAG-26 (1990) 1355.
[8] A. Thiaville, A. Hubert and R. Schäfer, J. Appl. Phys. 69 (1991) 4551.
[9] J.F. Dillon, Jr., E. Yi Chen, N. Giordano and W.P. Wolf, Phys. Rev. Lett. 33 (1975) 2.
[10] J.F. Dillon, Jr., E. Yi Chen and H.J. Guggenheim, Solid State Commun. 16 (1975) 371.
[11] R.B. Laughlin, Science 242 (1988) 525.
[12] B.I. Halperin, Proc. Intern. Sem. on Theory of High Temperature Superconductors (Dubna, USSR, 3-6 July 1990).
[13] X.G. Wen and Z. Zee, Phys. Rev. Lett. 62 (1989) 2873.
[14] K.B. Lyons, J. Kwo, J.F. Dillon, Jr., G.P. Espinosa, M. McGlashan-Powell, A.P. Ramirez and L.F. Schneemeyer, Phys. Rev. Lett. 64 (1990) 2949.
J.F. Dillon, Jr. and K.B. Lyons, in: New Horizons in Low Dimensional Electron Systems, ed. H. Aoki (Kluwer, Dordrecht, in press).
[15] H.J. Weber, D. Weitbrecht, D. Brach, A.L. Shelankov, H. Keiter, W. Weber, Th. Wolf, J. Geerk, G. Linker, G. Roth, P.E. Splittgerber-Hünnekes and G. Güntherodt, Solid State Commun. 76 (1990) 511.
[16] S. Spielman, K. Fesler, C.B. Eon, T.H. Geballe, M.M. Fejer and A. Kapitulnik, Phys. Rev. Lett. 65 (1990) 123.
[17] I.E. Dzyaloshinskii, Proc. Workshop on the Mathematics and Physics of Anyons (Houston, TX, 30 January-2 February 1991).
[18] G.W. Day and A.H. Rose, SPIE 985 (1988) 138.
[19] J.J. Krebs, W.G. Maisci, G.A. Prinz and D.W. Forester, IEEE Trans. Magn. MAG-16 (1980) 1179.
[20] M. Weber, to be published.
[21] Hitachi Cascaded Optical Isolator, Hitachi Metals Reports E-288.
[22] K. Shirashi and S. Kawakami, Opt. Lett. 12 (1987) 462.
[23] R. Wolfe, private communication (1989).
[24] E.g. R.M.A. Azzam and N.M. Bashara, Ellipsometry and Polarized Light (North-Holland, Amsterdam, 1977).
[25] S. Yamamoto, Y. Koyamada and T. Makimoto, J. Appl. Phys. 43 (1972) 5090.
[26] H. Dammann, E. Pross, G. Rabe and W. Tolksdorf, Appl. Phys. Lett. 56 (1990) 1302.
[27] R. Wolfe, J.F. Dillon, Jr., R.A. Lieberman and V.J. Fratello, Appl. Phys. Lett. 57 (1990) 960.
[28] F. Auracher and H.H. Witte, Opt. Commun. 13 (1975) 435.
[29] T. Mizumoto, H. Chihara, N. Tokui and Y. Naito, Electronics Lett. 26 (1990) 199.
[30] H. Inuzuka, Y. Okamura and W. Yamamoto, Electronics Commun. Japan 72 (1989) 107.
[31] V.J. Fratello, S.J. Licht and C.D. Brandle, US Patent No. 4981341, issues 1 January 1991.

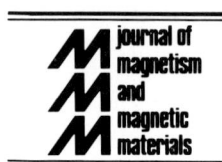

SMOKE

S.D. Bader

Materials Science Division, Argonne National Laboratory, Argonne, IL 60439, USA

Research activities associated with the surface magneto-optic Kerr effect (SMOKE) are reviewed. The purpose is to stimulate interest in the range of contemporary topics in surface magnetism and magneto-optics that SMOKE has addressed. These include the quest for monolayer magnetism, the nature of the surface magnetic anisotropy, trends in the Curie temperature with film thickness, and the critical magnetization exponent at the two-dimensional phase transition. Also, a macroscopic formalism that provides insights into magneto-optics in the ultrathin limit is examined and used to simulate the properties of a number of overlayer, sandwich and superlattice film configurations of interest. Comparisons are then made to experiment and the outlook for the future is examined.

1. Introduction

The surface magneto-optic Kerr effect (SMOKE) made its debut as an experimental technique to study surface magnetism in 1985. It was first used to detect ferromagnetic hysteresis loops from monolayer-range films of iron deposited epitaxially on Au(100) in ultrahigh vacuum [1]. Since that time SMOKE has emerged as an important surface magnetism technique. It has been applied to a variety of materials to search for magnetic ordering, to identify the dominant magnetic anisotropies, to correlate Curie-temperature trends with film thickness and to characterize the critical magnetization exponent at the two-dimensional phase transition. Additional interest in SMOKE stems from the conceptual and practical insights it provides into magneto-optics in the ultrathin limit. Our understanding of magneto-optical effects is historically rooted in the work of Michael Faraday and of the Rev. John Kerr [2]. They were the first to study the influence of magnetized media on the polarization of transmitted and reflected light, respectively. At present, magneto-optic effects are generally described within the context of macroscopic dielectric theory [3], although a microscopic basis also exists whereby the coupling between the electric field of the light and the magnetization of the spin system occurs through the spin–orbit interaction [4]. SMOKE highlights the importance of transcending macroscopic descriptions since continuum theory should break down in the atomic limit. Technological interest in SMOKE relates to the recent commercialization of magneto-optic information-storage media [5]. SMOKE enables applications-minded researchers to consider the ultimate limit of film thinness in their quest for miniaturization.

The present work surveys progress made in using SMOKE to address key issues in surface magnetism and in magneto-optics in the ultrathin limit. Table 1 illustrates the variety of materials that have been examined using the SMOKE technique. The materials are grouped into two categories: there are the ferromagnetic overlayers of Fe, Co and Ni deposited on single-crystal substrates [6–23], and there are the more exotically layered structures involving either two ferromagnetic layers in proximity or separated by an intervening layer, or a ferromagnetic layer sandwiched between nonmagnetic layers, or covered with controlled dosages of adsorbed gases [24–31]. The systems in table 1 are grown under state-of-the-art ultrahigh-vacuum conditions commonly referred to either as molecular beam epitaxy (MBE) or atomic layer epitaxy. The goal is to fabricate perfect films that grow in an ideal layer-by-layer

Table 1
Monolayer-range films studied via SMOKE

Fe	Co	Ni	Substrate	Refs.
A. Overlayers				
•			Ag(100)	[6,7]
	•			[8]
•			Ag(111)	[9]
	•			[8]
•			Au(100)	[1,10,11]
•			Cu(100)	[12,13]
	•			[14–18]
	•		Cu(111)	[18]
	•			[19]
•			Pd(100)	[20]
•			Ru(0001)	[21]
		•	W(110)	[22]
•			MgO(100)	[23]
B. Other coupled-layer systems				
			Co/Cu/Co/Cu(100)	[24]
			Fe/Cu/Fe/Ag(100)	[25]
			Fe/Cu/Fe/Cu(100)	[26]
			Fe/Cu/Fe/Pd(100)	[27]
			Ni/Fe/Ag(100)	[28]
			CrO_2/Co/Cu(100)	[29]
			Au/Co(0001)/Au(111)	[30]
			adsorbates/Fe/Cu(100)	[31]

fashion. Sometimes the goal is approached more closely than others. The art of film growth itself is a very active field, and the correlation between magnetic and structural properties forms the very basis of magnetic materials research. The scope of the present work, however, encompasses magnetism and magneto-optic much moreso than structural characterizations because the goal is to highlight SMOKE. Also, while magnetic superlattices will not be comprehensively covered, our interest in them stems from the view that ultrathin bilayers and sandwiches serve as the basic building block of the superlattice structure.

Background resources for the reader include an overview of magneto-optics [32], surveys of magneto-optical recording materials [33], a text on polarized-light optics [34], recent introductions to macroscopic theory for layered materials [35–38] and topical overviews of surface and thin-film magnetism [39,40], as well as other articles in this special, commemorative issue of the Journal of Magnetism and Magnetic Materials. The remainder of this paper is organized as follows: Section 2 provides background to the principles and experimental techniques. Section 3 is devoted to the contributions made by SMOKE in addressing issues in surface magnetism. Section 4 focuses on magneto-optical issues, including formalisms and simulations, as well as experimental results. Section 5 concludes the paper with a brief outline of future challenges and possible directions.

2. Background

In this section practical aspects of the measurement process are considered. First the underlying description of the magneto-optic interaction is qualitatively outlined, and then the operational approach of measuring signals is considered, including a discussion of the vacuum-chamber, magnet, optical-components and electronic signal-processing requirements.

2.1. Principles

The Kerr effect as applied to ferromagnetic films involves the change in polarization of light reflected from a magnetized medium. Linearly polarized incident light acquires a Kerr rotation and a Kerr ellipticity upon reflection. In fig. 1 the magneto-optic interaction schematically is shown to induce an orthogonal component in the electric field vector of the light upon reflection. The part of the induced response that is in-phase with the incident light gives rise to the rotation, while the out-of-phase part accounts for the ellipticity. If an external magnetic field is manipulated to reverse the magnetization direction of the sample, the magneto-optic rotation and ellipticity reverse sign. Macroscopically the effect can be described by off-diagonal terms in the dielectric tensor ϵ. In the case where the incident light and the magnetization direction are both along the surface normal:

$$\epsilon = N^2 \begin{pmatrix} 1 & iQ & 0 \\ -iQ & 1 & 0 \\ 0 & 0 & 1 \end{pmatrix},$$

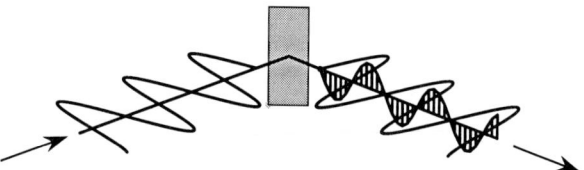

Fig. 1. Schematic representation of the magneto-optic interaction. The incident light is linearly polarized and the light reflected from the magnetized film has a Kerr rotation and ellipticity.

where N is the refractive index and Q is the magneto-optic (Voigt) constant of the medium. The light that enters the magnetized medium decomposes into beams that contain left- and right-circularly polarized modes. Different indices of refraction n are assigned to the two modes and for a general magnetization direction M:

$$n = N(1 \pm gQ),$$

where g is the direction cosine between the propagation vector of the light k and M. The two circular modes travel with different velocities and attenuate differently, due to the differences in the real and imaginary parts of the potential inside the medium. Upon emerging from the medium in reflection the two altered modes recombine to yield a rotated axis of polarization and an ellipticity.

There are three Kerr configurations that are of importance: polar, longitudinal and transverse. These terms correspond to the orientation of the magnetization with respect to the scattering geometry. In the polar Kerr effect the magnetization direction is perpendicular to the plane of the film (along the surface normal). In the longitudinal Kerr effect the magnetization is in the film plane and in the scattering plane of the light (plane of incidence). In the transverse Kerr effect the magnetization is also in the film plane, but it is perpendicular to the plane of incidence. In general, the magnetization can be in an arbitrary direction, but consideration of the above three configurations simplifies the algebra in describing the interaction. Simplifications are also realized by consideration of only p- or s-polarized incident light, for which the polarization vector is in the plane of incidence or perpendicular to it, respectively. In the transverse Kerr effect there is no change in the polarization of the reflected light because there is no component of the light propagating along the magnetization direction. The transverse Kerr effect involves only a small change in reflectivity upon magnetization reversal for p-poralized light. The polar Kerr effect usually gives rise to polarization changes that are an order of magnitude more pronounced than those associated with the longitudinal Kerr effect, as will be illustrated in section 4.

2.2. Experimental approaches

Operationally the detection of SMOKE signals can be as straightforward as is illustrated in fig. 2. The measurement scheme is basically the same as in conventional MOKE experiments. A polarized laser S serves as the light source. The beam enters into an ultrahigh-vacuum (UHV) surface science chamber through a window W to impinge on the magnetized target. The reflected light exits back through the window, passes through an analyzer A whose polarization axis is nearly crossed with that of the incident beam and is detected at D by a photodiode. The intensity of

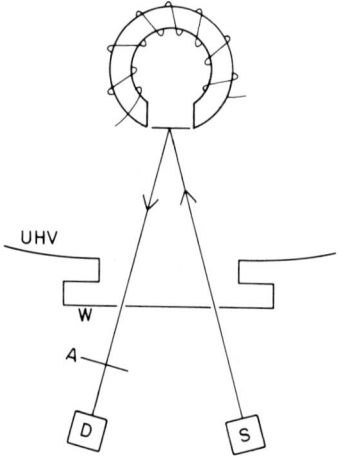

Fig. 2. Schematic of a SMOKE experiment, from ref. [1]. S is the polarized light source, A is the crystal-prism analyzer, D is the photodiode detector, W is the UHV window. The electromagnet is shown magnetizing the sample in-plane (longitudinal Kerr effect).

light that reaches the detector changes as the applied magnetic field is swept to reverse the magnetization of the target sample. Thus the output of the photodiode detector vs. the applied field H yields the hysteresis loop of the sample. Fig. 3 shows such a loop for 1.2 monolayers (ML) of Fe grown on Pd(100), as will be described further in section 3.

What is special about the SMOKE apparatus is that the sample chamber is maintained at a state-of-the-art vacuum level of 10^{-11} Torr to preserve the cleanliness of the film during and after growth. The chamber should be equipped with all the accouterments of a modern surface science laboratory, meaning typically that there are electron diffraction and spectroscopy capabilities to monitor the growth, cleanliness, epitaxy and electronic structure of the film, as well as provisions for film deposition and thickness monitoring, for heating and cooling the sample, and for eventually removing the overlayer and cleaning the substrate in preparation for further studies. Illustrative descriptions of the ancillary equipment and structural characterization methodologies appear in refs. [39,40].

The electromagnet used to magnetize the sample may be in the UHV chamber, as is shown in fig. 2 for the longitudinal Kerr configuration. Crossed magnet poles [12] or a rotation of the magnet [8,11] or sample by 90° make it possible to conveniently switch from the polar to longitudinal Kerr configuration. To achieve magnetic fields greater than a few kOe, water-cooled electromagnets can be mounted outside a pinched-down or tail section of the chamber [22], or the

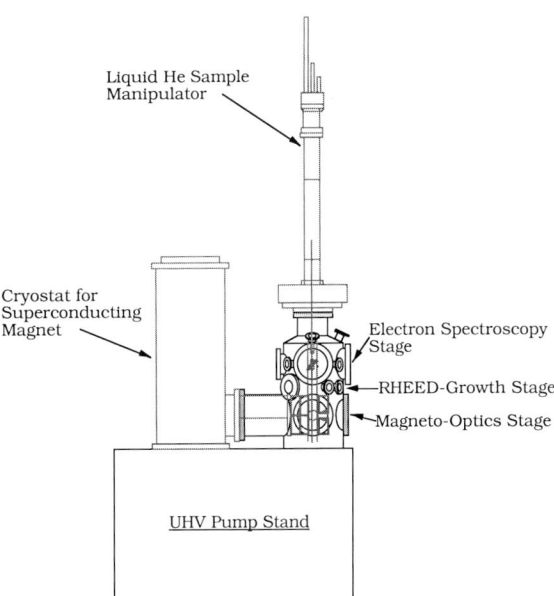

Fig. 4. Layout of a SMOKE system based on a superconducting magnet. The electron spectrometer, reflection high-energy electron diffraction (RHEED), MBE cells, thickness monitors, lasers and other ancillary equipment are not drawn in, except for their locations in the three-tiered UHV chamber. The sample holder has a long bellows for positioning the sample between tiers.

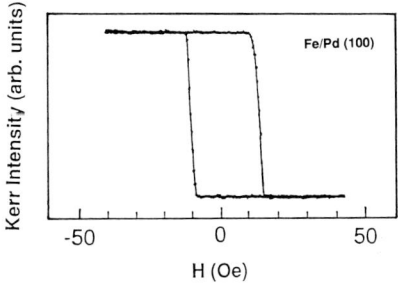

Fig. 3. Polar SMOKE signal from 1.2 ML Fe/Pd(100), as described in ref. [20].

magnet pole tips can be located in the uhv chamber with the rest of the magnet assembly outside [8]. In the most recent SMOKE system in the author's laboratory a split-coil superconducting magnet that is bakable and UHV compatible is introduced into the vacuum chamber (see fig. 4). It provides a ±20 kOe field with a charge time of a few minutes. It produces a low remanent field of < 10 Oe due to the use of twisted filamentary Nb–Ti wire. The sample can be rotated in the bore of the magnet to positions facing access ports for longitudinal or transverse and polar measurements, as well as for transmission studies. The access ports are windowless, but are surrounded with inserts anchored to a liquid-N_2 reservoir to limit the liquid-He boiloff.

The optical components shown in fig. 2 are outside the vacuum chamber, for the sake of convenience, and there are no mechanical adjustments made during the measurements. To improve the signal-to-noise ratio the hysteresis loop is swept multiple times and signal-averaged on a laboratory computer. The light source is typically a polarized He–Ne laser with a few mW output power. The analyzer could be a polaroid sheet, but a crystal prism has better extinction properties. Commercial mounts are available that can rotate the analyzer in steps as small as a few μrad, if desired. The detector is a photodiode covered with a filter to pass He–Ne light.

In magneto-optic studies the window of the vacuum chamber can introduce a stray birefringence that alters the polarization of the light. Compensators can be used to correct for this [41], and strain-free window mountings and careful choice of window materials can be used to minimize the unwanted effect. Also, the light can be directed perpendicular to the window and can enter and exit through the same region of the window to minimize these effects. Another optical problem is with drifting intensity of light. Using an intensity-stabilized laser offers great improvements [42], but sometimes the drift is due to mechanical instabilities. A detection scheme that consists of a polarization-splitting element, such as a Wollaston prism, followed by two matching photodiodes that record the intensity from the two orthogonal polarization components can yield a more stable signal because the ratio of the two intensities normalizes out the drift.

If the goal is to extract quantitative magneto-optic rotation and ellipticity values rather than merely to search for surface magnetism, there are various embellishments on the basic detection scheme that are valuable. A variety of modulation techniques have been used in magneto-optic studies that are also applicable in the SMOKE regime. Photoelastic modulators have been used to modulate the polarization of the incident light [43], and Faraday modulators [44] have been used to modulate the polarization of the reflected light. Also a phase-modulation approach to detect weak magneto-optic signals, based on a novel fiber-optic interferometer, was recently introduced [45]. In concluding this section, note that different strategies for separating rotation and ellipticity that do not rely on modulators have also been used [46,47].

3. Topics in surface magnetism

This section highlights the contributions made using SMOKE to elucidate key issues in surface magnetism. First, the quest for surface magnetism is considered. Then, for systems that exhibit magnetic order, we explore the nature of the surface magnetic anisotropy, trends in the Curie temperature and the critical behavior at the phase transition.

3.1. The quest for surface magnetism

Predictions of new magnetic materials, new structures and new couplings have motivated novel searches for surface magnetism. For example, band-structure calculations of the total energy have been used to predict that a monolayer of vanadium epitaxially deposited on Ag(100) or Au(100) is ferromagnetic, while V itself is not a magnetic element [48]. Similar predictions have been made for epitaxial monolayers of 4d-elements (T, Ru, Rh, Pd) grown on Ag(100) [49]. The top layers of Pd substrates (and Pt, by inference) in proximity to 3d-magnetic overlayers are predicted to have induced ferromagnetic moments of $0.32\mu_B$ for the first layer and $0.17\mu_B$ for the next [50]. Also, the surface layer [51] of antiferromagnetic Cr(100) and of monolayer [47] Cr grown on Au(100) have been predicted to be ferromagnetic with a giant moment of 2.7 and $3.7\mu_B$, respectively, compared to $0.6\mu_B$ for the bulk Cr moment. SMOKE is a convenient probe to search for new ferromagnets, but to date has not confirmed any of these dramatic predictions. However, SMOKE *has* been used to refine some of the predictions. For example, subsequent to SMOKE studies [52], monolayer Cr/Au(100) was calculated to be more stable as an antiferromagnet with giant moments of $3.5\mu_B$ [53].

Table 1 lists magnetic systems studied using SMOKE. While traditionally magnetic elements are the main species of interest in table 1, what is exciting is that epitaxy affords the possibility to grow these materials in lower dimensionality and with new or altered lattice constants. For example, while bulk Fe is bcc, the high-temperature fcc γ-phase is stabilized via epitaxy on Cu. Predictions of the magnetic properties of fcc Fe abound and demonstrate a multiplicity of magnetic states: fcc-Fe possesses ferromagnetic and antiferromagnetic (AF) phases at the same lattice constant [54–56]. The AF phase has a low moment and low Néel temperature, and should be paramagnetic at room temperature. SMOKE experiments have been used to discover how to prepare ferromagnetic and nonferromagnetic phases of ultrathin fcc-Fe on Cu(100) [12,13]. Other examples include the growth of Fe on Pd(100) [20] or Ru(0001) [21] substrates, for which Fe is not lattice-matched. The resultant structures are pseudomorphic – they take on the in-plane lattice spacing of the substrate – but have pronounced interlayer relaxations that yield tetragonal and trigonal distortions, respectively. In comparison to these cases, Fe epitaxially deposited on Au(100) yields relatively unstrained bcc-Fe, even though Au is an fcc 5d-element [1]. This is because the primitive unit cell of Au(100) is fortuitously lattice-matched to the bcc Fe square net upon 45° rotation. Thus, while the range of elements that have useful magneto-optical activities has been limited to date, there are vast possibilities to explore new and altered phases.

The sequential layering of magnetic films also has unveiled new and interesting magnetic couplings. Ferromagnetic layers of Fe (or Co) separated by spacer layers (i.e. Cr [57], Cu [58], Ru [59], Mo [60]) form multilayer structures that possess an oscillatory interlayer coupling as a function of spacer-layer thickness. The ferromagnetic layers either all align ferromagnetically with respect to each other or alternate in alignment and take on an AF structure. A modified RKKY-coupling gives rise to the oscillations in coupling [61]. The hysteresis loops for ferromagnetic films have high remanence, while those for AF films have low remanence. Fig. 5 shows a superlattice example wherein the Kerr effect was used to discover that Fe/Mo sputtered films belong to this new class of coupled-layer materials [60]. Plotted in fig. 5 is the ratio of the remanent-to-saturation-state Kerr intensity. The oscillations in fig. 5 are in good agreement with the magnetic-coupling behavior calculated by Levy et al. [62].

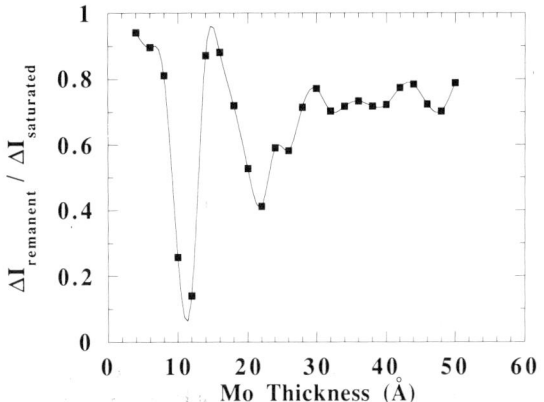

Fig. 5. Ratio of the remanent-to-saturation Kerr-loop intensities for sputtered Fe–Mo superlattices. The ferromagnetic films have a ratio close to unity, while the AF films have greatly reduced ratios.

3.2. Surface magnetic anisotropy

Néel [63] first introduced the idea that the surface anisotropy κ_s could give rise to perpendicular or vertical easy axes of magnetization [64]. This surface anisotropy would overwhelm the shape anisotropy of a thick Fe film, whose demagnetizing field, $-4\pi M_s \approx 20$ kG, would otherwise stabilize in-plane easy axes. Band-structure evaluations of the spin–orbit energetics for Fe confirm that large vertical anisotropies can exist [65,67] (although ref. [65] provides contradictory predictions for specific systems from refs. [66,67]). SMOKE is highly suitable for searching for vertical easy axes because the longitudinal and polar geometries can be used sequentially to determine if a film remanently magnetizes in-plane or vertically. To estimate the magnitude of the surface anisotropy field [68] note that the vertical easy axes will revert in-plane at a critical thickness d_c approximated by the expression:

$$2\kappa_s \approx 4\pi M_s d_c,$$

where $2\kappa_s$ is for the two interfaces (vacuum–overlayer and overlayer–substrate) and d_c is the number of monolayers thickness of the magnetic overlayer. For example, if $d_c = 5$ ML, the surface field, ≈ 50 kG, is indeed strong. The left panels of fig. 6 summarize SMOKE results vs. Fe film thickness for growth on four different substrates [69]. Strong polar Kerr intensity signals (the height of the hysteresis loop normalized by the total detected intensity) initially increase with thickness until d_c is reached, after which a relatively weak longitudinal signal is detected. The d_c-values in fig. 6 are in the 3–6 ML range. The right panels in fig. 6 show that the coercivities tend to peak in the monolayer range also. The films in fig. 6 of Fe on Cu(100), Au(100) and Pd(100), represent fcc, bcc and bct Fe phases, respectively, and were grown at low temperature (≈ 100 K) to suppress interdiffusion, while the Fe/Ru(0001) is a trigonally-compressed phase, grown at room temperature, whose first two monolayers do not exhibit ferromagnetism.

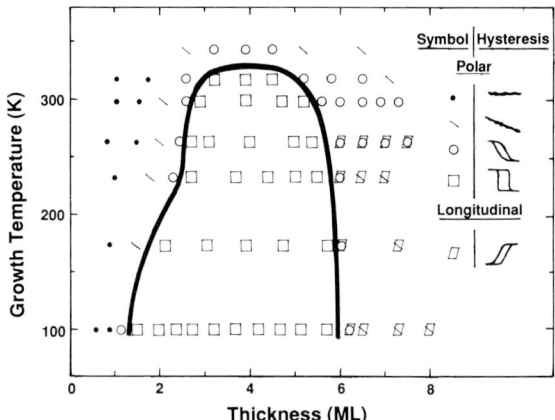

Fig. 7. The region of stability of vertical easy axes for fcc-Fe/Cu(100) is delineated by the heavy boundary. The measurements were made at the growth temperature and are described in ref. [13].

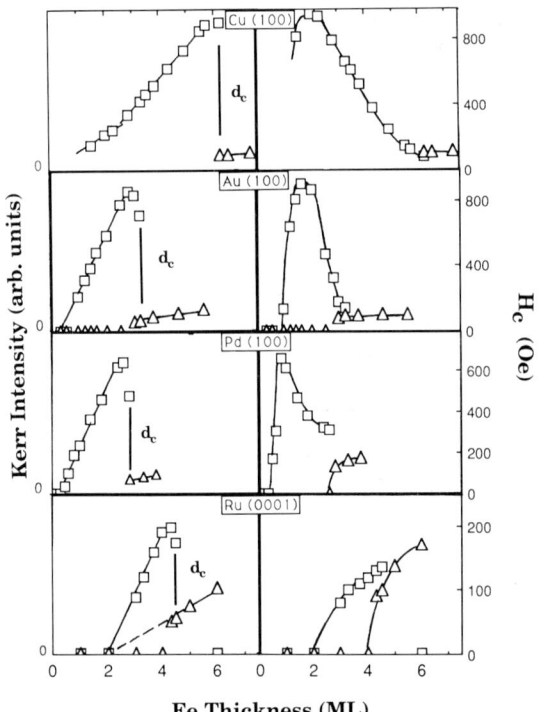

Fig. 6. Thickness dependence of the Kerr intensity (left panels) and coercivities H_c (right) for epitaxial Fe films grown on the indicated substrates, as described in the text and in ref. [69]. The Kerr intensity is the height of the hysteresis loop.

Remanent polar hysteresis loops, as in fig. 3 for Fe/Pd(100), verify the existence of vertical easy axes, but do not directly reveal the underlying stabilization mechanism of importance. The role of film growth and processing parameters in determining the easy axes of magnetization in ultrathin Fe/Cu(100) films are highlighted in fig. 7, which shows an 'anisotropy' phase diagram constructed from SMOKE hysteresis loops. The region of vertical easy axes are delineated with a heavy boundary in fig. 7. The d_c value is 6 ML, but the vertical easy axes are also destroyed for the thinnest films and for elevated-temperature growth due to intermixing.

3.3. Curie temperature vs. film thickness

The importance of magnetic anisotropy can be appreciated when one considers that the Mermin–Wagner theorem prohibits finite-temperature ordering in two dimensions (2D) for *isotropic* Heisenberg systems [70]. The strong surface anisotropies discussed above trigger the finite T_C-values and allow the thickness dependence of T_C to be studied. Bander and Mills [67] introduced a renormalization-group scaling relation between T_2 and T_3, the T_C-values in 2D and 3D, respectively, as a function of the effective anisotropy parameter $\kappa_{\text{eff}} = (2\kappa_s - 4\pi M_s d) > 0$:

$$T_2 = T_3 / \ln\left(\frac{3\pi}{4} \frac{T_3}{\kappa_{\text{eff}}}\right). \quad (1)$$

Their expression was derived for an N-component system evaluated in the large-N limit and applied to an anisotropic Heisenberg system ($N = 3$) that exhibits an Ising transition. Other estimates of the T_C-variation come from Monte Carlo simulations of Ising models. Binder and Hohenberg provided a classic treatment of the simple cubic lattice [71]. The difference is that eq. (1) predicts a nonmonotonic dip in T_C as d_c is approached, while the Monte Carlo simulations yield a monotonic increase from T_2 to T_3.

Fig. 8 shows the trend in T_C-values for Fe/Pd(100) obtained from SMOKE studies [20] compared to recent Monte Carlo simulations of Strandburg et al. [72] for a bcc lattice. The experimental data increase monotonically even though the thicker films tend to be magnetized in-plane while the thinner ones are vertically oriented. The Ising-model calculations associated with fig. 8 incorporate two interesting features that account for the good agreement with experiment. The strong T_C-variation in the submonolayer regime indicates that a 2D-island morphology would not work well, because it would lead to a T_C-value comparable to that of the monolayer, but with some finite-size broadening and shifting. Instead, a random-site-vacancy morphology was assumed, and the result of the dilute Ising model that T_C decreases with dilution was preserved, as also was found recently by Li and McQuistan [73]. The other interesting feature is that the interfacial Pd sites have induced moments when coupled to the Fe, as is indicated by the parameter $J_{\text{Fe-Pd}}$ in fig. 8. The coupling provides an increased effective number of nearest neighbors that raise T_C for small Fe thicknesses, but not in the bulk limit. A value of $J_{\text{Fe-Pd}}$ equal to 1/2 provides a good description of the experimental trend in fig. 8.

Note also that superparamagnetism can be of importance for ultrathin films. If islands form, the superparamagnetic blocking temperature can be mistaken for T_C, and can drop below the measurement range before the monolayer limit is reached. This does not occur for the Fe/Pd(100) system either because the random-site vacancy morphology yields continuous films, and/or because of the polarization of the underlying interfacial Pd. However, other systems may be suspect, like Co/Cu(100), where the monolayer T_C-value is not conclusively detected [14,16].

3.4. The critical magnetization exponent in 2D

For a second-order phase transition the order parameter increases continuously below the ordering temperature and exhibits scaling behavior in the critical regime wherein the universality class is determined by the dimensionality and the local symmetry. For a ferromagnet, as $T \to T_C$ from below:

$$M \propto (1 - T/T_C)^{\beta_c},$$

where β_c is the critical magnetization exponent. Values of β_c in 2D are illustrated in table 2, based on the previous tabulation of Einstein [74], and on the work of Suzuki [75] for the 4-state clock model, and that of Kosterlitz and Thouless [76] for the XY model. SMOKE has been used to measure the effective experimental exponent β_{eff} for comparison to the theoretical β_c, since the

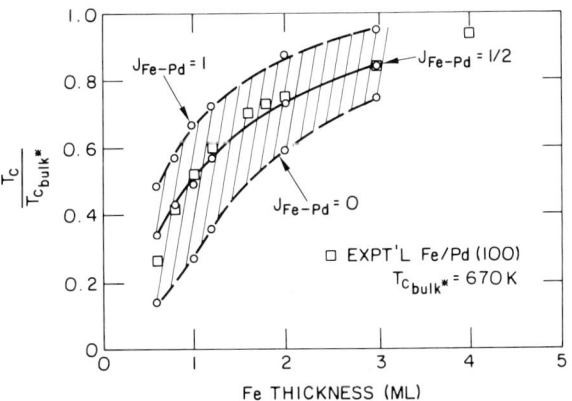

Fig. 8. The experimental T_C-trend for Fe/Pd(100) from ref. [20] plotted along with results of recent Monte Carlo calculations for 2D-Ising models. The model assumes a random-site occupancy of the Fe and permits coupling to interfacial Pd spins via the term $J_{\text{Fe-Pd}}$, as described in ref. [71].

Table 2
2D critical magnetization exponents

Universality class	β_c
Heisenberg (isotropic)	no transition
XY (isotropic)	Kosterlitz–Thouless transition
XY with cubic anisotropy	nonuniversal (depends on the strength of the anisotropy)
Ising (2-state Potts)	1/8
3-state Potts	1/9
chiral 3-state Potts	1/9
4-state Potts	1/12
4-state clock	1/8

Kerr intensity, or the height of the hysteresis loop, is proportional to the magnetization [19,20]. When the easy axes of magnetization are vertical the anisotropic 2D Heisenberg system belongs to the Ising universality class, and $\beta_c = 1/8 = 0.125$. Fig. 9 shows log–log plots of M vs. $(1 - T/T_C)$ near T_C for Fe/Pd(100) films, which indicate agreement with the 2D-Ising model at 1.2 ML Fe, and a gradual increase in β_{eff} in the submonolayer regime [20]. The T_C values are parameters

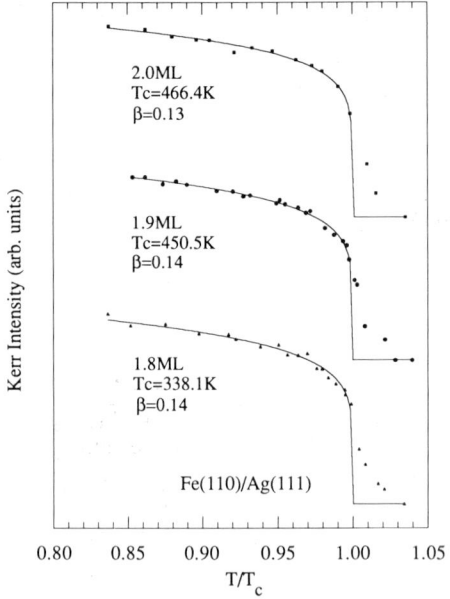

Fig. 10. Kerr intensity plots of the 2D phase transition for Fe(110)/Ag(111). The β_{eff}- and T_C-values from the least-squares fittings are indicated for each Fe thickness. These results are taken from the recent work of Qiu et al. [9].

chosen to maximize the linearity of the fit. The log–log plots show linearity over a broad T-range, which is characteristic of critical fluctuations in 2D. Note in fig. 9 that warming and cooling data were collected. It is important to demonstrate that the transition is reversible, and that temperature is not altering the physical properties due to irreversible changes in film morphology. The 3 ML film in fig. 9 is magnetized in-plane, but still exhibits an Ising-like exponent. Fig. 10 shows M vs. T plots for Fe(110)/Ag(111) magnetized in-plane, taken from the recent SMOKE study by Qiu et al. [9] using the apparatus outlined in fig. 4. The β_{eff} values in fig. 10, obtained from log–log plots, are again 2D Ising-like. The examples in figs. 9 and 10 are particularly interesting because of the controversy surrounding reports [19] of exponents with much larger β_{eff} values than expected from table 2.

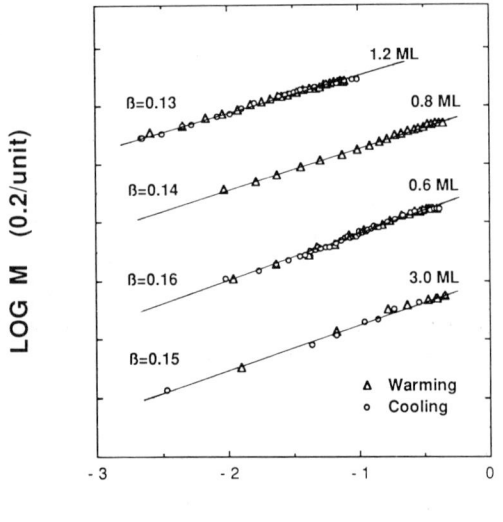

Fig. 9. Log–log plots of the temperature dependence of the Kerr-intensity data for Fe/Pd(100) used to obtain the critical exponent β that characterizes the phase transition. The data sets are shifted vertically for clarity. The effective β-exponents are obtained by linear least-squares fitting the slope of the curves, as described in ref. [20]. The results are in good agreement with the 2D-Ising value $\beta_c = 1/8$.

4. Topics in magneto-optics

In this section a relatively concise guide will be presented to the macroscopic Kerr-effect formalism of Zak et al. [35,36]. The choice of formalism is dictated by the interest and familiarity of the author. Insights will then be obtained by considering the ultrathin limit, or the SMOKE regime, and examining instructive simulations and measurements.

4.1. Formalism

A beam of light travelling from medium 1 to medium 2 conserves the tangential components of its electric E_x, E_y and magnetic H_x, H_y fields, where the xy-plane is the boundary between the two media. Expressed in terms of the electric fields of the incident (i) and reflected (r) waves we can write

$$F = \begin{pmatrix} E_x \\ E_y \\ H_x \\ H_y \end{pmatrix} = AP = A \begin{pmatrix} E_s^{(i)} \\ E_p^{(i)} \\ E_s^{(r)} \\ E_p^{(r)} \end{pmatrix},$$

where the 4×4 matrix A that connects column vectors F and P is called the *medium boundary matrix*. The matrix elements of A are constructed from the geometric angles of the problem and from the N and Q values of the medium. For a two-medium, one-boundary problem, the boundary matching condition becomes

$$A_1 P_1 = A_2 P_2.$$

If there is more than one boundary, the wave propagation inside the medium at depth z from the interface is described using the *medium propagation matrix* $\overline{D}$, where

$$P_2(z=0) = \overline{D}_2(z) P_2(z).$$

For a multilayer system, the light originates in the initial medium i, goes through the multilayer stack and ends up in the substrate or final medium f. The information of interest for l layers in the stack is contained in the expression

$$A_i P_i = \prod_{m=1}^{l} \left(A_m \overline{D}_m A_m^{-1} \right) A_f P_f.$$

If we put this expression in the form $P_i = MP_f$, where

$$M = A_i^{-1} \prod_m A_m \overline{D}_m A_m^{-1} A_f \equiv \begin{pmatrix} G & H \\ I & J \end{pmatrix},$$

then the 2×2 matrices G and I can be used to obtain the Fresnel transmission t and reflection coefficients, since

$$G^{-1} = \begin{pmatrix} t_{ss} & t_{sp} \\ t_{ps} & t_{pp} \end{pmatrix} \quad \text{and} \quad IG^{-1} = \begin{pmatrix} r_{ss} & r_{sp} \\ r_{ps} & r_{pp} \end{pmatrix}.$$

The Kerr rotation ϕ' and ellipticity ϕ'' for s- and p-polarized light are then expressed as:

$$\phi_s = \phi_s' + i\phi_s'' = r_{ps}/r_{ss} \quad \text{and}$$

$$\phi_p = -\phi_p' + i\phi_p'' = r_{sp}/r_{pp}.$$

Prescriptions for constructing the A and $\overline{D}$ matrices appear in ref. [36] and in the references cited therein.

4.2. The ultrathin limit

The formal expressions above simplify and provide useful insights in the ultrathin limit, which is defined by

$$\frac{2\pi}{\lambda} |N| d \ll 1,$$

where λ is the wavelength of the light and d is the thickness in Å of the magnetic layer [27,35]. In the ultrathin limit for a magnetic overlayer, characterized by N, Q and d, on a nonmagnetic substrate with refractive index N_{sub}, the polar (POL) and longitudinal (LON) Kerr effects become:

$$\phi^{POL} = -\left(\frac{4\pi}{\lambda}\right)\left(\frac{N^2}{1-N_{sub}^2}\right)Qd \quad \text{and}$$

$$\phi^{LON} = \left(\frac{4\pi}{\lambda}\right)\left(\frac{N_{sub}}{1-N_{sub}^2}\right)\theta Qd, \tag{2}$$

where θ, the angle of incidence measured from the surface normal, is assumed small. For a multilayer stack in the ultrathin limit there is an additivity law whereby the total ϕ^{POL} and ϕ^{LON} are represented by the appropriate expressions in eq. (2) summed over the magnetic layers: $\phi_{total} = \Sigma\phi$. Also, eq. (2) illustrates that $\phi^{POL} \gg \phi^{LON}$ because of the extra N factor and the lack of a θ factor in ϕ^{POL}. Note that ϕ^{LON} is independent of the refractive index of the magnetic layer. This implies that ϕ^{LON} can be enhanced by choosing a substrate with a suitable value of N_{sub}. The enhancement can be envisioned as due to the substrate acting as a mirror to reflect the transmitted beam back through the overlayer where it boosts the Kerr effect. Eq. (2) also demonstrates that the SMOKE signal is proportional to d, as in the Faraday effect, and unlike the bulk Kerr effect, which is independent of d. In this important respect SMOKE is distinct from MOKE.

Additional insights comes from consideration of superlattices (SL) of magnetic and nonmagnetic layers of thicknesses d_1 and d_2, respectively, that satisfy the ultrathin criterion. It can be shown that if $N_1 \approx N_2$, there is a remarkably simple relationship to the Kerr effect of bulk films [77]:

$$\frac{\phi_{SL}}{\phi_{bulk}} = \frac{d_1}{d_1 + d_2}.$$

4.3. Simulations

Fig. 11 shows simulations for Fe–Cu superlattices, overlayers and sandwiches as a function of Fe thickness, based on the full formalism of section 4.1. The quantity $\phi_m = [(\phi')^2 + (\phi'')^2]^{1/2}$ in fig. 11 is known as the magnitude of the complex rotation. Tabulated optical constants [78] and Q-values [79] were used in the simulation. The properties of the ultrathin regime are apparent in fig. 11. Comparisons between these types of simulations and experiment appear in the literature for Fe overlayers [80] on Au(100), and for Fe/Cr [81], Fe/Mo [60], Co/Pt [82] and related super-

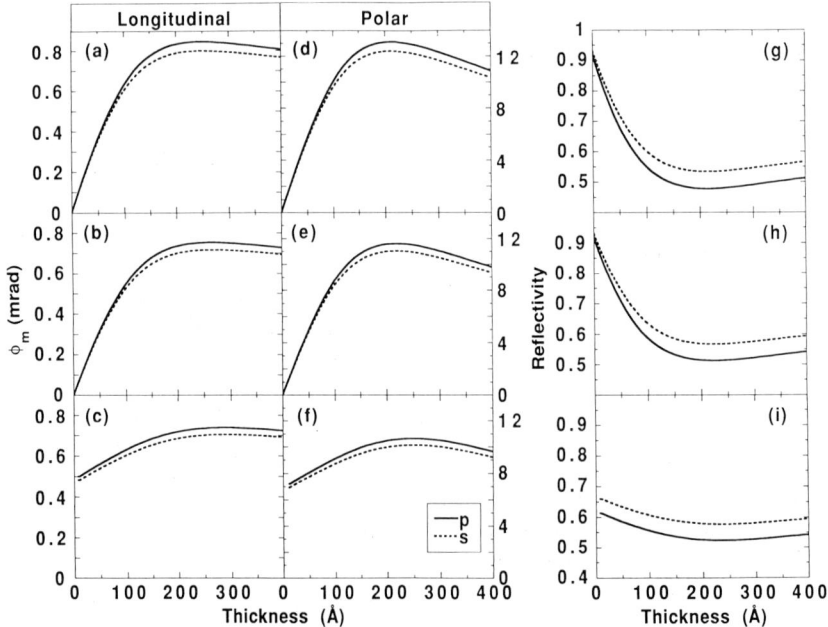

Fig. 11. Simulations for Fe overlayers, sandwiches and superlattices with Cu are shown in the panels from top to bottom, respectively, from ref. [35]. The sandwich structure consists of a 20 Å Cu overlayer and a Cu substrate, and the superlattice has 50 periods with equal Fe and Cu thicknesses and a 20 Å overlayer of Cu to protect the top Fe layer.

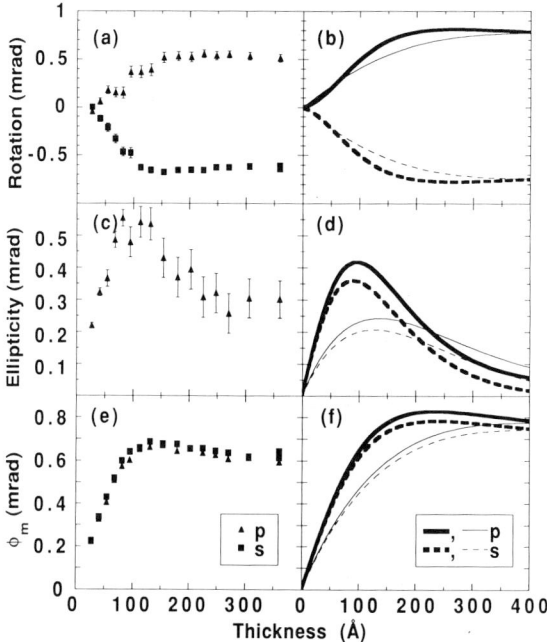

Fig. 12. Longitudinal magneto-optic measurements for Fe grown at elevated temperature on Au(100) are shown (left panels) along with simulations that use tabulated bulk data (right). The heavy curves on the right are for Fe/Au, while the thin curves assume $N_{sub} = N_{Fe}$. The difference between the two sets of curves in panel (f) illustrates the role of the substrate in enhancing the magneto-optic response of ultrathin films, as described in ref. [80].

lattices [83]. Fe/Au(100) results are shown in fig. 12, where the simulations were based on published expressions [84] for ϕ^{LON}. The simulations in the right panels of fig. 12 also include the case where $N_{sub} = N_{Fe}$, which causes the magneto-optic signal in the ultrathin limit to decrease, and demonstrates the substrate enhancement effect discussed in section 4.2 since $N_{Au} > N_{Fe}$. There are a number of controversies [85,86] in the literature surrounding magneto-optical enhancement effects that might become tractable with proper simulation capabilities.

The simulations in ref. [82] for the polar Kerr effect of Co/Pt superlattices are particularly all-inclusive, and were performed as a function of d_1 and d_2, the number of bilayers in the superlattice, the photon energy, angle of incidence, light polarization and the refractive index of the substrate. The role of antireflection ZnS coatings in enhancing the Kerr effect at select wavelengths was also explored. Calculated under these various conditions were the rotation, ellipticity, reflectivity R and ϕ_m, and the figure of merit $[(\phi_m)^2 R]$ which represents the rotating power of the film. Some of this information is reproduced in fig. 13. The versatility of the simulation scheme suggests the possibility of computer-designing materials with optimized magneto-optical properties. However, the experimental properties of Co/Pt superlattices [87] differ from the simulations because of the strong magneto-optic response from the interfacial Pt, due to its induced moments and large spin–orbit coupling [82].

Finally, it is interesting to see the macroscopically calculated trends for monolayer-range magnetic overlayers. Fig. 14 uses tabulated bulk quantities in a simulation of the SMOKE experiments in fig. 6. The discontinuities in fig. 14 between $\phi_m^{(POL)}$ and $\phi_m^{(LON)}$ at the d_c-values ticked in from experiment are reminiscent of the magnitude of those actually observed in fig. 6 when the magnetization reorients from vertical to in-plane.

5. Outlook

There are many challenges for the future. Of immediate interest are microscopic calculations of the Kerr effect in the monolayer regime, where the electronic, magnetic, structural and optical properties should be uniquely different than for bulk films. It is also important to quantify the Kerr rotation and ellipticity in the monolayer regime, and to do Kerr spectroscopy as a function of photon energy. Synchrotron and free-electron laser radiation at higher energies could be used to bridge the gap with information obtained from spin-polarized core-level spectroscopies. Additional directions include refinements that would enable Kerr microscopy [88] in the SMOKE regime to study magnetic switching dynamics and domain wall energetics. Kerr studies using second harmonic generation should increase surface sensitivity, so it might become possible to study surfaces of bulk magnetic films. In the materials realm there are seemingly endless film configurations that can be explored. It would be most

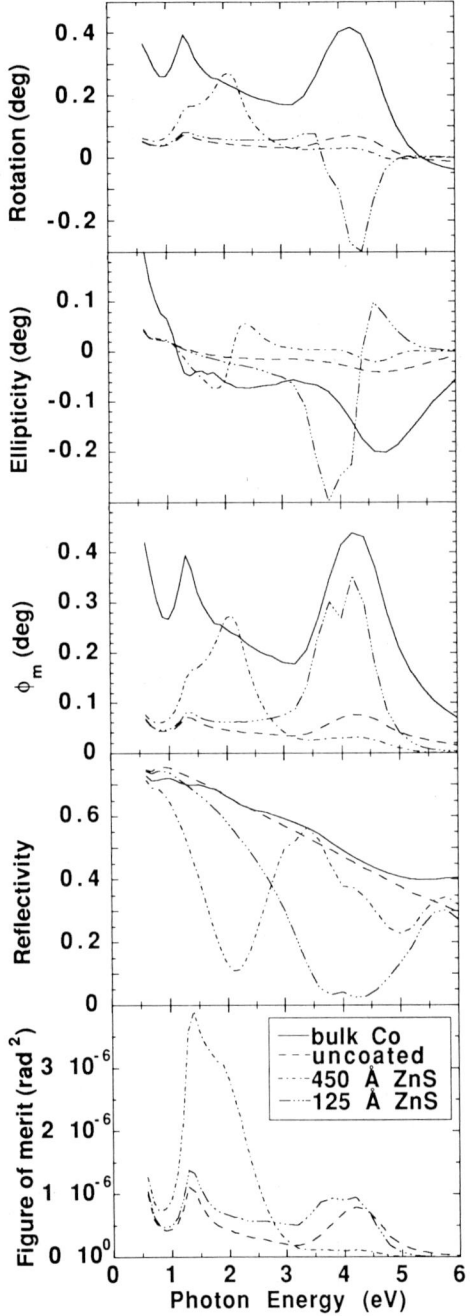

Fig. 13. Polar Kerr simulations for a Co/Pt superlattice with and without ZnS coating and for bulk Co, as described in ref. [82]. The superlattice is [Co(4.1 Å)/Pt(19 Å)]$_{25}$ on a Pt substrate.

interesting to obtain SMOKE signals from monolayer-range thicknesses of rare-earth metals. Spiral spin structures in heavy rare-earth metals

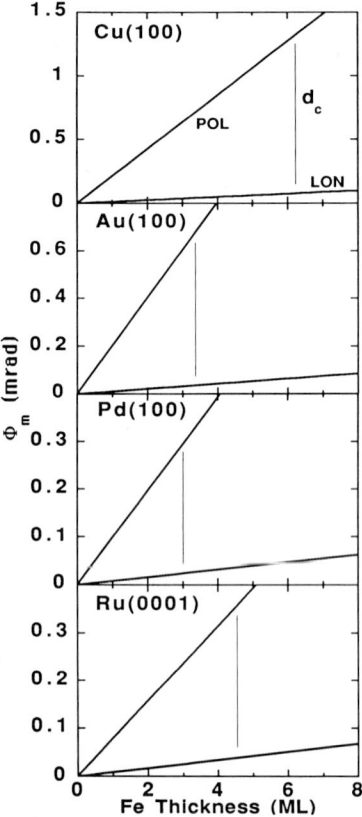

Fig. 14. SMOKE simulations for Fe overlayers based on tabulated bulk optical and magneto-optical constants. The experimental d_c-values from fig. 6 are indicated where the easy axes reorient from vertical to in-plane and remanent Kerr signals change from being polar (POL) to longitudinal (LON).

should be altered or suppressed when the film thickness approaches the spiral pitch length. While these are some ideas, the best future directions probably have not been dreamt of yet.

The purpose of this review was to acquaint the reader with SMOKE and to stimulate interest in the range of contemporary issues in surface magnetism and magneto-optics in the ultrathin limit that it has addressed.

Acknowledgements

This work was supported by the US Department of Energy, Basic Energy Sciences-Materials Sciences, under contract No. W-31-109-ENG-38.

The author thanks his coworkers M.E. Brubaker, C. Liu, E.R. Moog, J. Pearson, Z.Q. Qiu, K.J. Strandburg and J. Zak for their help, and those who sent preprints of their work.

References

[1] E.R. Moog and S.D. Bader, Superlattices Microstruct. 1 (1985) 543.
S.D. Bader, E.R. Moog and P. Grünberg, J. Magn. Magn. Mater. 53 (1986) L295.
S.D. Bader and E.R. Moog, J. Appl. Phys. 61 (1987) 3729.

[2] M. Faraday, Trans. Roy. Soc. (London) 5 (1846) 592.
J. Kerr, Phil. Mag. 3 (1877) 339, 5 (1878) 161.

[3] L.D. Landau and E.M. Lifshitz, Electrodynamics of Continuous Media (Pergamon, London, 1960).
C.C. Robinson, J. Opt. Soc. Am. 53 (1963) 681, 54 (1964) 1220, 58 (1968) 1342.
G. Metzger, P. Pluvinage and R. Torguet, Ann. de Phys. 10 (1965) 5.

[4] P.N. Argyres, Phys. Rev. 97 (1955) 343.
G.H.O. Daalderop, F.M. Mueller, R.C. Albers and A.M. Boring, J. Magn. Magn. Mater. 74 (1988) 211.

[5] D.S. Bloomberg and G.A.N. Connell, in: Magnetic Recording Handbook: Technology and Applications, eds. C.D. Mee and E.D. Daniel (McGraw-Hill, New York, 1990) p. 530.
M.H. Kryder, J. Appl. Phys. 57 (1985) 3913.

[6] J. Araya-Pochet, C.A. Ballentine and J.L. Erskine, Phys. Rev. B 38 (1988) 7846.

[7] Y. Suzuki, H. Kikuchi, T. Katayama and S. Yoshida, J. Appl. Phys. 67 (1990) 5394.

[8] C.A. Ballentine, R.L. Fink, J. Araya-Pochet and J.L. Erskine, Appl. Phys. A 49 (1989) 459.

[9] Z.Q. Qiu, J. Pearson and S.D. Bader, Bull. Am. Phys. Soc. 36 (1991) 677.
Z.Q. Qui, J. Pearson and S.D. Bader, Magnetic phase transition of ultrathin Fe films on Ag(111), preprint.

[10] C. Liu and S.D. Bader, J. Vac. Sci. Technol. A 8 (1990) 2727.

[11] J.-P. Qian and G.-C. Wang, J. Vac. Sci. Technol. A 8 (1990) 4117.

[12] P.A. Montano, G.W. Fernando, B.R. Cooper, E.R. Moog, H.M. Naik, S.D. Bader, Y.C. Lee, Y.N. Darici, H. Min and J. Marcano, Phys. Rev. Lett. 59 (1987) 1041.

[13] C. Liu, E.R. Moog and S.D. Bader, Phys. Rev. Lett. 60 (1988) 2422.

[14] T. Beier, H. Jahrreiss, D. Pescia, T. Woike and W. Gudat, Phys. Rev. Lett. 61 (1988) 1875.

[15] D. Kerkmann, Appl. Phys. A 49 (1989) 523.

[16] C.M. Schneider, P. Bressler, P. Schuster, J. Kirschner, J.J. de Miguel and R. Miranda, Phys. Rev. Lett. 64 (1990) 1059.

[17] M.T. Kief, G.J. Mankey and R.F. Willis, J. Appl. Phys. 69 (1991) 5000.

G.J. Mankey, M.T. Kief and R.F. Willis, J. Vac. Sci. Technol. A 9 (1991) 1595.

[18] Q. Chen, M. Onellion and A. Wall, Magnetic properties of Co/Cu(111) and Co/Cu(100) systems, preprint.

[19] C.A. Ballentine, R.L. Fink, J. Araya-Pochet and J.L. Erskine, Phys. Rev. B 41 (1990) 2631.

[20] C. Liu and S.D. Bader, J. Appl. Phys. 67 (1990) 5758.

[21] C. Liu and S.D. Bader, Phys. Rev. B 41 (1990) 553.

[22] M. Farle, A. Berghaus, Y. Li and K. Baberschke, Phys. Rev. B 42 (1990) 4873.

[23] G.P. Felcher, Y.Y. Huang and C. Liu, Bull. Am. Phys. Soc. 36 (1991) 677.

[24] A. Cebollada, R. Miranda, C.M. Schneider, P. Schuster and J. Kirschner, Experimental evidence of an oscillatory magnetic coupling in Co/Cu/Co epitaxial layers, preprint.

[25] B. Heinrich, Z. Celinski, J.F. Cochran, W.B. Muir, J. Rudd, Q.M. Zhong, A.S. Arrott and K. Myrtle, Phys. Rev. Lett. 64 (1990) 673.

[26] W.R. Bennett, W. Schwarzacher and W.F. Egelhoff, Jr., Phys. Rev. Lett. 65 (1990) 3169.

[27] J. Zak, E.R. Moog, C. Liu and S.D. Bader, J. Magn. Magn. Mater. 88 (1990) L261.

[28] Q.-M. Zhong, A.S. Arrott, B. Heinrich and Z. Celinski, J. Appl. Phys. 67 (1990) 4448.

[29] Q. Chen, M. Onellion and A. Wall, CrO_2/Co multilayers on Cu(100), preprint.

[30] J. Ferré, G. Pénissard, C. Marlière, D. Renard, P. Beauvillain and J.P. Renard, Appl. Phys. Lett. 56 (1990) 1588.

[31] M.T. Kief, G.J. Mankey and R.F. Willis, J. Appl. Phys. 67 (1990) 5416 [abstract].

[32] J.F. Dillon, Jr., in: Magnetic Properties of Materials, ed. J. Smith (McGraw-Hill, New York, 1971) p. 149.

[33] K.H.J. Buchow, J. Less-Common Met. 155 (1989) 307.

[34] D.S. Kliger, J.W. Lewis and C.E. Randall, Polarized Light in Optics and Spectroscopy (Academic Press, Boston, 1990).

[35] J. Zak, E.R. Moog, C. Liu and S.D. Bader, J. Magn. Magn. Mater. 89 (1990) 107.

[36] J. Zak, E.R. Moog, C. Liu and S.D. Bader, Phys. Rev. B 43 (1991) 6423.

[37] W.A. McGahan and J. Woollam, Appl. Phys. Commun. 9 (1989) 1.

[38] M. Mansuripur, J. Appl. Phys. 67 (1990) 6466.

[39] L.M. Falicov, D.T. Pierce, S.D. Bader, R. Gronsky, K.B. Hathaway, H.J. Hopster, D.N. Lambeth, S.S.P. Parkin, G. Prinz, M. Salamon, I.K. Schuller and R.H. Victora, J. Mater. Res. 5 (1990) 1299.

[40] S.D. Bader, Proc. IEEE 78 (1990) 909.

[41] See, for instance, E.R. Moog, C. Liu, S.D. Bader and J. Zak, Phys. Rev. B 39 (1989) 6949.

[42] J.A.C. Bland, M.J. Padgett, R.J. Butcher and N. Bett, J. Phys. E 22 (1989) 308.

[43] S.N. Jasperson and S.E. Schnatterly, Rev. Sci. Instr. 40 (1969) 761, erratum 41 (1970) 152.
K. Sato, Jpn. J. Appl. Phys. 20 (1981) 2403.
P.Q.J. Nederpel and J.W.D. Martens, Rev. Sci. Instr. 56 (1985) 687.

[44] J.F. Dillon, Jr., E.M. Gyorgy, F. Hellman, L.R. Walker and R.C. Fulton, J. Appl. Phys. 64 (1988) 6098.
K.B. Lyons, J. Kwo, J.F. Dillon, Jr., G.P. Espinosa, M. McGlashan-Powell, A.P. Ramirez and L.F. Schneemeyer, Phys. Rev. Lett. 64 (1990) 2949.

[45] S. Spielman, K. Fesler, C.B. Eom, T.H. Geballe, M.M. Fejer and A. Kapitulnik, Phys. Rev. Lett. 65 (1990) 123.

[46] M. Mansuripur, F. Zhou and J.K. Erwin, Appl. Opt. 29 (1990) 1308.

[47] E.R. Moog, J. Zak, M.L. Huberman and S.D. Bader, Phys. Rev. B 39 (1989) 9496.

[48] C.L. Fu, A.J. Freeman and T. Oguchi, Phys. Rev. Lett. 54 (1985) 2700.

[49] O. Eriksson, R.C. Albers and A.M. Boring, Phys. Rev. Lett. 66 (1991) 1350.

[50] S. Blügel, M. Weinert and P.H. Dederichs, Phys. Rev. Lett. 60 (1988) 1077.
S. Blügel, B. Drittler, R. Zeller and P.H. Dederichs, Appl. Phys. A 49 (1989) 547.

[51] R.H. Victora and L.M. Falicov, Phys. Rev. B 31 (1985) 7335.

[52] E.R. Moog, S.D. Bader, P.A. Montano, G. Zajac and T.H. Fleisch, Superlattices Microstruct. 3 (1987) 435.

[53] A.J. Freeman and C.L. Fu, J. Appl. Phys. 61 (1987) 3356.

[54] V.L. Moruzzi, Phys. Rev. Lett. 57 (1986) 2211.
V.L. Moruzzi, P.M. Marcus, K. Schwarz and P. Mohn, Phys. Rev. B 34 (1986) 1784.
V.L. Moruzzi, P.M. Marcus and P.C. Pattnaik, Phys. Rev. B 37 (1988) 8003.
V.L. Moruzzi, P.M. Marcus and J. Kübler, Phys. Rev. B 39 (1989) 6957.

[55] C.S. Wang, B.M. Klein and H. Krakauer, Phys. Rev. Lett. 54 (1985) 1852.

[56] F.J. Pinski, J. Staunton, B.L. Gyorffy, D.D. Johnson and G.M. Stocks, Phys. Rev. Lett. 56 (1986) 2096.

[57] G. Binasch, P. Grünberg, F. Saurenbach and W. Zinn, Phys. Rev. B 39 (1989) 4828.

[58] S.S.P. Parkin, R. Bhadra and K.P. Roche, Phys. Rev. Lett. 66 (1991) 2152.

[59] S.S.P. Parkin, N. More and K.P. Roche, Phys. Rev. Lett. 64 (1990) 2304.

[60] M.E. Brubaker, J.E. Mattson, C.H. Sowers and S.D. Bader, Appl. Phys. Lett. 58 (1991) 2306.

[61] Y. Wang, P.M. Levy and J.L. Fry, Phys. Rev. Lett. 65 (1990) 2732.

[62] P.M. Levy, J.L. Fry, and E.C. Ethridge, Interlayer magnetic coupling in transition-metal multilayered structures, preprint.

[63] L. Néel, Compt. Rend. 237 (1953) 1468.

[64] U. Gradmann, J. Magn. Magn. Mater. 54–57 (1986) 733.

[65] J.G. Gay and R. Richter, Phys. Rev. Lett. 56 (1986) 2728.

[66] W. Karas, J. Noffke and L. Fritsche, J. de Chimie Phys. 86 (1989) 861.

[67] C. Li, A.J. Freeman, H.J.F. Jansen and C.L. Fu, Phys. Rev. B 42 (1990) 5433.

[68] M. Bander and D.L. Mills, Phys. Rev. B 38 (1988) 12015.

[69] C. Liu and S.D. Bader, J. Magn. Magn. Mater. 93 (1991) 307.

[70] N.D. Mermin and H. Wagner, Phys. Rev. Lett. 17 (1966) 1133, erratum 17 (1966) 1307.

[71] K. Binder and P.C. Hohenberg, IEEE Trans. Magn. MAG-12 (1976) 66.
K. Binder, Thin Solid Films 20 (1974) 367.

[72] K.J. Strandburg, D.W. Hall, C. Liu and S.D. Bader, Monte Carlo simulations of the Curie temperature of ultrathin ferromagnetic films, preprint.

[73] J. Li and R.B. McQuistan, Submonolayer magnetization and the spin lattice gas model, preprint.

[74] T.L. Einstein, in: Chemistry and Physics of Solid Surfaces VII, eds. R. Vanselow and R.F. Howe (Springer, Berlin, 1988) p. 307.

[75] M. Suzuki, Prog. Theor. Phys. 37 (1967) 770.

[76] J.M. Kosterlitz and D.J. Thouless, J. Phys. C 6 (1973) 1181.
J.M. Kosterlitz, J. Phys. C 7 (1974) 1046.

[77] J. Zak, E.R. Moog, C. Liu and S.D. Bader, Appl. Phys. Lett. 58 (1991) 1214.

[78] CRC Handbook of Chemistry and Physics, 60th ed. (CRC Press, Boca Raton, 1988–89) p. E-387.

[79] G.S. Krinchik and V.A. Artemev, Sov. Phys. JETP 26 (1968) 1080.

[80] E.R. Moog, S.D. Bader and J. Zak, Appl. Phys. Lett. 56 (1990) 2687.

[81] J.E. Mattson, M.E. Brubaker, C.H. Sowers, M. Conover, Z. Qiu and S.D. Bader, Temperature dependence of the magnetoresistance of sputtered Fe/Cr superlattices, Phys. Rev. B (in press).

[82] E.R. Moog, J. Zak and S.D. Bader, J. Appl. Phys. 69 (1991) 880.

[83] E.R. Moog, J. Zak and S.D. Bader, J. Appl. Phys. 69 (1991) 4559.

[84] T. Yoshino and S. Tanaka, Jpn. J. Appl. Phys. 5 (1966) 989.

[85] H. Feil and C. Haas, Phys. Rev. Lett. 58 (1987) 65.
J. Schoenes and W. Reim, Phys. Rev. Lett. 60 (1988) 1988.
H. Feil and C. Haas, Phys. Rev. Lett. 60 (1988) 1989.

[86] T. Katayama, Y. Suzuki, H. Awano, Y. Nishihara and N. Koshizuka, Phys. Rev. Lett. 60 (1988) 11426.
R. Nies and F.R. Kessler, Phys. Rev. Lett. 64 (1990) 105.
T. Katayama, Y. Suzuki and Y. Nishihara, Phys. Rev. Lett. 64 (1990) 106.

[87] W.B. Zeper, F.J.A.M. Greidanus, P.F. Carcia and C.R. Fincher, J. Appl. Phys. 65 (1989) 4971.
S. Hashimoto and Y. Ochiai, J. Magn. Magn. Mater. 88 (1990) 49.

[88] F. Schmidt, W. Rave and A. Hubert, IEEE Trans. Magn. MAG-21 (1985) 1596.
D.A. Herman and B.E. Argyle, IEEE Trans. Magn. MAG-22 (1986) 772.
W.W. Clegg, N.A.E. Heyes, E.W. Hill and C.D. Wright, J. Magn. Magn. Mater. 95 (1991) 49.

Behavior of vertical-Bloch-line chains of hard domains in garnet bubble films

Bao Shan Han

Magnetism Laboratory, Institute of Physics, Chinese Academy of Sciences, Beijing 100080, China

As a main micromagnetic structure of domain walls in garnet bubble films, vertical-Bloch-line (VBL) chains play an important role in the characteristics of hard domains. In the 1970s, their study progressed rapidly during the period in which bubble devices were developed. When ultra-high-density Bloch line memory (BLM) was proposed in 1983, VBL chain behavior again attracted attention. This review will introduce the early achievements focusing primarily on the results of the last decade. A convenient method of forming VBL chains, a new classification scheme of hard domains and the unsolved "number effect" of VBLs will be discussed. Various behaviors of VBL chains under static compression, and in-plane magnetic fields will also be presented. Finally, the temperature stability of VBL chains will be reported.

1. Introduction

The potential of bubble domains for computer memory was first proposed by Bobeck in 1967 [1]. A rapid development of bubble technology lasted for more than a decade. In that period, the study of domain wall physics was promoted by the promise and achievements of bubble devices. In addition, the high quality of the transparent garnet bubble films was helpful in the experimental and theoretical study of domain walls. As a result, understanding of the micromagnetic internal structure of domain walls and domain-wall dynamics was greatly advanced. The book, "Magnetic Domain Walls in Bubble Materials" written by Malozemoff and Slonczewski in 1979 [2], is a comprehensive review of the achievements gained before and during this splendid period. The contents of this book will be quoted often in this review.

As discussed in ref. [2], garnet bubble films can contain Bloch domain walls and some complex micromagnetic structures, such as VBLs and Bloch points (BPs). A Bloch domain wall is the boundary region separating domains of opposite magnetic orientation. In this region the spontaneous magnetization M lies in the wall plane and rotates from up ($\theta = 0$) to down ($\theta = \pi$). A Bloch wall has two different, but energetically equal polarities, $\psi = 0$ or π, corresponding to two types of magnetization rotation in the wall. A VBL is the transition region separating two such different Bloch-wall polarities and running throughout the film thickness. In the same way that two polarities exist for Bloch walls a VBL can have two types of magnetic handedness. By analogy, we can consider a BP separating two sections of VBL with different magnetic handedness.

The static and dynamic properties of stripe-shaped (stripe) and round-shaped (bubble) domains in garnet bubble films are appreciably affected by the VBLs contained in their walls. Same signed (i.e. with a common twist sense) VBLs can link up together via magnetostatic and exchange interactions to form VBL chains. VBL chains give rise to interesting phenomena, such as hard bubbles [3–5], and dumbbell domains [6,7].

In the early 1970s, the incidental formation of "hard" bubbles with many VBLs in their walls brought a serious problem to bubble devices, and also generated enthusiasm for the study of hard domains and VBL chains. Fortunately, this prob-

0304-8853/91/$03.50 © 1991 – Elsevier Science Publishers B.V. All rights reserved

lem was soon solved. Hard bubbles can be suppressed by means of ion implantation. This produces a thin surface layer of in-plane magnetization to prevent VBL chains from forming. Afterwards, some attention was paid to hard domains and VBL chains.

Our Chinese group began to study hard domains in garnet bubble films in the early 1980s. By that time, due to declining interest in bubble technology, almost all the groups in this field had given up the study of hard domains. We began our studies because we had learned to form either normal (soft) bubbles or different hard domains, easily reproducible by means of the series bias pulse method [8–11]. This method enabled us to study their behavior statistically. In 1983, Konishi [12] proposed ultra-high-density Bloch line memory (BLM). With this application possibility, the significance of our studies became obvious. For instance, hard stripe domains with natural VBL chains provide a good simulation of the minor loops (information storage loops) of BLM. We were thus determined to use hard stripe domains to study the temperature stability of VBLs, the information carriers of BLM. This study led to the discovery of the critical temperature T_0 for the break-down of VBL chains [13]. We were also interested in the VBL capacity limit of BLM; how many VBLs could be contained in walls of stripe domains and how did the number of VBLs affect the characteristics of stripe domains? For this purpose, a new classification scheme of hard domains has been proposed [14–16]. On this basis, we have performed and are performing a series of experiments on various hard domains to further reveal the behavior of VBL chains.

The focus of our review will be the static behavior of VBL chains of hard domains in typical 5 μm liquid-phase-epitaxy (LPE) as-grown garnet bubble films. We will divide the review into five sections: formation of VBL chains; classification of hard domains; static compression of VBL chains; behavior of VBL chains subjected to an in-plane field; temperature dependence of formation and break-down of VBL chains. We will briefly summarize the early work, but the emphasis will be on the results of the last decade.

2. Formation of VBL chains

In contrast to soft bubbles with few VBLs, hard bubbles have many VBLs in their walls. As described above, these VBLs link up together to form VBL chains. Because VBL chains raise the wall energy, hard bubbles are characterized by high collapse fields, low mobility and propagation at an angle to a field gradient [3–5]. If the number of VBLs becomes even larger, the bubble diameter may become so large that the bubble will distort into a "dumbbell" shape. Dumbbell domains have their own characteristics [2,6,7]. The schematic wall structures of soft bubbles, hard bubbles and dumbbells are illustrated in fig. 1.

There are several methods to form hard domains in garnet bubble films. Nishida et al. [17] cut the stripe domains under parallel conductors with a single bias pulse to generate a few hard

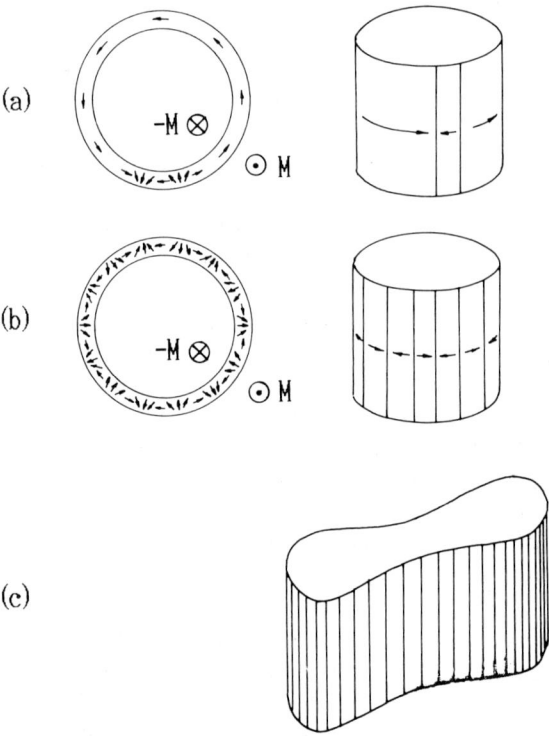

Fig. 1. Schematic wall structures with VBLs: (a) soft bubble with two negative VBLs (its winding number $S = 0$), (b) hard bubbles and (c) dumbbells.

bubbles, whose hardness can be controlled by changing the angle between the stripe domains and the conductors. Tabor et al. [3] proposed a method by which a single normal bubble is expanded by a strong bias field into a multi-branched domain which is then contracted into a hard bubble. A similar but less controlled method [18] is to rapidly demagnetize a sample from its saturated state to form a highly branched stripe pattern. However, the above three methods can only form a few hard bubbles.

The pulse-bias-field method to form a group of hard bubbles was first presented by Nishida et al. [4]. We adopted and developed their method. Hard domains were formed by applying a number (n_p) of rectangular pulse bias fields with the amplitude H_p and width τ_p, parallel to a fixed static bias field $(H_b)_0$, to stripe domains through a 10 turn pancake coil with an inner diameter of 1 mm [8]. With this arrangement, we further studied the formation of hard domains in garnet bubble films [8,9]. We found that hard bubbles can be formed only over a certain amplitude range of H_p, and there exists a critical amplitude $(H_p)_0$ above which hard bubbles are no longer formed, i.e. all bubbles generated are soft bubbles [8]. In fact, at the end of the pulses there is a group (several tens) of domain segments left inside the coil, as shown in fig. 2a. All of these segments can be contracted into soft bubbles and are thus referred to as "soft domain segments". Based on this, Han et al. [10] investigated the formation of hard bubbles by applying a single pulse bias field to soft domain segments. This work delimited the region of hard bubble formation. The authors found that an impressive fraction of soft segments may be hardened by applying just one pulse. This single-pulse method provides the simplest method to excite VBL chains in the walls of soft domain segments. Later, Nie et al. [11] showed how to use the series-pulse method to form hard bubbles and dumbbell domains. In sum, the above studies [8–11] provide a convenient way to control the formation and break-down of VBL chains.

The procedures to form both soft and hard domains have been described in detail in ref. [13]. First the fixed static bias field $(H_b)_0$ is carefully determined to avoid forming multi-branched domains with VBLs. Second, at the fixed $(H_b)_0$, the amplitude H_p is changed to determine the soft and hard bubble formation regions SB and HB. In fact, the regions SB and HB are functions of experimental variables, such as pulse width τ_p, in-plane field H_{ip} (see fig. 9 below), temperature T (see fig. 3 of ref. [13]). In our experiments, τ_p is usually selected as 0.5 µs with about 50 ns rise and fall times. Third, when the total bias field is set at the proper position in the region SB, a group of soft segments can be reproduced and easily formed in seconds, as shown for a typical sample in fig. 2a. The soft segments are then hardened by series bias pulses. The typical hardened domain pattern is shown in fig. 2b for the same sample. To achieve higher hardening fraction, the amplitude, H_p, is usually selected in the middle of the region HB. With an increase of n_p, hard bubbles and dumbbells can be successively formed. Once the appropriate $(H_b)_0$, H_p and n_p are decided, tens of hard domains can be easily

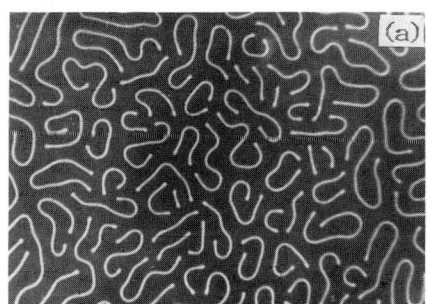 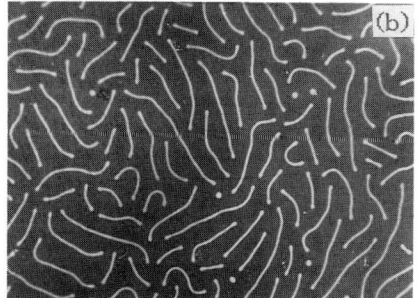

Fig. 2. Photos of (a) soft domain segments; (b) the domain segments having just undergone series bias pulses ($H_p = 44.6$ Oe, $\tau_p = 0.5$ µs and $n_p = 100$). These are a mixture of soft and hard segments. (Sample No. 2, $(H_b)_0 = 85.5$ Oe, RT.)

formed and changed. Although one group of hard domains is not exactly the same as another, their basic characteristics, such as the hardening fraction and the maximum collapse field $H_{\text{col,max}}$, are nearly the same. With experience using this method, we can select desired groups with the consistency in domain numbers. In other words, the series-pulse method enables us to study hard domains and the behavior of VBL chains statistically.

Now let us discuss the mechanism of VBL formation. We can imagine that during the rise time of the pulse, all soft segments, such as those in fig. 2a, shrink to some degree. They then undergo a rapid expansion and move to their new equilibrium positions during the fall time of the pulse. The non-uniform pulse field inside the pancake coil, and the mutual repelling forces between segments result in many types of soft segment "movements". Han et al. [10] developed a double-exposure photography method to statistically reveal the dependence of soft segment hardening on the movements caused by a single pulse. By analyzing more than 200 hard bubbles of sample No. 1 of nominal composition $(YSmCa)_3(FeGe)_5O_{12}$, they found that among the various types of movements, the two movements favoring for hard bubble formation are "medium head swing" and "bow contraction", as shown in fig. 2 of ref. [13].

In our cases, because there exists a non-uniform driving force needed for Bloch-line wind-up, although no cutting occurs, we also question the resulting preponderance of Bloch lines of one sign [2]. Despite the fact that the wind-up process of VBL chains is still unknown, ref. [2] has already provided one speculation perhaps, "one possibility is a Bloch-line annihilation in regions of high Bloch-line density by means of a generation of Bloch points, leading statistically to a preponderance of Bloch lines of one sign over the other". Considering that the role played by BPs in VBL annihilation is still not clear in some cases, as discussed in the following sections, we prefer to modify the above speculation by using "a process like the collapse of a hard bubble" to replace that of "a generation of Bloch points". Thus we postulate that only during the "medium head swing" and "bow contraction" can high Bloch-line density regions appear with high probability in the walls of soft segments, which are thus hardened. In the SB region, as described in ref. [10], the movements of soft segments are so intense, due to the large H_p, that favoring the formation of VBL chains, these movements no longer occur. Thus, they remain a group of soft segments.

3. Classification of hard domains

As introduced above, in the early 1970s hard bubbles and dumbbells were widely studied [3–7,2]. The typical studies regarding dumbbells were presented by Slonczewski et al. [6]. As one of their results, the shapes and sizes of hard domains vs. bias field in a 5.25 μm thick film of composition $(Tb_{0.04}Eu_{0.66}Y_{2.3})(Fe_{3.85}Ga_{1.15})O_{12}$ are shown in fig. 2 of ref. [6]. Actually this figure nicely outlined the essential features of all kinds of hard domains. They found that some dumbbells become elliptical and eventually turn into bubbles and collapse as the bias field is increased. But these dumbbells were classified under the same category as "hard bubbles". Thus their term "dumbbells" only denotes those dumbbells which undergo an instability at approximately the same field.

During our studies of hard domain formation, we were impressed by the substantial difference in the characteristics of various hard domains and considered that they should be better classified. In particular, we were interested in those dumbbells which appear to turn into bubbles. Han et al. [14] performed a so-called "exercise" experiment to check whether or not they remain "dumbbells" after all of them appear to turn into bubbles. The impressive fact is that for the films of composition $(YSmCa)_3(FeGe)_5O_{12}$, they remain dumbbells rather than become hard bubbles. Therefore, these dumbbells should be classified as a kind of dumbbell (ID) rather than hard bubbles. Later, Nie et al. [15,16] proposed that the family of hard domains in garnet bubble films can be divided into three types, the ordinary hard bubbles (OHB) and their corresponding stripe domains, the first, and the second kind of dumbbells (ID and IID). The classification criteria and formation conditions were also given [15,16].

We now briefly review the classification criteria [16]. In order to distinguish between OHBs and dumbbells better, we define $H'_{sb} = H_{sb} + 0.01 \times 4\pi M$ as the "standard stripe-to-bubble transition field", where H_{sb} is the maximum stripe-to-bubble transition field of soft bubbles and $4\pi M$ is the saturated magnetization. The conventional criterion for "hard" bubbles is that their collapse field H_{col} is greater than $H'_0 = H_0 + 0.01 \times 4\pi M$ [2], where H_0 is the minimum collapse field of soft bubbles. We define those hard domains, which can contract to a circular form as static bias field H_b is increased to H'_{sb} and whose collapse fields are distributed between H'_0 and the maximum collapse field, $H_{0,max}$, as "ordinary hard bubbles" (OHB). Furthermore, we define all of those hard domains, which exist in non-circular form, as H_b is increased to H'_{sb}, as "dumbbell domains", no matter whether they are in the forms of ellipses, dumbbells, S-shapes, C-shapes or other shapes [7]. Those dumbbells which eventually contract into circular form after H_b is greater than H'_{sb} and then collapse are referred to as "the first kind dumbbells" (ID). Their stripe-to-bubble transition fields are distributed over a range from H'_{sb} to a certain upper limit, $(H_{sb})_{ID,max}$. Finally, dumbbells which stably shrink in length but never contract to circular form and therefore eventually collapse in dumbbell form are called "the second kind of dumbbells" (IID).

After carefully examining about 20 as-grown samples with four compositions [15,16], we have found that OHBs, IDs and IIDs can all be formed in samples of (i) $(YSmCa)_3(FeGe)_5O_{12}$ and (ii) $(BiYLu)_3(FeGa)_5O_{12}$; but in the samples of (iii) $(YSmLuCa)_3(FeGe)_5O_{12}$ and (iv) $(YSmLuCa)_3(FeGeGa)_5O_{12}$, only OHBs and IDs can be produced. For the materials (i) and (ii), sample No. 1 with 7.6 μm thickness and composition $(YSmCa)_3(FeGe)_5O_{12}$ is typical. Its static parameters at room temperature (RT) are: The characteristic length $l = 0.84$ μm, $4\pi M = 145$ G, and the uniaxial anisotropy constant $K_u = 4.19 \times 10^3$ erg/cm^3. The typical histograms of H_{col} for OHBs, IDs and IIDs of sample No. 1 are plotted in fig. 3 [16]. For the materials with compositions (iii) and (iv), typical is the 6.6 μm thick sample No. 2 of $(YSmLuCa)_3(FeGe)_5O_{12}$. Its $l = 0.68$ μm, $4\pi M = 220$ G and $K_u = 6.51 \times 10^3$ at RT. The H_{col} histograms for OHBs and IDs in sample No. 2 are shown in fig. 4 [16]. Typical photos of each hard domain type for sample No. 1 have been shown in fig. 2 of ref. [16]. Here we present typical photos for sample No. 2 in fig. 5 [15].

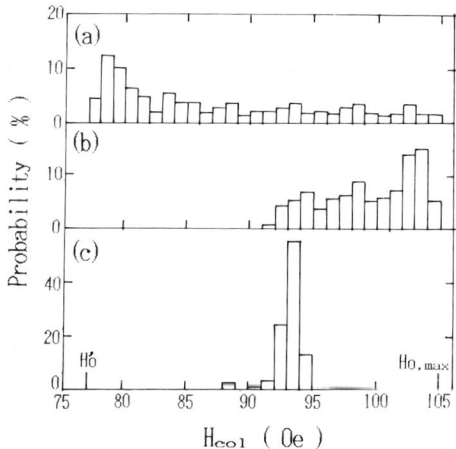

Fig. 3. Typical histograms of H_{col} for the hard domains of sample No. 1. (a) OHBs. Total number tested, $N = 215$, number of the series bias field pulses used, $n_p = 10$. (b) IDs. $N = 217$, $n_p = 100$. (c) IIDs. $N = 200$, $n_p = 600$. Other conditions used in formation: $(H_b)_0 = 53$ Oe, $H_p = 30$ Oe, $\tau_p = 0.5$ μs, RT (after Nie et al. [16]).

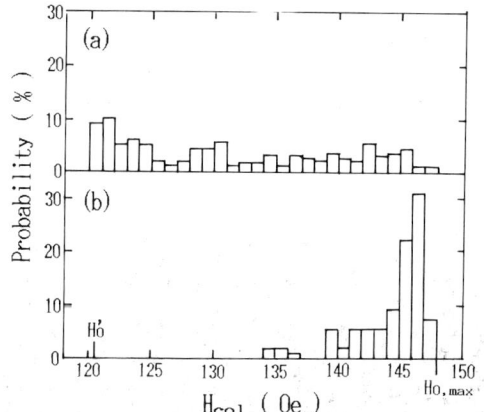

Fig. 4. Typical histograms of H_{col} for the hard domains of sample No. 2. (a) OHBs. Total number tested, $N = 199$; $n_p = 100$. (b) IDs. $N = 108$, $n_p = 5000$. $((H_b)_0 = 85.5$ Oe, $H_p = 44.6$ Oe, $\tau_p = 0.5$ μs, RT) (after Nie et al. [16]).

According to our classifications criteria, it is seen in fig. 5b and c that the only bubble is OHB, and the other five dumbbells, which will shrink into bubbles and then collapse, are IDs.

For the materials (i) and (ii), our experiments show that, for H_p and τ_p fixed, IIDs begin to be formed at larger n_p of bias pulses than IDs and dumbbells begin to be formed at larger n_p than OHBs [15]. For OHBs and IDs in the materials (iii) and (iv), the situation is the same [15]. This implies that the VBL number in their walls is successively increased from OHB to ID and to IID for the same sample. We believe that it is the change of VBL number of the VBL chains in hard domain walls that leads to the substantial variation of their static characteristics shown in figs. 3 and 4.

For OHBs, as is well-known, an increase of VBL number leads to an increase of their collapse fields [3–5,2]. As the VBL number exceeds that of the hardest bubbles, dumbbells appear, however, in contrast to OHBs, the collapse fields of IDs are decreased, as shown in fig. 3b. With this evidence, we have confirmed [19] that the peak of the H_{col} histograms of IDs for sample No. 1 moves down to lower bias fields with an increase of n_p, implying that for IDs, the more the VBL numbers, the lower their collapse fields. With a further increase of VBL number, IIDs appear; however, their walls can no longer contract into circular form, and will eventually collapse in dumbbell form within a narrow bias field range, as shown in fig. 3c. We believe that the same analysis is also valid for the OHBs and IDs of the sample No. 2 (see fig. 4), although no IIDs can be formed in it.

Following the new classification scheme, we have presented the "number effect" of VBL chains, i.e. substantial changes in static characteristics of hard domains arise from an increase of VBL number in VBL chains. Theoretically, this is still an unsolved problem in the field of micromagnetism. For BLM, negative VBL pairs in walls of stripe domains aligned in parallel ("minor loops") are used as information carriers [12]. Thus the following questions are raised: how many VBLs can be contained and how can these VBL pairs be arranged in minor loops without changing their static characteristics? This is actually a subject related to the chip organization and storage capacity limit of BLM.

In addition, it should be pointed out that the stability of IDs in static compression depends whether or not IIDs can be formed in the sample. Let us check the IDs of sample No. 1. Their H_{col} histograms (fig. 3b) are nearly complementary with their "exercise" curve (see fig. 4a of ref. [16]), indicating that the rapid fall of their "exercise" curve near $H_{0,max}$ is simply due to their collapse. Hence these IDs can contain more VBLs than the hardest bubbles in circular form and do not lose VBLs in static compression until collapse. Thus we will call the IDs of the materials (i) and (ii) "stable" IDs. In contrast, the "exercise" curve in fig. 4b of ref. [16] for the IDs of sample No. 2, shows that only about 40% of the IDs still remain "dumbbells" as the bias field reaches $(H_{sb})_{ID,max}$. Moreover, that curve is not complementary with the H_{col} histograms (fig. 4b). These facts show that with the increase of bias field, more and more IDs lose VBLs and become OHBs, and all IDs become OHBs before bias field reaches $H_{0,max}$. Thus, we will call the IDs of the materials (iii) and (iv) "less stable" IDs. Fi-

Fig. 5. The formation and typical photos of OHB and ID for sample No. 2 at RT. (a) $H_b = (H_b)_0$ (85.5 Oe). (b) $H_b = H'_{sb}$ (102.5 Oe). (c) $H_b = H'_0$ (120.5 Oe). The pulse conditions: $H_p = 44.6$ Oe, $\tau_p = 0.5$ μs, $n_p = 5000$ (after Nie et al. [15]).

nally, we should point out that our experiences from samples with ten compositions show that the formation of IIDs seems to depend only on composition. Unfortunately, limited sample resources restrict us from doing further study.

4. Static compression of VBL chains

The existence of VBL chains distinguishes hard domains from soft bubbles. In the high Q approximation, the equilibrium separation S_{eq} of VBLs was derived by Hubert [20] for a large finite VBL chain in an infinite plane wall:

$$S_{eq} = \sqrt{2}\,\pi\left[1 + (2Q)^{-1}\right]^{-1/2} \Lambda_0, \quad (1)$$

where Λ_0 is the width parameter of VBL, Q is the quality factor, $Q = H_k/4\pi M$, where $H_k = 2K_u/M$ is uniaxial anisotropy field. Hubert [20] also verified by computation that the structure of a periodic array of VBLs, like a VBL chain, is stable for the real separation $s \leqslant s_{eq}$, which is the case in the hard bubbles and dumbbells of garnet bubble films.

VBL chains break down under static compression during the collapse of hard bubbles. Apparently, the mechanism is different from the purely magnetostatic instability of normal bubbles [3,21,22]. A qualitative explanation [2] of the VBL compression behavior is as follows: As the bias field is increased above the normal collapse field H_0, the Bloch lines become squeezed together, increasing the Bloch wall energy far above its normal value $\sigma_0 = 4(AK_u)^{1/2}$, where A is exchange constant. Spontaneous annihilation of VBLs is assumed to occur at a critical local energy density w_c ($\approx 10^5$ erg/cm^3, which is almost an order of magnitude larger than the anisotropy energy in typical 5 μm garnet films). Bubble collapse occurs because once the Bloch

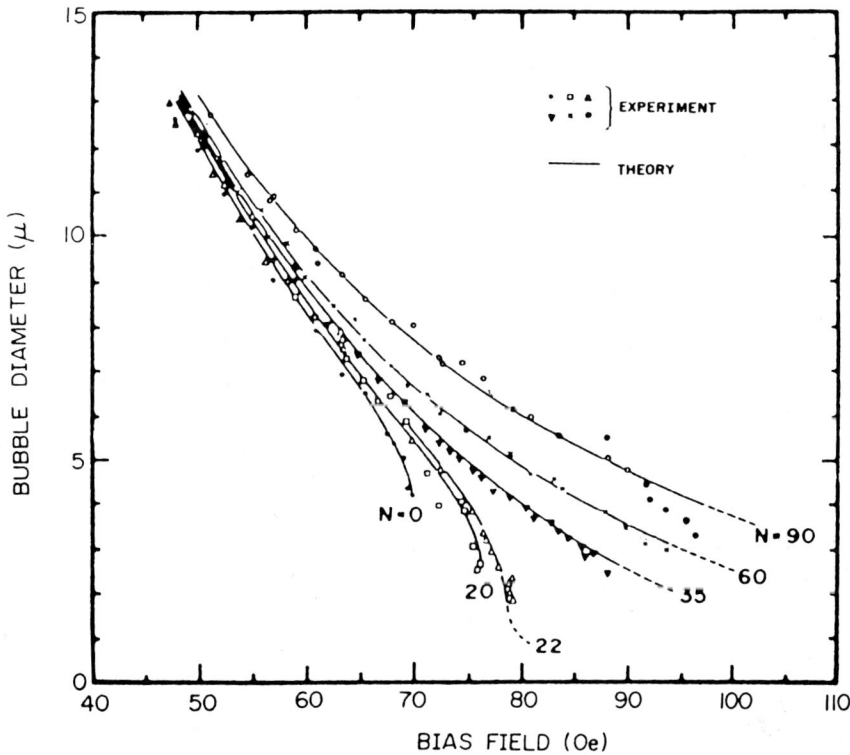

Fig. 6. Bubble diameter as a function of bias field for hard and normal ($N = 0$) bubbles in a EuErGaIG film. The solid lines are fits of eq. (2) to the experimental data. The curves are labeled by a winding number $N \equiv S - 1$ (after Kobayashi et al. [21]).

lines annihilate, the bubble is no longer stable above H_0. A quantitative description [3–5,2] involves the derivation of a new expression for the wall energy density with the VBL effect included to replace the normal one, σ_0, based on a uniform twist model. Thiele's normalized force equation [23] then becomes,

$$(\sigma_0/4\pi M^2 h)\left[1 + (2S\Delta_0/d)^2 + (1/2Q)\right]^{-1/2} + (d/h)(H_b/4\pi M) - F(d/h) = 0, \qquad (2)$$

where h is the thickness of the film, S is the winding number of the bubble [2], Δ_0 is the width parameter of the Bloch wall, d is the bubble diameter and F is the force function. By using eq. (2), the dependence of the diameter on bias field can be predicted for various bubbles. Then from the fit of experimental data with theoretical curves, the winding number S, i.e. the VBL numbers of hard bubbles can be determined. Fig. 6 is an example for a EuErGaIG sample [21]. In this figure, values of the winding number S up to 90 are found, corresponding to 180 VBLs in the wall of the hardest bubbles. Another example is shown in fig. 3 of ref. [10] for sample No. 1. From the fit, we found that there are 90 VBLs in its hardest bubbles whose diameters range from 2.2 (at $H_{0,\max}$) to 11 μm (at H_{sb}). By using eq. (1) and the data of sample No. 1, the calculated S_{eq} and s at H_{sb} and $H_{0,\max}$ are 0.79, 0.38 and 0.08 μm, respectively at RT. This work confirms that $s \leqslant S_{eq}$ and also gives us an idea how closely packed the VBL chains are just before collapse. Finally, we should point out that this quantitative work fails to predict the collapse fields and diameters of hard bubbles at collapse, and some other details [5].

On the collapse of dumbbells, the qualitative understanding is the same as that for hard bubbles [6]. In addition, the work done by Slonczewski et al. [6] should be introduced. They proposed a simple extension of Thiele's model [23] of elliptic and radial instability, by including the Bloch line energy term, to deal with the runout and collapse of the "hard bubbles" which lie above the d–H_b curve of the hardest bubble in fig. 2 of ref. [6]. According to our classification and discussion in the last section, these are "stable" IDs. The resultant theoretical curve of d/h vs. $H_b/4\pi M$ is in reasonable agreement with the experimental runout and collapse of these IDs. However, this effort did not establish a quantitative relation between the collapse field and VBL number for IDs. As described above, IIDs can stably shrink in length with an increase in H_b. During the shrink process, there is no sign to show the loss of VBLs. In contrast to IDs, the VBL chains of IIDs cannot be contained in circular walls, and all chains break down in a narrow range of bias fields. This feature undoubtedly arises from the very large number of VBLs in their chains. Unfortunately, the quantitative theoretical work, regarding IIDs is very sparse. We hope that the theory on the "number effect" of VBLs can explain the different behavior of VBL chains in collapse.

In connection with static compression there are two unique kinds of behavior of VBL chains: self-collapse of hard domains [24–26] and self-shrink of IDs [27]. The self-collapse of so-called superhard bubbles (most like OHBs and IDs with high collapse fields in our classification), i.e. their spontaneous shrink and then collapse with time, as the static bias field is set below, but close to their collapse field, was first investigated by Haisma et al. [24]. They proposed a qualitative explanation in terms of the creation and motion of BPs.

Considering that the "superhard" bubbles tested by Haisma et al. were formed under a large visible defect, we repeated the self-collapse experiments on "free" OHBs [25] and extended the experiments to IDs and IIDs [26]. The self-collapse percentage, R_{OHB}, of OHBs as a function of time with the treatment bias field $(H_b)_t$ as a parameter for sample No. 1 is plotted in fig. 7 [25]. It is seen in fig. 7 that self-collapse occurs only when $(H_b)_t > 100$ Oe. It other words, most OHBs with low H_{col} are unable to self-collapse, implying a sufficient high number of VBLs is necessarily for self-collapse. Moreover, when $(H_b)_t$ approaches $H_{0,\max}$ self-collapse occurs more easily and R_{OHB} increases faster, indicating that an increase of VBL density can speed up self-collapse. For the "stable" IDs and IIDs of

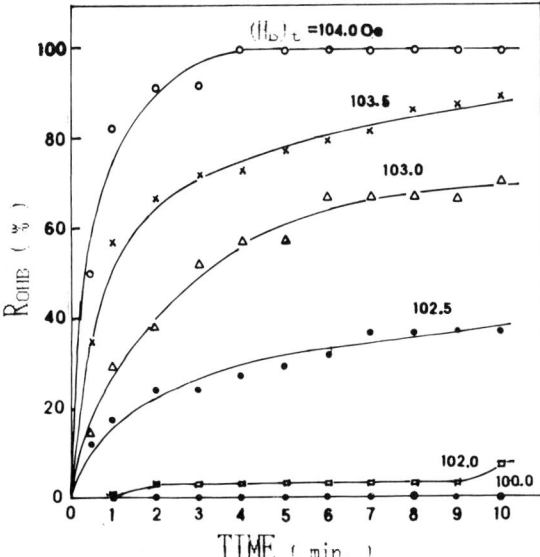

Fig. 7. The self-collapse percentage, R_{OHB}, of OHBs as a function of time with the treatment bias field, $(H_b)_t$, as a parameter for sample No. 1. ($H'_0 = 77$ Oe, $H_{0,max} = 105$ Oe at RT.)

sample No. 1, we found that their self-collapse region coincides with that of their collapse [26], showing that self-collapse can occur for any ID and IID formed in the same film. This is due to the large number of VBLs in their walls. For the "less stable" IDs, for instance, those formed in sample No. 2, the self-collapse behavior is similar to the sample's OHB behavior [25].

Finally, we should point out that, according to our calculation, the "insertion" energy, W_{bp}, [28] is usually 2 or more orders of magnitude higher than $k_B T$ for typical 5 μm bubble films even near 100°C (see the last section of this review). Hence, the thermal nucleation of BPs is unlikely, and the mechanism of self-collapse is still not clear.

Another unique behavior of VBL chains is so-called "self-shrink" of "less stable" IDs [27]. When the bias field is set below, but close to $(H_{sb})_{ID,max}$, we found that these IDs can spontaneously and gradually shrink into "bubbles". The results taken from sample No. 2 indicate that the closer the treatment bias field $(H_b)_t$ is to $(H_{sb})_{ID,max}$, the earlier self-shrink occurs and the faster the self-shrink percentage R_{ID} of IDs increases. A typical set of "time-lapse" photos for sample No. 2 is shown in fig. 8. From these photos, one can see that three of the four IDs in fig. 8a shrink into bubbles and only one is still in the dumbbell shape after 60 s. We also found [27] that the "bubbles" formed by ID self-shrink are a mixture of IDs (in circular form) and OHBs; IDs cannot entirely convert into OHBs by self-shrink. These results imply that the pressure exerted by $(H_b)_t$ ($< (H_{sb})_{ID,max}$) to squeeze IDs into circular form is "large" enough to make "less stable" IDs lose VBLs gradually and partially, giving rise to "self-shrink". Finally, we emphasize that no self-shrink occurs for "stable" IDs, showing again that the stability is different for two kinds of IDs in static compression.

5. Behavior of VBL chains subjected to an in-plane field

An in-plane magnetic field is an important factor affecting the behavior of VBL chains because it can move VBLs or change their structure. The effect of an in-plane field, H_{ip}, on an

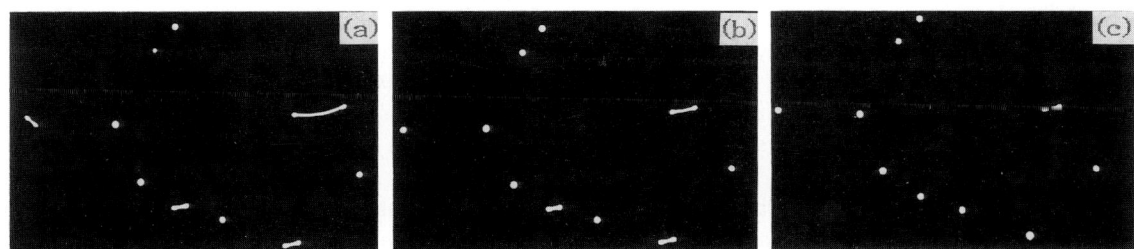

Fig. 8. A typical set of "time-lapse" photos of the self-shrink process of IDs for sample No. 2 with $(H_{sb})_{ID,max} = 132.5$ Oe at RT. $(H_b)_t = 130.5$ Oe. (a) $t = 0$ s. (b) $t = 20$ s. (c) $t = 60$ s (after Nie et al. [27]).

isolated VBL in a Bloch wall was described in ref. [2]: The component of the in-plane field parallel to the wall, H_x, favors one Bloch-wall polarity at the expense of the other. The force per unit length, $F_L = 2\pi\Delta_0 M H_x$, tends to displace the VBL which will continue to move until it gets to a point on the wall where the in-plane field is perpendicular to the wall or where it runs into other VBLs. The component of in-plane field perpendicular to the wall, H_y, will affect the static structure of the wall with VBLs, depending on the direction and magnitude of H_y. Because the position of VBLs can be controlled by H_{ip}, the application of H_{ip} plays an important role in the function operations of BLM [12,29].

The effect of an in-plane field, H_{ip}, on a bubble with two VBLs lying on opposite ends of the bubble diameter was summarized in ref. [2]. Because H_{ip} can stabilize the bubble states and lead to the movement of BPs, the application of H_{ip} is indispensable for the state transformation of bubbles with a few VBLs [30,2].

In ref. [2], only a short description was made for the cases of very large H_{ip}. Here we will discuss the effect of H_{ip} on the formation of hard bubbles [31] and the break-down of VBL chains at a critical in-plane field [32–34].

The simple single-pulse method [13,10] was used to study the influence of H_{ip} on the formation of VBL chains [31]. This experiment started from a group of soft segments at $(H_b)_0$ (see fig. 2a). When H_{ip} is applied, these segments prefer to align towards the direction of H_{ip}. Then the regions SB and HB were measured. The results for sample No. 1 are shown in fig. 9 [31]. It is seen in it that the region HB of hard bubble formation varies with the increase of H_{ip}. Above a critical in-plane field, H_{ip}^0, hard bubbles are no longer formed. For sample No. 1, as seen in fig. 9, $H_{ip}^0 = 380$ Oe and the normalized one $H_{ip}^0/H_k = 0.52$ at RT. To investigate these observations, we used double exposure photography to reveal statistically the dependence of soft segment hardening on their movements. With a great number of segments tested, we found that the movements favoring the formation of VBL chain are "small head swing" and "stretching", which are very different from the case with $H_{ip} = 0$ for the same

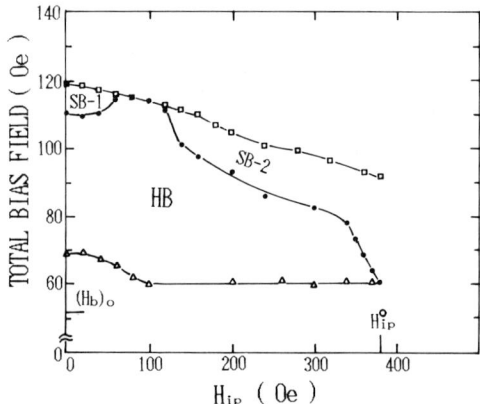

Fig. 9. The soft and hard bubble formation regions (SB and HB) vary with in-plane fields H_{ip} for sample No. 1 at RT (single-pulse method, $(H_b)_0 = 51.3$ Oe, $\tau_p = 0.5$ μs) (after Han et al. [31]).

sample [13,10]. By using the above statistical results, the variation of the region HB with H_{ip} can be qualitatively explained. On the critical in-plane field H_{ip}^0, it is believed that this in-plane field is so large as to suppress all kinds of movements favoring the formation of VBL chains.

Now let us turn to the behavior of VBL chains subjected to a very large H_{ip}. Obakata et al. [32] first observed that a sufficiently large in-plane field (a few hundred oersteds in a typical 5 μm garnet film) makes all hard bubbles turn into soft ones. Malozemoff and Slonczewski [2] pointed out that this is likely due to the field-nucleation of BPs. However, no precise threshold field for such a process had been known either experimentally or theoretically.

Later, an experimental study was performed by Li et al. [33]. They reduced the bias field to zero to elongate a group of hard domains formed by the series-pulse method [8,9]. Then an in-plane field, H_{ip}, was applied to them for 5 s. After H_{ip} was removed, the percentage ρ' of hard domains remaining was measured. With this procedure, the typical curve of ρ' vs. H_{ip} obtained is shown in fig. 10 [33]. It shows that there exists a critical in-plane field H_{ip}'. When H_{ip} is much lower than H_{ip}', the number of hard domains remains unchanged. When H_{ip} approaches H_{ip}', there is a transition region where more and more hard domains are softened. This region ends at H_{ip}',

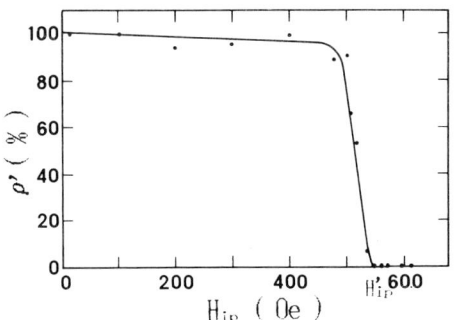

Fig. 10. The percentage ρ' of hard domain remaining vs. the in-plane field (extracted from Li et al. [33]).

above which all hard domains become soft, indicating the break-down of all VBL chains of these hard domains. For sample No. 1, H'_{ip} was measured to be 500 Oe at RT [34], giving $H'_{ip}/H_k = 0.69$, which is obviously greater than H^0_{ip}/H_k (0.52). In addition, it was observed that the value of H'_{ip} varies with the bias field, H_b, and hard domain segments can slowly contract into hard bubbles under appropriate H_b and H_{ip} [33]. Since the new classification scheme of hard domains was proposed [14–16], we can study the behavior of VBL chains subjected to an in-plane field for different types of hard domains. These experiments are now in process.

Concerning the mechanism of H'_{ip}, Yu et al. [34] measured H'_{ip} for eight films at RT and carried out a theoretical study to investigate the thermodynamic stability of hard walls under an in-plane field in the framework of micromagnetism by examining the positive definiteness of an appropriate 1-D homogeneous quadratic functional. They noted that H_{ip} makes hard stripe

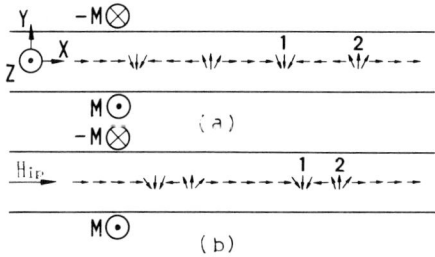

Fig. 11. (a) The equally spaced VBL structure when $H_{ip} = 0$. (b) The paired VBL structure when $H_{ip} > 0$ (after Yu et al. [34]).

domains, which tend to align along the field direction and it is assumed that H_{ip} makes the equally spaced VBLs of VBL chains paired, as shown in fig. 11 [34]. Their calculation, by means of finite element method, shows that when H_{ip} is greater than the critical value H'_{ip}, each pair becomes so close that the energy of this system is no longer positive definite. As a result, the VBL pair becomes unstable and the VBL chain breaks down. The derived theoretical H'_{ip}/H_k is uniquely determined by the quality factor Q of the films. However, an analysis of the topological stability of VBL structure [35] seems to be in conflict with the idea of ref. [34].

6. Temperature dependence of formation and break-down of VBL chains

The first study concerning the dependence of VBL formation on temperature was done by Henry et al. [36]. They found that there is a characteristic temperature, T_H, above which hard bubbles cannot be generated. However, they found it strange that the hard bubbles generated below T_H were still stable above T_H. This led to efforts to pursue "hard bubble free" garnet films [36,37]. Several years ago, Han et al. [13,38] studied the temperature dependence of the formation of VBL chains in garnet bubble films by means of the single-pulse method. The typical results were shown in fig. 3 of ref. [13]. In the experiment, they found that the region HB of hard bubble formation shrinks with temperature and finally vanishes at a critical temperature, T_{02}, above which hard bubbles are no longer formed, i.e. VBL chains cannot be excited. This T_{02} corresponds to T_H of ref. [35]. Meanwhile, due to the use of the single-pulse method they also found some details with regard to the formation of VBL chains and gave them a qualitative explanation.

In the early 1970s, certain bubble-lattice storage schemes were proposed. These devices relied on the long-term stability of VBLs because information was to be coded by means of the winding number S. This was the first time the question of VBL stability had been raised. However this question lost relevance when these bubble-lattice schemes were later given up. When the concept of BLM was proposed [12], this stability question was raised once again. Since a negative VBL pair is used as an information bit in BLM, a wide operating temperature range requires a study of the temperature stability for a number of VBL pairs in minor loops. As we know, the absence or presence of a pair of VBLs injected in the wall of a long stripe domain (a minor loop), acts as a binary "0" or "1", so that the two extreme cases for minor loops are all "0" and all "1". The former and the latter correspond to soft and hard stripe domains, respectively (if VBLs link up together). Despite the fact that VBL pairs are well separated and stabilized by "bit patterns" in minor loops of BLM, hard stripe domains with natural VBL chains are considered a good simulation of minor loops to investigate the temperature stability of VBLs before the model BLM is available.

For this purpose, hard stripe domains corresponding to OHBs were selected to study the temperature stability of VBL chains statistically. At zero bias field, a group of OHBs are elongated as stripe domains. The VBL chains in their walls are in a non-compressed equilibrium state. Temperature is then raised from RT to the testing temperature T and kept constant there. After two minutes, temperature is lowered to RT again. During this process, the softening fraction ρ and the maximum collapse field $H_{col,max}$ of this group of OHBs are measured. With this procedure, the curves of $H_{col,max}$ and ρ vs. T are obtained. From these two curves, the critical temperature T_0 for the break-down of VBL chains was first found in 1988 [13,39]. So far, tens of garnet films have been tested by this method. Fig. 12 shows typical curves for sample SEW-425 No. 11 [40,41]. The solid curves connect all the 2 min data together. It is seen in fig. 12 that when T is much lower than T_0, the VBL chains are unchanged after the temperature treatment. When T approaches T_0, there is a transition region where more and more hard stripe domains are softened. This region of several degrees of centigrade ends at T_0, above which all hard domains become soft, indicating the break-down of all VBL chains. It is important to point out that for all of our films tested, there

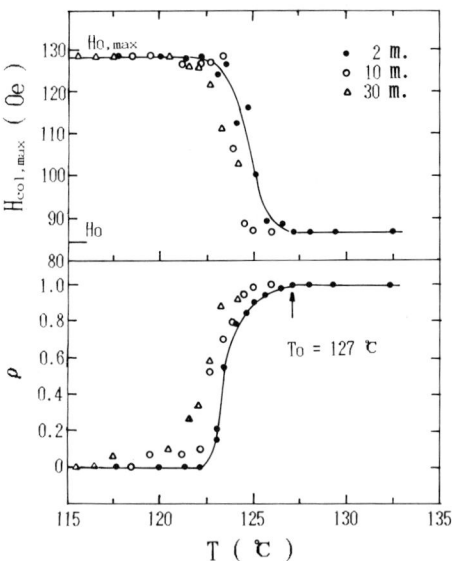

Fig. 12. For the different heat treatment duration, the maximum collapse field $H_{col,max}$ and the softening fraction ρ vs. the testing temperature T for the OHBs (86.1 Oe $< H_{col} \le$ 128.3 Oe) of sample SEW-425 No. 11 ((YSmLuCa)$_3$(FeGe)$_5$O$_{12}$, $l = 0.60$ μm, $4\pi M = 191$ Oe at RT = 29.9°C). The accuracy of temperature control of this experiment was ± 0.1°C over the range from 20 to 150°C (after Han et al. [41]).

is a relation: $T_{02} < T_0 < T_c$ (Curie temperature). For example, for the samples No. 1 and No. 2, these are 98, 120, 178 and 86, 94, 151°C, respectively. The fact that $T_{02} < T_0$ has apparently explained why the hard bubbles generated below T_{02} (T_H) are still stable above T_{02} [36]. On the other hand, the fact that $T_0 < T_c$ makes it necessary to study the mechanism of T_0 in order to improve the temperature stability of VBLs for developing BLM [12].

One possible mechanism for T_0 is the "Q-effect" [2], which says that "the Q factor usually decreases with temperature" and "low Q permits a larger wall width at the surface, thus giving a greater energy gain there for Bloch-point nucleation on energetically unfavored Bloch lines". However, among our 18 tested samples with seven compositions, 3 samples have Q values at T_0 greater than 4, 11 samples have Q values between 2 and 4 and only 3 samples have Q values between 1.2 and 2. This shows that the Q value is not too low for many of the samples.

It is simple to consider the break-down of VBL chains at T_0 as a consequence of BP thermal activation. To check this point, we have investigated the influence of the heat treatment duration on T_0 under constant testing temperature T. The results for different samples are similar. Typical results are shown in fig. 12 [40,41]. It is seen that for this sample $T_0 = 127\,°C$ and for longer duration (10 and 30 min) ρ begins to increase at lower T; near and within the transition region there is an obvious, but not very large increase and the increment for 30 min is greater than that for 10 min. As a result, T_0 is lowered by 2–3°C for the longer heat treatment duration. The above statistical results indicate the existence of a quasi-static athermal instability which leads to the break-down of VBL chains at T_0, with the thermal activation of BPs only of secondary importance. Afterwards, the "insertion" energy W_{bp} of a BP ($W_{bp} = 2\pi A^{3/2} K_u^{-1/2}(\ln Q + 1.90)$ [28]) at T_0 for this sample was calculated, giving $W_{bp} = 11 \times 10^{-12}$ erg, which is two and half orders of magnitude greater than the corresponding $k_B T_0$ (5.5×10^{-14} erg). If the calculated W_{bp} is correct, thermal nucleation of BPs is unlikely. However, fig. 12 shows that heat treatment duration does play a small role, indicating the existence of BP thermal activation to a certain degree. Hence the suitability of the above formular W_{bp} might be doubtful. Furthermore, if the activation of BPs were really the main reason causing the breakdown of VBL chains at T_0, T_0 would be appreciably decreased by the vibration of stripe domain walls, because the walls in vibration sweep over the micro-defects of the film and BPs are more easily activized near micro-defects [2]. However, the experimental data does not support this supposition [42].

The quasi-static athermal instability should be related to the variation of material parameters including their inhomogeneity with temperature. One possibility is that this instability may occur when the temperature approaches T_0 due to the formation of a capping layer with in-plane magnetization at the surface or interface of the film. If this is true, because of exchange coupling, hard stripe walls with VBL chains topologically demands a complex high-energy structure in the cap or interface that is strongly disfavored [2], i.e. VBL chains will break down. We hope that FMR experiments can be performed to precisely detect whether or not the capping layer exists around T_0. In addition, considering that VBL chains are always at a non-compressed equilibrium state in our experiments, and the reduction of VBL separation plays an important role in collapse and self-collapse of hard domains, the temperature dependence of the equilibrium separation S_{eq} [20] between VBLs has been calculated. It is interesting to note that T_0 seems to be the turning point for some samples; when $T > T_0$, the curve of S_{eq} falls faster or becomes flat [41,43]. This may be the first time that a relationship between T_0 and the material parameters has been found, which may contribute to the solution of the problem of the temperature stability of VBLs.

In summary, even though the behavior of VBL chains has been comprehensively studied and understood, many questions remain. The physical picture of the formation of VBL chains is still not clear. In addition, many observations, such as the "number effect" of VBLs, and the break-down of VBL chains in collapse and self-collapse, at the critical in-plane field, H'_{ip}, and the critical temperature, T_0, are far from explained. In particular, the role played by BPs is obscure. The answers to these questions await quantitative theories associated with the material parameters.

Acknowledgements

The author thanks all of his colleagues, collaborators and friends, in particular Prof. X.F. Nie, Prof. B.Z. Li, Mr. X.D. Xu and Mr. L. Tang for their assistance and support. The author acknowledges support from the National Natural Science Foundation of China. Thanks are also given to Prof. J. Engemann, Prof. A. Hubert, Dr. R. Dahlbeck, Dr. J. Ben Youssef, Dr. J.M. Desvignes and Dr. G. Vértesy for their support, collaboration and helpful discussions.

References

[1] A.H. Bobeck, Bell Syst. Tech. J. 46 (1967) 1901.
[2] A.P. Malozemoff and J.C. Slonczewski, Magnetic Domain Walls in Bubble Materials (Academic Press, New York, London, 1979).
[3] W.J. Tabor, A.H. Bobeck, G.P. Vello-Coleiro and A. Rosencwaig, Bell Syst. Tech. J. 51 (1972) 1427; AIP Conf. Proc. 10 (1973) 442.
[4] H. Nishida, T. Kobayashi and Y. Sugita, AIP Conf. Proc. 10 (1973) 493.
[5] R.F. Lacey, R.B. Clover, L.S. Cutler and R.F. Waites, AIP Conf. Proc. 10 (1973) 448.
[6] J.C. Slonczewski, A.P. Malozemoff and O. Voegeli, AIP Conf. Proc. 10 (1973) 458.
[7] F.G. West and D.C. Bullock, AIP Conf. Proc. 10 (1973) 483.
[8] Y.L. Liu, X.F. Nie and B.S. Han, Acta Phys. Sinica 29 (1980) 241 (in Chinese, with English abstract).
[9] T.D. Mao, J.Y. Li, S.Y. Fan and Y.L. Liu, Acta Phys. Sinica 32 (1983) 176 (in Chinese, with English abstract).
[10] B.S. Han, G.D. Tang, X.F. Nie and W. Xi, Acta Phys. Sinica 34 (1985) 1396 (in Chinese); English abstr., see Chin. Phys. 6 (1986) 232.
[11] X.F. Nie, G.D. Tang, J.W. Lin and B.S. Han, Acta Phys. Sinica 35 (1986) 338 (in Chinese); English abstr., see Chin. Phys. 6 (1986) 533.
[12] S. Konishi, IEEE Trans. Magn. MAG-19 (1983) 1838.
[13] B.S. Han, X.F. Nie, G.D. Tang and S.G. Huo, J. de Phys. 49 (1988) C8-1877.
[14] B.S. Han, X.F. Nie and D.P. Liu, Chin. Phys. Lett. 6 (1989) 329.
[15] X.F. Nie, G.D. Tang, X.D. Nie and B.S. Han, Acta Phys. Sinica 39 (1990) 296 (in Chinese, with English abstract).
[16] X.F. Nie, G.D. Tang, X.D. Niu and B.S. Han, J. Magn. Magn. Mater. 95 (1991) 231.
[17] H. Nishida, T. Kobayashi and Y. Sugita, IEEE Trans. Magn. MAG-9 (1973) 517.
[18] R. Wolfe and J.C. North, Bell Syst. Tech. J. 51 (1972) 1436.
[19] B.S. Han and X.F. Nie, private communication.
[20] A. Hubert, AIP Conf. Proc. 18 (1974) 178.
[21] T. Kobayashi, H. Nishida and Y. Sugita, J. Phys. Soc. Japan 34 (1973) 555.
[22] D.H. Smith and A.A. Thiele, AIP Conf. Proc. 18 (1974) 173.
[23] A.A. Thiele, Bell Syst. Tech. J. 48 (1969) 3287.
[24] J. Haisma, K.L.L. Van Mierloo, W.F. Druyvesteyn and U. Enz, Appl. Phys. Lett. 27 (1975) 459.
[25] G.D. Tang, X.F. Nie and B.S. Han, private communication.
[26] X.F. Nie, G.D. Tang and B.S. Han, Chin. Phys. Lett. (to be published).
[27] X.F. Nie, G.D. Tang and B.S. Han, ICM '91 (submitted).
[28] M. Margulies and J. Slonczewski, J. Appl. Phys. 49 (1978) 1912.
[29] Y. Hidaka, K. Matsuyama and S. Konishi, IEEE Trans. Magn. MAG-19 (1983) 1841.
[30] T.J. Beaulieu, B.R. Brown, B.A. Calhoun, T. Hsu and A.P. Malozemoff, AIP Conf. Proc. 34 (1976) 138.
[31] B.S. Han, J.W. Lin, B.Z. Li, X.F. Nie and G.D. Tang, Acta Phys. Sinica 35 (1986) 130 (in Chinese, with English abstract).
[32] T. Obakata, K. Yamaguchi and K. Asama, AIP Conf. Proc. 29 (1975) 74.
[33] J.Y. Li, Z.h. Yu, B.S. Han and S.G. Zhang, Acta Phys. Sinica 34 (1985) 24 (in Chinese, with English abstract).
[34] X.T. Yu, B.S. Han, B.Z. Li and F.C. Pu, J. Magn. Magn. Mater. 62 (1986) 233.
[35] F.L. Yan and B.Z. Li, Acta Phys. Sinica 37 (1988) 95; Chin. Phys. 9 (1989) 96.
[36] R.D. Henry, P.J. Besser, R.G. Warren and E.C. Whitcomb, IEEE Trans. Magn. MAG-9 (1973) 514.
[37] K. Hoshikawa, K. Minegishi, H. Nakanishi, N. Inoue, K. Takimoto and Y. Suemune, Jpn. J. Appl. Phys. 15 (1976) 387.
[38] B.S. Han, J.F. Xiu, X.F. Nie, H.L. Sun, F.L. Ouyang and G.D. Tang, Acta Phys. Sinica 37 (1988) 1527 (in Chinese, with English abstract).
[39] S.G. Huo, X.F. Nie and B.S. Han, Acta Phys. Sinica 37 (1988) 1703 (in Chinese, with English abstract).
[40] B.S. Han, R. Dahlbeck, Y. Yuan and J. Engemann, Chin. Phys. Lett. (to be published).
[41] B.S. Han, R. Dahlbeck, Y. Yuan and J. Engemann, ICM '91 (submitted).
[42] J. Ping, X.D. Nie and B.S. Han, Research Report (1990).
[43] J. Ben Youssef, J.M. Desvignes, R. Dahlbeck and B.S. Han, private communication.

Metastability in epitaxial magnetic metal films

G.A. Prinz

Naval Research Laboratory, Washington, DC 20375, USA

A brief review is provided which assesses the current status of the study of metastable phases of magnetic 3-d metals as epitaxial films on single crystal substrates. Three systems which have the largest literature are discussed in detail: fcc γ-Fe on Cu, fcc β-Co on Cu and bcc Co on GaAs.

1. Introduction

The growth of epitaxial magnetic metal films via vacuum deposition dates from 1936 when it was shown that Fe(100) could be successfully grown on a cleaved surface of NaCl(100) [1]. Since that time technological advances have enabled the experimental work to become increasingly more sophisticated. Remarkable progress was made in improved vacuum, so that by 1971 all-metal vacuum systems operating at 10^{-10} Torr could effectively address questions of the effect of surface cleanliness upon the nucleation and growth process [2]. In parallel with improved vacuum were advances in surface science techniques such as Auger electron spectroscopy. to characterize surface chemistry and electron diffraction techniques to characterize the surface crystallographic structure. More recent advances with bright synchrotron light sources have permitted frequency dependent photoemission studies of the electronic structure and, from the fine structure near the X-ray absorption edges, detailed structure determinations that are elementally specific. Finally, the most recent experimental work exploits the angular dependence of the forward scattering of Auger or photo-emitted electrons to provide an in situ elementally specific measure of the structure of films at an atomic monolayer level of growth.

Along with these experimental advances, the enormous strides in digital computing power have permitted condensed matter theorists to carry out total energy calculations of a large variety of new structures, not only as bulk materials (i.e. three-dimensional with infinite boundary conditions) but as two-dimensional slabs containing a countable number of atomic layers. These slab calculations are particularly important since they can be constructed to mimic actual thin films grown on a particular substrate. Unfortunately the cost of such calculations precludes their use for general surveys and for this the bulk calculations have proved to be indispensable.

One of the most interesting topics to emerge from this increasingly active field is the question of precisely what crystal structure results from the growth of a given metal on a given substrate. This is especially significant when one considers magnetic metals, because the magnetic metals often occur in several different crystallographic phases. Each of these different crystallographic structures will in general, possess different magnetic properties. This occurs because the differences in electronic energy which accompany a structural phase change are of the same magnitude as the energies associated with a magnetic phase change. This is nowhere seen to be more striking than in the thermodynamic phase diagram of Fe, shown in fig. 1 [3]. At ambient pressure the ferromagnetic bcc α phase persists up to 910°C, where it transforms to fcc γ-Fe which is antiferromagnetic. This phase obtains to 1390°C where it transforms to δ-Fe, a non-mag-

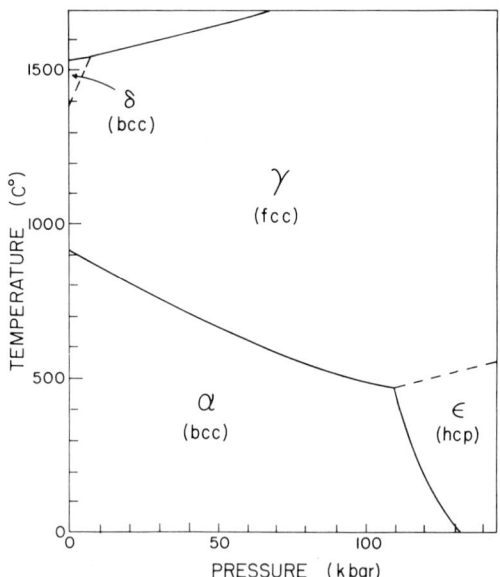

Fig. 1. Thermodynamic phase diagram of Fe [3].

netic bcc phase, which holds to the melting point of 1534°C. The application of high pressure can induce the non-magnetic hcp ε phase.

This phase diagram is of course established for bulk material under hydrostatic pressure. The physical situation for epitaxial growth is quite different, in that the epitaxial bond at the film/substrate interface is the only force exerted on the crystal film and that force may be either compressive or tensile. The system is free to relax in a direction perpendicular to the interface. It will, in general, unless there exists a bulk crystallographic phase whose total energy is at a minimum at a lattice constant defined by the interface. This phase need not be at a global minimum, as say bcc α-Fe is. It may be at a local, higher energy minimum, corresponding to a higher temperature phase such as fcc γ-Fe. Indeed, the probability that one can obtain a perfect lattice match to a bulk phase lattice constant is vanishingly small, considering the finite number of choices for substrate materials available. Consequently one should anticipate that most materials will grow with a uniaxial strain distortion perpendicular to the film plane as the unit cell attempts to minimize its energy by trying to maintain a volume close to a bulk phase.

To what thickness a uniaxially distorted crystalline film will ultimately grow depends upon the balance between the energy associated with the bonding at the interface, the energy necessary to introduce strain relieving defects and the accumulated energy associated with the strain distortion. This last energy accumulates because there is an energy per unit cell associated with its distortion away from some bulk minimum. Every added layer during growth contributes to this energy and finally when it exceeds the energy necessary to create defects or break the interface bonds, the system will relax to the bulk phase. This leads to the well-known concept of critical thickness t_c.

There is another physical basis for a critical thickness even when one is not growing a strained structure. If the substrate lattice spacing exactly matches that of a higher energy phase, as say fcc γ-Fe, then one will grow a metastable phase that is unstrained. The resulting film will be perfectly cubic but still will be limited in its ultimate thickness. This arises from again an accumulated energy because each unit cell (or layer) grown is of higher energy than a bcc α-Fe unit cell. This energy difference between the energy minima of the two structures will ultimately exceed the energy necessary to introduce defects or break interface bonds and the system will relax to the lower energy structure. The important distinction between these two cases is that the former is strain distorted while the latter is unstrained metastable.

There are few cases where either of these ideal cases occurs and one in general has a strain distorted crystal structure derived from either a stable or a metastable phase.

Finally, the metastable phases are not limited to those found on a three dimensional bulk thermodynamic equilibrium phase diagram, such as fig. 1. There are two important additional classes of metastable phases, neither of which were recognized before the advent of epitaxial film growth studies. The first class arises from the fact that via epitaxial growth one may expand a crystal structure by forcing its growth upon a substrate which imposes a larger near neighbor distance than in a bulk equilibrium phase. This is somewhat equivalent to a negative pressure experiment on a phase diagram such as fig. 1, except that one must always recognize that epitaxial growth represents a two-dimensional strain and calculations up to now are generally uniform three-dimensional variations in the lattice. Of course in general growth under tensile stress will merely result in a uniaxially strained structure of the sort described earlier. However there may exist true metastable phases with energy minima at an expanded lattice, which may be stabilized in this manner. We shall see that an example of this is bcc Co. A second class arises from the fact that a uniaxially distorted structure may itself be a metastable phase. For example growing upon a square net (i.e. a (100) face) may yield a film with a tetragonal unit cell. This cell may be merely a strained cubic cell or it may actually be a structure located at a local minimum in total energy. At first glance one would expect this to be unlikely on physical grounds since the electronic structure of the transition metals yields approximately spherical charge clouds and close packing of spheres is either hcp or fcc. Indeed the bcc structure of the lowest energy phase of V, Cr, Mn and Fe is already not the densest packing, and therefore indicates that the departure from sphericity is important, but it is still cubic. While there are metals which do exhibit low symmetry stable phases (e.g. Ga, In, Sn) no convincing report of any stable low-symmetry phases of 3d transition metals has been made. Whether or not there are metastable low symmetry phases of these metals is yet to be determined. Theoretical total energy calculations seemed to indicate a metastable tetragonal phase for Cu [4], but later work has discounted it [5]. The most recent theoretical work on Fe indicates that while one can move continuously from a minimum in total energy space for fcc to another for bcc through a continuous face centered tetragonal (fct) distortion, there are no local minima for an fct phase [6]. It should be noted however that incorporation of a small proportion of small interstitial ions is known to stabilize ordered tetragonal structures, as is well known from study of martensitic phase transitions of steel [7].

2. Stabilization of high temperature phases

In the discussion which follows we shall be concerned with films which attain a thickness sufficient to determine a three-dimensional structure. It is of course always possible to deposit a monolayer of metal on a single crystal substrate and, regardless of the lattice spacing of the substrate, establish at least one ordered atomic layer of adatoms. Generally, however, unless this lattice spacing is sufficiently close to some stable or metastable phase a critical thickness is reached before a three-dimensional structure (i.e. 3 monolayers (ML)) is obtained and the film structure collapses to a polycrystalline array of the lowest energy stable crystallographic phase. Examples of these pseudomorphs are Fe on W(110) [8] and Mn on Ru(0001) [9]. Although the magnetism of such two-dimensional monolayer structures is interesting in its own right, these structures do not constitute three-dimensionally stable or metastable phases.

The most commonly studied epitaxial film–substrate systems are listed in table 1 for thermodynamically stable phases of the 3d magnetic elements. Only two of the films listed in table 1

Table 1
Common substrate/film combinations for some of the thermodynamically stable phases of the 3d magnetic elements

	Substrate (a_0 [Å])			Film (a_0 [Å])	
fcc	Cu	(3.61)	fcc	Ni	(3.52)
				β-Co	(3.55)
				γ-Fe	(3.59)
	[45° rotation on (001)]				
fcc	LiF	(4.02)	bcc	α-Fe	(2.867)
	Al	(4.05)			$\times \sqrt{2}$
	Au	(4.07)			$\overline{5.733}$
	Ag	(4.09)			
	[$p(1/2 \times 1/2)$ on (001)]				
fcc	NaCl	(5.64)	bcc	α-Fe	(2.867)
	AlAs	(5.62)			$\times 2$
	GaAs	(5.65)			$\overline{5.733}$
	Ge	(5.66)			
	ZnSe	(5.67)			

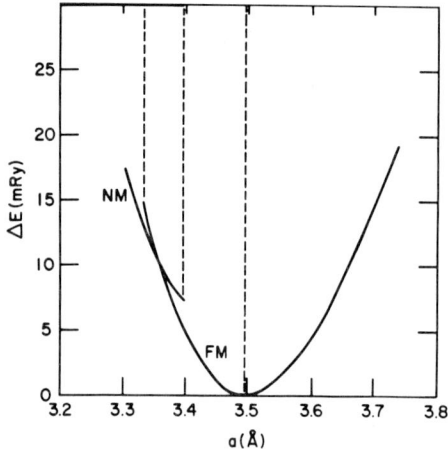

Fig. 2. Binding energy curves for fcc Co vs. cubic lattice constant a in Å. Energy changes per atom in mRy referenced to the minimum [10].

properly fall into the class of metastable systems: β-Co and γ-Fe. Both of these are thermodynamically stable high temperature phases, but may be formed at low temperature by epitaxial growth on Cu.

2.1. fcc-β-Co

Bulk Co exists in two ferromagnetic thermodynamically stable phases at ambient pressure, hcp-α-Co below 388°C and fcc-β-Co above 450°C, the transformation being sluggish between those temperatures. The total energy of bulk fcc Co as a function of the lattice constant has been calculated from first principles [10] and the result is shown in fig. 2. Although the accuracy in determining the lattice constant is only ≈ 2%, the minimum falls close to $a_0 = 3.5$ Å. The experimentally observed room temperature value is $a_0 = 3.54$ Å [11]. The growth of fcc Co on Cu has been studied and reported by several groups [12]. The lattice constant of Cu at 3.61 Å represents a 2% tensile strain in the film plane. From fig. 2 one can see that a *uniform* 2% strain would represent an energy increase of 5 mRy (68 meV). Although this estimate is for a uniform bulk expansion and its accuracy depends upon the accuracy of these calculations, it nevertheless is not a negligible change in the energy of the system. It is to be compared, for example, with the calculated energy difference between the bulk fcc phase (β-Co) and the bulk hcp phase (α-Co), at their respective energy minima, of 27 meV [13]. One expects however that in thin film form, the 2% mismatch is partially compensated by a relaxation of the out-of-plane spacing, although no detailed report of this appears to have yet been made. fcc films have been grown up to 1 μm, but in the vicinity of strain-relieving misfit dislocations the hcp phase is found [14]. More recent work has centered on thinner films and it is found that up to 20 Å no misfit dislocations are formed [15]. Growth at high temperature yields isolated nucleated islands which remain fcc up to a radius of 375 Å. This is an interesting result to compare with finely powdered Co, in which the surface free energy is sufficient to maintain the fcc structure in particles up to 10 μm in diameter! [16]. X-ray studies of those powders indicate they are single crystal with a lattice constant temperature (°C) dependence of [17]

$$a_0(T) = 3.54 \text{ Å}(1 + 12.3 \times 10^{-6}T + 2.0 \times 10^{-9}T^2).$$

Finally, fcc Co has been obtained as fine grain precipitates in Cu at the lattice constant of Cu [18]. Unlike epitaxial growth of films, this is a three-dimension epitaxial form which does approximate a uniform expansion of the lattice. At a critical grain diameter of ≈ 500 Å dislocations appear and the grains are no longer coherent structures. These grains were readily studied by ferromagnetic resonance [19]. There is literature for magnetic properties of Co on Cu(100), measured by a variety of techniques [12], but there is little agreement between them. Clearly surface preparation and growth conditions have proved to dramatically affect the reported results.

2.2. fcc-γ-Fe

The second example shown in table 1 is γ-Fe on Cu. fcc-γ-Fe has more magnetic phases competing near the minimum energy lattice constant than does fcc Co, as can be seen in fig. 3 [20]. Here we see the calculated low temperature local

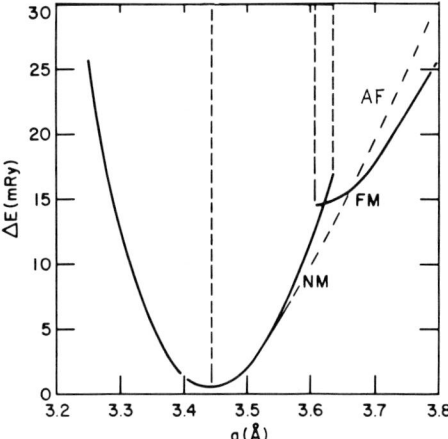

Fig. 3. Same as fig. 2, but for fcc Fe [20].

minimum is located at $a_0 = 3.44$ Å. This is 6% smaller than the observed high temperature (910°C) value of 3.64 Å [21] which in turn is only 1% larger than that of room temperature Cu (3.61 Å). Extrapolating to 20°C yields a_0 (γ-Fe) = 3.59 Å [22]. The energy difference between the fcc phase at minimum and bcc α-Fe is very small. Unfortunately the error in total energy calculations currently exceeds this energy difference, so it is impossible to estimate it from a first principles calculation. As in the case of fcc Co, however, the energy due to strain when grown on Cu is substantial and probably exceeds the difference between the two phases. If we assume that at room temperature the calculated theoretical lattice constant of fcc γ-Fe is 3.44 Å as shown in fig. 3, then the strain energy for uniform expansion would be ≈ 15 mRyd (200 meV). From fig. 3 however one sees that there is a bifurcation of the energy curve into non-magnetic and antiferromagnetic branches above the energy minimum, and at a_{Cu} these branches are crossed by yet a third ferromagnetic branch. It is this coincidence of three different magnetic fcc phases near the Cu lattice constant which has led to many conflicting accounts of the magnetic properties of fcc Fe grown on Cu [23]. Interestingly, the ferromagnetic metastable phase (which has no high temperature thermodynamic corrollory such as γ-Fe) is near a metastable minima at a_{Cu} and may explain why the ferromagnetic phase is found to exist in some experiments. The experimental history of the growth of Fe on Cu(001) dates back to 1956 [24], but reliable work meeting modern standards began in the late 60's. A remarkable study in 1980 demonstrated a clever display of lattice matchings [25]. Fe was grown on substrates made of Cu–Au alloys. This alloy system, which forms a continuous series of solid solutions in the fcc structure throughout its compositional range, offers a continuous choice of lattice constants from Cu (3.615 Å) to Au (4.018 Å). Very thin Fe films (≤ 3 ML) grown on these (111) substrates over an Au concentration range of 0 → 30 at% (corresponding to a lattice constant increase from 3.61 to 3.75 Å) show a dramatic increase in the magnetic moment per Fe atom from 0.6 to 2.6μ_B. These lattice constants are all larger than that for the stable high temperature fcc phase, γ-Fe, of 3.59 Å and these results support theoretical predictions of an increase in moment with increasing lattice spacing. More recent work has indicated, however, that the growth of Fe on Au and Cu is complicated by questions of diffusion at the interface. Studies of Fe growth on Au show that the intensity of Auger electrons emitted from both the Fe and Au follows a linear dependence on coverage with "breaks" in the slope at integral numbers of ML coverage. This is evidence of layer-by-layer growth. Unfortunately, a monolayer of Au appears to segregate to the top of the Fe film [26]. Growth of Fe on Cu, on the other hand, shows that unless the substrate temperature is maintained below 125°C during growth, Cu diffuses into the Fe to form an alloyed fcc structure [27,28]. Furthermore, careful LEED studies show that Fe on Cu(001) is not cubic, but is tetragonally distorted, the interplanar spacing perpendicular to the surface is contracted 2% with respect to the Cu spacing [22].

It should be noted that fcc Fe, like fcc Co, forms single crystal precipitated grains in Cu. These particles, 300 to 700 Å in diameter, have a room temperature lattice constant of 3.58 Å. Also, like Co, three dimensional nucleated islands form at high temperature on Cu. They remain single crystal up to a radius of 750 Å. The current status of Fe growth on Cu has been reviewed by Lu et al. [22].

3. Non-thermodynamic metastable phases

As mentioned in the introduction, in addition to the known thermodynamic phases, other phases not known from previous high temperature/high pressure studies may exist. An example of this is the ferromagnetic fcc phase of Fe predicted from total energy calculations discussed earlier. This is one example of many "expanded lattice" predictions which have come from such calculations. Before the advent of epitaxial growth studies, it was not possible to experimentally explore this world of "negative pressure" and there was little impetus to carry out theoretical calculations. The experimental work cited earlier of growing ferromagnetic fcc Fe on an expanded Cu–Au lattice, is an early example of this work and was not generally recognized as opening the door to undiscovered metastable phases. This may have finally been widely understood with the discovery of bcc Co [29]. Here was a crystallographic phase which did not appear on the known thermodynamic phase diagram for Co, but could be prepared by epitaxial growth. The first clue that such

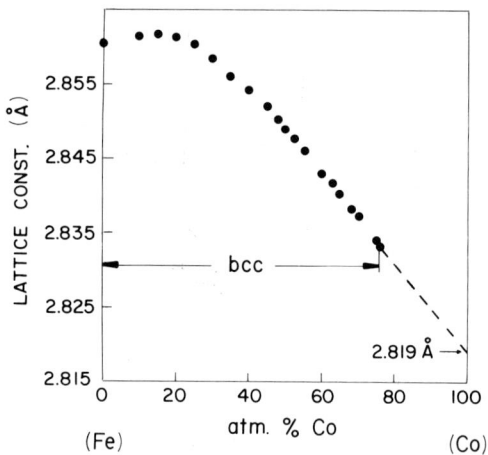

Fig. 4. Lattice constant vs. composition for the bcc Fe–Co alloy system [30].

a phase might exist was found in the Fe–Co alloy phase diagram [30] which supports the bcc phase up to 75% Co. An estimate of a lattice constant for a possible bcc phase of Co was obtained by extrapolation, as shown in fig. 4. This lattice constant of 2.819 Å is within 0.4% of that of

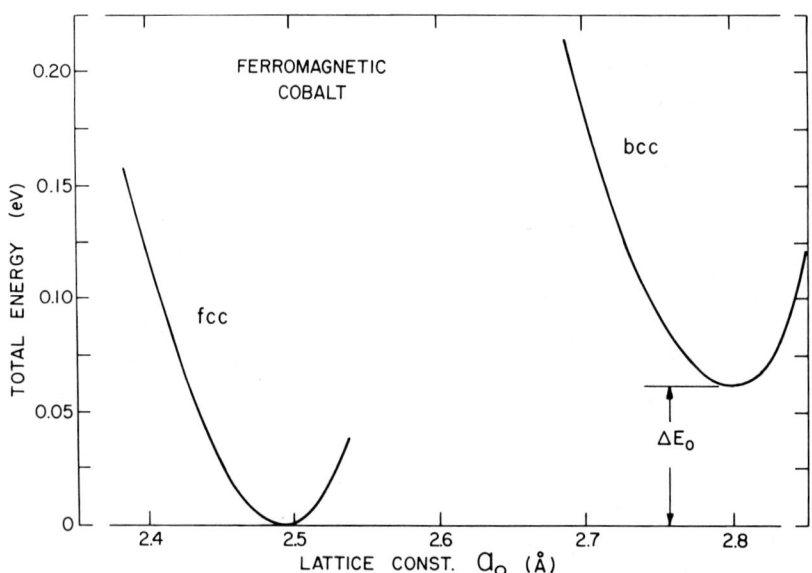

Fig. 5. Relative energies per atom of the two ferromagnetic cubic phases of Co as a function of lattice constant, adopted from ref. [32b]. To facilitate comparison, the fcc unit cell has been oriented to achieve a lattice match on a square net closest in size to the bcc cell, as commonly happens during epitaxial growth. This represents a 45° rotation on a (100) face, or a 90° rotation on a (110) face. Therefore the scale shown represents the true lattice constant of a bcc structure, but $(\sqrt{2})^{-1}$ times the true lattice constant of the fcc structure.

GaAs divided by two (i.e. 2.825 Å), which had already been shown to support the growth of bcc Fe ($a_0 = 2.867$ Å) [31]. The growth of bcc Co on GaAs has now been studied in a number of laboratories using a variety of techniques, establishing it as one of the most studied of the 3d magnetic metastable phases, and has stimulated much of the theoretical search for new phases. Total energy calculations for bcc Co yielded the curves shown in fig. 5, which compare fcc Co with bcc Co [32]. The energy minimum for bcc Co was found at $a_0 = 2.82$ Å, in agreement with the extrapolated value in fig. 4, and the energy difference between the two minima was 70 meV per atom.

A review of the studies carried out on bcc Co may serve to illustrate the tools available for characterizing a metastable phase, both structurally and magnetically. The first structural probe employed was Reflection High Energy Electron Diffraction (RHEED) during growth in the MBE chamber. Shown in fig. 6 along with the analogous RHEED patterns for the GaAs(110) substrate and a Fe(110) film, one immediately sees

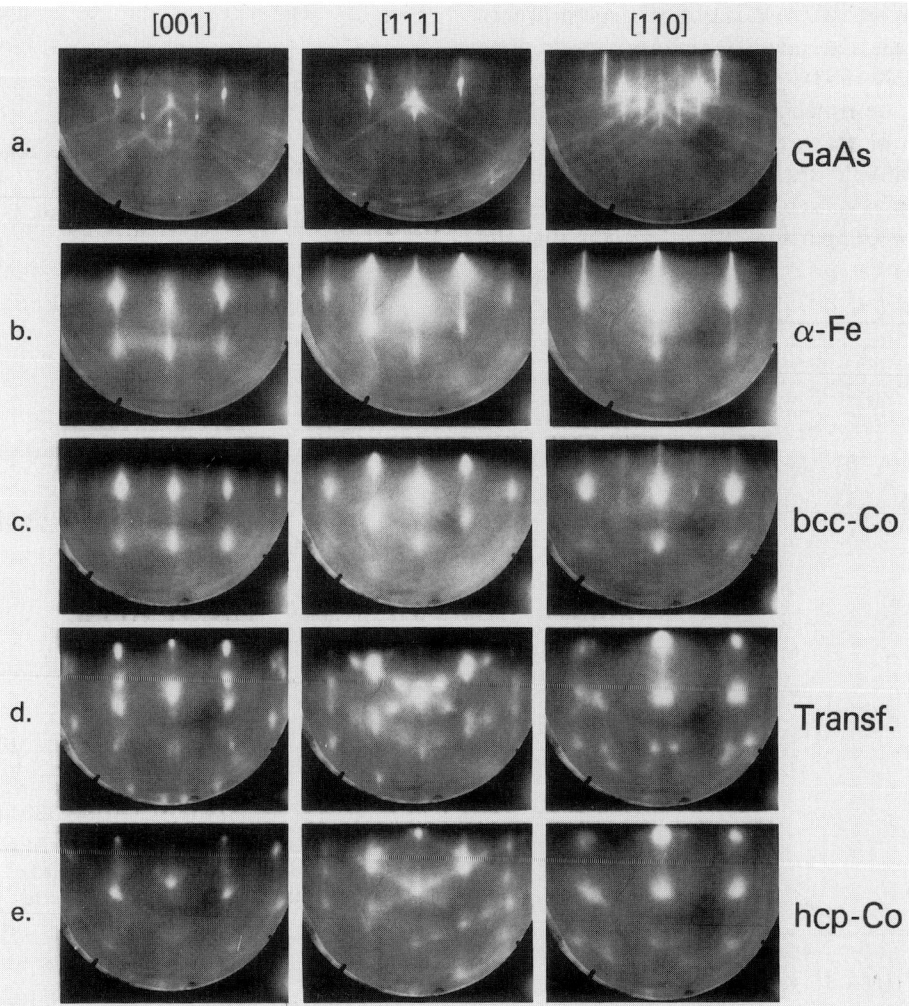

Fig. 6. Reflection high energy electron diffraction (RHEED) patterns along the princpal axes in the (110) face of (a) GaAs ;(b) α-Fe; (c) bcc Co; (d) crystallographic transformation; (e) hcp Co with (11·0)[00·1] Co ∥(110)[111] GaAs.

that the film is a single crystal, of approximately the same lattice spacing as bcc α-Fe and possessing the same in-plane symmetry and orientation. Also shown is the diffraction pattern obtained when a critical thickness is exceeded and the film transforms to the low energy stable phase α-Co (hcp). After removal from the growth chamber, X-ray diffraction measurements were made to obtain the inter-planar atomic spacing perpendicular to the film plane. Those results are shown in fig. 7 [33] and one sees that the (110) Co diffraction peak lies on the shoulder of the GaAs(220) exhibiting only a 0.1% contraction of the lattice planes when compared with a perfect 2-fold match to GaAs ($a_0 = 2.827$ Å). The next structural tool employed was Extended X-ray Absorption Fine Structure (EXAFS) exploiting an electron conversion technique especially suitable for thin films. The results of these measurements, which yield real space atomic distributions, are shown in fig. 8 where bcc Fe, hcp Co and fcc Cu data are provided for comparison [34]. It is clear that the

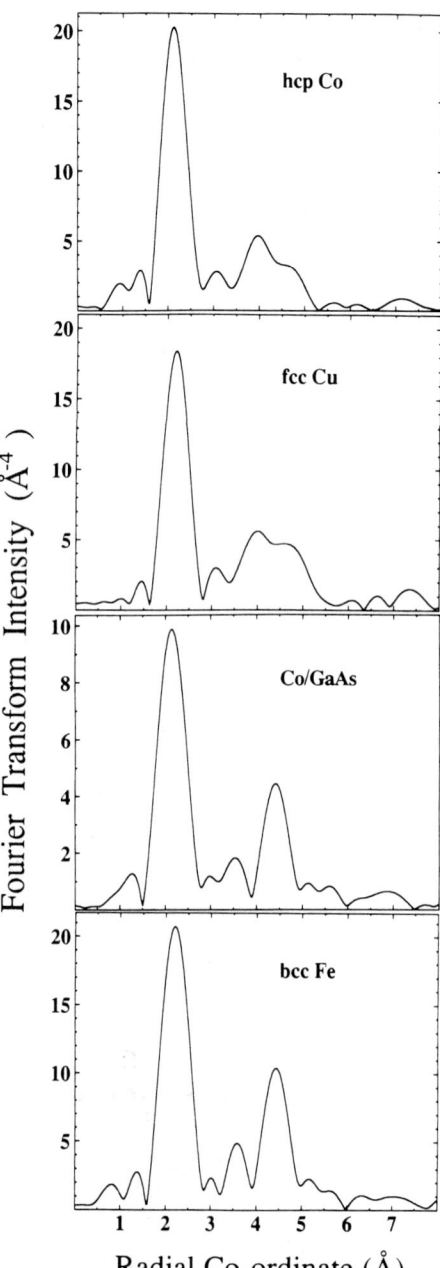

Fig. 8. Fourier transform of isolated fine structure for hcp Co rod; fcc Cu foil; Co film deposited on GaAs(110); a thick film of bcc Fe. Radial coordinate not corrected for phase shifts [34].

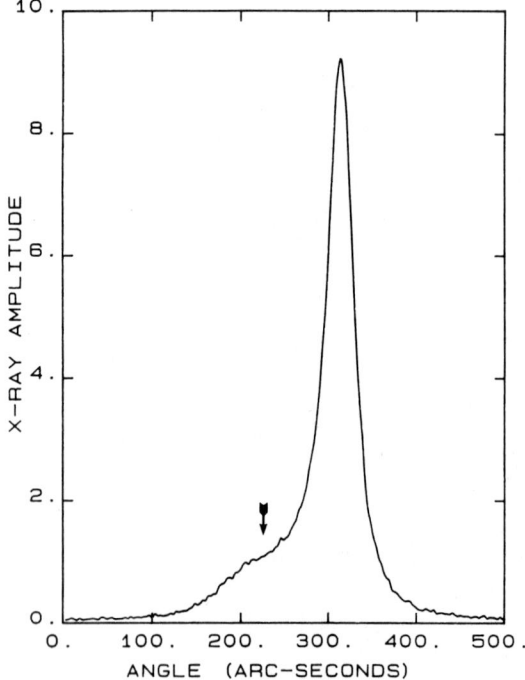

Fig. 7. Double-crystal X-ray rocking curve for a 357 Å Co film on a GaAs(110) substrate. The Co(110) peak forms a shoulder on the GaAs(220) peak [33].

Co/GaAs structure is bcc and the determined lattice constant of (2.82 ± 0.01) Å compares well with both the X-ray data and the theoretical calculations. Finally, the most recent structural work has exploited the relatively new technique of Auger Electron Diffraction (AED) or Auger Electron Forward Scattering (AEFS). This element-specific technique exploits the fact that Auger electrons emitted from atoms in a crystalline solid are strongly forward scattered along near-neighbor directions. Angle-resolved measurements therefore yield detailed measurements of the arrangements of atoms within the top few layers of a film. These results, shown in fig. 9 indicate that by the fourth atomic layer Co on GaAs (001) has already established the bcc structure with no indication of tetragonal distortion perpendicular to the plane of growth.

In addition to these structural studies, several magnetic characterizations have been carried out. The first of these was the magnetization. The value for the magnetic moment in a 357 Å thick film was found to be (1257 ± 10) emu/cm^3 or $1.53\mu_B$ per atom [29]. This value proved to be lower than the value of $1.7\mu_B$ predicted from first principles calculation of the bcc phase [32]. Subsequent nuclear magnetic resonance measurements made on the same sample obtained a broad resonance linewidth (75 MHz) centered at a frequency corresponding to the observed magnetic moment [33]. More recent NMR studies on (001) films confirm this low average value for the moment, but Polarized Neutron Reflection (PNR) studies of those films show a gradient in the moment. At the GaAs interface the moment falls to $1.0\mu_B$ but rises to $1.7\mu_B$ at the center of a 100 Å thick film [36]. This center value matches the theoretical prediction and confirms earlier speculation that chemical contamination near the GaAs interface decreases the magnetic moment in Co films as it does in Fe films. Detailed ferromagnetic resonance studies confirmed the prediction, obtained by extrapolation of Fe/Co alloy studies, that the cubic anisotropy constant K_1 in bcc Co should be opposite in sign and twice the magnitude of that found in bcc Fe [37]. Finally, recent Brillouin light scattering studies have yielded a value for the exchange stiffness constant in bcc Co of $D = 2.6\ (\pm 0.3) \times 10^{-9}$ Oe cm^2 [38]. This value can be compared to that obtained from hcp Co of $D = 4.04 \times 10^{-9}$ Oe cm^2. It is remarkable

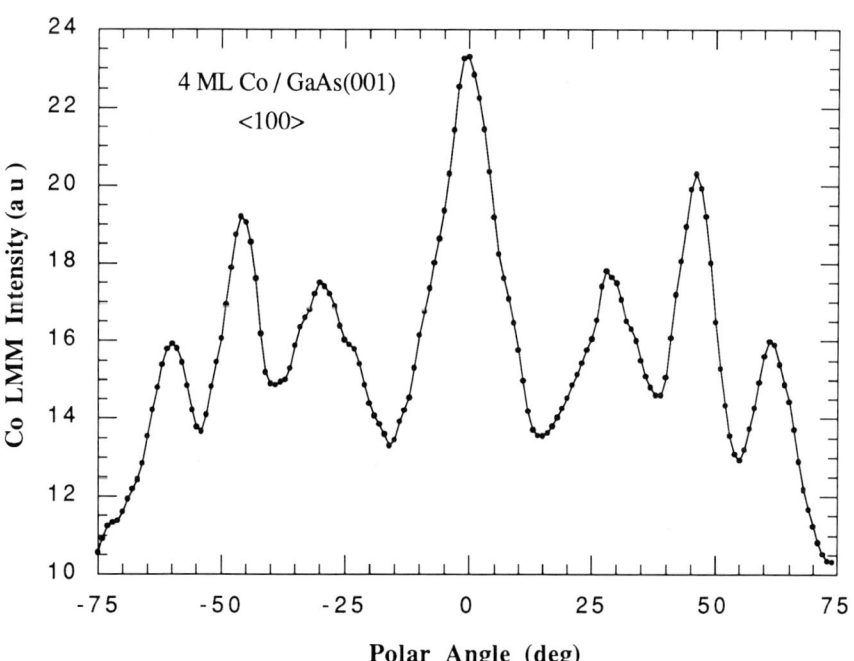

Fig. 9. Electron forward scattering pattern from 4 ML Co deposited on GaAs(001) at 175 °C [35].

that D scales from one structure to another exactly as the near-neighbor coordination. This suggests that for spin-wave excitations near $k = 0$ Co metal, a manifestly itinerant electron system, behaves very much like a local moment ferromagnet.

4. Other systems

In addition to these well documented cases, there are several overlayer systems which have been less thoroughly studied, but have nevertheless yielded interesting experimental and theoretical results. If we consider first the experimental cases, there are two groups. The first of these involves the curious element magnanese, which exhibits a rich variety of crystallographic and magnetic properties. The α-Mn phase is cubic with 58 atoms per unit cell, is antiferromagnetic below 100 K and paramagnetic at room temperature. The higher temperature phases are all nonmagnetic: β-Mn (20 atoms/unit cell) cubic ($T_β$ = 727°C); γ-Mn (4 atoms/unit cell) fcc ($T_γ$ = 1095°); δ-Mn (2 atoms/unit cell) bcc ($T_δ$ = 1133°C). Although there are reports by several authors for a few monolayers of growth of Mn on various substrates, the only two well-documented cases for many layers establishing a crystal structure are for Mn/Pd [39] and Mn/Ag [40]. For the former, Mn was grown to 21 monolayers (ML) on Pd {001} and assumed the γ-Mn fcc structure, with a tetragonal strain due to the lattice mismatch. For the latter, Mn was grown to 14 ML on Ag (001) and exhibited a body-centered-tetragonal structure with a c/a ratio of 1.13 ± 0.06. Neither of these structures was reported to be magnetic.

A second group consists of the pair Fe and Ni alternating roles as substrate and overlayer. Considering first Fe grown on Ni {001} [41], one obtains γ-Fe (fcc) which is tetragonally distorted (expanded perpendicular to the film) by the compressive stress at the interface. Above 6 ML domains of bcc α-Fe appear in the film. No magnetic measurement was reported. The alternative case of Ni on Fe(100) was reported by two laboratories, both of which agree that one obtains bcc Ni at the α-Fe lattice constant [42,43]. There remain discrepancies over the magnetic properties of these films and the thickness to which the bcc phase can be stabilized.

In addition to these experimental findings there are numerous theoretical predictions of metastable phases for bcc Ni on [Ag, Mn, Fe, Cr, V], bcc Mn on [Fe, Cr, V], fcc [Fe, Co, Pd] on [Co, Ni, Cu, Cr, Rh, Pd and Ag] [44,45]. A complete catalog of which of these candidates will, and will not, grow on these proposed substrates would be useful for understanding the phenomenon of metastable structures.

5. Conclusion

An attempt has been made, in this brief review, to give the reader a compact assessment of the status of metastable phases of the magnetic 3-d metals as films on single crystal substrates. This has been an area of intense interest in recent years as the references indicate. There remain several outstanding problems to be addressed, aside from the task of carefully investigating the remaining overlayer/substrate combinations to determine the resulting structures and their magnetic properties. One of the remaining problems is to determine if there actually are any metastable tetragonal phases. There are initial theoretical results for bct Cu which have been disputed and initial theoretical work on fct Fe which appears to show no energy minima. The apparent stability of some experimentally produced tetragonal systems may be misleading. All of these structures have been found to disorder above some critical film thickness. This leads to a second, and more difficult, outstanding problem, which is to determine the exact basis and mechanism for these structures to disorder. The critical thicknesses seem to depend very delicately upon conditions of film growth e.g. surface condition, temperature of substrate, rate of deposition and impurity incorporation. Some of these may be very subtle and difficult to control, such as surface roughness, step density, surface defects, surface contaminants, etc. Impurity incorporation may be from the substrate itself or residual contaminants in the vacuum system, some as subtle as hydrogen. Once all of those elements are under control, then one would like to know how much energy is stored in the strain field of the overlayer as compared to the accumulated energy difference in the overlayer arising from its being a higher energy phase, and how this energy deficit is balanced by the bonding forces at the interface or tension at the free surface. Finally, one would like to understand how the disorder mechanism is initiated and how it proceeds.

These are all difficult problems, both experimental and theoretical, but represent the challenge to be met before this field of study is to be considered understood.

Acknowledgement

This work was supported by the Office of Naval Research.

References

[1] L. Bruck, Ann. Phys. 26 (1936) 233.
[2] T.J. Patrician and C.M. Wyman, Phys. Stat. Sol. (a)6 (1971) 449.
[3] T. Takahashi and W.A. Bassett, Science 145 (1964) 483.
[4] I.A. Morrison, M.H. Kang and E.J. Mele, Phys. Rev. B 39 (1989) 1575.
[5] H. Li, S.C. Wu, D. Tian, J. Quinn, Y.S. Li, F. Jona and P.M. Marcus, Phys. Rev. B 40 (1989) 5841.
[6] S.S. Peng and H.J.F. Jansen, J. Appl. Phys. 69 (1991) 6132.
[7] L.H. van Vlack, Elements of Materials Science and Engineering, 6th ed. (Addison-Wesley, New York, Reading, 1989) p. 238.
[8] U. Gradmann and G. Waller, Surf. Sci. 116 (1982) 539.
[9] B. Heinrich, C. Liu and A.S. Arrott, J. Vac. Sci. Technol. B 3 (1985) 766.
[10] P.M. Marcus and V.L. Moruzzi, J. Appl. Phys. 63 (1988) 4045.
[11] J. Donohue, The Structures of the Elements (Krieger, Malabar, Florida, 1982). 208.
[12] a) L. González, R. Miranda, M. Salmerón, J.A. Verges and F. Yndurain, Phys. Rev. B 24 (1981) 3245.
b) R. Miranda, D. Chandesris and J. Lecante, Surf. Sci. 130 (1983) 269.
c) D. Pescia, G. Zampieri, M. Stampioni, G.L. Bona, R.F. Willis and F. Meier, Phys. Rev. Lett. 58 (1987) 933.
d) J.J. Miguel, A. Cebollada, J.M. Gallego, S. Ferrer, R. Miranda, C.M. Schneider, P. Bressler, J. Garbe, H. Bethke and J. Kirschner, Surf. Sci. 211 & 212 (1989) 732.
e) F.J. Lamelas, C.H. Lee, Hui He, W. Vavra and R. Clarke, Phys. Rev. B 40 (1989) 5837.
f) Hong Li and B.P. Tonner, Phys. Rev. B 40 (1989) 10241.
g) J.J. De Miguel, A. Cebollada, J.M. Gallego, R. Miranda, C.M. Schneider, P. Schuster and J. Kirschner, J. Magn. Magn. Mater. 93 (1991) 1.
[13] B.E. Min, T. Oguchi and A.J. Freeman, Phys. Rev. B 33 (1986) 7852.
[14] J. Goddard and J.G. Wright, Br. J. Appl. Phys. 15 (1964) 807.
[15] W..A. Jesser and J.W. Mathews, Phil. Mag. 17 (1968) 461.
[16] P. Reidi, St. Andrews, Scotland, private communication.
[17] E.A. Owen and D. Madoc Jones, Proc. Phys. Soc. B 67 (1954) 456.
[18] V.A. Phillips, Trans. Met. Soc AIME 230 (1964) 967.
[19] a) J.J. Becker, AIME Trans. 209 (1957) 59.
b) C.P. Bean, J.D. Livingston and D.S. Rodbell, Acta Met. 5 (1957) 682.
c) D.S. Rodbell, J. Appl. Phys. 29 (1958) 311.
d) J.D. Livingston, Trans. Met. Soc. AIME 215 (1959) 566.
[20] V.L. Moruzzi, P.M. Marcus and J. Kubler, Phys. Rev. B 39 (1989) 6957.
[21] J. Donohue, ref. [20], p. 205.
[22] S.H. Lu, J. Quinn, D. Tian, F. Jona and P.M. Marcus, Surf. Sci. 209 (1989) 364.
[23] a) W. Becker, H.-D. Pfannes and W. Keune, J. Magn. Magn. Mater. 35 (1983) 53.
b) R. Halbauer and U. Gonser, J. Magn. Magn. Mater. 35 (1983) 55.
c) A. Amiri Hezaveh, G. Jennings, D. Pescia, R.F. Willis, K. Prince, M. Surman and A. Bradshaw, Solid State Commun. 57 (1986) 329.
d) H.I. Starnberg, M.T. Johnson, D. Pescia and H.P. Hughes, Surf. Sci. 178 (1986) 336.
e) J.A.C. Bland, D. Pescia and R.F. Willis, Phys. Rev. Lett. 58 (1987) 1244.
f) D. Pescia, M. Stampanoni, G.L. Bona, A. Vaterlaus, R.F. Willis and F. Meier, Phys. Rev. Lett. 58 (1987) 2126.
g) P.A. Montano, G.W. Fernando, B.R. Cooper, E.R. Moog, H.M. Naid, S.D. Bader, Y.C. Lee, Y.N. Darici, H. Min and J. Marcano, Phys. Rev. Lett. 59 (1987) 1041.
[24] O. Haase, Z. Naturforsch. 11-A (1956) 862.
[25] U. Gradmann and H.O. Isbert, J. Magn. Magn. Mater. 15–18 (1980) 1109.
[26] S.D. Bader and E.R. Moog, J. Appl. Phys. 61 (1987) 3729.
[27] S.A. Chambers, T.J. Wagener and J.H. Weaver, Phys. Rev. B 36 (1987) 8992.
[28] D.A. Steigerwald, I. Jacob and W.F. Egelhoff, Jr., Surf. Sci. 202 (1988) 472.
[29] G.A. Prinz, Phys. Rev. Lett. 54 (1985) 1051.

[30] W.C. Ellis and E.S. Greiner, Trans. Am. Soc. Met. 29 (1941) 415.
[31] a) J.R. Waldrop and R.W. Grant, Appl. Phys. Lett. 34 (1979) 630.
b) G.A. Prinz and J.J. Krebs, Appl. Phys. Lett. 39 (1981) 397.
[32] a) D. Bagayoko, A. Ziegler and J. Callaway, Phys. Rev. B27 (1983) 7046.
b) V.L. Moruzzi, P.M. Marcus, H. Schwarz and P. Mohn, J. Magn. Magn. Mater. 54–57 (1986) 955.
c) J.I. Lee, C.L Fu and A.J. Freeman, J. Magn. Magn. Mater. 62 (1986) 93.
[33] P.C. Reidi, T. Dumelow, M. Rubinstein, G.A. Prinz and S.B. Qadri, Phys. Rev. B 36 (1987) 4595.
[34] Y.U. ldzerda, B.T. Jonker, W.T. Elam and G.A. Prinz, J. Vac. Sci. Technol. A 8 (1990)1572.
[35] S.T. Jonker, G.A. Prinz and Y.U. ldzerda, J. Vac. Sc. Technol. B 9 (1991) in press.
[36] J.A.C. Bland, R.D. Bateson, P.C. Reidi, R.G. Graham, H.J. Lauter, J. Penfold and C. Shackleton, J. Appl. Phys. 69 (1991) 4989.
[37] G.A. Prinz, C. Vittoria, J.J. Krebs and K.S. Hathaway, J. Appl. Phys. 57 (1985) 3672.
[38] J.M. Karanikas, R. Sooryakumar, G.A. Prinz and S.T. Jonker, J. Appl. Phys. 69 (1991) 6120.
[39] D. Tian, S.C. Wu, F. Jona and P.M. Marcus, Solid State Commun. 70 (1989) 199.
[40] a) B.T. Jonker, J.J. Krebs and G.A. Prinz, Phys. Rev. B39 (1989-I) 1399.
b) Y.U. ldzerda, B.T. Jonker, W.T. Elam and G.A. Prinz, J. Appl. Phys. 67 (1990) 5385.
[41] S.H. Lu, Z.Q. Wang, D. Tian, Y.S. Li, F. Jona and P.M. Marcus, Surf. Sci. 221 (1989) 35.
[42] B. Heinrich, A.S. Arrott, J.F. Cochran, C. Liu and K. Myrtle, J. Vac. Sci. Technol. 4 (1986) 1376.
[43] P.M. Marcus, V.L. Moruzzi, Z.Q. Wang, V.S. Li and F. Jona, Mater. Res. Soc. Proc. 83 (1987) 21.
[44] P.M. Marcus and V.L. Moruzzi, IBM Research Report RC 13159 (1987).
[45] S.C. Hong, A.J. Freeman and C.L. Fu, Phys. Rev. B 29 (1989) 5719.

Surface magnetism

U. Gradmann

Physikalisches Institut, Technische Universität Clausthal, W-3392 Clausthal-Zellerfeld, Germany

The main lines of surface magnetism research before and after 1975 are reported. The overall changes of magnetic properties in surfaces and interfaces were investigated by a wide variety of experimental and theoretical techniques. A softened temperature dependence of magnetization, enhanced moments in clean surfaces, the existence of magnetic surface anisotropies and the sensitive dependence of magnetic surface properties on the structural and physicochemical state of the surface are now generally accepted.

1. Introduction

The magnetic state of atoms or ions depends on their coordination, and so does the magnetic interaction with their surroundings. The surface as the most important and inevitable disturbance of translational symmetry in a solid therefore must show, for the case of magnetic crystals, strong magnetic pecularities. This applies for interfaces, too. Surface magnetism, including interface magnetism in what follows, therefore concerns basic questions both of magnetism and of surface physics. Amazingly, the field nevertheless was treated by a small group of experts only when the Journal of Magnetism and Magnetic Materials started 16 years ago, whereas it is now generally accepted as an important field of basic research with considerable importance for applications.

In discussing the steep rise of the field, we will show that the main body of magnetic surface phenomena had been addressed again 16 years ago. It is their rich enfolding, based on a wide variety of experimental and theoretical techniques, and the increasing awareness of the importance of magnetic surface and interface phenomena, which made surface magnetism such a fruitful and interesting field.

This paper intends a presentation of the main lines of the field and its evolution during the period since 1975, from the point of view of an experimentalist who enjoyed surface and ultrathin film magnetism for many years. A more complete review of surface magnetism from a theoretical point of view has been given by Mathon [1]. For a collection of experimental data compare ref. [2]; the state of the field at the beginning of this period has been reported by Gradmann [3]. We focus on the properties of well-defined single-crystal surfaces or interfaces, appropriate for an analysis in terms of basic idealized models. Strongly interdiffused and polycrystalline surfaces and interfaces are out of the scope of the review.

Surface magnetism is tightly connected with magnetism in ultrathin films, and a great deal of our knowledge on bulk magnetic surfaces comes from the theoretical analysis of ultrathin film slabs using first principles band structure calculations [4] or from experiments with ultrathin magnetic films [3]. In the present review, ultrathin magnetic films are reported only as an aid for surface magnetism. An extended report of ultrathin film magnetism is given elsewhere [5].

2. Temperature dependence of magnetic order

The loss of magnetic neighbours in a magnetic surface represents a weakening of exchange interaction which must result somehow in an enhanced thermal decrease of magnetic order. This had been addressed before 1975 in spinwave [6,7] and molecular field [8] models and in Monte Carlo simulations [9]. As usual, we discuss separately the limiting cases of critical behaviour near T_c and of low temperatures, $T \ll T_c$.

2.1. Critical behaviour

The following rough ideas on critical surface behaviour were available in 1975:

Molecular field calculations of Wolfram et al. [8] had indicated a critical surface magnetization exponent β_1 (exponent for the first layer) near 1.0, to be compared with $\beta = 1/3$ for bulk ferromagnets. Monte Carlo simulations of Binder and Hohenberg [9] had confirmed these rough predictions and had given a more accurate value of $\beta_1 = 0.75$. The only relating experimental evidence came from non-polarized electron scattering from antiferromagnetic NiO-surfaces [10], where the temperature-dependence of antiferromagnetic super-structure spots was interpreted in terms of $\beta_1 = 1$ (this experiment could never be reproduced).

Since 1975, the theoretical insight in surface critical behaviour has been promoted by treating the simple cubic (100) model surface using more extended Monte Carlo simulations [11,12], high temperature series expansions [13] and renormalization group [14,15] analysis, for various types of anisotropies. An excellent review of theoretical models has been given by Binder [16]. For critical exponents β_1 from different theoretical models compare table 1. All theoretical values are near $\beta_1 = 0.8$.

The experimental progress in critical surface behaviour is intimately connected with the evolution of powerful spin-polarized electron techniques, which were in a pioneering state at the beginning of the period and are now available in a broad variety of methods, which have been reviewed in the book edited by Feder [17]. All spin polarized electron methods which have been used for analysis of critical behaviour are emission type. What is observed as a magnetic signal is the polarization of emitted electrons,

$$P = \frac{n^\uparrow - n^\downarrow}{n^\uparrow + n^\downarrow}, \qquad (1)$$

where $n^\uparrow$ or $n^\downarrow$ is the number of electrons with magnetic moment parallel or antiparallel to the sample magnetization. Application to critical behaviour of bulk ferromagnetic surfaces has been reported for spin polarized angle resolved photoemission (SPARPES [18]), spin-polarized secondary electron emission [19], electron capture spectroscopy (ECS [20,21]) and spin polarized low energy electron diffraction [22] (SPLEED; the exchange asymmetry A_{ex} used in this method differs slightly from P, because of superimposed polarization by spin–orbit coupling, compare refs. [17,22]).

It is generally assumed in the evaluation of those experiments that P is proportional to magnetic long-range order, in the critical region. Using this assumption, critical exponents β_1 of the first monolayer could be determined, as shown in fig. 1 for the SPLEED-analysis of Ni(100) and Ni(110) [22]. Experimental results for β_1 for different surfaces are collected in table 2. Roughly, theory and experiment agree in $\beta_1 = 0.8$ (1); for a detailed discussion compare the original papers. The enhanced surface exponent is now a well established property of surface magnetic order.

Table 1
Critical surface exponents β_1 (exponent of the first atomic layer) for simple cubic (100) surfaces, from theoretical models

Model	Method	β_1	Ref.
Ising	Monte Carlo	0.78 (2)	[11]
	series expansion	0.77 (2)	[13]
	renormalization group	0.80 (2)	[14]
xy	Monte Carlo	0.84 (1)	[12]
	series expansion	0.79 (3)	[13]
Heisenberg isotropic	Monte Carlo	0.75 (10)	[19]
	series expansion	0.81 (4)	[13]
	renormalization group	0.84 (1)	[14]
Heisenberg anisotropic	renormalization group	0.84 (1)	[14]

2.2. Live layers

The Curie temperature is the macroscopic measure for exchange interaction. Modified exchange interactions must be considered near surfaces, too, and obviously sufficiently enhanced exchange interaction near the surface must finally result in some type of surface magnetic order even above the bulk Curie temperature, a so-called magnetic "live (surface) layer". It is easy to introduce such an enhanced exchange interaction in the surface layer in local theoretical models [14–16]. It turns out that an enhancement above roughly 50% results in live layers.

Table 2
Critical surface exponents β_1 (exponent of the first atomic layer) from experiments

Surface	Experimental method	β_1	Ref.
Ni(100)	SPLEED	0.81 (2)	[22]
Ni(110)	SPLEED	0.79 (2)	[22]
Ni(111)	ECS	1	[20]
Fe(100)	SPARPES	1	[18]
Fe(110)	spin polarized secondary electrons	0.83 (8)	[32]

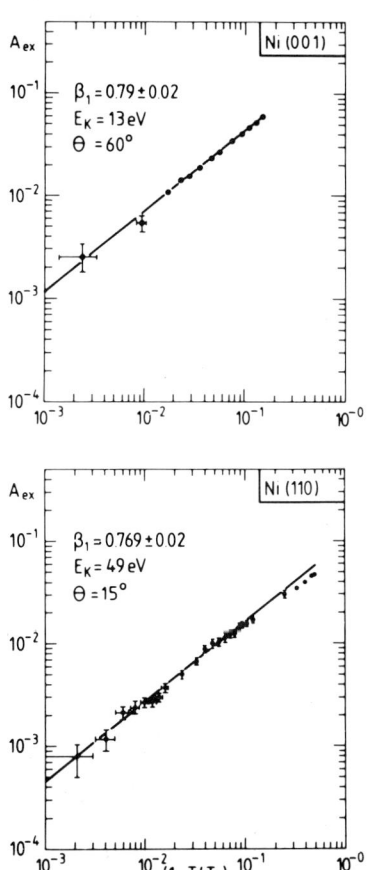

Fig. 1. Critical behaviour of Ni(100) and Ni(110) surfaces, by SPLEED [2]. Exchange asymmetry A_{ex} vs. $(1 - T/T_c)$ in a log–log plot to determine surface critical exponents β_1. Electron energies E and angles of reflexion θ as indicated (from Alvarado et al. [22]).

The first experimental indication of such a live layer came from unpolarized photoemission experiments of Klebanoff et al. [23] on Cr(100), where they observed some level splitting, which disappeared near 780 K, well above the Néel temperature $T_N = 312$ K of bulk antiferromagnetic Cr. With reference to a previous theoretical prediction of Grempel [24], Klebanoff et al. interpreted their observation as an exchange splitting and 780 K as a surface Curie temperature. However, previous spin-polarized photoemission experiments [25] had failed to detect this surface magnetic order on Cr.

For Gd(0001)-surfaces, Weller et al. [26] observed, using SPLEED, a surface live layer between the bulk Curie temperature $T_{c,b} = 293$ K and a surface critical temperature $T_{c,s}$ between 307 and 315 K. However, the phenomenon was observed only if the sample was cooled in a magnetic field.

Rau et al. [28,29] observed a similar live layer on Tb(0001) films on W(110), 50 Å thick, with $T_{c,s}/T_{c,b} = 1.14$, using ECS. They found a critical exponent $\beta = 0.35$ for this live layer. The only case where a magnetic live layer was reported on top of a crystal which is non-magnetic in bulk down to $T = 0$ is given by the V(100) surface, where Rau et al. [21] observed ferromagnetic order below $T_{c,s} = 540$ K, with $\beta \approx 1$, using ECS.

2.3. Thermal decrease of magnetic order at low temperatures and Bloch's law

For the case of temperatures well below the bulk Curie temperature $T_{c,b}$, that means in the regime of bulk spin-wave models, Rado [6] had noted again in 1957 that the surface or first layer magnetization m_i should follow, like the volume, the Bloch law

$$m_1 \sim \left(1 - b_1 T^{3/2}\right), \qquad (2)$$

however, with a spin-wave parameter of the first layer $b_1 = 2b_0$, enhanced twice in comparison with the bulk parameter b_0. It seems characteristic for the interest in the field at that time that Rado's result was published in the Bulletin of a conference only and not in a general Journal. Rado's prediction resulted from the appearance of spin-wave antinodes in the surface, which is expected for the case of missing or weak magnetic surface anisotropies. Mills and Maradudin [7] had substantiated Rado's statement by more rigorous spin-wave theory, and included surface modes.

An experimental confirmation of these ideas was performed first in 1982, for the case of the surface of a ferromagnetic glass, $Ni_{40}Fe_{40}B_{20}$, for which Pierce et al. [30], using SPLEED, could confirm a surface Bloch law (1), however with $b_1 = 3b_0$, enhanced beyond Rado's prediction. An explanation for this magnetic softening of the surface was given by Mathon et al. [1,31]. They

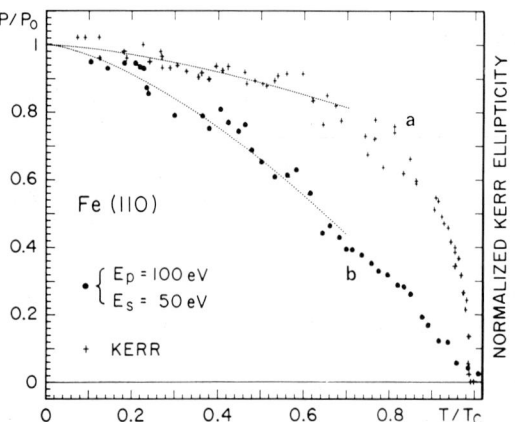

Fig. 3. Magnetic surface order versus temperature, for Fe(110), measured by Kerr ellipticity (+) for the bulk, and by the polarization P of scattered electrons (●) for the surface, respectively (from Taborelli et al. [32]).

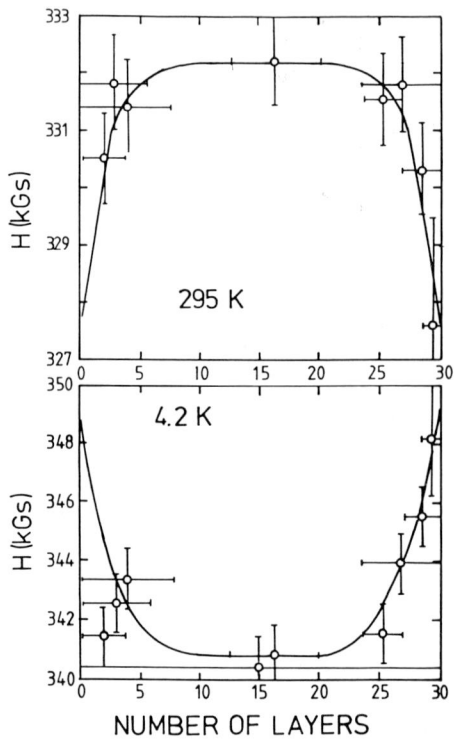

Fig. 4. Magnetic hyperfine field H versus position of ^{57}Fe-probe-layers in Fe(110)-films on Ag(111), consisting of 30 layers, coated by Ag. Near the Ag-coated surface (right), an increase of H towards the surface is observed in the ground state at 4.2 K. It is converted at 295 K to a decrease towards the surface by enhanced thermal decrease near the surface. Disturbances near the Ag(111)-substrate (left) indicate island-nucleation of the Fe-films (from Walker et al. [33]).

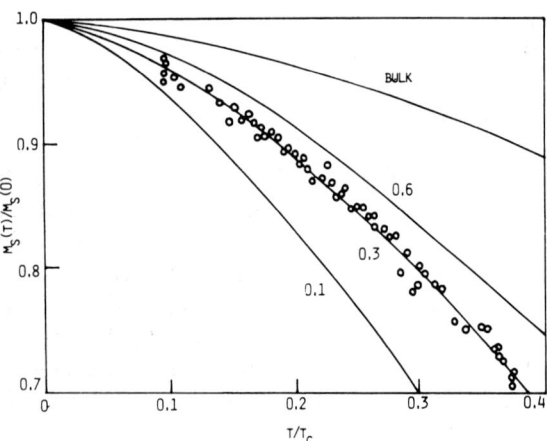

Fig. 2. Temperature dependence of (normalized) spin polarization of secondary electrons from a ferromagnetic glass $Ni_{40}Fe_{40}B_{20}$ [30], in comparison with the temperature dependence of surface magnetization $M_s(T)$, for a sc model, for different parameters J'/J, where J' and J are exchange parameters between the first and second layer and in the bulk material, respectively (from Mathon et al. [31]).

discussed the thermal decrease of surface magnetization for the case where the exchange interaction J' between the first and the second monolayer is reduced in comparison wit the bulk exchange J. They found that m_1 follows, in a wide range of temperatures, a pseudo-Bloch law, with increasing b_1 for decreasing J'. As shown in fig. 2, the experiments of Pierce et al. [30] can be interpreted in terms of a reduced surface exchange $J' = 0.3J$.

For the case of single-crystal Fe(110) surfaces, Taborelli et al. [32] compared the temperature dependence of scattered low energy electron polarization P, as a measure of surface magnetization, with Kerr rotation, as a measure of bulk magnetization, as shown in fig. 3, again resulting in an enhanced surface Bloch parameter $b_1 = 3b_0$.

A detailed analysis of thermal decrease of magnetic order and its local structure, near Fe-based surfaces or interfaces, was done using Mössbauer spectroscopy. The method is based on the observation that the magnetic hyperfine field $B_{hf}(T)$ follows the same temperature dependence as the magnetic moment μ of the electron shell, from which it originates. The proportionality between B_{hf} and $\mu(T)$ is known, from bulk materials, to hold to a very good approximation. Local analysis of $\mu(T)$ therefore becomes feasible if a probe layer of a Mössbauer isotope, e.g. ^{57}Fe for the case of Fe, is introduced by preparation near the surface of a crystal which consists otherwise of a non-Mössbauer isotope, e.g. ^{56}Fe. The method has been pioneered by the group of Walker [33] for Fe(110) films on Ag(111), using transmission Mössbauer spectroscopy. The enhancement of thermal decrease near Fe(110)/Ag interfaces could be clearly confirmed, as shown in fig. 4. However, true monolayer resolution could not be realized because of sensitivity problems of the transmission Mössbauer mode, and in situ structural testing of the Fe(110) surfaces before coating by Ag was not possible. Quantitative evaluation resulted, for a probe of 1.5 atomic layers of ^{57}Fe near the Ag-coated surface of a film of 30 atomic layers, in a Bloch law with $b_1 = 12(1) \times 10^{-6}$ K$^{-3/2}$ Compared with the bulk value $b_0 = 5.2 \times 10^{-6}$ K$^{-3/2}$ [34], this results in a surface enhancement $b_1/b_0 = 2.3$, only slightly above Rado's predictions.

In situ application and monolayer resolution of the Mössbauer method was realized by Korecki and Gradmann [35], who introduced the more

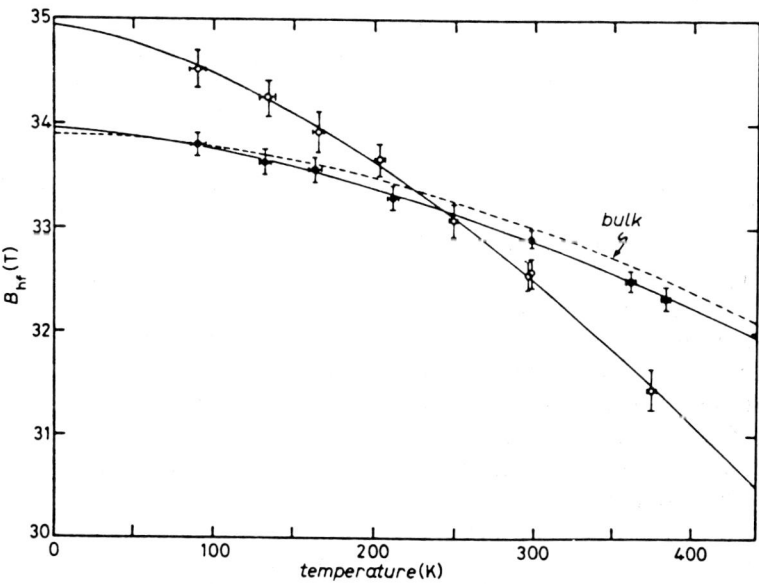

Fig. 5. Magnetic hyperfine field B_{hf} versus temperature, for the topmost atomic layer (○) and the central layer (●) of a 21 layers Fe(110)-film on W(110), coated by Ag (from Korecki et al. [35b]).

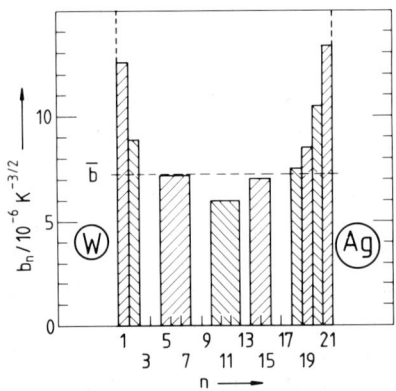

Fig. 6. Profile of single-layer spin-wave parameters b_n across a Fe(110)-film on W(110), coated by Ag, consisting of 21 atomic layers. Mean value $\bar{b}$ for comparison. The value in the center $b_{11} = 8.2 \times 10^{-6}$ K$^{-3/2}$) is only slightly enhanced with respect to the bulk value (5.2×10^{-6} K$^{-3/2}$) (from Przybylski et al. [36]).

sensitive detection of nuclear resonant absorption by the following emission of K-shell conversion electrons (Conversion Electron Mössbauer Spectroscopy, CEMS) to the field of surface magnetism. A detailed analysis of Fe(110) films on W(110), coated by Ag, again showed the enhanced thermal decrease of the first monolayer, as shown in fig. 5, whereas the center of the film follows nearly the bulk Bloch law. A profile of b-parameters across the film is symmetric, as shown in fig. 6.

Comparison of $b_1 = 13.3 \times 10^{-6}$ K$^{-3/2}$ in the Ag-coated surface with $b_1 = 6.2 \times 10^{-6}$ K$^{-3/2}$ from the film center results in an enhancement $b_1/b_{11} = 2.15$ very near Rado's prediction, and in agreement with Walkers results. Apparently, the enhancement of b_1 in Fe(110) beyond Rado's value, as shown in fig. 3, and the reduction of J', is a special feature of the free surface which is eliminated by metal coating.

There are indications that this difference between the free and the Ag-coated surface is dramatically enforced for Fe(100). For the case of the Ag-coated surface, Liu and Gradmann [37] obtained $b_1/b_0 = 2.7(4)$ using the CEMS method, corresponding to $J' \approx 0.5J$. In sharp contrast, Siegmann and Bagus [38] claimed a reduction of J' by the order of 10^{-5} for the free Fe(100) surface. This statement resulted from the interpretation of experiments of Allenspach et al. [39], who observed, using spin-polarized secondary electron emission, a surprisingly strong dependence of surface hysteresis loops on the information depth of the electrons used, interpreted in a tentative decoupling of the first monolayer by Siegmann and Bagus. A confirmation of this exciting result by alternative methods is missing.

3. Magnetic moments and hyperfine fields

The question of magnetic moments in surfaces of ferromagnetic crystals had been highlighted before 1975 by Liebermann et al. [40], who reported on non-magnetic "dead layers" even at $T = 0$ K on top of Ni surfaces. This was in contrast to the finding of a homogeneous ground state magnetization near the surface of 48Ni/52Fe(111) films in Cu [3]. Because Liebermann's experiments were not performed in UHV with clean surfaces, it was suggested [3] that his dead layers were an artifact of the preparation only. The magnitude of magnetic moments in free surfaces was an open question in 1975.

3.1. Magnetic moments in free surfaces

The progress in our knowledge on magnetic moments in surfaces and interfaces since 1975 is tightly connected with the development of theoretical methods of self-consistent, spin-dependent band-structure calculations in ultrathin film slabs, in particular the Full Potential Linearized Augmented Plane Wave (FLAPW) method [41], which is now widely used in several groups [42–44] and provides a powerful frame for calculating magnetic surface or interface moments. For extended reviews of theoretical results compare [42,44]. We concentrate our discussion on magnetic moments in the first monolayer for the cases where comparison with their experimental determination using SPLEED is possible. Results are collected in table 3 which we comment as follows:

It is now obvious that magnetic moments are enhanced in free surfaces of transition metal ferromagnets. There are definitely no dead layers in free surfaces. The agreement between different

Table 3
Magnetic moments in free surfaces of Ni and Fe. p_1 and p_b represent spin polarization (theory) or magneton number (experiment) for the first monolayer or bulk material, respectively, both in the ground state (extrapolated values for the experiments). Theoretical results from tight-binding and FLAPW-calculations, numbers of atomic layers in the calculated slabs are indicated. Experimental values from SPLEED on bulk or thick film surfaces

Surface	Method	$p_1 - p_b$	Ref.
Fe(100)	FLAPW, 7 layers	0.73	45
	FLAPW, 5 layers	0.71	43
Fe(110)	tight binding	0.43	46
	FLAPW, 9 layers	0.43	47
	SPLEED	0.8	48
Ni(100)	tight binding	0.18	49
	FLAPW, 5 layers	0.06	50
	FLAPW, 9 layers	0.12	51
	SPLEED	0.03 (3)	52
Ni(111)	tight binding	0.09	49
	FLAPW, 7 layers	0.05 (10)	53
	SPLEED	0.06 (12)	54

theoretical calculations, using FLAPW with different numbers of atomic layers in the slab or using tight binding approximations, is excellent for the case of Fe, it is moderate for the case of Ni. There is only qualitative agreement with experiments. For the case of SPLEED on Fe(110), the accuracy of the experimental result is somewhat uncertain because it was deduced from one spectral feature only, which reacts sensitively on p_1. For the case of Ni(100), the experimental result is definitely lower than any theoretical prediction.

3.2. Step moments

In simple terms, one may argue that the magnetic moments in surfaces are enhanced because surface atoms are something between free atoms, with $\mu = 4\mu_B$ for the case of Fe, and the bulk atoms with $\mu = 2.2\mu_B$. In agreement with this argumentation, the band-calculations show that the enhancement of the magnetic moment is connected with band narrowing at the surface. This interpretation suggests that atoms in surface steps or kinks might show even larger moments. Recently, Albrecht et al. [55] checked this suggestion by magnetometry of epitaxial Fe(110) films on W(110) in situ in UHV. Films with initially smooth surfaces were coated at lowered temperatures by some additional atomic layers of Fe, to create a high density of atomic steps which could be estimated from spot profile analysis LEED measurements. After measuring the magnetic moment (and anisotropies) of the film with the rough surface, the steps were eliminated by annealing and the magnetic measurements were repeated. The moment enhancement for step atoms in comparison with surface atoms could then be estimated to $\Delta\mu_{step} \approx 0.5 \mu_B$, in good qualitative agreement with model calculations using the theory of local magnetic moments of magnetic impurities in non-magnetic metals.

As a whole, the phenomenon of moment enhancement in surfaces and steps as a result of reduced magnetic coordination and resulting band narrowing became a well established phenomenon since 1975.

3.3. Moments in interfaces

Polarized electron methods are restricted to external surfaces, but interface magnetism is interesting as well. During the period, band-structure calculations of ultrathin film slabs became a powerful tool for interface magnetism, too, and it is now possible to calculate virtually any slab structure of up to roughly 10 layers, if a common two-dimensional translation symmetry is preserved. For a review compare ref. [42]. It is nice that ultrathin films, which thus became a tool of theoretical research in surface magnetism, have been used as a tool of experimental surface magnetism before [3] and increasingly since 1975. What can be measured easily, at least in principle, is the change of magnetic moments of ultrathin films during coating processes, giving directly the difference between free surface and interface moments. The method has been improved during the period by progress in methods of epitaxial growth and structural testing of ultrathin films and by the availability of torsion-oscillation magnetometry (TOM) with submonolayer sensitivity during growth processes in situ in UHV [56].

Application of the method to selected examples showed a tight connexion between interface and alloy magnetism, and good accordance with band calculations. We note the following examples:

1. In *NiPd-interfaces*, Pd is magnetically polarized [56] in tight analogy to its polarization in the well-known "giant moments" around Ni-impurities in Pd, the range of polarization in interfaces being in quantitative agreement with its range in the giant moments [57,1], compare fig. 7. The same polarization has been detected in FePd-interfaces using FMR [58] and by spin polarized photoelectron emission from ultrathin Pd on Fe(110) [59].

2. *Coating of Ni(111) films by Cu(111)* results in a strong drop of magnetic moment [60] induced by the first two layers of Cu(111), compare fig. 8, in full analogy to the loss of moments of Ni by Cu-impurities, by filling Cu-electrons in Ni-d-holes, in quantitative agreement with tight-binding band calculations [49].

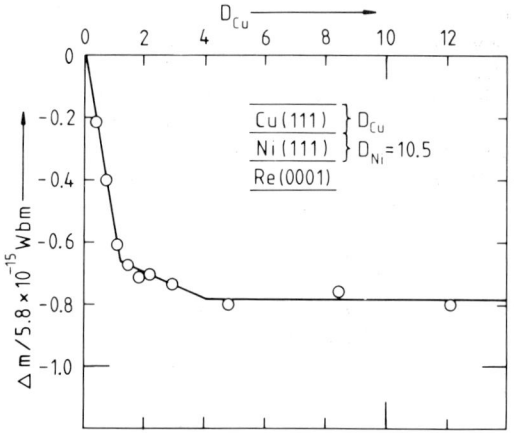

Fig. 8. Change of magnetic moment Δm, in units of the magnetic moment of one monolayer Ni(111), $m_{ML} = 5.8 \times 10^{-15}$ Wbm, caused by Cu(111) on a 10.5 layers Ni(111)-film, as a function of the number of Cu-layers, D_{Cu} (from Bergter et al. [60]).

3. For comparison a *coating of Fe(110) films by Cu, Ag or Au* does not change the magnetic moment in the limits of accuracy ($\approx 0.02\mu_B$/atom), in accordance with the more local character of Fe-moments in bulk Fe [61]. There is some discrepancy to FLAPW-calculations of Ohnishi et al. [62] who found in the first layer of a Fe(100)-surface, a moment enhancement of 0.7 μ_B, which was reduced by Ag-coating to 0.3 μ_B.

The combination of in situ interface magnetometry with band calculations became a powerful approach of surface magnetism in the period.

3.4. Magnetic hyperfine fields in Fe: Local analysis and Friedel-oscillations

An interesting aspect of spin-dependent band calculations is that they result in quantitative predictions on the main contribution of the magnetic hyperfine field B_{hf}, the Fermi-contact contribution which results from the s-electron polarization at the nucleus. For Fe, where B_{hf} can easily be measured by Mössbauer spectroscopy, this provides an interesting test of theoretical predictions.

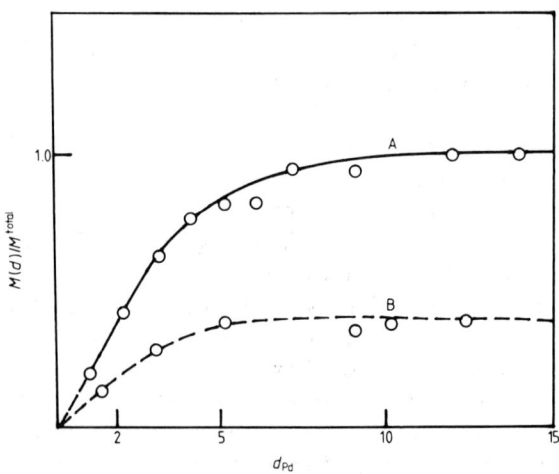

Fig. 7. Magnetic moment M of Pd-films on Ni(111)-substrates consisting of 30 (A) and 6 (B) atomic layers, versus the number of Pd-layers, d_{Pd}. M^{total} corresponds to $0.5\mu_B$ per interface Pd-atom, for (A). Experimental points from Gradmann and Bergholz [56]; full curve calculated by Mathon [57] (from Mathon [1]).

As pointed out above in section 2, the Mössbauer technique, in particular CEMS, provides a powerful tool for the local analysis of magnetic hyperfine field B_{hf} in Fe. Because B_{hf} is caused finally by the magnetic moments of the electron shell, there was originally some hope, persisting in 1975, to use B_{hf} for *local* analysis of surface magnetization. This is not true, as was shown by Ohnishi et al. [62] in their pioneering calculations of Fe(100) surfaces [62b] and Ag/Fe(001)-interfaces [62]. For the free Fe(100)-surface, an enhancement of the first layer moment by 32% was predicted in combination with a reduction of B_{hf} by 30%; the explanation is that B_{hf} contains a contribution from conduction electron polarization which of course depends on the magnetic state of neighbour atoms and therefore destroys the proportionality between B_{hf} and the electronic magnetic moment of the shell around the probe nucleus: B_{hf} cannot be taken as a local probe of the magnitude of magnetic moment.

One remarkable result of the analysis of Fe(100) by Ohnishi et al. [62a] was the prediction of a local oscillation of B_{hf} near the surface, resulting from an oscillation in s-electron polarization which in turn is a result of the combined cut-offs of the electron gas at the Fermi-level in k-space and at the surface in real space. Accordingly, the oscillation was predicted to appear near the free surface only and to disappear after Ag-coating. Experimental verification has not yet been performed for the case of Fe(100); for the case of Fe(110), however, it became possible using in situ CEMS [35] compare fig. 9; results are in qualitative agreement with calculations by Freeman and Fu [62b]. Note that the reduction of B_{hf} in the first monolayer is combined with an enhancement of the magnetic moment by at least 20%, as discussed above.

The disproportionality between magnetic moment and hyperfine field is even more expressed for a Ag-coated ferromagnetic monolayer Fe(110) on W(110) [63,64], for which a reduction of $B_{hf}(0)$ from 34 T (in bulk) to 12 T in the submonolayer as measured by CEMS [64,65], is connected with an increase of the magnetic moment from 2.22 to $2.53(12)\mu_B$, as measured by TOM [66], in excellent agreement with FLAPW-calculations by Hong et al. [67] ($B_{hf}(0) = 12.2$ T, $\mu(0) = 2.17\mu_B$.

4. Magnetic surface anisotropies

In general magnetic anisotropies are the result of structural anisotropies. It was Néel who noted

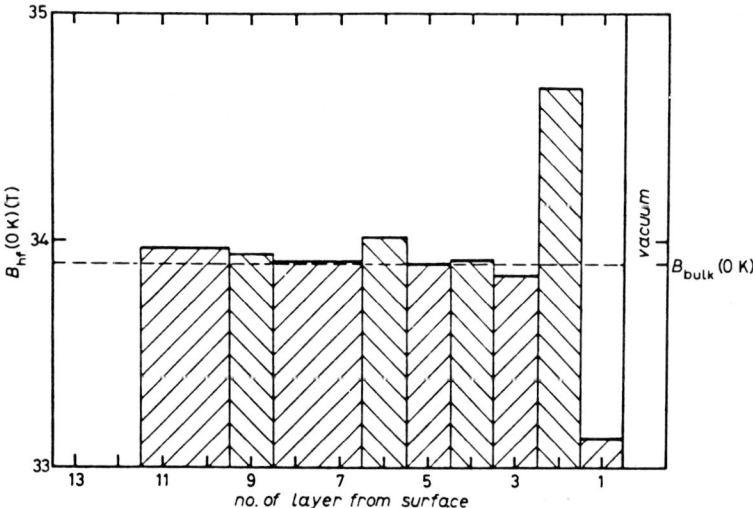

Fig. 9. Ground state magnetic hyperfine field, $B_{hf}(0)$, near the free surface of an Fe(110)-film on W(110), consisting of 21 atomic layers, versus number of layer from the free surface, determined by extrapolation of CEMS-measurements at finite temperatures (from Korecki and Gradmann [35]).

as early as 1954 [68] that the basic symmetry loss in a ferromagnetic surface must result in a magnetic surface anisotropy (MSA), that means a magnetic contribution to the surface energy σ the leading term of which is given by

$$\sigma = K_s \cos^2\vartheta, \qquad (3)$$

where ϑ is the angle between J_s of a homogeneously magnetized sample and the surface normal. In a phenomenological approach to magnetic anisotropies based on an anisotropic pair interaction

$$w = l \cos^2\phi \qquad (4)$$

for nearest neighbours, ϕ being the angle between magnetic moment and pair axis, Néel could connect the surface anisotropy constant K_s with magnetoelastic volume data and predict K_s of the order 0.1 to 1×10^{-3} Jm^{-2} for 3d-metal surfaces. This corresponds to surface anisotropy fields $\mu_0 H_s$ of the order of a few teslas, which could easily overcome in magnitude the shape anisotropy in ultrathin films, thus inducing perpendicular magnetization in ultrathin films, for appropriate sign of K_s, despite of shape anisotropy. The first experimental confirmation of MSA was performed in 1969 [69] using ultrathin NiFe(111)-films on Cu(111), where a linear dependence of the film anisotropy on $1/d$ (film thickness d) could be interpreted in terms of surface anisotropies, in quantitative agreement with Néel's predictions. This work included the first observation of perpendicular magnetization as a result of perpendicular MSA.

The basic concept of MSA as a result of broken symmetry in surfaces, its measurement using ultrathin films and its order of magnitude, able to induce perpendicular magnetization, was thus present before 1975.

The period since 1975 brought the general perception that magnetic surface anisotropy is not an esoteric special effect but a basic magnetic property of surfaces with immense importance for forthcoming magnetic thin film and multilayer devices.

Extended experimental research on MSA since 1975, mostly based on anisotropy measurements

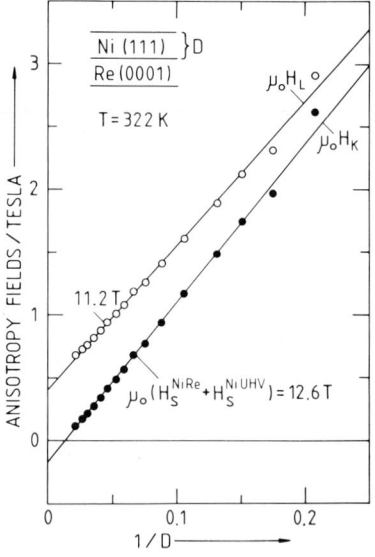

Fig. 10. Surface anisotropy in Ni(111). Total anisotropy field $\mu_0 H_L = \mu_0(2L/J_s)$, of anisotropies with respect to the surface normal, versus $1/D$, for Ni(111)-films on Re(0001), consisting of D layers. After subtraction of shape anisotropy J_s, which depends linearly on $1/D$, too, a crystalline contribution $\mu_0 H_K$ results, the slope of which gives a total surface anisotropy field of 12.6 T (from Gradmann [70]).

of ultrathin films using torsion oscillation magnetometry (TOM [70,71]), Brillouin light scattering (BLS [72]) and ferromagnetic resonance (FMR [72]) has been strongly promoted by the interest in perpendicular magnetization for recording devices. The method to determine MSA from a plot of film anisotropies versus $1/d$ has now been widely accepted, an example is shown in fig. 10. The results were collected in several reviews [5,70–73] and may be summarized as follows:

1. Free magnetic surfaces show surface anisotropy constants K_s with magnitudes of the order of 0.1 to 1×10^{-3} Jm^{-2}.
2. Both easy plane ($K_s > 0$) and easy axis (perpendicular, $K_s < 0$) MSA has been observed frequently. A fictitious preference of perpendicular MSA is pretended only because it is qualitatively indicated by perpendicular magnetization of ultrathin films. Such a qualitative indication is missing for $K_s > 0$, where MSA supports shape anisotropy. (Unfortunately, there is some confusion in sign of K_s, which

was changed by some authors with respect to Néel's original definition, which we prefer and retain.)

3. In lower symmetric surfaces, like Fe(110), in-plane anisotropies are observed in addition to the out-of-plane MSA according to eq. (2). Anisotropy constants and anisotropy fields are of the same order of magnitude as for the out-of-plane MSA [71]. The in-plane easy axis of the MSA (e.g. (110) in Fe(110)) can be at right angles to the bulk one, resulting in an in-plane magnetization switching (at ≈ 40 layers for Fe(110)).
4. MSA can depend sensitively on the chemical state of the surface. MSA tends to be maximum for free surfaces and to be reduced by coverage with any solid coating or reaction with gases.
5. MSA of free surfaces depends on roughness. Step anisotropies could be determined recently [74].
6. Néel's phenomenological pair-bonding model provides a useful connexion between different types of MSA (in-plane, out-of-plane, step) [66,67,74], and an order of magnitude connexion with bulk strain anisotropies.

The calculation of MSA by first principles band theories turned out to be an extremely difficult task. Even with restriction to such easy and unrealistic models like a free (100) Fe-monolayer the results of different groups differ in sign: Whereas Gay and Richter [75] predicted for the free Fe(100)-monolayer perpendicular magnetization, in agreement with experimental results from Fe(100) films on Ag(100) [72], a calculation of Karas et al. [43] suggested in-plane magnetization. Most recent calculations, in which noble metal substrates of Fe(001) monolayers were modelled very roughly only by noble metal monolayers [76] are consistent with some experimental observations. Calculations of magnetic anisotropies in Co Pd-multilayers [77] show encouraging correlation with experimental values. As a whole, however, a reliable theory of MSA is missing at present and it seems realistic to take surface anisotropy constants as empirical parameters, as is common use for bulk anisotropies.

5. Domain walls at surfaces

The surface has a profound effect on the structure of a domain wall. In the volume, the equilibrium wall structure is a Bloch wall, characterized by a constant magnetization component perpendicular to the wall plane, because this geometry is free from stray-fields. If such a Bloch wall would be terminated by a surface, a magnetization component perpendicular to the surface would result with energetically unfavourable stray fields. Micromagnetic calculations of walls near surfaces [78–80] had shown before 1975 that it is energetically favourable to form Néel-type closure surface structures, virtually free from stray field energies because the magnetization is rotated parallel to the surface. For thin films this results in the so-called asymmetric Bloch wall [78,79], for thick films or bulk surfaces in the asymmetric Néel-termination of a Bloch wall [80]. The experimental confirmation of this concept is a result of the period since 1975. The Néel-termination was detected first using magneto-optic Kerr-microscopy [81], however, a local resolution of the wall structure was not possible with the optical microscope. Great progress became possible by the introduction of scanning electron microscopy with polar-

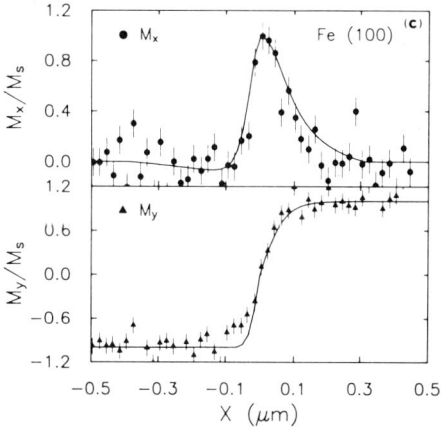

Fig. 11. Asymmetric Néel-termination of a Bloch wall on a Fe(100)-surface. Surface magnetization components in the surface plane M_x(●) perpendicular and M_y(▲) parallel to the wall plane, respectively, as measured by SEMPA [84], in comparison with micromagnetic calculation [86] (full lines); the component along the surface normal is zero both in theory and experiment (from Scheinfein et al. [86]).

ization analysis of secondary electrons (SEMPA) [82,83] which is now available with local resolution down to 40 nm [84]. This enabled a local resolution of the asymmetric spin-structure of the Néel termination [84,85] which agrees quantitatively with micromagnetic calculations [86] compare fig. 11. This quantitative confirmation of micromagnetic theory for a wall structure at a surface provides a strong support of theoretical results for wall structures in general, including walls in the volume where a comparable experimental test has not yet been possible. The surface wall structure is determined basically by bulk anisotropies and magnetostatic interactions. The agreement with the experiment could not be improved by inclusion of surface anisotropies in the calculations. Surface anisotropies have no detectable influence on the surface wall structure [86].

6. Surface magnetochemistry

The dependence of magnetic surface properties on the chemical state of the surface has been studied before 1975 in the pioneering work of Selwood [87] on chemical reactions at magnetic small particles. He found a decrease of magnetic moment in Ni-particles as a result of chemisorption of H_2, O_2 and CO.

The evolution of spin polarized electron techniques since 1975 opened a wide field to get information on magnetic phenomena connected with adsorption processes on single crystal surfaces and it seems that the inherent surface sensitivity of these techniques can be used much easier for following the magnetic aspects of reactions than for clean surfaces. The attentive reader may have missed the extended role of spin polarized techniques in the analysis of free surface magnetic properties which is expected from their inherent surface sensitivity. The problem is that up to now the separation of bulk and surface information in spin-polarized electron experiments remained an unsolved problem and most experiment e.g. in spin-polarized photoemission or inverse photoemission have been interpreted in terms of bulk band structures.

This is quite different if adsorption processes are followed by spin polarized techniques, where

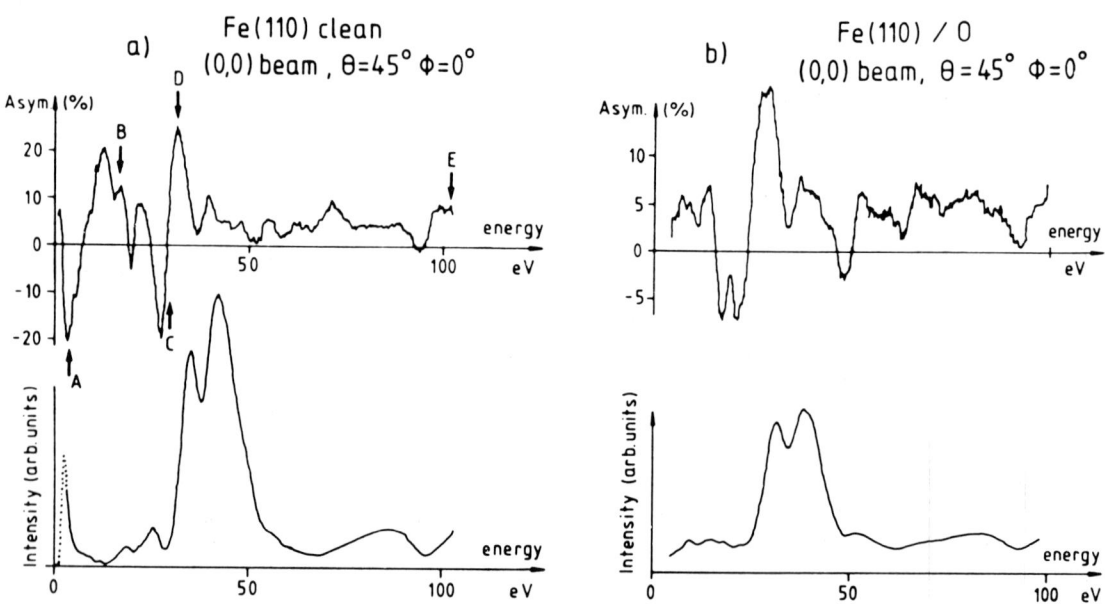

Fig. 12. Spin-polarized electron scattering from clean and oxygen-covered Fe(110)-surfaces. Scattering asymmetry $A_{sym} = (I^\uparrow - I^\downarrow)/(I^\uparrow + I^\downarrow)$ and reflected intensity of the specular beam versus electron energy, for (a) clean and (b) oxygen-covered Fe(110). A_{sym} reacts in far more sensitively on the reaction than the intensity (from Kirschner [88]).

the surface nature of any change in spin asymmetries is immediately evident. It is the general experience of the period since 1975 that spin asymmetries in low energy processes at surfaces depend sensitively on adsorption, as shown for an example in fig. 12.

As a further example, we show the result of a spin polarized inverse photoemission study of O_2-adsorption on Ni(110) in fig. 13 [88]. The peak for minority spin irradiation near the Fermi-level E_F indicates the absorption of electrons in empty minority-states just above E_F. Its rapid decrease with O_2-adsorption indicates that the d-holes are filled effectively by the adsorbed O_2. Similarly, in a spin-polarized photoemission study of Ni(100) [90] a reduction of spin-polarization was observed as a result of O_2-adsorption. Both experiments

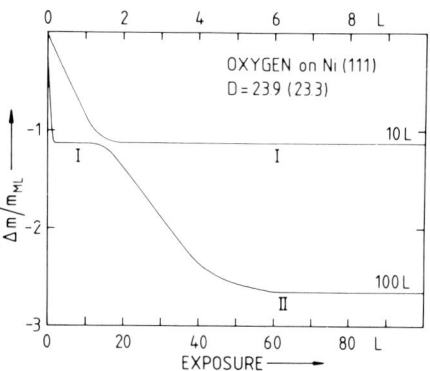

Fig. 14. Magnetometry of O_2-adsorption on Ni(111). Change of magnetic moment Δm, in units of the moment m_{ML} of one monolayer Ni(111), versus oxygen-exposure. Initial range (0–10 L) and extended range (100 L) with different exposure scales (from Elmers and Gradmann [91]).

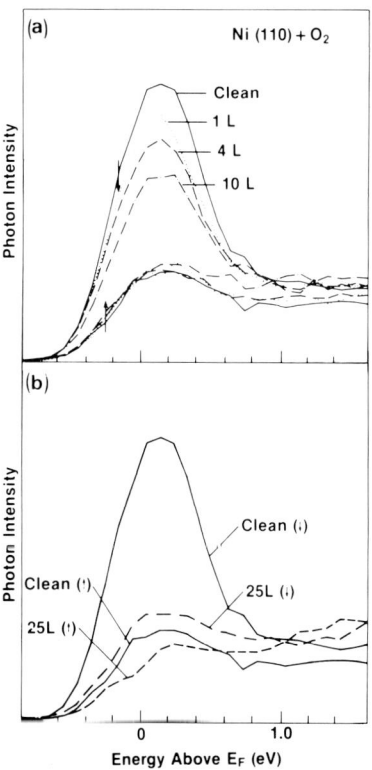

Fig. 13. O_2-adsorption in Ni(110) analyzed by spin-polarized inverse photoemission. Photon intensities at $\hbar\omega = 9.7$ eV, caused by incident electrons having their spin parallel (↑) or antiparallel (↓) to the spin of Ni majority electrons, versus final state electron energies. O_2-exposures in L (1 L ≡ 1 langmuir = 10^{-6} Torr s) (from Pierce et al. [89]).

indicate a reduction of magnetic moment in Ni(100) as a result of O_2-adsorption. A more direct measurement of this decrease of magnetic moment was done for the case of Ni(111) using in situ magnetometry of Ni(111) films on Re(0001) during oxygen adsorption [91], as shown in fig. 14. A rapid decrease caused by less than 2 L oxygen saturates in a chemisorption state which is followed above 10 L by a further decrease, caused by the formation of a NiO-layer. The initial rapid decrease corresponds to a giant loss of magnetic moment of $4.5\mu_{Ni}$ per O-atom, which is far from being understood; it is connected with a dramatic decrease of the surface anisotropy field from 10 T to 0 T, again far from being understood.

However, the loss of magnetic moment does by now mean result in a non-magnetic state of the Ni surface. This has been shown in a spin-polarized inverse photoemission study of Schönhense et al. [92], who detected an exchange splitting in oxygen-derived emission peaks from Ni(110)–(2 × 1)-O, compare fig. 15.

Similar exchange splittings in adsorbate induced bands have been observed on Fe(001) for p(1 × 1) oxygen and c(2 × 2) sulfur by use of spin-polarized angle-resolved photoemission [93], indicating magnetic moment in the adsorbate with ferromagnetic alignment to the substrate.

Chemisorption on bulk single-crystal Ni(110)-surface can even change the magnetization curves

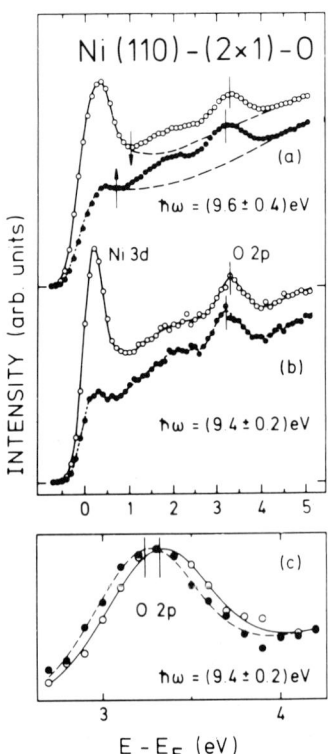

Fig. 15. Exchange splitting of an O-derived band at Ni(110). Photon intensities at (a) 9.6 and (b) 9.4 eV, respectively, caused by irradiation with electrons with minority (○) or majority (●) polarization, respectively. Dashed curves, in (a), correspond to clean Ni. In (c), the background asymmetry has been suppressed (from Schönhense et al. [92]).

of the bulk crystal as a whole [94]; this supports an early conjecture of Zijlstra [95] that surfaces may serve as nucleation centers for reversal domains in remagnetization processes.

The few examples where magnetic surface properties have been followed during adsorption processes at single crystal magnetic surfaces certainly show that surface magnetochemistry is a fascinating field for future research.

7. Summary and outlook

What are the main advantages in surface magnetism, what are the challenges for the future?

(1) Powerful experimental methods have been developed since 1975, and are now available for magnetic analysis of surfaces in UHV, in particular to analyze the spin-polarized electronic surface structure using spin-polarized electron techniques, to measure magnetic moments by ultrathin films magnetometry, to measure magnetic surface anisotropies using ferromagnetic resonance, Brillouin light-scattering and torsion magnetometry and to follow the local structure of magnetic hyperfine fields using Mössbauer spectroscopy.

(2) The temperature dependence of surface magnetization is basically understood. Sophisticated theoretical techniques have been developed to calculate both low-temperature and critical behaviour, by simulations and by analytical methods. Interpretation of surface sensitive experiments was comparatively easy, because most magnetic signals could be assumed to scale with the magnetization and important theoretical predictions could be confirmed. Bloch laws with an enhanced spin-wave parameter at low temperatures and critical power-laws with a surface magnetization exponent $\beta_s \approx 0.8$ are now generally accepted results.

(3) Powerful theoretical methods became available for first principles calculations of band structures, magnetic moments and hyperfine fields near surfaces and interfaces. Important results like the enhancement of moments in surfaces and oscillation of hyperfine fields could be confirmed. An experimental separation of surface band structures from bulk ones remains to be done.

(4) Magnetic surface anisotropy as a result of broken local symmetry in surfaces has now generally been accepted as an important basic surface property, and reliable experimental data became available for selected surfaces. Whereas early phenomenological theories remain useful for qualitative understanding of surface anisotropies, a microscopic theory remains a challenge for future research.

(5) The remarkable sensitivity of magnetic surface properties on the structural and physicochemical state of a surface has been shown up in some published examples; even more, it is the unpublished experience of many experi-

mentalists who know, from their effort to isolate magnetic properties of clean surfaces, that magnetic properties are frequently the most sensitive indicators of surface processes. Magnetochemistry of surface processes is a rich and mostly open field for future research.

(6) The combination of integrating experimental methods of surface magnetism with existing and forthcoming magnetic surface microscopies and micromagnetic analysis opens now a chance and challenge for a realistic and qualitative understanding of the role of surfaces in magnetic reversal processes.

References

[1] J. Mathon, Rep. Progr. Phys. 51 (1988) 1.
[2] U. Gradmann, in: Landolt–Börnstein, Neue Serie, ed. O. Madelung (to be published).
[3] U. Gradmann, Appl. Phys. 3 (1974) 161.
[4] A.J. Freeman and C.L. Fu, J. Appl. Phys. 61 (1987) 3356.
[5] U. Gradmann, Magnetism in Transition Metal Films, in: Ferromagnetic Materials, ed. K.H.J. Buschow (North-Holland, Amsterdam, 1992).
[6] G.T. Rado, Bull. Am. Phys. Soc. 2 (1957) 127.
[7] D.L. Mills and A.A. Maradudin, J. Phys. Chem. Solids 28 (1967) 1855.
[8] T. Wolfram, R.E. de Wames and W.F. Hall, Surf. Sci. 28 (1971) 45.
[9] K. Binder and P.C. Hohenberg, Phys. Rev. B 9 (1974) 2194.
[10] P.W. Palmberg, R.W. de Wames, L.A. Vredevoe and T. Wolfram, J. Appl. Phys. 40 (1969) 1158.
[11] K. Binder and D.P. Landau, Phys. Rev. Lett. 52 (1984) 318.
[12] D.P. Landau, R. Pandey and K. Binder, Phys. Rev. B 39 (1989) 12302.
[13] K. Ohno, Y. Okabe and A. Morita, Progr. Theor. Phys. 71 (1984) 714.
[14] H.W. Diehl and E. Eisenriegler, Phys. Rev. B 30 (1984) 300.
[15] H.W. Diehl and A. Nüsser, Phys. Rev. Lett. 56 (1986) 2834.
[16] K. Binder, in: Spin Polarized Electrons in Surface Physics, ed. R. Feder (World Scientific, Singapore, 1985).
[17] R. Feder, ed., Spin Polarized Electrons in Surface Physics (World Scientific, Singapore, 1985).
[18] E. Kisker, K. Schröder, W. Gudat and M. Campagna, Phys. Rev. B 31 (1985) 329.
[19] M. Landolt, R. Allenspach and D. Mauri, J. Appl. Phys. 57 (1986) 3626.
[20] C. Rau, K. Kuffner, J. Magn. Magn. Mater. 54–57 (1986) 767.
[21] C. Rau, C. Liu, A. Schmalzbauer and G. Xing, Phys. Rev. Lett. 57 (1986) 2311.
[22] S.F. Alvarado, M. Campagna, F. Ciccacci and H. Hopster, J. Appl. Phys. 53 (1982) 7920.
[23] L.E. Klebanoff, S.W. Robey, G. Liu and D.A. Shirley, Phys. Rev. B 30 (1984) 1048.
[24] D.R. Grempel, Phys. Rev. B 24 (1981) 3928.
[25] F. Meier, D. Pescia and T. Schriber, Phys. Rev. Lett. 8 (1982) 645.
[26] D. Weller, S.F. Alvarado, W. Gudat, K. Schröder and M. Campagna, Phys. Rev. Lett. 54 (1985) 1555.
[27] D. Weller and S.F. Alvarado, Phys. Rev. B 37 (1988) 9911.
[28] J. Rau and C. Jin, J. de Phys. 49 (1988) C8-1627.
[29] C. Rau, Appl. Phys. A 49 (1989) 579.
[30] D.T. Pierce, R.J. Celotta and J. Unguris, Phys. Rev. B 26 (1982) 2566.
[31] B.S. Ahmad, J. Mathon and M.S. Phan, J. de Phys. 49 (1988) C8-1639.
[32] M. Taborelli, O. Paul, O. Züger and M. Landolt, J. de Phys. 49 (1988) C8-1659.
[33] (a) A.H. Owens, J. Tysai, G. Bayreuther and J.C. Walker, Hyperfine Interactions 10 (1981) 805.
(b) J.C. Walker, R. Droste, G. Stern and J. Tyson, J. Appl. Phys. 55 (1984) 2500.
[34] M.A. Butler, G.K. Wertheim and D.N.E. Buchanan, Phys. Rev. B 5 (1972) 990.
[35] J. Korecki and U. Gradmann, Phys. Rev. Lett. 55 (1985) 2491.
[35a] J. Korecki and U. Gradmann, Europhys. Lett. 2 (1986) 651.
[36] M. Przybylski, U. Gradmann and J. Korecki, J. Magn. Magn. Mater. 69 (1987) 199.
[37] G. Liu and U. Gradmann, to be published.
[38] H.C. Siegmann and P.S. Bagus, Phys. Rev. B 38 (1988) 10434.
[39] R. Allenspach, M. Taborelli, M. Landolt and H.C. Siegmann, Phys. Rev. Lett. 56 (1986) 953.
[40] L.N. Liebermann, J. Clinton, D.M. Edwards and J. Mathon, Phys. Rev. Lett. 25 (1969) 539.
[41] E. Wimmer, H. Krakauer, M. Weinert and A.J. Freeman, Phys. Rev. B 24 (1981) 864.
[42] A.J. Freeman, A. Continenza and C. Li, in: Atomic-Level Properties of Interface Materials, eds. S. Yip and D. Wolf (Chapman and Hall, London, 1990).
[43] W. Karas, J. Noffke and L. Fritsche, J. de Chim. Phys. 86 (1988) 861.
[44] S. Blügel, B. Drittler, R. Zeller and P.H. Dederichs, Appl. Phys. A 49 (1989) 547.
[45] S. Ohnishi, A.J. Freeman and M. Weinert, Phys. Rev. B 28 (1983) 6741.
[46] R.H. Victoria, L.M. Falicov and S. Ishida, Phys. Rev. B 30 (1984) 3896.
[47] C.L. Fu and A.J. Freeman, J. Magn. Magn. Mater. 69 (1987) L1.

[48] E. Tamura, R. Feder, G. Waller and in. Gradmann, Phys. Stat. Sol. (b) 157 (1990) 627.
[49] J. Tersoff and L.M. Falicov, Phys. Rev. B 26 (1982) 6186.
[50] O. Jepsen, J. Madsen and O.K. Andersen, Phys. Rev. B 26 (1982) 2790.
[51] E. Wimmer, A.J. Freemann and H. Krakauer, Phys. Rev. B 30 (1984) 3113.
[52] R. Feder, S.F. Alvarado, E. Tamura and E. Kisker, Surf. Sci. 127 (1983) 83.
[53] C.L. Fu and A. Freeman, J. de Phys. 49 (1988) C8-1625.
[54] G.A. Mulhollan, A.R. Köymen, D.M. Lind, F.B. Dunning, G.K. Walters, E. Tamura and R. Feder, Surf. Sci. 204 (1988) 503.
[55] M. Albrecht, T. Furubayashi, U. Gradmann and W. Harrison, to be published.
[56] U. Gradmann and R. Bergholz, Phys. Rev. Lett. 52 (1984) 35.
[57] G. Mathon, J. Phys. F 16 (1986) 669.
[58] S. Celinski, B. Heinrich, J.F. Cochran, W.B. Muir and A.S. Arrott, Phys. Rev. Lett. 65 (1990) 1156.
[59] W. Weber, D.A. Wesner, G. Güntherodt and U. Linke, Phys. Rev. Lett. 66 (1991) 942.
[60] E. Bergter, U. Gradmann and R. Bergholz, Solid State Commun. 53 (1985) 565.
[61] H.J. Elmers and U. Gradmann, J. Appl. Phys. 64 (1988) 5328.
[62] (a) S. Ohnishi, M. Weinert and A.J. Freeman, Phys. Rev. B 30 (1984) 36.
(b) A.J. Freeman and C.L. Fu, in: Magnetic Properties of Low-Dimensional Systems, eds. L.M. Falicov and J.L. Moran-Lopéz (Springer, Berlin, 1986).
[63] U. Gradmann, M. Przybylski, H.J. Elmers and G. Liu, Appl. Phys. A 49 (1989) 563.
[64] U. Gradmann, G. Liu, H.J. Elmers and M. Przybylski, Hyperfine Interactions 57 (1990) 1845.
[65] M. Przybylski and U. Gradmann, Phys. Rev. Lett. 59 (1987) 1152.
[66] H.J. Elmers, G. Liu and U. Gradmann, Phys. Rev. Lett. 63 (1989) 566.
[67] S.C. Hong, A.J. Freemann and C.L. Fu, Phys. Rev. B 38 (1988) 12156.
[68] L. Néel, J. Phys. Radium 15 (1954) 225.
[69] U. Gradmann and J. Müller, Phys. Stat. Sol. 27 (1968) 313.
[70] U. Gradmann, J. Magn. Magn. Mater. 54–57 (1986) 733.
[71] H.J. Elmers and U. Gradmann, Appl. Phys. A 51 (1990) 255.
[72] B. Heinrich, K.B. Urquhart, J.R. Dutscher, S.T. Purcell, J.F. Cochran and A.S. Arrott, J. Appl. Phys. 63 (1988) 3863.
[73] P. Bruno and J.P. Renard, Appl. Phys. A 49 (1989) 499.
[74] M. Albrecht, T. Furubayashi, H. Przybybski, J. Korecki and U. Gradmann, to be published in J. Magn. Magn. Mater.
[75] J.G. Gay and R. Richter, Phys. Rev. Lett. 56 (1986) 2728.
[76] C. Li and A.J. Freeman, J. Appl. Phys. (1991) in press.
[77] G.H.O. Daalderop, P.J. Kelly and M.F.H. Schmurmans, Phys. Rev. B 42 (1990) 7270.
[78] A. Hubert, Phys. Stat. Sol. (b) 32 (1969) 519.
[79] A.E. La Bonte, J. Appl. Phys. 40 (1969) 2450.
[80] A. Hubert, Z. Angew. Phys. 32 (1971) 58.
[81] F. Schmidt, W. Rave and A. Hubert, IEEE Trans. Magn. MAG-21 (1985) 1596.
[82] K. Koike and K. Hayakawa, Jpn. J. Appl. Phys. 23 (1984) L187.
[83] J. Unguris, G.G. Hembree, R.J. Celotta and D.T. Pierce, J. Microsc. 139 (1985) SRP 1–2.
[84] H.P. Oepen and J. Kirschner, Phys. Rev. Lett. 62 (1989) 819.
[85] M.R. Scheinfein, J. Unguris, R.J. Celotta and D.T. Pierce, Phys. Rev. Lett. 63 (1989) 668.
[86] M.R. Scheinfein, J. Unguris, J.L. Blue, K.J. Coakley, D.T. Pierce, R.J. Celotta and P.J. Ryan, Phys. Rev. B 43 (1991) 3395.
[87] P.W. Selwood, Chemisorption and Magnetization (Academic Press, New York, London, 1975).
[88] J. Kirschner, Surf. Sci. 138 (1984) 191.
[89] D.T. Pierce, A. Seiler, C.S. Feigerle, J.L. Penna and R.J. Celotta, J. Magn. Magn. Mater. 54–57 (1986) 617.
[90] W. Schmitt, H. Hopster and G. Güntherodt, Phys. Rev. B 31, (1985) 4035.
[91] H.J. Elmers and U. Gradmann, Surf. Sci. 193 (1988) 94.
[92] G. Schönhense, M. Donath, U. Kolac and V. Dose, Surf. Sci. 206 (1988) L 888.
[93] P.D. Johnson, A. Clarke, N.B. Brookes, S.L. Hulbert, B. Sinkovic and N.V. Smith, Phys. Rev. Lett. 61 (1988) 2257.
[94] M. Donath, G. Schönhense, K. Ertl and V. Dose, Appl. Phys. A 50 (1990) 49.
[95] H. Zijlstra, IEEE Trans. Magn. MAG-15 (1979) 1246.

Electronic structure theory of surface, interface and thin-film magnetism

A.J. Freeman and Ru-qian Wu

Department of Physics and Astronomy, Northwestern University, Evanston, IL 60208-3112, USA

Low dimensional magnetic systems including surfaces, interfaces and thin-films, have attracted a great amount of attention in the past decade because, as expected, the lowered symmetry and coordination number offer a variety of opportunities for inducing new and exotic phenomena and so hold out the promise of new device applications. Local spin density functional (LSDF) ab initio electronic structure calculations played a key role in the development of this exciting field by not only providing a clearer understanding of the experimental observations but also predicting new systems with desired properties. Extensive calculated results reviewed here demonstrate that (1) weakened interatomic hybridization at clean surfaces or interfaces with inert substrates give rise to strong magnetic enhancement and (2) the strong interaction with nonmagnetic transition metals diminishes (entirely in some cases) the ferromagnetism and usually stabilizes the antiferromagnetic configuration. Surprisingly, experimentally observed surface (interface) magnetic anisotropy can be reproduced correctly in the theoretical calculations, although the anisotropy energy is only $\sim 10^{-4}$–10^{-5} eV.

1. Introduction

During the course of the last decade, low-dimensional magnetism of surfaces, interfaces and thin-films has matured into a major branch of modern condensed matter physics and is likely to open vast vistas for practical applications [1–3]. The abrupt termination of the lattice or composition in these systems leads to a variety of exotic phenomena such as localized electronic states, magnetic enhancement, magnetocrystalline anisotropy and complex magnetic ordering, etc. The goal of these fundamental studies is the strong expectation that the magnetic materials can be synthesized with expected properties in a controlled way. Indeed, as spelled out elsewhere in this volume, recent technical progress makes it now possible to synthesize high quality artificial ultrathin films, with stable or metastable lattice geometry in a layer by layer mode. Thus theoretical work is challenged not only to provide clear explanations for the experimental results but more importantly to predict new novel magnetic materials which have not yet been made experimentally.

Concurrently, several ab initio numerical methods (mainly the full-potential linearized augmented plane wave (FLAPW) method) based on the local spin density functional theory (LSD) were developed to calculate the electronic and magnetic structure of materials [4]. Great success has been achieved especially for lower-dimensional magnetic systems since the theoretical calculations predicted the large enhancement of the magnetic moment of 3d transition metal (TM) surfaces or TM ultrathin films deposited on inert substrates [5], even some normally nonmagnetic materials were also predicted to acquire a magnetic moment under certain conditions [6–8]. This predicted magnetic moment enhancement was recently verified experimentally for some systems such as the Fe(110) surface [9], Ag/Fe/W(110) [10] and a Cu/Co/Cu sandwich, etc. [11]. Stable magnetic structures, especially some antiferromagnetic (AFM) configurations, can now be predicted by comparing total energies and, in the same way, equilibrium atomic geometry and lattice relaxation (including surface and interface) can also be determined [12]. In particular, experimentally observed surface (interface) magnetic

0304-8853/91/$03.50 © 1991 – Elsevier Science Publishers B.V. All rights reserved

anisotropy, which is dominated by spin–orbit coupling, can be reproduced correctly using LDF calculations [13–16] although the anisotropy energy is only of the order of $\sim 10^{-4}$–10^{-5} eV.

The aim of the present review is to provide the main lines of important theoretical developments in this exciting area during the last decade with an emphasis on our own FLAPW calculations. Extensive examples presented here indicate that weakened interatomic hybridization at clean surfaces or interfaces with inert substrates is conductive to enhanced magnetization, while strong interactions with nonmagnetic transition metals diminishes the ferromagnetism and usually leads to AFM ordering. After a brief description of the local spin density and FLAPW method, predicted interfacial effects on the magnetization, the Fermi-contact hyperfine field and magnetic anisotropy etc. are surveyed.

2. Theoretical framework

2.1. Local spin density equations

Density functional theory provides an elegant cornerstone for the ab initio determination of the structural, electronic and magnetic properties of materials. Its basis is the proof by Hohenberg and Kohn [16] that the ground state energy of a many-body system is a unique functional of the charge density, $n(r)$ and is a minimum when evaluated for the true ground state density. Kohn and Sham [17] then showed how to obtain an equivalent single-particle equation by incorporating many-body effects into an exchange-correlation potential which can be acquired from the theory of the homogeneous electronic gas (local density approximation) [18–21]. By including relativistic effects, furthermore, the single-particle equations within the local spin density framework in the presence of an external magnetic field, B, can be written in the form [22–25]

$$[c\boldsymbol{\alpha} \cdot \boldsymbol{p} + \beta mc^2 + eV_{\text{eff}}(\boldsymbol{r}) - \mu_B \boldsymbol{\Sigma} \cdot \boldsymbol{U}_{\text{eff}}(\boldsymbol{r})]\Psi_i(\boldsymbol{r})$$
$$= \epsilon_i \Psi_i(\boldsymbol{r}), \quad (1)$$

where the effective potential $V_{\text{eff}}(\boldsymbol{r})$, the spin density operator $\boldsymbol{\Sigma}$ and magnetic potential $\boldsymbol{U}_{\text{eff}}(\boldsymbol{r})$ are given by

$$V_{\text{eff}}(\boldsymbol{r}) = \Phi(\boldsymbol{r}) + e\int \frac{n(\boldsymbol{r})\,\mathrm{d}\boldsymbol{r}'}{|\boldsymbol{r}-\boldsymbol{r}'|} + \frac{\delta E_{\text{xc}}[J_\mu]}{\delta J_0(\boldsymbol{r})}, \quad (2)$$

$$\boldsymbol{\Sigma} = \begin{bmatrix} \boldsymbol{\sigma} & 0 \\ 0 & -\boldsymbol{\sigma} \end{bmatrix}, \quad (3)$$

$$\boldsymbol{U}_{\text{eff}}(\boldsymbol{r}) = \boldsymbol{B} + \frac{\delta E_{\text{xc}}}{\delta \boldsymbol{m}(\boldsymbol{r})} \quad (4)$$

in the notation of Weinert and Freeman [25]. The electron density $n(\boldsymbol{r})$ and magnetization density $\boldsymbol{m}(\boldsymbol{r})$ can be obtained through a summation over all occupied states as

$$n(\boldsymbol{r}) = \sum_i \Psi_i^\dagger(\boldsymbol{r})\Psi_i(\boldsymbol{r}), \quad (5)$$

$$\boldsymbol{m}(\boldsymbol{r}) = \mu_B \sum_i \Psi_i^\dagger(\boldsymbol{r})\boldsymbol{\Sigma}\Psi_i(\boldsymbol{r}). \quad (6)$$

In the nonrelativistic limit (retaining the first order relativistic corrections), eq. (1) becomes the familiar Pauling equation

$$\left[\left(\frac{p^2}{2m} - \frac{p^4}{8m^3c^2}\right) - \mu_B \boldsymbol{\sigma} \cdot \left(\boldsymbol{B} - \frac{1}{2mc}(\nabla V \times \boldsymbol{p})\right)\right.$$
$$\left. + \left(eV + \frac{\hbar^2 e}{8m^2c^2}\nabla^2 V\right)\right]\Psi_i(\boldsymbol{r}) = \epsilon_i \Psi_i(\boldsymbol{r}), \quad (7)$$

where only spins ($\boldsymbol{\sigma}$) couple with the external magnetic field $\boldsymbol{B}$. V, the effective potential, includes the effects of exchange and correlation and depends on spin according to local spin density functional theory [19,20]. In practice, relativistic (for core states, eq. (1)) and semi-relativistic (for valence states, obtained by excluding the spin–orbit coupling term [26]) equations are solved self-consistently to obtain properties such as charge and magnetic (spin, for the nonrelativistic case) densities, single-particle eigenvalues and total energies, etc.

2.2. Thin-slab model and FLAPW method

The last decade has witnessed the development of one of the most accurate and powerful schemes to solve the local spin density single-particle equations: in the FLAPW method [27,28] for surface calculations a single slab geometry is used in which, as sketched in fig. 1, the real space is divided into three different regions, namely muffin-tin (MT) spheres around the nuclei, a vacuum region on each side of the slab and the remaining interstitial region. The single-particle wave function of each state, $\Psi_i(k, r)$, is expanded variationally in the reciprocal lattice

$$\Psi_i(k, r) = \sum_j c_{ij} \phi(K_j, r), \qquad (8)$$

where each basis function is a linearized augmented place wave (LAPW) given by

$$\phi(K_j, r) = \begin{cases} \Omega^{-1/2} \exp(\mathrm{i} K_j \cdot r), & r \in \text{interstitial}, \\ \sum_j \left[A_L(K_j) u_l(E_l, r) \right. \\ \left. + B_L(K_j) \dot{u}_l(E_l, r) \right] \\ \quad \times Y_L(\hat{r}), & r \in \text{MT sphere}, \\ \sum_q \left[A_q(K_j) u_q(E_v, z) \right. \\ \left. + B_q(K_j) \dot{u}_q(E_v, z) \right] \\ \quad \times \exp(\mathrm{i} K_q \cdot r_\parallel), & r \in \text{vacuum}, \end{cases} \qquad (9)$$

with $K_j = k + G_j$ and $K_q = K_{j\parallel}$. G_j is a vector of a three-dimensional reciprocal lattice defined in terms of the auxiliary periodicity domain, $\tilde{D}$. As shown in fig. 1, $\tilde{D}$ is chosen slightly larger than D to avoid an artificial zero slope at the vacuum boundary ($z = \pm D/2$) in the z direction perpendicular to the surface. Ω is the volume of the unit cell. The radial function, $u_l(E_l, r)$, and its energy derivative, $\dot{u}_l(E_l, r)$, are obtained by solving the radial Schrödinger equation with a spherical effective potential at a fixed energy E_l in each muffin-tin sphere. The coefficients $A_L(K_j)$ and $B_L(K_j)$ are determined by the matching conditions on the basis function and its derivative across the sphere boundary. Similarly, the z dependent wave function $u_q(E_v, z)$, and its energy derivative, $\dot{u}_q(E_v, z)$, are obtained by solving the

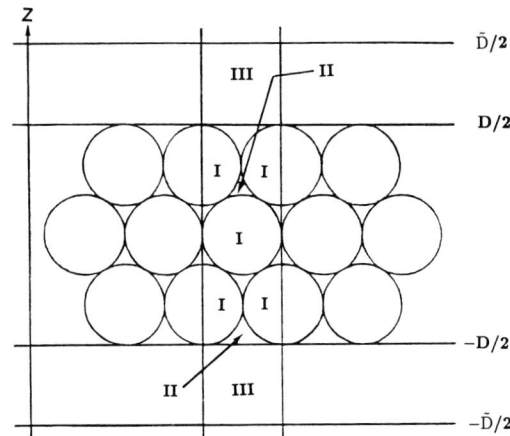

Fig. 1. Thin film geometry as used in the FLAPW method. I, II and III are the spheres, the interstitial and the vacuum regions, respectively. $\tilde{D}$ provides an auxiliary periodicity as discussed in the text.

one-dimensional Schrödinger equation in the vacuum region, while the continuity conditions on $\phi(K_j, r)$ and $\partial \phi(K_j, r)/\partial z$ across the vacuum boundary are used to determine the $A_q(K_j)$ and $B_q(K_j)$ coefficients.

In the FLAPW approach, the charge density, $n(r)$, is represented by a "natural" representation in each of the three spatial regions, namely

$$n(r) = \begin{cases} \sum_j n_j \exp(\mathrm{i} G_j \cdot r), & r \in \text{interstitial}, \\ \sum_L n_L(r) Y_L(\hat{r}), & r \in \text{MT sphere}, \\ \sum_q n_q(z) \exp(\mathrm{i} G_{j\parallel} \cdot r_\parallel), & r \in \text{vacuum}. \end{cases} \qquad (10)$$

The electrostatic potential is obtained by solving Poisson's equation using the technique developed by Weintert [29]. The exchange-correlation potential calculated from the charge density is added by a least squares fitting technique. Finally, the effective single-particle potential is represented in the form completely analogous to that for the charge density as given by eq. (10), without any shape approximation.

In the FLAPW framework, as stated by Weinert et al. [28], the Coulomb singularities in the kinetic and potential terms arising from the nuclear charge is explicitly canceled in the formula and thus can achieve a high precision for total energy calculations. Indeed, the FLAPW method has been successfully applied to solve such problems as surface and overlayer relaxation, surface reconstruction and, more importantly, appears to be the unique way today to determine theoretically the magnetic ground state among a limited set of magnetic configurations.

3. Magnetic moment enhancement on clean transition metal surfaces

Changes in symmetry and coordination number lead to a narrowed d band and localized electronic surface states (SS) or surface resonance (SR) states at a metal surface, and thus affect the magnetic properties in the surface region. While early studies [30] on nickel, iron and cobalt films stimulated interest in surface magnetism by suggesting that the surface layers were magnetically dead, later investigations traced these results to surface contamination. Now, it is well recognized that environments with few nearest neighbors and hence weaker interatomic hybridization, are conducive to enhanced magnetization [1,2,4]. Fortunately today, it is possible to synthesize and study surfaces of metastable phases and this had dramatically and importantly increased the range of materials that are magnetic and hence the challenges for understanding magnetism in low dimensional systems.

3.1. Fe, Co, Ni surfaces

We summarize the calculated results for the magnetic moment of 3d transition metal surface atoms and, for comparison, their bulk counterparts (taken from the center layer of the slab) in table 1. Both ground and metastable lattice structures are included. For most of the magnetic metals (Fe, Co and Ni), surface atoms achieve an obviously stronger magnetic polarization. Furthermore, atoms in a compacted lattice possess a smaller magnetic moment, e.g., the magnetic moment for a bulk-like atom in metastable fcc

Table 1
Magnetic moment (μ_B) in surface and center layer (bulk) muffin-tin spheres and the corresponding enhancement (in %) for magnetic metal surfaces

System	Surface	Center	Enhancement
bcc Fe(001) [32]	2.96	2.27	30
bcc Fe(110) [33]	2.65	2.22	19
bcc Fe(111) [34]	2.70	2.30	17
hcp Co(0001) [48]	1.76	1.64	7
fcc Ni(001) [41]	0.68	0.56	23
fcc Ni(110) [44]	0.63	0.56	13
fcc Ni(111) [42]	0.63	0.58	9
fcc Fe(001) [31]	2.85	1.99	43
bcc Co(001) [45]	1.95	1.76	11
bcc Co(110) [47]	1.82	1.76	3
fcc Co(001) [46]	1.86	1.65	13
bcc Cr(001) [60]	2.49	0.59	322
bcc V(001) [62]	0.00	0.00	–
fcc Pd(111) [69]	0.00	0.00	–
fcc Pd(001) [68]	0.00	0.00	–

Fe(001) [31] decreases to $1.99\mu_B$ versus a corresponding value of $2.27\mu_B$ for bcc Fe(001) [32].

The magnetism of Fe atoms depends most strongly on the environment. The calculated magnetic moments of bcc Fe(001) [32] and fcc Fe(001) [31] surface atoms are enhanced by as much as ~40% over their bulk values, but decrease drastically to only 19% for a closer packed surface, bcc Fe(110) [33]. Note that, although both have same number of nearest neighbors (4), the surface magnetic enhancement (17%) for bcc Fe(111) surface [34] is only about half of that for the bcc Fe(001) surface – indicating that the surface magnetism depends not only on the coordination number, but also on details of the atomic arrangement. Thus, for example, antiferromagnetic coupling between surface–subsurface or interior layers was found for fcc Fe(001) [31,35,36].

Ni(001) has been studied extensively since the report of magnetically dead layers on this surface. The first LSD calculation for Ni(001) was made by Wang and Freeman using the LCAO-DVM method [27]. With this early more approximate calculations, they found that the surface is not magnetically dead but the surface magnetic moment is reduced compared with that of the bulk-like atoms. Later calculations, by Jepsen et al. [38] and Krakauer et al. [39] using the more precise LAPW method and by Zhu et al. [40] using the SCLO technique reported a slight increase of the surface magnetism. By contrast, as listed in table 1, the highly precise FLAPW calculation by Wimmer et al. [41] obtained a large surface magnetic enhancement (23%) for this surface – demonstrating that the results are sensitive to the computational precision of the method employed. Again, because of the close packed character, the fcc Ni(111) surface [42,43] shows a much smaller magnetic enhancement (9%). Furthermore, an evident oscillation of the magnetic moment was found on going into the bulk; the magnetic moment is $0.64\mu_B$ for the subsurface layer and then converges to $0.58\mu_B$ for the interior layers. Similar to the Fe surfaces discussed above, by contrast to the Ni(001) results, the surface magnetic enhancement of Ni(110) is only 13%, although it has the same surface coordination number as Ni(001) [44].

By comparison, the magnetic moment at Co surfaces is more stable to the change of the environment. For the open surfaces, e.g., bcc Co(001) [45] and fcc Co(001) [46], the surface magnetic moment is only 12% larger than the bulk value, while for the close-packed surfaces, e.g., bcc Co(110) [47] and hcp Co(0001) [48], the magnetic enhancement is even as small as 3–7%.

The trend that the reduced number of the nearest neighbors on the surface leads to an enhancement of the surface magnetic moment, becomes even more obvious and a simple picture emerges when we include linear chains and free atoms into the consideration. For example, on going from bulk to a (001) surface, then to a linear chain [49] and finally to the free atom, the moments are 0.56, 0.68, 1.1 and 2.0 μ_B for Ni and 2.25, 2.96, 3.3 and 4.0 μ_B for Fe. Clearly, the magnetic moment approaches the free atom value as the dimensionality is decreased.

The spin densities of 7 layer Fe(001) and Ni(001) slabs are given in fig. 2. For Fe(001) [32], the positive spin density dominates most of the space around the Fe atoms, with only small pockets of negative spin density in the interstitial region; positive spin density (with sp like character) extends far into the vacuum. By contrast, Ni has a larger interstitial region with negative spin

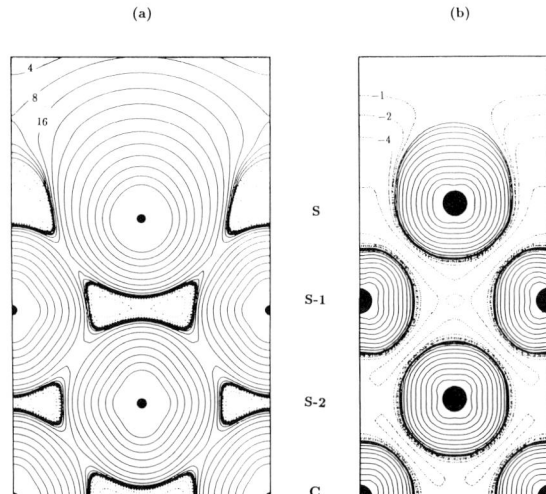

Fig. 2. Spin density of (a) a 7-layer bcc Fe(001) slab and (b) a 7-layer fcc Ni(001) slab in the vertical plane in units of 10^{-4} e/au^3. Dotted lines represent negative spin densities.

density [41]. Contours around the surface Ni atoms, as also suggested by the spin decomposition of the magnetic moment, possess almost pure d symmetry.

To clarify how strongly the surface relaxation may affect the magnetic enhancement, the calculated total energy and surface magnetic moment are plotted in fig. 3 as a function of the surface

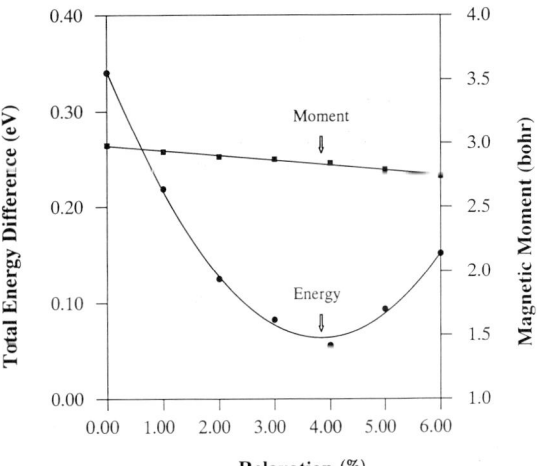

Fig. 3. Dependence of the total energy and surface magnetic moment of bcc Fe(001) on the surface relaxation. Arrows point out the equilibrium position.

layer relaxation for bcc Fe(001) [50]. The surface Fe atomic layer is found to relax downward by 4%. The surface magnetic moment varies slightly over the relaxation range shown and decreases to $2.80\mu_B$ at the equilibrium position – which is ~ 6% smaller than that for the unrelaxed case. That is to say, including the surface relaxation probably does not change the surface magnetic enhancement for most systems, but a careful check is needed for several close-packed surfaces – especially for bcc Co(110).

Unfortunately, few directly measured magnetic moments exist even today; a recent evaluation obtained by fitting SPLEED "rocking curves" [9] reported a 36–40% magnetic enhancement on Fe(110). As is well known, clean surfaces are very important for experimental observations since many adsorptions, e.g., H [51], S [52] and even Ag [53], may substantially reduce or eliminate the surface magnetic enhancement. Recent calculations for Al/Fe(001) [54] found that the Al coverage may reduce the surface magnetic moment of Fe(001) to $1.38\mu_B$ (~ 37% smaller than the bulk value).

3.2. Live surface layers

The ground state of bulk Cr is AFM with an incommensurate spin density wave along the (001) direction [55]. Thus, since the (001) plane contains atoms of one spin direction, Cr(001) is expected from this simple structure argument to be ferromagnetically ordered. This surface FM ordering was first reported experimentally by Ferguson [56] on polycrystalline Cr films, and was confirmed by subsequent observations to persist up to (780 ± 50) K, well above the bulk Néel temperature, $T_N = 312$ K [57]. Theoretically, Allan et al. [58] using the tight binding method, found that the Cr(001) surface is ordered ferromagnetically with an enhanced magnetic moment of $2.8\mu_B$. Victoria et al. [59] also using a tight binding technique, found that the Cr(001) surface magnetic moment is $3\mu_B$. The highly precise FLAPW calculations carried out by Fu et al. [60] confirmed the existence of the surface FM ordering while gave a smaller surface magnetic moment ($2.49\mu_B$) – which is still enhanced by ~ 300% compared to the local magnetic moment in Cr AFM bulk ($0.59\mu_B$). Furthermore, as was verified recently by FLAPW calculations [61], the $c(2 \times 2)$ AFM coupling in the Cr(001) surface plane was found to be unfavorable from the total energy point of view because it frustrates the magnetization of the subsurface layer.

To explore possible surface magnetism for nonmagnetic metals, FLAPW calculations were carried out for V(001) [62] including multilayer relaxation. A stable paramagnetic (PM) ground state and a 9% contraction of the topmost interlayer spacing was found for this surface. An analysis of the electronic structure of V(001) shows that the surface states, which are characteristic of the occurrence of surface magnetism in bcc Cr(001) [60] and Fe(001) [32,63], are located about 0.3 eV above the Fermi energy. In contrast to those theoretical results, Rau et al. [64] reported long range FM ordering for the V(001) surface at 300–540 K with local magnetic order persisting up to 650 K. The possibility exists that either a reduced coordination number for the surface V atoms due to adsorption of oxygen or surface reconstruction may be responsible for the discrepancy.

Pd is considered to be a "nearly magnetic" metal although neither its fcc bulk nor free atom possess magnetic moments. Large enhanced susceptibilities were observed for Pd either sandwiched by Au [65] or on a Re [66] substrate and thus the existence of a live surface Pd layer was suggested. Theoretically, a PM–FM phase transition (with ~ $0.3\mu_B$ moment) was found to occur in bulk Pd with ~ 6% lattice expansion [67]. For surfaces, paramagnetic calculations for Pd(001) [68] and Pd(111) [69] revealed that the DOS at E_F for the surface Pd layer is ~ 20% lower than that for the interior layers. Therefore, surface Pd atoms have a smaller magnetic susceptibility than to bulk atoms and thus, by the Stoner criterion, the possibility of surface ferromagnetic order should be ruled out. As a further check, we also performed spin polarized calculations for both orientations with an (anomalously) large starting moment of $2.0\mu_B$/atom. Upon carrying out the self-consistent procedure, this giant initial magnetic moment disappears gradually during the iterations, and finally converges to a nonmagnetic ground state, even for the geometry with a 5% artificial surface expansion [68,69].

3.3. Rare-earth metal surfaces: Gd(0001)

As a prototype of a rare-earth (RE) metal, Gd(0001) has attracted considerable attention in recent years. Most strikingly, as first found by Rau et al. [70] and Cerri et al. [71] on polycrystalline Gd thin films and later confirmed by Weller et al. [72] on epitaxially grown hcp Gd(0001), the ferromagnetic (FM) order in the surface layer may persist at temperatures up to 310 K, i.e., about 20 K above the bulk Curie temperature, $T_{C,b}$. To interpret the sharp minimum of the magnetic signal (detected by a magneto-optic Kerr effect (MOKE) experiment at a characteristic temperature T_{comp} = 289 K), Weller et al. [72,73] proposed that the surface layer on Gd(0001) couples antiferromagnetically to the underlying FM bulk. Electronic, magnetic and structural properties were recently determined using the FLAPW method [74]. As found by these total energy calculations, the surface AFM coupling was found to be the ground state – which supports the experimental interpretation [72]. Analysis of the charge redistribution revealed that the energy stability of the surface AFM coupling arises from the enhancement of the second-neighbor interactions between the surface and the third Gd layers. Surprisingly, the surface interlayer distance *expands* by about 6.3% outwardly at the equilibrium position. This appears to be the first theoretical prediction of an outward expansion for a pure metal surface. It can be attributed to the softness of the rare-earth metal lattice (cf., the small modulus of Gd metal [75]) and strong magnetic pressure effect [76] due to the giant 4f magnetic moment. Recently, such a magnetically driven surface expansion was reported for fcc Fe(001) [77] and even was used to explain the LEED result for Pd(001) [78].

For Gd, the giant magnetism ($7\mu_B$) of the 4f shell induces large and parallel spin-polarization of the valence (5d and 6s) electrons through a direct exchange coupling. As listed in table 2, the calculated valence electron magnetic moment

Table 2
Calculated conduction electron contributions to the magnetic moment for bulk Gd and the Gd(0001) surface [74]

		FM	AFM
Gd bulk	MT	0.47	–
	int./atom	0.12	–
	total	0.59	–
	expt.	0.63	–
Gd(0001)	S	0.58	−0.49
	S-1	0.44	0.41
	C	0.46	0.45

(M_v) for FM Gd bulk is $0.59\mu_B$/atom ($0.47\mu_B$ in the muffin-tin sphere and 0.12 in the interstitial region) – which agrees well with experiment, $0.63\mu_B$ [79]. For the FM state, the conduction electron magnetic moment in the surface muffin-tin sphere is enhanced by about 26% over that in the center layer atom. A slight Friedel oscillation was found for the magnetic moment upon going from the surface to the center layer indicating that the film is not thick enough to produce bulk properties in the center layer. The surface magnetic enhancement is markedly frustrated for the AFM state. The magnetic moments of the surface and subsurface layers decrease significantly because in following the 4f polarization, the spin density gets a node between these two atomic layers. Nevertheless, the magnetic moment of the surface atom is still enhanced by ~18% from that of the second layer and furthermore is still slightly larger that of the center layer. Fig. 4 gives the valence charge and spin density for AFM Gd(0001) (obtained from an 8 layer slab); here one can find some characteristic properties of the surface–vacuum interface, e.g., surface smoothness and no noticeable special bond-orientation, etc.

A $\bar{\Gamma}d_{z^2}$ SS was found to play a significant role for the surface magnetic enhancement [74]. It shows strong localization in the surface region in the vicinity around the $\bar{\Gamma}$ point. For the spin majority part, the SS lies just below E_F and thus is occupied. However, it is shifted up by the exchange interaction to lie ~1 eV above E_F for the spin minority part. The confirmation of this

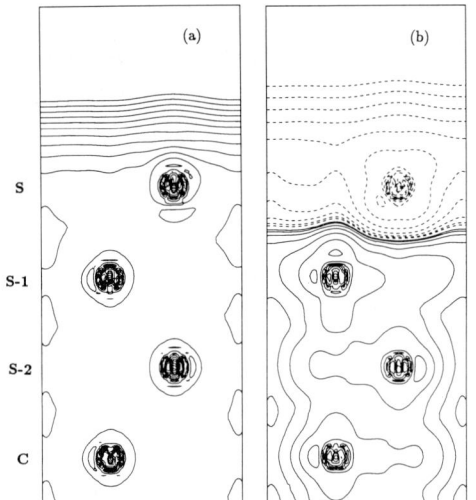

Fig. 4. (a) Total valence charge density and (b) spin density for AFM Gd(0001). Contours in panel (a) start from 5×10^{-4} e/au^3 and increase successively by a factor of $\sqrt{2}$. Contours in panel (b) start from $\pm 1.0\times 10^{-4}$ e/au^3 and increase by a factor of 2.

predicted surface state was reported recently by Li et al. [80].

4. 2D magnetism

An important development in the last decade is the ability to study 2D magnetism by the synthesis of ultrathin magnetic films on various kinds of substrates. The lively interplay of theory and experiment has already yielded unusual results, e.g., critical phenomena [81], etc. In table 3, we list some selected calculated magnetic properties of overlayer systems. In general, interfacial effects with the inert substrates (noble metals and ceramics) do not affect significantly the magnetism of free monolayers (except Ni), while strong d band interaction with transition metal substrates (except Pd) diminishes the overlayer ferromagnetism substantially and usually leads to some kind of AFM configuration.

4.1. Monolayer magnetism

As an ideal theoretical model system, a free standing monolayer is expected to exhibit even

Table 3
Magnetic moment (μ_B) and magnetic ground state of magnetic monolayer and adsorption systems

System	Monolayer		Overlayer	
	state	moment	state	moment
Fe/Ag(001) [82,87,88]	FM	3.20–3.4	FM	2.96–3.01
Fe/Au(001) [90]	FM	–	FM	2.97
Fe/MgO(001) [91]	FM	3.10	FM	3.07
Fe/Cu(001) [81,92]	FM	3.20	FM	2.69–2.85
Co/Cu(001) [92]	FM	–	FM	1.79
Co/Cu(111) [93]	FM	–	FM	1.63
Co/Ag(001) [82]	FM	2.20	FM	2.03
Ni/Cu(001) [94]	FM	–	FM	0.39
Ni/Ag(001) [82,95]	FM	1.02	FM	0.57–0.65
Ni/Cu(111) [42]	FM	–	FM	0.34
Mn/Ag(001) [82]	AFM	4.32	AFM	4.11
Cr/Ag(001) [82]	AFM	4.09	AFM	3.57
Cr/Au(001) [33]	AFM	3.84	AFM	3.48
V/Ag(001) [82]	FM	2.87	AFM	2.08
Ti/Ag(001) [82]	FM	1.72	FM	0.34
Pd/Ag(001) [84]	FM	0.40	PM	0.00
Rh/Au(001) [85]	FM	1.56	FM	1.09
Rh/Ag(001) [83]	FM	1.45	FM	0.95
Ru/Ag(001) [83]	FM	2.12	FM	1.57
Fe/Pd(001) [118]	FM	–	FM	3.19
Co/Pd(001) [82,118]	FM	–	FM	2.12
Co/Pd(111) [69]	FM	1.87	FM	1.88
Co/Pt(111) [69]	FM	1.89	FM	1.84
Ni/Pd(001) [118]	FM	–	FM	0.89
V/Pd(000) [118]	FM	–	AFM	1.39
Cr/Pd(001) [118]	AFM	–	AFM	3.46
Mn/Pd(001) [118]	AFM	–	AFM	4.05
Ti/Pd(001) [82]	FM	–	PM	0.00
Fe/Ni(111) [43]	FM	2.49	FM	2.33
Ni/Fe(001) [120]	FM	–	FM	0.83
Fe/W(110) [114]	Fm	2.98	FM	2.18
Fe/W(001) [113]	FM	3.10	AFM	0.93
Fe/Ru(0001) [111]	FM	2.90	AFM	2.23

stronger magnetic enhancement than the surface layer because of its more reduced atomic coordination number. Indeed, as listed in table 3, magnetic moments for Fe, Co and (especially) Ni monolayers increase greatly compared to those for their corresponding clean surfaces. Note, for example, that the magnetic moment of square Fe(001) monolayer remains in the range of 3.1–3.4μ_B (different radii of MT spheres employed may lead to some differences) for either contracted (on Cu) or expanded (on W and MgO) 2D lattice parameters. Therefore, the 2D lattice strain very likely does play a significant role on the magnetism.

Giant magnetic moments were also found for AFM ordered Mn [82] and Cr [33,82] monolayers. Strikingly, some nonmagnetic 3d metals like Ti [82] and V [62,82] and some 4d metals like Ru [83], Pd [84] and Rh [83,85] were calculated to possess ferromagnetism as monolayers. Physically, the large magnetic enhancement of the monolayers results from the narrowed energy bands which leads to a large density of states at E_F. For example, the DOS at E_F of Ru and Rh monolayers [83] (matching Ag(001)) are found to increase by 450% over their corresponding bulk values [86], which results in a large Stoner factor (1.45 and 1.89 for Ru and Rh monolayers, respectively) and thus an instability for the PM phase.

4.2. Overlayers on inert substrates

The nearly perfect match of the 2D lattice translational vectors along the (001) direction, for example, between bcc Fe (or Cr) and fcc Ag (or Au) with a 45° rotation, ensures epitaxial growth of the overlayers. Furthermore, noble metal or ceramic substrates lack electronic states at the Fermi energy, which minimizes the overlayer–substrate hybridization and thus enhances the 2D magnetism.

As seen from table 3, strongly enhanced magnetic moments are found for Fe/Ag(001) [82,87–89], Fe/Au(001) [90], Fe/MgO [91] and Co/Ag(001) [82]. They decrease only slightly from the values of the free-standing monolayers, and are close to or even larger than the magnetic moments of the corresponding clean surface – suggesting the weakness of the overlayer-substrate interaction. By comparison, because of the 2D lattice constant contraction ($\sim 11\%$) and the stronger sp–d band hybridization, Cu substrates affect the overlayer magnetism much more since the overlayer magnetic moment becomes obviously smaller than the value of clean surfaces, especially for Co [31,92,93].

The magnetism of Ni depends very sensitively on the environment. As seen in table 3, the magnetic moment of a Ni overlayer decreases drastically to only $\sim 0.35\mu_B$ ($\sim 40\%$ smaller than that in Ni bulk) on Cu(001) [94]. This decrease was found to be mainly due to a charge transfer from the Cu to the Ni sites which occupies the minority spin states [94]. A strong decrease of the magnetic moment was also found for Ni/Cu(111) [42]. Although the Ni–Ag interaction is believed to be much weaker, the magnetic moment of the Ni overlayer is also reduced markedly compared to the value for the free-standing monolayer. Hong et al. [95] found that the Ni layer lose its magnetism entirely when it is sandwiched by Ag – which appears to explain to ARUPS experiments [95]. A recent SMOKE observation reported that Ni loses its magnetism entirely on a Ag(111) surface [97], which may be due to the segregation of Ag atoms to the surface of Ni [95].

Mn and Cr monolayers exhibit AFM coupling of their giant magnetic moments for both the isolated and adsorbed cases. Surprisingly, the greatly enhanced moment of Cr/Au in only moderately reduced when Cr is itself covered by an additional Au layer [33]. Similar results were also obtained by Oguchi and Freeman [98] for a (1×1) Cr/Au superlattice, where the Cr magnetic moment is as large as $2.90\mu_B$ and the in-plane AFM coupling is more stable than the FM coupling.

Whereas an isolated V monolayer is ferromagnetically ordered, the coupling changes dramatically to antiferromagnetic ordering when deposited on a Ag(001) substrate [82]. As revealed in a recent study [99], this phase transition can be attributed primarily to the interfacial sp-d hybridization (with a charge transfer from both Ag and V sites to the interfacial region). While most of the recent experiments failed to detect unambiguous manifestations of ferromagnetism in V/Ag(001) [100,101], one group claims to have done so [102]. The reason for this difference is still a point of controversy.

4d magnetism still remains unconfirmed as yet. Although a Pd monolayer shows weak ferromagnetism [84], Pd/Ag(001) [8,84] and Pd/Au(001) [85] was found to be paramagnetic. Strikingly, Zhu et al. [85] (using norm-conserving pseudopotentials with a Gaussian-orbital expansion) and Eriksson et al. [8] (using a linear muffin-tin orbital film code with the fixed spin-moment method), obtained large ferromagnetic moments

(0.62–1.02μ_B) for Rh/Ag(001) and Rh/Au(001) systems. This prediction was confirmed by recent more precise FLAPW calculations for Rh/Ag(001) [83]. The Rh magnetic moment decreases by ~ 40% to 0.95μ_B from the value of 1.45μ_B for the free Rh monolayer – indicating the sensitivity of the magnetism of 4d metals to the environment. In contrast to the result of Eriksson et al. [8] FM ordering was found to be a stable ground state for Ru/Ag(001) in the FLAPW calculations [83]. The energy of magnetization is 5.5 mRy per Ru atom for Ru/Ag(001), while it is only 2.8 mRy per Rh atom for Rh/Ag(001). Moreover, Ru atoms possess a larger magnetic moment (2.12 and 1.57μ_B for the isolated Ru monolayer and Ru/Ag(001), respectively) than Rh atoms. Therefore, Ru/Ag(001) is a more appropriate system for experimental verification studies.

4.3. Transition metal substrates

Interfacial d band hybridization between magnetic overlayers and transition metal substrates has been found to lead to new significant phenomena. Experimentally, very unusual behavior was found in superlattice systems like Fe/Cr, Fe/Cu, Co/Cr and Co/Ru etc. [103–106], where the ferromagnetically ordered ultrathin Fe (Co) films couple antiferromagnetically via the magnetic or nonmagnetic intervening layers. The Fe(111)$_n$/Ru(0001) overlayer system [107] and the hcp Fe$_n$/Ru$_m$ [108] superlattice were reported to be "magnetically dead" when the number n of Fe layers is less than 2 (on the surface) or 4 (in the superlattice). Strikingly, the ferromagnetism of the first 2 (4) Fe layers cannot be activated by deposition of successive FM Fe layers, as deduced from a thickness-dependent extrapolation [2]. Co/Pd, Co/Pt and Co/Au multilayers show perpendicular magnetic anisotropy and 100% remanence for Co films thinner than a few monolayers [109] and appear to be promising candidates for high-density magneto-optical storage media [110].

In our recent FLAPW calculations [111], we found that in-plane AFM coupling becomes the stable ground state for Fe(111)/Ru(0001) – which

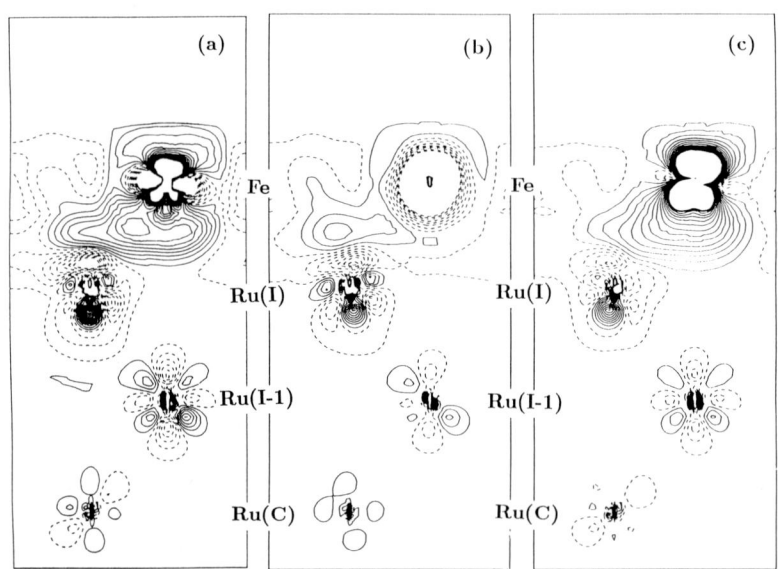

Fig. 5. Charge density difference between AFM Fe/Ru(0001) and the direct superposition of the free-standing AFM Fe monolayer and the clean Ru(0001) surface for (a) total charge, (b) majority spin, and (c) minority spin. Contours on the vertical (11$\bar{2}$0) plane start from $\pm 1 \times 10^{-3}$ e/au^3 and increase successively in steps of $\pm 2 \times 10^{-3}$ e/au^3. Solid and dashed lines represent positive and negative differences, respectively.

explains the origin of the observed "magnetically dead layers" [107,108]. As an example of hybridization with the transition metal substrates, valence charge density differences between AFM Fe/Ru(0001) and the direct superposition of the free-standing AFM Fe monolayer and the clean Ru(0001) surface are plotted in fig. 5. Strong Fe–Ru covalent bonding is obvious in panel (a) whereas both Fe and Ru atoms lose electrons to the bonding region. Furthermore, the spin decomposition in panels (b) and (c) indicates that the Fe–Ru bonding arises mainly from the minority spin states. The Fe magnetic moment was found to depend sensitively on the Fe–Ru interatomic distance and is diminished from a giant value of $2.90\mu_B$ for the isolated hexagonal Fe monolayer to $2.3\mu_B$ for Fe/Ru(0001) at the equilibrium position. A strong interplay between overlayer magnetism and relaxation was found: the 6% contraction of the nearest Fe–Ru interatomic distance for the PM case is entirely recovered (even with a 1% expansion) by introducing the magnetic polarization (in both the AFM and FM states). Suprisingly, recent LMTO-ASA calculations by Knab et al. [112] found that the Fe_n/Ru_m superlattice is magnetically dead when $n = 1$. For thicker Fe films, the antiferromagnetic coupling through the intervening Ru layers becomes more favorable.

For Fe/W(001) [113], the much stronger interfacial d band hybridization eliminates the ferromagnetic state (even as a metastable state) although the Fe monolayer has a large magnetic moment of $3.1\mu_B$. Instead, the AFM state is about 0.01 eV per Fe atom lower than that of the PM state in total energy and thus appears to be the ground state; the Fe moment of $0.93\mu_B$ is drastically reduced, but still observable. Very interestingly, in this case the magnetism of the "dead" Fe layer can be activated by an additional Fe overlayer; the magnetic moment in the Fe MT sphere in 2Fe/W(001) becomes 1.68 and $2.43\mu_B$ for the surface and subsurface layers, respectively. On W(110), the Fe overlayer remains FM coupled but the magnetic moment varies strongly with the overlayer relaxation [114]. At the equilibrium position, the reduced moment becomes $2.18\mu_B$ – about the value in bulk bcc Fe. An additional Ag overlayer alleviates the overlayer relaxation while not changing the Fe magnetic moment. By contrast, a recent experiment obtained a 14% magnetic enhancement on the Fe site over the bulk value for Ag/Fe/W(110) [10].

As an incipient ferromagnet, Pd is known to be easily magnetized by a small percentage of magnetic (Fe [115], Co [116] and Ni [117], etc.) atoms. As seen in table 3, as calculated by Blügel et al. using the FLAPW method, overlayer moments on a Pd(001) substrate are slightly enhanced for Fe, Co and Ni (FM) but reduced for Mn, Cr and V (AFM) compared to those on the Ag(001) substrate [82,118]. Pd(111) even keeps the magnetic moment of the Co monolayer intact as in the isolated case despite the strong Co–Pd d band interaction [69] – which indicates the value of using the Pd substrate to enhance overlayer ferromagnetism. However, the weak ferromagnetism of a Ti overlayer on an Ag substrate is destroyed when Ti is adsorbed on Pd(001) [82]. Magnetic overlayers induce a sizable and long ranged spin polarization in the Pd substrate; the magnetic moment is typically $0.3–0.4\mu_B$/atom on interfacial Pd sites but remains $0.2\mu_B$/atom on the subinterface Pd sites.

Fe–Ni hybridization in the interfacial region diminishes the magnetism of Fe while enhancing that of Ni slightly. Recently, metastable bcc Ni was grown on Fe(001) by MBE [119]. An FLAPW calculation [120] found a strongly enhanced Ni moment ($0.83\mu_B$ for 1Ni/Fe(001). The Fe induced enhancement is also evident in 2Ni/Fe(001) since the magnetic moment on the interfacial Ni site still remains as large as $0.69\mu_B$, only $0.03\mu_B$ smaller than that for the outermost Ni layer, $0.72\mu_B$. For both mono- and bilayer adsorptions, the moment of the interfacial Fe atom is $\sim 2.70\mu_B$, i.e., decreased by about 11% from that of the clean Fe(001) surface. Similar results were also found for Fe/Ni(111) [43], where the magnetic moment of the interfacial Ni atom increases by about 6% to $0.67\mu_B$ from that at the clean Ni(111) surface after the Fe absorption. By contrast, the Fe magnetism is significantly reduced compared to the isolated monolayer due to the interfacial exchange interaction – but is still larger than that of bulk Fe.

5. Fermi-contact hyperfine field

One of the striking successes of theory in the last decade has been the calculation of the Fermi contact hyperfine field at surfaces and interfaces. The magnetic hyperfine interaction, which describes the interaction of the electronic moment with the nuclear magnetic moment, provides an important tool for the study of surface and interface magnetism [121]. Quantitatively, the Fermi-contact part of this hyperfine field, H_{CF}, is proportional to the electronic spin density at the nucleus. Theoretically, great care has to be taken to adequately describe this sensitive quantity. In particular, it is essential to describe not only the wave function of the valence electrons at the nuclei accurately, but also to account for the dominant polarization of the core electrons. Thus an accurate all-electron method is needed which treats all electrons self-consistently. In a fully relativistic treatment, only orbitals with $j = 1/2$ (corresponding to $\kappa = 1$ in the Dirac equation) contribute to the spin density at the nucleus. The total hyperfine field can be conveniently decomposed into the contribution from the valence electrons, i.e. in the case of 3d transition metals,

Table 4
Calculated layer-by-layer Fermi contact hyperfine fields H_{CF} (in kgauss) broken down into valence ($H_{CF,v}$) and core ($H_{CF,c}$) contributions, magnetic moment M (in μ_B) and the ratio of $H_{CF,c}$ and M for selected systems

System	Layer	M	H_{CF}	$H_{CF,v}$	$H_{CF,c}$	$H_{CF,c}/M$
bcc Cr(001) [60]	S	2.49	−72	256	−328	131
	S-1	−1.29	23	−140	163	126
	S-2	0.89	−38	75	−113	127
	C	−0.89	−2	−115	113	127
bcc Fe(001) [32]	S	2.98	−252	143	−398	134
	S-1	2.35	−395	−89	−306	130
	S-2	2.39	−320	−16	−311	130
	C	2.25	−366	−75	−291	129
bcc Fe(110) [33]	S	2.65	−324	40	−364	138
	C	2.22	−350	−47	−305	137
Fe/Cu(001) [92]	Fe	2.69	−173	204	−376	140
Fe/Ag(001) [87]	Fe	2.96	−150	296	−419	141
Fe/Au(001) [90]	Fe	2.97	−213	200	−413	139
Fe/MgO(001) [91]	Fe	3.07	−42	396	−438	142
Ag/Fe(001) [53]	Fe	2.52	−335	−7	−328	131
Fe/Ru(0001) [111]	Fe	2.23	−48	278	−326	146
Fe/W(110) [114]	Fe	2.18	−148	158	−306	141
Ag/Fe/W(110) [114]	Fe	2.17	−177	127	−304	140
Fe/Ni(111) [43]	Fe	2.33	−288	35	−323	141
	Ni(I)	0.67	−205	−193	−102	152
	Ni(C)	0.59	−108	−19	−89	151
hcp Co(0001) [48]	S	1.76	−287	−32	−257	146
	C	1.64	−315	−76	−239	145
Co/Cu(001) [92]	Co	1.79	−96	161	−257	144
fcc Co(001) [46]	S	1.86	−211	55	−266	144
	C	1.65	−306	−69	−237	144
Co/Pd(111) [69]	Co	1.88	−207	75	−282	150
Co/Pt(111) [69]	Co	1.84	−210	66	−276	150
fcc Ni(001) [41]	S	0.68	−102	3	−105	154
	C	0.56	−119	−32	−87	155
FM Gd bulk [74]	Gd	0.47	−245	107	−352	−
AFM Gd(0001) [74]	S	−0.49	−425	−830	405	−
	S-1	−0.41	296	606	−309	−
	S-2	−0.45	−373	-15	−358	−

mainly from 4s-like components near a given nucleus, and into contributions from the core electrons, i.e. from 1s, 2s, $2p_{1/2}$, 3s and $3p_{1/2}$, etc. [122,123].

The standard interpretation of the Fe hyperfine field data for bulk systems is that H_{CF} scales with the magnetic moment since the dominant term is the (negative) core polarization contribution (the bulk valence electrons generally give an additional small negative contribution). Originally, Tyson et al. [124] used this interpretation to infer that there was an increase of the magnetic moment at the Fe surface. However, this interpretation is obviously not true as seen from the calculations for the Fe(001) surface [32] in table 4. The large and positive valence contribution $H_{CF,v}$ results in a reduction of the net H_{CF} at the surface compared to bulk, although the surface magnetic moment is enhanced by 32%. Nevertheless, the core contribution $H_{CF,c}$ was found to scale precisely with the local magnetic moment for a given system, with the ratio $H_{CF,v}/M$ staying about constant (in the range of 129–146 kG/μ_B) for all the Fe systems from clean FM surfaces to AFM overlayers or sandwiches. The calculated H_{CF} for Fe(110) [33] surface atoms agrees well with conversion electron Mössbauer spectroscopy measurement by Korecki and Gradmann [125].

For bcc Cr(001) [60], despite the 300% enhancement of the surface magnetic moment and the drastic oscillation of the valence contribution, the core hyperfine field, $H_{CF,c}$, again scales very precisely with the moment within each sphere. Increased $H_{CF,c}$ values are found for the surface atom corresponding to the large surface magnetic moment enhancement, and results in a net H_{CF} of -72 kG at the surface nuclei. The proportionality between $H_{CF,c}$ and M is also obvious for Co and Ni systems. However, the ratio of $H_{CF,c}/M$ increases upon going Cr to Ni (from ~ 130 to ~ 155 kG/μ_B) – as expected from the contraction of the 3d radial spin density [122].

For FM Gd bulk, the valence (6s) contribution is positive at the nucleus (different from the transition metals) and partly offsets the negative core contribution. Large oscillations of $H_{CF,v}$ in the first 2–3 layers is observed for both AFM and FM Gd(0001) surfaces. Nevertheless, the rule, that the core hyperfine field is proportional to the local moment, is still found to hold but one must separate the influence of the f electrons from that of the conduction electrons [74].

Finally, the importance of theoretical determinations is now clear because the hyperfine field is not proportional to the magnetization and so their interpretation is totally dependent on theory.

6. Magnetic anisotropy

As suggested by Néel [126], the lack of neighbors and the lowered symmetry at a surface or interface can give rise to a so-called *magnetocrystalline surface anisotropy*. The anisotropy energy at a surface or interface may reach as much as 10^{-3} eV/atom, which is two to three orders of magnitude larger than bulk values. Thus the demagnetizing energy can be overcome in ultrathin films, which leads to the spectacular result of spontaneous perpendicular magnetization, as observed experimentally in many cases [14,127].

As proposed by Van Vleck [128] more than 50 years ago, the magnetic anisotropy of 3d transition metals is thought to originate from the spin–orbit interaction. However, a theoretical understanding of the magnetic anisotropy in realistic systems remains a great challenge, because a highly precise determination of both the electronic structure and the total energy (for the latter up to $\pm 10^{-5}$ eV) is required. Brooks [129], and later Fletcher [130], used an itinerant-electron model to explain the magnetic anisotropy and the quenching of orbital angular momentum in cubic crystals by treating the spin–orbit coupling as a perturbation (fourth order for cubic lattices). First principles magnetic anisotropy calculations for bulk Fe, Co and Ni have been reported by several groups with varying results [131–134]. Recently, an LMTO-ASA calculation by Daalderop et al. [135] found that the easy axes for Co/X (along the [111] direction, X = Cu, Ag and Pd) multilayers are perpendicular to the multilayer plane and the magnetic anisotropy energy for Co/Pd superlattice decreases with increasing

Co thickness. For surfaces, a tight binding perturbative model was developed in the 1970's to study the magnetic anisotropy of a Ni(001) monolayer [136,137]. Because of a rather crude approximation and inaccurate knowledge of the band structure, these authors obtained only a reasonable order of magnitude for the magnetocrystalline anisotropy. Recently, Bruno [138] extended this treatment to include the orbital moment (which is largely quenched in transition metals) with the aim of providing qualitative trends. The calculated anisotropy energy, which has a strong connection to the orbital magnetic moment (0.1–0.3μ_B), depends sensitively on the crystal field parameter Δ, on surface or interface roughness, etc. [14].

Pioneering calculations for the spin anisotropy of ferromagnetic Fe, Co, Ni and V monolayers, and thick Fe slabs and Fe/Ag(001) [13,139] were carried out by Gay et al. using a self-consistent local orbital approach by incorporating the spin–orbit coupling as a perturbation. They found that the easy direction of magnetization is perpendicular to the plane of the monolayer for Fe and V, but in the plane for Ni and Co [13,139]. The result for Fe, i.e., perpendicular anisotropy, was used to explain the experimental observation by Jonker et al. for Fe/Ag(001) [140] of spin-split bands but no spin polarization. Recently, the effect of temperature was considered by Richter and Gay [141] by only smearing out of the states around the Fermi level according to the Fermi distribution function. The easy direction of the magnetization is not sensitive to the temperature in this approach.

Recently, we developed a so-called "second variation" method to treat the metal surface anisotropy based on the FLAPW approach [15], i.e., to solve the relativistic Dirac equation based on the charge and potential distribution obtained from a semi-relativistic self-consistent calculation [26,27]. By this means, relativistic eigenvalues, wave functions and total energy are obtained with respect to the spatial orientation of the spin polarization, which yields the magnetic anisotropy. Since the anisotropy energy has an extremely small value, a very careful numerical treatment (e.g., enough k points in the 2D BZ,

Table 5
The magnetic anisotropy energy ($E_\perp - E_\parallel$, in meV) for Fe [15] and Co systems [142]

Fe ML	Fe/Ag(001)	Fe/Au(001)	Fe/Pd(001)
0.03–0.04	−0.06	−0.57	−0.35
Co ML(squ)	Co/Cu(001)	Co ML(hex)	Co/Pt(111)
0.03	−0.05	0.65	−0.45

sufficient unoccupied states to construct the matrix elements of the spin–orbit coupling term of the Hamiltonian, etc.) is demanded. Because of the complexity of the calculations, the "second variation" method has been applied only to magnetic monolayers and to overlayer systems with only a monolayer of substrate atoms.

The calculated magnetic anisotropy energy for Fe and Co monolayer [142] is listed in table 5, where positive (negative) energy indicates in-plane (perpendicular) spin polarization. In all cases, the absolute value of the anisotropy energy is too small to be meaningful. Instead, it is the relative value that may yield the physics of the situation. Surprisingly, in-plane magnetization with a small anisotropy energy was obtained for the Fe monolayer (with 2D lattice constants corresponding to both Ag and Au substrates), i.e., different from the earlier Gay and Richter result [13]. However, when the effects of an Au, Ag or Pd substrate are treated, the direction of the spin polarization turns out of the plane. Interestingly, the anisotropy energy of Fe/Ag(001) is much smaller than that of Fe/Au(001), although the two substrates are believed to have the same influence in most cases (cf. the magnetic moment in table 3). Plots of the anisotropy energy vs. the angle of the spin polarization to the z axis are given in fig. 6 for Fe monolayer and adsorption systems. As expected, the theoretical data can be well fitted by $\cos^2(\theta)$ type curves.

For a square Co monolayer (with a lattice constant that of Cu(001)), we obtained a small in-plane anisotropy energy, 0.032 meV, which is close to the value for the Fe monolayer. By contrast, the anisotropy energy increases drastically for a hexagonal Co monolayer (at the Pt(111) lattice constant) to 0.65 meV, i.e., ∼ 20 times larger than that of the square lattice – demon-

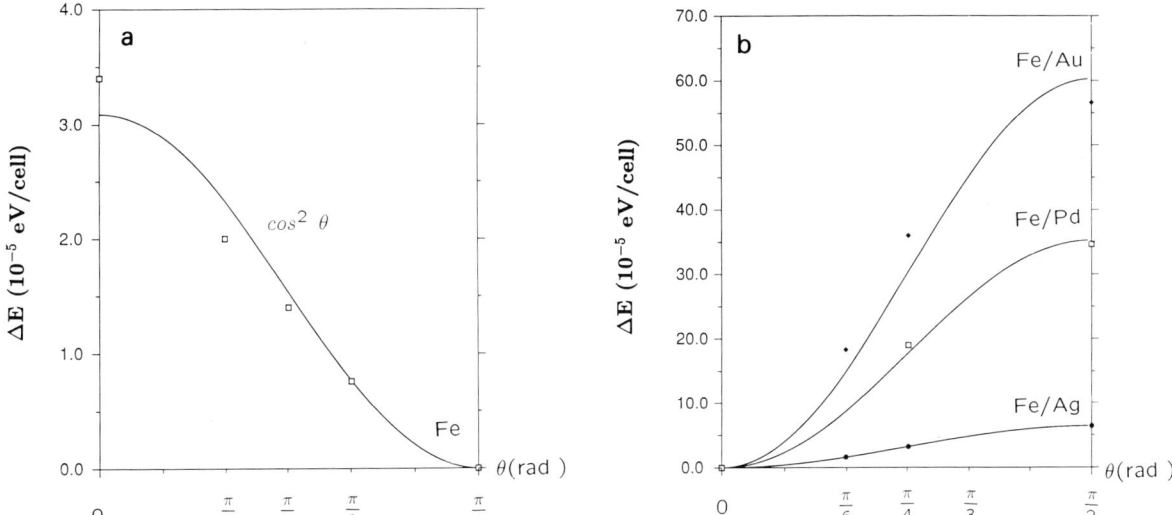

Fig. 6. Magnetic anisotropy energy (ΔE) for (a) Fe ML and (b) Fe Fe/Ag(001), Fe/Au(001) and Fe/Pd(001) vs. the angle (θ) between the direction of spin polarization and the z axis. Solid lines are $\cos^2(\theta)$ type fitting curves.

strating the important role of the lattice structure and symmetry. Such a role was also revealed for bulk Fe, Co and Ni [134], namely, the anisotropy energies of cubic Fe and Ni (~ 2 μeV) are much smaller than that of hcp Co (~ 60 μeV). Similar to Fe/Ag(001), the perpendicular anisotropy energy is very small (~ 0.049 μeV) for Co/Cu(001), and thus an in-plane magnetization is expected because of the demagnetization effect. This result is consistent with a recent experimental observation [143]. By contrast, Co/Pt(111) shows a strong perpendicular magnetic alignment with an anisotropy energy of 45 meV. The perpendicular orientation agrees with experimental conclusions [109].

7. Conclusions

In summary, state-of-the-art ab initio LSD electronic structure calculations have achieved great success in the exciting field of thin film magnetism, in both explaining existing phenomena and, more importantly, in predicting the properties of new systems. Illustrative results demonstrate that: (1) the lowered coordination number at clean metal surfaces leads to enhanced magnetic moments; (2) noble metal and MgO substrates do not affect the magnetism in most cases, but show significant effects on Ni and V overlayers; (3) the strong interaction (hybridization) with nonmagnetic transition metals diminishes (entirely in some cases) the ferromagnetism and usually leads to AFM ordering; (4) the core contribution to the Fermi contact hyperfine field (but not the total) is proportional to the local magnetic moment; and (5) the magnetic anisotropy can be predicted correctly for some systems. In the future, electronic structure theory is expected to continuously play a predictive role by considering more practical systems, by eliminating the limitation of the local spin density approximation and developing more efficient and precise methods to treat magnetic anisotropy, magneto-optic effects, effects of temperature, etc.

Acknowledgement

We are grateful to Drs. Chun Li, C.L. Fu and S.C. Hong for helpful discussions and collaboration. Work was supported by the National Science Foundation (Grant no. DMR 88-16126) and by the Office of Naval Research (Grant no. N00014-89-J-1290).

References

[1] L.M. Falicov, D.T. Pierce, S.D. Bader, R. Gronsky, K.B. Hathaway, H.J. Hopster, D.N. Lambeth, S.P. Parkin, G. Prinz, M. Salamon, I.K. Schuller and R.H. Victora, J. Mater. Res. 5 (1990) 1299.
[2] S.D. Bader, Proc. IEEE 78 (1990) 909.
[3] J. Mathon, Rep. Prog. Phys. 51 (1988) 1.
[4] A.J. Freeman, C.L. Fu, S. Ohnishi and M. Weinert, in: Polarized Electrons in Surface Physics, ed. R. Feder (World Scientific, Singapore, 1985).
[5] C.L. Fu, A.J. Freeman and T. Oguchi, Phys. Rev. Lett. 54 (1985) 2700.
[6] D. Wang, A.J. Freeman and M. Weinert, J. Magn. Magn. Mater. 31–34 (1983) 891.
[7] S.C. Hong, C.L. Fu and A.J. Freeman, J. Appl. Phys. 63 (1988) 3655.
[8] O. Eriksson, R.C. Albers and A.M. Boring, Phys. Rev. Lett. 66 (1991) 1350.
[9] E. Tamura, R. Feder, G. Waller and U. Gradmann, Phys. Stat. Sol. (b) 157 (1990) 627.
[10] H.J. Elmers, G. Liu and U. Gradmann, Phys. Rev. Lett. 63 (1989) 566.
[11] T. Beier, D. Pescia, M. Stampanoni, A. Vaterlaus and F. Meier, Appl. Phys. A 47 (1988) 73.
[12] M. Weinert, E. Wimmer and A.J. Freeman, Phys. Rev. B 26 (1982) 4571.
[13] J.G. Gay and R. Richter, Phys. Rev. Lett. 56 (1986) 2728.
[14] P. Bruno and J.P. Renard, Appl. Phys. A 49 (1989) 499.
[15] C. Li, A.J. Freeman, H.J.F. Jansen and C.L. Fu, Phys. Rev. B 42 (1990) 5433.
[16] P. Hohenberg and W. Kohn, Phys. Rev. B 136 (1964) 864.
[17] W. Kohn and L.J. Sham, Phys. Rev. A 140 (1965) 1133.
[18] J. Callaway and N.M. March, in: Solid State Physics, vol. 39, eds. F. Seitz, D. Turnbull and H. Ehrenreich (Academic Press, New York, London, 1983).
[19] U. von Barth and L. Hedin, J. Phys. C 5 (1972) 1629.
[20] O. Gunnarson, B.I. Lundqvist and S. Lundqvist, Solid State Commun. 11 (1972) 149.
[21] A.K. Rajagopal, in: Advances in Chemical Physics, vol. 41, eds. I. Prigogine and S.A. Rice (Wiley, New York, 1980).
[22] A.K. Rajagopal and J. Callaway, Phys. Rev. B 7 (1973) 1912.
[23] A.H. MacDonald and S.H. Vosko, J. Phys. C 12 (1979) 2977.
[24] A.K. Rajagopal, J. Phys. C 11 (1978) L943.
[25] M. Weinert and A.J. Freeman, Phys. Rev. B 28 (1983) 6262.
[26] D.D. Koelling and B.N. Harmon, J. Phys. C 10 (1977) 3107.
[27] E. Wimmer, H. Krakauer, M. Weinert and A.J. Freeman, Phys. Rev. B 24 (1981) 864, and references therein.
[28] M. Weinert, E. Wimmer and A.J. Freeman, Phys. Rev. B 26 (1982) 4571.
[29] M. Weinert, J. Math. Phys. 22 (1981) 2433.
[30] L.M. Liebermann, D.R. Fredkin and H.B. Shore, Phys. Rev. Lett. 22 (1969) 539.
L.M. Liebermann, J. Clinton, D.M. Edwards and J. Mathon, ibid. 25 (1970) 232.
[31] C.L. Fu and A.J. Freeman, Phys. Rev. B 35 (1987) 925.
[32] S. Ohnishi, A.J. Freeman and M. Weinert, Phys. Rev. B 28 (1983) 6741.
[33] A.J. Freeman and C.L. Fu, J. Appl. Phys. 61 (1987) 3356.
C.L. Fu and A.J. Freeman, J. Magn. Magn. Mater. 69 (1987) L1.
[34] R. Wu and C. Li, A.J. Freeman and G.B. Olson, to be published.
[35] G.W. Fernando and B.R. Cooper, Phys. Rev. B 38 (1988) 3016.
[36] W.A.A. Macedo and W. Keune, Phys. Rev. Lett. 61 (1990) 475.
[37] C.S. Wang and A.J. Freeman, Phys. Rev. B 21 (1980) 4585.
[38] O. Jepsen, J. Madsen and O.K. Andersen, J. Magn. Magn. Mater. 15–18 (1980) 867; Phys. Rev. B 26 (1982) 2790.
[39] H. Krakauer, A.J. Freeman and E. Wimmer, Phys. Rev. B 28 (1983) 610.
[40] X. Zhu, J. Hermanson, F.J. Arlingghaus, J.G. Gay, R. Richter and J.R. Smith, Phys. Rev. B 29 (1984) 4426.
[41] A.J. Freeman, D. Wang and H. Krakauer, J. Appl. Phys. 53 (1982) 1997.
E. Wimmer, A.J. Freeman and H. Krakauer, Phys. Rev. B 30 (1984) 3113.
[42] C.L. Fu and A.J. Freeman, J. de Phys. 49 (1988) C8-1625.
[43] R. Wu and A.J. Freeman, to be published.
[44] A.J. Freeman, C.L. Fu and T. Oguchi, Mat. Res. Soc. Symp. Proc. 63 (1985) 1.
[45] J.I. Lee, C.L. Fu and A.J. Freeman, J. Magn. Magn. Mater. 62 (1986) 93.
[46] C. Li, A.J. Freeman and C.L. Fu, J. Magn. Magn. Mater. 75 (1988) 53.
[47] J.I. Lee, C.L. Fu and A.J. Freeman, to be published.
[48] C. Li, A.J. Freeman and C.L. Fu, J. Magn. Magn. Mater. 94 (1991) 134.
[49] M. Weinert and A.J. Freeman, J. Magn. Magn. Mater. 38 (1983) 23.
[50] C. Li, Bull. Am. Phys. Soc. 36 (1991) 677.
[51] M. Weinert and J.W. Davenport, Phys. Rev. Lett. 54 (1985) 1547.
[52] C.L. Fu and A.J. Freeman, Phys. Rev. B 40 (1989) 5359.
[53] S. Ohnishi, M. Weinert and A.J. Freeman, Phys. Rev. B 30 (1984) 36.
[54] S.C. Hong, J.I. Lee and A.J. Freeman, J. Magn. Magn. Mater. 99 (1991) L45.
[55] W.C. Koehler, R.M. Moon, A.L. Trego and A.R. MacKintosh, Phys. Rev. 151 (1966) 405.
S.A. Werner, A.S. Arrott and H. Kendrick, ibid. 155 (1967) 528.

[56] P.E. Ferguson, J. Appl. Phys. 49 (1978) 2203.
[57] L.E. Klebanoff, S.W. Robey, G. Liu and D.A. Shirley, Phys. Rev. B 30 (1984) 1048.
G. Zajac, S.D. Bader and R.J. Friddle, Phys. Rev. 31 (1985) 4947.
L.E. Klebanoff and D.A. Shirley, Phys. Rev. B 33 (1986) 5301.
[58] G. Allan, Phys. Rev. B 19 (1979) 4774.
[59] R.H. Victora and L.M. Falicov, Phys. Rev. B 31 (1985) 7335.
L.E. Klebanoff, R.H. Victora, L.M. Falicov and D.A. Shirley, ibid. 32 (1985) 1997.
[60] C.L. Fu and A.J. Freeman, Phys. Rev. B 33 (1986) 1755.
[61] S. Blügel, D. Pescia and P.H. Dederichs, Phys. Rev. B 39 (1989) 1392.
[62] S. Ohnishi, C.L. Fu and A.J. Freeman, J. Magn. Magn. Mater. 50 (1985) 161.
[63] A.M. Turner, Y.J. Chang and J.L. Erskine, Phys. Rev. Lett. 48 (1982) 348.
A.M. Turner and J.L. Erskine, Phys. Rev. B 28 (1983) 5628, B 30 (1984) 6675.
[64] C. Rau, C. Liu, A. Schmalzbauer and G. Xing, Phys. Rev. Lett. 57 (1986) 2311.
[65] M.B. Brodsky and A.J. Freeman, Phys. Rev. Lett. 45 (1980) 133.
[66] U. Gradmann and R. Bergholz, Phys. Rev. Lett. 52 (1989) 771.
[67] H. Chen, N.E. Brener and J. Callaway, Phys. Rev. B 40 (1989) 1443.
[68] R. Wu and A.J. Freeman, unpublished.
[69] R. Wu, C. Li and A.J. Freeman, J. Magn. Magn. Mater. 99 (1991) 71.
[70] C. Rau and S. Eichner, in: Nuclear Method in Materials Research, eds. K. Bodge, H. Bauman, H. Jex and F. Rauch (Vieweg, Braunschweig, 1980) p. 354.
C. Rau, J. Magn. Magn. Mater. 31–59 (1983) 874.
C. Rau and S. Eichner, Phys. Rev. B 34 (1986) 6347.
C. Rau and M. Robert, Phys. Rev. Lett. 58 (1987) 2714.
[71] A. Cerri, D. Mauri and M. Landolt, Phys. Rev. B 27 (1983) 6526.
[72] D. Weller, S.F. Alvarado, W. Gudat, K. Schröder and M. Campagna, Phys. Rev. Lett. 54 (1985) 1555.
D. Weller and S.F. Alvarado, J. Appl. Phys. 59 (1986) 2908.
[73] D. Weller and S.F. Alvarado, Phys. Rev. B 37 (1988) 9911.
[74] R. Wu and A.J. Freeman, J. Magn. Magn. Mater. 99 (1991) 81.
R. Wu, C. Li, A.J. Freeman and C.L. Fu, Phys. Rev. B, submitted.
[75] See for example, C. Kittel, Introduction to Solid State Physics, 5th ed. (Wiley, New York, 1976) p. 85.
[76] J.F. Janak and A.r. Williams, Phys. Rev. B 14 (1976) 4199.
K.B. Hathaway, H.J.F. Jansen and A.J. Freeman, Phys. Rev. B 31 (1985) 7603.
[77] P.A. Montano, G.W. Fernando, B.R. Cooper, E.R. Moog, H.M. Naik, S.D. Bader, Y.C. Lee, Y.N. Darici, H. Min and J. Marcano, Phys. Rev. Lett. 59 (1987) 1041.
[78] J. Quinn, Y.S. Li, D. Tian, H. Li, F. Jona and P.M. Marcus, Phys. Rev. B 42 (1990) 11348.
[79] L.W. Roeland, G.J. Cock, F.A. Muller, A.C. Moleman, R.G. Jordan and K.A. McEwen, J. Phys. F 5 (1975) L233.
[80] D. Li, C.W. Hutchings, P.A. Dowben, C. Hwang, R.T. Wu, M. Onellion, A.B. Andrews and J.L. Erskine, J. Magn. Magn. Mater. 99 (1991) 85.
[81] See for example, C. Rau, Appl. Phys. A 49 (1989) 579.
[82] S. Blügel, B. Drittler, R. Zeller and P.H. Dederichs, Appl. Phys. A 49 (1989) 547.
[83] R. Wu and A.J. Freeman, to be published.
[84] M.J. Zhu, D.M. Bylander and L. Kleinman, Phys. Rev. B 42 (1990) 2874.
[85] M.J. Zhu, D.M. Bylander and L. Kleinman, Phys. Rev. B 43 (1991) 4007.
[86] J.F. Janak, Phys. Rev. B 16 (1977) 255.
[87] C.L. Fu and A.J. Freeman, J. Magn. Magn. Mater. 54–57 (1986) 777.
[88] R. Richter, J.G. Gay and J.R. Smith, Phys. Rev. Lett. 54 (1985) 2707.
[89] S. Blügel and P.H. Dederichs, Europhys. Lett. 9 (1989) 597.
[90] C. Li, A.J. Freeman and C.L. Fu, J. Magn. Magn. Mater. 75 (1988) 201.
[91] C. Li and A.J. Freeman, Phys. Rev. B 43 (1991) 780.
[92] C. Li, A.J. Freeman and C.L. Fu, J. Magn. Magn. Mater. 83 (1990) 51.
[93] R.H. Victora and L.M. Falicov, Phys. Rev. B 28 (1984) 5232.
[94] D.S. Wang, A.J. Freeman and H. Krakauer, Phys. Rev. B 26 (1982) 1340.
[95] S.C. Hong, A.J. Freeman and C.L. Fu, Phys. Rev. B 39 (1989) 5719.
[96] M.A. Thompson, M. Onellion and J.L. Erskine, Bull. Am. Phys. Soc. 31 (1986) 675.
[97] J. Araya-Pochet, C.A. Ballentine and J.L. Erskine, in: Magnetic Properties of Low-Dimensional Systems II, eds. L.M. Falicov, F. Mejia-Lira and J.L. Morán-López (Springer-Verlag, Berlin, 1990) p. 29.
[98] T. Oguchi and A.J. Freeman, J. Magn. Magn. Mater. 54–57 (1986) 797.
[99] R. Wu and A.J. Freeman, to be published.
[100] M. Stampanoni, A. Vaterlaus, D. Pescia, M. Aeschlimann, F. Meier, W. Dürr and S. Blügel, Phys. Rev. B 37 (1988) 10380.
[101] R.L. Fink, C.A. Ballentine, J.L. Erskine and J.A. Araya-Pochet, Phys. Rev. B 41 (1990) 10175.
[102] C. Rau, G. Xing, C. Liu and M. Robert, Phys. Lett. A 135 (1989) 227.
J.S. Moodera and R. Meservey, Phys. Rev. B 40 (1989) 8541.
[103] M.N. Baibich, J.M. Broto, A. Fert, F. Nguyen van Dau, F. Petroff, P. Etienne, G. Creuzet, A. Friederich and J. Chazelas, Phys. Rev. Lett. 61 (1988) 2472.

[104] G. Binasch, P. Grünberg, F. Saurenbach and W. Zinn, Phys. Rev. B 39 (1989) 4828.
[105] S.S.P. Parkin, N. More and K.P. Roche, Phys. Rev. Lett. 64 (1990) 2304.
[106] B. Heinrich, Z. Celinski, J.f. Cochran, W.B. Muir, J. Rudd, Q.M. Zhong, A.S. Arrott and K. Myrtle, Phys. Rev. Lett. 64 (1990) 673.
[107] C. Liu and S.D. Bader, Phys. Rev. B 41 (1990) 553.
[108] M. Maurer, J.C. Ousset, M. Piechuch, M.F. Ravet and J.P. Sanchez, Mat. Res. Soc. Symp. Proc. Vol. 151 (1989) 99.
M. Maurer, J.C. Ousset, M. Piechuch and M.F. Ravet, Europhys. Lett. 9 (1989) 803.
[109] P.F. Carcia, A.D. Meinholdt and A. Suna, Appl. Phys. Lett. 47 (1985) 178.
H.J.G. Draaisma, W.J.M. de Jonge and F.J.A. den Broeder, J. Magn. Magn. Mater. 66 (1987) 351.
P.F. Carcia, J. Appl. Phys. 63 (1988) 5066.
[110] S. Hashimoto, H. Matsuda and Y. Ochiai, Appl. Phys. Lett. 56 (1990) 1069.
F.J.A.M. Greidanus, W.B. Zeper, F.J.A. den Broeder, W.F. Godlieb and P.F. Carcia, Appl. Phys. Lett. 54 (1989) 5481.
W.B. Zeper, F.J.A.M. Greidanus, P.F. Carcia and C.R. Fincher, J. Appl. Phys. 65 (1989) 4971.
[111] R. Wu and A.J. Freeman, Phys. Rev. B, submitted.
[112] D. Knab and C. Koenig, Phys. Rev. B 43 (1991) 8370.
[113] R. Wu and A.J. Freeman, to be published.
[114] S.C. Hong, A.J. Freeman and C.L. Fu, Phys. Rev. B 38 (1988) 12156.
[115] B. Hillebrands, P. Baumgart and G. Güntherodt, Appl. Phys. A 49 (1989) 589.
Z. Celinski, B. Heinrich, J.F. Cochran, W.B. Muir, A.S. Arrott and J. Kirschner, Phys. Rev. Lett. 65 (1990) 1156.
[116] C.J. Nieuwenhuys, Adv. Phys. 24 (1975) 515.
R.M. Bozorth, P.A. Wolff, D.D. Davis, V.B. Compton and J.H. Wernick, Phys. Rev. 122 (1961) 1157.
A. Oswald, R. Zeller and P.H. Dederich, Phys. Rev. Lett. 56 (1986) 1419.
[117] A.T. Aldred, B.D. Rainford and M.W. Stringfellow, Phys. Rev. Lett. 24 (1970) 897.
[118] S. Blügel, M. Weinert and P.H. Dederichs, Phys. Rev. Lett. 60 (1988) 1077.
[119] B. Heinrich, A.S. Arrott, J.F. Cochran, C. Liu and K. Myrtle, J. Vac. Sci. Technol. A4 (1986) 1376.
B. Heinrich, S.T. Purcell, J.R. Dutcher, K.B. Urquhart, J.F. Cochran and A.S. Arrott, Phys. Rev. B 38 (1988) 12879.
[120] J.I. Lee, S.C. Hong and A.J. Freeman, Phys. Rev. B, submitted.
[121] T. Shinjo, Surf. Sci. Rep. 12 (1991) 51 and references therein to earlier work in the field.
[122] A.J. Freeman and R.E. Watson, in: Magnetism, vol. IIA, eds. G.T. Rado and H. Suhl (Academic Press, New York, 1965) p. 167.
[123] A.J. Freeman, C.L. Fu, M. Weinert and S. Ohnishi, Hyperfine Interactions 33 (1987) 53.
[124] J. Tyson, A.H. Owens, J.C. Walker and G. Bayreuther, J. Appl. Phys. 52 (1981) 2487.
[125] J. Korecki and U. Gradmann, Phys. Rev. Lett. 55 (1985) 2491.
[126] L. Néel, J. Phys. Radium 15 (1954) 225.
[127] C. Liu and S.d. Bader, in: Magnetic Properties of Low-Dimensional Systems II, eds. L.M. Falicov, F. Mejía-Lira and J.L. Morán-López (Springer-Verlag, Berlin, 1990) p. 22.
[128] J.H. van Vleck, Phys. Rev. 52 (1937) 1178.
[129] H. Brooks, Phys. Rev. 58 (1940) 909.
[130] G.C. Fletcher, Proc. R. Soc. London 67A (1954) 505.
[131] L. Fritsche, J. Noffke and H. Eckardt, J. Phys. F 17 (1987) 943.
[132] G.H. Daalderop, P.J. Kelly, M.F.H. Schuurmans and H.J.F. Jansen, J. de Phys. 12 (1989) C8–12.
[133] P. Strange, H. Ebert, J.B. Staunton and B.L. Gyorffy, J. Phys.: Condens. Matter 1 (1989) 3947.
[134] G.H. Daalderop, P.J. Kelly and M.F.H. Schuurmans, Phys. Rev. B 41 (1990) 11919.
[135] G.H. Daalderop, P.J. Kelly and M.F.H. Schuurmans, Phys. Rev. B 42 (1990) 7270.
[136] A.J. Bennett and B.R. Cooper, Phys. Rev. B 3 (1971) 1642.
[137] H. Takayama, K.P. Bohnen and P. Fulde, Phys. Rev. B 14 (1976) 2287.
[138] P. Bruno, Phys. Rev. B 39 (1989) 865.
[139] J.G. Gay and R. Richter, J. Appl. Phys. 61 (1987) 3362.
[140] B.T. Jonker, K.H. Walker, E. Kisker, G.P. Prinz and C. Carbone, Phys. Rev. Lett. 57 (1986) 142.
[141] R. Richter and J.G. Gay, Mat. Res. Soc. Symp. Proc. 151 (1989) 3.
[142] C. Li and A.J. Freeman, unpublished.
[143] F.J.A. den Broeder, D. Kuiper, A.P. van de Mosselaer and W. Hoving, Phys. Rev. Lett. 60 (1988) 2769.

The ferromagnetism of ultrathin films; from two to three dimensions

D.L. Mills

Department of Physics, University of California, Irvine, CA 92717, USA

We review the properties of ultrathin ferromagnetic films, with attention to simple physical arguments that outline their principal properties. A ferromagnetic monolayer clearly provides an illustration of two dimensional magnetism. It is argued that the crossover to three dimensional behavior occurs when only a few monolayers are present, both in the low temperature spin wave regime, and near the transition temperature. We discuss anisotropy and its role in inducing long range order in ultrathin films, and spin reorientation transitions.

1. Introduction

The study of magnetism in less than three (space) dimensions has proved most fundamental. In three dimensions, one expects long range order for virtually any spin system on a periodic lattice. Furthermore, if one wishes to obtain a broad overview of the behavior of any particular system, the various forms of mean field theory can work nicely. Spin wave theory then provides a quantitative account of the low temperature phase, and one can say a great deal about the critical regime as well, with modern theoretical methods.

In two dimensions, mean field theory proves inadequate as even a crude guide. It is well known that it predicts the onset of long range order for both the Ising and the Heisenberg model, while we know very well only the former picture describes a spin system which displays true long range order at low temperature. In systems of reduced dimensionality, the importance of fluctuations is enhanced, to the point where in one dimension neither the Ising nor the Heisenberg model displays long range order. One encounters also, for certain classes of two dimensional magnets the Kosterlitz–Thouless transition. The study of magnetic materials in less than three dimensionless thus allows one to explore basic questions in statistical mechanics.

Experimental studies of magnetism in one or two dimensions generally employ bulk crystals, within which one has a structure where linear chains of spins are separated by intervening molecular groups to the point where interchain coupling is weak [1] or where planes of spins are separated in a similar manner [2,3]. While such crystals indeed allow one to explore statistical mechanics in less than three dimensions, it remains the case that the interchain or interplanar couplings are present, and assert themselves at some level, in any regime of temperature or magnetic field.

In recent years, it has proved possible to synthesize in the laboratory ultrathin films of ferromagnetic transition metals on nonmagnetic metal substrates such as Cu(100), Ag(100) [4] or W(110) [5]. At least in principle, these systems can be true realizations of two dimensional magnetic materials, uncontaminated by couplings to nearby layers. The films can be as thin as one monolayer, and in fact can be grown one monolayer at a time. This allows the study of true two dimensional magnetism, along with the transition from two to three dimensions. At the time of this writing, we know rather little about the microscopic atomic arrangements in these ultrathin films. Also, there is ongoing discussion about the quality of the films realized on various substrates [5,6]. There is currently a high level of activity in this area, and we can expect these issues to be settled soon.

The purpose of this paper is to outline, with simple physical arguments, the basic magnetic properties of the ultrathin films, and the issues that may be addressed by further experimental study. We have fascinating glimpses of the physics of these new materials, but in fact many basic questions remain to be addressed within the laboratory and by theorists.

The ultrathin films of Fe and Ni order ferromagnetically, at Curie temperatures that can be well below the bulk Curie temperature but which are still substantial. We shall discuss a possible origin of the long range order in the next section; while these transition metal ferromagnets are not localized spin magnets, we expect them to belong to the same universality class as the two dimensional Heisenberg ferromagnet, which fails to exhibit long range order at any finite temperature.

Given the fact that these materials do order ferromagnetically, a simple argument allows us to outline how thin the film must be before its magnetic properties are two dimensional in nature. We begin by noting that below the ordering temperature, we may use spin wave theory to describe the materials.

Consider a film with $N_\perp$ atomic layers. The spin waves have the character of standing spin waves, with a quantized wave vector component $k_\perp^{(m)} \approx m\pi/N_\perp a_0$ perpendicular to the surfaces, where a_0 is the interplanar spacing and $m = 0, 1, \ldots$ The wave vector $k_\parallel$ parallel to the surface is continuous, so the spin wave spectrum of the film consists of branches whose frequencies may be labeled $\omega^{(m)}(k_\parallel)$. The spin waves that influence the thermodynamics of the film have excitation energies $\hbar\omega^{(m)}(k_\parallel)$ the order of $k_B T$, which is large compared to typical anisotropy and dipolar contributions to the excitation energy. Hence, we may write $\omega^{(m)}(k_\parallel) \approx D(k_\parallel^2 + (m\pi/N_\perp a_0)^2)$, with D the exchange stiffness.

If the film is so thin that $\hbar D(\pi/N_\perp a_0)^2 \gg k_B T$, then only the lowest $m = 0$ spin wave branch is thermally populated. In this limit, the thermodynamic properties of the film are those of a two dimensional magnet, even though the film may be a few atomic layers thick. In the opposite limit, $\hbar D(\pi/N_\perp a_0)^2 \ll k_B T$, many of the branches $\omega^{(m)}(k_\parallel)$ are thermally populated, sums over $k_\perp^{(m)}$ may be replaced by integrals to good approximation, and we are in the quasi-three dimensional limit.

This leads us to define a critical thickness $N_c = (\hbar D/k_B T)^{1/2}(\pi/a_0)$. If, at any temperature, $N_\perp < N_c$ we are in the quasi-two dimensional limit, while if $N_\perp > N_c$, the film is quasi-three dimensional. We do not know that value of D for ultrathin Fe films at present; this parameter is surely thickness dependent [7]. If we use D for bulk Fe (2.5×10^{-9} G cm^2 in magnetic field units) and $a_0 \approx 2A$, then $N_c \approx 3$ atomic layers. The monolayer is a true two dimensional magnet, and one makes the transition to three dimensional magnetism in only a few monolayers. Below, from a very different perspective, we will see once again that the transition takes place in rather few layers. Thus, the new ultrathin films offer us the possibility of studying both magnetism in two dimensions (the monolayer), and the transition to three dimensional magnetism as well.

There is experimental evidence that bears directly on this question, presented in ref. [5]. These authors measure the temperature variation and magnitude of the hyperfine field, at Fe sites in ultra thin Fe films on W(110). They find, at low temperatures and for hyperfine fields normalized to their value at $T = 0$,

$$\frac{H_{\text{hf}}(T)}{H_{\text{hf}}(0)} = 1 - bT^n, \tag{1}$$

where the exponent n lies in the range 1.3–1.5. Most importantly for the purposes of the present discussion, they find the prefactor $b \sim 1/N_\perp$, with $N_\perp$ the number of layers in the film. The data covers films for which $N_\perp$ ranges from 1 to 4.

The data are taken at temperatures sufficiently low that spin wave theory can be applied. The temperature dependence of the hyperfine field is that of $\langle S_z(l_\| l)\rangle_T$, which is the component parallel to the film magnetization of the spin associated with site $l_\|$, in the layer $l_\perp$. The Holstein–Primakoff transformation gives for all spins in layer $l_\perp$

$$\frac{H_{\text{hf}}(l_\perp, T)}{H_{\text{hf}}(l_\perp, 0)} = \frac{\langle S_z(l_\| l_\perp)\rangle_T}{\langle S_z(l_\| l_\perp)\rangle_0}$$

$$= 1 - \frac{1}{S}\langle a^+(l_\| l_\perp)a(l_\| l_\perp)\rangle, \quad (2)$$

where $a^+(l_\| l_\perp)$ is the spin wave creation operator for the site indicated.

Suppose we calculate the spin wave spectrum supposing only exchange couplings are present, representing anisotropy and dipolar effects by introducing an effective internal field parallel to the magnetization [8]. The normal modes have the standing wave character described above, and we may write

$$a(l_\| l_\perp) = \frac{1}{\sqrt{N_s}} \sum_{k_\|} \xi^m(k_\|, l_\perp)\, e^{ik_\| \cdot l_\|} a^{(m)}(k_\|), \quad (3)$$

where $a^m(k_\|)$ is the spin wave destruction operator for mode $(k_\|, m)$. We have a film in the form of $N_\perp$ planes of sites, each containing N_s spins. Then $\xi^{(m)}(k_\|, l_\perp)$ is an eigenvector normalized so that

$$\sum_{l_\perp = 1}^{N_\perp} \left|\xi^{(m)}(k_\| l_\perp)\right|^2 = 1. \quad (4)$$

We then have

$$\frac{\langle S_z(l_\| l_\perp)\rangle_T}{\langle S_z(l_\| l_\perp)\rangle_0} = 1 - \frac{1}{SN_s} \sum_{k_\|} \sum_m \left|\xi^{(m)}(k_\|, l_\perp)\right|^2$$

$$\times \bar{n}(\omega^{(m)}(k_\|)), \quad (5)$$

where $\bar{n}(\omega) = [\exp(\hbar\omega/k_B T) - 1]^{-1}$ is the Bose–Einstein function.

Suppose we are in the regime $N \ll N_c$, with N_c defined by the criterion given above. Then only the lowest branch $m = 0$ contributes to the right hand side of eq. (5). The lowest mode will be (in simple models) an acoustical mode, wherein for fixed $k_\|$, spins in each layer precesses in phase with the same amplitude. For this mode $|\xi^{(m)}(k_\|, l_\perp)|^2$ is independent of $l_\perp$, and eq. (4) then gives this quantity as $1/N_\perp$. Then eq. (5) becomes

$$\frac{\langle S_z(l_\| l_\perp)\rangle_T}{\langle S_z(l_\| l_\perp)\rangle_0} = 1 - \frac{1}{N_\perp}\left\{\frac{1}{S}\sum_{k_\|}\bar{n}(\omega^{(0)}(k_\|))\right\}. \quad (6)$$

The result in eq. (6), derived on the assumption that the thermodynamic properties of the film are quasi-two dimensional, show that the coefficient b in eq. (5) scales inversely with film thickness, *provided* the dispersion relation $\omega^{(0)}(k_\|)$ of the low lying spin wave branch is independent of thickness. In section 2, we shall discuss a scenario in which this is possible. If we accept this scenario, then the data reported in ref. [5] is clear experimental evidence that the thermodynamic properties of the ultrathin Fe films on W(110) is indeed quasi-two dimensional, at least for $N_\perp$ the order of four or less.

Note that N_c is temperature dependent, and decreases with increasing temperature. Thus a thicker film may have $N_\perp < N_c$ at low temperatures, but with increasing temperature one may pass into the domain $N_\perp \gtrsim N_c$ to enter the regime where two dimensional behavior breaks down. Thinner films will obey the constraint $N_\perp < N_c$ to higher temperatures. Thus, at higher temperatures, the scaling relation $b \sim 1/N_\perp$ should break down, with the deviations evident in data on thicker films. It would be interesting to see this point explored.

The above discussion assumes the presence of long range order, so spin wave theory may be applied at temperatures well below the ordering temperature. We comment next on one view of the origin of the long range order in ultrathin films.

2. Anisotropy and long range order in ultrathin ferromagnetic films

A monolayer of Fe atoms on a substrate may be viewed as a two dimensional Heisenberg ferromagnet, so far as its spin dynamics is concerned. But then there should be no long range order, by virtue of the Mermin–Wagner theorem. In these films, however, the presence of very large anisotropies may be inferred from the data; these anisotropies are in fact very much larger than those found in, say, bulk Fe. It has been argued that it is the presence of the large anisotropy that is responsible for the presence of the long range order [9,10]. We first comment on the nature of the anisotropy in these materials, then on its role in inducing long range order.

There are two possible origins of the anisotropy. One is the combination of spin–orbit coupling, and the fact that a large fraction of the spins reside on sites of low symmetry. There is in addition strain induced anisotropy, whose influence can be appreciable. We consider each in turn, with emphasis on simple, physical arguments.

Consider first the spin–orbit coupling, within a picture that assigns a spin S to each lattice site. The spin–orbit term in the Hamiltonian has the form $\lambda \mathbf{L} \cdot \mathbf{S}$, and we may treat the influence of this on the energy of the spin through use of perturbation theory. One does this by integrating over the coordinates of the electron, leaving the spin S as an operator that acts on the manifold of spin states. This procedure produces an effective Hamiltonian dependent on spin orientation referred to as the *spin Hamiltonian* [11].

The structure of the spin Hamiltonian may be analyzed from symmetry considerations. First, time reversal symmetry allows only even powers of S to appear in the result, so we have contributions from second order, fourth order, ... perturbation theory.

Suppose the spin S is in a site of cubic symmetry, as in bulk Fe. In the second order of perturbation theory, symmetry allows only the combination $(S_x^2 + S_y^2 + S_z^2)$, which is the constant $S(S+1)$. This contribution to the energy of the spin is thus independent of orientation. The fourth order contribution is made up of two invariants we write as

$$\Delta E^{(4)} = \Gamma_4 \{ \alpha_4 (S_x^4 + S_y^4 + S_z^4) \\ + \beta_4 (S_x^2 S_y^2 + S_x^2 S_z^2 + S_y^2 S_z^2) \} \\ = \Gamma_4 (\beta_4 - 2\alpha_4)(S_x^2 S_y^2 + S_x^2 S_z^2 + S_y^2 S_z^2) \\ + \Gamma_4 \alpha_4 [S(S+1)]^2. \quad (7)$$

Here β_4 and α_4 are dimensionless constants of order unity, while Γ_4 has units of energy.

The second line of eq. (7) describes a form which indeed depends on the orientation of the spin relative to the crystal axes. Thus, for a spin in a cubic site, the spin first senses the crystal axes through contributions to the energy fourth order in perturbation theory. If λ is the spin–orbit coupling constant and ΔE a typical energy denominator of perturbation theory, then $\Gamma_4 \sim \lambda^4/(\Delta E)^3$. In the bulk, cubic ferromagnetic metals such as Fe and Ni, the magnetic anisotropy is rather weak. From the measured anisotropy constants of bulk Fe ($\Gamma_4 \approx 5 \times 10^5$ erg/cm^2 or 5×10^{-2} cm^{-1}/atom) and the spin–orbit coupling constant ($\lambda \approx 3 \times 10^2$ cm^{-1}), one estimates $(\lambda/\Delta E) \approx 5 \times 10^{-2}$ or $\Delta E \approx 0.8$ eV.

A real calculation of the anisotropy of Fe must be based on a band picture of the magnetic moment bearing electrons, of course. The value of ΔE just estimated is surely reasonable, if we think of it as a typical energy denominator in a perturbation theoretical description of the influence of spin–orbit coupling; the total width of the collection of the five Fe 3d bands is close to 4 eV.

Now consider an Fe monolayer, with a square unit cell. This would be the case of Fe on Cu(100) or Ag(100). The z direction is now inequivalent to the x and y directions, with z normal to the plane of the monolayer. The second order terms now have the form

$$\Delta E^{(2)} = \Gamma_2 \{ \alpha_2 S_z^2 + \beta_2 (S_x^2 + S_y^2) \} \\ = \Gamma_2 (\alpha_2 - \beta_2) S_z^2 + \Gamma_2 \beta_2 S(S+1), \quad (8)$$

where $\Gamma_2 \approx \lambda^2/\Delta E$, an energy much larger than Γ_4, estimated earlier.

Thus, for the monolayer, the low site symmetry of the Fe "switches on" the second order terms, which are suppressed in the bulk by symmetry considerations. If $\lambda/\Delta E \approx 10^{-1}$, Γ_2 may be two orders of magnitude larger than Γ_4. The monolayer thus is characterized by uniaxial anisotropy very large compared to bulk Fe.

There are two possibilities. One is that, in the notation of eq. (8), $\Gamma_2(\alpha_2 - \beta_2) < 0$. The axis normal to the surface is then an easy axis. The magnetization "points" normal to the film. The other is $\Gamma_2(\alpha_2 - \beta_2) > 0$; the axis normal to the film is then a hard axis, and the magnetization lies in the plane. In this second case, the (weak) fourth order terms $(S_x^2 S_y^2 + \ldots)$ will orient the magnetic moments in a selected direction within the xy plane.

The very crude argument here cannot predict the sign of the anisotropy constants, of course. For this purpose, a full and complete calculation is required. Such full analyses have appeared in the recent literature [12].

In a film a few layers thick, we expect the spins in the surface layer, and nearest the interface with the substrate will experience very strong uniaxial anisotropy of the sort just described. As the film thickness increases, only spins in the outer layers, which sit in low symmetry sites, experience this anisotropy. The effective anisotropy per atom of the film thus decreases inversely with film thickness, when this mechanism is operative. Such behavior is indeed observed, and it is possible to extract explicit results for the magnitude and thickness dependence of the effective anisotropy from studies of spin waves in ultrathin films via light scattering spectroscopy [13].

It is also the case that the spins in the interior of an ultrathin film such as that discussed above sits in sites with symmetry less than cubic, if the interlayer spacing between films differs from that appropriate to the bulk crystal. For the (100) films, this effect may be small (though this is not clear at this writing), since at the (100) surfaces of the simple metals interlayer relaxations typically are small. However, for the Fe films grown epitaxially on W(110) studied by Gradmann et al. [5], the basic unit cell is both expanded, and distorted so that it is in fact rectangular. Under these circumstances, each atom sits in a site of low symmetry, and the second order contributions "switch on" for all atoms in the film.

For the Fe films on W(110), the magnetic moment lies in plane. We may view the anisotropy as strain-induced in character, where the distortion of the unit cell is induced by strains required to place the Fe film on the substrate. If this is the dominant source of anisotropy, then to good approximation, the effective anisotropy of the film, expressed again as the average anisotropy per atom, is independent of film thickness.

We now return to the discussion of section 1, and the experimental observation that the coefficient b in eq. (1) varies inversely with film thickness. We argued that this behavior is expected (i) if the thermodynamics of the film is quasi-two dimensional in nature *and* (ii) the dispersion of the acoustical spin wave does not vary with film thickness. If there is strong anisotropy only in the outer layers of an ultrathin film so, as discussed above the effective anisotropy per spin varies inversely with film thickness, there is no possibility that condition (ii) can be fulfilled. The dispersion relation of the long wavelength acoustical spin wave will vary with film thickness, and the relation $b \sim 1/N_\perp$ will not be obeyed. If the average anisotropy per spin is constant through the film, and if it is reasonable to presume the exchange contribution to the excitation energy of a spin wave varies little with film thickness (an assumption open to question, if one is interested in films down to the monolayer thickness, as one sees from our footnote [7]), then assumption (ii) is satisfied, and one expects $b \sim 1/N_\perp$.

The experimental observations in ref. [5] thus suggest that for Fe films on W(110), strain induced anisotropy acts throughout the film, and this is the dominant source of anisotropy in these examples. Clearly, for the samples explored in ref. [13], the low symmetry of surface and interface sites are responsible, since the effective anisotropy per spin scales inversely with thickness. Thus, one has evidence for both mechanisms described above.

We may summarize the discussion above by

saying that the ultrathin films may be described by a phenomenological Hamiltonian

$$H = -\tfrac{1}{2} \sum_l K(l_\perp) S_z^2(l)$$
$$-\tfrac{1}{2} \sum_{l,l'} J(l,l') S(l) \cdot S(l'). \qquad (9)$$

Here $l_\perp$ is a layer index, which ranges from 1 to $N_\perp$, the number of layers. The z direction is the easy direction. If $K(l_\perp)$ is nonzero in only the outer layer or both outer layers of the film, we are modeling the circumstance where the film anisotropy is dominated by the contributions from the low symmetry sites in the surfaces. If $K(l_\perp)$ acts on all layers, we are modeling the case of Fe on W(110) (as described above), where strain induced anisotropy influences all spins. We have not yet mentioned the role of magnetic dipole–dipole interactions, whose influence will be discussed in section 3. For the moment, we shall assume they are incorporated into an effective anisotropy term, with a form in the end similar to that displayed in eq. (9). We shall elaborate on this issue later.

The ultrathin film is thus not a simple two dimensional Heisenberg model, but rather a two dimensional Heisenberg model perturbed by the uniaxial anisotropy. While, for reasons given above, the strength of the anisotropy $K(l_\perp)$ is large compared to that found in bulk Fe or Ni, in fact it remains the case that it is still weak compared to the very strong exchange described by the second term in eq. (9). Hence, we have a weakly perturbed Heisenberg film.

The presence of the uniaxial anisotropy leads to long range order. A recent analysis of the weakly perturbed two dimensional Heisenberg model [9] leads to a rather simple expression for the transition temperature. The calculation is based on a renormalization group approach. One considers a monolayer, with spins S of length unity, and models the influence of exchange within a continuum theory appropriate to the limit where long wavelength fluctuations dominate the critical behavior of the film. If T_2 is the transition temperature of the two dimensional monolayer, and T_3 that of the three dimensional array of exchange coupled films, one finds [9]

$$T_2 = 2T_3 / \ln\left[\frac{\pi^2 J}{K}\right]. \qquad (10)$$

The expression shows that as $K \to 0$, $T_2 \to 0$, as required. However, the fact that the ratio J/K appears in a logarithm means that even when K/J is very small, T_2 can be an appreciable fraction of T_3. For instance, if $K/J = 10^{-2}$, eq. (10) gives $T_2 = 0.30 T_3$. While the two dimensional Heisenberg ferromagnet fails to display long range order at any nonzero temperature, in fact it is very close to ordering, so to speak. A weak, symmetry breaking perturbation will drive the transition temperature up to a rather appreciable fraction of T_3. The analysis shows [9], as one might expect, that the transition has Ising character.

The expression given in eq. (9) is in fact approximate. It is based on use of a large N expansion, where for the moment N is the number of degrees of freedom in the order parameter. One extrapolates the result obtained for the large N limit down to $N = 3$ to generate eq. (9). One is then led to question the accuracy of this expression. Erickson and the present author [10] have explored this question within the framework of classical Monte Carlo calculations based on the full Hamiltonian in eq. (9). We find the expression works remarkably well.

We have also explored the dependence of the transition temperature on film thickness for this model [10], for the case where the spins in one outer layer experience the easy axis anisotropy. For $K/J = 0.1$, we reproduce our results in fig. 1. The transition temperature rises quite dramatically with increasing thickness, to assume a value quite close to the bulk value by the time six monolayers are present. We expect that if spins in *each* surface layer were to experience the anisotropy, the transition temperature would rise to the bulk value more quickly. For the model explored in fig. 1, in fact the spins in one surface of the film are exchange coupled to fewer neighbors than spins in the bulk, so they must be

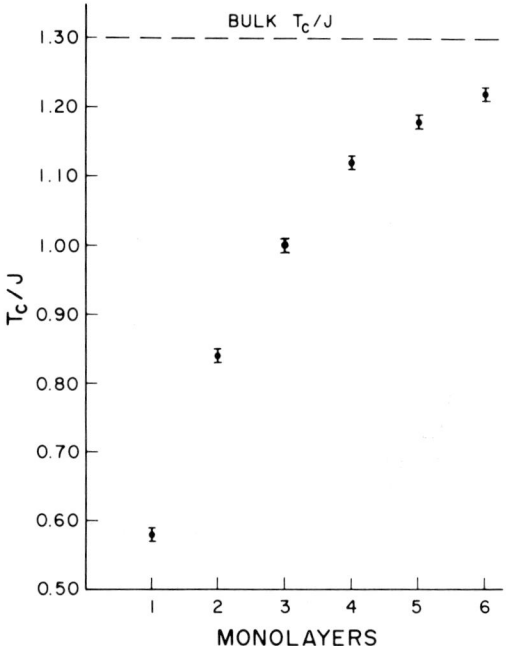

Fig. 1. The transition temperature of a model film described by the Hamiltonian in eq. (9), as a function of film thickness. The calculations assume that the easy axis anisotropy acts only on one outer layer of spins, and that for these spins, $K/J = 0.1$.

pulled into order by those in the remaining sample.

Fig. 1 shows the transition from true two dimensional behavior to bulk behavior occurs by the time only a few monolayers are present. The simple argument given in section 1 suggests that well below the ordering temperature, the crossover from 2D to 3D behavior also occurs on this length scale. Thus, the monolayer is an ideal laboratory for the study of real two dimensional magnetism, and by adding only a few layers, one may study the transition to three dimensions. For this reason, in the view of the present author, the study of magnetism in ultrathin films is of fundamental interest.

3. Spin reorientation transitions in ultrathin films; magnetic field induced and spontaneous

So far, the role of magnetic dipole interactions between spins has been ignored, save for a brief remark in section 3 that these may be incorporated into the anisotropy term in a phenomenological manner.

The dipolar terms in the spin Hamiltonian in fact play a strong role in controlling the behavior of the magnetization of the ultrathin ferromagnetic films. Again in this paper, our attention is confined to a simple physical model within which the essential aspects emerge. In recent papers, full microscopic studies of the properties of ultrathin ferromagnetic films have been carried out, in the presence of both dipolar coupling and uniaxial anisotropy [8,14].

Consider a thin film, with magnetization per unit volume M directed at an angle θ with respect to the film normal, which coincides with the z axis. It is an elementary matter to see that the spins generate within the films an internal demagnetizing field $H_d = -4\pi \hat{z} M \cos\theta$. All the spins are assumed locked together by very strong exchange, and thus are parallel to each other. Let the film have $N_\perp$ layers, each separated by the distance a_0. The magnetostatic energy per unit area may then be written as $2\pi M^2 \cos^2\theta N_\perp a_0$. If this energy is the only contribution to the energy per unit area dependent on the orientation of M, then the lowest energy state of the film is that where $\theta = \pi/2$. In very thick films of a ferromagnet like Fe or Ni, where the bulk anisotropy is weak for reasons discussed in section 2, then since $4\pi M \approx 20$ kG, very much larger than the effective internal anisotropy fields (≈ 500 G), then what is realized is the configuration where the magnetization is parallel to the film surfaces.

We have seen that in the ferromagnetic ultrathin films (Fe on Cu(100), for instance) one encounters very strong uniaxial anisotropy of the form given in eq. (9), where the easy axis is normal to the surface. If the anisotropy is confined to spins in either surface layer only, the energy per unit area is given by $-n_s K S^2 \cos^2\theta$, where n_s is the number of spins per unit area in the surface layers. The total energy per unit area associated with orientation of the magnetic moment is then

$$E(\theta) = -\left(n_s K S^2 - 2\pi M^2 N_\perp a_0\right)\cos^2\theta. \quad (11)$$

It is convenient to rewrite eq. (11). Let the lattice consist of entities whose magnetic moments are $g\mu S$, with μ the Bohr magneton. Then $M = g\mu S n_v$ for the material, where n_v is the number of spins per unit volume; we let $n_v = n_s/a_0$. If one examines the equations of motion of spin wave theory, the combination $KS/g\mu$ acts as an effective magnetic field in the surface, the surface anisotropy field H_s, which acts on the spins there. In terms of the quantities just introduced, eq. (11) may be arranged to read

$$E(\theta) = -g\mu n_s S(H_s - 2\pi M N_\perp) \cos^2\theta. \quad (12)$$

For the monolayer, it should be remarked, eq. (12) should be replaced by $([H_s/2] - 2\pi M)$, since we have assumed in our derivation that spins in each of two outer layers experience the anisotropy field H_s; there is only one layer of spins when $N_\perp = 1$.

Suppose $H_s > 4\pi M$, for the monolayer. Then the favored state has $\theta = 0$, and the easy direction is normal to the surface. The role of the dipole interactions, accounted for here by introducing the macroscopic field H_d, is to just renormalize the anisotropy from H_s to $\tilde{H}_s = H_s - 4\pi M$.

Now as layers are added, when $N_\perp > (H_s/2\pi M)$, the orientation $\theta = \pi/2$ is the low energy state. The magnetism lies in the plane for all films thicker than $H_s/2\pi M$. In practice, for monolayers of Fe which have $H_s > 4\pi M$, and which are thus magnetized normal to the surface, one finds the critical value of $N_\perp$ lies in the range from 3 to 4 layers. Hence, H_s lies in the range of 50 to 100 kG. If one converts this to the language of section 2, the crude arguments there account for the magnitude of H_s.

Application of a magnetic field parallel to the surface causes the magnetization to cant away from the normal to the surfaces, for a film with easy axis normal to the surface, and a sufficiently strong field will tip it into the plane. Upon application of a field H_0 parallel to the surface, eq. (12) is replaced by, letting $\tilde{H}_s = H_s - 2\pi M N_\perp$,

$$E(\theta) = -g\mu n_s S\left[\tilde{H}_s \cos^2\theta - H_0 \sin\theta\right], \quad (13)$$

so the canting angle is given by the condition

$$\cos\theta\left[H_0 - 2\tilde{H}_s \sin\theta\right] = 0 \quad (14)$$

so we have

$$\sin\theta = \frac{H_0}{2\tilde{H}_s}, \quad H_0 < 2\tilde{H}_s \quad (15a)$$

and

$$\theta = \pi/2. \quad (15b)$$

Now the magnetic moment parallel to the surface is just $M_\parallel = M \sin\theta$. Hence for $H_0 < 2\tilde{H}_s$, $M_\parallel = (MH_0/2\tilde{H}_s)$, while for $H_0 > 2\tilde{H}_s$, $M_\parallel = M$ independent of H_0.

A plot of $M_\parallel$ vs. H_0 thus exhibits a *discontinuity in slope* at the critical field $H_0^{(c)} = 2\tilde{H}_s$. This means the free energy of the system is a nonanalytic function of H_0, at $H_0^{(c)}$; we have a magnetic induced phase transition. The phase transition is of second order.

The phase transition just described has in fact been studied in experimentally, very beautiful light scattering experiments reported by the Vancouver group [15]. By means of light scattering spectroscopy (Brillouin scattering), these authors study the magnetic field dependence of the low frequency acoustic spin wave of a 3 monolayer Fe film, as H_0 is swept through the critical field $H_0^{(c)} = 2\tilde{H}_s$. They find the frequency of the mode drops to zero at $H_0^{(c)}$, to stiffen again as H_0 is increased beyond the critical field. This is the direct observation of the "soft mode" associated with field induced phase transition. They also see a dramatic increase in the Brillouin intensity, as H_0 is swept through $H_0^{(c)}$; this is scattering from critical fluctuations.

The spin reorientation transition, or an analogue of it, can be induced in any material with an easy axis, of course. For example, one may consider the garnet films which a few years ago were of great interest from the point of view of magnetic bubble technology. These films, macroscopic in thickness, can undergo an identical magnetic field induced spin reorientation transition to that reported in ref. [15].

What is of fundamental interest, from the point of view of the present author, is that we have in the ultrathin films a very interesting field induced phase transition in a two dimensional magnetic material. In contrast, the thermodynamics of the macroscopic garnet film is fully three dimensional. There are then very substantial differences in the thermodynamic behavior of the material near the transition, in the two cases.

Consider first the ultrathin film, with $N_\perp$ layers, where $N_\perp$ is small enough that we are in the two dimensional limit, in the sense of section 1. At low temperatures where spin wave theory applies, the thermodynamic properties of the film are then dominated by a single spin wave branch. As the critical field $H_0^{(c)}$ is approached from either above or below, this one low lying branch "goes soft". Right at the critical field, in the notation of section 1, the dispersion relation of the one low lying branch has the form $\omega^{(0)}(k_\parallel) = Dk_\parallel^2$ at long wavelengths, identical to the two dimensional isotropic Heisenberg ferromagnet. Thus, at H_c, the long range order in the film disappears. We have large amplitude, low frequency spin fluctuations that destroy long range order.

At long wavelengths, and near the critical field, the acoustic spin wave dispersion relation is well approximated by [13] $\omega^{(0)}(k_\parallel) = \omega_0(H_0) + Dk_\parallel^2$, where the gap $\omega_0(H_0)$ vanishes as the external field approaches the critical field $H_0^{(c)}$ from either above or below. For this form, eq. (6) is easily evaluated to give

$$\frac{\langle S_z(l_\parallel l_\perp)\rangle_T}{S(l_\parallel l_\perp)_0} = 1 - \frac{A_c k_B T}{4\pi\hbar DS} \ln\left[\frac{1}{1-\exp[-\hbar\omega_0(H_0)/k_B T]}\right], \quad (16)$$

where A_c is the area of the two dimensional unit cell of the structure. We see a logarithmic divergence in eq. (16) as $H_0 \to H_0^{(c)}$; spin wave theory breaks down in this regime, and a more complete treatment must be used.

The discussion which leads to eq. (16) is somewhat incomplete. The experiments reported in ref. [14] were in fact carried out on a three layer

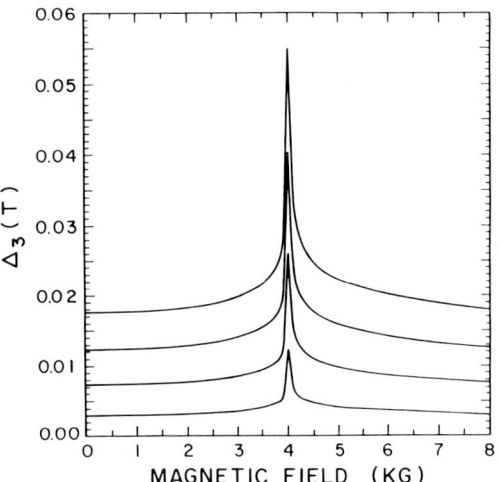

Fig. 2. For a model of a three monolayer Fe film with easy axis perpendicular to the film surfaces, we show for various temperatures the magnetic field dependence of the quantity $\Delta(H_0, T)$. The divergence occurs at the critical magnetic field discussed in section 3. The quantity Δ has been evaluated for the outermost layer of the film, though little layer dependence is encountered in the calculations reported in ref. [7], in which a full description of these calculations is presented.

film, and the discussion in sections 1 and 2 shows the transition from two to three dimensions to take place when only a few atomic layers are present. Also, it is necessary to take explicit account of dipolar couplings between spins, to discuss the behavior of the acoustic spin wave branch of a few layer film. This modifies not only the dispersion relation, but also the form of eq. (6).

In fig. 2 we reproduce calculations completed by Erickson and the present author [8], for a three monolayer film within which a full microscopic account of the dipolar couplings has been incorporated. We may write

$$\frac{\langle S_z(l_\parallel l_\perp)\rangle_T}{\langle S_z(l_\parallel l_\perp)\rangle_0} = 1 - \Delta(H_0, T) \quad (17)$$

and the figure displays $\Delta(H_0, T)$. The divergence at the critical field is evident, for all temperatures. It should be remarked that in low fields, the temperature variation of $\Delta(H_0, T)$ is rather well fitted by the form $\Delta(0, T) \approx bT^n$, where $n = 1.3$. This result is compatible with the data reported by Gradmann and his co-workers in ref.

[5], though no particular effort was made to choose parameters appropriate to these films.

One may inquire about the behavior of the thick garnet films near the field induced spin reorientation transition; in these films, one is very far from the quasi-two dimensional limit. To follow through with our simple model, we now incorporate a near continuum of spin wave "mini bands" by choosing $\omega^{(m)}(k_\parallel) = \omega_0(H_0) + D(k_\parallel^2 + [\pi m/N_\perp a_0]^2)$ where now we assume, in the language of section 1 that $N_\perp \gg N_c$. Thus for $\omega_0(H_0)$ of eq. (16) we have

$$\Delta(H_0, T) = \frac{1}{N_s N_\perp S} \sum_m \sum_{k_\parallel} \bar{n}\big(\omega^{(m)}(k_\parallel)\big). \quad (18)$$

We replace the sum on m by an integral, keeping the first correction term to this procedure provided by the Euler–MacLauren formula [16]:

$$\sum_{n=n_m}^{n_M} f(n) = \int_{n_m}^{n_M} dn f(n) + \tfrac{1}{2}[f(n_m) + f(n_M)] + \ldots \quad (19)$$

This gives, at low temperatures,

$$\Delta(T, H_0) = C_\infty T^{3/2} \left\{ \frac{1}{\zeta(\tfrac{3}{2})} \sum_{n=1}^\infty \frac{1}{n^{\tfrac{3}{2}}} \times \exp[-n\hbar\omega_0(H_0)/k_B T] \right\}$$
$$+ \frac{\pi A_c k_B T}{N_\perp \hbar D} \ln\left[\frac{k_B T}{\hbar\omega_0(H_0)}\right] + \ldots \quad (20)$$

In this expression, C_∞ is the coefficient in the famous Bloch $T^{3/2}$ law of spin wave theory, for the infinite ferromagnet; $\zeta(3/2)$ is the Riemann zeta function. When $\hbar\omega_0(H_0) \ll k_B T$, the quantity in curly brackets in eq. (17) is very close to unity in value. Thus, the first term exhibits a negligible dependence on field, near the critical field $H_0^{(c)}$.

There is a logarithmic divergence in eq. (15), quite similar to that in eq. (15). However, we see the factor $1/N_\perp$ in the prefactor. As $N_\perp \to \infty$, the singularity becomes vanishingly weak.

Hence, the singular behavior of the ferromagnetic film near the spin reorientation phase transition is confined to the quasi-two dimensional regime $N_\perp < N_\perp^{(c)}$. In these materials, it would be fascinating to see further experimental studies, with this issue in mind.

We direct the reader's attention to refs. [8,14], for the full spin wave formalism for a film of $N_\perp$ layers, with magnetic dipole couplings, layer dependent anisotropies and an external magnetic field possibility noncolinear with the average magnetization. Spin wave theory in such systems can be formulated in standard and well known manners. However, the proper link between the spin wave equations of motion, and the thermodynamic properties involve rather subtle mathematical issues not covered fully in the literature. These equations are addressed in ref. [8], where the nature of spin correlations in the films is explored as well, for fields near $H_0^{(c)}$.

The above discussion concerns spin reorientation transitions induced by an externally applied field, perpendicular to an easy axis. Most remarkably recent experiments along with theory show that such transitions occur *spontaneously*, in zero external field, with increasing temperature.

So far, the discussion presented above was phrased in terms of parameters that characterize the fully saturated film, at the absolute zero of temperature. At finite temperatures, the effective anisotropy constants and the effective dipolar contributions to the free energy of the film will be temperature dependent. There is no reason to believe that the temperature variation of these two different pieces will be similar. It is thus possible to envision a film for which the effective anisotropy field $\tilde{H}_s = H_s - 4\pi M$ favors an easy axis normal to the surface at low temperatures but then as the temperature is raised, the effective value of H_s "softens" more readily than $4\pi M$. With increasing temperature, $\tilde{H}_s$ can then change sign. Thus, there will be a spontaneous spin reorientation transition with increasing temperature; the magnetization will rotate from perpendicular, to parallel as temperature is increased.

In very interesting theoretical work, Pescia and Pokrovskii [17] proposed that precisely this behavior occurs, providing the initial system parameters are such that $\tilde{H}_s$ changes sign below the ordering temperature of the film. By means of a renormalization group calculation, these authors show that the effective perpendicular anisotropy decreases faster with temperature than the magnetization. Then they suggest the possibility of spontaneous spin reorientation transitions.

At virtually the same time the theoretical work of Pescia and Pokrovskii was completed, Pappas, Kämper and Hopster [18] directly observed a spontaneous spin reorientation transition, in a few monolayer Fe films on Cu(100). They monitor the direction of the substrate spins by detecting all three Cartesian components of spins of electrons photoemitted from the Fe 3d shell; one assumes the spin of the photoemitted electrons to be parallel to those in the ferromagnetic film. At low temperatures, they infer the films to have easy axis character. They see the substrate magnetization shrink to zero, while it remains always normal to the surface. There is then a temperature interval where no long range order is discerned, and at higher temperatures the magnetization grows in the substrate, parallel to the surface.

From the discussion in the paper by Pescia and Pokrovskii, it is not clear this behavior is the one expected. Their theory suggests the magnetization rotates from perpendicular to parallel to the surface at the spin reorientation temperature, but they make no mention of the continuous disappearance, and regrowth of the magnetization found by Pappas et al. However, in the experiments, domain effects may be responsible for this behavior. Also, large fluctuations present near transition may obscure the details of the picture put forward in the theoretical work. This possibility was noted by Pescia and Pokrovskii.

4. Concluding remarks

It is the view of the present author that ultrathin magnetic films are very rich in their properties, and can provide an excellent arena to explore the influence of reduced dimensionality on magnetism. It is the hope that the remarks in this paper will stimulate further experimental and theoretical studies.

References

[1] For an interesting experimental study of the dynamics of a quasi-one dimensional magnet, see J.K. Kjems and M. Steiner, Phys. Rev. Lett. 41 (1978) 1137.
Related theoretical issues are discussed by H.J. Mikeska, J. Phys. C 11 (1978) L29 and K.M. Leung, D. Hone, D.L. Mills, P. Riseborough and S.E. Trullinger, Phys. Rev. B 21 (1980) 4017.

[2] For data on the spin dynamics of quasi-two dimensional magnetic materials, and references to the related theoretical and experimental literature, see P.A. Fleury and H.J. Guggenheim, Phys. Rev. Lett. 24 (1970) 1346 and R.R.P. Singh, P.A. Fleury, K.B. Lyons and P.E. Sulewski, Phys. Rev. Lett. 62 (1989) 2736.

[3] S.O. Demokritov et al., Soviet Phys. JETP 68 (1989) 1277. These authors explore the Kosterlitz–Thouless transition in a quasi-two dimensional easy plane material, along with the influence of a magnetic field perpendicular to the film plane.

[4] B. Heinrich, A.S. Arrott, J.F. Cochran, S.T. Purcell, K.B. Urguhart and K. Myrtle, J. Crystal Growth 81 (1987) 562.
B. Heinrich, K.B. Urguhart, A.S. Arrott, J.F. Cochran, K. Myrtle and S.T. Purcell, Phys. Rev. Lett. 59 (1987) 1756.
J.F. Dutcher, J.F. Cochran, B. Heinrich, and J.S. Arrott, J. Appl. Phys. 64 (1988) 6095.
N.C. Koon, B.T. Jonker, F.A. Volkening, J.J. Krebs and G.A. Prinz, Phys. Rev. Lett. 59 (1987) 2463.

[5] M. Przybylski, I. Kaufmann and U. Gradmann, Phys. Rev. B 40 (1989) 8631.

[6] H. Li, Y.S. Li, J. Quinn, D. Tian, J. Sokolov, F. Jena and P.M. Marcus, Phys. Rev. B 42 (1990) 9195.

[7] One sees this easily for an array of Heisenberg films, with nearest neighbor ferromagnetic coupling. Consider films of square symmetry, stacked to form an ideal fcc film. For the monolayer, D is half the value of the infinite crystal, if one assumes (naively of course) that the value of the exchange is the same for both cases.

[8] The full theory is considerably more complex. See the preprint "Thermodynamics of Thin Ferromagnetic Films in the Presence of Anisotropy and Dipolar Coupling", R.P. Erickson and D.L. Mills, Phys. Rev. B, to be published. In the end, the final results can be described by the picture under discussion, with reasonable accuracy.

[9] M. Bander and D.L. Mills, Phys. Rev. B 38 (1988) 12015.

[10] R.P. Erickson and D.L. Mills, Phys. Rev. B 43 (1991) 11527.

[11] See the chapter by K.W.H. Stevens, Magnetism, vol. 1, eds. H. Suhl and G. Rado (Academic Press, New York, 1963) p. 1.

[12] Chun Li, A.J. Freeman, H.J.F. Jansen and C.L. Fu, Phys. Rev. B 42 (1990) 5433.

[13] B. Hillebrands, P. Baumgart and G. Guntherodt, Phys. Rev. B 36 (1987) 2450.
[14] R.P. Erickson and D.L. Mills, Phys. Rev. B 43 (1991) 10715.
[15] J.R. Dutcher, J.F. Cochran, I. Jacob and W.F. Egelhoff, Jr., Phys. Rev. B 39 (1989) 10430.
[16] Handbook of Mathematical Functions, eds. M. Abramovitz and I.A. Stegun (Dover, New York, 1965) p. 806.
[17] D. Pescia and V.L. Pokrovskii, Phys. Rev. Lett. 65 (1990) 2599.
[18] D.P. Pappas, K.P. Kämper and H. Hopster, Phys. Rev. Lett. 64 (1990) 3179.

Theory of magnetic multilayers. Exchange interactions and transport properties

J. Mathon

Dept. of Mathematics, City University, London EC1V 0HB, UK

The temperature dependence of the local magnetization in the spin-wave regime is discussed and it is shown that it can be used as a sensitive probe of local exchange interactions in magnetic surfaces and overlayers. The present status of the theory of oscillatory exchange interactions through transition and noble metal spacer layers is reviewed. The theory of the giant magnetoresistance observed recently in some antiferromagnetically coupled multilayers is also discussed. It is shown that both the oscillatory exchange and giant magnetoresistance effects can be understood in terms of different distributions of up and down spin d holes in different parts of the layer structure. The total energy and hence the exchange coupling is determined directly by the d band owing to its high density of states and the transport properties more indirectly through the Mott mechanism in which conduction electrons scatter into d band.

1. Introduction

During last few years it has become possible to prepare magnetic layer structures based on transition metals with interfaces that are sharp on an atomic scale. This development means that one can engineer materials exhibiting new magnetic phenomena with great potential for novel devices [1–4]. To exploit these opportunities to the full a theoretical understanding on new multilayer systems is essential.

Perhaps the most exciting systems are sandwiches and superlattices with magnetic layers separated by nonmagnetic spacer layers. Typical examples are Fe/Cr, Co/Cr, Co/Ru and Co/Cu layer structures. It is found experimentally [5] that the magnetic moments of the magnetic layers are aligned ferromagnetically or antiferromagnetically depending on the thickness of the spacer layer. This implies the existence of an exchange coupling mediated by the spacer layer which oscillates as a function of its thickness. Furthermore, the resistance of the structure is much higher in the antiferromagnetic configuration than in the ferromagnetic one. The relative change of the resistance can be very large, the current record is over 100% for Co/Cu structures. Change of the magnetic configuration from antiferromagnetic to ferromagnetic, and hence a change of the resistance, can be effected by an applied magnetic field and this mechanism can be exploited to read information from a magnetic disc. Such magnetoresistive sensors have the great advantage over the conventional inductive reading heads that they sense the strength of the magnetic field rather than its rate of change. Consequently, one can store data at the same high density throughout the magnetic disc without the need of reduced density near the slow moving centre.

Magnetic properties of transition metal systems are dominated by holes in the d band and both the exchange coupling and giant magnetoresistance effects can be understood in terms of different distributions of up and down spin d holes in different parts of the layer structure. The total energy and hence the exchange coupling is determined directly by the d band owing to its high density of states and the transport properties more indirectly through the Mott mechanism in which conduction electrons scatter into d band.

Historically, the influence of layering on local

exchange interactions was first studied for magnetic surfaces and overlayers. In section 2 I shall, therefore, describe how measurements of the temperature dependence of the local magnetisation can be combined with spin wave analysis to extract information about local exchange in magnetic surfaces and overlayers.

In section 3, the current status of the theory of oscillatory exchange interactions through transition and noble metal spacers is reviewed.

Finally, in section 4 the giant magnetoresistance effect observed in multilayers with antiferromagnetic exchange coupling is discussed.

2. Temperature dependence of the magnetization as a probe of local exchange interactions

Although a spectacular progress has been made over past ten years in calculating from first principles the ground state moment for magnetic surfaces and interfaces [6], it is not always easy to verify the theoretical predictions since it is very difficult to measure the *absolute* value of the local moment $M(0)$ (see ref. [7]). On the other hand, the *relative* change of the local moment, i.e. its temperature dependence $M(T)/M(0)$ can be determined very accurately by several methods such as photoemission, SPLEED, spin-polarized secondary electrons, Mössbauer spectroscopy, SMOKE, ECS (see e.g., refs. [7–10]). Since it is well known that $M(T)$ in bulk ferromagnets is a direct measure of exchange interactions, it is natural to ask whether we can exploit the high spatial resolution (on atomic scale) of modern experimental methods to determine from the measured local $M(T)$ the strength of local exchange interactions in magnetic surfaces and overlayers.

To extract information about local exchange from $M(T)$, we need a good theory of the temperature dependence of the magnetisation. One region where we have a good theory is the spin wave regime $0 < T < T_C/3$, where T_C is the Curie temperature. It has, therefore, been proposed [10,11] to use the following strategy to access local exchange via spin waves.

First, a model exchange Hamiltonian H for a layer structure is assumed. Next the density of spin-wave states $N_i(E)$ in any atomic plane of the structure is determined from the local spin-wave Green function $G = (E - H)^{-1}$

$$N_i(E) = (1/\pi)\text{Tr Im } G_{ii}(E, q_\parallel), \qquad (1)$$

where the trace is over the wave vector $q_\parallel$ parallel to the structure surface and i labels atomic planes in the structure. Since spin waves are bosons, the temperature-induced reduction of the local magnetisation in any atomic plane i is given by

$$M_i(0) - M_i(T)$$
$$= \int_0^\infty 2\mu_B N_i(E)[\exp(E/kT) - 1]^{-1} \, dE, \qquad (2)$$

where μ_B is the Bohr magneton.

Finally, the exchange parameters in the trial Hamiltonian H are deduced from the best fit of eq. (2) to the observed $M_i(T)$.

To calculate the local spin-wave Green function we have developed a method which can be viewed as the theoretical equivalent of an MBE machine [12,13].

One starts with a homogeneous semi-infinite ferromagnetic substrate with exchange J and spin S whose surface spin-wave Green function (density of states) is known from the classical spin wave theory [14,15]. The next step is the "deposit" one-by-one all the atomic layers of the structure we wish to study. With every new adlayer the surface Green function is updated using the Dyson equation. At every stage of the deposition process we have, therefore, the exact surface Green function G^s expressed in terms of the known substrate Green function G^0. When the layer structure is complete we stop the deposition process and evaluate the local Green function in ever atomic plane from the exact G^s (see ref. [13]). We have demonstrated elsewhere [12,13] that the method is so fast and accurate that the local spin-wave Green function for magnetic multilayers of 10^3–10^4 at. planes can be calculated essentially exactly. It follows that one can determine from eqs. (1) and (2) the local temperature dependence $M_i(T)$ for any magnetic multilayer with an arbitrary distribution of exchange integrals and with an arbitrary distribution of local spin.

The whole strategy outlined above is based on the hypothesis that a variation of local exchange on an atomic scale is reflected in the local $M(T)$. This can be easily tested for a ferromagnetic surface.

Consider first a perfect surface with no changes to the surface exchange and spin. Classical spin wave theory [14] predicts that the surface magnetisation $M_s(T)$ obeys a Bloch law

$$M_s(T)/M_s(0) = 1 - kCT^{3/2} \qquad (3)$$

with $k = 2$ (C is the bulk prefactor), i.e. the surface magnetisation should decrease twice as fast as in the bulk. Measurements of $M_s(T)$ for ferromagnetic metals using SPLEED [16], Mössbauer spectroscopy [17] and the polarization of secondary electrons [18,19] confirm that M_s obeys a Bloch law but with a prefactor which is different for different surfaces and can be as large as $k = 5.7$ [20].

It is natural to assume that a faster decrease of M_s is due to a softening of surface exchange. Such a softening can be clearly modelled by an overlayer with a suitable distribution of exchange deposition on a homogeneous ferromagnetic substrate.

The method of adlayers [12] gives the exact *initial* surface density of spin wave states (DOS) for an overlayer consisting of N at. layers with arbitrary (ferromagnetic) exchange interactions within and between the layers and with an arbitrary local spin S_i in each layer

$$N_s(E) = (S_N/S_{N-1}) \cdots (S_1/S) 2N_B(E)$$
$$= (S_N/S) 2N_B(E) + \mathcal{O}(E^{3/2}), \qquad (4)$$

where $N_B(E) \propto E^{1/2}$ is the bulk DOS and S_N is the surface spin.

The most remarkable feature of the DOS in eq. (4) is that it is quite *independent* of the exchange in the overlayer. Moreover, normalisation to $M_s(0)$ in eq. (3) means that the surface spin is cancelled and we recover from eqs. (2) and (4) the classical result $k = 2$. The spin wave theory, therefore, predicts that the surface magnetisation should *initially* always decrease twice as fast as in the bulk irrespective of the surface exchange and spin. Not only this exact result

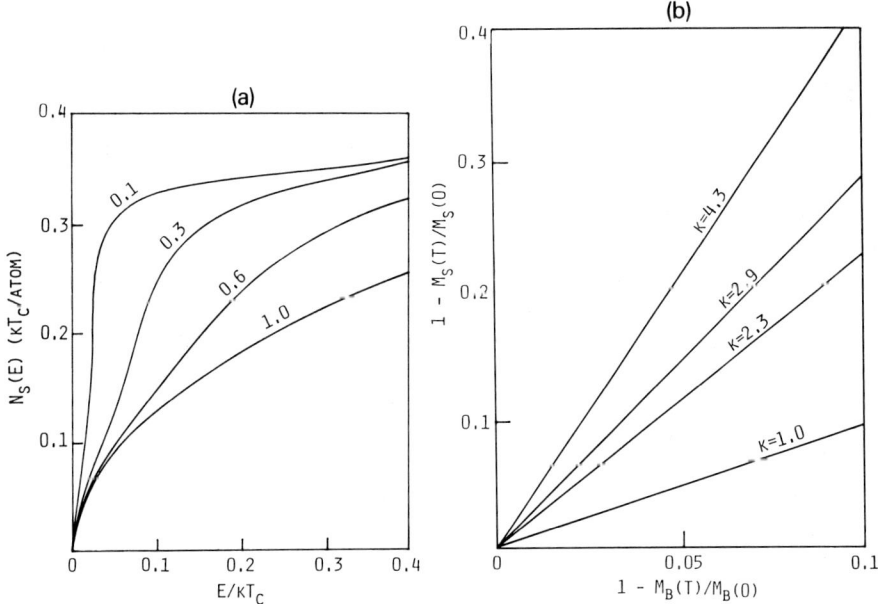

Fig. 1. Surface density of spin wave states (a) and the plots of $M_s(T)$ against the bulk $M_B(T)$ (b) for an overlayer with a weaker surface to bulk exchange $J_\perp/J = 1.0, 0.6, 0.3, 0.1$.

contradicts the experiment but it also seems to indicate that no information about local exchange can be gained from the Bloch law for $M_s(T)$.

Fortunately, quite a different picture emerges [15] when the surface DOS is computed over a wider energy range. This is illustrated in fig. 1a for an overlayer consisting of a single atomic layer which is coupled to the substrate by a weaker exchange $J_\perp/J = 1.0, 0.6, 0.3, 0.1$ (J is the bulk exchange). It can be seen that the classical result (4) breaks down at energies as low as $0.01 kT_C$. The surface DOS at experimentally relevant energies is determined entirely by the surface to bulk exchange $J_\perp$ and it increases rapidly as the exchange $J_\perp$ weakens. We expect, therefore, that $M_s(T)$ should decrease more rapidly with increasing temperature. This is illustrated in fig. 1b, where $M_s(T)$ calculated from eqs. (2) and (4) is plotted against the bulk magnetisation $M_B(T)$ for the same range of values of $J_\perp/J$ as in fig. 1a.

The results shown in fig. 1b allow us to make two firm theoretical predictions:

(i) $M_s(T)$ for a ferromagnetic surface with a weaker surface to bulk exchange should always obey a good $T^{3/2}$ law;

(ii) the prefactor k in this "second" $T^{3/2}$ law should increase with decreasing $J_\perp$.

These predictions of the spin-wave theory were tested by Mauri et al. [20] using secondary electrons. They first deposited on a clean Permalloy surface a thin spacer layer of nonmagnetic Ta and then covered it with a very thin Permalloy film. The Ta spacer layer thus formed a weak exchange link between the bulk and surface of Permalloy whose strength could be controlled by varying the thickness of Ta. The experimental results of Mauri et al. [20] are reproduced in fig. 2, where the curves (a), (b) and (c) correspond to a progressively weaker exchange link (thicker Ta interlayer; 0.5, 1.0 and 1.5 at. layers) and the curve 0 is for clean Permalloy surface. The curves (a)–(c) demonstrate clearly that $M_s(T)$ decreases faster with decreasing $J_\perp$ (following in all cases a good $T^{3/2}$ law with prefactors $k = 2.5, 4.1$ and 5.7.

This pioneering work on magnetic overlayers shows that surface exchange interactions on the

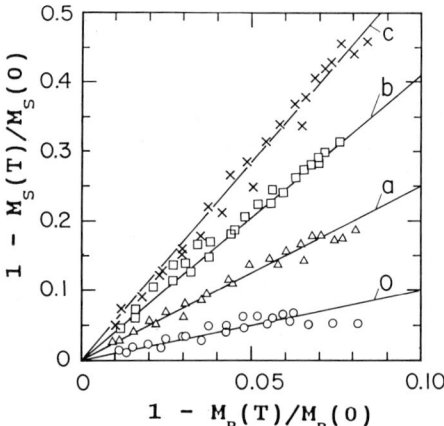

Fig. 2. Experimental results of ref. [20] for the temperature dependence of the surface magnetisation in FeNi/Ta/FeNi overlayers.

path perpendicular to the surface can be engineered to be weaker than in the bulk by the introduction of a suitable spacer layer. On the other hand, the surface exchange in a magnetic overlayer can be also made much stronger than in the bulk. When a Permalloy surface is covered with a submonolayer of Fe, it is found [21], using again spin wave analysis of $M_s(T)$, that the exchange in the surface layer $J_\parallel$ is approximately three times stronger than the bulk J.

To determine the local exchange in magnetic overlayers by the method I have described, we require sophisticated experimental probes of the local magnetization with a resolution on an atomic scale and a careful spin wave analysis of $M(T)$ to separate the surface and bulk contributions. This problem does not arise for magnetic sandwiches and superlattices because we do not have here the bulk background. Sandwiches and superlattices with magnetic layers separated by nonmagnetic spacer layers are, therefore, most promising candidates for studying interlayer exchange.

3. Oscillatory exchange interactions through a nonmagnetic spacer layer

The work on magnetic overlayers has demonstrated that interlayer exchange can be reduced significantly when a nonmagnetic metallic spacer is inserted between the surface and the bulk of a ferromagnetic metal. Most recent results for sandwiches and superlattices show that a nonmagnetic metallic spacer can even change the sign of the exchange coupling between ferromagnetic layers separated by the spacer [5]. Such antiferromagnetic coupling has been observed by several experimental techniques for Fe/Cr/Fe sandwiches and superlattices [2,3,22]. Typically, the antiferromagnetic coupling was seen in these experiments when the thickness of the chromium layer was less than 20–30 Å. Recently, Parkin et al. [5] reported oscillations in the exchange coupling as a function of the thickness d of the nonmagnetic spacer layer in Co/Ru, Co/Cr and Fe/Cr superlattices. Light scattering experiments of Grünberg et al. [23] confirm that the exchange coupling of iron layers oscillates as a function of chromium thickness d and the observed oscillations are reproduced in fig. 3. The oscillations are reminiscent of RKKY oscillations but the period (≈ 15–18 Å for Cr) is much longer than one would expect from the conventional RKKY theory. The amplitude of the oscillations decreases as $1/d^2$ at low temperatures but the decrease for Fe/Cr is much faster at room temperature [24]. The maximum strength of the antiferromagnetic coupling for Co/Ru with Ru spacer of ≈ 3 Å is ≈ 4 erg cm^{-2}. The strength of the coupling in Fe/Cr at short distances is about half of this value. Such large values of the coupling rule out the possibility of magnetostatic interaction. One is, therefore, lead to the conclusion that the origin of the coupling must be electronic.

The most direct way to calculate the coupling is to evaluate the total energy different $\Delta E = E_{\uparrow\uparrow} - E_{\uparrow\downarrow}$ between the ferromagnetic and antiferromagnetic configurations of the magnetic layers. This has been tried using LSDF [25] and tight-binding [26,27] methods but all such calculations have so far failed to reproduce the observed *long-period* oscillations. The most likely reason for the failure of such "brute force" calculations will be discussed later.

Edwards and Mathon [28] and Edwards et al. [29] have proposed a model of the coupling in which the total energy difference ΔE can be evaluated numerically for all thicknesses d of the spacer layer and analytically in the asymptotic limit of large d. The model is based on the assumption that the coupling is mediated by holes in the d band of the spacer layer. Edwards and Mathon [28] and Edwards et al. [29] consider a sandwich consisting of two infinitely thick magnetic layers separated by a nonmagnetic spacer containing $N-1$ atomic planes. The ferromagnetic metal is assumed to have a full majority spin d band and a partially occupied minority spin d band, which is satisfied for a strong ferromagnet such as Co and is close to the situation in Fe. The

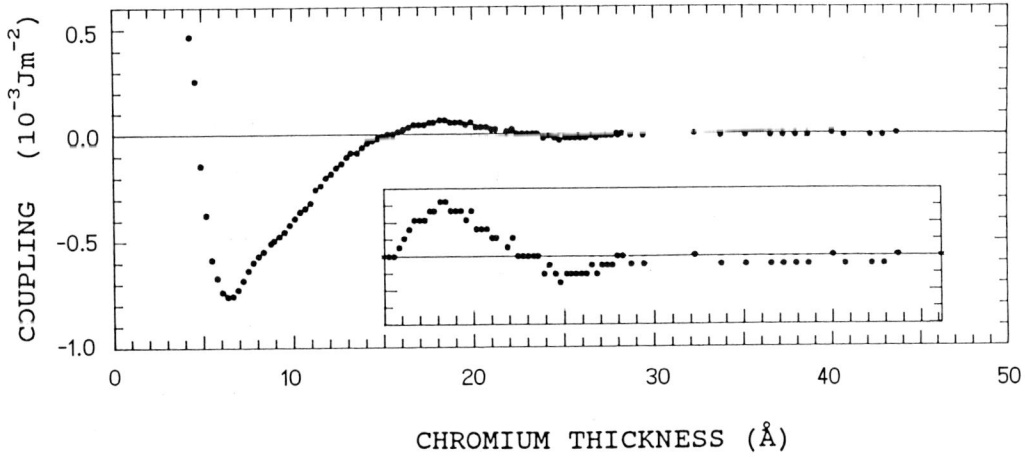

Fig. 3. Oscillations of the exchange coupling between two iron layers in an Fe/Cr sandwich as a function of the thickness of the chromium spacer layer (after Grünberg et al. [23]).

nonmagnetic metal has equal numbers of holes in each spin sub-band (Cr or Ru). The densities of states in the up and down spin sub-bands in the ferromagnetic and nonmagnetic spacer layers, together with the spatial variation of the hole densities $\rho_\uparrow$, $\rho_\downarrow$ are shown schematically in figs. 4a and b for the parallel and antiparallel orientations of the layer moments.

Two further simplifying assumptions which do not alter the basic physics are now made: (i) the number of d holes per atom of each spin in the bulk nonmagnetic metal is equal to the number of holes in the bulk ferromagnetic metal; (ii) the exchange splitting in the ferromagnetic layers is so large that holes of the wrong spin are completely excluded. It follows that the intra-atomic Coulomb interaction between holes of opposite spin plays no role; holes of opposite spin completely avoid each other except in the spacer layer where the holes are assumed to be noninteracting. As a result, the model depicted in fig. 4 is exactly solvable in the Hartree–Fock approximation and the total energies of the two configurations may be calculated as sums of one-electron energies.

To understand the physical origin of the coupling, it is instructive to examine how d holes move across the sandwich structure. It is clear from fig. 4a that in the ferromagnetic configuration there are no available states for down spin holes in the two ferromagnets and, therefore, they become confined in the spacer layer by the exchange potential. The energy of the down spin holes trapped in a quantum well of width $d = Na$ (a is the interatomic distance) is thus size quantized. Up spin holes, on the other hand, can move freely across the whole structure.

In the antiferromagnetic configuration, up spin holes can cross the left interface but encounter a high potential barrier at the right interface which causes their density to go to zero in the right ferromagnet. Similarly, the density of the down spin holes goes to zero at the left interface.

Quantum confinement of d holes costs energy. In the antiferromagnetic configuration, we have a constant surface energy associated with the con-

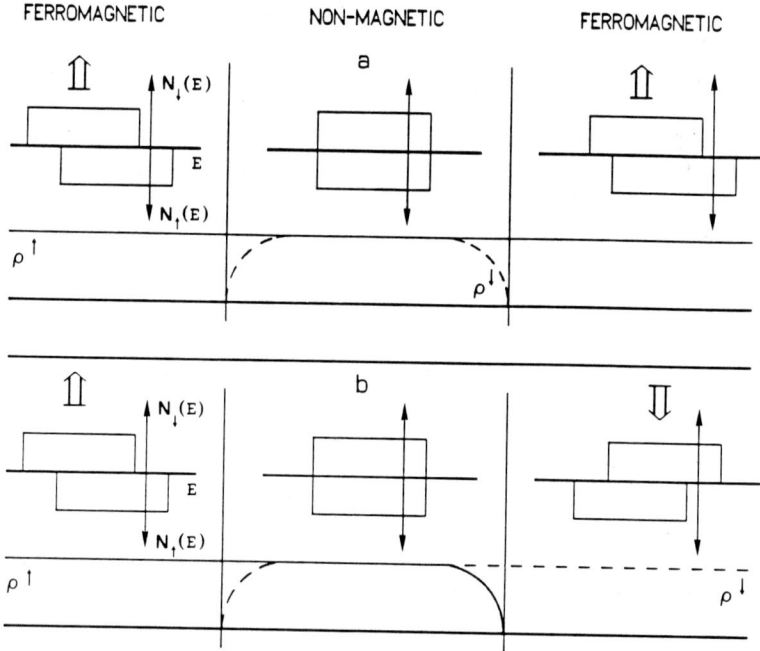

Fig. 4. Schematic representation of the densities of states of holes in the up and down spin bands of a sandwich in its ferromagnetic (a) and antiferromagnetic (b) configurations. $\rho_\uparrow$ and $\rho_\downarrow$ are the corresponding variations of the local hole densities across the sandwich.

finement of up and down spin holes to two semi-infinite regions. This energy is *independent* of the spacer thickness. It follows that the dependence of the exchange coupling on the spacer thickness, which is given by the energy difference between the ferromagnetic and antiferromagnetic configurations, must be determined entirely by the confinement of down spin holes in the spacer layer which occurs in the ferromagnetic configuration. It is clear that the energy difference $\Delta E(N-1) = E_{\uparrow\uparrow} - E_{\uparrow\downarrow}$ is given by [29]

$$\Delta E(N-1) = E(N-1) - E(\infty), \quad (5)$$

where $E(N-1)$ is the kinetic energy of the down-spin holes measured relative to an $N-1$ plane reference state with bulk density.

To conserve the number of particles (overall charge neutrality) we must consider instead of the total energy of holes in the spacer layer their thermodynamic potential

$$\Omega(N-1) = E(N-1) - E_F n(N-1), \quad (6)$$

where E_F is the Fermi energy and $n(N-1)$ is the number of holes in the spacer measured again relative to the bulk reference state. The change in energy between the two configurations conserving the number of particles is, therefore, given by

$$\Delta\Omega(N-1) = \Omega(N-1) - \Omega(\infty). \quad (7)$$

Following Parkin et al. [5], one can finally define an exchange coupling constant per unit area for a spacer layer of $N-1$ at. planes by

$$J(N-1) = \Delta\Omega(N-1)/A. \quad (8)$$

The physical argument leading to eq. (8) is based solely on the availability of states in the phase space and is, therefore, quite general. In particular, eq. (8) applies to an arbitrary band structure of holes in the spacer layer. However, to illustrate the origin of oscillations in the exchange coupling, consider first the simplest case of a parabolic band. The difference between the densities of states of holes in the ferromagnetic and antiferromagnetic configurations for a parabolic band is shown in fig. 5. The density of states is step-like due to size quantization and oscillates

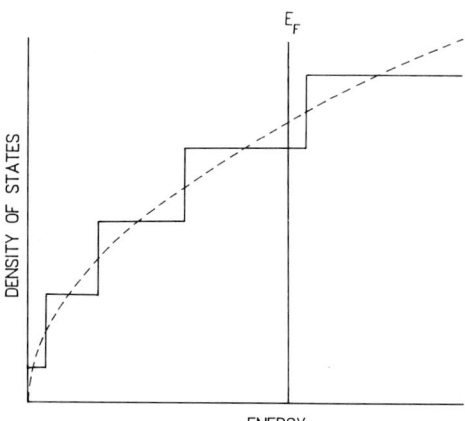

Fig. 5. Difference between the densities of states of holes in the ferromagnetic and antiferromagnetic configurations of a sandwich assuming a parabolic band (solid curve); the density of states for free particles (broken curve). E_F is the Fermi level.

about the bulk density of states (broken curve). It is clear that the distribution of steps in fig. 5 changes as the thickness of the spacer layer increases, and every time a step passes through the Fermi level an oscillation in the exchange coupling occurs. Such oscillations are quite analogous to oscillations in the de Haas–van Alphen effect and it is straightforward to formalize the correspondence between the two effects [29].

To determine the coupling from eq. (8), we require the thermodynamic potential Ω of holes confined in the spacer layer. At finite temperatures Ω is clearly given by

$$\Omega = -T \sum_{k,r} \ln\left[1 + \exp\frac{\mu - E(k,r)}{T}\right], \quad (9)$$

where μ is the chemical potential, $E(k,r)$ is the hole energy, k is the vector parallel to the sandwich and r labels the discrete energy levels of holes ($k_B = 1$).

Since the exchange potentials in the present model are equivalent to two infinitely high potential barriers, we simply require the dependence of the thermodynamic potential Ω of size-quantised holes in a layer of $N-1$ at. planes on the thickness of the layer $d = Na$. This is a familiar problem in the theory of de Haas–van Alphen effect [30] where two-dimensional quantisation of the carrier energy in a magnetic field takes place in a plane perpendicular to the field. It is well known that oscillation periods in the de Haas–van Alphen effect are determined by extremal cross sections of the Fermi surface. Since, in the present problem the quantization is one-dimensional rather than two-dimensional, we expect the period of oscillations in the exchange coupling to be determined by extremal diameters of the Fermi surface (calliper measurements) in the direction perpendicular to the layers. This is confirmed by the calculation of Edwards et al. [29]. They determined the exchange coupling J from the thermodynamic potential (9) using a simple cubic tight-binding model for the spacer layer

$$\epsilon(k_x, k_y, k_z) = -[\cos(k_x a) + \cos(k_y a) + \cos(k_z a)], \quad (10)$$

where the energy is measured in units of the hopping $2|t|$. For such a band and a sandwich parallel to the x, y plane, the quantisation of k_z is given by

$$k_z = \frac{r\pi}{Na}, \quad r = 1, 2, \ldots, N-1, \quad (11)$$

where a is the distance between neighboring planes. One can easily show that the same quantisation is obtained for any single band and any orientation of the sandwich. Adapting the conventional theory of the de Haas–van Alphen effect, it is then found [29] that the coupling $J(N-1)$ is given by the following general asymptotic formula valid for large spacer thickness N:

$$J(N-1) = \frac{1}{4\pi Na} \mathrm{Re} \sum_{s=1}^{\infty} \frac{\sigma}{s^2} \left| \frac{\partial k_z^2}{\partial k_x^2} \frac{\partial k_z^2}{\partial k_y^2} \right|^{-1/2}$$

$$\times \frac{\exp[2isNak_z^0(\mu)]}{T^{-1} \sinh[2\pi sNaT \, \partial k_z/\partial \epsilon]},$$

$$\sigma = \begin{cases} i, & \text{both 2nd derivatives} > 0, \\ -i, & \text{both derivatives} < 0, \\ 1, & \text{one derivative} > 0, \text{ the other} < 0. \end{cases}$$
$$(12)$$

Here, $k_z^0(\mu)$ is an extremal radius of the Fermi surface in the direction perpendicular to the layers and all the derivatives in eq. (12) are taken at the stationary point $k_z^0 \equiv k_z^0(\mu, k_x^0(\mu), k_y^0(\mu))$.

The consequences of the general asymptotic formula for $J(N)$ are:

(i) The exchange coupling J oscillates with a period $N_p = \pi/ak_z^0(\mu)$ determined by the calliper measurement of the Fermi surface;
(ii) the amplitude contains a factor determined by the curvature of the Fermi surface at its extremal points;
(iii) the temperature dependence of the oscillations is governed by the velocity of carriers at the extremal points;
(iv) the asymptotic decay of the amplitude at $T = 0$ is proportional to $1/N^2$.

The most important question is whether this theory can explain the observed long oscillation periods. Since the sum over s in eq. (12) reflects just the effect of higher harmonics, it is sufficient to consider the term $s = 1$. The fundamental oscillation $s = 1$ is governed by the factor

$$\exp[2iNak_z^0(\mu)]. \quad (13)$$

For an electron gas, eq. (13) leads to the conventional RKKY result for the oscillation period if we set $k_z^0(\mu)$ equal to the radius of the Fermi sphere and let the distance between the two ferromagnets $d = Na$ vary *continuously*. However, it will be seen that the *discreteness* of the spacer layer leading to Bragg scattering of electrons is crucial for obtaining long oscillation periods. This was first shown by Edwards et al. [29] for the tight-binding band (10) and (001) orientation of the layers but their argument based on eq. (13) is valid quite generally for any band and any orientation of the layers.

Oscillations in J are determined by extrema of the bulk Fermi surface in the spacer layer in the direction perpendicular to the layers. These can be seen in the cross section of the Fermi surface by a plane perpendicular to the (001) plane and passing through the centre of the Brillouin zone. Such cross sections are shown in fig. 6 for three different fillings of the tight-binding band: $E_F =$

−2.5 (a), −1.05 (b) and −0.95 (c) (the band edges are at ±3 in our energy units).

The case (b) when the Fermi surface almost touches the zone boundary is most interesting since it leads to a long period. In fact, if ϵ is the distance from the extremum to the zone boundary, then $k_z^0(\mu) + \epsilon = \pi/a$ ($2\pi/a$ being a reciprocal lattice vector) and the exponential factor in eq. (13) becomes

$$\exp[2i Na k_z^0(\mu)] \equiv \exp[-2i Na\epsilon]. \quad (14)$$

It is clear from eq. (14) that the oscillation period tends to infinity as the Fermi surface approaches the zone boundary ($\epsilon \to 0$). This is a completely new feature which is absent in the conventional RKKY for an electron gas where the period decreases monotonically with increasing radius of the Fermi sphere.

For the tight-binding band considered, long period oscillations of J with a period N_p of about ten interatomic distances (as observed) are obtained for $E_F = -1.05$ and they are shown in fig. 7 (squares). Note that the initial sign of J is antiferromagnetic, which is as observed.

Long period oscillations are obtained not only for Fermi surfaces close to a zone boundary but also for Fermi surfaces with necks. For a tight-binding band with $E_F = -0.95$ and (001) orientation of the layers, the Fermi surface develops

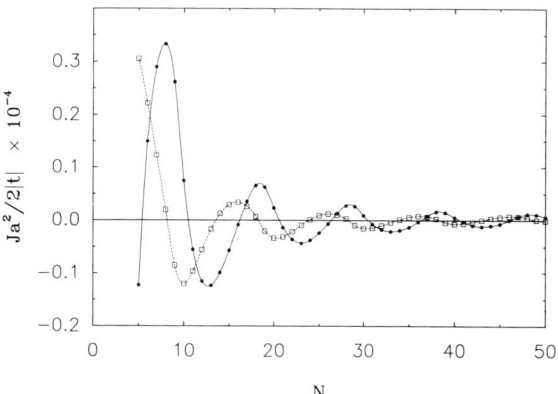

Fig. 7. Exchange coupling $J(N)$ as a function of the number of atomic planes N in the spacer layer for two differents values of the Fermi energy close to the Brillouin zone boundary: $E_F = -0.95$ (circles) and -1.05 (squares).

four necks of a small diameter at $(0, \pm\pi/a)$ and $(\pm\pi/a, 0)$. The necks on the opposite faces of the Brillouin zone combine into two saddle points which lead again to long period oscillations. These are shown also in fig. 7 (circles). It is interesting to note that the initial sign of the oscillations for $E_F = -0.95$ is ferromagnetic. This comes about because the factor σ in eq. (12) has value $\sigma = 1$ for a saddle point, which leads to a phase shift in the oscillations.

The dependence of the amplitude of oscillations in J on the curvature of the Fermi surface is illustrated in fig. 8 where the oscillations for $E_F = -1.05$ (case b in fig. 6) are compared with the oscillations for $E_F = -2.5$ (case a in fig. 6). In the case of $E_F = -2.5$ the Fermi surface is almost spherical, its curvature is small, and therefore, the oscillation amplitude is much larger than for the long period oscillations at $E_F = -1.05$. Nevertheless, direct evaluation of $\Delta\Omega$ from eq. (9) shows [28,29] that the strength of the exchange coupling for $N = 2$ is $J \approx 1$ erg cm^{-2} in the case of long period oscillations ($E_F = -1.05$). This is in qualitative agreement with the largest value of the exchange $J \approx 5$ erg cm^{-2} observed for $N = 2$ in Co/Ru.

The temperature dependence of the oscillations in J is controlled by the sinh factor in eq. (12) which has exactly the same form as in the de

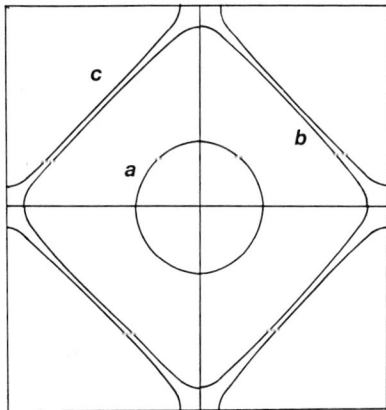

Fig. 6. Cross sections of the Fermi surface in the spacer layer by a plane perpendicular to the layers and passing through the center of the Brillouin zone for three fillings of the band: $E_F = -2.5$ (a), -1.05 (b) and -0.95 (c).

Haas–van Alphen effect provided the following correspondence is made:

$$N\frac{\partial k_z a}{\partial \epsilon} \to \frac{1}{\hbar \omega_c} \quad (15)$$

where ω_c is an effective "cyclotron" frequency. Substituting in eq. (15) typical values of E_F, it is easy to see that the effective cyclotron frequency in the present problem is $\hbar \omega_c \approx 10^3$ K for small N. The dominant temperature dependent factor in eq. (12) is $T/\sinh(2\pi T/\hbar\omega_c)$ and hence the temperature dependence of the exchange coupling is on the scale ≈ 100 K. It is important to note that, at $T = 0$, eq. (12) predicts $J \propto 1/N^2$ for the dependence of the exchange on the spacer thickness. However, at $T > 0$, the attenuation of the oscillations becomes exponential. Such a crossover from a slow attenuation at $T = 0$ to a fast attenuation at room temperature has been observed [24].

Finally, the exact asymptotic formula (12) can be used to assess the feasibility of direct numerical calculations of J (see ref. [29]). Every such calculation is either for a slab [25,26] or for a cluster of atoms [27]. In slab calculations, Brillouin zone summations over a two-dimensional zone have to be done numerically and it turns out that these summations are the main factor that limits the accuracy of such calculations.

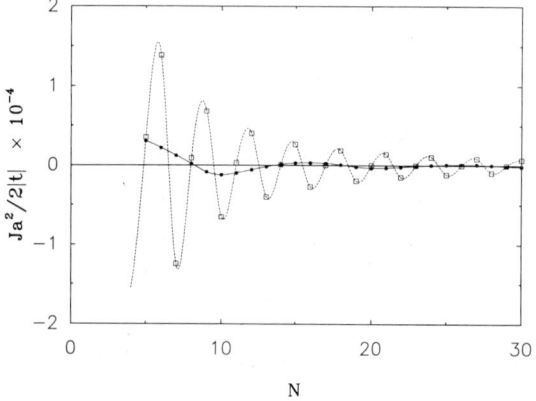

Fig. 8. Comparison of the amplitudes and periods of oscillations in the exchange coupling $J(N)$ as a function of the number of atomic planes N in the spacer layer for two different fillings of the band: $E_F = -2.5$ (squares) and -1.05 (circles).

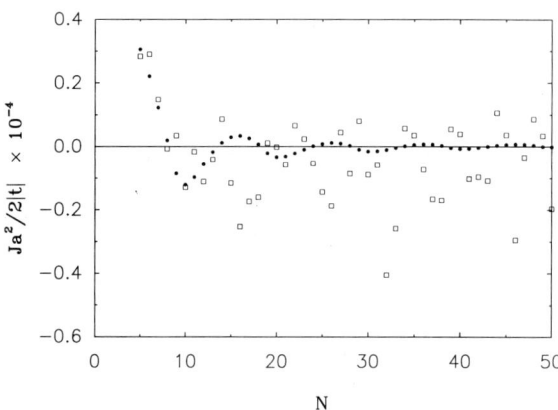

Fig. 9. Exchange coupling $J(N)$ obtained by direct evaluation of eq. (7) using different numbers of Cunningham points in the two-dimensional Brillouin zone. Squares are for 16 640 k points and circles are for 1 050 624 k points. Both results are for $E_F = -1.05$.

In the approach of Edwards et al. [29], "brute-force" calculation of J means direct evaluation of the thermodynamic potential $\Delta\Omega$ of magnetic carriers trapped in the spacer layer. This involves a two-dimensional BZ sum over k_x, k_y. Any discrete summation means that the wave vectors k_x, k_y effectively become quantized as if periodic (or other) boundary conditions were imposed over a region with transverse dimension L. It is clear that to obtain reliable results for a layer structure, the length L must be much larger than the thickness of the structure d. Taking $L/d \approx 100$ and the number of atomic planes in the spacer $N \approx 50$, which is needed to see long period oscillations, the required number of k points in the two-dimensional zone is $\approx 2.5 \times 10^7$. However, even using this huge number of ordinary k-space points the computed results for J amount to more or less random noise. The situations improves somewhat when one uses special Cunningham points [31] and the results are shown in fig. 9 for 16 640 and 1 050 624 Cunningham points in the two-dimensional Brillouin zone.

It is clear from fig. 9 that well-defined oscillations are obtained only for $\approx 10^6$ k points and this number is several orders of magnitude larger than the number of k points used in conventional first-principle band calculations. We believe that this is the most likely reason why such calculations have failed to reproduce the observed long period oscillations in J.

The theory of Edwards and Mathon [28] and Edwards et al. [29] is so far the only total energy calculation that explains the physical origin of long-period oscillations in the exchange coupling and, at the same time, gives a qualitatively correct estimate of the overall strength of the interaction. A weakness of the original calculation [29] is that the coupling has only been worked out quantitatively for a simple tight-binding band. However, the calculation has now been extended [32] to a realistic band structure for Cr(001) spacer using canonical d bands and 10^6 Cunningham points in the two-dimensional BZ. This high-precision calculation without any adjustable parameters confirms the existence of long-period oscillations with a period ≈ 9 atomic planes and, in addition, it shows that an oscillation with a period of two atomic planes is superimposed over the long-period oscillations. Such short-period oscillations were seen recently in experiments on high-quality Fe/Cr samples [24,33].

An alternative to total energy calculations is to use perturbation theory. The advantage of using perturbation theory is that one avoids accuracy problems involved in subtracting two large energies $E_{\uparrow\uparrow}$, $E_{\uparrow\downarrow}$. The disadvantage is that the overall strength of the coupling cannot be obtained by this method since it is just a parameter in all such calculations. It is also unlikely that perturbation theory remains valid for a small thickness of the spacer layer. However, the period and asymptotic decay should be correctly reproduced provided one uses realistic band structure of the spacer layer.

The simplest method is to adapt RKKY to the layer geometry [34,35]. The coupling is then determined by the one-dimensional Fourier transform of the wave vector-dependent susceptibility

$$\frac{1}{2\pi}\int_{-\infty}^{+\infty}\chi(q_\| = 0, q_z)\, e^{izq_z}\, dq_z,$$

where $q_\|$ and q_z are the components of the wave vector parallel and perpendicular to the layers. This method has been applied by Yafet [34] to interpret the observed oscillations of the exchange coupling in Gd/Y superlattices.

To explain the initially antiferromagnetic coupling through chromium, Lacroix et al. [36] have proposed to include in the perturbation expansion the fourth-order superexchange-type term. Levy et al. [37] used the same idea but their calculation of the susceptibility was based on a realistic band structure of the chromium spacer. Using perturbation theory with two adjustable parameters, they find long-period oscillations of the coupling with superimposed short-period oscillations having a period of two atomic planes.

Concerning the coupling through chromium, one can, therefore, conclude that there is an agreement between the perturbation calculation of Levy et al. [37] and the total energy calculation [32] in that they both lead to two oscillation periods in Cr, which is as observed.

All the results discussed so far concern exchange coupling through transition metal spacers. It is difficult to calculate the coupling for such spacers because d as well as s–p electrons are involved. In the case of coupling through noble metals, we have only a simple s band and an RKKY-type theory should be applicable (at least in the asymptotic limit of large spacer thickness).

An oscillatory coupling with a period ≈ 10–12 Å was observed [38,39] in polycrystalline Co/Cu structures with predominant (111) orientation of the layers. There is also some experimental evidence [40] for a shorter oscillation period ≈ 5–7 Å in Co,Fe/Cu(100) system.

When one applies conventional RKKY for a free electron gas to Cu, the observed long periods seem puzzling since the radius of the Fermi sphere in Cu would appear to lead to a period $\pi/k_F \approx 2.3$ Å. However, such a conclusion is incorrect since it ignores discreteness of the spacer layer. One must use again eq. (13) and examine the extrema of the Cu Fermi surface in relation to the Brillouin zone boundary. It is only necessary to extend the argument of Edwards et al. [29] from sc to fcc lattice. This was done recently by Coehoorn [41] (see also ref. [42]) and he shows, assuming a spherical Fermi surface for Cu, that proximity of the Fermi surface to the zone boundary can lead to long-period oscillations.

However, his discussion ignores the effect of Cu Fermi surface necks parallel to the (111) plane. Such necks can be easily reproduced with a tight-binding model for an fcc lattice and eq. (12) shows that they lead to oscillations in J with a period determined by the neck diameter. Taking the known neck diameter for Cu, we obtain a period $\approx 10-12$ Å. This mechanism seems, therefore, to be the natural explanation of the oscillations observed in Co/Cu(111).

4. Giant magnetoresistance in magnetic multilayers

Metallic multilayer structures consisting of magnetic layers (Fe, Co) separated by nonmagnetic spacer layers (Cr, Ru, Au, Cu) exhibit unexpectedly high values of saturation magnetoresistance when the current flows in a direction parallel to the layers [2,3,5,38,39]. It is found that the resistance is high when the neighboring magnetic layers have antiparallel magnetic moments and it is low when the moments are aligned parallel in a saturating magnetic field. The relative reduction of the resistance $\Delta R/R$ can be as large as 100% (see refs. [38,39]). The effect is referred to as giant magnetoresistance but this is somewhat misleading. The main role of the magnetic field is not to influence electron motion directly as in the conventional magnetoresistance but merely to align the magnetic layer moments. The direction of the magnetic field within the plane of the layer is found to be unimportant [43]. The effect is of particular interest in view of its applications in magnetic sensors and magnetic storage technology.

It was suggested by Baibich et al. [3] that spin-dependent scattering at the interfaces is responsible for the magnetoresistance. A detailed theory of the spin-dependent interfacial scattering based on the Boltzmann equation was worked out by Camley and Barnás [44] and more recently it was extended to include spin-dependent scattering in the bulk of the ferromagnetic layers [43]. Levy et al. [45] considered also spin-dependent interfacial and bulk scattering but used a quantum-mechanical approach based on the Kubo formula to calculate the magnetoresistance. Both Levy et al. [45] and Barnás et al. [43] conclude that to explain the observed giant magnetoresistance a strong interfacial scattering is required. Edwards et al. [46] have proposed an explanation of the giant magnetoresistance effect for a superlattice in terms of an equivalent network of resistors. In their approach, the magnetoresistance is attributed to bulk scattering in the ferromagnetic layers and they show that the magnetoresistance of Co/Cu structures can be well explained by this mechanism.

Apart from different emphasis attached to interfacial and bulk scattering, the basic assumptions in all these calculations are similar. However, the method of Edwards et al. [46] has the advantage that one can derive in two physically important limiting cases a very simple analytic formula for $\Delta R/R$ in terms of the resistivities and thicknesses of the magnetic and nonmagnetic layers in the superlattice unit cell. To review the theory of the magnetoresistance I shall, therefore, begin with the equivalent resistor network model of Edwards et al. [46] and then show how it can be modified to include interfacial scattering.

The main assumptions of the model are:

(i) The conduction electron spin is conserved over macroscopic distances [47] much longer than a superlattice unit cell (no spin flip scattering). This means that the current is carried separately by $\uparrow$ and $\downarrow$ spin conduction electrons (two channels in parallel);

(ii) conduction electrons do not see any potential barriers when crossing the interfaces, i.e., they travel in the superlattice as ordinary plane waves. This is a good approximation when the bottoms of the conduction bands of the two metals separated by an interface are well aligned on the scale of the Fermi energy;

(iii) the electron mean-free path l_σ in a transition metal ferromagnet is spin dependent ($l_\uparrow \neq l_\downarrow$). This arises because the densities of states (DOS) at the Fermi level E_F available for scattering are different for up and down spins [48–50].

It follows from the assumptions (i)–(iii) that an electron of a given spin travelling in a superlattice

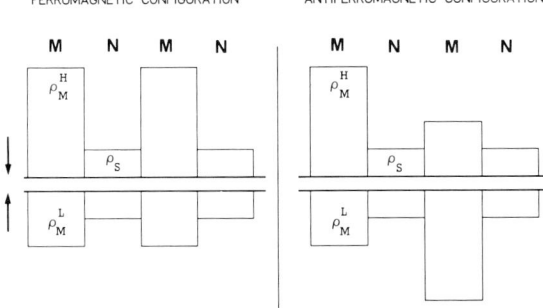

Fig. 10. Schematic representation of the distribution of local resistivities ρ in the magnetic unit cell for the ferromagnetic and antiferromagnetic configurations of the magnetic layers. Both the resistivities in the spin up ($\uparrow$) and down ($\downarrow$) channels are shown. M and N denote, respectively, the magnetic and nonmagnetic layers.

sees regions of different local resistivities. The resistivity is high in those regions where there is a high DOS at E_F available for scattering of conduction electrons. Generally speaking, a high DOS is obtained whenever the Fermi level intersects a d band. The distribution of such regions in a superlattice cell is shown schematically in fig. 10. The magnetic superlattice cell in fig. 10 is defined as twice the conventional unit cell and consists of two ferromagnetic layers containing M atomic planes and two nonmagnetic layers containing N atomic planes.

There are three different resistivities in the superlattice cell shown in fig. 10: the resistivity of the nonmagnetic spacer layer ρ_S, which is the same for both spin orientations, and the high ρ_M^H and low ρ_M^L resistivities for the two different spin orientations in the ferromagnet. To estimate the magnitude of all these resistivities, we assume that the conduction electrons, of mainly s and p character, are scattered predominantly into the d band (Mott scattering mechanism [47,48]), i.e., the resistivity in each spin channel σ is proportional to the density of states (DOS) at the Fermi level in the d band

$$\rho_\sigma \propto N_\sigma(E_F). \qquad (16)$$

As an example, consider a Co/Cu type superlattice. For a strong ferromagnet such as Co there are no d states at E_F in the majority spin band (spin up) and the resistivity in this channel should be comparable with the resistivity of Cu whose d band lies well below E_F. It follows that both the resistivities ρ_S and ρ_M^L for a Co/Cu superlattice should be low. On the other hand, the resistivity ρ_M^H should be much higher than ρ_M^L and ρ_S since the Fermi level in the minority band of Co lies in the region where the DOS is high.

Let us now return to the calculation of the magnetoresistance. It is clear from fig. 10 that the magnetic superlattice cell is equivalent to a system of eight resistors, with four resistors in each spin channel. To determine the magnetoresistance, we first need a rule for combining the four resistors in a given spin channel. Once the total resistances of both spin channels are known they can be simply combined as resistors in parallel to give the total resistance of the superlattice cell. This needs to be done both for the ferromagnetic ($\uparrow\uparrow$) and antiferromagnetic ($\uparrow\downarrow$) configurations of the magnetic layers. The magnetoresistance $\Delta R/R$ as defined by Parkin et al. [5] is then given by

$$\Delta R/R = (R_{\uparrow\downarrow} - R_{\uparrow\uparrow})/R_{\uparrow\uparrow}, \qquad (17)$$

where $1/R_{\uparrow\uparrow} = [1/R_\uparrow + 1/R_\downarrow]_{\uparrow\uparrow}$, $1/R_{\uparrow\downarrow} = [1/R_\uparrow + 1/R_\downarrow]_{\uparrow\downarrow}$ and R_σ is the total resistance of the superlattice cell in the spin channel σ.

It is now necessary to determine the rule for adding up the four resistors in the same spin channel. It is clear from fig. 10 that for the ferromagnetic configuration the problem reduces to the calculation of the resistance of a two-component superlattice with alternating regions of thicknesses a and b having resistivities ρ^a and ρ^b. For the antiferromagnetic configuration a four-component superlattice needs to be considered. It is nontrivial to evaluate the total resistance of such superlattices because electrons may cross freely from one layer to the other and, in general, the conventional rules for a network of resistors do not apply.

Consider first the case studied experimentally when the current is parallel to the layers (parallel magnetoresistance). There are two simple physical limits in which the total resistance can be easily evaluated without any microscopic calculation.

A) The mean-free path in each layer of a superlattice cell is much shorter than the thickness of the layer. In this limit electrons from different layers do not "mix" and flow in their respective separate resistor channels. All resistors then behave as resistors in a conventional resistor network and should be added in parallel. Since there are exactly the same numbers of resistors of each type in the ferromagnetic and antiferromagnetic configurations (see fig. 10), we have $R_{\uparrow\uparrow} = R_{\uparrow\downarrow}$ and there is no parallel magnetoresistance in this limit.

B) The mean-free path in each layer is much longer than the thickness of the layer. For a metallic superlattice with a small number of atomic planes in each layer we are clearly always close to this limit. It is, therefore, the limit which should be applicable to magnetic superlattices exhibiting the giant magnetoresistance.

Conduction electrons, which behave as plane waves, sample equally layers with low and high resistivity (uniform mixing of channels) and, therefore, experience an average resistivity. For a two-component superlattice this is clearly given by

$$\bar{\rho} = \frac{a\rho^a + b\rho^b}{a+b}, \quad (18)$$

where a, b are the thicknesses of the layers and ρ^a, ρ^b the corresponding resistivities. The generalization of eq. (18) to the four-component superlattice is obvious. The results for cases A and B obtained here from simple physical considerations are derived rigorously in ref. [46] as limits of a more general approach based on the Boltzmann equation.

It is now easy to apply eq. (18) and the corresponding result for the four-component superlattice to evaluate $\Delta R/R$. Because the length L of the sample in the direction of the current and the transverse area S per atomic plane are the same for $(_{\uparrow\uparrow})$ and $(_{\uparrow\downarrow})$ configurations they cancel out in the calculation of $\Delta R/R$. It is, therefore, sufficient to determine for each spin channel the reduced resistance R_σ per unit L and unit S. It is then clear from fig. 10 that

$$[1/R]_{\uparrow\uparrow}$$
$$= 1/R_\uparrow + 1/R_\downarrow = 2(M+N)^2$$
$$\times \left\{ (M\rho_M^L + N\rho_S)^{-1} + (M\rho_M^H + N\rho_S)^{-1} \right\},$$
$$[1/R]_{\uparrow\downarrow} = 2/R_\downarrow = 2/R_\uparrow = 8(M+N)^2$$
$$\times (M\rho_M^L + M\rho_M^H + 2N\rho_S)^{-1}. \quad (19)$$

It follows from eqs. (17) and (19) that the parallel magnetoresistance is given by

$$(\Delta R/R)_{\text{bulk}} = \frac{(\alpha - \beta)^2}{4(\alpha + N/M)(\beta + N/M)}, \quad (20)$$

where $\alpha = \rho_M^H/\rho_S$ and $\beta = \rho_M^L/\rho_S$ are the only parameters in this simple resistor network model and the subscript "bulk" indicates that the magnetoresistance is due to bulk scattering from the magnetic layers. Because M and N, originally defined as the numbers of atomic planes, appear both in the numerator and denominator they can be regarded in eq. (20) as the actual thicknesses of the magnetic and nonmagnetic layers.

We can use eq. (20) to determine favourable values of α, β for a large $\Delta R/R$. Clearly $\Delta R/R$ is a function of two variables, α/β and $M\beta/N$. The most important requirement for large $\Delta R/R$ is that the spin asymmetry ratio $\alpha/\beta = \rho_M^H/\rho_M^L$ in the ferromagnetic layer should be large. If there is no spin asymmetry, i.e. if $\alpha = \beta$, $\Delta R/R = 0$. For a given α/β, $\Delta R/R$ increases with increasing $M\beta/N$ but saturates for large values of this parameter. As a function of N, $\Delta R/R$ decreases monotonically and falls off as $1/N^2$ when $N \gg \alpha M$.

It is appropriate to ask how long the mean free path should be for eq. (20) to be applicable. By solving numerically the Boltzmann equation for a superlattice, Edwards et al. [46] showed that the long mean free path limit B is reached very rapidly and eq. (20) is a very good approximation already for a mean free path comparable with the thickness of the superlattice unit cell. Clearly the main restriction is that the thickness of the magnetic layer should be less than the shorter mean free path in that layer. For a superlattice with very thin magnetic layers, eq. (20) can, therefore, be safely used to discuss the dependence of $\Delta R/R$ on the thickness of the nonmagnetic spacer. It was applied in ref. [46] to a Co/Cu superlattice and the calculated dependence of $\Delta R/R$ on Cu thickness is shown in fig. 11a together with the experimental results of Parkin et al. [38]. It should be noted that oscillations in the observed magnetoresistance arise because it is measured only for those values of N for which the Co layers are coupled antiferromagnetically. On the other hand, eq. (20) gives the relative difference between the resistances of the antiferromagnetic and ferromagnetic configurations for all values of N. It is, therefore, only meaningful to compare the calculated results with the envelope of the observed $\Delta R/R$. A good fit to the observed $\Delta R/R$ is obtained for $\alpha = 22$ and $\beta = 2$. It is interesting that the scattering asymmetry $\alpha/\beta = 11$ is very close to the ratio of the calculated densities of states at E_F in the majority and minority spin bands in bulk Co [51] and the value $\beta = 2$ is close to the ratio of the conductivity of bulk Cu to bulk Co at room temperature. This gives support to the claim made in ref. [46] that the bulk scattering mechanism is dominant in Co/Cu systems.

Eq. (20) is not appropriate to discuss the dependence of $\Delta R/R$ on the Co thickness M since the short mean free path in Co soon becomes shorter than the thickness of Co layers. Full solution of the Boltzmann equation is then needed [52] and it reproduces well with the same values of the parameters $\alpha = 22$, $\beta = 2$ the observed dependence of $\Delta R/R$ on M (fig. 11b). One additional parameter is required in this calculation and this is the mean free path in Cu, which was taken to be ≈ 250 Å.

When eq. (20) is applied to Fe/Cr, it is found [46] that an asymmetry α/β larger than that obtained from bulk densities of states for Fe is needed to explain the observed $\Delta R/R$. This gives support to the claim made in refs. [43–45] that interfacial scattering is important in Fe/Cr.

Using the resistor network model, it is easy to understand the effect of interfacial scattering. Assuming the extreme case of no bulk scattering [44], we can use again fig. 10 to model a superlattice with interfacial scattering by a system of resistors. The only regions with a spin dependent resistivity ρ_I^H, ρ_I^L are now thin interfacial layers of thickness I. The resistivities of the magnetic layers are assumed to have the same values for both spin orientations and, for Fe/Cr system, it is a good approximation to set the resistivity of Fe equal to the resistivity of Cr. This means that Fe

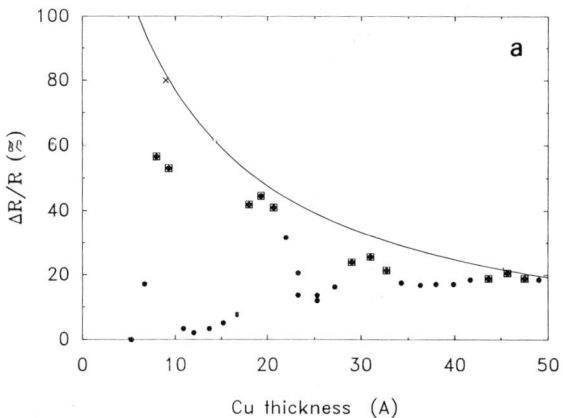
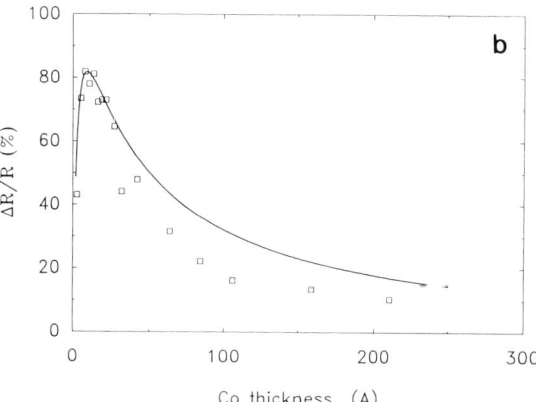

Fig. 11. Dependence of the magnetoresistance $\Delta R/R$ of a $Co_{10}Cu_N$ superlattice on the thickness of the Cu spacer N (a); dependence of $\Delta R/R$ of a $Co_M Cu_9$ superlattice on the thickness of the Co layer M (b). The experimental points are from ref. [38]; explanation of different symbols in (a) can be found in ref. [38].

layers now play the same role as nonmagnetic Cr layers. Using again eqs. (18) and (19), it is easy to show that $(\Delta R/R)_{\text{interface}}$ is given by

$$(\Delta R/R)_{\text{interface} } = \frac{(\alpha - \beta)^2}{4[\alpha + (N+M)/2I][\beta + (N+M)/2I]}. \quad (21)$$

Eq. (21) reduces to eq. (20) if the following correspondence is made: $(N+M) \to N$ and $I \to M$. This simply reflects the fact that magnetic layers are now inactive in the spin-dependent scattering and their thickness should, therefore, be added to the thickness of the nonmagnetic layers. On the other hand, the role of magnetic layers in eq. (20) is taken over by the interfacial layers.

It is clear from these results that it is difficult to distinguish between the bulk and interfacial scattering. One may, therefore, conclude that there is general consensus on the mechanism that causes the giant magnetoresistance effect, i.e. that it is spin-dependent scattering but more theoretical work is needed to decide where the scattering takes place. A more microscopic theory of the scattering is also needed so that we can link the scattering rate to the local band structure and thus predict which particular combinations of magnetic and nonmagnetic layers are likely to lead to a large magnetoresistance. Such a theory will have to include the effect of interface roughness which is known experimentally to influence $\Delta R/R$.

Finally, I would like to mention an interesting result that follows directly from the resistor network model. We have discussed only parallel magnetoresistance and it is natural to ask what happens when the current flows in the direction perpendicular to the layers. In the limit where the mean free path in each layer is much shorter than the thickness of the layer the system behaves as conventional resistors in series. It is then easily shown that $\Delta R/R$ is again given by eq. (20) or (21), the result obtained for current parallel to the layers in the opposite limit of long mean free paths.

Acknowledgements

Most of the theoretical ideas and calculations I have described are the product of collaboration with my colleagues David Edwards, Bechara Muniz and Son Phan and without their contribution this paper could not have been written. I have also greatly benefited from numerous discussions with Stuart Parkin at IBM Almaden Research Center.

References

[1] S. Iwasaki, IEEE Trans. Magn. MAG-20 (1984) 657.
[2] P. Grünberg, R. Schreiber, Y. Pang, M.B. Brodsky and H. Sower, Phys. Rev. Lett. 57 (1986) 2442.
[3] M.N. Baibich, J.M. Broto, A. Fert, F. Nguyen Van Dau, F. Petroff, P. Etiene, G. Creuset, A. Friederich and J. Chazelas, Phys. Rev. Lett. 61 (1988) 2472.
[4] G. Prinz, J. Magn. Magn. Mater. 100 (1991) 469.
[5] S.S.P. Parkin, N. More and K.P. Roche, Phys. Rev. Lett. 64 (1990) 2304.
[6] A.J. Freeman and R.Q. Wu, J. Magn. Magn. Mater. 100 (1991) 497.
[7] J. Mathon, Rep. Prog. Phys. 51 (1988) 1.
[8] L.M. Falicov and J.L. Moran-Lopez, eds., Magnetic Properties of Low-Dimensional Systems 1 (Springer, Berlin, 1986).
[9] R. Feder, ed., Polarized Electrons in Surface Physics (World Scientific, Singapore, 1985).
[10] H.C. Siegmann, D. Mauri, D. Scholl and E. Kay, J. de Phys. 49 (1988) C8–9.
[11] J. Mathon, in: Magnetic Properties of Low-Dimensional Systems II, eds. L.M. Falicov and J.L. Moran-Lopez (Springer, Berlin, 1990) p. 167.
[12] J. Mathon, Physica B 149 (1988) 31.
[13] J. Mathon, J. Phys.: Condens. Mater. 1 (1989) 2505.
[14] G.T. Rado, Bull. Am. Phys. Soc. 2 (1957) 127.
D.L. Mills and A.A. Maradudin, J. Phys. Chem. Solids 28 (1967) 1855.
[15] J. Mathon and S.B. Ahmad, Phys. Rev. B 36 (1988) 660.
[16] D.T. Pierce, R.J. Celotta, J. Unguris and H.C. Siegmann, Phys. Rev. B 26 (1982) 2566.
[17] J.C. Walker, R. Droste, G. Stern and J. Tyson, J. Appl. Phys. 55 (1984) 2500.
[18] D. Mauri, D. Scholl, H.C. Siegmann and E. Kay, Phys. Rev. Lett. 61 (1988) 758.
[19] M. Landolt, in: Polarized Electrons in Surface Physics, ed. R. Feder (World Scientific, Singapore, 1985).
[20] D. Mauri, D. Scholl, H.C. Siegmann and E. Kay Phys. Rev. Lett. 62 (1989) 1900.
[21] D. Scholl, M. Donath, D. Mauri, E. Kay, J. Mathon, R.B. Muniz and H.C. Siegmann, Phys. Rev. B (in press).

[22] C. Carbone and S.F. Alvarado, Phys. Rev. 36 (1987) 2433.
[23] P. Grünberg, S. Demokritov, A. Fuss, M. Vohl and J.A. Wolf, J. Appl. Phys. (in press).
[24] S. Demokritov, M. Vohl, J.A. Wolf, P. Grünberg and W. Zinn, Proc. MRS Spring Meeting 1991 (to be published).
[25] P.M. Levy, K. Ounadjela, S. Zhang, Y. Wang, C.B. Sommers and A. Fert, J. Appl. Phys. 67 (1990) 5914.
[26] H. Hasegawa, Phys. Rev. B 42 (1990) 2368.
[27] D. Stoeffler, K. Ounadjela and F. Gautier, J. Magn. Magn. Mater. 93 (1991) 386.
[28] D.M. Edwards and J. Mathon, J. Magn. Magn. Mater. 93 (1991) 85.
[29] D.M. Edwards, J. Mathon, R.B. Muniz and M.S. Phan, J. Phys.: Condens. Matter 3 (1991) 4941.
[30] A.A. Abrikosov, Introduction to the Theory of Normal Metals (Academic Press, New York, 1972).
[31] S.L. Cunningham, Phys. Rev. B 10 (1974) 4988.
[32] J. Mathon, D.M. Edwards, R.B. Muniz and M.S. Phan, Proc. ICM '91 (to be published).
[33] S.T. Purcell, W. Folkerts, M.T. Johnson, N.W.E. McGee, K. Jager, J. Van de Stegge, W.B. Zeper, W. Hoving and P. Grünberg, (preprint).
I. Unguris, R.J. Celotta and D.T. Pierce, (preprint).
[34] Y. Yafet, J. Appl. Phys. 61 (1987) 4058.
[35] W. Baltensberger and J.S. Helman, Appl. Phys. Lett. 57 (1990) 2954.
[36] C. Lacroix, B. Dieny, J.P. Gavignan and D. Givord, Thin Solid Films (in press).
[37] Y. Wang, P.M. Levy and L.L. Fry, Phys. Rev. Lett. 65 (1990) 2732.
[38] S.S.P. Parkin, R. Bhadra and K.P. Roche, Phys. Rev. Lett. 66 (1991) 2152.
[39] D.H. Mosca, F. Petroff, A. Fert, P.A. Schroeder, W.P. Pratt, Jr. and R. Laloee, J. Magn. Magn. Mater. 94 (1991) L1.
[40] A. Cebolada, J.L. Martínez, J.M. Gallego, J.J. de Miguel, R. Miranda, S. Ferrer, F. Batallan, G. Fillion and J.P. Rebouillat, Phys. Rev. B 39 (1989) 9726.
D. Pescia, D. Kerkmann, F. Schumann and W. Gudat, Z. Phys. B 78 (1990) 475.
B. Heinrich, Z. Celinski, J.F. Cochran, W.B. Muir, J. Rudd, Q.M. Zhong, A.S. Arrott, K. Myrtle and J. Kirschner, Phys. Rev. Lett. 64 (1990) 673.
[41] R. Coehoorn, (preprint).
[42] C. Chappert and J.P. Renard, Europhys. Lett. (in press).
[43] J. Barnás, A. Fuss, R.E. Camley, P. Grünberg and W. Zinn, Phys. Rev. B 42 (1990) 8110.
[44] R.E. Camley and J. Barnás, Phys. Rev. Lett. 63 (1989) 664.
[45] P.M. Levy, S. Zhang and A. Fert, Phys. Rev. Lett. 65 (1990) 1643.
[46] D.M. Edwards, R.B. Muniz and J. Mathon, IEEE Trans. Magn. (in press).
[47] J.M. Ziman, Electrons and Phonons (Oxford University Press, London, 1960).
[48] N.F. Mott, Adv. Phys. 13 (1964) 325.
[49] G.G. Cabrera and L.M. Falicov, Phys. Stat. Sol. (b) 61 (1974) 539, (1974) 217.
[50] A. Fert and I.A. Campbell, J. Phys. F 6 (1976) 849.
[51] V.L. Moruzzi, J.F. Janak and A.R. Williams, Calculated Electronic Properties of Metals (Pergamon Press, Oxford, 1978).
[52] D.M. Edwards, J. Mathon, R.B. Muniz and S.S.P. Parkin, Proc. ICM '91 (to be published).

Nonlinear dynamics of coupled systems near magnetic phase transitions of the "order–order" type

V.I. Ozhogin
Kurchatov Institute of Atomic Energy, 123182 Moscow, USSR

and

V.L. Preobrazhenskii
Moscow Institute of Radioengineering, Electronics and Automation, Moscow 117454, USSR

The results are reviewed of the 15-year development of the concept of effective anharmonicity of the normal modes that describe the vibrations of a continuous medium having two or more mutually coupled subsystems. Magnetoelastic interaction near a spin reorientation phase transition leads to a giant acoustic nonlinearity and its magnetic field sensitivity in antiferromagnets such as α-Fe_2O_3, $FeBO_3$, $TmFeO_3$, etc.

1. Introduction

The time period which has passed since the publication of the 1st volume of the Journal of Magnetism and Magnetic Materials (JMMM) turned out to be the period of an intensive development of the physics of nonlinear phenomena in continuous media (plasma, solid state, liquids, polymers and biological media). This general tendency reverberated in the physics of magneto-ordered substances as well. Strong nonlinearity and complex dispersion inherent to spin systems have attracted and are still attracting attention to magnets not only as to objects of self-dependent scientific interest but also as to model media for a study of general problems of nonlinear dynamics of a continuous medium.

The nonlinearity of an electron spin system manifests itself most vividly (even under relatively weak excitations) in the vicinity of the second kind phase transitions of the "order–order" type (so-called spin-reorientation transitions, SRT). At the phase transition a spin system looses its stability, while on nearing it the activation energy of one of the branches ("soft mode") of the spin spectrum anomalously shrinks. At the same time the extraordinary increase of the contribution of nonlinearity into the energy of long-wave "soft" excitations of the finite amplitude can be noted. A principal distinction of SRT in magnets is a rather narrow fluctuation region for thermally activated fluctuations of the order parameter (see below), which gives a good opportunity to study the anomalies of linear and nonlinear properties of magnetic crystals outside the region, the close proximity to the critical point included.

In real crystals the excitations of an electron spin system are to a certain extent coupled with excitations of a different physical nature, such as crystal lattice oscillations, nuclear spin precession, electromagnetic waves or electric polarization oscillations in ferroelectromagnetic substances. Excitations of a small amplitude in this case exist in the form of normal mixed (hybridized) modes. Generally speaking all the degrees of freedom of interacting "partial" subsystems are excited simultaneously in any normal mode of coupled vibrations. If even one of the

0304-8853/91/$03.50 © 1991 – Elsevier Science Publishers B.V. All rights reserved

subsystems or a coupling between them is nonlinear in amplitude the normal modes will also be nonlinear. At resonance conditions, i.e. at conditions of intersecting the partial subsystem spectra (dispersion curves $\omega(k)$), the excitation energy is equally distributed between them. Obviously, the main contribution to nonlinearity of coupled waves is expected here from the subsystem where strong nonlinearity is realized with less energy densities. At a distance off the resonance each of the normal modes retains to a great extent the physical individuality of the subsystem which is the closest to it in frequency. Nevertheless the contribution of the interaction of vibrations to the nonlinearity of a specific mixed mode may turn out to be decisive. A glaring example of such a situation is the interaction of a magnetic and an elastic subsystem in magnets near SRT.

A detailed discussion of this example is still more important because in many cases or SRT it is not a spin (or rather a quasispin) mode but an activationless acoustic (quasisound) magnetoelastic mode that plays the role of a soft mode with the critical behavior, while in the quasispin branch spectrum there remains an energy gap [1–4] resulting from the magnetostriction coupling. In ref. [3] it was noted that this situation is similar to spontaneous symmetry breaking [43].

Atom displacements of a relatively small amplitude in a wave of the acoustic branch may cause substantial deviations of magnetic moments from the equilibrium direction in the region of the spin system stability loss. Magnetostriction stresses turn out to be nonlinearly dependent on the strains amplitude in an acoustic wave. The nonlinearity of the stress dependence on deformations is nothing else but elastic anharmonicity which is introduced in this case into acoustic oscillations by the interaction of a sound with a spin subsystem of a crystal. Such "effective" anharmonicity which is a reflection of the nonlinearity of mixed modes is often of a giant value and hence is a main mechanism of a number of nonlinear wave processes.

The concept of the effective anharmonicity of mixed modes, first formulated in ref. [5] and analyzed in detail in ref. [6], appeared to be very fruitful not only for the nonlinear magnetoacoustics and the kinetics of mixed excitations near SRT [8–10] but also for acousto-optics of magnets [11–13]. The admixture of spin excitations in acoustic oscillations near SRT was found to be capable to increase considerably the amplitudes of the light-on-sound scattering (acoustomagneto-optical interaction [11]) as compared to conventional photoelasticity – similarly to the increase of the Raman sound-on-sound scattering amplitudes near SRT [6,14].

The increase of nonlinearity near the point of the spin system stability loss reduces considerably the applicability of anharmonic expansions of the magnetic energy density by dynamic variables of the critical soft mode. Thus, for a quasisound wave the anharmonic approximations turn invalid near SRT already for dynamic deformations ($\hat{u}_-$) of the order of magnitude of spontaneous magnetostriction $|\hat{u}_0| \approx 10^{-5}$, while outside the SRT region a wave with such deformations remains very weakly nonlinear. It suggests that magnets near SRT are objects of special interest for experimental and theoretical investigations of strongly nonlinear wave processes in a mixed wave system (in particular, of the processes of formation and propagation of mixed solitons of a topological and a dynamic type [16–22] and nonlinear periodic mixed waves [12,23]. They are also interesting for a study of the stability problems in strongly nonlinear mixed excitations [20,12]. A particular place in the investigations of nonlinear dynamics of coupled waves near SRT belongs to the phenomena of strong phase nonlinear effects of double bistability over the threshold of the parametric instability of acoustic waves [24].

Taking into account a specific character of the jubilee edition of the journal this article is aimed at summarizing the results of a 15-year development period of the physics of mixed nonlinear excitations of the acoustic (hydrodynamic) type in a magnet near SRT. We have dwelt on a number of original papers initially published in editions almost inaccessible for JMMM readers. We also wanted to draw the readers' attention to some results in the field of nonlinearity of coupled waves obtained within the last three years which passed since the publication of our previous review [7].

2. Nonlinear magnetoelastic dynamics of an antiferromagnet with the "Easy Plane" type Anisotropy (EPAF)

Crystals of EPAF are the most convenient and interesting objects for studying nonlinear dynamics of coupled modes. In crystals with a negligibly small in-plane anisotropy the situation, which is completely similar to the region of the second kind SRT, is realized at weak magnetic fields (required only for rendering a single-domain in the sample). The antiferromagnetic character of spin ordering weakens the long-range dipole–dipole interactions which only complicate the nonlinear dynamics pattern near SRT. The decrease of the Zeeman energy due to the AFM exchange widens the field interval for the observation of critical anomalies. Finally, such EPAFs as α-Fe_2O_3 and $FeBO_3$ have a high temperature of magnetic ordering ($T_N = 960$ and 348 K, respectively) which not only facilitates the experimental conditions but also enables one to hope for practical applications of the phenomena under investigation in solid-state electronics.

The magnetoelastic vibrations in an antiferromagnet are described phenomenologically in the two-suitable model by a system of the coupled nonlinear equations for the sublattice magnetization vectors M_1 and M_2 and the equations for elasticity:

$$-\gamma^{-1}\dot{M}_n = [M_n \times H_n] \quad (n = 1, 2), \tag{1}$$

$$\rho \ddot{U}_i = \frac{\partial \sigma_{ij}}{\partial x_j} \quad (i, j = 1, 2, 3). \tag{2}$$

Here the effective fields are

$$H_n \equiv -\delta \int F \, dV / \delta M_n,$$

the displacement vector is U_i, and the tensor of the mechanical stresses is

$$\sigma_{ij} = \delta F / \delta(\partial U_i / \partial x_j).$$

The energy density of the crystal is

$$F\left(M_1, M_2, \frac{\partial M_1}{\partial x_i}, \frac{\partial M_2}{\partial x_i}, \frac{\partial U_i}{\partial x_j}\right).$$

In its magnetic component F_m we shall take account of the energy of intersublattice exchange interaction, the Dzyaloshinskii interaction, the uniaxial anisotropy energy (with the effective fields respectively of H_E, H_D and H_A), and the energy of interaction of the magnetic moments with the external field H:

$$F_m = 2M_0 \left[\tfrac{1}{2} H_E m^2 - H_D [m \times l]_z \right.$$

$$\left. + \tfrac{1}{2} H_A l_z^2 - (m \cdot H) - \frac{\alpha}{2} \left(\frac{\partial l_i}{\partial x_j}\right)^2 \right].$$

Here $m \equiv (M_1 + M_2)/2M_0$ is the ferromagnetic vector, $l \equiv (M_1 - M_2)/2M_0$ is the antiferromagnetic vector, $\alpha \equiv v_m^2/4H_E\gamma^2$, and v_m is the so-called "limiting" velocity of the spin waves (magnons), which is proportional to the exchange field and the square of the lattice constant.

The effective exchange field $H_E \sim kT_N/\mu_B$ in crystals having a high enough Néel temperature T_N considerably exceeds the fields of the relativistic (H_A) and exchange-relativistic (H_D) interactions, just as it does the external fields (H) used in experiments. For example, in crystals of α-Fe_2O_3 the characteristic values of the fields are the following: $H_E \approx 10^7$ Oe, $H_D \approx 2 \times 10^4$ Oe and $H_A \approx 2 \times 10^2$ Oe. In many cases, taking account of the actually existing hierarchy of interactions in the spin system enables one substantially to simplify the description of the nonlinear dynamics of antiferromagnets. In the first approximation in the parameter $\tilde{H}/H_E \ll 1$, where $\tilde{H} = H_D$, H_A or H, it proves possible to reduce the system of precession equations to the equation of motion for the antiferromagnetic vector l alone, since $l^2 = 1 - m^2 \approx 1$ and $m \ll l$ [6,25,26]:

$$[l \times L] = 0,$$

$$L \equiv \gamma^{-2}(\ddot{l} - v_m^2 \nabla^2 l) - \gamma^{-1}\{2[H \times \dot{l}] + [\dot{H} \times l]\}$$

$$+ H(H \cdot l) + H_D [H \times z]$$

$$+ (2H_E H_A + H_D^2) l_z z - 2H_E H_{me}. \tag{3}$$

In this approximation it suffices to restrict the treatment in the expression for the magnetoelastic energy density to taking account of invariants of the type

$$F_{ml}^{(l)} = l(\hat{B}_l \hat{u}) l. \quad (4)$$

Here $\hat{B}_l$ is the tensor of magnetoelastic constants corresponding to the antiferromagnetic vector.

The spin-wave vector of an EPAF, which can be found from the linearized eq. (3) or, as is usual, directly from the linearized system (1), contains two branches. Their frequencies *without allowance* for the magnetoelastic interaction are given by the relationships

$$\tilde{\omega}_{ak}^2 = \gamma^2[2H_E H_A + H_D(H + H_D)] + v_m^2 k^2, \quad (5)$$

$$\tilde{\omega}_{fk}^2 = \gamma H(H + H_D) + v_m^2 k^2. \quad (6)$$

One of the branches, the "antiferromagnetic" one, $\tilde{\omega}_{ak}$ has the relatively high activation energy $\gamma(2H_E H_A)^{1/2}$. For example, for α-Fe$_2$O$_2$ $\tilde{\omega}_{a0}$ lies in the millimeter UHF range [27]. The other ("quasiferromagnetic") branch $\tilde{\omega}_{fk}$ amounts to the "soft mode" of the spin system, whose activation energy is small, in line with the smallness of the external field intensity. The activation branch of the spectrum $\tilde{\omega}_{ak}$ corresponds to vibrations with departure of the vector l from the basis plane and with change of the angle inclination of the magnetic sublattices. The soft mode $\tilde{\omega}_{fk}$ corresponds to rocking of l in the base plane and precession of the ferromagnetic vector m about the equilibrium vector m_s, as in a ferromagnet. The interaction of the elastic subsystem with the soft spin-wave mode gives rise to very strong coupling. It determines the fundamental features of the magnetoacoustic properties of EPAFs. One can show that, under ordinary conditions of not too high frequencies $\tilde{\omega} \ll \omega_{ak}$ and weak magnetostrictive fields $H_{me} \ll H_A$, where

$$H_{me} = -\frac{1}{2M_0} \frac{\partial F_{me}}{\partial l},$$

we can neglect the departure of the antiferromagnetic vector from the base plane in describing the magnetoelastic dynamics ($l_z \approx 0$). Then the equation of motion for l is reduced to the form

$$\gamma^{-2}[l \times (\ddot{l} - v_m^2 \nabla^2 l)]_z$$
$$= (H \cdot l)([H \times l]_z + H_D)$$
$$+ 2H_E[l \times H_{me}]_z + \gamma^{-1} \dot{H}_z. \quad (7)$$

When we take account of the condition of conservation of the modulus $|l| = 1$, the only dynamical magnetic variable in eq. (7) proves to be the angle φ of rotation of the vector l in the base plane. Transforming to this variable allows us to write the energy density of spin excitations w_m and the magnetoelastic excitation F_{me} in the form

$$w_m = \frac{M_0}{2H_E}\left[\gamma^{-2}(\dot{\varphi})^2 + \gamma^{-2} v_m^2 (\nabla \varphi)^2 \right.$$
$$\left. - (H\cos\varphi + H_D)^2\right], \quad (8)$$

$$F_{me} = (\hat{B}_1(\alpha)\cos 2\varphi + \hat{B}_2(\alpha)\sin 2\varphi)\hat{u}. \quad (9)$$

The explicit form of the components of the tensor of magnetoelastic constants $B_n(\alpha)$ ($n = 1, 2$) depends on the specific symmetry of the crystal and the angle α of the external field orientation with respect to the marked crystal axis in the basis plane. For rhombohedral symmetry, which many of the experimentally studied EPAFs possess (α-Fe$_2$O$_3$, FeBO$_3$, MnCO$_3$, CoCO$_3$, having the symmetry D_{3d}^6 and the marked axis $x \parallel U_2$), the following relationships hold:

$$\hat{B}_1(\alpha) = \hat{B}_1(0)\cos 2\alpha + \hat{B}_2(0)\sin 2\alpha,$$
$$\hat{B}_2(\alpha) = -\hat{B}_1(0)\sin 2\alpha + \hat{B}_2(0)\cos 2\alpha,$$
$$\hat{B}_1(0)\hat{u} = -\tfrac{1}{2}(B_{11} - B_{12})(u_{xx} - u_{yy}) - 2B_{14} u_{yz},$$
$$\hat{B}_2(0)\hat{u} = -(B_{11} - B_{12})u_{xy} - 2B_{14} u_{xz}.$$
$$(10)$$

Thus it proves possible to describe phenomenologically the magnetoelastic dynamics of EPAFs by using the four-vector magnetoelastic displacement (U, φ), which satisfies the system of equations of motion [16]

$$\rho \ddot{U}_i = \frac{\partial}{\partial x_j} \left(\hat{C}^{(2)} \hat{u}_\sim + \hat{B}_1(\alpha) \cos 2\varphi \right.$$
$$\left. + \hat{B}_2(\alpha) \sin 2\varphi \right)_{ij} \equiv \frac{\partial}{\partial x_j} \sigma_{ij}, \quad (11)$$

$$\gamma^{-2} \left(v_m^2 \nabla^2 \varphi - \ddot{\varphi} \right)$$
$$= \tfrac{1}{2} \left(\gamma^{-2} \omega_{f0}^2 - H H_D \right) \sin 2\varphi + H H_D \sin \varphi$$
$$+ \frac{2 H_E}{M_0}$$
$$\times \left| \hat{B}_2(\alpha) \hat{u} \cos 2\varphi - \hat{B}_1(\alpha) \hat{u}_\sim \sin 2\varphi \right|, \quad (12)$$

here $\hat{u}_\sim = \hat{u} - \hat{u}_0$ and $\hat{u}_0 = -[\hat{C}^{(2)}]^{-1} \hat{B}_1(\alpha)$ is the tensor of spontaneous magnetostrictive deformations,

$$\omega_{f0} = \gamma \left[H(H + H_D) + 2 H_E H_{mS} \right]^{1/2}$$

is the frequency of the ferromode of the antiferromagnetic resonance (AFMR), $H_{mS} = -(2/M_0) \hat{B}_1(\alpha) \hat{u}_0$ is the magnitude of the effective field of the spontaneous striction; and we use

$$\hat{u} \equiv \frac{1}{2} \left(\frac{\partial U_i}{\partial x_j} + \frac{\partial U_j}{\partial x_i} \right).$$

We should note that using eqs. (11) and (12) does not presuppose any restrictions on the amplitude of the spin oscillations. They are suitable for describing both weakly and strongly nonlinear effects to which anharmonic approximations are inapplicable. We note also that one can derive eq. (12) directly from the relationships (8) and (9) by variation methods. It has the form of the well known double sine-Gordon equation [28] supplemented by nonlinear dynamical couplings. It is encountered in the theory of nonlinear spin excitations in liquid ^{3}He and in the theory of self-induced optical transparency. Antiferromagnets are another example of physical systems whose description can be reduced to an equation of this type. Here the nonlinear dynamical couplings considerably enrich the pattern of possible wave processes.

3. Coupled magnetoacoustic modes in EPAFs in the SRT-like region

Let us take up in greated detail in properties of coupled magnetoelastic excitations of small amplitude. Their spectrum Ω_k is determined by the well-known dispersion equation, which can be easily derived from the system (11), (12) linearized over small deviations $\varphi \ll 1$:

$$\left(\omega_{fk}^2 - \Omega_k^2 \right) + \sum_{S=1}^{3} \frac{\zeta_{Sk}^2 \omega_{fk}^2 \omega_{Sk}^2}{\Omega_k^2 - \omega_{Sk}^2} = 0. \quad (13)$$

Here ω_{Sk} is the partial frequency of the concrete ($S = 1, 2, 3$) elastic mode having the wave vector k and the polarization $e^{(S)}$, and ω_{fk} is the partial frequency of the ferromode. The adjective "partial" is essential – it means that the frequencies ω_{Sk} and ω_{fk} that enter into eq. (13) are calculated for a fixed partner subsystem. For example, ω_{fk} is calculated for a "frozen" lattice (for more details on this, see ref. [3]) and hence it contains a so-called magnetoelastic gap: $\omega_{fk} = (\tilde{\omega}_{kf}^2 + 2\gamma^2 H_E H_{mS})^{1/2}$. Recurring to eq. (13), now we have:

$$\zeta_{Sk}^2 = \frac{H_E}{M_0} \frac{\left(\hat{2} B_2 \hat{u}_S \right)^2}{(\omega_{fk}/\gamma)^2 \hat{C}^{(2)} \hat{u}_S \hat{u}_S}, \quad (14)$$

$$(\hat{u}_S)_{ij} = \frac{1}{2k} \left(e_i^{(S)} k_j + e_j^{(S)} k_i \right).$$

The asymptotic linearity in k (i.e., when $k \gg k_m \equiv \omega_{f0}/v_m$, fig. 1) of the spin-wave spectrum in an antiferromagnet determines the qualitative differences of the spectra of magnetoelastic excitations in crystals having a high and a low Néel temperature T_N, to which the "limiting" magnon velocity v_m is related by direct proportionality. In low temperature EPAFs (such as $MnCO_3$, $CoCO_3$ and $CsMnF_3$), T_N is lower than the Debye temperature T_D and the velocity of spin waves is smaller than the velocity of sound (see fig. 1b). Here the spectra of the "pure" (partial) spin and elastic excitations ω_{fk} and ω_{Sk} intersect. The conditions for strongest interactions are realized in the resonance region near the intersection. In high-temperature EPAFs ($T_N > T_D$), for which $v_m > v_S$, intersection of the "partial" spectra is not realized. In this case the spin waves for any k are unambiguously separated into acoustic (more exactly – soundlike) and spin (magnonlike) waves, while the strongest coupling corresponds to the long-wavelength region of the spectrum ($k \ll \omega_{f0}/v_m$). We note for the discussion below that, in an EPAF of the type of α-Fe$_2$O$_3$ or FeBO$_3$, the relation holds that $v_S^2 \ll v_m^2$, and the frequencies of the acoustic waves satisfy the condition $\Omega_k^2 \ll \omega_{fk}^2$ for any wave vectors k.

Magnetoelastic coupling renormalizes the velocities of the acoustic waves and causes them to depend on the field intensity. The acoustic waves contain a substantial admixture of "nonresonance" spin excitations caused by the deformations. Their amplitude can be found from the linearized equation (11) with allowance for the condition $\partial/\partial t \ll \omega_{fk}$:

$$\varphi_k = -\frac{2H_E}{M_0}\left(\frac{\gamma}{\omega_{fk}}\right)^2 \hat{B}_2 \hat{u}_k. \qquad (15)$$

It is precisely these excitations, participating in nonlinear interactions inherent in the spin system, that introduce anharmonicity into the acoustic vibrations. We can easily convince ourselves that in weak magnetic fields, and namely when $H(H + H_D) \lesssim 2H_E H_{mS}$, the amplitudes of the nonresonance excitations prove to be large ($\varphi \approx 1$) even at deformations of the order of the spontaneous deformations ($u_k \approx u_0$). The nonlinearity of the acoustic modes under these conditions *cannot be considered to be weak*.

When $u_k \ll u_0$ and $\varphi \ll 1$ the elastic-stress tensor can be linearized and reduced to usual form $\hat{\sigma}_k = C_{\text{eff}}^{(2)}(k)\hat{u}_k$, where the effective second-order elastic moduli, i.e., those renormalized by the interaction, are determined by the relationships

$$\hat{C}_{\text{eff}}^{(2)}(k) \equiv \hat{C}^{(2)} + \Delta\hat{C}^{(2)}(k), \qquad (16)$$

$$\Delta\hat{C}^{(2)}(k) = -\frac{H_E}{M_0}\left(\frac{2\gamma\hat{B}_2}{\omega_{fk}}\right)^2. \qquad (17)$$

The spectrum and polarization of the magnetoelastic waves of the acoustic branch are found here, as in ordinary elasticity theory, from an equation like the Green–Christoffel equation:

$$\left[\rho\Omega_k^2\hat{I} - \hat{\Gamma}_{\text{eff}}(k)\right]e_k = 0, \qquad (18)$$

where we have

$$\hat{\Gamma}_{\text{eff}} = \hat{C}_{\text{eff}}^{(2)} k \cdot k.$$

The field dependence of the moduli $\hat{C}_{\text{eff}}^{(2)}(H)$ and of the corresponding velocities of sound bear

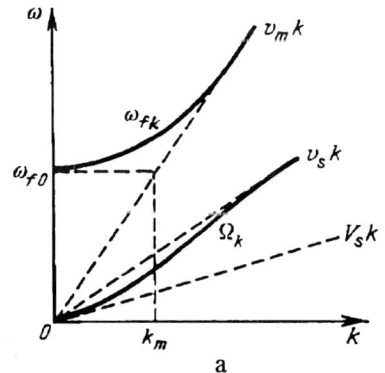
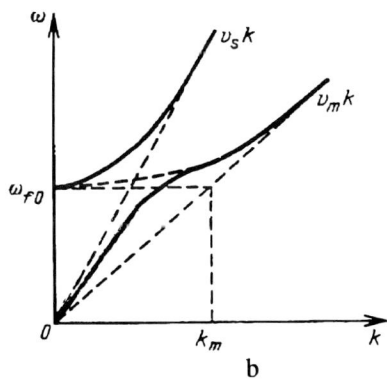

Fig. 1. Spectra of coupled magnetoelastic waves in crystals of an EPAF (solid lines) – high-temperature ($T_N > T_D$) (a) and low-temperature ($T_N < T_D$) (b).

direct information on the magnitude of the linear coupling of the elastic and spin waves. In the simplest case of interaction of one elastic mode (S) with the spin system, the field dependence of the velocity of sound is described by the relationship

$$V_{Sk}(H) = v_{Sk}\left(1 - \zeta_{Sk}^2(H)\right)^{1/2}. \qquad (19)$$

The specific participation of the exchange reaction, which is characteristic specifically of antiferromagnets, in the formation of the amplitudes of the interaction of the excitations and the magnetoelastic activation of the spectrum of spin waves has the result that the coupling coefficient proves to be of the order of unity over a broad interval of magnetic fields $0 < H(H + H_D) \lesssim 2H_E H_{mS}$ that considerably exceeds the monodomainization field of the crystal. For hematite the characteristic field is $H^* = 2H_E H_{mS}/H_D \approx 0.5$ kOe.

The acoustic modes that satisfy the condition of limiting strong coupling $\zeta_{S,k \to 0}(H \to 0) \to 1$ are of fundamental physical and practical interest. The existence of such modes in a crystal that is isotropic in its magnetic properties (though not necessarily in its magnetoelastic properties) involves losses of stability of the equilibrium state with respect to slow rotation of the magnetization on the xy plane by an arbitrary angle (with a corresponding change in the spontaneous deformation).

Measurement of the field dependence of the velocity of sound is one of the fundamental methods of experimental study of the coupling of acoustic with spin waves. Early experiments on standing waves in α-Fe$_2$O$_3$ and FeBO$_3$ crystals were carried out in refs. [29,30]. Fig. 2 shows the data of measurements of the velocities of certain types of running bulk and surface acoustic waves in hematite. Also shown there are the results of measurements of the intrinsic frequency of acoustic vibrations of the "contour-shear" mode for a resonator made of α-Fe$_2$O$_3$ in the form of a disk cut in the basis plane [31]. The results of calculations performed by the methods of elasticity the-

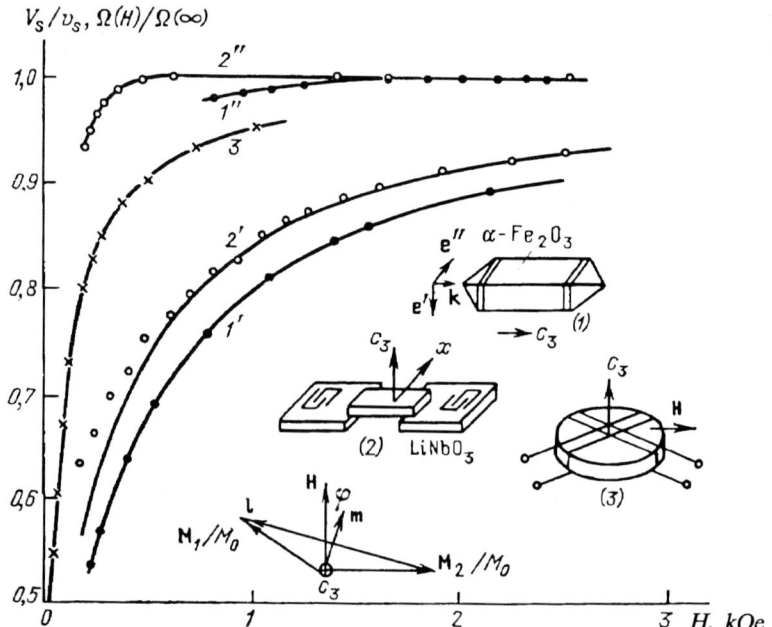

Fig. 2. Dependence of the normalized velocities (curves 1 and 2) and acoustic-resonance frequencies (curve 3) on the magnetic field intensity in α-Fe$_2$O$_4$. 1 – volume transverse waves (1 – $e' \parallel x \parallel H$, 1" – $e'' \perp x \parallel H$ [15]); 2 – surface waves (2' – $H, x = \pi/4$, 2" – $H \parallel x$ [60]; 3 – resonance of the shear mode over the contour of the thin disk. Dots – experiment [31], lines – calculation, insets – geometry of excitation and detection of magnetoelastic waves.

ory using the effective moduli $\hat{C}^{(2)}_{\text{eff}}(k \to 0, H)$ are presented at the same time. The differing symmetries of the tensors $\hat{C}^{(2)}$ and $\hat{C}^{(2)}_{\text{eff}}$ lead to removal of the degeneracy of the spectrum of transverse waves propagating along the trigonal axis of rhombohedral EPAFs. One of the normal modes is characterized by strong coupling, whereas the other one does not interact linearly with the spin system (see the curves 1a and b in fig. 2).

The interest in the properties of the contour-shear mode arises from the fact that, according to the calculations, specifically it satisfies the criterion of limiting strong coupling. The field dependence of its frequency is described by a relationship analogous to that derived in refs. [4,6] for running waves:

$$\Omega(H) = \Omega(\infty)\left(1 - \frac{2H_E H'_{mS}}{\tilde{\omega}^2_{f0}\gamma^{-2}}\right)^{1/2}.$$

Here the frequency

$$\tilde{\omega}_{f0} = \gamma[H(H + H_D) + 2H_E H'_{mS}]^{1/2}$$

differs from the experimentally measured antiferromagnetic resonance frequency by the amount of the magnetoelastic "gap". The data presented in fig. 2 imply that the magnetoelastic interaction leads to a variation in the frequencies and velocities of the bulk acoustic waves by a factor of practically two (i.e., fourfold for the corresponding dynamical moduli), while the variation in the velocities of surface acoustic waves reaches 35%. Such a substantial renormalization of the acoustic parameters is direct experimental proof of the strong coupling of the elastic and spin waves in α-Fe$_2$O$_3$ crystals near SRT. Magnetoelastic waves in FeBO$_3$ possess analogous properties. Similar strongly coupled modes are the fundamental objects of experimental studies in the nonlinear magnetoelastics of high-temperature EPAFs.

The orientation phase transition in EPAFs can be caused by a combined action of a magnetic field and of applied mechanical stresses.

Relatively weak elastic deformations comparable in value with spontaneous magnetostriction induce a magnetoelastic anisotropy in the basis plane of an EPAF. Its energy may exceed the energy of the crystalomagnetic in-plane anisotropy as it takes place in a hematite. This induced anisotropy manifests itself, in particular, in the variation of the antiferromagnetic resonance (AFMR) frequency. When the "hard axis" of the induced anisotropy is oriented in parallel to the applied magnetic field in the basis plane the deformation lowers the AFMR frequency down to a complete compensation of the contribution of the Zeeman interaction to the spectrum activation. The point of compensation is correlated with the reorientation phase transition of the second kind [4]. Further build-up of deformations results in the deviation of the magnetic moment *m* off the field direction and in the corresponding reorientation of the AFM vector *l*. The AFMR frequency begins to increase. Such nonmonotonic dependence of the AFMR frequency on the deformation stresses in α-Fe$_2$O$_3$ was experimentally observed in refs. [30,32] and the corresponding reorientation of vector *l* was studied using the magneto-optical method in ref. [33].

A detailed experimental study of the behavior of an acoustic wave spectrum near SRT when a field and external stresses ($\hat{\sigma}$) are applied to an α-Fe$_2$O$_3$ crystal was first carried out in ref. [34]. On compressing the crystal ($\sigma_{xx} > 0$) along the magnetic field in the basis plane the effective field of the external pressure is directed in antiparallel to the applied magnetic field and compensates the latter on reaching the critical value of $H_{pc} = H(H + H_D)/2H_E$. The coupled magnetoacoustic wave velocity in this case, similarly to the case of $H \to 0$, has its minimal value. The results of measuring the dependence of shear coupled waves propagating along the crystal threefold axis on elastic stresses are plotted in fig. 3 for different values of the magnetizing field. (The experiment geometry is shown in the insert). What draws our attention is a noticeable limitation of the sound velocity in the proximity to the SRT. It should be noted that the acoustic mode under investigation in a rhombohedral EPAF is not critical and the wave velocity should not turn to zero within the limit of $k \to 0$ in the critical point. However the observed limitation of the sound velocity can be ascribed quantitatively neither to the specific elastic and magnetoelastic anisotropy in α-Fe$_2$O$_3$ nor to the sound velocity dispersion in the transition point. The comparison of experimental data with the calculated curve (see fig. 3, curve 1') suggests a substantial difference between them near the transition. In ref. [34] an explanation of the experiment was proposed that the SRT is affected by the defects of the "random field"-type [35]. The thermal fluctuations are known to be insignificant at reorientation transitions, on the contrary the defects distorting the translation symmetry of a crystal seem to play the decisive role in critical behavior at SRT. The correlation radius of spin fluctuations $r_c = v_m / (\omega_{f0}^2 - 2\gamma^2 H_E H_{mS})^{1/2}$ for high-temperature EPAFs turns out to be macroscopically large in a wide range of fields $\Delta H \approx 2 H_E H_{mS}/H_D$ ($\approx 10^3$ Oe for hematite) around the SRT field.

4. Mixed mode anharmonicity near SRT

In the crystal lattice dynamics the weak nonlinearity is described by anharmonic terms in the expansion of the potential energy density F_e of a crystal in a power series in small deformations:

$$F_e = \frac{1}{2!} \hat{C}^{(2)} \hat{u}\hat{u} + \frac{1}{3!} \hat{C}^{(3)} \hat{u}\hat{u}\hat{u} + \cdots .$$

The feasibility of this expansion is justified by usually small displacements of the lattice atoms

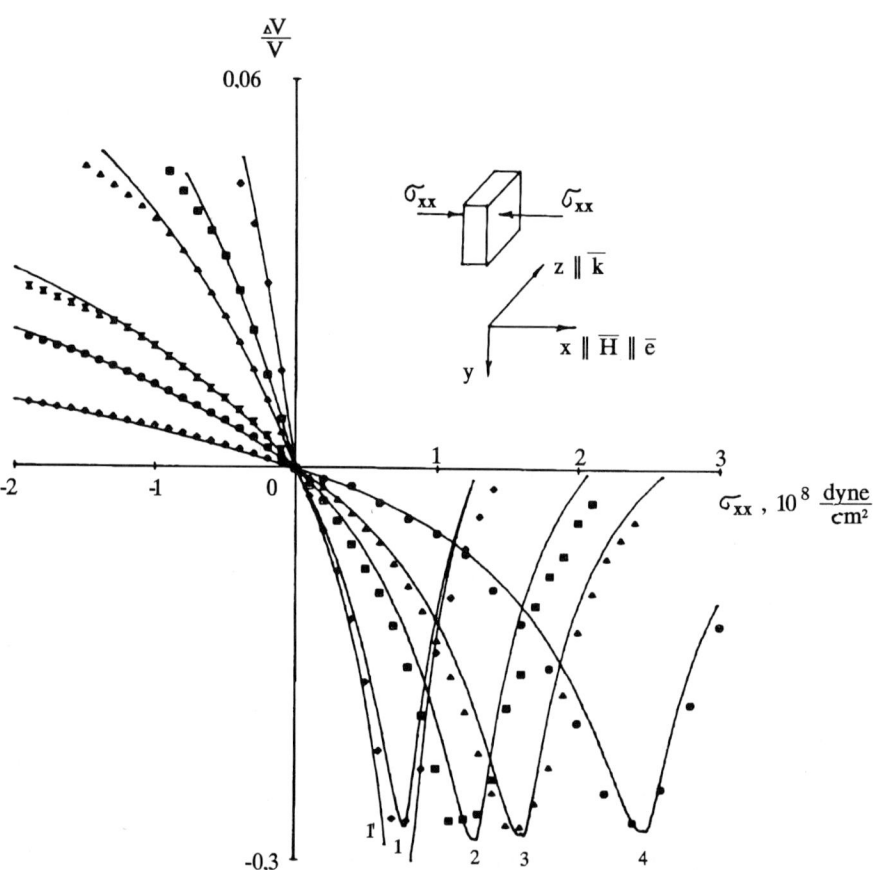

Fig. 3. Relative variation of the magnetoelastic wave velocity vs. elongation-compression stresses at the field H, kOe: 1 – 0.5; 2 – 0.8; 3 – 1.0; 4 – 1.5, Solid lines – calculation with allowance for the influence of the "random field"-type defects [34], 1' – calculation without allowance for defects.

as compared to interatomic distances: $|\hat{u}| \ll 1$, and by relatively weak variations of the magnitude of anharmonic moduli with the increase of their order [36]: $C^{(n+1)}/C^{(n)} \approx 1$–$10$. This ratio of constants is in agreement with widely spread ideas of the mechanism of interionic interactions in crystals with different types of chemical bond. In the acoustic experiment conditions the deformations are, as a rule, sufficiently small for the nonlinearity of elastic waves to be considered weak ($|\hat{C}^{(3)}\hat{u}|/|\hat{C}^{(2)}| \ll 1$).

In the lowest order of perturbation theory, the anharmonicity of the acoustic branch excitations arises from the nonlinearity of the magnetoelastic interaction, as well as from pure elastic nonlinearity [6]. The expansion of the magnetoelastic energy density of (9) in a power series in the amplitudes φ of the spin oscillations contains the anharmonic term

$$F_{me}^{(3)} = -2\hat{B}_1 \hat{u}\varphi^2. \tag{20}$$

Eq. (20) describes the processes of interaction of one sound wave and two spin waves. Upon taking account of the spin oscillations of (15) that accompany the sound wave, we can easily pick out the contribution to the energy density of the acoustic excitation, which is proportional to the cube of the deformation. One can correlate this contribution with the tensor of the anharmonic elastic moduli $\Delta \hat{C}^{(3)}$ [6]:

$$\Delta w^{(3)} = \frac{1}{3!}\Delta\hat{C}^{(3)}\hat{u}\hat{u}\hat{u},$$

$$\Delta\hat{C}^{(3)} = -6\left(\frac{H_E}{M_0}\right)^2 \frac{(2\hat{B}_1)(2\hat{B}_2)^2}{(\omega_{f0}/\gamma)^4}. \tag{21}$$

For simplicity we have restricted the treatment to the wavelength region of the spectrum, i.e., $\omega \ll \omega_{f0} \approx \omega_{fk}$. Using the characteristic parameters $B \approx 10^7$ erg/cm^3, $H_E/M_0 \approx 10^4$, and $(\omega_{f0}/\gamma)^2 \approx 10^7$ erg/cm^3, we obtain for α-Fe$_2$O$_3$ at $H \approx 1$ kOe the estimate presented above of $\Delta C^{(3)} \approx 10^4 C^{(2)} \approx 10^{16}$ erg/cm^3.

Eq. (21) describes the processes of interaction of three acoustic (soundlike) waves. A number of experimentally observable nonlinear acoustic phenomena requires account to be taken of a higher-order effective nonlinearity to describe them. The magnetic and magnetoelastic energies of the crystal contain anharmonic terms of all orders in the amplitudes of the magnetoelastic excitations. The effective fourth-order elastic moduli are formed by three fundamental mechanisms: four-wave interaction of non-resonance-excited spin waves, magnetoelastic interaction with participation of sound wave and three non-resonance spin waves, and the interaction of (20) in second-order perturbation theory. The first two mechanisms corresponds to the following terms in the energy density:

$$w^{(4)} = -\frac{M_0}{2H_E}\left[\frac{1}{3}\left(\frac{\omega_{f0}}{\gamma}\right)^2 - \tfrac{1}{4}HH_D\right]\varphi^4$$
$$-\tfrac{4}{3}\hat{B}_2\hat{u}\varphi^3. \tag{22}$$

Taking into account all the contributions cited above, the anharmonic terms of the expansion of the energy density of interest to us are reduced to the standard form [12,37]:

$$\Delta w^{(4)} = \frac{1}{4!}\Delta \hat{C}^{(4)}\hat{u}\hat{u}\hat{u}\hat{u}.$$

Here we have

$$\Delta \hat{C}^{(4)} = 12\left(\frac{H_E}{M_0}\right)^3 \frac{(2\hat{B}_2)^4}{(\omega_{f0}/\gamma)^6}\left(1 + \frac{1}{4}\frac{\gamma^2 HH_D}{\omega_{f0}}\right)$$
$$- 48\left(\frac{H_E}{M_0}\right)^3 \frac{(2\hat{B}_2)^2(2\hat{B}_1)^2}{(\omega_{f0}/\gamma)^6}. \tag{23}$$

For α-Fe$_2$O$_3$ with $H \approx 0.5$ kOe we have $\Delta C^{(4)} \approx 10^{20}$ erg/cm^3.

A steep increase of effective moduli with the growth of their order necessitates the discussion of the problem of the anharmonic expansion parameter. It is easy to notice that it is $\chi \equiv \zeta^2 |u/u_0|$ (where u is the dynamic deformation amplitude and u_0 is the spontaneous magnetostriction deformation) that plays the role of this parameter.

As it has already been mentioned, a distinctive feature of the behaviour of the coupling coefficient of a critical mode near SRT is its tendency to the limit $\zeta^2 \to 1$ which is independent of the crystal constants and environment (fields, pressure, temperature, etc.). As a result, near SRT the anharmonic expansion parameter turns out to be not the amplitude of dynamic deformation proper but its ratio to the magnitude of spontaneous striction. Thus, at the reorientation point the magnetostriction-introduced elastic nonlinearity increases with the decrease of the magnetostriction constants of a crystal (it is true only if $H_{mS} > H_a$, where H_a is the in-plane anisotropy field). This, on the face of it, paradoxal result conforms to experimental data which demonstrate a much stronger effective acoustic nonlinearity near "in-plane" SRT of α-Fe_2O_3 crystals with a conventional weak striction ($u_0 \approx 10^{-5}$) in comparison with recently investigated [38] effective nonlinearity of the rare-earth compound $Tb_{0.3}Dy_{0.7}Fe_2$ with the compensated magnetic anisotropy and the giant magnetostriction ($u_0 \approx 10^{-3}$) [39]. The peculiarity under consideration, i.e. the increase of nonlinearity of activationless (hydrodynamic) modes near the critical point with the decrease of the constants of their interaction with the order parameter, is, in all probability, of general character for the second kind phase transitions of the "order–order" type.

5. Experimental nonlinear magnetoacoustics near SRT

The anomalously large magnitude, specific symmetry and strong dependence of the effective elastic moduli on the magnitude and direction of the magnetic field facilitate the experimental detection and identification of the magnetoelastic machanisms of many nonlinear acoustic processes.

Direct experimental measurements [40] of the effective third-order anharmonic moduli have been performed on synthetic single crystals α-Fe_2O_3 of dimensions 5×4 mm^2 in the basis plane and 13 mm in length along the C_3 axis. The quality of the crystal was sufficiently high: for the soundlike wave being studied having $\alpha_k \approx 600$ cm^{-1}, the quality factor $Q_k = \omega_k / \Delta\omega_k = k/\alpha_k \approx 3 \times 10^3$ for a field $H \approx 3$ kOe which practically "turns off" the magnetoelastic coupling (α_k is the decay coefficient for the intensity of the wave). The nonlinear moduli were determined from the change in the velocities of the acoustic waves under a relatively weak static deformation of the crystal. We note that the anomalously high tensosensitivity of the velocities of sound in EPAFs had been prediced also in ref. [30b] – on the basis of analysis of the linear effects of static elastic stresses on the antiferromagnetic resonance frequency and the magnetoelastic coupling. In ref. [40] the geometry of experiment was chosen taking into account the strong magnetoelastic anisotropy of hematite, owing to which the anharmonic moduli ΔC_{455} and ΔC_{155} have the largest magnitude (with H parallel to the twofold axis $U_2 \| x$). In determining these components of the tensor $\Delta C^{(3)}$, one can use a transverse wave with polarization $e \| x$ and wave vector $k \| z$. Here one must create static stresses of two types that are homogeneous throughout the crystal: tensile (σ_{yy}) and shear (σ_{yz}). An acoustic wave at the frequency 30 MHz was excited and detected with piezotransducers. The variation of the velocity of this transverse wave δv_t upon deforming the crystal was measured from the change in the phase of

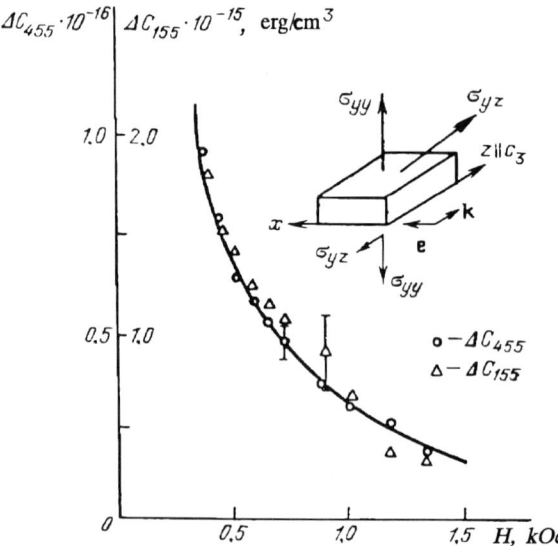

Fig. 4. Dependences of the effective third-order elastic moduli on the magnetic field intensity (solid line – calculation by eqs. (21)) [40].

the signal in the receiving transducer. The results were processed by using the relationships

$$\Delta C_{155} = \tfrac{1}{2}\bigl[C_{44}(C_{11} - C_{12})S_{\|}(H) + C_{14}C_{44}S_{\perp}(H)\bigr], \qquad (24)$$
$$\Delta C_{455} = \tfrac{1}{2}\bigl[C_{44}^{2}S_{\perp}(H) + 2C_{14}C_{44}S_{\|}(H)\bigr].$$

Here we have

$$S_{\|} = \frac{\rho(\delta v_{t}^{2})_{\|}}{C_{44}\sigma_{yy}}, \quad S_{\perp} = \frac{\rho(\delta v_{t}^{2})_{\perp}}{C_{44}\sigma_{yz}}.$$

The subscripts $\|$ and $\perp$ corresponds to tensile and shear deformations. The results of the measurements and of calculation of the modulus $\Delta C_{455}(H)$ are presented in fig. 4.

The use of static deformation for determining the dynamical fourth-order elastic moduli (e.g., from its influence on frequency doubling of sound) involves a fundamental difficulty. The static deformations that cause the change in the velocity of the magnetoelastic wave owing to fourth-order anharmonicity alter the direction of the equilibrium magnetization. Since the magnetostrictive field that determines the gap in the spectrum of spin waves does no hinder static remagnetization, in contrast to high-frequency remagnetization, the deviation of the magnetic moments (and the

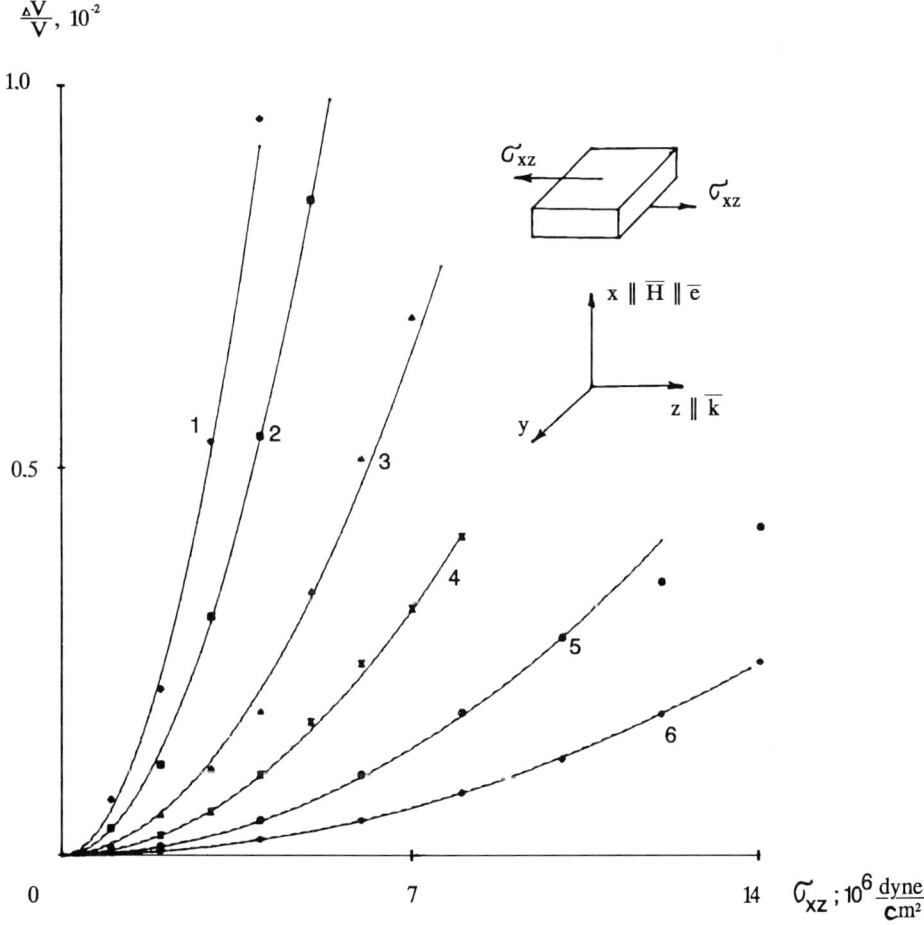

Fig. 5. Dependence of the relative variation of the magnetoelastic wave velocity on shear stresses [34] H, kOe: 1 – 0.5; 2 – 0.7; 3 – 1.0; 4 – 1.3; 5 – 1.7; 6 – 2.5.

changes in the velocity of sound associated with them) in static and dynamic deformation can differ substantially. Accordingly the need arises of distinguishing the dynamic and quasistatic moduli. As $H \to 0$ and near orientational phase transitions, the latter diverge proportionally to $1/(\omega_{f0}^2 - 2\gamma^2 H_E H_{mS})^n$, where n is the order of the expansion of the energy density in the components of the static deformation tensor that are associated with the change in the direction of the vector l in the linear approximation.

Measurements of the quasistatic fourth-order moduli in α-Fe$_2$O$_3$ using static deformations were performed in ref. [34]. Fig. 5 shows the measurement results of the mixed shear acoustic mode velocity, propagating along the threefold axis (z) of a crystal, vs. the magnitude of shear stresses σ_{xz}. In this figure the result of the square approximation of experimental dependences is plotted in solid lines. While processing the experimental data presented in the figure, an allowance was made for the fact that shear stresses σ_{zx} (for $H \parallel x$) give rise to the variation not only of velocities but also of polarizations of transversal waves propagating in the direction of the threefold axis. Rotation of polarization vectors of normal waves results from the rotation of vector l in the basis plane under the σ_{xz} stresses, the effective field of anisotropy of the latter being oriented at an angle of $\pi/4$ to the x-axis in the basis plane. Transverse wave polarization degeneracy peculiar to rhombohedral crystals in the basis plane is elimi-

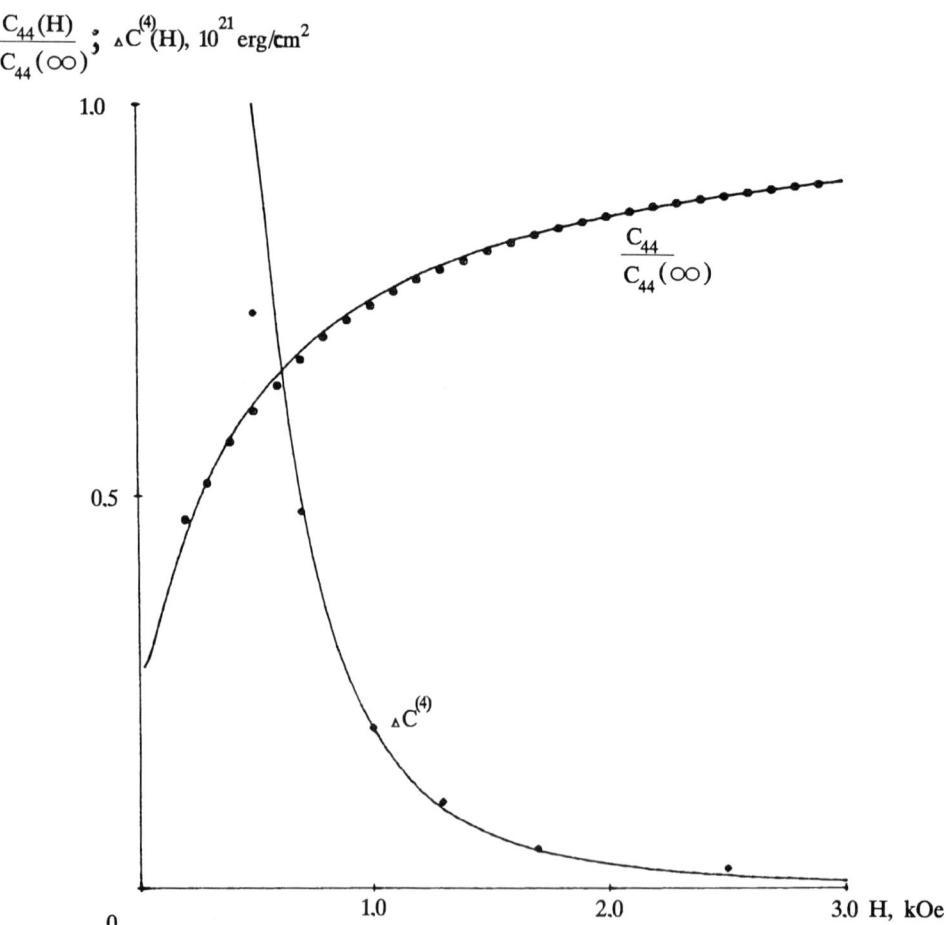

Fig. 6. Second- and fourth-order elasticity moduli vs. magnetic field intensity (dots – experiment, lines – calculation) [34].

nated by the magnetoelastic interaction. A linear magnetoelastic coupled and incoupled waves have different velocities and their polarizations are directed at angles $2\varphi_0$ and $(2\varphi_0 + \pi/2)$, respectively, to the twofold axis (x), where φ_0 is the angle of the vector l deviation from axis y. The polarization rotation appears in the sound propagation description in terms of effective elasticity moduli because of the existence of nondiagonal (in variable deformation indexes) components of quasistatic fourth-order moduli $C^{(3)}_{(5)45}$ where indexes in brackets relate to the static component of deformation, u^σ_{xz}. Allowing for their contribution to the renormalization of wave velocities in the second order of perturbation theory it can be proved that it is the combination of effective moduli $\Delta C^{(4)}_{(55)44} + \Delta C^{(4)}_{(55)55}$ which is responsible for the transverse wave velocity variation ($\Delta v_{\rm t}$) following the square law in relation to static deformations u^σ_{xz}. Here we have

$$\rho(\Delta v_{\rm t}^2) = 2\left[\Delta C^{(4)}_{(55)44} + \Delta C^{(4)}_{(55)55}\right](u^\sigma_{xz})^2. \quad (25)$$

For the combination of moduli in eq. (25) the following expression can be written:

$$\Delta C^{(4)}_{(55)55} + \Delta C^{(4)}_{(55)44}$$

$$= \left(\frac{H_{\rm E}}{M_0}\right)^3 \left(\frac{2B_{14}\gamma^2}{\omega_{\rm f0}\omega_H}\right)^2 (7HH_{\rm D} + H^2), \quad (26)$$

where $\omega_H \equiv \gamma\sqrt{H(H + H_{\rm D})}$. The solid line in fig. 6 indicates the result of the calculation of elasticity moduli vs. field dependences using eq. (26). The dots in the same figure show the results of processing the measurement results according to eq. (25). Point of reduction is chosen at the field strength $H = 1$ kOe. It is easy to notice the qualitative conformity between calculated and experimental curves of the field dependence at $H > 0.5$ kOe. The quantitative calculation of moduli in the point of reduction with allowance for independently determined magnetoelastic parameters of a crystal provides the value very close to that in fig. 6.

These results verify theoretically predicted anomalous rise of nonlinear moduli with the growth of the nonlinearity order as well as a giant value (10^{20}–10^{21} erg/cm^2) and strong field sensitivity of effective fourth-order elasticity moduli of EPAFs near SRT.

Measurements of the field dependence of the dynamic moduli ΔC_{5555} and ΔC_{6666} have been performed while using the effect of nonlinear frequency shift (NFS) of the magnetoelastic oscillations of a thin plate. In the experiment the acoustic vibrations of a monocrystalline resonator in the form of a disk 0.35 mm thick and 5.5 mm in diameter cut in the basis plane were studied. The NFS was measured from the distortion of the shape of the resonance line upon increasing the

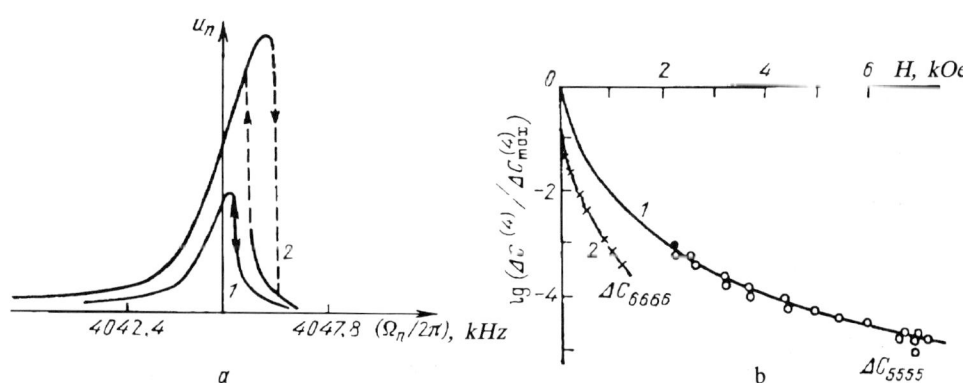

Fig. 7. Nonlinear frequency shift and hysteresis of the amplitude–frequency characteristics of acoustic resonance of the thickness-shear mode observed for different values of the amplitude of the alternating field ($h_2/h_1 = 2$, $H = 2$ kOe). (b) Field dependences of the effective fourth-order elastic moduli (curves – calculation by eqs. (27)) [37].

amplitude of the vibrations (a recording of a characteristic shape of an acoustic resonance line is shown in fig. 7a). The magnitude of the NFS of the acoustic vibrations is proportional to the fourth-order elastic moduli: ΔC_{5555} for the thickness shear mode and ΔC_{6666} for the contour shear mode. Fig. 7b compares the results of measurements and calculations of the field dependence of the parameters $\Delta C^{(4)}(H)$. The calculations were performed by using the relationships

$$\Delta C_{5555} = 12 \left(\frac{H_E}{M_0} \right)^3 \frac{(2B_{14})^4}{(\omega_{f0}/\gamma)^6} \left(1 + \frac{\gamma^2 HH_D}{4\omega_{f0}^2} \right), \quad (27a)$$

$$\Delta C_{6666} = \frac{3}{2} \frac{\left[(C_{11} - C_{12})C_{44} - 2C_{14}^2 \right]^2}{M_0 H'_{ms} C_{44}^2}$$

$$\times \left(\frac{2H_E H'_{ms}\gamma^2}{\tilde{\omega}_{f0}^2} \right)^3 \left(1 + \frac{\gamma^2 HH_D}{4\tilde{\omega}_{f0}^2} \right). \quad (27b)$$

In agreement with the theoretical ideas on the features of the effective elastic anharmonicity, the experiment performed on hematite demonstrates the anomalously large magnitude and strong field dependence of the anharmonic elastic moduli of EPAFs.

One of the first confirmations of the concept of effective elastic anharmonicity was the strong field dependence discovered in α-Fe$_2$O$_3$ of the amplitude of the second acoustic harmonic generated by a running elastic wave [41]. A wave at the frequency of 37 MHz was excited at one end of the specimen. At the opposite end the acoustic signal was received with a resonance piezotransducer tuned to the doubled frequency. Fig. 8 shows the results of measuring the dependence of the power of the second-harmonic signal on the power of the pump wave and the field dependence of the efficiency of conversion, which is proportional to $\omega_{f0}^{-8}(H)$, in agreement with the theory. The decline in the efficiency of conversion with decreasing field in weak fields is explained by the increase in damping of the acoustic waves as the homogeneity of the magnetization distribution over the volume of the specimen breaks down – primarily owing to crystal-structure defects, since the disorientation of the magnetization near a defect increases with decreasing field.

At the same time an effect was discovered in hematite of acoustic detection by running waves.

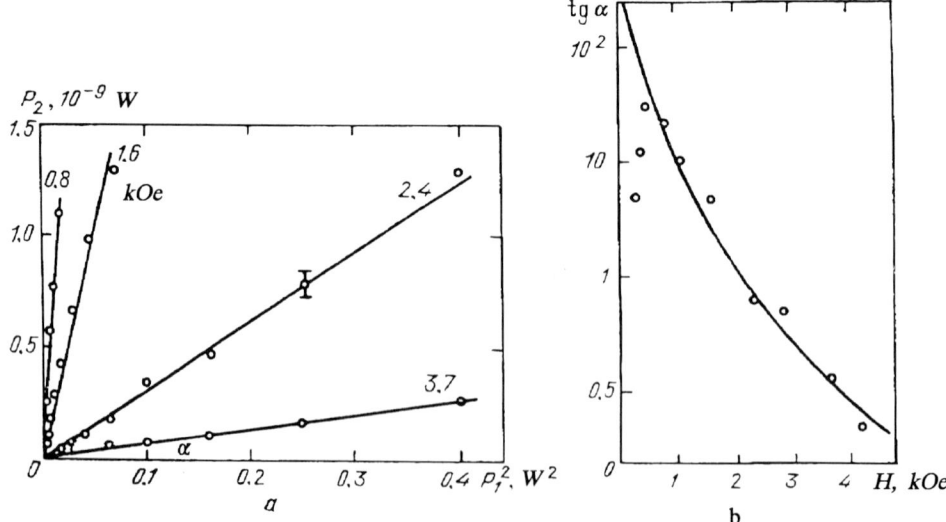

Fig. 8. (a) Dependence of the power of the second acoustic harmonic on the square of the power of the pump wave in α-Fe$_2$O$_3$. (b) Field dependence of the efficiency of conversion of sound into the second harmonic (line – calculation) [41].

The essence of the effect consists in generation of a sound wave at the frequency of the envelope of the amplitude-modulated acoustic signal. The field dependence of the power of the detected signal that was found was also proportional to $\omega_{f0}^{-8}(H)$. Generation of the second acoustic harmonic of a surface magnetoelastic wave was found later – also for α-Fe$_2$O$_3$ [42].

A convincing demonstration of the strong acoustic nonlinearity of hematite was the observation of the effect predicted in ref. [6] of stimulated combination scattering (SCS) of running acoustic waves [14]. Upon propagation of a "pure" transverse pump wave of frequency ω_p and wave vector k parallel to the trigonal axis, a threshold process was observed of generation of backward-running magnetoelastic waves at the combination frequencies ω_1 and ω_2, namely such that $\omega_1 + \omega_2 = \omega_p$.

The conditions for space–time synchronization for this process are illustrated in fig. 9a. Since the velocities of magnetoelastic waves (with polarization $e_{1,2} \parallel H \parallel x$) substantially depend on the field intensity (see fig. 2), the combination frequencies corresponding to the synchronization condition also depend on the field. Fig. 9b shows the data of the measurements and the results of calculation of the field dependences of the generation frequencies. Calculation [6] of the magnitude of the threshold deformation within the framework of the theory of effective anharmonicity for the process being discussed yields the following relationships:

$$u_p^c = \pi \frac{2|B_{14}|}{C_{44}} \frac{(1-\zeta^2)^{3/2}}{k_p L \zeta^5}, \qquad (28)$$

$$\zeta^2 \equiv \frac{H_E}{M_0} \frac{(2B_{14}\gamma)^2}{\omega_{f0}^2 C_{44}}.$$

Here L is the length of the specimen.

A calculated estimate of the threshold deformation $(u_p^c) \approx (3.5 \pm 1.5) \times 10^{-7}$ for $H \approx 0.5$ kOe agrees in order of magnitude with the result of measurement of $(u_p^c) \approx (8 \pm 4) \times 10^{-7}$. In the same geometry and also for α-Fe$_2$O$_3$ under syn-

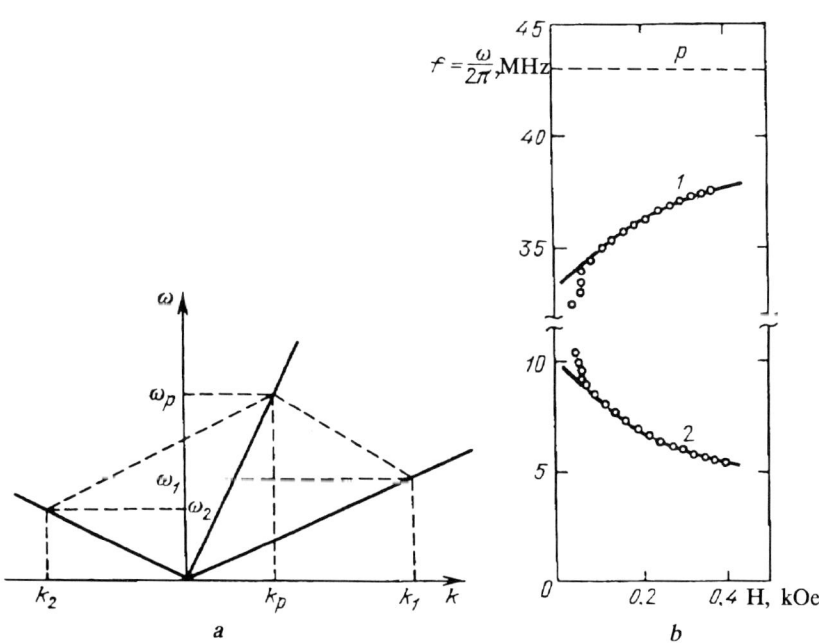

Fig. 9. (a) Diagram of the conditions of synchronization for acoustic stimulated combination scattering [6]. Field dependence of the combination frequencies of transverse magnetoelastic waves in α-Fe$_2$O$_4$ (p – ω_p, 1 – ω_1, 2 – ω_2, lines – calculation) [14].

chronization conditions, a threshold-free process conditions, a threshold-free process was observed of merger of magnetoelastic waves into a running pure-sound wave that was the reverse of the SCS [44].

When a high-frequency magnetic field $h(t)$ *homogeneous* throughout the specimen acts on a crystal of α-Fe$_2$O$_3$, a number of parametric acoustic phenomena can occur [45]. The parametric coupling of sound with an alternating field, just like the effective interactions of elastic waves, is mediated by the excitation of the spins. In an alternating field parallel to the magnetizing field (the so-called parallel pumping ($h_\parallel \ll H$), the following term in the energy density is responsible for the parametric coupling:

$$w_{p\parallel} = \frac{M_0}{2H_E}(2H + H_D)h_\parallel \varphi^2. \quad (29)$$

Upon substituting into this the amplitudes of the nonresonance spin oscillations of (15), we obtain the following expression for the coupling of the acoustic waves with the field:

$$w_{p\parallel} = \left(\frac{2H_E}{M_0}\right)\frac{2H + H_D}{(\omega_{f0}/\gamma)^4}(\hat{B}_2 \hat{u})^2 h_\parallel. \quad (30)$$

The magnetoelastic interaction of (20) also determines the coupling of the sound with the transverse pump field $h_\perp(t)$. Taking into account the nonresonance excitations of the spin system caused by the transverse field, e.g. ($\varphi_h = (H + H_D)H_\perp(\omega_{f0}/\gamma)^{-2}$, and by the sound (15), we find

$$w_{p\perp} = 8\frac{H_E}{M_0}(H + H_D)(\omega_{f0}/\gamma)^{-4}$$
$$\times (\hat{B}_1 \hat{u})(\hat{B}_2 \hat{u})h_\perp. \quad (31)$$

The parametric interactions (30) and (31) allow a graphic physical interpretation. The effective elastic moduli $\Delta C^{(2)}(H)$ (and this implies also the frequencies of the acoustic spectrum and velocities of sound) depend on the magnitude and direction of the external magnetic field. Modulation of the external field in magnitude or in direction leads to modulation of the elastic pa-

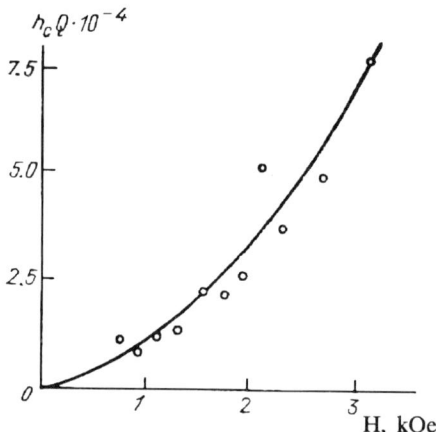

Fig. 10. Dependence of the amplitude of the reduced threshold field on the constant magnetic bias field; the reduction point is $H = 1.6$ kOe [45].

rameters of the crystal and parametric couplings with the energy $w_p = w_{p\parallel} + w_{p\perp}$, where

$$w_p = -\left(\frac{\partial \Delta \hat{C}^{(2)}}{\partial H}\right)h \frac{\hat{u}\hat{u}}{2}. \quad (32)$$

When the amplitude of the ac field with the frequence ω exceeds a critical value, e.g., $h_\parallel > h_c = Q^{-1}\Omega/(\partial\Omega/\partial H)$ (Q is the Q-factor of the quasisound mode), a parametric instability arises in α-Fe$_2$O$_3$ crystals of the magnetoelastic acoustic modes with the frequency $\Omega = \omega/2$. An analogous effect has been observed earlier in the ferrite garnet Eu$_3$Fe$_5$O$_{12}$ [46]. Under the conditions of the experiment the amplitude of the parametric interactions characterized by the quantity $\partial \Delta C^{(2)}/\partial H$ is substantially larger for α-Fe$_2$O$_3$ than for the ferrite. Moreover, in hematite the effect is observed over a considerably broader range of magnetic fields.

Fig. 10 shows the field dependence of the threshold field for the thickness shear mode of a disc acoustic resonator made of α-Fe$_2$O$_3$ (the plane of the disk is parallel to the basis plane). The pumping and recording of the instability were performed by an induction method. The line in fig. 10 shows the result of calculation of the field dependence of h_c from the data of independent measurement of $Q(H)$ and $\Omega(H) = \Omega(\infty)(1 - \zeta^2(H))^{1/2}$ of the studied specimen. For

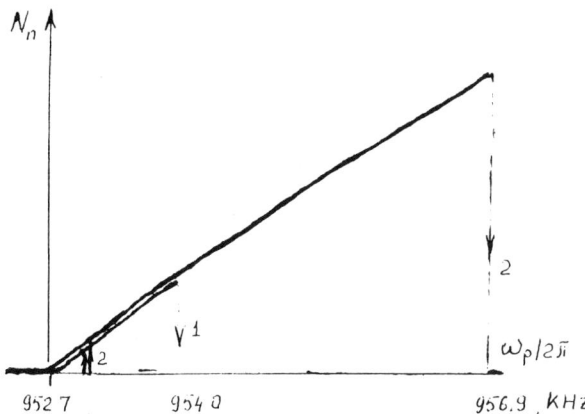

Fig. 11. Dependence of the amplitude of parametric magnetoelastic vibrations on the pump frequency for different degrees of supercriticality $h_p/h_{pc} = 1.25$ (1); 2 (2).

the characteristic values $Q \approx 10^4$, $H \approx 0.5$ kOe, and $\Omega/(\partial\Omega/\partial H) \approx 0.2$ kOe, the threshold amplitude has the magnitude $h_c = 0.5$ Oe.

When the resonance conditions are satisfied and with an intense enough external agent (pump), one observes the so-called nondegenerate threshold processes of parametric generation of magnetoelastic modes – at the combination frequencies $\Omega_n + \Omega_m = \omega_p$ [45]).

Past the threshold of parametric excitation, the effects of nonlinear-self-action of acoustic modes manifest themselves. In particular they are found from the difference of the values of the frequency detuning ($\Delta\omega = \omega_p - 2\Omega_n$) with respect to the parametric resonance at which sound generation arises and disappears. A characteristic recording of the hysteresis dependence of the amplitude of a parametrically excited oscillation on the detuning is shown in fig. 11.

The states with detuning $\Delta\omega_n > Q_n^{-1}[(h_p^2/h_c^2) - 1]^{1/2}$ are bistable. Small fluctuations of the quasisound intensity relax towards to stationary metastable state while oscillating with the characteristic frequency (Ω_M) depend on the stationary

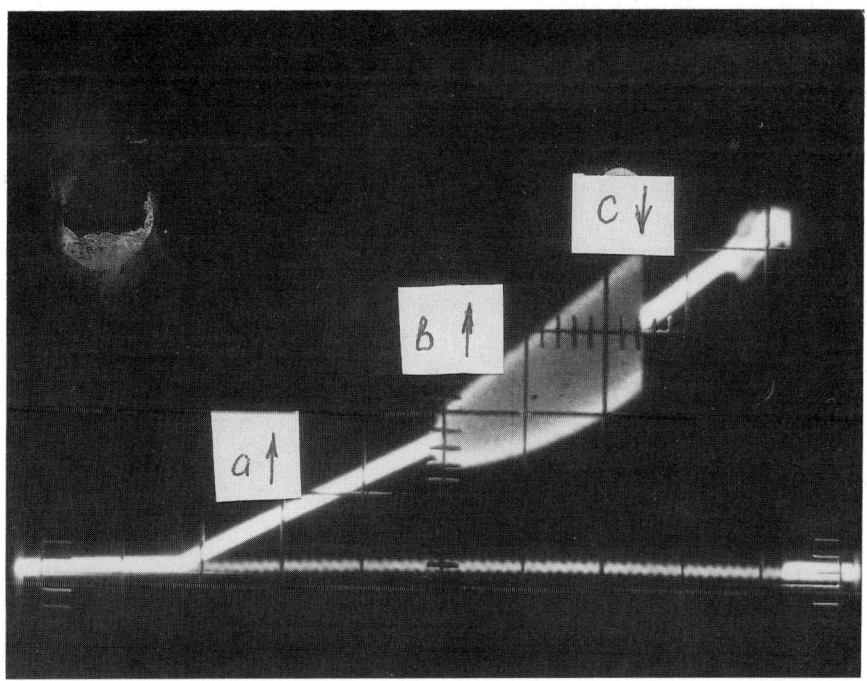

Fig. 12. Oscillogram of sound intensity dependence on pumping frequency ω_p. Line widening corresponds to "collective oscillations". Arrow a indicates the frequency at which the "primary" parametric quasi-sound vibrations arise while ω_p decreases; arrow b – the same for collective oscillations. Arrows c and d show the quenching points for primary vibrations and collective oscillations while ω_p increases.

state parameters. These oscillations can be excited by the parametric parallel low frequency (LF) pumping $h_M \cos \omega_M t \parallel H$ that modulates the quasisound spectrum with frequency $\omega_M \approx 2\Omega_M$. The effects observed here [24] are to a certain extent similar to earlier investigated processes of a double parametric resonance under collective oscillation excitations of parametric magnons in ferrites and antiferromagnets [47,48]. The most interesting result of the acoustic experiment [24] proved to be the discovered bistability of oscillations of the quasi-sound intensity against the background of the bistability of an over-threshold stationary state (a double over threshold bistability). Fig. 12 shows an oscillogram of an observed dependence of the quasisound intensity (associated with the critical contour-shift mode with frequency $\omega_n/2\pi = 0.5$ MHz) on the frequency detuning of HF pumping $\Delta\omega_n = \omega_p - 2\omega_n$ under LF modulation $\omega_M \approx 1$ kHz. Beam trace spreading corresponds to the intensity oscillations. The oscillation bistability can be clearly seen in the dependence of their intensity on the modulation frequency (fig. 13). The double bistability effect can be explained in terms of the theory of the parametric excitation of quasisound modes with allowance for the effective elastic fourth-order anharmonicity which is responsible for the nonlinear frequency shift of the acoustic mode. Complex amplitude b_n of a parametric quasi-sound is described by equation

$$\dot{b}_n + \tfrac{1}{2}\omega_n Q_n^{-1} b_n + \tfrac{1}{2}i\,\Delta\omega_n b_n - i\psi_n |b_n|^2 b_n$$
$$= ih_p V_n b_n^*, \qquad (33)$$

where $V_n \equiv \tfrac{1}{2}\partial\omega_n/\partial H$ is the parametric coupling amplitude, ψ_n is an amplitude of four quasi-phonon interaction and Q_n is the Q-factor of a mode. A stationary overthreshold state possesses the following normalized sound intensity $N_n \equiv 2\psi_n |b_n|^2/\omega_n$:

$$N_n = \frac{\Delta\omega_n}{\omega_n} + Q_n^{-1}\left(\frac{h_p^2}{h_{pc}^2} - 1\right)^{1/2}. \qquad (34)$$

The linear dependence of intensity on detuning

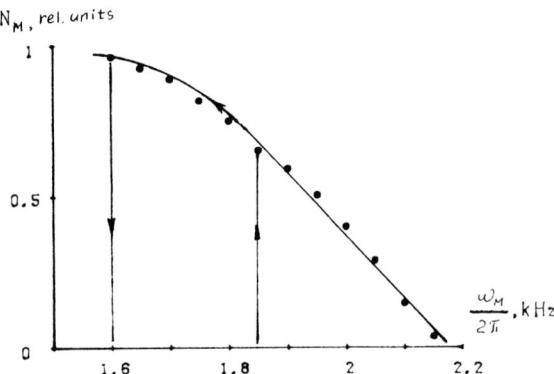

Fig. 13. Collective oscillations intensity vs. modulation field frequency. The frequency $\Omega_M/2\pi$ of collective oscillations is 1 kHz.

agrees well with the experiment (fig. 11). The oscillation frequency of small perturbations of the stationary state is deduced from eq. (33) and has the form of the following relationship:

$$\Omega_M = \omega_n\left[N_n\left(N_n - \frac{\Delta\omega_n}{\omega_n}\right) - \tfrac{1}{4}Q_n^{-2}\right]^{1/2}. \qquad (35)$$

As it follows from eq. (33) the parametric quasisound phase $\theta_n = 2\arg(b_n)$ is described by the expression:

$$\ddot{\theta}_n + \omega_n Q_n^{-1}\dot{\theta}_n + \left(\omega_n Q_n^{-1} - 2h_p V_n \sin\theta_n\right)$$
$$\times\left(\Delta\omega_n - 2h_p V_n \cos\theta_n\right) = 0, \qquad (36)$$

where detuning (under a slow modulation of the quasisound spectrum) is equal to

$$\Delta\omega_n = \omega_p - 2\omega_n - 2\frac{\partial\omega_n}{\partial H}h_M \cos\omega_M t.$$

The phase equation is strongly nonlinear and for small deviations from the stationary state it allows the expansion (to an accuracy of cubic members) into phase deviation from the stationary value $\theta_n - \theta_n^0 = d\exp(\tfrac{1}{2}i\omega_M t) + $ c.c. The equation being formed for complex amplitudes $d(t)$ is completely identical to the initial one. Detuning, a relaxation rate, a threshold modulation field, amplitudes of parametric coupling and of nonlinear self-action in this case have the forms of

$$\Delta\omega_M = \omega_M - 2\Omega_M, \quad \tfrac{1}{2}\Omega_M Q_M^{-1} = \tfrac{1}{2}\omega_n Q_n^{-1},$$

$$h_{Mc} = \Omega_M \left\{ \frac{\partial \omega_n}{\partial H} \left[(h_p/h_{pc})^2 - 1 \right]^{1/2} \right\}^{-1},$$

$$V_M = -\frac{\partial \omega_n}{\partial h} \left(\frac{\omega_n}{2\Omega_M} \right) Q_n^{-1} \left[(h_c/h_{pc})^2 - 1 \right]^{1/2}$$

and

$$\psi_M = -\frac{\Omega_M}{N_n} \left\{ \frac{\Delta\omega_n}{4\omega_n} + Q_n^{-1} \left[(h_p/h_{pc})^2 - 1 \right]^{1/2} \right\},$$

respectively. In conformity with the experimental data (fig. 13) the nonlinear shift of oscillation frequency appears to be negative ($\psi_M < 0$) in contrast to the nonlinear shift of the initial quasisound frequency ($\psi_n > 0$). Detunings

$$\Delta\omega_M > \Omega_M Q_M^{-1} \left[(h_M/h_{Mc})^2 - 1 \right]^{1/2} \equiv \Delta_M$$

correlate with metastable excited states. After the quenching of oscillations they restore on increasing the pumping frequency ω_M in the region of absolute instability $|\Delta\omega_M| < \Delta_M$ and this is an explanation of the experimentally observed hysteresis of the oscillation amplitude.

Let us note that the above approach towards the double bistability description can be applied to a newly formed metastable state, etc. It gives good ground to believe that in case of sufficiently high Q-factor and strong nonlinearity it is possible to realize not only double but also cascaded bistabilities near SRT.

A manifestation of self-action and nonlinear intermode interactions proves to be the automodulation of the over-threshold amplitude of deformations observed in α-Fe$_2$O$_3$ in the parametric generation of sound. The effect arises at values of the magnetic field, supercriticality, and frequency detuning defined for the given mode [45].

The parametric effects in running acoustic waves also include the frequency shift discovered in hematite of a wave in a non-steady-state, monotonically varying magnetic field.

Phenomena of parametric amplification and front reversal of running magnetoelastic waves in a high frequency magnetic field were observed in α-Fe$_2$O$_3$ [49].

The experimentally observable phenomena of generation of long-wavelength nonresonance excitations of the spin system in a field of nonlinearly interacting acoustic waves are associated with the interactions of (30), (31) and (32) [50]. The acoustic waves give rise in the crystal to oscillations of the magnetization $\mu(t)$ that depend nonlinearly on the elastic deformations. The quadratic components of the ac magnetizations can be calculated by using (32): $\mu = -\partial w_p/\partial h$. When harmonic sound waves with waves vectors identical in magnitude propagate in opposite directions ($k_2 = -k_1$), spatially homogeneous oscillations of the magnetization are excited in the crystal at the sum frequency. In the case of amplitude-modulated waves, the time envelope of the integral of the magnetization over the length L of the interaction region amounts to the convolution function of the envelopes of the interacting waves:

$$\mu(t) \propto \int u_1(\xi) u_2(\xi - 2vt) \, d\xi.$$

(We assume that L is larger than the spatial extent of the acoustic pulses.) The effect of acoustic convolution was observed in a crystal of α-Fe$_2$O$_3$ [50] upon interaction of transverse bulk waves propagating along the trigonal axis. Sound waves of frequencies ≈ 30 MHz were excited at opposite ends of the crystal by piezotransducers made of LiNbO$_3$. The spatially integrating detection of the signal of magnetization oscillations was performed by an induction method. In agreement with the theoretical concepts of mediated interactions of the type of (30) and (31), intramode interactions of coupled waves ($e_1 \parallel e_2 \parallel x \parallel H$) were observed in the experiment with excitation of a longitudinal component of the magnetization $\mu_\parallel \propto u_{xz}^2$ and intermode interactions of the coupled and the pure sound waves ($e_1 \parallel x \parallel H; e_2 \perp e_1$) with excitation of a transverse component $\mu_\perp \propto u_{xz} u_{yz}$. The efficiency of conversion (for fixed polarizations of the emitters) shows a sharp dependence on the magnitude of the external field and on its direction (see ref. [50]). Here the character of the interaction is described by the discussed mechanisms of the interactions.

Effects of acoustic convolution in α-Fe_2O_3 have been observed also in the nonlinear interaction of surface magnetoelastic waves [51,52]. The internal bilinear factor of the processes, i.e., the ratio of output electromagnetic power to the product of the acoustic powers of the interacting signals, in fields $H \approx 0.5$ kOe amounted to about -30 dB m (-10 dB m corresponds to 0.1 mW at the output for 1 mW at each of the two inputs).

In closing this section we note the experimental results in the field of nonlinear magnetoacoustics obtained with another high-temperature EPAF – $FeBO_3$. Under transverse UHF pumping

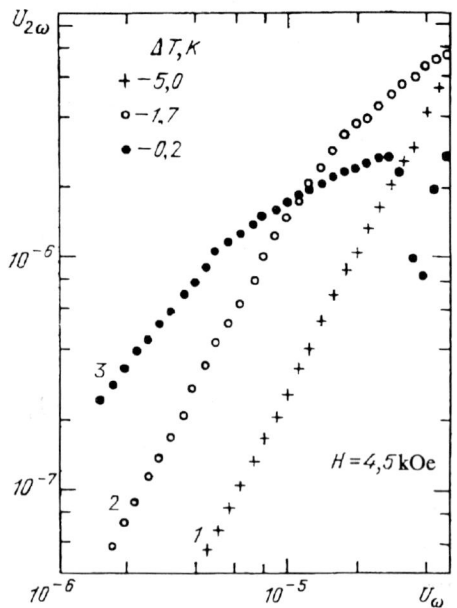

Fig. 15. Dependence of the amplitude of the second acoustic harmonic on the amplitude of the pump wave in $TmFeO_3$ ($\Delta T \equiv T - T_1$).

under conditions of antiferromagnetic resonance in iron borate, parametric instability of sound was observed [53,54]. The effect was treated theoretically in ref. [55]. The excitation of a parametric acoustic echo by a high-frequency magnetic field was discovered in $FeBO_3$ [56]. It was noted in an excellent review on iron borate [57], and this viewpoint cannot but be shared, that $FeBO_3$, just like α-Fe_2O_3, will serve in many regards as an extremely convenient object for studying numerous effects engendered by the very strong dynamic coupling of the elastic and magnetic subsystems of this EPAF.

Experimentally the appearance of strong acoustic nonlinearity near an orientational phase transition (of second-order) has been demonstrated by observing an effect of generation of the second acoustic harmonic in the orthoferrite $TmFeO_3$ [58]. The $TmFeO_3$ crystal has an orthorhombic structure (the b axis is the "hard axis" for the magnetic moments of the sublattices), and the temperature of antiferromagnetic ordering is $T_N = 630$ K. In the absence of an

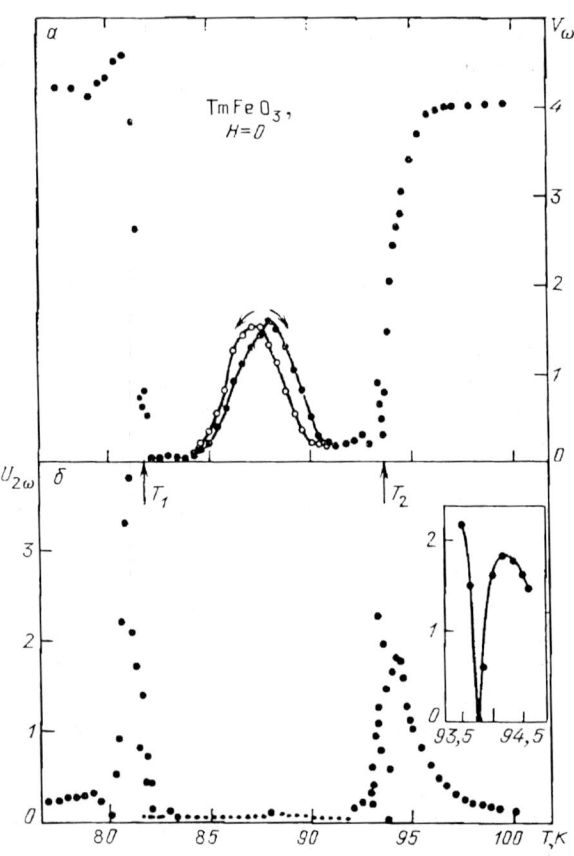

Fig. 14. Temperature dependence of the amplitudes of the first and second harmonics of sounde at the output of the $TmFeO_3$ crystal near the SRT points T_1 and T_2.

external magnetic field, changing temperature leads to reorientation of the spins from the a axis at $T \geq T_1 = 94$ K to the c axis at $T \leq T_2 = 82$ K. At $T_1 < T < T_2$ the vector l rotates smoothly in the ac plane from one axis to the other. The temperatures T_1 and T_2 are second-order orientational-transition points, near which the activation energy of the spin-wave modes is depressed (the mode becomes "soft"), as is confirmed experimentally by measuring the antiferromagnetic resonance frequency. In the vicinity of the temperatures T_1 and T_2 a sharp increase has been observed in the efficiency of generation of the second acoustic harmonic (figs. 14, 15). With increasing power of the pump wave, the output of the second harmonic is saturated and even declines owing to conversion of the energy of the incident wave, not only into the second harmonic, but also into a multitude of higher harmonics.

6. On the quantum theory of nonlinear interactions of coupled excitations

The interaction of elementary excitations determines many thermodynamics and kinetic properties of crystals. As a rule, description of these properties requires one to apply quantumstatistical methods. The coupling of excitations gives shape to the effective nonlinear interactions, whose amplitudes can have an anomalously large magnitude and unusual dispersion properties. The selection rules for such interactions are often controlled by the external conditions, e.g., the orientation of the external field with respect to the crystallographic axes. All this introduces certain specifics into a number of phenomena that arise in crystals in the presence of coupling of subsystems.

The concept of the effective nonlinearity as being the nonlinearity of mixed elementary excitations is the basis for constructing its quantum theory [8–10]. In a second-quantization scheme the transition to a representation of mixed modes is performed by diagonalizing the bilinear component of the Hamiltonian operator taking into account the coupling of the subsystems [59]. This has the form

$$\hat{H}^{(2)} = \sum_{k,s} \left[\omega_{Sk} b^+_{Sk} b_{Sk} + \omega_{fk} c^+_k c_k \right.$$
$$\left. + (G_{Sk} b^+_{Sk} c_k + \text{H.c.}) \right] = \sum_{\lambda=0}^{3} \Omega_{\lambda k} d^+_{\lambda k} d_{\lambda k}. \quad (37)$$

Here b^+_{Sk}, b_{Sk}, c^+_k and c_k are the operators for creation and annihilation of phonons and magnons, and we have $G_{Sk} = \zeta_{Sk}(\omega_{Sk}\omega_{fk})^{1/2}$. The operators for creation and annihilation of mixed excitations $d^+_{\lambda k}$ and $d_{\lambda k}$ are connected to the operators for the "pure" excitations by the unitary transformation:

$$c^+_k \pm c_{-k} = \sum_{\lambda=0}^{3} P_{\lambda k}(d^+_{\lambda k} \pm d_{\lambda,-k}), \quad (38)$$

$$b_{Sk} \pm b^+_{S,-k} = \sum_{\lambda=0}^{3} R_{\lambda k}(d^+_{\lambda k} \pm d_{\lambda,-k}). \quad (39)$$

Here $P_{\lambda k}$ and $R_{\lambda k}$ are the transformation coefficients given in ref. [8]. Far from the intersection point of the "pure" spectra, the coupling of the terms with $\lambda = 1, 2, 3$ with the operators c^+_k and c_{-k} in eq. (38), which is the quantum analog of eq. (15), governs the nonresonance (virtual) excitation of magnons by quasiphonons. Therefore every nonlinear process in the system of magnons is also a source of nonlinearity in the system of quasiphonons. Thus, an interaction of the type of (20) with participation of one phonon and two magnons studied phenomenologically is described by a contribution to the Hamiltonian having the form [8]

$$H^{(3)} = N^{-1/2} \sum_{k,q} \Psi^{(3)}_{m-\text{ph}}(k,q)(b_{pk} + b^+_{p,-k})$$
$$\times (c^+_q + c_{-q})(c^+_{-k-q} + c_{k+q}). \quad (40)$$

When we take account of the transformation (38), (39), this interaction is the source of three-particle processes in the system of quasiphonons:

$$\Delta H^{(3)} = N^{-1/2} \sum_{\substack{k,q \\ l,\lambda,\nu}} \Psi^{(3)}_{l\lambda\nu}(k,q)(d^+_{\lambda k} + d_{\lambda,-k})$$
$$\times (d^+_{\nu q} + d_{\nu,-q})(d_{l(k+q)} + d^+_{l(-k-q)}). \quad (41)$$

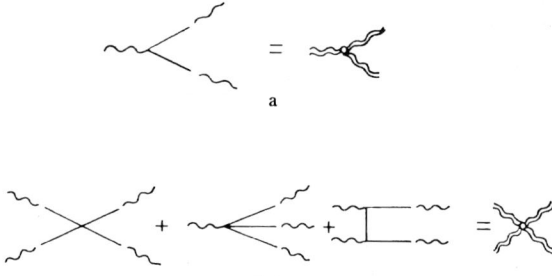

Fig. 16. Diagrams for vertices arising from effective anharmonicity of third (a) and fourth (b) order [8].

Here l, λ and ν are the polarization indices of the quasiphonons, while the amplitude $\Psi^{(3)}_{l\lambda\nu}$ of the interaction is expressed in terms of the parameters of the quasiphonons and the effective elastic moduli, just as in ordinary lattice dynamics:

$$\Psi^{(3)}_{l\lambda\nu}(k,q)$$
$$= -\frac{1}{2^{3/2}} \left(\frac{\Omega_{\lambda k}\Omega_{\nu q}\Omega_{l(k+q)}}{M_v^3 V_\lambda^2 V_\nu^2 V_l^2} \right)^{1/2}$$
$$\times \frac{J_0^2}{\omega_{fk}^2 \omega_{fq}^2} (\hat{\beta}_2 \hat{u}_{\lambda k})(\hat{\beta}_2 \hat{u}_{\nu q})(\hat{\beta}_1 \hat{u}_{l,k+q}). \quad (42)$$

Here J_0 is the exchange energy; $V_\lambda = \Omega_{\lambda k}/k$; $\hat{\beta}_{1,2} = B_{1,2} v_0$; and v_0 and M_v are the volume and mass of the unit cell.

We can make the construction perspicuous by using the graphic representation for the vertex parts of the Feynman diagrams (fig. 16). The diagram of fig. 16a illustrates the interaction (40). The straight and wavy lines correspond to magnons and phonons. We can conveniently treat the transformation to the representation of the coupled waves, with formation of an effective vertex of interaction of quasiphonons (which corresponds to the double line), as the result of joining the phonon and the magnon lines. This merger is put into correspondence with the factor (G_{Sk}/ω_{fk}) in the analytic expression for the vertex with subsequent replacement of the phonon parameters (frequencies and polarizations) by the quasiphonon parameters. The description of any

interaction of quasiphonons mediated by spin interactions is constructed analogously. The construction of the amplitude of the interaction of four quasiphonons corresponding to the effective fourth-order anharmonicity of (23) is illustrated in fig. 16b. A quantum theory of the EPAF acoustic nonlinearity based on the diagram technique for spin operators [61–63] has been developed in refs. [9–10].

A specific spatial dispersion of the interaction amplitudes (i.e., their dependence on k) caused b the dispersion of magnons in characteristic for effective anharmonicity. Dispersion restricts the phase volume of the interacting excitations to the region of small wave vectors $k_m \lesssim \omega_{f0}/v_m$, since phonons with wave vectors of the order of the Debye value practically do not interact linearly with magnons ($\zeta_{k \to k_D} \to 0$). In this regard the effective anharmonicity of the elastic subsystem, which is giant in the long-wavelength region of the spectrum, is weak in the short-wavelength region and hence gives rise to relatively small energy losses of acoustic waves in processes of their scattering by short-wavelength (thermal) excitations.

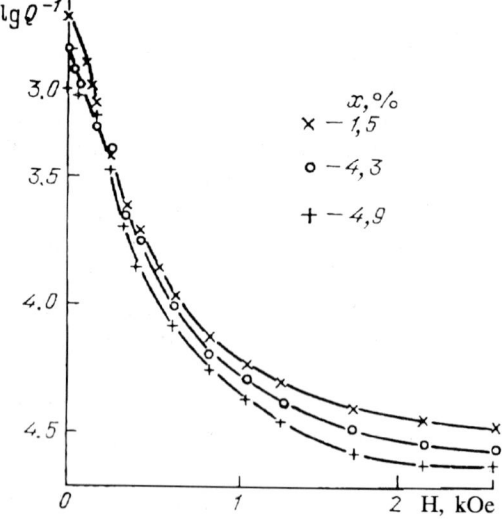

Fig. 17. Field dependence of the damping of magnetoelastic vibrations at the frequency 0.5 MHz in crystals of α-(Fe$_{1-x}$Al$_x$)$_2$O$_3$.

The contributions of effective anharmonicity to the damping γ_k of sound is determined by the relationship [8]

$$\frac{\gamma_k}{\Omega_k} \sim \frac{TT_N \beta^6}{\omega_{f0}^5 (M_v v_S^2)^3}.$$

For hematite an estimate of the acoustic Q-factor $Q \equiv \Omega_k/2\gamma_k$ yields $Q \approx 10^5$–10^6 for $k \ll k_m$. Here it is pertinent to note that the experimentally measurable acoustic losses in crystals of α-Fe$_2$O$_3$ are substantially higher [12] (fig. 17). Apparently this is explained by the defect content of real crystals, which corresponds to the results of comparing the losses upon passage of sound through a specimen of α-Fe$_2$O$_3$ and relaxation times of the sonic field in experiments of reversal of the wave front of sound. We should note that crystals of hematite containing an admixture of Al^{3+}, which is isomorphous to trivalent iron, have a relatively higher Q-factor, owing to the decreased concentration of Fe^{2+} ions usually present in hematite in an amount of about 1%.

7. Outside the anharmonic approximations

The arguments presented in section 1 imply that, for magnetics having near SRT a strong enough magnon–phonon coupling ($\zeta \approx 1$), the region of applicability of anharmonic expansions is limited to elastic deformations close in magnitude to the deformations of spontaneous magnetostriction. Such deformations are usually far from the limits for breakdown of real crystals and are relatively easily attained under the conditions of acoustic experimentation. In this regard it is of interest to study magnetoelastic excitations whose description lies outside the framework of anharmonic approximation and which involves the need for exact solution of strongly nonlinear systems of the dynamic equations (11) and (12). Under certain conditions the nonlinearity and dispersion introduced by the spin system into the magnetoelastic excitations can compensate one another. Consequently the possibility arises of formation of isolated coupled magnetoelastic waves – magnetoacoustic solitons. Such waves as applied to EPAFs were first studied in ref. [16]. Their different modifications problems of stability and the character of evolution of soliton solutions have been studied in refs. [17–21].

We shall analyze the conditions of formation of magnetoacoustic solitons by using eq. (7) with the example of waves propagating in rhombohedral EPAFs along the trigonal axis. If the wave is a steady-state one, i.e., $U = U(\xi)$ and $\varphi = \varphi(\xi)$, where $\xi \equiv z - Vt$, then we can eliminate the dynamic deformations from the system of equations (11), (12) and reduce it to the "double sine-Gordon" steady-state equation:

$$m_p \frac{\partial^2 \varphi}{\partial \xi^2} = A \sin \varphi + \tfrac{1}{2} D \sin 2\varphi. \tag{43}$$

Here we have

$$m_p \equiv \gamma^{-2}(v_m^2 - V^2), \quad A = HH_D,$$

$$D = \gamma^{-2} \left(\omega_{f0}^2 \frac{V_{St}^2 - V^2}{v_{St}^2 - V^2} - HH_D \gamma^2 \right),$$

$V_{St} \equiv v_{St}(1 - \zeta^2)^{1/2}$ is the velocity of the sound-like wave. We can easily establish the intervals of velocities V of waves corresponding to soliton solutions and the qualitative features of the motion of the magnetic moments by using the evident analogy of eq. (32) with the equation of motion of a particle of mass m_p and momentum $p = m_p \partial \varphi / \partial \xi$ in a force field having the potential

$$\tilde{f}(\varphi) = A(\cos \varphi - 1) + \tfrac{1}{4} D (\cos 2\varphi - 1).$$

In crystals of EPAfs having a high Néel temperature (α-Fe$_2$O$_3$, FeBO$_3$), for which $v_m > v_{St}$, the attainable velocities of motion of solitons in the z direction must satisfy the conditions: $V^2 < V_{St}^2$ or $v_m^2 > V^2 > v_{St}^2$. In these cases the effective potential has the characteristic form shown in fig. 18a. An isolated wave on the background of the equilibrium state of the crystal ($\varphi \equiv 0$) corresponds to motion of the "particle" in the potential well $F(\varphi)$ from the point ($\varphi|_{\xi \to -\infty} = 0$ for $p|_{\xi \to -\infty} = 0$) to the point ($\varphi|_{\xi \to +\infty} = 2\pi$ for $p|_{\xi \to +\infty} = 0$), i.e., continuous rotation of the magnetic moment in the wave by the angle 2π. A solution of the given type amounts to the well

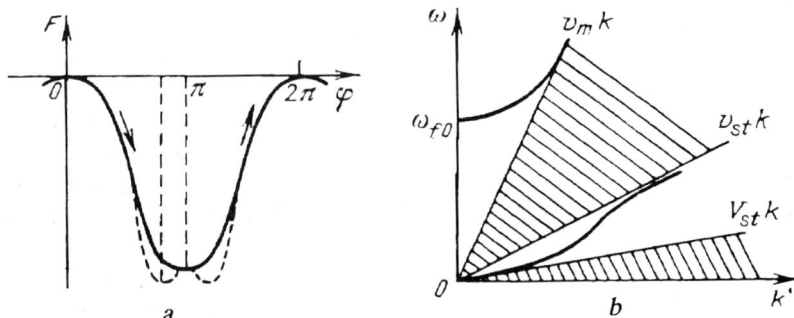

Fig. 18. (a) Dependence of the effective potential on the angle φ of deviation of the vector l from the equilibrium direction. (b) Regions of existence for the velocities of solitons in a high-temperature EPAF ($v_M > v_{St}$).

known "2π kink" of the "double sine-Gordon" equation [16]:

$$\operatorname{tg}\frac{\varphi}{2} = \left(\frac{D+A}{A}\right)^{1/2} \left(\operatorname{sh}\frac{\xi}{\xi^0}\right)^{-1}. \quad (44)$$

Here the characteristic dimension of a soliton is $\xi_0 = [m_p/(D+A)]^{1/2}$. For the α-Fe$_2O_3$ crystal with $H \approx 0.5$ kOe and $V \ll V_{St}$ the magnitude of ξ_0 is $\sim 10^{-3}$–10^{-4} cm, while the maximum deformations in the wave are of the order of 10^{-5}. We should pay attention to the fact that the velocities of solitons take on values that do not coincide with the phase velocities of linear magnetoelastic waves. In the ωk plane each value $v_{\mathrm{ph}} \equiv \omega/k$ of the harmonic wave can be put into correspondence with a straight line passing through the coordinate origin with a slope equal to $\tan \alpha = v_{\mathrm{ph}}$. The regions in which these straight lines can lie are left uncrosshatched in fig. 18b. In turn, the crosshatched regions correspond to the attainable velocities of solitons ($V = \tan \alpha$). Such a segmentation of the ωk plane is valid for any directions of propagation of magnetoelastic waves, if we do not take into account relaxation, which amounts to a generalized result of the action of the nonlinear interactions not taken into account in (11), (12).

In low-temperature EPAFs for which $v_m < v_{St}$ (MnCO$_3$, CoCO$_3$), solitons correspond to velocities $v_{St} > V > V_{St}$ or $V < v_m$ (fig. 19b). In the former case the solitons also have the form of a

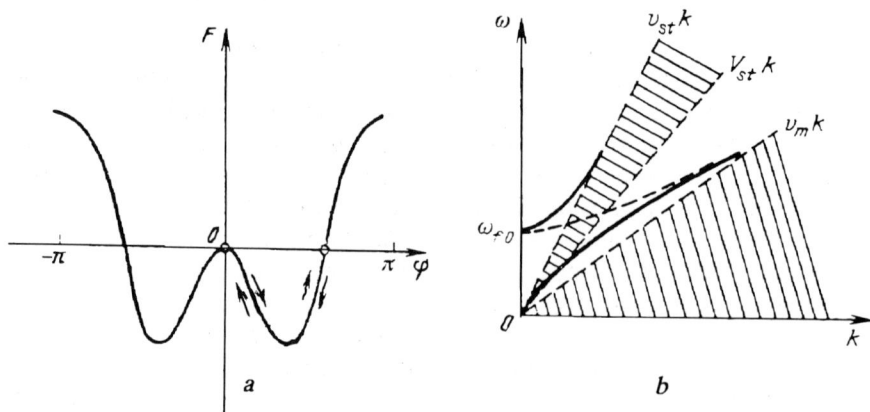

Fig. 19. (a) Dependence of the effective potential on the angle φ of deviation of the vector l from the equilibrium direction. (b) Region of existence for the velocities of solitons in a low temperature EPAF ($v_m < v_{St}$).

2π kink, and in the latter case their structure proves qualitatively different. The characteristic form of the effective potential a of "particle" of mass $|m_p|$ for the range of near-sonic velocities $v_{St} > V > V_{St}$ is shown in fig. 19a. A soliton corresponds to motion of the "particle" from the point $\varphi = 0$ (for $p = 0$) to the point φ_{max} with subsequent return to the point $\varphi = 0$. Here the solution of eq. (32) has the form [16]

$$\text{tg}\frac{\varphi}{2}\left(\frac{|D+A|}{A}\right)^{1/2}\left(\text{ch}\frac{\xi}{\xi_0}\right)^{-1}. \quad (45)$$

In contrast to a 2π kink, the given type of solitons can be realized for any deviations from equilibrium ($\varphi_{max} \ll 1$). In this case the result goes over into the known soliton solution of the modified Korteweg–de Vries equation. From the experimental standpoint it seems interesting to estimate the time τ during which such a soliton evolves from the initial steplike magnetoelastic perturbation moving with the velocity V_{St} [17]. For the typical low-temperature EPAF $MnCO_3$ we find $\tau = \xi_0/|V - V_{St}| \approx 10^{-6}$, and the length of specimen necessary for a pulse experiment does not exceed 1 cm. Estimates of τ for an elastic soliton with a nonlinear elastic lattice or for a magnetoelastic soliton in a ferromagnet having a moderate magnetoelastic interaction yield $\tau \approx 0.1$ s, thus demonstrating the advantages of nonconductive EPAFs for direct observation of solitary magnetoelastic waves. We must note that soliton solutions of the original system (11), (12) for magnetoelastic waves of small amplitude with near-sonic velocities also exist in high-temperature EPAFs, but for other directions of propagation [17]. However, in these crystals such excitations prove to be unstable with respect to transverse perturbations of their front [18,20]. One of the variants of the development of this instability turns out to be self-focusing of the magnetoelastic excitations [18]. Its physical cause is the decrease in velocity of the excitation with increase in its amplitude, which causes an accumulating deflection of the wave front in a direction opposite to the direction of propagation. A detailed analysis of the initial distributions of the amplitude from which the self-focusing wave is formed has been performed in ref. [20].

The evident qualitative differences between solitons of the types (44) and (45) (see figs. 18 and 19) allow a certain topological treatment. The solitons of (44) have a nonzero topological charge $(1/2\pi)\oint \nabla\varphi \cdot dl = 1$, where the integration is performed over a contour closed at infinity and penetrating the basis plane. By analogy with the vortex states in extended Josephson structures, such excitations can be classified as vortex excitations. For the solitons of (45) the topological charge is zero.

In the case of the excitations (45) a crystal with fixed boundary conditions ($\varphi|_{z=\pm L} = 0$) can be converted by continuous transformation to the equilibrium state. Under the same conditions for excitations of the form of (44) the transition to equilibrium involves overcoming a finite energy barrier due to the exchange interaction. Accordingly, the solitons (44) are topologically stable – in contrast to the solitons (45). A calculation [12] of the spectra of their localized excitations shows that the given type of solitons is stable not only topologically, but also dynamically, which allows us to expect a possible experimental observation.

We note that strongly nonlinear spin excitations are substantially magnetoelastic even in crystals with relatively weak magnon–phonon coupling if their velocity of propagation is close the sound velocity. In particular, such a situation arises in the orthoferrites in the motion of domain boundaries with near-sonic velocities [21,22].

Recently experimental results have been obtained [64,23,74] on $FeBO_3$ which confirm the conclusions of the theory on magnetoelastic gaps [65,66] in the spectrum of the stationary motion velocity of domain boundaries in EPAFs. Also it was obtained the evidence of dynamic SRTs in the sound wave field [67].

8. Conclusion

As we see it, the idea presented in the review on the nonlinearity of mixed modes are highly general in character. Not only magnetoelastic, but also electron–nuclear-spin and electron–

nuclear-magnetoelastic waves, ferroelectromagnetic, and ferroelectric–magnetoelastic waves and other types of coupled oscillations can play the role of mixed excitations. Nonlinear processes in such systems have been intensively studied, but mainly theoretically [68–73] and all of them are particularly prominent in the vicinity of spin reorientation transitions – from a corrugated instability of a magnetoelastic wave front [74] up to shock waves [73c] and even strains-spouts (to be published). An expansion of the experimental studies of strong dynamic interactions in coupled systems in the vicinity of SRT would facilitate the further development of our views of the mechanisms of formation of the dynamical properties of solids and their functional potentialities in technical applications.

References

[1] A.S. Borovik-Romanov and E.G. Rudashevskii, Zh. Eksp. Teor. Fiz. 47 (1964) 2095 [Sov. Phys. JETP 20 (1965) 1407].
[2] E.A. Turov and V.G. Shavrov, Fiz. Tverd. Tela (Leningrad) 7 (1965) 217 [Sov. Phys. Solid State 7 (1965) 166].
[3] E.A. Turov and V.G. Shavrov, Usp. Fiz. Nauk 140 (1983) 429 [Sov. Phys. Usp. 26 (1983) 593].
[4] I.E. Dikshtein, V.V. Tarasenko and V.G. Shavrov, Zh. Eksp. Teor. Fiz. 67 (1974) 816 [Sov. Phys. JETP 40 (1975) 404].
[5] V.I. Ozhogin and V.L. Preobrazhenskii, Physica B 86–88 (1977) 979.
[6] V.I. Ozhogin and V.L. Preobrazhenskii, Zh. Eksp. Teor. Fiz. 73 (1977) 988 [Sov. Phys. JETP 46 (1977) 523].
[7] V.I. Ozhogin and V.L. Preobrazhenskii, Sov. Phys. Usp. 31 (1988) 713.
[8] V.S. Lutovinov, V.L. Preobrazhenskii, and S.P. Semin, Zh. Eksp. Teor. Fiz. 74 (1978) 1159 [Sov. Phys. JETP 47 (1978) 609].
[9] L.V. Panina and V.L. Preobrazhenskii, Fiz. Met. Metalloved. 57 (1984)39 [Phys. Met. Metallogr. (USSR) 57 (1984)30].
[10] L.V. Panina, V.L. Preobrazhenskii, V.N. Shumilov and N.A. Ekonomov, Akust. Zh. 30 (1984) 566 [Sov. Phys. Acoust. 30 (1984) 335].
[11] N.N. Evitkhiev, V.V. Moshkin, V.L. Probrazhenskii and N.A. Ekonomov, Pis'ma Zh, Eksp. Teor. Fiz. 35 (1982) 31 [JETP Lett. 35 (1982) 38].
[12] V.L. Preobrazhenskii, Abstract of dissertation for doctorate in physical-mathematical sciences, Institute of Radio Engineering, Electronics, and Automation, Moscow, (1986) in Russian.
[13] E.A. Turov, JETP 98 (1990) 655.
[14] A.Yu. Lebedev, V.I. Ozhogin and A.Yu. Yakubovskii, Pis'ma Zh. Eksp. Teor. Fiz. 34 (1981) 22 [JETP Lett. 34 (1981) 19].
[15] V.V. Berezhnov, N.N. Evtikhiev, V.L. Preobrazhenskii and N.A. Ekonomov, Radiotekh. Electron 28 (1983)].
[16] V.L. Preobrazhenskii and M.A. Savchenko, Proc. 20th Congr. AMPERE. Tallinn (1978) p. 410.
[17] V.I. Ozhogin, and A.Yu. Lebedev, J. Magn. Magn. Mater. 15–18 (1980) 617.
[18] V.I. Ozhogin, D.Yu. Manin, V.I. Petviashvili and A.Yu. Lebedev, IEEE Trans. Magn. MAG-19 (1983) 1977.
[19] V.D. Buchel'nikov and V.G. Shavrov, Fiz. Tverd. Tela (Leningrad) 25 (1983) 90 [Sov. Phys. Solid State 25 (1983) 49].
[20] S.K. Turitsyn and G.E. Fal'kovich, Zh. Eksp. Teor. Fiz. 89 (1985) 258 [Sov. Phys. JETP 62 (1985) 146].
[21] A.K. Zvezdin and A.F. Popkov, Fiz. Tverd. Tela (Leningrad) 21 (1979) 1334 [Sov. Phys. Solid State 21 (1979) 771].
[22] V.G. Bar'yakhtar, B.A. Ivanov and V.V. Chetkin, Usp. Fiz. Nauk 146 (1985) 417 [Sov. Phys. Usp. 28 (1985) 563].
[23] M.V. Chetkin and V.V. Lykov, JETP Lett. 52 (1990) 869.
[24] A.O. Kiselev, V.I. Ozhogin and V.L. Preobrazhenskii, JETP Lett. 51 (1990) 82.
[25] V.G. Bar'yakhtar, B.A. Ivanov and A.L. Sukstanckii, Zh. Eksp. Teor. Fiz. 78 (1980) 1509 [Sov. Phys. JETP 51 (1980) 757].
[26] A.F. Andreev and V.I. Marchenko, Usp. Fiz. Nauk 130 (1980) 39 [Sov. Phys. Usp. 23 (1980) 21].
[27] L.V. Velikov, S.V. Mironov and E.G. Rudashevskii, Zh. Eksp. Teor. Fiz. 75 (1987) 1110 [Sov. Phys. JETP 48 (1978) 559].
[28] R.K. Bullough, P.J. Caudrey and H.M. Gidds, in: Solitons, eds. R.K. Bullough and P.J. Caudrey (Springer-Verlag, Berlin, 1980) p. 107 [Russ. transl. (Mir, Moscow, 1983) p. 122].
[29] M.H. Seavey, Solid State Commun. 10 (1972) 219.
[30] a) V.I. Ozhogin and P.P. Maksimenkov, Digests of Intermag Conf., Kyoto, (1972) 49-4; b) IEEE Trans. Magn. MAG-8 (1972) 130; c) Zh. Eksp. Teor. Fiz. 65 (1973) 657 [Sov. Phys. JETP 38 (1974) 342].
[31] E.F. Andruschak, N.N. Evtikhiev, S.A. Pogozhev, V.L. Preobrazhenskii and N.A. Ekonomov, Akoust. Zh. 27 (1981) 170 [Sov. Phys. Acoust. 27 (1981)].
[32] A. Tasaki and S. Iida, J. Phys. Soc. Japan 18 (1963) 1148.
[33] V.S. Merkulov, E.G. Rudashevskii, H. LeGall and C. Leycuras, Zh. Eksp. Teor. Fiz. 75 (1978) 628.
[34] A.F. Ermolov, V.V. Moshkin, V.L. Preobrazhenskii and N.A. Ekonomov, Fiz. Tverd. Tela 30 (1988) 65.
[35] A.P. Levanyuk, V.V. Osipov, A.S. Sigov and A.A. Sobyanin, Zh. Eksp. Teor. Fiz. 76 (1979) 345.
[36] L.K. Zarembo and V.A. Krasil'nikov, Usp. Fiz. Nauk 102 (1970) 549 [Sov. Phys. Usp. 13 (1971) 778].

[37] V.L. Preobrazhenskii, M.A. Savchenko and N.E. Ekonomov, Pis'ma Zh. Eksp. Teor. Fiz. 28 (1978) 93 [JETP Lett. 28 (1978) 87].

[38] A.M. Zaikin and V.L. Preobrazhenskii, 5th All-Union Conf. on Precisive Alloys, Rostov-Velikii (September 1991) Abstracts.

[39] A.E. Clark, H.S. Belson, N. Tamagave and E. Callen, Proc. ICM '73, 4 (1973) 335.

[40] V.V. Berezhnov, N.N. Evitkhiev, V.L. Probrazhenskii and N.A. Ekonomov, Fiz. Tverd. Tela (Leningrad) 25 (1982) 1870 [Sov. Phys. Solid State 25 (1982) 1079].

[41] V.I. Ozhogin, A.Yu. Lebedev and A.Yu. Yakubovskii, Pis'ma Zh. Ekskp. Teor. Fiz. 27 (1978) 333 [JETP Lett. 27 (1978) 313].

[42] V.A. Krasil'nikov, T.A. Mamatova and V.G. Prokoshev, Pis'ma Zh. Tekh. Fiz. 10 (1984) 1196 [Sov. Tech. Phys.. Lett. 10 (1984) 506].

[43] D.A. Kirzhnits, Usp. Fiz. Nauk 124 (1978) 487.

[44] V.V. Berezhnov, V.L. Preobrazhenskii, N.A. Ekonomov and D.E. El'yashev, Abstracts of reports at the 16th All-Union Conf. on Physics of Magnetic Phenomena, Tula (1983) p. 23.

[45] N.N. Evtikhiev, V.L. Preobrazhenskii, M.A. Savchenko and N.A. Ekonomov, Vopr. Radioelektron. Ser. Obshchetekhn. 5 (1978) 124.

[46] R.L. Comstock and R.C. LeCraw, Phys. Rev. Lett 10 (1963) 219.

[47] V.V. Zautkin, V.S. L'vov, B.I. Orel and S.S. Starobinets, Zh. Eksp. Teor. Fiz. 72 (1977) 272.

[48] V.I. Ozhogin, A.Yu. Yakubovskii, A.V. Abryutin, S.M. Sileimanov, J. Magn. Magn. Mater. 15–18 (1980) 757.

[49] V.A. Krasil'nikov, T.A. Mamatova and V.G. Prokoshev, Fiz. Tverd. Tela (Leningrad) 28 (1986) 615 [Sov. Phys. Solid State 28 (1986) 346].

[50] V.V. Berezhnov, N.N. Evtikhiev, V.L. Preobrazhenskii and N.A. Ekonomov, Akust. Zh. 26 (1980) 328 [Sov. Phys. Acoust. 26 (1980) 180].

[51] M.K. Gubkin, T.A. Mamatova and V.G. Prokoshev, ibid. 31 (1985) 678 [Sov. Phys. Acoust. 31 (1985) 410].

[52] V.A. Ermolov, A.I. Alekseev and V.G. Pankratov, Pis'ma Zh. Tekh. Fiz. 11 (1985) 377 [Sov. Tech. Phys. Lett. 11 (1985) 156].

[53] W. Wettling, W. Jantz and C.E. Patton, J. Appl. Phys. 50 (1979) 2030.

[54] B.Yu. Kotyuzhanskii and L.A. Prozorova, Zh. Eksp. Teor. Fiz. 83 (1982) 1567 [Sov. Phys. JETP 56 (1982) 903].

[55] V.S. Lutovinov and M.A. Savchenko, Abstracts of reports at the All-Union Conf. on the Physics of Magnetic Phenomena, Khar'kov (1979) p. 85.

[56] M.V. Petrov, A.P. Paugurt, I.V. Pleshakov and A.V. Ivanov, Pis'ma Zh. Tekh. Fiz. 11 (1985) 1204 [Sov. Tech. Phys. Lett. 11 (1985) 498].

[57] R. Dichl, W. Jantz, B.I. Nolang and W. Wettling, Current Topics in Materials Science, vol. II, ed. E. Kaldis, (1986) p. 242.

[58] A.Yu. Lebedev, V.I. Ozhogin, V.L. Safonov and A.Yu. Yakubovskii, Zh. Eksp. Teor. Fiz. 85 (1983) 1059 [Sov. Phys. JETP 58 (1983) 616].

[59] M.A. Savchenko, Fiz. Tverd. Tela (Leningrad) 6 (1964) 864 [Sov. Phys. Solid State 6 (1964) 666].

[60] R.I. Kukhtin, V.L. Preobrazhenskii and N.A. Ekonomov, Fiz. Tverd. Tela (Leningrad) 26 (1984) 884 [Sov. Phys. Solid State 26 (1984) 356].

[61] V.G. Vaks, A.I. Larkin and S.A. Pikin, Zh. Eksp. Teor. Fiz. 53 (1967) 1089 [Sov. Phys. JETP 28 (1969) 734].

[62] E.M. Pikalev, M.A. Savchenko and J. Shoyom, Zh. Eksp. Teor. Fiz. 55 (1968) 1404 [Sov. Phys. JETP 28 (1969) 734].

[63] Yu.A. Izyumov, S.A. Kasan-Ogly and Yu.N. Skryabin, Field Methods in the Theory of Ferromagnetism (in Russian), (Nauka, Moscow, 1984) p. 224.

[64] M.V. Chetkin, V.V. L'vov and V.D. Tereshchenko, Fiz. Tverd. Tela (Leningrad) 32 (1990) 939.

[65] A.K. Zvezdin, V.V. Kostuchencko and A.A. Mukhin, Preprint No. 209, Lebedev Phys. Inst. of Ac. Sci (1983).

[66] L.V. Panina and V.L. Preobrazhenskii, Fiz. Tverd. Tela (Leningrad) 60 (1985) 449.

[67] A.F. Kabychenkov and V.G. Shavrov, Acta Phys. Polon. 4 (1988) 531.

[68] A.V. Andrienko, V.I. Ozhogin, V.L. Safonov and A.Yu. Yakubovskii, Sov. Phys. Usp. (October 1991).

[69] M.A. Savchenko and V.P. Sobolev, Abstracts of reports at the All-Union Conf. on the Physics of Magnetic Phenomena, Donetsk (1985) p. 318.

[70] M.A. Savchenko and M.A. Khabakhpashev, Fiz. Tverd. Tela (Leningrad) 18 (1976) 2699; 20 (1978) 1845 [Sov. Phys. Solid State 18 (1976) 1573, 20 (1978) 1065].

[71] G.A. Farias and A.A. Maradudin, Phys. Rev. Ser. B 28 (1983) 1870.

[72] V.V. Men'shenin, I.F. Mirsaev and G.G. Taluts, Fiz. Met. Metalloved. 56 (1983) 1078 [Phys. Met. Metallogr. 56 (1983) 1983].

[73] A.F. Kabychenkov, V.G. Shavrov and A.L. Shevchenko, Fiz Tverd. Tela (Leningrad) a) 31 (1989) 193, b) 32 (1990) 1182, c) 32 (1990) 2010.

[74] M.V. Chetkin and V.V. Lykov, J. Magn. Magn. Mater. (1991) to be published.

Properties, units and constants in magnetism

Lydon J. Swartzendruber

National Institute of Standards and Technology, Gaithersburg, MD 20234, USA

The accompanying table (table 1) consists of data and numbers designed to be useful as a pocket data card for workers in the fields of magnetism and magnetic materials. They have been selected from a number of sources [1–6]. The symbols used in the table are:

T_c the ferromagnetic Curie temperature;

μ_{at} the magnetic moment of each atom in Bohr magnetons;

σ_s the specific saturation magnetization. The saturation magnetization $4\pi M_s$ in gauss can be obtained by multiplying σ_s by 4π and by the density;

ν_r the ferromagnetic resonance frequency. There are two isotopes for Gd: ^{155}Gd (14.9% abundance) and ^{157}Gd (15.7%). Other isotopes are: ^{57}Fe (2.14%), ^{61}Ni (1.1%) and ^{59}Co (100%);

ρ the density. Densities given are calculated densities from X-ray results. Actual densities can vary a few percent from the values given, even for very pure materials;

A the atomic weight, or the number of grams per mole;

χ_v the volume susceptibility;

χ_g the mass susceptibility;

μ_0 the permeability of free space;

μ_B the Bohr magneton;

μ_N the nuclear magneton;

μ_B/k the ratio of the Bohr magneton to the Boltzmann constant. This is useful because the ration $\mu_B H/kT$ appears quite often;

$N_A\mu_B$ the product of Avogadro's constant and the Bohr magneton;

Φ_0 the magnetic flux quantum.

The value of T_c for hcp Co given in table 1 is estimated. The ferromagnetic nuclear magnetic resonance (FNR) frequencies listed are for ^{57}Fe, ^{59}Co, ^{61}Ni, ^{155}Gd and ^{157}Gd. The two relationships noted are useful for converting the magnetic induction σ in emu/g to the magnetic moment per atom in Bohr magnetons or to the magnetic induction in gauss. The susceptibility values and their temperature coefficients for several NIST standard reference materials are included since many workers have these standards in their laboratory but may no longer have the certificate. Further, if great care is used in the selection of materials, one can probably use these numbers to fabricate special standards at the 1% accuracy level. The temperature coefficients have a limited range of validity but can be considered valid for normal variations in room temperature.

Conversions between cgs Gaussian units and mks (SI) units seem to have been designed to torment both the novice and the seasoned professional alike. Both unit systems have their advantages and both

Table 1
Data and numbers in the field of magnetism and magnetic materials

Ferromagnetic elements:	Fe (bcc)	Co (fcc)	Co (hcp)	Ni (fcc)	Gd (hcp)
T_0 [K]	1044	1388	1360	627.4	293.4
μ_{at} at 0 K [μ_B]	2.217	1.753	1.721	0.6157	7.56
σ_s at 0 K [Tm^3/kg $4\pi\ 10^{-7}$] [a]	221.7	166.1	163.1	58.57	268.4
σ_s at 298 K [Tm^3/kg $4\pi\ 10^{-7}$] [a]	217.2	164.8	161.8	55.09	0
NMR ν_r at 0 K [MHz]	46.64	228	217	28.46	48.6, 64.2
ρ at 298 K [g/cc]	7.875	8.793	8.804	8.912	7.898
A [g/mol]	55.847	58.933	58.933	58.71	157.25

NIST calibration samples, SI units, $T = 298$ K

SRM 772 (Ni sphere) [b]	$\sigma = 54.95 \times 4\pi \times 10^{-7} (1 - 955/H)$	-0.03%/K	($\pm 0.4\%$)
SRM 766 (MnF$_2$)	$\chi_g = 123.0 \times 4\pi \times 10^{-9}$	-0.28%/K	($\pm 0.5\%$)
SRM 764 (Pt)	$\chi_g = 0.991 \times 4\pi \times 10^{-9}$	-0.07%/K	($\pm 0.5\%$)
SRM 765 (Pd)	$\chi_g = 5.24 \times 4\pi \times 10^{-9}$	-0.19%/K	($\pm 0.5\%$)
SRM 763 (Al)	$\chi_g = 0.604 \times 4\pi \times 10^{-9}$	-0.06%/K	($\pm 0.5\%$)
SRM 763 (Al) @ 77.7 K	$\chi_g = 0.696 \times 4\pi \times 10^{-9}$	-0.07%/K	($\pm 0.5\%$)

Gaussian to SI:	multiply no. of	by	to obtain no. of
Flux density, B	G	10^{-4}	T
Field strength, H	Oe	$10^3/4\pi$	A/m
Magnetization, M	G or Oe	10^3	A/m
Magnetization, $4\pi M$	G	10^{-4}	T
Susceptibility (M/H), χ_v [c]	emu cm^{-3} Oe^{-1}	4π	(no dim.)
Mass susceptibility, χ_g [c]	emu g^{-1} Oe^{-1}	$4\pi \cdot 10^{-3}$	m^3/kg
Magnetic moment, μ [c]	erg/G ($\equiv$ emu)	10^{-3}	J/T ($\equiv$ A m^2)
Specific magnetization, σ [c]	emu/g	$4\pi \cdot 10^{-7}$	Tm^3/kg
Demagnetizing factor, N	(no dim.)	$1/4\pi$	(no dim.)
Flux, Φ	maxwell	10^{-8}	Wb
Anisotropy constant, K	erg/cm^3	10^{-1}	J/m^3

Useful constants:	Gaussian	SI
μ_0	1 (no dim.)	$4\pi \times 10^{-7}$ H/m ($\equiv$ Tm/A)
μ_B	9.274015×10^{-21} erg/G	9.274015×10^{-24} J/T
μ_N	5.050787×10^{-24} erg/G	5.050787×10^{-27} J/T
μ_B/k	0.671710×10^{-4} K/G	0.671710 K/T
$N_A \mu_B$	5.584939×10^3 erg G^{-1} mol^{-1}	5.584939 J T^{-1} mol^{-1}
Φ_0 ($= h/2e$)	2.067835×10^{-7} maxwell	2.067835×10^{-15} Wb

[a] Note for σ in emu/g: μ_{at} (in μ_B) $= \sigma A / N_A \mu_B$; B_s (in G) $= 4\pi \sigma_s \rho$.
[b] Valid for H between 280 and 795 kA/m.
[c] G cm^3 is often substituted for emu (which is not a unit). In the Gaussian system $B = H + 4\pi M$, in SI (mks) $B = \mu_0 (H + M)$.

will no doubt be used for some time to come. The conversions given are the ones we have found most often useful. The often used "unit", emu, is not really a unit to which dimensional analysis can be applied. As an aid to overcoming this, emu is sometimes replaced by G cm^3. (It could also be replaced by erg/G.) Susceptibilities in the Gaussian system, such as χ_g, alternately appear with the units emu/g, emu/g Oe, emu/g G, G cm^3/g and cm^3/g. Although volume susceptibility is dimensionless in both systems, it is often assigned a unit in the Gaussian system. For further discussion of magnetic units see ref. [5].

References

[1] G.C. Carter, L.H. Bennett and D.J. Kahan, Metallic Shifts in NMR (Pergamon, New York, 1977).
[2] J.J. Rhyne, Bull. Alloy Phase Diagrams 3 (1982) 401.
[3] E.R. Cohen and B.N. Taylor, Rev. Mod. Phys. 59 (1987) 1121.
[4] E.P. Wohlfarth, Ferromagnetic Materials, vol. 1 (North-Holland, Amsterdam, 1980).
[5] L.H. Bennett, C.H. Page and L.J. Swartzendruber, AIP Conf. Proc. 29 (American Institute of Physics, New York, 1976); J. Res. Nat. Bur. Stds. 83 (1978) 9.
[6] H.P.J. Wijn, Landolt–Börstein Numerical Data, Group III, vol. 19, subvol. A, (Springer, Berlin, 1986).

Author index

Adroja, D.T. and S.K. Malik, Valence fluctuation and heavy fermion behaviour in rare earth and actinide based compounds — 126
Aeppli, G., see G.H. Lander — 151
Akai, H., see P.H. Dederichs — 241

Bader, S.D., SMOKE — 440
Bate, G., Magnetic recording materials since 1975 — 413
Belanger, D.P. and A.P. Young, The random field Ising model — 272
Borsje, H.R., see R.J.H. Kappert — 363
Budnick, J.I., see Y.D. Zhang — 13
Buschow, K.H.J., Permanent magnet materials based on tetragonal rare earth compounds of the type $RFe_{12-x}M_x$ — 79

Croat, J.J., see J.F. Herbst — 57

Das, B.N., see N.C. Koon — 173
De Jonge, W.J.M. and H.J.M. Swagten, Magnetic properties of diluted magnetic semiconductors — 322
De Visser, A. and J.J.M. Franse, Uranium-based heavy-fermion superconductors: an experimental survey — 204
Dederichs, P.H., R. Zeller, H. Akai and H. Ebert, Ab-initio calculations of the electronic structure of impurities and alloys of ferromagnetic transition metals — 241
Dhar, S.K., see K. Ikeda — 292
Dillon Jr., J.F., Magnetooptics — 425

Ebert, H., see P.H. Dederichs — 241

Ford, J.C., see Y.D. Zhang — 13
Franse, J.J.M., see A. de Visser — 204
Freeman, A.J. and R.-Q. Wu, Electronic structure theory of surface, interface and thin-film magnetism — 497
Fuggle, J.C., see R.J.H. Kappert — 363

Gignoux, D. and D. Schmitt, Rare earth intermetallics — 99
Gradmann, U., Surface magnetism — 481
Gschneidner Jr., K.A., see K. Ikeda — 292

Hakonen, P., O.V. Lounasmaa and A. Oja, Spontaneous nuclear magnetic ordering in copper and silver at nano - and picokelvin temperatures — 394
Han, B.S., Behavior of vertical-Bloch-line chains of hard domains in garnet bubble films — 455
Hasegawa, R., Amorphous magnetic materials – a history — 1
Herbst, J.F. and J.J. Croat, Neodymium–iron–boron permanent magnets — 57
Hines, W.A., see Y.D. Zhang — 13

Ikeda, K., S.K. Dhar, M. Yoshizawa and K.A. Gschneidner Jr., Quenching of spin fluctuations by high magnetic fields — 292

Johnston, D.C., Normal state magnetism of the high T_c cuprate superconductors — 218

Kappert, R.J.H., H.R. Borsje and J.C. Fuggle, High energy spectroscopies and magnetism — 363
Koon, N.C., C.M. Williams and B.N. Das, Giant magnetostriction materials — 173

Lacroix, C., Magnetic properties of the Kondo lattice — 90
Lander, G.H. and G. Aeppli, Neutron scattering studies of magnetic properties of actinide systems — 151
Lounasmaa, O.V., see P. Hakonen — 394

Malik, S.K., see D.T. Adroja — 126
Mathon, J., Theory of magnetic multilayers. Exchange interactions and transport properties — 527
Mills, D.L., The ferromagnetism of ultrathin films; from two to three dimensions — 515
Moon, R.M. and R.M. Nicklow, Neutron scattering of lanthanide materials — 139
Moriya, T., Theory of itinerant electron magnetism — 261

Nicklow, R.M., see R.M. Moon — 139

Oja, A., see P. Hakonen — 394
Ozhogin, V.I. and V.L. Preobrazhenskii, Nonlinear dynamics of coupled systems near magnetic phase transitions of the "order–order" type — 544

Preobrazhenskii, V.L., see V.I. Ozhogin 544
Prinz, G.A., Metastability in epitaxial magnetic metal films 469

Schmitt, D., see D. Gignoux 99
Steglich, F., Experimental study of Ce-based heavy-fermion compounds 186
Strnat, K.J. and R.M.W. Strnat, Rare earth–cobalt permanent magnets 38
Strnat, R.M.W., see K.J. Strnat 38
Swagten, H.J.M., see W.J.M. de Jonge 322
Swartzendruber, L.J., Properties, units and constants in magnetism 573

Wassermann, E.F., The Invar problem 346
Williams, C.M., see N.C. Koon 173
Wu, R.-Q., see A.J. Freeman 497

Yoshizawa, M., see K. Ikeda 292
Young, A.P., see D.P. Belanger 272

Zeller, R., see P.H. Dederichs 241
Zhang, Y.D., J.I. Budnick, J.C. Ford and W.A. Hines, Some application of NMR to the study of magnetically-ordered materials with emphasis on the short-range order in (Fe–B)-based crystalline and amorphous alloys 13

Subject index

Actinides	126, 151, 186, 204, 292
Amorphous systems	1, 13, 57
– alloys	13, 57
– glassy	1
– metals	1
– Metglas	13
Anharmonicity	544
– normal modes	544
Anisotropy	38, 79, 99, 173, 218, 440, 481, 497, 515
– energy	497
– magnetocrystalline surface	497
– RE compounds	38, 79, 99, 173, 218
– strain induced	515
– surface	440, 481, 497
– theory	497
– uniaxial	515
Antiferromagnetism	99, 126, 204, 218, 261, 346, 425, 497, 527, 544
– weak itinerant	261
Band narrowing	481
Band structure	322, 481, 497, 527
Birefringence	425
– gradient related	425
Bloch $T^{3/2}$ hw	527
Bloch tight-binding $T^{3/2}$ law	481
Boltzmann equation	527
Bragg scattering	527
Bubble films	455
Charge transfer	497
Coercivity	38, 57, 79, 413
– magnetic particles	413
Conduction-electron polarization	481, 497
Conduction-electron scattering	527
Critical phenomena	151, 272, 481
– critical exponents	272
– exponents	139, 440
– Heisenberg model	515
– Ising model	272, 515
Crystal field	57, 79, 99, 173, 497
– RE compounds	57, 79, 99, 173
Crystal structure	57, 79, 218
De Haas–van Alphen effect	186, 527
Density functional formalism	497

Dilute alloys	241
Dimensionality	515
Dipole coupling	481, 515
Domain structure	425, 455, 481
– bubbles	455
– classification	455
Domain walls	481
Electrical resistivity	126, 186, 204, 218, 292
Electronic structure	241
Electroplated films	413
Enhanced surface Bloch parameter	481
Enhanced surface exponent	481
Epitaxial metal films	469
Exchange	57, 79, 90, 173, 322, 481, 527
– anisotropic	322, 527
– asymmetry	481
– 3d-conduction electrons	497
– 4f-conduction	173, 497
– magnetic semiconductors	322
– RKKY	90, 322, 527
– splitting	481, 527
Excitons	515
Fermi liquids	186, 292
Fermi surface	527
Ferrimagnetism	1
Ferrites	413
Ferromagnetic alloys	241
Ferromagnetism	1, 261, 322
carrier induced	322
– soft	1
– weak itinerant	261
Field induced transition	515
Finite temperature properties	261
FLAPW method	481, 497
FMR	481
Friedel	497
Friedel-oscillations	481
Garnets	455, 515
Generalized dynamical susceptibility	261
Giant moments	497
Green function	527

Hard magnetic materials	38, 57, 79	Magnetostriction	173, 346
Heavy fermion	126, 151, 186, 204	Magnetothermal effect	292
Heisenberg ferromagnet	515	Magnetovolume effects	346
– two-dimensional	515	MBE	527
High pressure	126	Mean field theory	272, 515
Hyperfine fields	13, 481, 497, 515	Mean free path	527
Hyperfine fields of impurities	241	Metamagnetism	90, 99, 292
		Micromagnetics	481
Impurities in magnetic metals	241	Molecular-field theory	481
Impurity effects	218	Monolayer	469, 481, 497, 515, 527
Interface magnetism	481	Monolayer magnetism	440
Interface roughness	527	Monte Carlo simulation	481
Intermediate valence	126, 292	Mössbauer spectroscopy	13, 481, 527
Ising model	272, 515	– linear chains	515
– antiferromagnetism	497	– transition-metal alloys	13
– ground state	497		
– paramagnetism	497	Neutron diffraction	99, 139, 151, 186, 204, 218, 394
– surfaces	497	Neutron scattering	139, 151, 173, 204, 272, 394
Isolators	425	– elastic	139, 151, 204, 394
– bulk	425	– inelastic	139, 151, 173, 204
– film	425	NMR	13, 394
Itinerant electrons	218, 261, 292, 363	– Knight shift	13
– antiferromagnetism	218	– spin echo	13
– ferromagnetism	292, 363	Nonmagnetic spacer layers	527
– spin fluctuations	292	Nuclear magnetism	139, 394
Kerr effect	425, 481	One-dimensional magnetism	515
Kondo effect	1, 90, 186, 204	One-dimensional systems	527
		Oscillation	497, 527
Light scattering	481	– long-period	527
Live layers	481	– short-period	527
Local moments	1, 186, 241, 481, 497	Overlayers	440
Localized surface state	497		
Long-range order	515	Paramagnons	292
Longe-range order	272	Particulate coatings	413
		Permanent magnets	38, 57, 79
Mach–Zehnder interferometer	425	Phase transition	99, 186, 346
Magnetic multilayers	527	– crystallographic	346
Magnetic overlayer	527	– field induced	99, 515
Magnetic phase diagram	99, 151, 204, 394	– metamagnetic	99
Magnetic phase transitions	99	Phases	469
Magnetic recording	413	– high temperature	469
– materials	413	– metastable	469
– performance	413	Photoemission	363
Magnetic semiconductors	322	Potential barrier	527
Magnetic structures	99, 139, 218, 455, 481	Pressure effect	186, 346
Magnetite	413		
Magnetization patterns	38, 57, 99, 204, 218, 346	Random field systems	272
Magnetization processes	99	Rare earth alloys	139
Magneto-optics	1, 425, 481, 527	Rare earth compounds	38, 126, 151, 173, 186, 218, 292, 363
Magnetoelasticity	173, 544		
– magnetoelastic modes	544	– intermetallic	38, 126, 173, 186, 292, 363
Magnetoresistance	204, 218, 292, 527		
– RE alloys	292	– pnictides	151
– transition-metal alloys	204, 481	Rare earth metals	99, 139, 363, 497
Magnetostatics	515, 527		

Rare earth–transition metal compounds	38, 57, 79, 90, 99, 126, 173	Superconductivity	126, 186, 204, 218, 261, 425
Recording	425	– anions	425
– magneto-optical	425	– high temperature	218
Relativistic effects	241	Superexchange	322, 527
Relaxation	497	Superlattice	440
Remanence	38, 57, 413	Superstructure	481
Renormalization group	515	Surface magnetism	363, 440, 469, 481, 497
		– enhancement	497
Sandwiches	440	Surface magneto-optic Kerr effect (SMOKE)	440
Scanning microscopy	481		
Self-consistent renormalization theory	261	Surface magnetochemistry	481
		Susceptibility	497
Semiconductors	322	– magnetic	497
– diluted magnetic	322		
Sensors	1	Thermal excitations	261
Short-range order	13	Thermal expansion	186, 346
Slater–Pauling curves	241	Thin films	413, 481, 497, 515, 527
SMOKE	527	– coatings	413
Soft modes	515	– recording media	413
Specific heat	99, 126, 186, 204, 272, 292, 346	– ultra	481, 515
		Transition metal alloys	241, 346, 363
Sperimagnetism	1	– concentrated	241
Spin density	497	Transition metal compounds	151, 346, 363
Spin-dependent interfacial scattering	527	Transition metals	363, 481, 497, 515, 527
		Transport properties	527
Spin fluctuations	204, 261, 292	Two-dimensional magnetism	515
– antiferromagnetic	261	Two-dimensional systems	497, 527
Spin glass transition	322		
Spin–orbit coupling	481, 515	Vertical Bloch line chains	455
Spin-polarized electron technques	481	– formation energy	455
Spin reorientation	57, 99, 173, 515, 544	– temperature dependence	455
Spin waves	139, 481, 515, 527	Voigt effect	425
– surface modes	481		
– theory	481, 515	Weak ferromagnetism	292
– two-dimensional	515		
Step moments	481	X-ray photoelectron spectroscopy	363